1990 ASHRAE HANDBOOK

REFRIGERATION
Systems and Applications

Inch-Pound Edition

American Society of Heating, Refrigerating and Air-Conditioning Engineers, Inc.
1791 Tullie Circle, N.E., Atlanta, GA 30329
404-636-8400

Copyright © 1990 by the American Society of Heating, Refrigerating and Air-Conditioning Engineers, Inc. All rights reserved.

DEDICATED

TO THE ADVANCEMENT OF

THE PROFESSION

AND ITS ALLIED INDUSTRIES

No part of this book may be reproduced without permission in writing from ASHRAE, except by a reviewer who may quote brief passages or reproduce illustrations in a review with appropriate credit; nor may any part of this book be reproduced, stored in a retrieval system, or transmitted in any form or by any means — electronic, photocopying, recording, or other — without permission in writing from ASHRAE.

Although great care has been taken in compilation and publication of this volume, no warranties, express or implied, are given in connection herewith and no responsibility can be taken for any claims arising herewith.

Comments, criticisms, and suggestions regarding the subject matter are invited. Any errors or omissions in the data should be brought to the attention of the Editor. If required, an errata sheet will be issued at approximately the same time as the next Handbook. Notice of any significant errors found after that time will be published in the ASHRAE *Journal*.

ISBN 0-910110-69-7
ISSN 1041-2344

CONTENTS—1988 EQUIPMENT VOLUME

AIR-HANDLING EQUIPMENT

Chapter
1. Duct Construction
2. Air-Diffusing Equipment
3. Fans
4. Evaporative Air-Cooling Equipment
5. Humidifiers
6. Air-Cooling and Dehumidifying Coils
7. Sorption Dehumidification and Pressure Drying Equipment
8. Forced-Circulation Air Coolers
9. Air-Heating Coils
10. Air Cleaners for Particulate Contaminants
11. Industrial Gas Cleaning and Air Pollution Control

REFRIGERATION EQUIPMENT

Chapter
12. Compressors
13. Absorption Cooling, Heating, and Refrigeration Equipment
14. Air-Cycle Equipment
15. Condensers
16. Liquid Coolers
17. Liquid Chilling Systems
18. Component Balancing in Refrigeration Systems
19. Refrigerant-Control Devices
20. Cooling Towers
21. Factory Dehydrating, Charging, and Testing

HEATING EQUIPMENT

Chapter
22. Automatic Fuel-Burning Equipment
23. Boilers
24. Furnaces
25. Residential In-Space Heating Equipment
26. Chimney, Gas Vent, and Fireplace Systems
27. Unit Ventilators, Unit Heaters, and Makeup Air Units
28. Radiators, Convectors, Baseboard, and Finned-Tube Units
29. Infrared Heaters

GENERAL COMPONENTS

Chapter
30. Centrifugal Pumps
31. Motors and Motor Protection
32. Engine and Turbine Drives
33. Pipes, Tubes, and Fittings
34. Air-to-Air Energy-Recovery Equipment

UNITARY EQUIPMENT

Chapter
35. Retail Food Store Refrigeration Equipment
36. Food Service and General Commercial Refrigeration
37. Household Refrigerators and Freezers
38. Drinking Water Coolers and Central Systems
39. Bottled Beverage Coolers and Refrigerated Vending Machines
40. Automatic Ice Makers
41. Room Air Conditioners and Dehumidifiers
42. Unitary Air Conditioners and Unitary Heat Pumps
43. Applied Packaged Equipment

GENERAL

Chapter
44. Solar Energy Equipment
45. Codes and Standards

CONTENTS—1987 HVAC SYSTEMS AND APPLICATIONS

AIR-CONDITIONING AND HEATING SYSTEMS

Chapter
1. Air-Conditioning System Selection and Design
2. All-Air Systems
3. Air-and-Water Systems
4. All-Water Systems
5. Unitary Refrigerant-Based Systems for Air Conditioning
6. Heat Recovery Systems
7. Panel Heating and Cooling Systems
8. Cogeneration Systems
9. Applied Heat Pump Systems
10. Air Distribution Design for Small Heating and Cooling Systems
11. Steam Systems
12. Heating and Cooling from a Central Plant
13. Basic Water System Design
14. Chilled and Dual-Temperature Water Systems
15. Medium and High Temperature Water Heating Systems
16. Infrared Radiant Heating

COMFORT AIR-CONDITIONING AND HEATING APPLICATIONS

Chapter
17. Residences
18. Retail Facilities
19. Commercial and Public Buildings
20. Places of Assembly
21. Domiciliary Facilities
22. Educational Facilities
23. Health Facilities
24. Surface Transportation
25. Aircraft
26. Ships
27. Environmental Control for Survival

INDUSTRIAL AND SPECIAL AIR CONDITIONING AND VENTILATION

Chapter
28. Industrial Air Conditioning
29. Enclosed Vehicular Facilities
30. Laboratories
31. Engine Test Facilities
32. Clean Spaces
33. Data Processing System Areas
34. Printing Plants
35. Textile Processing
36. Photographic Materials
37. Environment for Animals and Plants
38. Drying and Storing Farm Crops
39. Air-Conditioning of Wood and Paper Product Facilities
40. Heating, Ventilating, and Air Conditioning for Nuclear Facilities
41. Ventilation of the Industrial Environment
42. Underground Mine Air Conditioning and Ventilation
43. Industrial Exhaust Systems
44. Industrial Drying Systems

GENERAL

Chapter
45. Geothermal Energy
46. Thermal Storage
47. Solar Energy Utilization
48. Energy Management
49. Owning and Operating Costs
50. Control of Gaseous Contaminants
51. Automatic Control
52. Sound and Vibration Control
53. Water Treatment
54. Service Water Heating
55. Snow Melting
56. Evaporative Air Cooling
57. Testing, Adjusting, and Balancing
58. Fire and Smoke Control
59. Mechanical Maintenance
60. Computer Applications
61. Codes and Standards

CONTENTS

Contributors

Technical Committees and Task Groups

Preface

REFRIGERATION SYSTEM PRACTICES

Chapter
1. **Engineered Refrigeration Systems** (Industrial Design Practices)(TC 10.1, Custom Engineered Refrigeration Systems)
2. **Liquid Overfeed Systems** (TC 10.1)
3. **System Practices for Halocarbon Refrigerants** (TC 10.3, Refrigerant Piping)
4. **System Practices for Ammonia** (TC 10.3)
5. **Secondary Coolants in Refrigeration Systems** (TC 10.1)
6. **Refrigerant System Chemistry** (TC 3.2, Refrigerant System Chemistry)
7. **Moisture and Other Contaminant Control in Refrigerant Systems** (TC 3.3, Contaminant Control in Refrigerating Systems)
8. **Lubricants in Refrigerant Systems** (TC 3.4, Lubrication)

FOOD REFRIGERATION

Chapter
9. **Commercial Freezing Methods** (TC 11.6, Prepared Food Products)
10. **Microbiology of Foods** (TC 11.6)
11. **Methods of Precooling Fruits, Vegetables, and Cut Flowers** (TC 11.5, Fruits, Vegetables, and Other Products)
12. **Meat Products** (TC 11.1, Meat, Fish, and Poultry Products)
13. **Poultry Products** (TC 11.1)
14. **Fishery Products** (TC 11.1)
15. **Dairy Products** (TC 11.3)
16. **Deciduous Tree and Vine Fruits** (TC 11.5)
17. **Citrus Fruits, Bananas, and Subtropical Fruits** (TC 11.5)
18. **Vegetables** (TC 11.5)
19. **Fruit Juice Concentrates** (TC 11.6)
20. **Precooked and Prepared Foods** (TC 11.6)
21. **Bakery Products** (TC 11.6)
22. **Candies, Nuts, Dried Fruits, and Vegetables** (TC 11.5)
23. **Beverage Processes** (TC 11.7, Fermented and Carbonated Beverages)
24. **Eggs and Egg Products** (TC 11.1)
25. **Refrigerated Warehouse Design** (TC 10.5, Refrigerated Distribution and Storage Facilities)
26. **Commodity Storage Requirements** (TC 10.5)
27. **Refrigeration Load** (TC 10.8, Refrigeration Load Calculations)

DISTRIBUTION OF CHILLED AND FROZEN FOOD

Chapter 28. **Trucks, Trailers, and Containers** (TC 10.6, Transport Refrigeration)
29. **Railroad Refrigerator Cars** (TC 10.6)
30. **Marine Refrigeration** (TC 10.6)
31. **Air Transport** (TC 10.6)
32. **Retail Food Store Refrigeration** (TC 10.7, Commercial Food Display and Storage Equipment)

INDUSTRIAL APPLICATIONS OF REFRIGERATION

Chapter 33. **Ice Manufacture** (TC 10.2, Automatic Icemaking Plants and Skating Rinks)
34. **Ice Rinks** (TC 10.2)
35. **Concrete Dams and Subsurface Soils** (TC 10.1)
36. **Refrigeration in the Chemical Industry** (TC 10.1)

LOW TEMPERATURE APPLICATIONS

Chapter 37. **Environmental Test Facilities** (TC 9.2, Industrial Air Conditioning)
38. **Cryogenics** (TC 10.4, Ultra-Low Temperature Systems and Cryogenics)
39. **Low Temperature Metallurgy** (TC 10.4)
40. **Biomedical Applications** (TC 10.4)

GENERAL

Chapter 41. **Codes and Standards**

ADDITIONS AND CORRECTIONS

1987 HVAC Systems and Applications
1988 Equipment
1989 Fundamentals

INDEX

Composite index to the 1987 HVAC Systems and Applications, 1988 EQUIPMENT, 1989 FUNDAMENTALS, and 1990 REFRIGERATION Systems and Applications volumes.

CONTRIBUTORS

In addition to the Technical Committees, the following individuals contributed significantly to this volume. The appropriate chapter numbers follow each contributor's name.

William J. Buck (1)
Engineered Refrigeration Systems, Inc.

Sam P. Soling (2)
St. Onge, Ruff and Associates, Inc.

R.A. Jones (3)
Nordyne

W. Barbier (3)
Concept Technology Inc.

J.D. Gaines (3)
Hussmann Co.

William F. Stoecker (4)
University of Illinois

Donald K. Miller (4, 5)
MDK Engineering Corporation

Ronald A. Cole (5, 27)
Ron A. Cole and Associates, Inc.

S.R. Dunne (7)
UOP Research and Molecular Sieve Technology

Kenneth W. Manz (7)
Robinair Division, SPX Corporation

K.S. Sanvordenker (8)
Tecumseh Products

John R. Polky (8)
Calumet Refining Company

Carl F. Speich (8)
The Trane Company

Ward D. Wells (8)
E.I. duPont de Nemours

Ivor N. Morgan (9)

Douglas R. Maschin (9)
Air Products and Chemicals

C.D. Baird (11)
University of Florida—Gainesville

Ronald P. Vallort (12)
Epstein Engineering, Inc.

Benjamin C. Smith (14)
National Sea Products Limited

Rick Vowels (15)
Flav-O-Rich

John M. Harvey (16)
USDA (retired)

William M. Miller (17)
University of Florida

G. Eldon Brown (17)
Florida Department of Citrus

Robert E. Hardenburg (18, 26)
USDA (retired)

Ivor N. Morgan (20, 21)

Harold R. Bolin (22)
Western Regional Research Center (USDA)

Norman C. Grisewood (23)
F.X. Matt Brewing Company

Evans J. Lizardos (23)
Lizardos Engineering Associates, P.C.

George J. Lizardos (23)
Lizardos Engineering Associates, P.C.

Richard R. Nelson (23)
Canandaigua Wine Company

N.F. Norkaitis (23)
Anheuser-Busch Companies, Inc.

Peter J. Hoey (24)
Emtek, Inc.

Daniel J. Schlachter (25)
Hixson

George R. Smith (27)
Industrial Air Conditioning Co.

David H. Taylor (28)

Clifford A. Woodbury, III (28, 29)
LTK Engineering Services

E. Thomas Harley (29)
LTK Engineering Services

Michael Levin (29)
LTK Engineering Services

W. F. Anderson (30)
General Marine Refrigeration Corporation

W.J. Hannett (30)
Carrier Transicold Company

D.W. Sergius (30)
Polar Industries, Ltd.

James J. Bushnell (31)
General Dynamics, Convair

Elio Battista (32)
Hussmann Refrigeration

T.A. Niblack, Jr. (32)
Furr's Inc. Planning

Martin L. Timm (34)
Henry Vogt Machine Co.

Frederick H. Belz (35)
Kaiser Engineers Hanford

Brent Allardyce (36)
Startec Refrigeration Services Ltd.

Earl M. Clark (36)
E.I. Dupont

W.L. Holladay (37)
Consulting Engineer (retired)

W.R. Bruskrud (37)
Bemco, Inc.

R.W. Haines (37)
Consulting Engineer

Ram K. Shukla (38)
Westinghouse Electric Corporation West Valley Nuclear Service Company Division

Hsien-Sheng Pei (40)
Digital Equipment Corporation

ASHRAE HANDBOOK COMMITTEE
Benford G. Liebtag, Chairman

1990 Refrigeration Volume Subcommittee: **Peter J. Hoey**, Chairman

Jerome Gaffney   **John M. Harvey**   **Donald K. Miller**   **Arnett S. Smiley**

ASHRAE HANDBOOK STAFF

Robert A. Parsons, Editor   **Claudia Forman**, Associate Editor

Andrea S. Andersen, Editorial Assistant

Ron Baker, Production Manager

Gene A. Sweigart and **Nancy F. Thysell**, Typography   **Susan M. Boughadou**, Graphics

W. Stephen Comstock, Publishing Director

ASHRAE TECHNICAL COMMITTEES AND TASK GROUPS

SECTION 1.0—FUNDAMENTALS AND GENERAL
1.1 Thermodynamics and Psychrometrics
1.2 Instruments and Measurements
1.3 Heat Transfer and Fluid Flow
1.4 Control Theory and Application
1.5 Computer Applications
1.6 Terminology
1.7 Operation and Maintenance Management
1.8 Owning and Operating Costs
1.9 Electrical Systems

SECTION 2.0—ENVIRONMENTAL QUALITY
2.1 Physiology and Human Environment
2.2 Plant and Animal Environment
2.3 Gaseous Air Contaminants and Gas Contaminant Removal Equipment
2.4 Particulate Air Contaminants and Particulate Contaminant Removal Equipment
2.5 Air Flow Around Buildings
2.6 Sound and Vibration Control
TG Safety
TG Halocarbon Emission
TG Seismic Restraint Design

SECTION 3.0—MATERIALS AND PROCESSES
3.1 Refrigerants and Brines
3.2 Refrigerant System Chemistry
3.3 Contaminant Control in Refrigerating Systems
3.4 Lubrication
3.5 Sorption
3.6 Corrosion and Water Treatment
3.7 Fuels and Combustion

SECTION 4.0—LOAD CALCULATIONS AND ENERGY REQUIREMENTS
4.1 Load Calculation Data and Procedures
4.2 Weather Information
4.3 Ventilation Requirements and Infiltration
4.4 Thermal Insulation and Moisture Retarders
4.5 Fenestration
4.6 Building Operation Dynamics
4.7 Energy Calculations
4.8 Energy Resources
4.9 Building Envelope Systems
TG Indoor Environment Calculations
TG Cold Climate Design

SECTION 5.0—VENTILATION AND AIR DISTRIBUTION
5.1 Fans
5.2 Duct Design
5.3 Room Air Distribution
5.4 Industrial Process Air Cleaning (Air Pollution Control)
5.5 Air-to-Air Energy Recovery
5.6 Control of Fire and Smoke
5.7 Evaporative Cooling
5.8 Industrial Ventilation
5.9 Enclosed Vehicular Facilities

SECTION 6.0—HEATING EQUIPMENT, HEATING AND COOLING SYSTEMS AND APPLICATIONS
6.1 Hydronic and Steam Equipment Systems
6.2 District Heating and Cooling
6.3 Central Forced Air Heating and Cooling Systems
6.4 In Space Convection Heating
6.5 Radiant Space Heating and Cooling
6.6 Service Water Heating
6.7 Solar Energy Utilization
6.8 Geothermal Energy Utilization
6.9 Thermal Storage

SECTION 7.0—PACKAGED AIR-CONDITIONING AND REFRIGERATION EQUIPMENT
7.1 Residential Refrigerators, Food Freezers and Drinking Water Coolers
7.2 Beverage Coolers
7.5 Room Air Conditioners and Dehumidifiers
7.6 Unitary Air Conditioners and Heat Pumps

SECTION 8.0—AIR-CONDITIONING AND REFRIGERATION SYSTEM COMPONENTS
8.1 Positive Displacement Compressors
8.2 Centrifugal Machines
8.3 Absorption and Heat Operated Machines
8.4 Air-to-Refrigerant Heat Transfer Equipment
8.5 Liquid-to-Refrigerant Heat Exchangers
8.6 Cooling Towers and Evaporative Condensers
8.7 Humidifying Equipment
8.8 Refrigerant System Controls and Accessories
8.9 Valves
8.10 Pumps and Hydronic Piping
8.11 Electric Motors—Open and Hermetic
TG Unitary Combustion-Engine-Driven Heat Pumps

SECTION 9.0—AIR-CONDITIONING SYSTEMS AND APPLICATIONS
9.1 Large Building Air-Conditioning Systems
9.2 Industrial Air Conditioning
9.3 Transportation Air Conditioning
9.4 Applied Heat Pump/Heat Recovery Systems
9.5 Cogeneration Systems
9.6 Systems Energy Utilization
9.7 Testing and Balancing
9.8 Large Building Air-Conditioning Applications
TG Building Commissioning
TG Laboratory Systems
TG Tall Buildings

SECTION 10.0—REFRIGERATION SYSTEMS
10.1 Custom Engineered Refrigeration Systems
10.2 Automatic Icemaking Plants and Skating Rinks
10.3 Refrigerant Piping
10.4 Ultra-Low Temperature Systems and Cryogenics
10.5 Refrigerated Distribution and Storage Facilities
10.6 Transport Refrigeration
10.7 Commercial Food Display and Storage Equipment
10.8 Refrigeration Load Calculations

SECTION 11.0—REFRIGERATED FOOD TECHNOLOGY AND PROCESSING
11.1 Meat, Fish and Poultry Products
11.3 Dairy Products
11.5 Fruits, Vegetables and Other Products
11.6 Prepared Food Products
11.7 Fermented and Carbonated Beverages
11.9 Thermal Properties of Food

PREFACE

The Refrigeration Handbook covers refrigeration equipment or systems in a particular application, process, or cold storage facility and describes current designs for specific applications. Also covered in this book are industrial applications of refrigeration and an introduction to low temperature refrigeration. While the Refrigeration Handbook is primarily a reference for the practicing engineer, it is also a useful reference for anyone involved in the storage and handling of most foods.

Volunteers, working in conjunction with the society's Handbook Committee and Technical Committees, spend thousands of hours revising the Handbook series. ASHRAE sponsored research has also been an important source of technical information for the Handbook series. Independently funded by members, their colleagues, and businesses, ASHRAE research defines many of the design concepts used by the profession.

All chapters in the 1986 Refrigeration Handbook were revised to some extent for this volume. Some of the topics that were significantly revised include:

- liquid overfeed systems, halocarbon refrigerants, ammonia systems, and secondary coolants (Chapters 2, 3, 4, and 5)
- fish products, dairy products, breweries, and retail food store refrigeration (Chapters 14, 15, 23, and 32)
- truck, rail, air, and marine transport (Chapters 28, 29, 30, and 31)

New topics include:

- Chapter 27, "Refrigeration Load Calculations" (previously in the 1985 Fundamentals Handbook)
- information on stabilization methods of permanently frozen earth (Chapter 35)

In addition, the data in the food refrigeration chapters were reviewed and information from the previous chapter on supplements to refrigeration was integrated into appropriate chapters for specific commodities.

A chapter listing additions and corrections for the 1987, 1988, and 1989 volumes precedes the index. This Handbook is published in two editions. One edition contains Inch-Pound (I-P) units of measurement and the other, the International System of Units (SI).

The Handbook Committee welcomes reader input. If you have suggestions and comments on improving a chapter or would like more information on helping review a chapter, write to Handbook Editor, ASHRAE, 1791 Tullie Circle, Atlanta, GA 30329.

Robert A. Parsons
Handbook Editor

CHAPTER 1

ENGINEERED REFRIGERATION SYSTEMS (INDUSTRIAL DESIGN PRACTICES)

Refrigeration Load 1.1
System Selection 1.2
Refrigerants 1.3
Equipment 1.3
Accessory Equipment 1.5
Piping 1.8
Control 1.9
Moisture in Refrigerant Circuit 1.10
Insulation and Vapor Barriers 1.10

REFRIGERATION is the process of moving heat from one location to another by use of refrigerant in a closed refrigeration cycle. Oil management, gas and liquid separation, subcooling, superheating, and piping of refrigerant liquid, gas, and two-phase flow are all part of the refrigeration discipline. Application areas for the use of refrigeration include air conditioning, commercial refrigeration, and industrial refrigeration.

For the purposes of this chapter, refrigeration is the use of mechanical refrigerating machinery for applications other than human comfort. Engineered refrigeration systems are unique in that they require specific engineering analysis of all components to select balanced elements to accomplish the desired results.

Human comfort air-conditioning systems usually work over a small, well-defined range of evaporating temperatures, typically 35 to 55 °F. Because of these limited temperatures and the large market for air-conditioning equipment, manufacturers can standardize their equipment and package it for volume production at lower cost. This is particularly true in water chilling and unitary air-conditioning equipment.

Conversely, custom-engineered refrigeration systems often have design conditions that span a wide range of evaporating and condensing temperatures, e.g., (1) an industrial process heat pump with 100 °F evaporator and 230 °F condensing temperature; (2) a candy storage requiring 60 °F dry bulb with precise humidity control; (3) a beef chill room at 28 to 30 °F with high humidity; (4) a distribution warehouse requiring multiple temperatures for storage of ice cream, frozen food, meat, produce, docks; (5) cooling combustion air to −90 °F for jet engine testing; or (6) a chemical process requiring multiple temperatures ranging from +60 to −120 °F.

Typical considerations for a refrigeration system include some of the following:

- Year-round operation regardless of outdoor ambient conditions
- Possible wide load variations (0 to 100% capacity) during short periods without seriously disrupting the required temperature levels
- Frost control for continuous performance applications
- Oil management for different refrigerants under varying load and temperature conditions
- A wide choice of heat exchange methods, e.g., dry expansion, liquid overfeed, or flooded feed of the refrigerants, and the use of secondary coolants such as salt brines, alcohol, and glycol
- System efficiency, maintainability, and operating simplicity
- Operating pressures and pressure ratios that might require multistaging, cascading, and so forth

The preparation of this chapter is assigned to TC 10.1, Custom Engineered Refrigeration Systems.

REFRIGERATION LOAD

An analysis of the refrigeration load is vital to proper equipment selection and satisfactory system performance. All load components and often the time profile of the load components must be considered.

Typical factors to consider are as follows:

1. Heat leakage of either latent or sensible heat into the space or product. This flow is a function of the temperature difference and insulating efficiency of the container walls, including insulation and any heat introduced by ventilation or infiltration.
2. Product load, which is the heat that must be removed to bring the product to the desired state. This load includes both sensible and latent heat associated with a change of state.
3. Internal sensible load can include heat from motors (including evaporator fan motors), lights, and other heat sources.

A plot of load versus time is important in selecting equipment that will function well at minimum load as well as at peak load. Minimum load conditions are often overlooked in design so that severe operating problems such as inefficiency, short-cycling, oil migration, shortage of hot gas for defrost, or other adverse conditions arise. Often small- and large-capacity components must be selected and properly controlled to match operating conditions. Industrial processes are particularly prone to abrupt interruptions, giving from 100 to 0% load and then back to 100% over a short time. Improperly designed systems will not function well under these conditions.

Maximum and minimum ambient temperatures during varying loads must be considered in the design of a system with capacity reduction features, particularly when selecting the condenser.

In food freezing or storage, the effect of frost accumulation on heat transfer surface efficiency must be considered and sufficient defrost time allocated. If exact conditions have to be continuously maintained, dual evaporators may have to be used and alternated to permit defrosting. When selecting evaporators for air cooling (as in cold storage warehouses), the defrost system should be based on the following tabulation of cooling hours:

Defrost Method	Maximum Hours of Operation Per Day
Air (above 34 °F)	16 to 18
Water or brine	20 to 22
Hot gas	20 to 22
Electrical	20 to 22

This table is based on average frost conditions. For higher frost conditions and lower evaporating temperatures, more time must be allowed for defrosting.

1990 Refrigeration Handbook

During defrost, heat is absorbed by the frost, the metal of the air-cooling unit, and the room air. When the system returns to cooling, all of the heat added to the room (except that in the condensate that was drained away) must be added back to the load because it must also be removed. The designer and the owner must agree on an economical balance of system capacity versus load.

SYSTEM SELECTION

In selecting an engineered refrigeration system, several design decisions that are not part of typical HVAC system design must be considered. Decisions include whether to use: (1) single-stage compression, (2) economized compression, (3) compound compression, (4) a cascade system, (5) dry expansion feed, (6) flooded feed, (7) liquid recirculation feed, and (8) secondary coolants (brines).

Single-Staged Systems

The basic single-stage system consists of evaporators, a compressor, a condenser, a refrigerant receiver (if used), and a refrigerant control device (expansion valve, float, etc.). Chapter 1 of the 1989 ASHRAE *Handbook—Fundamentals* discusses the compression refrigeration cycle. Piping practices and accessories are discussed in Chapters 2 through 4.

Economized Systems

Economized systems are frequently used with centrifugal or rotary screw compressors. Figure 1 shows an arrangement of the basic components. Subcooling the liquid refrigerant before it reaches the evaporator reduces its enthalpy, resulting in a higher net refrigerating effect. Economizing is beneficial since the vapor generated during the subcooling is taken in by the compressor partway through its compression cycle and must be compressed only from the inlet port pressure (which is higher than suction pressure) to the discharge pressure. This produces additional refrigerating capacity with less increase in unit energy input. Economizing is most dramatic at high-pressure ratios with refrigerants having high mass flows and high liquid specific heat (mostly halocarbons), but it is also useful with ammonia. Under certain conditions, economizing can nearly provide the operating efficiency of a two-staged compound system, with less complexity and simpler maintenance.

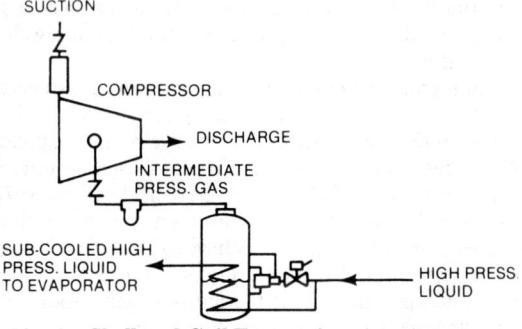

Fig. 1 Shell and Coil Economizer Arrangement

Multistage Systems

A system with more than one stage of compression can be either a compound or cascade system. As pressure ratios increase, single-staged systems encounter problems including: (1) high discharge temperatures causing the oil to deteriorate; (2) loss of volumetric efficiency as high pressure leaks back to the low-pressure side through compressor clearances; and (3) excessive stresses on compressor moving parts.

For these reasons, manufacturers usually limit the maximum pressure ratios for multicylinder reciprocating machines to

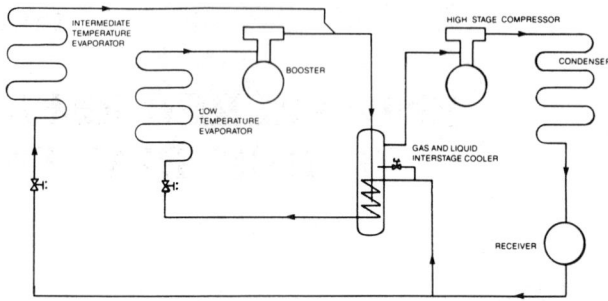

Fig. 2 Two-Stage Compound System with Side Load Evaporator at Intermediate Pressure

approximately 9, and for screw compressors to about 19. (Screw compressors usually incorporate some form of oil cooling.) When the overall system pressure ratio (absolute discharge pressure divided by absolute suction pressure) begins to exceed these limits for a particular refrigerant and system, the pressure ratio occurring across the compressor must be reduced. This is usually accomplished by multistaging. A properly designed two-staged system exposes each of the two compressors to a pressure ratio equal to about the square root of the overall pressure ratio. In a three-staged system, each compressor is exposed to a pressure ratio equal to about the cube root of the overall ratio.

Another advantage to multistaging is that successively subcooling the liquid at each stage of compression increases the overall system operating efficiency. Additionally, multistaging can be used to accommodate multiple side loads at different suction pressures and temperatures in the same central refrigeration system. Figure 2 shows a compound system with a side load at the interstage pressure.

Compound Systems

A compound system is a multistage application in which compressors are interconnected in series in the same refrigerant system. In some cases, two or more stages of compression can be contained in a single compressor such as in a multiwheel centrifugal or an internally compounded reciprocating compressor as shown in Figure 3. In these units, one or more cylinders are isolated from the others so that they can act as an independent stage of compression.

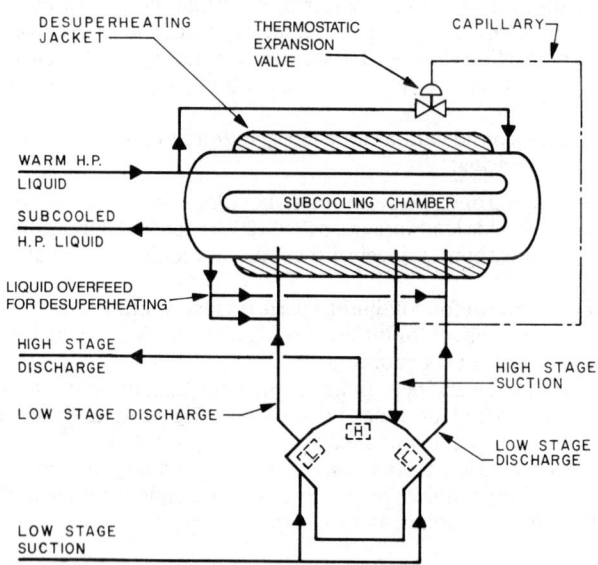

Fig. 3 Internally-Compounded Two-Stage Compressor with Intercooler

Engineered Refrigeration Systems (Industrial Design Practices)

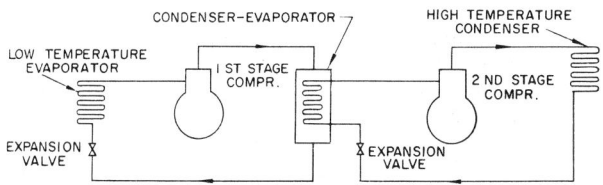

Fig. 4 Basic Elements of a Cascade System

Cascade Systems

A cascade system is a multistaged application in which different refrigerants are used in the various stages. In this case, the evaporator of the higher temperature stage is the condenser for the lower temperature stage. Figure 4 shows an elementary cascade system.

The cascade system is most often used where (1) the range of pressures from the evaporating temperature to the condensing temperature of a single refrigerant is greater than commercial compressors or vessels can handle, or (2) operation of a cascade system is more economical because it requires less displacement of the low-stage compressor and smaller line sizes. The reason for this is that the refrigerants commonly used in the low-temperature stage of a cascade system have much higher suction pressures and densities at a given temperature than the refrigerants used in most conventional systems. A cascade system also reduces the problem of balancing the inventory of oil between high- and low-stage compressors. An inherent inefficiency of the cascade system is the temperature difference required between the high-stage evaporator and the low-stage condenser. This inefficiency is offset to some extent by the proper choice of refrigerants.

Each circuit in a cascade system may be either single-stage or compound. An example is the use of R-13 in a single-stage compressor with a condenser that rejects heat to the evaporator of a two-staged R-22 system.

In the following sections, multistage refers to both compound and cascade systems.

REFRIGERANTS

Many refrigerants can be used in multistage systems. Figure 5 shows an approximate range of temperatures within which the more common refrigerants are used. The most frequently used refrigerants shown are Refrigerants 12, 13, 13B1, 22, 502, 503, and hydrocarbons will probably decline due to their ozone depletion potential. Industrial systems down to about −55 °F frequently use ammonia or R-22 in single- or two-staged systems. R-503 is commonly used with systems in the −100 to −150 °F range. R-503 is commonly used in the low stage, whereas R-22, 12, or 502 are used in the high stage.

R-503 has a higher suction pressure and density at a given temperature than R-13. For this reason, at −120 °F suction and −20 °F condensing, R-503 has approximately 57% more capacity for a given size compressor than R-13. Also R-503 maintains a positive suction pressure at lower temperatures. R-503 has 0 psig suction at approximately −125 °F, while R-13 is near 0 psig at about −110 °F. The disadvantage of R-503 is that it has a higher operating pressure at ambient temperature—nearly 560 psig at 60 °F for R-503 versus 402 psig for R-13. These refrigerants require special design and handling with an expansion tank to hold the refrigerant during periods of shutdown or power failure.

Factors that should be considered in selecting a particular refrigerant include operating and standby pressures, operating efficiencies, system size, size of available compressors, toxicity, flammability, cost, and qualifications of operating and maintenance personnel.

If a refrigerant is extremely flammable (*e.g.*, propane), it should be limited to pressures above atmospheric to prevent drawing air into the system, which would possibly create an explosion hazard. Special seal designs can overcome this problem by maintaining a slight positive pressure in the seal area.

For safety and minimum design criteria for common refrigerants, refer to the *Safety Code for Mechanical Refrigeration* (ANSI/ASHRAE 15-1989) or to IIAR Bulletin 109, *Minimum Safety Criteria for a Safe Ammonia Refrigeration System* (1988).

EQUIPMENT

Compressors

Various compressors available for multistage applications include the following:

Reciprocating
  Single-stage (booster or high-stage)
  Internally compounded
  Oil-free/dry cylinder
Rotary vane (primarily low-stage)
Helical rotary screw (booster or high-stage)
Centrifugal
  Single-stage
  Multiwheel

The reciprocating compressor is the most common compressor used in small, single-staged or multistaged systems. The screw compressor is used in small and large systems, both in single- and multistaged systems. Various combinations of compressors may be used in multistage systems. Rotary vane, screw, or centrifugal compressors are frequently used for the low-pressure stage, where large volumes of gas must be moved. The high-pressure stage usually uses a reciprocating or screw compressor.

Many factors must be considered in selecting a compressor, including:

- system size and capacity requirements
- location, such as indoor or outdoor installation at ground level or on the roof
- equipment noise
- part- or full-load operation
- winter and summer operation
- pulldown time required to reduce the temperature to desired conditions for either initial or normal operation. The temperature must be pulled down frequently for some applications for a process load, while a large cold storage warehouse may require pulldown only once in its lifetime.

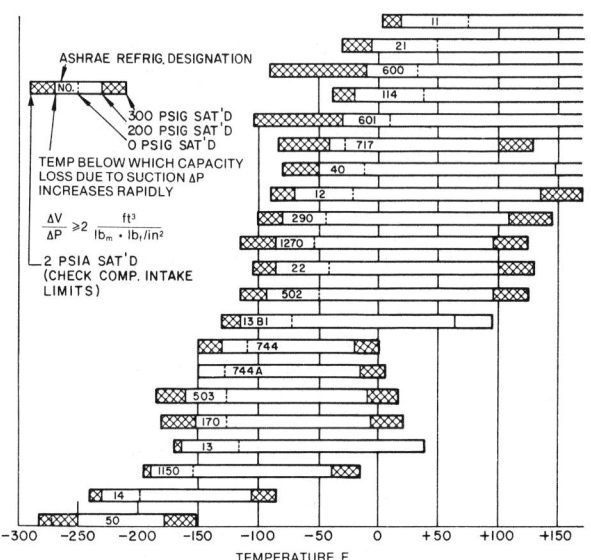

Fig. 5 Equivalent Temperature Limitations for Reciprocating Compressors with Several Refrigerants

Oil Cooling

Oil for screw compressors is usually cooled by one of three methods: (1) direct injection of the liquid refrigerant into the compressor to cool the discharge gas; (2) external cooling by a heat exchanger using water or brine; or (3) external cooling using refrigerant (dry expanded or thermosyphon). Thermosyphon oil cooling uses refrigerant from the condenser in a flooded shell-and-tube heat exchanger to cool the oil. Direct liquid refrigerant injection may shorten compressor life by thinning out the oil, reduce efficiency, and contribute to increased oil carryover through the oil separator. Water-cooled oil coolers are subject to corrosion and freezing.

Compressor Drivers

The correct electric motor(s) size(s) for a multistaged system is determined by the pulldown load. When the final low-stage operating level is $-100\,°F$, the pulldown load can be three times the operating load. Positive displacement compressor motors are usually selected for about 150% of operating power requirements for 100% load. The compressor's unloading mechanism can be used to prevent motor overload. Electric motors should not be overloaded, even when a service factor is indicated, to avoid overloading in low-voltage conditions.

Gasoline, natural gas, propane, or diesel internal combustion engines are used when electricity is unavailable, or if the selected energy source is cheaper. Sometimes they are used in combination with electricity to reduce peak demands. The power output of a given size engine can vary as much as 15% depending on the fuel selected.

The standard power rating of an engine is the absolute maximum, not the recommended power available for continuous use. Also, torque characteristics of internal combustion engines and electric motors differ greatly. The proper engine selection is at 75% of its maximum power rating. For longer life, the full-load speed should be at least 10% below maximum engine speed.

Internal combustion engines, in some cases, can reduce operating cost below that for electric motors. Disadvantages include (1) higher initial cost of the engine; (2) additional safety and starting controls; (3) higher noise levels; (4) larger space requirements; (5) air pollution; (6) requirement for heat dissipation; (7) higher maintenance costs; and (8) higher levels of vibration than with electric motors.

Steam turbine drives for refrigerant compressors are usually limited to very large installations where steam is already available at moderate to high pressure. In all cases, a torque analysis is required to determine what coupling must be used to dampen out any pulsations transmitted from the compressor. For optimum efficiency, a turbine should operate at a high speed that must be geared down for reciprocating and most rotary compressors. Since the gear reducer as well as the turbine cannot tolerate a pulsating backlash from the driven end, torque analysis and special couplings are essential.

Advantages of turbines include variable speed for capacity control and low operating and maintenance costs. Disadvantages include higher initial costs and possible high noise levels. The turbine must be started manually to bring the turbine housing up to temperature slowly and to prevent excess condensate from entering the turbine. Chapter 32 of the 1988 ASHRAE *Handbook—Equipment* has further information on compressor drives.

Condensers

Ordinarily, the condensers used in multistage systems are the same as those used in single-stage systems. However, since the heat rejection per unit capacity of refrigeration varies widely depending on operating conditions, a condenser should not be selected on the basis of refrigeration capacity, but on the basis of total heat rejection. The cascade condenser, which operates between two separate refrigerant circuits, is different in that it must function as both an evaporator for the high side and a condenser for the low side. This unit must conform to both condenser and evaporator design practices.

Often the heat rejected at the start of pulldown is several times the amount rejected at normal, low-temperature operating conditions. Some means, such as low-temperature compressor unloading, can be used to limit the maximum amount of heat rejected during pulldown. If the condenser is not sized for pulldown conditions and compressor capacity cannot be limited during this period, condensing pressure might increase enough to shut down the system. In other words, a system that works satisfactorily and economically at the design operating low temperature may be designed, but, without external aid, it cannot reach the condition under normal ambient conditions. This problem is discussed in the section on controls.

Evaporators

Several types of evaporators can be used in multistage systems. A tubular, direct expansion evaporator returns oil easily and requires the smallest refrigerant charge. Where direct expansion is impractical, a flooded system or a recirculated system may be used, but these methods compound oil return problems. For more details, see the section on oil return.

Some problems that can become more acute in low-temperature systems than in high-temperature systems include oil transport properties, loss of capacity caused by static head from the depth of the pool of liquid refrigerant in the evaporator, deterioration of refrigerant boiling heat transfer coefficients, and higher specific volumes for the vapor.

The effect of pressure losses in the evaporator and suction piping is more acute in low-temperature systems because of the large change in saturation temperatures and specific volume in relation to pressure changes at these conditions. Systems that operate near zero absolute pressure are particularly affected by pressure loss. For example, with R-12 and R-22 at 20 psia suction and 80 °F liquid feed temperature, a 1 psi loss increases the volume flow rate by about 5%. At 5 psia suction and 20 °F liquid feed temperature, a 1 psi loss increases the volume flow rate by about 25%.

The depth of the pool of boiling refrigerant in a flooded evaporator causes a liquid head or static pressure that is exerted on the lower part of the heat transfer surface. Therefore, the saturation temperature at this surface is higher than the pressure in the suction line, which is not affected by the static head. Although tubular dry expanded evaporators do not have appreciable static liquid head, gas pressure drops from the inlet to the outlet of the evaporator create a velocity head that causes a similar condition.

The liquid depth penalty for the evaporator can be eliminated if the pool of liquid is below the heat transfer surface and a refrigerant pump sprays the liquid over the surface. Of course, the pump energy is an additional heat load to the system, and more refrigerant must be used to provide the Net Positive Suction Head (NPSH) required by the pump. The pump is also an additional item to be maintained.

Another type of low-temperature evaporator is the flash cooler in which liquid refrigerant is cooled by boiling off some vapor. The remaining cold liquid can then be pumped from the flash cooler to the evaporator. There it is either top or bottom fed at a rate greater than the evaporation rate to ensure wetting of the entire evaporator surface for maximum heat transfer without an appreciable static head penalty. This liquid overfeed system is frequently used in large refrigerated warehouses with many evaporators.

Another less frequently used system pumps the liquid refrigerant as a secondary coolant from the flash cooler at low temperature. As the coolant passes through a secondary cooler or coil, heat transfers to it from the material being cooled. The liquid tem-

Engineered Refrigeration Systems (Industrial Design Practices)

perature rises to develop a temperature range, but because pressure is maintained sufficiently above saturation by the liquid pump, the coolant does not evaporate until it returns via a restriction to the flash cooler. Sufficient refrigerant must be circulated to accommodate the temperature range. The flash cooler in this system is an accumulator receiver similar to that used in a liquid overfeed system, except that no excess refrigerant is fed to the remote heat transfer surface.

In both types of liquid recirculation systems, the cold liquid can be moved by mechanical pumps or by pressure from the compressor discharge.

ACCESSORY EQUIPMENT

Receivers

Almost all large systems have a refrigerant receiver to store temporary excess quantities of refrigerant when parts of the system are pumped down for repairs, some evaporators are defrosted, or some evaporators are shut off by their thermostat. The receiver should be large enough to hold the entire system charge, but in large systems this may not be possible. To size the receiver, determine whether the entire system will ever be pumped down at the same time. In cases where systems have grown too large for the originally installed receiver, a separate storage tank may be installed to hold the excess refrigerant when a large portion of the system is shut down. If additional receiver capacity is needed during normal operation, extreme caution should be exercised in the design. Designers usually remove the inadequate receiver and replace it with a larger one rather than install an additional receiver in parallel. This procedure is best because even slight differences in piping pressure or temperature can cause the refrigerant to migrate to only one receiver and not to the other.

Subcoolers

As previously mentioned, to take full advantage of the two-stage system, the refrigerant liquid should be cooled to a temperature near the interstage temperature to reduce the amount of flash gas handled by the low-stage compressor. The net result is a reduction in total system power requirements. The amount of gain from cooling to near interstage conditions varies among refrigerants.

Figure 6 illustrates an open or flash type cooler. This is the simplest and least costly type of cooler, which has the advantage of cooling liquid to the saturation temperature of the interstage pressure. Its disadvantage is that the pressure of the cooled liquid is reduced to interstage pressure leaving less pressure available for liquid transport. Although the liquid temperature is reduced, the pressure is correspondingly reduced, and the expansion device controlling flow to the cooler must be large enough to pass all of the liquid refrigerant flow. Failure of this valve could allow a large flow of liquid to the upper stage compressor suction, which could seriously damage the compressor.

Liquid from a flash cooler is saturated, and liquid from a cascade condenser usually has little subcooling. In both cases, the liquid temperature is usually lower than the temperature of the surroundings. Thus it is important to avoid heat input and pressure losses that would cause flash gas to form in the liquid line to the expansion device or to recirculating pumps. The cold liquid lines should be insulated, as expansion devices are usually designed to feed liquid, not vapor.

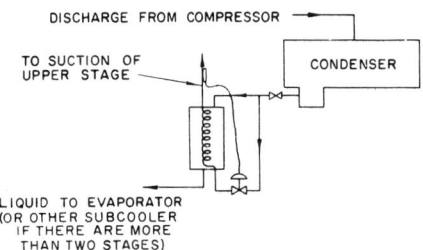

Fig. 7 Exchange-Type Subcooler

Figure 7 shows the closed or heat exchanger type of subcooler. It should have sufficient heat transfer surface to transfer heat from the liquid to the evaporating refrigerant with a small final temperature difference. The pressure drop should be small, so that full pressure is available for feeding liquid to the expansion device at the low-temperature evaporator. The subcooler liquid control valve should be sized to supply only the quantity of refrigerant required for the subcooling. This prevents a tremendous quantity of liquid from flowing to the upper stage suction in the event of a valve failure.

Gas Intercooler (Desuperheater)

The discharge gas from a low-stage compressor is normally too hot for the suction of a high-stage compressor. The pressure-enthalpy diagram shows that the booster discharge should be cooled to within 10 to 25 °°F of saturation for the suction gas to the high-stage compressor. The most efficient intercooler would reject heat of the low-stage discharge to the environment instead of passing it through to the high-stage compressor, but this is usually not possible.

One practice directly injects liquid refrigerant into the low-stage discharge gas. Care must be taken to ensure that all of the injected liquid is evaporated in the process or that a separation chamber is available to remove excess liquid from the vapor before it reaches the high-stage compressor. Figure 8 shows an arrangement that promotes thorough mixing of the liquid and vapor.

Combined Subcooler and Desuperheater

In most cases, the subcooler and desuperheater are combined into a single unit with one expansion device. The subcooling function requires evaporating liquid on the surface of a pipe carrying the liquid to be subcooled, and the desuperheating function requires the liquid to be in contact with the gas to be desuperheated. The design of the unit must include these features along with protection from liquid carryover into suction, which would result in immediate compressor damage.

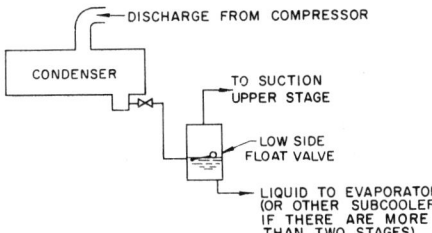

Fig. 6 Flash-Type Cooler

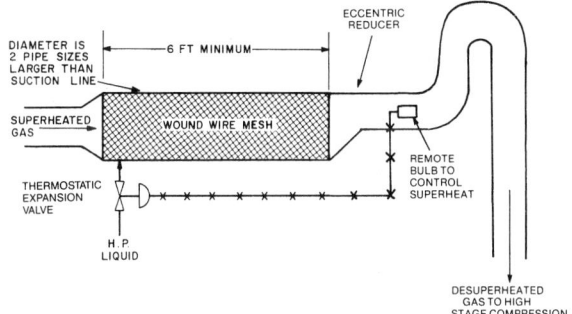

Fig. 8 Mixing Desuperheater for Interstage Gas

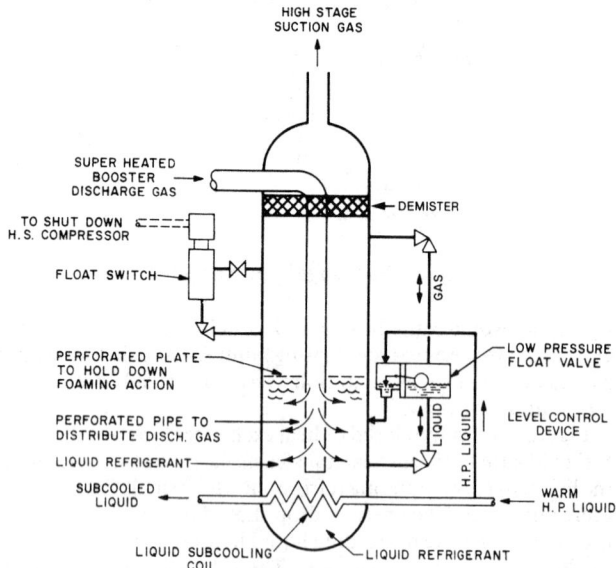

Fig. 9 Intercooler

Vertical shell and coil intercoolers with float valve feed (Figure 9) perform well in many applications using ammonia as the refrigerant. Halocarbon refrigerants require a different design to handle the oil. The valve and shell must be sized properly to separate the liquid from the vapor that is returning to the high-stage compressor. The superheated gas inlet pipe should extend below the liquid level and have perforations or slots to distribute the gas evenly in small bubbles. The addition of a perforated baffle across the area of the vessel slightly below the liquid level protects against violent surging. A float switch that shuts down the high-stage compressor when the liquid level gets too high should always be used with this type of intercooler in case the feed valve fails to control properly.

Heat Exchanger (Liquid to Suction)

Chapter 3 discusses liquid to suction heat exchangers. This section pertains especially to low-temperature applications.

For some compressors, it may be necessary to raise the temperature of the low-stage suction gas above a predetermined point to maintain lubricity of the oil in the compressor or, in the case of very low-temperature systems, to keep the metal of the compressor casting from becoming brittle. Other compressor components such as seals, bearings, valves, or rings can be damaged by overheating. This is common in low-temperature compressors because there is minimal mass flow of refrigerant to carry heat away. Some form of external cooling may be required.

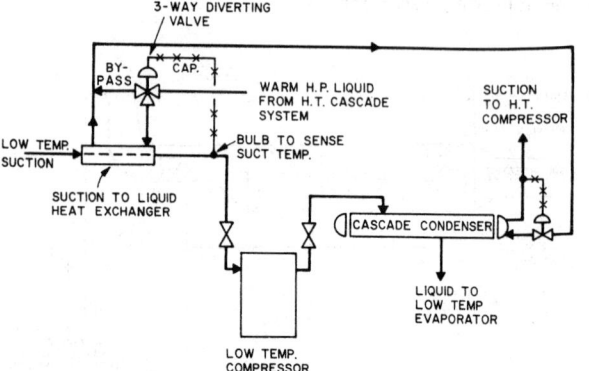

Fig. 10 Application of Heat Exchanger to Cascade System

A good heat source for the low-temperature suction of a cascade system is the warm, high-pressure liquid from the high-temperature system as shown in Figure 10. A three-way temperature actuated valve can be used to control the suction temperature to the low-temperature compressor. Temperature control may be needed (1) if the low-temperature evaporator has a fluctuating load, such as in a batch operation; (2) if the full flow of warm liquid might cause the suction to become highly superheated; and (3) to prevent condensation in the cylinders during compression of hydrocarbons, such as a butane refrigerant.

The advantages of a liquid-to-suction heat exchanger must be balanced with the pressure drop in the gas as it passes through the heat exchanger. Liquid-to-suction heat exchangers should be carefully sized to handle the various operating conditions.

Oil Return Equipment

One of the most difficult problems in low-temperature refrigeration systems using halocarbon refrigerants is returning lubrication oil from the evaporator to the compressors. With the exception of most centrifugal compressors and with rarely used nonlubricated compressors, refrigerant continuously carries oil into the discharge line from the compressor. Much of this oil can be removed from the stream by an oil separator and returned to the compressor. Coalescing oil separators are far better than separators using only mist pads or baffles; however, they are not 100% effective. The oil that finds its way into the system must be managed.

Oil is immiscible in R-717 (ammonia) and separates out of the liquid easily when flow velocity is low or when temperatures are lowered. This oil can be easily drained from the system (see Chapter 4). Oil mixes well with halocarbon refrigerants at higher temperatures. As the temperature lowers, the miscibility is reduced, and some of the oil separates to form an oil-rich layer near the top of the liquid level in a flooded evaporator. If the temperature is very low, the oil becomes a gummy mass that prevents refrigerant controls from functioning, blocks flow passages, and fouls the heat transfer surfaces. Proper oil management is often the key to a properly functioning system.

In compound systems, proper design usually calls for oil separators on both the high-stage and booster compressors. A properly designed coalescing oil separator can remove almost all of the oil that is in droplet or fog form. Oil that reaches its saturation vapor pressure and becomes a vapor cannot be removed by an oil separator. Separators equipped with some means of cooling the discharge gas will condense much of the oil vapor for consequent separation. Care must be exercised, however, to be sure refrigerant is not condensed and fed back into the compressor crankcase where it can lower lubricity and cause compressor damage.

An oil separator is not used on the booster of a compound system where the intercooler is designed not to trap oil. If oil is trapped, a float arrangement is used to return oil to the booster from the high-stage separator.

In general, direct expansion and liquid overfeed system evaporators have fewer oil return problems than do flooded system evaporators since refrigerant flows continuously at good velocities to sweep oil from the evaporator. Low-temperature systems using hot-gas defrost can also be designed to sweep oil out of the circuit each time the system defrosts. This reduces the possibility of oil coating the evaporator surface and hindering heat transfer.

Flooded evaporators can promote oil contamination of the evaporator charge since they may only return dry refrigerant vapor back to the system. In ammonia systems, the oil is simply drained from the surge drum. In halocarbon systems, however, skimming systems must sample the oil-rich layer floating in the drum, a heat source must distill the refrigerant, and the oil must be returned to the compressor. Because flooded halocarbon systems can be elaborate, some designers avoid them.

Engineered Refrigeration Systems (Industrial Design Practices)

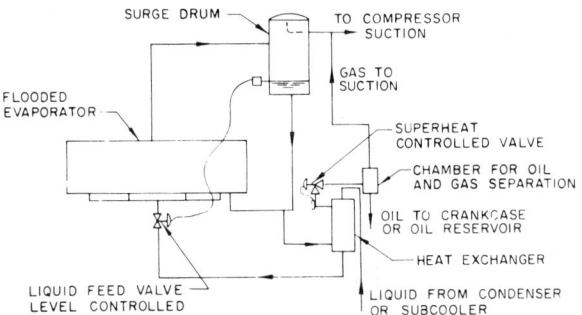

Fig. 11 Flooded Evaporator with Oil Recovery Arrangement for Refrigerants of Greater Density than Oil

Figure 11 illustrates an evaporator, a surge drum, and an oil recovery arrangement that automatically returns oil to the compressor for a halocarbon system. In this system, the level in the flooded evaporator is maintained sufficiently high to guarantee a continuous return of percolated liquid to the surge drum along with the vapor. Since the liquid is agitated by boiling, the oil cannot stratify and is mixed with the refrigerant. Under low load conditions, the oil will form an oil-rich layer at some level in the pool of refrigerant.

The refrigerant and oil mixture that falls to the bottom of the surge drum is sampled as it recirculates to the evaporator; the sample then goes to the heat exchanger for distillation. Heat from the warm liquid line boils off much of the refrigerant, leaving mostly oil to return to the compressor crankcase. This arrangement depends on gravity to flow oil back to the compressor. If system geometry does not permit this, a small oil pump may be used to move the distilled oil. In multistaged systems, the oil is sometimes drained to a central oil reservoir for feed to booster or high-stage compressors as needed. For a liquid overfeed system with multiple evaporators fed from a central recirculator, the oil recovery system can be installed on the one recirculator rather than on each evaporator.

Expansion Tanks

A cascade refrigeration system is often used for refrigeration systems operating below approximately −85 °F. One drawback to the cascade system is that a loss of cooling on the high stage causes the pressure in the low stage to increase rapidly above the normal design pressures of common refrigerant vessels. Thus, if a power failure on a cascade system using R-13 causes it to warm to 75 °F, the pressure will approach 500 psig if the safety valves do not open. Expansion tanks are used to prevent loss of valuable refrigerant in such a system.

In a cascade system charged with a critical charge of R-13, an expansion tank is piped into the system with sufficient volume, so that the entire refrigerant charge can evaporate and then expand due to superheat without creating a high-pressure condition. Figure 12 shows a typical cascade system with an expansion tank. While the system is operating, the expansion tank contains only the mass of refrigerant that corresponds to its volume of vapor at suction pressure. When the system is shut down, the volume of the expansion tank prevents an overpressure condition. The pressure relief valve from the discharge side of the cascade loop should be independent of the pressure downstream of the relief valve. On larger systems, a downstream pressure regulator should be used in the line from the expansion tank to the low-side suction. This prevents the compressor from overloading during pulldown.

The size of the expansion tank may be determined by:

$$V_f = M_s v_2 \quad \text{(where the tank must hold the total charge)}$$

or

$$V_f = M_s v_2 - V_s \quad \text{(where the tank and system hold the total charge)}$$

where

M_s = total refrigerant charge, lb
V_f = volume of expansion tank, ft^3
V_s = volume of the rest of the refrigeration system, ft^3
v_2 = specific volume of the refrigerant gas at the maximum shutdown pressure and temperature, ft^3/lb

Relief Valves

Safety valves must be provided in conformance with *The Safety Code for Mechanical Refrigeration* (ANSI/ASHRAE 15-1989) and Section VIII, Division I of the ASME *Boiler and Pressure Vessel Code*. For ammonia systems, Bulletin No. 109-1988, IIAR, *Minimum Safety Criteria for a Safe Ammonia Refrigeration System*, published by the International Institute of Ammonia Refrigeration, also addresses the subject of safety valves.

Isolated Line Sections

Sections of piping that can be isolated between hand valves or check valves can be subjected to extreme hydraulic pressures if cold liquid refrigerant is trapped in them and subsequently warmed. Additional safety valves for such areas should be provided.

Suction Accumulators

Under certain conditions, any refrigeration system can boil liquid refrigerant into the suction line in quantities sufficient to damage the compressor. A suction accumulator (also known as a knockout drum or suction trap) is used to remove the liquid. Both vertical and horizontal accumulators are used, and separation may be enhanced by the tangential nozzles or demister pads.

The liquid trapped in the accumulator must be returned to the system. The simplest method boils out the excess liquid by pass-

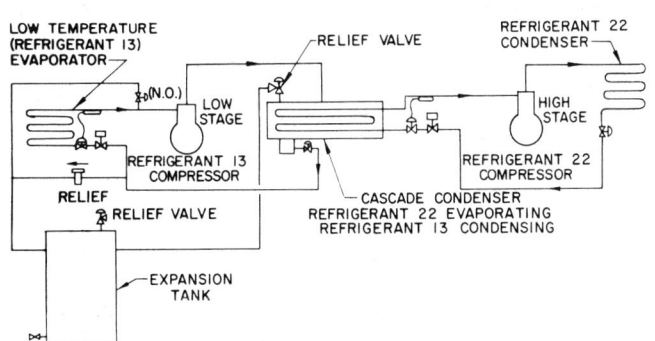

Fig. 12 Cascade System Using Refrigerants 13 and 22

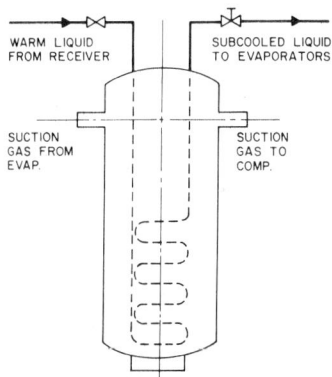

Fig. 13 Warm Liquid Coil

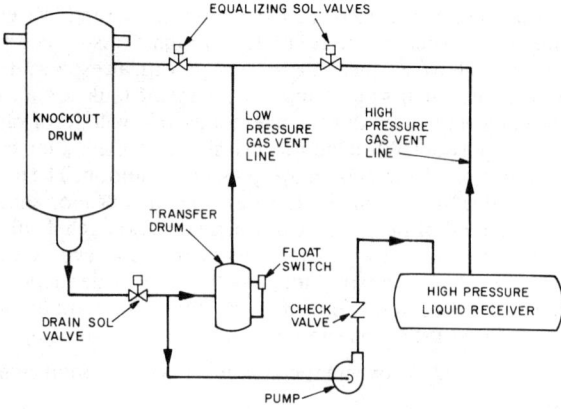

Fig. 14 Equalized Pressure Pump Transfer System

ing warm receiver liquid through a coil in the accumulator as it goes to the evaporator (Figure 13). This method has no moving parts to maintain and requires no external energy source to evaporate excess liquid. External heat sources are used only if no better solution can be found.

If proper connections on the accumulator are provided, the excess liquid can be transferred by pumps or discharge pressure to other parts of the system where it can be managed. High head pumps or special gravity configurations (Figures 14 and 15) can be used to transfer to the high-pressure receiver or to feed the liquid to the evaporators where it can be used by the system. Chapters 3 and 4 include more information on suction traps for halocarbons and ammonia, respectively.

Refrigerant Liquid Pumps

As mentioned in the section on evaporators, liquid refrigerant that has been flashed to saturation is often pumped. Positive displacement gear type or centrifugal-type pumps either of open drive or sealed construction are used. Any pump that moves a liquid near its saturation pressure must have a net liquid column above the pump centerline to provide the necessary pressure for liquid flow into the pump suction without allowable flashing. This liquid column provides the NPSH of the pump. When the liquid in the tank is saturated, the NPSH equals the liquid height above the pump centerline minus the pressure loss from pipe, fittings, or valves. The pressure equivalent of any temperature rise and the effect of any change in velocity must also be subtracted. The NPSH represents the net equivalent liquid height required to overcome the dynamic losses for the liquid entering the pump.

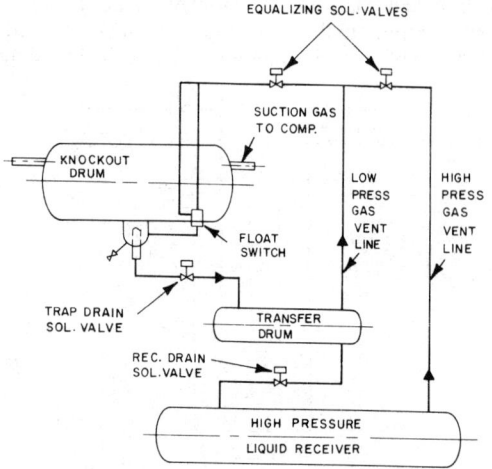

Fig. 15 Gravity Transfer System

If the net liquid column above the pump is less than the NPSH requirements, pure liquid will not flow into the impeller. Because of pressure drop and some heat gain, some liquid will evaporate and form gas bubbles. This can diminish, if not totally eliminate, the pump's ability to move liquid. Cavitation from collapsing gas bubbles can destroy pump seals and impellers. The problem is compounded in low-temperature refrigerant pumps because heat gain or a sudden drop in system pressure will form gas bubbles in the column entering the pump suction reducing the available suction head. Variations due to system transients are compensated for by having as much liquid column as possible with a minimum of 1 ft net extra above NPSH requirements for the worst conditions.

Liquid Level Indicators and Sight Glasses

A liquid sight flow indicator in the liquid line indicates whether sufficient charge is in the system to maintain liquid flow without bypassing high-pressure gas. For halocarbon systems, the flow indicator usually includes a moisture indicator.

Level indicators for vessels at low temperatures may include:

1. A bare pipe connected as a level column with the frost line indicating liquid level. A variation of this indicator uses an insulated column with short, horizontal closed-end pipes protruding through the insulation at various heights; the frost then indicates liquid level.
2. An insulated column with reflex sight glass and transparent plastic frost shield allows direct viewing of the liquid refrigerant level.

PIPING

Pipe Sizing

Since low-temperature systems may work with low suction pressures, suction lines must be sized for the minimum practical pressure drop. Liquid lines are generally sized for friction loss limits and to avoid excessive velocities, particularly where quick shutoff valves such as solenoid valves are used. Discharge and suction lines are generally sized for reasonable friction losses in horizontal or downward flow pipes. Risers may require different treatment for oil management.

Pipe friction data are available for Refrigerants 12, 22, and 502 in Chapter 3, and for R-717 in Chapter 4. Hydrocarbon data are available from petroleum industry handbooks. Other refrigerants must be estimated on the basis of Moody Curves, if the Reynolds Number and relative roughness are known. The suction line sizing tables are based on dry gas flow only. Generally, liquid overfeed suction lines with two-phase flow are oversized one pipe size above that required for the vapor itself. Liquid line sizing tables are based on warm liquid flow at a rate equal to that used by the evaporators. When liquid overfeed systems call for circulating at a rate greater than 1:1, the evaporator rate must be multiplied by the liquid overfeed rate before selecting the liquid line size.

To avoid liquid slugging, the compressor piping arrangement should prevent any liquid accumulation during the compressor off cycle or low load. Otherwise, when the load increases, increased gas velocity could pull liquid into the compressor, causing considerable damage. Chapters 3 and 4 illustrate recommended design practices for halocarbon and ammonia systems. Included are piping practices that encourage oil return and recommended allowable pressure drops.

Riser Sizing for Oil Return

In halocarbon systems where oil is transported throughout the system, special care must be taken to ensure that the system does not become oil logged because the oil is unable to travel up a vertical section of pipe. Frequently, risers must be trapped and reduced in size to cause sufficient friction on the oil droplets to lift

Engineered Refrigeration Systems (Industrial Design Practices)

them through the riser and into the horizontal pipe run. Chapter 3 includes tables for the minimum upward velocities for Refrigerants 12, 22, and 502; data on other refrigerants are not available. Table 1 lists pressure drop sizing data for oil returns.

Table 1 Pressure Drop Sizing Data for Oil Return in Discharge or Suction Lines with Flow Vertically Upward

Saturation Temperature, °F	Line Size	
	2 in. and less	Above 2 in.
50 and above	0.25 psi/100 ft	0.15 psi/100 ft
0	0.35 psi/100 ft	0.20 psi/100 ft
−50	0.45 psi/100 ft	0.25 psi/100 ft

Friction values, at minimum operating flow rate, should not be less than these values. (Adapted from *System Design and Installation Manual*, Worthington Corp., p. A 37a-5.)

Piping Codes

The *Pressure Code for Refrigeration Piping* (ANSI B31.5-1986) lists piping materials and thicknesses used for refrigerant piping. Ammonia systems use steel piping systems, as ammonia is corrosive to copper. Depending on their size, halocarbon systems may use copper or steel refrigerant piping. This code also outlines acceptable methods of refrigerant pipe supports and fabrication.

CONTROL

Refrigerant flow controls are discussed in Chapter 19 of the 1988 ASHRAE *Handbook—Equipment*. The following precautions are necessary in the application of certain controls in low-temperature or low absolute pressure systems.

Liquid Feed Control

Many types of controls available for single-staged, high-temperature systems may be used with some discretion on multistaged low-temperature systems. If the liquid level is controlled by a low-side float valve with the float in the chamber where the level is controlled, low pressure and temperatures have no appreciable effect on their operation. External float chambers, however, must be thoroughly insulated to prevent heat influx that might cause boiling and an unstable level affecting the float response. Equalizing lines to external float chambers, particularly the upper line, must be sized generously so that liquid can reach the float chamber, and gas resulting from any evaporation may be returned to the vessel without appreciable pressure loss. A loss of 0.1 psi corresponds to a 2-in. column height of liquid R-12, R-22, or R-13, and a 4-in. column height of R-717 liquid, depending on temperature.

The superheat-controlled expansion (thermostatic) valve is generally used in direct-expansion evaporators. This valve operates on the pressure difference between the bulb, which is responsive to the suction temperature, and the pressure below the diaphragm, which is the actual suction pressure.

Many fluids have been developed for charging the power element of the superheat-controlled valve to adapt it to the varied operating conditions of refrigeration systems. Manufacturers' literature details the operating characteristics of a valve with a specific charge. Superheat-controlled expansion valves are available for a wide range of evaporator requirements.

An evaporator with more than one section must have a separate expansion valve and suction line for each section. The suction lines must not join until after each expansion valve bulb location, and care must be taken to ensure that flow conditions from one circuit do not affect the bulb for another circuit.

The thermostatic expansion valve is designed to maintain a preset superheat in the suction gas. Although the pressure-sensing part of the system responds almost immediately to a change in conditions, the temperature-sensing bulb must overcome thermal inertia before its effect is felt on the power element of the valve. For this reason, when compressor capacity decreases suddenly, the expansion valve may overfeed before the bulb senses the presence of liquid in the suction line and reduces the feed. Therefore, a suction accumulator must be installed on direct-expansion low-temperature systems with multiple expansion valves.

Controlling Load During Pulldown

System transients during pulldown can be managed by controlling compressor capacity. Proper load control reduces the compressor capacity so that the energy requirements stay within the capacities of the motor and the condenser. On larger systems using screw compressors, a current sensing device reads motor amperage and adjusts the capacity control device appropriately. Cylinders can be unloaded on reciprocating compressors for similar control. Some expansion valves prevent overload by limiting the amount of refrigerant they feed in response to a maximum allowed suction pressure. These pressure limiting expansion valves are usually found on relatively small installations.

Alternatively, a downstream pressure regulator can be installed in the suction line to throttle the suction flow, should the pressure exceed a preset limit. This regulator limits the compressor's suction pressure during pulldown. The disadvantage of this device is the extra pressure drop it causes when the system is at the desired operating conditions. To overcome some of this pressure drop, the designer can use external forces to drive the valve, causing it to be held fully open when the pressure is below the maximum allowable. Systems incorporating downstream pressure regulators and compressor unloading must be carefully designed so that the two controls complement each other.

Operation at Varying Loads and Temperatures

Compressor and evaporator capacity controls are similar for multi- and single-stage systems. Control methods include compressor capacity control, hot gas bypass, or evaporator pressure regulators. Low pressure can affect control systems by significantly increasing specific volume of the refrigerant gas and pressure drop. A small pressure reduction can cause a large percentage capacity reduction.

System load usually cannot be reduced to near zero, since this would result in little or no flow of gas through the compressor and consequent overheating. Additionally, high pressure ratios would be detrimental to the compressor if it is required to run at very low loads. If the compressor cannot be allowed to cycle off during low load, an acceptable alternative is a hot gas bypass. The high-pressure gas is fed into the low-pressure side of the system through a downstream pressure regulator. The gas should be desuperheated by injecting it at a point in the system where it will be in contact with expanding liquid. Otherwise, extremely high discharge temperatures can result. The artificial load supplied by the high-pressure gas can fill the gap between the actual load and the lowest stable compressor operating capacity. Figure 16 shows such an arrangement. Chapter 3 discusses hot gas bypass capacity control in further detail.

Electrical and Electronic Control

Microprocessor and computer-based control systems are becoming the norm for control systems on individual compressors as well as for entire system control. Almost all screw compressors use microprocessor control systems to monitor all safety functions and operating conditions. These machines are frequently linked together with a programmable controller or computer for sequencing multiple compressors so that they load and unload in response to system fluctuations in the most economical manner. Programmable controllers are also used to replace multiple defrost time clocks on larger systems for more accurate and economical defrosting. Communications and data logging

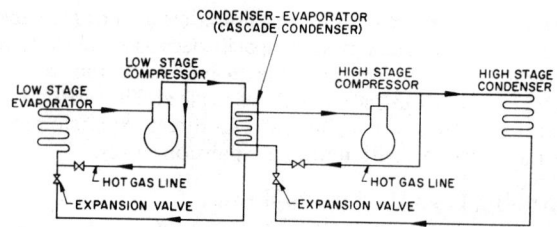

Fig. 16 Hot-Gas Injection Evaporator for Operations at Low (Even Zero) Load with a Cascade System

permit systems to operate at optimum conditions under transient load conditions even when operatiors are not in attendance.

MOISTURE IN REFRIGERANT CIRCUIT

Water in a refrigeration system should be avoided, since it can cause the refrigerant and oil to deteriorate and cause damage to the system. Testing and drying procedures for halocarbon refrigeration systems are discussed in Chapters 6 and 7. The moisture problem is often compounded in low-temperature systems as it is often necessary to operate the system below atmospheric pressure where a leak on the suction side of the system will draw in humid air.

Liquid line sight glasses with moisture indicators are used to detect moisture in low-temperature halocarbon systems. To prevent moisture from entering a system, do not open a system under vacuum for maintenance; use dry, clean oil, refrigerant, and piping when charging a system; and keep the system tight, especially when it operates below atmospheric pressure.

Dehydrators. To keep moisture, acids, dirt, and sludge levels as low as possible, install a liquid line drier and moisture indicator on all halocarbon refrigeration systems; a bypass drier can be used for partial filtration on systems with larger than 2-in. nominal liquid line size.

Piping. Piping should be kept clean and dry during the installation of the system. A pressure test for leaks should be made after the system is completed. The system should then be evacuated and held to below 1 mm of mercury absolute pressure to boil out all residual water before the system is charged with refrigerant.

INSULATION AND VAPOR BARRIERS

Chapters 20 and 21 of the 1989 ASHRAE *Handbook—Fundamentals* cover insulation and vapor barriers. Insulation and effective vapor barriers on low-temperature systems are very important. At low temperatures, the smallest leak in the vapor barrier can allow ice to form inside the insulation, which can totally destroy the integrity of the entire insulation system. The result can cause a significant load and power increase on a low-temperature system.

CHAPTER 2

LIQUID OVERFEED SYSTEMS

Operation of Mechanical Pump System 2.1	Evaporator Design................................. 2.6
Operation of Gas Pump System 2.2	Top Feed/Bottom Feed 2.6
Distribution 2.2	Refrigerant Charge 2.6
Oil in System 2.3	Start-up and Operation 2.6
Circulating Rate 2.3	Operating Costs 2.7
Type of Pump..................................... 2.5	Line Sizing 2.7
Installing and Connecting Mechanical Pumps 2.5	Low-Pressure Receiver Sizing...................... 2.7
Controls 2.5	

OVERFEED systems are those in which excess liquid is forced, mechanically or by gas pressure, through organized-flow evaporators, separated from the vapor and returned to the evaporators. Terms commonly used with overfeed systems include:

Low-Pressure Receiver. Sometimes referred to as an accumulator, this vessel acts as the separator for the mixture of vapor and liquid returning from the evaporators. The pumping unit is located below the low-pressure receiver. A constant refrigerant level is usually maintained by conventional control devices.

Pumping Unit. Consists of one or more mechanical pumps or gas-operated liquid circulators arranged to pump the overfeed liquid to the evaporators.

Wet Returns. The connections between the evaporator outlets and low-pressure receiver through which the mixture of vapor and overfeed liquid is drawn.

Liquid Feeds. The connections between the pumping unit outlet and the evaporator inlets.

Flow Control Regulators. Devices used to regulate the overfeed flow into the evaporators. They may be needle-type valves, fixed orifices, calibrated manual regulating valves, or automatic valves designed to provide a fixed liquid rate.

The main advantages of liquid overfeed systems are high system efficiency and reduced operating expenses. These systems have lower energy cost and fewer operating hours because:

1. The evaporator surface is used efficiently through good refrigerant distribution and completely wetted internal tube surfaces.
2. The compressors are protected. Liquid slugs resulting from fluctuating loads or malfunctioning controls are separated from suction gas in the low-pressure receiver.
3. Low-suction superheats are achieved where the suction lines between the low-pressure receiver and the compressors are short. This condition causes a minimum discharge temperature, preventing lubrication breakdown and minimizing condenser fouling.
4. With simple controls, evaporators can be hot-gas defrosted with little disturbance to the system.
5. Refrigerant feed to evaporators is unaffected by fluctuating ambient and condensing conditions. The flow control regulators do not need to be adjusted after the initial setting, since the overfeed rates are not generally critical.
6. Flash gas, resulting from refrigerant throttling losses, is removed at the low-pressure receiver before entering the evaporators. This gas is drawn directly to the compressors and is eliminated as a factor in the design of the system low side. It does not contribute to increased pressure drops in the evaporators or overfeed lines.

7. Refrigerant level controls, level indicators, refrigerant pumps and oil drains are generally located in the equipment rooms, which are under operator surveillance.
8. Because of ideal entering suction gas conditions, compressors last longer. There is less maintenance and fewer breakdowns. The oil circulation rate to the evaporators is reduced as a result of the low compressor discharge superheat (Scotland 1963).
9. Overfeed systems have convenient automatic operation.

Possible disadvantages are:

1. In some cases, refrigerant charges are greater than those used in other systems.
2. Higher refrigerant flow rates to and from evaporators cause the size of the liquid feed and wet return lines to be larger than the high-pressure liquid and suction lines for other systems.
3. Piping insulation is costly and is generally required on all feed and return lines to prevent moisture or frost formation.
4. The installed cost may be greater, particularly for small systems or those with fewer than three evaporators.
5. The operation of the pumping unit requires added expenses that are offset by the increased efficiency of the overall system.
6. The pumping units may require more maintenance.

Generally, the more evaporators that are used, the more favorable the initial costs are for the liquid overfeed compared to a gravity recirculated or flooded system (Scotland 1970). Liquid overfeed systems also compare favorably with thermostatic valve feed systems for the same reason. For small systems, the initial cost for liquid overfeed may be higher than for direct expansion.

Easy operation and less maintenance are attractive features for even small ammonia systems. However, for ammonia systems operating at below 0°F evaporating temperatures, some manufacturers do not supply direct expansion evaporators because of unsatisfactory refrigerant distribution and control problems.

OPERATION OF MECHANICAL PUMP SYSTEM

Figure 1 shows the basic system in which a constant liquid level is maintained in a low-pressure receiver. A mechanical pump circulates liquid through the evaporator(s). The two-phase return mixture is separated in the low-pressure receiver. The vapor is directed to the compressor(s). The makeup refrigerant enters the low-pressure receiver by means of a refrigerant metering device.

Figure 2 shows a horizontal low-pressure receiver with a minimum pump head, service valves in place, and a strainer on the suction side of the pump. The strainer protects hermetic pumps when oil is miscible with the refrigerant. It should have a free area twice the transverse cross-sectional area of the line in which it is

---

The preparation of this chapter is assigned to TC 10.1, Custom Engineered Refrigeration Systems.

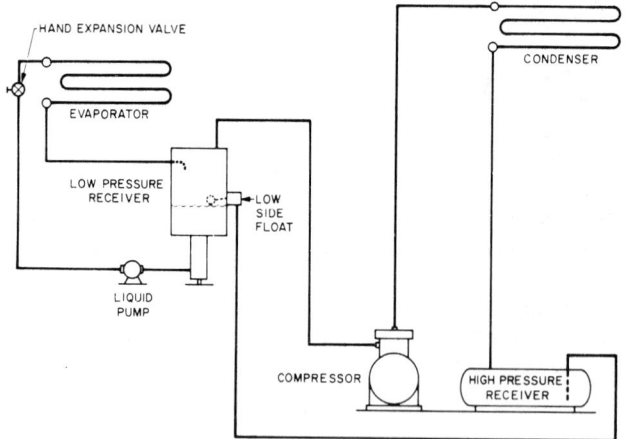

Fig. 1 Liquid Overfeed with Mechanical Pump

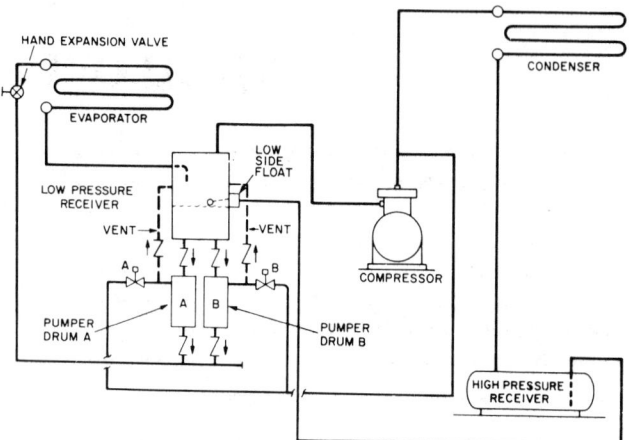

Fig. 3 Double Pumper Drum System

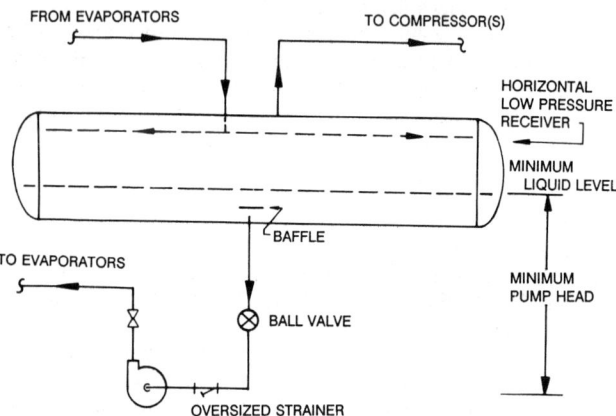

Fig. 2 Pump Circulation, Horizontal Separator

installed. A suction strainer should not be used with ammonia, and open drive pumps do not require strainers.

The minimum pump head should be at least double the net positive suction head (NPSH) to avoid cavitation. The liquid velocity to the pump should not exceed 3 ft/s.

OPERATION OF GAS PUMP SYSTEM

Figure 3 shows a liquid overfeed system in which the pumping power is supplied by gas at condenser pressure. In this system, a level control maintains the liquid level in the low-pressure receiver. There are two pumper drums; one is filled by the low-pressure receiver, while the other is drained as hot gas pushes liquid from the pumper drum to the evaporator. Pumper drum B drains when hot gas enters the drum through Valve B. To function properly, the pumper drums must be correctly vented so they can fill during the fill cycle.

Another common arrangement is shown in Figure 4. In this system, the high-pressure liquid is flashed into a controlled-pressure receiver, which maintains constant liquid pressure at the evaporator inlets, resulting in continuous liquid feed at constant pressure. The flash gas is drawn into the low-pressure receiver through a receiver pressure regulator. Excess liquid drains into a liquid-pump trap from the low-pressure receiver. Check valves and a three-way equalizing valve transfer the liquid into the controlled-pressure receiver during the dump cycle. Refinements of this system are used for multistage systems.

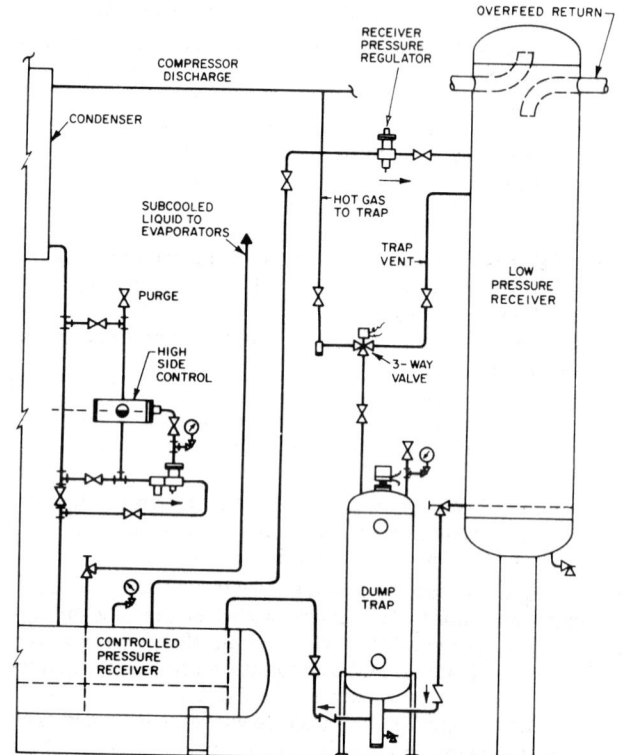

Fig. 4 Constant-Pressure Liquid Overfeed System

DISTRIBUTION

To prevent underfeeding and excessive overfeeding of refrigerants, metering devices regulate the liquid feed to each evaporator and/or evaporator circuit. An automatic regulating device continually controls refrigerant feed to the design value. Other devices used are hand expansion valves, calibrated regulating valves, orifices, and distributors.

It is time-consuming to adjust hand expansion valves to achieve ideal flow conditions. However, they have been used with some success in many installations prior to the availability of more sophisticated controls. One factor to consider is that standard hand expansion valves are designed to regulate flows caused by the relatively high pressure differences between condensing and evaporating pressure. In overfeed systems, large differences do not

Liquid Overfeed Systems

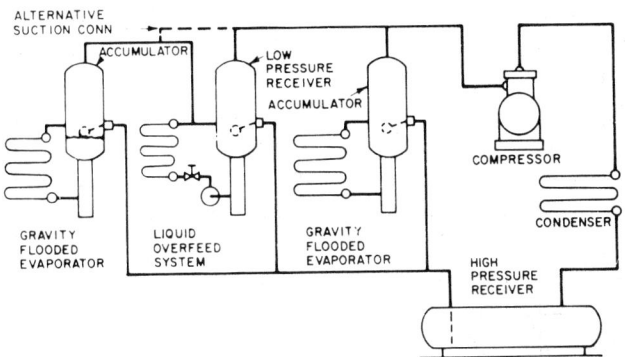

Fig. 5 Liquid Overfeed System Connected on Common System with Gravity-Flooded Evaporators

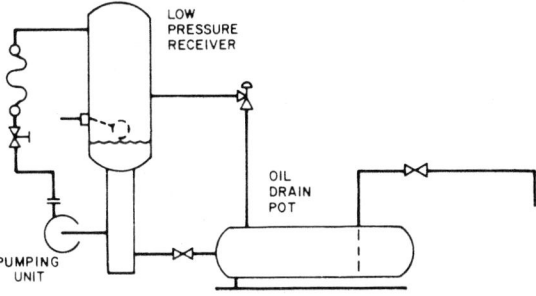

Fig. 6 Oil Drain Pot Connected to Low-Pressure Receiver

exist, so valves with larger orifices may be needed to cope with the combination of the increased quantity of refrigerant and the relatively small pressure differences—approximately 10 to 15 psi.

Calibrated, manually operated regulating valves reduce some of the uncertainties involved in using conventional hand expansion valves. To be effective, the valves should be adjusted to the manufacturer's recommendations. Because the refrigerant in the liquid feed lines is above saturation pressure, the lines should not contain flash gas. However, liquid flashing can occur if excessive heat gains by the refrigerant and/or high pressure drops build up in the feed lines.

Orifices should be carefully designed and selected. These generally are used only for top- and horizontal-feed multicircuit evaporators. Foreign matter and congealed oil globules can cause flow restriction; a minimum orifice of 0.1 in. is recommended. With the small liquid volume of ammonia normally circulated, the rate of circulation may need to be increased beyond that needed for the minimum orifice size. Pumps and feed and return lines larger than minimum may be needed. This does not apply to halocarbons because of the greater liquid volume circulated as a result of fluid characteristics.

Conventional multiple outlet distributors with capillary tubes of the type usually paired with thermostatic expansion valves have been used successfully in liquid overfeed systems. Capillary tubes may be installed downstream of a distributor with oversized orifices to achieve the required pressure reduction and efficient distribution.

Existing gravity-flooded evaporators with accumulators can be connected to liquid overfeed systems. Changes may be needed only for the feed to the accumulator, with suction lines from the accumulator connected to the system wet return lines. An acceptable arrangement is shown in Figure 5. Generally, gravity-flooded evaporators have different circuiting arrangements from overfeed evaporators. In many cases, the circulating rates developed by thermal-syphon action are greater than the circulating rates used in conventional overfeed systems.

Example 1. Find the orifice diameter of an ammonia overfeed system with a refrigeration load per circuit of 1.27 tons and a circulating rate of 7. The evaporating temperature is $-30°F$ and the pressure drop across the orifice is 8 psi. The circulation per circuit is 0.528 gpm.

Solution: Orifice diameter may be calculated by Equation (1) as:

$$d = \left[\frac{Q}{a\,C_d\sqrt{p/S}}\right]^{0.5} \quad (1)$$

where

a = units conversion, 29.81

Q = discharge through orifice, 0.528 gpm
p = pressure drop through orifice, 8 psi
S = specific gravity of fluid relative to water at $-30°F$
  = 5.701/8.336 = 0.6839
C_d = Coefficient of discharge for orifice = 0.61

$$d = \left[\frac{0.528}{29.81 \times 0.61\,\sqrt{8/0.6839}}\right]^{0.5} = 0.0922 \text{ in.}$$

OIL IN SYSTEM

In spite of reasonably efficient compressor discharge oil separators, oil finds its way into the system low-pressure sides. In the case of ammonia overfeed systems, the bulk of this oil can be drained from the low-pressure receivers with suitable oil drainage facilities. In low-temperature systems, a separate valved and pressure-protected noninsulated oil drain pot can be placed in a warm space at the accumulator. The oil/ammonia mixture will flow into the pot and the refrigerant will evaporate. This arrangement is shown in Figure 6. Because of the low solubility of oil in liquid ammonia, thick oil globules circulate with the liquid and can restrict flow through strainers, orifices and regulators. To maintain high efficiency, oil should be removed from the system by regular draining.

Because halocarbons are miscible with oil, positive oil return to the compressor must be assured. Many methods are used, including oil stills using both electric heat and heat exchange from high-pressure liquid or vapor. Some arrangements are discussed in Chapter 3.

Excessive oil must not be allowed to build up in evaporators, because efficiency will rapidly decrease. This is particularly critical in evaporators with high heat transfer rates associated with low volumes, such as flake-type icemakers, ice cream freezers, and scraped-surface heat exchangers. Because high refrigerant flow rate occurs through such evaporators, excessive oil can accumulate and rapidly deteriorate efficiency.

CIRCULATING RATE

In a liquid overfeed system, the mass ratio of liquid pumped to the amount of vaporized liquid is the circulating number or rate. The amount of liquid vaporized is based on the latent heat for the refrigerant at the evaporator temperature. Overfeed rate is the ratio of liquid to vapor returning to the low-pressure receiver. When vapor leaves an evaporator at saturated vapor conditions with no excess liquid, the circulating rate is 1 and the overfeed rate is 0. With a circulating rate of 4, the overfeed rate at full load is 3; at no load, it is 4. Most systems are designed for steady flow conditions. With few exceptions, the load conditions may vary, causing fluctuating temperatures outside and within the evaporator. Evaporator capacities vary considerably; with constant refrigerant flow to the evaporator, the overfeed rate fluctuates.

For each evaporator, there is an ideal circulating rate for every loading condition that will result in the minimum temperature

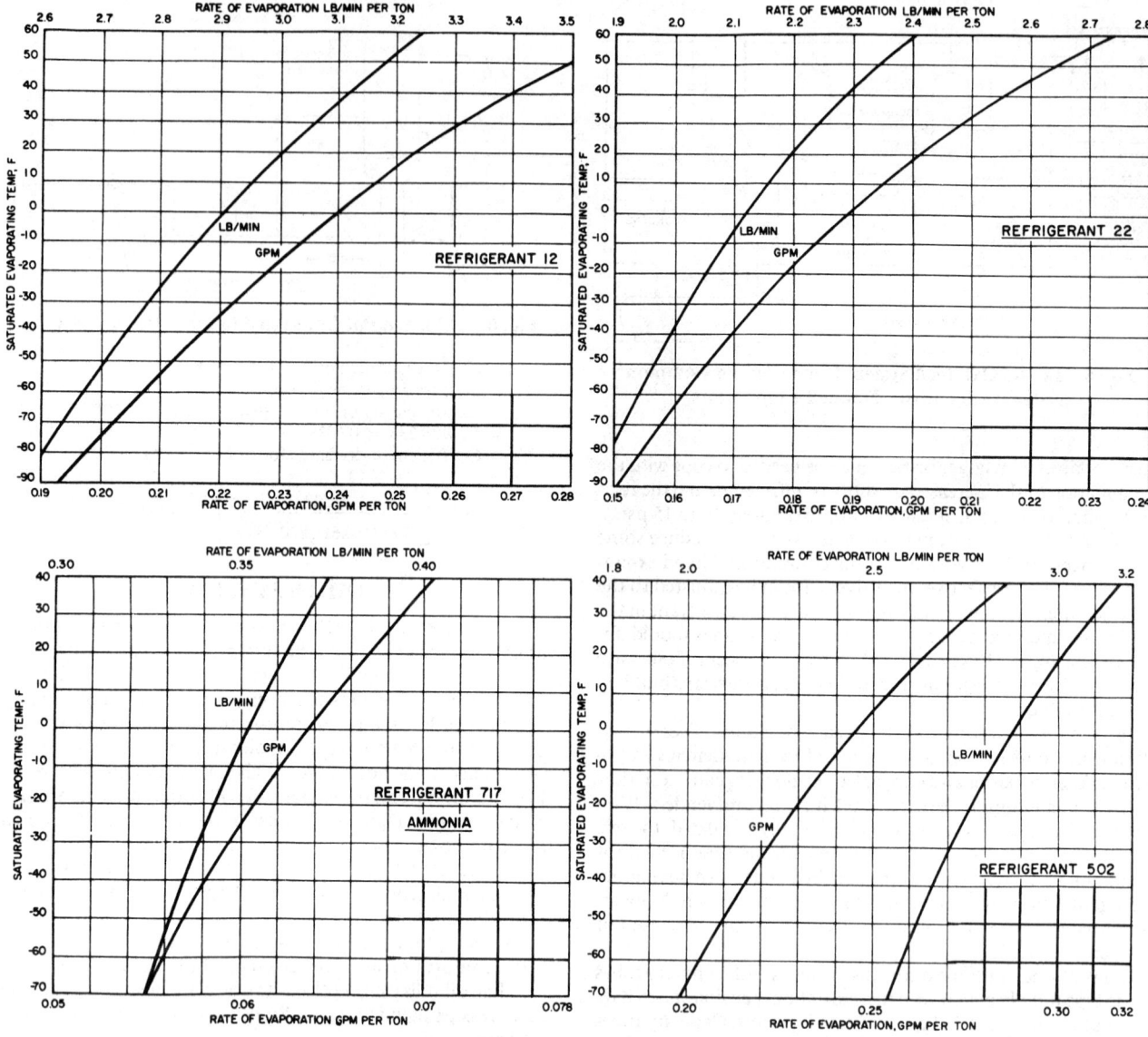

Fig. 7 Charts for Determining Rate of Refrigerant Feed (No Flash Gas)

difference and the best evaporator efficiency (Lorentzen 1968, Lorentzen and Gronnerud 1967). With few exceptions, it is impossible to predict ideal circulating rates or to design a plant for automatic adjustment of the rates to suit fluctuating loads. The optimum rate can vary with heat load, pipe diameter, circuit length, and number of parallel circuits to achieve the best performance. High circulation rates can cause excessively high pressure drops through evaporators and wet return lines. The sizing of these return lines, discussed in a separate section, can have a bearing on the ideal rates. Many evaporator manufacturers specify recommended circulating rates for their equipment. The rates shown in Table 1 agree with these recommendations.

Because of distribution considerations, higher circulating rates are common with top feed evaporators. In multicircuit systems, the refrigerant distribution must be adjusted to provide the best possible results. Incorrect distribution can cause excessive overfeed in some circuits, while others may be starved. Manual or automatic regulating valves can be used to control flow for the optimum or design value.

Halocarbon densities are about twice that of ammonia. If halocarbons R-12, R-22, and R-502 are circulated at the same rate as ammonia, the halocarbons will require 6 to 8.3 times more energy for pumping to the same height as the less dense ammonia. Since this pumping energy must be added to the system load, halocarbon circulating rates are usually lower than those for ammonia. Because ammonia has a relatively high latent heat of vaporization, for equal heat removal, much less ammonia mass must be circulated compared to halocarbons.

Even though halocarbons circulate at lower rates than ammonia, the wetting process in the evaporators is still efficient because of the liquid and vapor volume ratios. For example, at −40 °F evaporating temperatures, with constant flow conditions in the wet return connections, similar ratios of liquid and vapor are experienced with a circulating rate of 4 for ammonia, 2.5 for R-22 and R-502, and 2 for

Table 1 Recommended Minimum Circulating Rate

Refrigerant	Circulating Rate
Ammonia (R-717)	
Downfeed (large diameter tubes)	6 to 7
Upfeed (small diameter tubes)	2 to 4
R-12, R-502 - upfeed	2
R-22 - upfeed	3

Liquid Overfeed Systems

R-12. With halocarbons, some additional wetting is also experienced because of the solubility of the oil in these refrigerants.

When bottom feed is used for multicircuit coils, a minimum feed rate per circuit is not necessary since orifices or other distribution devices are not required. The circulating rate for top feed and horizontal feed coils may be determined by the minimum rates from the orifices or other distributors in use.

Figure 7 provides a method for determining the liquid refrigerant flow (Niederer 1963). The charts indicate the amount of refrigerant vaporized in a 1-ton system with circulated operation having no flash gas in the liquid feed line. The value obtained from the chart may be multiplied by the desired circulating rate and by the total refrigeration to determine total flow.

The pressure drop through the flow control regulators is usually 10 to 50% of the available feed pressure. The pressure at the outlet of the flow regulators is higher than the vapor pressure at the low-pressure receiver by an amount representing the total pressure drop of the two-phase mixture through the evaporator, any evaporator pressure regulator, and wet return lines. This outlet pressure could be 5 psi in a typical system. When using recommended liquid feed sizing practices, assuming a single-story building, the frictional pressure drop from the pump discharge to the evaporators is about 10 psi. Therefore, a pump for 20 to 25 psi should be satisfactory in this case, depending on the lengths and sizes of feed lines, the quantity and types of fittings, and the vertical lift involved.

TYPE OF PUMP

Mechanical pumps, gas pressure pumping systems, and injector systems are available for liquid overfeed systems.

Types of mechanical pumps include open, semihermetic, magnetic clutch, and hermetic rotor arrangements with positive rotary, centrifugal, or turbine vane construction. Positive rotary and gear-type pumps are generally operated at slow speeds from 90 to 300 rpm. Whatever type of pump is used, care should be taken to prevent flashing at the pump suction and/or within the pump itself.

Centrifugal pumps are preferred for larger volumes, while semihermetic pumps are best suited for halocarbons at or below atmospheric refrigerant saturated pressure.

Open-type pumps are fitted with a wide variety of packing or seals. For continuous duty, a mechanical seal with an oil reservoir or a liquid refrigerant supply to cool, wash, and lubricate the seals is commonly used. Experience with the particular application or the recommendations of an experienced pump supplier are the best guides for selecting the packing or seal. A small immersion-type electric heater within the oil reservoir can be used with low temperature systems to assure that the oil will remain fluid. Motors are selected that have a service factor that compensates for drag on the pump if the oil is cold or stiff.

Application considerations should include ambient temperatures, heat leakage, fluctuating system pressures from compressor cycling, internal bypass of liquid to pump suction, friction heat, and motor heat conduction. They should also include dynamic conditions, cycling of automatic evaporator liquid and suction stop valves, action of regulators, gas entrance with liquid, and loss of subcooling by pressure drop. Other factors to consider are the time lag caused by the heat capacity of pump suction, cavitation, and NPSH factors (Lorentzen 1963).

The motor and stator of hermetic pumps are separated from the refrigerant by a thin nonmagnetic membrane. The metal membrane should be strong enough to withstand system design pressures. Normally, the motors are cooled and the bearings lubricated by liquid refrigerant bypassed from the pump discharge. It is good practice to use two pumps, one operating and one standby.

INSTALLING AND CONNECTING MECHANICAL PUMPS

Because of the sensitive suction conditions of mechanical pumps operating on overfeed systems, the manufacturer's application and installation specifications must be followed closely. Suction connections should be as short as possible, without restrictions, valves, or elbows. Angle or fullflow ball valves should be used. Using valves with horizontal valve spindles eliminates possible traps. Gas binding is more likely with high evaporating pressures.

Installing discharge check valves prevents back flow. Relief valves should be used, particularly for positive displacement pumps. Strainers are not usually installed in ammonia pump suction lines because they plug with oil. Strainers, although a poor substitute for a clean installation, protect halocarbon pumps from damage by dirt or pipe scale.

Pump suction connections to liquid legs (vertical drop legs from low-pressure receivers) should be made above the bottom of the legs to allow collection space for solids and sludge. Vortex eliminators should be considered, particularly when submersion of the suction inlet is insufficient to prevent the intake of gas bubbles. Lorentzen (1963, 1965) gives more complete information.

Sizing the pump suction line is important. The general velocity should be about 3 ft/s. Small lines cause restrictions, while oversized lines can cause bubble formation during decreasing evaporator temperature conditions because of the heat capacity of the liquid and piping. Oversized lines also impose increased heat gain from the ambient spaces. Oil heaters for the seal lubrication system keep the oil fluid, particularly during subzero operation. Thermally insulating all cold surfaces of pumps, lines, and receivers increases efficiency.

CONTROLS

The liquid level in the low-pressure receiver can be controlled by conventional devices such as low-pressure float valves, combinations of float switch and solenoid valve with manual regulator, thermostatic level controls, or other proven automatic devices. High level float switches are useful to stop compressors and/or operate alarms. Solenoid valves should be installed on liquid lines (minimum sized) feeding low-pressure receivers so that positive shutoff is automatically achieved with system shutdown. This prevents excessive refrigerant from collecting in low-pressure receivers, which can cause spill-over at start-up.

To prevent pumps from operating without liquid, low level float switches can be fitted on liquid legs. An alternative device, a differential pressure switch connected across pump discharge and suction connections, causes the pump to stop without interrupting liquid flow. In extreme cases, cavitation can also cause this control to operate. When hand expansion valves are used to control the circulation rate to evaporators, the orifice should be sized for operation between system high and low pressures. Occasionally, with reduced inlet pressure conditions, these valves can starve the circuit. Calibrated, manually-adjusted regulators are available to meter the flow according to the design conditions. An automatic-flow regulating valve is available specifically for overfeed systems.

Liquid and suction solenoid valves are selected for refrigerant flow rates by mass or volume, not by refrigeration ratings from capacity tables. Evaporator pressure regulators should be sized according to the manufacturer's ratings for overfeed systems. When ordering valves, tell the manufacturer they are for overfeed application, since slight modifications may be required. When evaporator pressure regulators are used on overfeed systems for controlling air-defrosting of cooling units (particularly when fed with very low-temperature liquid), the refrigerant heat gain may be achieved by sensible effect, not by latent effect. In such cases, investigate other defrosting methods. The possibility of connecting the units directly to high-pressure liquid should be considered, especially if the loads are minor.

When a check valve and a solenoid valve are paired on an overfeed system liquid line, the check valve should be downstream from the solenoid valve. When the solenoid valve is closed, dangerous hydraulic pressure can build up from the expansion of the trapped liquid as it is heated. When evaporator pressure regulators are used, the pressure of the entering liquid should be high enough to cause flow into the evaporator.

Multicircuit systems have a bypass relief valve in the pump discharge. When some of the circuits are closed, the excess liquid is bypassed into the low-pressure receiver, rather than forced through the evaporators still in operation. This prevents higher evaporating temperatures from pressurizing evaporators and reducing capacities of operating units. Where low-temperature liquid feeds can be isolated manually or automatically, relief valves can be installed to prevent damage from excessive hydraulic pressure.

EVAPORATOR DESIGN

There is an ideal refrigerant feed and flow system for each evaporator design and arrangement. An evaporator designed for gravity-flooded operation cannot always be converted to an overfeed arrangement, or vice versa, nor can systems always be designed to circulate the optimum flow rate. When top feed is used to ensure good distribution, a minimum quantity per circuit must be circulated, generally about 0.5 gpm. Distribution in bottom feed evaporators is less critical than for top or horizontal feed, since each circuit fills with liquid to equal the pressure loss in other parallel circuits.

Circuit length in evaporators is determined by allowable pressure drop, load per circuit, tubing diameter, overfeed rate, type of refrigerant, and heat transfer coefficients. The most efficient circuiting is resolved in most cases through laboratory tests conducted by the evaporator manufacturers. Their recommendations should be followed when designing systems.

TOP FEED/BOTTOM FEED

System design must determine whether evaporators are to be top or bottom fed, although both feed types can be installed in a single system. Each feed type has advantages; no best arrangement is common to all systems. Top feed advantages include (1) smaller refrigerant charge, (2) possibly smaller low-pressure receiver, (3) possible absence of static head penalty, (4) better oil return and (5) quicker, simpler defrost arrangements. For halocarbon systems with greater fluid densities, the refrigerant charge, oil return, and static head are very important.

Bottom feed is advantageous in that (1) distribution considerations are less critical, (2) relative locations of evaporators and low-pressure receivers are less important, and (3) the system design and layout are simpler. The top feed system is limited by the relative location of components. Because this system sometimes requires more refrigerant circulation than bottom feed systems, it has greater pumping load, possibly larger feed and return lines, and increased line pressure drop penalties. In bottom feed evaporators, multiple headers with individual inlets and outlets can be installed to reduce static head penalties. For high lift of return overfeed lines from the evaporators, dual-suction risers eliminate static head penalties (Miller 1974, 1979).

Distribution must be considered when a vertical refrigerant feed is used because of the static head variations in the feed and return header circuits. For example, for equal circuit loadings in a horizontal airflow unit cooler, using gradually smaller orifices for the bottom feed circuits than for the upper circuits can compensate for pressure differences.

When the top feed free-draining arrangement is used for air cooling units, liquid solenoid control valves can be used during the defrost cycle. This applies in particular to air, water, or electric defrost units. Any remaining liquid in the coils will rapidly evaporate or drain to the low-pressure receiver. Defrost is faster than in bottom feed evaporators.

REFRIGERANT CHARGE

Overfeed systems need more refrigerant than dry expansion systems. Top feed arrangements have smaller charges than bottom feed systems. The amount of charge also depends on the evaporator volume, the circulating rate, the sizes of flow and return lines, the operating temperature differences, and the heat transfer coefficients. Generally, top feed evaporators operate with the refrigerant charge occupying about 25 to 40% of the evaporator volume. The refrigerant charge for the bottom feed arrangement occupies about 60 to 75% of the evaporator volume with corresponding variations in the wet returns. Under certain no-load conditions in up-feed evaporators, the charge may occupy 100% of the evaporator volume. In this case, the liquid surge volume from full to no-load condition must be considered in sizing the low-pressure receiver (Miller 1971, 1974).

Evaporators with high heat flux, such as flake icemakers and scraped-surface heat exchangers, have small charges because of small evaporator volumes. The amount of refrigerant in the low side has a major effect on the size of the low-pressure receiver, especially in horizontal vessels. The cross-sectional area for vapor flow in horizontal vessels is reduced with increasing liquid level. It is important to ascertain the evaporator refrigerant charge with fluctuating loads for correct vessel design, particularly for a low-pressure receiver that does not have a constant level control but is fed through a high-pressure control.

START-UP AND OPERATION

All control devices should be checked prior to start-up. If mechanical pumps are used, the direction of operation must be correct. System evacuation and charging procedures are similar to those for other systems. The system must be operating under normal conditions to determine the total required refrigerant charge. Liquid height is established by liquid level indicators in the low-pressure receivers.

Calibrated, manually-operated regulators should be set for the design conditions and adjusted for better performance when necessary. When hand expansion valves are used, start the system by opening the valves about one-quarter to one-half turn. When balancing is necessary, cut back the regulators on those circuits not starved of liquid to force the liquid through the underfed circuits. The outlet temperature of the return line from each evaporator should be the same as the saturation temperature of the main return line, allowing for pressure drops. Starved circuits are indicated by higher temperatures than those for adequately fed circuits. Excessive feed to a circuit increases the evaporator temperature because of excessive pressure drop.

The relief bypass from the liquid line to the low-pressure receiver should be adjusted and checked to ensure that it is functioning. During operation, follow the pump manufacturer's recommendations regarding lubrication and maintenance. Regular oil draining procedures should be established for ammonia systems; a comparison should be made between the oil quantities added to and drained from each system. This comparison will determine if oil is accumulating in systems. Oil should not be drained in halocarbon systems, but drain valves should be opened, particularly after start-up, to blow out any foreign matter that may accumulate. Due to the miscibility of oil with halocarbons, it may be necessary to add oil to the system until an operating balance is achieved (Stoecker 1960, Soling 1971).

Liquid Overfeed Systems

OPERATING COSTS

Operating costs for overfeed systems generally are less than for other systems. Operating costs may not be less in all cases because of the variety of inefficiencies that exist from system to system and from plant to plant. However, in cases where existing dry expansion plants were converted to liquid overfeed, the operating hours, power, and maintenance costs were reduced. The efficiencies of the early gas pump systems have been improved by using flash gas to circulate the overfeed liquid. One flash gas system is indicated in the controlled pressure system shown in Figure 4. Refinements of the double pumper drum arrangement (shown in Figure 3) have also been developed.

Gas pumped systems, which use refrigerant gas to pump liquid to the evaporators or to the controlled-pressure receiver, require additional compressor volume, from which no useful refrigeration is obtained. These systems consume 4 to 10% or more of the compressor power to maintain the gas flow.

If the condensing pressure is reduced as much as 10 psig, the compressor power per unit of refrigeration will drop by about 7%. Where outdoor dry- and wet-bulb conditions allow, a mechanical pump can be used to pump the gas with no effect on evaporator performance. Gas-operated systems must, however, maintain the condensing pressure within a much smaller range to pump the liquid and maintain the required overfeed rate.

LINE SIZING

The liquid feed line to the evaporator and the wet return line to the low-pressure receiver cannot be sized by the method described in Chapter 33 of the 1989 ASHRAE *Handbook—Fundamentals*. Figure 7 can be used to size liquid feed lines. The circulating rate is multiplied by the evaporating rate. For example, an evaporator forming vapor at a rate of 5 lb/min and with a circulating rate of 4 needs a feed line sized for $4 \times 5 = 20$ lb/min. Alternative methods that may be used to design wet returns include:

1. Use one pipe size larger than calculated for vapor flow alone.
2. Use a velocity selected for dry expansion reduced by the factor $(1/\text{Circulating Rate})^{0.5}$. This method suggests that the wet return velocity for a circulating rate of 4 is $(1/4)^{0.5} = 0.5$ that of the acceptable dry vapor velocity.
3. Use the design method described by Chaddock et al. (1972). The report includes tables of flow capacities at 2°F drop per 100 ft of horizontal lines for R-717 (ammonia), R-12, R-22, and R-502.

When sizing refrigerant lines, the following design precautions should be taken:

1. Carefully size overfeed return lines with vertical risers, since more liquid will be held in risers than in horizontal pipe. This holdup increases with reduced vapor flow, and increases pressure loss because of gravity and two-phase pressure drop.
2. Use double risers with halocarbons to maintain velocity at partial loads and to reduce liquid static head loss (Miller 1979).
3. Add the equivalent of a 100% liquid static height penalty to the pressure drop allowance to compensate for liquid holdup in ammonia systems that have unavoidable vertical risers.
4. As alternatives in severe cases, provide traps and a means of pumping liquids, or use dual-duct risers.
5. Use low loss coefficient valves installed so the stems are horizontal or nearly so (Chisolm 1971).

LOW-PRESSURE RECEIVER SIZING

Low-pressure receivers are also called liquid separators, suction traps, accumulators, liquid-vapor separators, flash-type coolers, gas and liquid coolers, intercoolers, surge drums, knock-out drums, slop tanks, or low-side pressure vessels, depending on their function and the preferred term of the user.

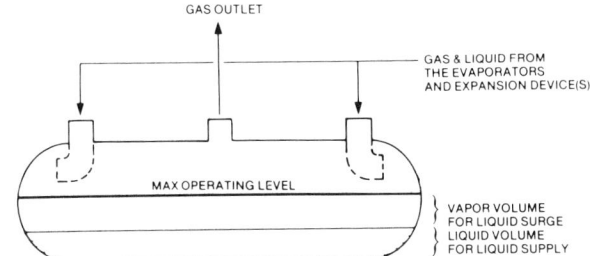

Fig. 8 Basic Horizontal Gas and Liquid Separator

The sizing of low-pressure receivers is determined by the required liquid holdup volume and the allowable gas velocity. The volume must accommodate the fluctuations of liquid in the evaporators and overfeed return lines as a result of load changes. It must also handle the swelling and foaming of the liquid charge in the receiver, caused by boiling during temperature or pressure reduction. At the same time, a liquid seal must be maintained on the supply line for continuous circulation devices. A separating space must be provided for gas velocity low enough to cause a minimum entrainment of liquid drops into the suction outlet. Space limitations and design requirements result in a wide variety of configurations (Miller 1971; Stoecker 1960; Lorentzen 1966; Niemeyer 1961; Scheiman 1963, 1964; Sonders and Brown 1934; Younger 1955).

In selecting a gas and liquid separator, adequate volume for the liquid supply and a vapor space above the minimum liquid height for liquid surge must be provided. This requires an analysis of operating load variations. This, in turn, determines the Maximum Operating Liquid Level. Figures 8 and 9 identify these levels and the important parameters of vertical and horizontal gravity separators.

Vertical separators maintain the same separating area with level variations, while separating areas in horizontal separators change with level variations. Horizontal separators should have inlets and outlets separated horizontally by at least the vertical separating distance. A useful arrangement in horizontal separators distributes the inlet flow into two or more connections to reduce turbulence and the horizontal velocity without reducing the residence time of the gas flow within the shell (Miller 1971). In horizontal separators, as the horizontal separating distance is increased beyond the vertical separating distance, the residence time of the

Table 2 Maximum Effective Separation Velocities for R-717, R-22, R-12, and R-502, with Steady Flow Conditions

Temp., °F	Vertical Separation Distance, in.	Maximum Steady Flow Velocity, fpm			
		R-717	R-22	R-12	R-502
+50	10	29	13	16	11
	24	125	62	70	50
	36	139	77	85	62
+20	10	42	19	22	15
	24	172	86	96	69
	36	195	102	115	83
−10	10	61	27	32	22
	24	253	120	135	97
	36	281	141	159	116
−40	10	95	41	47	33
	24	392	173	198	140
	36	428	205	230	165
−70	10	158	65	72	50
	24	649	267	303	212
	36	697	310	351	247

Adapted from Miller (1971)

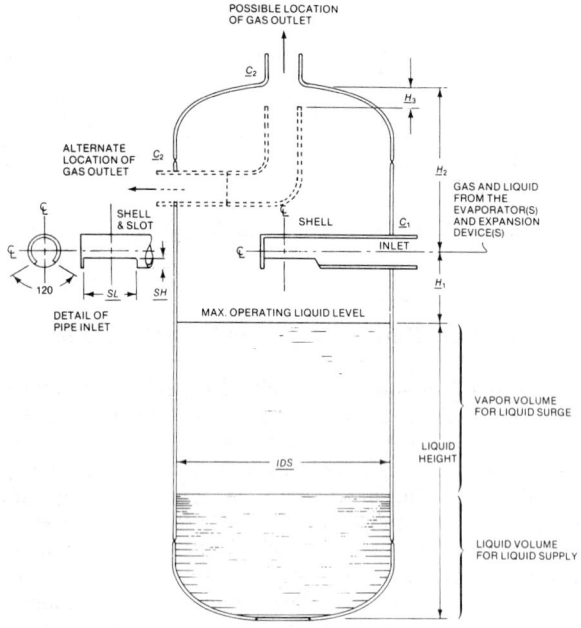

C_1 = inlet pipe diameter, OD, in.
C_2 = outlet pipe diameter, OD, in.
SH = slot height = $C_1/4$, in.
SL = slot length = $3\,C_1$, in.
H_1 = height of C_1 above maximum liquid level, in.; for pseudo SD = 24 in.
H_1 = $\sqrt{7.33\ \text{cfm}/SV}$ in. (at SD = 24 in.)
cfm = maximum gas flow in the shell at maximum sustained operating conditions
SV = separation velocity, fpm
H_2 = location of C_1 from inside top of shell, in.
H_2 = SD + 0.5 depth of curved portion of head = $C_1/4$
SD = vertical separation distance, in. actual
H_3 = location of gas exit point for alternate location of C_2 measured from inside top of shell, in.
H_3 = 0.5 depth of curved portion of shell or 2 in., whichever is greater
IDS = internal diameter of shell, in.
IDS = $\sqrt{183\ \text{cfm}/SV}$, in.

Fig. 9 Basic Vertical Gravity Gas and Liquid Separator

vapor passing through is longer so that higher velocities than allowed in vertical separators can be tolerated.

As the separating distance is reduced, the amount of liquid entrainment from gravity separators increases. Table 2 shows the gravity separation velocities. For surging loads or pulsating flow associated with large step changes in capacity, reduce the maximum steady flow velocity to a value achieved by a suitable multiplier such as 0.75.

The gas and liquid separator may be designed with baffles or eliminators to separate liquid from the suction gas returning from the top of the shell to the compressor. More often, sufficient separation space is allowed above the liquid level for this purpose. Such a design usually is of the vertical type, with a separation height above the liquid level of from 24 to 36 in. The shell diameter is sized to keep the suction gas velocity at a value low enough to allow the liquid droplets to separate and not be entrained with the returning suction gas off the top of the shell.

Although separators are made with length to diameter (L/D) ratios of 1/1 increasing to 10/1, the least expensive separators usually have L/D ratios between 3/1 and 5/1. Vertical separators are normally used for systems with reciprocating compressors. Horizontal separators may be preferable where vertical height is critical and/or where large volume space for liquid is required. The procedures for designing vertical and horizontal separators are different.

The vertical gas-and-liquid separator is shown in Figure 9.

The end of the inlet pipe C_1 is capped so that flow dispersion is directed downward toward the liquid level. The opening is four times the transverse internal area of the pipe. The height H_1 with a 120° dispersion of the flow reaches to approximately 70% of the internal diameter of the shell.

An alternate inlet pipe with a downturned elbow or mitered bend can be used. However, the jet effect of entering fluid must be considered to avoid undue splashing. The outlet of the pipe must be a minimum distance of $IDS/5$ above the maximum liquid level in the shell. H_2 is measured from the outlet to the inside top of the shell. It equals SD + 0.5, the depth of the curved portion of the head.

For the alternate location of C_2, determine IDS from Equation (2):

$$IDS = \sqrt{(183\ \text{cfm}/SV) + C_2^2} \qquad (2)$$

The maximum liquid height in the separator is a function of the type of system in which the separator is being used. With some systems this can be estimated, while in others previous experience provides the only means of selecting the proper liquid height. The accumulated liquid must be returned to the system by a suitable means at a rate comparable to the rate at which it is being collected.

When using a horizontal separator, the vertical separation distance used is an average value. The top part of the horizontal shell restricts the gas flow so that the maximum vertical separation distance cannot be used. If H_t represents the maximum vertical distance from the liquid level to the inside top of the shell, the average separation distance as a fraction of IDS is as follows:

H_t/IDS	0.1	0.2	0.3	0.4	0.5
SD/IDS	0.068	0.140	0.215	0.298	0.392
H_t/IDS	0.6	0.7	0.8	0.9	1.0
SD/IDS	0.492	0.592	0.693	0.793	0.893

The suction connection(s) for refrigerant gas leaving the horizontal shell must be located at or above the location established by the average distance for separation. The maximum cross-flow velocity of gas establishes the residence time for the gas and any entrained liquid droplets in the shell. The most effective removal of entrainment occurs when the residence time is at a maximum practical value. Regardless of the number of gas outlet connections for uniform distribution of gas flow, the cross-sectional area of the gas space is:

$$A_x = \frac{288\ SD \cdot \text{cfm}}{SV \cdot L} \qquad (3)$$

where

A_x = minimum transverse net cross-sectional area or gas space, in.^{2}
SD = average vertical separation distance, in.
cfm = total quantity of gas leaving the vessel
L = inside length of shell, in.
SV = separation velocity for the separation distance used, fpm

For nonuniform distribution of gas flow in the horizontal shell, determine the minimum horizontal distance for gas flow from point of entry to point of exit as follows:

$$RTL = \frac{144\ \text{cfm} \cdot SD}{SV \cdot A_x} \qquad (4)$$

where

RTL = residence time length, in.
cfm = maximum flow for that portion of the shell

All connections must be sized for the flow rates and pressure drops permissible and must be positioned to minimize liquid splashing.

Liquid Overfeed Systems

Internal baffles or mist eliminators can reduce the diameter of vessels; however, test correlations are necessary for a given configuration and placement of these devices.

An alternate formula for determining separation velocities that can be applied to separators is:

$$v = k[(\rho_l - \rho_v)/\rho_v]^{0.5} \quad (5)$$

where

- v = velocity of vapor, fps
- ρ_l = density of liquid, lb/ft^3
- ρ_v = density of vapor, lb/ft^3
- k = factor based on experience without regard to vertical separation distance and surface tension for gravity separators

In gravity liquid/vapor separators that must separate heavy entrainment from vapors, use a k of 0.1. This gives velocities equivalent to those used for 12 to 14 in. vertical separating distance (*SD*) for R-717, and 14 to 16 in. vertical *SD* for halocarbons. In knockout drums that separate light entrainment, use a k of 0.2. This gives velocities equivalent to those used for 36 in. vertical *SD* for R-717 and for halocarbons.

REFERENCES

Chaddock, J.S., H. Lau, and E. Skuchas. 1976. Two-phase pressure drop in refrigerant liquid overfeed systems—Experimental measurements. ASHRAE *Transactions* 82(2):134-50.

Chaddock, J.B., D.P. Werner, and C.G. Papachristou. 1972. Pressure drop in the suction lines of refrigerant circulation systems. ASHRAE *Transactions* 78(2):114-23.

Chisholm, D. 1971. Prediction of pressure drop at pipe fittings during two-phase flow. *Proceedings* I.I.R., Washington, D.C.

Lorentzen, G. 1963. Conditions of cavitation in liquid pumps for refrigerant circulation. *Progress Refrigeration Science Technology* I:497.

Lorentzen, G. 1965. How to design piping for liquid recirculation. *Heating, Piping & Air Conditioning* (June):139.

Lorentzen, G. 1966. On the dimensioning of liquid separators for refrigeration systems. *Kaltetechnik* 18:89.

Lorentzen, G. and R. Gronnerud. 1967. On the design of recirculation type evaporators. *Kulde* 21(4):55.

Miller, D.K. 1971. Recent methods for sizing liquid overfeed piping and suction accumulator-receivers. *Proceedings* I.I.R., Washington, D.C.

Miller D.K. 1974. Refrigeration problems of a VCM carrying tanker. ASHRAE *Journal* 11.

Miller, D.K. 1979. Sizing dual suction risers in liquid overfeed refrigeration systems. *Chemical Engineering* 9.

Niederer, D.H. 1964. Liquid recirculation systems—What rate of feed is recommended. *The Air Conditioning & Refrigeration Business* (December).

Niemeyer, E.R. 1961. Check these points when designing knockout drums. *Hydrocarbon Processing and Petroleum Refiner*, June.

Scheiman, A.D. 1964. Horizontal vapor-liquid separators. *Hydrocarbon Processing and Petroleum Refiner* (May).

Scheiman, A.D. 1963. Size vapor-liquid separators quicker by nomograph. *Hydrocarbon Processing and Petroleum Refiner*, October.

Scotland, W.B. 1963. Discharge temperature considerations with multi-cylinder ammonia compressors. *Modern Refrigeration* (February).

Scotland, W.B. 1970. Advantages, disadvantages and economics of liquid overfeed systems. ASHRAE *Symposium Bulletin* KC-70-3, Liquid overfeed systems.

Soling, S.P. 1971. Oil recovery from low temperature pump recirculating hydrocarbon systems. ASHRAE *Symposium Bulletin* PH-71-2, Effect of oil on the refrigeration system.

Sonders, M. and G.G. Brown. 1934. Design of fractionating columns, entrainment and capacity. *Industrial & Engineering Chemistry* (January).

Stoeker, W.F. 1960. How to design and operate flooded evaporators for cooling air and liquids. *Heating, Piping & Air Conditioning* (December).

Younger, A.H. 1955. How to size future process vessels. *Chemical Engineering* (May).

BIBLIOGRAPHY

Chaddock, J.B. 1976. Two-phase pressure drop in refrigerant liquid overfeed systems—Design tables. ASHRAE *Transactions* 82(2):107-33.

Geltz, R.W. 1967. Pump overfeed evaporator refrigeration systems. *Air Conditioning, Heating & Refrigeration News* (January 30, February 6, March 6, March 13, March 20, March 27).

Lorentzen, G. and A.O. Baglo. 1969. An investigation of a gas pump recirculation system. *Proceedings* of the Xth International Congress of Refrigeration, p. 215. International Institute of Refrigeration, Paris.

Lorentzen, G. 1968. Evaporator design and liquid feed regulation. *Journal of Refrigeration* (November-December): 160.

Richards, W.V. 1959. Liquid ammonia recirculation systems. *Industrial Refrigeration* (June):139.

Richards, W.V. 1970. Pumps and piping in liquid overfeed systems. ASHRAE *Symposium Bulletin* KC-70-3, Liquid overfeed systems.

Slipcevic, B. 1964. The calculation of the refrigerant charge in refrigerating systems with circulation pumps. *Kaltetechnik* 4:111.

Thompson, R.B. 1970. Control of evaporators in liquid overfeed systems. ASHRAE *Symposium Bulletin* KC-70-3, Liquid overfeed systems.

Watkins, J.E. 1956. Improving refrigeration systems by applying established principles. *Industrial Refrigeration* (June).

CHAPTER 3

SYSTEM PRACTICES FOR HALOCARBON REFRIGERANTS

Piping 3.1	*Refrigeration Accessories* 3.19
Refrigerant Line Sizing 3.1	*Head Pressure Control for Refrigerant Condensers* 3.22
Air-Cooled Condensers 3.14	*Keeping Liquid from Crankcase During Off-Cycles* 3.23
Piping at Multiple Compressors 3.15	*Hot-Gas Bypass Arrangements* 3.24
Piping at Various System Components 3.16	

PIPING

A SUCCESSFUL refrigeration system depends on good piping design and an understanding of the accessories required to perform its functions.

The fundamentals of piping and accessories applications in halocarbon refrigerant systems are discussed in this chapter. Ammonia piping applications are covered in Chapter 4. Refer to Chapter 33 of the 1989 ASHRAE *Handbook—Fundamentals* for general refrigeration piping information. Hydrocarbon refrigerant pipe friction data can be found in petroleum industry handbooks. Use the refrigerant properties and information in Chapters 2, 16, and 17 of the 1989 ASHRAE *Handbook—Fundamentals* to calculate friction losses.

Basic Principles

Refrigerant piping systems are designed and operated to:

1. Assure proper refrigerant feed to evaporators.
2. Provide practical refrigerant line sizes without excessive pressure drop.
3. Prevent excessive amounts of lubricating oil from being trapped in any part of the system.
4. Protect the compressor at all times from loss of lubricating oil.
5. Prevent liquid refrigerant or oil slugs from entering the compressor during operating and idle time.
6. Maintain a clean and dry system.

REFRIGERANT LINE SIZING

In sizing refrigerant lines, cost considerations favor keeping line sizes as small as possible. However, suction and discharge line pressure drops cause loss of compressor capacity and increased power. Excessive liquid line pressure drops can cause flashing of the liquid refrigerant, resulting in faulty expansion valve operation.

Refrigeration systems are designed so that friction pressure losses do not exceed a pressure differential equivalent to a corresponding change in the saturation boiling temperature. The primary measure for determining pressure drops is a given change in saturation temperature.

Pressure Drop Considerations

Liquid Lines. Pressure drop should not be so large as to cause gas formation in the liquid line, insufficient liquid pressure at the liquid feed device, or both. Systems are normally designed so that the pressure drop in the liquid line, due to friction, is not greater than that corresponding to about 1 to 2 °F change in saturation temperature. See Tables 1 through 4 for liquid line sizing information. Liquid pressure losses for a change of 1 °F saturation change at 100 °F condensing pressure is approximately:

Refrigerant	psi
R-12	1.8
R-22	2.9
R-502	3.1

The velocity of liquid leaving a partially filled vessel (such as a receiver or shell-and-tube condenser) is limited by the height of the liquid above the point at which the liquid line leaves the vessel, whether or not the liquid at the surface is subcooled. Since the liquid in the vessel has a very low (or zero) velocity, the velocity V in the liquid line (usually at the vena contracta) is $V^2 = 2gh$, where h is the height of the liquid in the vessel. Gas pressure does not add to the velocity without gas flowing in the same direction. As a result, both gas and liquid flow through the line, limiting the rate of liquid flow. If this factor is not considered, excess operating charges in receivers and flooding of shell-and-tube condensers may result.

No specific data are available to precisely size a line leaving a vessel. If the height of the liquid level above the vena contracta produces the desired velocity, the liquid will leave the vessel at the expected rate. Thus, if the level in the vessel falls to one pipe diameter above the bottom of the vessel from which the liquid line leaves, the capacity of lines for R-22 at 3 lb/min per ton of refrigeration is approximately as follows for copper with OD sizes:

Inches OD	Tons
$1\frac{1}{8}$	14
$1\frac{3}{8}$	25
$1\frac{5}{8}$	40
$2\frac{1}{8}$	80
$2\frac{5}{8}$	130
$3\frac{1}{8}$	195
$4\frac{1}{8}$	410

The whole liquid line need not be as large as the leaving connection. After the vena contracta, the velocity is about 40% less. If the line continues down from the receiver, the value of h increases. For a 200-ton capacity with R-22, the line from the bottom of the receiver should be about $3\frac{1}{8}$ in. After a drop of 1 ft, a reduction to $2\frac{1}{8}$ in. is satisfactory.

The preparation of this chapter is assigned to TC 10.3, Refrigerant Piping.

Friction pressure drops in the liquid line are caused by accessories such as solenoid valves, strainer-driers, hand valves, and so forth, as well as by the actual pipe and fittings, from the receiver outlet to the refrigerant feed device at the evaporator.

Suction Lines. A pressure drop in the suction line reduces a system's capacity because it forces the compressor to operate at a lower suction pressure to maintain desired evaporating temperature in the coil. The suction line is normally sized to have a pressure drop from friction no greater than the equivalent of about 2°F change in saturation temperature. See Tables 1 through 7 for suction line sizing information.

The equivalent pressure loss at 40°F saturated suction temperature is approximately:

Refrigerant	Suction Loss, °F	Pressure Loss, psi
R-12	2	1.81
R-22	2	2.94
R-502	2	3.14

At suction temperatures lower than 40°F, the pressure drop equivalent to a given degree change decreases. For example, at −40°F suction with R-22, the pressure drop equivalent to 2°F change in saturation temperature is about 0.8 psi. Therefore, low-temperature lines must be sized for a very low pressure drop or higher equivalent temperature losses, with resultant loss in equipment capacity, must be accepted. For very low pressure drops, any suction or hot *gas risers* must be sized properly to assure oil entrainment up the riser so that the oil is always returned to the compressor.

Where pipe size must be reduced to provide sufficient gas velocity to entrain oil up vertical risers at partial loads, greater pressure drops are imposed at full load. These can usually be compensated for by oversizing the horizontal and downcomer lines.

Discharge Lines. Pressure loss in hot gas lines increase the required compressor power per unit of refrigeration and decrease the compressor capacity. Pressure drop is kept to a minimum by generously sizing the lines for low friction losses, but still maintaining refrigerant line velocities to entrain and carry oil along at all loading conditions. Pressure drop is normally designed not to exceed the equivalent of 2°F change in saturation temperature.

Table 1 Suction, Discharge, and Liquid Line Capacities in Tons for Refrigerant 12 (Single or High-Stage Applications)

Line Size Type L Copper, OD	Suction Lines (Δt = 2°F) Saturated Suction Temperature, °F					Discharge Lines (Δt = 1°F), Δp = 1.9 psi Saturated Suction Temp., °F			Line Size Type L Copper, OD	Liquid Lines	
	−40 Δp=0.49	−20 Δp=0.72	0 Δp=1.01	20 Δp=1.38	40 Δp=1.82	−40	0	40		Vel. = 100 fpm[a]	Δt=1°F Δp = 1.9 psi[b]
½	—	—	—	0.20	0.30	0.38	0.43	0.46	½	1.9	2.2
⅝	—	0.16	0.25	0.38	0.56	0.72	0.80	0.87	⅝	3.0	4.1
⅞	0.25	0.42	0.67	1.0	1.5	1.9	2.10	2.30	⅞	6.2	10.8
1⅛	0.51	0.86	1.4	2.1	3.0	3.8	4.3	4.6	1⅛	10.5	21.9
1⅜	0.90	1.5	2.4	3.6	5.3	6.7	7.4	8.1	1⅜	16.0	38.3
1⅝	1.4	2.4	3.8	5.7	8.3	10.6	11.7	12.8	1⅝	22.7	60.7
2⅛	3.0	5.0	7.8	11.8	17.3	21.9	24.2	26.4	2⅛	39.5	126.4
2⅝	5.3	8.8	13.9	21.0	30.5	38.6	42.8	46.8	2⅝	60.9	223.8
3⅛	8.4	14.1	22.1	33.4	48.6	61.6	68.2	74.5	3⅛	86.9	357.7
3⅝	12.6	20.9	32.9	49.7	72.2	91.4	101.2	110.6	3⅝	117.6	532.3
4⅛	17.8	29.5	46.5	70.1	101.9	128.8	142.6	155.8	4⅛	152.9	751.7
Steel									**Steel**		
IPS / SCH									IPS / SCH		
½ 40	—	—	0.30	0.44	0.64	0.8	0.9	1.0	½ 80	3.0	3.4
¾ 40	0.25	0.40	0.63	0.93	1.3	1.7	1.9	2.0	¾ 80	5.5	7.7
1 40	0.47	0.76	1.2	1.8	2.5	3.2	3.5	3.8	1 80	9.2	15.1
1¼ 40	0.97	1.6	2.4	3.6	5.2	6.5	7.2	7.9	1¼ 80	16.4	32.5
1½ 40	1.5	2.4	3.7	5.5	7.8	9.8	10.9	11.9	1½ 80	22.6	49.5
2 40	2.8	4.6	7.1	10.6	15.1	18.9	20.9	22.9	2 40	42.8	115.0
2½ 40	4.5	7.3	11.3	16.8	24.1	30.1	33.4	36.5	2½ 40	61.1	183.6
3 40	8.0	13.0	20.0	29.7	42.6	53.2	59.0	64.4	3 40	94.3	324.3
4 40	16.3	26.4	40.8	60.6	86.8	108.3	119.9	131.0	4 40	162.5	662.4

Notes:
1. Table capacities are in tons of refrigeration.
   Δp = pressure drop due to line friction, psi per 100 ft of equivalent line length
   Δt = corresponding change in saturation temperature per 100 ft, °F
2. Line capacity for other saturation temperatures Δt and equivalent lengths L_e

$$\text{Line capacity} = \text{Table capacity} \left(\frac{\text{Table } L_e}{\text{Actual } L_e} \times \frac{\text{Actual } \Delta t}{\text{Table } \Delta t} \right)^{0.55}$$

3. Saturation temperature Δt for other capacities and equivalent lengths L_e

$$\Delta t = \text{Table } Dt \frac{\text{Actual } L_e}{\text{Table } L_e} \left(\frac{\text{Actual capacity}}{\text{Table capacity}} \right)^{1.8}$$

4. Values in the table are based on 105°F condensing temperature. Multiply table capacities by the following factors for other condensing temperatures.

Condensing Temperature, °F	Suction Line	Discharge Line
80	1.11	0.81
90	1.06	0.89
100	1.02	0.97
110	0.98	1.05
120	0.94	1.13
130	0.87	1.20
140	0.82	1.26
150	0.76	1.33

[a] The sizing shown is recommended where any gas generated in the receiver must return up the condensate line to the condenser without restricting condensate flow. Water-cooled condensers, where the receiver ambient temperature may be higher than the refrigerant condensing temperature, fall in this category.

[b] The line pressure drop Δp is conservative; if subcooling is substantial or the line is short, a smaller size line may be used. Applications with very little subcooling or very long lines may require a larger line.

System Practices for Halocarbon Refrigerants

Refrigerant Line Capacity Tables

Tables 1 to 4 show capacities for R-12, 22, and 502 at specific pressure drops. The capacities shown in the tables are based on the refrigerant flow that develops a friction loss, per 100 ft of equivalent pipe length, corresponding to a 1°F change in the saturation temperature Δt for discharge and liquid lines. Suction lines are based on a 2°F change. Tables 5, 6, and 7 show suction line capacities for a 0.5 and 1°F change of the saturation suction temperature. Pressure drops are given in degrees because this pipe sizing method is convenient and accepted throughout the industry. Corresponding pressure drops are also shown.

The refrigerant line sizing capacity tables are based on the Darcy-Weisbach relation and friction factors as computed by the Colebrook function (1938, 1939). Tubing roughness factors were 0.000005 for copper and 0.00015 for steel pipe. Viscosity extrapolations and adjustments for pressures other than 1 atm were based on correlation techniques as presented by Keating and Matula (1969). Discharge gas superheat was 80°F for R-12 and R-502, and 105°F for R-22.

The refrigerant cycle for determining capacity is based on saturated gas leaving the evaporator. The calculations neglect the presence of oil and assume nonpulsating flow.

Equivalent Lengths of Valves and Fittings

Refrigerant line capacity tables are based on unit pressure drop per 100 ft length of straight pipe; or a combination of straight pipe, fittings, and valves with friction drop equivalent to a 100 ft length of straight pipe.

Generally, pressure drop through valves and fittings is determined by establishing the equivalent straight length of pipe of the same size with the same friction drop. Line sizing tables can then be used directly. Tables 8, 9, and 10 give equivalent lengths of straight pipe for various fittings and valves, based on nominal pipe sizes.

The following example illustrates the use of various tables and charts to size refrigerant lines.

Example 1. Determine the line size and pressure drop equivalent (in degrees) for the suction line of a 30 ton R-22 system, operating at 40°F suction and 100°F condensing temperatures. The suction line is copper tubing, with 50 ft of straight pipe and 6 long radius elbows.

Solution: Add 50% to the straight length of pipe to establish a trial equivalent length. Trial equivalent length is $50 \times 1.5 = 75$ ft. From Table 2 (for 40°F suction, 105°F condensing), 33.1 tons capacity in 2-1/8 in. OD results in a 2°F loss per 100 ft equivalent length. Referring to Note 4,

Table 2 Suction, Discharge, and Liquid Line Capacities for Refrigerant 22 (Single or High-Stage Applications)

Line Size Type L Copper, OD	Suction Lines ($\Delta t = 2°F$) Saturated Suction Temperature, °F					Discharge Lines ($\Delta t = 1°F$, $\Delta p = 1.9$ psi) Saturated Suction Temp., °F		Line Size Type L Copper, OD	Liquid Lines	
	−40 $\Delta p = 0.79$	−20 $\Delta p = 1.15$	0 $\Delta p = 1.6$	20 $\Delta p = 2.22$	40 $\Delta p = 2.91$	−40	40		Vel. = 100 fpm[a]	$\Delta t = 1°F$ $\Delta p = 3.05$ psi[b]
½	—	—	—	0.40	0.6	0.75	0.85	½	2.3	3.6
⅝	—	0.32	0.51	0.76	1.1	1.4	1.6	⅝	3.7	6.7
⅞	0.52	0.86	1.3	2.0	2.9	3.7	4.2	⅞	7.8	18.2
1⅛	1.1	1.7	2.7	4.0	5.8	7.5	8.5	1⅛	13.2	37.0
1⅜	1.9	3.1	4.7	7.0	10.1	13.1	14.8	1⅜	20.2	64.7
1⅝	3.0	4.8	7.5	11.1	16.0	20.7	23.4	1⅝	28.5	102.5
2⅛	6.2	10.0	15.6	23.1	33.1	42.8	48.5	2⅛	49.6	213.0
2⅝	10.9	17.8	27.5	40.8	58.3	75.4	85.4	2⅝	76.5	376.9
3⅛	17.5	28.4	44.0	65.0	92.9	120.2	136.2	3⅛	109.2	601.5
3⅝	26.0	42.3	65.4	96.6	137.8	178.4	202.1	3⅝	147.8	895.7
4⅛	36.8	59.6	92.2	136.3	194.3	251.1	284.4	4⅛	192.1	1263.2
Steel								**Steel**		
IPS / SCH								IPS / SCH		
½ / 40	—	0.38	0.58	0.85	1.2	1.5	1.7	½ / 80	3.8	5.7
¾ / 40	0.50	0.8	1.2	1.8	2.5	3.3	3.7	¾ / 80	6.9	12.8
1 / 40	0.95	1.5	2.3	3.4	4.8	6.1	6.9	1 / 80	11.5	25.2
1¼ / 40	2.0	3.2	4.8	7.0	9.9	12.6	14.3	1¼ / 80	20.6	54.1
1½ / 40	3.0	4.7	7.2	10.5	14.8	19.0	21.5	1½ / 80	28.3	82.6
2 / 40	5.7	9.1	13.9	20.2	28.5	36.6	41.4	2 / 40	53.8	192.0
2½ / 40	9.2	14.6	22.1	32.2	45.4	58.1	65.9	2½ / 40	76.7	305.8
3 / 40	16.2	25.7	39.0	56.8	80.1	102.8	116.4	3 / 40	118.5	540.3
4 / 40	33.1	52.5	79.5	115.9	163.2	209.5	237.3	4 / 40	204.2	1101.2

Notes:
1. Table capacities are in tons of refrigeration.
   Δp = pressure drop due to line friction, psi per 100 ft of equivalent line length
   Δt = corresponding change in saturation temperature per 100 ft, °F
2. Line capacity for other saturation temperatures Δt and equivalent lengths L_e
   $$\text{Line capacity} = \text{Table capacity} \left(\frac{\text{Table } L_e}{\text{Actual } L_e} \times \frac{\text{Actual } \Delta t}{\text{Table } \Delta t} \right)^{0.55}$$
3. Saturation temperature Δt for other capacities and equivalent lengths L_e
   $$\Delta t = \text{Table } \Delta t \frac{\text{Actual } L_e}{\text{Table } L_e} \left(\frac{\text{Actual capacity}}{\text{Table capacity}} \right)^{1.8}$$
4. Values in the table are based on 105°F condensing temperature. Multiply table capacities by the following factors for other condensing temperatures.

Condensing Temperature, °F	Suction Line	Discharge Line
80	1.11	0.79
90	1.07	0.88
100	1.03	0.95
110	0.97	1.04
120	0.90	1.10
130	0.86	1.18
140	0.80	1.26

[a] The sizing shown is recommended where any gas generated in the receiver must return up the condensate line to the condenser without restricting condensate flow. Water-cooled condensers, where the receiver ambient temperature may be higher than the refrigerant condensing temperature, fall in this category.

[b] The line pressure drop Δp is conservative; if subcooling is substantial or the line is short, a smaller size line may be used. Applications with very little subcooling or very long lines may require a larger line.

Table 1, the capacity at 40/100 is 1.03 × 33.1 = 34.1 tons. This trial size is used to evaluate actual equivalent length.

Straight pipe length	= 50 ft
Six 2-in.-long-radius elbows at 3.3 ft each (Table 8)	= 19.8 ft
Total Equivalent Length	= 69.8 ft

The actual pressure loss in degrees (Note 3, Table 2)

$$\Delta t = 2 \times \frac{69.8}{100} \left(\frac{30}{34.1}\right)^{1.8} = 1.1°F \text{ or } 1.9 \text{ psi}$$

Liquid Lines

Liquid lines should be designed so that slightly subcooled liquid reaches the liquid feed device at a pressure high enough for proper operation. Two factors must be considered:

1. The liquid line and its valves and accessories must be sized for a pressure drop to prevent flash gas because of friction losses.
2. Precautions should be taken to prevent flash gas in the liquid line, or to handle it if prevention is not possible.

Suction Lines

Design Considerations. Suction lines are more critical than liquid and discharge lines from a design and construction standpoint. Refrigerant lines should be sized to (1) provide a minimum pressure drop at full load, (2) return oil from the evaporator to the compressor under minimum load conditions, and (3) prevent oil from draining from an active evaporator into an idle one.

Oil Circulation. Some lubricating oil is lost from all compressors during normal operation. Since oil inevitably leaves the compressor with the discharge gas, systems using halocarbon refrigerants must return this oil at the same rate at which it leaves (Cooper 1971).

Oil that leaves the compressor or oil separator reaches the condenser and dissolves in the liquid refrigerant, enabling it to pass readily through the liquid line to the evaporator. In the evaporator, the refrigerant evaporates and the liquid phase becomes enriched in oil. The concentration of refrigerant in the oil depends on the evaporator temperature and types of refrigerant and oil used. The viscosity of the oil/refrigerant solution is determined by the system parameters. Oil separated in the evaporator is returned to the compressor by gravity or by the drag forces of the returning gas.

The effect of oil on pressure drop is large, increasing the pressure drop by as much as a factor of ten in some cases (Alofs et al. 1990).

System Capacity Reduction. The use of automatic capacity control on compressors requires careful analysis and design. The compressor is capable of loading and unloading as it floats with the system load requirements through a considerable range of capacity variation. A single compressor can unload down to 25% of full-load capacity, while multiple compressors connected in parallel can unload to a system capacity of 12.5% or lower. System piping must be designed to return oil at the lowest loading, yet not impose excessive pressure drops in the piping and equipment at full load.

Oil Return Up Suction Risers. Many refrigeration piping systems contain a suction riser because the evaporator is at a lower level than the compressor. Oil circulating in the system can return up gas risers only by being transported by the returning gas or by auxiliary means such as a trap and a pump. The minimum conditions for oil transport correlate with buoyancy forces, *i.e.* the density difference between the liquid and the vapor, and the momentum flux of the vapor (Jacobs et al. 1976).

The principal criteria determining the transport of oil are gas velocity, gas density, and pipe inside diameter. The density of the

Table 3 Suction, Discharge, and Liquid Line Capacities in Tons for Refrigerant 502 (Single or High-Stage Applications)

Line Size Type L Copper, OD	Suction Lines ($\Delta t = 2°F$)					Discharge Lines ($\Delta t = 1°F$, $\Delta p = 1.9$ psi)			Line Size Type L Copper, OD	Liquid Lines	
	Saturated Suction Temperature, °F					Saturated Suction Temp., °F				Vel. = 100 fpm[a]	$\Delta t = 1°F$ $\Delta p = 3.15$ psi[b]
	-40 $\Delta p = 0.92$	-20 $\Delta p = 0.133$	0 $\Delta p = 1.84$	20 $\Delta p = 2.45$	40 $\Delta p = 3.18$	-40	0	40			
½	0.08	0.14	0.22	0.33	0.49	0.56	0.63	0.70	½	1.5	2.4
⅝	0.16	0.27	0.42	0.63	0.91	1.0	1.2	1.3	⅝	2.3	4.5
⅞	0.43	0.70	1.1	1.7	2.4	2.7	3.1	3.4	⅞	4.9	11.8
1⅛	0.87	1.4	2.2	3.4	4.8	5.5	6.3	7.0	1⅛	8.3	24.1
1⅜	1.5	2.5	3.9	5.8	8.4	9.6	10.9	12.1	1⅜	12.6	42.0
1⅝	2.4	4.0	6.2	9.2	13.3	15.2	17.2	19.1	1⅝	17.9	66.4
2⅛	5.0	8.2	12.8	19.1	27.5	31.4	35.6	39.5	2⅛	31.1	138.0
2⅝	8.8	14.5	22.6	33.7	48.4	55.3	62.8	69.5	2⅝	48.0	243.7
3⅛	14.1	23.2	36.0	53.7	77.0	87.9	99.8	110.5	3⅛	68.4	389.3
3⅝	21.0	34.4	53.5	79.7	114.3	130.5	148.1	164.0	3⅝	92.6	579.0
4⅛	29.7	48.5	75.4	112.3	161.0	183.7	208.4	230.9	4⅛	120.3	816.9
5⅛	53.2	86.7	134.6	200.3	287.1	327.3	371.3	411.3	—	—	—
6⅛	85.6	139.5	216.2	321.3	460.6	525.2	595.9	660.1	—	—	—

Notes:
1. Table capacities are in tons of refrigeration.
   Δp = pressure drop due to line friction, psi per 100 ft of equivalent line length
   Δt = corresponding change in saturation temperature per 100 ft, °F
2. Line capacity for other saturation temperatures Δt and equivalent lengths L_e

   $$\text{Line capacity} = \text{Table capacity} \left(\frac{\text{Table } L_e}{\text{Actual } L_e} \times \frac{\text{Actual } \Delta t}{\text{Table } \Delta t}\right)^{0.55}$$

3. Saturation temperature Δt for other capacities and equivalent lengths L_e

   $$\Delta t = \text{Table } \Delta t \frac{\text{Actual } L_e}{\text{Table } L_e} \left(\frac{\text{Actual capacity}}{\text{Table capacity}}\right)^{1.8}$$

4. Values in the table are based on 105°F condensing temperature. Multiply table capacities by the following factors for other condensing temperatures.

Condensing Temperature, °F	Suction Line	Discharge Line
80	1.20	0.83
90	1.12	0.91
100	1.04	0.97
110	0.96	1.02
120	0.88	1.08
130	0.80	1.16

[a] The sizing shown is recommended where any gas generated in the receiver must return up the condensate line to the condenser without restricting condensate flow. Water-cooled condensers, where the receiver ambient temperature may be higher than the refrigerant condensing temperature, fall in this category.

[b] The line pressure drop Δp is conservative; if subcooling is substantial or the line is short, a smaller size line may be used. Applications with very little subcooling or very long lines may require a larger line.

System Practices for Halocarbon Refrigerants

oil-refrigerant mixture plays a somewhat lesser role, since it is almost constant over a wide range. In addition, at temperatures somewhat lower than −40°F, oil viscosity may be significant.

Greater gas velocities are required as the temperature drops and the gas becomes less dense. Higher velocities are also necessary if the pipe diameter increases. Table 11 translates these criteria to minimum refrigeration capacity requirements for oil transport.

Suction risers must be sized for minimum system capacity. Oil must be returned to the compressor at the operating condition corresponding to the *minimum* displacement and *minimum* suction temperature at which the compressor will operate. When suction or evaporator pressure regulators are used, suction risers must be sized for actual gas conditions in the riser.

For a single compressor with capacity control, the minimum capacity is the lowest capacity at which the unit can operate. For multiple compressors with capacity control, the minimum capacity is the lowest at which the last operating compressor can run.

Riser Sizing. Example 2 describes the use of Tables 11 and 12 in establishing maximum riser sizes for satisfactory oil transport down to minimum partial loading.

Example 2. Determine the maximum size suction riser that will transport oil at the minimum loading, using R-22 with a 40-ton compressor with a capacity in steps of 25, 50, 75, and 100%. Assume the minimum system loading is 10 tons at 40°F suction and 105°F condensing temperatures with 15°F superheat.

Solution: From Table 11, a 2.125-in. OD pipe at 40°F suction and 90°F condensing temperature has a minimum capacity of 7.57 tons. When corrected to 105°F condensing, the minimum capacity becomes 7.3 tons. Therefore, the 2.125-in. OD pipe is suitable.

Based on Table 11, the next smaller line size should be used for marginal suction risers. When vertical riser sizes are reduced to provide satisfactory minimum gas velocities, the pressure drop at full load increases considerably; horizontal lines should be sized to keep the total pressure drop within practical limits. As long as the horizontal lines are level or pitched in the direction of the compressor, oil can be transported with normal design velocities.

Because most compressors have multiple capacity reduction features, gas velocities required to return oil up through vertical suction risers under all load conditions are difficult to maintain. When the suction riser is sized to permit oil return at the minimum operating capacity of the system, the pressure drop in this portion of the line may be too great when operating at full load. If a correctly sized suction riser imposes too great a pressure drop at full load, a double-suction riser should be used.

Table 4 Suction, Discharge, and Liquid Line Capacities in Tons for Intermediate or Low-Stage Duty for Refrigerants 12 and 22

Refrigerant and Δt Equivalent of Friction Drop[a]	Line Size Type L Copper OD	Suction Lines							Discharge Lines[a]	Line Size Type L Copper OD	Liquid Lines
		Saturated Suction Temperature, °F									
		−90	−80	−70	−60	−50	−40	−30			
Refrigerant 12	⅞				0.23	0.30	0.40	0.51	0.9	⅞	
	1⅛	0.17	0.25	0.34	0.46	0.62	0.81	1.1	1.9	1⅛	
	1⅜	0.31	0.44	0.60	0.81	1.1	1.4	1.8	3.3	1⅜	
	1⅝	0.49	0.69	0.95	1.3	1.7	2.3	2.9	5.3	1⅝	
	2⅛	1.0	1.4	2.0	2.7	3.6	4.7	6.1	11.0	2⅛	
2°F Δt per 100 ft equivalent length	2⅝	1.8	2.6	3.6	4.8	6.4	8.4	10.8	19.5	2⅝	See Table 1
	3⅛	3.0	4.2	5.7	7.7	10.3	13.4	17.2	31.1	3⅛	
	3⅝	4.4	6.2	8.5	11.6	15.3	20.0	25.6	46.2	3⅝	
	4⅛	6.2	8.8	12.1	16.3	21.7	28.2	36.2	65.2	4⅛	
	5⅛	11.3	15.8	21.8	29.4	38.9	50.7	65.0	116.7	5⅛	
	6⅛	18.2	25.6	35.2	47.4	62.8	81.6	104.8	188.0	6⅛	
Refrigerant 22	⅝								0.7	⅝	
	⅞	0.18	0.25	0.34	0.46	0.61	0.79	1.0	1.9	⅞	
	1⅛	0.36	0.51	0.70	0.94	1.2	1.6	2.1	3.8	1⅛	
	1⅜	0.6	0.9	1.2	1.6	2.2	2.8	3.6	6.6	1⅜	
	1⅝	1.0	1.4	1.9	2.6	3.4	4.5	5.7	10.5	1⅝	
	2⅛	2.1	3.0	4.1	5.5	7.2	9.3	11.9	21.7	2⅛	
2°F Δt per 100 ft equivalent length	2⅝	3.8	5.3	7.2	9.7	12.7	16.5	21.1	38.4	2⅝	See Table 2
	3⅛	6.1	8.5	11.6	15.5	20.4	26.4	33.8	61.4	3⅛	
	3⅝	9.1	12.7	17.3	23.1	30.4	39.4	50.2	91.2	3⅝	
	4⅛	12.9	18.0	24.5	32.7	43.0	55.6	70.9	128.6	4⅛	
	5⅛	23.2	32.3	43.9	58.7	77.1	99.8	126.9	229.5	5⅛	
	6⅛	37.5	52.1	71.0	94.6	124.2	160.5	204.2	369.4	6⅛	

Notes:
1. Table capacities are in tons of refrigeration.
   Δp = pressure drop due to line friction, psi per 100 ft of equivalent line length
   Δt = corresponding change in saturation temperature per 100 ft, °F
2. Line capacity for other saturation temperatures Δt and equivalent lengths L_e
   $$\text{Line capacity} = \text{Table capacity} \left(\frac{\text{Table } L_e}{\text{Actual } L_e} \times \frac{\text{Actual } \Delta t}{\text{Table } \Delta t} \right)^{0.55}$$
3. Saturation temperature Δt for other capacities and equivalent lengths L_e
   $$\Delta t = \text{Table } \Delta t \frac{\text{Actual } L_e}{\text{Table } L_e} \left(\frac{\text{Actual capacity}}{\text{Table capacity}} \right)^{1.8}$$
4. Refer to the refrigerant thermodynamic property tables (Chapter 17, 1989 ASHRAE *Handbook—Fundamentals*) for the pressure drop corresponding to Δt.
5. Values in the table are based on 105°F condensing temperature. Multiply table capacities by the following factors for other condensing temperatures. Flow rates for discharge lines are based on −50°F evaporating temperature.

Condensing Temp., °F	Refrigerant 12		Refrigerant 22	
	Suction	Discharge	Suction	Discharge
−30	1.12	0.55	1.09	0.58
−20	1.07	0.70	1.06	0.71
−10	1.03	0.85	1.03	0.85
0	1.00	1.00	1.00	1.00
10	0.96	1.25	0.97	1.20
20	0.93	1.50	0.94	1.45
30	0.90	1.80	0.90	1.80

[a] See section titled "Pressure Drop Considerations."

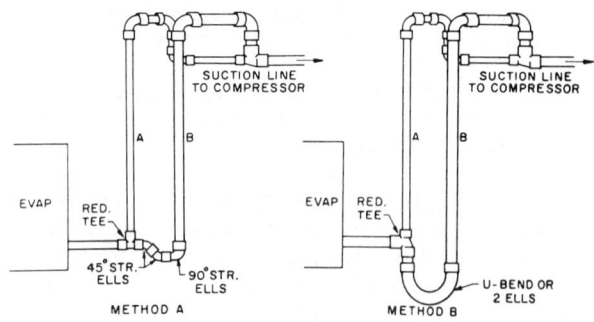

Fig. 1 Double Suction Riser Construction

Oil Return Up Suction Risers—Multistage Systems. The movement of oil in the suction lines of multistage systems requires the same design approach as in single-stage systems. In low-temperature systems where high oil viscosity prevents flow, apply viscosity-reducing additives to keep the oil fluid.

When refrigerants other than those listed in Tables 11 and 12 are used, follow the recommendations listed in Table 13. For oil to flow up along a pipe wall, a certain minimum drag of the gas flow is required. This can be represented by the friction gradient. Table 13 shows values proven satisfactory for minimum friction gradients.

Double Suction Risers. Figure 1 shows two methods of double suction riser construction. While oil return in this arrangement is accomplished at minimum loads, it does not cause excessive pressure drops at full load. The sizing and operation of a double-suction riser is described as follows:

1. Riser A is sized to return oil at the minimum load possible.
2. Riser B is sized for satisfactory pressure drop through both risers at full load. The usual method is to size riser B so that the combined cross-sectional area of A and B is equal to or slightly greater than the cross-sectional area of a single pipe, which would be sized for an acceptable pressure drop at full load without regard for oil return at minimum load. The combined cross-sectional area, however, should not be greater than the cross-sectional area of a single pipe that would return oil in an up-flow riser under maximum load conditions.
3. A trap is introduced between the two risers, as shown in both methods. During partial load operation, the gas velocity is not sufficient to return oil through both risers and the trap gradually fills up with oil until the second riser B is sealed off. The gas then travels up riser A only, and now has enough velocity to carry oil along with it back into the horizontal suction main.

The trap's oil holding capacity is limited to a minimum by close-coupling the fittings at the bottom of the risers. If this is not done, the trap can accumulate enough oil on partial load operation to lower the compressor crankcase oil level. Note in Figure 1 that Riser B forms an inverted loop and enters the horizontal suction line from the top. This prevents oil drainage into this riser, which may be idle during partial load operation. The same purpose can be served by going horizontally into the main, providing it is larger in size than either riser.

Often, double suction risers are essential on low-temperature systems that can tolerate very little pressure drop. Any system using these risers should include a suction trap (accumulator) and a means of returning oil gradually.

Table 5 Suction Line Capacities in Tons for Refrigerant 12 (Single or High-Stage Applications) for Pressure Drops of 1.0 and 0.5 °F/100 ft Equivalent

Line Size Type L Copper, OD		Saturated Suction Temperature, °F								
		−40		−20		0		20		40	
		$\Delta t = 1°F$ $\Delta p = 0.243$	$\Delta t = 0.5°F$ $\Delta p = 0.122$	$\Delta t = 1°F$ $\Delta p = 0.358$	$\Delta t = 0.5°F$ $\Delta p = 0.179$	$\Delta t = 1°F$ $\Delta p = 0.506$	$\Delta t = 0.5°F$ $\Delta p = 0.253$	$\Delta t = 1°F$ $\Delta p = 0.689$	$\Delta t = 0.5°F$ $\Delta p = 0.345$	$\Delta t = 1°F$ $\Delta p = 0.910$	$\Delta t = 0.5°F$ $\Delta p = 0.455$
1/2		0.03	0.02	0.06	0.04	0.09	0.06	0.14	0.09	0.20	0.14
5/8		0.06	0.04	0.11	0.07	0.17	0.12	0.26	0.18	0.39	0.26
3/4		0.11	0.07	0.18	0.12	0.29	0.19	0.44	0.30	0.64	0.44
7/8		0.17	0.11	0.29	0.19	0.46	0.31	0.69	0.47	1.02	0.70
1 1/8		0.35	0.24	0.58	0.40	0.93	0.63	1.41	0.97	2.07	1.42
1 3/8		0.61	0.42	1.02	0.70	1.63	1.11	2.47	1.69	3.62	2.48
1 5/8		0.97	0.66	1.63	1.11	2.58	1.76	3.92	2.68	5.73	3.93
2 1/8		2.03	1.38	3.39	2.32	5.38	3.68	8.14	5.59	11.90	8.18
2 5/8		3.61	2.47	6.03	4.13	9.54	6.54	14.42	9.92	21.04	14.49
3 1/8		5.79	3.96	9.65	6.62	15.24	10.45	23.08	15.85	33.60	23.18
3 5/8		8.64	5.91	14.37	9.87	22.70	15.57	34.34	23.63	49.95	34.48
4 1/8		12.22	8.36	20.36	13.95	32.06	22.02	48.39	33.33	70.46	48.69
Steel											
IPS	SCH										
3/8	80	0.03	0.02	0.05	0.03	0.07	0.05	0.11	0.08	0.16	0.11
1/2	80	0.06	0.04	0.09	0.06	0.15	0.10	0.22	0.15	0.32	0.22
3/4	80	0.13	0.09	0.21	0.15	0.33	0.23	0.50	0.35	0.71	0.50
1	80	0.26	0.18	0.42	0.29	0.65	0.46	0.97	0.68	1.40	0.93
1 1/4	40	0.68	0.47	1.11	0.77	1.72	1.20	2.56	1.80	3.68	2.58
1 1/2	40	1.02	0.71	1.67	1.16	2.58	1.81	3.84	2.70	5.52	3.88
2	40	1.98	1.37	3.23	2.25	5.00	3.50	7.42	5.22	10.65	7.50
2 1/2	40	3.15	2.20	5.15	3.60	7.96	5.59	11.84	8.33	16.96	11.95
3	40	5.60	3.91	9.11	6.38	14.08	9.91	20.91	14.72	30.04	21.14
4	40	11.46	8.01	18.60	13.05	28.75	20.22	42.72	30.09	61.21	43.09
5	40	20.71	14.51	33.56	23.59	51.87	36.57	77.02	54.25	110.23	77.76
6	40	33.55	23.54	54.38	38.22	83.97	59.27	124.67	87.71	178.47	126.06
8	40	68.83	48.33	111.47	78.53	171.92	121.34	254.97	179.81	365.59	257.79
10	40	124.77	87.58	201.84	142.17	311.34	219.74	461.88	326.03	660.26	466.25
12	ID	199.85	140.44	322.82	227.67	498.57	351.39	737.68	520.11	1056.24	745.84

Δp = pressure drop due to line friction, psi per 100 ft equivalent length of line
Δt = corresponding change in saturation temperature with the pressure drop, °F/100 ft

System Practices for Halocarbon Refrigerants

Table 6 Suction Line Capacities in Tons for Refrigerant 22 (Single or High-Stage Applications)

Line Size Type L Copper, OD	Saturated Suction Temperature, °F									
	−40		−20		0		20		40	
	Δt = 1°F Δp = 0.393	Δt = 0.5°F Δp = 0.197	Δt = 1°F Δp = 0.577	Δt = 0.5°F Δp = 0.289	Δt = 1°F Δp = 0.813	Δt = 0.5°F Δp = 0.406	Δt = 1°F Δp = 1.104	Δt = 0.5°F Δp = 0.552	Δt = 1°F Δp = 1.455	Δt = 0.5°F Δp = 0.727
½	0.07	0.05	0.12	0.08	0.18	0.12	0.27	0.19	0.40	0.27
⅝	0.13	0.09	0.22	0.15	0.34	0.23	0.52	0.35	0.75	0.51
¾	0.22	0.15	0.37	0.25	0.58	0.39	0.86	0.59	1.24	0.85
⅞	0.35	0.24	0.58	0.40	0.91	0.62	1.37	0.93	1.97	1.35
1⅛	0.72	0.49	1.19	0.81	1.86	1.27	2.77	1.90	3.99	2.74
1⅜	1.27	0.86	2.09	1.42	3.25	2.22	4.84	3.32	6.96	4.78
1⅝	2.02	1.38	3.31	2.26	5.16	3.53	7.67	5.26	11.00	7.57
2⅛	4.21	2.88	6.90	4.73	10.71	7.35	15.92	10.96	22.81	15.73
2⅝	7.48	5.13	12.23	8.39	18.97	13.04	28.19	19.40	40.38	27.84
3⅛	11.99	8.22	19.55	13.43	30.31	20.85	44.93	31.00	64.30	44.44
3⅝	17.89	12.26	29.13	20.00	45.09	31.03	66.81	46.11	95.68	66.09
4⅛	25.29	17.36	41.17	28.26	63.71	43.85	94.25	65.12	134.81	93.22

Steel											
IPS	SCH										
⅜	80	0.06	0.04	0.10	0.07	0.15	0.10	0.21	0.15	0.30	0.21
½	80	0.12	0.08	0.19	0.13	0.29	0.20	0.42	0.30	0.60	0.42
¾	80	0.27	0.18	0.43	0.30	0.65	0.46	0.95	0.67	1.35	0.95
1	80	0.52	0.36	0.84	0.59	1.28	0.89	1.87	1.31	2.64	1.86
1¼	40	1.38	0.96	2.21	1.55	3.37	2.36	4.91	3.45	6.93	4.88
1½	40	2.08	1.45	3.32	2.33	5.05	3.55	7.38	5.19	10.42	7.33
2	40	4.03	2.81	6.41	4.51	9.74	6.85	14.22	10.01	20.07	14.14
2½	40	6.43	4.49	10.23	7.19	15.56	10.93	22.65	15.95	31.99	22.53
3	40	11.38	7.97	18.11	12.74	27.47	19.34	40.10	28.23	56.52	39.79
4	40	23.24	16.30	36.98	26.02	56.12	39.49	81.73	57.53	115.24	81.21
5	40	42.04	29.50	66.73	47.05	101.16	71.27	147.36	103.82	207.59	146.38
6	40	68.04	47.86	108.14	76.15	163.77	115.21	238.29	168.07	335.71	236.70
8	40	139.48	98.06	221.17	155.78	334.94	236.21	488.05	344.19	686.71	484.74
10	40	252.38	177.75	400.53	282.05	606.74	427.75	881.59	622.51	1243.64	876.79
12	ID	403.63	284.69	639.74	451.09	969.02	683.22	1410.30	995.80	1987.29	1402.63

Table 7 Suction Line Capacities in Tons for Refrigerant 502 (Single or High-Stage Applications)

Line Size Type L Copper, OD	Saturated Suction Temperature, °F										
	−60		−40		−20		0		20		40
	Δt = 1°F Δp = 0.307	Δt = 0.5°F Δp = 0.153	Δt = 1°F Δp = 0.462	Δt = 0.5°F Δp = 0.231	Δt = 1°F Δp = 0.666	Δt = 0.5°F Δp = 0.333	Δt = 1°F Δp = 0.919	Δt = 0.5°F Δp = 0.460	Δt = 1°F Δp = 1.227	Δt = 0.5°F Δp = 0.614	Δt = 1°F Δp = 1.590 Δt = 0.5°F Δp = 0.795
½	0.03	0.02	0.06	0.04	0.10	0.06	0.15	0.10	0.23	0.16	0.33 0.23
⅝	0.06	0.04	0.11	0.07	0.18	0.12	0.29	0.19	0.43	0.29	0.63 0.43
¾	0.10	0.07	0.18	0.12	0.30	0.21	0.48	0.32	0.72	0.49	1.04 0.71
⅞	0.17	0.11	0.29	0.20	0.48	0.33	0.76	0.52	1.14	0.78	1.65 1.13
1⅛	0.34	0.23	0.59	0.40	0.98	0.67	1.54	1.05	2.30	1.58	3.34 2.29
1⅜	0.59	0.40	1.04	0.71	1.71	1.17	2.68	1.84	4.02	2.76	5.81 4.00
1⅝	0.94	0.64	1.64	1.12	2.71	1.86	4.25	2.91	6.36	4.38	9.19 6.33
2⅛	1.96	1.34	3.43	2.34	5.64	3.87	8.81	6.06	13.19	9.09	19.00 13.14
2⅝	3.48	2.39	6.08	4.17	9.98	6.85	15.39	10.74	23.30	16.07	33.55 23.21
3⅛	5.59	3.83	9.72	6.67	15.94	10.97	24.90	17.17	37.16	25.69	53.48 37.03
3⅝	8.32	5.71	14.48	9.93	23.72	16.33	37.02	25.55	55.20	38.21	79.46 55.06
4⅛	11.77	8.08	20.45	14.06	33.49	23.07	52.22	36.08	77.93	53.89	111.96 77.70

Steel												
IPS	SCH											
⅜	80	0.03	0.02	0.05	0.03	0.08	0.05	0.12	0.08	0.18	0.12	0.25 0.18
½	80	0.05	0.04	0.09	0.07	0.15	0.11	0.23	0.16	0.35	0.24	0.49 0.35
¾	80	0.12	0.09	0.21	0.15	0.34	0.24	0.53	0.37	0.78	0.55	1.11 0.78
1	80	0.24	0.17	0.42	0.29	0.67	0.47	1.04	0.73	1.53	1.07	2.17 1.53
1¼	40	0.65	0.45	1.10	0.77	1.77	1.24	2.72	1.91	4.01	2.82	5.70 4.02
1½	40	0.97	0.68	1.66	1.16	2.67	1.87	4.09	2.87	6.03	4.23	8.56 6.03
2	40	1.88	1.81	3.20	2.24	5.14	3.62	7.88	5.55	11.60	8.17	16.49 11.61
2½	40	3.01	2.10	5.10	3.58	8.19	5.77	12.55	8.84	18.49	13.02	26.25 18.50
3	40	5.33	3.73	9.04	6.34	14.50	10.20	22.22	15.64	32.67	23.00	46.39 32.69
4	40	10.81	7.64	18.43	12.98	29.59	20.84	45.28	31.87	66.61	46.94	94.49 66.64
5	40	19.66	13.80	33.30	23.42	53.33	37.65	81.66	57.52	119.95	84.62	170.60 120.29
6	40	31.81	22.36	53.84	37.95	86.35	60.82	132.02	93.12	193.98	136.82	275.83 194.52
8	40	65.18	45.92	110.37	77.79	176.81	124.69	270.40	190.70	396.78	280.20	563.70 397.90
10	40	118.01	83.13	200.13	140.34	319.78	225.48	488.43	344.91	718.87	506.82	1019.75 719.75
12	ID	189.00	133.11	319.25	225.57	511.53	360.66	781.35	551.73	1148.74	810.78	1629.08 1151.46

Δp = pressure drop due to line friction, psi per 100 ft equivalent length of line
Δt = corresponding change in saturation temperature with the pressure drop, °F

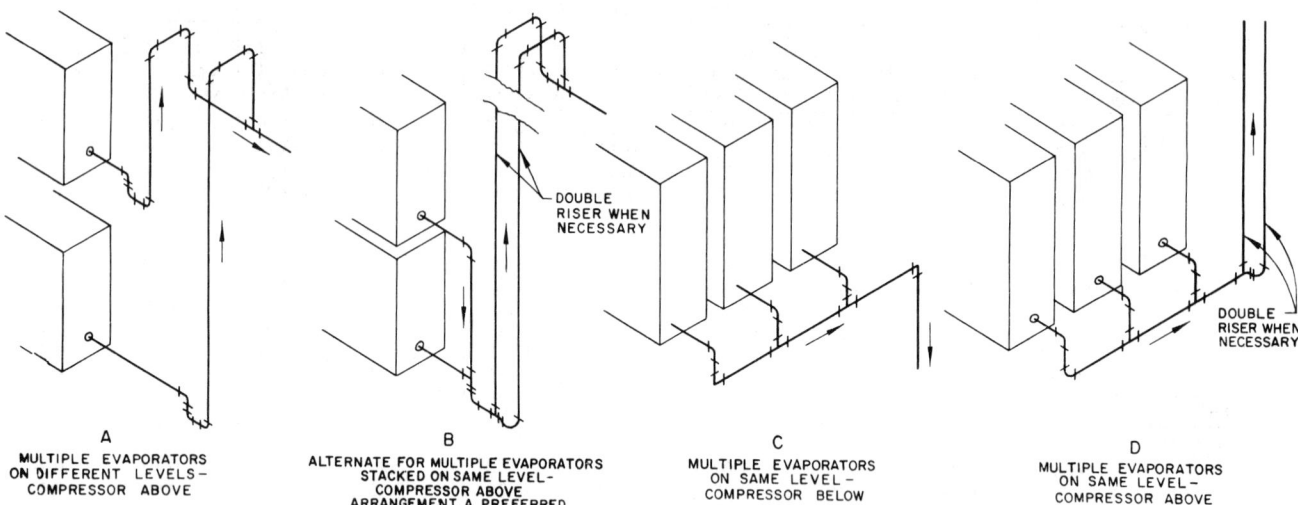

Fig. 2 Suction Line Piping at Evaporator Coils

Note: All arrangements should include a pumpdown cycle.

For systems operating at higher suction temperatures, such as for comfort air conditioning, single suction risers can be sized for oil return at minimum load. Where single compressors are used with capacity control, minimum capacity will usually be 25 or 33% of maximum displacement. With this low ratio, pressure drop in single suction risers designed for oil return at minimum load are rarely serious at full load.

When multiple compressors are used, one or more may shutdown while another continues to operate, and the maximum to minimum ratio becomes much larger. This may make a double suction riser necessary.

The remaining portions of the suction line are sized to permit a practical pressure drop between the evaporators and compressors, since oil is carried along in horizontal lines at relatively low gas velocities. It is good practice to give some pitch to these lines toward the compressor. Traps should be avoided, but when that is impossible, the risers from them are treated the same as those leading from the evaporators.

Preventing Oil Trapping in Idle Evaporators. Suction lines should be designed so that oil from an active evaporator does not drain into an idle one. Diagram A of Figure 2 shows multiple evaporators on different floor levels and the compressor above. Each suction line is brought upward and looped into the top of the common suction line to prevent oil from draining into inactive coils.

Diagram B shows multiple evaporators stacked with the compressor above. Oil cannot drain into the lowest evaporator because the common suction line drops below the outlet of the lowest evaporator before entering the suction riser.

Diagram C shows multiple evaporators on the same level, with the compressor located below. The suction line from each evaporator drops down into the common suction line so that oil cannot drain into an idle evaporator. An alternate arrangement is shown in Diagram D for cases where the compressor is above the evaporators.

Figure 3 illustrates typical piping for evaporators above and below a common suction line. All horizontal runs should be level or pitched toward the compressor to assure oil return.

The traps shown in the suction lines after the coil suction outlet are recommended by various thermal expansion valve manufacturers to prevent erratic operation of the thermal expansion valve. The expansion valve bulbs are located in the suction lines between the coils and these traps. The traps serve as drains and help prevent liquid from accumulating under the expansion valve bulbs during compressor off-cycles. They are useful only where straight runs or risers are encountered in the suction line leaving the coil outlet.

Discharge (Hot-Gas) Lines

Hot-gas lines should be designed (1) to avoid trapping oil at partial load operation; (2) to prevent condensed refrigerant and oil in the line from draining back to the head of the compressor, either during shutdown or during operation at low ambients when long outdoor discharge lines are required; (3) with carefully selected connections from a common line to multiple compressors; and (4) to avoid developing excessive noise or vibration from hot-gas pulsations, compressor vibration, or both.

Oil Transport Up Risers at Normal Loads. Even though a low pressure drop is desired, oversized hot-gas lines can reduce gas velocities to a point where the refrigerant will not transport oil. Therefore, when using multiple compressors with capacity control, hot-gas risers must transport oil at all possible loadings.

Minimum Gas Velocities for Oil Transport in Risers. Minimum capacities for oil entrainment in hot-gas line risers are shown in Table 12. On multiple compressor installations, the lowest possible system loading should be calculated and a riser size selected to give at least the minimum capacity indicated in the table for successful oil transport.

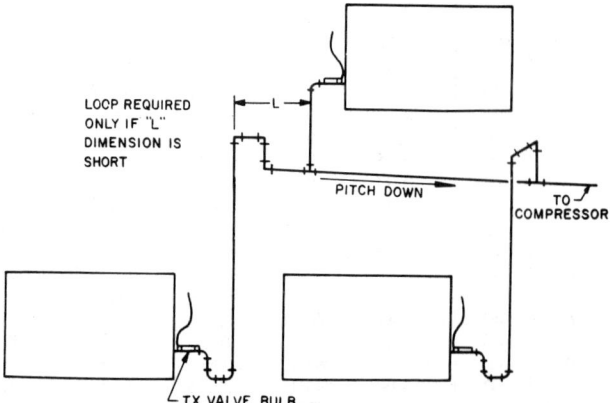

Fig. 3 Typical Piping from Evaporators Located Above and Below Common Suction Line

System Practices for Halocarbon Refrigerants

Table 8 Fitting Losses in Equivalent Feet of Pipe
(Screwed, Welded Flanged, Flared, and Brazed Connections)

Nominal Pipe or Tube Size (in.)	Smooth Bend Elbows							Smooth Bend Tees			
	90° Std[a]	90° Long Rad.[b]	90° Street[a]	45° Std[a]	45° Street[a]	180° Std[a]	Flow Through Branch	Straight-Through Flow		
								No Reduction	Reduced 1/4	Reduced 1/2
3/8	1.4	0.9	2.3	0.7	1.1	2.3	2.7	0.9	1.2	1.4
1/2	1.6	1.0	2.5	0.8	1.3	2.5	3.0	1.0	1.4	1.6
3/4	2.0	1.4	3.2	0.9	1.6	3.2	4.0	1.4	1.9	2.0
1	2.6	1.7	4.1	1.3	2.1	4.1	5.0	1.7	2.2	2.6
1 1/4	3.3	2.3	5.6	1.7	3.0	5.6	7.0	2.3	3.1	3.3
1 1/2	4.0	2.6	6.3	2.1	3.4	6.3	8.0	2.6	3.7	4.0
2	5.0	3.3	8.2	2.6	4.5	8.2	10.0	3.3	4.7	5.0
2 1/2	6.0	4.1	10.0	3.2	5.2	10.0	12.0	4.1	5.6	6.0
3	7.5	5.0	12.0	4.0	6.4	12.0	15.0	5.0	7.0	7.5
3 1/2	9.0	5.9	15.0	4.7	7.3	15.0	18.0	5.9	8.0	9.0
4	10.0	6.7	17.0	5.2	8.5	17.0	21.0	6.7	9.0	10.0
5	13.0	8.2	21.0	6.5	11.0	21.0	25.0	8.2	12.0	13.0
6	16.0	10.0	25.0	7.9	13.0	25.0	30.0	10.0	14.0	16.0
8	10.0	13.0	—	10.0	—	33.0	40.0	13.0	18.0	20.0
10	25.0	16.0	—	13.0	—	42.0	50.0	16.0	23.0	25.0
12	30.0	19.0	—	16.0	—	50.0	60.0	19.0	26.0	30.0
14	34.0	23.0	—	18.0	—	55.0	68.0	23.0	30.0	34.0
16	38.0	26.0	—	20.0	—	62.0	78.0	26.0	35.0	38.0
18	42.0	29.0	—	23.0	—	70.0	85.0	29.0	40.0	42.0
20	50.0	33.0	—	26.0	—	81.0	100.0	33.0	44.0	50.0
24	60.0	40.0	—	30.0	—	94.0	115.0	40.0	50.0	60.0

[a] R/D approximately equal to 1.
[b] R/D approximately equal to 1.5.

Table 9 Special Fitting Losses in Equivalent Feet of Pipe

Nominal Pipe or Tube Size, in.	Sudden Enlargement, d/D			Sudden Contraction, d/D			Sharp Edge		Pipe Projection	
	1/4	1/2	3/4	1/4	1/2	3/4	Entrance	Exit	Entrance	Exit
3/8	1.4	0.8	0.3	0.7	0.5	0.3	1.5	0.8	1.5	1.1
1/2	1.8	1.1	0.4	0.9	0.7	0.4	1.8	1.0	1.8	1.5
3/4	2.5	1.5	0.5	1.2	1.0	0.5	2.8	1.4	2.8	2.2
1	3.2	2.0	0.7	1.6	1.2	0.7	3.7	1.8	3.7	2.7
1 1/4	4.7	3.0	1.0	2.3	1.8	1.0	5.3	2.6	5.3	4.2
1 1/2	5.8	3.6	1.2	2.9	2.2	1.2	6.6	3.3	6.6	5.0
2	8.0	4.8	1.6	4.0	3.0	1.6	9.0	4.4	9.0	6.8
2 1/2	10.0	6.1	2.0	5.0	3.8	2.0	12.0	5.6	12.0	8.7
3	13.0	8.0	2.6	6.5	4.9	2.6	14.0	7.2	14.0	11.0
3 1/2	15.0	9.2	3.0	7.7	6.0	3.0	17.0	8.5	17.0	13.0
4	17.0	11.0	3.8	9.0	6.8	3.8	20.0	10.0	20.0	16.0
5	24.0	15.0	5.0	12.0	9.0	5.0	27.0	14.0	27.0	20.0
6	29.0	22.0	6.0	15.0	11.0	6.0	33.0	19.0	33.0	25.0
8	—	25.0	8.5	—	15.0	8.5	47.0	24.0	47.0	35.0
10	—	32.0	11.0	—	20.0	11.0	60.0	29.0	60.0	46.0
12	—	41.0	13.0	—	25.0	13.0	73.0	37.0	73.0	57.0
14	—	—	16.0	—	—	16.0	86.0	45.0	86.0	66.0
16	—	—	18.0	—	—	18.0	96.0	50.0	96.0	77.0
18	—	—	20.0	—	—	20.0	115.0	58.0	115.0	90.0
20	—	—	—	—	—	—	142.0	70.0	142.0	108.0
24	—	—	—	—	—	—	163.0	83.0	163.0	130.0

Note: Enter table for losses at smallest diameter d.

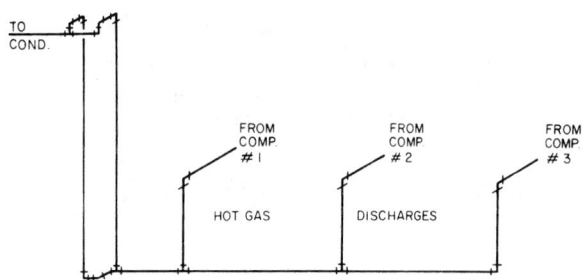

Fig. 4 Double Hot-Gas Riser

In some installations with multiple compressors with capacity control, a vertical hot-gas line sized to transport oil at minimum load has excessive pressure drop at maximum load. When this problem exists, either a double riser or a single riser and an oil separator can be used.

Double Hot-Gas Risers. A double gas riser can be used in the same way as it is used in a suction line. Figure 4 shows the double riser principle applied to a hot-gas line. Its operating principle and sizing technique is described in the section Double Suction Risers.

Single Riser and Oil Separator. As an alternative, an oil separator located in the discharge line just before the riser permits sizing the riser for a low pressure drop. Any oil draining back down the riser accumulates in the oil separator. With multiple compressors, it may be necessary to use individual oil separators for each discharge line. Horizontal lines should be level or pitched downward in the direction of gas flow to facilitate travel of oil through the system and back to the compressor.

Piping to Prevent Liquid and Oil from Draining to the Compressor Head. Each discharge line of a multiple compressor arrangement should have a check valve to prevent discharge gas from active compressors from condensing on the heads of idle compressors.

Whenever the condenser is located above the compressor, the hot-gas line should loop to the floor near the compressor before rising to the condenser, especially when the hot-gas riser is a long one. This minimizes the possibility that refrigerant, condensed in the line during off-cycles, will drain back to the head of the compressor. Any oil traveling up the pipe wall also will not drain back to the compressor head.

The loop in the hot-gas line (Figure 5) serves as a reservoir and traps liquid resulting from condensation in the line during shutdown, thus preventing gravity drainage of liquid and oil back to the compressor head. Hot-gas lines discharging straight across to a condenser directly adjacent to the compressor do not require this loop at the compressor.

Whenever the condenser and receiver are located where the ambient temperature can be higher than that of the compressor, a tightly closing check valve should be installed in the hot-gas line. This valve prevents refrigerant from boiling off in the condenser or receiver and condensing on the compressor heads during off-cycles. For compressors equipped with water-cooled oil coolers, a water solenoid and water-regulating valve should be installed in the water line so that the regulating valve maintains adequate cooling during operation, and the solenoid stops flow during the off-cycle to prevent localized condensing of the refrigerant.

Hot-Gas (Discharge) Mufflers. Mufflers can be installed in hot-gas lines to dampen the discharge gas pulsations, reducing vibration and noise. Mufflers should be installed in a horizontal or downflow portion of the hot-gas line, immediately after it leaves the compressor, not in a riser.

Because gas velocities through the muffler are substantially lower than they are through the hot-gas line, the muffler may form an oil trap if it is installed in a riser or horizontal line so that oil is not drained at the outlet.

Receivers. Refrigerant receivers require additional piping considerations, so they should not be installed in a refrigerant system unless they are required. Receivers used in through-type arrangements (Figure 6) also require a minimum operating refrigerant charge, which adds to the overall system charge. This minimum charge should be listed in the manufacturers' literature.

Table 10 Valve Losses in Equivalent Feet of Pipe

Nominal Pipe or Tube Size, in.	Globe[a]	60° − Y	45° − Y	Angle[a]	Gate[b]	Swing Check[c]	Lift Check
3/8	17	8	6	6	0.6	5	
1/2	18	9	7	7	0.7	6	Globe and
3/4	22	11	9	9	0.9	8	vertical
1	29	15	12	12	1.0	10	lift
1 1/4	38	20	15	15	1.5	14	same as
1 1/2	43	24	18	18	1.8	16	globe
2	55	30	24	24	2.3	20	valve[d]
2 1/2	69	35	29	29	2.8	25	
3	84	43	35	35	3.2	30	
3 1/2	100	50	41	41	4.0	35	
4	120	58	47	47	4.5	40	
5	140	71	58	58	6.0	50	
6	170	88	70	70	7.0	60	
8	220	115	85	85	9.0	80	
10	280	145	105	105	12.0	100	Angle lift
12	320	165	130	130	13.0	120	same as
14	360	185	155	155	15.0	135	angle
16	410	210	180	180	17.0	150	valve
18	460	240	200	200	19.0	165	
20	520	275	235	235	22.0	200	
24	610	320	265	265	25.0	240	

Note: Losses are for valves in fully open position and with screwed, welded, flanged, or flared connections.

[a] These losses do not apply to valves with needle point seats.
[b] Regular and short pattern plug cock valves, when fully open, have same loss as gate valve. For valve losses of short pattern plug cocks above 6 in., check with manufacturer.
[c] Losses also apply to the in-line, ball-type check valve.
[d] For Y pattern globe lift check valve with seat approximately equal to the nominal pipe diameter, use values of 60° − Y valve for loss.

System Practices for Halocarbon Refrigerants

Receivers offer the following benefits:

1. Receivers can provide pumpdown storage capacity when another part of the system must be serviced or the system must be shut down for an extended time. (In some water-cooled condenser systems, the condenser also serves as a receiver if the total refrigerant charge does not exceed its storage capacity.)
2. A receiver is required to handle the excess refrigerant charge that occurs with air-cooled condensers using the coil-flooding type condensing pressure control (see Head Pressure Control for Refrigerant Condensers).
3. On systems where the operating charge in the evaporator and/or condenser varies for different loading conditions, receivers accommodate for the fluctuating charge in the low side and drain the condenser of liquid to prevent reducing the effective condensing surface. When an evaporator is fed with a thermal expansion valve, hand expansion valve or low pressure float, the operating charge in the evaporator varies considerably depending on the loading. During low load, the evaporator requires a larger charge since the boiling is not as intense. When the load increases, the operating charge in the evaporator decreases, and the receiver must store excess refrigerant.
4. On systems with multicircuit evaporators that shut off the liquid supply to one or more circuits during reduced load and pump out the idle circuit, receivers are used to hold the full charge of the idle circuit.
5. On systems with air-cooled condensers, receivers provide enough refrigerant to fill the lines on start-up to prevent starvation of the evaporator and subsequent shutdown by the low pressure cutout.

Table 11 Minimum Refrigeration Capacity in Tons for Oil Entrainment up Suction Risers Type L Copper Tubing

Refrigerant	Saturated Temp., °F	Suction Gas Temp., °F	Pipe OD, in.											
			0.500	0.625	0.750	0.875	1.123	1.375	1.625	2.125	2.625	3.125	3.625	4.125
			Area, in^2											
			0.146	0.233	0.348	0.484	0.825	1.256	1.780	3.094	4.770	6.812	9.213	11.970
12	−40.0	−30.0	0.045	0.061	0.133	0.201	0.391	0.662	1.02	2.04	3.51	5.48	7.99	11.1
		−10.0	0.044	0.078	0.130	0.196	0.381	0.645	0.997	1.99	3.42	5.34	7.78	10.8
		10.0	0.044	0.080	0.132	0.199	0.388	0.655	1.01	2.02	3.47	5.42	7.91	11.0
	−20.0	−10.0	0.059	0.106	0.175	0.264	0.513	0.868	1.34	2.68	4.60	7.19	10.5	14.5
		10.0	0.058	0.103	0.171	0.258	0.503	0.850	1.31	2.62	4.51	7.04	10.3	14.2
		30.0	0.059	0.105	0.173	0.262	0.510	0.863	1.33	2.66	4.57	7.14	10.4	14.4
	0.0	10.0	0.077	0.139	0.229	0.345	0.673	1.14	1.76	3.51	6.03	9.42	13.7	19.1
		30.0	0.075	0.134	0.221	0.334	0.650	1.10	1.70	3.39	5.82	9.09	13.3	18.4
		50.0	0.075	0.135	0.223	0.337	0.657	1.11	1.72	3.43	5.89	9.19	13.4	18.6
	20.0	30.0	0.094	0.169	0.279	0.421	0.820	1.39	2.14	4.28	7.35	11.5	16.7	23.2
		50.0	0.095	0.170	0.280	0.423	0.825	1.39	2.16	4.30	7.39	11.5	16.8	23.4
		70.0	0.095	0.170	0.281	0.425	0.828	1.40	2.17	4.32	7.42	11.6	16.9	23.4
	40.0	50.0	0.121	0.217	0.358	0.541	1.05	1.78	2.76	5.50	9.45	14.8	21.5	29.8
		70.0	0.117	0.210	0.347	0.524	1.02	1.73	2.67	5.33	9.16	14.3	20.8	28.9
		90.0	0.117	0.211	0.348	0.526	1.02	1.73	2.68	5.34	9.18	14.3	20.9	29.0
22	−40.0	−30.0	0.067	0.119	0.197	0.298	0.580	0.981	1.52	3.03	5.20	8.12	11.8	16.4
		−10.0	0.065	0.117	0.194	0.292	0.570	0.963	1.49	2.97	5.11	7.97	11.6	16.1
		10.0	0.066	0.118	0.195	0.295	0.575	0.972	1.50	3.00	5.15	8.04	11.7	16.3
	−20.0	−10.0	0.087	0.156	0.258	0.389	0.758	1.28	1.98	3.96	6.80	10.6	15.5	21.5
		10.0	0.085	0.153	0.253	0.362	0.744	1.26	1.95	3.88	6.67	10.4	15.2	21.1
		30.0	0.086	0.154	0.254	0.383	0.747	1.26	1.95	3.90	6.69	10.4	15.2	21.1
	0.0	10.0	0.111	0.199	0.328	0.496	0.986	1.63	2.53	5.04	8.66	13.5	19.7	27.4
		30.0	0.108	0.194	0.320	0.484	0.942	1.59	2.46	4.92	8.45	13.2	19.2	26.7
		50.0	0.109	0.195	0.322	0.486	0.946	1.60	2.47	4.94	8.48	13.2	19.3	26.8
	20.0	30.0	0.136	0.244	0.403	0.608	1.18	2.00	3.10	6.18	10.6	16.6	24.2	33.5
		50.0	0.135	0.242	0.399	0.603	1.17	1.99	3.07	6.13	10.5	16.4	24.0	33.3
		70.0	0.135	0.242	0.400	0.605	1.18	1.99	3.08	6.15	10.6	16.5	24.0	33.3
	40.0	50.0	0.167	0.300	0.495	0.748	1.46	2.46	3.81	7.60	13.1	20.4	29.7	41.3
		70.0	0.165	0.296	0.488	0.737	1.44	2.43	3.75	7.49	12.9	20.1	29.3	40.7
		90.0	0.165	0.296	0.488	0.738	1.44	2.43	3.76	7.50	12.9	20.1	29.3	40.7
502	−40.0	−30.0	0.051	0.092	0.152	0.230	0.447	0.756	1.17	2.33	4.01	6.26	9.13	12.7
		−10.0	0.053	0.095	0.157	0.237	0.461	0.779	1.21	2.41	4.13	6.45	9.41	13.1
		10.0	0.055	0.098	0.163	0.246	0.476	0.809	1.25	2.50	4.29	6.39	9.76	13.5
	−20.0	−10.0	0.068	0.122	0.201	0.303	0.591	0.999	1.54	3.08	5.30	8.27	12.1	16.7
		10.0	0.070	0.125	0.207	0.312	0.608	1.03	1.59	3.17	5.45	8.51	12.4	17.2
		30.0	0.072	0.129	0.213	0.322	0.627	1.06	1.64	3.27	5.62	8.78	12.8	17.8
	0.0	10.0	0.087	0.157	0.259	0.391	0.761	1.29	1.99	3.97	6.82	10.6	15.5	21.5
		30.0	0.089	0.160	0.264	0.399	0.777	1.31	2.03	4.05	6.96	10.9	15.9	22.0
		50.0	0.092	0.165	0.273	0.412	0.802	1.36	2.10	4.19	7.19	11.2	16.4	22.7
	20.0	30.0	0.110	0.197	0.325	0.491	0.957	1.62	2.50	4.99	8.58	13.4	19.5	27.1
		50.0	0.112	0.201	0.331	0.501	0.975	1.65	2.55	5.09	8.74	13.6	19.9	27.6
		70.0	0.115	0.207	0.342	0.516	1.01	1.70	2.63	5.25	9.02	14.1	20.5	28.5
	40.0	50.0	0.136	0.243	0.401	0.606	1.18	2.00	3.09	6.16	10.6	16.5	24.1	33.4
		70.0	0.138	0.247	0.408	0.616	1.20	2.03	3.14	6.28	10.8	16.8	24.5	34.0
		90.0	0.142	0.254	0.420	0.634	1.23	2.09	3.23	6.44	11.1	17.3	25.2	35.0

Note: The capacity in tons is based on 90°F liquid temperature and superheat as indicated by the listed temperature. For other liquid line temperatures, use correction factors in the table below.

Refrigerant	Liquid Temperature, °F								
	50	60	70	80	100	110	120	130	140
12	1.17	1.13	1.09	1.04	0.96	0.91	0.87	0.81	0.76
22	1.17	1.14	1.10	1.06	0.98	0.94	0.89	0.85	0.80
502	1.24	1.18	1.12	1.06	0.94	0.87	0.81	0.74	0.67

If a through-type receiver is used, the liquid must always flow from the condenser to the receiver. Therefore, either the pressure in the receiver must be lower than that in the condenser outlet, or the elevation and piping between the two must allow a refrigerant column high enough to overcome the pressure difference and friction loss.

The receiver and its associated piping provide free flow of liquid from the condenser to the receiver by equalizing the pressures between the two so that the receiver cannot build up a higher pressure than the condenser. This occurs by either of two methods:

1. The piping between condenser and receiver (condensate) is sized so liquid flows in one direction and gas flows in the opposite direction. Sizing the condensate line for 100 ft/min liquid velocity is usually adequate to attain this flow. Piping should slope at least 0.25 in/ft and eliminate any natural liquid traps. See Figure 6 for this configuration.

2. The piping between the condenser and receiver can be equipped with a separate vent (equalizer) line to allow receiver and condenser pressures to equalize. This external vent line can be piped either with or without a check valve in the vent line (see Figures 8 and 9). When piping without a check valve in the vent line, prevent the discharge gas from flowing directly into the vent, or install a gas bypass around the condenser. When the piping configuration is unknown, install a check valve in the vent (with the direction of flow toward the condenser). The condensate line should be sized so the velocity does not exceed 150 fpm.

Connections for a Surge-Type Receiver. When a receiver is required for variation in liquid requirements during normal opera-

Table 12 Minimum Refrigeration Capacity for Oil Entrainment up Hot-Gas Risers Type L Copper Tubing

Refrigerant	Saturated Temp., °F	Suction Gas Temp., °F	Pipe OD, in.											
			0.500	0.625	0.750	0.875	1.123	1.375	1.625	2.125	2.625	3.125	3.625	4.125
			Area, in²											
			0.146	0.233	0.348	0.484	0.825	1.256	1.780	3.094	4.770	6.812	9.213	11.970
12	80.0	110.0	0.161	0.289	0.478	0.721	1.41	2.38	3.67	7.33	12.6	19.7	28.7	39.8
		140.0	0.150	0.270	0.443	0.672	1.31	2.21	3.42	6.83	11.7	16.3	26.7	37.1
		170.0	0.143	0.256	0.423	0.638	1.24	2.10	3.25	6.49	11.1	17.4	25.4	35.2
	90.0	120.0	0.167	0.299	0.494	0.745	1.45	2.46	3.80	7.58	13.0	20.3	29.6	41.1
		150.0	0.155	0.278	0.459	0.694	1.35	2.29	3.53	7.05	12.1	18.9	27.6	38.3
		180.0	0.147	0.264	0.436	0.639	1.28	2.17	3.36	6.70	11.5	18.0	26.2	36.3
	100.0	130.0	0.171	0.307	0.506	0.765	1.49	2.52	3.89	7.77	13.4	20.8	30.4	42.2
		160.0	0.159	0.285	0.470	0.710	1.38	2.34	3.62	7.22	12.4	19.4	28.2	39.2
		190.0	0.151	0.271	0.448	0.677	1.32	2.23	3.43	6.88	11.8	18.4	28.9	37.3
	110.0	140.0	0.174	0.312	0.515	0.778	1.52	2.56	3.96	7.91	13.6	21.2	30.9	42.9
		170.0	0.162	0.290	0.479	0.724	1.41	2.38	3.69	7.36	12.6	19.7	28.8	39.9
		200.0	0.153	0.274	0.452	0.683	1.33	2.25	3.49	6.95	11.9	18.6	27.2	37.7
	120.0	150.0	0.175	0.314	0.518	0.782	1.52	2.58	3.96	7.95	13.7	21.3	31.1	43.2
		180.0	0.162	0.291	0.480	0.725	1.41	2.39	3.69	7.37	12.7	19.8	28.8	40.0
		210.0	0.153	0.274	0.452	0.682	1.33	2.25	3.47	6.93	11.9	18.6	27.1	37.6
22	80.0	110.0	0.235	0.421	0.695	1.05	2.03	3.46	5.35	10.7	18.3	28.6	41.8	57.9
		140.0	0.223	0.399	0.659	0.996	1.94	3.28	5.07	10.1	17.4	27.1	39.6	54.9
		170.0	0.215	0.385	0.635	0.960	1.87	3.16	4.89	9.76	16.8	26.2	38.2	52.9
	90.0	120.0	0.242	0.433	0.716	1.06	2.11	3.56	5.50	11.0	18.9	29.5	43.0	59.6
		150.0	0.226	0.406	0.671	1.01	1.97	3.34	5.16	10.3	17.7	27.6	40.3	55.9
		180.0	0.216	0.387	0.540	0.956	1.88	3.18	4.92	9.82	16.9	26.3	38.4	53.3
	100.0	130.0	0.247	0.442	0.730	1.10	2.15	3.83	5.62	11.2	19.3	30.1	43.9	60.8
		160.0	0.231	0.414	0.884	1.03	2.01	3.40	5.26	10.5	18.0	28.2	41.1	57.0
		190.0	0.220	0.394	0.650	0.982	1.91	3.24	3.00	9.96	17.2	26.8	39.1	54.2
	110.0	140.0	0.251	0.451	0.744	1.12	2.19	3.70	5.73	11.4	19.6	30.6	44.7	62.0
		170.0	0.235	0.421	0.693	1.05	2.05	3.46	3.35	10.7	18.3	28.6	41.8	57.9
		200.0	0.222	0.399	0.658	0.994	1.94	3.28	5.06	10.1	17.4	27.1	39.5	54.8
	120.0	150.0	0.257	0.460	0.760	1.15	2.24	3.78	5.85	11.7	20.0	31.3	45.7	63.3
		180.0	0.239	0.428	0.707	1.07	2.08	3.51	5.44	10.8	18.6	29.1	42.4	58.9
		210.0	0.225	0.404	0.666	1.01	1.96	3.31	5.12	10.2	17.6	27.4	40.0	55.5
502	80.0	110.0	0.192	0.344	0.567	0.857	1.67	2.82	4.36	8.71	15.0	23.4	34.1	47.3
		140.0	0.180	0.323	0.534	0.806	1.57	2.66	4.11	8.20	14.1	22.0	32.1	44.5
		170.0	0.173	0.310	0.512	0.773	1.50	2.54	3.94	7.85	13.5	21.1	30.7	42.8
	90.0	120.0	0.194	0.348	0.574	0.867	1.69	2.85	4.41	8.81	15.1	23.6	34.5	47.8
		150.0	0.182	0.326	0.538	0.813	1.58	2.68	4.14	8.26	14.2	22.2	32.3	44.8
		180.0	0.169	0.303	0.501	0.756	1.47	2.49	3.85	7.69	13.2	20.6	30.1	41.7
	100.0	130.0	0.194	0.348	0.575	0.869	1.69	2.86	4.42	8.83	15.2	23.7	34.5	47.9
		160.0	0.182	0.326	0.539	0.813	1.58	2.68	4.14	8.27	14.2	22.2	32.3	44.9
		190.0	0.170	0.304	0.503	0.739	1.48	2.50	3.87	7.71	13.3	20.7	30.2	41.9
	110.0	140.0	0.170	0.305	0.504	0.761	1.48	2.51	3.87	7.73	13.3	20.7	30.2	42.0
		170.0	0.162	0.291	0.481	0.726	1.41	2.39	3.70	7.38	12.7	19.8	28.9	40.1
		200.0	0.152	0.273	0.450	0.680	1.33	2.24	3.46	6.92	11.9	18.5	27.0	37.5
	120.0	150.0	0.170	0.305	0.503	0.760	1.48	2.50	3.87	7.73	13.3	20.7	30.2	41.9
		180.0	0.153	0.275	0.453	0.683	1.33	2.26	3.49	6.96	12.0	18.7	27.2	37.8
		210.0	0.149	0.267	0.440	0.665	1.30	2.19	3.39	6.76	11.6	18.1	26.4	36.7

Note: The capacity in tons is based on a saturated suction temperature of 20°F with 15°F superheat at the indicated saturated condensing temperature with 15°F subcooling. For other saturated suction temperatures with 15°F superheat, use the following correction factors:

Sat. suction temperature, °F	−40	−20	0	40
Correction factor	0.88	0.95	0.96	1.04

System Practices for Halocarbon Refrigerants

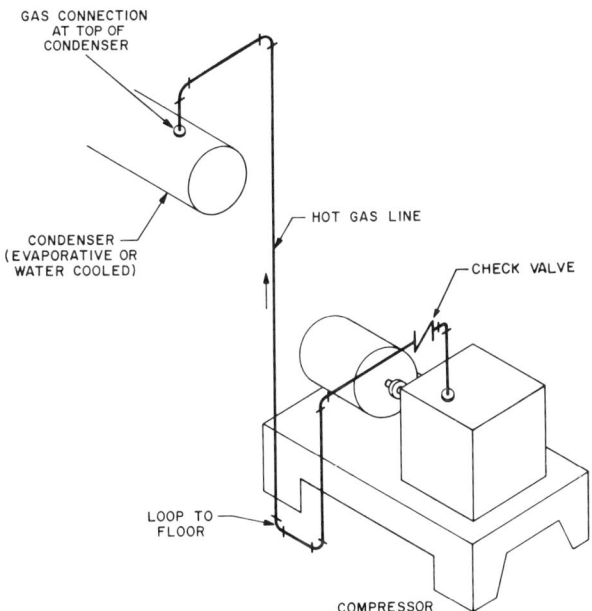

Fig. 5 Hot-Gas Loop

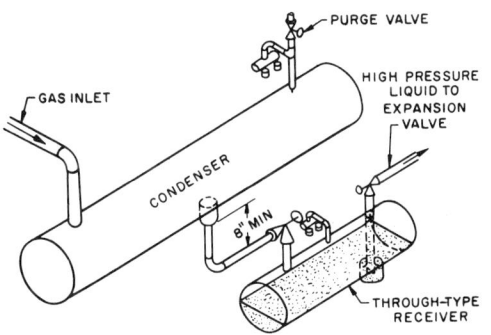

Fig. 6 Shell-and-Tube Condenser to Receiver Piping
(Through-Type Receiver)

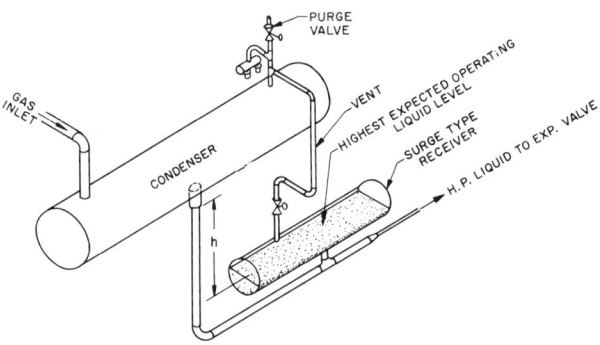

Max. Velocity of Drain Line, fpm	Type Valve between Condenser and Receiver	h Required, in.
150	None	14
150	Angle	16
150	Globe	28
100	Angle or globe	14

Fig. 7 Shell-and-Tube Condenser to Receiver Piping
(Surge-Type Receiver)

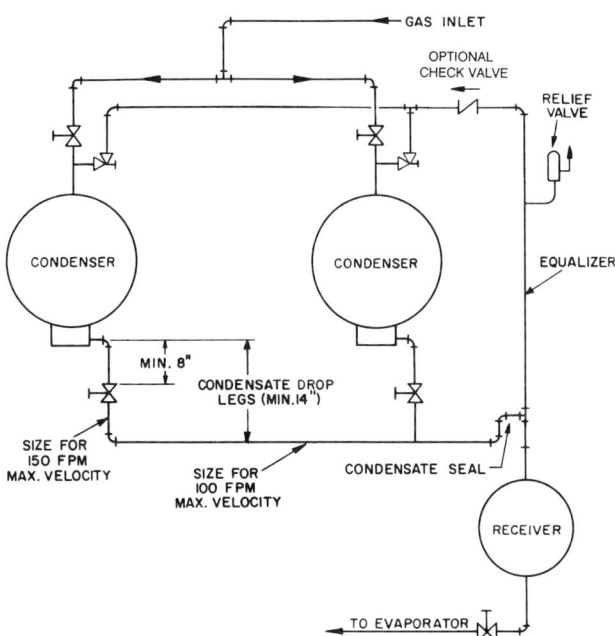

Fig. 8 Parallel Condensers with Through-Type Receiver

Table 13 Sizing Data for Oil Return in Discharge or Suction Lines with Flow Vertically Upward

Saturation Temp., °F	Line Size	
	2 in. or less	Above 2 in.
0	0.35 psi/100 ft	0.20 psi/100 ft
−50	0.45 psi/100 ft	0.25 psi/100 ft

tion, it can be connected as shown in Figure 7. The full receiver volume is available for liquid that is to be removed from the circuit. Also, because liquid flows to the expansion valve without exposure to gas in the receiver, it can remain subcooled.

The height h must be adequate for a liquid head at least as large as the pressure loss through the condenser, the liquid line and the vent line at the maximum temperature difference between the receiver ambient and the condensing temperature. The condenser pressure drop at the greatest expected heat rejection should be obtained from the manufacturer. The minimum value of h can then be calculated and a decision made as to whether or not the available height will permit the surge-type receiver.

The vent line flow is from receiver to condenser when the receiver temperature is warmer than the condensing temperature. Flow is from condenser to receiver when the air temperature around the receiver is below the condensing temperature. The rate of flow depends on this temperature difference as well as on the amount of receiver surface. Vent size can be calculated from this flow rate.

Multiple Condensers. Two or more condensers connected in series or in parallel can be used in a single refrigerant circuit.

When connected in series, the pressure losses through each condenser must be added. Condensers are more often arranged in parallel. The pressure loss through any one of the parallel circuits is always equal to that through any of the others, even if it results in filling much of one circuit with liquid while gas passes through another.

Figure 8 shows a basic arrangement for multiple condensers with a through-type receiver. The condensate drop legs must be long enough to allow liquid levels to adjust in them to equalize pressure losses between condensers at all operating conditions. The drop legs should be 6 to 12 in. higher than calculated to assure that liquid outlets remain free-draining. This height will provide

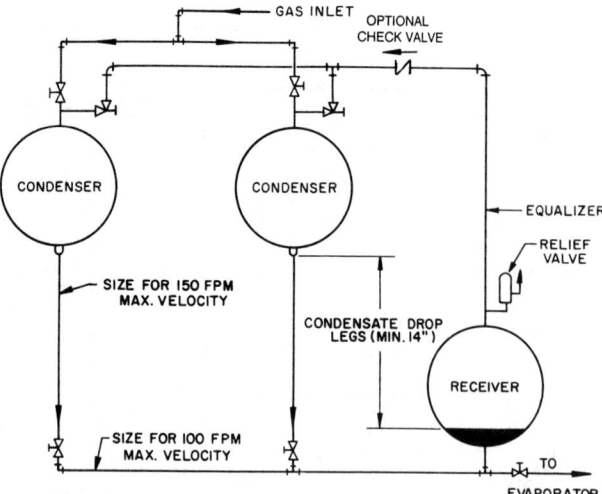

Fig. 9 Parallel Condensers with Surge-Type Receiver

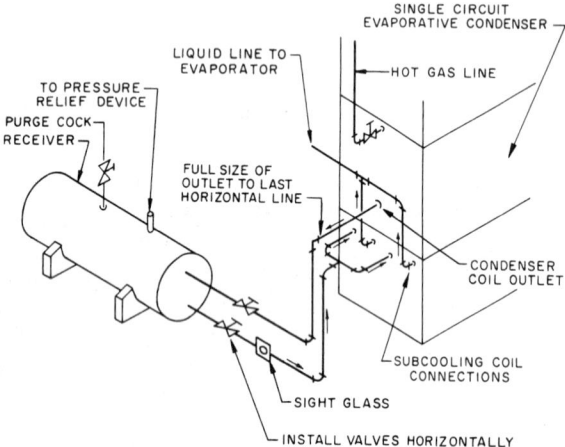

Fig. 10 Single-Circuit Evaporative Condenser with Receiver and Liquid Subcooling Coil

a liquid head to offset the largest condenser pressure loss. The liquid seal will prevent gas blow-by between condensers.

Large single condensers with multiple coil sections should be piped as if the independent sections were parallel condensers. As an example, assume the left condenser in Figure 8 has 2 psi more pressure drop than the right condenser. The liquid level on the left side will be about 4 ft higher than on the right. If the condensate lines do not have enough vertical height for this level difference, the liquid will rise in the condenser until the pressure drop is the same through both circuits. Enough surface may be covered in this event to reduce the condenser capacity significantly.

The condensate drop legs should be sized based on 150 fpm velocity. The main condensate lines should be based on 100 fpm velocity. Depending on prevailing local and/or national safety codes, a relief device may have to be installed in the discharge piping.

Figure 9 shows a piping arrangement for multiple condensers with a surge-type receiver. When only one or two compressors operate, the flow paths may not be symmetrical through the discharge lines. Small pressure differences would not be unusual, and the liquid line junction should be about 2 or 3 ft below the bottom of the condensers. The exact amount can be calculated from pressure loss through each path at all possible operating conditions.

When condensers are water-cooled, a single automatic water valve for the condensers in one refrigerant circuit should be used because individual valves for each condenser are difficult or impossible to set alike.

With evaporative condensers (Figure 10), the pressure loss may be high. If parallel condensers are alike and all are operated, the differences may be small, and the height of the condenser outlets above the liquid line junction need not be more than 2 or 3 ft. If the fans of any condenser are not operated while the fans on one condenser are, the level above the junction must equal the loss through any condenser at its operating condition.

When the level difference between condenser outlets and the liquid line junction is sufficient, the receiver may be vented to the condenser inlets. In that case, the surge-type receiver should be used. The level difference must then be at least equal to the greatest loss through any condenser circuit plus the greatest vent line loss when the air temperature is greater than the condensing temperature. In this type of connection (Figure 11), the height of drop leg H should equal 6 to 8 ft.

AIR-COOLED CONDENSERS

The refrigerant pressure drop through air-cooled condensers must be obtained from the supplier for the particular unit at the specified load. If the refrigerant pressure drop is low enough and it is practical to so arrange the equipment, parallel condensers can be connected to allow for capacity reduction to zero on one unit without causing liquid backup in other active units (Figure 12).

A single air-cooled condenser with any pressure drop can be connected to a receiver without an equalizer and without trapping height if the condenser outlet and the line from it to the receiver can be sized for sewer flow without a trap or restriction, using a maximum velocity of 100 fpm. A single condenser can be connected with an equalizer line to the hot-gas inlet, if the vertical drop leg is sufficient to balance the refrigerant pressure drop through the condenser and the liquid line to the receiver.

Multiple condensers with high pressure drops can be connected as shown in Figure 12, provided that: (1) the receiver is located in an ambient equal to or lower than the inlet air temperature to the

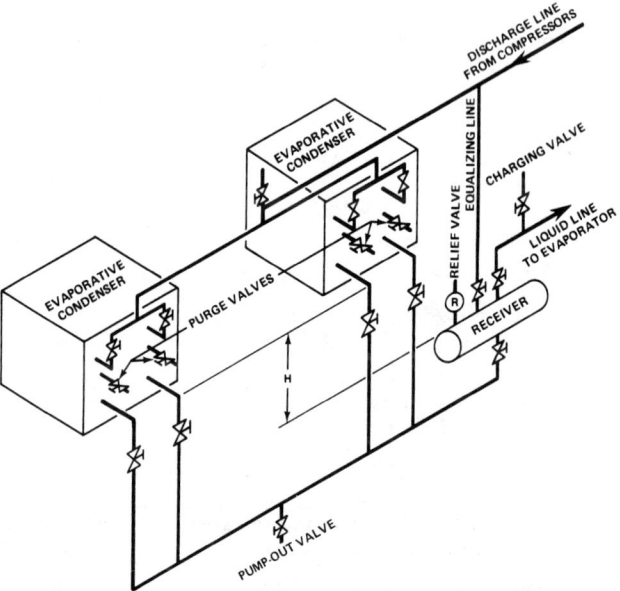

All outlets must be trapped. The height of the trap leg must be such that when one or more units are idle (fan or pump stopped), liquid may rise in the leg of the operating unit so that the static head equals the pressure drop in the operating unit at all conditions. The trap height should be 6 to 12 in. greater than the calculated height to assure that the liquid outlets remain free-draining.

Fig. 11 Multiple Evaporative Condensers with Equalization to Condenser Inlets

System Practices for Halocarbon Refrigerants

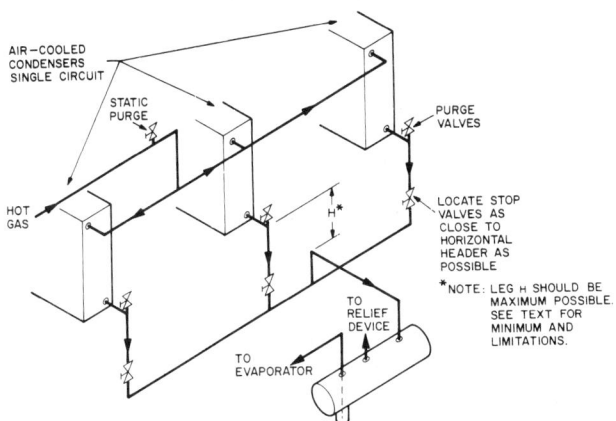

Fig. 12 Multiple Air-Cooled Condensers

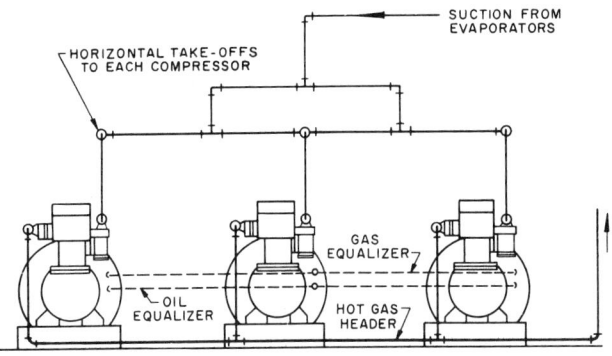

NOTE: GAS EQUALIZER MUST BE LARGE ENOUGH TO APPROXIMATE THE SAME CRANKCASE PRESSURE IN ALL COMPRESSORS WITH ANY COMBINATION OF IDLE AND OPERATING COMPRESSORS (ANY PRESSURE DIFFERENCE IS REFLECTED BY DIFFERENCE IN OIL LEVEL.)

Fig. 13 Suction and Hot-Gas Headers for Multiple Compressors

condenser, (2) capacity control affects all units equally, (3) all units operate when one operates, unless valved off at both inlet and outlet, and (4) all units are of equal size. A through-type receiver is recommended for this application.

If unit sizes are unequal, additional liquid height H equivalent to the difference in full-load pressure drop is required. Usually, condensers of equal size are used in parallel applications.

If the receiver cannot be located in an ambient below the inlet air for all operating conditions, sufficient extra height of drop leg H is required to overcome the equivalent differences in saturation pressure of the receiver and the condenser. The subcooling formed by the liquid leg tends to condense vapor in the receiver to reach a balance of rate of condensation, at an intermediate saturation pressure, with heat gain from ambient to the receiver. A relatively large liquid leg is required to balance a small temperature difference; therefore, this method is probably limited to marginal cases. In any case, the liquid leaving the receiver will be saturated, and any subcooling to prevent flashing in the liquid line must be obtained downstream of the receiver. If the temperature of the receiver ambient is above the condensing pressure only at part-load conditions, it may be acceptable to back liquid into the condensing surface, sacrificing the operating economy of lower part-load head pressure for a lower liquid leg requirement. The receiver must be adequately sized to contain a minimum of the backed-up liquid so that the condenser can be fully drained when full load is required. If a low ambient control system of backing liquid into the condenser is used, consult the system supplier for proper piping.

PIPING AT MULTIPLE COMPRESSORS

Multiple compressors operating in parallel must be carefully piped to assure proper operation.

Suction Piping

Suction piping should be designed so that all compressors run as nearly as possible at the same suction pressure, and so that oil is returned in equal proportions. All suction lines should be brought into a common suction header in order to return the oil to each crankcase as uniformly as possible. It may be best to return oil through a suction trap (accumulator).

The suction header should be run above the level of the compressor suction inlets so that oil can drain into the compressors by gravity. Branch suction lines to the compressors should be connected to the side of the header.

The return mains from the evaporators should not be connected into the suction header so as to form crosses with the branch suction lines to the compressors.

The suction header should be run full size. The horizontal takeoffs to the compressors should be the same size as the suction header. The branch suction lines to the compressors should not be reduced until the vertical drop is reached. With welded construction, the branch can be taken off eccentric on the bottom of the header.

Figure 13 shows a pyramidal or yoke-type suction header to maximize pressure equalization at each of three compressor suction inlets, piped in parallel. An acceptable alternative, although not as good with regard to equal pressure drops to all compressors, is to have the suction line from the evaporators enter at one end of the header instead of using the yoke arrangement.

Takeoffs to each compressor from the common suction header should be horizontal from the side to distribute oil equally and to prevent accumulating liquid refrigerant in an idle compressor in case of slopover.

Suction traps are recommended wherever the following are used: (1) parallel compressors, (2) flooded evaporators, (3) double suction risers, (4) long suction lines, (5) multiple expansion valves, (6) hot-gas defrost, (7) reverse cycle operation, and (8) suction pressure regulators.

The suction trap must be sized for effective gas and liquid separation. Adequate liquid volume and a means of disposing of it must be provided. A liquid transfer pump or electric heater may be used (see Chapter 2).

An oil receiver equipped with an electric heater effectively evaporates liquid refrigerant accumulated in the suction trap.

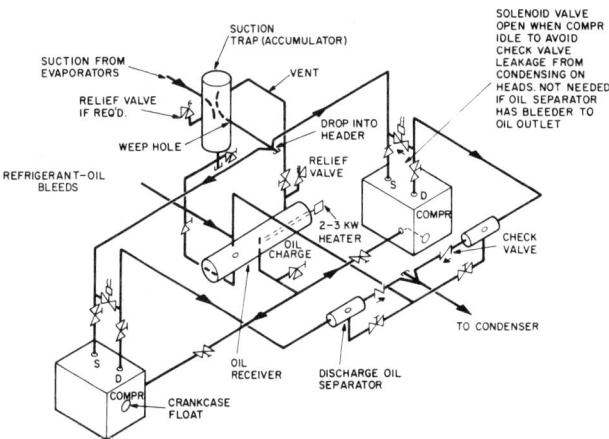

Fig. 14 Parallel Compressors with Gravity Oil Flow

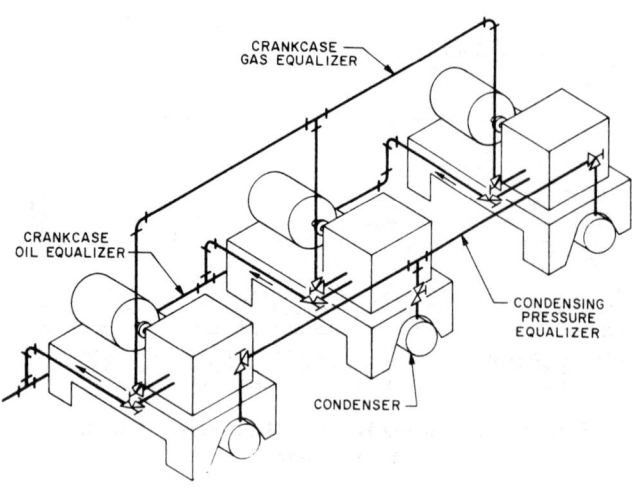

Fig. 15 Interconnecting Piping for Multiple Condensing Units

It also operates on the assumption that each compressor receives its share of oil. Either crankcase float valves or external float switches and solenoid valves can be used to control the oil flow to each compressor.

The oil receiver should be elevated to overcome the pressure drop between it and the crankcase. The oil receiver should be sized so that a malfunction of the oil control mechanism cannot overfill an idle compressor.

Figure 14 shows a recommended hookup of multiple compressors, suction trap (accumulator), oil receiver, and discharge line oil separators. The oil receiver also provides a reserve supply of oil for the compressors where the oil in the system external to the compressor varies with system loading. The heater element should always be submerged.

Discharge Piping

The piping arrangement shown in Figure 13 is suggested for discharge piping. Using a bullheaded tee at the junction of two compressor branches and the main discharge header causes increased turbulence, increased pressure drop, and possible hammering in the line and should therefore be avoided.

Interconnection of Crankcases

When two or more compressors are to be interconnected, they should be placed on foundations so that all oil equalizer tappings are exactly level. If crankcase floats (as shown in Figure 14) are not used, an oil equalization line should connect all of the crankcases to maintain uniform oil levels. The oil equalizer may be run level with the tappings, or, for convenient access to the compressors, it may be run at the floor (Figure 15). It should *never* be run at a level higher than that of the tappings.

For the oil equalizer line to work properly, equalize the crankcase pressures by installing a gas equalizer line above the oil level. This line may be run to provide head room (Figure 15) or run level with the tappings on the compressors. It should be piped so that oil or liquid refrigerant will not be trapped.

Both lines should be the same size as the tappings on the largest compressor and should be valved so that any one machine can be taken out for repair. The piping should be arranged to absorb vibration.

PIPING AT VARIOUS SYSTEM COMPONENTS

Flooded Fluid Coolers

For a description of flooded fluid coolers, see Chapter 16 of the 1988 ASHRAE *Handbook—Equipment*.

Shell-and-tube flooded coolers designed to minimize liquid entrainment in the suction gas require a continuous liquid bleed line installed at some point in the cooler shell below the liquid level to remove trapped oil. This continuous bleed of refrigerant liquid and oil prevents the oil concentration in the cooler from getting too high. The location of the liquid bleed connection on the shell depends on the refrigerant and oil used. Refrigerants 11 and 12 can use a connection anywhere below the liquid level because they are highly miscible with mineral oil.

Refrigerants 22 and 502 are less miscible with oil, however, and a separate oil-rich phase floating on a refrigerant-rich layer may exist. This becomes possible as the evaporating temperature drops. When Refrigerants 22 and 502 are used with mineral oil, the bleed line usually is taken off the shell just slightly below the liquid level, or there may be more than one valved bleed connection at slightly different levels so that the optimum point can be selected during operation. With alkyl benzene lubricants (see Chapter 8), miscibility between the oil and R-22 and R-502 is sufficient so that the oil bleed connection can be anywhere below the liquid level.

Where the flooded cooler design requires an external surge drum to separate the liquid carryover from the suction gas off the tube bundle, the richest oil concentration may or may not be in the cooler. In some cases, the surge drum will have the highest concentration of oil. In this instance, the refrigerant and oil bleed connection is taken from the surge drum. The refrigerant and oil bleed from the cooler by gravity. The bleed sometimes drains into the suction line so that the oil can be returned to the compressor with the suction gas after the accompanying liquid refrigerant is vaporized in a liquid-suction heat interchanger. A better method drains the refrigerant-oil bleed into an electrically heated receiver that boils the refrigerant off to the suction line and drains the oil back to the compressor.

Refrigerant Feed Devices

For further information on refrigerant feed devices see Chapter 19 of the 1988 ASHRAE *Handbook—Equipment*.

The *pilot-operated low-side float control* (Figure 16) is sometimes selected for flooded systems using halocarbon refrigerants. Except for small capacities, direct-acting low-side float valves are impractical for these refrigerants. The displacer float controlling

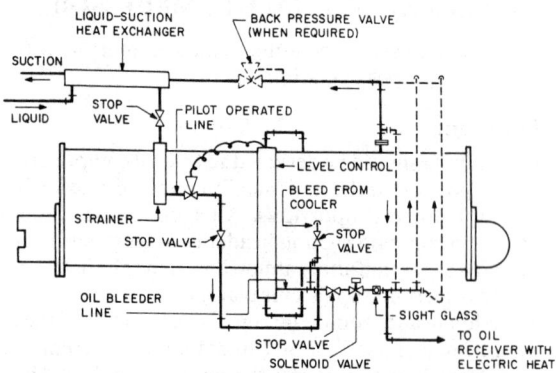

NOTE: WHEN OIL BLEED IS TAKEN TO A HEATED OIL RECEIVER THE DOUBLE SUCTION RISER SERVES NO PURPOSE AND THE LIQUID-SUCTION HEAT EXCHANGER IS NOT REQUIRED.

Fig. 16 Typical Piping at Flooded Fluid Center

System Practices for Halocarbon Refrigerants

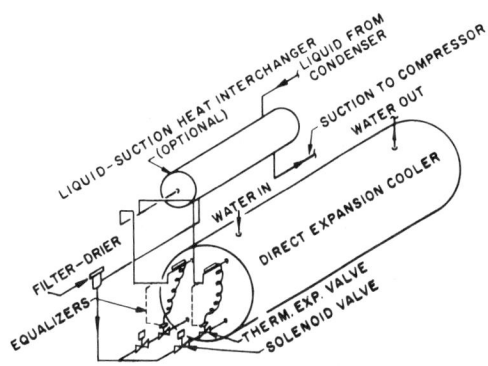

Fig. 17 Two-Circuit Direct Expansion Cooler Connections
(for Single-Compressor System)

a pneumatic valve works well for low-side liquid level control. It allows the cooler level to be adjusted within the instrument without disturbing the piping.

High-side float valves are practical only in systems having one evaporator, because distribution problems result when multiple evaporators are used.

Float chambers should be located as near to the liquid connection on the cooler as possible, since a long length of liquid line, even though insulated, can pick up room heat and give an artificial liquid level in the float chamber. Equalizer lines to the float chamber must be amply sized to minimize the effect of heat transmission toward creating false liquid levels in the chamber. The float chamber and its equalizing lines must be insulated.

Each flooded cooler system must have a way to keep oil concentration in the evaporator low, both to minimize the bleedoff needed to keep oil concentration in the cooler low and to reduce system losses from large stills. A highly efficient discharge gas/oil separator can be used for this purpose.

At low temperatures, periodic warm-up of the evaporator permits recovery of oil accumulation in the chiller. If continuous operation is required, dual chillers may be needed to permit deoiling an oil-laden evaporator, or an oil-free compressor may be used.

Direct Expansion Fluid Chillers

For further information on this chiller type, see Chapter 17 of the 1988 ASHRAE *Handbook—Equipment*.

Figure 17 shows typical piping connections for a multicircuit direct expansion chiller. Each circuit contains its own thermal and solenoid valves. One of the solenoid valves can be wired to close at reduced system capacity. The thermal valve bulbs should be located between the cooler and the liquid-suction interchanger, if used. Locating the bulb downstream from the interchanger can

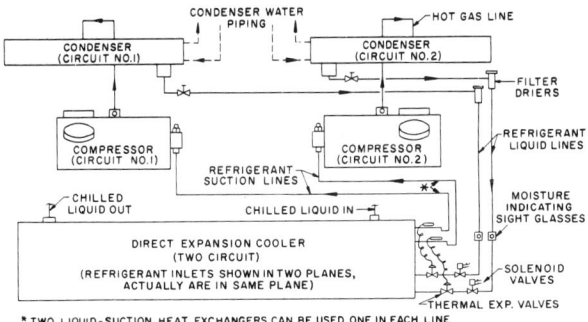

Fig. 18 Typical Refrigerant Piping in Liquid Chilling
Package Having Two Completely Separate Circuits

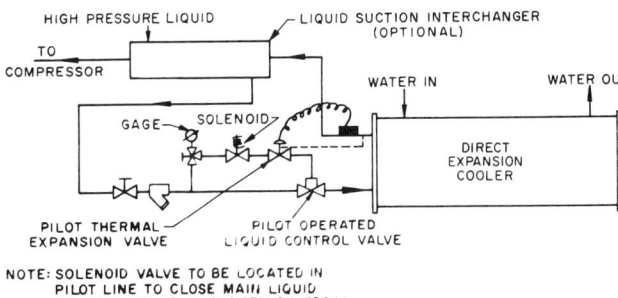

Fig. 19 Direct-Expansion Cooler with Pilot-Operated
Control Valve

cause bad cycling of the thermal valve, since the high-pressure liquid flow through the interchanger ceases when the thermal valve closes; consequently, no heat source is available from the high-pressure liquid and the cooler must starve itself to obtain the superheat necessary to open the valve. When the valve does open, excessive superheat causes it to overfeed until the bulb senses liquid downstream from the interchanger. Therefore, position the remote bulb between the cooler and the interchanger.

Figure 18 shows a typical piping arrangement that has been successful in packaged water chillers having direct expansion coolers. With this arrangement, automatic recycling pumpdown is needed on the lag compressor to prevent leakage through compressor valves, permitting migration to the cold evaporator circuit. Liquid will slug the compressor at startup if this is not done.

On larger systems, the limited size of thermal valves may require the use of a pilot-operated liquid valve controlled by a small thermal expansion valve (Figure 19). The small thermal valve pilots the main liquid control valve. The equalizing connection and the thermal bulb of the pilot thermal valve should be treated as a direct-acting thermal expansion valve. A small solenoid valve in the pilot line will shut off the high side from the low during shutdown. However, the main liquid valve will not open and close instantaneously.

Direct Expansion Air Coils

For further information on this type of coil, see Chapter 6 of the 1988 ASHRAE *Handbook—Equipment*.

The most common ways of arranging direct expansion coils are shown in Figures 20 and 21. The method shown in Figure 21 provides the necessary superheat to operate the expansion valve and is very effective for heat transfer because the leaving air contacts the coldest evaporator surface. This arrangement is highly advantageous on low-temperature applications where the coil pressure drop represents an appreciable change in evaporating temperature.

Direct expansion air coils can be located in any position provided proper refrigerant distribution and continuous oil removal facilities are provided.

Figure 20 shows top feed, free-draining piping at a vertical up airflow coil. In Figure 21, which illustrates a horizontal airflow coil, the suction is taken off the bottom header connection, providing free oil draining. Many coils are supplied with connections at each end of the suction header so that a free-draining connection can be used regardless of which side of the coil is up; the other end is then capped.

In Figure 22, a refrigerant upfeed coil is used with a vertical downflow air arrangement. Here, the coil design must provide for sufficient gas velocity to entrain oil at lowest loadings and carry it into the suction line.

Pumpdown compressor control is desirable on all systems using downfeed evaporators to protect the compressor against a liquid slugback in cases where liquid can accumulate in the suction

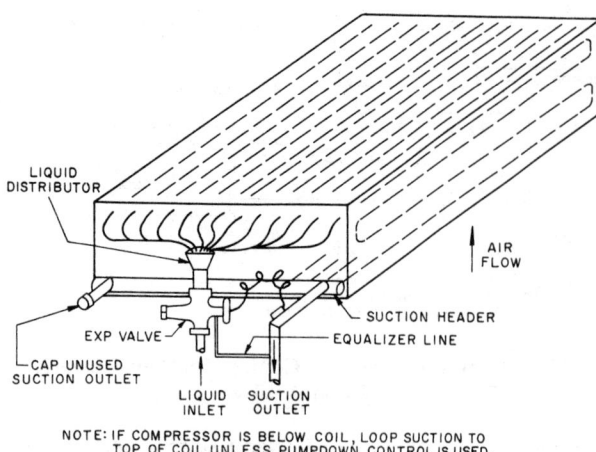

Fig. 20 Direct Expansion Evaporator (Top Feed, Free-Draining)

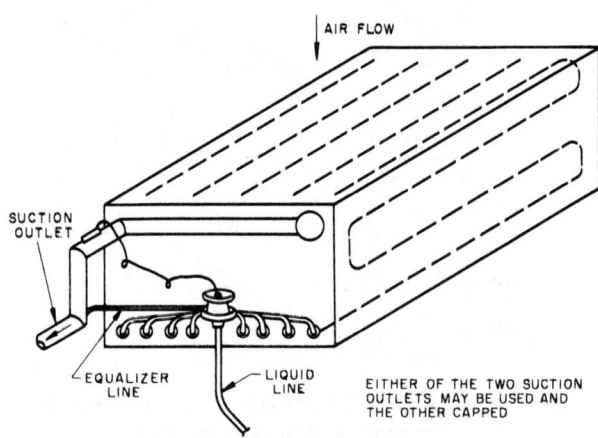

Fig. 22 Direct Expansion Evaporator (Bottom Feed)

header on system off-cycles. Pumpdown compressor control is described later in this chapter.

Thermal expansion valve operation and application are described in Chapter 19 of the 1988 ASHRAE *Handbook—Equipment*.

Thermal expansion valves should be sized carefully to avoid undersizing at full load and oversizing at partial load. The refrigerant pressure drops through the system (distributor, coil, condenser, and refrigerant lines, including liquid lifts) must be properly evaluated to determine the correct pressure drop available across the valve on which to base the selection. Variations in condensing pressure greatly affect the pressure available across the valve, and hence its capacity.

Oversized thermal expansion valves result in a cycling condition that alternates flooding and starving the coil. This occurs because the valve is attempting to throttle at a capacity below its capabilities, which causes periodic flooding of the liquid back to the compressor and wide temperature variations in the air leaving the coil. Reduced compressor capacity further aggravates this problem. Systems having multiple coils can use a solenoid valve located in the liquid line feeding each evaporator or group of evaporators to close them off individually as compressor capacity is reduced.

Flooded Coils

Flooded coils may be desirable where a small temperature differential is required between the refrigerant and the medium being cooled, an advantageous feature on low-temperature applications.

In a flooded evaporator, the coil is kept full of refrigerant when cooling is required. The refrigerant level is generally controlled by using a high- or low-side float control. Figure 23 represents a typical arrangement showing a low-side float control, oil return line, and heat interchanger.

Circulation of refrigerant through the evaporator depends on gravity and a thermosyphon effect. A mixture of liquid refrigerant and vapor returns to the surge tank, and the vapor flows into the suction line. A baffle installed in the surge tank helps prevent foam and liquid from entering the suction line.

A liquid refrigerant circulating pump (Figure 24) provides a more positive way to obtain a high rate of circulation.

Where the suction line is taken off the top of the surge tank, difficulties arise if no special provisions are made for oil return. For this reason, the oil return lines shown in Figure 23 should be installed. These lines are connected near the bottom of the float

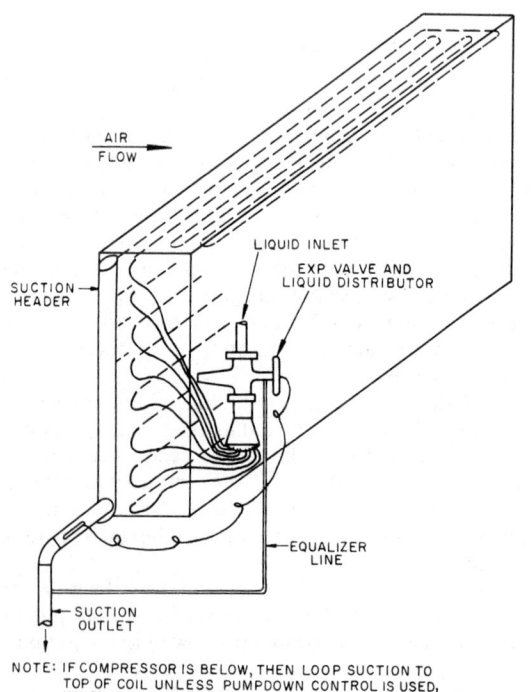

Fig. 21 Direct Expansion Evaporator (Horizontal Airflow)

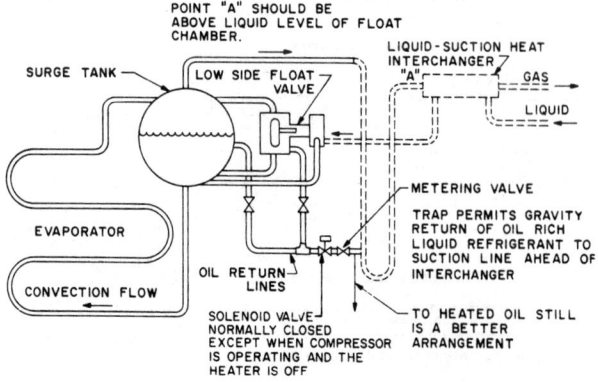

Fig. 23 Flooded Evaporator (Gravity Circulation)

System Practices for Halocarbon Refrigerants

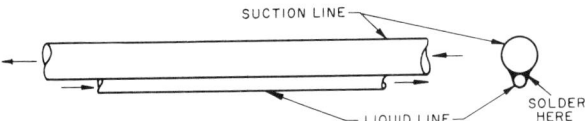

Fig. 25 Soldered-Tube Heat Interchanger

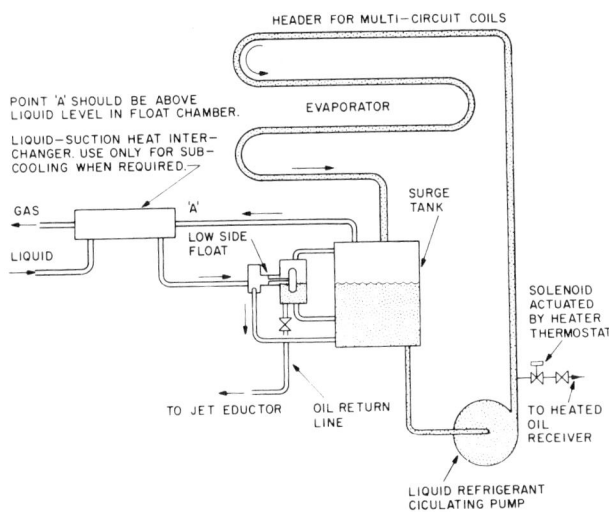

Fig. 24 Flooded Evaporator (Forced Circulation)

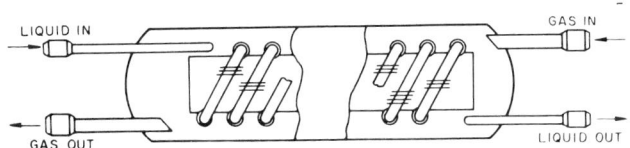

Fig. 26 Shell-and-Finned-Coil Heat Interchanger

chamber and also just below the liquid level in the surge tank (where an oil-rich liquid refrigerant exists). They extend to a lower point on the suction line to allow gravity flow. Included in this oil return line is a solenoid valve that is open only while the compressor is running and a metering valve that is adjusted to allow a constant but small volume return to the suction line. A liquid line sight glass may be installed downstream from the metering valve to serve as a convenient check on the liquid being returned.

Oil can be returned satisfactorily by taking a bleed of refrigerant and oil from the pump discharge (Figure 24) and feeding it to the heated oil receiver. If a low-side float is used, a jet eductor can be used to remove oil from the quiescent float chamber.

REFRIGERATION ACCESSORIES

Liquid-Suction Heat Interchangers

Generally, liquid suction heat exchangers subcool the liquid refrigerant and superheat the suction gas. They are used for one or more of the following functions:

1. *To increase the efficiency of the refrigeration cycle.* The efficiency of the thermodynamic cycle of certain halocarbon refrigerants can be increased when the suction gas is superheated by removing heat from the liquid. This increased efficiency must be evaluated against the effect of pressure drop through the suction side of the exchanger, which forces the compressor to operate at a lower suction pressure.

    Liquid-suction interchangers are most beneficial at low suction temperatures. The increase in cycle efficiency for systems operating in the air-conditioning range (down to about 30°F evaporating temperature) usually does not justify their use. The interchanger can be located wherever convenient.

2. *To subcool the liquid refrigerant to prevent flash gas at the expansion valve.* The interchanger should be located near the condenser or receiver to achieve subcooling before pressure drop occurs.

3. *To evaporate small amounts of expected liquid refrigerant returning from evaporators in certain applications.* Many heat pumps incorporating reversals of the refrigerant cycle include a suction line accumulator and liquid-suction heat exchanger arrangement to trap liquid floodbacks and vaporize them slowly between cycle reversals.

If the design of an evaporator makes a deliberate slight overfeed of refrigerant necessary either to improve evaporator performance or to return oil out of the evaporator, a liquid-suction heat interchanger is needed to evaporate the refrigerant.

A flooded water cooler usually incorporates an oil-rich liquid bleed from the shell into the suction line for returning oil. The liquid-suction heat exchanger boils the liquid refrigerant out of the mixture in the suction line. Exchangers used for this purpose should be placed in a horizontal run near the evaporator.

Several types of liquid-suction heat exchangers are used.

1. *Liquid and Suction Line Soldered Together.* The simplest form of heat exchanger is obtained by strapping or soldering the suction and liquid lines together to obtain counterflow and then insulating the lines as a unit. To obtain the greatest capacity, the liquid line should always be on the bottom of the suction line, since liquid in a suction line runs along the bottom, as shown in Figure 25. This arrangement is limited by the amount of suction line available.

2. *Shell-and-Coil or Shell-and-Tube Interchangers* (Figure 26). These units are usually installed so that the suction outlet drains the shell. When used to evaporate liquid refrigerant returning in the suction line, the free-draining arrangement is not recommended. Liquid refrigerant can run along the bottom of the interchanger shell, having very little contact with the warm liquid coil, and drain into the compressor. By installing the interchanger at a slight angle with the horizontal (Figure 27), and with the gas entering at the bottom and going out at the top, any liquid returning in the line is trapped in the shell and held in contact with the warm liquid coil, where most of it is vaporized. An oil return line, with a metering valve and solenoid valve (open only when the compressor is running), is required to return oil that collects in the trapped shell.

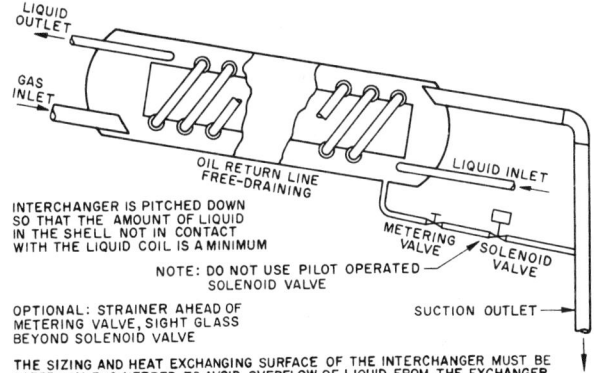

Fig. 27 Shell-and-Finned-Coil Interchanger Installed to Prevent Liquid Floodback

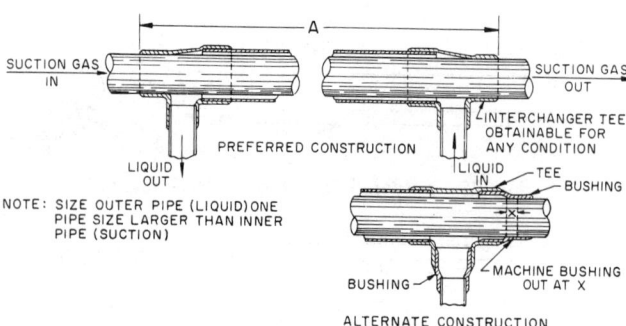

Fig. 28 Tube-in-Tube Heat Exchanger

3. *Concentric Tube-in-Tube Interchangers*. The tube-in-tube interchanger is not as efficient as the shell-and-finned-coil type. It is, however, quite suitable for cleaning up small amounts of excessive liquid refrigerant returning in the suction line. Figure 28 shows typical construction with available pipe and fittings.

For *air-conditioning ranges* in systems using Refrigerants 12 or 502, interchangers should be used only for liquid subcooling or cleaning up excess liquid in the suction line. For *refrigeration ranges* in R-12 and R-502 systems, interchangers are frequently used to increase cycle efficiency, as well as for liquid subcooling and removing small amounts of excess liquid in the suction line. Interchangers should always be used in R-502 systems to maximize performance.

In R-22 systems, liquid suction interchangers are usually used only to evaporate excess liquid in the suction line. The tube-in-tube type is most suitable. Excessive superheating of suction gas in R-22 systems should be avoided.

Discharge Line Oil Separators

Oil is always in circulation in systems using halocarbon refrigerants. Refrigerant piping is designed to assure this oil passes through the entire system and returns to the compressor as fast as it leaves. Although well-designed piping systems can handle the oil in most cases, an oil separator can have certain advantages in some applications. For further information, see Chapter 19 of the 1988 ASHRAE *Handbook—Equipment*.

Some of the applications where discharge line oil separators can be useful are:

1. In systems where it is impossible to prevent substantial absorption of refrigerant in the crankcase oil during shutdown periods. When the compressor starts up with a violent foaming action, oil will be thrown out at an accelerated rate, and the separator will immediately return a large portion of this oil to the crankcase. Normally, however, the system should be designed with pumpdown control or crankcase heaters to minimize liquid absorption in the crankcase.
2. In systems using flooded evaporators, where refrigerant bleedoff is necessary to remove oil from the evaporator. Oil separators reduce the amount of bleedoff from the flooded cooler needed for operation.
3. In direct expansion systems using coils or tube bundles that require bottom feed for good liquid distribution and where refrigerant slopover from the top of the evaporator is essential for proper oil removal.
4. In low-temperature systems, where it is advantageous to have as little oil as possible going through the low side.

In applying oil separators in refrigeration systems, the following potential hazards must be recognized and dealt with:

1. Oil separators are not 100% efficient; it is therefore necessary to design the complete system for oil return to the compressor.
2. Oil separators tend to condense out liquid refrigerant during compressor off-cycles and on compressor start-up. This is true if the condenser is in a warm location, such as an evaporative condenser and receiver on a roof. During the off-cycle, the oil separator cools down and acts as a condenser for refrigerant that evaporates in warmer parts of the system. A cool oil separator may condense discharge gas and, on compressor start-up, automatically drain it into the compressor crankcase. To minimize this possibility, the drain connection from the oil separator can be connected into the suction line. This line should be equipped with a shutoff valve, a fine filter, hand throttling and solenoid valves and a sight glass. The throttling valve should be adjusted so that the flow through this line is only a little greater than would normally be expected to return oil through the suction line.
3. The float valve is a mechanical device that may stick open or closed. If it sticks open, hot gas will be continuously bypassed to the compressor crankcase. If the valve sticks closed, no oil is returned to the compressor. To minimize this problem, the separator can be supplied without an internal float valve. A separate external float trap can then be located in the oil drain line from the separator preceded by a filter. Shutoff valves should isolate the filter and trap. The filter and traps are also easy to service without stopping the system.

The solenoid valve should be wired so that it is open when the compressor is running. When the compressor starts up with liquid refrigerant in the oil separator, the valve prevents this liquid from dumping into the crankcase, but allows it to be bled slowly into the suction line, where it can be taken care of normally, similar to oil returning from the system.

Another method returns oil from an oil separator to the crankcase through an oil reservoir that is heated, boiling off the refrigerant into the suction line.

A solenoid valve can be installed in the drain line from the separator (to the crankcase) and can be controlled by a thermostat at the bottom of the oil separator set high enough so that the solenoid valve will not open until the separator temperature is higher than the condensing temperature. A superheat-controlled expansion valve can perform the same function. If a discharge line check valve is used, it should be downstream of the oil separator.

Surge Drums or Accumulators

These vessels are described in Chapter 2. A surge drum is required on the suction side of almost all flooded evaporators to prevent liquid slopover to the compressor. The exceptions are shell-and-tube coolers and similar shell-type evaporators, which provide ample surge space above the liquid level or contain eliminators to separate gas and liquid. A horizontal surge drum is sometimes used where headroom is limited.

The drum can be designed with baffles or eliminators to separate liquid from the suction gas. More often, sufficient separation space is allowed above the liquid level for this purpose. Usually the design is vertical, with a separation height above the liquid level of from 24 to 30 in. and with the shell diameter sized to keep the suction gas velocity at a value low enough to allow the liquid droplets to separate. Since these vessels are also oil traps, it is necessary to provide oil bleed.

Although separators may be fabricated with length to diameter (L/D) ratios of 1/1 increasing to 10/1, the lowest cost separators will usually be for L/D ratios between 3/1 and 5/1.

Compressor Floodback Protection

Certain systems periodically flood the compressor with excessive amounts of liquid refrigerant. When periodic floodback through the suction line cannot be controlled, the compressor must be protected against it.

System Practices for Halocarbon Refrigerants

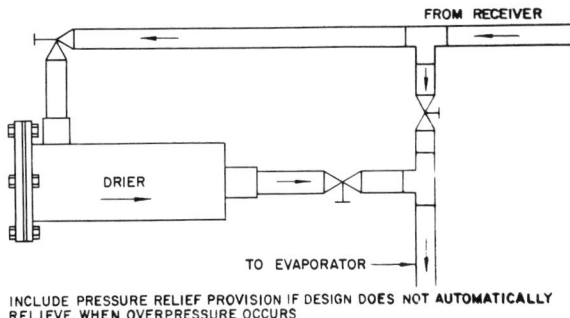

Fig. 30 Drier with Piping Connections

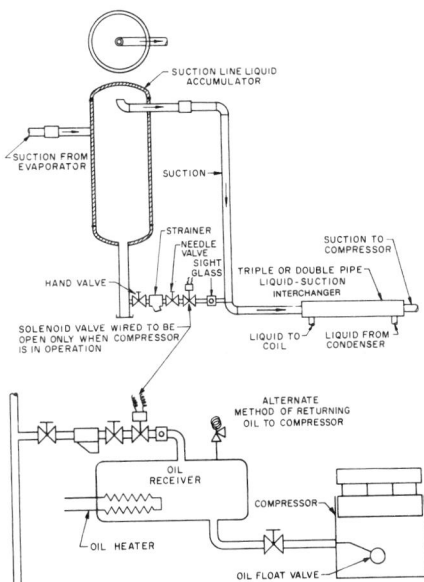

Fig. 29 Compressor Floodback Protection Using Accumulator with Controlled Bleed

The most satisfactory method appears to be a trap arrangement that catches the liquid floodback and may do one of the following: (1) meter it slowly into the suction line where the floodback is cleaned up with a liquid-suction heat interchanger; (2) evaporate the liquid 100% in the trap itself by using a liquid coil or electric heater, and then automatically return oil to the suction line, or (3) return it to the receiver or one of the evaporators.

Figure 27 illustrates an arrangement that handles moderate liquid floodbacks, getting rid of the liquid by a combination of boiling off in the exchanger, plus a limited bleedoff into the suction line. This device, however, would not have sufficient trapping volume for most heat pump jobs or hot-gas defrost systems employing reversal of the refrigerant cycle.

For heavier floodbacks, a larger volume is required in the trap. The arrangement shown in Figure 29 has been applied successfully in reverse cycle heat pump jobs using halocarbon refrigerants. It consists of a suction line accumulator with sufficient volume to hold the maximum expected floodback and of large enough diameter to separate liquid from suction gas. The trapped liquid is slowly bled off through a properly sized and controlled drain line into the suction line, where it is boiled off in a liquid-suction heat interchanger, between cycle reversals.

With the alternate arrangement shown, liquid-oil mixture is heated to evaporate the refrigerant, and the remaining oil is drained into the crankcase or the suction line.

Refrigerant Driers and Moisture Indicators

The effect of moisture in refrigeration systems is discussed in Chapters 6 and 7.

Using a permanent refrigerant drier is recommended on all systems and with all refrigerants. It is especially important on low-temperature systems to prevent ice forming at expansion devices.

A *full-flow drier* is always recommended in hermetic compressor systems to keep the system dry and to prevent the products of decomposition from getting into the evaporator in the event of a motor burnout.

Replaceable element filter-driers are preferred for large systems, since the drying element can be replaced without breaking any refrigerant connections. The drier is usually located in the liquid line near the liquid receiver. It may be mounted horizontally or vertically with the flange at the bottom but should never be mounted vertically with the flange on top, because any loose material would then fall into the line when the drying element is removed.

A three-valve bypass is usually used, as shown in Figure 30, to provide a means for isolating the drier for servicing. The refrigerant charging connection should be located between the receiver outlet valve and the liquid line drier so that all refrigerant added to the system will pass through the drier.

Reliable moisture indicators can be installed in refrigerant liquid lines to provide a positive indication of when the drier cartridge should be replaced.

Strainers

Strainers should be used in both liquid and suction lines to protect automatic valves and the compressor from foreign material such as pipe and welding scale, rust, and metal chips. The strainer should be mounted in a horizontal line with the flanged end down, so that the screen can be replaced without loose particles falling into the system.

A liquid line strainer should be installed before each automatic valve to prevent particles from lodging on the valve seats. Where multiple expansion valves with internal strainers are used at one location, a single main liquid line strainer will protect all of these. The liquid line strainer can be located anywhere in the line between the condenser (or receiver) and the automatic valves, preferably near the valves for maximum protection. Strainers should trap the particle size that could affect valve operation. In pilot-operated valves, a very fine strainer should be installed in the pilot line ahead of the valve.

Filter-driers perform the dual function of refrigerant drying and filtering out particles far smaller than those trapped by mesh strainers. No other strainer is needed in the liquid line if a good filter-drier is used.

Refrigeration compressors are usually equipped with a built-in suction strainer, which is adequate for the usual system with copper piping. The suction line should be piped at the compressor so that the built-in strainer is accessible for servicing.

Both liquid and suction line strainers should be adequately sized to assure sufficient foreign material storage capacity without excessive pressure drop. In steel piping systems, an external suction line strainer is recommended in addition to the compressor strainer.

Liquid Indicators

Every refrigeration system should have a way to check for sufficient refrigerant charge. The common devices used are liquid line sight glass, liquid level test cock on condenser or receiver, and an external gauge glass with equalizing connections and shutoff valves. A properly-installed sight glass will show bubbling when the charge is insufficient.

Liquid indicators should be located in the liquid line as close as possible to the receiver outlet, or condenser outlet if no receiver

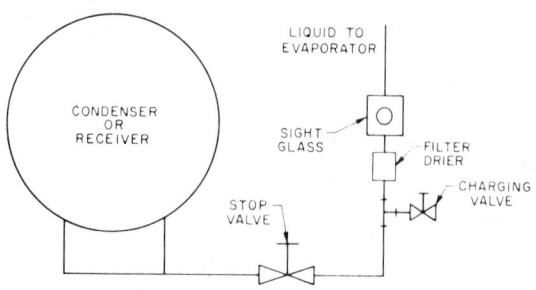

Fig. 31 Sight Glass and Charging Valve Locations

is used (Figure 31). The sight glass is best installed in a vertical section of the line a sufficient distance downstream from any valve so that the resulting disturbance does not appear in the glass. If it is installed too far away from the receiver, the line pressure drop may be sufficient to cause flashing and bubbles in the glass, even though the charge is sufficient for a liquid seal at the receiver outlet.

When sight glasses are installed near the evaporator, often no amount of system overcharging will give a solid liquid condition at the sight glass because of pressure drop in the liquid line or lift. Subcooling is required here. An additional sight glass near the evaporator may be needed to check on the condition of the refrigerant at that point.

Sight glasses should be installed full size in the main liquid line. In very large liquid lines this may not be possible, and the glass can then be installed in a bypass that is arranged so that any gas in the liquid line will tend to go to it.

A sight glass with double ports (for back lighting) and seal caps, which provide added protection against leakage, is preferred. Moisture-liquid indicators large enough to be installed directly in the liquid line serve the dual purpose of liquid line sight glass and moisture indicator.

Oil Receivers

Oil receivers serve as reservoirs for replenishing crankcase oil pumped by the compressors and provide the means to remove refrigerant dissolved in the oil. They are selected when:

1. Flooded or semiflooded evaporators with large refrigerant charges are used
2. Two or more compressors are operated in parallel
3. Long suction and discharge lines are installed
4. Double suction line risers are installed

A typical hookup is shown in Figure 29. Outlets are arranged to prevent oil from draining below the heater level to avoid heater burnout and to prevent scale and dirt from being returned to the compressor.

Purging

Noncondensable gas separation using a purge unit is useful on most large refrigeration systems where suction pressure may fall below atmospheric (see Figure 11 of Chapter 4).

HEAD PRESSURE CONTROL FOR REFRIGERANT CONDENSERS

For further information regarding head pressure control, see Chapter 15 of the 1988 ASHRAE *Handbook—Equipment*.

Water-Cooled Condensers

With water-cooled condensers, head pressure controls are used both for maintaining the condensing pressure and conserving water. On cooling tower applications, they are used only where it is necessary to maintain condensing temperatures.

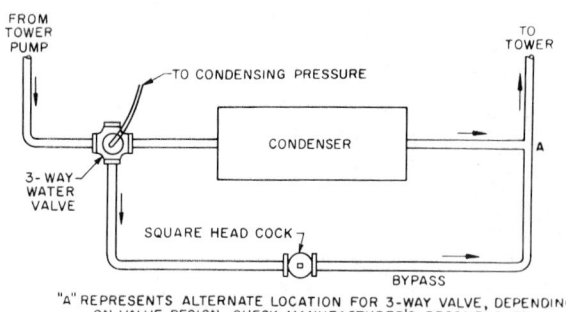

Fig. 32 Head Pressure Control for Condensers Used with Cooling Towers (Water Bypass Modulation)

Condenser Water Regulating Valves

The shutoff pressure of the valve must be set slightly higher than the saturation pressure of the refrigerant at the highest ambient temperature expected when the system is not in operation. This ensures the valve will not pass water during off-cycles. These valves are usually sized to pass the design quantity of water at about a 25 to 30 psi difference between design condensing pressure and valve shutoff pressure. Chapter 19 of the 1988 ASHRAE *Handbook—Equipment* has further information.

Water Bypass

In cooling tower applications, a simple bypass can also be used to maintain condensing pressure, using a manual or automatic valve responsive to head pressure change. Figure 32 shows an automatic three-way valve arrangement. The valve divides the water flow between the condenser and the bypass line to maintain the desired condensing pressure. This maintains balanced flow of water on the tower and pump.

Evaporative Condensers

Methods used for condensing pressure control with evaporative condensers are: (1) cycling the spray pump motor, (2) cycling both fan and spray pump motors, (3) throttling the spray water, (4) bypassing air around duct and dampers, (5) throttling air via dampers, either on inlet or discharge, and (6) combinations of these methods. For further information, see Chapter 15 of the 1988 ASHRAE *Handbook—Equipment*.

In water pump cycling, a pressure control at the gas inlet starts and stops the pump in response to head pressure changes. The

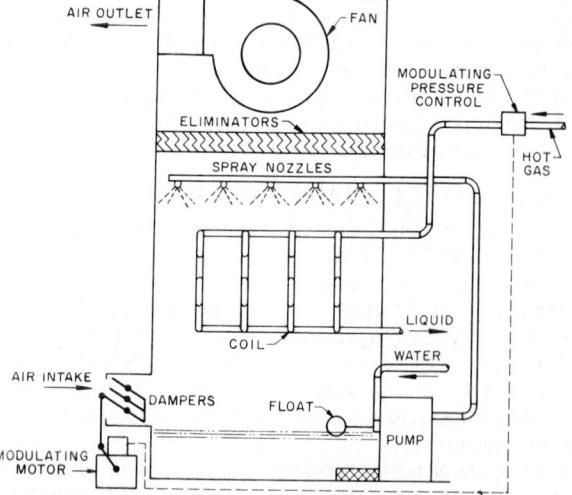

Fig. 33 Head Pressure Control for Evaporative Condenser (Air Intake Modulation)

System Practices for Halocarbon Refrigerants

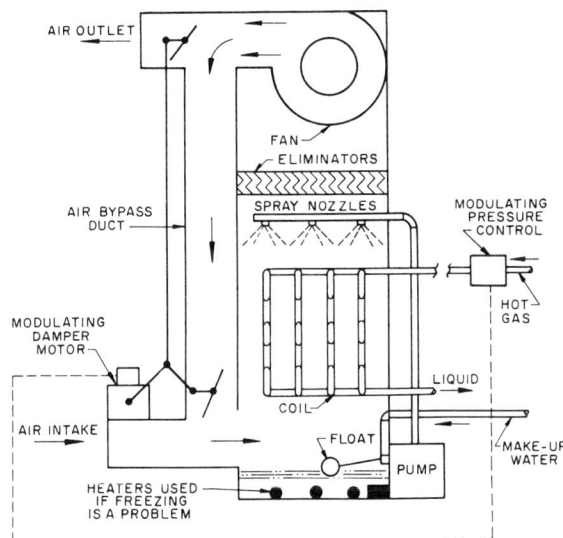

Fig. 34 Head Pressure for Evaporative Condenser (Air Bypass Modulation)

pump sprays water over the condenser coils. As the head pressure drops, the pump stops and the unit becomes an air-cooled condenser.

It is difficult to maintain a constant head pressure with coils of prime surface tubing, because as soon as the pump stops, the pressure goes up and the pump starts again. This occurs because this type of coil has insufficient capacity when operating as an air-cooled condenser. The problem is not as acute with extended surface coils. Short cycling results in excessive deposits of mineral and scale on the tubes and decreases the life of the water pump.

One method of regulating head pressure is to control the amount of air the damper admits (Figure 33). The modulating head pressure control located on the incoming hot-gas line operates the damper motor at the inlet air duct. With a rise of head pressure, the damper opens and allows increased airflow over the condenser. One of the drawbacks of this control in outdoor condensers is that ice forms on the dampers under certain rain or snow conditions.

Figure 34 incorporates an air bypass arrangement for controlling head pressure. A modulating motor, acting in response to a modulating pressure control, positions dampers so that the mixture of recirculated and cold inlet air maintains the desired pressure. In extremely cold weather, most of the air is recirculated.

Air-Cooled Condensers

Methods used for condensing pressure control with aircooled condensers are: cycling fan motor, air throttling or bypassing, coil flooding, and fan motor speed control.

The first two methods are described in the section on evaporative condensers. The third method holds the condensing pressure up by backing liquid refrigerant up in the coil to cut down on effective condensing surface.

When the head pressure drops below the setting of the modulating control valve, it opens, allowing discharge gas to enter the liquid drain line. This restricts liquid refrigerant drainage and causes the condenser to flood enough to maintain the condenser and receiver pressure at the control valve setting. A pressure difference must be available across the valve to open it. Although the condenser would impose sufficient pressure drop at full load, this may practically disappear at partial loading. Therefore, a positive restriction must be placed parallel with the control valve. Systems using this type of control require extra refrigerant charge.

In multiple fan air-cooled condensers, it is common to cycle fans off down to one fan, then to apply air throttling to that section or modulate the fan motor speed. Consult the manufacturer before using this method, as not all condensers are properly circuited for it.

Using ambient temperature change (rather than condensing pressure) to modulate air-cooled condenser capacity prevents rapid cycling of condenser capacity means, which in turn upsets thermal expansion valve operation.

KEEPING LIQUID FROM CRANKCASE DURING OFF-CYCLES

The control of reciprocating compressors should prevent excessive accumulation of liquid refrigerant in the crankcase during off-cycles. Any one of the following control methods will accomplish this.

Automatic Pumpdown Control (Direct Expansion Air-Cooling Systems)

The most effective way to keep liquid out of the crankcase during system shutdown is to operate the compressor on automatic pumpdown control. The recommended arrangement involves the following devices and provisions:

1. A liquid line solenoid valve in the main liquid line or in the branch to each evaporator.
2. Compressor operation through a low-pressure cutout providing for pumpdown whenever this device closes, regardless of whether or not the balance of the system is operating.
3. Electrical interlock of the liquid solenoid valve with the evaporator fan, so that the refrigerant flow will be stopped when the fan is out of operation.
4. Electrical interlock of the refrigerant solenoid valve with the safety devices (such as the high-pressure cutout, oil safety switch, and motor overloads), so that the refrigerant solenoid valve closes when the compressor stops, through the action of these safety devices.
5. Low-pressurestat settings such that the cut-in point will correspond to a saturated refrigerant temperature lower than any expected compressor ambient air temperature. If the cut-in setting of the low-pressure switch is higher than this, then liquid refrigerant can accumulate and condense in the crankcase at a pressure corresponding to the ambient temperature. In this case, the crankcase pressure would not rise high enough to reach the cut-in point and effective automatic pumpdown would not be obtained.

Crankcase Oil Heater (Direct Expansion Systems)

A crankcase oil heater with or without single (nonrecycling) pumpout at the end of each operating cycle does not keep liquid refrigerant out of the crankcase as effectively as automatic pumpdown control, but many compressors equalize too fast after stopping to use automatic pumpdown control. Crankcase oil heaters will maintain the crankcase oil at a temperature higher than that of other parts of the system, minimizing the absorption of the refrigerant by the oil.

Operation with the single pumpout arrangement is as follows. Whenever the temperature control device opens the circuit, or the manual control switch is opened for shutdown purposes, the crankcase heater is energized and the compressor keeps running until it cuts off on the low-pressure switch. Because the crankcase heater remains energized during the complete off-cycle, it is important that a continuous live circuit be available to the heater during the off time. The compressor cannot start again until the temperature control device or manual control switch closes, regardless of the position of the low-pressure switch.

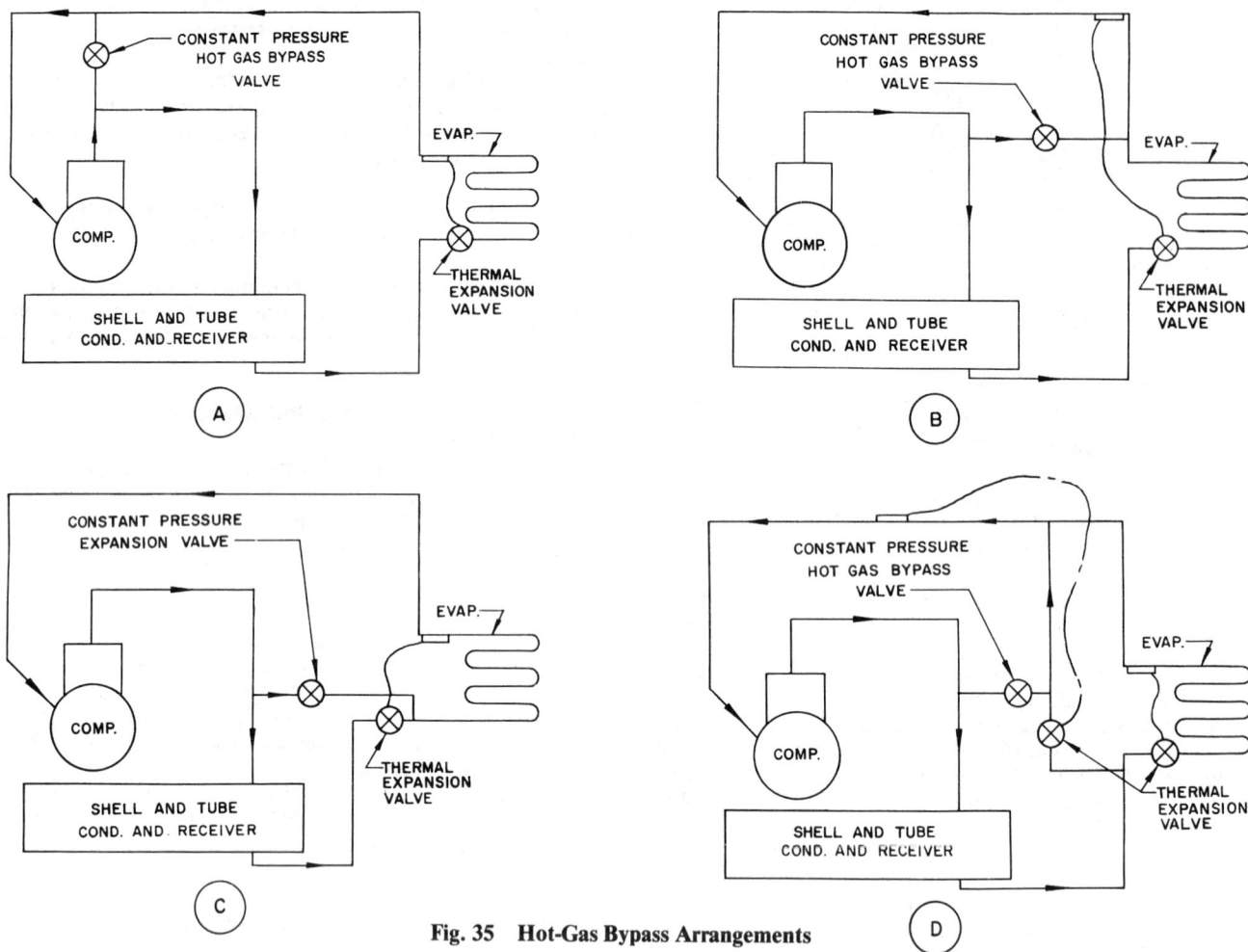

Fig. 35 Hot-Gas Bypass Arrangements

This control method requires:

1. A liquid line solenoid valve in the main liquid line or in the branch to each evaporator.
2. Using a relay or the maintained contact of the compressor motor auxiliary switch to obtain a single pumpout operation before stopping the compressor.
3. A relay or auxiliary starter contact to energize the crankcase heater during the compressor off-cycle and deenergize it during the compressor on-cycle.
4. Electrical interlock of the refrigerant solenoid valve with the evaporator fan, so that the refrigerant flow will be stopped when the fan is out of operation.
5. Electrical interlock of the refrigerant solenoid valve with the safety devices (such as the high-pressure cutout, oil safety switch and motor overloads) so that the refrigerant flow valve will close when the compressor stops, through the action of one of these safety devices.

Control for Direct Expansion Water Chillers

Automatic pumpdown control is not desirable for direct expansion water chillers because freezing is possible if excessive cycling occurs. A crankcase heater is the best solution, with a solenoid valve in the liquid line that closes when the compressor stops.

Effect of Short Operating Cycle

With reciprocating compressors, oil will leave the crankcase at an accelerated rate immediately after starting. Therefore, each start should be followed by a sufficiently long operating period to permit the regain of the oil level. Thermostats used for compressor control should have differentials wide enough so that the running cycles will not be less than 7 or 8 min.

HOT-GAS BYPASS ARRANGEMENTS

Most large reciprocating compressors are equipped with unloaders, which allow the compressor to start with most of its cylinders unloaded. However, it may be necessary to further unload the compressor to (1) reduce starting torque requirements so that the compressor can be started both with low starting torque prime movers and on low current taps of reduced voltage starters, and (2) permit capacity control down to 0% load conditions without stopping the compressor.

Full (100%) Unloading for Starting

Loadless starting can be done with a manual or automatic valve in a bypass line between the hot-gas and suction lines at the compressor.

To prevent overheating, this valve is open only during the starting period and closed after the compressor is up to full speed and full voltage is applied to the motor terminals.

The sequence of control is as follows:

The unloading bypass valve is energized on demand of the control calling for compressor operation. This equalizes pressures across the compressor. After an adequate time delay, a timing relay closes a pair of normally open contacts to start the compressor. After a further time delay, a pair of normally closed timing relay contacts open, deenergizing the bypass valve.

Full (100%) Unloading for Capacity Control

Where full unloading is required for capacity control, hot-gas bypass arrangements can be used in ways that will not overheat the compressor. In using these arrangements, hot gas should not be bypassed until after the last unloading step.

A hot-gas bypass should: (1) give acceptable regulation throughout the range of loads, (2) not cause excessive superheating of the suction gas, (3) not cause any slopover to the compressor, and (4) maintain an oil return to the compressor.

A hot-gas bypass for capacity control is an artificial loading device that maintains a minimum evaporating pressure during continuous compressor operation, regardless of the evaporator load. This is usually done by an automatic or manual pressure reducing valve, which establishes a constant pressure on the downstream side.

Four of the more common methods of using hot-gas bypass are shown in Diagrams A, B, C, and D of Figure 35. Diagram A illustrates the simplest type of hot-gas bypass. It will dangerously overheat the compressor if used for protracted periods of time. Diagram B shows the use of hot-gas bypass to the exit of the evaporator. The expansion valve bulb should be placed at least 5 ft downstream from the bypass point of entrance, and preferably further, to assure good mixing.

In Diagram D, the hot-gas bypass enters after the evaporator thermostatic expansion valve bulb. Another thermostatic expansion valve supplies liquid directly to the bypass line for desuperheating purposes. It is always important to install the hot-gas bypass far enough back in the system to maintain sufficient gas velocities in suction risers and other components to assure oil return at any evaporator loading.

Diagram C shows the most satisfactory hot-gas bypass arrangement. Here the bypass is connected into the low side between the expansion valve and entrance to the evaporator. If a distributor is used, the gas enters between the expansion valve and distributor. Refrigerant distributors are commercially available with side inlet connections that can be used for hotgas bypass duty to a certain extent. Pressure drop through the distributor tubes must be evaluated to determine how much gas bypassing can be effected. This arrangement provides good oil return.

Solenoid valves should be used before the constant pressure bypass valve and before the thermal expansion valve used for liquid injection desuperheating, so that these devices cannot function until they are required.

The control valves for the hot gas should be close to the main discharge line because the line preceding the valve will usually fill with liquid when closed.

The hot-gas bypass line should be sized so that its pressure loss is only a small percentage of the pressure drop across the valve. Usually, it is the same size as the valve connections. When sizing the valve, consult a control valve manufacturer to determine the minimum compressor capacity that must be offset, the refrigerant used, the condensing pressure, and the suction pressure.

When unloading as in Diagram C, the head pressure control requirements increase considerably, because the only heat delivered to the condenser is that caused by the motor power delivered to the compressor. The discharge pressure should be kept high enough so that the hot-gas bypass valve can deliver gas at the required rate. The condenser head pressure control must be capable of meeting this condition.

REFERENCES

Alofs, D.J., M.M. Hasan, and H.J. Sauer, Jr. 1990. Influence of oil on pressure drop in refrigerant compressor suction lines. ASHRAE *Transactions* 96:1.

Colebrook, D.F. 1938, 1939. Turbulent flow in pipes. *Journal of Institute of Engineers* 11.

Cooper, W.D. 1971. Influence of oil-refrigerant relationships on oil return. ASHRAE Symposium Bulletin, PH-71-2.

Jacobs, M.L., F.C. Scheideman, F.C. Kazem, and N.A. Macken. 1976. Oil transport by refrigerant vapor. ASHRAE *Transactions* 81(2):318-29.

Keating, E.L. and R.A. Matula. 1969. Correlation and prediction of viscosity and thermal conductivity of vapor refrigerants. ASHRAE *Transactions* 75(1).

CHAPTER 4

SYSTEM PRACTICES FOR AMMONIA

Piping Requirements 4.1	Evaporator Piping 4.9
Oil in Ammonia Systems 4.3	Suction Traps 4.12
Compressor Piping 4.3	Liquid Ammonia Pumped Recirculation Systems 4.13
Ammonia Compressor Cooling 4.5	Compound Compression Systems 4.14
Condenser and Receiver Piping 4.6	Cascade Systems 4.15

AMMONIA (Refrigerant 717) is an economical choice for industrial systems. While ammonia has superior thermodynamic properties, at low concentration levels of 35 to 50 ppm, it is considered toxic. Large quantities of ammonia should be used to reduce hazards. Cold liquid refrigerant should not be vented to enclosed areas near open flames or heavy sparks. A proportion of 16 to 25% by volume in air in the presence of an open flame burns rapidly and can explode.

The importance of ammonia piping is sometimes minimized when the main emphasis is on selecting major equipment pieces. Mains should be sized carefully to provide low pressure drop and avoid capacity or power penalties caused by inadequate piping.

Rusting pipes and vessels in older systems containing ammonia can create a safety hazard. Oblique X-ray photographs of welded pipe joints and ultrasonic inspection of vessels should be used to disclose defects where of concern. Only vendor certified parts for pipe valving and pressure-containing components according to designated assembly drawings should be used to reduce hazards. Cold liquid refrigerant should not be confined between closed valves in a pipe where the liquid can warm and expand to burst piping components. Rapid multiple pulsations of ammonia liquid in piping components, such as those developed by cavitation forces or hydraulic hammering from compressor pulsations with massive slugs of liquid carryover to the compressor, must be avoided to prevent equipment and piping damage and injury to personnel.

Most service problems are caused by inadequate precautions during design, construction, and installation, (ANSI/IIAR 1984, ANSI/ASHRAE 1989). Ammonia is a powerful solvent that removes dirt, scale, sand, or moisture remaining in the pipes, valves, and fittings during installation. These substances are swept along with the suction gas to the compressor, becoming a menace to the bearings, pistons, cylinder walls, valves, and lubricating oil. Most compressors are equipped with suction strainers and/or additional disposable strainer liners for the large quantity of debris that can be present at initial start-up.

Moving parts are often scored when a compressor is run for the first time. Damage starts with minor scratches, which increase progressively until they seriously affect the operation of the compressor or render it inoperative.

Serious cylinder wall and ring wear results if an appreciable quantity of liquid returns with the suction gas. Relatively large masses of liquid (called *slugs*) break the valves or cause more serious damage. Liquid returning steadily, at rates too low to cause immediately evident damage, reduces the cylinder lubrication, causing extremely rapid cylinder wall and ring abrasion. With double-acting machines, which do not deliver the connecting rod thrust to the piston, cylinder lubrication is less important. However, with single-acting machines, particularly at the prevalent higher speeds, the thrust forces are high, and a steady return of even a mist of liquid can cause cylinder wall wear, requiring replacement after less than 500 h of operation.

A system that has been carefully and properly installed with no foreign matter or liquid entering the compressor will operate satisfactorily for many years. As piping is installed, it should be power rotary wire brushed and blown out with compressed air. Then the piping system should be blown out with compressed air or nitrogen before evacuation and charging. Discharge and liquid piping should be tested to 300 psig air pressure, and suction and low-side piping to 150 psig.

PIPING REQUIREMENTS

The following recommendations are given for ammonia piping. Local codes or ordinances governing ammonia mains should also be consulted.

Recommended Material

Since copper and copper-bearing materials are attacked by ammonia, they are not used in ammonia piping systems. Iron and steel piping, fittings, and valves of the proper pressure rating are suitable for ammonia gas and liquid.

Ammonia piping should conform to ANSI/ASME, *Code for Pressure Piping,* which states:

1. Liquid lines 1.5 in. and smaller shall be not less than schedule 80 carbon steel pipe.
2. Liquid lines 2 in. through 6 in. shall be not less than schedule 40 carbon steel pipe.
3. Vapor lines 6 in. and smaller shall be not less than schedule 40 carbon steel pipe.

Fittings

Couplings, elbows, and tees for threaded pipe are for a minimum of 2000 psi design pressure and constructed of forged steel. Fittings for welded pipe should match the type of pipe used; i.e., full weight fittings for full weight pipe and extra heavy fittings for extra heavy pipe. For welded 1.5 in. and smaller pipe, use socket-weld fittings.

Tongue and groove or ANSI flanges should be used in ammonia piping. Welded flanges for low-side piping can have a 150 psi design pressure rating. The high side should be 300 psi.

Pipe Joints

Joints between lengths of pipe or between pipe and fittings can be threaded if the pipe size is 1.25 in. or smaller. Sizes of 1.5 in. or larger should be welded. An all-welded piping system is superior.

Threaded Joints. Many sealants and compounds are available for sealing threaded joints. The manufacturer's instructions cover compatibility and application method. Do not use excessive amounts or apply on female threads, because any excess can contaminate the system.

Welded Joints. Pipe should be cut and beveled before welding. Backup rings (chill rings) keep slag and metallic particles from entering the pipe. Use pipe alignment guides to align the pipe and provide a proper gap between pipe ends so that a full penetration weld is obtained. The weld should be made by a qualified welder, using proper procedures such as the *Standard Procedure Specifications for Welding of Pipe, Fittings and Flanges,* prepared by the Mechanical Contractors Association of America.

The preparation of this chapter is assigned to TC 10.3, Refrigerant Piping.

Gasketed Joints. When using flanges, a 0.063-in. fiber gasket should be used. Before tightening flange bolts to valves, controls, or flange unions, properly align the pipe and bolt holes. If flanges are used to straighten pipe, they may put stress on adjacent valves, compressors, and controls, causing the operating mechanism to bind. To prevent leaks, flange bolts are drawn up evenly when connecting the flanges.

Union Joints. Steel (2000 psi) ground joint unions are used for gage and pressure control lines with screwed valves and for joints up to 0.75 in. When tightening this type of joint, the two pipes must be axially aligned. To be effective, the two parts of the union must match almost perfectly.

Pipe Location

Piping should be at least 7.5 ft above the floor where possible. Locate pipes carefully in relation to other piping and structural members, especially when the lines are to be insulated. The distance between insulated lines should be at least three times the thickness of the insulation for screwed fittings, and four times for flange fittings. The space between the pipe and adjacent surfaces should be three-fourths of these amounts.

Hangers located close to the vertical risers to and from compressors keep the piping weight off the compressor. Pipe hangers should be placed not more than 8 to 10 ft apart and within 2 ft of a change in direction of the piping. Hangers should be designed to bear on the outside of insulated lines. Sheet metal sleeves on the lower half of the insulation are usually sufficient.

Piping to and from compressors and to other components must provide for expansion and contraction. Sufficient flange or union joints should be located in the piping so that components can be assembled easily during initial installation, and also disassembled for servicing.

Pipe Sizing

In addition to the refrigerant line sizing data and tables in Chapter 33 of the 1989 ASHRAE *Handbook—Fundamentals*, Table 1 presents *practical* suction line sizing data based on 0.25 °F and 0.50 °F differential pressure drop equivalent per 100 ft total equivalent length of pipe. For data on equivalent lengths of valves and fittings, refer to Tables 8, 9, and 10 in Chapter 3. Table 2 lists data for sizing suction and discharge lines at 1 °F differential pressure drop equivalent per 100 ft equivalent length of pipe, and for sizing liquid lines at 100 fpm. Charts prepared by Wile (1977) present pressure drops in saturation temperature equivalents for important refrigerants.

Valves

Stop Valves. These valves should be placed in the inlet and outlet lines to all condensers, vessels, evaporators, and long lengths of pipe so they can be isolated in case of leaks and to facilitate "pumping out" by evacuation.

Table 1   Suction Line Capacities (Tons) for Ammonia with Pressure Drops of 0.25 and 0.50 °F/100 ft Equivalent

Steel Line Size		Saturated Suction Temperature, °F					
		−60		−40		−20	
IPS	SCH	$\Delta t = 0.25\,°F$ $\Delta p = 0.046$	$\Delta t = 0.50\,°F$ $\Delta p = 0.092$	$\Delta t = 0.25\,°F$ $\Delta p = 0.077$	$\Delta t = 0.50\,°F$ $\Delta p = 0.155$	$\Delta t = 0.25\,°F$ $\Delta p = 0.123$	$\Delta t = 0.50\,°F$ $\Delta p = 0.245$
3/8	80	0.03	0.05	0.06	0.09	0.11	0.16
1/2	80	0.06	0.10	0.12	0.18	0.22	0.32
3/4	80	0.15	0.22	0.28	0.42	0.50	0.73
1	80	0.30	0.45	0.57	0.84	0.99	1.44
1-1/4	40	0.82	1.21	1.53	2.24	2.65	3.84
1-1/2	40	1.25	1.83	2.32	3.38	4.00	5.80
2	40	2.43	3.57	4.54	6.59	7.79	11.26
2-1/2	40	3.94	5.78	7.23	10.56	12.50	18.03
3	40	7.10	10.30	13.00	18.81	22.23	32.09
4	40	14.77	21.21	26.81	38.62	45.66	65.81
5	40	26.66	38.65	48.68	70.07	82.70	119.60
6	40	43.48	62.83	79.18	114.26	134.37	193.44
8	40	90.07	129.79	163.48	235.38	277.80	397.55
10	40	164.26	236.39	297.51	427.71	504.98	721.08
12	ID	264.07	379.88	477.55	686.10	808.93	1157.59

Steel Line Size		Saturated Suction Temperature, °F					
		0		20		40	
IPS	SCH	$\Delta t = 0.25\,°F$ $\Delta p = 0.184$	$\Delta t = 0.50\,°F$ $\Delta p = 0.368$	$\Delta t = 0.25\,°F$ $\Delta p = 0.265$	$\Delta t = 0.50\,°F$ $\Delta p = 0.530$	$\Delta t = 0.25\,°F$ $\Delta p = 0.366$	$\Delta t = 0.50\,°F$ $\Delta p = 0.582$
3/8	80	0.18	0.26	0.28	0.40	0.41	0.53
1/2	80	0.36	0.52	0.55	0.80	0.82	1.05
3/4	80	0.82	1.18	1.26	1.83	1.87	2.38
1	40	1.62	2.34	2.50	3.60	3.68	4.69
1-1/4	40	4.30	6.21	6.63	9.52	9.76	12.42
1-1/2	40	6.49	9.34	9.98	14.34	14.68	18.64
2	40	12.57	18.12	19.35	27.74	28.45	36.08
2-1/2	40	20.19	28.94	30.98	44.30	45.37	57.51
3	40	35.87	51.35	54.98	78.50	80.40	101.93
4	40	73.56	105.17	112.34	160.57	164.44	208.34
5	40	133.12	190.55	203.53	289.97	296.88	376.18
6	40	216.05	308.62	329.59	469.07	480.96	609.57
8	40	444.56	633.82	676.99	962.47	985.55	1250.34
10	40	806.47	1148.72	1226.96	1744.84	1786.55	2263.99
12	ID	1290.92	1839.28	1964.56	2790.37	2862.23	3613.23

Note: Capacities are in tons of refrigeration resulting in a line friction loss (Δp in psi) per 100 ft equivalent pipe length, with corresponding change (Δt in °F per 100 ft) in saturation temperature.

System Practices for Ammonia

Installing globe-type stop valves with the valve stems horizontal lessens the chance for dirt or scale to lodge on the valve seat or disk and cause it to leak, or for liquid or oil to pocket in the area below the seat.

Welded flanged or weld-in-line valves are desirable for all line sizes; however, screwed valves may be used for 1.25 in. and smaller sizes. Ammonia globe and angle valves should have the following features:

1. Soft seating surfaces for positive shutoff (no copper or copper alloy)
2. Back seating to permit repacking the valve stem while in service
3. Arrangement that allows packing to be tightened easily
4. Body of steel, cast iron, or nodular iron
5. Bolted bonnets above 1 in., threaded bonnets for 1 in. and smaller

Control Valves. Pressure regulators, solenoid valves, and thermostatic expansion valves should be flanged for easy assembling and removal. Valves 1.5 in. and larger should have welding companion flanges. Smaller valves can have threaded companion flanges.

A strainer should be used in front of self-contained control valves to protect them from pipe construction material and dirt. A ceramic filter, installed in the pilot line to the power piston, will protect the close tolerances from foreign material when pilot-operated control valves are used.

Solenoid Valves. Solenoid valve stems should be located upright and their coils protected from moisture. They should have flexible conduit connections, where allowed by codes, and have an electric pilot light, wired in parallel, to indicate when the coil is energized. A manual opening stem is useful for emergencies.

Solenoid valves for high-pressure liquid feed to evaporators should have soft seats for positive shutoff.

Solenoid valves for other applications, such as in suction lines, hot gas lines, or gravity feed lines, should be selected for the pressure and temperature of the fluid flowing and for the pressure drop available.

OIL IN AMMONIA SYSTEMS

Oil is miscible with liquid ammonia only in very small proportions. The proportion decreases with the temperature, and the oil separates with this decrease. The evaporation of ammonia increases the oil ratio, causing more oil to separate. The increased density causes the oil (saturated with ammonia at the existing pressure) to form a separate layer below the ammonia liquid.

Unless oil is removed periodically or continuously from the point where it collects, it can cover the heat transfer surface in the evaporator, reducing performance. If gage lines or branches to level controls are taken from low points (or oil is allowed to accumulate), these lines will contain oil. The higher oil density will be at a lower level than the ammonia liquid. Draining oil from a properly located collection point is not difficult unless the temperature is so low that the oil does not flow readily. In this case, maintaining the oil receiver at a higher temperature may be beneficial.

Oil in the system is saturated with ammonia at the existing pressure. (See Table 5 in Chapter 8.) When the pressure is reduced, the ammonia vapor separates, causing foaming.

COMPRESSOR PIPING

Figures 1 and 2 show a typical piping arrangement for two compressors operating in parallel off the same suction main. Suction

Table 2 Refrigerant Suction and Discharge Line Capacities (Tons) for Ammonia (Single or High-Stage Applications) for a Pressure Drop of 1°F Equivalent

Steel Line Size		Suction Lines $\Delta t = 1°F$					Discharge Lines	Liquid Lines			
		Saturated Suction Temperature, °F						Steel Line Size			
IPS	SCH	-40 $\Delta p = 0.31$	-20 $\Delta p = 0.49$	0 $\Delta p = 0.73$	20 $\Delta p = 1.06$	40 $\Delta p = 1.46$	$\Delta t = 1°F$ $\Delta p = 2.95$	IPS	SCH	Velocity/ 100 fpm	$1°F \Delta p = 2.0$ psi $\Delta t = 0.7°F$
3/8	80	—	—	—	—	—	—	3/8	80	8.6	12.1
1/2	80	—	—	—	—	—	3.1	1/2	80	14.2	24.0
3/4	80	—	—	—	2.6	3.8	7.1	3/4	80	26.3	54.2
1	80	—	2.1	3.4	5.2	7.6	13.9	1	80	43.8	106.4
1 1/4	40	3.2	5.6	8.9	13.6	19.9	36.5	1 1/4	80	78.1	228.6
1 1/2	40	4.9	8.4	13.4	20.5	29.9	54.8	1 1/2	80	107.5	349.2
2	40	9.5	16.2	26.0	39.6	57.8	105.7	2	40	204.2	811.4
2 1/2	40	15.3	25.9	41.5	63.2	92.1	168.5	2 1/2	40	291.1	1292.6
3	40	27.1	46.1	73.5	111.9	163.0	297.6	3	40	449.6	2287.8
4	40	55.7	94.2	150.1	228.7	333.0	606.2	4	40	774.7	4662.1
5	40	101.1	170.4	271.1	412.4	600.9	1095.2	5	40	—	—
6	40	164.0	276.4	439.2	667.5	971.6	1771.2	6	40	—	—
8	40	337.2	566.8	901.1	1366.6	1989.4	3623.0	8	40	—	—
10	40	611.6	1027.2	1634.3	2474.5	3598.0	—	10	40	—	—
12	ID	981.6	1644.5	2612.4	3963.5	5764.6	—	12	ID	—	—

Notes:
1. Table capacities are in tons of refrigeration resulting in a line friction loss per 100 ft equivalent pipe length, with corresponding change in saturation temperature, where

    Δp = pressure drop due to line friction, psi per 100 ft of line
    Δt = change in saturation temperature, °F

2. Line capacity for other saturation temperatures Δt and equivalent lengths L_e

    $$\text{Line capacity} = \text{Table capacity} \left(\frac{\text{Table } L_e}{\text{Actual } L_e} \times \frac{\text{Actual } \Delta t}{\text{Table } \Delta t} \right)^{0.55}$$

3. Saturation temperature Δt for other capacities and equivalent lengths L_e

    $$\Delta t = \text{Table } \Delta t \ \frac{\text{Actual } L_e}{\text{Table } L_e} \left(\frac{\text{Actual capacity}}{\text{Table capacity}} \right)^{1.8}$$

4. Values in the table are based on 90°F condensing temperature. Multiply table capacities by the following factors for other condensing temperatures.

Condensing Temperature, °F	Suction Lines	Discharge Lines
70	1.05	0.78
80	1.02	0.89
90	1.00	1.00
100	0.98	1.11

5. Discharge and liquid line capacities are based on 20°F suction. Evaporator temperature is 0°F. The capacity is affected less than 3% when applied from $-40°F$ to $+40°F$ extremes.

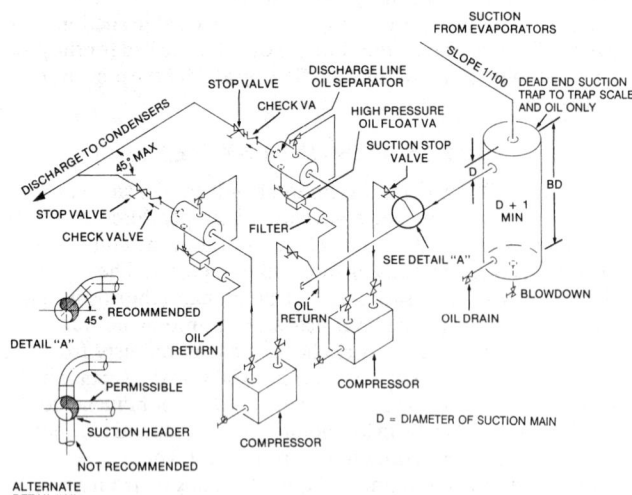
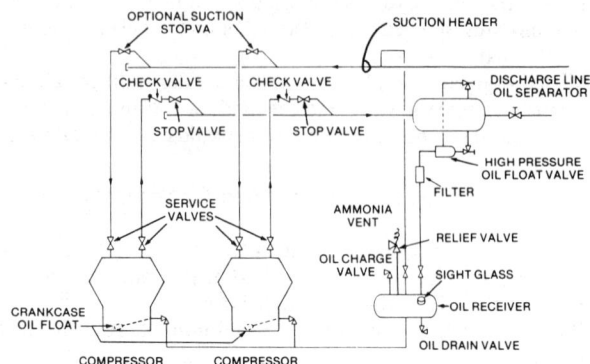

Fig. 1 Piping for Two Compressors in Parallel Operation

Fig. 2 Parallel Compressors with Common Oil Separator

mains should be laid out with the objective of returning only clean, dry gas to the compressor. This usually requires a suction trap sized adequately for gravity gas and liquid separation based on permissible gas velocities for specific temperatures (see Chapter 2, Table 2). A dead-end trap can usually only trap scale and oil.

In sizing the suction mains and the takeoffs from the mains to the compressors, consider how the pressure drop in the selected piping affects the compressor size required. First costs and operating costs for compressor and piping selections should be optimized.

Good suction line systems have a total friction drop of 2 to 5 °F pressure drop equivalent. Practical suction line friction losses should not exceed 0.5 °F equivalent per 100 ft equivalent length.

A well-designed discharge main has a total friction loss of 3 to 5 psi. Generally, a slightly oversized discharge line is desirable to hold down discharge pressure and, consequently, discharge temperature.

High- and low-pressure cutouts and gages are installed on the compressor side of the stop valves to protect the compressor.

The service valves on the compressor open and close frequently and sometimes do not seat tightly. Therefore, an additional stop valve can be used in the suction takeoff to each compressor. However, the additional pressure drop should be considered.

The main suction line should slope toward the trap with a vertical drop 1% of the horizontal length as a minimum.

Oil Separators

Oil separators are located in the discharge line of each compressor (Figure 1). A high-pressure float valve drains the oil back into the compressor crankcase or oil receiver. Place the oil separator as far from the compressor as possible, so that the extra pipe length can be used to cool the discharge gas before it enters the oil separator. This reduces the temperature of the ammonia vapor and makes the oil separator more effective. The external float return valve, as shown, is preferable for automatic oil return.

Liquid ammonia must not reach the crankcase. A valve (preferably automatic) in the drain from the oil separator, open only when the temperature at the bottom of the separator is higher than the condensing temperature, is recommended to prevent this.

A filter is recommended in the drain line on the downstream side of the high-pressure float valve. The minimum size of the oil return line is 0.75 in.

Oil Receivers

Figure 2 illustrates two compressors on the same suction line with one discharge line oil separator. The oil separator float drains into an oil receiver, which maintains a reserve supply of oil for the compressors. Compressors should be equipped with crankcase floats to regulate the oil flow to the crankcase.

Discharge Check Valves

Discharge check valves on the downstream side of each oil separator prevent high-pressure gas from flowing into an inactive compressor and causing condensation (Figure 1). Screw compressors must be equipped with two check valves to prevent reverse rotation; one must be on the compressor suction and one downstream of the oil separator.

If the engine room temperature is less than the condensing temperature, gas may leak past the check valve and condense in the discharge line and oil separator. When this occurs, liquid ammonia drains into the compressor from the float under the oil separator. In this situation, a positive method of equalizing between the discharge line and suction line is recommended; equalization is always recommended when compressors are outside or in unheated spaces. The equalizing can be done automatically, either by a 0.25-in. solenoid valve between the discharge line and suction line that is open when the compressor motor stops, or by a compressor oil pressure-actuated valve that opens whenever the compressor stops. It can be done manually by opening a 0.25-in. hand valve between the discharge and suction lines whenever the compressor stops.

The discharge line from each compressor should enter the discharge main at a 45° maximum angle in the horizontal plane so that the gas flows smoothly. Crankcase oil heaters should always be used to hold the oil temperature well above condensing to prevent refrigerant condensation and to boil off liquid ammonia entrained in the oil. These heaters are energized only when the compressor is off.

Unloaded Starting

Unloaded starting is frequently needed to stay within the torque or current limitations of the motor. Most compressors are unloaded either by holding the suction valve open or by external bypassing. Control can be manual or automatic.

Many slow-speed, heavy-duty compressors are equipped with a manual bypass manifold with starting bypass valves. Suction and discharge of compressors without manual starting bypass can be cross-connected externally to accomplish the same purpose (Figure 3).

Automatic unloading for starting is also used with slow speed heavy-duty compressors and consists of an external bypass posi-

System Practices for Ammonia

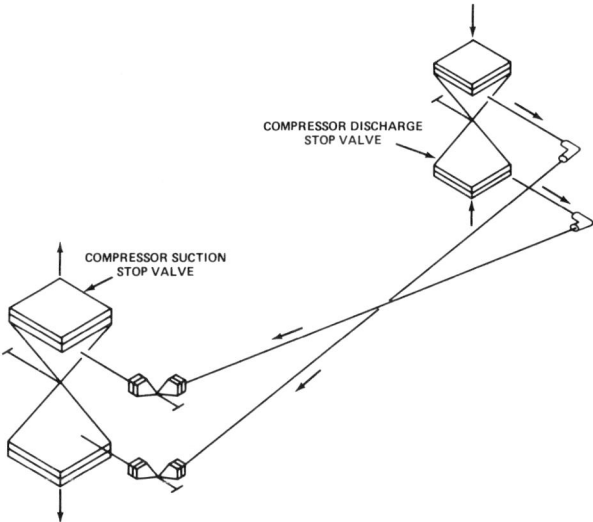

Close the compressor suction and discharge valves and open all cross-connection valves. The refrigerant flows as the arrows indicate in the above sketches; *i.e.*, from the discharge line into the top of the discharge valve, across the bottom of the suction valve, through the compressor into the bottom of the discharge valve, across to the top of the suction valve, and out into the suction line. The size of the cross-connection lines varies depending on the size and number of cylinders on the machine.

Fig. 3 External Cross Connections for Unloaded Starting

tioned on the compressor or on external piping on the compressor side of the discharge line check valve (Figure 4).

Suction Gas Conditioning

Suction main piping should be insulated, especially outside the cooled space, to limit superheat at the compressor. Additional superheat results in increased discharge temperatures and reduces compressor capacity. Low discharge temperatures in ammonia plants are important to reduce oil carryover and because the compressor lubricating oil can carbonize at higher temperatures, which can cause cylinder wall scoring and oil sludge throughout the system. Discharge temperatures above 300 °F should be avoided at all times.

AMMONIA COMPRESSOR COOLING

Generally ammonia compressors are constructed with internally cast cooling passages along the cylinders and/or in the top heads. These passages provide space for circulating a heat transfer medium, which minimizes heat conduction from the hot discharge gas to the incoming suction gas and oil in the compressor's crankcase. Water is usually the medium circulated through these

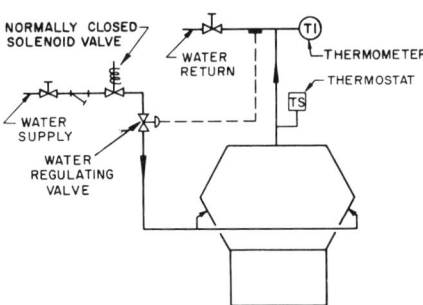

Fig. 5 Jacket Water Cooling for Ambient Temperatures Above Freezing

passages, or *water jackets*. About 0.1 gpm/ton of refrigeration is circulated. The oil in the crankcase (depending on type of construction) is about 120 °F. Temperatures above this level reduce the oil's lubricating properties.

For compressors operating in ambients above 32 °F, water flow sometimes is controlled entirely by hand valves, although a solenoid valve in the inlet line is desirable to make the system automatic. Water flow *must* be stopped when the compressor stops to keep the residual gas from condensing and to conserve water. A water regulating valve, installed in the water supply line with the sensing bulb in the water return line, is also recommended. This type of cooling is shown in Figure 5.

The thermostat in the water line leaving the jacket serves as a safety cutout to stop the compressor if the temperature becomes too high.

For compressors installed where ambient temperatures below 32 °F may exist, provide a means for draining the jacket on shutdown to prevent freeze-up. One method is shown in Figure 6. Water flow is through the normally closed solenoid valve, which is energized when the compressor starts. The water then circulates through the jacket and out through the water return line, which contains a check valve. When the compressor stops, the solenoid valve in the water inlet line is deenergized and stops the water flow to the compressor. At the same time, the solenoid valve opens to drain the water out of the low point of the jacket to the sewer. A check valve in the air vent line opens when pressure is relieved and allows the jacket to be drained. This flapper check valve is installed so that water pressure will close it, but absence of water pressure allows it to swing open.

When compressors are installed in spaces below 32 °F or where water quality is very poor, cooling is best handled by using an antifreeze solution or other suitable fluid in the jackets and by cooling with a secondary heat exchanger. Halocarbon or chlorinated fluids that leak into the ammonia side will cause a chemical reaction and precipitate solids.

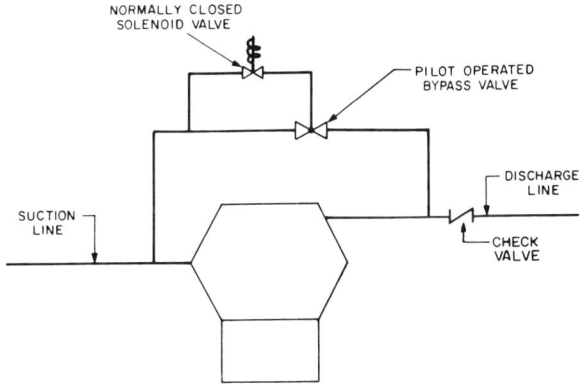

Fig. 4 External Bypass for Automatic Unloading

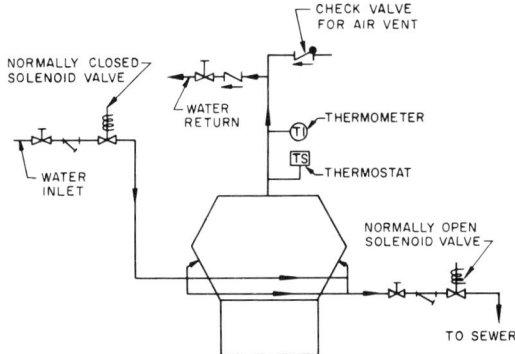

Fig. 6 Jacket Water Cooling for Ambient Temperatures Below Freezing

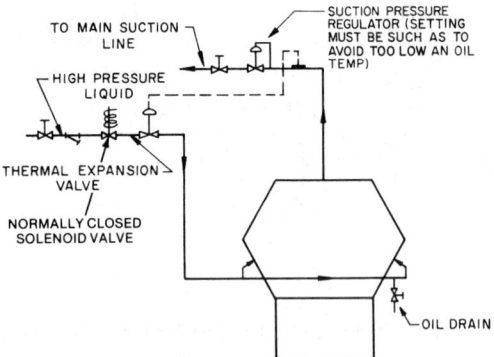

Fig. 7 Jacket Cooling with Ammonia

Another method of cooling the compressor on systems operating below 32°F evaporator temperature is to expand ammonia through the jacket at temperatures safely above 32°F and to control this temperature with a back pressure regulator. Some systems use a thermostatic expansion valve to control liquid flow through the jacket; others use a hand expansion valve. In either case, a solenoid valve opens in the liquid line when the compressor is started. When using a thermostatic expansion valve, suction gas from the jacket can be taken directly into a suction line leading to the compressors. Figure 7 illustrates this type of jacket cooling. The method using the hand expansion valve and solenoid valve requires the branch suction line to enter the suction main before a suction trap or low-pressure receiver.

The jacket heat probably will require about 0.2 ton of refrigeration for each hp of compressor friction.

In rotary vane, low-stage compressors, even temperatures should be maintained around the cylinder for even expansion and contraction. Water or oil circulation through the jacket has been used for cooling rotary compressors; oil is recommended (see Figure 8).

CONDENSER AND RECEIVER PIPING

Properly designed piping around the condensers and receivers keeps the condensing surface at its highest efficiency by draining the ammonia out of the condenser immediately as it condenses and keeping air and other noncondensables purged.

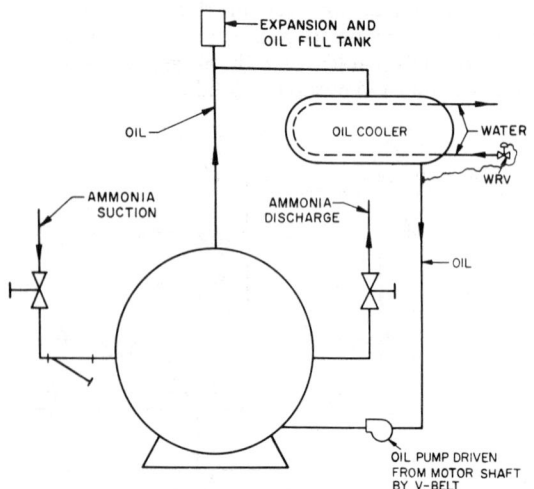

Fig. 8 Rotary Booster Compressor Cooling with Oil

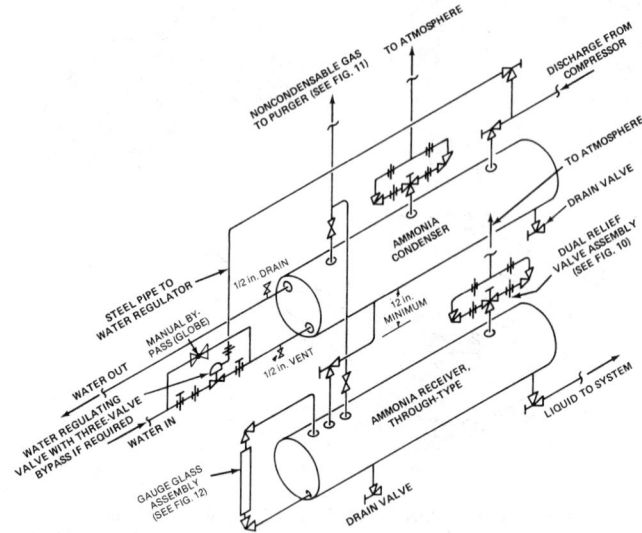

Fig. 9 Horizontal Condenser and Through Receiver Piping

Horizontal Shell-and-Tube Condenser and Through Receiver

Figure 9 shows a horizontal water-cooled condenser draining into a through-type receiver. Ammonia plants do not require controlled water flow to maintain head pressure. Usually, pressure is adequate to force the ammonia to the various evaporators without water regulation. Each situation should be evaluated by comparing water costs with input power cost savings at lower head pressures.

Water piping should be arranged so that condenser tubes are always filled with water. Air vents should be provided on condenser heads and have hand valves for manual purging.

Receivers must be below the condenser so that the condensing surface is not flooded with ammonia. The piping should provide (1) free drainage from the condenser and (2) static height of

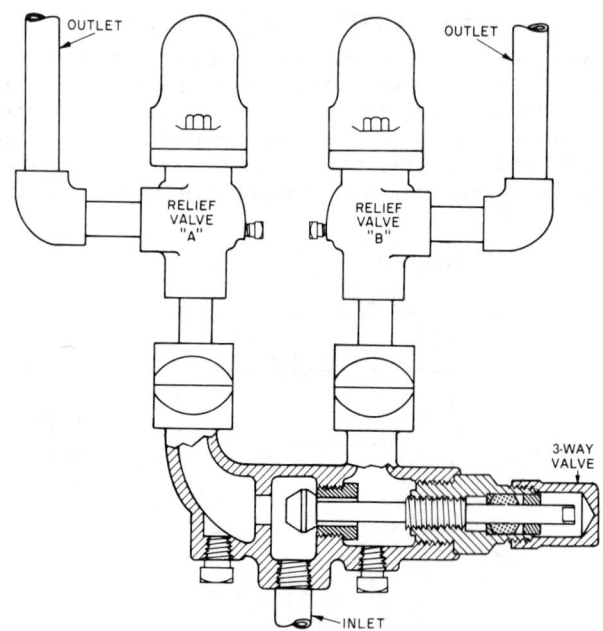

Fig. 10 Dual Relief Valve Fitting for Ammonia Condenser

System Practices for Ammonia

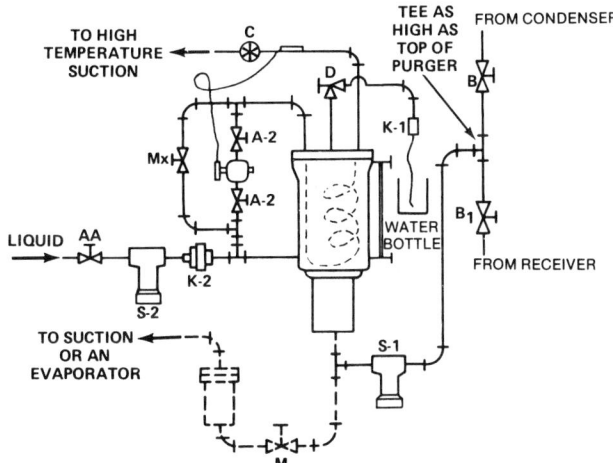

Fig. 11 Purge Unit and Piping for Noncondensable Gas

ammonia above the first valve out of the condenser greater than the pressure drop through the valve.

The drain line from condenser to receiver is designed on the basis of 100 fpm maximum velocity to allow gas equalization between condenser and receiver. Refer to Chapter 33 of the 1989 ASHRAE *Handbook—Fundamentals* for sizing criteria.

Accessories

Any ammonia vessel that can be valved off must have an automatic relief valve. Valve sizing should conform to ANSI/ASHRAE 15-1989 *Safety Code for Mechanical Refrigeration*. The valve should automatically close.

Dual relief valve arrangements enable testing of the relief valves (Figure 10). The three-way stop valve is constructed so that it is always open to one of the relief valves if the other is removed to be checked or repaired.

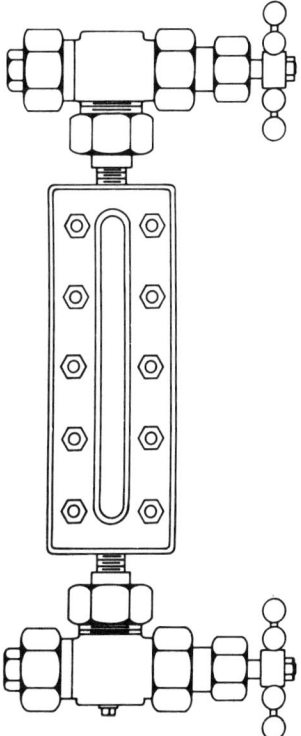

Fig. 12 Gage Glass Assembly for Ammonia

A noncondensable gas separator (purge unit) is useful in most plants, especially when suction pressure is below atmospheric pressure. Purge units on ammonia systems are piped to carry noncondensables (air) from the receiver and condenser to the purger, as shown in Figure 11. High-pressure liquid expands through a coil in the purge unit, providing a cold spot in the purge drum. The suction from the coil should be taken into one of the high-temperature suction mains. Ammonia vapor and noncondensable gas are drawn up into the purge drum, and the ammonia condenses on the cold surface. When the drum fills up with air and other noncondensables, a float valve within the purger opens and permits them to leave the drum and pass into the open water bottle.

A receiver should be equipped with an armored gage glass so that the quantity of ammonia in the receiver can be determined. The valves on each side of the gage glass should be constructed with a check valve so that, if the glass breaks, the check valve will prevent the receiver from losing refrigerant. Figure 12 shows a typical gage glass assembly.

Field or factory fabricated sight columns constructed of pipe with multiple "level eyes" or reflex lenses have also proven satisfactory. In this case, standard valves are usually used to isolate the column from the vessel.

Sufficient liquid ammonia has to be charged into the system to seal the outlet pipe of the receiver and to prevent hot gas from blowing over to the low side. The outlet pipe of a top outlet receiver should extend down into the receiver to a distance of one pipe diameter from the bottom. The lowest operating liquid level in the receiver should be equivalent to two diameters of the outlet pipe for both top and bottom outlet receivers.

A through-type receiver is piped so that all of the ammonia from the condenser passes through on the way to the low side. It should be adequate to handle fluctuations in liquid inventory of the low side. Maintenance is easier if the receiver is sized to hold the entire plant charge. On very large systems the receiver should be sized for at least one-half the total charge.

Parallel Horizontal Shell-and-Tube Condensers

Figure 13 shows two condensers operating in parallel with one through-type receiver. The length of horizontal liquid drain lines

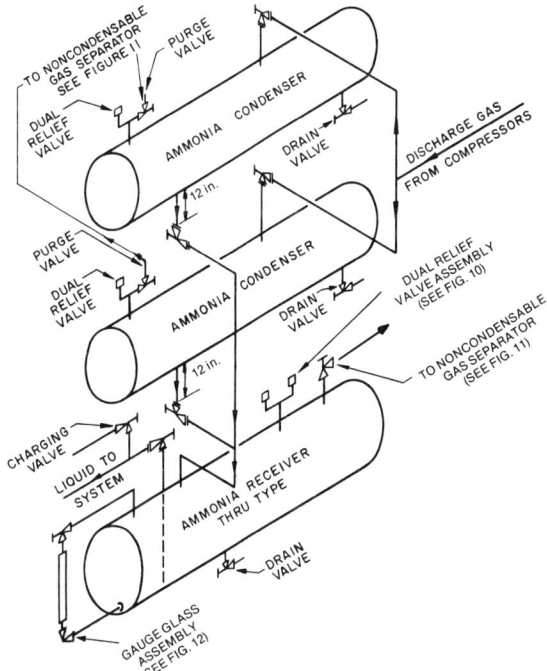

Fig. 13 Parallel Condensers with Through Receiver

Table 3 Equalizing Line Size (Surge Receivers)

Maximum Plant Capacity, tons	Equalizing Line Size, in.
50	0.50
100	0.75
170	1.00
310	1.25
425	1.50
650	2.00

to the receiver should be minimized and with no traps permitted. Equalization between the shells is achieved by keeping the liquid velocity in the drain line less than 100 fpm. The drain line can be sized from Chapter 33 of the 1989 ASHRAE *Handbook—Fundamentals*.

Vertical Shell-and-Tube Condensers and Surge Receiver

Figure 14 illustrates the piping of two vertical condensers and a surge receiver. An equalizing connection relieves the gas pressure above the liquid level to keep the condenser drained when the receiver is in an atmosphere warmer than the condensing temperature. This connection also provides gas pressure on the liquid in the receiver when the receiver is in an atmosphere colder than the condensing temperature. Sizing the equalizing line depends on the exposed surface in the receiver and on the greatest temperature difference between the receiver ambient temperature and the condensing temperature. Table 3 gives equalizing line sizes used for general applications.

The liquid line from the condenser to the surge receiver can be based on a velocity up to 200 fpm and can be sized according to the data in Chapter 33 of the 1989 ASHRAE *Handbook—Fundamentals*. This velocity is higher (compared with 100 fpm on through-type receivers) because the line can be run full of liquid. It is unnecessary to allow room for gas (which might flash in a warm receiver) to travel up the same line to the condenser without restricting liquid drainage from the condenser.

An oil drain leads from the lowest part of each condenser into an oil receiver. Oil, which is heavier than ammonia liquid, settles out at the lowest point.

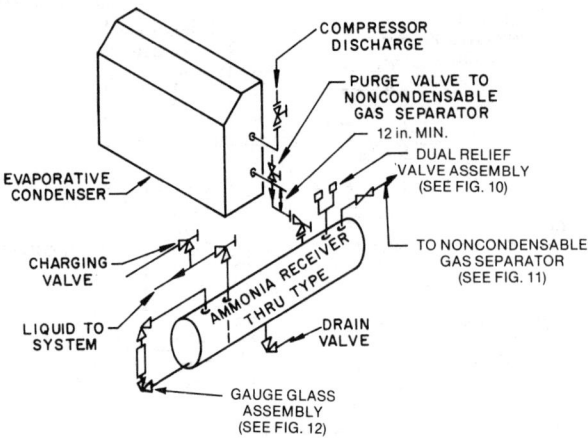

Fig. 15 Piping for Single Evaporative Condenser and Through Receiver

Evaporative Condenser

A single evaporative condenser used with a through-type receiver can be connected as shown in Figure 15. The receiver *always* must be at a pressure lower than the condensing pressure (except that a liquid column height provides about 1 psi for each 4 ft). As indicated in Chapter 33 of the 1989 ASHRAE *Handbook—Fundamentals*, the receiver must be cooler than the condensing temperature.

In areas having ambient temperatures below 32°F, the water must be kept from freezing at light plant loads. When the temperature is at freezing, the evaporative condenser can operate as

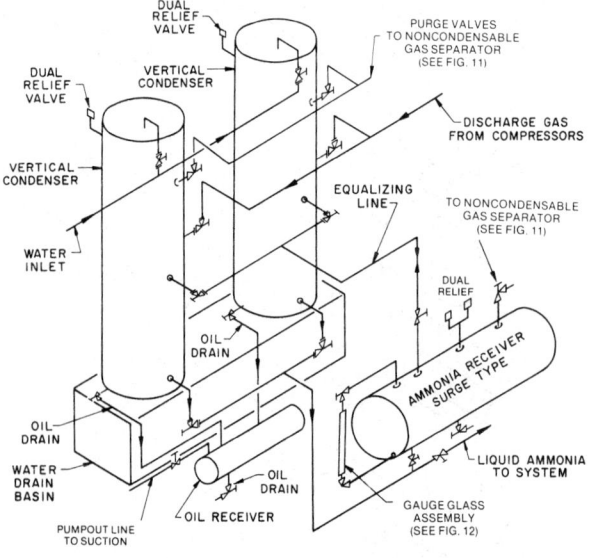

Fig. 14 Parallel Vertical Condensers with Surge Receiver

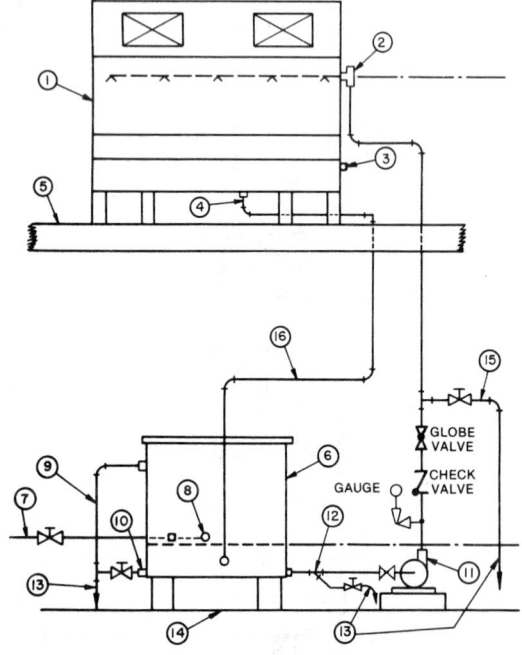

1. Evaporative condenser
2. Spray header connection
3. Pan overflow
4. Pan drain
5. Roof
6. Tank
7. Make-up water inlet
8. Float
9. Tank overflow
10. Tank drain
11. Pump
12. Strainer with blow-off valve
13. To drain
14. Floor
15. Bleed off and drain line
16. Return from evaporative condenser

Fig. 16 Evaporative Condenser with Inside Water Tank

System Practices for Ammonia

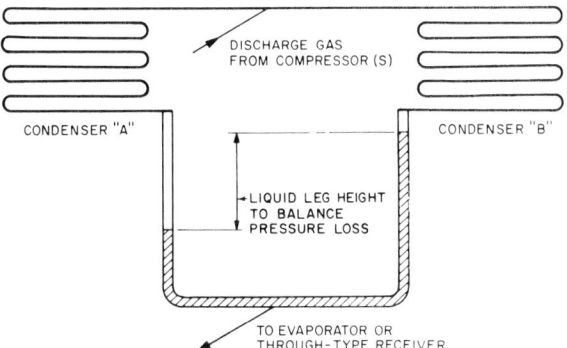

Fig. 17 Piping Through Condensers with Unequal Pressure Loss

a dry-coil unit, and the water pump(s) can be drained and secured for the season.

Another method of preventing the water from freezing is to place the water tank inside and install it as illustrated in Figure 16. When the outdoor temperature drops, the head pressure drops, and a pressure switch, with its sensing element in the discharge pressure line, stops the water pump; the water is then drained into the tank. Another method is to use a thermostat that senses the water temperature or outdoor ambient temperature and stops the pump at low temperatures. The exposed piping and any trapped water headers in the evaporative condenser should be drained.

Air volume capacity control methods include inlet, outlet, or bypass dampers; two-speed fan motors; or fan cycling in response to pressure controls.

Parallel Operation of Evaporative Condensers

When two or more condensers (any type) discharge into a common liquid line, the individual liquid line must be sufficiently high before the junction to permit a liquid level difference that balances any difference in pressure loss of the condensers without filling part of a condenser. Figure 17 illustrates conditions where condenser B has a greater pressure drop than does A.

All operating conditions must be considered. Suppose, at design, A rejects half the heat with half the pressure loss of B. During lower outside temperature, B does not operate and A rejects all heat. The pressure loss through A will be four times its design loss. The loss through B will be near zero. The required liquid leg height will then be four times the original height difference.

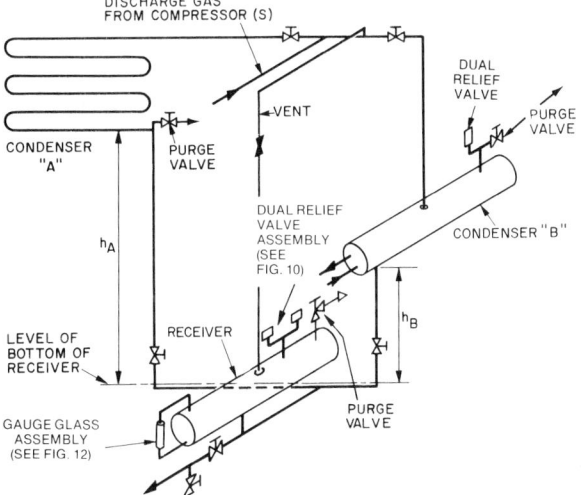

Fig. 18 Parallel Condensers with Surge Receiver

With the conditions given, if B rejected the same total heat while A was not operating, the difference would be about eight times the original liquid leg height.

If condenser A is the shell-and-tube type, its pressure loss will be near zero at normal heat rejection. The level required in the liquid lines is calculated as already described.

A separate high-side float valve in the liquid line from each condenser provides control without any level difference when the receiver operates at a lower pressure than the condensers.

When a surge-type receiver is used, a vent line installed to the condenser inlet prevents loss of liquid subcooling from the condenser. Figure 18 shows such an arrangement. At the greatest difference between the air temperature around the receiver and the condensing temperature, the minimum value of h_A must be about 50 in. for each psi loss through condenser A, plus through the discharge line branch, the liquid line to the receiver branch and the vent. The minimum value of h_B must be the same with relation to condenser B and its branch lines. For both h_A and h_B, this must be based on the pressure loss when the heat rejected through the particular condenser is greatest.

The size of the vent (or equalizer) is based on the surface of the receiver and the temperature difference between the air around the receiver and the condensing temperature. The sizes given in Table 3 are generous.

Figure 19 shows three methods of connecting evaporative condensers in parallel. The common liquid header in each case has sufficient volume to fill the liquid legs to the outlet of the condensers. The height H is commonly assumed to be 4 ft minimum for ammonia, but more generous heights are useful to accommodate a variety of sizes or makes of condensers, particularly for low-head pressure operation. Always provide an adequate difference in elevation between the bottom of a condenser and the liquid level in the receiver.

EVAPORATOR PIPING

Proper evaporator piping and control are necessary to keep the cooled space at the desired temperature and also to adequately protect the compressor from surges of liquid ammonia out of the evaporator. The evaporators illustrated in this section show some methods used to accomplish these objectives. In some cases, combinations of details shown on several illustrations have been used.

Pipe Coils

Figure 20 illustrates a gravity air circulation coil fed by a thermostatic expansion valve. The relative height of the coil and suction mains should allow the coil suction to drop vertically into the suction main. This prevents liquid that may be in the main from affecting the expansion valve bulb. The amount the expansion valve opens is governed by refrigerant pressure on the expansion valve outlet and by the suction temperature on the coil outlet. To operate properly, the suction gas from the coil must be superheated. Therefore, the pipe coil design should provide about 15% extra coil surface to superheat the suction gas. If the refrigerant pressure drop through the coil is excessive, use an externally equalized expansion valve.

The solenoid valve shown in the line ahead of the thermostatic expansion valve is controlled by a room thermostat. This solenoid valve should be wired so that it can be energized only when the compressor is operating. This prevents the coil from filling with liquid at low loads, and the resulting surge of liquid to the suction line when the load is suddenly increased.

When a solenoid valve is used only in the liquid line, it is necessary to pump out the evaporator coil whenever the room thermostat is satisfied. This requires adequate storage capacity in the high-pressure receiver. If desired, the refrigerant can be bottled up in the evaporator by using a suction line solenoid valve wired in parallel with the liquid line solenoid valve.

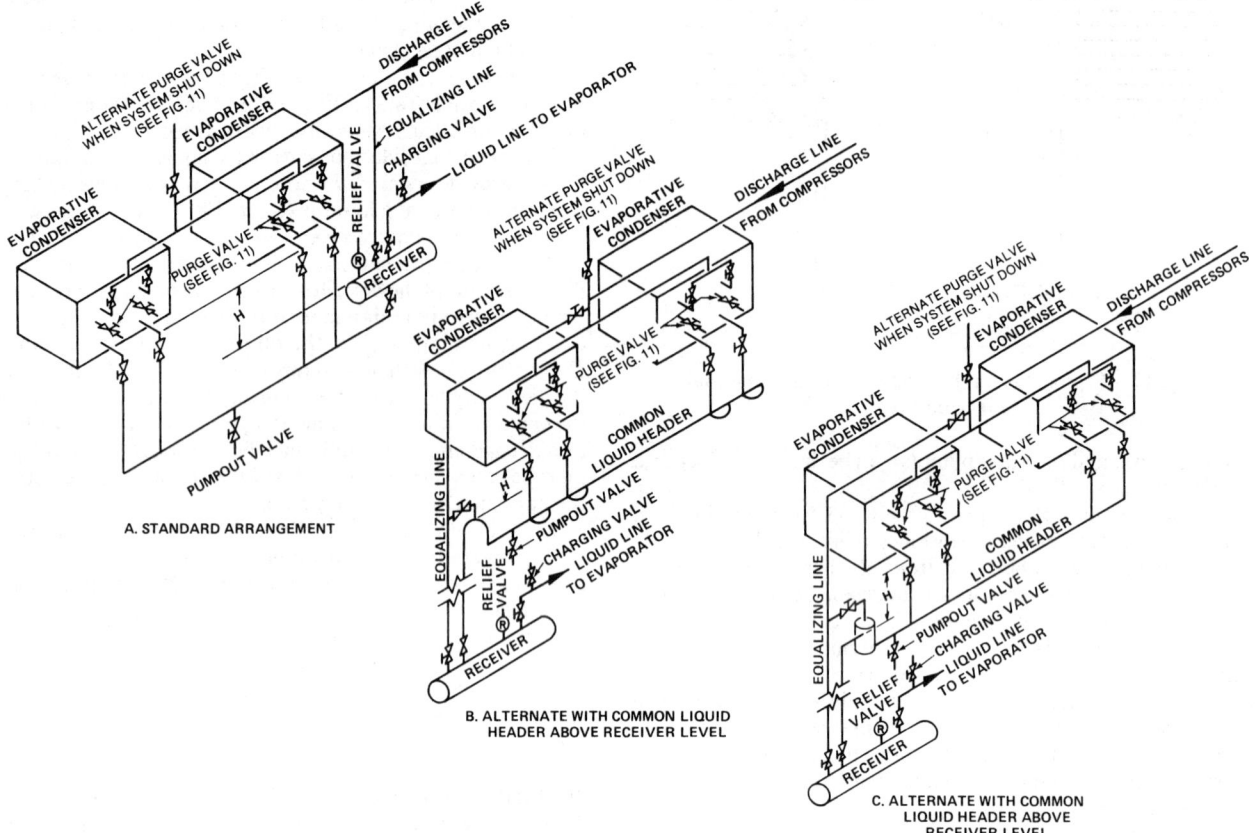

Fig. 19 Paralleling Evaporative Condensers (Any Size) with One or More Compressors

When evaporator temperatures are above 32°F, condensate drips from the coils. Drain troughs collect the condensate for discharge to a drain.

When temperatures are below 32°F, frost accumulates on the coils, and they must be defrosted. Pipe coils are usually defrosted by hot gas. This can be done manually since pipe coils are used only where defrosting is required at infrequent intervals. Defrosting can be done by a valve off the discharge line in the engine room. This prevents hot gas from condensing in the hot-gas pipe during times when no defrosting is taking place. At the end of defrosting, the coil contains a considerable amount of liquid ammonia, which surges out of the evaporator as soon as the suction valve is opened. Therefore, a suction trap is required. Oil that accumulates during normal operation of the coil is heated during the defrost cycle and surges out of the coil when the suction valve is open. A pumpout connection is useful when it is necessary to open the coil for modifications.

Hand expansion valves feed a constant amount of liquid into the coil at all times at any set position. As the load changes, the valve setting must be changed or the coil will operate with highly superheated leaving gas (room temperature rises) at increasing loads; or the liquid will surge through the coil into the suction line at decreasing loads. It is imperative for this type of coil control to use a suction trap to protect the compressor from liquid surges

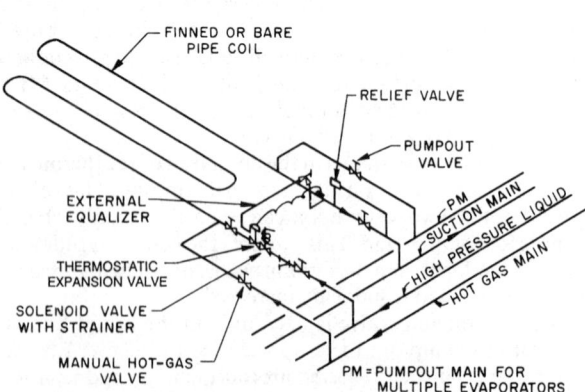

Fig. 20 Piping for Gravity Air Circulation Coil Fed by a Thermal Expansion Valve

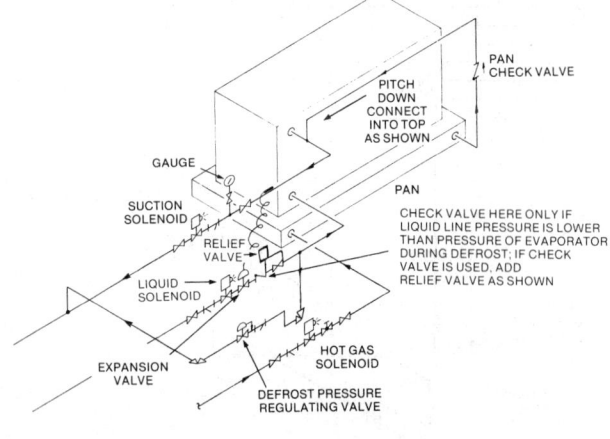

Fig. 21 Piping for Thermostatic Expansion Valve Application for Automatic Defrost on Air Blower

System Practices for Ammonia

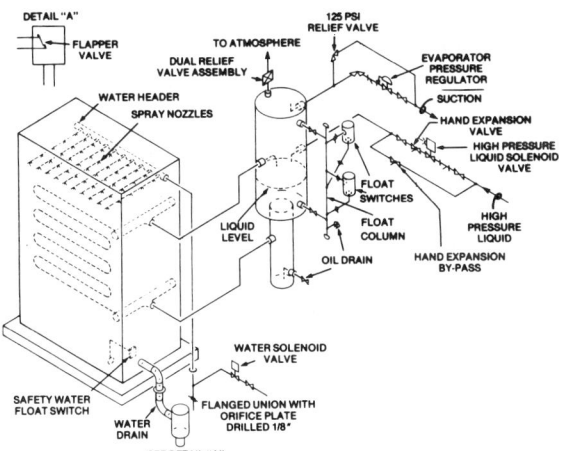

Fig. 22 Arrangement for Automatic Defrost of Air Blower with Flooded Cell

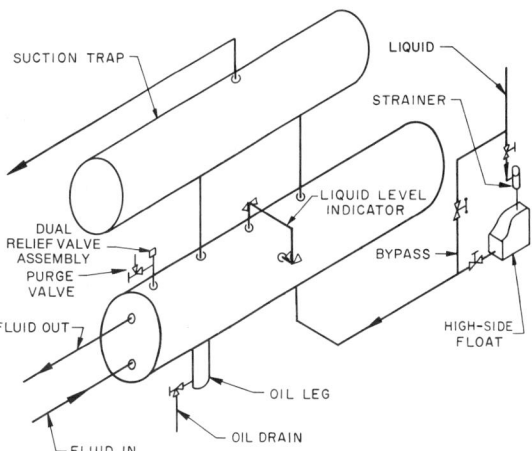

Fig. 23 Arrangement for Horizontal Liquid Cooler and High-Side Float

and to size the suction line liberally to offset the effect of ammonia liquid in the suction main.

Air Blowers

Figure 21 illustrates a thermostatic expansion valve on an air blower using hot gas for automatic defrosting.

When using hot gas or electric heat for defrosting, the drain pan and drain line must be heated to prevent the condensate from refreezing. When hot gas is used for defrosting, a heating coil is imbedded in the drain pan. The hot gas first flows through this coil and then into the evaporator coil. When using electric heat for defrosting, an electric heating coil is used under the drain pan.

Wrap-around electric heating cables are used on the condensate drain line when the room temperature is below 32°F.

Since this is an automatic defrosting arrangement, hot gas must always be available at the hot-gas solenoid valve near the unit. This usually means that the system must contain multiple evaporators so that the compressor will be running when the evaporator to be defrosted is shut down. The hot-gas header must be kept in a space where ammonia will not condense in the pipe. Otherwise, the coil receives liquid ammonia at the start of defrosting and is unable to take full advantage of the latent heat of hot-gas condensation entering the coil. Also, an extra amount of liquid ammonia will be handled by the suction trap when defrosting. If this cannot be arranged, the insulated hot gas main must be kept drained by a high-pressure float back to the suction trap.

The liquid line and suction line solenoid valves are open during normal operation only and are closed during the defrost cycle. When the defrost cycle starts, the hot-gas solenoid valve is open. A pressure regulator in the condensed liquid main maintains a gage pressure of about 70 psi in the coil. A hand valve may also be used to maintain pressure in the main.

Air Blower—Flooded Operation

Figure 22 illustrates a flooded evaporator with a close coupled low-pressure vessel for feeding ammonia into the coil and automatic water defrost.

The lower float switch on the float column at the vessel controls the opening and closing of the liquid line solenoid valve, regulating ammonia into the unit to maintain a liquid level. The hand expansion valve downstream of the solenoid valve should be adjusted so that it will not feed ammonia into the vessel at a rate higher than the vessel can take without raising the suction pressure of gas from the vessel more than 1 or 2 psig.

The static height of liquid in the vessel should be sufficient to flood the coil with liquid under normal loads. The higher float switch should be wired into an alarm circuit in the engine room to indicate when the liquid level in the vessel is too high. With flooded coils having horizontal headers, distribution between the multiple circuits is accomplished without distributing orifices.

A combination evaporator pressure regulator and stop valve is used in the suction line from the vessel. During operation, the regulator maintains a nearly constant back pressure in the vessel. A solenoid coil in the regulator mechanism closes it during the defrost cycle. The liquid solenoid valve should also be closed at this time. One of the best methods to control room temperature is a room thermostat that controls the effective setting of the evaporator pressure regulator.

A spring-loaded relief valve is used around the suction pressure regulator and is set for a differential so that the vessel is kept below 125 psig.

A solenoid valve unaffected by downstream pressure is used in the water line to the defrost header. The defrost header is constructed so that it is drained at the end of the defrost cycle and the downstream side of the solenoid valve is drained through a fixed orifice.

Unless the room is maintained above 32°F, the drain line from the unit should be wrapped with a heater cable or provided with another heat source and then insulated to prevent the defrost water from refreezing in the line.

The length of the water line within the space leading up to the header and the length of the drain line in the cooled space should be kept to a minimum. A flapper or pipe trap on the end of the drain line prevents warm air from flowing up the drain pipe and into the unit.

An air outlet damper should be closed during defrosting to prevent thermal circulation of air through the unit during the defrost cycle, which otherwise would affect the temperature of the cooled space. The fan is stopped during defrost.

This type of defrosting requires a drain pan float switch for safety control. If the drain pan fills with water, the switch overrides the time clock to stop the flow into the unit by closing the water solenoid valve.

There should be a 5-min. delay at the end of the water spray part of the defrosting cycle so that the water can drain from the coil and pan. This limits the ice buildup in the drain pan and on the coils after the cycle is completed.

On completion of the cycle, the pressure in the low-pressure vessel may be about 75 psig. When the unit is opened to the suction main at a much lower pressure, some liquid surges out into the main; therefore, it may be necessary to gradually bleed off this pressure before fully opening the suction valve, to prevent thermal

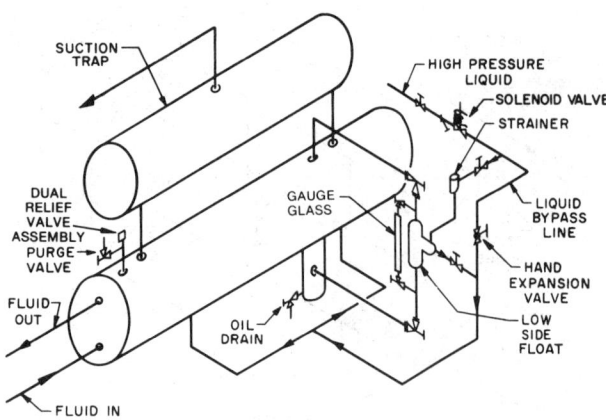

Fig. 24 Piping for Evaporator and Low-Side Float with Horizontal Liquid Cooler

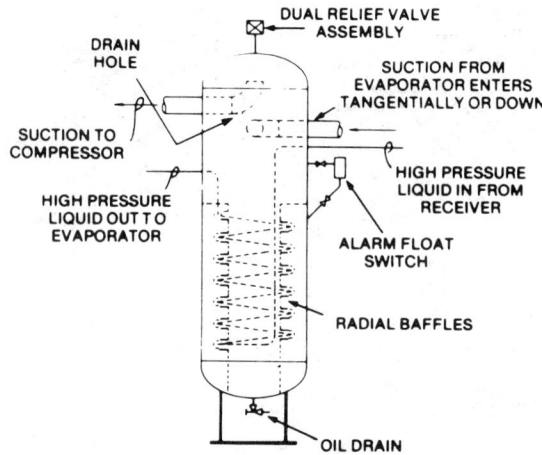

Fig. 25 Vertical Suction Trap with High-Pressure Liquid Coil

shock. Generally, a suction trap in the engine room removes this liquid before the gas stream enters the compressors.

The type of refrigerant control shown in Figure 22 can be used on "brine spray" type units where brine is sprayed over the coil at all times to pick up the condensed water vapor from the airstream. The brine concentration is reconcentrated continually to remove the water absorbed from the airstream.

High-Side Float Control

When a system has only one evaporator, a high-pressure float control can be used to keep the condenser drained and to provide a liquid seal between the high side and the low side. Figure 23 illustrates a brine or water cooler with this type of control. The high-side float should be located near the evaporator to avoid insulating the liquid line.

The amount of ammonia in this type of system is critical, because the charge must be limited so that, under the highest loading in the evaporator, liquid will not surge into the suction line. Some type of suction trap should be used. One method is to place a horizontal shell above the cooler, with the suction gas piped into the bottom and out of the top.

Coolers should include a liquid indicator. Use a reflex glass lens with a large liquid chamber and vapor connections for boiling liquids and a plastic frost shield to determine the actual level. A refrigeration thermostat measuring chilled fluid out temperature should be wired into the compressor starting circuit to prevent freezing.

A flow switch or differential pressure switch should prove flow before the compressor starts. The fluid to be cooled should be piped into the lower portion of the tube bundle and out of the top portion.

Low-Side Float Control

For multiple evaporator systems, low-side float valves are used to control the refrigerant level in flooded evaporators. The low-pressure float shown in Figure 24 has an equalizer line from the top of the float chamber to the space above the tube bundle, and an equalizer line out of the lower side of the float chamber to the lower side of the tube bundle.

For positive shutoff of liquid feed when the system stops, a solenoid valve in the liquid line is wired so that it is only energized when the brine or water pump motor is operating and the compressor is running.

Use a reflex glass lens with large liquid chamber and vapor connections for boiling liquids and with a plastic frost shield to determine the actual level with front extensions as required.

If this type of brine cooler is subjected to fluctuating loads, a suction trap is used in the suction line to intercept liquid flowing in the gas stream before it arrives at the compressor.

If the ammonia contains large quantities of water, there is no satisfactory, economical way to remove it. Water contamination can cause black deposits in reciprocating compressors with subsequent mechanical failures. Where excessive water contamination is evident, the entire ammonia charge and the oil charge for the plant should be replaced.

If the ammonia contains small quantities of water, capacity loss and excessively high temperatures can result. A separate "still", off the oil leg, removes the water accumulation.

SUCTION TRAPS

Every system should have a properly designed and adequate suction trap to assure that the suction gas will be free of liquid at all operating conditions. Such a trap intercepts most foreign matter as well as liquid. The vapor velocity limit for effective gas and liquid separation should not be exceeded (see Chapter 2, Table 2).

A simple suction trap is a vertical vessel with a tangential entrance. Additional elimination of fine droplets and mist can be obtained by adding knitted wire mesh pads. Figure 25 shows a vertical trap with a liquid coil in the lower part to evaporate trapped liquid, while simultaneously subcooling the high-pressure liquid to the evaporator.

Vertical Suction Trap and Pump

Figure 26 shows the piping of a vertical suction trap that uses a high-head ammonia pump to transfer liquid from the system's

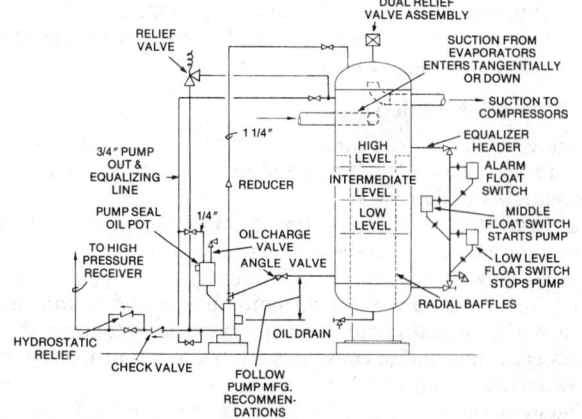

Fig. 26 Piping for Vertical Suction Trap and High-Head Pump

System Practices for Ammonia

low-pressure side to the high-pressure receiver. Float switches piped on a float column on the side of the trap can start and stop the liquid ammonia pump, sound an alarm in case of excess liquid and, sometimes, stop the compressors.

When the liquid level in the suction trap reaches the setting of the middle float switch, the liquid ammonia pump starts and reduces the liquid level to the setting of the lower float switch, which stops the liquid ammonia pump. A check valve in the discharge line of the ammonia pump prevents gas and liquid from flowing backward through the pump when it is not in operation. Depending on the type of check valve used, some installations have two valves in a series as an extra precaution against pump "back spin."

Compressor controls adequately designed for starting, stopping, and capacity reduction result in minimal agitation, which aids in separating the vapor and liquid in the suction trap. Increasing the compressor capacity slowly and in small increments reduces boiling liquid in the trap, caused by the refrigeration load of cooling the refrigerant and metal mass of the trap. If another compressor is started when plant suction pressure increases, it should be brought on line slowly to prevent a sudden pressure change in the suction trap.

A high level of liquid in a suction trap should activate an alarm or stop the compressors. Although eliminating the cause is the most effective way to reduce a high level of excess surging liquid, a more immediate solution is to stop part of the compression system and raise the plant suction pressure slightly. Continuing high levels indicate insufficient pump capacity or suction trap volume.

Pressurized Return Systems

Liquid can be transferred from the suction trap to the high side by draining the main suction trap into an auxiliary suction trap below it. When the auxiliary suction trap is partially filled with ammonia liquid, it is valved off. High-pressure gas is then directed into the auxiliary trap to increase the pressure sufficiently so that it can be returned either by gravity or by a low-head ammonia pump to the high-pressure receiver. Adequate pressure relief valve protection must be provided.

LIQUID AMMONIA PUMPED RECIRCULATION SYSTEMS

The following discussion gives an overview of liquid recirculating (sometimes called liquid overfeed) systems. For additional engineering details on liquid overfeed systems, refer to Chapter 2 or Chapter 8 in Stoecker (1988).

In a liquid ammonia recirculating system, a pump circulates the ammonia from a low-pressure receiver to the evaporators. The low-pressure receiver is a shell for storing refrigerant at low-pressure and is used to supply evaporators with refrigerant, either by gravity or by a low-head pump. It also takes the suction from the evaporators and separates the gas from the liquid. Since the amount of liquid fed into the evaporator is usually several times the amount that actually evaporates there, liquid is always present in the suction return to the low-pressure receiver. Frequently, three times the evaporated amount is circulated through the evaporator.

Generally, the liquid ammonia pump is sized by the flow rate required and a head differential of about 25 psi. This is satisfactory for most single-story installations. If there is a static lift on the pump discharge, the differential is increased accordingly.

Size the low-pressure receiver by the cross-sectional area required to separate liquid and gas, and by the volume between the normal liquid level and the alarm liquid level in the low-pressure receiver. This volume should be sufficient to contain the maximum fluctuation in liquid from the various load conditions.

The liquid at the pump discharge is in the subcooled region. A total pressure drop of about 5 psi in the piping can be tolerated.

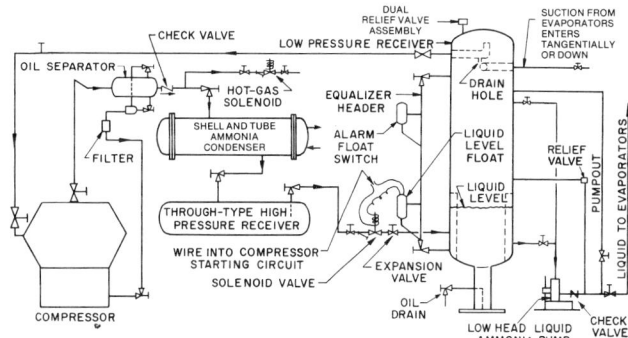

Fig. 27 Piping for Single-Stage System with Low-Pressure Receiver and Liquid Ammonia Circulation

The remaining pressure is expended through the control valve, coil, and suction return line. The pressure drop and heat pickup in the liquid supply line should be low enough to prevent any flashing in the liquid supply line.

Provisions for liquid relief are required from the liquid main back to the low-pressure receiver, so that when the liquid line solenoid valves at the various evaporators are closed, either for defrosting or for temperature control, the excess liquid can be relieved back to the low-pressure receiver. Generally, relief valves used for this purpose are set at about 40 psi differential.

The suction header between the evaporators and the low-pressure receiver should be pitched 1% to permit excess liquid flow back to the low-pressure receiver. The header should be designed to avoid traps.

Liquid Recirculation in Single-Stage System

Figure 27 shows the piping of a typical single-stage system with a low-pressure receiver and liquid ammonia recirculating feed. While some of the details illustrated in previous figures are omitted, references are made to them in the following text.

Liquid Recirculation and Evaporator Piping

Figures 28 and 29 show the typical piping of an evaporator designed for liquid ammonia recirculation and equipped with hot-gas defrosting. Liquid is fed through the liquid line into the header and distributed to the various coils through orifices in the header. Suction is taken from the suction header through a strainer and an automatic flow stop valve.

During the defrost cycle, the suction and liquid line solenoid valves are closed and the hot-gas solenoid valve is open. The hot gas flows through a coil located underneath the unit pan to keep the condensate from refreezing on the pan. The gas then flows through a check valve into the suction header and is distributed

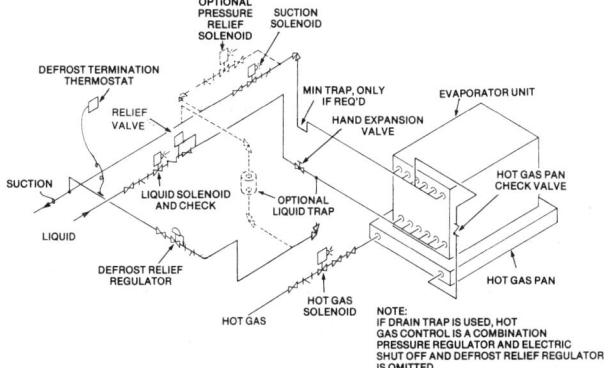

Fig. 28 Piping for High-Temperature Evaporator with Liquid Recirculation and Hot-Gas Defrost

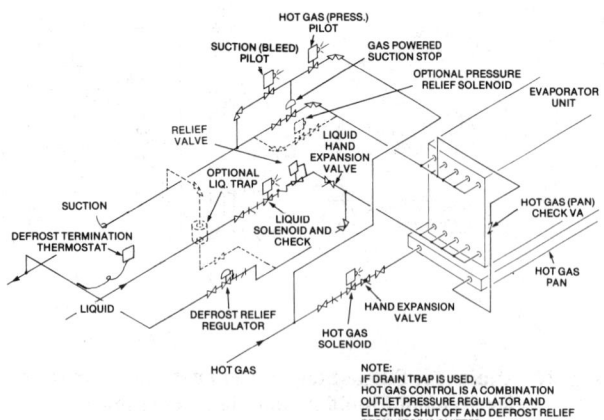

Fig. 29 Piping for Low-Temperature Evaporator with Liquid Recirculation and Hot-Gas Defrost

to the various coils where defrosting takes place. The condensed liquid is maintained at a temperature above 32°F by a pressure relief regulator or liquid drain trap that delivers condensed liquid to the suction line and then to the suction trap.

COMPOUND COMPRESSION SYSTEMS

Compound compression systems compress the gas from the evaporator to the condenser in several stages (see Chapter 1). They are used to produce temperatures of −15°F and below, where this is not economical with single-stage compression.

Single-stage reciprocating compression systems are generally limited to between 5 and 10 psig suction pressure. With oil-injected rotary screw compressors, where the discharge temperatures are lower because of the oil cooling, the low-suction temperature limit is −40°F. Two-stage systems are used down to about −65°F evaporator temperatures. Below this temperature, use three-stage systems.

Two-stage systems consist of one or more compressors that operate at low suction pressure and discharge at intermediate pressure, and have one or more compressors that operate at intermediate pressure and discharge to the condenser.

Where either single- or two-stage compression systems can be used, two-stage systems require less power and have lower operating costs, but can have a higher initial equipment cost.

Gas and Liquid Intercoolers

One reason for using an intercooler in a compound system is to cool the discharge gas between stages to prevent overheating the higher-stage compressor. This is done by bubbling the discharge gas from the low-stage compressor through a bath of liquid refrigerant at intermediate pressure and corresponding temperature. The heat removed from the discharge gas is absorbed by the evaporation of part of the liquid in the bath and eventually passes through the high-stage compressor to the condenser.

For plant operating cost economy, the liquid for the evaporators on the low-stage compressor is subcooled after leaving the condenser, increasing the refrigerating effect per unit mass of ammonia. This lowers the low-stage compressor displacement per unit of refrigeration capacity and reduces its operating power.

Two types of intercoolers for compound compression systems are illustrated. Figure 30 shows a vertical type shell-and-coil intercooler. The liquid level is maintained in the intercooler by a float that controls the solenoid valve feeding liquid into the shell side of the intercooler. Gas from the first-stage compressor enters the lower head of the intercooler, is distributed by a perforated plate, and cooled to the saturation temperature corresponding to intermediate pressure.

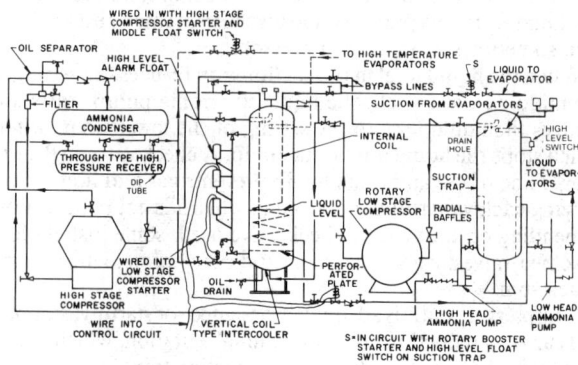

1. For suction trap and high-head pump details, see Figure 26.
2. For rotary booster compressor cooling details, see Figure 8.
3. For high-stage compressor cooling, see Figures 5, 6, or 7.
4. For condenser-receiver details, see Figure 9.
5. For low-head liquid ammonia pump piping detail, see Figure 27.

Fig. 30 Arrangement for Compound System with Vertical Intercooler and Suction Trap

High-pressure liquid from the receiver flows through a coil immersed in the lower part of the intercooler, where it is subcooled to within approximately 10°F of an intermediate temperature before it flows to the evaporator. This liquid is kept at a high pressure, which reduces the required size of the liquid control valve at the evaporators.

Figure 31 shows a horizontal flash intercooler. A float switch maintains the level in the intercooler by controlling the flow of ammonia liquid from the high-pressure receiver and injecting it into the intercooler, where it flashes to the intermediate pressure and temperature. Liquid at intermediate pressure flows from the intercooler to the low-pressure receiver through a solenoid valve controlled by a float switch, to maintain the level in the low-pressure vessel (on the float column). Discharge gas from the low-stage compressor is piped to an internal slotted pipe in the intercooler, distributed up through the liquid, and cooled to the saturation temperature corresponding to intermediate pressure.

Two-Stage System Piping

Figure 30 shows a two-stage compound compression system that uses a vertical coil intercooler. A vertical suction trap in the suction line from the evaporators intercepts any liquid present.

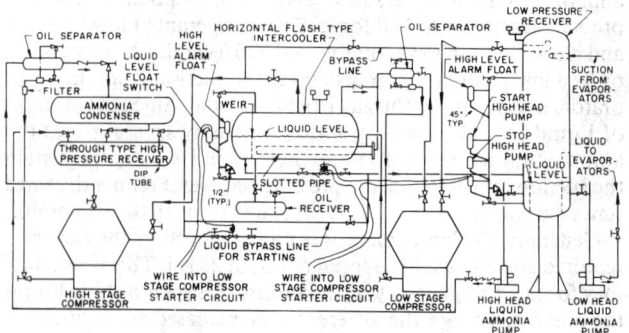

1. For low-pressure receiver and low-head liquid ammonia pump details, see Figure 27.
2. For high-head pump details, see Figure 26.
3. For condenser-receiver details, see Figure 9.
4. For high-stage compressor cooling, see Figures 5, 6, or 7.

Fig. 31 Arrangement for Compound System with Flash Intercooler

System Practices for Ammonia

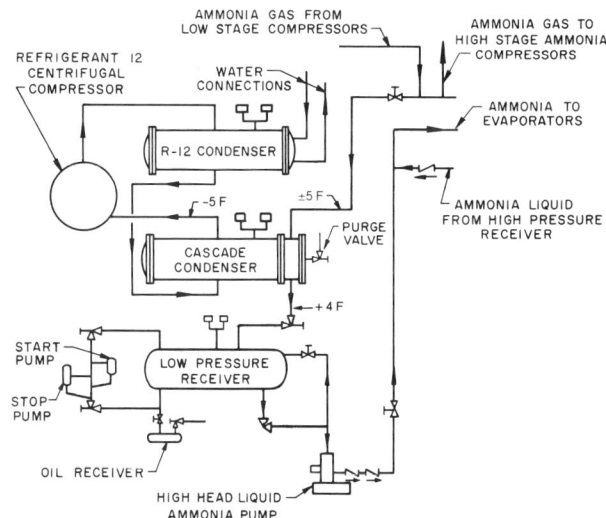

Fig. 32 Cascade System Using a Centrifugal Refrigerant 12 Compressor for Condensing Ammonia

A low-stage compressor takes suction from the vertical suction trap and discharges into the vertical coil intercooler. It has a valved bypass line around the rotary compressor and intercooler so that the plant suction pressure can be lowered with the high-stage compressor before starting the low-stage. This prevents overloading the motor on the low-stage compressor. In case of shutdown, a check valve in the discharge line of the rotary compressor minimizes backflow of gas from the intercooler to the compressor and into the suction trap and evaporators. Similarly, a check valve from the gas space at the top of the intercooler to the top of the rotary compressor discharge line prevents the intermediate pressure in the intercooler from backing liquid up the discharge line and through the check valve into the compressor.

Suction gas from high-temperature evaporators can be taken into the intercooler, where any liquid present can be separated before going to the high-stage compressor.

The intercooler has a float column for mounting the float switches. The lowest float switch maintains the normal liquid level in the intercooler by controlling the flow of high-pressure liquid from the receiver to the intercooler.

If the intercooler receives excess liquid from the high-temperature evaporators, the liquid level rises to the setting of the middle float switch. This switch opens a solenoid valve in a drain line from the intercooler at intermediate pressure to the suction trap, where it can be pumped to the high-pressure receiver by the high-head pump.

The highest float switch is an alarm and/or high level cut out switch for the compression system in the event of excess liquid.

The high-stage compressor takes its suction from the top of the intercooler and discharges its gas into the condenser.

Figure 31 illustrates a two-stage system with a low-pressure receiver, a low-head liquid ammonia pump for circulating to the evaporators, and a horizontal flash intercooler. A high-head liquid ammonia pump can also take excess liquid from the low-pressure receiver and pump it back to the high-pressure receiver during unusual operation of the evaporators.

An oil separator, used on both the low- and high-stage compressors, returns oil to the crankcase.

A check valve on the line from the top of the intercooler to the top of the low-stage compressor's discharge piping prevents liquid from flowing back through the discharge line when the low-stage compressor is shut down. This check valve is required because the discharge and suction sides of the compressor are equalized when it stops.

When starting a system with only the high-stage compressor in operation, the liquid feed to the low-pressure receiver should bypass the intercooler. This is required because, under startup conditions, stored cold liquid at the intercooler is at a lower pressure than the liquid at low-pressure receiver and liquid from the intercooler is not able to flow against the higher pressure.

Converting Single-Stage into Two-Stage Systems

When plant refrigeration capacity must be increased and the system is operating below about 10 psig suction pressure, it is usually more economical to increase capacity by adding a compressor to operate as the low-stage compressor of a two-stage system. The existing single-stage compressor then becomes the high-stage compressor of the two-stage system. Some items to consider when converting are:

1. The motor on the existing single-stage compressor may have to be increased in size when used at a higher suction pressure.
2. The suction trap should be checked for sizing at the increased gas flow rate.
3. An intercooler should be added to cool the low-stage compressor discharge gas and to cool high-pressure liquid.
4. A condenser may have to be added to handle the increased condensing load.
5. A means to purge air should be added if plant suction pressure is below 0 psig.
6. A means to automatically reduce compressor capacity should be added so that the system will operate satisfactorily at reduced system capacity points.

Piping of two-stage systems is shown in Figures 30 and 31.

CASCADE SYSTEMS

Cascade systems use two different refrigerant circuits to produce the desired temperature (see Chapter 1).

A series cascade system could be illustrated by an ammonia system operating from −40°F evaporator temperature to a condensing temperature of 5°F with an R-22 system providing an evaporator temperature of −5°F for condensing the ammonia.

A parallel cascade system could be an ammonia system that operates at an evaporator temperature of 5°F, with the evaporated gas condensed at a slightly lower pressure in a cascade condenser by an R-22 system at an evaporator temperature of −5°F.

In an added parallel cascade system, only part of the ammonia at the 5°F temperature is recondensed by the R-22 system. The remaining ammonia is taken into an ammonia compression system and discharged into the condensers.

Figure 32 is a cascade system using a centrifugal R-22 compressor to condense ammonia at 5°F.

Noncondensables from the low-pressure receiver usually are purged by a small purge compressor that takes its suction from the drain-end pass of the cascade condenser and discharges into a water-cooled ammonia condenser and receiver, from which the air can be purged as shown in Figures 9 and 11.

REFERENCES

ANSI/ASHRAE. 1989. Standard 15-1989, Safety code for mechanical refrigeration.
ANSI/ASME. 1987. Code for pressure piping, B31.5-1987. American Society of Mechanical Engineers, New York.
ANSI/IIAR 2-1984. Equipment, design, and installation of ammonia mechanical refrigeration systems. International Institute of Ammonia Refrigeration, Chicago.
Stoecker, W.F. 1988. Industrial refrigeration, Chapter 8. Business New Publishing Company, Troy, MI.
Wile, D.D. 1977. Refrigerant line sizing. ASHRAE.

BIBLIOGRAPHY

Bradley, W.E. 1984. Piping evaporative condensers. Proceedings of IIAR meeting, International Institute of Ammonia Refrigeration, Chicago.

Cole, R.A. 1986. Avoiding refrigeration condenser problems. Heating, Piping and Air-Conditioning, Parts I and II, 58(7 and 8).

Evaporative condenser engineering manual. 1983. Baltimore Aircoil Company, Inc., Baltimore, MD.

Nuckolls, A.H. "The comparative life, fire, and explosion hazards of common refrigerants". Miscellaneous Hazard No. 2375. Underwriters Laboratory, Northbrook, IL.

CHAPTER 5

SECONDARY COOLANTS IN REFRIGERATION SYSTEMS

Coolant Selection 5.1
Design Considerations 5.2
Applications 5.6

A secondary coolant is a liquid that is used as a heat transfer fluid and which changes temperature as it gains or loses heat energy without changing into another phase. For the lower temperatures of refrigeration, this requires a coolant with a freezing point below that of water.

In this chapter, the design considerations for components, system performance requirements, and applications for secondary coolants are discussed. Related information can be found in Chapters 2, 3, 17, 18, and 33 of the 1989 ASHRAE *Handbook—Fundamentals*.

COOLANT SELECTION

A secondary coolant must be compatible with the other materials in the system at the pressures and temperatures encountered for maximum component reliability and operating life. The coolant should also be compatible with the environment and the applicable safety regulations, and it should be economical to use and replace.

The coolant should have a minimum freezing point of 5 °F and preferably 15 °F below the lowest temperature to which it will be exposed. When subjected to the lowest temperature in the system, the viscosity of the coolant should be low enough to allow satisfactory heat transfer and reasonable pressure drop.

The vapor pressure of the coolant should not exceed that allowed at the maximum temperature encountered. To avoid a vacuum in a low vapor pressure secondary coolant system, the coolant can be pressurized with pressure-regulated dry nitrogen in the expansion tank. However, recognize that some special secondary coolants such as those used for computer circuit cooling have a high solubility for nitrogen and must therefore be isolated from the nitrogen with a suitable diaphragm.

Load Versus Flow Rate

The secondary coolant pump is usually in the return line upstream of the chiller. Therefore, to be accurate, the pumping rate in gpm is based on the density at the return temperature. The mass flow rate for a given heat load is based on the desired temperature range and required coefficient of heat transfer at the average bulk temperature.

To determine heat transfer and pressure drop, the specific gravity, specific heat, viscosity, and thermal conductivity are based on the average bulk temperature of the coolant in the heat exchanger, noting that film temperature corrections are based on the average film temperature. Trial solutions of the secondary coolant side coefficient compared to the overall coefficient and the total LMTD (log mean temperature difference) determine the average film temperature. Where the secondary coolant is cooled, the more viscous film reduces the heat transfer rate and raises the pressure drop compared to what can be expected at the bulk temperature. Where the secondary coolant is heated, the less viscous film approaches the heat transfer rate and pressure drop expected at the bulk temperature.

The greater the amount of turbulence and mixing of the bulk and film, the better the heat transfer and the higher the pressure drop. Where secondary coolant velocity in the tubes of a heat transfer device results in laminar flow, the heat transfer can be improved by inserting spiral tapes or spring turbulators that promote mixing the bulk and film. This usually increases pressure drop. The inside surface can also be spirally grooved or augmented by other devices. Since the state of the art of heat transfer is constantly improving, use the most cost-effective heat exchanger to provide optimum heat transfer and pressure drop. Energy costs for pumping the secondary coolant must be considered when selecting the fluid to be used and the heat exchangers to be installed.

Pumping Cost

Pumping costs are a function of the secondary coolant selected, the load and temperature range where energy is transferred, the pump head required by the system pressure drop (including that of the chiller), the mechanical efficiencies of the pump and driver, and the electrical efficiency and power factor where the driver is an electric motor. Small centrifugal pumps, operating in the range of approximately 50 gpm at 80 ft head to 150 gpm at 70 ft head, for 60 Hz applications, typically have 45 to 65% efficiency, respectively. Larger pumps, operating in the range of 500 gpm at 80 ft of head to 1500 gpm at 70 ft of head, for 60 Hz applications, typically have 75 to 85% efficiency, respectively.

A pump should operate near its peak operating efficiency for the flow rate and head that usually exist. The secondary coolant temperature increases slightly from the energy expended at the pump shaft. If a semihermetic electric motor is used as the driver, the motor inefficiency is added as heat to the secondary coolant, and the total kilowatt input to the motor must be considered in establishing load and temperatures.

Performance Comparisons

Assuming that the total refrigeration load at the evaporator includes the pump motor input and brine line insulation heat gains, as well as the delivered beneficial cooling, tabulating typical secondary coolant performance values assists in the coolant selection. A 1.06 in. ID smooth steel tube evaluated for pressure drop and internal heat transfer coefficient at the average bulk temperature of 20 °F and a temperature range of 10 °F for 7 fps tubeside velocity provides comparative data (see Table 1) for some typical coolants. Table 2 ranks the same coolants comparatively, using data from Table 1.

For a given evaporator configuration, load, and temperature range, select a secondary coolant that gives satisfactory velocities, heat transfer, and pressure drop. At the 20 °F level, hydrocarbon and halocarbon secondary coolants must be pumped at a rate of 2.3 to 3.0 times the rate of water-based secondary coolants for the same temperature range.

Higher pumping rates require larger coolant lines to keep the head and brake horsepower requirement for the pump within reasonable limits. Table 3 lists approximate ratios of pump power for secondary coolants. The heat transferred by a given secondary coolant affects the cost and perhaps the configuration and pressure drop of a chiller and other heat exchangers in the system; therefore, Tables 2 and 3 are only guides of the relative merits of each coolant.

The preparation of this chapter is assigned to TC 10.1, Custom-Engineered Refrigeration Systems.

Table 1 Secondary Coolant Performance Comparisons

Secondary Coolant	Concentration (by Weight), %	Freeze Point, °F	gpm/ton[c]	Pressure Drop,[a] psi	Heat Transfer Coeff.,[b] h_i Btu/h·ft²·°F
Propylene glycol	39	−5.1	2.56	2.91	205
Ethylene glycol	38	−6.9	2.76	2.38	406
Methanol	26	−5.3	2.61	2.05	473
Sodium chloride	23	−5.1	2.56	2.30	558
Calcium chloride	22	−7.8	2.79	2.42	566
Aqua-ammonia	14	−7.0	2.48	2.44	541
Trichlorethylene	100	−123	7.44	2.11	432
d-Limonene	100	−142	6.47	1.48	321
Methylene chloride	100	−142	6.39	1.86	585
R-11	100	−168	7.61	2.08	428

[a] Based on one length of 16-ft tube with 1.06-in. ID and use of Moody Chart (1944) for 7 fps velocity. I/O losses equal one Vel. H_D for 7 fps velocity. Evaluations are at a bulk temperature of 20°F and a temperature range of 10°F.
[b] Based on a curve fit equation for Kern's adaptation (1950) of Sieder & Tate heat transfer equation (1936) using a 16-ft tube for L/D = 181 and a film temperature of 5°F lower than average bulk temperature with 7 fps velocity.
[c] Based on inlet secondary coolant temperature at the pump of 25°F.

Table 2 Comparative Ranking of Heat Transfer Factors at 7 fps[a]

Secondary Coolant	Heat Transfer Factor	Secondary Coolant	Heat Transfer Factor
Propylene glycol	1.000	Methanol	2.307
d-Limonene	1.566	Aqua-ammonia	2.639
Ethylene glycol	1.981	Sodium Chloride	2.722
R-11	2.088	Calcium chloride	2.761
Trichlorethylene	2.107	Methylene chloride	2.854

[a] Based on Table 1 values using 1.06-in. ID tube 16 ft long. The actual ID and length vary according to the specific loading and refrigerant applied with each secondary coolant, tube material, and surface augmentation.

Table 3 Relative Pumping Energy Required[a]

Secondary Coolant	Energy Factor	Secondary Coolant	Energy Factor
Aqua-ammonia	1.000	Calcium chloride	1.447
Methanol	1.078	d-Limonene	2.406
Propylene glycol	1.142	Methylene chloride	3.735
Ethylene glycol	1.250	Trichlorethylene	4.787
Sodium chloride	1.295	R-11	5.022

[a] Based on the same pump head, refrigeration load, 20°F average temperature, 10°F range, and freezing point (for water-based secondary coolants) 20 to 23°F below the lowest secondary coolant temperature.

Other Considerations

The corrosion that occurs when a secondary coolant contacts system materials must be considered when selecting the coolant, an inhibitor, and the system components. The effect of secondary coolant and inhibitor toxicity on the health and safety of plant personnel or consumers of food and beverages must be considered. The flash point and explosive limits of secondary coolant vapors must also be evaluated.

Examine the secondary coolant stability for anticipated moisture, air, and contaminants at the temperature limits of materials used in the system. The skin temperatures of the hottest elements determine the secondary coolant stability.

If defoaming additives are necessary, their effect on the thermal stability and toxic properties of the coolant must be considered for the application.

DESIGN CONSIDERATIONS

The secondary coolant vapor pressure at the lowest operating temperature determines whether a vacuum could exist in the secondary coolant system. To keep air and moisture out of the system, pressure-controlled dry nitrogen can be applied to the top level of secondary coolant (e.g., in the expansion tank or a storage tank). The gas pressure over the coolant plus the pressure created at the lowest point in the system by the maximum vertical height of coolant determines the minimum internal pressure for design purposes. The coincident highest pressure and lowest secondary coolant temperature dictate the design working pressure (DWP) and material specifications for the secondary coolant system components.

To select proper relief valve(s) with settings based on the system DWP, the highest temperatures to which the secondary coolant could be subjected should be considered. This temperature would occur in case of heat radiation from a fire in the area or the normal warming of the valved-off sections. Normally, a valved-off section is relieved to an unconstrained portion of the system and the secondary coolant can expand freely without loss to the environment.

Safety considerations for the system are found in the Safety Code for Mechanical Refrigeration (ASHRAE 1978). The design standards for pressure piping can be found in ANSI (1974), and the design standards for pressure vessels can be found in ASME Pressure Vessel Code (1986).

Piping and Control Valves

Piping should be sized for reasonable pressure drop using the calculation methods in Chapters 2 and 33 of the 1989 ASHRAE

Secondary Coolants in Refrigeration Systems

Handbook—Fundamentals. Balancing valves or orifices in each of the multiple feed lines help distribute the secondary coolant. A reverse-return piping arrangement balances the flow. Control valves that vary the flow are sized for 20 to 80% of the total friction pressure drop through the system for proper response and stable operation. Valves sized for pressure drops smaller than 20% may respond too slowly to a control signal for a flow change. Valves sized for pressure drops in excess of 80% can be too sensitive, causing control cycling and instability.

Storage Tanks

Storage tanks can shave peak loads for brief time periods, limit the size of the refrigeration equipment, and reduce energy costs considerably. In off-peak hours, a relatively small refrigeration plant cools a secondary coolant stored for later use. A separate circulating pump sized for the maximum flow needed by the peak load is started to satisfy the peak load. The amount of secondary coolant stored determines the length of time the peak load can be sustained while running the refrigeration equipment. Energy cost savings are enhanced if the refrigeration equipment is used to cool secondary coolant at night, when the cooling medium for heat rejection is generally at the lowest temperature.

The load profile over 24 h and the temperature range of the secondary coolant determine the minimum net capacity required for the refrigeration plant, the sizes of the pumps, and the minimum amount of secondary coolant to be stored. For maximum use of the storage tank volume at the expected temperatures, choose inlet velocities and locate the connections and the tank for maximum stratification. Note, however, that maximum use will probably never exceed 90% and, in some cases, may equal only 75% of the tank volume.

Example 1. Figure 1 depicts the load profile and Figure 2 shows the arrangement of a refrigeration plant with storage of a 23% (by weight) sodium chloride secondary coolant at a nominal 20°F. During the peak load of 50 tons, a range of 8°F is required. At an average temperature of 24°F, with a range of 8°F, the specific heat of the coolant c_p is 0.791 Btu/lb·°F. At 28°F, the weight per unit volume of coolant at the pump (ρ_L) is 1.183 [(62.4 lb/ft^3)/(7.48 gal/ft^3)]; at 20°F, the ρ_L is 1.185 [(62.4 lb/ft^3)/(7.48 gal/ft^3)]. Determine the minimum size storage tank for 90% use, the minimum capacity required for the chiller, and the sizes of the two pumps. The chiller and the chiller pump run continuously. The secondary coolant storage pump runs only during the peak load. A control valve to the load source diverts all coolant to the storage tank during a zero load condition, so that the initial temperature of 20°F is restored in the tank. During the low load condition, only the required flow rate for a range of 8°F at the load source is used; the balance returns to the tank and restores the temperature to 20°F.

Solution: If x is the minimum capacity of the chiller, determine the energy balance in each segment by subtracting the load in each segment from x. Then multiply the result by the time length of the respective segments, and add as follows:

$$(x - 0) \times 6 + (x - 50) \times 4 + (x - 9) \times 14 = 0$$
$$6x + 4x - 200 + 14x - 126 = 0$$
$$24x = 326$$
$$x = 13.58 \text{ tons}$$

Calculate the secondary coolant flow rate (W) at peak load:

$$W = (50 \times 200)/(0.791 \times 8) = 1580.3 \text{ lb/min.}$$

For the chiller at 15 tons, the secondary coolant flow rate is:

$$W = (15 \times 200)/(0.791 \times 8) = 474.1 \text{ lb/min.}$$

Therefore, the coolant flow rate to the storage tank pump is 1580.3 − 474.1 = 1106.2 lb/min. The chiller pump size is determined by:

$$474.1/[(1.183 \times 62.4)/7.48] = 48 \text{ gpm}$$

Calculate the storage tank pump size as follows:

$$1106.2/[(1.185 \times 62.4)/7.48] = 111.9 \text{ gpm}$$

Using the concept of stratification in the storage tank, the interface between warm return and cold stored secondary coolant falls at the rate

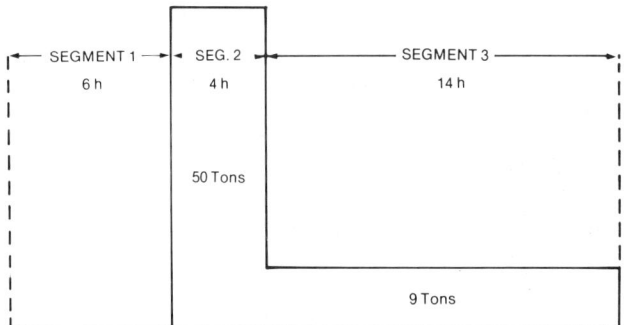

Fig. 1 Load Profile of a Refrigeration Plant where Secondary Coolant Storage Can Save Energy

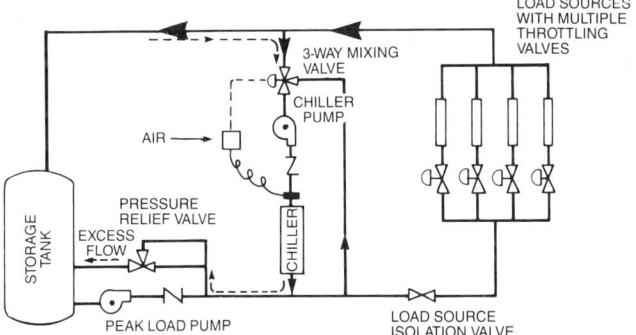

Fig. 2 Arrangement of a System with Secondary Coolant Storage

pumped from the tank. Since the time segments fix the total amount pumped and the storage tank pump operates only in segment 2 (see Figure 1), the minimum tank volume (V) at 90% use is determined as follows:

Total mass = [(1106.2 lb/min.) (60 min./h) (4 h)]/0.90 = 294,987 lb

and

$$V = 294,987/[(1.185 \times 62.4)/7.48] = 29,840 \text{ gal}$$

A larger tank (*e.g.*, 50,000 gal) provides flexibility for longer segments at peak load and accommodates potential mixing. It may be desirable to insulate and limit heat gains to 8000 Btu/h for the tank and lines. Energy use for pumping can be limited by designing for 46-ft head. With the smaller pump operating at 51% efficiency and the larger pump at 52.5% efficiency, the pump heats added to the secondary coolant would be 3300 Btu/h and 7478 Btu/h, respectively.

For cases with various time segments and their respective loads, the maximum load for segment 1 or 3 with the smaller pump operating, cannot exceed the net capacity of the chiller minus insulation and pump heat gain to the secondary coolant. For various combinations of segment time lengths and cooling loads, the recovery or restoration rate of the storage tank to the lowest temperature required for satisfactory operation should be considered.

Figure 2 depicts a system arrangement with secondary coolant storage as described in Example 1.

As load source circuits shut off, the excess flow is bypassed back to the storage tank. The temperature setting of the 3-way valve is the normal return temperature for full flow through the load sources.

When only the storage tank requires cooling, the flow is as shown by the dotted lines with the load source isolation valve closed. When the storage tank temperature is at the desired level, the load isolation valve can be opened to allow cooling of the piping loops to and from the load sources for full restoration of storage cooling capacity.

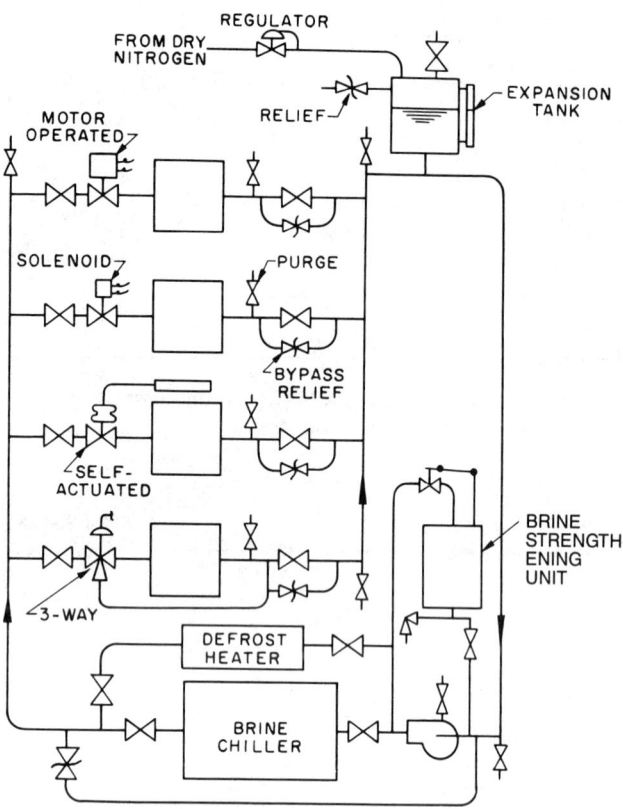

Fig. 3 Typical Closed Salt Brine System

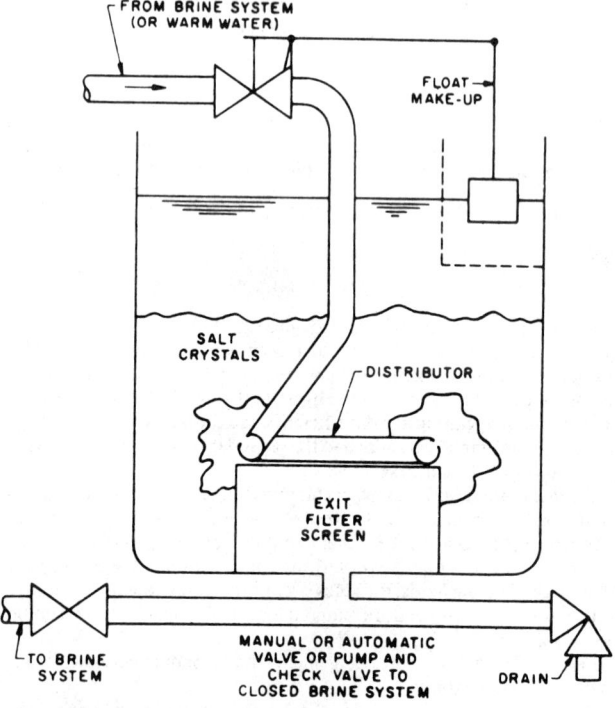

Fig. 4 Brine Strengthening Unit for Salt Brines Used as Secondary Coolants

Expansion Tanks

Figure 3 shows a typical closed secondary coolant system without a storage tank; it also illustrates the different control strategies that might be employed. The reverse-return piping assists flow balance. Figure 4 shows a secondary coolant strengthening unit for salt brines. The secondary coolant expansion tank volume is determined by considering the total coolant inventory and the differences in coolant density at the lowest temperature of coolant pumped to the load location (t_1) and the maximum temperature. The expansion tank is sized to accommodate a residual volume with the system coolant at t_1, plus an expansion volume and vapor space above the coolant. A vapor space equal to 20% of the expansion tank volume should be adequate. A level indicator, used to prevent overcharging, is calibrated at the residual volume level versus lowest system secondary coolant temperature.

Example 2. Assume a 50,000 gal charge of 23% sodium chloride secondary coolant at t_1 of 20°F in the system. If 100°F is the maximum temperature, determine the size of the expansion tank required. Assume that the residual volume is 10% of the total tank volume and that the vapor space at the highest temperature is 20% of the total tank volume.

$$ETV = \frac{V_S[(SG_1/SG_2) - 1]}{1 - (R_F + V_F)}$$

where

ETV = expansion tank volume
V_S = system secondary coolant volume at t_1 temperature
SG_1 = specific gravity at t_1
SG_2 = specific gravity at maximum temperature
R_F = residual volume of tank liquid (low level) at t_1, expressed as a fraction
V_F = volume of vapor space at highest temperature, expressed as a fraction

If the specific gravity of the secondary coolant is 1.185 at 20°F and 1.155 at 100°F, the tank volume is:

$$ETV = \frac{50,000\,[(1.185/1.155) - 1]}{1 - (0.10 + 0.20)} = 1855 \text{ gal}$$

Pulldown Time

Example 1 is based on a static situation of secondary coolant temperature at two different loads—normal and peak.

The length of time for pulldown from 100°F to the final 20°F may need to be calculated. For a graphical solution, required heat extraction versus secondary coolant temperature is plotted. Then, by iteration, the pulldown time is solved by finding the net refrigeration capacity for each increment of coolant temperature change. A mathematical method may also be used.

The 15-ton refrigeration system cited in the examples has a 30.03-ton capacity at a maximum of 50°F saturated suction temperature (STP). For pulldown, a compressor suction pressure regulator (holdback valve) is sometimes used in the system. The maximum secondary coolant temperature must be determined when the holdback valve is wide open, and the STP is at 50°F. For the above example, this is at 70°F coolant temperature. As the coolant temperature is further reduced with a constant 48.1 gpm, the capacity of the refrigeration system is gradually reduced until a 15-ton capacity is reached with 26°F coolant in the tank. Further cooling to 20°F will be at reduced capacity.

Temperatures of the secondary coolant mass, storage tanks, piping, cooler, pump, and insulation must all be reduced. In Example 1, as the coolant is reduced in temperature from 100 to 20°F, the total heat removed from these items is as follows:

Secondary Coolants in Refrigeration Systems

Brine Temperature, °F	Total Heat Removed, Million Btu
100	31.54
80	23.62
70	19.67
60	15.73
40	7.85
20	0

From a secondary coolant temperature of 100 to 70°F, the refrigeration system capacity is fixed at 30.03 tons, and the time for pulldown is essentially linear (system net tons for pulldown is less than the compressor capacity because of heat gain through insulation and added pump heat). In Example 1, the pump heat was not considered. When recognizing the variable heat gain for a 95° ambient, and the pump heat as the secondary coolant temperature is reduced, the following net capacity is available for pulldown at the various secondary coolant temperatures:

Brine Temperature, °F	Net Capacity, Tons
100	29.86
80	29.58
70	29.44
60	25.28
40	17.80
26	14.10
20	12.70

A curve fit shows capacity is a straight line between the values for 100 and 70°F. Therefore, the pulldown time for this interval is:

$$\theta = \frac{[(31.54 \times 10^6) - (19.67 \times 10^6)]}{12,000[(29.86 + 29.44)/2]} = 33.4 \text{ h}$$

From 70 to 20°F, the capacity curve fits a second-degree polynomial equation as follows:

$$q = 9.514809086 + 0.1089883647t + 0.002524039t^2$$

where

t = secondary coolant temperature, °F
q = tons capacity for pulldown

Rearranging to another form:

$$q = 0.002524039\,(t^2 + 43.1801429t + 3769.675938)$$
$$= 0.002524039\,[(t + 21.59007145)^2 + (57.4764713)^2]$$

Using the arithmetic average pulldown net capacity from 70 to 20°F, the time interval would be:

$$\theta = \frac{19.67 \times 10^6}{12,000\,[(29.44 + 12.7)/2]} = 77.8 \text{ h}$$

If the ln mean average net capacity for this temperature interval is used, the time is:

$$\theta = \frac{19.67 \times 10^6}{(19.91 \times 12,000)} = 82.4 \text{ h}$$

This is a difference of over 4.5 h and neither solution is correct. A more exact calculation uses a graphical analysis or calculus. When using a mathematical approach, determine the heat removed per degree of secondary coolant temperature change per ton capacity. Since the coolant's heat capacity and the heat leakage change as the temperature is reduced, this is best determined by a curve fit of the data for total heat removed versus secondary coolant temperature, using a series of iterations for secondary coolant temperature ± 1°F as the temperature is reduced.

A curve fit of the total heat removed from a given secondary coolant temperature to 20°F gives the following second-degree polynomial:

$$Q = -7832831.847 + 391110.316t + 26.45203369t^2$$

where

t = secondary coolant temperature, °F
Q = Btu

By a series of iterations for 1°F temperature change, the polynomial that represents the heat removed per hour for one degree change in secondary coolant temperature per ton is developed from the following relationship:

$$\frac{1}{\Delta \text{Btu}} = \frac{1}{\text{h} \cdot \text{ton}} = \frac{\Delta 1\,°F}{\text{h} \cdot \text{ton}}$$
$$\text{°F range} \times \frac{12{,}000 \text{ Btu}}{\text{h} \cdot \text{ton}} \qquad \text{°F Range}$$

or

$$\frac{\Delta 1°}{\text{h} \cdot \text{ton}} = 0.0306818755 - 0.0000041498t$$

where

t = secondary coolant temperature, °F

The differential equation for the differential change in coolant temperature versus time in hours is (use minus sign since coolant temperature is being reduced):

$$\frac{dt}{d\theta} = \left(\frac{-\Delta 1°F}{\text{h} \cdot \text{ton}}\right) \text{tons}$$

Integrating this equation determines time t in hours:

$$\int_{70°}^{20°} d\theta = \theta = \frac{-1}{\left(\frac{\Delta 1°F}{\text{h} \cdot \text{ton}}\right)} \times \int_{70°}^{20°} \frac{dt}{(\text{tons})}$$

The polynomial equations are then substituted for the parenthesized terms (rounding to four significant numbers):

$$\theta = \frac{-1}{(0.03068 - 0.00000415t)}$$
$$\times \int_{70°}^{20°} \frac{dt}{0.002524[(t + 21.59)^2 + (57.48)^2]}$$

These are in the general form of:

$$\theta = \text{constant} \times \int \frac{dv}{v^2 + a^2} = \frac{\text{constant}}{a} \times \tan^{-1}\frac{v}{a} + C$$

where

$dv = dt$
$v = t + 21.59$
$a = 57.48$
v/a = radians

Substituting:

$$\theta = \frac{-1}{[0.002524(0.03068 - 0.000004157t)]}$$
$$\times \frac{1}{57.48} \times \tan^{-1}\frac{(t + 21.59)}{57.48}\bigg|_{70°}^{20°} + C$$

Subtracting values for the two temperatures, C drops out:

$$\theta_{20} - \theta_{70} = -141.1 - (-229.2) = 88.1 \text{ h}$$

The correct answer is 88.1 h, which is 7% greater than the ln mean average capacity and 13% greater than the arithmetic average capacity over the temperature range.

Therefore, total time for temperature pulldown from 100 to 20°F is:

$$\theta = 33.4 + 88.1 = 121.5 \text{ h}$$

Averaging methods would be 5.7 to 10.3 h in error on total pulldown time. Graphical methods have less error than averaging methods, but calculus produces the greatest accuracy.

In applying calculus to the problem, note that the general form of the equation may develop as:

$$\theta = \text{constant} \times \int \frac{dv}{v^2 - a^2}$$

Therefore:

$$\theta = \frac{1}{2a} \times l_n \left| \frac{v - a}{v + a} \right| + C$$

The polynomial equations may be solved by computer or a handheld calculator and a suitable program or spreadsheet. Computers can also be used to solve the equations by iterative methods.

The time for pulldown will be less if supplemental refrigeration is available for pulldown or if less secondary coolant is stored.

System Costs

Various alternatives may be evaluated to justify a new project or system modification. Means (1988) lists the installed cost of various projects. Various references, such as Parks and Jackson (1984) and NBS (1978), discuss engineering and life-cycle cost analysis. Using the various time value of money formulas, the payback for storage tank handling of peak loads compared to large refrigeration equipment and higher energy costs can be evaluated. The trade-offs in these costs—initial, maintenance, insurance, increased secondary coolant, loss of space, and energy escalation—all must be considered.

Corrosion Prevention

Corrosion prevention requires choosing proper materials and inhibitors, routine testing for pH, and eliminating contaminants. Because potentially corrosive calcium chloride and sodium chloride salt brine secondary coolant systems are widely used, test and adjust the brine solution monthly. To replenish salt brines in a system, a concentrated solution may be better than a crystalline form, because it is easier to handle and mix.

A refrigerating salt brine should not be allowed to change from an alkaline to an acid condition. Acids rapidly corrode the metals ordinarily used in refrigeration and ice-making systems. Calcium chloride usually contains sufficient alkali to render the freshly prepared brine slightly alkaline. When any brine is exposed to air, it gradually absorbs carbon dioxide and oxygen, which neutralize the alkalinity and eventually make the salt brine slightly acid. Dilute salt brines dissolve oxygen more readily and generally are more corrosive than concentrated brines. One of the best preventive measures is to make a closed rather than open system, using a regulated inert gas over the surface of a closed expansion tank (see Figure 2). However, many systems, such as icemaking tanks, brine spray unit coolers, and brine spray-type carcass chill rooms, cannot be closed.

The degree of alkalinity or acidity of a salt brine or other secondary coolant can be expressed in pH value. In the pH scale, pH 7 represents the neutral point; values from 7 down represent increasing acidity, whereas values greater than 7 represent increasing alkalinity. A salt brine pH of 7.5 for a sodium or calcium chloride system is ideal, since it is safer to have a slightly alkaline rather than a slightly acid brine. Various chemical and electronic pH indicators are available. Every salt brine system operator should use an indicator regularly.

If a salt brine is acid, the pH can be raised by adding caustic soda that has been dissolved in warm water. If a salt brine is alkaline (indicating ammonia leakage into the brine), carbonic gas or chromic, acetic, or hydrochloric acid should be added. Ammonia leakage must be stopped immediately so that the brine can be neutralized.

In addition to controlling the pH, an inhibitor should be used. Generally, sodium dichromate is the most effective and economical for salt brine systems. The dichromate has a bright orange color, a granular form, and readily dissolves in warm water. Since it dissolves very slowly in cold brine, it should be dissolved in warm water and added to the brine far enough ahead of the pump so that only a dilute solution reaches the pump. The quantities recommended are: 125 lb/1000 ft^3 of calcium chloride brine, and 200 lb/1000 ft^3 of sodium chloride brine.

Adding sodium dichromate to the salt brine does not make it noncorrosive immediately. The process is affected by many factors, including water quality, specific gravity of the brine, amount of surface and kind of material exposed in the system, age, and temperature. Corrosion stops only when protective chromate film has built up on the surface of the zinc and other electrically positive metals exposed to the brine. No simple test is available to determine the chromate concentration. Since the protection afforded by the sodium dichromate treatment depends greatly on maintaining the proper chromate concentration in the brine, brine samples should be analyzed anually. The proper concentration for calcium chloride brine is 7.58 gr/gal (as $Na_2Cr_2—2H_2O$); for sodium chloride brine, it is 12.128 gr/gal (as $Na_2Cr_2O—2H_2O$).

Since crystals and concentrated solutions of sodium dichromate can cause severe skin rash, avoid contact. If contact does occur, wash the skin immediately. *Warning: Sodium dichromate should not be used for brine spray decks, spray units, or immersion tanks where food or personnel may come in contact with the spray mist or the brine itself.*

Polyphosphate-silicate and orthophosphate-boron mixtures in water-treating compounds are useful for sodium chloride brines in open systems. However, where the rate of spray loss and dilution is very high, any treatment other than density and pH control is not economical. For the best protection of spray unit coolers, housings and fans should be of a high quality, hot-dipped galvanized construction. Stainless steel fan shafts and wheels, scrolls, and eliminators are desirable.

While the nonsalt secondary coolants described in this chapter are generally noncorrosive when used in systems for long periods, recommended inhibitors should be used, and a pH check should be performed occasionally.

Steel, iron, or copper piping should not be used to carry the salt brines. Use copper nickel or suitable plastic. Use all-steel and iron tanks if the pH is not ideal. Similarly, calcium chloride systems usually have all-iron and steel pumps and valves to prevent electrolysis in the presence of acidity. Sodium chloride systems usually have all-iron or all-bronze pumps. When the pH can be controlled in a system, brass valves and bronze fitted pumps may be satisfactory. A stainless steel pump shaft is desirable. Consider salt brine composition and temperature to select the proper rotary seal or, for dirtier systems, the proper stuffing box.

APPLICATIONS

Applications for secondary coolant systems are extensive (see Chapters 9 through 36). A glycol coolant prevents freezing in solar collectors and outdoor piping. Secondary coolants heated by solar collectors or by other means can be used to heat absorption-cooling equipment, to melt a product such as ice or snow, or to heat a building. Process heat exchangers can use a number of secondary coolants to transfer heat between locations at various temperature levels. Using secondary coolant storage tanks for the application increases the availability of cooling and heating and reduces peak demands for energy.

Each supplier of refrigeration equipment that uses secondary coolant flow has specific ratings. Flooded and direct expansion coolers, dairy plate heat exchangers, food processing, and other air, liquid, and solid chilling devices come in various shapes and sizes. Refrigerated secondary coolant spray wetted-surface cooling and humidity control equipment has an open system that absorbs moisture while cooling and then continuously regenerates the secondary coolant with a concentrator. Although this assists the cooling, dehumidifying, and defrosting process, it is not strictly a secondary coolant flow application for refrigeration, unless the secondary coolant also is used in the coil. Heat transfer coefficients can be determined from vendor rating data or by methods described in Chapter 3 of the 1989 ASHRAE *Handbook—Fundamentals* and accompanying references.

A primary refrigerant may be used as a secondary coolant in a system by being pumped at a flow rate and pressure high enough that the primary heat exchange occurs without evaporation. But the refrigerant is then subsequently flashed at a low pressure, with the resulting flash gas being drawn off to a compressor in the conventional manner.

REFERENCES

Moody, L.F. 1944. Frictional factors for pipe flow. ASME *Transactions* (November): 672-73.

Kern, D.Q. 1950. *Process heat transfer*. McGraw-Hill Book Co., Inc., New York, p. 134.

Sieder, E.N. and G.E. Tate. 1936. Heat transfer and pressure drop of liquids in tubes. *Industrial and Engineering Chemistry* 28(12):1429.

ASHRAE. 1978. Safety code for mechanical refrigeration. ASHRAE *Standard* 15-1978.

ASME. 1987. Refrigeration piping. *Standard* B31.5. American Society of Mechanical Engineers, New York.

ASME. 1986. Boiler and pressure vessel code, Section VIII, Recommended Guidelines for the care of power boilers (non-interfiled). American Society of Mechanical Engineers, New York.

Means. 1988. *Means mechanical cost data,* 11th ed. Section 2, U.C.I. Division 15.5. Robert Snow Means Co., Inc., Kingston, MA.

Park, W.R. and D.E. Jackson. 1984. *Cost engineering analysis,* 2nd ed. John Wiley & Sons, New York.

NBS. 1978. National Bureau of Standards Building Science Series 113, Life cycle costing. SD Catalog Stock No. 003-003-01980-1, U.S. Government Printing Office, Washington, D.C.

CHAPTER 6

REFRIGERANT SYSTEM CHEMISTRY

Evaluating Chemical Problems 6.1
Material Properties 6.2
Chemical Reactions 6.4

REFRIGERATION systems can fail as a result of the chemical breakdown of the materials used. The breakdown can be *direct,* as caused by the reaction of the lubricant and refrigerant in the presence of metals, or *indirect,* where the decomposition products from one set of reactions initiate other chemical reactions. This chapter deals primarily with these complex chemical aspects, apart from the physical and mechanical actions of a refrigeration system. The physical aspects (*i.e.,* measurement and contaminant level control, including moisture) are discussed in Chapter 7. Similarly, reactions with lubricants are covered in this chapter, while lubricant reactivity as a function of composition is discussed in Chapter 8. Contaminant reactions, thermal decomposition, and other effects of high temperature; the behavior of motor insulation and other polymeric materials; hydrolysis (reactions with moisture); and copper plating phenomena are also covered in this chapter.

A chemical reaction can be defined as a reaction in which the identity of the materials involved changes. Usually, but not necessarily, more than one material enters the reaction. Thermal decomposition is a familiar example of a reaction involving only one material. In refrigerant systems, reactions involving two or more materials are hydrolysis, oxidation and reduction, reaction with oils, and direct reaction with metals.

EVALUATING CHEMICAL PROBLEMS

When unexpected chemical problems occur, they can usually be attributed to inadequate testing of new material, improper application of a previously tested material, or inadvertent introduction of contaminants into the system. Three fundamental laboratory techniques are used to chemically evaluate materials: (1) materials tests in sealed tubes, (2) component tests, and (3) accelerated life and system tests.

Sealed Tube Tests

The glass sealed tube test, as described by ASHRAE *Standard* 97 (1989), is the method most widely used to assess the stability of refrigerant system materials. It is also an excellent tool for identifying the chemical reactions likely to occur in operating units.

Generally, glass tubes are charged with refrigerant, oil, metal strips, and other materials to be tested. The tubes are then sealed and aged at elevated temperatures for a specified time. The tubes are visually inspected and compared to a "control" tube, which is processed identically to the specimen tube but does not contain the test material; it can contain a known reference material. A record of observations of the color of the liquids and the appearance of the metals and other specimens provides the simplest means of judging the results from the sealed tube test. At times,

---

The preparation of this chapter is assigned to TC 3.2, Refrigerant System Chemistry.

depending on the objective of the test, visual observations are adequate. But more often, after the aging period, the contents are analyzed for changes in the test material and for refrigerant decomposition by gas analysis or by wet chemical methods.

Originally, the sealed tube test was designed to compare oils, but the technique is effective in testing other materials as well. For example, Huttenlocher (1972) evaluated zinc die castings, Elsey and Flowers (1956) studied the degradation of cellulosic ground insulation by R-22, and Mays (1962) has reported the decomposition of R-22 in the presence of 4A-type molecular sieve desiccants. Although the glass sealed tube test is useful, it has some pitfalls. Since the sealed tube test greatly magnifies chemical reactions likely to occur in a refrigeration system, results can be misinterpreted; *i.e.,* polyester film is perfectly adequate as ground insulation in actual practice, though early test results were negative.

The sealed tube test, in spite of its proven utility, is only a screening tool, and *not* a simulation of a refrigerant system. While the data obtained are valuable, experience and judgment are required to predict field behavior. For this reason, material selection for refrigerant systems requires follow-up with component or system tests or both.

Component Tests

Component tests carry materials evaluation a step beyond the sealed tube tests in that materials are not only tested in the proper environment, but also under dynamic conditions. Motorette tests, used to evaluate hermetic motor insulation, are a good example of this type of test. Component tests are conducted in large pressure vessels or autoclaves (enameled wire, ground insulation, and other motor materials assembled into a simulated motor) in the presence of oil and refrigerant. Unlike glass sealed tubes, where temperature and pressure are the only means of accelerating the aging process, the autoclave tests can involve mechanical vibrations, on-off electrical voltages, and refrigerant liquid floodback as external stresses that accelerate the phenomena likely to occur in an operating system.

System Tests

System tests can be divided into two major categories:

1. Test a sufficient number of systems under a broad spectrum of operating conditions to obtain a good, statistical reference base. The failure rates of two populations of units can be compared: (a) those containing the new materials, and (b) those containing previously proven materials.
2. Test under extremely well-controlled conditions. Temperatures, pressures, and other operating parameters are continuously monitored, and chemical analysis of the refrigerant and oil is made during and after the test is completed.

In most cases, the tests are conducted under severe operating conditions so that results can be obtained quickly. Analyzing the oil and refrigerant samples during the test and inspecting the components after the teardown can yield valuable information on (1) the nature and rate of chemical reactions taking place in the system, (2) the products formed by the reactions, and (3) the possible effect on system life and performance. Accurate interpretation of these data determines operating limits for the system that keep chemical reactions at an acceptable level.

MATERIAL PROPERTIES

Refrigerants

Halocarbons. These refrigerants are derivatives of methane, ethane, and other gases that contain fluorine, chlorine, hydrogen, and, in a few instances, bromine, in various combinations. Although numerous compounds in this category are listed as refrigerants (ANSI/ASHRAE *Standard* 34), only a few are commercially important. R-12, R-22, and R-502 are the most widely used, and R-11, R-13B1, and R-500 are used in many specialized applications. All of these refrigerants are characterized by low toxicity and are classified as nonflammable by Underwriters Laboratories.

Sand and Andrjeski (1982) studied the flammability of R-11, R-12, and R-22 under extreme conditions and found that at pressures above 200 psig, mixtures of R-22/air containing a minimum of 50 mole percent air, can explode in the presence of a high-temperature ignition source (greater than 1800°F). Attempts to ignite R-11 or R-12 under similar conditions were not successful.

Ammonia. Used extensively in large, open-type compressors for industrial and commercial applications, ammonia has high refrigerating capacity per unit displacement, low pressure losses in connecting piping, and low reactivity with refrigeration oils.

The toxicity and flammability of ammonia offset its advantages. Ammonia is such a strong irritant to the human nose (detectable below 5 ppm) that people automatically avoid exposure to it. Ammonia-air mixtures are flammable, but only within a relatively narrow range of 16 to 25% by volume of ammonia. These mixtures can explode, but are difficult to ignite since they require a high intensity ignition source. The apparent ignition temperature is 1200°F, and the maximum explosion pressure is relatively low, *i.e.,* 50 psi. Therefore, ammonia is not a serious fire or explosion hazard.

The thermal stability of ammonia is often subject to confusion. At atmospheric pressure, ammonia starts to dissociate into nitrogen and hydrogen at about 570°F in the presence of active catalysts such as nickel and iron. However, since such high temperatures are unlikely to occur in open type compressors, thermal stability is not a problem in ammonia systems.

Ammonia attacks copper in the presence of even small amounts of moisture; therefore, except for some specialty bronzes, copper-bearing materials are meticulously excluded in ammonia systems. A corollary of this design practice is that the copper plating common to halocarbon refrigerant systems is nonexistent in ammonia systems. Since ammonia is not used in hermetically sealed units, interactions of motor insulation and ammonia are not discussed in this chapter.

Other Refrigerants. Sulfur dioxide is thermally stable to very high temperatures. Methyl chloride is thermally stable, but reacts with metals such as aluminum, magnesium, and zinc. Hydrocarbons present no stability problems in refrigeration use.

Oils

The physical and chemical properties of refrigeration oils and additives are covered in Chapter 8.

Electrical Insulation

Electrical insulating materials used in hermetic motors include the magnet wire covering, materials for ground and phase insulation, impregnating varnish, and miscellaneous minor applications such as tie cords and lead insulation. Since hermetic motors operate in the refrigerant-oil environment, the following questions should be considered:

1. Does the insulation material contribute contaminants to the hermetic system?
2. Does the refrigeration system environment affect the performance of the insulation material with respect to its physical, chemical, or electrical properties?

Test procedures for commonly used hermetic motor insulating materials and their performance characteristics are reviewed in this section.

Magnet Wire Insulation. The wire covering is the primary insulation and plays an important role in the service life of hermetic motors. The chemical structure must be known to predict the behavior of the insulation in the particular hermetic environment. In small hermetic motors, the magnet wire is coated with an organic resin, which is subsequently cured by heating. Many large reciprocating compressors using random-wound motors and hermetic centrifugal systems with form-wound motors are insulated with composites of organic resin films and fiber wrapping. Both types of wire coverings are qualified according to their mechanical, electrical, chemical, and thermal properties, as measured by traditional magnet wire test methods (NEMA 1987).

These tests are useful, but most of them are conducted in air, so they are not valid for hermetic refrigerant applications. For example, wire enamels absorb R-22 up to 15 to 30% by mass (Hurtgen 1971a) and at different rates depending on their chemical structure, degree of cure, and the conditions of exposure to the refrigerant. Refrigerant permeation is shown by changes in electrical, mechanical, and physical properties of the wire enamels.

Wire enamels in refrigerant vapor typically exhibit dielectric loss with increasing temperature, as shown in Figure 1. Depending on the atmosphere and degree of cure, each wire enamel or enamel/varnish combination exhibits a characteristic temperature t_{max} above which dielectric losses increase sharply. Table 1 shows values of t_{max} for several hermetic enamels. Continued heating above t_{max} causes aging, evidenced by the irreversible alteration of dielectric properties and an increase in the conductance of the insulating material.

Spauschus and Sellers (1969) suggest that the change rate in conductance is a quantitative measure of aging in a refrigerant environment. They give the aging rate for varnished and unvarnished enamels at two levels of R-22 pressure typical of high- and low-side hermetic motor operation.

Apart from the effects on long-term aging, R-22 can also affect the short-term insulating properties of certain wire enamels. Beacham and Divers (1955) demonstrated that the resistance of

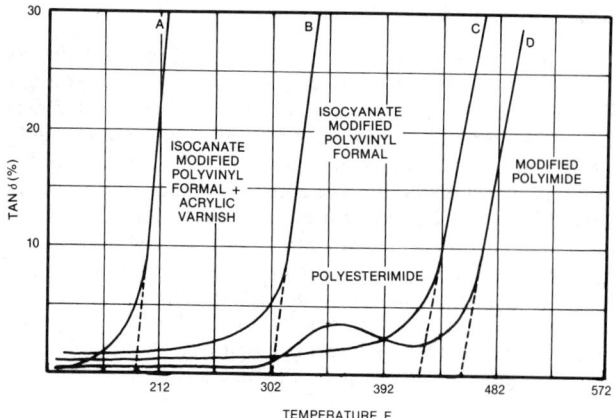

Fig. 1 Loss Curves of Various Insulating Materials

Refrigerant System Chemistry

Table 1 Maximum Temperature t_{max} for Hermetic Wire Enamels in R-22 at 65 psia

Enamel Type	t_{max}, °F
Acrylic	226
Polyvinyl formal	277
Isocyanate modified polyvinyl formal	304
Polyamide-imide	361
Polyester-imide	419
Polyimide	450

polyvinyl formal drops drastically when the material is submerged in liquid R-22. A parallel experiment using R-12 showed a much smaller drop, followed by quick recovery to the original resistance. The relatively rapid permeation of R-22 into polyvinyl formal, coupled with the low-volume resistivity of liquid R-22 and other electrical properties of the two refrigerants, explains the phenomenon.

Softening of the wire coating, which can cause the insulation to fail, occurs with certain combinations of coatings and refrigerants. Table 2 shows data on softening measured in terms of abrasion resistance for a number of wire enamels exposed to R-22. At the end of the shortest soaking period, the urethane-modified polyvinyl formal had lost all its abrasion resistance. All the other insulations, except polyimide, lost their abrasion resistance at a slower rate, which over a 3-month period approached the rate of the urethane-polyvinyl formal. Only the polyimide showed a minimal effect even though its abrasion resistance was lower originally than most of the others.

Because of the time dependency on softening, which is related to the permeation rate of R-22 into the enamel, Sanvordenker and Larime (1971) proposed that comparative tests on magnet wire be made only after the enamel is completely saturated with refrigerant, so that the effect of long-term exposure to R-22 on the enamel properties can be evaluated.

The second consequence of R-22 permeation is blistering, caused by the rapid change in pressure and temperature of a wire enamel when exposed to R-22. Heating greatly increases the internal pressure as the dissolved R-22 expands; and, since the polymer film has already been softened, portions of the enamel lift up in the form of blisters. While blistered wire has a poor appearance, field experience indicates that mild blistering is not a cause for concern, as long as the blisters do not break and the enamel film remains flexible. Currently used wire enamels have the above characteristics and maintain dielectric strength even after blistering. However, hermetic wire enamel with strong resistance is preferred.

Table 2 Effect of Liquid R-22 on Abrasion Resistance

	Average Cycles to Failure			
		After Time in Liquid R-22		
Magnet Wire Insulation	As Received	7 - 10 Days	One Month	Three Months
Urethane-polyvinyl formal batch 1	40	3	2	2
Urethane-polyvinyl formal batch 2	42	2	2	7
Polyester-imide batch 1	44	15	18	6
Polyester-imide batch 2	24	10	5	6
Dual coat amide-imide top coat, polyester base	79	35	23	11
Dual coat, polyester	35	5	5	9
Polyimide	26	25	23	21

From Sanvordenker and Larime (1971).

The extractables in wire enamels can contribute to the contaminants in an operating system. Their effects are covered in Chapter 7.

Harris and Sonnino (1959) and Emmons (1967) describe test procedures for determining abrasion resistance, blistering, and extractables. NEMA (1987) provides modified test procedures for blistering and extractables that reflect the behavior of modern insulation systems. Petry (1971) described tests that involve motorettes in autoclaves, and Hurtgen (1971b) used production compressors to simulate field behavior.

Because of cost, an enamel with superior resistance to refrigerants may not be the first choice for hermetic applications. For example, polyimide, which has the highest t_{max} and is not softened by R-22 (see Tables 1 and 2), is acceptable only in specialized applications. Polyamide-imide, which has very good blister resistance, is widely accepted, but only as a fraction of the total insulation thickness, i.e., as the top coat. This is partly because the resin requires a special solvent (n-methyl-pyrrolidone) for the wire coating process.

Because of their optimum hermetic properties and cost, the two wire enamels most widely used are the polyesterimide and the two-resin construction, which has a polyester undercoat and a polyamide-imide top coat. Variations of these systems include a polyesterimide base coat under the polyamide-imide top coat, and a polyester-amide-imide, which is a one-resin system (Bich *et al.* 1971). Current development efforts are aimed at polyamide-imide structures; these require no special solvent, yet satisfy requirements of hermetic application (Okada *et al.* 1981).

Varnishes. The stators of many hermetic motors, particularly those above 3 hp, are treated with a varnish. Varnish treatment immobilizes the enameled wires in the end turns of stators when subjected to electromechanical forces prevalent in rotating machinery. Varnishes are usually applied by dipping the stator into a solvent-thinned mixture, allowing the varnish to drip-dry. The varnish film is then cured.

Methods have been developed for testing the bond strength between varnish and the wire enamel. Although bond strength tests in air at room temperature have some significance, data at elevated temperatures and in the refrigerant environment are much more meaningful. In addition to using laboratory bench tests, bond strengths should be evaluated by repeated start-stop-reversing tests of finished motors operating in the refrigerant environment. Felger and Baciu (1971) describe a method and equipment for determining varnish/magnet wire bond strengths in refrigerants at elevated temperatures and pressures. A unique mechanical fixture measures the force in the exposure vessel. The equipment allows the use of either varnished splints or coils.

Many different chemical types of varnishes can be used in hermetic motors. The commonly used varnishes include acrylics, phenolics, epoxies, and modified polyimides, with numerous formulations of each chemical type available. In addition to bond strength, other important considerations include chemical compatibility with the wire enamel; low extractables; good electrical properties and adhesion; service conditions, such as the nature of the wire enamel; motor operating temperatures; and type of refrigerant.

Ground Insulation. Sheet insulation material is used in slot liners, phase insulation, wedges, and for tie point insulation in hermetic motors. Some of the fractional horsepower motors intended for R-12 systems continue to use a quality electrical insulation grade paper for this service, but the vast majority use a polyethyleneterephthalate (Mylar) film. For special applications involving very high temperatures, either an aramid fiber mat or an aromatic polyimide film is used. Each of these materials is characterized by unique properties that must be considered for successful application.

Paper is comprised largely of cellulose, which yields moisture, noncondensable gases, and tars on thermal degradation. In addition, paper can be degraded by acidic contaminants sometimes found in hermetic systems, leading to darkening and loss of strength, particularly in systems that use R-22. On the other hand, electrical grade paper is more economical than synthetic insulation materials and will not melt if local hot spots develop in a motor.

Polyethyleneterephthalate films possess excellent dielectric properties and are characterized by good chemical resistance to refrigerants and oils. Their nonhygroscopic nature relative to paper permits shorter and more effective dehydration cycles. Also, these films do not release significant amounts of moisture to the refrigeration system, even under abnormal operating conditions.

The polyethyleneterephthalate film selected must contain a minimum of the low molecular weight polymers that exhibit a temperature-dependent solubility in mineral oils and tend to precipitate as noncohesive granules at temperatures lower than those experienced by the motor. Another limitation of this film is that, like most polyesters, it is susceptible to degradation by hydrolysis; but the quantity of water required is in excess of that generally found in refrigerant systems. However, this film should not be exposed to alcohols that are sometimes used as antifreeze in servicing refrigeration systems.

Elastomers

Refrigerants, oils, or mixtures of both can, at times, extract enough filler or plasticizer from an elastomer to change its physical or chemical properties. This extracted material can harm the refrigeration system by increasing its chemical reactivity or by clogging screens and expansion devices. Data on swelling of elastomers in most of the common halocarbon refrigerants are presented in Chapter 16 of the 1989 ASHRAE *Handbook—Fundamentals*.

Plastics

The effect of refrigerants on plastics usually decreases as the amount of fluorine in the molecule increases. For example, R-12 has less effect than R-11, while R-13 is almost entirely inert. Test each type of plastic material for compatibility with the refrigerant before it is used. Two samples of the same type of plastic might be affected differently by the refrigerant because of differences in polymer structure, molecular weight, and plasticizer. Some characteristics of different plastics are:

Polystyrene is dissolved by R-11, R-21, and R-22, but is only slightly affected by R-12 and R-113. As a general rule, it should not be used.

Methacrylate polymers, such as Lucite or Plexiglas, craze and crack on long exposure to most refrigerants and dissolve in R-22 and R-21.

Fluorinated polymers tend to swell slightly in highly fluorinated refrigerants and may be unsuitable in some cases.

Fluorocarbon resins like TFE and Teflon do not swell when submerged in the liquid phase of refrigerants, but both R-12 and R-22 can diffuse through Teflon film.

Polychlorotrifluoroethylene swells slightly, but generally is suitable for use with fluorocarbon refrigerants.

Polyvinylalcohol, although unaffected by fluorocarbon refrigerants, is very sensitive to water.

Vinyl includes many types of vinyl elastomers with good resistance to the fluorocarbon refrigerants in most cases. However, test each formulation before it is used.

Acrylic fiber generally is suitable for use with fluorocarbon refrigerants.

Nylon generally is suitable for use with fluorocarbon refrigerants, but may tend to become brittle at high temperatures in the presence of air, water, or alcohol.

Polyethylene is suitable for some applications at moderate temperatures. Many variations exist, so each type should be thoroughly tested with the refrigerant before use.

Phenolic resins are usually unaffected by refrigerants.

Epoxy resins are resistant to most solvents and suitable for use with refrigerants unless plasticized.

Cellulose acetate or nitrate is suitable for use with refrigerants under most conditions.

Acetal resin is suitable for use with refrigerants under most conditions.

Chapter 16 of the 1989 ASHRAE *Handbook—Fundamentals* discusses the effect of refrigerants on plastic materials.

CHEMICAL REACTIONS

Halocarbon Refrigerants

Thermal stability. All the common halocarbon refrigerants have excellent thermal stability, as shown in Table 3.

Reactions with Water. The halogenated refrigerants are susceptible to reaction with water (hydrolysis), but the rates of reaction are so slow that they are negligible (see Table 4).

Reactions with Metals. Chloro-fluorocarbon refrigerants can be used satisfactorily with most common metals under normal conditions (see Chapter 16 of the 1989 ASHRAE *Handbook—Fundamentals*). Under extreme conditions, such as above red heat or molten metal temperatures, these refrigerants react exothermically with metals and produce metal halides and carbon. Extreme temperatures may occur in applications such as centrifugal compressors if the impeller rubs against the housing when the system malfunctions. Using R-12 as the test refrigerant, Eiseman (1963) found that aluminum is the most reactive, followed by iron and stainless steel. Copper is relatively unreactive. Also using aluminum as the reactive metal, Eiseman reports that R-14 causes the most vigorous reaction, followed by R-22, R-12, R-114, R-11, and R-113, in that order.

Table 3 Inherent Stability of Refrigerants

Refrigerant	Formula	Decomposition Rated at 400 °F in Steel, % per yr[a]	Temperature Where Decomposition Is Readily Observed in Laboratory,[b] °F	Temperature Where 1%/Year Decomposes in Absence of Active Materials, °F	Major Gaseous Decomposition Products[c]
22	$CHClF_2$	—	800	480	CF_2CF_2,[d] HCL
11	CCl_3F	2	1100	570[e]	R-12, Cl_2
114	$CClF_2\text{-}CClF_2$	1	1100	710	R-12
115	$CClF_2CF_3$	—	1160	740	R-13
12	CCl_2F_2	Less than 1	1400	930	R-13, Cl_2
13	$CClF_3$	—	1550	1000[f]	R-14, Cl_2, R-116

Data from Norton (1957), duPont (1959, 1969), and Borchardt (1975).
[a] Data from Underwriters Laboratories.
[b] Decomposition rate is about 1% per min.
[c] Data from Borchardt (1975).
[d] A variety of side products are also produced, here and with the other refrigerants, some of which may be quite toxic.
[e] Conditions were not found where this reaction proceeds homogeneously.
[f] Rate behavior too complex to permit extrapolation to 1% per year.

Refrigerant System Chemistry

Table 4 Rate of Hydrolysis in Water
Grams per Litre of Water per Year

Refrigerant	Formula	1 Atm Pressure 86°F		Saturation Pressure at 122°F
		Water Alone	With Steel	With Steel
113	CCl_2FCClF_2	< 0.005	50	40
11	CCl_3F	< 0.005	10	28
12	CCl_2F_2	< 0.005	1	10
21	$CHCl_2F$	< 0.01	5	9
114	$CClF_2\text{-}CClF_2$	< 0.005	1	3
22	$CHClF_2$	< 0.01	0.1	—

Data from Du Pont (1959, 1969).

Ammonia

Reactions involving ammonia, oxygen, oil degradation acids, and moisture are common factors in the formation of ammonia compressor deposits. Sedgwick (1966) suggested ammonia or ammonium hydroxide reacts with organic acids produced by the oxidation of the compressor oil to form ammonium salts (soaps), which can decompose further to form amides (sludge) and water. This reaction may be described as follows:

$$NH_3 + RCOOH \rightleftharpoons RCOONH_4 \rightleftharpoons RCONH_2 + H_2O$$

Water may be consumed or released during the reaction, depending on system temperature, metallic catalysts, and chemical state (acidic or basic condition). Compressor deposits can be minimized by keeping the system clean and dry, preventing entry of air, and maintaining proper compressor temperatures.

Oils

Oils in the refrigeration system may: (1) react with the refrigerant, (2) oxidize, and (3) coke. Oxidation is not a problem in normal systems because no oxygen is available to react with the oil. However, when a system is inadequately evacuated or air is allowed to leak into the system, organic acids and sludges can be formed. These materials are detrimental to refrigeration systems and can cause their failure.

Coking can be caused by thermal decomposition of the oil as a result of high operating temperatures.

Oil-Refrigerant Reactions

The reactions of the present halocarbon refrigerants with different oils have been studied, using the glass sealed tube and, to some extent, the Phillip Test, which is more popular in Europe (Steinle 1964). Some oils are more reactive with a given refrigerant than others and, likewise, some refrigerants are more stable with a given oil than with others. For example, R-14 is almost inert, while R-11 is highly reactive. General relationships between molecular structure and reactivity with oils in the presence of metals are as follows:

1. Substitution of a chlorine atom by a fluorine atom decreases reactivity. Examples in order of decreasing reactivity are: R-11, R-12, R-13, R-14, R-113, R-114, and R-115.
2. Substitution of a hydrogen atom by a fluorine atom decreases reactivity. For example, R-12 is less reactive than R-21, and R-13 is less reactive than R-22.
3. Substitution of a hydrogen atom by a chlorine atom increases reactivity. For example, R-22 is less reactive than R-12, and R-21 is less reactive than R-11.

The reactivity of oils as a function of their chemical structure is more complicated since, unlike refrigerants, lubricating oils are not pure compounds. However, reactivity of oils, based on predominant components (saturates, aromatics, and noncarbons), has been investigated (see Chapter 8).

Regardless of the type of oil and refrigerant, the reaction accelerates greatly in the presence of metals and contaminants, compared to oil and refrigerant alone. The major construction materials in refrigerant systems, steel, copper, and aluminum, differ considerably in their catalytic activity. Borchardt (1972, 1973, 1978), in studies of the effects of these metals (individually and in combinations) on R-12-oil reaction, found that steel is the most reactive while copper is essentially inert. Aluminum has a small catalytic effect unless the protective layer of aluminum oxide is broken down by the chlorides or hydrochloric acid formed by the oil-refrigerant reaction.

Parmelee (1965) studied the effects of contaminants likely to occur in refrigerant systems. He concluded that, as a general rule, the simpler (chemically) the system, the longer its life; every added material makes more chemical reactions possible. Walker *et al.* (1960, 1962) investigated the effect of combinations of variables (refrigerants, oils, metals, air, and moisture) and reported the effects of each variable on an individual and combined basis.

Elsey *et al.* (1952) postulated that the carbonaceous sludges noticed in sealed tube tests occur when a chlorine atom splits off from the refrigerant and combines with a hydrogen atom from the oil to form hydrochloric acid. They also found a correlation between the amount of chloride ion formed and the color development in the oil. They suggested that the darkening of the oil was a much simpler and adequate criterion for judging the extent of the oil-refrigerant reaction, except when the refrigerant itself was unstable at the test temperature. Based on this darkening, they recorded the reactivity of several halocarbon refrigerants, including R-12 and R-22.

Spauschus and Doderer (1961), using gas analysis procedures, showed that R-12 exchanges one of its chlorine atoms for a hydrogen atom from the oil, forming R-22. This reaction occurs in other chlorinated refrigerants; for example, R-22-oil reaction forms R-32 (Spauschus and Doderer 1964), and R-115 forms R-125 (Parmelee 1965). The proposed chemical reactions are as follows:

$$\text{R-12} + \text{oil} \rightarrow \text{R-22} + \text{chlorinated oil}$$

or written as a chemical formula:

$$CCl_2F_2 + R\text{—}CH_2\text{—}CH_3 \rightarrow CHClF_2 + R\text{—}\underset{\underset{Cl}{|}}{CH}\text{—}CH_3$$

Then the chlorinated oil reacts to become an unsaturated hydrocarbon as follows:

$$R\text{—}\underset{\underset{Cl}{|}}{CH}\text{—}CH_3 \rightarrow RCH\text{=}CH_2 + HCl$$

Finally, polymerization converts the unsaturated hydrocarbon to carbonaceous sludge.

This work provides a more sensitive criterion for measuring the extent of the oil-refrigerant reaction, which previously could only be determined from wet chemical analysis of the chloride ion formed.

Sanvordenker (1985) presented further details of oil-R12 reactions in glass sealed tubes. This study shows that iron, a participant rather than a catalyst, dissolves in the oil as the reaction proceeds. The dissolved iron shows a structural change in the oil as if the oil were oxidized. Further on, the iron-oil molecule along with the refrigerant can react with the glass, and form extraneous products that should not be considered a part of hermetic chemistry. A criterion to identify the entry of glass has been defined.

Figures 2 and 3 show the sealed tube test data for the reaction rates of R-22 and R-12 with oil in the presence of copper and mild steel. Formation of the chloride ion is the criterion of the refrigerant decomposition. As expected, temperature greatly accelerates reaction rates. Also, R-22 is considerably less reactive than R-12. Again, note that the sealed tube is an accelerated test and that the extent of reaction shown in Figures 2 and 3 can

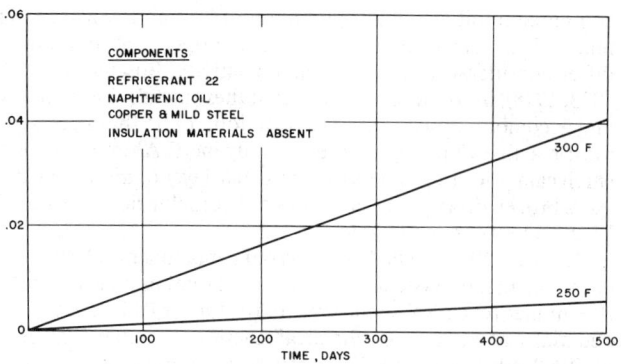

Fig. 2 Stability of Refrigerant 22 Control System
(Kvalnes and Parmelee 1957)

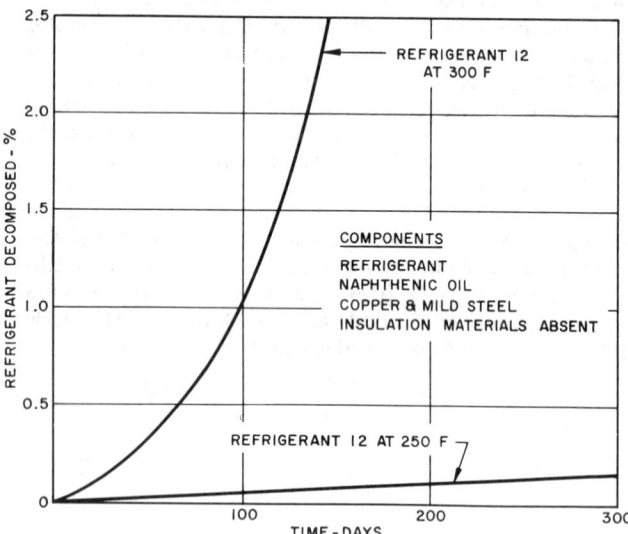

Fig. 3 Stability of Refrigerant 12 Control Systems
(Kvalnes and Parmelee 1957)

exceed the reaction in the actual system by several orders of magnitude.

Copper Plating

Copper plating is the formation of a copper film on metal surfaces of refrigeration compressors. A blush of copper is often discernible on compressor bearing and valve surfaces when machines that operate at high temperatures are cut apart. After several hours of exposure to air, this thin film becomes invisible, probably because metallic copper is converted to copper oxide. In more severe cases, the copper deposit can build up to substantial thickness and interfere with compressor operation. Extreme copper plating can cause compressor failure.

Physical or mechanical copper entrapment can occur when very fine copper crystals or particles are removed from improperly annealed copper tubing and are transported to the bearings, pistons, and other surfaces where mechanical forces plate the copper onto the other metals.

Chemical copper plating involves complex reactions. While some details of the mechanism remain obscure, the principal causes of copper plating are understood well enough to avoid catastrophic field problems. Although water was once considered the major factor in copper plating, Walker *et al.* (1960, 1962) have shown that water below the saturation level has no significant effect on copper plating.

Theory also stresses the importance of oil-refrigerant reaction products, in which copper plating takes place in two steps. First, copper is dissolved in the oil or oil-refrigerant solution. The quantity that dissolves is established by the nature of the oil, temperature, refrigerant-oil reaction products, oxygen, and other impurities or contaminants such as residual processing chemicals. In a subsequent step, which can involve an electrochemical reaction or thermal decomposition, dissolved copper is deposited on metallic surfaces. Products of refrigerant-oil reactions can be precursors for gross copper plating. Therefore, selecting a stable halogenated refrigerant and a quality oil, removing air and water, and operating the system at moderate temperatures greatly reduces or eliminates copper plating problems.

Contaminant Generation by High Temperature

Hermetic motors can overheat well beyond the design levels under adverse conditions. These include line voltage fluctuations, operation of a central air-conditioning system during an electrical storm, or inadequate airflow over the condenser coils. Under these conditions, the motor windings can exceed 300 °F. Prolonged exposure to these thermal excursions can damage the motor insulation, depending on the thermal stability of the insulation materials, their reactivity with the refrigerant and oil that surround them, and the temperature levels encountered.

Thermal decomposition of organic insulation materials produces noncondensable gases such as carbon monoxide and carbon dioxide. These liquids can be as simple as water, as is the case with cellulosic materials, or they can be highly complex materials similar to coal tar in appearance. Noncondensable gases that circulate with the refrigerant vapors increase the discharge pressure and lower the unit efficiency. At the same time, the compressor temperature and rate of insulation degradation increase. Liquid products circulate with the lubricating oil either in true solutions or as colloidal suspensions. Dissolved and suspended decomposition products circulate throughout the refrigeration system, where they can clog oil passages, interfere with the operation of expansion, suction, and discharge valves, or plug capillary tubes.

Appropriate control mechanisms in refrigerant system design minimize exposure to such thermal excursions. Identifying potential reactions, performing adequate laboratory tests to qualify the materials before using them in the field, and finding the means to remove contaminants generated by high-temperature excursions are equally important (see Chapter 7).

REFERENCES

ASHRAE. 1989. ANSI/ASHRAE *Standard* 97-1989. Sealed glass tube method to test the chemical stability of material for use within refrigerant systems.

Beacham, E.A. and R.T. Divers. 1955. Some practical aspects of the dielectric properties of refrigerants. *Refrigerating Engineering* (July):33.

Bich, G.J., J.R. Kwiecenski, and F.A. Satter. 1971. Effect of composition of polyester-amide-imide wire enamels on Refrigerant 22 resistance. *Proceedings of the 10th Electrical Insulation Conference*, 126.

Borchardt, H.J. 1972. The reaction of aluminum with R-12 and oil. ASHRAE *Bulletin* NA72-5.

Borchardt, H.J. 1973. Effect of metals on the reaction of R-12 with oil. ASHRAE *Bulletin* LO-73-5.

Borchardt, H.J. 1975. *DuPont innovation*, Vol. G, No. 2 (Spring).

Borchardt, H.J. 1978. The reaction of Refrigerant 12 with oils. *Freddo*, No. 6:488.

Du Pont. 1959, 1969. Properties and application of the "Freon" fluorinated hydrocarbons, and Stability of several "Freon" compounds at high temperatures. *Technical Bulletins* B-2 and X1A, Freon Products Division, E.I. du Pont de Nemours & Co., Inc.

Refrigerant System Chemistry

Eiseman, B.J., Jr. 1963. Reactions of chlorofluoro-hydrocarbons with metals. ASHRAE *Journal* 5(5):63.

Elsey, H.M. and L.C. Flowers. 1956. The reaction between electrical insulating paper and monochlorodifluoromethane. *Refrigerating Engineering* (April): 31.

Elsey, H.M., L.C. Flowers, and J.B. Kelley. 1952. A method of evaluating refrigerator oils. *Refrigerating Engineering* (July):737.

Emmons, H.L. 1967. Latest technology for hermetic motor insulation testing: Part I, Varnishes, Insulation (August):42; Part II, Magnet wire overload, *Insulation* (September):57; Part III, Sheet and integral insulation cut-through, *Insulation* (November):53; Part IV, Motorette-autoclave for systems evaluation, *Insulation* (December):53.

Felger, M.M. and N. Baciu. 1971. The determination of varnish bond strength in refrigerant atmospheres. *Insulation/Circuits* (December):30.

Harris, J.F. and C.B. Sonnino. 1959. New insulation media for hermetic motors require new tests. ASHRAE *Journal* 1(3):49.

Hurtgen, J.R. 1971a. R-22 blister testing of magnet wire. *Proceedings of the 10th Electrical Insulation Conference*, Chicago, 183.

Hurtgen, J.R. 1971b. Hermetic motor life tests. *Proceedings of the 10th Electrical Insulation Conference*, Chicago, 113.

Huttenlocher, D.F. 1972. Accelerated sealed-tube test procedure for Refrigerant 22 reactions. *Proceedings of the 1972 Purdue Compressor Technology Conference*, 469.

Kvalnes, D.E. and H.M. Parmelee. 1957. Behavior of Freon-12 and Freon-22 in sealed-tube tests. *Refrigerating Engineering* (November):40.

Mays, R.L. 1962. Molecular sieve and gel-type desiccants for refrigerants. ASHRAE *Transactions* 68:330.

NEMA. 1987. Magnet wire. *Standard* MW 1000-87. National Electrical Manufacturer's Association, Washington, D.C.

Norton, F.J. 1957. Rates of thermal decomposition of $CHClF_2$ and Cl_2F_2. *Refrigerating Engineering* (September):33.

Okada et al. 1981. A new class 200C enamelled wire. *15th Electrical Electronics Insulation Conference*, 121.

Parmelee, H.M. 1965. Sealed-tube stability tests on refrigeration materials. ASHRAE *Transactions* 71(1):154.

Petry, R.D. 1982. Hermetic insulation tests using motorettes in autoclave. *Proceedings of the 10th Electrical Insulation Conference*, 116.

Sand, J.R. and D.L. Andrjeski. 1982. Combustibility of chlorodifluoromethane. ASHRAE *Journal* 24(5):38.

Sanvordenker, K.S. 1985. Mechanism of oil-R-12 reactions—The role of iron catalyst in glass sealed tubes. ASHRAE *Transactions* 91(1A):356.

Sanvordenker, K.S. and M.W. Larime. 1971. Screening tests for hermetic magnet wire insulation. *Proceedings of the 10th Electrical Insulation Conference*, 122.

Sedgwick, N.V. 1966. *The organic chemistry of nitrogen, 3rd ed.* Clarendon Press, Oxford, England.

Spauschus, H.O. and G.C. Doderer. 1961. Reaction of Refrigerant 12 with petroleum oils. ASHRAE *Journal* 3(2):65.

Spauschus, H.O. and G.C. Doderer. 1964. Chemical reactions of Refrigerant 22. ASHRAE *Journal* 6(10).

Spauschus, H.O. and R.A. Sellers. 1969. Aging of hermetic motor insulation. IEEE *Transactions*, EI-4(4):90.

Steinle, H. 1950. Chemical reactions between refrigerants and oils in refrigerating machines. *Kaltetechnik* 2(7):174.

Steinle, H. 1964. Development and testing of lubricants for refrigerating machines. ASHRAE *Transactions* 70:195.

Walker, W.O., S. Rosen, and S.L. Levy. 1960. A study of the factors influencing the stability of mixtures of Refrigerant 22 and refrigerating oils. ASHRAE *Transactions* 66:445.

BIBLIOGRAPHY

Chemical Evaluation

Armstrong, T.D., Jr. 1965. Chloride analysis as a measure for the evaluation of sealed-tube tests. ASHRAE *Transactions* 71(1):150.

Doderer, G.C. and H.O. Spauschus. 1966. A sealed-tube gas chromatograph method for measuring reaction of Refrigerant 12 with oil. ASHRAE *Transactions* 72(2):IV.4.1.

Eiseman, B.J., Jr. 1949. Effect on elastomers of "Freon" and other halocarbons. *Refrigerating Engineering* (December):1171.

Kvalnes, D.E. 1965. The sealed-tube test for refrigeration oils. ASHRAE *Transactions* 71(1):138.

Olsen, R.S. 1971. Sealed-tube test techniques for hermetic enamel evaluation. *Proceedings of the 10th Electrical Insulation Conference*, 119.

Sanvordenker, K.S. 1972. Chemical aspects of refrigerant systems. *Proceedings* of the 1972 Purdue Compressor Technology Conference, p. 456.

Spauschus, H.O. and R.S. Olesen. 1959. Gas analysis—A new tool for determining the chemical stability of hermetic systems. *Refrigerating Engineering* (February):31.

Walker, W.O. 1965. Sealed-tube tests—A comparison of methods. ASHRAE *Transactions* 71(1):134.

Walker, W.O., S. Rosen, and S.L. Levy. 1960. A study of the factors influencing the stability of mixtures of Refrigerant 22 and refrigerating oils. ASHRAE *Transactions* 66:445.

Material Properties

Bushouse, C.J. 1961. Degradation of polyester film by alcohols. ASHRAE *Journal* 3(9):61.

Ditzler, J.L. 1960. Better methods for evaluating hermetic motor insulation. *Air Conditioning, Heating and Ventilating* 57(April):64.

Harrington, J.P. and R.J. Ward. 1959. Polyester film insulation for hermetic motors. ASHRAE *Journal*, 1(4):75.

Hoffman, J.E. 1971. Alcohol at work. *Refrigeration Service and Contracting* (August):18.

McMahon, W., et al. 1959. Degradation studies of polyethylene teraphthalate. *Journal of Chemical and Engineering Data* 4(1):57.

Spauschus, H.O. and R.A. Sellers. 1967a. Apparatus and procedures for evaluating dielectric properties of wire enamels in controlled environments. *Proceedings of the 7th Electrical Insulation Conference*, Chicago.

Spauschus, H.O. and R.A. Sellers. 1967b. Electrical properties of wire enamels in refrigerant environments. *Proceedings of the 12th International Congress of Refrigeration*, Madrid.

Walker, W.O. and K.S. Willson. 1937. The action of methyl chloride on aluminum. *Refrigerating Engineering* (August):89.

Chemical Reactions

McGovern, E.W. 1942. Chemical effect in refrigerating systems. *Refrigerating Engineering* (May):276.

Spauschus, H.O. 1963. Copper transfer in refrigerant-oil solutions. ASHRAE *Journal* 5(6):89.

Steinle, H. 1955. Tests on copper plating in refrigerating machines. *Kaltetechnik* 7(4):101.

Steinle, H. and W. Seeman. 1953. The cause of copper plating in refrigerating machines. *Kaltetechnik* 5(4):90. See also H. Steinle and W. Seeman, 1951, *Kaltetechnik* 3(8):194.77

CHAPTER 7

MOISTURE AND OTHER CONTAMINANT CONTROL IN REFRIGERANT SYSTEMS

MOISTURE 7.1	OTHER CONTAMINANTS 7.6
Sources of Moisture 7.1	Metallic Contaminants and Dirt 7.6
Effects of Moisture 7.1	Organic Contaminants—Sludge, Wax, and Tars 7.6
Moisture Indicators 7.3	Residual Solvents 7.7
Drying Methods 7.3	Antifreeze Agents (Methyl Alcohol) 7.7
Moisture Measurement 7.3	Noncondensable Gases 7.7
Desiccants 7.4	Motor Burnouts 7.8
Desiccant Applications 7.5	Field Assembly 7.8
Driers 7.5	APPENDIX 7.8
Drier Selection 7.6	System Cleanup Procedure After Hermetic Motor
Testing and Rating 7.6	Burnout 7.8

MOISTURE, an important and universal contaminant in refrigerant systems, is discussed in the first part of this chapter. Other contaminants, including dirt, waxes, and metal particles introduced during manufacture or installation, and organic materials such as acids, sludges, and other products of chemical reaction from system operation are covered in the second part.

MOISTURE

The amount of moisture in a refrigerant system must be kept below an allowable maximum to provide satisfactory operation. Moisture must be removed from refrigerant system components during manufacture and assembly to guard against moisture entering the system during installation or servicing. Any moisture that enters the refrigerant system should be removed as quickly as is practicable.

SOURCES OF MOISTURE

Moisture in a refrigerant system results from:

1. Faulty equipment drying in factories and service operations
2. Introduction during installation or service operations in the field
3. Low-side leaks, resulting in entrance of moisture-laden air
4. Leakage of the water-cooled condenser
5. Oxidation of certain hydrocarbons of oil to produce moisture
6. Wet oil, refrigerant, or both
7. Decomposing cellulose insulation in hermetically sealed units
8. Moisture entering a nonhermetic refrigerant system via permeation of nonmetallic hoses and seals

Moisture vapor transmission rates (MVTR) for nonmetallic air-conditioning hoses have been measured by a vacuum/cold trap method, where the hose is placed in a controlled temperature/humidity chamber and the ingressed moisture is collected in vacuum cold trap maintained at approximately $-90\,°F$. Moisture vapor transmission rates of 0.02 to 0.12 $g/cm^2 \cdot year$ have been measured for hose materials currently used in nonhermetic systems.

The preparation of this chapter is assigned to TC 3.3, Contaminant Control in Refrigerating Systems.

Drying equipment in the factory is discussed in Chapter 21 of the 1988 ASHRAE *Handbook—Equipment*. Proper installation and service procedures eliminate the sources listed in items 2, 3, and 4. Lubricants are discussed in Chapter 8 of this volume. If purchased refrigerants and oils meet specifications and are properly handled, the moisture content generally remains satisfactory. See the section on hermetic motor insulation in Chapter 6 and the section on motor burnouts in this chapter.

EFFECTS OF MOISTURE

Excess moisture in a refrigerating system can cause one or all of the following undesirable effects:

1. Ice formation in expansion valves, capillary tubes, or evaporators
2. Corrosion of metals
3. Copper plating
4. Chemical damage to insulation in hermetic compressors or other system materials

Ice or solid hydrate separates from refrigerants if the water concentration is high enough and the temperature low enough. Solid hydrate, a complex molecule of refrigerant and water, can form at temperatures higher than those required to separate ice. Liquid water forms at temperatures above those required to separate ice or solid hydrate. Ice forms during refrigerant evaporation when the relative saturation of vapor reaches 100% at temperatures of 32°F or below.

The separation of water as ice or liquid also is related to the solubility of water in a refrigerant. This solubility varies for different refrigerants and with temperature. Data for the more important refrigerants are presented in Table 1. The greater the solubility of water in a refrigerant, the less the possibility that ice or liquid water will separate in a refrigerating system. The solubility of water in ammonia, carbon dioxide, and sulfur dioxide is so high that ice or liquid water separation is not a problem.

Elsey and Flowers (1949) recognized that the concentration of water by mass at equilibrium is greater in the gas phase than in the liquid phase of Refrigerant 12. This same relation holds for all commercial halocarbon refrigerants, with the exception of Refrigerants 22 and 502. The ratio of weight concentrations differs for

Table 1 Solubility of Water in the Liquid Phase of Certain Refrigerants

Temperature, °F	Solubility, ppm by weight							
	R-11	R-12	R-13	R-21	R-22	R-113	R-114	R-502
160	460	700	—	4700	4100	460	450	1780
150	400	560	—	4200	3600	400	380	1580
140	340	440	—	3600	3150	344	320	1400
130	290	350	—	3200	2750	290	270	1220
120	240	270	—	2700	2400	240	220	1080
110	200	210	—	2350	2100	200	180	930
100	168	165	—	1980	1800	168	148	810
90	140	128	—	1700	1580	140	120	690
80	113	98	—	1410	1350	113	95	580
70	90	76	—	1150	1140	90	74	490
60	70	58	44	960	970	70	57	400
50	55	44	—	800	830	55	44	335
40	44	32	26	645	690	44	33	278
30	34	23.3	—	512	573	34	25	225
20	26	16.6	14	398	472	26	18	180
10	20	11.8	—	314	384	20	13	146
0	15	8.3	7	260	308	15	10	115
−10	11	5.7	—	205	244	11	7	90
−20	8	3.8	3	158	195	8	5	69
−30	6	2.5	—	118	152	6	3	53
−40	4	1.7	1	90	120	—	2	40
−50	3	1.1	—	66	91	—	1.5	30
−60	2	0.7	—	49	68	—	1	22
−70	1	0.4	—	35	50	—	0.6	16
−80	0.8	0.3	—	25	37	—	0.4	11
−90	0.5	0.1	—	17	27	—	0.2	8
−100	0.3	0.1	—	12	19	—	0.1	5

Data adapted from E. I. duPont de Nemours & Company, Inc., and Industrial Chemicals Division, Allied Corporation. Used by permission.

Table 2 Distribution of Water Between the Vapor and Liquid Phases of Certain Refrigerants

Temperature, °F	H$_2$O in Vapor, mass % ÷ H$_2$O in Liquid, mass %					
	R-11	R-12	R-22	R-113	R-114	R-502
−40	—	17.1	0.203	—	52.2	0.40
−20	71.0	15.3	0.251	172.6	37.4	0.47
0	62.4	13.1	0.301	140.9	32.1	0.53
20	57.9	11.9	0.351	120.7	29.5	0.61
30	—	11.2	0.378	111.4	26.9	0.65
40	50.5	9.89	0.390	99.3	24.7	0.66
50	—	9.0	0.391	89.3	21.7	0.66
60	43.1	8.2	0.401	79.0	19.7	0.66
70	—	7.5	0.404	68.6	17.6	0.66
80	34.8	6.3	0.405	61.4	16.0	0.65
90	—	6.1	0.397	54.6	14.4	0.64
100	30.1	5.5	0.400	50.2	13.3	0.63

Data adapted from E. I. duPont de Nemours & Company, Inc. Used by permission.

each refrigerant; it also varies with temperature. Table 2 shows the distribution ratios of water in the vapor phase to water in the liquid phase for common refrigerants. It can be used to calculate the equilibrium water concentration of the liquid phase refrigerant if the gas phase concentration is known, and vice versa. Water content in the vapor phase is determined by:

$$W = [P_w/(P_w)^0][(d_w)^0/(d_R)^0] \quad (1)$$

where

P_w = partial pressure of water vapor
$(P_w)^0$ = partial pressure of water vapor at saturation
$(d_w)^0$ = density of water vapor at saturation
$(d_R)^0$ = density of refrigerant vapor at saturation

Freezing at expansion valves or capillary tubes can occur when excessive moisture is present in a refrigerating system. Formation of ice or hydrate in evaporators can partially insulate the evaporator. Walker *et al.* (1962) showed that excess moisture can cause corrosion and enhance copper plating. Other factors affecting copper plating are discussed in Chapter 6.

The moisture required for freeze up is a function of the amount of flash gas formed during expansion and the distribution of water between the liquid and gas phases downstream of the expansion device. For example, in an R-12 system with a 110°F liquid temperature and a −20°F evaporator temperature, the refrigerant after expansion is 41.3% vapor and 58.7% liquid by weight. The percentage of vapor formed is determined by:

$$\% \text{ Vapor} = 100 \frac{h_{L(\text{liquid})} - h_{L(\text{evap})}}{h_{fg(\text{evap})}} \quad (2)$$

where

$h_{L(\text{liquid})}$ = saturated liquid enthalpy for refrigerant at liquid temperature
$h_{L(\text{evap})}$ = saturated liquid enthalpy for refrigerant at evaporating temperature
$h_{fg(\text{evap})}$ = latent heat of vaporization of refrigerant at evaporating temperature

Table 1 lists the saturated water content of the liquid phase at −20°F as 3.8 ppm. Table 2 is used to determine the saturated vapor phase water content as:

$$3.8 \times 15.3 = 58 \text{ ppm}$$

When the vapor contains more than the saturation quantity (100% rh), free water will be present as a third phase. If the temperature is below 32°F, ice will form. Using the saturated moisture values and the liquid-vapor ratios, the critical water content of the circulating refrigerant can be calculated as:

$$3.8 \times 0.587 = 2.2$$
$$58.0 \times 0.413 = \underline{24.0}$$
$$26.2 \text{ ppm}$$

Moisture and Other Contaminant Control in Refrigerant Systems

Table 3 Survey Results on Moisture in Refrigeration Systems

Application	Max Tolerable Moisture, ppm by Weight		
	R-12	R-22	R-502
Domestic refrigerators and freezers	—	—	—
Food display cases	—	—	—
Frozen food cases	10[a]	50 to 100	25
Room air conditioners	15	50 to 200	—
Residential and package air conditioners	20	50 to 150	—
Automotive air conditioners	60	—	—

[a]Or lower for low-temperature applications.

Maintaining moisture levels below critical value keeps free water from the low side of the system.

The above analysis can be applied to all refrigerants and applications. An R-22 system with 110°F liquid and −20°F evaporating temperatures reaches saturation when the moisture circulating is 139 ppm. Note that this value is less than the liquid solubility, 195 ppm at −20°F.

Excess moisture causes paper or polyester motor insulation to become brittle, which can cause premature motor failure. However, not all motor insulations are affected adversely by moisture. The amount of water present in a refrigerant system must be small enough to avoid ice separation, corrosion, and insulation breakdown.

Exact experimental data on the maximum permissible moisture level in refrigerant systems are not known because so many factors are involved. Table 3 shows the results of an opinion survey on moisture. Where no values are shown, the refrigerant normally is not used for the given application.

MOISTURE INDICATORS

Moisture-sensitive elements that change color according to moisture content can gauge the moisture level in the system; the color changes at a low enough level to be safe. Manufacturer's instructions must be followed since the color change point is also affected by the liquid line temperature and the refrigerant used.

DRYING METHODS

Factory dehydration of refrigeration and air-conditioning units is discussed in Chapter 21 of the 1988 ASHRAE *Handbook—Equipment*. Field systems are dried by evacuation and driers. After installation or extensive service work, all field systems are evacuated to remove the air. Evacuation should be done with a high vacuum pump that reduces the absolute pressure to 500 μm of mercury or less. When the internal pressure is reduced below the boiling point of water at the existing temperature, any water inside the system vaporizes and is carried to the vacuum pump. Frequently, external heat is required to vaporize the water. Even with these procedures, small amounts of moisture trapped under an oil film, adsorbed by the motor windings, or located far from the vacuum pump will be difficult to remove.

It is good practice to install a drier. On larger systems a drier with a replaceable element is frequently used, and may need to be changed several times before the proper degree of dryness is obtained. A moisture indicator in the liquid line can indicate when the system has been dried satisfactorily.

If large quantities of water from a burst tube or water chiller leak enters a refrigeration or air-conditioning system, parts of the system may have to be disassembled and the water drained from system low points. If the system contains a semihermetic compressor, it should be cleaned frequently by disassembling and hand wiping the various parts. After reassembly, it should be dried further by heating and evacuation, and by changing driers during the initial operating period.

When conditions in the field do not permit heating of the equipment or proper drainage, anhydrous methanol (methyl alcohol) can be used to flush out the moisture from a seriously contaminated system (Hoffman 1971). The water in the system is dissolved by the circulating alcohol. The alcohol-water solution has a lower boiling point than water alone, and the residue can be removed more readily than water by a stream of dry air, carbon dioxide, oil-pumped nitrogen, or by the application of a vacuum. Because flushing with methyl alcohol is a fire hazard, this procedure is used only under extreme conditions. Alcohol should never be used in hermetic compressors with polyester insulations.

MOISTURE MEASUREMENT

Techniques for measuring the amount of moisture in a compressor, or in an entire system, are discussed in Chapter 21 of the 1988 ASHRAE *Handbook—Equipment*. The following methods are used to measure the moisture content of various halocarbon refrigerants. The moisture content to be measured is generally in the ppm range, and the procedures require special laboratory equipment and techniques.

The *Gravimetric Method* is used extensively for measuring the moisture content of refrigerants and is used in ASHRAE *Standards* 63-79 and 35-1983. In this method, a measured amount of refrigerant vapor is passed through two tubes in series, each containing phosphorous pentoxide (P_2O_5). Any moisture present in the refrigerant reacts chemically with the P_2O_5 and appears as a weight increase in the first tube. The second tube is used as a tare. This method is satisfactory when the refrigerant is pure, but the presence of oil will produce inaccurate results, since it will be weighed as moisture. Approximately 200 g of refrigerant are required for accurate results. Since the refrigerant must pass slowly through the tube, analysis requires many hours of elapsed time. In spite of its limitations, this method is considered the primary standard for laboratory refrigerant moisture analysis.

The *Karl Fischer Method* is suitable for measuring the moisture content of a refrigerant, even if it contains oil. The refrigerant sample is bubbled through predried methyl alcohol in a special sealed glass flask; any water present remains with the alcohol. Karl Fischer reagent is added, and the solution is immediately titrated to a "dead stop" electrometric end point. The Karl Fischer reagent reacts with any moisture present. The amount of water in the sample can be calculated from a previous calibration of the Karl Fischer reagent. This method, considered among the most accurate, is also suitable for measuring the moisture content of pure oil or other liquids. Special instruments designed for this particular analysis are available from laboratory supply companies. Haagen-Smit *et al.* (1970) describe improvements in the equipment and technique that significantly reduce analysis time.

DeGeiso and Stalzer (1969) discuss the *electrolytic moisture analyzer*. Refrigerant vapor is passed over a thin hygroscopic film of phosphoric acid on an electrometric probe. A meter measures the electrical conductivity of this film, which varies according to the moisture content. Oil must not coat the detector; if the detector is contaminated with oil, it must be carefully cleaned and recoated. The instrument requires at least 0.0176 ft^3 of vapor for an analysis. The analysis with this instrument is so rapid that it can be used to monitor moisture changes within a system. Brisken (1955) used this method in a study of moisture migration in hermetic systems. His work describes the sampling technique and running time necessary to obtain reproducible samples from an operating refrigeration system.

Another method, the infrared spectrophotometer, is used for moisture analysis, but requires a large sample for precise results and is subject to interference if oil is present in the refrigerant. In applications where a simple indication of "wet" or "dry" is satisfactory, commercial moisture indicators work satisfactorily.

DESICCANTS

Desiccants used in refrigeration systems adsorb or react chemically with the moisture contained in a liquid or gaseous refrigerant-oil mixture. Solid desiccants, used widely as dehydrating agents in refrigerant systems, remove moisture from both new or field-installed equipment. The desiccant is contained in a mechanical device called a drier or receiver-drier and can be installed in either the liquid or the suction line of a refrigeration system.

Desiccants must remove water and not react unfavorably with any other materials in the system. Activated alumina, silica gel, and molecular sieves are the most widely used desiccants acceptable for refrigerant drying. These materials remove moisture from refrigeration systems by physical adsorption. They are available in granular, bead, and block forms. The manufacturer should be consulted in selecting any desiccant.

Table 4 Reactivation of Desiccants

Desiccant	Temperature, °F
Activated alumina	400 to 600
Silica gel	350 to 600
Molecular sieves	500 to 660

Combinations of desiccants can be used in a single drier and may have certain advantages over a single desiccant because they can adsorb a greater variety of refrigeration contaminants. Two combinations are activated alumina with molecular sieves and silica gel with molecular sieves. Activated carbon is also used in some mixtures.

Solid core desiccants, or block forms, consist of desiccant granules held together by a binder (Walker 1963). The binder is usually a nondesiccant material. Suitable filtering action, adequate contact of the desiccant with the refrigerant, and low pressure drop are obtained by properly sizing the desiccant particles used to make up the core, and by the proper geometry of the core with respect to the flowing refrigerant.

Desiccants that take up water by chemical reaction are not recommended. Calcium chloride reacts with water to form a corrosive liquid. Barium oxide is known to cause explosions. Magnesium perchlorate and barium perchlorate are powerful oxidizing agents, which are potential explosion hazards in the presence of oil. Phosphorous pentoxide is an excellent desiccant, but its fine powdery form makes it difficult to handle and produces a high resistance to gas and liquid flow. A mixture of calcium oxide and sodium hydroxide has limited use, but is not recommended as a desiccant.

Desiccants are extremely sensitive to moisture and must be protected against it at all times until ready for use. The desiccant should be obtained in a factory-packed unit and handled under moisture-proof conditions. If a desiccant has picked up moisture, it can be reactivated by heating for about 4 h at a suitable temperature, preferably with a dry air purge or in a vacuum oven (see Table 4). Only adsorbed water is driven off at the temperatures listed, and the desiccant is returned to its initial reactive state. This does not include alumina and silica gels, where removing the water of constitution destroys the desiccants. Molecular sieves have no water of constitution, but can be destroyed by excessive temperature. A desiccant on a refrigerating equipment drier should not be reactivated for reuse, because oil and other contaminants are present.

Equilibrium Conditions of Desiccants

Desiccants function on the equilibrium principle. If a freshly activated desiccant contacts a moisture-laden refrigerant, the water is adsorbed from the refrigerant-water mixture onto the desiccant surface until the vapor pressures of the adsorbed water (*i.e.*, at the desiccant surface) and the water remaining in the refrigerant are in equilibrium. Conversely, if the vapor pressure of the water on the desiccant surface is higher than that in the refrigerant, water is released into the refrigerant-water mixture, and the same equilibrium point will be established if the temperature is constant.

Adsorbent desiccants function by holding (adsorbing) moisture on their internal surfaces. The amount of water adsorbed from a refrigerant by an adsorbent at equilibrium is determined by pore volume, pore size, surface characteristics, and the temperature and moisture content of the refrigerant.

Equilibrium curves for the various adsorbent desiccants with R-12 and R-22 are presented in Figures 1, 2, and 3. These curves (adsorption isotherms) are based on the technique developed by Gully, Tooke, and Bartlett (1954), as modified by ASHRAE *Standard* 35-1983. ASHRAE *Standards* 35-1983 and 63-79 define the moisture content of the refrigerant as End Point Dryness (EPD), and the moisture held by the desiccant as water capacity. The curves show that for any specified amount of water in a particular refrigerant, the desiccant holds a corresponding specific quantity of water.

Figure 1 shows moisture equilibrium curves for three common adsorbent desiccants in drying R-12 at 75 °F. Figure 2 presents data for drying R-22 with these desiccants also at 75 °F. As shown, the capacity of a desiccant can vary widely for different refrigerants when the same EPD is required. Generally, a refrigerant in which moisture is more soluble requires more desiccant for adequate drying than one with less solubility requires.

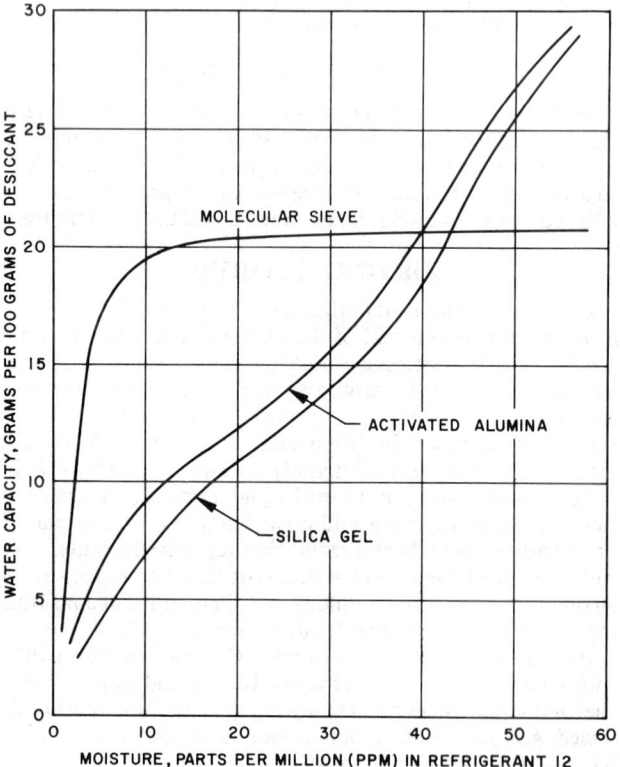

Fig. 1 Moisture Equilibrium Curves for R-12 and Three Common Desiccants at 75 °F

Moisture and Other Contaminant Control in Refrigerant Systems

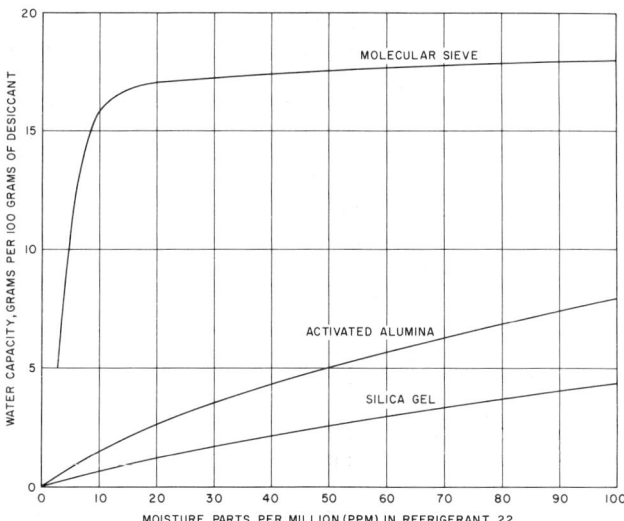

Fig. 2 Moisture Equilibrium Curves for R-22 and Three Common Desiccants at 75 °F

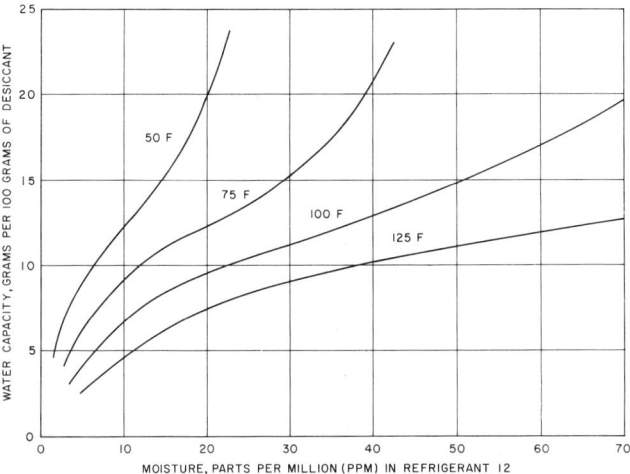

Fig. 3 Moisture Equilibrium Curves for Activated Alumina at Various Temperatures in R-12

Figure 3 shows the effect of temperature on moisture equilibrium capacities of activated alumina and R-12. Much higher water capacities are obtained at lower temperatures, demonstrating the advantage of locating the alumina driers at relatively cool spots in the system. When using molecular sieves, the effect of temperature on water capacity is much less because of the shape of the equilibrium curve. ARI *Standard* 710-86 requires determining the water capacity for R-12 at an EPD of 15 ppm, and for R-22 at 60 ppm. Each determination must be made at 75 °F (see Figures 1 and 2) and 125 °F.

Although the figures show that molecular sieves have higher capacities than activated alumina or silica gel at the indicated EPD, all three desiccants are suitable if sufficient quantities are used. Cost, operating temperature, other contaminants present, and equilibrium capacity at the desired EPD must be considered when choosing a desiccant for refrigerant drying. The desiccant manufacturer has information and equilibrium curves for any specific desiccant-refrigerant system.

DESICCANT APPLICATIONS

In addition to removing water, desiccants may be capable of adsorbing or reacting with acids, dyes, chemical additives, refrigerant oil reaction products, and colloids.

Acids

The quantity of acid that a refrigerant system tolerates depends in part on the size, the mechanical design and construction of the system, the type of motor insulation, the type of acid, and the quantity of water in the system. Generally, acids can harm refrigerant systems. Desiccants remove acids by adsorption and/or chemical reaction.

The acid capacity of desiccants in a refrigeration system is difficult to determine because the environment is complex. Refrigerant systems can have different refrigerants, oils, and concentrations of refrigerant and oil circulating through the drier, all of which affect the rate and capacity of desiccants to remove acids. In addition, acids formed in these systems can be inorganic, such as HCL and HF, or a mixture of organic acids. All factors must be considered to establish acid capacities of desiccants.

Hoffman and Lange (1962), and Mays (1962) showed that acids are removed from refrigerants and oils by adsorption and/or chemical reaction. Hoffman also showed that the concentration of water in the desiccant, the type of desiccant, and the type of acid play a major role in a desiccant's ability to remove acids from refrigerant systems.

Colors

Colored materials frequently are adsorbed by activated alumina and silica gel and occasionally by calcium sulfate and molecular sieves. Leak detector dyes may lose their effectiveness when used in systems containing desiccants. The interaction of the dye and drier should be evaluated before putting a dye in the system.

Oil Deterioration Products

Oils in refrigerant systems can react chemically to produce substances that are adsorbed by desiccants. Some of these are hydrophobic and, when adsorbed by the desiccant, reduce the rate at which it can adsorb liquid water. The rate and capacity of the desiccant to remove water dissolved in the refrigerant are not significantly impaired, however (Walker *et al.* 1955). Frequently, the reaction products are sludges or powders that can be filtered out mechanically by the drier.

Chemicals

Refrigerants that can be adsorbed by desiccants cause the drier temperature to rise considerably when the refrigerant is first admitted. This temperature rise is not the result of moisture in the refrigerant, but the adsorption heat of the refrigerant. Oil additives may be adsorbed by silica gel and activated alumina. Methanol is coadsorbed with moisture and competes with it for adsorption capacity in desiccants. Because of small pore size, molecular sieves generally do not adsorb additives or the oil.

DRIERS

A drier is a mechanical device containing a desiccant. It collects and holds moisture, but it also acts as a moisture adsorber, filter, and adsorber of acids and other contaminants.

To prevent moisture from freezing in the expansion valve or capillary tube, a drier is installed in the liquid line close to these devices. Hot locations should be avoided. Driers can function on the low side of the metering devices, but this is not the preferred location (Jones 1969).

Moisture is reduced to a low level by one pass of the liquid refrigerant through a drier. The moisture is usually distributed throughout the entire system, and time is required for the circulating oil-refrigerant mixture to carry the moisture to the drier. Krause *et al.* (1960) showed that considerable time is required to produce moisture equilibrium in a refrigeration unit.

Driers are also used effectively to clean up hermetic motor burnouts. The use of both liquid and suction line driers can prevent repeat burnouts. Replaceable element-type driers have been used in the suction line after a hermetic motor burnout as a "system cleaner." Service personnel prefer to use sealed driers in the suction line to protect the new compressor. This results in a lower overall cost, since the drier can be left in the system and no callback is required.

DRIER SELECTION

The drier manufacturer selection chart lists the amount of desiccants, flow capacity, filter area, water capacity, and a specific recommendation covering the type and refrigeration capacity of the drier for various applications.

The manufacturer must consider the following factors when designing and producing a drier:

1. The *desiccant* is the heart of the drier and its selection is most important. The section on desiccants has further information.
2. The drier's *water capacity* is measured by methods described in ARI *Standard* 710-80. The reference points are set arbitrarily to prevent confusion arising from determinations made at other points. The specific refrigerant, the amount of desiccant, and the effect of temperature are all considered in the statement of water capacity.
3. The *liquid line flow capacity* is listed at 2 psi pressure drop across the drier by the official procedures of ASHRAE *Standard* 63-79 and ARI *Standard* 710-80. Jones (1964) developed a gravity flow method for determining flow capacities that uses R-113 and converts the results to R-12 and R-22. Rosen *et al.* (1965) described a closed-loop method for evaluating filtration and flow characteristics of liquid line refrigerant driers. The flow capacity of suction line filters and filter-driers is determined according to ASHRAE *Standard* 78-74 and ARI *Standard* 730-80. The latter standard gives recommended pressure drops for selecting suction line filter-driers for permanent and temporary installations. Flow capacity may reduce quickly when critical quantities of solids and semisolids are filtered out by the drier. The amount or time of occurrence cannot be predicted. Whenever flow capacity drops below the machine's requirements, the drier should be replaced.
4. Although limits for particle size vary with refrigerant system size and design, and with the geometry and hardness of the particles, manufacturers publish *filtration capabilities* in the low micrometre range, which is acceptable to the industry.

TESTING AND RATING

Desiccants and driers are tested according to the procedures of ASHRAE *Standards* 35-1983 and 63-79, respectively. Driers are rated under ARI *Standard* 710-80. Minimum standards for listing of refrigerant driers can be found in nonelectrical UL 207, *Standard for Refrigerant Containing Components*. Filtration ratings and test standards have not been developed.

OTHER CONTAMINANTS

Refrigerant filter-driers are the principal devices used to remove contaminants from refrigeration systems. The filter-drier is not a substitute for poor workmanship or design, but a maintenance tool necessary for continued and proper system performance. Contaminants removed by filter-driers include moisture, acids, high-molecular-weight hydrocarbons, oil decomposition products, and insoluble material, such as metallic particles and copper oxide.

METALLIC CONTAMINANTS AND DIRT

Small contaminant particles frequently left in refrigerating systems during manufacture or servicing include chips of copper, steel, or aluminum; copper or iron oxide; copper or iron chloride; welding scale; brazing or soldering flux; sand; and other dirt. Some of these contaminants, such as copper chloride, develop from normal wear or chemical breakdown during system operation. Solid contaminants vary widely in size, shape, and density.

Solid contaminants create problems by:

1. Scoring cylinder walls and bearings
2. Lodging in the motor insulation of a hermetic system, where they act as conductors between individual motor windings or abrade the wire coating when flexing of the windings occurs
3. Depositing on terminal blocks and serving as a conductor
4. Plugging expansion valve screen or capillary tubing
5. Depositing on suction or discharge valve seats, significantly reducing compressor efficiency
6. Plugging oil holes in compressor parts, leading to improper lubrication
7. Increasing the rate of chemical breakdown in the system. At elevated temperatures, R-12 and R-22 decompose more readily when in contact with iron powder, iron oxide, or copper oxide (Norton 1957).

Liquid line filter-driers, suction filters, and strainers isolate contaminants from the compressor and expansion valve. Filters minimize the return of particulate matter to the compressor and expansion valve; but the capacity of permanently installed liquid and/or suction filters must accommodate this particulate matter without causing excessive, energy-consuming pressure losses in the refrigeration system. Equipment manufacturers should use the following procedures to ensure proper operation during the system's design life:

1. Develop system cleanliness specifications for production systems that include a practical value for maximum residual matter. Some manufacturers specify allowable quantities in terms of internal surface area.
2. Multiply the factory contaminant level by a factor of five to allow for solid contaminants that will be added during installation. The safety factor depends on the type of system and the previous experience of the installers, among other considerations.
3. Determine a value for maximum pressure drop to be incurred by the suction or liquid filter when loaded with the quantity of solid matter calculated in Step 2.
4. Conduct pressure drop tests by adding contaminants. The chosen test fluid and its flow rates should be predictable in terms of typical system flow rates and refrigerant. Contaminants in the test system should be graded for particle size and distribution and held constant from test to test.
5. Select driers for each system according to its capacity requirements and test data. In addition to contaminant removal capacity, tests can evaluate filter efficiency, maximum escaped particle size, and average escaped particle size.

Very small particles passing through filters tend to accumulate in the crankcase. Most compressors tolerate a small quantity of these particles without allowing them into the oil pump inlet, where they can damage running surfaces. If the compressor will not tolerate these particles, or if they become excessive, the particles can be reduced by changing the oil.

ORGANIC CONTAMINANTS—SLUDGE, WAX, AND TARS

Organic contaminants in a refrigerating system can appear when organic materials, such as oil, insulation, varnish, gaskets,

Moisture and Other Contaminant Control in Refrigerant Systems

and adhesives decompose. As opposed to inorganic contaminants, these materials are mostly carbon, hydrogen, and oxygen. Organic materials may be partially soluble in the refrigerant-oil mixture or may become so through the action of heat. They then circulate in the refrigerating system and can plug small orifices.

Some organic contaminants remain in a new refrigerating system during manufacture or assembly. For example, excessive solder paste introduces a wax-like contaminant into the refrigerant stream. Certain cutting oils, corrosion inhibitors, or drawing compounds frequently contain wax-like material. Normal fabricating methods can leave very small amounts of these oils deposited on the copper tubing. These contaminants are a mixture of substances that usually have no definite melting point. Organic contamination also results during the normal method of fabricating return bends. The die used during forming is lubricated with oil or other organic materials, and afterwards the return bend is brazed to the tubes to form a condenser. During brazing, residual lubricant inside the tubing and bends is baked to a resinous deposit.

If organic materials are handled improperly, certain contaminants remain in a system. Resins used in varnishes, wire coating, or casting sealers may not be cured properly and can dissolve in the refrigerant-oil mixture. Solvents used in washing stators may be adsorbed by the wire film and later, during compressor operation, carry chemically reactive organic extractables. Chips of varnish, insulation, or fibers can detach and circulate in the system. Portions of improperly selected or cured rubber parts or gaskets can dissolve in the refrigerant.

Refrigeration grade oil decomposes under adverse conditions to form a resinous liquid or a solid frequently found on refrigeration filter-driers. These oils decompose noticeably when exposed for as little as 2 h to temperatures as low as 250°F in an atmosphere of air or oxygen. The compressor manufacturer should perform all high-temperature dehydrating operations on the machines prior to adding the oil charge. In addition, equipment manufacturers should not expose compressors to processes requiring high temperatures unless the compressors contain refrigerant or inert gas.

The result of organic contamination is frequently noticed at the expansion device. Resinous or wax-like material dissolved in the refrigerant-oil mixture, under liquid line conditions, may precipitate out at the lower temperature in the expansion device, resulting in restricted or plugged capillary tubes or sticky expansion valves (Zahorsky 1967). Wax-like material, weighing less than a gram with a volume less than that of a few grains of rice, can render a system inoperative. These materials have physical properties that range from a fluffy powder to a solid resin entraining inorganic dirt. If the contaminant is dissolved in the refrigerant-oil mixture in the liquid line, it usually will not be removed by a filter-drier.

Chemical identification of these organic contaminants is very difficult. Infrared spectroscopy can characterize the type of organic groups present in contaminants. Materials found in actual systems vary from wax-like aliphatic hydrocarbons to resin-like materials containing double bonds, carbonyl groups and carboxyl groups. In some cases, organic compounds of copper or iron have been identified.

These contaminants can be eliminated by carefully selecting refrigeration system materials and strictly controlling cleanliness during manufacture and assembly. Because heat degrades most organic materials and enhances chemical reactions, operating conditions with excessively high discharge temperatures must be avoided to prevent formation of degradation products.

RESIDUAL SOLVENTS

Solvents used for cleaning compressor parts are likely contaminants in refrigerating systems. Solvents in this category are considered pure liquids without additives. If additives are present, they are reactive materials and should not be in a refrigerating system. Some solvents are relatively harmless to the chemical stability of the refrigerating system, while others initiate or accelerate degradation reactions. For example, the common mineral spirits solvents, such as *Stoddard Solvent,* are considered harmless. Carbon tetrachloride, on the other hand, reacts rapidly with hydrocarbon lubricating oils (Elsey *et al.* 1952). Parmelee (1965) reported the effect of residual solvents, using the sealed tube method containing R-12, oil, and 1% of solvent in addition to aluminum and/or copper and/or iron. General ratings are:

R-113 < R-11 < trichloroethylene < methyl chloroform
(Least harmful)   (Most harmful)

Generally, even the least harmful chlorinated solvents should be removed from the system.

During manufacture, solvents are used to degrease or clean parts and are removed in the regular process operations. Extreme care should be used in washing porous or absorbent parts to avoid carrying residual solvents into the system. Such parts as motors are very absorbent, and the safest solvents are mineral spirits, naphtha, and R-113. If residual solvents are entrapped, only the least reactive and the most similar to mineral spirits should be used. Residues from solvents, emulsion cleaners, or rustproofing agents could contain materials that are chemically very reactive.

Flushing the parts with a safe solvent and a final cleaning with the refrigerant is recommended for field or installation cleaning. In these and other operations, only minimal residues of even the more stable solvents should remain. Ideally, refrigerant and oil are the only liquids in refrigeration equipment.

ANTIFREEZE AGENTS (METHYL ALCOHOL)

Antifreeze agents are sometimes added to refrigerating systems to act as cosolvents for the small quantities of water present. They prevent ice formation at the expansion device but add to the contaminants already present in the system.

Methyl alcohol has shown no adverse effects when used in the proportion of 3 cm^3 per pound of R-12. Larger concentrations of methyl alcohol can cause corrosion problems, especially where aluminum is present, because it will be attacked and hydrogen gas will form.

If similar systems, one dry and one with 0.05 to 1% methyl alcohol, are test-run side by side, the alcohol system will always have more stain, corrosion, copper plating, and debris on fine filter screens than the dry system. However, these systems still perform as well as those containing no methyl alcohol.

NONCONDENSABLE GASES

Gases, other than the refrigerant, are a contaminant frequently found in refrigerating systems. These gases result: (1) from incomplete evacuation, (2) when functional materials release sorbed gases or decompose to form gases at an elevated temperature during system operation, (3) through low side leaks, and (4) from chemical reactions during system operation. Chemically reactive gases, such as hydrogen chloride, attack other components in the refrigerating system; in extreme cases, the refrigerating unit fails.

Chemically inert gases in the system, which do not liquefy in the condenser, reduce the cooling efficiency. The quantity of inert, noncondensable gas that is harmful depends on the design and size of the refrigerating system and the nature of the refrigerant. Its presence contributes to higher than normal head pressures and resultant higher discharge temperatures. Higher temperatures speed up undesirable chemical reactions.

Gases found in hermetic refrigeration units include nitrogen, oxygen, carbon dioxide, carbon monoxide, methane, and hydrogen. The first three gases listed originate from incomplete air evacuation or a low side leak in the system. Carbon dioxide and carbon monoxide usually form when organic insulation materials are overheated. Hydrogen has been detected when a compressor is experiencing serious bearing wear. Only trace amounts of these gases are present in well-designed, properly functioning equipment.

Spauschus and Olsen (1959), Doderer and Spauschus (1966), and Gustafsson (1977) developed sampling and analytical techniques for establishing the quantities of contaminant gases present in refrigerating systems. Parmelee (1965), Spauschus and Doderer (1961, 1964), and Kvalnes (1965) applied gas analysis techniques to sealed tube tests to yield information on stability limitations of refrigerants, in conjunction with other materials used in hermetic systems.

MOTOR BURNOUTS

Motor burnout is the final result of hermetic motor insulation failure. During burnout, high temperatures and arc discharges severely deteriorate the insulation, producing large amounts of carbonaceous sludge, acid, water, and other contaminants, and some deterioration of the refrigerant and oil. The products of burnout escape into the system, causing severe cleanup problems. If decomposition products are not removed from the system, replacement motors fail with increasing frequency.

To clean a system after a burnout, differentiate between mild and severe burnout. A rapid burn from a spot failure in the motor winding results in a mild burnout with little oil discoloration and no carbon deposits. A severe burnout occurs when the compressor remains on line and burns over a longer period, resulting in highly discolored oil, carbon deposits, and acid formation.

Since the condition of the oil indicates the amount of system contamination, examine the oil during the cleanup process. Wojtkowski (1964) has shown that acid in the oil should not exceed 0.05 acid number. Commercial acid test kits can be used for this analysis.

Various methods are recommended for cleaning a system after hermetic motor burnout (RSE 1978). However, the suction filter-drier method is commonly used (see Appendix).

FIELD ASSEMBLY

Proper field assembly and maintenance are essential for contaminant control in refrigerating systems and to prevent undesirable refrigerant emissions to the atmosphere. Refrigeration components that are adequate for intended design limits can be misapplied or mismatched. Connecting piping may be too large or too small to carry the necessary gas and still circulate oil properly. Driers may be too small, or carelessly handled so that drying capacity is lost. Improper tube-joint soldering is a major source of water, flux, and oxide scale contamination. Copper oxide scale from improper brazing is one of the most frequently found contaminants. Careless tube cutting and handling can introduce excessive quantities of dirt and metal chips.

Because dehydration of the assembled system is not easy in the field, oversized driers are recommended. Even if manufacturers' components are delivered sealed and bone-dry, the weather and the open-time during assembly can introduce considerable quantities of moisture that must be removed. Inaccurate control settings, dirt-fouled air condensers, scaled or corroded heat exchangers, and improper evacuation can cause any system to break down.

APPENDIX

SYSTEM CLEANUP PROCEDURE AFTER HERMETIC MOTOR BURNOUT

Introduction

A. This procedure is limited to positive displacement (reciprocating or rotary) hermetic compressors. Centrifugal compressor systems are highly specialized and are frequently designed for a particular application. A centrifugal system should be cleaned according to the manufacturer's recommendations.

B. After a hermetic motor burnout, the system must be cleaned thoroughly to remove all contaminants. Otherwise, a repeat burnout will occur. Failure to follow these minimum cleanup recommendations as quickly as possible increases the potential for repeat burnout.

C. Flushing the system with R-11 or similar refrigerants has been used to some extent. Although this method is satisfactory under certain conditions, it has many limitations. Therefore, the flushing method is seldom used and is not recommended here.

Procedure

A. *Make sure a burnout has occurred.* Although a motor that will not start appears to be a motor failure, the problem may be improper voltage, starter malfunction, or a compressor mechanical fault.

1. To check for proper voltage, turn off the main disconnect switch so that all power is off. Remove the compressor leads at the compressor side of the starter. Close the disconnect switch to energize the control circuit and check for voltage on all lines at both the line and load side of the starter.

2. Before checking the compressor motor, make sure the compressor is cool to the touch. Otherwise, a false indication can be obtained since the internal motor protectors are open.

3. Check the compressor motor to see if it is electrically grounded or open. A 500 megohmmeter or an ohmmeter can be used for this test. If no fault is found and if the normal values for winding resistance are known, check motor resistance with a precision ohmmeter to determine if turn-to-turn shorts exist. Winding resistance varies 2.25% for each 10°F difference from the 77°F at which published values are usually obtained.

Before pronouncing the hermetic motor unfit, determine whether a ground or open circuit exists in the external leads from the starter to the compressor, in the compressor terminals, or in the internal leads to the starter.

4. Purge a small quantity of refrigerant gas from the compressor and smell it cautiously. A motor burnout is usually indicated by a characteristic burned odor.

B. *Safety.* In addition to electrical hazards, service personnel should be aware of the hazard of acid burns.

1. When testing for odor, release a small amount of gas and smell it cautiously to avoid inhaling toxic decomposition products.

2. If it is necessary to touch the oil or sludge in a burned-out compressor, wear rubber gloves to avoid a possible acid burn.

C. *Determine the severity of the burnout.* Classify burnouts as *mild* or *severe,* and use the severity as a guide for the cleanup procedure to be followed. The severity can be determined by the following procedure:

1. Obtain a small sample of the oil from the burned-out compressor and analyze it using an acid test kit. Excessive acidity (over 0.05 acid number) in the oil indicates a severe burnout. Discoloration of the oil may also indicate a severe burnout.

2. Release a small amount of refrigerant and smell it. A characteristic burned odor indicates a severe burnout.

3. Inspect the suction line at the compressor and the liquid line drier. Carbon deposits indicate a severe burnout.

4. If none of the above indications of severe contamination are found, the burnout can be classified as mild.

D. *Cleanup after a mild burnout.* When the burnout is mild, the contaminants can be removed by changing the liquid line filter-drier, or by installing one if the system does not have one. The procedure to follow is:

Moisture and Other Contaminant Control in Refrigerant Systems

1. System refrigerant should be isolated within the system or recovered into an external storage container to avoid discharge into the atmosphere.

2. Remove the burned-out compressor and install the replacement.

3. Remove the liquid line filter-drier and install an oversized replacement drier.

4. Evacuate the system or portion opened to the atmosphere according to the manufacturer's recommendations.

5. Recharge the system and begin operation according to the manufacturer's startup instructions.

6. If the system has service valves, the refrigerant can be saved by closing the service valves and pumping down the system with the new compressor to change the liquid line filter-drier.

E. *Cleanup after a severe burnout.* The system must be completely cleaned. Using a suction line filter-drier in addition to a liquid line filter-drier is recommended as follows:

1. System refrigerant should be isolated within the system or recovered into an external storage container to avoid discharge into the atmosphere.

2. Remove the burned-out compressor and install the replacement. Install a filter-drier in the suction line to protect the new compressor from any contaminants remaining in the system. Leaving a permanent filter-drier in the suction line allows service personnel to complete the cleanup at one time. Install a pressure tap upstream of the filter-drier. This tap permits measuring the pressure drop from the tap to the service valve during the first hours of operation to determine if the suction line filter-drier becomes plugged because of excessive contaminants remaining in the system. Follow the manufacturer's recommendations on pressure drop.

3. Remove the old liquid line filter-drier, if one exists, and install a replacement drier of the next larger capacity than is normal for this system.

4. Check the expansion device and clean or replace it.

5. Evacuate the system or portion opened to the atmosphere according to the manufacturer's recommendations.

6. Recharge the system and begin operation according to the manufacturer's startup instructions. Check the pressure drop across the suction line filter-drier during the first hours of operation, changing the cores if necessary.

F. *Additional suggestions*

1. If sludge or carbon has backed up into the suction line, swab it out or replace that section of the line.

2. Install a moisture indicator in the liquid line to make sure the system is dry.

3. If a change of cores in the suction line filter-drier is required, change the oil in the compressor each time the cores are changed, if the compressor design permits.

4. If the system has service valves, save the refrigerant by the following procedure: Close the service valves, change the compressor, evacuate the new compressor, open the service valves, and use the new compressor to pump down the system; change the liquid line filter-drier, and install a suction line filter-drier. This procedure can be used on a mild or severe burnout.

5. Evaluate an oil sample from the system after one day's operation and again after two weeks of operation. Provide some means for taking oil samples. Observe the oil color and test it with an acid test kit to measure the degree of acidity. If the oil is either dirty or acidic, change the suction line filter-drier and change the oil in the system before checking an additional sample. For best possible results, the cleanup is not considered complete until the oil is clean and free of acid (less than .05 acid number).

6. Remove the suction line filter-drier after several weeks of system operation to avoid excessive pressure drop in the suction line. This problem is particularly significant on commercial refrigeration systems.

7. In some cases, noncondensable gases are produced during the burnout. Compare the measured heat pressure with the pressure equivalent to the condensing temperature. If the head pressure is excessive, the system should be purged.

Special System Characteristics and Procedures

Because of unique system characteristics, the procedures described here may require adaptations.

A. If an oil sample cannot be obtained from the new compressor, determine another method to get a sample from the system.

1. Install a tee and a trap in the suction line. An access valve at the bottom of the trap permits easy oil drainage. Only 0.5 oz of oil is required for an acid analysis. Be certain the oil sample represents oil circulating in the system. It may be necessary to drain the trap and discard the first amount of oil collected, before collecting the sample to be analyzed.

2. Make a trap from 1 3/8-in. copper tubing and valves. Attach this trap to the suction and discharge gauge port connections with a charging hose. By blowing discharge gas through the trap and into the suction valve, enough oil will be collected in the trap for analysis. This trap becomes a tool that can be used repeatedly on any system that has suction and discharge service valves.

B. On semihermetic compressors, remove the cylinder head to determine the severity of burnout. Dismantle the compressor for solvent cleaning and hand wiping to remove contaminants. Consult the manufacturer's recommendations on compressor rebuilding and motor replacement.

C. In rare instances on a close-coupled system, where it is not feasible to install a suction line filter-drier, the system can be cleaned by repeated changes of the cores in the liquid line filter-drier and repeated oil changes.

D. On heat pump systems, the four-way valve and the compressor should be carefully inspected after a burnout. In cleaning a heat pump after a motor burnout, it is essential to remove any drier originally installed in the liquid line. These driers may be replaced for cleanup, or a reversible filter-drier may be installed in the common reversing liquid line.

E. Systems with a critical charge require a particular effort for proper operation after cleanup. If an oversized liquid line filter-drier is installed, an additional charge must be added. However, no additional charge is required for the suction line filter-drier that may be added to the system.

F. The new compressor should not be used to pull a vacuum. Refer to the manufacturer's recommendations for evacuation. Normally one of two methods is used, after determining that there are no refrigerant leaks in the system.

1. Pull a high vacuum to an absolute pressure of less than 500 microns of mercury for several hours. Allow the system to stand several hours to be sure the vacuum is maintained. This requires a good vacuum pump and an accurate high vacuum gauge.

2. Use a vacuum pump to pull a vacuum to an absolute pressure of at least 27 in. of mercury, vacuum. Break this vacuum with refrigerant and allow the system to stand for 1 h. Reevacuate the system, and repeat the procedure two more times for a triple evacuation. If a fully halogenated CFC refrigerant is used in the system, a different medium such as R-22 should be used prior to evacuation, and the fully halogenated CFC refrigerant should be used for final charging. The refrigerant used in intermediate evacuations should be recovered.

REFERENCES

ARI. 1986. Flow-capacity rating and application of suction-line filters and filter-driers. *Standard* 730-86. Air-Conditioning and Refrigeration Institute, Arlington, VA.

ARI. 1986. Liquid-line driers. Standard 710-86. Air-Conditioning and Refrigeration Institute, Arlington, VA.

ASHRAE. 1979. Method of testing liquid line refrigerant driers. ASHRAE *Standard* 63-79.

ASHRAE. 1983. Standard method of testing desiccants for refrigerant drying. ASHRAE *Standard* 35-83.

Brisken, W.R. 1955. Moisture migration in hermetic refrigeration systems as measured under various operating conditions. *Refrigerating Engineering* (July):42.

DeGeiso, R.C. and R.F. Stalzer. 1969. Comparison of methods of moisture determination in refrigerants. ASHRAE *Journal* (April).

Doderer, G.C. and H.O. Spauschus. 1966. A sealed tube-gas chromatograph method for measuring reaction of Refrigerant 12 with oil. ASHRAE *Transactions* 72(2):IV,4.1.

Elsey, H.M. and L.C. Flowers. 1949. Equilibria in Freon-12—Water systems. *Refrigerated Engineering* (February):153.

Elsey, H.M., L.C. Flowers, and J.B. Kelley. 1952. A method of evaluating refrigerator oils. *Refrigerating Engineering* (July):737.

Gully, A.J., H.A. Tooke, and L.H. Bartlett. 1954. Desiccant-refrigerant moisture equilibria. *Refrigerating Engineering* (April):62.

Gustafsson, V. 1977. Determining the air content in small refrigeration systems. Purdue Compressor Technology Conference.

Haagen-Smit, I.W., P. King, T. Johns, and E.A. Berry. 1970. Chemical design and performance of an improved Karl Fischer titrator. *American Laboratory* (December).

Hoffman, J.E. 1971. Caution: Alcohol at work. *Refrigeration Service and Contracting* (August).

Hoffman, J.E. and B.L. Lange. 1962. Acid removal by various desiccants. ASHRAE *Journal* (February):61.

Jones, E. 1964. Determining pressure drop and refrigerant flow capacities of liquid line driers. ASHRAE *Journal* (February):70.

Jones, E. 1969. Liquid or suction line drying? *Air Conditioning and Refrigeration Business* (September).

Krause, W.O., A.B. Guise, and E.A. Beacham. 1960. Time factors in the removal of moisture from refrigerating systems with desiccant type driers. ASHRAE *Transactions* 66:465.

Kvalnes, D.E. 1965. The sealed tube test for refrigeration oils. ASHRAE *Transactions* 71(1):138.

Mays, R.L. 1962. Molecular sieve and gel-type desiccants for Refrigerants 12 and 22. ASHRAE *Journal* (August):73.

Norton, F.J. 1957. Rates of thermal decomposition of $CHCIF_2$ and CF_2Cl. *Refrigerating Engineering* September):33.

Parmelee, H.M. 1965. Sealed tube stability tests on refrigeration materials. ASHRAE *Transactions* 71(1):154.

Rosen, S., *et al.* 1965. A method of evaluating filtration and flow characteristics of liquid line driers. ASHRAE *Transactions* 71(1):200.

Spauschus, H.O. and G.C. Doderer. 1961. Reaction of Refrigerant 12 with petroleum oils. ASHRAE *Journal* (February):65.

Spauschus, H.O. and G.C. Doderer. 1964. Chemical reactions of Refrigerant 22. ASHRAE *Journal* (October):54.

Spauschus, H.O. and R.S. Olsen. 1959. Gas analysis—A new tool for determining the chemical stability of hermetic systems. *Refrigerating Engineering* (February):25.

Standard procedure for servicing hermetic motor burnouts. 1978. *Service Manual*, Section 91:9101. Refrigeration Service Engineers Society, Chicago.

Walker, W.O. 1963. Latest ideas in use of desiccants and driers. *Refrigerating Service & Contracting* (August):24.

Walker, W.O., J.M. Malcolm, and H.C. Lynn. 1950. Hydrophobic behavior of certain desiccants. *Refrigerating Engineering* (April):50.

Walker, W.O., S. Rosen, and S.L. Levy. 1962. Stability of mixtures of refrigerants and refrigerating oils. ASHRAE *Journal* (August):59.

Wojtkowski, E.F. 1964. System contamination and cleanup. ASHRAE *Journal* (June):49.

Zahorsky, L.A. 1967. Field and laboratory studies of wax-like contaminants in commercial refrigeration equipment. ASHRAE *Transactions* 73(1):II,1.1.

BIBLIOGRAPHY

Dunne, S.R. and T.J. Clancy. 1984. Methods of testing desiccant for refrigeration drying. ASHRAE *Transactions* 90(1A):164.

Zhukoborshy, S.L. 1984. Application of natural zeolites in refrigeration industry. Proceedings of the International Symposium on Zeolites, Portoroz, Yugoslavia (September).

CHAPTER 8

LUBRICANTS IN REFRIGERANT SYSTEMS

Tests for Boundary Lubrication 8.1	*Oil Return from Evaporators* 8.13
Refrigeration Oil Requirements 8.2	*Wax Separation (Floc Tests)* 8.17
Mineral Oil Composition 8.2	*Solubility of Hydrocarbon Gases* 8.19
Component Characteristics 8.3	*Solubility of Water in Oils* 8.19
Synthetic Oils 8.3	*Solubility of Air in Oil* 8.19
Oil Additives 8.4	*Foaming and Antifoam Agents* 8.19
Oil Properties 8.4	*Oxidation Resistance* 8.20
Oil-Refrigerant Solutions 8.7	*Chemical Stability* 8.20

THE primary function of a lubricant is to reduce friction and minimize wear. A lubricant achieves this by interposing a film between moving surfaces that reduces direct solid-to-solid contact and is itself easily sheared.

Understanding the role of a lubricant requires an analysis of the surfaces to be lubricated. While bearing surfaces and other machined parts may appear and feel smooth, close examination reveals microscopic peaks (asperities) and valleys. With a sufficient quantity of lubricant, a layer is provided that has a molecular thickness greater than the maximum height of the mating asperities, so that moving parts ride on a lubricant cushion.

Ideal conditions are not always easily attained. For example, when the shaft of a horizontal journal bearing is at rest, the static loads squeeze out the lubricant, producing a discontinuous film with metal-to-metal contact at the bottom of the shaft. When the shaft begins to turn, there is no layer of liquid lubricant separating the surfaces.

As the shaft picks up speed, the lubricating fluid is drawn into the converging clearance between the bearing and the shaft, generating a hydrodynamic pressure that eventually can support the load on an uninterrupted fluid film.

Various regimes or conditions of lubrication can exist when surfaces are in motion with respect to one another. Regimes of lubrication are defined as follows:

Full Fluid Film. Mating surfaces are completely separated by the lubricant film.
Boundary. Gross surface-to-surface contact occurs because the bulk lubricant film is too thin to separate the mating surfaces.
Mixed Film. Occasional or random surface contact occurs.

A wide variety of materials can be used to separate and lubricate contacting surfaces. Separation can be a boundary layer on a metal surface, a fluid film, or a combination of both. The function of a lubricant extends beyond preventing surface contact. It also removes heat, provides a seal to keep out contaminants or retain pressures, prevents corrosion, and disposes of debris created by wear. Lubricating oils are best suited to meet these various requirements.

Viscosity is the most important factor to consider in choosing a lubricating oil under full fluid film conditions. Under boundary conditions, the asperities are the contact points and support much, if not all, of the load. The contact pressures are usually sufficient to cause welding and surface deformation. However, wear can be controlled effectively with nonfluid, multimolecular films formed on the surface. These films must be strong enough to resist rupturing, yet have acceptable frictional and shear characteristics. These films reduce surface fatigue, adhesion, abrasion, and corrosion, which are the four major sources (either singularly or together) of rapid wear under boundary conditions. The slightly active constituents left in commercially refined mineral oils give them their natural film-forming properties.

Additives have also been developed to improve lubrication under boundary conditions. These materials are characterized by terms such as oiliness agents, lubricity improvers, and film strength enhancers. They form a film on the metal surface through polar attraction or chemical action. These films or coatings yield or flow better under the shear stress imposed by boundary conditions. In chemical action, the frictional heat between the contacting surfaces energizes the reaction between the additive and the metal surface. Films such as iron sulfide and iron phosphate can be formed depending on the additives and the energy available for the reaction. In some instances, organic phosphates and phosphites are used in refrigeration oils to improve boundary lubrication. The nature of the metal and the condition of the metal surfaces are more important. Refrigeration compressor designers often treat ferrous pistons, shafts, and wrist pins with phosphating processes that impart a crystalline, discontinuous film of metal phosphate to the surface. This film aids boundary lubrication during the initial break-in period.

TESTS FOR BOUNDARY LUBRICATION

Film strength or load carrying ability are terms often used to describe oil lubricity characteristics under boundary conditions. Laboratory tests that measure the degree of scoring, welding, or wear have been developed to evaluate oils. Some of these tests have been standardized by the American Society of Testing and Materials (ASTM) and other organizations. The following tests evaluate lubricant performance.

In the *Four-ball* method (D-2783), the antiwear property is determined from the average scar diameter on the stationary balls and is stated in terms of a load-wear index. The smaller the scar, the better the load-wear index. The maximum load carrying capability is defined in terms of a weld point; *i.e.*, the load at which welding by frictional heat occurs. The *Falex* method (D-2670) allows measurement of wear during the test itself, and the scar

---
The preparation of this chapter is assigned to TC 3.4, Lubrication.

width on the V-blocks and/or the mass loss of the pin can be used as a measure of the antiwear properties. The load carrying capability is determined from a failure, which can be caused by excess wear or extreme frictional resistance. The *Timken* method (D-2782) determines the load at which rupture of the oil film occurs, and the *Alpha LFW-1* machine (D-2714) measures frictional force and wear.

However, since all these machines operate in air, available data may not apply to a refrigerant environment. Divers (1958) questioned the validity of tests in air because several of the oils that performed poorly in Falex testing have always been used successfully in refrigerant systems. Murray *et al.* (1956) suggest that halocarbon refrigerants can aid in boundary lubrication. Refrigerant 12, for example, when run hot in the absence of oil, reacted with steel surfaces to form a lubricating film. Studies emphasize the need for laboratory testing in a simulated refrigerant environment.

In Huttenlocher's (1969) method of simulation, refrigerant vapor is bubbled through the oil reservoir before the test to displace the dissolved air. It is continued during the test to maintain a blanket of refrigerant on the oil surface. Using the Falex tester, Huttenlocher has shown the beneficial effect of Refrigerant 22 on the load carrying capability of the same lubricant compared with air or nitrogen. Sanvordenker and Gram (1974) describe a further modification of the Falex test using a sealed sample system. Both R-12 and R-22 atmospheres had beneficial effects on an oil's boundary lubrication characteristics when compared with tests in air.

Test parameters must simulate as closely as possible the system conditions in the base material from which the test specimens are made, their surface condition, the processing methods, and the operating temperature. There are several bearings or rubbing surfaces in a refrigerant compressor, each of which may use different materials and may operate under different conditions. A different test may be required for each bearing. Moreover, hermetic compressors are precision devices with internal and bearing clearances often controlled to within several ten-thousandths of an inch. Permissible bearing wear is minimal because wear debris remains in the system and can cause other problems even if the clearances stay within working limits. Compressor system mechanics must be understood to perform and interpret simulated tests.

Some aspects of compressor lubrication are not suitable for laboratory simulation. One aspect, the return of liquid refrigerant to the compressor, can cause the oil to dilute or wash away from the bearings, creating conditions of boundary lubrication. Tests using operating refrigerant compressors have also been considered, and one such wear test has been proposed as a German Standard (DIN 8978). The test is a functional for a given compressor system and may permit comparison of lubricants within that class of compressors. However, it is not designed to be a generalized test for the boundary lubricating capability of an oil. Other tests using radioactive tracers in refrigerant systems have given useful results (Rembold and Lo 1966).

REFRIGERATION OIL REQUIREMENTS

Refrigeration compressors are classified as continuous or dynamic and positive displacement. Dynamic types such as the centrifugal compressor depend on energy transfer from a rotating set of blades to the gas. Momentum imparted to the gas is converted to useful pressure by decelerating the gas. Positive displacement compressors can be either reciprocating or rotary. Both designs confine discrete volumes of gas within a closed space and then elevate pressure by reducing volume.

Refrigeration systems require oil to do more than lubricate. Oil seals compressed gas between the suction and discharge sides, and acts as a coolant to remove heat from the bearings and transfer heat from the crankcase to the compressor exterior. Oil also reduces noise generated by moving parts inside the compressor.

Although the compression components of centrifugal compressors require no internal lubrication, rotating shaft bearings, seals, and couplings must be adequately lubricated. Turbine or other types of oils can be used when the oil is not in contact or circulated with refrigerant gas.

Chapter 12 of the 1988 ASHRAE *Handbook—Equipment* describes how reciprocating and rotary compressors are lubricated. The lubricant lubricates moving parts, helps seal internal clearances, and cools the refrigerant.

Hermetic systems, where the motor is exposed to the oil, require oil with electrical insulating properties. The refrigerant gas carries some oil with it into the condenser and evaporator. This oil must return to the compressor within a reasonable time and must have adequate fluidity at low temperatures. The oil should remain miscible with the refrigerant for good heat transfer in the evaporator and for good oil return. The oil must be free of suspended matter or components such as wax that might clog the expansion tube or deposit in the evaporator and interfere with heat transfer. In hermetic refrigeration systems, the oil is charged only once, so it must function for the lifetime of the compressor. The chemical stability required of the oil in the presence of refrigerants, metals, motor insulation, and extraneous contaminants is perhaps the most important characteristic distinguishing refrigeration oils from those used for all other applications (see Chapter 6).

As expected, an ideal oil does not exist; a compromise must be made to balance the requirements. A high viscosity oil seals the gas pressure best, but may offer more frictional resistance. Slight foaming can reduce noise, but excessive foaming can carry too much oil into the cylinder and cause structural damage. Oils that are most stable chemically are not necessarily good lubricants. The oils should not be considered alone, since the functions of the lubricant are performed by oil-refrigerant mixtures, not by pure oils.

The behavior of an oil in refrigeration systems depends on composition and structure. It is difficult to define the composition and to relate it to the oil's performance in refrigerant systems. Understanding refrigeration oil behavior requires information on its composition, how it is processed, and how its composition affects its properties.

An oil used in refrigeration applications must be evaluated on the basis of its physical and chemical properties. Many of these properties can be determined by standard procedures available from ASTM publications. Among these are: (1) viscosity, (2) viscosity index, (3) color, (4) specific gravity, (5) refractive index, (6) pour point, (7) aniline point, (8) oxidation resistance, (9) dielectric breakdown voltage, (10) foaming tendency in air, (11) moisture content, and (12) wax separation. Other properties, particularly those involving interactions with refrigerants, must be determined by special tests described in refrigeration literature. Among these nonstandard properties are: (1) mutual solubility with various refrigerants, (2) chemical stability in the presence of refrigerants and metals, (3) chemical effects of contaminants or additives that may be in the oils, (4) boundary film forming ability, and (5) solubility of air.

MINERAL OIL COMPOSITION

For practical purposes, the numerous compounds in refrigeration oils of mineral origin can be grouped into the following structures: (1) paraffins, (2) naphthenes (cycloparaffins), (3) aromatics, and (4) nonhydrocarbons. *Paraffins* consist of all straight chain and branched carbon chain saturated hydrocarbons. N-pentane and isopentane are examples of hydrocarbons. *Naphthenes* are also completely saturated but consist of cyclic or ring structures; cyclopentane is a simple example. *Aromatics* are unsaturated cyclic hydrocarbons containing one or more rings characterized by alter-

Lubricants in Refrigerant Systems

nate double bonds; benzene is a typical example. The *nonhydrocarbons* are molecules containing atoms such as sulfur, nitrogen, or oxygen in addition to carbon and hydrogen.

The preceding structural components do not necessarily exist in pure states. In fact, a paraffinic chain frequently is attached to a naphthenic or aromatic structure. Similarly, a naphthenic ring to which a paraffinic chain is attached may in turn be attached to an aromatic molecule. Because of such complications, mineral oil composition is usually described by carbon-type and molecular analysis.

In carbon-type analysis, the number of carbon atoms on the paraffinic chains, naphthenic structure, and aromatic rings is determined and represented as the percentage of the total. Thus, $\%C_P$, or the percentage of carbon atoms having a paraffinic configuration, includes not only the free paraffins but also those paraffinic chains attached to naphthenic or to aromatic rings. Similarly, $\%C_N$ includes the carbon atoms on the free naphthenes as well as those on the naphthenic rings attached to the aromatic rings, and $\%C_A$ represents the carbon atoms on the aromatic rings. Carbon analysis describes an oil in its fundamental structure and correlates and predicts many physical properties of the oil. However, direct methods of determining carbon composition are laborious. Therefore, it is common practice to use a correlative method, such as the one based on the refractive index-density-molecular weight (n-d-m) (Van Nes and Weston 1951), or one standardized by ASTM D-2140 or ASTM D-3288. Other valuable methods are ASTM D-2008, which uses ultraviolet absorbence and a rapid method employing infrared spectrophotometry and calibration from known oils.

Molecular analysis is based on methods of separating the structural molecules. For refrigeration oils, important structural molecules are: (1) saturates or nonaromatics, (2) aromatics, and (3) nonhydrocarbons. All the free paraffins and naphthenes (cycloparaffins), as well as mixed molecules of paraffins and naphthenes are included in the saturates. However, any paraffinic and naphthenic molecules attached to an aromatic ring are classified as aromatics. This representation of oil composition is less fundamental than carbon analysis. However, many properties of the oil relevant to refrigeration can be explained with this analysis, and the chromatographic methods of analysis are fairly simple (ASTM D-2549, ASTM D-2007, Mosle and Wolf 1963, Sanvordenker 1968).

The traditional classification of oils as paraffinic or naphthenic refers to the preponderance of paraffinic or naphthenic molecules in refined oil. This terminology is used since paraffinic crudes usually have a lower aromatic content than naphthenic crudes.

COMPONENT CHARACTERISTICS

Saturates have excellent chemical stability, but poor solubility with polar refrigerants, such as R-22; they are also poor boundary lubricants. Aromatics are somewhat more reactive but have very good solubility with refrigerants and good boundary lubricating properties. Nonhydrocarbons are the most reactive but are beneficial for boundary lubrication, although the amounts needed for that purpose are small (Steinle 1964). The reactivity, solubility, and boundary lubricating properties of a refrigeration oil are affected by the relative amounts of these components in the oil.

The saturate and aromatic fractions separated from an oil do not have the same viscosity as the parent oil. The saturate fraction is much less viscous, while the aromatic fraction is much more viscous than the parent oil. Both fractions have the same boiling range. Thus, for this range, the aromatics have a higher viscosity than the saturates. For the same viscosity, the aromatics have a lower boiling range, *i.e.*, higher volatility, than the saturates. Also, the saturate fraction has a lower density and a lower refractive index, but higher viscosity index and molecular mass, than the aromatic fraction of the same oil.

Among the saturates, the free, straight chain paraffins are undesirable for refrigeration applications because they precipitate as wax crystals when the oil is cooled to its pour point. The branched chain paraffins and naphthenes are less viscous at low temperatures and have extremely low pour points.

Nonhydrocarbons are mostly removed during refining of refrigeration oils and have little effect on the physical properties of the oil, except perhaps on the color. However, since all the nonhydrocarbons (*e.g.*, sulfur compounds) are not dark, even a colorless oil does not necessarily guarantee the absence of nonhydrocarbons. Kartzmark *et al.* (1967) and Mills and Melchoire (1967) found indications that nitrogen-bearing compounds cause or act as catalysts toward the deterioration of oils. The sulfur and oxygen compounds are thought to be less reactive, with some types considered to be natural inhibitors.

Solvent refining is often used to remove more thermally unstable aromatic and unsaturated compounds from the base stock by a liquid-liquid extraction process using a solvent that selectively dissolves such compounds.

The properties of the components naturally are reflected in the parent oil. An oil with a very high saturate content, as is frequently the case with paraffinic oils, also has a high viscosity index, low specific gravity, high molecular weight, low refractive index, and low volatility. In addition, it would have a high aniline point and would be less miscible with polar refrigerants. The reverse is true of naphthenic oils. Table 1 lists typical properties of several refrigeration oils.

SYNTHETIC OILS

The limited solubility of mineral oils with R-13, R-22, and R-502 has led to the investigation of synthetic oils for refrigeration use. Of the available types, alkylbenzenes perform satisfactorily.

There are two basic types of alkylbenzenes—branched and linear. The products are synthesized by reacting an olefin or chlorinated paraffin with benzene in the presence of a catalyst. Catalysts commonly used for this reaction are aluminum chloride and hydrofluoric acid.

In addition to the type of olefin or chlorinated paraffin used, the alkylbenzene structure is affected by the catalyst ($AlCl_3$ or HF) as well as other processing variables. After the catalyst is removed, the product is distilled into fractions. The relative size of these fractions can be changed by adjusting the molecular weight of the side chain (olefin or chlorinated paraffin) and by changing other variables. Therefore, the amounts of the middle and heavy fractions produced depend largely on demand and economics. Alkylbenzene-based refrigeration oils are available and sold commercially in the United States. In addition to good solubility with highly fluorinated refrigerants, these lubricants have better high temperature and oxidation stability than mineral oil-based refrigeration oils. The quality of alkylbenzene refrigeration oil varies, depending on the type (branched or linear) and manufacturing scheme. Typical properties for a branched alkylbenzene are shown in Table 1.

Gunderson and Hart (1962) describe other commercially available synthetic oils. Many have properties suited to refrigeration purposes, such as: synthetic paraffins, polyglycols, dibasic acid esters, neopentyl esters, silicones, silicate esters, and fluorinated compounds. Sanvordenker and Larime (1972) describe the properties of these synthetic oils, alkylbenzenes, and phosphate esters in regard to refrigeration application. The phosphate esters are unsuitable for refrigeration use because of their poor thermal stability. Although very stable and compatible with refrigerants, the fluorocarbon oils are expensive. Among others, only the synthetic paraffins have poor miscibility relations with R-22. Dibasic acid esters, neopentyl esters, silicate esters, and polyglycols all have excellent viscosity temperature relations and remain miscible with

Table 1 Typical Properties of Refrigeration Oils

Property	ASTM[a]	Naphthenic Oils					Paraffinic Oils			Alkylbenzene
Viscosity, cs (SSU) at 100°F	D-445	33.1(155)	32.2(151)	36.2(169)	61.9(287)	68.6(318)	34.2(160)	44.1(205)	112.8(523)	31.7(149)
Viscosity index	D-2270	0	35	60	0	46	95	91	92	27
Density	D-1298	0.913	0.892	0.889	0.917	0.900	0.862	0.869	0.879	0.872
Color	D-1500	0.5	1.0	1.5	1.0	1.0	1.0	1.0	1.0	0.5
Refractive index	D-1747	1.5015	1.4883	1.4850	1.5057	1.4918	1.4752	1.4793	1.4836	—
Molecular weight	D-2503	300	318	332	321	345	378	394	474	320
Pour point, °F	D-97	−45	−45	−40	−40	−35	0	0	0	−50
Floc point, °F	—	−68	−60	−26	−60	−60	−31	−39	−18	−100
Flash point, °F	D-92	340	345	356	360	400	395	420	485	350
Fire point, °F	D-92	390	375	—	400	450	450	490	550	365
Carbon-type composition	n-d-m									
%C_A	Van Nes and	14	7	5	16	7	3	5	4	24
%C_N	Weston (1951)	43	46	45	42	46	32	31	32	None
%C_P		43	47	50	42	47	65	64	64	76
Molecular composition	D-2549									
% Saturates		62	78	88	59	78	87	85	83	None
% Aromatics		38	22	12	41	22	13	15	17	100
Aniline point, °F	D-611	160	188	207	165	197	220	220	244	125
Critical solution temperature with R-22, °F	—	25	48	65	35	74	81	110	140	−100

[a] All designated ASTM methods refer to the current version.

R-22 and R-502 to very low temperatures. These and the alkylbenzenes are considered suitable for low-temperature applications.

Extreme caution must be exercised in considering a synthetic oil for refrigeration purposes. In mineral oils, characteristics such as hydrolytic stability or thermal decomposition are taken for granted and are only cursorily examined. Within the same chemical class, the properties of industrial synthetic compounds can vary widely depending on the compound's structure. Every conceivable property should be determined for each formulation, and extensive accelerated system tests should be made prior to actual field use. Otherwise, unforeseen field problems can occur, as was the case of polybutyl silicate described by Downing and Cooper (1972).

OIL ADDITIVES

Additives can enhance certain oil properties or impart new characteristics depending on their chemical composition and the nature of the base oil. They generally fall into three groups: polar compounds, polymers, and compounds containing active elements such as sulfur or chlorine. Additive types include: (1) pour point depressants, (2) floc point depressants, (3) viscosity index improvers, (4) thermal stability improvers, (5) extreme pressure and antiwear additives, (6) rust inhibitors, (7) antifoam agents, (8) metal deactivators, (9) dispersants, and (10) oxidation inhibitors. Some additives offer performance advantages in one area but are detrimental in another. Some additives work best when combined with other additives. They must be compatible with system materials and be present in the optimum concentration; too little may be ineffective, while too much can be detrimental or offer no incremental improvement.

In general, additive-type oils are not required to lubricate a refrigerant compressor. However, oils that contain additives give highly satisfactory service. Their use is justifiable as long as the user knows of their presence; they offer performance advantages, and their composition remains unaltered.

An additive-type oil should only be used after thorough testing to determine whether the additive material (1) is removed by system driers, (2) is inert to system components, (3) is soluble in refrigerants at low temperatures so as not to cause deposits in capillary tubes or expansion valves, and (4) is stable at high temperatures to avoid adverse chemical reactions such as harmful deposits. This can best be done by sealed tube and compressor testing using the actual additive/base oil combination intended for field use.

OIL PROPERTIES

Viscosity and Viscosity Grades

Viscosity defines a fluid's resistance to shearing force. It is expressed as absolute viscosity (centipoises, cps), kinematic viscosity (centistokes, cs), or Saybolt Seconds Universal viscosity (SSU). Conversion from SSU to centistokes can be made from tables contained in ASTM D-2161, but density must be known to convert kinematic viscosity to absolute viscosity. Refrigeration oils are sold in viscosity grades, and ASTM has proposed a system of standardized viscosity grades for industry-wide use (D-2422).

In selecting a viscosity grade, consider the environment to which the oil will be exposed. The viscosity of the lubricating fluid decreases if temperatures rise or if the refrigerant dissolves appreciably in the oil.

A large reduction in viscosity can affect the lubricating function or, more likely, the sealing function of the lubricant, depending on the nature of the machinery. The design of some types of hermetically sealed units, such as the single-vane rotary type, requires the lubricating fluid to act as an efficient sealing agent. In reciprocating compressors, the lubricant film is spread over the

Lubricants in Refrigerant Systems

entire area of contact between the piston and the cylinder wall, and there is a very large area to resist leakage from the high to the low-pressure side. In a single-vane rotary type, however, the critical sealing area is a line contact between the vane and a roller. In this case, viscosity reduction is serious.

The oil with the lowest viscosity that gives the necessary sealing properties with the refrigerant used for the entire range of temperatures and pressures encountered should be chosen. A practical method for determining the minimum safe viscosity is to calculate the total volumetric efficiency of a given compressor system, using several oils of widely varying viscosities. The oil of lowest viscosity that gives satisfactory volumetric efficiency should be selected. Tests should be run at a number of ambient temperatures, for example, 70, 90, and 110°F. As a guideline, Table 2 lists the viscosity ranges recommended for various refrigeration systems.

The International Standards Organization (ISO) has established a series of definite viscosity levels as a standard for specifying or selecting lubricant for industrial applications. This system, covered in the United States by ASTM D-2422, is designed to eliminate intermediate or unnecessary viscosity grades while providing enough viscosity grades for operating equipment. The system reference point is kinematic viscosity at 104°F, and each viscosity grade with suitable tolerances is identified by kinematic viscosity at this temperature in centistokes (cs). Therefore an ISO VG 32 grade oil would identify an oil with a viscosity grade of 32 cs at 104°F. Table 3 is a chart of the various standardized viscosity grades of lubricants.

Table 2 Recommended Viscosity Ranges

Small Systems

Refrigerant	Type of Compressor	Oil Viscosities at 100°F SSU	cs
Ammonia	Screw	280-300	60-65
Ammonia	Reciprocating	150-300	32-65
Carbon dioxide	Reciprocating	280-300[a]	60-65[a]
Refrigerant 11	Centrifugal	280-300	60-65
Refrigerant 12	Reciprocating	150-300	32-65[b]
Refrigerant 12	Centrifugal	280-300	60-65[b]
Refrigerant 12	Rotary	280-300	60-65[b]
Refrigerant 22	Reciprocating	150-300	32-65
Other halogenated refrigerant types	Reciprocating	150-300[a]	32-65[a]
	Centrifugal	280-300	60-65
	Rotary	280-300	60-65
Halogenated refrigerants	Screw	150-4000	32-800

Industrial Refrigeration

(Ammonia and carbon dioxide compressors with splash, force-feed, or gravity circulating systems)

Type of Compressor	Oil Viscosities at 100°F SSU	cs
Where oil may enter refrigeration system or compressor cylinders	150-300	32-65
Where oil is prevented from entering system or cylinders:		
In force-feed or gravity systems	500-600	108-129
In splash systems	150-160	32-34
Steam-driven compressor cylinders when condensate is reclaimed for ice-making	High viscosity oil with 140-165 SSU at 210°F	30-35 at 210°F

[a]Some applications may require lighter oils of 75-85 SSU (14-17 cs); others, heavier oils of 500-600 SSU (108-129 cs).
[b]Automotive applications using R-12 may require 100 cs (450-550 SSU) oil.

Table 3 Viscosity System for Industrial Fluid Lubricants (ASTM-D2422)

Viscosity System Grade Identification	Midpoint Viscosity cs at 104°F	Kinematic Viscosity Limits cs at 104°F Minimum	Maximum	Approximate Equivalents, SSU Units
ISO VG 2	2.2	1.98	2.42	32
ISO VG 3	3.2	2.88	3.52	(a)
ISO VG 5	4.6	4.14	5.06	40
ISO VG 7	6.8	6.12	7.48	(a)
ISO VG 10	10	9.00	11.00	60
ISO VG 15	15	13.50	16.50	75
ISO VG 22	22	19.80	24.20	105
ISO VG 32	32	28.80	35.20	150
ISO VG 46	46	41.40	50.60	215
ISO VG 68	68	61.20	74.80	315
ISO VG 100	100	90	110	465
ISO VG 150	150	135	165	700
ISO VG 220	220	198	242	1000
ISO VG 320	320	288	352	1500
ISO VG 460	460	414	506	2150
ISO VG 680	680	612	748	3150
ISO VG 1000	1000	900	1100	4650
ISO VG 1500	1500	1350	1650	7000

[a]The 36 and 50 SSU grades are not currently considered standardized grades in the United States.

Viscosity Index

Mineral oil viscosity decreases as the temperature increases and increases as the temperature decreases. The relationship between temperature and kinematic viscosity is represented by:

$$\log \log (v + 0.7) = A + B \log T$$

where

v = kinematic viscosity, cs
T = thermodynamic temperature, K or °R
A, B = constants for each oil

This relationship is the basis for the viscosity-temperature charts published by ASTM and permits a straight line plot of viscosity over a wide temperature range. Figure 1 shows a plot for two different mineral oils. This plot is applicable over the temperature range in which the oils are homogenous liquids.

The slope of the vicosity-temperature lines is different for different oils. The viscosity-temperature relationship of an oil is described by an empirical number called the Viscosity Index (VI)

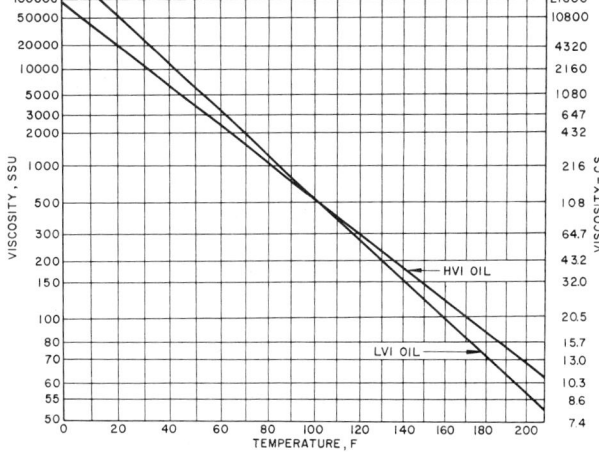

Fig. 1 Viscosity-Temperature Chart for 108 cs (500 SSU) HVI and LVI Oils

(ASTM D-2270). An oil with a high viscosity index (HVI) shows less change in viscosity over a given temperature range than an oil with a low viscosity index (LVI). In the example shown in Figure 1, both oils possess equal viscosities [109 cs (504 SSU)] at 100°F. However, the viscosity of the LVI oil, as shown by the steeper slope of the line, decreases to 7.9 cs (52.4 SSU) at 210°F, whereas the HVI oil viscosity drops only to 11.1 cs (62.5 SSU) at the same temperature.

The viscosity index is related to the composition of the oil. Generally, an increase in cyclic structure, aromatic and naphthenic, decreases the viscosity index. Paraffinic oils usually have a high viscosity index and low aromatic content. Naphthenic oils, on the other hand, have a lower viscosity index and are usually higher in aromatics. For the same base oil, the viscosity index decreases with increasing aromatic content.

Superficially, the HVI oils appear to be advantageous for refrigeration. Except for the effect on viscosity by the dissolved refrigerant, this would be true. Under some circumstances, such as where the oil is prevented from entering the evaporator, an HVI oil may be preferred to an LVI oil. However, when the oil circulates through the system, the usual practice is to use an LVI oil because the dissolved refrigerant affects viscosity more than the temperature does.

Density

To convert viscosity units *e.g.*, from centistokes (kinematic viscosity) to centipoises (absolute viscosity), the density of the lubricating oil must be known. Figure 2 shows published values for pure oil densities over a range of temperatures. These density temperature curves all have approximately the same slope and appear merely to be displaced from one another. If the density of a particular oil is known at one temperature but not over a range of temperatures, a reasonable estimate at other temperatures can be obtained by drawing a line paralleling those in Figure 2.

Density indicates the composition of a mineral oil. Naphthenic oils are usually more dense than paraffinic oils. Also, the higher the aromatic content, the higher the density. For equivalent compositions, higher viscosity oils have higher densities, but the change in density with aromatic content is greater than it is with viscosity.

Molecular Mass

In refrigeration applications, the molecular mass of an oil is often needed. Albright and Lawyer (1959) showed that, on a molar basis, Refrigerants 22, 115, 13, and 13B1 have about the same viscosity reducing effects on a paraffinic oil.

For most mineral oils, a reasonable estimate of the average molecular mass can be obtained by the standard ASTM test D-2502, based on kinematic viscosities at 100 and 210°F; or from viscosity-gravity correlations of Mills *et al.* (1946). Direct methods (ASTM D-2503, D-2224) can also be used when greater precision is needed or when the correlative methods are not applicable.

Pour Point

Any oil intended for low-temperature service should be able to flow at the lowest potential temperature. This requirement is usually met by specifying a suitable low pour point. The pour point of an oil is defined as the lowest temperature at which it will pour or flow, when tested according to the standard method prescribed in ASTM D-97.

The loss of fluidity at the pour point may manifest itself in two ways. Naphthenic oils usually approach the pour point by a steady increase in viscosity. Paraffinic oils, unless heavily dewaxed, tend to separate out a rigid network of wax crystals, which may prevent the flow while still retaining unfrozen liquid in the interstices. Although not recommended for halogenated refrigerants, pour points can be lowered by adding chemicals known as pour point

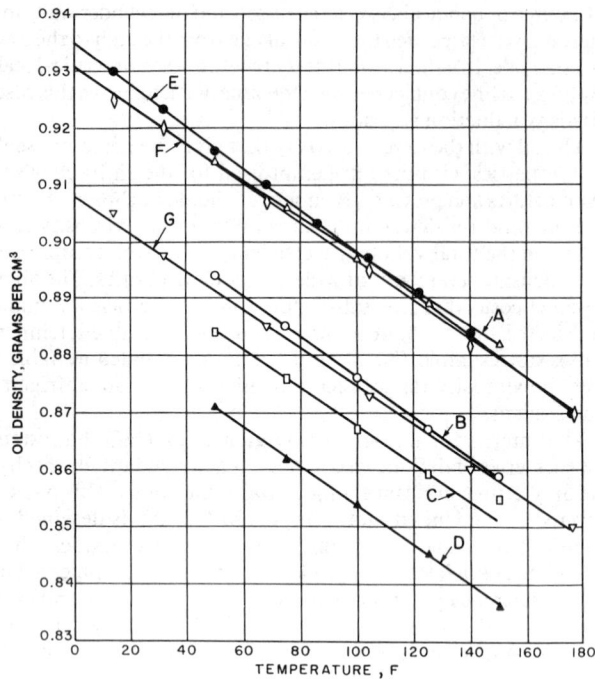

Oil	Base	Viscosity cs 100°F	SSU	C_A	C_N	C_P	Ref.
A	Naphthene	64.7	300	—	—	—	1
B	Naphthene	15.7	80	—	—	—	1
C	Paraffin	64.7	300	—	—	—	1
D	Paraffin	32.0	150	—	—	—	1
E	Paraffin	51.7	240[a]	—	—	—	2
F	Naphthene	33.1	155[a]	12	44	44	3
G	Naphthene	45.2	210[a]	4	44	52	3

References:
1. Albright and Lawyer (1959)
2. Bambach (1955)
3. Loffler (1959)

[a]German oils, base as stated by the author; viscosities converted to SSU and interpolated at 100°F

Fig. 2 Variation of Refrigeration Oil Densities with Temperature

depressors. These chemicals are believed to modify the wax structure, possibly by depositing a film on the surface of each wax crystal, so that the crystals no longer adhere to form a matrix, and do not interfere with the oil's ability to flow.

Standard pour test values are significant in the selection of oils for ammonia and carbon dioxide systems, and any other system in which refrigerant and oil are almost totally immiscible. In such a system, any oil that gets into the low side is essentially refrigerant free; therefore, the pour point of the oil itself determines whether loss of fluidity, congealment, or wax deposition will occur at low-side temperatures.

Since oil in the low side of fluorocarbon systems contains significant amounts of dissolved refrigerant, the pour point test, which is conducted on pure oils and in air, is of little significance. The viscosity of the oil-refrigerant solutions at the low-side conditions, and the wax separation or the floc test, are important considerations.

Volatility—Flash and Fire Points

Since the boiling ranges and vapor pressure data on lubricating oils are not readily available, an indication of the volatility of an oil is obtained from the flash and fire points (ASTM D-92), properties that are not significant in refrigerant systems. However,

Lubricants in Refrigerant Systems

some refrigerants, such as sulfur dioxide, ammonia, and methyl chloride, have a high ratio of specific heats (c_p/c_v) and consequently have a high adiabatic compression temperature. These refrigerants frequently produce carbonization of oils with low flash and fire points when operating in high ambient temperatures. Even with the present fluorocarbon refrigerants, in some applications requiring high compression ratios, such as a domestic refrigerator-freezer operating in high ambient temperatures, carbonization of oils sometimes occurs. Because such carbonization or coking of the valves is not necessarily accompanied by the general oil deterioration, this characteristic is often referred to as thermal stability, in contrast to chemical stability. Some manufacturers circumvent such problems by using paraffinic oils, which in comparison to naphthenic oils, have higher flash and fire points. Others prevent them through appropriate system design.

Vapor Pressure

Vapor pressure is the pressure at which the vapor phase of a substance is in equilibrium with the liquid phase at a specified temperature. The composition of the vapor and liquid phases (when not pure) influences the equilibrium pressure. With refrigeration oils, the type, boiling range, and viscosity are also factors influencing vapor pressure; naphthenic oils of a specific viscosity grade will generally show higher vapor pressures than paraffinic oils.

The vapor pressure of an oil increases with increasing temperature, as shown in Table 4. In practice, the vapor pressure of a refrigeration oil at an elevated temperature is negligible compared with the refrigerant at that temperature. The vapor pressure of narrow boiling petroleum fractions can be plotted as straight line functions. If the oil's boiling range and its type are known, standard tables may be used to determine the oil's vapor pressure up to 760 mm Hg at any given temperature (API 1970).

Table 4 Increase in Vapor Pressure and Temperature

Temperature, °F	Vapor Pressure 32 cs (150 SSU) Oil	
	Alkylbenzene, mm Hg	Naphthene Base, mm Hg
300	0.72	0.93
325	1.58	1.92
350	3.36	3.78
375	6.70	7.15
400	13.0	13.1
425	24.3	23.0
450	43.8	39.4

Aniline Point

Aniline, a ring-structured aromatic compound, is more soluble in oils containing a greater quantity of similar compounds. Therefore, measure the aromaticity of an oil by its solubility in aniline. The temperature at which an oil and aniline are mutually soluble is the oil's aniline point (ASTM D-611). In comparing oils, the lower the aniline point, the more naphthenic or aromatic is the oil.

Aniline point also has practical significance in predicting an oil's effect on elastomer seal materials. Generally, a highly naphthenic oil swells a specific elastomer material more than a paraffinic oil. This is caused by the greater solvency of aromatic and naphthenic compounds present in a naphthenic-type oil. However, aniline point gives only a general indication of oil-elastomer compatibility. Within a given class of elastomer material, oil resistance varies widely because of differences in compounding practiced by the elastomer manufacturer. Therefore, test for oil-elastomer compatibility under conditions anticipated in actual service.

Table 5 Absorption of Low Solubility Refrigerant Gases in Oil

Ammonia[a] (Percent by Mass)					
Absolute Pressure, psi	Temperature, °F				
	32	68	149	212	302
14.2	0.246	0.180	0.105	0.072	0.054
28.4	0.500	0.360	0.198	0.144	0.108
42.7	0.800	0.540	0.304	0.228	0.166
57.0	—	0.720	0.398	0.300	0.222
142.0	—	—	1.050	0.720	0.545

Carbon Dioxide[b] (Percent by Mass)				
Absolute Pressure, psi	Temperature, °F			
	32	68	140	212
14.7	0.26	0.19	0.13	0.10

[a] Type of oil: Not given (Steinle 1950)
[b] Type of oil: HVI oil, 34.8 cs (163 SSU) at 100°F (Baldwin and Daniel 1953)

Solubility of Refrigerants in Oils

All gases are soluble to some extent in mineral oils, and many of the refrigerant gases are highly soluble. The amount dissolved depends on the pressure of the gas and the temperature of the oil; on the nature of the gas; and on the nature of the oil. Since refrigerants are much less viscous than oils, any appreciable amount in solution causes a marked reduction in viscosity.

Two refrigerants usually regarded as poorly soluble in mineral oil are ammonia and carbon dioxide. Data showing the slight absorption of these gases by mineral oil are given in Table 5. The amount absorbed increases with increasing pressure and decreases with increasing temperature. In ammonia systems, where pressures are moderate, the 1% or less refrigerant that dissolves in the oil should have little, if any, effect on oil viscosity. However, operating pressures in CO_2 systems tend to be much higher (not shown in Table 5), and in that case, the quantity of gas dissolving in the oil may be enough to substantially reduce viscosity. At 390 psig, for example, Beerbower and Greene (1961) observed a 69% reduction when a 32 cs (150 SSU) oil (HVI) was tested under CO_2 pressure at 80°F.

OIL-REFRIGERANT SOLUTIONS

Many of the halogenated refrigerants used today are highly soluble in mineral oils at any temperature likely to be encountered. R-11 and R-12 are examples of such refrigerants. The only limit to the amount of these refrigerants that the oil can dissolve is established by the refrigerant pressure at a given temperature. Other halogenated refrigerants such as R-22 and R-114 may show limited solubilities with oil at evaporator temperatures (exhibited in the form of phase separation) and unlimited solubilities in the higher temperature regions of a refrigerant system.

As a result of the high solubilities of halogenated refrigerants, the lubricating fluid can no longer be treated as a pure oil, but rather as an oil-refrigerant solution whose properties are markedly different from those of pure oil. The amount of refrigerant dissolved in an oil depends on the pressure and temperature. Therefore, the composition of the lubricating fluid is different in different sections of a refrigeration system operating at steady state and changes from the time of startup until the system attains the steady state. The most pronounced effect is on viscosity.

Consider the crankcase of a compressor as an example. If the system at start-up was at 77°F, the viscosity of a pure 32 cs (150 SSU) naphthenic oil would be about 66.9 cs (310 SSU). Under operating conditions, it is not unusual for the oil temperature to be 130°F and the viscosity of the pure oil to be about 16.9 cs (85 SSU). If the system is operating with R-22 as the refrigerant, and the pressure in the crankcase is 94.7 psia, according to Little (1952), the viscosity of the lubricating fluid at start-up would be 15.7 cs (80 SSU) rather than 66.9 cs, and this would decrease to 9.7 cs (58 SSU) at 130 °F.

If only oil properties are considered, an erroneous picture of the system is obtained. When the oil circulates through the system and returns from the evaporator to the compressor, a similar situation exists. The highest viscosity does not occur at the lowest temperature, because the oil contains a large amount of dissolved refrigerant. As the temperature increases, the oil loses some of the refrigerant and the viscosity reaches a maximum at a point away from the coldest spot in the system.

Similar to the lubricating fluid, the properties of the working fluid (the refrigerant) are also affected. The vapor pressure of an oil-refrigerant solution, for example, is markedly lower than that of the pure refrigerant. One result of this property is that the evaporator temperature is higher than if the refrigerant is pure. Another result is the so-called flooded start-up. When the crankcase and the evaporator are at about the same temperature, the fluid in the evaporator, which is mostly refrigerant, has a higher vapor pressure than the fluid in the crankcase, which is mostly oil. This difference in vapor pressures acts as the force driving the refrigerant to the crankcase to be absorbed in the oil, until the pressures are equalized. At times, the moving parts in the crankcase may be completely immersed in this oil-refrigerant solution. At startup, the change in pressure and the turbulence can cause excessive amounts of liquid to enter the cylinders, causing damage to the valves and creating oil-starvation in the crankcase. The use of crankcase heaters to prevent such problems caused by highly soluble refrigerants is discussed in Chapter 3 and in Neubauer (1958).

The occurrence of the solutions of oils and refrigerants results in a somewhat different set of rules for the lubricating fluid and for the working fluid, requiring a detailed discussion. Much of the remaining chapter deals with the properties of oil-refrigerant solutions.

Density

For a rough estimate of the density or specific gravity of an oil-refrigerant solution, assume that the solution is ideal so that the specific volumes of the components are additive. The formula for calculating the ideal density (D_{id}) is:

$$D_{id} = d_o/[1 + W(d_o/d_R - 1)] \qquad (1)$$

where

d_o = density of pure oil at the solution temperature
d_R = density of refrigerant liquid at the solution temperature
W = mass fraction of refrigerant in solution

Depending on the refrigerant, the actual density of an oil-refrigerant solution may deviate from the ideal by as much as 8%. The solutions are usually more dense than calculated, but sometimes are less. For example, R-11 forms ideal solutions with oils, whereas R-12 and R-22 show significant deviations. Density correction factors for R-12 and R-22 solutions are depicted in Figure 3. The corrected densities can be obtained from the relation:

$$\text{Mixture Density} = D_m = D_{id}/A \qquad (2)$$

where A is the density correction factor read from Figure 3 at the desired temperature and refrigerant concentration.

Thermodynamics and Transport Phenomena

Dissolving oil in liquid refrigerant affects the thermodynamic properties of the working fluid. The vapor pressures of refrigerant-oil solutions at a given temperature are always less than the vapor pressure of pure refrigerant at that temperature. Therefore, dissolved oil in an evaporator leads to lower suction pressures and higher evaporator temperatures than expected from pure refrigerant tables. Bambach (1955) gives an enthalpy diagram for R-12/oil solutions over the range of compositions from 0 to 100% oil and temperatures from −40 to 240 °F. Spauschus (1963) has developed general equations for calculating thermodynamic functions of refrigerant-oil solutions and has applied them to the special case of R-12/mineral oil solutions.

Pressure-Temperature-Solubility Relations

When a refrigerant is in equilibrium with an oil, a fixed and definite amount of the refrigerant is present in the oil at a given temperature and pressure. This is evident if the phase rule is applied to the system, which is basically a two-phase, two-component system. The oil, although a mixture of several compounds, may be considered one component, and the refrigerant the other. The two phases are the liquid phase and the vapor phase. The phase rule defines this system as having two degrees of freedom.

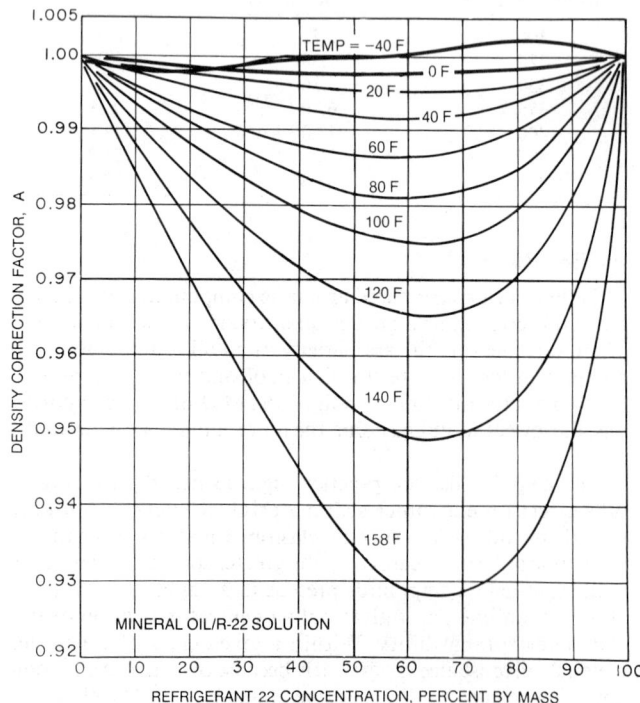

Fig. 3   Density Correction Factors
(Loffler 1959)

Lubricants in Refrigerant Systems

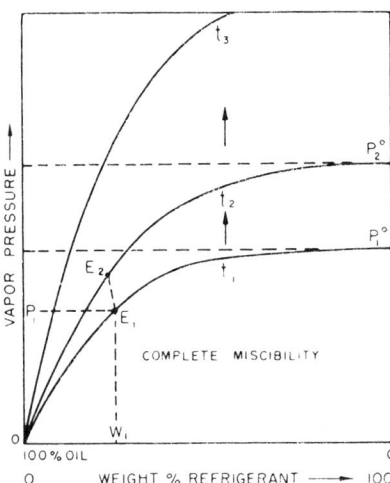

Fig. 4 P-T-S Diagram for Completely Miscible Refrigerant-Oil Solutions

Normally, the variables involved are pressure, temperature, and the compositions of the liquid and of the vapor phase. Since the vapor pressure of the oil is negligible compared with that of the refrigerant, the vapor phase is essentially pure refrigerant, and only the composition of the liquid phase needs to be considered. If the pressure and temperature are defined, it makes the system invariant, i.e., there can be only one composition of the liquid phase. This is a different but more precise way of stating that an oil-refrigerant mixture having a known composition will exert a certain vapor pressure at a certain temperature. If the temperature is changed, the vapor pressure will also change.

Pressure-temperature-solubility relations are usually presented in the form shown in Figure 4. On this graph, P_1^o and P_2^o represent the saturation pressures of the pure refrigerant at temperatures t_1 and t_2, respectively. Point E_1 represents an equilibrium condition, where one and only one composition of the liquid, represented by W_1, is possible at the pressure P_1. If the temperature of this system is increased to t_2, some of the liquid refrigerant evaporates, and the equilibrium point shifts to E_2, corresponding to a new pressure P_2. In either case, the oil-refrigerant solution exerts a vapor pressure less than that of the pure refrigerant at the same temperature. Example 1 illustrates the utility of the p-t-s diagrams.

Example 1. A refrigerant system with 1.9 lb of a 32 cs (150 SSU) naphthenic oil and an R-12 charge of 2 lb with a gas space of 1 ft^3 is allowed to come to equilibrium with ambient temperature at 77°F. Determine the pressure existing in the system and the amount of R-12 dissolved in the oil.

Solution: The problem can best be solved by the method of successive approximations.

1. First approximation (all R-12 considered as vapor).
   - R-12 vapor density = 2 lb ft^3
   - R-12 pressure at 77°F and 2 lb/ft^3 = 83 psia
   - Solubility limit at 77°F and 83 psia from Figure 18 = 48 wt %
2. Second approximation (assume actual solubility is one-half of above value).
   - R-12 dissolved in oil at 24 wt % = 0.6 lb
   - R-12 vapor density, 2 − 0.6 lb ft^3 = 1.4 lb/ft^3
   - Solubility limit at 77°F and 61 psia = 28 wt %
3. Third approximation (assume mean value between 24 and 28 %).
   - R-12 dissolved in oil at 26 wt % = 0.66 lb
   - R-12 vapor density, 2 − 0.66 lb/ft^3 = 1.34 lb/ft^3
   - R-12 pressure at 77°F and 1.34 lb/ft^3 = 59 psia
   - Solubility limit at 77°F and 59 psia = 25 wt %

While pure R-12 would have exerted a pressure of 83 psia, the presence of the oil has dropped the pressure to 59 psia, and almost one-fourth of the refrigerant is absorbed by the oil. The absolute viscosity of the oil, which without the refrigerant would be about 60 cps (300 SSU), has been reduced to about 8 cps (see Figure 20).

Table 6 Mutual Solubility of Refrigerants and Mineral Oil

Completely Miscible	Partially Miscible			Immiscible
	High Miscibility	Intermediate Miscibility	Low Miscibility	
R-11	R-13B1	R-22	R-13	Ammonia
R-12	R-501	R-114	R-14	Carbon Dioxide
R-21	R-123		R-115	R-134a
R-113			R-152a	
R-500			R-C318	
			R-502	

Mutual Solubility

In dealing with the lubricating and sealing problems in a compressor, the lubricating fluid is a solution of refrigerant dissolved in oil. In other parts of the refrigerant system, the problem may involve a solution of oil in liquid refrigerant. In both instances, either the oil or the refrigerant could exist alone as a liquid if the other were not present; therefore, any distinction between the dissolving and dissolved component merely reflects a point of view. Either of the liquids can be considered as dissolving the other. This relationship is termed mutual solubility.

Refrigerants are classified as completely miscible, partially miscible, or immiscible, according to their mutual solubility relations with mineral oils. Since several commercially important refrigerants are partially miscible, further designation as having high, intermediate, or low miscibility is shown in Table 6.

Completely miscible refrigerants and oils are mutually soluble in all proportions at any temperature. This type of mixture always forms a single liquid phase under equilibrium conditions, no matter how much refrigerant or oil is present. R-11 and R-12 with mineral oil are examples.

Partially miscible refrigerant-oil systems are mutually soluble to a limited extent. Above the critical solution temperature (CST) or consolate temperature, oil-refrigerant mixtures in this class are completely miscible, and their behavior is identical to that described in the preceding section. Below this temperature, however, the liquid may separate into two phases. Such phase separation does not mean that the oil and the refrigerant are insoluble in each other. Each liquid phase is a solution; one is oil-rich and the other refrigerant-rich, depending on the predominant component. Each phase may contain substantial amounts of the leaner component in the mutual solution, and these two solutions are themselves immiscible.

The importance of this concept is best illustrated by R-502, which is considered a low miscibility refrigerant exhibiting a high CST as well as a broad immiscibility range. However, even at 0°F, the oil-rich phase contains about 20 wt % of dissolved refrigerant. Examples of partially miscible systems, in addition to R-502, are R-22, R-114, and R-13, with mineral oils.

The basic properties of the immiscible region can be recognized by applying the phase rule to the system. With three phases (two liquid and one vapor) and two components, there can be only one degree of freedom. Therefore, either the temperature or the pressure automatically determines the composition of both liquid phases. If the system pressure is changed, the temperature of the system changes and the two liquid phases assume somewhat different compositions determined by the new equilibrium conditions.

Figure 5 illustrates the behavior of partially miscible mixtures. Point C on the graph represents the critical solution temperature t_3. There are three separate regions below this temperature on the diagram. Reading from left to right, a family of the smooth solid curves represents a region of completely miscible oil-rich solutions. These are followed by a wide break representing a region of partial miscibility in which there are two immiscible liquid phases. On the right side, the partially miscible region disappears into a

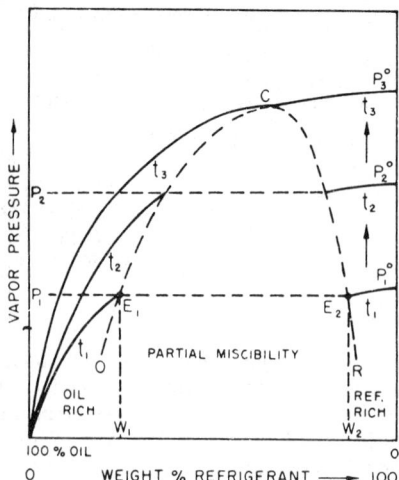

Fig. 5 P-T-S Diagram for Partially Miscible Refrigerant Oil Solutions

second completely miscible region of refrigerant-rich solutions. A dome-shaped envelope (broken line curve OCR) encloses the partially miscible region; everywhere outside this dome the refrigerant and oil are completely miscible. In a sense, Figure 5 is a variant of Figure 4 in which the partial miscibility dome (OCR) blots out a substantial portion of the continuous solubility curves. Under the dome, *i.e.*, in the immiscible region, if one were to follow the line at temperature t_1, the two points E_1 and E_2 represent the two phases coexisting in equilibrium.

These two phases differ considerably in composition (W_1 and W_2) but have the same refrigerant pressure P_1. The solution pressure P_1 lies not far below the saturation pressure of pure refrigerant P_1°. It is not uncommon for refrigerant-oil solutions near the partial miscibility limit to show less reduction in refrigerant pressure than observed at the same oil concentration with completely miscible refrigerants.

Totally immiscible oil-refrigerant systems are defined in this chapter as only very slightly miscible. In such mixtures, the immiscible range is so broad that mutual solubility effects can be ignored. Critical solution temperatures are seldom found in mixtures of the totally immiscible type. Examples are ammonia and oil, and carbon dioxide and oil.

Effects of Partial Miscibility in Refrigerant Systems

Evaporator. The evaporator is the coldest part of the system and the most likely part in which immiscibility or phase separation will occur. If the evaporator temperature is below the critical solution temperature, phase separation is likely to occur in some part of the evaporator. The fluid entering the evaporator is mostly liquid refrigerant containing a small fraction of oil, whereas the liquid leaving the evaporator is mostly oil, since the refrigerant is in vapor form. No matter how little oil the entering refrigerant carries, the liquid phase, as it progresses through the evaporator, passes through the critical composition which usually lies in 15 to 20% oil in the total liquid phase.

Phase separation in the evaporator can sometimes cause problems. In a dry-type evaporator, there is usually enough turbulence to cause the phases to emulsify. In this case, the heat transfer characteristics of the evaporator may not be significantly affected. In flooded-type evaporators, however, the working fluid may separate into layers, and the oil-rich phase may float on top of the boiling liquid. In addition to the heat transfer, partial miscibility may also affect the oil's return from the evaporator to the crankcase. Usually the oil is moved by high velocity suction gas transferring momentum to the droplets of oil on the return line walls.

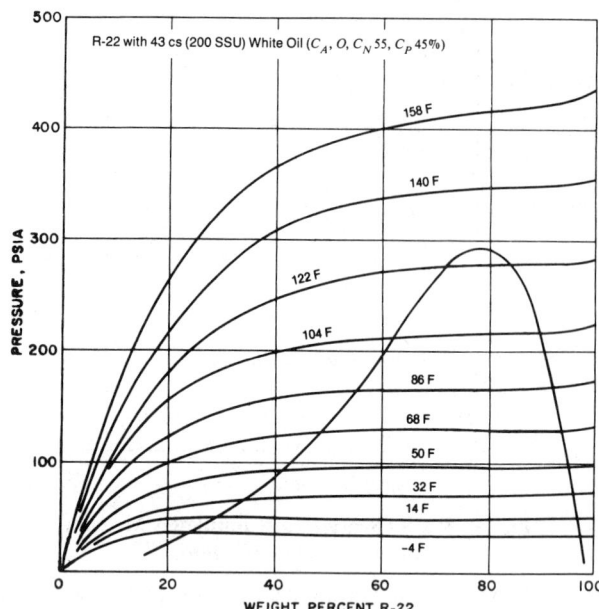

Fig. 6 P-T-S Relations of R-22 with White Oil
(Spauschus 1964)

If an oil-rich layer separates at the evaporator temperatures, this viscous, nonvolatile liquid can migrate and collect in pockets or blind passages not easily reached by the high velocity suction gas. The oil return problem may be magnified and, in some cases, an oil-logging can occur. The design of the system should take into account all these possibilities, and evaporators should be designed to promote entrainment (see Chapter 3). Oil separators are frequently required in the discharge line to minimize oil circulation when refrigerants of poor solvent power are used or in systems involving very low evaporator temperatures (Soling 1971).

Crankcase. With certain refrigerant systems, such as R-502 and mineral oil, or even with R-22 in applications such as heat pumps, phase separation sometimes occurs in the crankcase when the system is shut down. When this happens, the refrigerant-rich layer settles to the bottom, often completely immersing the pistons, bearings, and other moving parts. At start-up, the fluid that lubricates these moving parts is mostly refrigerant with little lubricity, and severe bearing damage may result. The turbulence at start-up may cause the liquid refrigerant to enter the cylinders, carrying large amounts of oil with it. Precautions in design will prevent such problems in partially miscible systems.

Condenser. Partial miscibility is not a problem in the condenser, since the liquid flow lies in the turbulent region and the temperatures are relatively high. Even if phase separation occurs, there is little danger of separation of layers, the main obstacle to efficient heat transfer.

Solubility Curves and Miscibility Diagrams

Figure 6 shows mutual solubility relations of partially miscible refrigerant-oil mixtures. More than one curve of this type can be plotted on a miscibility diagram. Each single dome then represents the immiscible ranges for one oil and one refrigerant. Miscibility curves for Refrigerants 13, 13B1, 502 (Parmelee 1964), and for Refrigerant 22 and mixtures of R-12 and R-22 (Walker *et al.* 1957), are shown in Figure 7. Miscibility curves for R-13, R-22, R-502, and R-503 in an alkylbenzene refrigeration oil are shown in Figure 8. A comparison with Figure 7 illustrates the greater solubility of refrigerants in this type of oil.

Effect of Oil Type on Solubility and Miscibility

When compared on a mass basis, low viscosity oils absorb more refrigerant than high viscosity oils. Also, naphthenic oils absorb

Lubricants in Refrigerant Systems

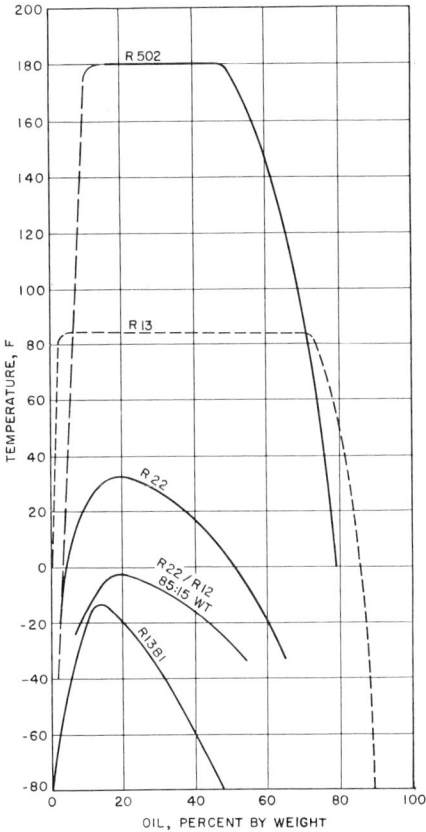

Fig. 7 Critical Solubilities of Refrigerants with a 32 cs (150 SSU) Naphthenic Oil (C_A 12, C_N N 44, C_P 44%)

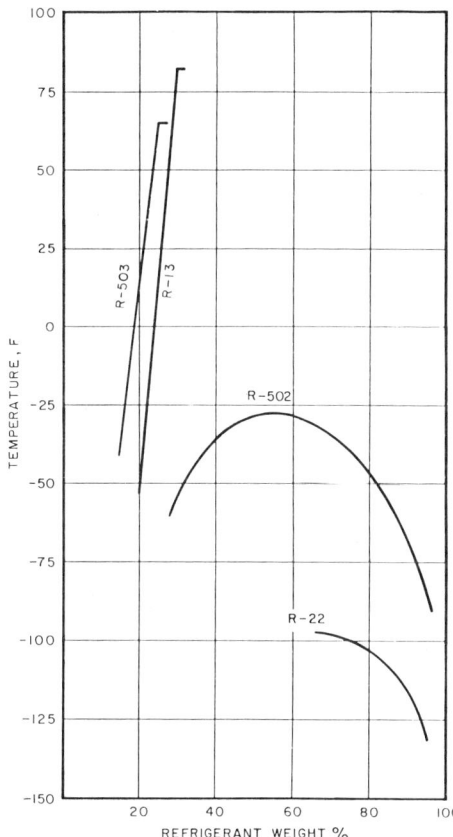

Fig. 8 Critical Solubilities of Refrigerants with a 32 cs (150 SSU) Alkylbenzene Oil

more than paraffinic oils. However, when compared on a mol basis, some confusion arises. Paraffinic oils absorb more refrigerant than naphthenic oils, i.e., reversal of the mass basis, and there is little difference between a 15.7 cs (80 SSU) and a 64.7 cs (300 SSU) naphthenic oil (Albright and Lawyer 1959, Albright and Mandelbaum 1956). The differences on either basis are small, i.e., within 20% of each other. Comparisons of oils by their carbon type analyses are not available, but in view of the data on naphthenic and paraffinic types, differences between oils with different carbon-type analyses, except perhaps for extreme compositions, are unlikely.

The effect of oil-type and composition on miscibility is better defined than solubility. When the critical solution temperature (CST) is used as the criterion of miscibility, oils with higher aromatic contents show a lower CST. Higher viscosity grade oils show a higher CST than lower viscosity grade oils, and paraffinic oils show a higher CST than naphthenic oils (see Figures 9 and 21). When the entire dome of immiscibility is considered, a similar result is noticeable. Oils that exhibit a lower CST, usually show a narrowing of the immiscibility range, i.e., the mutual solubility is greater at any given temperature.

Miscibility of R-22 With Oils

Parmelee (1964) showed that polybutyl silicate improves miscibility with R-22 (and also R-13) at low temperatures. Alkylbenzenes, by themselves or mixed with mineral oils, also have better miscibility with R-22 than do mineral oils alone (Seemann and Shellard 1963).

Among the mineral oils, several reports deal with miscibility differences resulting from different types and viscosity grades. Walker *et al.* provide detailed miscibility diagrams of 12 brand name oils commonly used for refrigeration systems. Walker's data show that in every case, higher viscosity oil of the same base and type has a higher critical solution temperature.

Loffler (1957) provides complete miscibility diagrams of R-22 and 18 oils. A portion of the properties of the oils used and the critical solution temperatures are summarized in Table 7. Although precise correlations are not evident in the table, certain trends are clear. For the same viscosity grade and base, the effect of aromatic carbon content is seen in oils 2, 3, 7, and 8, and between 4 and 6. Similarly, for the same viscosity grades, the effect of paraffinic structure (with essentially the same % C_A) is noticeable between oils 6 and 17 and between oils 8 and 18.

According to Loffler, the most pronounced effect on the critical solution temperature is exerted by the aromatic content of the oil; the table indicates that the paraffinic structure reduces the miscibility compared with naphthenic structures. Sanvordenker (1968) reports the miscibility relations of the saturated fractions and the aromatic fractions of mineral oils as a function of their physical properties. The critical solution temperatures with R-22 increase with increasing viscosities for the saturates, as well as for the aromatics. For equivalent viscosities, aromatic fractions with naphthenic linkages show lower critical solution temperatures than aromatics with only paraffinic linkages.

Generally, adding R-12 to R-22 systems improves the miscibility by reducing the critical solution temperature and narrowing the immiscible range. The former effect appears as a drop of about 40 or 50 °F when the mixture contains 85% R-22 and 15% R-12 (Walker *et al.* 1957, Loffler 1960). The narrowing of the immiscible range also improves the fluidity of the oil-rich phase, so that the minimum temperature at which the oil-rich phase can still flow drops by about 25 °F (Loffler 1960).

Table 7 Critical Miscibility Values of R-22 with Different Oils

Oil No.	Oil Base Type[a]	Approximate Viscosity Grade		Viscosity at 122°F Converted		Carbon-Type Composition			Critical Solution Temperature, °F
		SSU	cs	to SSU	to cs	%C_A	%C_N	%C_P	
2	N	75	15	63	11.2	23	34	43	−35
3	N	75	15	60	10.2	2.5	48.5	49	21
1	N	150	32	92	18.5	13	43	44	27
8	N	200	46	118	24.6	0.6	45	55	75
7	N	200	46	127	26.7	2.8	44	54	68
5	N	200+	46+	132	27.9	22	30	47	3
4	N	250	46	135	28.6	26	28	46	−4
6	N	250	46	140	29.7	4	45	51	62
13	N	500	100	253	54.5	1.9	41	56	None[b]
12	N	500	100	282	60.7	4	41	55	None[b]
11	N	700	150	320	69.1	7	40	53	None[b]
10	N	1000	220	434	93.2	21	27	52	61
9	N	1200	220	502	109.0	27	24	50	48
18	P	200+	46+	138	29.3	0.5	33	67	None[b]
17	P	250	46	148	31.6	3.5	34	63	None[b]
16	P	300	68	164	35.2	6.4	30	63	None[b]
15	P	350	68	210	45.2	14.3	25	61	None[b]
14	P	400	100	232	50.0	18.1	22	60	111[c]

[a] P = Paraffinic, N = Naphthenic
[b] Never completely miscible at any temperature
[c] A second (inverted) miscibility dome was observed above 136°F. Above this temperature, the oil/R-22 mixture again separated into two immiscible solutions.

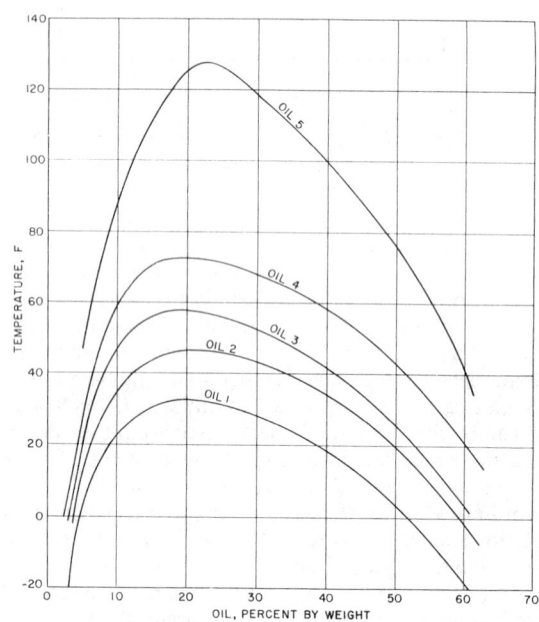

Oil No.	Viscosity		Compositions, %			Ref.
	cs 100°F	SSU	C_A	C_N	C_P	
1	34.0	159	12	44	44[a]	1
2	33.5	157	7	46	47[a]	1
3	63.0	292	12	44	44[a]	1
4	67.7	314	7	46	47[a]	1
5	41.3	192	0	55	45	2

References:
1. Walker et al. (1957)
2. Spauschus (1964)

[a] Estimated composition, not in original reference.

Fig. 9 Effect of Oil Properties on Miscibility with R-22

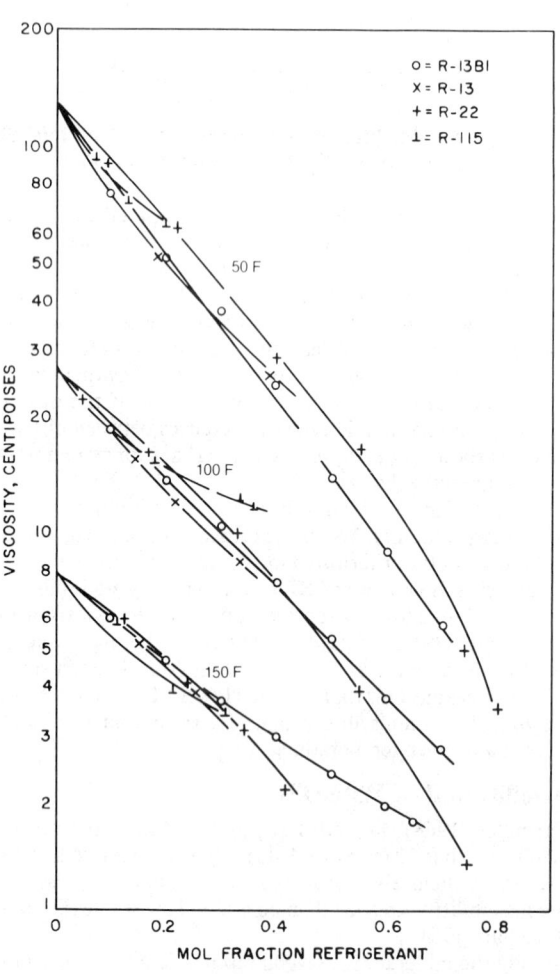

Fig. 10 Viscosity of Mixtures of Various Refrigerants and 32 cs (150 SSU) Paraffinic Oil
(Albright and Lawyer 1959)

Lubricants in Refrigerant Systems

Solubilities and Viscosities of Oil-Refrigerant Systems

Although the differences are small on a mass basis, naphthenic oils are better solvents than paraffinic oils. When considering the viscosity of oil-refrigerant mixtures, naphthenic oils show greater viscosity reduction than paraffinic oils for the same mass percent of dissolved refrigerant. When the two effects are compounded, under the same conditions of temperature and pressure, a naphthenic oil in equilibrium with a given refrigerant shows a significantly lower viscosity than a paraffinic oil.

Refrigerants also differ in their viscosity-reducing effects when the solution concentration is measured in mass percent. However, when the solubility is plotted in terms of mol percent, the reduction in viscosity is approximately the same, at least for Refrigerants 13, 13B1, 22, and 115 (Figure 10).

Spauschus (1964) reports numerical vapor pressure data on a R-22/white oil system; solubility-viscosity graphs on naphthenic and paraffinic oils have been published by Loffler (1960), Little (1952), and Albright and Mandelbaum (1956). Some discrepancies, particularly at high R-22 contents, have been shown in data on viscosities, which apparently could not be attributed to the properties of the oil and remain unexplained. However, generalized plots reported by the abovementioned authors are satisfactory for engineering and design purposes.

References of solubility and viscosity data, as compiled by Speaker and Spauschus (1986), are listed in the Bibliography.

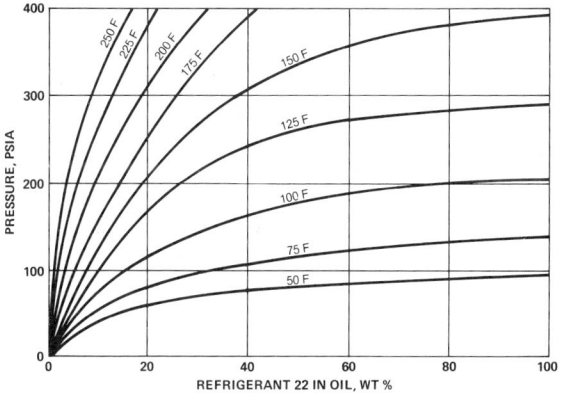

Fig. 11 Solubility of R-22 in 32 cs (150 SSU) Naphthenic Oil
(Witco)

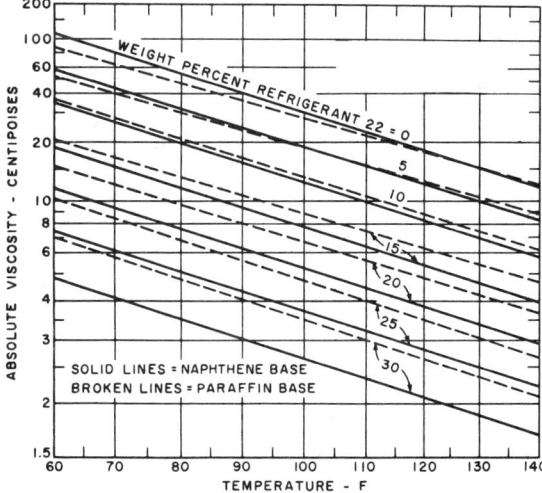

Fig. 12 Viscosity-Temperature Chart for Solutions of Refrigerant 22 in 32 cs (150 SSU) Naphthene and Paraffin Base Oils

Selected solubility-viscosity data are summarized in Figures 6 and 11 through 23.

Wherever possible, solubilities have been converted to weight percent to provide consistency among the various charts. Figures 6 and 11 through 15 contain data on R-22 and oils, Figure 16 on R-502, Figures 17 and 18 on R-11, Figures 19 and 20 on R-12, and Figures 21 and 22 on R-114. Figure 23 contains data on the solubility of various refrigerants in alkylbenzene oil. Viscosity-solubility characteristics of mixtures of R-13B1 and lubricating oils have been investigated by Albright and Lawyer (1959). Similar studies on R-13 and R-115 are covered by Albright and Mandelbaum (1956).

OIL RETURN FROM EVAPORATORS

Regardless of the miscibility relations of oil with refrigerants, for a refrigeration system to function properly, the oil must return

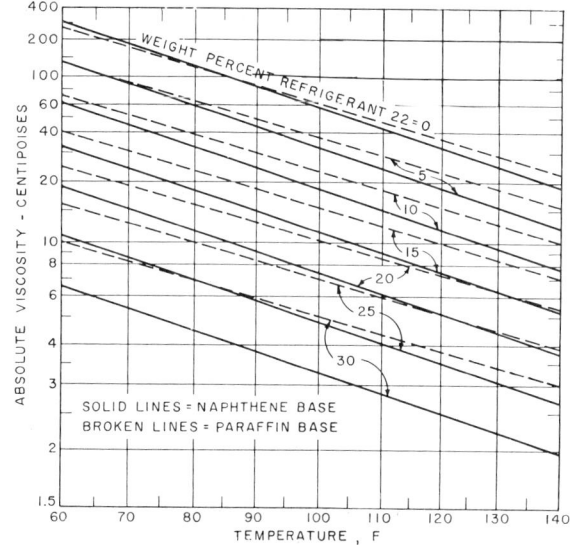

Fig. 13 Viscosity-Temperature Chart for Solutions of Refrigerant 22 in 65 cs (300 SSU) Naphthene and Paraffin Base Oils

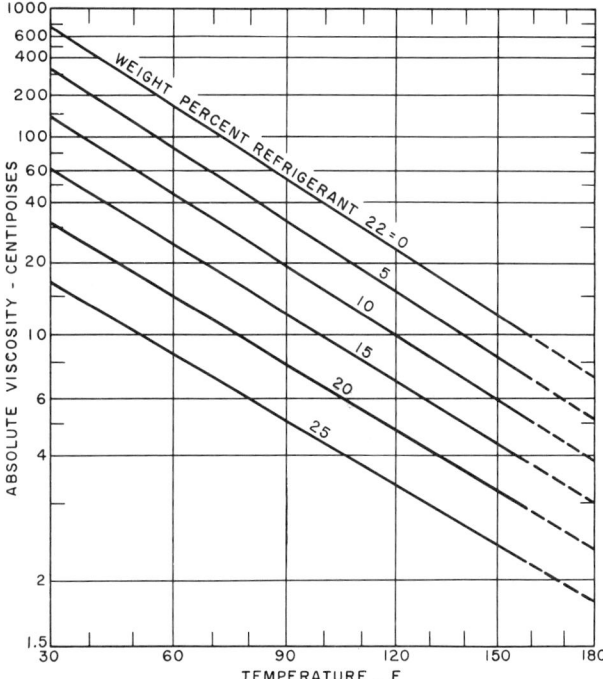

Fig. 14 Viscosity-Temperature Chart for Solutions of Refrigerant 22 in a 42.5 cs (210 SSU) German Naphthene Base Oil
(Loffler 1960)

adequately from the evaporator to the crankcase. Parmelee (1964) showed that the viscosity of the oil, saturated with refrigerant under low pressure and low-temperature conditions, is important in providing good oil return. The viscosity of the oil-rich liquid that accompanies the suction gas changes as it sees rising temperatures on its way back to the compressor. Two opposing factors then come into play. First, the increasing temperature tends to decrease the viscosity of the fluid. Second, since the pressure remains unchanged, the increasing temperature also tends to drive off some of the dissolved refrigerant from the solution, thereby increasing its viscosity.

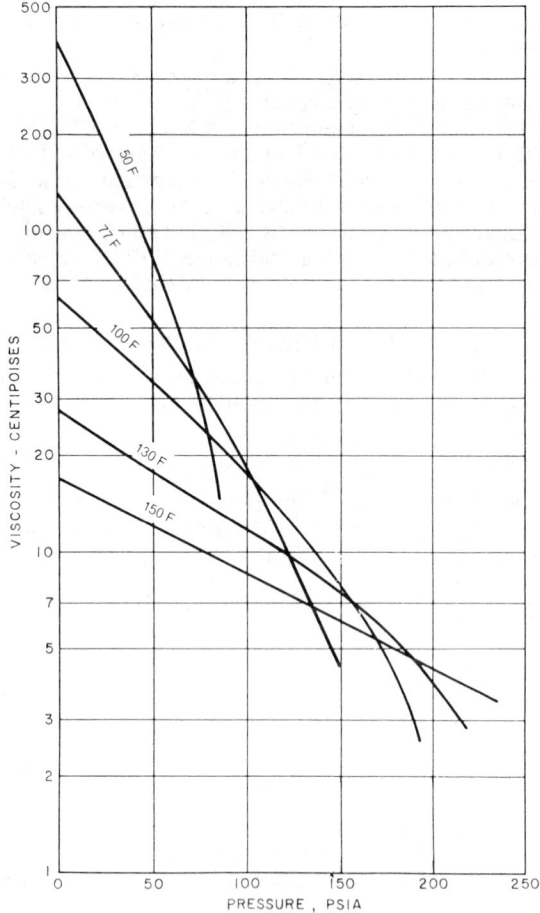

Fig. 15 Viscosity of Mixtures of 65 cs (300 SSU) Paraffin Base Oil and Refrigerant 22
(Albright and Mandelbaum 1956)

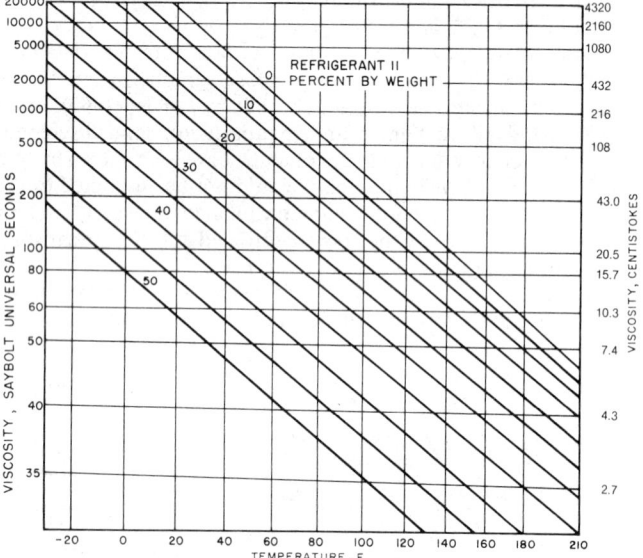

Fig. 17 Viscosity-Temperature Curves for Solutions of Refrigerant 11 in 65 cs (300 SSU) Naphthene Base Oil

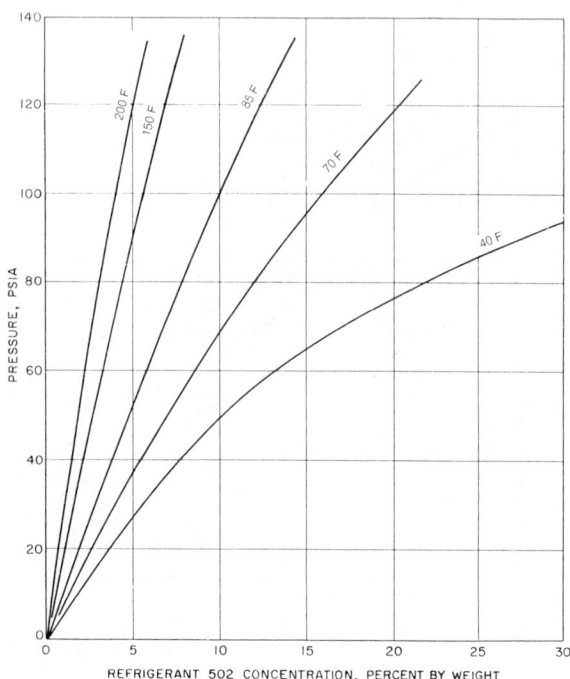

Fig. 16 Solubility of R-502 in a 32 cs (150 SSU) Naphthenic Oil (C_A 12, C_N 44, C_P 44%)

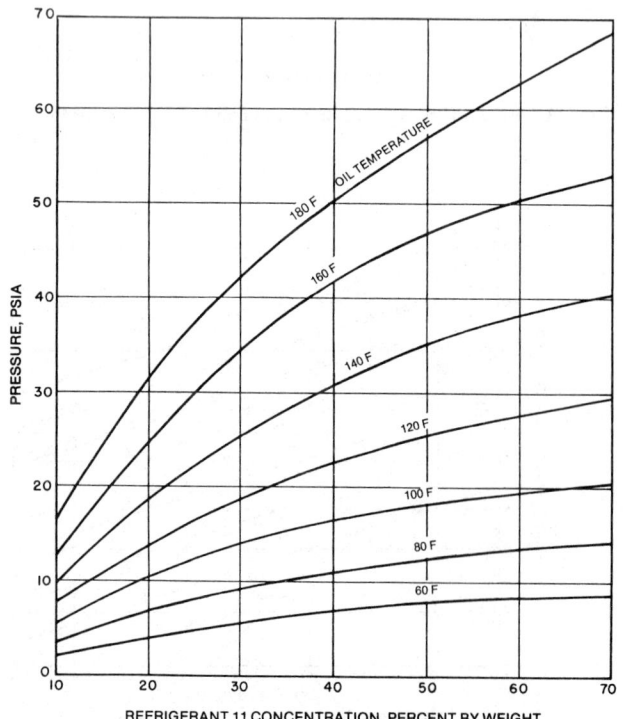

Fig. 18 Solubility of Refrigerant 11 in 65 cs (300 SSU) Oil

Lubricants in Refrigerant Systems

Figures 24 through 28 show the variation in viscosity with temperature and pressure for 5 oil-refrigerant solutions ranging from −100 to 70 °F. In all cases, the viscosities of the solutions passed through maximum values as the temperature was changed at constant pressure, a finding that was also consistent with previous data obtained by Bambach (1955) and Loffler (1960). According to Parmelee, the existence of a viscosity maximum is significant, because the oil-rich solution becomes most viscous, not in the coldest regions in the evaporator, but at some intermediate point where much of the refrigerant has escaped from the oil. This is possibly in the suction line. The velocity of the return vapor, which might

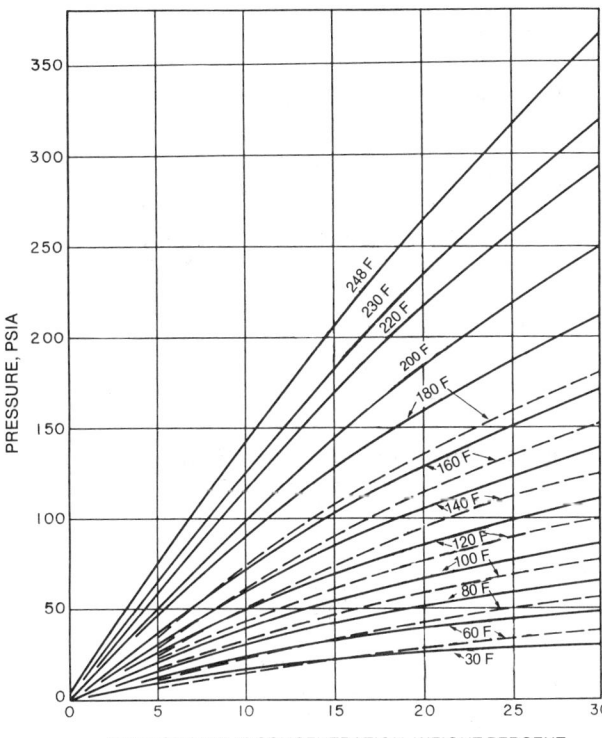

Solid lines: 32 mm²/s (32 cs or 150 SSU) Naphthene Base also 52 mm²/s (52 cs or 240 SSU) German Base (Bambach 1955).
Broken lines: 32 mm²/s (32 cs or 150 SSU) and 70 mm²/s (70 cs or 325 SSU) Mixed Base Oils.

Fig. 19 Solubility of Refrigerant 12 in Refrigeration Oils

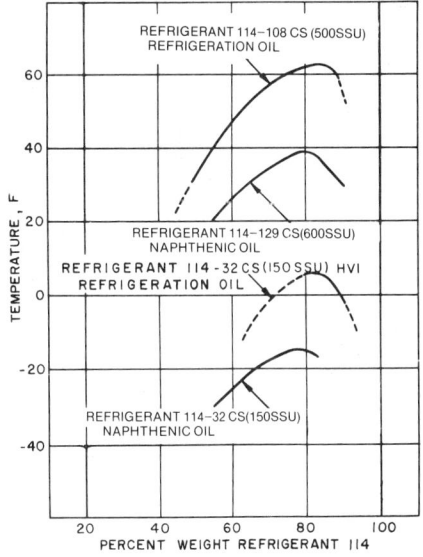

Fig. 21 Critical Solution Temperatures of Refrigerant 114/Oil Mixtures

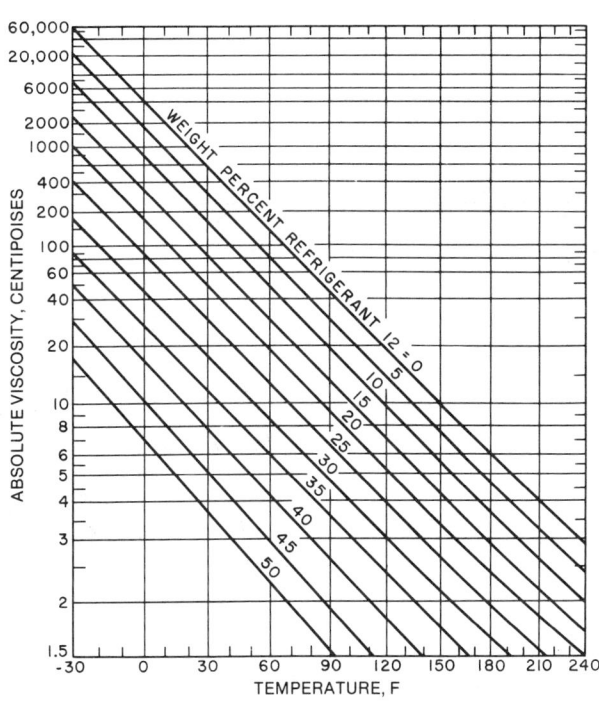

Fig. 20 Viscosity-Temperature Chart for Solutions of Refrigerant 12 in 32 cs (150 SSU) Naphthene Base Oil
(Loffler 1960)

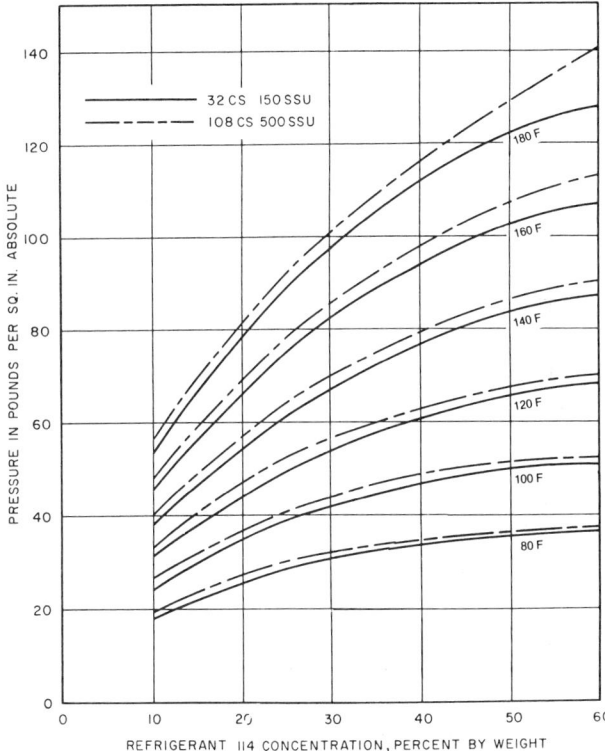

Fig. 22 Solubility of Refrigerant 114 in HVI Oils

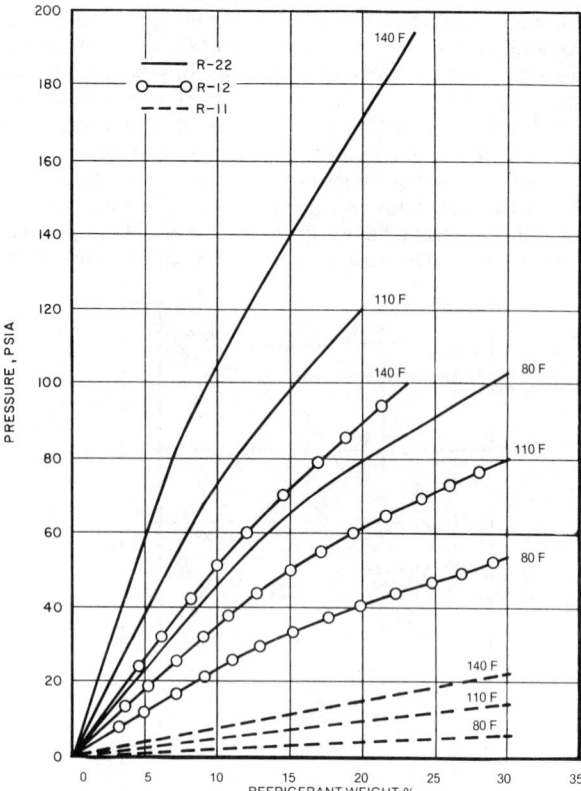

Fig. 23 Solubility of Refrigerants in 32 cs (150 SSU) Alkylbenzene Oil

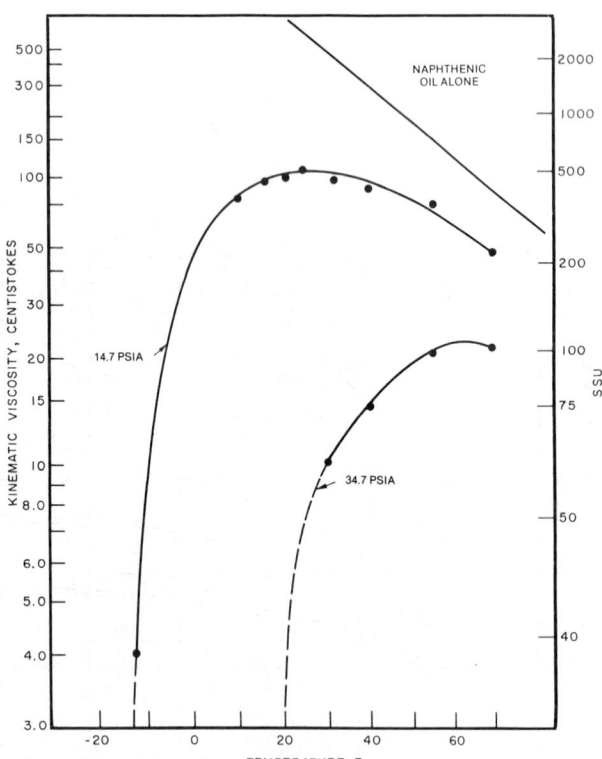

Fig. 24 Vicosity of Refrigerant 12/Oil Solutions at Low Side Conditions
(Parmelee 1964)

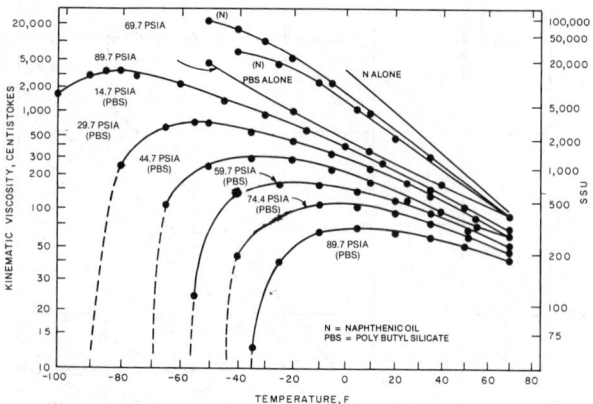

Fig. 25 Viscosity of Refrigerant 13/Oil Solutions at Low Side Conditions
(Parmelee 1964)

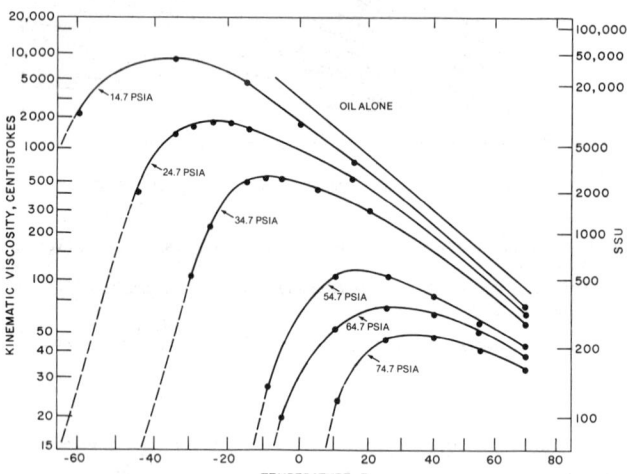

Fig. 26 Viscosity of Refrigerant 13B1/Oil Solutions at Low Side Conditions
(Parmelee 1964)

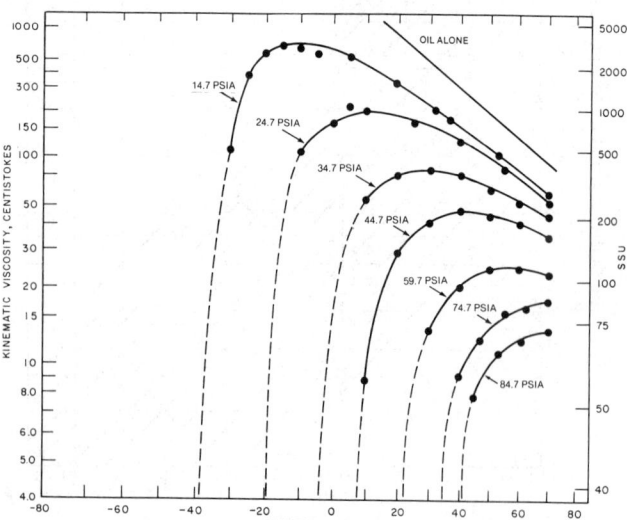

Fig. 27 Viscosity of Refrigerant 22/Naphthenic Oil Solutions at Low Side Conditions
(Parmelee 1964)

Lubricants in Refrigerant Systems

be high enough to move the oil-refrigerant solution in the colder part of the evaporator, might be too low to achieve the same result at the point of maximum viscosity. The designer must consider this factor to minimize any oil return problems. Chapters 3 and 4 have further information on velocities in return lines.

Another aspect of viscosity data at the evaporator conditions is shown in Figure 29, which compares a synthetic alkylbenzene oil with a naphthenic mineral oil. The two oils are the same viscosity grade, but the highly aromatic alkylated benzene oil has a much lower viscosity index in the pure state and shows a higher viscosity at low temperatures. However, at 19.7 psia or approximately $-40\,°$F evaporator temperature, the viscosity of the oil/R-502 mixture is considerably lower for alkylbenzene than for naphthenic oil. In spite of the lower viscosity index, alkylated benzene returns more easily than naphthenic oil.

Estimated viscosity-temperature-pressure relationships for an alkylated benzene oil with R-12, R-22, and R-502 are shown in Figures 30, 31, and 32.

WAX SEPARATION (FLOC TESTS)

Petroleum derived lubricating oils are mixtures of large numbers of chemically distinct hydrocarbon molecules. At low temperatures in the low-pressure side of refrigeration units, some of the larger molecules separate from the bulk of the oil, forming wax-like deposits. This wax can clog capillary tubes and cause expansion valves to stick and is, therefore, undesirable in refrigeration systems. Bosworth (1952) describes other wax separation problems in various systems.

In selecting an oil to use with completely miscible refrigerants such as R-11 or R-12, the wax forming tendency of the oil can be determined by the floc test. The test evaluates the wax precipitation tendency of a 90% by volume refrigerant and 10% by volume oil mixture. The floc point is the highest temperature at which wax-

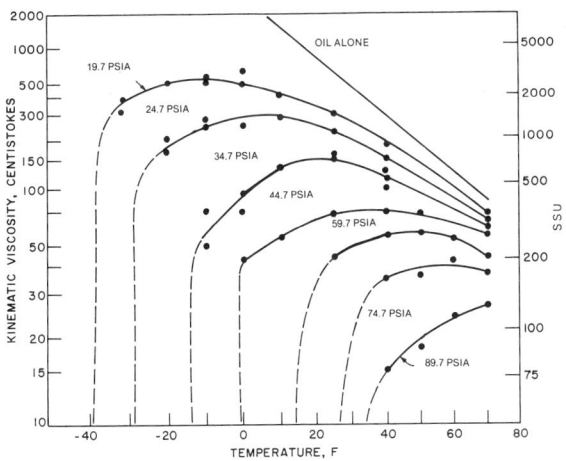

Fig. 28 Viscosity of Refrigerant 502/Naphthenic Oil Solutions at Low Side Conditions

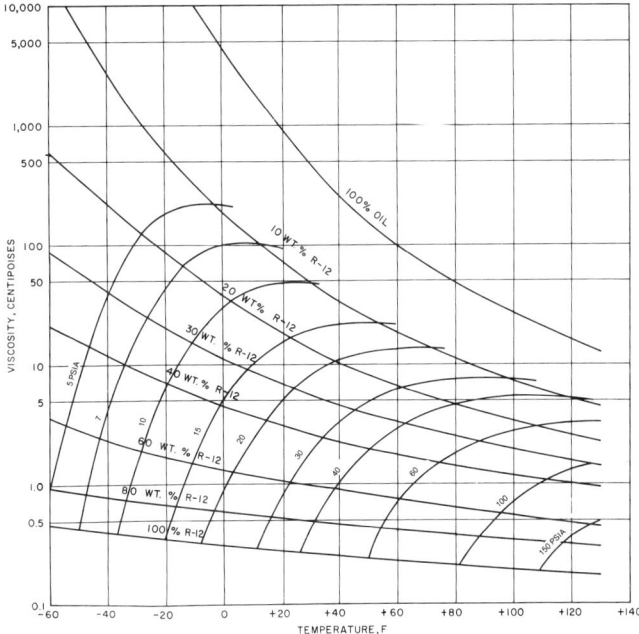

Fig. 30 Estimated Viscosity-Temperature-Pressure Relationships 32 cs (150 SSU) Alkylbenzene Oil With R-12

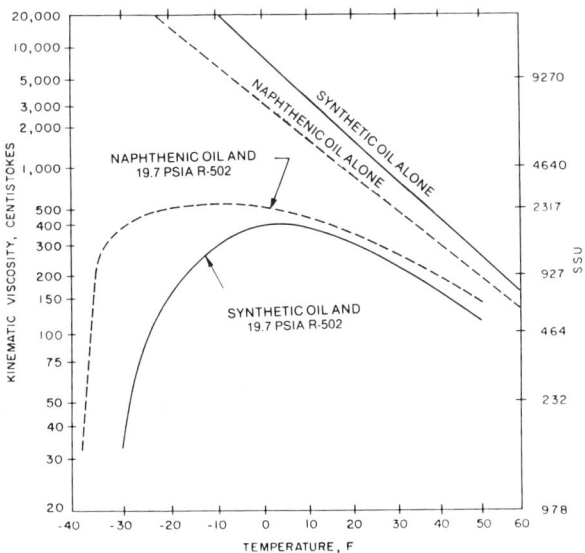

Fig. 29 Viscosities of Solutions of R-502 with a 32 cs (150 SSU) Naphthenic Oil (C_A 12, C_N 44, C_P 44%) and a 32 cs (150 SSU) Synthetic Alkylbenzene Oil (see Table 1)

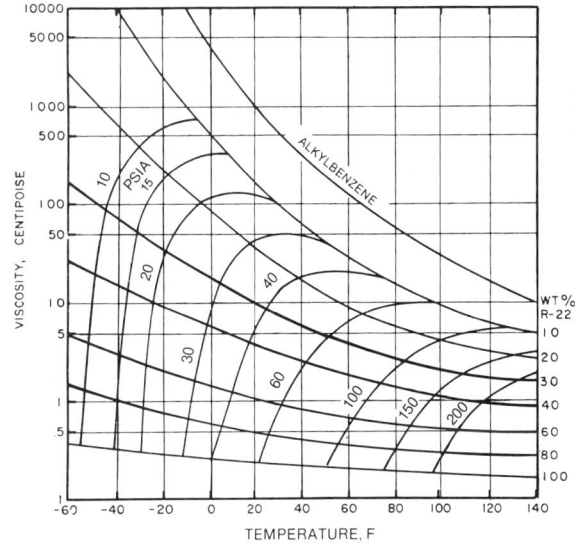

Fig. 31 Estimated Viscosity-Temperature-Pressure Relationships of 32 cs (150 SSU) Alkylbenzene Oil With R-22

like materials or other solid substances precipitate when a mixture of 10% oil and 90% Refrigerant 12 is cooled under specific conditions. The test can be used on other oils as well. Since different refrigerant and oil concentrations are encountered in actual systems, test results cannot be used directly to predict actual system performance. The oil concentration in the expansion devices of most refrigeration and air-conditioning systems is considerably less than 10%, resulting in significantly lower temperatures at which wax separates from the oil-refrigerant mixture.

ASHRAE *Standard* 86-1983, *Method of Testing the Floc Point of Refrigeration Grade Oils*, describes a standardized method of determining the floc characteristics of refrigeration oils in the presence of R-12.

Attempts to develop a test for the floc print of partially miscible refrigerants for use with R-22 have not been successful. The solutions being cooled often separate into two liquid phases. Once phase separation occurs, the components of the oil distribute themselves into the oil-rich phase and the refrigerant-rich phase in such a way that the highly soluble aromatics concentrate into the refrigerant phase, while the less soluble saturates concentrate into the oil phase. The waxy materials stay dissolved in the refrigerant-rich phase only to the extent of their solubility limit. On further cooling, any wax that separates out from the refrigerant-rich phase migrates into the oil-rich phase. Therefore, a significant floc point cannot be obtained with partially miscible refrigerants once phase separation has occurred. However, lack of flocculation does not mean lack of wax separation. Wax may separate in the oil-rich phase causing it to congeal. Parmelee (1964) has reported such phenomena with a paraffinic oil and R-22 system.

Floc point determinations might not be reliable when applied to used oils. A part of the original wax may already have been deposited, and the used oil may contain extraneous material from the operating system.

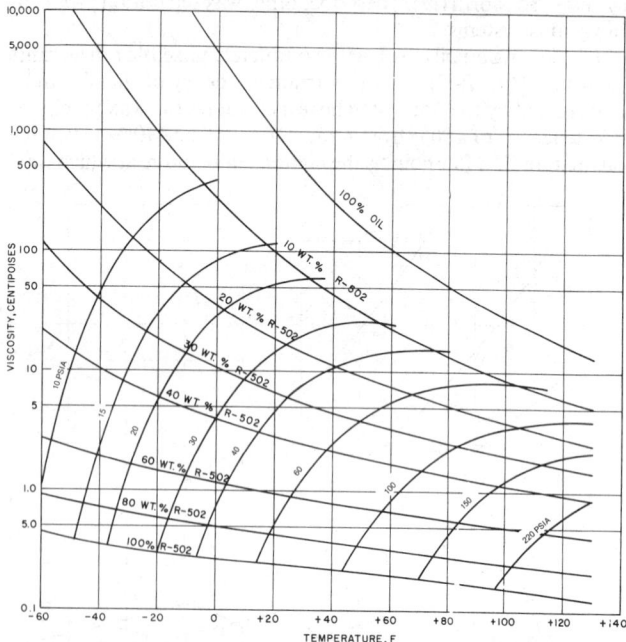

Fig. 32 Estimated Viscosity-Temperature-Pressure Relationships of 32 cs (150 SSU) Alkybenzene Oil With R-502

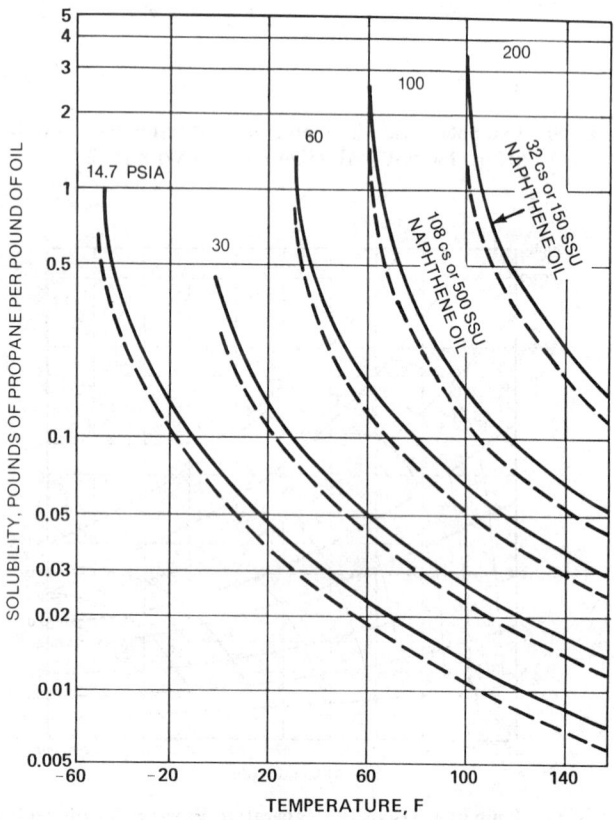

Fig. 33 Solubility of Propane in Oil

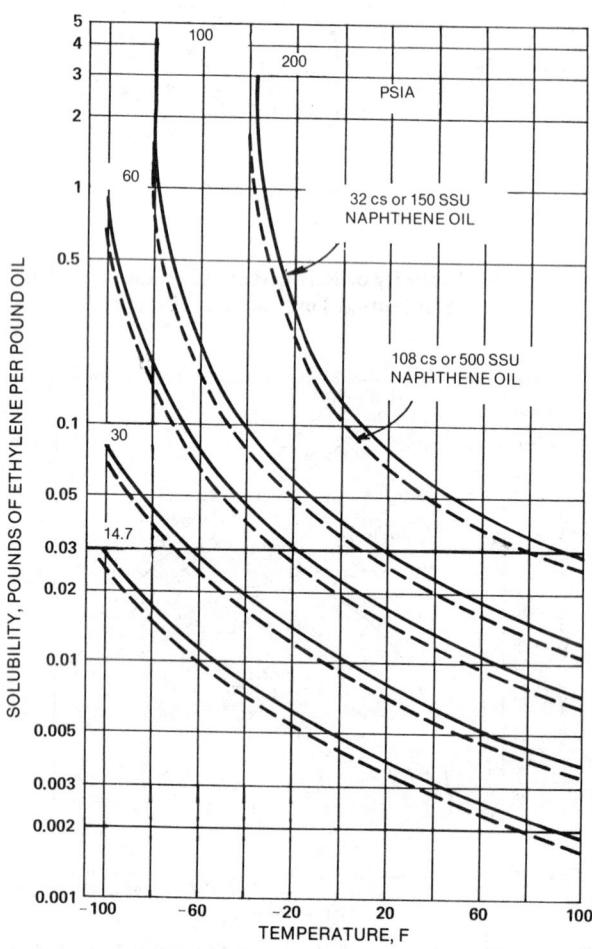

Fig. 34 Solubility of Ethylene in Oil
(Witco)

Lubricants in Refrigerant Systems

Good design practice suggests that oils that do not deposit wax in low sides, regardless of single-phase or two-phase refrigerant-oil solutions, be selected.

Mechanical design affects how susceptible an equipment design is to wax deposition. Wax is deposited from refrigerant solutions at points where sharp bends occur, and the suspended wax particles build up on the tubing walls by impingement. Careful design avoids bends and materially reduces the tendency to deposit wax.

SOLUBILITY OF HYDROCARBON GASES

Hydrocarbon gases such as propane (R-290) and ethylene (R-1150) can be used as refrigerants. These gases are miscible with the compressor lubricating oil and are absorbed by the oil. The solubility of the gas depends on its structure. The lower the boiling point or critical temperature, the less soluble the gas—all other values being equal. Gas solubility increases with decreasing temperature and increasing pressure. These relationships for propane and ethlyene are illustrated in Figures 33 and 34. As with other oil miscible refrigerants, absorption of the hydrocarbon gas reduces oil viscosity.

SOLUBILITY OF WATER IN OILS

Refrigerant systems must be dry internally for trouble-free performance (see Chapter 7). As with other components, it is essential that the refrigeration oil contains the least practicable amount of moisture. Normal refinery handling practices result in a moisture content for refrigeration oils of about 30 ppm. However, this amount may increase between the time of shipment from the refinery and the time of actual usage, unless proper preventive measures are taken. Small containers are usually sealed, but it is not common practice to pressure seal bulk tank cars. During transit, changes in ambient temperatures cause oil to expand and contract, and therefore, breathe humid air from outside. Depending on the extent of such cycling, oil may have significantly higher moisture content than at the time of shipment.

Users of large quantities of refrigeration oils frequently dry the oil before usage. Chapter 21 of the 1988 ASHRAE *Handbook—Equipment* discusses the methods of drying oils.

Spot checks show that data taken by Clark (1940) on transformer oils apply to refrigeration oils as well (Figure 35).

A simple method, widely used in industry to detect free water in refrigeration oils, is the dielectric breakdown voltage (ASTM D-877), which is designed to control moisture and other contaminants in electrical insulating oils. According to Clark (1940), the dielectric breakdown voltage decreases with increasing moisture content at the same test temperature and increases with temperature for the same moisture content. At room temperature, 80°F, when the solubility of water in a 32 cs (150 SSU) naphthenic oil is between 50 to 70 ppm, a dielectric breakdown voltage of about 25,000 V indicates that no free water is present in the oil. However, the oil may contain dissolved water up to the solubility limit. Therefore, it is not uncommon to specify a dielectric breakdown voltage of 35,000 V to indicate that the moisture content is well below saturation. The ASTM D-877 test is not sensitive below about 60% saturation. General practice has been to measure directly the total moisture content by procedures such as the Karl Fischer (ASTM D-1533) method for low moisture levels.

SOLUBILITY OF AIR IN OIL

Refrigerant systems should not contain excessive amounts of air or other noncondensable gases. Because oxygen present in air can react with oil, all air should be removed. Also, the air's 79% nitrogen content (which does not react with oil) constitutes a noncondensable gas that, even in amounts small enough to not pose the threat of oil oxidation, can interfere with refrigerating machine performance. In some systems, the tolerable volume of noncondensables is very low, and if the oil is added after the system is evacuated, it must not contain an excessive amount of dissolved air or any other noncondensable gas. As mentioned in the preceding section, dissolved air is removed when a vacuum process is used to dry the oil. However, if the deaerated oil is subsequently stored under dry air pressure, it will reabsorb air in proportion to the pressure. Figure 36 illustrates the volume of air under standard conditions that can be absorbed (*i.e.*, dissolved) in mineral oil over a range of pressures. The data in Figure 36 were determined at 68°F but apply to other temperatures since variation in solubility between 20 and 120°F is minor.

FOAMING AND ANTIFOAM AGENTS

Excessive foaming of the lubricant is undesirable in refrigeration systems. Brewer (1951) suggests that abnormal refrigerant foaming reduces the oil's effectiveness in cooling the motor windings and removing heat from the compressor. Too much foaming also can cause too much oil to pass through the pump and enter the low side. Foaming in a pressure oiling system can result in starved lubrication under some conditions.

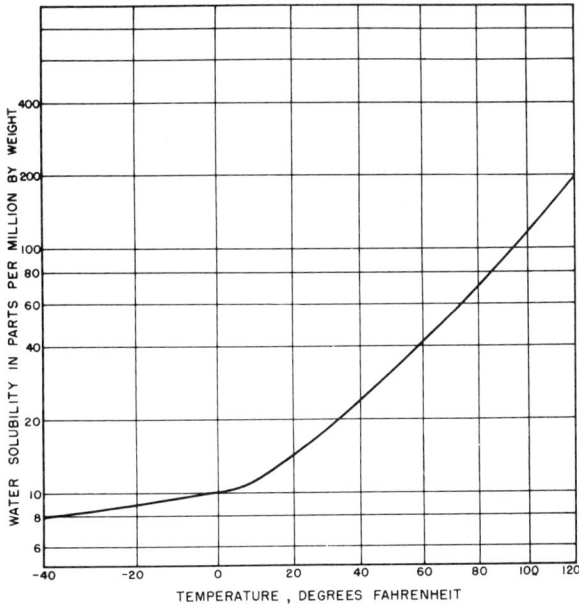

Fig. 35 Solubility of Water in Mineral Oil

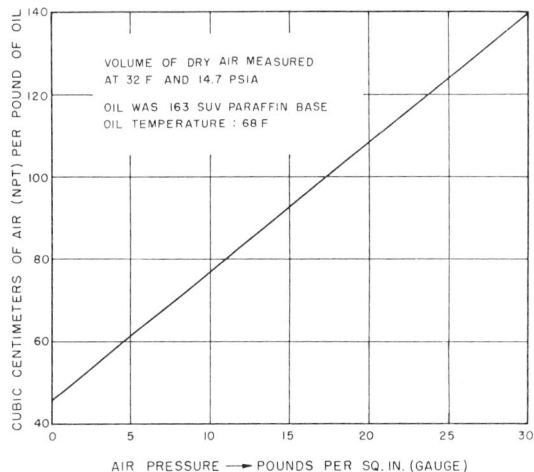

Normally refined transformer oil. Redrawn from Clark (1940).

Fig. 36 Effect of Air Pressure on Solubility of Air in Mineral Oil
(Baldwin and Daniel 1953)

However, moderate foaming is beneficial in refrigeration systems, particularly for noise suppression. A foamy layer on top of the oil level dampens the noise created by the moving parts of the compressor. There is no general agreement on what constitutes excessive foaming or how it should be prevented. Some manufacturers add small amounts of an antifoam agent, such as silicone fluid, to refrigerator oils. Others believe that foaming difficulties are more easily corrected by equipment design.

OXIDATION RESISTANCE

Refrigeration oils used in hermetic systems are seldom exposed to oxidizing conditions. Handling and manufacturing practices include elaborate care to protect refrigeration oils against air, moisture, or any other contaminant. Oxidation resistance by itself is rarely included in refrigeration oil specifications.

Nevertheless, oxidation tests are justified, since oxidation reactions are generally free radical in nature, and chemically similar to the reactions between oils and refrigerants. An oxygen test, using the power factor as the measure, correlates with established sealed-tube tests. However, such oxidation resistance tests are not used as primary criteria of chemical reactivity, but rather to support the claims of chemical stability determined by sealed-tube and other tests.

Oxidation resistance may become a prime requirement during manufacture. The small amount of oil used during compressor assembly and testing is not always completely removed before the system is dehydrated. If the subsequent dehydration process is carried out in a stream of hot dry air, as is frequently the case, the hot oxidizing conditions can cause the residual oil to become gummy, leading to stuck bearings, overheated motors, and other operating difficulties. For these purposes, the oil should have high oxidation resistance. However, the oil used under such extreme conditions should be classed as a specialty process oil rather than a refrigeration oil. Any harmful effects occurring because of such processing is a contamination of refrigeration oils.

Once a refrigerant system is sealed against air and moisture, the oxidation resistance of an oil is not significant unless it reflects the chemical stability.

CHEMICAL STABILITY

Refrigeration oils must have excellent chemical stability. Within the enclosed refrigeration environment, the lubricating oil must resist chemical attack by the refrigerant in the presence of all the materials encountered, including the various metals, motor insulation, and any unavoidable contaminants trapped in the system. A lubricating oil reacts with the refrigerant at elevated temperatures, and the reaction is catalyzed by metals. Methods for evaluating the chemical stability of oil-refrigerant mixtures are covered in Chapter 6.

Various phenomena in an operating system, such as sludge formation, carbon deposits on valves, gumming, and copper plating of bearing surfaces have been attributed to oil decomposition in the presence of the refrigerant. In addition to the direct reactions of the oil and refrigerant, the lubricating oil may also act as a medium for reactions between the refrigerant and the motor insulation, particularly when the refrigerant extracts the lighter components of the insulation. Factors affecting the stability of various components such as wire insulation materials in hermetic systems are also covered in Chapter 6.

Copper plating is one visible sign of reactions between oils and refrigerants in an operating system. Often, after an accelerated life test and teardown of a compressor, films of copper are seen on the ferrous parts such as bearings, pistons, and valves, which operate at temperatures higher than the rest of the system and are in contact with oil. Usually, these films are too thin to affect system operation. In more severe cases, the buildup can be substantial and may interfere with the compressor's functioning.

The mechanism of dissolution and copper plating in refrigerant-oil systems, as well as methods to minimize the phenomena, are discussed in Chapter 6.

Effect of Oil Type

Mineral oils differ in their ability to withstand chemical attack by a given refrigerant. In an extensive laboratory sealed tube test program, Walker *et al.* (1960, 1962) showed that color darkening, corrosion of metals, deposits, and copper plating occur less in paraffinic oils than in naphthenic oils. Using gas analysis, Spauschus and Doderer (1961) and Doderer and Spauschus (1965) show that a white oil containing only saturates and no aromatics is considerably more stable in the presence of R-12 and R-22 than a medium refined oil. Steinle (1950) has reported the effect of oleoresin (nonhydrocarbons) and sulfur content on the reactivity of the oil, using the Philipp test. A decrease in the oleoresin content, accompanied by a decrease in sulfur and aromatic content, showed an improvement in the chemical stability with R-12, while the oil's lubricating properties became poorer. Schwing's (1968) study on a synthetic polyisobutyl benzene oil reports that it is not only chemically stable but also has good lubricating properties.

REFERENCES

Albright, L.F. and J.D. Lawyer. 1959. Viscosity-solubility characteristics of mixtures of Refrigerant 13B1 and lubricating oils. ASHRAE *Journal* (April):67.

Albright, L.F. and A.S. Mandelbaum. 1956. Solubility and viscosity characteristics of mixtures of lubricating oils and "Freon-13 or -115". *Refrigerating Engineering* (October):37.

API Technical data book—*Petroleum refining*, 2nd ed. 1970.

ASTM D-2549, Standard test method for separation of representative aromatics and nonaromatics fractions of high-boiling oils by elusion chromatography, or ASTM D-2007, Standard test method for characteristic groups in rubber extender and processing oils and other petroleum derived oils by the clay-gel absorption chromatographic method. American Society for Testing and Materials, Philadelphia.

Baldwin, R.R. and S.G. Daniel. 1953. *Journal of the Institute of Petroleum* 39:105.

Brewer, A.F. 1951. Good compressor performance demands the right lubricating oil. *Refrigerating Engineering* (October):965.

Divers, R.T. 1958. Better standards are needed for refrigeration lubricants. *Refrigeration Engineering* (October):40.

Downing, R.C. and W.D. Cooper. 1972. Operation of a three stage cascade system. ASHRAE *Transactions* 78(2):44.

Gunderson, R.C. and A.W. Hart. 1962. *Synthetic lubricants*. Reinhold Publishing Corp., New York.

Huttenlocher, D.F. 1969. A bench scale test procedure for hermetic compressor lubricants. ASHRAE *Journal* (June):85.

Kartzmark, R., J.B. Gilbert, and L.W. Sproule. 1967. Hydrogen processing of lube stocks. *Journal of the Institute of Petroleum* 53:317.

Mills, I.W., *et al.* 1946. Molecular weight-physical property correlations for petroleum fractions. *Industrial and Engineering Chemistry* 38:442.

Mills, I.W. and J.J. Melchoire. 1967. Effect of aromatics and selected additives on oxidation stability of transformer oils. *Industrial and Engineering Chemistry*, Product Research and Development 6:40.

Mosle, H. and W. Wolf. 1963. *Kaltetechnik* 15:11.

Murray, S.F., R.L. Johnson, and M.A. Swikert. 1956. Difluoro-dichloromethane as a boundary lubricant for steel and other metals. *Mechanical Engineering* 78(3):233.

Rembold, U. and R.K. Lo. 1966. Determination of wear of rotary compressors using the isotope tracer technique. ASHRAE *Transactions* 72:VI.1.1.

Sanvordenker, K.S. 1968. Separation of refrigeration oil into structural components and their miscibility with R-22. ASHRAE *Transactions* 74, Part I:III.2.1.

Sanvordenker, K.S. and W.J. Gram. 1974. Laboratory testing under controlled environment using a falex machine. Compressor Technology Conference, Purdue University.

Sanvordenker, K.S. and M.W. Larime. 1972. A review of synthetic oils for refrigeration use. Paper presented at ASHRAE Symposium, Lubricants, Refrigerants and Systems—Some Interactions. Nassau, Bahamas.

Schwing, R.C. 1968. Polyisobutyl benzenes and refrigeration lubricants. ASHRAE *Transactions* 74(1):III.1.1.

Spauschus, H.O. and G.C. Doderer. 1961. Reaction of Refrigerant 12 with petroleum oils. ASHRAE *Journal* (February):65.

Speaker, L.M. and H.O. Spauchus. 1986. A study to increase state-of-the-art solubility and viscosity relationships for oil/refrigerant mixtures. Final report prepared for ASHRAE for RP 444.

Steinle, H. 1950. *Kaltemaschinenole*. Springer-Verlag, Berlin, Germany, 81.

Van Nes, K. and H.A. Weston. 1951. *Aspects of the constitution of mineral oils*. Elsevier Publishing Co., Inc., New York.

Walker, W.O., S. Rosen, and S.L. Levy. 1960. A study of the factors influencing the stability of the mixtures of Refrigerant 22 and refrigerating oils. ASHRAE *Transactions* 66:445.

Walker, W.O., S. Rosen, and S.L. Levy. 1962. Stability of mixtures of refrigerants and refrigerating oils. ASHRAE *Transactions* 68:360.

Witco. Sonneborn Division, *Bulletin* 8846.

BIBLIOGRAPHY

ASTM. *Hydrocarbon analysis*. ASTM STP 389. American Society for Testing and Materials, Philadelphia.

ASTM. 1968. *Manual on hydrocarbon analysis*, 2nd ed. ASTM STP 332A. American Society for Testing and Materials, Philadelphia.

Bowden, F.P. and D. Tabor. 1956. *Friction and lubrication*. John Wiley and Sons, Inc., New York.

Daniel, G., M.J. Anderson, W. Schmid, and M. Tokumitsu. 1982. Performance of selected synthetic lubricants in industrial heat pumps. *Journal of Heat Recovery Systems* 2(4):359-68.

Elsey, H.M., L.C. Flowers, and J.B. Kelley. 1952. A method of evaluating refrigeration oils. *Refrigerating Engineering* (July):737.

Freon Products Division, E.I. du Pont de Nemours and Co., Inc. 1977. *Bulletin* RT-56 (Feburary).

Heckmatt, H. 1960. Some lubrication problems of refrigerating machines using F12. *Journal of Refrigeration* 3(1):5.

Heide, R. 1970. A contribution to the viscosity of oil-refrigerant mixtures. *Luft-und Kaltetechnik* 6:308-10.

Heide, R. 1977. About the viscosity of refrigerant-refrigerating machine oil mixtures. *Luft-und Kaltetechnik* 10(3):34-7.

Heide, R., B. Herre, Ch. Staege, and H. Finger. 1981-82. Lufrigol XK 35—A new refrigerating machine oil. *Luft-und Kaltetechnik* 17(2):75-7.

Heide, R., H. Lippold, and G. Hackstein. 1974. Physical properties of a new refrigerating machine oil and its mixtures with R22 and R502. *Luft-und Kaltetechnik* 10(1):44-8 (German).

Hirschberg, H.G. 1966. Determining the viscosity of mixtures of mineral oil and refrigerants. *Modern Refrigeration*, 711-16.

Hughes, T.P., Jr. 1960. Application of refrigerator lubrication system to heat pumps. *Journal of Refrigeration* 3(1):12.

Jaeger, H.P. 1969. Thermodynamic properties of binary mixtures of refrigeration oils and R11, Part II. *Kaltetechnik und Klimatisierung* 21(12):367-69 (German).

Jaeger, H.P. 1972. Empirical methods for predicting thermodynamic properties of oil-refrigerant mixtures. Dissertation, Technical University, Braunschweig (German).

Jaeger, H.P. 1975. Calculating the viscosity of liquid oil-refrigerant mixtures. *Die Kalte* 9:33240 (German).

Jaeger, H.P. and H.J. Loffler. 1970. Thermodynamic properties of oil-refrigerant mixtures. *Kaltetechnik-Klimatisierung* 22(8):246-56.

Jaeger, H.P. and H.J. Loffler. 1971. Concerning the behavior of oil/R22 mixtures. *KaltetechnikKlimatisierung* 239(10):305-9.

Kriebel, M., H.J. Loffler, and H. Matthias. 1966. Thermodynamic properties of binary mixtures of tetrafluorodichloroethane (R114) and refrigerating machine oils. *Kaltetechnik-Klimatisierung* 7(18):261-67 (German).

Kruse, H.H. and M. Schroeder. 1984. Fundamentals of lubrication in refrigerating systems and heat pumps. ASHRAE *Transactions* 90:763-83.

Kruse, H.H. and M. Schroeder. 1985. Modified version of Kruse and Schroeder (1984). Rev. Int Froid 8 (November):347-55.

Loffler, H.J. 1956. Influence of physical properties of mineral oils on the solubility with Refrigerant 22. Dissertation, University of Karlsruhe (German).

Loffler, H.J. 1957. Review of refrigerants and mixtures of oils and refrigerants. *Kaltetechnik* 9(11):358.

Loffler, H.J. 1959. Fluidity of oil-Frigen 22 mixtures at low temperatures. *Kaltetechnik* 11(9):258.

Loffler, H.J. 1967. Thermodynamic properties of binary mixtures of Refrigerant 502 and refrigerating machine oils. *Kaltetechnik-Klimatisierung* 19(7):201-7 (German).

McKenzie, K.G. 1951. Tests for lubricating oils. *Refrigerating Engineering* (April):375.

Ross, E.S. Selection and use of oils for refrigeration units. *Air Conditioning, Heating, and Ventilating* 55(10):97.

Rutledge, O.C. 1938. Viscosity of oils diluted with refrigerants. *Refrigerating Engineering* (January):31.

Sakaki, U., K. Mifaji, and M. Taunemi. 1979. Physical properties of refrigerator oil under Freon atmosphere. *Nisseki Rebyu* 21(2):129-33 (Japanese).

Schroeder, M. 1985. Development and testing of a low cost screw compressor for Refrigerants 12 NS 114 and their mixtures. Dissertation, University of Hannover (German).

Schubert, R. 1957. Determination of fluidity of refrigeration oils at low temperatures. *Kaltetechnik* 9(2):40.

Standard handbook of lubrication engineering. 1968. American Society of Lubrication Engineers, New York: McGraw-Hill.

Steinle, H. 1960. The surface tension of refrigerants, of lubricants, and of their mixtures. *Kaltetechnik* 12(11):334.

Stirling, R. and F.G. Drakesmith. 1979. Refrigerant 114: Viscosity and solubility measurements in hydrocarbon and synthetic lubricants. Paper Bl-38, Proceedings of the 15th International Congress of Refrigeration, Venice.

Symposium on composition of petroleum oils, determination and evaluation. ASTM STP No. 224. University Microfilms, Ann Arbor, MI.

The Texas Company. 1957. *Lubricants for refrigeration systems*.

Thelen, A. and H.J. Loffler. 1960. Lubrication of cylindrical bearings with oil-refrigerant mixtures. Abhandlung des Deutschen Kaltetechnischen Vereins, Verlag C.F. Muller, Karlsruhe (German).

CHAPTER 9

COMMERCIAL FREEZING METHODS

Freezing Equipment 9.1
Freezer Applications 9.6

FREEZING is a cost-effective method of maintaining the quality, nutritional value, and sensory properties of food for extended periods. As shown in Table 1, canning has a greater energy requirement than freezing.

Table 1 Approximate Energy Cost Expressed as Index

Process	Energy Index Canning	Energy Index Freezing
Freezing	—	1
Sterilization	2	—
Cooling	2	—
Packaging	33	13.5
Store 6 months	0.5	5
Transport 300 miles	1	1.1
Total	38.5	20.6

Freezing is the process that changes a product's water content to ice and reduces the product temperature from ambient to storage level. *Frozen storage* is the holding of a product at a constant temperature, generally 0°F or lower. In Europe, food products are stored at −10°F or lower, which reduces the rate of deterioration and increases the storage life. Fish should be stored at −15°F or lower. Table 2 shows a comparison of equivalent age versus temperature for typical frozen fruits and vegetables. For example, storage at 20°F may only maintain fruit at an acceptable quality for a maximum of 7 days, but storage at 0°F will maintain the fruit at the equivalent quality for 365 days.

Table 2 Equivalent Age Versus Temperature of Frozen Fruits and Vegetables

Temperature, °F	Equivalent Age, Days	Biological Age[a] Chronological Age
0	365	1.0
5	152	2.4
10	42	8.7
15	21	17.4
20	7	52.1
25	2	183
30	1	365

Source: *Commodity Storage Manual*. The Refrigeration Research Foundation. Bethesda, MD.
[a] Biological Age/Chronological Age ≈ 1.2173^t, where t = product temperature (from 0 to 30°F).

FREEZING EQUIPMENT

Freezing equipment can be grouped by the basic method for extracting heat from food products as follows:

Air-blast Freezing. Air at relatively high velocities circulates over the product. The air picks up heat, then it is recooled by an air/refrigerant heat exchanger before being recirculated.

Contact Freezing. Food (packed or unpacked) is placed on or between metal plates or surfaces. The heat is extracted by direct conduction through the metal surface that is refrigerated by a circulating medium.

Immersion Freezing. Food is immersed in a low-temperature brine that is cooled by evaporators in a conventional refrigeration system.

Cryogenic Freezing. Food is exposed to an environment below −76°F, which is achieved by spraying liquid nitrogen or liquid carbon dioxide into the freezing chamber.

All these methods are used in the food processing industry; however, the favored systems are those that can be operated in-line with the preceding processing and preparation operation and the subsequent packaging function.

Air-Blast Systems

Because air is the most common freezing medium, equipment designs vary widely. These designs include storage room, blast room, stationary tunnel, push-through tunnel, automatic tunnel, belt freezer, and fluidized bed freezer.

Storage Room. A storage room, which is usually a sectioned-off portion of a cold storage, should not be considered freezing equipment, although it is sometimes used for this purpose. Freezing in a storage room has so many disadvantages that it should be used only in emergencies. The freezing is slow, and the quality of almost all products suffers. If products are already being stored in the room, their quality is jeopardized because flavors may be transferred from the warm products yet to be frozen.

Because a storage room is not designed for freezing, the cooling coils may frost up so quickly that the total refrigeration capacity is reduced below the level required for proper storage temperature. The temperature of products already frozen may rise considerably, affecting their quality. Generally, freezing in a storage room leads to lower quality and higher cost than using an in-line freezer specially designed for the application.

Blast Room. This sectioned-off portion of a cold storage is usually equipped with more forced-air cooling units or a larger evaporator surface than normal. The air coolers are equipped with fans that create some turbulence in the air. Products can be laid on trays, which are loaded in a freezing rack that is moved in and out of the blast zone by a forklift truck. The airspace between trays is at least 50% of the product thickness. Because air circulation is not controlled, the resulting heat transfer at the product/air interface is less effective. Therefore, the blast room offers acceptable conditions for a limited range of products, such as those with large cross sections (*e.g.*, carcass meat).

Stationary Freezing Tunnel. This simple freezer (Figure 1) can produce satisfactory results for the majority of products. It can be located in the processing line area rather than in a separately located cold storage facility. It is an insulated enclosure equipped with refrigeration coils and fans that circulate the air in a controlled pattern over the product. The air-circulation system design influences the freezing rate and the resulting loss of product mass.

The preparation of this chapter is assigned to TC 11.6, Prepared Food Products.

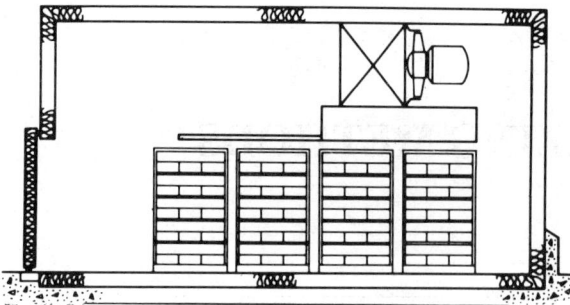

Fig. 1 Stationary Freezing Tunnel

Products are placed on trays that are then placed into a rack. Racks are arranged to provide an airspace above and below each level of trays and are moved in and out of the tunnel manually. The human element becomes important when situating the racks inside the tunnel. Preventing air bypass is the key to a more effective freezing process.

Practically all products can be frozen in a stationary freezing tunnel. Whole, sliced or diced vegetables may be frozen in cartons or unpacked in a 1.2 to 1.5-in. deep layer on trays. Spinach, broccoli, meat patties, fish fillets, and prepared foods are often frozen in packages in this type of equipment. By using different rack designs, thick packages and whole meat carcasses can also be frozen. System capacity depends on product thickness and composition as well as the existence of packaging.

The flexibility of this type of freezer makes it suitable during the initial development stage of a new frozen food market, but it also requires a heavy outlay of manpower and can cause considerable weight loss or quality impairment if improperly used.

Push-Through Tunnel. This more mechanized version of a stationary freezing tunnel (see Figure 2) has racks fitted with casters or wheels. The racks or trolleys are usually moved on rails by a pushing mechanism that is often hydraulically powered. This freezer has the same basic advantages and disadvantages as the stationary tunnel. However, the labor costs can be decreased, although with a corresponding loss of flexibility. Products with different freezing times should have separate tracks or rails.

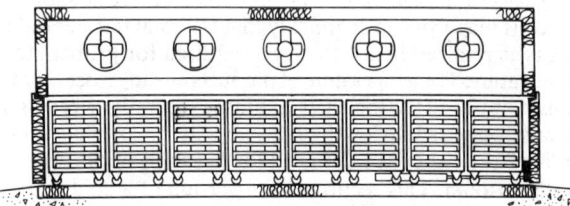

Fig. 2 Push-Through Tunnel

Automatic Carton Freezing Tunnel. The demand for automatic operation of freezers has led to a great variety of freezing tunnels with more sophisticated mechanization. Types of freezers include: sliding tray freezer, traveling tray freezer, and carrier freezer.

The sliding tray freezer consists of one great rack that accommodates many big trays on each tier. At one end of the system, an elevating mechanism lifts entering trays to the top tier. Here they are pushed inward, forcing all the other trays in the tier to advance one step. The tray at the far end is pushed onto an elevator, lowered one tier, and pushed in the return direction; in every odd tier, the trays advance, and on every even tier, they return. For each tray that enters, all trays advance one step.

In another version of the automatic freezing tunnel, the trays move on only one tier at a time, which gives almost the same result. This version is sometimes equipped with a plate freezer on 20 to 30% of each tier. This eliminates the bulging of packages but requires more space.

Usually, all mechanisms are powered hydraulically. Outside the freezer enclosure, trays must be automatically loaded and unloaded. Because each tray is exposed to considerable mechanical stress that limits both width and length, this type of freezer is suitable for handling intermediate size packages at moderate capacity.

The carrier freezer (Figure 3) may be regarded as two push-through tunnels on top of each other. In the top section, the carriers are pushed forward; in the lower section, they are returned. Both ends contain elevating mechanisms. A carrier is similar to a bookcase. When it is indexed up at the loading end of the freezer, frozen product is pushed from the carrier, one shelf at a time, onto a discharge conveyor. As the carrier is indexed upward, unfrozen product is transferred from an infeed conveyor to the emptied shelf.

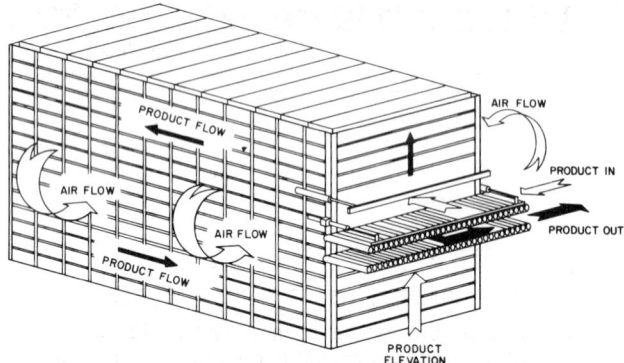

Fig. 3 Carrier Freezer

The carriers can be designed for almost any pitch between shelves and for variable lengths and widths. This allows maximum compactness. Loading and unloading may be manual or fully automatic. If fully automatic, speeds of over three packages per second can be achieved.

The automatic freezers described are primarily for packaged products, specifically cases of food such as fruit pies or 50-lb red meat portions. Attempts to freeze unpackaged fish fillets, meat patties, or similar products individually on trays have been only moderately successful. Products stick to the trays, causing damage and weight losses when mechanically removed. Although heating the trays will release the products, this requires complex equipment and reduces capacity. Furthermore, the trays must be washed after removing the frozen product to maintain hygienic conditions. The handling of the trays from outfeed to infeed of the freezer is costly, whether it is done manually or automatically.

Belt Freezers. The first belt freezer consisted of a wire mesh belt conveyor in a blast room and satisfied the need for continuous product flow. But this design suffered from the disadvantages of poor heat transfer and mechanical problems.

Modern belt freezer designs have improved contact between air and product, which is achieved by a vertical airflow pattern. Uniform distribution of product over the entire belt surface is a prerequisite for effective freezing. A thin or nonexistent product layer offers less resistance to the airflow. This causes a concentration of air in less densely loaded areas as the air bypasses the deeper product layer. This *channeling* of air may result in poorly frozen products. Therefore, the product must be spread uniformly over the total belt surface area.

The single-belt freezer, the multitier belt freezer, and the spiral belt freezer are the main types of belt freezers used. The simplest is the single-belt freezer, consisting of a single belt exposed to an updraft of air. It is suited for deep fried or relatively dry

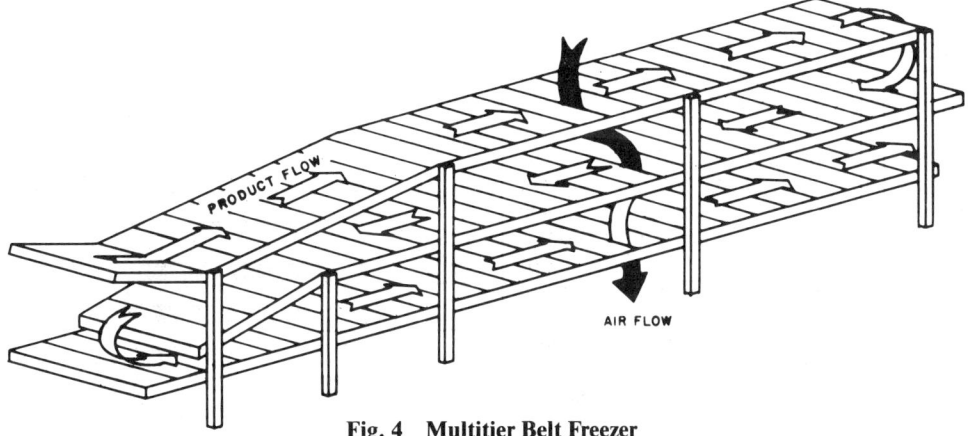

Fig. 4 Multitier Belt Freezer

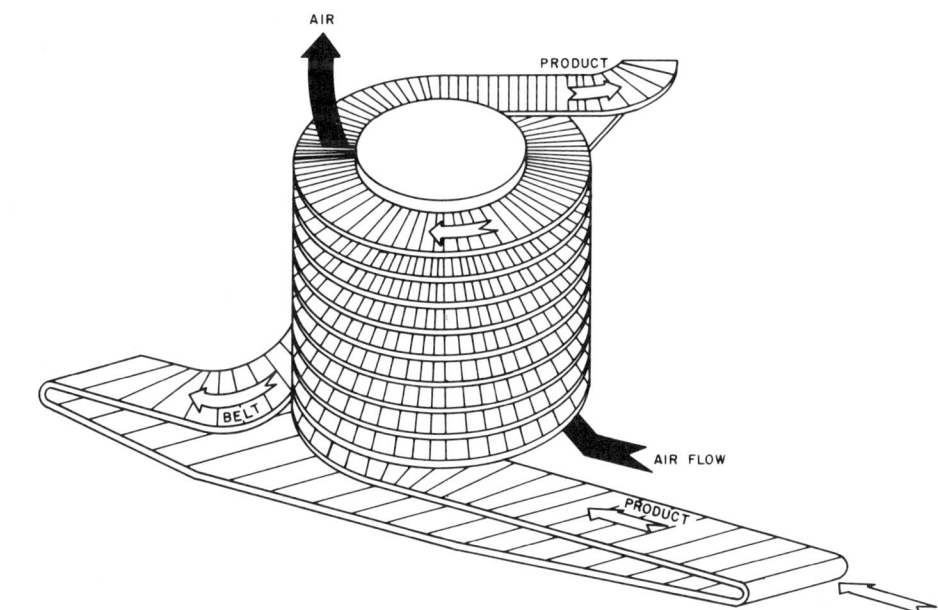

Fig. 5 Spiral Belt Freezer

products that do not tend to freeze to each other or form clumps (*e.g.,* fish sticks, French fried potatoes, or bakery products). Wet products tend to form clumps and ice buildup that can jeopardize belt life and give the product a less attractive appearance. This can be avoided by installing one of a variety of "clump-breakers" at the point where the surface of the product begins to freeze.

For large capacities, the single-belt freezer requires ample floor space. This requirement is reduced by building several tiers of belt above each other, an arrangement that has another advantage: the product, after being surface frozen on the first belt, may be stacked in a rather deep bed on the lower belts. Therefore, total belt area required is reduced. A freezer design combining the advantages of the single-belt and multitier types consists of two in-line tandem belts. Product is crust-frozen in a thin layer on the first belt, then bulk-frozen to final temperature on the slower-moving second belt.

Multitier belt freezers (Figure 4) are suitable for individual quick freezing of fried fish sticks, fish portions, bakery items, and other products. However, in most cases, the longer freezing times for packaged products make these freezers uneconomical in an individual quick freezing (IQF) application.

The spiral belt freezer (Figure 5) maximizes the belt surface area in a given floor space by using a product belt that can bend laterally around a rotating drum. By stacking up to 40 tiers of belt, a minimum of floor space is occupied by the freezer.

A traditional spiral freezer design features a low tension drive system. The special collapsible belt can be wound around one or more drums in a continuous circuit. The belt is sometimes supported by spiraling rails and driven by the friction developed between the belt and the rotating drum. One manufacturer provides articulated side plates on the belt, which makes the circular belt tiers self-stacking.

The continuous belt eliminates product transfer points internal to the system. Product is transferred only at the infeed and outfeed ends of the freezer. Products are placed on the belt outside the freezer, where the transfer can be monitored. Because the belt is continuous, product occupies the same area throughout the freezing process. No significant product movement occurs with respect to the belt. By employing a single belt, a continuous cleaning process can be installed for those products that warrant it. The flexibility of the belt allows for more than one infeed and outfeed using the same continuous belt. Furthermore, the relative locations of the infeed and outfeed can be arranged in various configurations.

Spiral belt freezers are suitable for unpackaged meat, fish, and poultry products, including meat patties, meatballs, fish fillets, and cut chicken portions. The versatility of the spiral design also makes it suitable for packaged products, such as prepared foods. The spiral belt freezer is one of the few conventional freezing tunnels that can be used for freezing soft formed products such as raw meatballs or patties.

Fluidized Bed Freezers. In freezing, particles of similar shape and size are subjected to an upward stream of low-temperature air. At a certain air velocity, the particles float in the airstream; each particle is separated from the other but surrounded by air and free to move. In this state, the particle mass assumes the properties of a liquid or fluid. The fluidization principle, when applied to an in-line freezer, corresponds to the concept of a dam with spillway. If the particles are retained in a cubical area that is designed with an open end (the outfeed end) slightly lower than the other three walls, product fed in at the infeed will displace product in the fluidized bed and cause a flow through the freezer bed. The low-temperature air achieves the fluidization, freezes the product, and simultaneously conveys it without the aid of a mechanical belt.

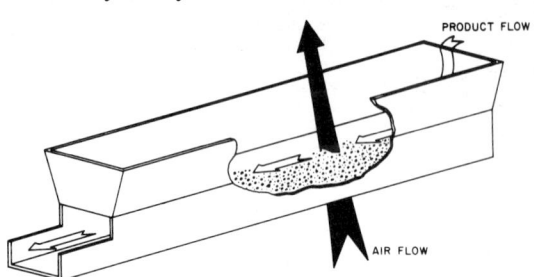

Fig. 6 Fluidized Bed Freezer

Using the fluidization principle for freezing (Figure 6) provides several advantages over the more conventional belt. Most products that tend to stick together (*e.g.,* sliced green beans, sliced carrots, and sliced cucumbers) are individually quick-frozen. The fluidized bed freezer is effective and reliable when freezing wet products, and it can accept products with high surface water content. Another advantage of the fluidized bed freezer is its complete independence from product flow variations. Even when running at reduced capacity, the same evenly distributed air pattern is maintained without channeling or air bypass.

Fluidization achieves effective air/product contact, which produces heat transfer rates higher than those considered normal for conventional air-blast freezing tunnels. The heat-removal efficiency can be seen in the physical dimensions of the freezer, which are generally one-third of the comparable capacity belt freezer. The fluidized bed freezer is suitable for vegetables, fruits, berries, and processed products, such as French fried potatoes, peeled cooked shrimps, diced meat, or cooked meatballs.

Contact Freezers

Most contact freezing is accomplished in manual or automatic plate freezers or on continuous, solid stainless steel moving belts. Plate freezers (Figure 7) are suited for freezing packaged products, although they are sometimes used to freeze fish fillets and various liquids such as coffee, soup, and gravies.

Plate Freezers. In a plate freezer, the product is pressed firmly between top and bottom metal plates. The refrigerant is circulated in channels housed inside the plates. This ensures good heat transfer and relatively short freezing times if certain design and operating criteria are met. Plates should be flat and free of distortion and packages must be tightly filled with product. It is also an advantage if the product itself is a good conductor. All of these factors have a positive influence on heat transfer by conductance, the primary heat removal mode. Because heat transfer at the surface is gradually reduced with increasing thickness of the product, package thickness is often limited to a maximum of 2 in.

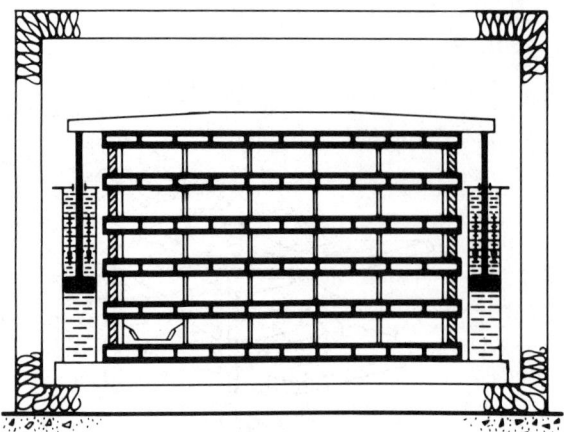

Fig. 7 Plate Freezer

Pressure from the plates has a secondary positive influence during the freezing process. It eliminates bulging in packages, a common occurrence in air-blast tunnels. Therefore, packages discharge with straight sides that are within close tolerances.

Plate freezers are horizontal or vertical, referring to the orientation of the plates. Either type can be manual or automatic. Generally, manual horizontal plate freezers have 15 to 20 plates. Product is placed on metal trays at the end of the packaging line, loaded in a rack or trolley, and transported to the freezer. Trays are loaded manually between the plates.

A horizontal plate freezer can be automated by designing the entire battery of plates to move up and down in an elevating system. At the level of the loading conveyor, the plates separate and packages that have accumulated on an infeed conveyor are pushed between the plates. This action simultaneously discharges a row of frozen packages at the opposite end of the plates. The cycle repeats until all frozen packages are replaced. Plates are then closed and all plates are indexed upwards.

The vertical plate freezer, developed originally for freezing fish at sea and now popular for freezing offal from slaughtering plants, is comprised of a series of vertical freezing plates that form partitions in a container. Products are fed from the top; the finished block of frozen product is discharged to either side, top, or bottom. Usually this operation is mechanized. Block thickness varies from 2 to 6 in.

Immersion Freezers. For irregularly shaped products such as chicken, high heat transfer can be achieved in an immersion freezer, which normally consists of a tank that houses a refrigerated brine. The brine is most often a glycol or sodium chloride solution. Product is immersed in the brine or sprayed while it is conveyed through the tank.

Immersion freezers are commonly used for plastic-bagged turkeys. Another use is surface freezing of poultry. Final freezing is accomplished in a separate blast tunnel or during cold storage. The latter, however, may impair quality because of slow core freezing. The product is protected from contact with many of the brines by high quality packaging that gives an absolutely tight seal. Brine residue on the packages is washed off with water at the freezer exit; this brine consumption can be costly.

High quality products commonly require double handling in an immersion freezer, which is unnecessary in air-blast freezers. The immersion freezer has been improved to satisfy the poultry industry's special demands concerning the color setting of birds.

Another type of brine freezer, which uses a sodium chloride solution as the brine, is frequently applied in the fishing industry to freeze unpackaged crab sections or whole dressed fish. The brine is refrigerated in a heat exchanger by conventional refrigeration equipment. However, brine in this system can contaminate the product.

Commercial Freezing Methods

Cryogenic Freezing

Cryogenic freezing exposes food products to an environment below $-76°F$. Liquid nitrogen (LN_2) or liquid carbon dioxide (LCO_2) are the primary cryogens and are generally used so that they make direct contact with the food in either their liquid or solid phase.

Freezer designs offered are straight belt, flighted belt, spiral belt, multitier belts, and immersion designs. These designs may differ based on the cryogen used. The size and mobility of cryogenic freezers offer flexibility when designing or redesigning a processing line in a food plant.

Key attributes of cryogenic freezers are high heat transfer rates, low initial cost, rapid installation and start-up, ability to fit in highly mechanized food-processing lines, and mechanical simplicity.

Because of their high heat transfer rates, cryogenic freezers can be used for chilling, firming, or crusting products such as chicken or beef patties, candy, other foodstuffs, plastic, and rubber. They typically are sized to perform specific refrigeration functions based on process requirements. Commonly, several specialty freezers are placed in series, each fulfilling a different refrigeration requirement for the process.

A cryogen freezing system is uniquely set apart from other systems in a food plant. This allows good cost control because the cost components are easily identified. The cryogen cost is normally identified per unit mass of cryogen delivered. The cryogenic storage vessel is designed for the CO_2 or LN_2 and has its associated cost. The standard CO_2 storage system has no standby loss, if a supplemental refrigeration system is employed. If not used, the standard LN_2 system loss is 0.25 to 1% per day. The processing plant does not require a separate refrigeration installation with cryogenic freezing systems.

The cost of operating a cryogenic freezer is influenced by equipment design and operation methods. Outside air should not be allowed to infiltrate into the freezing chamber because it reduces cryogen cooling efficiency and freezer capacity. Typical cooling capacity is 120 Btu/lb for CO_2 and 160 Btu/lb or more for LIN. Figures 8 and 9 combine to give the total available cooling capacity of LIN. The higher the exhaust gas temperature, the greater the cooling capacity and subsequent efficiency.

Of this available cooling capacity, 70 to 80% should be transferred to the product in a well-designed and operated system. Two-shift operations may be up to 5% more efficient. This capacity includes all normal losses in storage and transfer, as well as in the freezer.

The large temperature differences between cryogen, product, and ambient conditions can result in large design errors if improper assumptions or data are used. For example, food exiting any freezer is much colder on the outside than it is in the center. The product takes 10 to 20 min before its temperature equalizes, depending on the temperature differential, mass, and surface area. Therefore, probing individual pieces with a thermometer before that time will give erroneous results.

Liquid Nitrogen Cryogen

In a typical straight belt LIN cryogenic freezer, liquid nitrogen at $-320°F$ is sprayed into a single-belt freezer in which the atmosphere is circulated with turbulators. As the liquid nitrogen contacts the product, it quickly vaporizes, and cold nitrogen vapors are directed over the incoming product in a counterflow arrangement. Typically, vapor is discharged to the atmosphere at -50 to $20°F$.

Liquid nitrogen provides a fast freezing cycle and very high reserve freezing capacity (often referred to as "turn up"). In some instances, the freezing capacity can increase by 30% but with the penalty of higher cryogen consumption (exit vapor temperatures below $-100°F$) and corresponding higher freezing cost. Also,

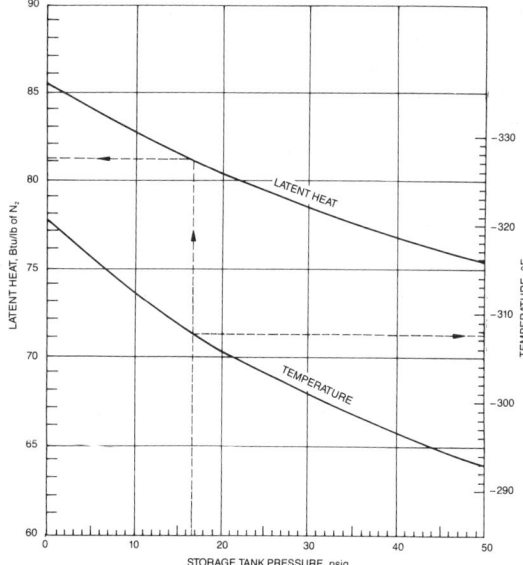

Fig. 8 Liquid Nitrogen Latent Heat Versus Tank Pressure

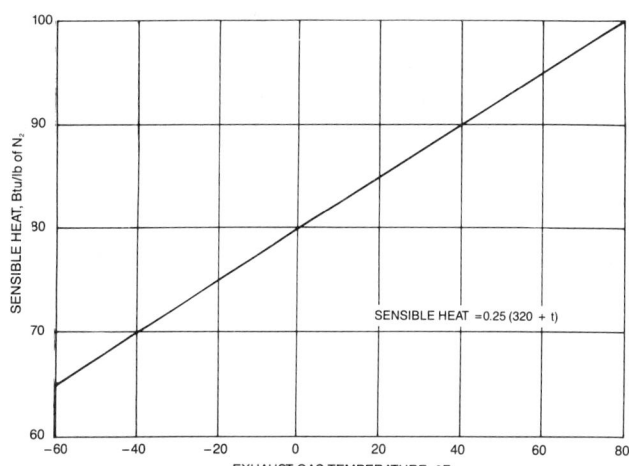

Fig. 9 Liquid Nitrogen Sensible Heat Versus Exhaust Gas Temperature

some food products may not withstand the thermal stress imposed by this "turn up" method of operation, as its surface may crack.

Liquid Halocarbon Cryogen

Liquid halocarbon freezers are seldom manufactured today because of the environmental consequences of releasing R-12 into the atmosphere; however, they are mentioned because many units are still operating. In this type of freezer, the product advances on the conveyor belt and is sprayed with food-grade liquid R-12 at $-22°F$. While the product quickly freezes, the vaporized freezant is liquified in a $-45°F$ reflux condenser mounted at the top of the enclosure. It then drains to a collection tank at the bottom of the unit, from which it is repumped to the spray headers.

Carbon Dioxide Cryogen

Carbon dioxide is normally stored as a liquid at 300 psig and $0°F$. At atmospheric pressure, it exists only as a solid or a gas. When liquid is released to the atmosphere, 46% becomes dry ice snow at $-109°F$, and 54% becomes $-109°F$ vapor, with approximately 95% of its refrigeration in the snow portion. Because of these unusual properties, CO_2 freezer designs vary widely. Food product characteristics usually influence the design decision. The highest heat transfer rates are achieved by direct contact of the food product with the snow via a tumbling and scrubbing action. Very high heat transfer rates are also achieved by direct impingement of a jet of snow particle and vapor on the food product. High velocity vapor machines are also used as required by some food product characteristics.

FREEZER APPLICATIONS

When matching freezer design with product and process requirements, such factors as production capacity, quality, and economics should be considered.

Production Capacity

The user of freezing equipment is mainly concerned with its capacity to meet the expected production rate. Generally, the following equation applies to every freezer.

$$C = Q/F = (Vq)/F$$

where

- C = capacity of freezer, lb/h
- Q = quantity of product that can be kept in freezer at one time, lb
- F = freezing time of the product in the freezer, h
- V = volume of product that can be kept in freezer, ft^3
- q = density of product, lb/ft^3

The freezing time is the time required for the entire temperature change, normally from ambient temperature to 0°F. Infeed temperatures typically range from 176 to 26°F and outfeed temperatures from 32 to −40°F.

Some commercial freezer manufacturers have model freezers to determine experimentally the freezing time of a customer's product. During its life, however, a freezer is generally required to freeze different products and/or other sizes of food containers than those originally specified. At least one manufacturer uses a computer model, based on the original, experimentally determined performance parameters, to predict the new performance.

Input data include physical properties of the product, belt size, produce bed depth, circulating air volume, evaporator temperature, coil area, heat transfer rate, product initial and discharge temperature, and nonproduct tunnel loads such as air infiltration and electrical motor loads. Output information includes product freezing rate, freezing time, total refrigerating load, and temperature profiles of the product and air along the tunnel. Computer programs are available for (1) single-belt freezing tunnels, (2) two-tiered belt tunnels, and (3) inline tandem-belt tunnels.

Quality

Quality of processed food products is affected by physical changes and the speed of microbiological and chemical reactions. Each of these is influenced by the rate of temperature change. The freezing process physically changes the food product. The *rate* of physical change or freezing time determines the size of ice crystals produced. At a slow freezing rate, initially formed ice crystals can grow to a relatively large average size. Fast freezing forces more crystals with a smaller average size to be formed. In practice, however, different-sized ice crystals are formed because the product's surface freezes faster than its inner parts.

Large ice crystals may puncture the cell walls of the product, usually increasing drip loss of juices during thawing. Depending on the product, this can greatly affect the texture and flavor of the remaining product tissue.

The influence of freezing time is more apparent on some products than others. For strawberries, the drip loss is reduced from more than 20% to nearly zero with shorter freezing time. A difference in drip loss from 20% for strawberries frozen in 12 h to 8% for strawberries frozen in 15 min is significant. Cryogenic systems perform the same freezing function in 8 min or less, reducing the drip loss to less than 5%.

A modern mechanical freezer, which is well applied, or a cryogenic freezer can crust products rapidly, minimizing loss of natural juices, aromatics, and flavor essences, and locking in the qualities that make the product more marketable. Quality is also affected by storage temperature and temperature fluctuations during storage. Lower storage temperature and fewer less-severe temperature fluctuations tend to preserve quality better.

Economics

Ironically, freezing equipment is considered both the most expensive and the least expensive link in the modern processing chain. Although the freezer frequently represents the single largest investment in a line, its operating costs are usually only 3 to 5% of the total. Packaging costs may vary widely, but normally are several times greater than total freezing cost.

One essential factor to consider when choosing freezing equipment is the loss in product mass that occurs during freezing. This loss may be about the same as the operating cost of the freezer. This applies to inexpensive products like peas, and is even more significant for expensive products like seafoods.

Weight loss during freezing may be caused by mechanical losses, downgrading, and dehydration. Mechanical losses refer to products dropping to the floor, sticking to conveyor belts, or juice dripping, all of which are very specific for each plant. A modern freezer should have almost no losses in this category. Downgrading losses refer to damaged product, breakage, and similar occurrences that render the product unsalable at the top quality price.

Dehydration losses will always be present in any freezing system. The evaporation of water vapor from unpackaged products during freezing becomes evident as frost builds up on evaporator surfaces. This frost is also caused by excessive infiltration of warm, moist air into the freezer. Still air inside the diffusion-tight carton often creates larger dehydration losses than the unpackaged products frozen in an IQF freezer. Heat transfer is poor because no circulation of air occurs within the package. The result is an evaporation of moisture that can be significant; however, the frost stays inside the carton.

A poorly designed freezing tunnel easily operates with dehydration losses of 3 to 4%, while a well-designed tunnel can be built to operate with losses of 0.25 to 1.5%. Liquid nitrogen tunnels normally operate with a dehydration loss of about 0.2 to 1.25%. This loss occurs when the nitrogen gas is circulated over the product at the infeed end of the freezer. Infeed-end circulation is sometimes necessary to temper the product and to use the heat capacity in the nitrogen most efficiently. Nitrogen immersion freezers tend to have lower dehydration losses but use more liquid nitrogen. CO_2 freezer operation using jet impingement operates with a dehydration loss of about 0.2 to 1.25%. The vapor circulation system has much less heat transfer to accomplish in a CO_2 freezer.

CHAPTER 10

MICROBIOLOGY OF FOODS

Foods Refrigerated at Temperatures above Freezing..................................10.1
Frozen Foods..10.5

THE need for extending the storage life of foods after processing or harvesting for distribution to the consumer has furthered the development of the refrigeration industry. While refrigeration is applied for purposes other than food preservation, its largest overall application is for the prevention or retardation of microbial, physiological, and chemical changes in foods.

Even at temperatures near the freezing point, foods may deteriorate through the growth of microorganisms, through changes caused by the action of enzymes, or through chemical reactions. Holding foods at low temperatures merely reduces the rate at which deteriorative changes take place.

This chapter outlines the bacteriology of refrigerated, unfrozen, and frozen foods. Methods of applying refrigeration to foods may be found in other chapters in this volume.

FOODS REFRIGERATED AT TEMPERATURES ABOVE FREEZING

Fruits, vegetables, flesh-type foods, eggs, and dairy products are all subject to microbial deterioration when refrigerated in the unfrozen state. However, loss of quality in fruits and vegetables held under such conditions is primarily because of physiological changes.

Spoilage of these foods occurs primarily through the growth of bacteria. Minimum growth temperatures for some spoilage-type bacteria and for some food-borne disease causing bacteria are listed in Table 1.

Table 1 Minimum Growth Temperatures for Some Bacteria Which May Be Found in Foods

Organism or Type	Possible Significance	Approx Min. Growth Temp, °F
Staphylococcus aureus	food-borne disease	44
Salmonella spp	food-borne disease	44
Clostridium botulinum, types A and B	food-borne disease	44
Clostridium botulinum, type E	food-borne disease	38
Clostridium botulinum, type F	food-borne disease	38
Lactobacillus and *Leuconostoc*	spoilage of cooked sausage	38
Pseudomonas fluorescens	spoilage of fish	31
Acinetobacter spp	spoilage of precooked foods	31
Pseudomonas spp	spoilage of fish, meats, and dairy products	31

Meat Products

Source of Contamination. The main source of the bacteria present on meat surfaces immediately after slaughter is the hide or skin of the animal. These bacteria are primarily water and soil types. During dressing operations, the clothes and hands of the worker and skinning knives contact the outer surface of the cattle's hide and the surface of the skinned animal. Thus, the hide may become the chief source of contamination, because bacteria are transferred from it to meat surfaces either directly or indirectly. Additionally, contact with the contents of the animal's rumen or bung, and with airborne contamination may be sources of contamination but are less likely to be the origin of bacteria on meat surfaces. Usually, the esophagus and the bung are tied off to prevent contamination with intestinal contents. Cutting knives and saws usually do not carry such large numbers of bacteria as do skinning knives, clothes, and hands, which contact the outside of the hide.

Internal tissue contamination of hogs is less significant because short-time cures are used for pork products. When 30- to 40-day cures are used for bacon, and 60- to 90-day cures are used for hams, bacteria contaminate internal tissues and cause spoilage in cured pork products. Internal tissue contamination may be an important cause of spoilage of canned (pasteurized) hams, under conditions of mishandling. Also, commercial sterility in canned meat products not held under refrigeration depends in part on low-level contamination of the meat prior to heat processing.

Spoilage of Meats. Preslaughter conditions may affect the susceptibility of meat to bacterial spoilage. During *rigor mortis* in animals, muscle glycogen is changed into lactic acid. This causes the pH of the tissues of properly handled animals to reach values, after slaughter, ranging between 5.1 and 6.2 for beef, 5.4 and 6.7 for lamb, and 5.3 to 6.9 for pork. However, if an animal is fatigued prior to slaughter, the muscle glycogen may be depleted. Under this condition, normal amounts of lactic acid are not formed, resulting in meat with high pH, sometimes as high as 7.4. The bacteria that cause spoilage of meat grow best at a pH of 7.0, or somewhat above, and do not grow nearly as fast at a pH of 6.0 or below. Meat with a high pH may, therefore, be subject to comparatively rapid spoilage, even when held at refrigerator temperatures above freezing.

During beef cooling and holding, bacteria grow on surfaces, and if sufficient growth occurs, slime formation will be evident. This requires a concentration of bacteria between 0.5×10^6 and 1×10^6 cells per square millimetre (Ayres 1955). Sliming occurs mostly in places such as fissures in the neck muscle, made by removing the head, and the folds of flesh under the foreleg, because these are the areas where surface moisture in quantities sufficient to support rapid bacterial growth is frequently found.

Surface moisture on uncut meat is affected by the fat covering on the side or quarter, the relative humidity of the air in coolers and storages where the meat is held, and the rate of air movement over the product during holding. Relative humidities of 90 to 95% are ordinarily employed. Temperatures should be held as near 32 °F as possible without freezing the meat after cooling, although starting temperatures may be as low as 25 °F. High air movements and low relative humidities dry surfaces and cause a loss of weight, whereas low air movements and high relative humidities promote bacterial growth, causing excessive trimming losses. The operator must regulate and control cooler and storage conditions so that

---

The preparation of this chapter is assigned to TC 11.6, Prepared Food Products. The chapter last received a major revision in 1967.

meat surfaces are neither too moist nor excessively dehydrated. Usually a sodium chloride solution is used for spraying to prevent spoilage and maintain moisture.

When meat is cut into whole or retail portions, the new surfaces are contaminated with bacteria. Cut surfaces are moist and, under ordinary circumstances, cannot be dried. Such areas, therefore, provide a better environment for the growth of bacteria than does the surface of the side or quarter. Portions such as loins or ribs, and especially retail cuts, have a shorter storage life than sides, mainly because of the greater areas of moist surfaces available for bacterial growth. The larger portions may be trimmed when excessive bacterial growth occurs.

Ground beef generally has a shorter refrigerated storage life than that of unground retail portions, because surface areas increase greatly and are contaminated with bacteria when meat is ground.

Pork that is cut into portions is subject to the same surface moisture-bacterial growth relationships as beef. Pork is cut into the larger portions, such as loins, hams, and so forth, in rooms at temperatures much higher than those used in storage rooms. Thus, low relative humidities must be established in cutting rooms to prevent condensation of air moisture on meat surfaces (sweating), which facilitates bacterial growth.

Bacterial Spoilage of Fresh Meats. The bacteria that grow at relatively low temperatures are called psychrophiles and are aerobic, gram-negative, rod-shaped organisms of both pigmented and nonpigmented types. Generally, the preponderance of bacteria causing spoilage of fresh meats belong to the genus *Pseudomonas*. Both the proteolytic and nonproteolytic types of *Pseudomonas* may grow on flesh-type foods; however, proteolytic types of pseudomonads are predominant when meat spoilage occurs.

Bacteria of various genera, including *Pseudomonas, Acinetobacter, Micrococcus,* and *Flavobacterium* have been found on the surface of freshly slaughtered and dressed beef. *Acinetobacter* may grow to some extent on beef, but the predominating bacteria present on spoiled beef are pseudomonads.

Sides or quarters of beef are sometimes held at low temperatures (32 °F or slightly below) for as long as 30 days, for purposes of special aging (special tenderizing). In such cases, the surface may become entirely covered with mold growth. While such mold growth does not spoil the meat, and some consumers may prefer products handled in this manner, this kind of aging may cause costly trimming losses. The molds which grow on meat surfaces belong to the genera *Mucor, Thamnidium,* and *Rhizopus*.

Bacteria of the family *Lactobacillaceae* may cause spoilage of fresh meat. These may include species of the *Lactobacillus, Leuconostoc, Pediococcus,* or *Streptococcus* genera. When ground beef is packed in material which excludes the transfer of oxygen, lactobacillaceae grow while pseudomonads do not. Under such conditions, relatively high numbers of lactobacillaceae may eventually be present. In packaged meat, even that in which oxygen is available for bacterial growth, the lactobacillaceae are often present in large concentrations after the meat has been held for some time at refrigerator temperatures above freezing. However, under conditions in which oxygen is available and relatively low temperatures are applied, the pseudomonads grow so much faster in fresh meat that they will ordinarily cause spoilage before significant growth of the lactobacillaceae occurs.

Spoilage (Greening) of Processed Meats. Processed meat products, such as frankfurters, may undergo bacterial spoilage. Such spoilage usually involves the formation of green surface spots, green rings, green cores, or surface sliming. Greening of any type is the result of oxidation of nitrosohemochromogen (the color complex of heated nitrite-cured meat) with hydrogen peroxide produced by the growth of bacteria. Species of the lactobacillaceae cause green spots and green cores in frankfurters and, in some cases, greening of cured and cooked ham. Grown aerobically or anaerobically and exposed to oxygen for only a few hours, these organisms produce hydrogen peroxide which causes greening.

The lactobacillaceae are not normal contaminants of meat or meat products, but are brought into processing plants by humans, by indirect contamination of meat with bacteria from the contents of animal intestines, and by contamination from other sources. These bacteria will grow at temperatures used for the refrigeration of meat products, although more slowly than many of the psychrophilic bacteria. When frankfurters or ham are cooked during plant processing, some of the lactobacillaceae may survive, while all of the ordinary psychrophiles which grow on fresh meat are destroyed. This relatively high heat resistance (for vegetative types) and contamination during handling subsequent to cooking are responsible for spoilage of cooked-cured meat products caused by these bacteria.

Public Health Aspects. Fresh meat and meat products have sometimes been associated with food poisoning because they are a source of bacteria that cause food infections; sometimes toxin-producing bacteria have grown in these products, causing food intoxication. A relatively high percentage of fresh pork sausage contains bacteria of the *Salmonella* group. Leistner *et al.* (1961) found that only 3% of pig feces on farms contained *Salmonellae*, while, at the time of slaughter, 59% of the feces samples examined contained these bacteria. The greater incidence of *Salmonella* in feces at the time of slaughter appeared to be because of contamination or infection occurring during the holding of animals in pens used at the terminal market. Deposits in these pens were found to be highly contaminated with *Salmonellae*.

Salmonellosis usually occurs in the very young or the very old. People of other ages contract the disease only after ingestion of relatively large numbers of the causative bacteria. Since, in freshly slaughtered meats, only small numbers of these organisms are expected to be present, and since *Salmonellae* do not grow at temperatures below 40 °F, their presence in fresh meat would have little significance if meat were always adequately refrigerated. *Salmonellae* are also destroyed at the temperatures to which all parts of pork should be brought during cooking. Proper refrigeration and cooking have been important in limiting outbreaks of salmonellosis because of meat consumption.

Assuming meat to be a source of *Salmonella* infections, control measures which might be applied to decrease the incidence of infection include: (1) rigid inspection of plants that produce ingredients for animal feeds, such as meat and bone meal and fish meal, which may be a source of these organisms in animals; (2) testing of animal feeds; (3) better sanitary conditions in, or the elimination of, holding pens in handling animals prior to slaughter; and (4) the adoption of methods of disinfecting dehairing machines in slaughtering plants. Refrigeration of all products at temperatures below 40 °F at all times is another means to control infections of this type.

Perfringens poisoning has been caused by the consumption of meat or meat products, including meat pies, stews, meats, cold meat sandwiches, ham, bacon, and ham pies, which after cooking had been allowed to cool slowly at ambient temperatures. The organism causing this type of gastroenteritis, *Clostridium perfringens*, is normally present in the human intestine, yet the disease is considered to be an infection. The strains of this organism causing the disease are comparatively heat resistant.

Food intoxication caused by ingestion of meat products has mostly been of staphylococcal origin. Fresh meat products are almost never involved. Usually processed meat products such as hams, meat loaves, and certain types of fermented sausage are implicated. During processing, essentially all bacteria in these meats are killed; during later handling and processing, the product becomes contaminated with enterotoxic staphylococci, the chief source of which is the human being. Growth of these micro-

Microbiology of Foods

organisms is not inhibited by high concentrations of salt. The salt content prevents the multiplication of many types of bacteria which might otherwise grow and interfere with the proliferation of staphylococci. As a result, the staphylococci increase to reach a concentration of millions of cells per gram, producing a toxin which causes gastroenteritis.

Staphylococcus aureus, which produces the enterotoxin causing Staphylococcus poisoning, will not increase in numbers at temperatures below 44 °F. Since the organism causing staphylococcal poisoning is present in the nose and throat of approximately 50% of all humans, complete elimination of these bacteria from foods is difficult. Methods of controlling staphylococcal poisoning caused by meats include adequate refrigeration, application of good sanitary procedures during processing and subsequent handling, and the exclusion of food handlers with respiratory and pustulating infections.

Fish and Marine Products

Bacterial Spoilage of Fish. Bacterial spoilage of fish held at refrigerator temperatures above freezing occurs mainly in fish held on ice in the eviscerated or filleted state. However, bacterial spoilage may also occur in such fish as ocean perch and flounders which are held on ice without evisceration.

The surface skin and body cavity lining of freshly caught and eviscerated fish contain comparatively large proportions of *Pseudomonas, Acinetobacter,* and *Corynebacterium* species, *Flavobacterium* and *Micrococcus* species being present in lower concentrations. As the fish are held in ice, *Pseudomonas* species grow faster and eventually become the predominating types, although there may be instances in which *Acinetobacter* species are present in large numbers. The growth of pseudomonads may be the chief cause of fish spoilage.

Spoilage of fish because of bacterial growth is primarily found on the surface. The bacteria grow on the lining of the visceral cavity, the skin, or even on the boards lining fish hold pens, and produce enzymes which decompose chemical compounds in fish, forming products which effuse into the flesh and cause off-flavors and odors.

Spoilage Mechanism from Bacterial Growth. General types of spoilage occur either under aerobic conditions or under anaerobiosis. Salt water fish species contain about 0.1 to 1.0% of trimethylamine oxide, an end product of their nitrogenous metabolism. Bacterial enzymes activate this compound, subjecting it to reduction by cell dehydrogenases and causing the formation of trimethylamine. Trimethylamine oxide has essentially no flavor or odor, but trimethylamine has a disagreeable odor usually described as fishy. If trimethylamine accumulates in fish flesh to any extent, the fish becomes a second- or third-grade product. This type of spoilage occurs under aerobic conditions.

Under situations in which air is present, the enzymes of aerobic bacteria act on sulfur-containing amino acids to produce hydrogen sulfide, which has an offensive odor but is quickly oxidized to sulfate, which has no odor. Fish may be held under anaerobic conditions, which is what happens when they are pressed tightly against the boards of hold pens. While there may be little further growth of bacteria under these conditions, the enzymes of the bacteria present on board and on fish surfaces produce hydrogen sulfide from sulfur-containing amino acids present in fish tissue. This hydrogen sulfide cannot be oxidized to sulfate, since oxygen is not available. Eventually the hydrogen sulfide effuses into the fish tissues where, together with other compounds of decomposition, it produces the extremely bad odor called bilginess.

Since at the time of spoilage small amounts of formic acid, otherwise not present, accumulate in cod, haddock, ocean perch, and whiting, a test for this compound may be used to determine spoilage.

Fish fillets, when held in the fresh state, may go through several spoilage phases. At first, fresh-water, nonproteolytic pseudomonads grow, causing fruity, vegetable, or garbage-like odors. This type of deterioration does not make the product inedible, or even untasty. As the fillet is held, freshwater, and especially saltwater, proteolytic pseudomonad types eventually take over, producing trimethylamine and other off-flavor compounds which, with sufficient accumulation, cause the product to be considered inedible. If held for extended times, fillets would eventually become putrid, from anaerobiosis established by the growth of aerobes which consume the limited amounts of available oxygen.

Cured Fish Products. Pickled fish, which should be held under refrigeration, are not subject to bacterial spoilage unless held at temperatures above 40 to 50 °F, since they are packed in a solution of salt and acetic acid which inhibits the growth of psychrophiles. Lightly salted and lightly smoked fish, such as smoked fillets and kippered herring, are subject to spoilage when held at temperatures between 34 and 50 °F. Because of some destruction of psychrophilic bacteria during smoking, the refrigerated storage life is slightly extended. However, in most cases this extension of holding time is minimal.

Marine Foods and Food-Borne Diseases. Salmonellosis and staphylococcus poisoning have not been associated to any extent with the consumption of fresh fish.

Clostridium botulinum, type E, has been shown to grow and produce toxin in a nutrient medium when incubated for extended times at temperatures as low as 38 °F. This organism in fresh fish or fish products has not become a greater public health hazard because under the conditions of holding, bacterial types such as pseudomonads grow best. It is only after these bacteria have grown extensively and caused spoilage that conditions suitable for the growth of *Clostridium botulinum* are established. The product by that time would be considered inedible by most consumers.

Shellfish such as clams and oysters have caused salmonellosis, typhoid fever, cholera, and even infectious hepatitis. Shellfish are harvested close to shore, where waters are sometimes polluted and may contain pathogenic bacteria or viruses. The shellfish's bivalves filter out and retain bacteria or organisms present in the water. Since these shellfish are often eaten raw, if infectious germs are present, disease may result.

Poultry

Bacteria Causing Spoilage. Bacteria found on the skin of eviscerated, cut-up poultry, immediately after processing, are mostly species of the genus *Pseudomonas (Ps.)*. At spoilage, the bacteria in the pseudomonad group were mostly nonpigmented types which would probably be classified as *Ps. fragi, Ps. putrida,* and *Ps. ambuiga,* but a pigmented type, *Ps. fluorescens,* was included, as were some species of *Xanthomonas*. A buildup of bacteria on the skin can be prevented if adequate sanitary precautions are observed.

Changes Indicative of Bacterial Spoilage. In poultry distributed in the eviscerated or eviscerated, cut-up state, bacteria grow on chicken surfaces, causing slime on lean or skin sides, accompanied by odors described as tainted, acid, or sour. The product starts to develop an off-odor when the number of bacteria on the skin reaches a concentration of $10^7/cm^2$, and slime formation occurs at a concentration of about $10^8/cm^2$.

Transmitted Food-Borne Illness. Poultry and poultry products have often been associated with outbreaks of salmonellosis, because poultry of all types are quite often infected with *Salmonellae*. However, salmonellosis usually occurs from eating cooked fresh poultry products which have been mishandled. An example: stuffed birds not heated enough during cooking, thereby providing conditions in the dressing that allows bacteria to multiply. Outbreaks of salmonellosis have also occurred from ingestion of cooked poultry products which have not been adequately refrigerated. In these cases, contamination of the food after cooking are more probable than survival of these organisms.

Outbreaks of staphylococcus poisoning have occurred from consumption of poultry products. Contamination with the causative organism by those handling the food, and holding after preparation at temperatures well above 32°F, have probably been the reasons for such incidents. Food poisoning from the growth of *Clostridium perfringens* in cooked poultry is usually associated with inadequate refrigeration.

Eggs

Shell eggs (distributed as storage eggs) are held for a long time at temperatures near 32°F and 80-85% humidity. About 95% of eggs contain no bacteria when laid. The shell of the freshly laid egg is covered with a thin layer of proteinaceous material (keratin), which may seal the pores in the shell and tend to prevent the entrance of microorganisms. However, the inner shell membrane is much more effective than either the shell or the outer shell membrane in preventing bacteria from penetrating the egg. The washing of eggs or their contact with moisture or moist droppings may remove the cuticle. Then, after the egg is about 3 weeks old, the cuticle becomes brittle and is easily chipped off. When the cuticle is removed, bacteria or even molds may penetrate the shell and grow.

Dirty eggs are regarded as low quality because of the greater probability of contamination with bacteria and subsequent spoilage. Shell eggs may be washed under controlled conditions, for sale as such, or they may be washed prior to breaking, for production of frozen or dried products.

Spoilage from Growth of Microorganisms. Ayres (1960) observed that about 80% of spoilage in shell eggs is from the growth of pigmented pseudomonads. The types of decomposition found in eggs are: (1) *Green rots*—caused by pseudomonads; the white takes on a greenish appearance, but the yolk is not affected; (2) *Red rots*—the white may be either liquefied or still viscous and contains patches or spots of red material (probably patches of microorganisms), caused by species of *Pseudomonas*; (3) *Black rots*—the white is liquefied and the yolk is solidified and turns black, caused by *Proteus malanovogenes*; (4) *Miscellaneous rots*—indicated by many different decomposition changes including, upon occasion, the formation of gas, caused by different bacterial species, including *Pseudomonas, Serratia, Micrococcus,* and *Bacillus*; and (5) *Fungal rots*—the white may be either liquefied or coagulated and the egg contents have musty odors and flavors, caused by the growth of *Penicillium, Mucor, Cladisporium,* or *Sporotrichum* species.

In addition to strains of *Pseudomonas fluorescens,* which are the most common cause of egg spoilage, *Ps. putrefaciens, Ps. ovalis, Ps. geniculata, Ps. convexa* and *Ps. aeruginosa* have been shown to cause spoilage of shell eggs.

Salmonellosis and Egg Products. Salmonellosis has not often been associated with foods prepared from shell eggs, while dried and frozen egg products are frequently incriminated. However, *Salmonella* bacteria are sometimes present in freshly laid eggs.

Prior to freezing or drying, liquid whole egg magma or even liquid egg whites are usually pasteurized by heating at 140 to 144°F for 180 to 240 s to destroy *Salmonellae*. Small amounts of citric acid and aluminum or iron salts may be added to egg whites prior to pasteurization to prevent coagulation during heating and deterioration of the functional characteristics of the egg proteins.

Dairy Products

Even when milk is drawn aseptically from the udder, it contains small numbers of bacteria. When milk is drawn under the usual conditions, some bacteria from the air, milking equipment, or other sources enter the product. It is from these sources that the bacteria that cause natural lactic acid-type fermentations become present in milk.

Milk freshly drawn under good sanitary conditions contains at most only a few thousand bacteria per millilitre (mL). Under unsanitary conditions, the immediate count may approach 400 000 per mL.

Streptococcus lactis types, and *Pseudomonas, Bacillus, Enterobacter,* and *Escherichia* species may be present in freshly-drawn milk. If milk is held at suitably low temperatures and is collected promptly, the psychrophilic bacteria, especially pseudomonads and possibly some types of coliforms, will not have increased significantly by the time the raw product arrives at the pasteurization plant.

Bacterial Spoilage of Raw Milk. When raw milk is mishandled prior to pasteurization, it may merely become sour, because of the growth of *Streptococcus lactis* or similar types, or it may spoil through a variety of other changes. In addition to souring, raw milk may become spoiled: (1) because of malty, butyric, or other flavors; (2) through the development of ropiness or coagulation; (3) through yeasty (gassy) fermentations; or (4) through the development of blue, yellow, or red colors. All of these changes are from the growth of microorganisms, usually bacteria. The presence of large numbers of bacteria or yeasts, usually more than 50×10^6 per mL, is required for flavor, color, or consistency changes in milk, but some bacteria may cause sweet curdling at concentrations between 1.25×10^6 and 4.9×10^6 cells per mL and *Ps. graveolens* have been reported to cause off-odors in milk at concentrations of 5.5×10^6 cells per mL.

Pasteurization. Pasteurization treatment to destroy pathogenic bacteria may be of the low-temperature, long-time type (heated at least 0.5 h at a temperature not below 145°F, then cooled), or of the high-temperature, short time type (heated at a temperature not below 162°F for at least 15 s, then cooled). Pasteurization ordinarily destroys about 95% of the bacteria present in milk, although the proportion surviving will depend on the relative number of thermoduric bacteria originally present. Spore-forming bacteria, thermophilic and nonthermophilic, thermoduric strains of *Lactobacillus, Streptococcus, Micrococcus,* and *Sarcina* usually survive pasteurization. Some coliform bacteria may also survive this treatment, but these organisms either do not grow, or grow only very slowly at low temperatures. Psychrophilic bacteria are mostly destroyed during pasteurization. When milk is held at about 40°F, pasteurization extends storage life by approximately 10 to 13 days.

Spoilage of Pasteurized Milk. When pasteurized milk is mishandled by holding it at high temperatures, spoilage is usually indicated by curdling, with or without souring. When held at adequate refrigeration temperatures for long periods, sweet curdling without souring occurs more often. The lactic acid streptococci grow only very slowly at low temperatures, and under such conditions, bacteria better suited to growth at low temperatures develop and cause curdling without the production of acid.

Food-Borne Diseases and Dairy Products. *Salmonellae* have been isolated from a significant number of samples of nonfat dried milk, while there have been few, if any, outbreaks of salmonellosis from fluid milk. This has probably been because of pasteurization. Major outbreaks appear to have occurred from the eating of cheese, but dairy products seem to provide few public health problems concerning the transmission of *Salmonellae*.

While fluid milk is not involved in outbreaks of staphylococcal poisoning, cheese, and even dried milk, may present a problem in this respect. In one instance involving dried milk and one involving cheese, staphylococcal poisoning occurred even though the microorganisms originally producing the toxin had been destroyed during processing.

Microbiology of Foods

Ready-to-Eat Refrigerated Foods

Heat-pasteurized, ready-to-eat foods which are refrigerated at temperatures above freezing include many types, such as caterer items and foods dispensed in vending machines. Such products should be held at temperatures below 40 °F at all times to prevent growth of food-borne pathogenic disease bacteria.

Food Preservation by Refrigeration

Several factors determine the length of time during which flesh-type foods or milk may be held at refrigerator temperatures above freezing. The level of contamination with psychrophilic bacteria, and whether or not the contaminating bacteria are in the logarithmic (exponential) phase of growth at comparatively low temperatures are most important. Two of the factors governing spoilage time are the duration of the lag phase and the exponential growth rate of the contaminating bacteria. As the holding temperature is decreased from 40 °F to a point near 32 °F, the lag phase of psychrophilic bacteria increases markedly, and the exponential growth rate (increase in generation time) decreases considerably.

Most industries recognize the advantages of good sanitation and the use of temperatures near the freezing point for holding products. To obtain the maximum storage life of refrigerated foods, the use of relatively low temperatures is as important during transportation and retail holding as at any other time. Many of the advantages obtained with good processing procedures may be lost by poor handling procedures during transportation and at the retail level. The same sanitation and holding temperatures should be applied at transportation and retail levels for products of this kind.

FROZEN FOODS

Microorganisms do not grow in foods held below 7 to 10 °F. At temperatures between 31 and 38 °F, psychrophilic bacteria can grow, but food poisoning organisms cannot. Public health problems are chiefly associated with handling either type of organism prior to freezing, or after defrosting, or both. Most frozen foods are cooked before they are eaten, which is helpful because cooking may destroy harmful bacteria or bacterial end products. Still, it is not certain that all harmful materials or microorganisms will be destroyed by cooking. Only good quality foods should be frozen. They should be prepared, processed, and packaged under sanitary conditions and frozen at a rate which will prevent deterioration because of bacterial growth. After freezing, foods should be held at suitable frozen storage temperatures at all points, at 0 °F or lower. After defrosting, suitable low refrigerated temperatures should be used until the product is consumed.

Microorganisms in Foods

On freezing, considerable numbers of bacteria present are quickly destroyed, followed by a slower death rate during storage. However, bacteria survive freezing better when suspended in food than when suspended in water. Bacteria also survive freezing in neutral foods better than in acid foods. Because all bacteria present are not destroyed by freezing, the food is subject to bacterial decomposition when it is defrosted.

Significance of Bacterial Counts

Bacterial counts are not accurate; in addition, the bacterial counts of most foods give little indication of edible quality until sufficient bacterial action has produced off-flavors. However, the bacterial count may be used as a check to improve the sanitary condition of foods.

The presence of a certain level of coliform bacteria is not adequate evidence of pollution, so total bacteria count may indicate quality better than tests for coliform bacteria alone. But fecal coliform counts or tests for enterococci should be made on frozen food, although few fecal coliform bacteria are generally present. Thatcher and Clark (1968) discuss the significance of microorganisms in foods, including total numbers, indicator organisms, and pathogens.

The following factors help keep the bacterial count of frozen vegetables below 100 000 per gram: (1) blanching the product at a high temperature, (2) quickly and adequately cooling the product, (3) maintaining a sanitary packing plant, (4) using clean packaging materials, and (5) keeping the product from thawing after freezing.

Public Health Aspects

Any food, if grossly mishandled, may cause an epidemic of food infection or food intoxication. Neither *Clostridia Salmonellae* nor *Staphylococci* are all destroyed by freezing and storage. As a rule, large numbers of *Salmonellae* must be ingested to cause salmonellosis, although cases have been reported where normal, healthy children have been infected by less than 10 viable cells of *Salmonella* in chocolate candy. These bacteria may grow at relatively low temperatures (above 43 °F). Since cases of salmonellosis are rarely reported as having been caused by the consumption of frozen foods, they are not likely to have been defrosted and held at temperatures above 43 °F for long periods.

Since toxigenic staphylococci will not grow at temperatures below 44 °F, staphylococcal gastroenteritis from frozen foods is no less probable than salmonellosis. In either case, frozen foods appear to be relatively safe, compared with certain other types of food products. Unless they are grossly mishandled, it is unlikely that the consumption of such products will cause food infections or food intoxications.

When certain foods are defrosted and stored at temperatures that permit bacterial development, conditions favor the growth of acid-forming organisms that prohibit the growth of putrefactive anaerobes. The danger of toxin production from *Clostridium botulinum* may be lessened by the growth of these acid-forming organisms, unless foods are held above 38 °F for an extended period. Experience with commercial foods indicates that this danger is very small.

Some doubt has been expressed regarding the use of hermetically-sealed cans for frozen food packaging. Tests with frozen vegetables indicate the probability that *Clostridium botulinum* will develop and grow in foods packed in hermetically-sealed containers, but not before the product is grossly decomposed.

Plant Sanitation

Many food industries apply the sanitary principles developed by public health officials. The Association of Food and Drug Officials of the United States has published a code for the handling of frozen foods. This code considers the type of equipment and handling practices to be applied in the processing, transportation, warehousing, and retailing of frozen foods. In this code, the type of construction materials for equipment and the particular design and construction of equipment are discussed and recommendations made.

Personnel in plants producing products that may be eaten after defrosting, without further cooking, or without cooking to high temperatures, should pay special attention to plant sanitation. All plants preparing frozen foods should have an adequate supply of potable water. In-plant chlorination of water supplies is a desirable feature for food plants, and such an installation may help simplify cleanup.

Waste-elimination systems that do not allow accumulations around or near the plant are essential. Such wastes held near the plant become breeding places for insects which eventually find their way into the plant. Plants should be constructed in a manner that will prevent the entrance of insects and rodents. This involves ratproof building construction and adequate screening. For doors opened frequently, special measures must be taken, such as air curtains, to prevent the entrance of insects.

Plants should also be constructed so that steam may be properly vented to the outside. Accumulation of steam in the plant causes condensation and dripping, which may contaminate the food products produced. Hoods used to eliminate steam should be constructed so that condensed moisture does not run back and drip within the hooded area.

Floors should be constructed of impervious material such as concrete or tile and should be easy to clean. In areas where waste materials may accumulate during processing, floors should be sloped to drains at a minimum of 0.25 in. per ft. All equipment contacting foods should be constructed of a smooth metal which is not corroded by the material with which it comes in contact. Such construction provides for easy cleaning and generally improves sanitary conditions.

Machinery should either be mounted on a solid base or raised above the floor in such a manner that cleaning is easy. All machinery, belts, containers, and other equipment contacting foods should be freed of resident product, thoroughly cleaned with hot water and detergents, rinsed with cold water, then sanitized with some bactericidal agent such as chlorinated water. The floors in processing parts of the plant should also be thoroughly cleaned periodically.

While cleanup itself may be carried out by production personnel, cleanup procedures and the detergents and sanitizing agents used should be specified by the quality control group. Moreover, quality control representatives should inspect all areas after cleanup and determine whether or not a suitable job has been done. It is also the function of the quality control group to make frequent bacteriological line checks on the product as it is processed, to see whether or not a significant buildup of microorganisms is taking place on equipment located at any particular point.

Bacteriological counts should be made a number of times daily on the finished product or on the packaged product just prior to freezing, as part of a good quality assurance program. If batch-type production is used, such as might be the case with certain precooked products, at least one package from each batch should be examined bacteriologically.

In any food plant operation, the success of a quality control program depends on the cooperation and vigilance of management officials, as well as that of personnel carrying out the actual tests.

REFERENCES

Ayres, J. C. 1955. Microbiology of meat animals. *Advances in Food Research* 6:109.

Ayres, J. C. 1960. The relationship of organisms of the genus *Pseudomonas* to the spoilage of meat, poultry and eggs. *Journal of Applied Bacteriology* 23:471.

Leistner, L. *et al.* 1961. The occurrence and significance of *Salmonella* in meat animals and animal by-product feeds. *Proceedings of the 13th Research Conference, American Meat Institute Foundation* p. 9.

CHAPTER 11

METHODS OF PRECOOLING FRUITS, VEGETABLES, AND CUT FLOWERS

Product Requirements................................11.1
Calculation Methods................................11.1
Cooling Methods....................................11.3
Hydrocooling.......................................11.3
Forced-Air Cooling.................................11.4
Package Icing......................................11.6
Vacuum Cooling.....................................11.6
Selecting a Cooling Method.........................11.9
Cooling Cut Flowers................................11.9

COOLING is generally considered the removal of field heat from freshly harvested products in time to inhibit spoilage and to maintain preharvest freshness and flavor. The term "precooling" implies the removal of heat *before* the product is shipped to a distant market, processed, or stored. Some products are *room cooled* in the same room in which they are stored. In this case, cooling is accomplished over a day or more. Precooling is generally done in a separate facility within a few hours or even minutes. Therefore, room cooling is not considered precooling.

PRODUCT REQUIREMENTS

Product physiology, in relation to harvest maturity and ambient temperature at harvest time, largely determines cooling requirements and methods. Some products are highly perishable and must be cooled as soon as possible after harvest. Vegetables in this category include: asparagus, snap beans, broccoli, cauliflower, sweet corn, cantaloupes, summer squash, vine-ripened tomatoes, leafy vegetables, globe artichokes, brussels sprouts, cabbage, celery, carrots, snowpeas, and radishes. Vegetables such as white potatoes, sweet potatoes, winter squash, pumpkins, and mature green tomatoes need to be cured or ripened at some temperature higher than desirable for holding more perishable produce. Cooling of these products is not as important; however, some cooling is necessary if ambient temperature is high during harvest. Vegetables not listed may or may not be cooled because of lack of economic importance or susceptibility to chilling injury—for example, cucumbers.

Commercially important fruits that need to be precooled immediately after harvest include: apricots, avocados, all of the berries except cranberries, tart cherries, peaches and nectarines, plums and prunes, and tropical and subtropical fruits such as guavas, mangos, papayas, and pineapples. The tropical and subtropical fruits of this group are susceptible to chilling injury and thus need to be cooled according to individual temperature requirements. Sweet cherries, grapes, pears, and citrus fruit have a longer post-harvest life than the former fruits, yet prompt cooling is essential to high quality retention during the holding period. Bananas require special ripening treatment and therefore are not precooled. Because of their keeping quality, apples generally do not need to be precooled. Early varieties, harvested when the ambient temperature is high, are normally more perishable and may benefit by precooling. Other varieties, particularly those that are stored for several months, may benefit by precooling when the apples cannot be cooled to 32°F in storage within seven days after harvest.

CALCULATION METHODS

Heat Load

The refrigeration capacity needed for precooling is much greater than that required for holding a product at a constant temperature or for slow cooling of a product. Therefore, the heat load on a precooling system should be determined as accurately as possible. While it is imperative to have an adequate amount of refrigeration for effective precooling, it is uneconomical to have more refrigerating capacity available than is normally needed.

On jobs where refrigeration is needed only during a regular 8- to 10-h workday, ice builders make it possible to reduce the amount of refrigeration equipment. Equipment size can be cut in half, or more, depending on the hours of off time in relation to the hours of precooler operation.

The total heat load comes from the product, surroundings, air infiltration, containers, and heat-producing devices such as motors, lights, fans, and pumps. Product heat accounts for the major portion of total heat load on a precooling system. Product heat load depends on product temperature and cooling rate, amount of product cooled in a given time, and specific heat of the product. Heat from respiration is part of the product heat load, but only when the cooling time exceeds a few hours.

Mass Average Temperature. An accurate determination of product temperature is essential for accurate heat load calculations. Due to the rapid heat transfer, a temperature gradient develops within the product, with faster cooling causing larger gradients. This gradient is a function of product properties, surface heat transfer parameters, and cooling rate. For example, initially hydrocooling removes heat from the exterior of a product. Consequently, temperature in this area changes rapidly while the center temperature may not change at all. The major product mass constituency lies in the exterior of the product. Thus, it is possible, based on center temperature, to calculate a small or negligible heat load while, in fact, substantial heat has been extracted. For this reason, the product mass-average temperature must be used for product heat load calculations (Smith and Bennett 1965). A mass-average temperature denotes the single value from the transient temperature distribution that would become the uniform product temperature when held for a period under adiabatic conditions.

Figure 1 can be used to determine the mass average temperature t_{ma} of peaches during hydrocooling. Subsequently, the product cooling load can be calculated by Equation 1.

$$q = mc_p(t_i - t_{ma}) \qquad (1)$$

The nomograph can be applied to products other than peaches, where temperature ratio with respect to cooling time and product size is known. Figure 2 illustrates the comparative relationship of fractional temperature difference, or temperature ratio, Y, to cooling time for 2.5 in diameter peaches, 2.88 in. diameter apples and citrus fruit, and 2 in. diameter sweet corn when hydrocooled under ideal conditions of negligible surface resistance to heat transfer.

Cooling Rate

The terms *cooling coefficient*, C, and *half-cooling time*, Z, are perhaps the most meaningful in presenting cooling rate data. The

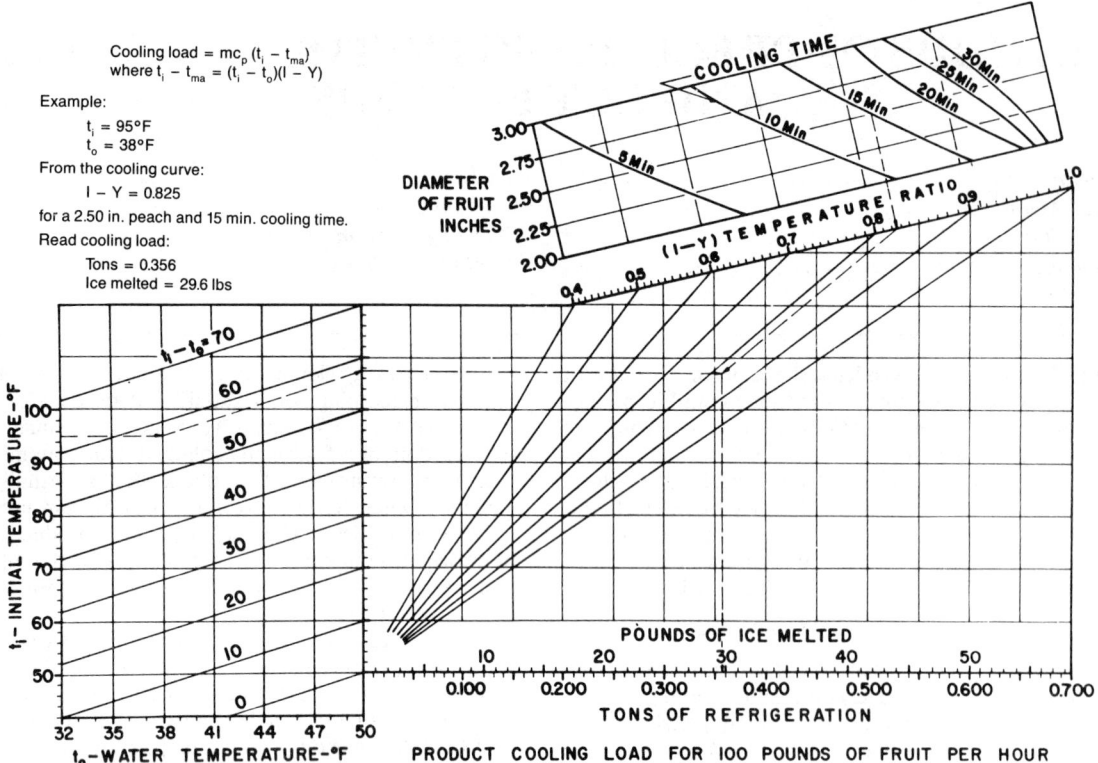

Fig. 1 Nomograph to Determine Product Heat Load of Hydrocooled Peaches

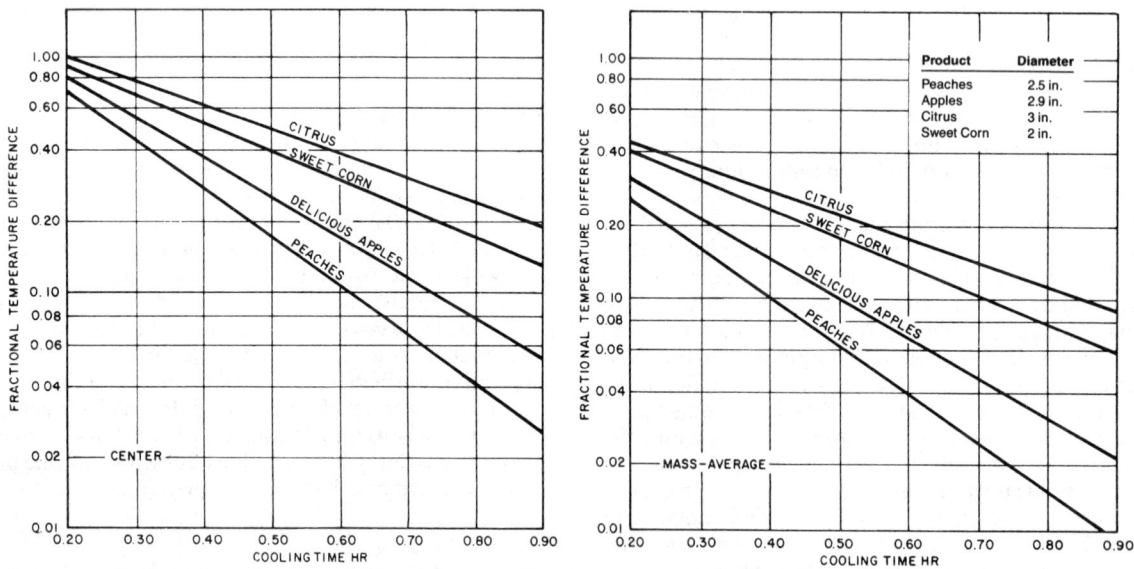

Fig. 2 Time-Temperature Response for Hydrocooled Produce

temperature ratio Y is the unaccomplished temperature change at any time in relation to the total temperature change possible for the cooling condition. It is calculated by the equation,

$$Y = (t - t_o)/(t_i - t_o) \qquad (2)$$

The inverted temperature ratio denotes the percent accomplished temperature change. The value of the cooling coefficient is the same in either case.

The cooling coefficient denotes the change in product temperature per unit change of cooling time for each degree temperature difference between the product and its surroundings. In most cases, a logarithmic transformation of the temperature ratio values provides a fairly accurate straight line fit to the data points. If Newtonian heat transfer occurs (a negligible temperature gradient within the product during cooling) and the straight line asymptote intercepts the temperature ratio at unity for zero time,

Methods of Precooling Fruits, Vegetables, and Cut Flowers

the cooling coefficient may be calculated by the simplified formula

$$C = \ln Y/\theta \qquad (3)$$

When produce is rapidly cooled, the conditions for Newton's law are seldom satisfied. A considerable temperature gradient often develops within the product, depending on its properties and rate of surface heat transfer. Also, the zero asymptote intercepts the temperature ratio at some value either greater or less than unity. In this case, the constant cooling coefficient is calculated by the equation

$$C = (\ln Y_1 - \ln Y_2)/(\theta_1 - \theta_2) \qquad (4)$$

where the subscripts 1 and 2 denote points along the straight line asymptote to the cooling curve.

The half cooling time is the time required to reduce the temperature difference between the product and its surroundings by one-half. Mathematically, it is expressed as

$$Z = \ln(0.5)/C \qquad (5)$$

where C is a negative value.

If certain properties and surface parameters are known, cooling coefficients may be predicted by the equation

$$C = \alpha M_1^2/l^2 \qquad (6)$$

where l is the characteristic length.

$M_1^2 = G\pi^2$ if there is no resistance to heat transfer at the surface interface. Smith *et al.* (1967) derived Equation (6) and the geometry index G. When there is resistance to heat transfer at the surface, usually the case in air cooling, the transcendental function M_1^2 becomes dependent on the Biot number in relation to the product geometry. Smith *et al.* (1968) present a nomograph to obtain M_1^2 as a function of the geometry index and the Biot number.

Cooling coefficients predicted by means of Equation (6) have physical meaning only if the value of the intercept is known or can be estimated. Employing the approach of Ball and Olson (1957), Pflug *et al.* (1965) outlined a procedure for developing time-temperature response charts for objects that can be approximated by a sphere, an infinite plate, or a cylinder, and gave examples for the corresponding solution of practical problems involving use of the intercept and Biot number as functions of the surface heat transfer characteristics. The geometry analysis developed by Smith *et al.* (1967) permits extension of these solutions to anomalous shapes.

These solutions treat the product as an individual unit in a specified environment and isolated from the influence of surrounding material, rounding material, which is rarely the case. Often, products are cooled in bulk or in packages of several layers, where the surrounding material influences the cooling response, particularly in air-cooling systems. The individual approach usually gives poor estimates for cooling of bulk loads.

Baird and Gaffney (1976) developed a numerical technique for calculating cooling rates in bulk loads of fruits or vegetables. This procedure, applicable to spherical products, uses finite difference solutions of the differential equations that describe heat transfer both within individual fruits and at different levels in bulk loads. Eshleman *et al.* (1976) extended this development to provide solutions for heat transfer in irregularly shaped objects. These models can be used to calculate temperatures at various points within individual products and at different depths in bulk loads at regular time intervals during cooling. Required inputs to the models include product size and shape, thermal properties of the product, the convective coefficient, and flow rate of the cooling medium.

COOLING METHODS

The principal methods of precooling are hydrocooling, forced-air cooling, package icing, and vacuum cooling. Most cooling is done at the packinghouse or in central cooling facilities. Some products can be cooled by any of these methods without suffering any adverse effects. For these products, the cooling method chosen is often determined more by such factors as economy, convenience, relation of the cooling equipment to the total packing operation, and personal preference.

HYDROCOOLING

Because of its simplicity, economy, and effectiveness, hydrocooling is a popular precooling method. The rate of heat transfer Q is directly proportional to the surface heat transfer coefficient h, the total surface area A, and the difference in temperature between the surface and its surroundings Δt. When a film of cold water flows briskly and uniformly over the surface of a warm substance, the surface temperature of the substance becomes essentially equal to that of the water. Rate of internal cooling is limited by the size and shape (volume in relation to surface area) and thermal properties of the substance being cooled.

If the fluid velocity is sufficient, as in gravity flow or forced convection over fruit and vegetable products, the resistance to heat transfer at the surface is negligible. In this case, the heat is removed as fast as it comes to the surface, and the temperature difference across the surface boundary layer is small (1 °F or less). In ideal hydrocooling, the optimum average film coefficient of heat transfer is roughly 120 Btu/(h · ft² · °F); the average temperature difference across the boundary layer is about 0.8 °F. On this basis, the rate of heat transfer per unit surface area is $h = 120 \times 0.8 = 96$ Btu/h per square foot of product area.

The value of h varies substantially, depending on many conditions. However, because of limitations in water temperature and product thermal properties and similarity among products, this is about as fast as any fruit or vegetable can be cooled in water. The ideal (maximum h) time-temperature response curves for hydrocooling select fruit and vegetable products are shown in Figure 2, based on both center and mass-average temperatures.

Commercial Hydrocooling

Freestone peaches, clingstone peaches hauled more than 100 miles for canning, tart cherries, and cantaloupes are hydrocooled. Few apples and citrus are hydrocooled. Hydrocooling is not popular for citrus fruit because it has a long marketing season and good postharvest holding ability. It is also susceptible to increased peel injury and to decay and loss of quality and vitality after hydrocooling. Apples are usually cooled in the storage rooms.

Many vegetables are successfully hydrocooled. The more important of these are sweet corn, celery, radishes, and carrots. Cooling rate data for hydrocooling are given by Grierson (1957), Ford (1956), and Perry and Perkins (1968).

Types. Hydrocooling is accomplished by flooding, spraying, or by immersing the product in an agitated bath of chilled water.

Commercial hydrocoolers for freestone peaches are *flood-type* and *bulk-type*. The flood-type hydrocooler cools the packaged product by flooding as it is conveyed through a cooling tunnel. Adaptations consist of conveying the product through the cooling tunnel in loose bulk or in bulk bins. The bulk-type cooler uses combined immersion and flood cooling. Loose fruit, dumped into cold water, remains immersed through half of its travel through the cooling tunnel. An inclined conveyor gradually lifts the fruit out of the water and moves it through an overhead shower. The bulk-type cooler permits greater packaging flexibility than the flood-type.

Most vegetables are moved by fork truck in unit loads of as many as 40 packages. In one system, chilled water sprays on stacks of packed vegetables placed in a refrigerated room. Water is collected in floor drains leading to a sump and recirculated over the product. One nozzle spraying approximately 90 gpm of water is located over each stack, which may contain from 30 to 40 crates (Grizzell and Bennett 1966).

Other unit load systems convey the stacks through cooling tunnels. Either spray nozzles or a flood pan may be used to deliver up to 400 gpm of chilled water to each stack. Most commercial hydrocoolers provide overhead showering at a rate of 10 to 15 gpm per square foot of top face area. This type of hydrocooler requires less space than the batch type, but is not as flexible. With the batch system, chilled water can be sprayed on the product indefinitely, depending on the season and the incoming product temperature. Also, the crates can be left in place after cooling until shipped.

Hydraircooling uses a mixture of refrigerated air and water in a fine mist spray that is circulated around and through the stack by forced convection (Henry et al. 1976). It has the advantage of reduced water requirements, the potential for improved sanitation, and the capability of adapting to fiberboard containers of the type that cannot be used in conventional hydrocooling systems. Cooling rates equal to, and in some cases better than, those obtained in conventional unit load hydrocoolers are possible.

Design and Operation. Flooded ammonia systems are used to cool the water for hydrocoolers in large packinghouses. Some use a secondary coolant. The cooling coils are contained in a tank through which water is rapidly circulated. Refrigerant temperature inside the cooling coils is approximately 28 °F, except in systems with ice building coils where it may be slightly lower.

To determine refrigeration requirements, the product mass-average temperature should be used. Because the mass-average temperature denotes the amount of heat within a product (above some reference level) no matter what the thermal gradient is, it is the only temperature value that will yield an accurate heat load computation.

Insolation, radiation from hot surfaces, convection from ambient air, or conduction from surroundings can affect the load. Protecting the hydrocooler from these sources of heat gain will enhance efficiency. Refrigeration capacity in excess of needs and cooling to a temperature below that required also reduces efficiency.

Cooling rate decreases in proportion to cooling time, provided water temperature is constant. Cooling time should be governed according to initial and final product temperature, product size, and thermal characteristics of the product with respect to surface parameters. Showalter and Grierson (1970) showed a substantial difference in half-cooling time for sizes 36 and 45 cantaloupes. A weighted average of temperatures taken at different depths showed that 20 min was required to half-cool size 36 melons and only 10 min for size 45.

When hydrocooler water is recirculated, as is usually the case in mechanically refrigerated units, decay-producing organisms accumulate. Mild disinfectants such as chlorine or approved phenol compounds will reduce buildup bacteria and fungus spores, but they will not kill infections already in the products or sterilize either the water or the product surfaces. Brown rot and rhizopus decay spores on the surface and under the skin of peaches can be destroyed by soaking the fruit in water at 125 °F for 2 to 3 min. (Smith and Redit 1968), or by treating the fruit with a chemical fungicide, which is usually put into the hydrocooling water. The principal problem with chemicals is maintenance of uniform concentrations, particularly in ice-refrigerated equipment because of the constant dilution from melting ice. In addition, hydrocooler shower pans and/or trash screens need regular (daily) cleaning to provide maximum efficiency.

FORCED-AIR COOLING

Air cooling at a rate comparable to hydrocooling can be theoretically obtained by providing certain conditions of product exposure and air temperature.

Considering the classic heat transfer correlation of Nusselt-Reynolds-Prandtl numbers,

$$\frac{2hl}{k} = C \left(\frac{2lG'}{\mu}\right)^m \left(\frac{\mu c_p}{k}\right)^n \quad (7)$$

for a specified air temperature and product characteristics length, the average product film coefficient of heat transfer h is a function of the interstitial mass velocity and may be written in the form:

$$h = f(G') \quad (8)$$

The value of h, therefore, depends on the volume rate of airflow and the physical characteristics of the product.

In air cooling, the optimum value of h is considerably smaller than when cooling with water. However, Pflug et al. (1965) showed that apples moving through a cooling tunnel on a conveyer belt cool faster with air at 20 °F approaching the fruit at a velocity of 600 fpm than they would in a water spray at 35 °F. For this condition of air cooling, they calculated an average film coefficient of heat transfer of 7.3 Btu/(h·ft²·°F). They note that the advantage of air is because of the lower air temperature and that if the water were reduced to 34 °F, time for water-cooling would be less. In tests to evaluate film coefficients of heat transfer for anomalous shapes, Smith et al. (1970) obtained an experimental value of 6.66 Btu/(h·ft²·°F) for a single Red Delicious apple in a cooling tunnel with air approaching at 1570 fpm. At this rate of air flow, the logarithmic mean surface temperature of a single apple cooled for 0.5 h in air at 20 °F is approximately 35 °F. The average temperature difference across the surface boundary layer is, therefore, 15 °F and the rate of heat transfer per square foot of surface area is:

$$q/A = 6.66 \times 15 = 100 \text{ Btu}/(h \cdot ft^2)$$

For these conditions, the cooling rate compares favorably with that obtained in ideal hydrocooling. However, these coefficients are based on single specimens isolated from surrounding fruit. Had the fruit been in a packed bed at equivalent flow rates, the values would have been less because less surface area would have been exposed to the cooling fluid. Also, the rate of evaporation from the product surface significantly affects the cooling rate.

Because of physical characteristics, mostly geometry, various fruit and vegetable products respond differently to similar treatments of airflow and air temperature. Perry et al. (1968) found that peaches cool faster than potatoes when they are cooled in a packed bed under similar conditions of airflow and air temperature, for example.

Surface coefficients of heat transfer are sensitive to the physical conditions involved among objects and their surroundings. Experimental surface coefficients ranging from 9 to 12 Btu/(h·ft²·°F) were obtained by Soule et al. (1966) for bulk lots of Hamlin oranges and Orlando tangelos with air approaching the mass of fruit at velocities ranging from 225 to 350 fpm. Bulk bins containing 1000 lb of 2.85 in. diameter Hamlin oranges were cooled from 80 °F to a final mass-average temperature of 46.5 °F in 1 h with air approaching the fruit at 330 fpm (Bennett et al. 1966). Surface heat transfer coefficients for these tests averaged slightly above 11 Btu/(h·ft²·°F). On the basis of a log mean air temperature of 44 °F, the calculated half-cooling time was 0.27 h.

By correlating data from experiments on cooling 2.8 in. diameter oranges in bulk lots with results of a mathematical model, Baird and Gaffney (1976) found surface heat transfer coefficients of 1.5 and 9 Btu/(h·ft²·°F) for approach velocities of 11 and 412 fpm, respectively. A Nusselt-Reynolds heat transfer correlation representing data from six experiments on air cooling of 2.8 in. diameter oranges and seven experiments on 4.2 in. diameter grapefruit, with approach air velocities ranging from 5 to 412 fpm, gave the relationship Nu = 1.17 Re$^{0.529}$, with a correlation coefficient of 0.996.

Ishibashi et al. (1969) constructed a stage-type forced-air cooler that exposed bulk fruit to air at a progressively declining temper-

Methods of Precooling Fruits, Vegetables, and Cut Flowers

ature (50, 32, and 14°F) as the fruit was conveyed through the cooling tunnel. Air approached the fruit at 700 fpm. With this system, 2.5 in. diameter citrus fruit cooled from 77 to 41°F in 1 h. Their half-cooling time of 0.32 h compares favorably with a half-cooling time of 0.30 h for similarly cooled Red Delicious apples at an approach air velocity of 400 fpm (Bennett et al. 1969). Perry et al. (1968) obtained a half-cooling time of 0.5 h for potatoes in a bulk bin with air approaching at 250 fpm, as compared to 0.4 h for similarly treated peaches and 0.38 h for apples. Optimum approach velocity for this type of cooling is in the range of 300 to 400 fpm, depending on conditions and circumstances.

Commercial Methods

Produce can be satisfactorily cooled (1) with air circulated in refrigerated rooms adapted for that purpose; (2) in rail cars or highway vans using special portable cooling equipment that cools the load before it is transported; (3) with air forced through the voids of bulk products moving through a cooling tunnel on continuous conveyors; (4) on continuous conveyors in wind tunnels; or (5) by the forced-air method of passing air through the containers by pressure differential. Each of these methods is used commercially, and each is suitable for certain commodities when properly applied.

In circumstances where air cannot be forced directly through the voids of products in bulk, a container type and a load pattern that permits air to circulate through the container and reach a substantial part of the product surface is beneficial. Examples of this are: (1) small products such as grapes and strawberries that offer appreciable resistance to airflow through the voids in bulk lots; (2) delicate products that cannot be handled in bulk; and (3) products that are packed in the shipping containers before they are precooled.

Forced air or pressure cooling involves definite stacking patterns and the baffling of stacks so that the cooling air is forced through, rather than around, the individual containers. Successful use of the method requires a container with vent holes placed in the direction in which the air will move and a minimum of packaging materials that would interfere with free movement of air through the containers. Under these conditions, a relatively small pressure differential between the two sides of the containers results in good air movement and excellent heat transfer. Differential pressures in use are about 0.25 to 3 in. H_2O, with airflows ranging from 1.0 to 3.0 cfm/lb of product.

Because the cooling air comes in direct contact with the product being cooled, cooling is much faster than with conventional room cooling. This gives the advantage of rapid product movement through the cooling plant, and the size of the plant is one-third to one-fourth that of an equivalent cold room type of plant.

Mitchell et al. (1972) observed that forced-air cooling usually cools in one-fourth to one-tenth the time needed for conventional room cooling but that it still takes two to three times longer than hydrocooling or vacuum cooling. Guillou (1963) reported a half-cooling time of 1 h for forced-air pressure cooling of cup-packed peaches in lidded lugs as compared to 6 h for similarly packed peaches in conventional room cooling.

A proprietary direct contact heat exchanger cools air and maintains high humidities using chilled water as a secondary coolant and a continuously wound polypropylene monofilament packing. It contains about 2000 linear feet of filament per cubic foot of packing section. Air is forced up through the unit while chilled water flows downward.

The dew-point temperature of the air leaving the unit equals the entering water temperature. Chilled water can be supplied from coils submerged in a tank. Buildup of ice on the coils provides an extra cooling effect during peak loads. This design also allows an operator to add commercial ice during long periods of mechanical equipment outage. A general diagram of the system is shown in Figure 3.

Effects of Containers and Stacking Patterns

The accessibility of the product to the cooling medium, essential to rapid cooling, may involve both access to the product within the container and to the individual container in a stack.

This effect is evident in the cooling rate data of various commodities in various types of containers reported by Mitchell et al. (1972). Parsons et al. (1972) developed a corrugated paperboard container venting pattern for palletized unit loads that produced cooling rates equal to those from conventional register stacked patterns. Fisher (1960) demonstrated that spacing apple containers on pallets reduced cooling time by 50% as compared with pallet loads stacked solidly.

Computer Solution

Baird et al. (1988) developed an engineering economic model for designing forced-air cooling systems. Figure 4 is an example of information that can be obtained from the model. By selecting a set of input conditions (which varies with each application), as indicated in Table 1, and varying the approach air velocity, the entering air temperature, or some other variable, the optimum (minimum cost) value can be determined. The selection of air velocity for cartons is critical, whereas the selection of entering air temperature is not as critical until the desired final product temperature of 40°F is approached. As indicated in Table 1, the results are for four cartons deep with a 4% vent area in the direction of airflow, and they would be quite different if the carton vent area was changed. Other design parameters that can be optimized using this program are the depth of product in direction of airflow and the size of evaporator and condenser.

Table 2 gives additional information on the economics of forced-air cooling, based on the same set of input conditions in Table 1. Each unit cost is varied from zero to twice the nominal

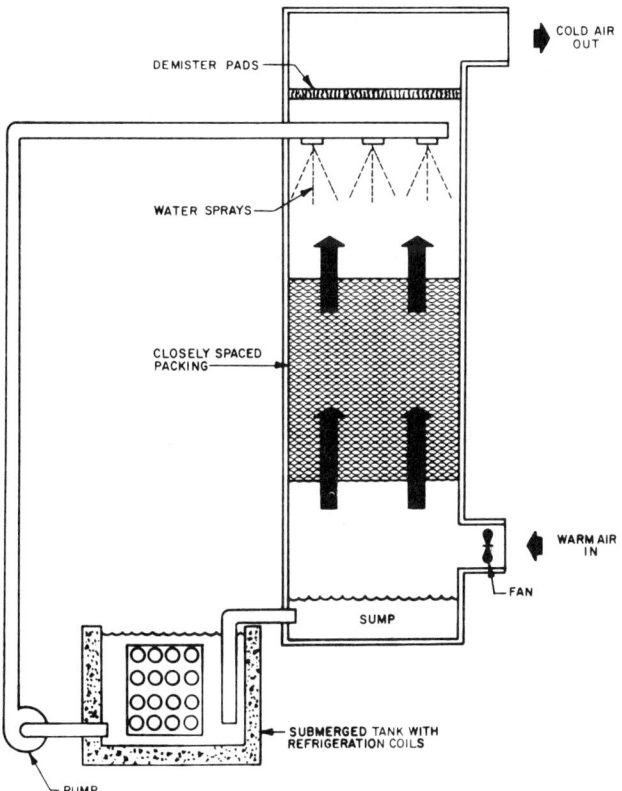

Fig. 3 Packed Tower for Forced-Air Cooling System

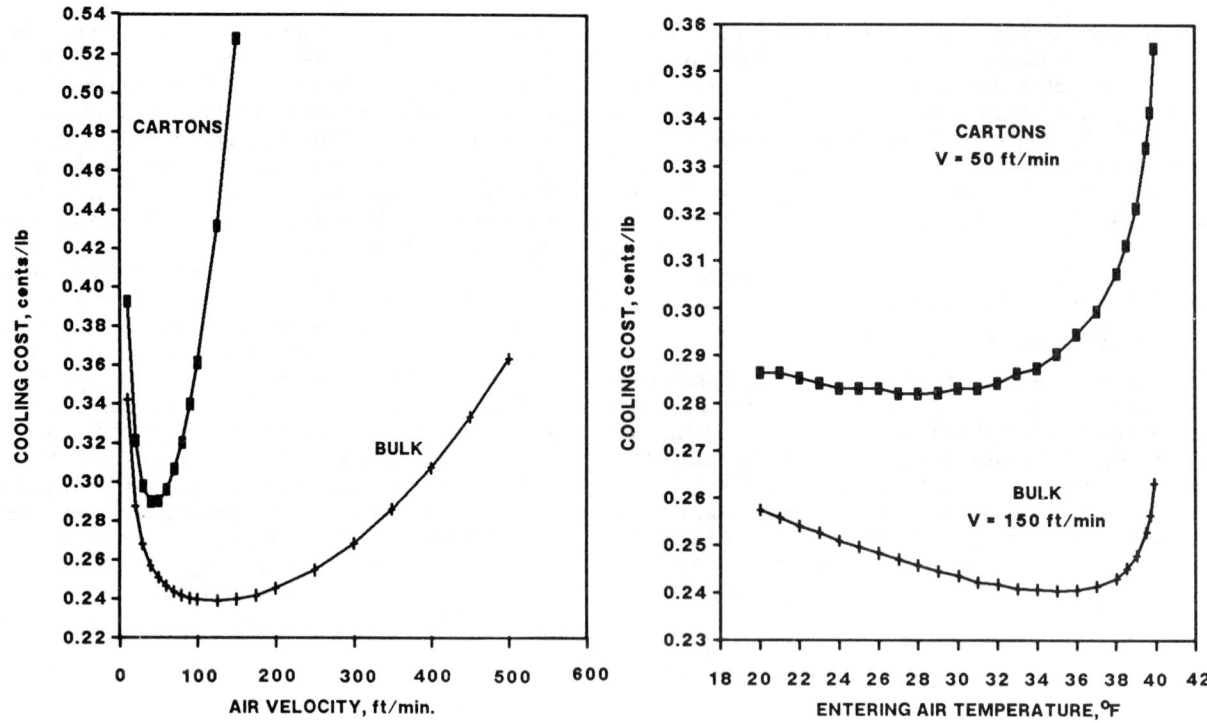

Fig. 4 Engineering-Economic Model Output for a Forced-Air Cooler
(Based on input values in Table 1)

Table 1 Input Values to an Engineering/Economic Model

Fixed Values	
Density	62.4 lb/ft^3
Porosity	0.39
Bulk density	38.1 lb/ft^3
Thermal conductivity	0.30 Btu/h·ft^2·°F
Specific heat	0.90 Btu/lb·°F
Transpiration coefficient	0.005 lb/h·ft^2·psi

Nominal Values	
Product diameter	2 in.
Air approach velocity (Cartons)	50 fpm
(Bulk)	150 fpm
Cooling air temperature	35°F
Product depth	4 ft
Ambient temperature	90°F
Refrigerant	R-12
Condenser Δt	20°F
Evaporator Δt	10°F
Product throughput	20,000 lb/h
Annual operating time	800 h
Power cost	$0.10/kwh
Building cost	$50/ft^2
Heat exchanger cost	$0.50/$UA$(Btu/h·°F)
Compressor and accessories cost	$750/hp
Air handling system cost	$150/hp + $100
Interest rate	10%
Carton configuration	4 cartons deep
Carton vent area	4%
Final product temperature	40°F
Initial product temperature	80°F
Labor cost	$10/h
Labor required	1 man-hour/20,000 lb

Output results in Figure 4 are based on these input values.

(considered to be average cost) value with the unit cooling cost given for each product. For these specific inputs, no one unit cost dominates; in fact, they all have about the same effect.

PACKAGE ICING

Finely crushed ice placed in shipping containers can effectively cool products that are not harmed by contact with ice. Spinach, collards, kale, brussels sprouts, broccoli, radishes, carrots, and onions are commonly packaged with ice (Hardenburg et al. 1986). Cooling a product from 95 to 35°F requires melting ice equal to 38% of the product's mass. Additional ice must melt to remove heat leaking into the packages and to remove heat from the container. In addition to removing field heat, package ice can keep the product cool during transit.

Top icing, or placing ice on top of packed containers, is used occasionally to supplement another cooling method. Because corrugated containers have largely replaced wooden crates, the use of top ice has decreased in favor of forced-air and hydrocooling. Wax-impregnated corrugated containers, however, have allowed the use of icing and hydrocooling of products after packaging.

Pumping *slush ice* or *liquid ice* into the shipping container through a hose and special nozzle that connect to the package is another method used for cooling some products. Some systems can ice an entire pallet at one time.

VACUUM COOLING

Vacuum cooling of fresh produce, mostly vegetables having a high ratio of surface area to volume, by the rapid evaporation of water from the product, is a particular application of vacuum refrigeration. In vacuum refrigeration, water, as the primary refrigerant, vaporizes in a flash chamber under low pressure. The pressure in the chamber is lowered to the saturation point corresponding to the lowest required temperature of the water.

Vacuum cooling is a batch process. The product to be cooled is loaded into the flash chamber, the system is put into operation,

Methods of Precooling Fruits, Vegetables, and Cut Flowers

Table 2 Relationships Among Unit Costs and Cost per Unit of Product Cooled
(Other Parameters are Shown in Table 1.)

	Unit Cost for Carton Cooling				
	−100%	−50%	Nominal	+50%	+100%
Building, $/ft^2	0	25	50	75	100
(Cooling cost, ¢/lb)	(0.236)	(0.263)	(0.290)	(0.317)	(0.344)
Heat exchangers, $ per Btu/h · °F	0	0.25	0.50	0.75	1.00
(Cooling cost, ¢/lb)	(0.233)	(0.261)	(0.290)	(0.319)	(0.347)
Compressor, $/hp	0	375	750	1125	1500
(Cooling cost, ¢/lb)	(0.223)	(0.256)	(0.290)	(0.323)	(0.357)
Fan, $/hp	0	75	150	225	300
(Cooling cost, ¢/lb)	(0.286)	(0.288)	(0.290)	(0.292)	(0.294)
Labor, $/h	0	5	10	15	20
(Cooling cost, ¢/lb)	(0.240)	(0.265)	(0.290)	(0.315)	(0.340)
Interest rate, %	0	5	10	15	20
(Cooling cost, ¢/lb)	(0.201)	(0.241)	(0.290)	(0.345)	(0.404)
Power cost, $/kWh	0	0.05	0.10	0.15	0.20
(Cooling cost, ¢/lb)	(0.233)	(0.261)	(0.290)	(0.319)	(0.347)

and the product is cooled by reducing the pressure to the corresponding saturation temperature desired. The system is then shut down, the product removed, and the process repeated. Since the product is normally at ambient temperature before it is cooled, vacuum cooling can be thought of as a series of intermittent operations of a vacuum refrigeration system where the water in the flash chamber is allowed to come to ambient temperature before each start. The functional relationships for determining refrigerating capacity are the same in each case.

Cooling is achieved by boiling water, mostly off the surface of the product to be cooled. The heat of vaporization required to boil the water is furnished by the product, which is cooled accordingly. As the pressure is further reduced, cooling continues to the desired temperature level. The saturation pressure for water at 212 °F is 760 mm Hg. At 32 °F, the saturation pressure is 4.58 mm Hg. Commercial vacuum coolers normally operate in this range.

Although the cooling rate of lettuce could be increased without danger of freezing, by reducing the pressure to 3.8 mm Hg, corresponding to a saturation temperature of 27 °F, most operators do not reduce the pressure below the freezing temperature of water because of the extra work involved and the freezing potential.

Pressure, Volume, and Temperature

In a vacuum cooling operation, the thermodynamic process is assumed to take place in two phases. In the first phase, the product is assumed to be loaded into the flash chamber at ambient temperature, and the temperature in the flash chamber remains constant until saturation pressure is reached. At the onset of boiling, the small remaining amount of air in the chamber is replaced by the water vapor, the first phase ends, and the second phase begins simultaneously. The second phase continues at saturation until the product has cooled to the desired temperature.

If the ideal gas law is applied for an approximate solution in a commercial vacuum cooler, the pressure-volume relationships are:

$$\text{Phase 1} \quad pv = 29{,}318 \text{ ft·lb/lb}$$
$$\text{Phase 2} \quad pv^{1.056} = 66{,}370 \text{ ft·lb/lb}$$

where

p = absolute pressure, lb/ft^2
v = specific volume, ft^3/lb

The pressure-temperature relationship is determined by the value of ambient and product temperature. Based on 90 °F for this value, the temperature in the flash chamber theoretically remains constant at 90 °F as the pressure is reduced from atmospheric to saturation, after which it declines progressively along the saturation line. These relationships are illustrated in Figure 5. The product temperature would respond similarly but would vary depending on where temperature is measured in the product, the physical characteristics of the product, and the amount of product surface water available. While it is possible for some vaporization to occur within the intercellular spaces beneath the product surface, most of the water is vaporized off the surface. The heat required to vaporize this water is also taken off the product surface where it flows by conduction under the thermal gradient produced. Thus, the rate of cooling depends on the relation of surface area to volume of the product and the rate at which the vacuum is drawn in the flash chamber.

Since water is the sole refrigerant, the amount of heat removed from the product depends on the amount of water W_v vaporized

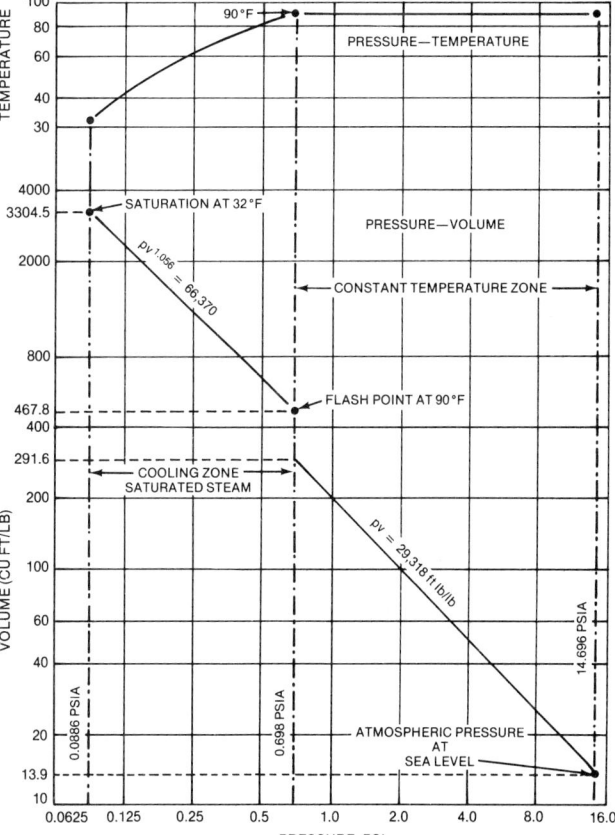

Fig. 5 Pressure, Volume, and Temperature in a Vacuum Cooler Cooling Product from 90 to 32 °F

and its latent heat of vaporization L. Assuming an ideal condition, with no heat gain from surroundings, total heat Q removed from the product is:

$$Q = W_N L_v L \qquad (9)$$

The amount of moisture removed from the product during vacuum cooling, then, is directly related to the specific heat of the product and the amount of temperature reduction accomplished. A product with a specific heat capacity of 0.95 Btu/(lb · °F) would theoretically lose 1% moisture for each 11°F reduction in temperature. In a study of vacuum cooling of 16 different vegetables, Barger (1963) showed that cooling of all products was proportional to the amount of moisture evaporated from the product. Temperature reductions averaged 9 to 10°F for each 1% of weight loss, regardless of the product cooled.

Commercial Systems

The four types of vacuum refrigeration systems that use water as the refrigerant are: (1) steam ejector, (2) centrifugal, (3) rotary, and (4) reciprocating. A schematic of the vacuum-producing mechanism of each is illustrated in Figure 6.

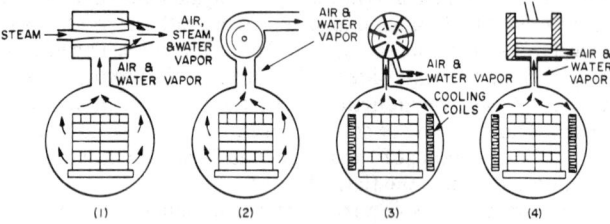

Fig. 6 Schematic Cross Sections of Vacuum-Producing Mechanisms

Of these, the steam ejector type is best suited for displacing the extremely high volumes of water vapor encountered at the low pressures needed in vacuum cooling. It also has the advantage of having few moving parts, thus requiring no compressor to condense the water vapor. High-pressure steam is expanded through a series of jets or ejectors arranged in series and condensed in barometric condensers mounted below the ejectors. Cooling water for condensing is accomplished by means of an induced-draft cooling tower. In spite of these advantages, few steam ejector vacuum coolers are used today, due to the inconvenience of using steam and the lack of portability. Instead, vacuum coolers mounted on semitrailers are used to follow the seasonal crops.

The centrifugal compressor is also a high-volume pump and can be adapted to water vapor refrigeration. However, its use in vacuum cooling is limited because of inherent mechanical difficulties at the high rotative speeds required to produce the low pressures needed.

Both rotary and reciprocal vacuum pumps are capable of producing the low pressures needed for vacuum cooling, and they also have the advantage of portability. Being positive displacement pumps, however, they have low volumetric capacity; therefore, vacuum coolers using rotary or reciprocating pumps have separate refrigeration systems to condense much of the water vapor that evaporates off the product, thus substantially reducing the volume of water vapor passing through the pump. Ideally, where it can be assumed that all of the water vapor is condensed, the required refrigeration capacity is equal to the amount of heat removed from the product during cooling.

The condenser must contain adequate surface to condense the large amount of vapor removed from the produce in a few minutes. Refrigeration is furnished from cold brine or a direct-expansion system. A very large peak load occurs from rapid condensing of so much vapor. Best results are obtained if the refrigeration plant is equipped with a large brine or icemaking tank having enough stored refrigeration to smooth out the load. A standard three-tube plant, with capacity to handle three cars per hour, will have a peak refrigeration load of at least 250 tons.

To increase cooling effectiveness and reduce product moisture loss, the product is sometimes wetted before cooling begins. However, lettuce is rarely prewetted. A propietary modification of vacuum cooling continuously circulates chilled water over the product throughout the vacuum cooling process. Among the chief advantages are increased cooling rates and residual refrigeration that is stored in the chilled water following each vacuum process. It also prevents water loss from products that show objectionable wilting after conventional vacuum cooling.

Applications

Because vacuum cooling is generally more expensive, particularly in first cost, than other cooling methods, its use is primarily restricted to products for which vacuum cooling is much faster or more convenient. Lettuce is ideally adapted to vacuum cooling. The numerous individual leaves provide a large surface area and the tissues release moisture readily. It is possible to freeze lettuce in a vacuum chamber if pressure and condenser temperatures are not carefully controlled. However, even lettuce does not cool entirely uniformly. The fleshy core, or butt, releases moisture more slowly than the leaves. Temperatures as high as 43°F have been recorded in core tissue when leaf temperatures were down to 33°F (Barger 1961).

Other leafy vegetables such as spinach, endive, escarole, and parsley are also suitable for vacuum cooling. Vegetables that are less suitable but adaptable by wetting are asparagus, snap beans, broccoli, brussels sprouts, cabbage, cauliflower, celery, green peas, sweet corn, leeks, and mushrooms. Of these vegetables, only cauliflower, celery, cabbage, and mushrooms are commercially vacuum cooled in California. Fruits are generally not suitable, except some of the berries, especially strawberries. Cucumbers, cantaloupes, tomatoes, dry onions, and potatoes cool very little because of their low surface-to-mass ratio and relatively impervious surface. The final temperature of various vegetables when vacuum cooled under similar conditions is illustrated in Figure 7.

The rate of cooling and the final temperature attained by vacuum cooling are largely affected by the ratio of the surface area of

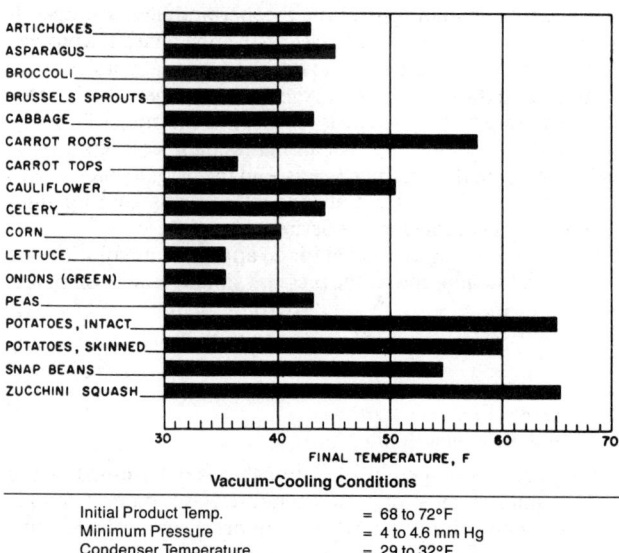

Fig. 7 Comparative Cooling of Vegetables Under Similar Vacuum Conditions

the commodity to its mass and the ease with which the product gives up water from its tissues. Consequently, the adaptability of fruits and vegetables varies tremendously for this method of precooling. For products that have a low surface-to-mass ratio, high temperature gradients occur. To prevent the surface from freezing before the desired mass-average temperature is reached, a procedure referred to as "bouncing" is practiced. This is accomplished by switching the vacuum pump off and on to keep the saturation temperature above freezing.

Mechanical vacuum coolers have been designed in several sizes. Most installations use cylindrical or rectangular retorts, sized to hold either a half- or a quarter-car of produce. However, a few are large enough to hold an entire refrigerator car-load.

SELECTING A COOLING METHOD

Packinghouse size and operating procedures, response of product to the cooling method, and market demands largely dictate the type of cooling method used. Other factors considered are whether or not the product is packaged in the field or in a packing-house, the product mix being cooled, length of cooling season, and comparative costs of dry versus water-repellant cartons. In some cases, there is little question about the type of cooling to be used. For example, vacuum cooling is most effective on lettuce and other similar vegetables. Peach packers in the southeastern United States, and some vegetable and citrus packers, are satisfied with hydrocooling. Air (room) cooling is used for apples, pears, peaches, plums, nectarines, sweet cherries, strawberries, and apricots. In other cases, choice of cooling method is not so clearly defined. Celery and sweet corn are usually hydrocooled, but they may be vacuum cooled as effectively. Cantaloupes may be satisfactorily cooled by several methods.

When more than one method can be used, cost becomes a major consideration. While rapid, forced-air cooling is more costly than hydrocooling, if the product does not require rapid cooling, a forced air system can operate almost as economically as hydrocooling. In a study to evaluate costs of hypothetical precooling systems for citrus fruit, Gaffney and Bowman (1970) found that the cost for forced-air cooling in bulk lots was 20% more than that for hydrocooling in bulk and forced-air cooling in cartons costs 45% more than hydrocooling in bulk.

COOLING CUT FLOWERS

Because of their high rates of respiration and low tolerance to heat, deterioration in cut flowers is rapid at field temperatures. Refrigerated highway vans do not have the capacity to remove the field heat in sufficient time to prevent some deterioration from occurring (Farnham *et al.* 1979). Forced-air cooling is commonly used by the flower industry. As with most fruits and vegetables, the cooling rate of cut flowers varies substantially among the various types. Rij *et al.* (1979) found that the half-cooling time for packed boxes of gypsophila was about 3 min compared to about 20 min for chrysanthemums at airflows ranging from 80 to 260 cfm per box. Within this range, cooling time was proportional to the reciprocal of airflow but varied less with airflow than with flower type.

SYMBOLS

Y = temperature ratio $(t - t_o)/(t_i - t_o)$
t = temperature of any point in product, °F
t_i = initial uniform product temperature, °F
t_o = surrounding temperature, °F
t_{ma} = mass average temperature, °F
θ = cooling time, h
h = surface heat transfer coefficient, Btu/(h·ft²·°F)
A = product surface area, ft²
q = cooling load, Btu/h
Q = total heat, Btu
W_v = mass, lb
L = heat of vaporization, Btu/lb
C = cooling coefficient, reciprocal of hours
Z = half-cooling time, h
α = thermal diffusivity, ft²/h
M_1 = first root of transcendental function
l = characteristic length, ft
G = geometry index
k = thermal conductivity, Btu/[h·ft²(°F/ft)]
μ = dynamic viscosity, lb/(h·ft)
G' = mass rate of airflow, lb/(h·ft²)
c_p = specific heat, Btu/(lb·°F)
p = pressure, lb/ft²
v = volume, ft³

REFERENCES

Baird, C.D. and J.J. Gaffney. 1976. A numerical procedure for calculating heat transfer in bulk loads of fruits or vegetables. ASHRAE *Transactions* 82(2):525.

Baird, C.D., J.J. Gaffney, and M.T. Talbot. 1988. Design criteria for efficient and cost effective forced air cooling systems for fruits and vegetables. ASHRAE *Transactions* 94(1):1434.

Ball, C.O. and F.C.W. Olson. 1957. *Sterilization in food technology.* McGraw Hill Book Company, New York.

Barger, W.R. 1961. Factors affecting temperature reduction and weight loss of vacuum-cooled lettuce. USDA, *Marketing Research Report.*

Barger, W.R. 1963. Vacuum precooling—A comparison of cooling of different vegetables. USDA, *Marketing Research Report* No. 600.

Bennett, A.H., J. Soule, and G.E. Yost. 1966. Temperature response of citrus to forced-air precooling. ASHRAE *Journal* 8(4):48.

Bennett, A.H., J. Soule, and G.E. Yost. 1969. Forced-air precooling for Red Delicious apples. USDA, ARS 52-41.

Eshleman, W.D., C.D. Baird, and J.J. Gaffney. 1976. A numerical simulation of transient heat flow in irregular shaped foods. ASAE Paper No. 76-6504.

Farnham, D.S., *et al.* 1979. Comparison of conditioning, precooling, transit method, and use of a floral preservative on cut flower quality. Proceedings, *Journal of American Society of Horticultural Science* 104(4):483.

Fisher, D.V. 1960. Cooling rates of apples packed in different bushel containers and stacked at different spacing in cold storage. ASHRAE *Journal* (July):53.

Ford, K.E. 1956. Hydrocooling cantaloupes. *Proceedings* of the Florida State Horticultural Society 69:138.

Gaffney, J.J. and E.K. Bowman. 1970. An economic evaluation of different concepts for precooling citrus fruits. ASHRAE Symposium on Precooling Fruits and Vegetables, San Francisco (January).

Grierson, W. 1957. Preliminary studies for cooling Florida oranges prior to packing. *Proceedings* of the Florida State Horticultural Society 70:264.

Grizell, W.G. and A.H. Bennett. 1966. *Hydrocooling stacked crates of celery and sweet corn.* USDA, Agricultural Research Service, ARS 52-12.

Hardenburg, R.E., A.E. Watada, and C.Y. Wang. 1986. The commercial storage of fruits, vegetables, and florist and nursery stocks. USDA *Agricultural Handbook* No. 66.

Henry, F.E. and A.H. Bennett. 1973. Hydraircooling vegetable products in unit ads. *Transactions* of the ASAE 16(4):731.

Henry, F.E. A.H. Bennett, and R.H. Segall. 1976. Hydraircooling—A new concept for precooling pallet loads of vegetables. ASHRAE *Transactions* 82(2):541.

Ishibashi, S., R. Kojima, and T. Kaneko. 1969. Studies on the forced-air cooler. JSAM, Japan 31(2), September.

Mitchell, F.G., R. Guillou, and R.A. Parsons. 1972. *Commercial cooling of fruits and vegetables.* Manual 43, Division of Agricultural Sciences, University of California, Berkeley.

Parsons, R.A., F.G. Mitchell, and G. Mayer. 1972. Forced-air cooling of palletized fresh fruit. *Transactions* of the ASAE 15(4):729.

Perry, J.S., A.H. Bennett, and T.V. Minh. 1968. Experiments with a prototype commercial forced-air precooler on peaches, potatoes, apples, and strawberries. Unpublished data, University of Georgia.

Perry, R.L., and R.M. Perkins. 1968. Hydrocooling sweet corn. ASAE Paper No. 68-880 (December), Chicago.

Pflug, I.J., J.L. Blaisdell, and I.J. Kopelman. 1965. Developing temperature-time curves for objects that can be approximated by a sphere, infinite plate, or infinite cylinder. ASHRAE *Transactions* 71(1):238.

Rij, R.E., J.F. Thompson, and D.S. Farnham. 1979. *Handling, precooling, and temperature management of cut flower crops for truck transportation*. USDA-SEA Western Series No. 5 (June).

Smith, R.E. and A.H. Bennett. 1965. Mass-average temperature of fruits and vegetables during transient cooling. *Transactions* of the ASAE 8(2):249.

Smith, R.E., A.H. Bennett, and A.A. Vacinek. 1970. Convection film coefficients related to geometry for anomalous shapes. *Transactions* of the ASAE, Paper No. 69-373.

Smith, R.E., G.L. Nelson, and R.L. Henrickson. 1967. Analyses on transient heat transfer from anomalous shapes. *Transactions* of the ASAE 10(2):236.

Smith, R.E., G.L. Nelson, and R.L. Henrickson. 1968. Applications of geometry analysis of anomalous shapes to problems in transient heat transfer. *Transactions* of the ASAE 11(2):296.

Smith, W.L. and W.H. Redit. 1968. Postharvest decay of peaches as affected by hot-water treatments, cooling methods, and sanitation. USDA, *Marketing Research Report* No. 807.

Soule, J., G.E. Yost, and A.H. Bennett. 1966. Certain heat characteristics of oranges, grapefruit and tangelos during forced-air precooling. *Transactions* of the ASAE 9(3):355.

BIBLIOGRAPHY

Ansari, F.A. and A. Afaq. 1986. Precooling of cylindrical food products. *International Journal of Refrigeration* 9(3):161-63.

Arifin, B.B. and K.V. Chau. 1988. Cooling of strawberries in cartons with new vent hole designs. ASHRAE *Transactions* 94(1):1415-26.

Bennett, A.H. 1962. Thermal characteristics of peaches as related to hydrocooling. USDA, *Technical Bulletin* No. 1292.

Bennett, A.H., W.G. Chace, Jr., and R.H. Cubbedge. 1969. Heat transfer properties and characteristics of Appalachian area Red Delicious apples. ASHRAE *Transactions* 75(2):133.

Bennett, A.H., R.E. Smith, and J.C. Fortson. 1965. Hydrocooling peaches—A practical guide for determining cooling requirements and cooling times. USDA, *Agriculture Information Bulletin* No. 298 (June).

Bennett, A.H., W.G. Chace, Jr., and R.H. Cubbedge. 1970. Thermal properties and heat transfer characteristics of marsh grapefruit. USDA, *Technical Bulletin* No. 1413.

Burton, K.S., C.E. Frost, and P.T. Atkey. 1987. Effect of vacuum cooling on mushroom browning. *International Journal of Food Science & Technology* 22(6):599-606.

Freeman, C.D. 1984. Cost reducing technologies in cooling fresh vegetables. ASAE Paper No. 841074.

Gaffney, J.J. and C.D. Baird. 1977. Forced-air cooling of bell peppers in bulk. ASAE *Transactions* 20(6):1174-80.

Gaffney, J.J., C.D. Baird, and W.D. Eshleman. 1976. Temperature response of avocados during cooling with chilled water. ASAE Paper No. 76-6017.

Gariepy, Y., G.S.V. Raghavan, and R. Theriault. 1987. Cooling characteristics of cabbage. *Canadian Agricultural Engineering* 29(1):45-50.

Hackert, J.M., R.V. Morey, and D.R. Thompson. 1987. Precooling of fresh market broccoli. *Transactions* of the ASAE 30(5): 1489-93.

Hayakawa, K. 1978. Computerized simulation for heat transfer and moisture loss from an idealized fresh produce. *Transactions* of the ASAE 21(5):1015-24.

Hayakawa, K. and J. Succar. 1982. Heat transfer and moisture loss of spherical fresh produce. *Journal of Food Science* 47(2):596-605.

Isenberg, F.M.R., R.F. Kasmire, and J.E. Parson. 19??. Vacuum cooling vegetables. *Information Bulletin* 186, Cornell Cooperative Extensive Service.

Jiang, H., D.R. Thompson, and R.V. Morey. 1987. Finite element model of temperature distribution in broccoli stalks during forced-air precooling. *Transactions* of the ASAE 30(5):1473-77.

Lentz, C.D. and L. van den Berg. 1977. Cabbage precooling study. *J. Inst. Can. Sci., Technol. Aliment.* 10(4):265.

Misener, G.C. and G.C. Shove. 1976. Simulated cooling of potatoes. *Transactions* of the ASAE 19(5):954.

Morey, R.V., S.A. Sargent, C.D. Baird, and M.R. Talbot. 1988. ASAE Paper No. 88-7539.

Rohrbach, R.P., R. Ferrell, E.O. Beasley, J.R. Fowler. 1984. Precooling blueberries and muscadine grapes with liquid carbon dioxide. *Transactions* of the ASAE 27(6):1950-55.

Shaw, J. and C. Kuo. 1987. Vacuum precooling green onion and celery. ASAE Paper No. 87-5522.

Stewart, J.K. and H.M. Couey. 1963. Hydrocooling vegetables—A practical guide to predicting final temperatures and cooling times. USDA, *Marketing Research Report* No. 637.

Thompson, J.F. and Y.L. Chen. 1986. Energy use in hydrocooling stone fruit. ASAE Paper No. 866556.

Thompson, J.F. and R.F. Kasmire. 1979. Evaporative cooling of chilling sensitive vegetable crops. ASAE Paper No. 79-6516.

Thompson, J.F., Y.L. Chen, and T.R. Rumsey. 1987. Energy use in vacuum coolers for fresh market vegetables. *Applied Engineering in Agriculture* 3(2):196-99, American Society of Agricultural Engineers, St. Joseph, MI.

CHAPTER 12

MEAT PRODUCTS

CARCASS CHILLING AND HOLDING 12.1	Bacon Slicing and Packaging Room 12.12
Refrigeration Systems for Coolers 12.2	Sausage Dry Rooms 12.13
Modified Atmospheres 12.3	Lard Chilling 12.14
Beef Cooler Layout and Capacity 12.3	Blast and Storage Freezers 12.15
Boxed Beef 12.7	Direct Contact Meat Chilling 12.15
Hog Chilling and Tempering 12.7	FROZEN MEAT PRODUCTS 12.15
Pork Trimmings 12.9	Prefreezing Quality of Meat 12.15
Fresh Pork Holding 12.10	Effect of Freezing on Quality 12.15
Calf and Lamb Chilling 12.10	Storage and Handling 12.16
Chilling and Freezing Variety Meats 12.10	Packaging 12.16
Packaging and Storage 12.11	SHIPPING DOCKS 12.16
Refrigeration Load Computations 12.11	ENERGY CONSERVATION 12.17
PROCESSED MEATS 12.11	

SOUND sanitary practices should be used at all stages of food processing, not only to protect public health but to meet aesthetic requirements as well. In this respect, meat processing plants are no different from other food plants; the same principles apply regarding sanitation of buildings and equipment; provision of sanitary water supplies and wash facilities; disposal of waste materials; insect and pest control; and proper use of sanitizers, germicides, and fungicides. All U.S. meat plants operate under regulations set forth in inspection service orders. For detailed sanitation guidelines to be followed in all plants producing meat under federal inspection, refer to *Agriculture Handbook No. 570,* available from the U.S. Department of Agriculture, FSIS.

Proper safeguards should mimimize bacterial contamination and growth. This involves using clean raw materials, clean water and air, sanitary handling throughout, good temperature control (particularly in coolers and freezers), and scrupulous between-shift cleaning of all surfaces in contact with the product.

Precooked products present additional problems because favorable conditions for bacterial growth exist after the product has cooled to below 130°F. Any delay in processing at this stage allows surviving microorganisms to multiply, especially where the cooked and cooled meat is handled and packed into containers prior to processing and freezing. Creamed products afford especially favorable conditions for bacterial growth. Filled packages should be removed immediately on filling and quickly chilled. Fast chilling not only reduces the time for growth, but can also reduce the number of bacteria.

It is even more important during processing to avoid any opportunity for the growth of pathogenic bacteria (such as *Salmonellae, Clostridium perfringens, Staphylococci,* or *Streptococci*) that may have entered the product (Thatcher and Clark 1968). While these organisms do not grow at temperatures below 40°F, they can survive freezing and prolonged frozen storage.

Storage at a temperature of about 25°F will permit the growth of psychrophillic spoilage bacteria, but at 14°F these, as well as all other bacteria, are dormant. Even though some cells of all bacteria types die off during storage, activity of the survivors is quickly renewed with rising temperature. It should be the duty of the processor to recommend safe preparation practices to the consumer. The best procedure is to provide workable instructions for cooking the food without preliminary thawing. In the freezer, sanitation is confined to keeping physical cleanliness and order and preventing access of foreign odors.

CARCASS CHILLING AND HOLDING

A hot carcass cooler removes live animal heat as rapidly as possible, with consideration given to side effects such as cold shortening in beef. Rapid temperature reduction is important in reducing the growth rate of microorganisms that may exist on carcass surfaces. Conditions of temperature, humidity, and air motion must be considered to attain desired meat temperatures within the time limit and to prevent excessive shrinkage, bone taint, sour rounds, surface slime, mold, or discoloration. The carcass must be delivered with a bright, fresh appearance.

Although certain basic principles are identical, beef and hog carcass chilling differs substantially. The massive beef carcass is only partially chilled (although shippable) at the end of the standard overnight period; the average hog carcass may be fully chilled (but not ready for cutting) in 8 to 12 h, while the balance of the period accomplishes only temperature equalization.

The beef carcass surface retains a large amount of wash and shroud water, which provides much evaporative cooling in addition to that derived from actual shrinkage; but evaporative cooling of the hog carcass, which retains little wash water and is not shrouded, occurs only through actual shrinkage. A beef carcass, without skin and destined largely for sale as fresh cuts, must be chilled in air temperatures sufficiently high to avoid freezing and damage to appearance. Although it must subsequently be well tempered for cutting and scheduled for in-plant processing, a hog carcass, including the skin, can tolerate a certain amount of surface freezing. Shrouded beef carcasses can be chilled with an overnight shrinkage of 0.75 to 1.25%, whereas equally good practice on hog carcasses will result in 1.25 to 2% shrinkage.

The bulk (16 to 20 h) of beef chilling is done overnight in chilling rooms with a large refrigeration and air circulation capacity. The balance of the chilling and temperature equalization occurs during a subsequent holding or storage period that averages one day, but frequently extends to 2 or 3 days, usually in a separate holding room with a low refrigeration and air circulation capacity.

Some packers load for shipment the day after slaughter, since some refrigerated transport vehicles have ample capacity to

---

The general responsibility for this chapter is assigned to TC 11.1, Meat, Fish, and Poultry Products.

remove the balance of the internal heat in round or chuck beef during the first two days in transit. This practice is most important in rapid delivery of fresh meat to the marketplace. Carcass beef that is not shipped the day after slaughter should be kept in a beef-holding cooler at temperatures of 34 to 36°F with minimum air circulation to avoid excessive color change and weight loss.

REFRIGERATION SYSTEMS FOR COOLERS

Refrigeration systems commonly used in carcass chilling and holding rooms are operated with ammonia as the primary refrigerant and are of three general types: chilled brine spray, sprayed coil, and dry coils.

Chilled brine spray systems are generally being abandoned in favor of other systems due to such a system's large required building space, inherent low capacity, brine carryover tendencies, and difficulty of control.

Sprayed coil systems consist of unit coolers equipped with coils, brine spray banks, eliminators to prevent brine carryover, and fans for air-vapor circulation. The units are usually mounted (without ductwork) either on the floor or overhead on converted brine spray decks. Refrigeration is supplied by the primary refrigerant in the coils. Chilled or nonchilled recirculated brine is continually sprayed over the coils, thus eliminating ice formation and the need for periodic defrosting.

The brine predominantly used is sodium chloride, with caustic soda or another additive for controlling pH. Because sodium chloride brine is corrosive, bare-pipe coils (without fins) generally see service. The brine is also highly corrosive to the rail system and other cooler equipment.

Propylene glycol with added inhibitor complexes is another coil spray solution used in place of sodium chloride. As with sodium chloride brine, propylene glycol is constantly diluted by moisture condensed out of the spaces being refrigerated and must be concentrated by evaporating water from it. The reconcentration process requires special equipment designed to minimize glycol losses. Sludging in the concentrator may become an operating problem; to avoid it, additives must be selected and pH closely controlled. Finned coils are usually used with propylene glycol.

Because of its noncorrosiveness in comparison to sodium chloride, propylene glycol greatly reduces the cost of unit cooler construction as well as maintenance of space equipment.

Dry coil systems comprise most chilling and holding room installations. Dry coil systems usually include unit coolers equipped with coils, defrosting equipment, and fans for air-vapor circulation. Because the coils are operated without continuous brine spray, eliminators are not required. Coils are usually finned, with fins limited to 3 or 4 per inch or with variable fin spacing to avoid icing difficulties. The units may be mounted on the floor, overhead on the rail beams, or overhead on converted brine spray decks.

Dry coil systems operated at surface temperatures below 32°F build up a coating of frost or ice, which ultimately reduces the airflow and cooling capacity. Coils must therefore be defrosted periodically, normally every 4 to 24 h for coils with 3 or 4 fins per inch, to maintain capacity. The rate of buildup, and hence the defrosting frequency, decreases with large coil capacity and high evaporating pressure.

Defrosting may be done either manually or automatically by the following:

- *Hot gas defrost* is accomplished, with fans either on or off, by introducing hot gas direct from the system compressors into the evaporator coils. The evaporator suction is throttled to maintain a coil pressure of about 60 to 75 psig (at approximately 40 to 50°F). The coils then act as condensers and supply the heat for melting the ice coating. Other evaporators in the system must supply the load for the compressors during this period. Hot gas defrost is rapid, normally requiring 5 to 15 min. for completion.

- *Coil spray defrost* is accomplished (with the fans turned off) by spraying the coil surfaces with water, which supplies the heat required to melt the ice coating. Suction and feed lines are closed off, with pressure relief from the coil to the suction line to minimize refrigeration effect. Enough water at 50 to 75°F must be used to avoid freezing on the coils, and care must be taken that freezeup does not occur in the drain lines. The sprayed water tends to produce some fog in the refrigerated space. Coil spray defrost is equally as rapid as hot gas defrost.

- *Room air defrost* is accomplished with the fans running while suction and feed lines are closed off (with pressure relief from coil to suction line), to permit buildup of coil pressure and melting of the ice coating by transfer of heat out of the air flowing across the coils. Refrigeration therefore continues during the defrosting period, but at a drastically reduced rate. Room air defrost is slow; the time required may vary from 30 min. to several hours if the coils are undersized for dry coil operation.

- *Electric defrost* is accomplished with electric heaters with fans either on or off. During defrost, refrigerant flow is interrupted.

Unit coolers may be defrosted by any one or combinations of the first three methods. All methods involve a reduction in chilling capacity, which varies with time loss and heat input. Hot gas and coil spray defrost interrupt the chilling only for short periods, but they introduce some heat into the space. Room air defrost severely reduces the chilling rate for long periods, but the heat required to vaporize the ice is obtained entirely from the room air. This method is generally considered impractical in air temperatures below 40°F because of the long defrosting periods. However, it is practicable in air temperatures as low as 35°F, if the coils are large enough to avoid excessive icing.

Evaporator controls customarily employed in carcass chilling and holding rooms include refrigerant feed controls, evaporator pressure controls, and air circulation control.

Refrigerant feed controls are designed to maintain under varying loads as high a liquid level in the coil as can be carried without excessive liquid spillover into the suction line. This is accomplished by using an expansion valve that throttles the liquid from supply pressure (typically 150 psig) to evaporating pressure (usually 20 psig or higher). The throttling of the liquid flashes some of it to gas, which chills the remaining liquid to saturation temperature at the lower pressure. If it does not bypass the coil, the flashed gas tends to reduce flooding of the interior coil surface, thus lowering coil efficiency.

The valve used may be a hand-controlled expansion valve supervised by operator judgment alone, a thermal expansion valve governed by the degree of superheat of the suction gas, or a float valve (or solenoid valve operated by a float switch) governed by the level of feed liquid in a surge drum placed in the coil suction line. This surge drum suction trap permits the ammonia flashed to gas in the throttling process to flow directly to the suction line, bypassing the coil. The trap may be small and placed just high enough so that its level governs that in the coils by gravity transfer. Or, as in the ammonia recirculation system, it may be placed below coil level so that the liquid is pumped mechanically through the coils in much greater quantity than is required for evaporation. In the latter case, the trap is sized large enough to carry its normal operating level plus all the liquid flowing through the coils, thus effectively preventing liquid spillover to the compressors. Nevertheless, it is necessary in all cases to provide further protection at the compressors' liquid return.

Present practice strongly favors liquid ammonia recirculation systems, mainly because of the greater coil heat transfer rates with the resultant greater refrigerating capacity over other systems (see Chapter 4). Some have coils mounted above the rail beams with 4 to 6 ft of ceiling head space. Air is forced through the coils, sometimes using two-speed fans.

Meat Products

Manual and thermal expansion valves do not provide good coil flooding under varying loads and do not bypass the flashed feed gas around the coils. As a result, evaporators so controlled are usually rated 15 to 25% less in capacity than those controlled by float valve or ammonia recirculation.

Evaporator pressure controls regulate coil temperature, and thereby the rate of refrigeration, by varying evaporating pressure within the coil. This is accomplished by using a throttling valve in the evaporator suction line downstream from the surge drum suction trap. All such valves impose a definite loss on the refrigeration system, and the amount of the loss varies directly with the pressure drop through the valve. This increases the work of compression for a given refrigeration effect.

The valve used to control evaporating pressure may be a manual suction valve set solely by operator judgment, a back pressure valve actuated by coil pressure or temperature, or a back pressure valve actuated by a temperature-sensing element somewhere in the room. Manual suction valves require excessive attention when loads fluctuate. The coil-controlled back pressure valve seeks to hold a constant coil temperature but does not control room temperature unless the load is constant. Only the room-controlled compensated back pressure valve responds to room temperature.

Air circulation control is frequently used when an evaporator must handle separate load conditions differing greatly in magnitude, such as the load in chilling rooms that are also used as holding rooms or for the negligible load on weekends. The use of two-speed fan motors (operated at reduced speed during the periods of light load) or turning the fans off and on can control air circulation to a degree.

Considerable attention is being directed to system designs that will reduce the amount of evaporative cooling at the time of entrance into the cooler and eliminate ceiling rail and beam condensation and drip. Good results have been achieved by using low-temperature blast chill tunnels before entrance into the chill room. The volume of ceiling condensate is reduced because the rate of evaporative cooling is reduced in proportion to the degree of surface cooling. Room condensation has been reduced by the addition of heat above carcasses (out of the main airstream), fans, minimized hot water usage during cleanup, better dry cleanup, timing of cleanup, and using wood rail supports.

Grade and yield sorting, with its simultaneous filling of several rooms, has shortened the chilling time available if refrigeration is kept off during the filling cycle. Its effect has to be offset by more chill rooms and more installed refrigeration capacity. If full refrigeration is kept at the start of filling, the peak load is reduced to the rooms being filled. Recently, hot carcass cutting has been started with only a short prechilling time. Cryogenic chilling has also been tested for hot carcass chilling.

MODIFIED ATMOSPHERES

Carbon dioxide gas may be used as a supplement to refrigeration in the storage and transportation of meat. The gas inhibits the growth of many bacteria that cause rapid deterioration. The storage life of meat refrigerated at 28.5 to 29.5 °F is about 40 days, but in a 10 to 20% CO_2 atmosphere, it remains in a salable condition for 60 to 70 days. CO_2 may prevent rancidity in pork and bacon. Fresh pork keeps for over 60 days in CO_2 at 32 °F, where it spoils in about 17 days at 32 °F in normal atmosphere.

An atmosphere with 20% CO_2 inhibits most bacteria and molds that are injurious to meat. Higher levels cause loss of color and flavor and may accelerate *Staphylococcus* bacteria, which cause food poisoning. Good sanitation is essential, since the length of successful storage, even with a CO_2 atmosphere, depends largely on the initial bacterial load.

Ozone prevents mold on dressed meat. A concentration of 2.5 to 3 ppm for 2 h twice each day is effective for initially clean meat stored at 34 to 37 °F and 90% rh. Higher concentrations of ozone oxidize the fats and cause rancidity and bad flavors to develop.

BEEF COOLER LAYOUT AND CAPACITY

Carcass halves or sides are supported by hooks suspended from one-wheel trolleys running on overhead rails. The trolleys are generally pushed from the dressing floor to the chilling room by powered conveyor chains equipped with fingers that engage the trolleys, which are then distributed manually over the chilling and holding room rail system. Chilling and holding room rails are commonly placed on 3 to 4 ft centers in the holding rooms, with pullout or sorting rails between them. The rails must be placed a minimum of 2 ft from the nearest obstruction, such as a wall or building column, and the tops of the rails must be at least 11 ft above the floor. The supporting beams should be placed a minimum of 6 ft below the ceiling for optimum air distribution. Applicable to new construction in plants engaged in interstate commerce, regulations for some of these dimensions are issued by the Meat Inspection Division of the USDA.

To assure effective air circulation, carcass sides should be placed on the rails in both chilling and holding coolers so that they do not touch each other. Required spacing varies with the size of the carcass and averages 2.5 ft per two sides of beef. In practice, however, sides are often more crowded.

A chilling room should be of such size that the last carcass loaded into it does not materially retard the chilling of the first carcass. While size is not as critical as in the case of the hog carcass chill room (because of the slower chill), it is desirable to limit chill cooler size to hold not more than 4 h of the daily kill to better control shrinkage and condensation. Holding coolers may be as large as desired because of their ability to maintain more uniform temperature and humidity.

While overall plant chilling and holding room capacities vary widely, chilling coolers generally require a capacity equal to the daily kill; holding coolers require 1 to 2 times the daily kill.

Beef Carcasses

Dressed beef carcasses, each split into two sides, range in weight from approximately 300 to 1000 lb, averaging about 550 lb per head. Specific heats of carcass components range from 0.50 Btu/(lb·°F) for fat to 0.8 Btu/(lb·°F) or more for lean muscle, averaging about 0.75 Btu/(lb·°F) for the carcass as a whole.

The body temperature of an animal at slaughter is about 102 °F. After slaughter, changes occur that generate heat and tend to increase carcass temperature, while heat loss from the surface tends to lower it.

The largest part of the carcass is the round, and at any given stage of the chilling cycle its center has the highest temperature of all carcass parts. This *deep round* temperature (about 105 °F when the carcass enters the chilling cooler) is therefore universally used as a measure of chilling progress. If it is to be significant, the temperature must be taken accurately. Incorrect techniques, in common use, will show temperatures as much as 10 °F lower than actual deep round temperature. An accurate technique that yields consistent results is shown in Figure 1. The technique applies a fast-reacting, easily read stem dial thermometer, calibrated before and after tests, inserted upward to the full depth through the hole in the aitchbone.

At the time of slaughter, the water content of beef muscle is approximately 300% that of its dry weight (75% of the total weight). Thereafter, a gradual drying of the surface takes place, resulting in weight loss or shrinkage. Shrinkage and its measurement are greatly affected by the final operations of the dressing process: weighing, washing, and shrouding. Weighing must be done prior to washing and shrouding if the weights are to reflect actual product shrinkage.

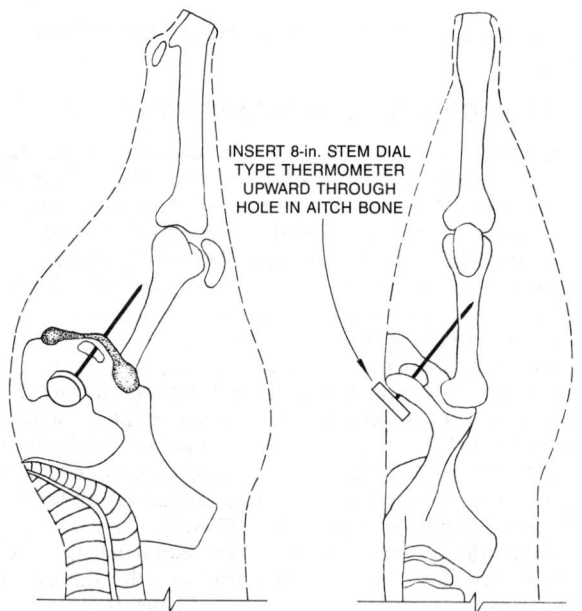

Fig. 1 Deep Round Temperature Measurement in Beef Carcass

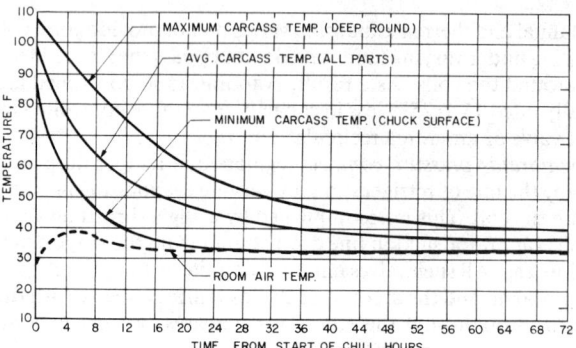

Fig. 2 Beef Carcass Chill Curves

A beef carcass retains large amounts of wash and shroud water on its surface, which it carries into the cooler. The loss of this water, occurring in the form of vapor, does not constitute actual product loss. However, it must be considered when estimating the system capacity since the vapor must be condensed on the coils, thus constituting an important part of the refrigeration load.

The amount of wash water retained by the carcass depends on its condition and on washing techniques. A carcass typically retains 8 lb, part of which is lost by drip and part by evaporation. Water pressures used in washing vary from 50 to 300 psig, and temperatures from 60 to 115 °F.

The shrouds—cloths saturated in warm (120 to 130 °F), weak (under 20 salometer) brine, and applied to outer surfaces of better grades of beef to enhance appearance—remain in place throughout the chill period but are removed before the carcasses are transferred to holding rooms. The shrouds weigh about 10 lb wet and 4 lb dry, per carcass (two sides). They lose about 6 lb of water per carcass in the chilling cooler, partially by drip but mostly by evaporation.

To minimize spoilage, a carcass should be reduced to a uniform temperature of about 35 °F as rapidly as possible. In practice, deep round temperatures of 60 °F (measured as in Figure 1), with surface temperatures of 35 to 45 °F, are common at the end of the first day's chill period.

To prevent formation of surface slime, a carcass surface needs to be a certain dryness during storage. Exposed beef muscle chilled to an actual temperature of 36 °F will not slime readily if dried at the surface to a water content of 90% of dry weight (47.4% of total weight). Such a surface is in vapor pressure equilibrium with a surrounding atmosphere at the same temperature (36 °F) and 96% rh. In practice, a room atmosphere at 32 to 34 °F and approximately 90% rh will maintain a well-chilled carcass in nonsliming condition (Thatcher and Clark 1968).

Chilling-Drying Process

Curves of carcass temperature during a chilling-holding cycle are shown in Figure 2. Note that some heat loss occurs before a carcass enters the chilling cooler. The evaporative cooling of surface water dominates in the initial stages of hot carcass chilling. As chilling progresses, the rate of losses by evaporative surface cooling diminishes and the sensible transfer of heat from the carcass surface increases. Note that the time-temperature rates of change are subject to variations between summer and winter ambient conditions, which influence system capacity.

The rate of transfer is increased both by more rapid circulation of air and lower air temperature, but these are limited by the necessity of avoiding surface freezing.

Estimated differences in vapor pressure between surface water (at average surface temperature) and atmospheric vapor during a typical chilling-holding cycle, and the corresponding shrinkage curve for an average carcass, is shown in Figure 3. Note the tremendous vapor pressure differences during the early part of the chill cycle when the carcass is warm. The evaporative loss could be reduced by beginning the chill with room temperature high, then lowering it slowly to minimize the pressure difference between carcass surface water and room vapor at all times. However, this slows the chill and prolongs the period of rapid evaporation. The quick chill practice is favored; but the cold shortening effect and bacterial growth must be considered in carcass quality and keeping time.

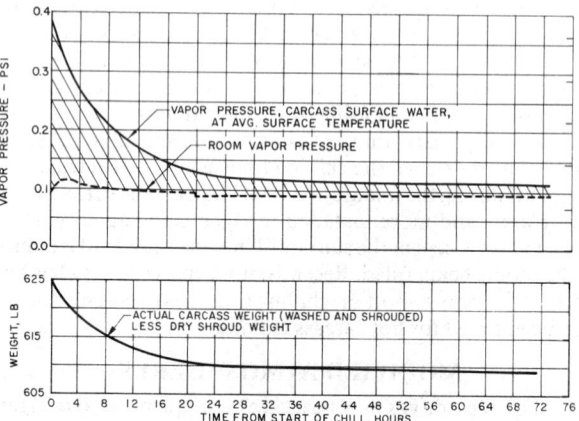

Fig. 3 Beef Carcass Shrinkage Rate Curves

Evaporation from the warm carcass in cool air is nearly independent of room relative humidity, because the warm carcass surface generates a much higher vapor pressure than the cooler vapor surrounding the carcass. If the space surrounding a warm carcass is saturated, evaporation forms fog, which can be observed at the beginning of any chill.

Evaporation from the well-chilled carcass with surface temperature at or near room temperature is different. The spread between surface and room vapor pressures approaches zero when room air is near saturation. Evaporation proceeds slowly, without forming fog. Evaporation does not cease when a room is saturated; it ceases only if the carcass is chilled through to room temperature, and no heat transfer is taking place.

Meat Products

The ultimate disposition of water condensed on the coils depends on the temperature of the coil surface and the method of coil operation. In continuous defrost (sprayed coil) operation, condensed and trapped water dilutes the solution sprayed over the coil. In nonfrosting dry-coil operation, condensed water falls to the evaporator pan and drains to the sewer. Water frozen on the coil is lost to the sewer if removed by hot gas or coil spray defrost. Periodic room air defrost, however, vaporizes part of the ice and returns it to room atmosphere, while losing the remainder to drain. This method of defrost is not normally used in beef chill or holding coolers because temperatures are not suitable and thus the defrost period is excessive, resulting in abnormal room temperature variations. Most chill and holding evaporators are automatically defrosted with water or hot gas on preselected time cycles. The weight changes that take place in an average beef carcass are given in Table 1.

Table 1 Weight Changes in Beef Carcass

In Chilling Cooler	Weight, lb	Percent
Initial dry weight	615	
Wash water pickup	8	
Initial wet weight	625	
Shroud water pickup[a]	6	
Drip loss, not vaporized	4	
Weight at start of chill[a]	625	
Weight loss in chill cooler	14.3	
Weight at end of 20-h chill[a]	610.7	
Net loss through chill, wet basis[b]	12.3	2.0
Net loss through chill, dry basis[c]	4.3	0.7
In Holding Cooler		
Initial weight (shroud removed)	610.7	
Weight loss during 48-h hold	1.8 +	
Weight loss per day	1.8 +	0.3
Final weight after 48-h hold	608.9	

[a]Exclusive of 4-lb weight of the dry shroud cloth.
[b]Carcass weighed in after washing, but before shrouding.
[c]Carcass weighed in before washing and shrouding.

Chilling of the beef carcass is not completed in the chill cooler but continues at a reduced rate in the holding cooler. A carcass well chilled when it enters the holding cooler shows minimum holding shrinkage; a poorly chilled one shows high holding shrinkage.

If shrinkage values are to have any significance, they must be carefully derived. Actual product loss must be determined by first weighing the dry carcass prior to washing and then weighing it out of the cooler with the shroud removed. In-motion weights are not sufficiently precise; carcasses must be weighed at rest. Scales must be accurate, and, if possible, the same scale should be used for both weighings. If shrinkage is to have any comparison value, it must be measured on carcasses chilled to the same temperature, since chilling occurs largely by evaporative weight loss.

Design Conditions and Refrigeration Load

Equipment selection should be based on conditions at peak load, when product loss is greatest. Room losses, equipment heat, and carcass heat add up to a total load that varies greatly—not only in magnitude but in proportion of sensible to total heat (sensible heat ratio)—throughout the chill. As the chill progresses, the vapor load decreases and the sensible load becomes more predominant.

Under peak chilling load, excess moisture condenses into fog—enough to warm the air-vapor-fog mixture to the sensible heat ratio of the heat removal process. The heat removal process of the coil therefore underestimates the actual rate of water removal by the amount of vapor condensed to fog (Table 2).

Table 2 Load Calculations for Beef Chilling

Cooler size, ft: 62 × 74 × 18.5
Cooler capacity: 476 carcasses
Average carcass weight: 625 lb
Assumed chill rate: 50°F in 20 h
Assumed air circulation: 167,400 cfm
Loading time: 3.3-h maximum
Assumed air to coil: 33°F, 100% rh
Assumed fan motive power: 36 hp
Specific heat of beef: 0.75 Btu/(lb·°F)

	Loads, Btu/h		
Heat Gain—Room Load	Sensible Heat	Latent Heat	Total Heat
1. Transmission, infiltration, personnel, fan motor, lights, and equipment heat	162,000	4100	166,100
2. Product heat (average, first 4 h):			
a. 476 × 625 × 0.75 × 50 × 0.1	1,115,700		
b. 476 × 14.3 × 0.13 × 1070[a]	−947,000	947,000	1,115,700
3. Total heat gain (room load), kW (Items 1 + 2a + 2b)	332,700	951,100	1,283,800
Heat Removal—Coil Load			
4. Air circulation, dry air, lb/h 167,400 × 0.08 × 60 = 803,500	—	—	—
5. Heat removed per lb of dry air, total heat (Item 3)/(Item 4) = 1.6 Btu/lb	—	—	—
6. Air-vapor enthalpy, Btu/lb dry air:			
a. Air to coil, 33°F, 100% rh	7.927	4.242	12.17
b. Btu removed, temperature drop 3.7°F	−0.927	−0.672	−1.60
c. Air from coil, 29.3°F, 100% rh	7.000	3.570	10.57
7. Coil air-vapor heat removal, Btu (Item 4)(Item 6b)	744,000	539,800	1,283,800
8. Room vapor condensed to fog (Item 7) − (Item 3)	411,300	−411,300	
9. Water (ice) removed by coil 476 × 14.3 × 0.13 × 144[b]		128,000	128,000
10. Total heat removal (coil load), Btu/h (Items 3 + 8 + 9)	744,000	667,800	1,411,800

[a]Heat of vaporization
[b]Heat of fusion

Fog does not generally form under later chilling room loads and all holding room loads, although it may form locally and then vaporize. Sensible heat ratios of air vapor heat gain and air vapor heat removal are then equal (Table 3).

Beef chilling rooms generally have evaporator capacity sufficient to hold room temperature under load approximately as shown in Figure 2. This results in an increase in room temperature to 35 to 40°F, with gradual reduction to 32 to 34°F. Many installations provide greater capacity, however, particularly dry coil systems, which thereby avoid excessive coil frosting. In batch-loaded coolers, room temperature may be as low as 25°F under peak load, provided it is raised to 30°F as the chill progresses, without surface freezing of the beef. The shrinkage improvement affected by these lower temperatures, however, tends to be less than expected (in beef chilling) because of the relatively small part played by sensible transfer of heat.

It is standard practice in the holding room to provide evaporator capacity to keep the room temperature at 32 to 34°F at all times. Holding room coils sized at peak load, low air vapor circulation rate, and a coil temperature 10°F below room temperature tend to maintain the approximately 90% rh that avoids excessive shrinkage and prevents surface sliming.

From the average temperature curve shown in Figure 2 and the shrinkage curve in Figure 3, certain generalizations useful in calculating carcass chilling load may be made. In the chilling cooler,

Table 3 Load Calculations for Beef Holding

Cooler size, ft: 100 × 136 × 18.5
Cooler capacity, one day's kill: 1120 carcasses
Average carcass weight: 610 lb
Assumed chill rate: 7.5 °F in 24 h
Assumed air circulation: 91,200 cfm
Assumed air to coil: 34 °F, 96% rh
Assumed fan motive power: 24 hp
Specific heat of beef: 0.75 Btu/(lb·°F)

Heat Gain—Room Load	Sensible Heat	Latent Heat	Total Heat
1. Transmission, infiltration, personnel, fan motor, lights, and equipment heat	245,000	10,000	255,000
2. Product heat, Btu/h:			
a. 1120 × 610 × 0.75 × 7.5 × 0.5	192,000		
b. 1120 × 1.8 × 0.06 × 1070[a]	−131,000	131,000	192,000
3. Total heat gain (room load), Btu/h (Items 1 + 2a + 2b)	306,000	141,000	447,000
Heat Removal—Coil Load			
4. Air circulation, lb/h dry air 91,200 × 0.08 × 60 = 437,800	—	—	—
5. Heat removed per lb of dry air, Btu (Item 3)/(Item 4) = 1.023	—	—	—
6. Air-vapor enthalpy, Btu/lb dry air:			
a. Air to coil, 34 °F, 96% rh	8.167	4.227	12.394
b. Btu removed, temperature drop 2.9 °F	−0.700	−0.323	−1.027
c. Air from coil, 31.1 °F, 100% rh	7.467	3.904	11.367
7. Coil air-vapor heat removal, Btu/h	306,000	141,00	447,000
8. Room vapor condensed to fog	—	—	—
9. Water (ice) removed by coil 1120 × 1.8 × 0.06 × 144[b]	—	17,400	17,400
10. Total heat removal (coil load), Btu/h (Items 3 + 8 + 9)	306,000	158,400	464,400

[a] Heat of vaporization
[b] Heat of fusion

Table 4 Sample Evaporator Installations for Beef Chilling[a]

Cooler size, ft: 62 × 74 × 18.5
Cooler capacity: 476 carcasses
Deep-round chill: to 50 °F in 20 h
Design load: 1,411,800 Btu
Coil operation: liquid recirculation
Loading time: 3.3 h
Average carcass weight: 625 lb
Assumed air to coil: 33 °F, 100% rh
Sensible heat ratio: 53%

	Dry Coil
Coil Description:	
Type of coil	Finned
Fin spacing fins/in.	4
Coil depth, number of pipe rows	8
Coil face area, ft²	20.4
Coil surface area, ft² total	2238
Fan Description, Airflow:	
Type of fan	Centrifugal
Flow through coil, cfm	9300
Flow, cfm/ft² coil face area	455
Fan motive power, hp	2
Unit Rating[b] (Total Heat):	
TD for capacity rating, °F[c]	10
Chilling capacity, Btu/h	81,000
Temperature drop, air through coil, °F	3.7
Equipment for 520 Carcasses:	
Number of units required	18
Total motive power, fans and pumps, hp	36
Coil surface per carcass, ft²	88
Airflow per carcass, cfm	350

[a] While data describe actual successful installations, other successful installations may be different.
[b] Ratings shown are estimated from performance of actual systems. Dry coil ratings are at average frost conditions, with airflow reduced by frost obstruction. While this example describes actual installations, it is not to be interpreted as an accepted standard. Other installations, employing both more and less equipment, are also successful.
[c] TD is temperature difference between refrigerant and air.

the average carcass temperature is reduced approximately 50 °F, from about 97 °F to about 47 °F, in 20 h. Simultaneously, about 14.3 lb of water is vaporized for each 625 lb carcass entering the chill; only 4.3 lb of this is actual shrinkage. The losses of sensible heat and water occur at about the same rate. In the sample load calculations, this is calculated at an average of 10% for the first 4 h of chill for sensible heat and 13% for the evaporation of moisture, which roughly agrees with the curves of Figures 2 and 3. This is the maximum rate of chill and is used for sizing the refrigeration equipment and piping.

In the holding cooler, the average carcass temperature is reduced from 47 to 39.5 °F in 24 h. Simultaneously, about 1.8 lb of water is vaporized per carcass (all actual shrinkage). Here also, the losses of sensible heat and water occur at about the same rate. The sample load calculations are figured at a 5% average for the first 4 h for sensible heat and 6% for latent heat.

Under peak chilling and holding room conditions, water trapped and condensed out by the coils imposes a further latent load on the evaporators. This occurs in the form of heat extracted to freeze condensed water into ice, or of heat removed to chill the returning warmed and strengthened spray solution. In the absence of a more complex evaluation, this load may be considered equal to the latent heat of fusion (144 Btu/lb) of the water removed.

Based on the data just mentioned, cooler loads may be calculated as illustrated in Tables 2 and 3. Transmission, infiltration, personnel, and equipment loads are estimated by standard methods.

The complete calculation is made to illustrate the heat removal process associated with the chilling-drying of the carcass, and in particular to illustrate that the sensible heat ratio of the heat transfer in the coil cannot be used to measure the amount of water removed from the space when fog is involved.

Evaporator Selection

Evaporator selection is a procedure of approximation only, because of the inaccuracies of load determination on the one hand and of predicting sustained field performance of coils on the other. Furthermore, there is rarely complete freedom of specification; for example, the air vapor circulation rate for a given coil may be limited to avoid spray solution carryover or excessive fan horsepower.

Sprayed and dry coil systems perform equally well with respect to shrinkage, provided compressor capacity is adequate and the evaporators are correct for the system selected. Evaporator requirements vary widely with the type of system. Comparative evaporator data on a typical, successful flooded coil installation in the chilling cooler is presented in Table 4.

The coil U-value (overall heat transfer coefficient) and airflows shown describe sustained field performance under actual chilling conditions and loads; they should not be confused with clean coil test ratings. The U-value varies greatly with the character of the coil and its operation and is influenced by such variables as the ratio of extended-to-prime surface, which may, for example, range from 7-to-1 to 21-to-1 in standard dry coils; coil depth, which typically ranges from 8 to 12 rows in sprayed coils and from 4 to 10 rows in dry coils; fin spacing, which may be 3 or 4 per inch in typical dry coils; condition of the surface, either continuously defrosted or generally coated with frost; and airflow, which may vary from 250 to 750 cfm/ft² coil face area.

Meat Products

Greater temperature differences (TD) than those shown are sometimes used, but a higher TD is valid only at high room temperatures. The lower TD (10 °F) shown for dry coils is desirable to limit frosting. Many dry coil evaporators have higher ratios of extended-to-prime surface and higher airflows per unit face area than shown.

The difficulties involved in obtaining accurate shrinkage figures on carcasses chilled to a specified degree cause wide differences of opinion as to the coil capacity required for good chilling. While data describe actual successful installations, other successful installations may differ.

BOXED BEEF

An increasing proportion of the output of beef slaughterhouses is in the form of prefabricated sections of the carcass, vacuum-packed in plastic bags and shipped in corrugated boxes. Standard cuts can be sold at cost savings to the market. The shipping density is much greater, with easier material handling, and the bones and fat are removed where their value as a byproduct is greater. Customers purchase only the sections they need, and the trim loss at final processing to primal cuts is minimized.

Vacuum-packaging or gas flush packaging with either CO_2 or N_2 has the following advantages:

- Creates anaerobic conditions, preventing the growth of mold (which is aerobic and requires the presence of oxygen for growth)
- Provides more sanitary conditions for carcass breaking
- Retains moisture, retards shrinkage
- Excludes bacteria entry, extends shelf life
- Retards bloom until opened

After normal chilling, a carcass is broken into primal cuts, vacuum packed, and boxed for shipment. Temperatures must be held under 40 °F to prevent the development of pathogenic organisms. Aging of the beef continues after vacuum-packaging and during shipment, because the exclusion of oxygen does not slow enzymatic action on the meat.

Freezing Times of Boneless Meat

The cooling of boneless meat from 50 to 10 °F requires the removal of about 133 Btu/lb of lean meat (74% water), most of which is latent heat liberated when the liquid water in the meat changes into ice. Most of the latent heat is produced as the meat is cooled from 30 to 25 °F. Accordingly, most of the time needed to freeze meat is spent in cooling through this range.

For boneless meat in cartons, the rate of freezing depends on the temperature and velocity of the surrounding air and on the thickness and thermal properties of the carton and the meat itself. Figure 4 shows the effects of the first three factors on cooling times for lean meat in two carton types.

For example, the chart shows a total cooling time of 30.75 h for solid cardboard cartons 5-in. thick at an air temperature of −18 °F and a velocity of 400 fpm.

The corresponding air temperatures and cooling times may be found for any specified thickness of carton and air velocity. Conversely, the chart can be used to find combinations of air velocity and temperature needed to freeze cartons of a particular thickness in a specified time (see Figure 5).

Accuracy of the estimated freezing times is about 3% for air velocities greater than 400 fpm. Calculations are based on a modified Plank's Equation. Latent heat of 107 Btu/lb and average freezing point of 28 °F is assumed for lean meat.

Increasing the fat content of meat reduces the water content and hence the latent heat load. Thermal conductivity of the meat is reduced at the same time, but the overall effect is for freezing times to drop as the percentage of fat rises. Actual cooling times for mixtures of lean and fatty tissue should therefore be somewhat less than the times obtained from the chart. For meat with 15% fat, the reduction is about 17%.

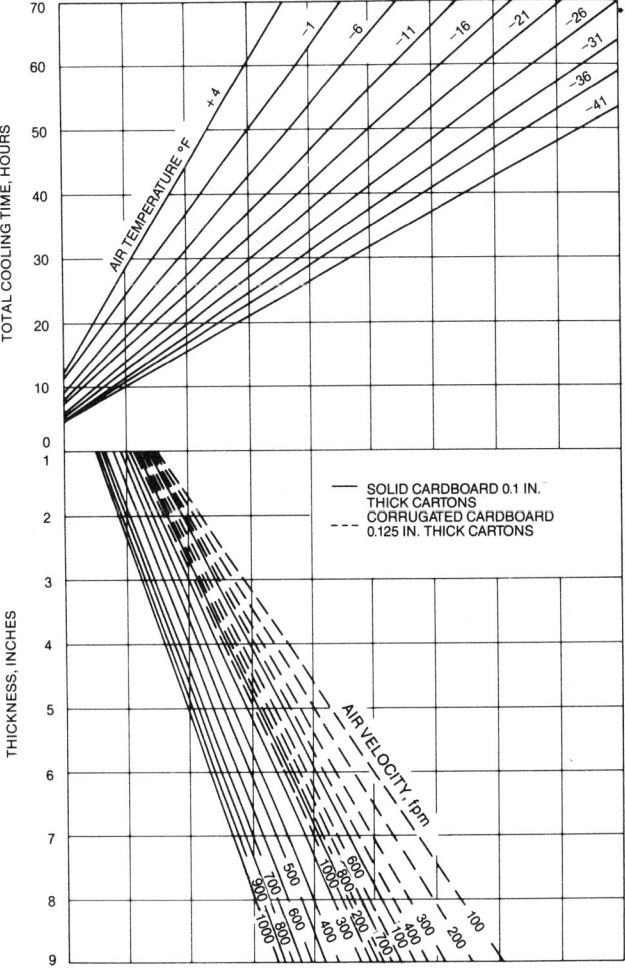

Fig. 4 Freezing Times of Boneless Meat

HOG CHILLING AND TEMPERING

The internal temperature of hog carcasses entering the chill coolers from the killing floor varies from 100 to 106 °F. The specific heat shown in the 1989 ASHRAE *Handbook—Fundamentals* is 0.62 Btu/(lb·°F), but in practice 0.7 to 0.75 Btu/(lb·°F) is used because changed feeding techniques have created leaner hogs. The dressed weight varies from 90 to 450 lb approximately, the average being near 180 lb. Present practice requires dressed hogs to be chilled and tempered to an internal ham temperature of 37 to 39 °F on an overnight basis. This limits the chilling and tempering time to 12 to 18 h.

Cooler and refrigeration equipment must be designed to chill the hogs thoroughly with no frozen parts at the time the carcasses are moved to the cutting floor. Carcass crowding, reducing exposure to circulated chilled air, and excessively high peak temperatures are all detrimental to proper chilling.

The following hog cooler design details will provide:

- Sufficiently quick chilling to retard bacterial development and prevent deterioration.
- A cooler shrinkage from 1.2 to 1.5%. A lower shrinkage could be obtained with additional evaporator surface, but the added cost often offsets the savings realized by reduction in shrinkage. Also, shrinkage below 1.2% could require all fresh cuts to be rechilled before shipping to prevent off-condition resulting from

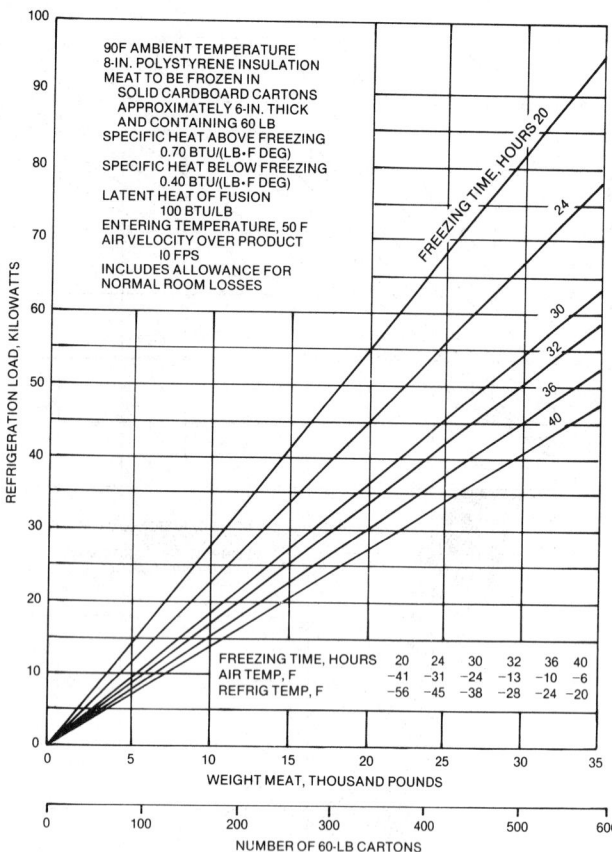

Fig. 5 Blast Freezer Loads

excessive moisture in the product. This extra handling could result in added cost that, again, would offset the shrinkage savings.
- Firm carcasses that are dry and bright without frozen surface or internal frost, suitable for efficient cutting.

Hog Cooler Design

The capacity of hog coolers is set by the dressing rate of hogs and the planned hours of operation. However, on a one-shift basis it is economically sound to provide cooler hanging capacity for 10-h dressing in order to properly handle the chilling of large sows that require more than 24-h exposure in the chill room; handle increased dressing volumes when market conditions warrant overtime operation; and have some flexibility in unloading and loading the cooler during normal operations. On a two-shift basis, extra cooler capacity for overtime operation is not necessary.

The rail height should be 9 ft to provide both good air circulation and adequate clearance between the floor and the largest dressed hog. (This is a requirement of the USDA Meat Inspection Division and most state regulations.) Rails should be spaced at a minimum of 30 in. on centers to provide sufficient clearance for the hanging hogs and to prevent contact between carcasses.

The spacing of hogs on the rail varies according to the size of the hog. Hogs should be spaced so that there is at least 1.5 to 2 in. between the flank of one carcass and the back of the carcass immediately in front of it. The rail spacing of 13 in. on centers is normal for 180-lb dressed hogs.

Many meat-packing refrigeration engineers maintain that several hog chill coolers with a capacity of 2-h loading for 300 to 600 kill/h or 4-h loading for lower killing capacities is more economical than one large chill cooler.

The hog cooler should be designed on the following basis:
1. Total amount of hanging rail should be equal to:
   a. (One-shift operation) (10h × rate of kill × 13 per 12 ft).
   b. (Two-shift operation) (16h × rate of kill × 13 per 12 ft).
   c. For combination carcass loading, hogs and cattle, calves, or sheep, the figures should be modified accordingly.
2. The rail height should be 9 ft. It may be 11 ft for combination beef and hog coolers.
3. The rail spacing should be a minimum of 30 in.
4. The inside building height will vary depending on the type of refrigeration equipment installed. A clear height of 6 ft above the rail support is adequate for space to install units, piping, and controls; it provides sufficient plenum over the rails to ensure even air distribution over the hog carcasses.

In Europe, prechilled, intensive batch, in-line chilling is practiced, resulting in smaller shrinkages with the variations in chilling systems.

Selecting Refrigeration Equipment

Both floor units and units installed above the rail supports are used. Floor units, with a top discharge outlet equipped with a short section of duct to discharge air in a space over the rail supports, are used by a few pork processors. A few brine spray units equipped with ammonia coils, and water or hot gas defrost units are being used.

Two types of units are available for installation above the rail supports. One uses a blower fan to force air below and through a horizontally placed coil with the air discharging horizontally from the front or top of the unit. The other consists of a vertical coil with axial fans or blowers to force the air through the unit. Both units are designed for various types of liquid feed control. The horizontal units are equipped for both hot gas and water defrosting. All finned surfaces designed for hog chill cooler operation. Evaporator controls should be provided as described in the previous section on beef carcass coolers.

Careful selection of units and use of automatic controls, including liquid recirculation, provides an air circulation, temperature, and humidity balance that chills hogs with minimum shrinkage in the quickest time.

The temperature control should be set to provide an opening room temperature of 26 to 28 °F. As the cooler is loaded, the suction temperature decreases to provide the additional refrigeration effect required to handle increased refrigeration load and maintain room temperature below 34 to 36 °F. Ample compressor and unit capacity is required to achieve these results.

The selection of dry coil units based on 10 to 12 °F temperature difference (refrigerant to air) at peak operation provides adequate coil surface and a TD of 1 to 5 °F prior to opening and about 10 h after closing the cooler. This low temperature difference results in maximum economical high humidity conditions throughout the entire chilling cycle. New cutting practices use higher initial chilling TD with lower TD at the end of the chilling cycle.

Sample Calculation

The Hog Chilling Time-Temperature Curves (Figure 6) are composite curves developed from several operation tests. The relation of the room temperature and ammonia suction gas temperature curves show that the refrigeration load decreased about 9 h after closing the cooler. After about 9 h, the room temperature is increased and the hog is tempered to an internal ham temperature of 37 to 39 °F.

Table 5 was prepared using empirical calculations to coordinate product and unit refrigeration loads. A shortcut method for determining hog chill cooler refrigeration loads is presented in Table 6. The latent heat of the product has been neglected, since the latent heat of evaporation is equal to the reduction of sensible heat load of the product. Total sensible heat was used in all calculations.

Meat Products

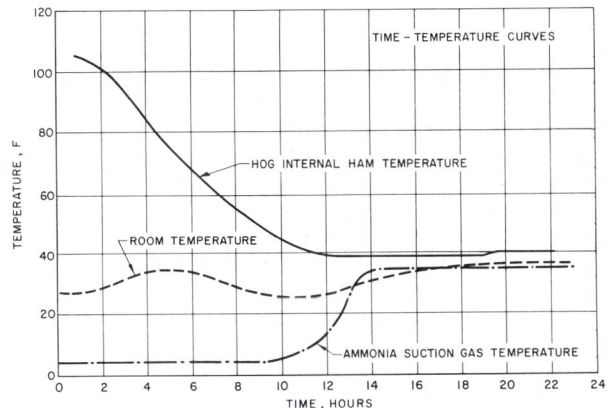

Fig. 6 Composite Hog Chilling Time-Temperature Curves

Table 5 Product Refrigeration Load, Tons

Hours	Cooler Loading Time, h			
	1	2	4	8
1	7.20[a]	7.20	7.20	7.20
2	6.80	14.00[a]	14.00	14.00
3	6.49	13.29	20.49	20.49
4	6.19	12.68	26.88[a]	26.68
5	5.94	12.13	25.42	32.62
6	5.71	11.65	24.33	38.33
7	5.56	11.27	23.40	43.89
8	5.44	11.00	22.65	49.33[a]
9	5.38	10.82	22.09	47.51
10	—	5.38	16.38	39.71
11	—	—	10.82	34.22
12	—	—	5.38	28.03
13	—	—	—	22.09
14	—	—	—	16.38
15	—	—	—	10.82
16	—	—	—	5.38

[a]Values are for peak load:
100 hogs/h at 180 lb dressed weight average.
Chilled from 102 to 38 °F in 16 h.
Based on operating test, not laboratory standards.

Table 6 Average Chill Cooler Loads Exclusive of Product

Cooler Capacity	Room Dimensions, ft	Floor Area, ft²	Room Volume, ft³	Refrigeration, tons[a]
1200 Hogs	40 × 100 × 17	4000	68,000	11.3
2400 Hogs	80 × 100 × 17	8000	136,000	22.6
3600 Hogs	120 × 100 × 17	12,000	204,000	33.9
4800 Hogs	100 × 160 × 17	16,000	272,000	45.2
6000 Hogs	100 × 200 × 17	20,000	340,000	56.5

[a]Based on 6000 ft³ room volume/ton (or use detailed calculations for building heat gains, infiltration, people, lights, and average unit cooler motors from Chapter 26 in the 1989 ASHRAE *Handbook—Fundamentals*).

Example:
600 hogs/h at 180 lb average dressed weight using 2-h loading time cooler.
Four coolers minimum requirement; five desirable.

Each cooler:
Capacity 1200 Hogs = 11.3 tons (Table 6)
Product Peak Load = 6 × 14 = 84.0 tons (Table 5)
Total = 95.3 tons

Select 18 units of 5.3 tons at 10.3 °F temperature difference per cooler
Approximately 198,000 cfm
Air changes per min. = 198,000/68,000 = 2.9.

The refrigeration of the hog cutting room, where the carcass is cut up into its primal parts, is an important factor in maintaining the quality of the product. A maximum dry-bulb temperature of 50 °F should be maintained. This level is low enough to prevent excessive rise in temperature of the product during its relatively short stay in the cutting room and also complies with USDA-FSIS requirements.

Chilled carcasses entering the room may have a surface temperature as low as 30 °F. Unless the dew point of the air in the cutting room is maintained at a temperature below 30 °F, moisture will condense on the surface of the product, providing an excellent medium for bacterial growth.

Floor, walls, and all machinery in the cutting floor must be thoroughly cleaned at the end of each day's operation. Cleaning releases a large amount of vapor in the room that, unless quickly removed, will condense on the walls, ceiling, floor, and machinery surfaces. When the outside dew-point temperature is less than the room temperature, vapor removal can be accomplished by installing fans to continuously exhaust the room during cleanup.

These fans should operate only during the cleanup period and as long as required to remove the vapor produced by cleaning. Exhaust facilities equivalent to five air changes per hour should be satisfactory. When the room temperature is lower than the outside dew-point temperature, this method cannot be used. The water vapor must then be condensed out on the evaporator surfaces.

The hog cutting room is highly populated. Attention should be given to the heat given off by personnel, normally 1000 Btu/h per person. Heat given off by electric motors must also be included in the refrigeration load.

Special consideration should also be given to the latent heat load caused by knife boxes, wash water, workers, and infiltration. This high latent heat load must be offset by reheat at the refrigeration units in order to maintain the desired low relative humidity if the sensible heat load is not sufficient.

The quantity of air circulated is influenced by the amount of sensible heat to be picked up and the relative humidity to be maintained, usually between 7 and 12 complete air changes each hour. The air distribution pattern requires careful attention to prevent drafts on the workers.

Forced air units are satisfactory for refrigerating cutting floors. Ceiling height must be sufficient to accommodate the units. A wide selection of forced air units may be applied to these rooms. They can be floor or ceiling mounted, with either dry-coil or wetted-surface units arranged for flooded, recirculated, or direct expansion refrigerant systems.

Suction pressure regulators should be provided with both the flooded and the direct expansion units. Automatic dry- and wet-bulb controls are essential for best operation.

PORK TRIMMINGS

Pork trimmings come from the chilled hog carcass, principally from the primal cuts: belly, plate, fatback, shoulder, and ham. Trimmings per hog average 4 to 8 lb. Only trimmings used in sausage or canning operations are discussed here.

In the cutting or trimming room, trimmings are usually at a temperature between 38 and 45 °F; an engineer must design for the higher temperature. The product requires only moderate chilling to be in proper condition for grinding if it is to be used locally in sausage or canning operations. If it is to be stored or shipped elsewhere, hard chilling is required. Satisfactory final temperature for local processing is 34 °F. This is the average temperature after tempering and should not be confused with surface temperature immediately after chilling. It is possible for the trimmings to have a much lower surface temperature than internal temperature immediately after chilling, especially if they have been quick-chilled.

Good operating practice requires rapid chilling of pork trimmings as soon as possible after removal from the primal cuts. This retards enzymatic action, which is responsible for poor flavor, rancidity, loss of color, and excessive shrinkage.

The choice of chilling method depends largely on local conditions and consists of a variation of air temperature, air velocity, and method of achieving contact of air and meat. Continuous belt equipment using low-temperature air fluids or one of the cryogenics to obtain lower shrinkage is available.

Truck Chilling

Economic conditions may require existing chilling or freezer rooms to be used. Some may require an overnight chill, others less than an hour. The following methods make use of existing facilities:

- Trimmings are put on truck pans to a depth of 2 to 4 in. and held in a suitable cooler kept at 32°F. This method requires a short chill time and results in a near uniform temperature (34°F) of the trimmings.
- Trimmings are spread 4 or 5 in. deep on truck pans in a 0°F freezer and held for 5 to 6 h, or until the meat is well stiffened with frost. Use of temperatures lower than 0°F (with or without fans) will expedite chilling if time is limited. After trimmings are hard-chilled, they are removed from the metal pans and tightly packed into suitable containers. They are held in a 26 to 28°F room until shipped or used. Average shrinkage using this system is 0.5% up to the time they are put in the containers.
- Trimmings from the cutting or trimming room are put in a meat truck and held in a cooler with a temperature from 28 to 32°F. This method usually requires an overnight chill, and is not likely to reach a temperature of 34°F in the center of the load.
- If trimmings are not to be used within one week, they should be frozen immediately and held at −10°F or lower.

FRESH PORK HOLDING

Fresh pork cuts are usually packed on the cutting floor. If they are not shipped the same day that they are cut and packed, they should be held in a cooler with a temperature of 26 to 28°F.

Forced air cooling units are frequently used for holding room service because they provide better air circulation and more uniform temperatures throughout the room, minimize ceiling condensation caused by air entering doorways from adjacent warmer areas (because of traffic), and eliminate the necessity of coil scraping or drip troughs if hot gas defrost is used.

Cooling units may be the dry type with hot gas defrost, or wetted surface with brine spray. Units should have air diffusers to prevent direct air blast on the products. Unless the room shape is unusually odd, discharge ductwork should not be necessary.

Since the product is boxed and wrapped and the holding period is short, humidity control is not too important. Various methods of automatic control may be used. CO_2 has been applied within boxes of pork cuts. Care must be taken to maintain the ratio of pounds of CO_2 to pounds of meat for the retention period. The enclosures must be relieved and ventilated in the interest of life safety.

CALF AND LAMB CHILLING

Dry-coils are typically used for calf and lamb chilling. These can be either the between-the-rail type, the suspended type above the rail, or floor units. The same type of refrigerating units used for pork may be used for lamb, with some modifications. However, in the chilling of lambs and calves, it is desirable to use reduced air changes over the carcass. This is accomplished by two-speed motors, using the higher speed for the initial chill and reducing the rate of air circulation when the carcass temperatures are reduced, approximately 4 to 6 h after the cooler is loaded.

Lambs usually weigh 40 to 80 lb, with an approximate average dressed carcass weight of 50 lb. Sheep weigh up to an average of approximately 125 lb and readily take refrigeration. Adequate coil surface should be installed to maintain a room temperature below 38°F and a relative humidity of 90 to 95% in the loading period. The evaporating capacity should be based on an average 10°F temperature differential between refrigerant and room air temperature, with an opening room temperature of 34°F.

The use of compensating back pressure regulating valves, which vary the evaporator pressure as the room temperature changes, is recommended. As the room and carcass temperatures drop, the temperature differential is reduced, thus holding a high relative humidity. At the end of a 4 to 6-h chill period, the air over the carcass may be reduced to help keep the bloom and color of the product.

Carcasses should not touch each other. They enter the cooler at 98 to 102°F, with the carcass temperature taken at the center of the heavy section of the rear leg. The specific heat of a carcass is 0.7 Btu/(lb·°F). Air circulation for the first 4 to 6 h should be approximately 50 to 60 changes per hour, reduced to 10 to 12 changes per hour. The carcass should reach 34 to 36°F internally in about 12 to 14 h and should be held at that point with 85 to 90% rh room air until shipped or otherwise processed. This gives the least possible shrinkage and prevents excess surface moisture.

In calf chilling coolers, approximately the same procedure is acceptable, with carcasses hung on 12 to 15 in. centers. The dressed weight varies at different locations, with an approximate 85 to 90 lb average in dairy country and a 200 to 350 lb (sometimes heavier) average in beef-producing localities. The same time and temperature relationship and air velocities previously given for chilling lambs are used for chilling calves, except when calves are chilled with the hide on. The time may be extended for air circulation. Air circulation need not be curtailed in hide-on chilling in this application because rapid cooling gives better color to these carcasses after they are skinned.

The refrigerating capacity for lamb and calf chill coolers is calculated the same as for other coolers but additional capacity should be added to permit reduced air circulation and maintain close temperature differential between room air and refrigerant.

CHILLING AND FREEZING VARIETY MEATS

The basic requirement for chilling variety meats is the rapid lowering of the product temperature to 32 to 34°F. From this initial chill point, the particular end use of the product by the individual establishment will determine the best refrigeration method. Weights are average, with temperatures averaging 100°F for design purposes. Specific heats of variety meats vary with the percent of fat and moisture in each. For design purposes, a specific heat of 0.75 Btu/(lb·°F) should be used.

Quick Chilling

A better and more widely used method consists of quick chilling at lower temperatures and higher air velocities, using the same type of truck equipment as in the overnight chilling method. This method is also used for chilling trimmings. Care must be exercised in the design of the quick chilling cabinet or room to provide for the refrigeration load imposed by the hot product. One industry survey shows that approximately 50% of large establishments are using quick-chill in their variety meat operations.

The quick-chilling cabinet or room should be designed to operate at an air temperature of approximately −5°F with velocities over the product of 500 to 1000 fpm. Curing initial loading, the air temperature may rise to 10°F. In the design of quick-chilling units, refrigeration coils are used with axial flow fans for air circulation.

The recommended defrosting method is with water and/or hot gas, except where units with continuous defrost are used. The product is chilled to the point where the outside is frosted or frozen

Meat Products

Table 7 Storage Life of Meat Products

	Months			
	Temperature, °F			
Product	10	0	−10	−20
Beef	4 to 12	6 to 18	12 to 24	12+
Lamb	3 to 8	6 to 16	12 to 18	12+
Veal	3 to 4	4 to 14	8	12
Pork	2 to 6	4 to 12	8 to 15	10
Chopped beef	3 to 4	4 to 6	8	10
Pork sausage	1 to 2	2 to 6	3	4
Smoked ham and bacon	1 to 3	2 to 4	3	4
Uncured ham and bacon	2	4	6	6
Beef liver	2 to 3	2 to 4		
Cooked foods	2 to 3	2 to 4		

and a temperature of 32 to 34 °F is obtained when the product later reaches an even temperature throughout in the packing or tempering room. The time required to chill the product by this method depends on the depth of product in the pans, the size of individual pieces, air temperature, and velocity. Normally, 0.5 to 4 h is a satisfactory chill period to attain the required 32 to 34 °F temperature. In addition to the obvious savings in time and space, an important advantage of this method is the low total shrinkage, averaging only 0.5 to 1%. These values were obtained in the same survey as those in Table 7.

Another method of handling variety meats involves packaging the product before chilling as near as possible to the killing floor. The packed containers are placed on platforms and frozen in a freezer. Separators should permit air circulation between packages.

This method is used in preparing products for frozen shipment or freezer storage. The internal temperature of the product should reach 25 °F for prompt transfer to a storage freezer. For immediate shipment, the internal temperature of the product must be reduced to 0 °F; this may be done by longer retention in the quick freezer. Here package material and size, particularly package thickness, largely determine the rate of freezing. For example, a 5 in. thick box will take from 16 h upward to freeze, depending on the type of product, package material, size, and loading method.

The dry-bulb air temperature in these freezers is maintained at −40 to −20 °F, with air velocities over the product at 500 to 1000 fpm. The time required to reach the desired internal temperature is a combination of refrigeration capacity, size of largest package, insulating properties of package material, and so forth. A generous safety factor should be used in sizing evaporator coils for this type of service. These freezers are best incorporated within refrigerated rooms. Defrosting is by the water or hot gas method, except where units with continuous defrost are used. Shrinkage varies in the range of only 0.5 to 1%.

Cryogenic chilling is being used to obtain lower shrinkages; again, CO_2 addition to boxes is being used for quick processing and lower shrinkage.

PACKAGING AND STORAGE

Packages for variety meats do not yet have any standard sizes or dimensions. Present requirements are a package that will stand shipping, with sizes to suit individual establishments. The importance of the package is brought more into focus in the case of the hot pack freezing method. Standardization of sizes and package materials promotes faster chilling and more economical handling.

Storage of variety meats depends on its end use. For short storage (under one week) and local use, 32 to 34 °F is considered a good internal product temperature. If stored for shipping, the internal temperature of the product should be kept at 0 °F or below. Recommended length of storage is controversial; type of package, freezer temperature and relative humidity, amount of moisture removed in original chill, and the variations of the products themselves all affect storage life.

Packers' recommendations regarding storage time vary from 2 to 6 months and longer, since variety meats pick up rancidity on the surface and soft muscle tissue dehydrates while freezing. More rapid freezing and vapor-proof packaging are important in increasing storage life.

REFRIGERATION LOAD COMPUTATIONS

The average evaporator refrigerating load for a typical chilling process above freezing may be computed by:

$$H_r = W_m s_m (t_1 - t_2) + W_t s_t (t_1 - t_2) + H_w + H_i + H_m \quad (1)$$

where
H_r = refrigeration load, Btu/h
W_m = weight of meat, lb/h
W_t = weight of trucks, lb/h
s_m = specific heat of meat, Btu/(lb·°F)
s_t = specific heat of truck, containers, or platforms (0.12 for steel), Btu/(lb·°F)
t_1 = average initial temperature, °F
t_2 = average final temperature, °F
H_w = heat gain through room surfaces, Btu/h
H_i = heat gain from infiltration, Btu/h
H_m = heat gain from equipment and lighting, Btu/h

The following example illustrates the method of computing the refrigeration load for a quick-chill operation.

Example 1. Find the refrigeration load for chilling six trucks of offal from a maximum temperature of 100 to 34 °F in 2 h. Each truck weighs 400 lb empty and holds 720 lb of offal. The specific heat is 0.12 for the truck and 0.75 for the offal. The room temperature is to be held at 0 °F, with an outdoor temperature of 40 °F and a rh of 70%. The walls, ceiling, and floor gain 72 Btu/ft² in 24 h and have an area of 947 ft². The volume of the room is 1881 ft³ and 12 air changes in 24 h are assumed.

Solution: Values for substitution in Equation 1 are obtained as follows:

W_m = 6 × 720/2 = 2160 lb/h
s_m = 0.75 Btu/(lb·°F)
$t_1 - t_2$ = 100 − 34 = 66 °F
W_t = 6 × 400/2 = 1200 lb/h
s_t = 0.12 Btu/(lb·°F)
H_w = 72 × 947/24 = 2841 Btu/h
H_i = (1.12 × 1881 × 12)/24 = 1053 Btu/h where
1.12 = heat removed per ft³ of air entering, Btu
H_m = (10 × 2545) + (200 × 3.4) = 26,140 Btu/h
10 = assumed horsepower
2545 = Btu per horsepower hour
200 = lights, W
3.4 = Btu/w·h

Substituting in Equation 1:

H_r = (2160 × 0.75 × 66) + (1200 × 0.12 × 66) + 2841
+ 1053 + 26,140 = 146,534 Btu/h = 12.2 tons

It is good practice to add 10 to 25% to the computed refrigeration load.

PROCESSED MEATS

Prompt chilling, handling, and storage under controlled temperatures help in the production of mild and rapidly cured and smoked meats. The product is usually transferred directly from the smokehouse to a refrigerated room, but sometimes a drop in temperature of 10 to 30 °F can occur if the transfer time is appreciable.

Since the day's production is not usually removed from the smokehouses at one time, the refrigeration load is spread over a large portion of the 24 h. Table 8 outlines temperatures, relative humidities, and time required in refrigerated rooms used in handling smoked meats.

Smoked meat prechilling results in a reduction of drips of moisture and fat, thus increasing the yield. This can be done at higher temperatures, with air velocities of up to 500 fpm (Table 8). At lower temperatures, air velocities of 1000 fpm and higher

Table 8 Room Temperatures and Relative Humidities for Smoking Meats

	Room Conditions			
	°F Dry-Bulb	% Relative Humidity	Final Product Temp, °F	Time, h
Prechill method				
Hams, picnics, etc.				
High temperature	38 to 40	80	60	8 to 10
Low temperature	26 to 28	80	60	2 to 3
Derind bacon				
Normal	26 to 28	80	28	8 to 10
Blast	0 to 10	80	26	2 to 3
Hanging or tempering				
Ham, picnics, etc.	45 to 50	70	50 to 55	
Derind bacon	26 to 28	70	26 to 28	
Wrapping or packaging				
Hams, picnics, etc.	45 to 50	70		
Storage	28 to 45	70		

are used. Chilling in the hanging room or in the wrapping and packaging rooms results in slow chilling and high temperatures when packing. This is not desirable for a product that is to be stored or shipped a considerable distance.

Meats handled through smoke and into refrigerated rooms are hung or racked on cages that are moved on an overhead track or mounted on wheels. Sometimes the product is transferred from suspended cages to wheel-mounted cages between smoking and subsequent handling.

Smoked hams and picnic meats must be chilled as rapidly as possible through the incubation temperature range of 105 to 50°F. Product requiring cooking before eating is brought to a minimum internal temperature of 140°F to destroy possible live trichinae, while product not requiring cooking before eating is brought to a minimum internal temperature of 155°F.

A maximum storage room temperature of 40°F dry bulb should be used where delivery from the plant to retail outlets is made within a short time. A room dry-bulb temperature of 28 to 32°F is desirable where delivery is to points considerably removed from the plant and transfer is made through controlled low humidity rooms, docks, cars, or trucks, keeping the dew point below that of the product.

Bacon usually reaches a maximum temperature of 125°F in the smokehouse. Since most smoked bacon in a sliced form is packaged, it may be transferred directly to the chill room if it has been skinned before smoking. If the bacon is to be skinned after smoking, it is usually allowed to hang in the smokehouse vestibule for 2 to 4 h, until it drops to a temperature of 90°F before skinning.

Bacon is usually molded and sliced at temperatures just below 28°F. Chill rooms are usually designed to reduce the internal temperature of the bacon to 26°F in 24 h or less, requiring a room dry-bulb temperature of 18 to 20°F. A tempering room (which also serves as storage for stock reserve), held at the exact temperature at which bacon is sliced, is often used.

Bacon molding can be done either after tempering, in which case it is moved directly to the slicing machines, or after the initial hardening, and then be transferred to the tempering room. In the latter case, care should be taken that none of the slabs is below 24°F so that the product will not crack during molding. Bacon cured by the pickle injection process generally shows less pickle pockets if it is molded after hardening, placed no more than eight slabs high on pallets, and held in the tempering room.

In any of the rooms mentioned, air distribution must be uniform. To minimize shrinkage, the air supply from floor-mounted unit coolers should be delivered through slotted ducts or by means of closed ducts supplying properly spaced diffusers directed so that no high velocity airstreams impinge on the product itself. The exception would be in blast chill rooms where high air velocities are needed, but in reality the product is subjected to the condition for only a short time.

Refrigeration may be supplied by floor or ceiling-mounted dry or wet coil units. If the latter are selected, water, hot gas, or electric defrost must also be used.

In recent years, many processors have begun using three different methods of chilling smoked meats: rapid blast chilling, brine chilling by direct contact spraying, and cryogenic chilling. Direct contact spraying is especially emphasized, because it affords minimization of shrinkage, longer shelf life, and more uniform chilling. This method is usually carried out in enclosures specially designed to combat the detrimental effects of salt brine. Color and salt taste may need close monitoring in contact spraying.

BACON SLICING AND PACKAGING ROOM

Bacon should enter the slicing room chilled to a uniform internal temperature from 26 to 32°F depending on the individual packer's temperature standard. Internal temperatures below 26°F tend to cause shattering of the product during slicing. Internal bacon temperatures above 32°F cause improper shingling from the slicing machine.

The slicing and packaging room temperature and air movement are usually the result of a compromise between the physical comfort demands of the operating personnel and the requirements of the product. The design room dry-bulb temperature is below 50°F, according to USDA-FSIS regulations.

An objectionable amount of condensation on the product may occur. To guard against this, the coil temperature should be maintained below the temperature of the bacon entering the room, thus keeping the dew point of the room air below the product temperature. The product should be exposed to the room air for the shortest possible time.

Exhaust ventilation should remove smoke and fumes from the sealing and packaging equipment and comply with OSHA occupancy regulations. Again, heat exchangers should be considered for reducing the resulting increased refrigeration load.

Refrigeration for this room may be supplied by forced air units—floor or ceiling mounted, dry or wetted coil—or finned-tube ceiling coils.

Dry coils should have defrost facilities if coil temperatures are to be kept at more than several degrees below freezing. Air discharge and return should be evenly distributed, using ductwork if necessary. To avoid drafts on personnel, air velocities in the occupied zone should be in the range of 25 to 35 fpm. The temperature differential between primary air and room air should not exceed 10°F to assure personnel comfort. To provide optimum comfort conditions and dew-point control, reheat coils are necessary.

Where ceiling heights are adequate, multiple ceiling units can be used to minimize the amount of ductwork. Automatic temperature and humidity controls are desirable in cooling units.

To provide draft-free conditions, drip troughs with suitable drainage should be added to finned-type ceiling coils. However, it is difficult, if not impossible, to maintain a room air dew point low enough to approach the product temperature. Some installations operating with relative humidities of 60 to 70% do not have product condensation problems.

One control method consists of individual coil banks connected to common liquid and suction headers. Each bank is equipped with a thermal expansion valve. The suction header has an automatically operated back-pressure valve. Thermostatically controlled dual back-pressure regulator and liquid-header solenoid are both controlled by a single thermostat. This arrangement provides a simple automatic defrosting cycle.

Meat Products

Another system uses fin coils with glycol sprays. Humidity is controlled by varying the concentration strength of the glycol and the refrigerant temperature.

SAUSAGE DRY ROOMS

Refrigeration or air conditioning is an integral part of year-round sausage dry rooms. The purpose of these systems is to produce and control the air conditions for proper removal of moisture from the sausage.

A variety of dry sausage is manufactured, for the most part uncooked. Its keeping qualities depend on curing ingredients, spices, and removal of moisture from the product by drying. This sausage is generally of two distinct types—smoked and not smoked.

FSIS regulates the minimum temperature and amount of time that the product must be held after stuffing and prior to release, depending on the method of production. The dry room temperature shall not be lower than 45 °F and the length of time the product shall be held in the dry room prior to release is dependent on the diameter of the sausage after stuffing and the method of preparation used.

After stuffing, the sausages are held at a temperature of 60 to 75 °F and 75 to 95% rh in the sausage greenroom to develop the cure. The sausages are suspended from sticks at the time of stuffing and may be held on the trucks or railed cages or be transferred to racks in the greenroom. Sausages in 3.5 to 4-in. diameter casings are generally spaced about 6 in. on centers on the sausage sticks. The length of time the sausages are held in the greenroom depends on the method of preparation, the type and dimensions of the sausage, the operator, and the judgment of the sausage maker regarding proper flavor, pH, and other characteristics.

Those sausage varieties that are not smoked are then transferred to the sausage dry room; those that are to be smoked are transferred from the greenroom to the smokehouse and then to the dry room.

In the dry room, approximately 30% of the moisture is removed from the sausage, to a point at which the sausage will keep for a long time, virtually without refrigeration. The drying period required depends on the amount of moisture to be removed to suit trade demand, type of sausage, and type of casing. The moisture transmission characteristics of synthetic casings vary widely and greatly influence the rate of drying. The diameter of the sausage is probably the most important factor influencing the drying rate.

Small-diameter sausages, such as pepperoni, have more surface in proportion to the weight of material than do large-diameter sausages. Furthermore, the moisture from the sausage interior has to travel a much shorter distance to reach the surface, where it can evaporate. Thus, the drying time for small-diameter sausages is much shorter than that for large-diameter sausages.

Typical conditions in the dry room are approximately 45 to 55 °F and 60 to 75% rh. Some sausage makers favor the lower range of temperatures for unsmoked varieties of dry sausage and the higher range for smoked varieties.

In processing dry sausage, moisture can only be removed from the product at the rate at which the moisture comes to the surface of the casing. Any attempt to speed the drying rate results in overdrying the surface of the sausage, a condition known as case hardening. This condition is identified by a dark ring inside the casing, close to the surface of the sausage, which precludes any further attempt to remove moisture from the interior of the sausage. On the other hand, if sausage is dried at too slow a rate, excessive mold occurs on the surface of the casing, sometimes leading to an unsatisfactory appearance. An exception is the Hungarian Salami, which requires a high humidity so that a prolific mold growth can occur and flourish.

As with any other cool or refrigerated space, sausage dry rooms should be properly insulated to prevent temperatures in adjoining spaces from influencing the temperature in the dry room. Ample insulation is especially important in the case of dry rooms located adjacent to rooms of much lower temperature or rooms on the top floor, where the ceiling may be exposed to relatively high temperatures in summer and low temperatures in winter.

Insulation should be adequate to prevent the inner surface of the walls, floor, and ceiling of the dry room from differing more than a degree or two from the average temperature in the room. Otherwise, there is the possibility of condensation due to the high relative humidity in these rooms, which leads to mold growth on the surfaces themselves and, in some cases, on the sausages as well.

The sticks of sausage are generally supported on permanent racks built into the dry room. In the past, these were frequently made of wood; however, sanitary requirements have virtually outlawed the use of wood for this purpose in new construction. The uprights and rails for the racks are now made of either galvanized pipe, hot dip galvanized steel, or stainless steel. The rails for supporting the sausage sticks should be spaced vertically at a distance that leaves ample room for air circulation below the bottom row and between the top row and the ceiling. Generally, a spacing of not less than 1 or 2 ft between rails depends on the length of sausage stick used by the individual manufacturer.

The horizontal spacing between sausages should be such that they do not touch at any point, to prevent mold formation or improper development of color. Generally, with the large 4-in. diameter sausages, a spacing of 6 in. on centers is adequate.

Dry-Room Equipment

In general, two types of refrigeration equipment are in current use for attaining the required conditions in a dry room. The most common is a refrigeration-reheat system, in which the room air is circulated either through a brine spray or over a refrigerated coil and sufficiently cooled to reduce the dew point to the temperature required in the room. The other type, involves the use of a hygroscopic liquid sprayed over a refrigerating coil in the dehumidifier, thus condensing the moisture from the air without the severe overcooling usually required by the refrigeration-reheat type of system. The chief advantage of this arrangement is that refrigeration and heating loads are greatly reduced.

The use of any type of liquid, brine or hygroscopic, requires periodic tests and adjusting the pH to minimize corrosion of the equipment. Although most systems depend on a type of liquid spray to prevent frost buildup on the refrigerating coils, some successful rooms use dry coils with hot gas or water defrost.

The air for conditioning the dry room is normally drawn through the refrigerating and dehumidifying systems by a suitable blower fan (or fans) and discharged into the distribution ductwork.

Rooms used exclusively for small-diameter products with a rapid drying rate may actually have air leaving the room to return to the conditioning unit at a lower dry-bulb temperature and greater density than the point at which it is introduced. A dry-room designer needs to know what the room will be used for to determine the natural circulation of the room air. Supply and return ducts can then be arranged to take advantage of and accelerate this natural circulation to provide thorough mixing of incoming dry air with the air in the room.

Regardless of the location of the supply and return ducts, care should be taken to prevent strong drafts or high velocity airstreams from impinging on the product; this will result in local overdrying and unsatisfactory products.

A study of air circulation within the product racks will show that as air passes over the sausages and moisture evaporates from them, this air becomes cooler and heavier, and thus has a tendency to drop toward the bottom of the room, creating a vertical downward air movement within the sausage racks. This natural tendency must be considered in designing the duct installation if uniform conditions are to be achieved.

An example of the calculation involved in designing a sausage dry room follows. These calculations apply to a room used for an assortment of sausages, with an average drying time of approximately 30 days. They would not be directly applicable to a room used primarily for very large salami (which has a much longer drying period) or a room used primarily for small-diameter sausage.

In the latter case, use of the air-circulating rate shown in this example will allow the air to absorb so much moisture in passing through the room that it will be difficult to obtain uniform conditions throughout the space. Furthermore, the amount of refrigeration required to lower the air temperature enough to produce the required low inlet air dew point would be excessive. An air circulation rate of 12 air changes per hour should therefore be considered average for use in average rooms. The actual circulating rate should be adjusted to obtain the best compromise of refrigeration load and air uniformity for the particular type of product handled.

Example 2. Air conditioning for sausage drying room

Room Dimensions:

40 ft, 2 in. × 33 ft, 6 in. × 11 ft, 6 in.
Floor space: 1350 ft^2
Volume: 15,600 ft^3
Outdoor wall area: 980 ft^2
Partition wall area: 770 ft^2

Hanging Capacity:

Number of racks: 12
Length of racks: 27 ft
Number of rails high: 5
Spacing of sticks: 2/ft of rail
Number of pieces of sausage per stick: 7
Average weight per sausage: 4 lb
12 × 27 × 5 × 2 × 7 × 4 = 90,720 lb of product
Assume 90,000 lb green weight hanging capacity
Loading per day: 1500 lb

Outdoor Conditions Assumed (Summer):
95°F db; 74.5°F wb; 39% rh; h = 37.8 Btu/lb; 66°F dp; 96 gr/lb

Dry-Room Conditions Desired:
55°F db; 50°F wb; 70% rh; h = 20.2 Btu/lb; 46°F dp; 46 gr/lb

Sensible Heat Calculations:

Walls (2 in. insulation):	
980 (95 − 55) × 0.10	= 3920 Btu/h
Partition (4 in. insulation):	
770 (95 − 55) × 0.067	= 2060 Btu/h
Floor and ceiling:	
2700 (55 − 55) × 0.10	= none
Infiltration: 0.5 × 15,600 (95 − 55)	
× 0.243 × 0.075	= 5700 Btu/h
Lights: 600 W × 3.415	= 2050 Btu/h
Motors: 5 × 1 hp × 2546	= 12,730 Btu/h
Product: 1500 × 0.8 (95 − 55)/24	= 2000 Btu/h
Total sensible heat	= 28,460 Btu/h

Latent Heat Calculations:

Product	
90,000 × 0.30 × 1000/(60 × 24)	= 18,720 Btu/h
(18,720 × 7000)/(60 × 1000)	= 2184 gr/min.
Infiltration	
(0.5 × 15,600 × 0.075 (37.8 − 20.2)	= 10,300 Btu/h
(10,300 − 5700) 7000/(60 × 1000)	= 537 gr/min.
Total grains of moisture = 2184 + 537	= 2721 gr/min.

Assume 12 air changes per hour with an empty room volume of 15,600 ft^3 = (15,600)(12/60)(0.76) = 237 lb of air per minute. Then each lb of air must absorb 2721/237 = 11.5 gr of moisture. Since air at the desired room condition carries 46 gr/lb, the air must enter the room with only 46 − 11.5 = 34.5 gr/lb, corresponding to 41.8°F db and 40°F wb (h = 15.3).

Temperature rise due to sensible heat gain (air specific heat = 0.243 Btu/lb·°F).

28,460/(237 × 0.243 × 60) = 8.2°F db.

Temperature drop due to evaporative cooling from latent heat of product only.

18,720/(237 × 0.243 × 60) = 5.4°F

Net temperature rise in the room = 8.2 − 5.4 = 2.8°F, or the air entering the room must be 55 − 2.8 = 52.2°F db at 38.5°F dp (34.5 gr).

Refrigerating load = 237(20.2 − 15.3) × 60 = 69,700 Btu/h
Reheat load = (17.8 − 15.3) × 60 = 35,600 Btu/h
Room load = 237(20.2 − 17.8) × 60 = 34,100 Btu/h

LARD CHILLING

In federally inspected plants, the USDA FSIS designates the types of pork fats which, when rendered, are classified as *lard*. Other pork fats, when rendered, are designated as *rendered pork fats*. The rendering process requires considerable heat, and the subsequent temperature of the lard at which refrigeration is to be applied may be as high as 120°F. The following data for refrigeration requirements may be used for either product type.

The fundamental requirement of the FSIS is good sanitation through all phases of handling. The use of copper or copperbearing alloys that come in contact with lard should be avoided, because minute traces of copper lower the stability of the product.

Specific gravity of lard varies from about 0.99 at 0°F to 0.88 at 160°F. Specific gravity at 70°F is about 0.93. Heat of solidification of lard is approximately 48 Btu/lb. Melting begins at −35 to −40°F and ends at 110 to 115°F. The point of half fusion is around 40°F. Specific heat of lard in a completely solid form varies from 0.28 Btu/(lb·°F) at −110°F to 0.34 Btu/(lb·°F) at −40°F. Specific heat in the liquid state varies from 0.50 Btu/(lb·°F) at 110°F to 0.52 at 212°F.

In the production of lard, refrigeration is applied so that the final product will have enough texture and a firm consistency. The finest possible crystal structure is desired.

Calculations for chilling 1000 lb of lard per hour are:

Initial temperature: 120°F
Final temperature: 80°F
Heat of solidification: 48 Btu/lb
Specific heat: 0.50 Btu/(lb·°F)

$$S_f = 100 \frac{t_e - t_f}{t_e - t_b} = 100 \frac{115 - 80}{115 - (-40)} = (35/155)100 = 22.6\%$$

where

S_f = percent solidification at final temperature
t_e = temperature at which melting ends
t_f = final temperature
t_b = temperature at which melting begins

Latent heat of solidification:
48(22.6/100) = 10.8 Btu/lb

Sensible heat removed:
0.50(120 − 80) = 20 Btu/lb

Total heat removed:
1000(10.8 + 20) = 30,800 Btu/h, = 2.57 tons refrigeration

Assuming a 15% loss because of radiation, for example, in the process, the required refrigeration for application to chill 1000 lb lard per hour would be 1.15 × 2.57 = 2.96 tons.

Filtered lard at 120°F can be chilled and plasticized in compact internal swept surface chilling units, which use either ammonia or halogenated hydrocarbons. A refrigerating capacity of about 36,000 Btu/h per 1000 lb of lard handled per hour for the product only should be provided. Additional refrigeration for the requirements of heat equivalent to the work done by the internal swept surface chilling equipment will also be needed.

When operating this type of equipment, it is essential to keep the refrigerant free of oil and other impurities so that the heat

Meat Products

transfer surface will not have a film of oil on it to act as insulation and cut down the unit's capacity. Some installations have oil traps connected to the liquid refrigerant leg on the floor below to provide an oil accumulation drainage space.

The safety requirements for this type of chilling equipment are described in the *Safety Code for Mechanical Refrigeration* (ANSI/ASHRAE 15-1978). Note that such units are pressure vessels and, as such, require properly installed and maintained safety values.

The recommended storage temperature for packaged refined lard is 31 to 33°F. The storage temperature required for prime steam lard in metal containers is 40°F or below. These conditions are for up to a 6-month storage period. Lard stored for a year or more should be kept at 0°F.

BLAST AND STORAGE FREEZERS

The standard method of sharp-freezing a product destined for storage freezers comprises freezing the product directly from the cutting floor in a blast freezer until its internal temperature reaches the holding room temperature. The product is then transferred to holding or storage freezers.

The product to be sharp-frozen may be bagged, wrapped, or boxed in cartons. Individual loads are usually placed on pallets or dead skids or in wire basket containers. In general, the larger the ratio of surface exposed to blast air to the volume of either the individual piece or the container of product, the greater the rate of freezing. Product loads should be placed in a blast freezer to assure that each load is well exposed to the blast air and to minimize possible short circuiting of the airflow. Each layer on a load should be separated by spacers to give the individual pieces as much exposure to the blast air as possible.

The most popular types of blast equipment are the self-contained air handling or cooling units, that consist of a fan, evaporator, and other elements in one package. These units are usually used in multiples and placed in the blast freezer to provide the optimum blast air coverage. The unit fans should be capable of high air velocity and volumetric flow; two air changes per minute is the accepted minimum.

The coils of the evaporator may have either a wetted or a dry surface. Dry coils can be defrosted with hot gas or ethylene glycol, which is used as a wetting agent on wetted coil surfaces. When a wetting agent is used, concentrators must be installed to remove moisture and maintain concentration of the wetting agent. Evaporator coils are best operated flooded. Temperature differential between coil and room air should be in the range of 10 to 15°F.

Blast chill design temperatures vary throughout the industry. Most designs are within a -20 to $-40°F$ range. For low temperatures, booster compressors that discharge through a desuperheater into the general plant suction system are used.

Blast freezers require sufficient insulation and good vapor barriers. If possible, a blast freezer should be located so that temperature differentials between it and adjacent areas are minimized in order to decrease insulation costs and refrigeration losses.

Blast freezer entrance doors should be power operated and suitable vestibules provided as air locks to decrease infiltration of outside air.

Besides normal losses, heat calculations for a blast freezer should include the loads imposed by such material handling equipment as electric trucks, skids, and spacers, together with the packaging materials for the product.

Storage freezers are usually maintained at 0 to $-15°F$. Pipe coils are generally sufficient as evaporators and can be operated as direct expansion, flooded, or recirculated liquid systems. If the plant operates with several high and low suction pressures, the evaporators can be tied to a suitable plant suction system. The evaporators can also be tied to a booster compressor system; if the booster system is operated intermittently, provisions must be made to switch to a suitable plant suction system when the booster system is down.

Storage freezer coils can be defrosted by hot gas, electricity, or water. Emphasis should be placed on not defrosting too fast with hot gas (because of pipe expansion) and on providing well-insulated, sloped, heat traced drains and drain pans to prevent freezeups.

DIRECT CONTACT MEAT CHILLING

Continuous processes for smoked and cooked wieners use direct sodium chloride brine tanks or deluge tunnels to chill the meat as soon as it comes out of the cooker. The brine is prepared fresh every day in 2 to 13% solutions depending on the chilling temperature and the salt content of the meat.

Cooling is usually done on sanitary stainless steel surface coolers, which are either refrigerated coils or plates in cabinets. Using this type of unit allows coolant temperatures near the freezing point without damaging the cooler; damage may occur when brine is confined in a tubular cooler. The brine is circulated in quantities necessary to fully wet the surface cooler and fill the distribution troughs of the deluge.

Another type of continuous process uses a brine tank into which the wieners drop out of a conveyor that has carried them through the smoking and cooking process. The pumped brine moves the product to the end of the tank, where it is removed by hand and inserted into peeling and packaging lines.

FROZEN MEAT PRODUCTS

The handling and selling of consumer portions of frozen meats has many potential advantages compared with merchandising fresh meat. The preparation and packaging can be done at the packing house, allowing economies of mass production, byproduct savings, lower transportation costs, and flexibility in meeting market demands. At the retail level, frozen meat products reduce space and investment requirements and labor costs.

PREFREEZING QUALITY OF MEAT

After an animal is slaughtered, physiological and biochemical reactions continue in the muscle until the complex system supplying energy for work has run down and the muscle goes into rigor. These changes continue for up to 32 h postmortem in major beef muscles. Hot boning with electrical stimulation renders meat tender on a continuous basis without conventional chilling. Freezing meat or cutting carcasses for freezing before the completion of these changes cause cold-shortening and thaw-shortening, which render meat tough. The best time to freeze meat is either after rigor has passed or later, when natural tenderization is more or less complete. Natural tenderization is completed during seven days of aging in most major beef muscles. Where flavor is concerned, freezing as soon as tenderization is complete is desirable.

For frozen pork, the age of the meat before freezing is even more critical than it is for beef. Pork loins aged 7 days before freezing deteriorate more rapidly in frozen storage than loins aged 1 to 3 days. In tests, a difference could be detected between 1- and 3-day-old loins, favoring those only 1-day old. With frozen pork loin roasts from carcasses chilled for 1 to 7 days, the flavor of lean and fat in the roasts was progressively poorer with longer holding time after slaughter.

EFFECT OF FREEZING ON QUALITY

Freezing affects the quality—including color, tenderness, and amount of drip—of meat.

Color

The color of frozen meat depends on the rate of freezing. Tests in which prepackaged, steak-size cuts of beef were frozen by immersion in liquid or exposure to an airblast at between 20 and −40°F revealed that airblast freezing at −20°F produced a color most similar to that of the unfrozen product. An initial meat temperature of 32°F was necessary for best results (Lentz 1971).

Flavor and Tenderness

Flavor does not appear to be affected by freezing per se, but tenderness may be affected depending on the condition of the meat and the rate and end-temperature of freezing (Jul 1969). Faster freezing to lower temperatures was found to increase tenderness; however, consensus on this effect has not been reached.

Drip

The rate of freezing generally affects the amount of drip, and meat nutrients, such as vitamins, are lost from cut surfaces after thawing. Faster freezing tends to reduce the amount of drip, although many other factors, such as the pH of meat, also have an effect on drip.

Changes in Fat

Fat of pork changes significantly in 112 days at −5°F, whereas beef shows no change within 260 days at this temperature. At −20 and −30°F, no measurable change occurs in either meat in one year.

The relationship of fat rancidity and oxidation flavor has not been clearly established for frozen meat, and the usefulness of antioxidants in reducing flavor changes during frozen storage is doubtful.

STORAGE AND HANDLING

Pork remains acceptable for a shorter storage period than beef, lamb, and veal because of higher oxidative rancidity associated with higher salt content. Storage life is also related to storage temperature. Because animals within a species vary greatly in nutritional and physiological backgrounds, their tissues differ in susceptibility to change when stored. As a result of the differences between meat animals, packaging methods, and acceptability criteria, a wide range of storage lives is reported for any one type of meat (Table 7). Jul (1969) discusses the storage lives of frozen meats and the factors that affect them.

Appreciable changes in the color and flavor of frozen beef occur at storage temperatures down to −40°F in 1 to 90 days (depending on temperature) for samples held in the dark. Changes were much more rapid (1 to 7 days) for samples exposed to light. Color changes were less pronounced after thawing than when frozen (Lentz 1971).

Reports on the effect of different storage temperatures on fat oxidation and palatability of frozen meats indicate that a temperature of 0°F or lower is desirable. Cuts of pork back held at 20, 10, 0, and −10°F show increases in peroxide value; free fatty acid is most pronounced at the two higher temperatures. For a storage period of 48 weeks, 0°F or lower is essential to avoid fat changes. Pork rib roasts of 0°F showed little or no flavor change up to 8 months, whereas at 10°F fat was in the early stages of rancidity in 4 months. Ground beef and ground pork patties stored at 10, 0, and −10°F indicate that meats must be stored at 0°F or lower to retain good quality for 5 to 8 months. For longer storage, −20°F is desirable.

When storage temperatures fluctuate, the effect on meat appears to be about the same as a steady temperature midway between the extremes. At storage temperatures below 0°F, variable temperature is not important. The desirability of flavor in pork loin roasts stored at −6 to −8°F with maximum fluctuations of 5 to 8 degrees all decreased slightly, apparently without significant difference between treatments. Fluctuations from 0 to 10°F did not harm quality.

Storage temperature is perhaps more critical with meat in frozen meals because of the differing stability of the various individual dishes included. Frozen meals show marked deterioration of most of the foods after 3 months at 13 to 15°F.

Storage and Handling Practices

In view of the effect of temperature on frozen foods, surveys of practices in the industry indicate why a certain amount of product reaches the consumer in poor condition.

One unpublished survey indicated that 10% of frozen foods may be found to be at 6°F or higher in warehouses, 16°F or higher in assembly rooms, 21°F or higher during delivery, and 17°F or higher in display cases. All these temperatures should be maintained at 0°F for complete protection of the product.

PACKAGING

At the time of freezing, a package or packaging material serves to hold the product and prevent it from losing moisture. Other functions of the wrapper or box become important as soon as the storage period begins. Ideal packaging material in direct contact with meat should have: low moisture vapor transmission rate; low gas transmission rate; high wet strength; grease proofness; flexibility over a temperature range including subfreezing; freedom from odor, flavor, and any toxic substance; easy handling and application characteristics adaptable to hand or machine use; easy strippability; and reasonable price.

Individually or collectively, these properties are desired for good appearance of the package, protection against handling, preventing dehydration, which is unsightly and damages the product, and keeping oxygen out of the package.

Desiccation through use of unsuitable packaging material is one of the major problems with frozen foods. Another problem is that of distorted or damaged containers due either to lack of expansion space for the product in freezing or to selection of low-strength box material.

Whenever free space is present in a container of frozen food, ice sublimes and condenses on the film or package. temperature fluctuation increases the severity of frost deposition.

SHIPPING DOCKS

A refrigerated shipping dock can eliminate the need for assembling orders on the nonrefrigerated dock or other area, or using a more valuable storage space for this purpose. This is especially true in the case of freezer operations. Some businesses do not really need a refrigerated order assembly area. One example is a packing plant that ships out whole carcasses or sides in bulk quantities and does not need a large area in which to assemble orders. Many such plants are constructed without any dock at all, simply having the load-out doors lead directly into the carcass-holding cooler, which is satisfactory provided adequate refrigerating capacity is installed in the vicinity of the shipping doors to prevent undue temperature rise in the coolers during shipping.

A refrigerated shipping dock can perform a second function of reducing the refrigeration load, which is most important in the case of freezers; although it serves almost as valuable a function with coolers. Even with cooler operations, the installation of a refrigerated dock greatly reduces the load on the cooler's refrigerating units and assures a more stable temperature within the cooler. At the same time, it is possible to only provide refrigeration to maintain dock temperatures on the order of 40 to 45°F, so that the refrigerating units can be designed to operate with a wet coil. In this way, frost buildup on the units is avoided and the capacity of the units themselves substantially increased,

Meat Products

making it unnecessary to install as many or as large units in this area.

In the case of freezers, the units should be designed and selected to maintain a dock temperature slightly above freezing, usually about 35 °F. With this dock temperature, orders may be assembled and held prior to shipment without the risk of defrosting the frozen product, and the workers can assemble orders in a much more comfortable space than the freezer. The design temperature should be low enough so that the dew point of the dock atmosphere is below the product temperature. Condensation on the surface of the products is one step in developing off-condition product.

With a dock temperature of 35 °F, the temperature difference between the freezer itself and the outdoor summer condition is split roughly in half, and since the airflow through the loading doors or other openings is proportional to the square root of the temperature difference, this results in an approximate 30% reduction in airflow through the doors—both the doors into the dock itself and the doors from the dock into the freezer. At the same time, by cooling the outdoor air to approximately 35 °F, in most cases about 50% of the total heat in the outdoor air is removed by the refrigerating units on the dock.

Since using a refrigerated dock reduces the airflow through the door into the freezer by approximately 30%, and 50% of the heat in the air that does pass through this door is removed, the net effect is to reduce the infiltration load on the units within the freezer itself by about 65%. This is not a net gain; since an equal number of these units operate at a much higher temperature, the power required to remove the heat on the dock is substantially lower than it would be if the heat were allowed to enter the freezer.

The infiltration load from the shipping door, whether it opens directly into a cooler or freezer or into a refrigerated dock, is extremely high. Even with well-maintained foam or inflatable door seals, a great deal of warm air leaks through the doors whenever they are open. Such air infiltration may be calculated approximately by:

$$V = 1.4 H W \theta (H)^{0.5} (t_1 - t_2)^{0.5}$$

where

V = air volume, ft^3/h at the higher temperature condition
1.4 = an arbitrary constant selected to convert the airflow into ft^3/h, after allowance for the contraction of the airstream as it passes through the door and allowing for the obstruction created by a truck parked at the door, with only nominal sealing
H = height of the door, ft
W = width of the door, ft
θ = time the door is open, min.
t_1 = outdoor air temperature or the air at the higher temperature, °F
t_2 = temperature of air in the dock or cooler, °F

Time, θ, is estimated, based on the time the door is assumed to be obstructed or partially obstructed. If the doors are equipped with good seals and these are well-maintained so that the average truck will be tightly sealed to the building, this time would be assumed as only the time necessary to spot the truck at the door and complete the air seal.

The unit cooler providing refrigeration for the dock area should be ceiling suspended with a horizontal air discharge. Each unit should be aimed toward the outer wall and above each of the truck loading doors, if possible, so that cold air strikes the wall and is deflected downward across the door. This downward airflow just inside the door tends to oppose the natural airflow of the entering warm air, thus helping reduce the total amount of infiltration.

In general, a between the rails unit cooler has proved most successful for this purpose, since it distributes the air over a fairly wide area and at low outlet velocity. Such an airflow pattern does not create severe drafts in the working area and is more acceptable to employees working in the refrigerated space. The above comments and equation for determining air infiltration also apply to shipping doors, which open directly into storage or shipping coolers.

ENERGY CONSERVATION

Water, a utility previously considered free, frequently has the most rapid rate increases. Coupled with high sewer rates, it is the largest single-cost item in some plants. If fuel charges are added to the hot water portion of water usage, water is definitely the most costly utility. Costs can be reduced by better dry cleanup, use of heat exchangers, use of filters and/or settling basins to collect solids and greases, use of towers and/or evaporative condensers, elimination of water as a means of product transport, and an active conservation program.

Air is needed for combustion in steam generators, sewage aeration, air coolers or evaporative condensers, and blowing product through lines. Used properly in conjunction with heat exchangers, air can reduce other utility costs—fuel, sewage, water, and electricity. Nearly all plants need close monitoring of valves either leaking through or left open in product conveying. Low-pressure blowers are frequently used in place of high-pressure air, reducing initial investment and operating costs of driving equipment.

Steam generation is a source of large savings through efficient boiler operation (fuel and water sides). Reduced use of hot water and sterilizer boxes, and proper use of equipment in plants with electric and steam drives, should be promoted. Sizable reductions can be realized by scavenging heat from process-side steam and hot water and by systematically checking steam traps. In some plants, excess hot water and low energy heat can be recovered using heat exchangers and better heat balances.

Electrical energy needs can be reduced by:

- Properly sizing, spacing, and selecting light fixtures and an energy program of keeping lights off (lights comprise 25 to 33% of an electric bill)
- Monitoring and controlling the demand portion of electric usage
- Checking and sizing motors to their actual loads for operation within the more efficient ranges of their curves
- Adjusting the power factor to reduce initial costs in transformers, switchgear, and wiring
- Good lubricating to cut power demands
- Operating refrigeration equipment to reduce the horsepower per ton requirements (i.e., low condenser and high suction pressures while still obtaining desired product temperatures)

Although refrigeration is not a direct utility, it involves all or some of the factors just mentioned. Energy use in refrigeration systems can be reduced by:

- Operating with reduced head and higher compressor suctions
- Properly removing oil from the system
- Purging noncondensable gases from the system
- Adequately insulating floors, ceilings, walls, and hot and cold lines
- Using energy exchangers on exhaust and air makeup
- Keeping doors closed to cut humidity or prevent an infusion of warmer air

Utility savings are also possible when usage is considered with product line flows and storage space. A strong energy conservation program not only saves total energy but frequently results in greater product yields and product quality improvements, and therefore increased profits. Prerigor or hot processing of pork and beef products is being used to greatly reduce the energy required for postmortem chilling. Removal of waste fat and bone prior to chilling reduces the amount of chilling space by 30 to 35% per beef carcass.

REFERENCES

ASHRAE. 1978. Safety code for mechanical refrigeration. ANSI/ASHRAE *Standard* 15-1978.

Jul, M. 1969. "Quality and stability of frozen meats." In *Quality and Stability in Frozen Foods*, Chapter 8. Wiley-Interscience, New York.

Lawrie, R.A. 1966. *Meat science,* Pergamon Press, New York.

Lentz, C.P. 1971. "Effect of light and temperature on color and flavor of prepackaged frozen beef." *Canadian Institute of Food Technology Journal* 4:166.

Thatcher, F.S. and D. S. Clark. 1968. *Microorganisms in foods: Their significance and methods of enumeration.* University of Toronto Press, Toronto, Canada.

CHAPTER 13

POULTRY PRODUCTS

Slaughtering, Feather Removal, and Evisceration 13.1
Chilling 13.1
Tenderness Control 13.2
Raw Poultry Processing and Packaging 13.3
Distribution and Retail Holding Refrigeration 13.3
Freezing 13.4
Preserving Quality in Storage and Marketing 13.6
Thawing and Use 13.8

POULTRY products may be ice-chilled (33 to 35 °F), deep chilled (around 28 °F), or frozen (0 °F or lower). Means of refrigeration include ice, mechanically cooled water or air, dry ice (carbon dioxide sprays), and liquid nitrogen sprays. Continuous chilling and freezing systems, with various means for conveying the product, are common. This chapter discusses the temperature-sensitive qualities of poultry products and the refrigeration and related requirements needed to maintain these qualities.

SLAUGHTERING, FEATHER REMOVAL, AND EVISCERATION

Slaughtering on a continuous conveyer line involves automatic electrical stunning to prevent struggling and to facilitate cutting and bleeding the birds with mechanical devices (Ingling 1978). After 90 to 120 s bleeding periods, birds are scalded for periods up to 120 s to loosen the feathers, usually by immersion in a tank of warm water. However, steam-spray scalders are used to improve sanitation.

Scalding water temperatures can range from 120 to 160 °F and higher. In commercial practice, two temperature ranges are used—semiscalding (124 to 130 °F) and subscalding (138 to 140 °F). Semiscalding removes the feathers without loss of the outer pigmented skin layer (cuticle) and is used for most broilers marketed in the chilled state. Subscalding provides complete feather removal, but the cuticle is removed during the feather picking process. Subscalding is used for all turkeys and mature chickens, and for some broilers to be further processed and frozen. Steam scalders have been developed successfully for subscalding, but have had only limited success for semiscalding.

Compared to a semiscalded skin surface, the exposed surface resulting from subscalding is darker and more susceptible to dehydration and further darkening during chilling and storage. Therefore, this product must be chilled at high relative humidity and all subsequent operations must be performed in a manner that prevents moisture loss from the surface. Tight, moistureproof packaging is recommended.

Feather-picking machines vary in design, depending on the speed of the picking line and specific type of feathers to be picked. All machines remove feathers through the beating action of long rubber fingers mounted on rotating double drums or the rubbing action of rotating short stiff rubber fingers mounted in a series of discs along the length of the picker. Proper clearance between rubber fingers and bird prevents excessive skin abrasion and beating of the carcass and muscles, which may toughen the meat.

The technology of evisceration and related steps essential to the establishment of a ready-to-cook carcass have changed from hand to completely automatic machine operation. Thorough washing of the eviscerated carcass, inside and out, prior to chilling, is essential. The extent to which the body temperature of the bird is lowered when it reaches the chiller depends on the temperature of ambient air and processing water, and on the extent of washing. Internal temperatures may range from 75 to 95 °F.

During processing, sanitary practices in design, use, and cleaning of equipment are essential. Chlorinated water for processing plant use is desirable.

CHILLING

Chilling retards deteriorative changes, principally those caused by microbial growth. According to USDA regulations (1972), poultry carcasses weighing less than 4 lb should be chilled to 40 °F or below in less than 4 h; carcasses of 4 to 8 lb in less than 6 h; and carcasses of more than 8 lb in less than 8 h.

Chilling procedures and equipment have evolved to meet the changing needs of the poultry processing industry. Slow air chilling was considered adequate for semiscalded uneviscerated poultry in the past. But with the transformation to eviscerated, ready-to-cook poultry, sometimes subscalded, air chilling was replaced by chilling in tanks of slush ice. Immersion chilling is more rapid than air chilling, and more important, it prevents dehydration and affects a net absorption of water. Slush ice tank chilling, a batch process, was replaced in turn by mechanical continuous immersion slush ice chillers, which are fed automatically from the end of the evisceration conveyer line. In general, tanks are only used to hold chilled carcasses in an iced condition prior to cutting up, or to age prior to freezing.

Continuous chillers vary greatly in their mechanical design. They include: *continuous drag chillers,* in which suspended carcasses are pulled through troughs containing agitated cool water and ice slush; *slush ice chillers,* through which the carcasses are pushed by a continuous series of power-driven rakes; *concurrent tumble systems,* which involve passage of the free-floating carcasses through horizontally rotating drums suspended in tanks of successively cool water and ice slush (the movement of the carcasses is regulated by the flow rate of recirculated water in each tank); *counterflow tumble systems,* in which the carcasses are carried through tanks of cool water and ice slush by horizontally rotating drums with helical flights on the inner surface of the drums; and *rocker vat systems,* in which carcasses are conveyed by the recirculating water flow, and agitation is accomplished by an oscillating, longitudinally oriented paddle. Carcasses are removed automatically from the tanks by continuous elevators. These chillers are capable of reducing the internal temperature of broilers from 90 to 40 °F in 20 to 40 min, at processing speeds of 5000 to 10,000 birds/h. USDA regulations require a minimum overflow of 2 quarts of water per broiler from continuous immersion chillers and 1 gal per turkey. Chilling water must not be higher than 65 °F in the warmest part of the chilling system.

Adjuncts and replacements for continuous immersion chilling should be used if available because immersion chilling is believed to be a major cause of bacterial contamination. Water spray

---

The preparation of this chapter is assigned to TC 11.1, Meat, Fish, and Poultry Products. This chapter last received a major revision in 1974.

chilling, air blast chilling, and use of carbon dioxide snow or liquid nitrogen spray are alternatives but are limited, as the following comparisons indicate.

Liquid water has a much higher heat transfer coefficient than any gas at the same temperature of cooling medium, so water immersion chilling is more rapid and efficient than gas chilling. Water spray chilling, without recirculation, requires much greater amounts of water than immersion chilling. Product appearance should be equivalent to water immersion or spray chilling, but inferior to air blast or carbon dioxide or nitrogen chilling, due to dehydration. Gas chilling without packaging could cause a 1 to 2% loss of moisture, while water immersion chilling permits from 4 to 15% moisture uptake, and water spray chilling up to 4% moisture uptake. Carpenter *et al.* (1979) also reported on the effect of salt brine chilling on dripless ice-packed broiler carcasses.

With adequately washed carcasses and with adequate chiller overflow in a direction counterflow to the carcasses, the bacterial count on carcasses should be reduced by continuous water immersion chilling. However, incidence of a particular low level contaminant, such as *Salmonella*, may increase during continuous water immersion chilling (Morris and Wells 1970, Surkiewicz *et al.* 1969). Such transfer of microbes can be controlled by chlorinating the chill water (Mast and MacNeil 1972, Lillard 1980). Spray chilling without recirculation has reduced bacterial surface counts 85 to 90% (Peric *et al.* 1971). Microbe transfer by spray chilling is unlikely. Chilling with air, carbon dioxide, or nitrogen presents no obvious microbiological hazards, although good sanitary practices are essential. If crust freezing is accomplished as a part of the chilling process, a reduction in bacterial load, possibly as much as 90%, can be expected.

Many chilling methods involving low-temperature air blast, carbon dioxide snow, or liquid nitrogen spray are used. Cryogenic gases are generally used in long insulated tunnels through which the product is conveyed on an endless belt. Some freezing of the outer layer (crust freezing) usually occurs, and an equilibration of temperature in outer and inner portions is allowed to occur to reach the final, intended chill temperature. In some plants, a combination of continuous water immersion chilling to reach 35 to 40 °F and a cryogenic gas tunnel to finally reach 28 °F is used. The water-chilled poultry, either whole or cut up, is generally packaged before gas chilling, to prevent dehydration.

Water absorption by the carcass in water immersion or spray chilling, and possible loss of water during gas chilling, are important economic and quality factors. USDA regulations set graded maximum levels of water absorption and retention varying with weight and type of product and specify testing procedures and basis for rejection or retention of product. The set values range from 4.3% weight increase for 27 lb and over consumer-packaged whole turkeys to 12.0% for ice-packed poultry at the end of the drip line. Of the water absorbed, very little appears in the muscle, most being held in the connective tissue in the skin and between skin and flesh (Klose *et al.* 1960, Osner and Shrimpton 1966).

For example, increases in percentage of moisture because of water immersion chilling have been reported from 0.5 to 2% for muscle and 10 to 12% for skin. Kotula *et al.* (1960) compared moisture uptake and drainage loss for air-agitated slush ice tank chilling, and for four mechanical continuous immersion slush ice chillers. Immediately after chilling, the total moisture uptake including 2 to 3% by prechill washing ranged from 7% for 2 to 4 h of tank chilling to 17% for a counterflow tumble continuous immersion water-ice chiller. After draining, shipping, and a final 30 min drain, all uptakes were in the range of 5 to 7%. Initial uptakes were much greater in a counterflow tumble system for carcasses with thigh skin area cut compared to uncut, *i.e.*, 18 versus 12%. After shipping and draining, the residual uptakes were 7 and 5%.

Although soluble solids are lost from the carcass to the chill water in immersion chilling, no evidence exists that, under accepted chilling practices, any appreciable loss of flavor or other desirable qualities occurs (Pippen and Klose 1955, Janky *et al.* 1978).

Ice requirements per bird for continuous immersion chilling depends on entering carcass temperatures and weight, entering water temperature, and exit water and carcass temperature. For a counterflow system, 60 °F entering water and 65 °F exit water, 0.25 lb of ice per pound of carcass is a reasonable estimate. This may be compared to a requirement of 0.5 to 1 lb of ice per pound of poultry for static ice slush chilling in tanks. For continuous counterflow water immersion chillers, if the plant water temperature is considerably above 65 °F, it may be economical to use a heat exchanger between incoming plant water and exiting (overflow) chill water.

Ice production for chilling is usually a complete in-plant operation, with large piping and pumps to convey the small crystalline ice or ice slush to the point of use. To reduce the use of ice, some immersion chillers are double-walled and depend on circulating refrigerant to chill the water within the chiller. The chiller has an ammonia or refrigerant oil between the outer and inner jacket with the inner jacket serving as the heat transfer medium. Agitation or a defrost cycle must be provided during periods of slack production to prevent the chiller from freezing up.

Chilling and holding to about 28 °F, the point of incipient freezing, gives the product a much greater shelf life compared with a product held at ice-pack temperatures (Stadelman 1970).

TENDERNESS CONTROL

Tenderness in cooked poultry meat is a prerequisite to acceptability. Relative tenderness decreases as birds mature, and this toughness has always been considered in the recommendations for cooking birds of various ages. Sadayoski *et al.* (1979) studied the effects of cysteamine in relation to the tenderness of meat obtained from mature chickens. However, another type of toughness depends primarily on the length of time that the carcass is held in an unfrozen state before cooking. Birds cooked before they have time to pass through rigor are very tough. Normal tenderization after slaughter is arrested by freezing. For birds held at 40 °F, complete tenderization occurs for all muscles within 24 h and for many muscles in a much shorter time.

Other factors that interfere with normal tenderization are immersion in 140 °F water and cutting into the muscle. Formerly, birds were held unfrozen for sufficient time in the normal channels of processing and utilization to permit adequate tenderization. Shorter chilling periods, more rapid freezing, and cooking without a preliminary thawing period have shortened the period during which tenderization can occur to such an extent that toughness has become a potential consumer complaint. Hanson *et al.* (1942) observed a rapid increase in tenderness within the first 3 h of holding and a gradual increase thereafter.

Shannon *et al.* (1957), working with hand-picked stewing hens, demonstrated an increased toughness because of increased scalding temperature or increased scalding time in the ranges of 120 to 195 °F and 5 to 160 s. However, the differences in toughness that occurred within the limits of temperature and time, necessary or practical in commercial plants, were quite small.

Klose *et al.* (1959 a,b, 1961) and Pool *et al.* (1959) showed that, in addition to the beneficial effect of increasing holding time above freezing, tenderness is increased by lower scalding temperatures, shorter scalding time, and particularly by reducing the extent of beating received by the birds during picking operations. In addition, the data obtained support the following conclusions. Turkey fryers should be held at least 12 h above freezing to develop optimum tenderness. Holding fryers at 0 °F for 6 months and longer has no tenderizing effect, but holding in a thawed state (35 °F) after frozen storage has as much tenderizing effect as an equal period of chilling before freezing. Turkeys frozen 1 h after slaughter were adequately tenderized by holding for 3 days at 28°F, a temperature at which the carcass is firm and no important quality loss occurs for the period involved. Behnke et al. (1973) confirmed this effect for Leghorn hens.

Machine feather picking may result in meat twice as tough as that obtained for a theoretical, hand-picked control. Toughening induced by the beating action of the mechanical feather pickers is not resolved appreciably by otherwise adequate chilling periods. This toughening action is accumulative in that increasing the number of picking machines to which the bird is exposed increases the degree of toughness. Thus, optimum tenderness can be attained through careful attention to chilling (aging) periods, feather picking conditions, and scalding practices. Exposure to unusually high temperatures during the aging period may also lead to irreversible toughening.

Overall processing efficiency could be impoved by cutting up the carcass directly from the end of the eviscerating line, packaging the parts, and then chilling the still-warm packaged product in a low-temperature air blast or cryogenic gas tunnel. Several evaluations have been made of the relative tenderness of such a hotcut product. Webb and Brunson (1972) reported that cutting the breast muscle and removing a wing at the shoulder joint before chilling significantly decreased tenderness of treated muscles, though cut carcasses were aged in ice slush before cooking. Klose et al. (1972) found that under commercial plant conditions, making an eight-piece hotcut before chilling and aging significantly reduced tenderness of breast and thigh muscles, compared to cutting after chilling.

No difference in shear score was found between the outer 0.6 in. slice and the second 0.6 in. slice of breast muscle for the chillcut birds, but a higher shear value was observed for the second slice from hotcut birds (Treat and Goodwin 1973). Cutting the muscle from the sternum on one breast piece resulted in a higher shear value than was recorded for the other side of the breast that was left intact. Only in hotcut birds was this found to be significant. Aging for 24 h eliminated the difference in shear value between hotcut and chillcut broilers (Wyche and Goodwin 1968).

Smith et al. (1966) indicated that too-rapid chilling of poultry might have a toughening effect, similar to cold shortening observed in red meats.

RAW POULTRY PROCESSING AND PACKAGING

Chickens and turkeys, for both chilled and frozen distribution, can be cut up in the processing plant. A popular cut for broilers is a nine-piece cut, usually two drumsticks, two thighs, two wings, two breast halves, and back. However, there are at least eight different cutting patterns, involving greater subdivision of breast and leg portion. Backs and necks are often mechanically deboned, giving a comminuted slurry which is frozen in rectangular flat cartons containing about 60 lb. Turkey breasts, legs, and drumsticks are available as separate film-packaged parts, and turkey thigh meat is marketed as a ground product resembling high quality hamburger. Partial cooking and breading and battering of broiler parts is done in some poultry processing plants.

Most packaged poultry is now tray packed, either for frozen or chilled distribution. All-plastic packages have been developed, and automated packaging lines using plastic film have been engineered. Changes in packaging methods and materials are so rapid that the best sources of information on this subject are the companies that fabricate films and packages and distribute the materials. They are listed in the most recent Encyclopedia Issue of *Modern Packaging*.

Many available packaging films are satisfactory concerning relative moisture and gas permeability, but they sometimes lack sufficient strength to withstand the rough handling encountered in normal marketing operations. Trays used in the packaging of cut-up chilled poultry need to contribute a certain amount of rigidity and form to the package and, with a blotter liner, provide absorbency to pick up the drip which exudes from the meat during storage. Plastic film overwraps should be tight fitting and reasonably moisture- and airproof. Wells et al. (1958) reported that partial evacuation of the air from packages wrapped in impermeable films tended to inhibit bacterial growth by reducing oxygen tension.

Ideally, packages for frozen poultry should possess low moisture-vapor permeability and should be strong, protective, attractive, and suitable for rapid freezing and prolonged storage of the product. Packages for precooked frozen poultry products must in addition be adaptable for reheating the product. In general, packaging requirements increase from whole to cut-up to precooked forms.

Packages for frozen, whole, and ready-to-cook poultry consist principally of plastic film bags that are tough and reasonably impermeable to moisture vapor and air. The commonly used polyvinylidene chloride, polyethylene, and polyester films are sufficiently water vaporproof and airproof to give adequate protection for normal commercial times and temperatures. Turkeys, ducks, and geese are packaged mostly in the whole, ready-to-cook form, while frozen chickens appear whole and in packaged, cut-up form.

Large fiberboard cartons or containers for holding and shipping from 2 to 12 individually packaged birds should be rectangular in shape to facilitate palletizing and be strong enough to support 16-ft high stacked loads common in refrigerated warehouses. If freezing is to be accomplished for material such as fryer turkeys that need to be frozen rapidly, holes or cutaway sections in the sides and ends are needed to permit a rapid airflow across the poultry surfaces in the air blast freezer.

DISTRIBUTION AND RETAIL HOLDING REFRIGERATION

Chilled poultry handled under proper conditions is an excellent product. However, there are limitations in its marketability because of the relatively short shelf life caused by bacterial deterioration. Bacterial growth on poultry flesh, as on other meats, has a high temperature coefficient. Studies based on total bacterial counts have shown that birds held at 36°F for 14 days are equivalent to those held at 50°F for 5 days or 75°F for 1 day. Spencer and Stadelman (1955) found that birds at 31°F had 8 days additional shelf life over those at 38°F. The magnitude of this temperature coefficient is illustrated by Farrell and Barnes (1964) on generation times (time for population to double) for a species of *Pseudomonas*: 36 h at 29°F, 14 h at 32°F, 7 h at 41°F, and less than 1 h at 77°F. This indicates the advantage of deep chill over ice chill.

The generation time of psychrophilic organisms isolated from chickens was 10 to 35 h at 32°F, depending on the species studied (Ingraham 1958). Raising the temperature to 36°F reduced generation time to 8 to 14 h, again depending on the species.

The bacterial flora in poultry meat include the genera *Pseudomonas, Achromobacter, Micrococcus,* and *Flavobacterium*. Nagel et al. (1960) found that the distribution pattern of bacterial genera (mostly *Pseudomonas* and *Achromobacter-Alcaligenes*) was unaffected by geographical location of processing or cutting operations, or by the use of antibiotics. Ayres et al. (1950) observed that the growth and coalescence of colonies of the above microorganisms on the surface of chilled poultry led to a characteristic off-odor, accompanied by the development of bacterial slime. The shelf life, or time before off-odor occurred, depended not only on the holding temperature but on the initial amount of bacterial contamination. This emphasizes the importance of adequate sanitation in processing, which minimizes initial bacterial contamination and thus prolongs shelf life. Frequent cleaning of processing equipment, as well as thorough washing of the eviscerated carcasses, is necessary. Goresline *et al.* (1951) reported a substantial decrease in bacterial contamination and an increase in shelf life by the use of 20 ppm of chlorine in processing and chilling water.

Since shelf life is limited considerably by bacterial growth (slime formation) on the skin layer, it is reasonable to assume that drastic changes in the skin surface, such as the removal of the epidermal layer by high temperature scalding, might appreciably affect shelf life. Ziegler and Stadelman (1950) reported approximately 1 day more chilled shelf life for 128°F scalded birds than for 140°F scalded ones.

Chickens, principally broilers, are sold as whole ready-to-cook or cut-up ready-to-cook. Poultry is shipped in wax-coated corrugated fiberboard boxes with crushed ice, or fiberboard boxes with dry-ice snow or solid particulate dry ice (carbon dioxide). Comparisons of dry-ice and water-ice packing in several types of boxes were made by Risse and Thomson (1971). While overall costs for packaging materials, refrigerant, and labor did not differ much between alternative systems, dry ice costs significantly less than water ice packing because of the greater shipping weight of 20 lb of water ice compared to 2 lb of dry-ice snow.

The limited number of warehouse receivers and retail receivers involved in this study preferred dry-ice packed poultry over water-ice packed poultry because the boxes were lighter in weight, had less drainage, and the surface appearance was superior for the dry-ice packed product. However, in the six commercial-type shipments of this study, the percentage weight loss of poultry from processing plant to receiver's warehouse was significantly higher for dry-ice packed (6.3%) than for water-ice packed (5.5%), and the warehouse arrival temperature was significantly higher for dry-ice packed (37°F) than for water-packed poultry (33°F). This suggests the desirability of having 28 to 30°F truck and warehouse temperatures if dry-ice packing is used.

FREEZING

Effect on Product Quality

The behavior of muscle tissue during freezing represents the combined effects of the soluble constituents that produce a freezing point lowering and a freezing range, the size and location of the ice crystals formed, and the translocation of cellular constituents during freezing. Koonz and Ramsbottom (1939) pointed out that very slowly frozen poultry appears dark, with very large ice crystals of the extra-fiber type. When poultry is frozen at a temperature that assures a natural bloom, the ice crystals are smaller and more numerous, but are still principally of the extra-fiber type. Ice crystals are small and of the intra-fiber type when freezing is rapid enough to assure a white appearance. Ice crystals become progressively larger in poultry as the time between slaughter and freezing is increased.

Rapid freezing is essential to obtain satisfactorily light appearance in certain types of frozen poultry (Van den Berg and Lentz 1958). Birds scalded at 140°F and above completely lose their outer layer of skin during normal machine picking and become particularly susceptible to the development of a dark frozen appearance if not frozen rapidly. Also, immature fryer-roaster turkeys, which have a thin, practically fat-free skin, are naturally dark and require a rapid rate of freezing to assure a light, pleasing frozen appearance.

Increased rates of freezing were provided by use of air blast tunnels operating at air temperatures ranging from −20 to −40°F and at air velocities of 500 fpm or more. In many cases, packaged birds freeze better on open shelves in the air blast rather than after packing the packaged birds in cartons. In an evaluation of various factors contributing to freezing rate and frozen appearance, Klose and Pool (1956) found that freezing on open shelves versus freezing in boxes, lower air blast temperature, and higher air blast velocity were important, in that order. At an air blast temperature of −20°F or below, increasing air blast velocity beyond 600 fpm had little beneficial effect. Also at 1300 fpm, decreasing air-blast temperature from −20 to −30°F produced almost no additional improvements in frozen appearance. Birds placed in the blast freezer immediately after evisceration and packaging and while still warm did not develop an appreciably darker frozen appearance than those chilled in ice slush before packaging and blast freezing. However, this procedure has the serious disadvantage that the duration of holding above freezing temperatures may be reduced to an amount inadequate for optimum tenderization. The factor of finish, or amount of fat in the skin layer, can exert a much greater beneficial effect on frozen appearance than any possible changes in processing or freezing practices.

The freezing rate of diced cooked chicken meat does affect the quality of the frozen meat. Hamre and Stadelman (1967a) reported that cryogenic freezing procedures were desirable as the resulting color was lighter, but too rapid a freezing rate resulted in a shattering of the meat cubes. The freeze-drying rates for rapidly frozen material were slower than for products frozen by slower methods. Hamre and Stadelman (1967b) indicated that tenderness of freeze-dried chicken after rehydration was affected by freezing rate prior to drying. Liquid nitrogen spray or carbon dioxide snow freezing were selected as preferred methods for overall quality of diced cooked chicken meat to be freeze-dried.

Freezing Methods

Poultry may be frozen between refrigerated double-walled plates in a blast of refrigerated air, by immersion in a refrigerated liquid, or by a shower of liquefied gas such as nitrogen. Rectangularly packaged, cut-up poultry can be frozen in multiplate freezers within several hours. A detailed description of the available commercial equipment may be found in Chapter 9 of this volume.

The major proportion of whole, ready-to-cook birds is frozen in air blast tunnel freezers, with air temperatures ranging from −10 to −40°F and air velocities of 300 to 1000 fpm and up. It is desirable to have air temperatures at −30°F or below during operation and air velocities over the product surfaces of at least 600 fpm. To obtain high air velocity over the product, the blast tunnel should be completely loaded across its cross section, with proper spacing of the units of the product to assure airflow around all sides and no large openings to permit bypassing of the airstream. In some cases, the whole bird may be packed into cartons or boxes, and the cartons stacked on pallets in the blast tunnel, with spacing between layers and between cartons in the same layer to permit adequate airflow and freezing rate.

However, as mentioned, freezing rates required to give a sufficiently light frozen appearance are so high for birds scalded at elevated temperatures and immature turkeys that modified freezing methods are necessary. Boxes may be left open during freezing, or they may be constructed with holes or cutouts on the ends, or, best of all, the individual, bagged birds may be frozen on open racks or shelves to provide maximum airflow and freezing rate. If freezing rates in excess of those possible in air blast tunnels are required, immersion freezing may be considered.

Freezing by direct contact with low-temperature brine or other liquids such as glycols has occasionally been proposed. Esselen *et al.* (1954) found that the freezing time per pound of packaged ready-to-cook birds immersed in −20°F brine was in the range of 20 to 30 min. In −20°F calcium chloride brine, times for the interior to reach 15°F for warm eviscerated, packaged broilers, 12-lb turkeys, and 25-lb turkeys were approximately 1.5, 5, and 7 h, respectively. Lentz and Van den Berrg (1957) have evaluated factors affecting freezing time and frozen appearance of poultry. Immersion freezing at −20°F produced a lightness of appearance that could be matched by air blast freezing only at an air blast

Poultry Products

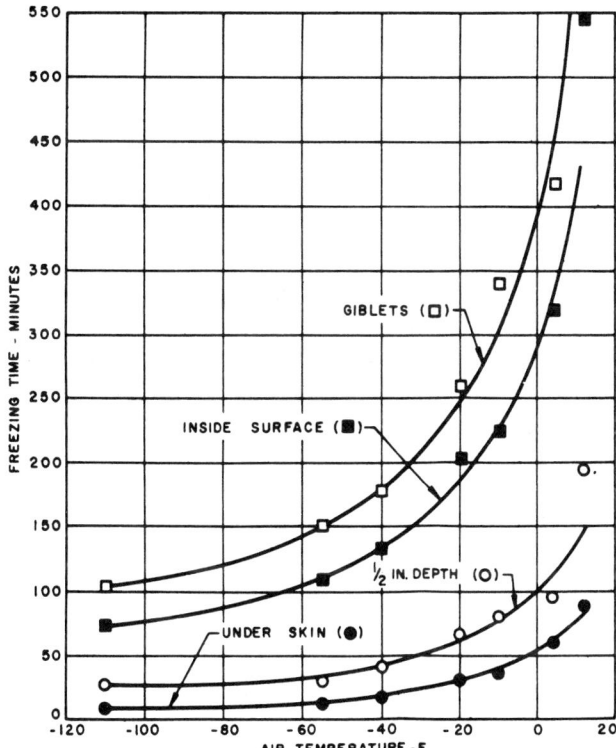

Note: Freezing time is the time required for temperature to fall from 32 to 25 °F. The values are for 5- to 8-lb chickens with initial temperature of 32 to 35 °F and with air velocity of 450 to 550 fpm.

Fig. 1 Relation between Poultry Freezing Time and Air Temperature
(Van den Berg and Lentz 1958)

temperature of −100 °F. Combinations of immersion freezing and air blast freezing were tested. A minimum immersion time at −20 °F of 20 min for chickens and 40 min for turkeys was necessary to obtain a white appearance, typical of immersion freezing, in birds subsequently frozen on open shelves in a −20 °F, 300 to 500 fpm air blast. The light-colored chalk-like appearance of immersion frozen poultry was maintained indefinitely in −20 °F storage, but darkened after 6 weeks at 0 °F or 2 weeks at 20 °F.

Comparison of various liquids (methanol, salt brines, and glycols) that might be used in immersion freezing revealed that freezing times depended mainly on viscosity of the liquids and were maximum in glycerol and minimum in calcium chloride brine. Increased agitation decreased freezing times in the surface layers by as much as 50%. Spraying the liquid was as effective in reducing freezing times as the highest rate of agitation used. Advantages of immersion freezing are lighter, more pleasing appearance; high rate of heat transfer that makes a line operation feasible; lower initial and upkeep costs than air blast; and possible use in combination with subsequent air freezing. Possible disadvantages include lack of uniformity of frozen appearance, development of leaks due to pinpoint holes or cracks in packaging film, the question of acceptance of the chalk-white birds, and lack of versatility for freezing all types of material.

Typical freezing rate curves are shown in Figures 1 to 5, which illustrate the effect of bird size and depth below skin surface, temperature, and velocity of the cooling medium, and type of medium (air versus liquid). Pflug (1957) reviews the relative efficiencies and requirements for air and liquid immersion freezing and the critical importance of the conductance of the coolant-product interfacial film, citing a film heat transfer coefficient of 3

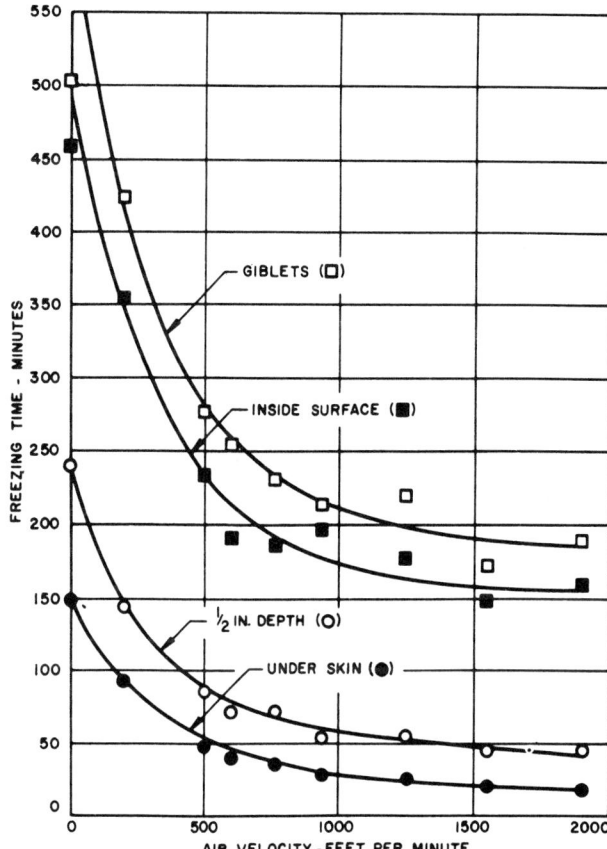

Note: Freezing time is the time required for temperature to fall from 32 to 25 °F. The values are for 5- to 8-lb chickens with initial temperature of 32 to 35 °F and with air temperature of −20 °F.

Fig. 2 Relation between Freezing Time and Air Velocity
(Van den Berg and Lentz 1958)

Btu/(h·ft²·°F) for air at −20 °F and 500 fpm, compared to 30 Btu/(h·ft²·°F) for sodium chloride brine at 0 °F and a velocity of 5 fpm. Table 1 lists physical properties of poultry related to refrigeration.

A liquid nitrogen shower system has been used to freeze the outer shell of products to −100 °F or below, after which the entire product may be equilibrated in a holding room to a temperature around −10 °F. Temperature in the freezer box where the liquid nitrogen is released in a shower may be as low as −250 °F. IQF (Individually Quick Frozen) freezing with CO_2 is also used, particularly in preparing for freeze-drying.

The freezing rate of precooked chicken affects the quality of product. Breaded precooked drumsticks frozen with liquid nitrogen are susceptible to cracking and separation of the meat from the bone. Precooked chicken that is lightly breaded (7 to 10% breading) and frozen by cryogenic procedures is susceptible to developing small white freezer burn areas next to the surface. These areas may or may not show up immediately after freezing, but will show up almost immediately after reconstituting with hot oil. In some instances, the white areas will cover the entire surface of the piece of chicken (Yingst and Goodwin 1971). Altering the fat content of thighs does not reduce the severity of this problem.

Automated units may be designed to handle packages, cartons of birds, or unwrapped pieces of chicken or turkey. Product may be transported through the freezing chamber on belts or trays. One such system is adaptable to all sizes of whole birds, packages, or cartons. The system automates the freezing operation from the

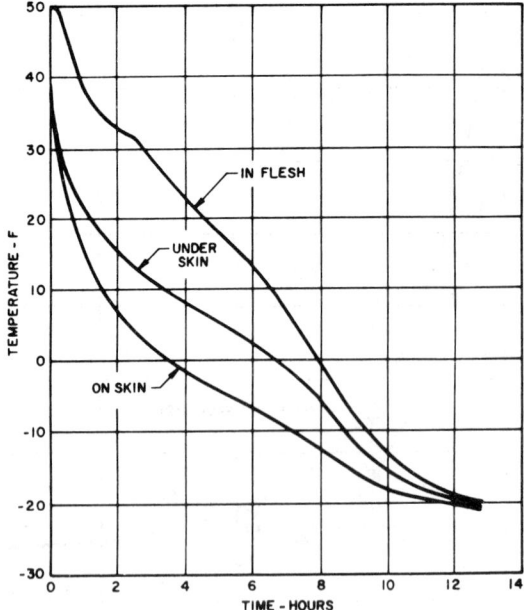

Note: For 21-lb, bronze tom turkeys on shelves in air blast.

Fig. 3 Temperature during Freezing of Packaged,
Ready-to-Cook Turkeys
(Klose and Pool 1956)

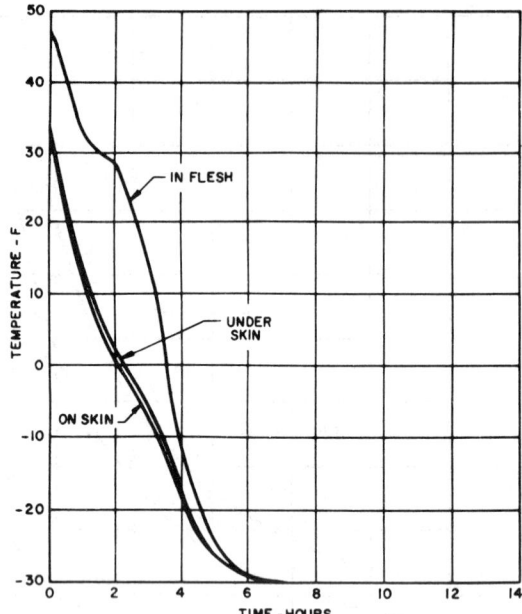

Note: For 7-lb, bronze tom turkeys on shelves in air blast.

Fig. 4 Temperature during Freezing of Packaged,
Ready-to-Cook Turkeys
(Klose and Pool 1956)

Table 1 Thermal Properties of Ready-to-Cook Poultry

Property	Value	Reference
Specific heat, above freezing	0.70 Btu/(lb·°F)	Pflug (1957)
Specific heat, below freezing	0.37 Btu/(lb·°F)	Pflug (1957)
Latent heat of fusion	106 Btu/lb	Pflug (1957)
Freezing point	27°C	Pflug (1957)
Average Density		
Poultry muscle	67 lb/ft^3	
Poultry skin	64 lb/ft^3	
Thermal conductivity [Btu/(h·ft·°F)]		
Broiler breast muscle[b]	0.24 at 80°F	Walters and May (1963)
Broiler breast muscle[a]	0.29 at 68°F	Sweat et al. (1973)
Broiler breast muscle[a]	0.80 at −4°F	Sweat et al. (1973)
Broiler breast muscle[a]	0.87 at −40°F	Sweat et al. (1973)
Broiler dark muscle[a]	090 at −40°F	Sweat et al. (1973)
Turkey breast muscle[a]	0.73 at −4°F	Sweat et al. (1973)
Turkey breast muscle[b]	0.93 at −4°F	Sweat et al. (1973)
Turkey leg muscle[a]	0.83 at −4°F	Lentz (1961)

[a] and [b] indicate heat flow perpendicular and parallel, respectively, to the direction of the muscle fibers.

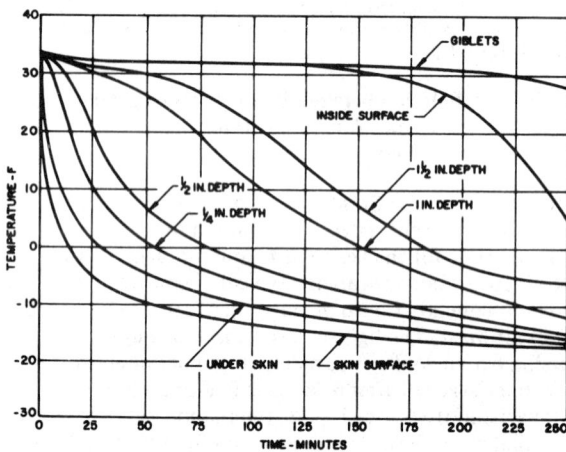

Fig. 5 Temperatures at Various Depths in Breast of
15-lb Turkeys during Immersion Freezing at −20°F
(Lentz and Van den Berg 1957)

point birds are placed in cartons until they are frozen and ready for a carton top to be put on. A typical design handles nearly 1 bird/s with about 150,000 lb total capacity. Refrigeration coils and fans are located at the side of the machine to give a high-velocity two-pass airflow that applies the coldest air to the warmest product. Frost or ice buildup is minimized since the shelves never come outside the freezer.

A tray system is available that automatically loads trays from a moving belt, conveys them through the air blast freezing chamber, empties the trays into a holding bin within the freezing chamber, and conveys the product onto a belt, which carries it out of the freezer to a packaging station. This system is particularly useful for cut-up chicken parts. Belt systems, using refrigerated air blast or cryogenic gases, are used for small parts, packages, and particle size poultry such as precooked diced chicken.

PRESERVING QUALITY IN STORAGE AND MARKETING

Important qualities of frozen poultry include appearance, flavor, and tenderness. If these are not optimum, the convenience and flexibility of use of the frozen product will not be sufficient inducement to the consumer. Optimum quality requires care in every phase of the marketing sequence, from the frozen storage warehouse, through transportation facilities, wholesaler, retailer, and finally to the frozen food case or refrigerator in the home.

Darkening of the bones is a condition that occurs in immature chickens and has become more prevalent as broilers are marketed at younger and younger ages. During chilled storage or during the freezing and defrosting process, some of the heme pigment normally contained in the interior of the bones of particularly young chickens leaches out through spongy areas and discolors the adjacent tissues. This in no way affects the palatability of the product, because the pigment is a natural tissue pigment, which, after being cooked, is known as hematin and is responsible for the cooked appearance of all meats. The freezing rate has no marked influence in preventing this discoloration. Also, the time between slaughter and freezing has no significant effect on the color of poultry bones or on the displacement of pigment from the interior of the bones to adjacent tissues.

Brant and Stewart (1950) found that development of dark bones was greatly reduced by a combination of freezing and storage at −30 °F and immediate cooking after rapid thawing. Aside from this combination, freezing rate, temperature and length of storage, and temperature fluctuations during storage were not found to have a significant effect. Further research suggested that freezing and thawing not only liberated hemoglobin from the bone marrow cells but modified the bone structure to permit penetration by the released pigment. Roasting pieces of chicken 0.5 h prior to freezing reduced discoloration of the bone. Ellis and Woodroof (1959) found that heating legs and thighs to 180 °F before freezing effectively controlled meat darkening. Methods of preheating, in order of preference, included microwave oven, steam, radiant heat oven, and deep fat frying.

During storage, poultry may become dehydrated, yielding a condition known as *freezer burn*. Dehydration can be controlled by humidification, lowering of storage temperatures, or by packaging the product adequately. Aside from adversely affecting the appearance of the product, dehydration, unless severe, does not impair quality. When freezer burn is extensive, quality is decreased because of toughening and development of oxidative rancidity of the affected area. If storage temperatures of 0 °F or lower are maintained, freezer burn is usually the factor limiting the length of time that poultry can be held in storage without adequate packaging.

Wills et al. (1948) found that the appearance of poultry suffered greatly when stored at 20 °F. The most serious defects were microbiological changes, desiccation, and development of old odors. Serious changes in flavor and juiciness occurred in poultry that had been frozen 3 to 9 months at 10 °F. Swanson and Sloan (1953) studied chemical changes in the proteins of ready-to-cook fowl held at −5 °F over a period of 40 weeks. Proteolysis was indicated during storage by increases in soluble nitrogen and nonprotein nitrogen of leg and breast muscle.

The effect of variations in storage temperature on the quality of frozen turkeys was studied by Klose et al. (1955) for moisture losses, chemical changes, and palatability. Evaluations after 6-, 12-, and 18-month storage times indicated that, under average commercial conditions of frozen storage (moisture impermeable package and temperature range of −10 to 10 °F), the only factor for which a periodic temperature fluctuation is inferior to the mean temperature is the accumulation of frost in the package. Frost formation, which influences appearance but not eating quality, increased with storage temperature, and for the −10 to 10 °F fluctuation was considerably greater than for the highest (10 °F) constant temperature. Results after 12-month storage indicated a definite superiority to −10 °F storage over 10 and 0 °F but no detectable organoleptic superiority of −30 over −10 °F.

Poultry fat becomes rancid during very long storage periods or at extremely high storage temperatures. Rancidity in frozen, eviscerated whole poultry stored for 12 months is not a serious problem if the bird is packaged in essentially impermeable film and held at 0 °F or below. Danger of rancidification is greatly increased when poultry is cut up before freezing and storage, because of the increased surface exposed to atmospheric oxygen.

Quality losses in frozen, packaged, and cut-up frying chickens were studied by Klose et al. (1959) over temperatures of −30 to 20 °F and storage periods from 1 month to 2 years. All commercial-type samples examined were acceptable after a storage period at 0 °F of at least 6 months, and some were stable for more than a year. In a comparison of a superior (moisture-vapor-proof) commercial package with a fair commercial package, increased adequacy of packaging resulted in as much extension in storage life as a decrease in storage temperature of about 20 °F. The results indicate that no statement on storage life can have general value unless the packaging condition is accurately specified.

Frozen storage tests by Klose et al. (1960) on commercial packs of ready-to-cook ducklings and ready-to-cook geese established that these products have frozen storage lives similar to other commercial forms of poultry. Ducks and geese should be stored at 0 °F or below to maintain their original high quality for 8 to 12 months.

Incorporation of polyphosphates into poultry meat by adding it to the chilling water has been shown to increase shelf life in frozen or refrigerated storage and to control loss of moisture in refrigerated storage and during thawing and cooking.

Storage studies by Hanson et al. (1959) on frozen fried chicken indicated that precooking produces a product much less stable than a raw product. Rancidity development is the limiting factor and, surprisingly enough, it is detected in the meat slightly sooner than in the skin and fatty coating of the fried product. The marked beneficial effect of oxygen (air)-free packaging was demonstrated in tests in which detectable off-flavors were observed at 0 °F in air-packed samples after 2 months, while nitrogen-packed samples developed no off-flavors for periods exceeding 12 months.

Cooling the precooked parts in ice water prior to breading was found to reduce the TBA values of precooked parts (Webb and Goodwin 1970). In this study, no difference in rancidity was noted for chicken stored 6, 8, or 10 months. By removing the skin from precooked broilers, the TBA values were lower, but yield and tenderness were reduced (Wyche et al. 1972). No difference was detected in the TBA values of the thighs frozen in liquid refrigerant with or without skin (Surkiewicz et al. 1969). The chicken parts that had the skin removed and were blast frozen were less rancid than parts with skin and frozen the same way. Precooked frozen chicken parts browned for 120 s at 400 °F were less rancid than those parts browned at 300 °F (Love and Goodwin 1974).

The susceptibility of frozen fried chicken to storage deterioration indicates the need for low temperature (0 or −10 °F) and short storage times (6 months or less) unless it is packed in an oxygen-free atmosphere. To prevent cracking and peeling of the batter coating from frozen fried chicken, Hansen and Fletcher (1963) shrunk the parts by cooking before dipping in batter. Cooling the parts to a temperature of 35 °F before dredging improved the texture and adhesion of the breading (Hale and Goodwin 1968).

In contrast to a loosely packed product, such as the frozen fried chicken mentioned above, Hanson and Fletcher (1958) reported that a solid-pack product such as chicken and turkey pot pies, in which the cooked poultry is surrounded by sauce or gravy, with consequent exclusion of air, had a storage life at 0 °F of at least 1 year. As is the case with raw poultry, turkey products have less fat stability than chicken products, but the stability can be increased by substituting more stable fats in the sauces or by using antioxidants. A quality defect found in precooked frozen products containing a sauce or gravy is a liquid separation and curdled appearance of the sauce or gravy when thawed for use. This separation is extremely sensitive to storage temperature. Sauces can be stored at least five times as long at 0 °F as at 10 °F before separation takes place. Hanson et al. (1951) established that the flour in the sauce was the cause of the separation, and found among a large number of alternative thickening agents that waxy rice flour produced superior stability. Sauces and gravies prepared with waxy rice flour are completely stable for about a year at 0 °F.

Since precooked frozen foods are not apt to be sterilized in the reheating process in the home, the processor has an added responsibility to keep bacterial counts in the product well below hazardous levels. Extra precautions should be taken in general plant sanitation, in rapid chilling and freezing of the cooked products, and in seeing that the products do not reach a temperature at any time during storage or distribution that will permit bacterial growth.

THAWING AND USE

Under ordinary conditions, poultry should be kept frozen until shortly before its consumption. The general procedure is to defrost in air or in water. Hoffert *et al.* (1952) found no significant differences in palatability between thawing in oven, refrigerator, room, and water.

For turkeys that have been scalded at high temperatures and fast frozen to give a light appearance, the temperature in retail storage and display must be kept as low as possible (0 °F is reasonable) to prevent darkening. Thawing in the package will minimize darkening.

Detailed instructions for thawing turkeys should be given on the bag used for turkey packaging. The safest procedure is to hold the turkey in the refrigerator (35 to 40 °F) for 2 to 4 days depending on the size of the bird. Other methods are immersion in cool water in the bag for 4 to 6 h or holding in a paper bag or styrofoam chest for 12 to 36 h at room temperature. When using these nonrefrigerated thawing techniques, care must be taken to keep the bird's surface cool to inhibit microbiological growth (Cunningham *et al.* 1979).

Some retail stores allow frozen poultry to start thawing in the chilled (33 to 38 °F) section of the meat display case where poultry is for sale on the particular day. This is an advantage to the consumer who wants to cook the poultry that night. However, this is a safe practice only if careful, constant control is maintained over the chilled inventory so that the product is not held beyond its overall shelf life in store and home. Freezing and thawing in itself does not reduce the refrigerated shelf life of the product. Elliott and Straka (1964) found that frozen-thawed chicken had a shelf life at 36 °F about equal to unfrozen counterparts at 36 °F, as measured by total counts of psychrophilic bacteria and by odor tests. Sauter *et al.* (1978) also reported a comparison of the microflora of fresh and thawed frozen fryers.

Ready-to-cook turkeys in a frozen, prestuffed raw form have been marketed. Extreme care should be exercised in producing and consuming this type of product to assure that the original bacterial count in the birds and stuffing is at a minimum and that, in roasting, the internal temperature reaches a value high enough to provide a safe product.

REFERENCES

Ayres, J.C., W.S. Ogilvy, and G.F. Stewart. 1950. Postmortem changes in stored meats. Part I, Microorganisms associated with development of slime on eviscerated and cut-up poultry. *Food Technology* (May):199.

Behnke, J.R., O. Fennema, and R.W. Haller. 1973. Quality changes in prerigor poultry at -3 °C *Journal of Food Science* (February):275.

Brant, A.W. and G.F. Stewart. 1950. Bone darkening in frozen poultry. *Food Technology* (May):168.

Carpenter, M.D., D.M. Janky, A.S. Arafa, J.L. Oblinger, and J.A. Koburger. 1979. The effect of salt brine chilling on drip loss of icepacked broiler carcasses. *Poultry Science* (March):369.

Cunningham, F.E., D.R. Suderman, and M.H. Wu. 1979. Composition of drainage from thawed poultry carcasses. *Poultry Science* (March):365.

Elliott, R.P. and R.P. Straka. 1964. Rate of microbial deterioration of chicken meat at °C after freezing and thawing. *Poultry Science* (January):81.

Ellis, C. and J.G. Woodroof. 1959. Prevention of darkening in frozen broilers. *Food Technology* (September):533.

Esselen, W.B., *et al.* 1954. Brine immersion cooling and freezing of packaged ready-to-cook poultry. *Refrigerating Engineering* (July):61.

Farrell, A.J. and E.M. Barnes. 1964. Bacteriology of chilling procedures in poultry processing plants. *British Poultry Science* (January):89.

Goresline, H.E., *et al.* 1951. In-plant chlorination does a 3-way job. *U.S. Egg and Poultry Magazine* (April):12.

Hale, K.K., Jr. and T.L. Goodwin. 1968. Breaded fried chicken. Effects of precooking, batter composition, and temperature of parts before breading. *Poultry Science* (May):739.

Hamre, M.L. and W.J. Stadelman. 1967a. Effect of various freezing methods on frozen diced chicken. *Quick Frozen Foods* (April):78.

Hamre, M.L. and W.J. Stadelman. 1967b. The effect of the freezing method on tenderness of frozen and freeze dried chicken meat. *Quick Frozen Foods* (August):50.

Hanson, H.L. and L.R. Fletcher. 1958. Time-temperature tolerance of frozen foods. Part XII, Turkey dinners and turkey pies. *Food Technology* (January):40.

Hanson, H.L. and L.R. Fletcher. 1963. Adhesion of coatings on frozen fried chicken. *Food Technology* (June):115.

Hanson, H.L., A. Campbell, and H. Lineweaver. 1951. Preparation of stable frozen sauces and gravies. *Food Technology* (October):432.

Hanson, H.L., L.R. Fletcher, and H. Lineweaver. 1959. Time-temperature tolerance of frozen foods. Part XVII, Frozen fried chicken. *Food Technology* (April):221.

Hanson, H.L., G.F. Stewart, and B. Lowe. 1942. Palatability and histological changes occurring in New York dressed broilers held at 1.7 °C (35 °F). *Food Research* (March-April):148.

Hoffert, E., *et al.* 1952. The defrosting method and palatability of poultry. *Food Technology* (September):337.

Ingling, A.L. 1978. Electrical terminology, measurements and units associated with the stunning technique in poultry processing plants. *Poultry Science* (January):127.

Ingraham, J.L. 1958. Growth of psychrophilic bacteria. *Journal of Bacteriology* (January):75.

Janky, D.M., A.S. Arafa, J.L. Oblinger, J.A. Koburger, and D.L. Fletcher. 1978. Sensory, physical, and microbiological comparison of brine-chilled, water-chilled, and hot-packaged (no chill) broilers. *Poultry Science* (March):417.

Klose, A.A., *et al.* 1959a. Poultry tenderness. Part I, Influence of processing on tenderness of turkeys. *Food Technology* (January):20.

Klose, A.A., *et al.* 1959b. Time-temperature tolerance of frozen foods. Part XIX, Ready-to-cook cut-up chicken. *Food Technology* (September):477.

Klose, A.A., *et al.* 1960. Effect of laboratory scale agitated chilling of poultry on quality. *Poultry Science* (September):1193.

Klose, A.A., *et al.* 1961. Turkey tenderness in relation to holding in and rate of passage through thawing range of temperature. *Poultry Science* (November):1633.

Klose, A.A., *et al.* 1972. Effect of hot cutting and related factors in commercial broiler processing on tenderness. *Poultry Science* (March):634.

Klose, A.A. and M.F. Pool. 1956. Effect of freezing conditions on appearance of frozen turkeys. *Food Technology* (January):34.

Klose, A.A., A.A. Campbell, and H.L. Hanson. 1960. Stability of frozen ready-to-cook ducks and geese. *Poultry Science* (September):1136.

Klose, A.A., M.F. Pool, and H. Lineweaver. 1955. Effect of fluctuating temperatures on frozen turkeys. *Food Technology* (August):372.

Koonz, C.H. and J.M. Ramsbottom. 1939. A method for studying the histological structure of frozen products, Part I. *Poultry Food Research* (March-April):117.

Kotula, A.W., J.E. Thomson, and J.A. Kinner. 1960. Water absorption by eviscerated broilers during washing and chilling. USDA, Agricultural Marketing Service, *Marketing Research Report* No. 438 (October).

Lentz, C.P. 1961. Thermal conductivity of meats, fats, gelatin, gels, and ice. *Food Technology* (May):243.

Lentz, C.P. and L. van den Berg. 1957. Liquid immersion freezing of poultry. *Food Technology* (April):247.

Lillard, H.S. 1980. Effect on broiler carcasses and water of treating chiller water with chlorine or chlorine dioxide. *Poultry Science* (August):1761.

Love, B.E. and T.L. Goodwin. 1974. Effects of cooking methods and browning temperatures on yields of poultry parts. *Poultry Science* (July):1391.

Mast, M.G. and J.H. MacNeil. 1972. Use of glutaraldehyde as a disinfectant in immersion. *Poultry Science* (May):681.

Morris, G.K. and J.G. Wells. 1970. Salmonella contamination in a poultry processing plant. *Applied Microbiology* (May):795.

Nagel, W.C., *et al.* 1960. Microorganisms associated with spoilage of refrigerated poultry. *Food Technology* (January):21.

Osner, R.L. and D.H. Shrimpton. 1966. Influence of chilling and thawing procedures on the origin of constituents of fluids lost from the chicken carcasses. *British Poultry Science* 7(4):301.

Poultry Products

Peric, M., E. Rossmanith, and L. Leistner. 1971. Verbesserung der microbiologischen qualitat von schlachthanchen durch die sprunhkuhling. *Die Fleischwirtschaft* (April):574.

Pflug, I.J. 1957. Immersion freezing found to improve poultry appearance. *Frosted Food Field* (June):17.

Pippen, E.L. and A.A. Klose. 1955. Effects of ice water chilling on flavor of chicken. *Poultry Science* (September):1139.

Pool, M.F., *et al*. 1959. Poultry tenderness. Part II, Influence of processing on tenderness of chickens. *Food Technology* (January):25.

Risse, L.A. and J.E. Thomson. 1971. Comparative performance and costs of dry ice and water ice in shipping fresh poultry. *Marketing Research Report*, No. 906 (February). Agricultural Research Service, USDA.

Sauter, E.A., C.F. Peterson, and J.F. Parkinson. 1978. Microfloral comparison of fresh and thawed frozen fryers. *Poultry Science* (March):422.

Sekoguchi, S., R. Nakamura, and Y. Sato. 1979. Cysteamine induced changes in the properties of intramuscular collagen and its relation to the tenderness of meat obtained from mature chickens. *Poultry Science* (September):1213.

Shannon, W.G., W.W. Marion, and W.J. Stadelman. 1957. Effect of temperature and time of scalding on the tenderness of breast meat of chicken. *Food Technology* (May):284.

Smith, M.C., Jr., M.D. Judge, and W.J. Stadelman. 1966. A cold shortening effect in avian muscle *Journal of Food Science* (May):450.

Spencer, J.V. and W.J. Stadelman. 1955. Effect of certain holding conditions on shelf life of fresh poultry meat. *Food Technology* (July):358.

Stadelman, W.J. 1970. 28 to 32 °F temperature is ideal for preservation, storage and transportation of poultry. *ASHRAE Journal* (March):61.

Surkiewicz, B.F., *et al*. 1969. A bacteriological survey of chicken eviscerating plants. *Food Technology* (August):80.

Swanson, M.H. and H.J. Sloan. 1953. Some protein changes in stored frozen poultry. *Poultry Science* (July):643.

Sweat, V.E., C.G. Haugh, and W.J. Stadelman. 1973. Thermal conductivity of chicken meat at temperatures between -75 and 20 °C. *Journal of Food Science* (January):158.

Treat, D.W. and T.L. Goodwin. 1973. Effects of sex, size and time of cutting on processing yields and tenderness of broilers. *Poultry Science* (July):1348.

USDA/FSIS. 1990. Poultry products inspection regulations. Chapter 3, Sub-Chapter C, Part 381. Washington, D.C.

Van den Berg, L. and C.P. Lentz. 1958. Factors affecting freezing rate and appearance of eviscerated poultry frozen in air. *Food Technology* (April):183.

Walters, R.E. and K.N. May. 1963. Thermal conductivity and density of chicken breast, muscle and skin. *Food Technology* (June):808.

Webb, J.E. and C.C. Brunson. 1972. Effects of eviscerating line trimming on tenderness of broiler breast meat. *Poultry Science* (January):200.

Webb, J.E. and T.L. Goodwin. 1970. Precooked chicken. Effect of cooking methods and batter formula on yields and storage conditions on 2-thiobarbituric acid values. *British Poultry Science* (January):171.

Wells, F.E., J.V. Spencer, and W.J. Stadelman. 1958. Effect of packaging materials and techniques on shelf life of fresh poultry meat. *Food Technology* (August):425.

Willis, R., B. Lowe, and G.F. Stewart. 1948. Poultry storage at subfreezing temperatures—Comparisons at -10 and $+10$ °F. *Refrigerating Engineering* (September):237.

Wyche, R.C. and T.L. Goodwin. 1974. Hot-cutting of broilers and its influence on tenderness and yield. *Poultry Science* (September):1668.

Wyche, R.C., B.E. Love, and T.L. Goodwin. 1972. Effect of skin removal, storage time and freezing methods on tenderness and rancidity of broilers. *Poultry Science* (March):655.

Yingst, L.D. and T.L. Goodwin. 1971. Freezing methods influence on fat and moisture composition of precooked thighs. *Poultry Science* (May):957.

Ziegler, F. and W.J. Stadelman. 1955. The effect of different scald water temperatures on the shelf life of fresh, non-frozen fryers. *Poultry Science* (January):237.

CHAPTER 14

FISHERY PRODUCTS

FRESH FISHERY PRODUCTS 14.1	FROZEN FISHERY PRODUCTS 14.4
Care Aboard Vessels 14.1	Packaging 14.4
Shore Plant Procedure and Marketing 14.2	Freezing Methods 14.5
Packaging Fresh Fish 14.3	Storage of Frozen Fish 14.8
Fresh Fish Storage 14.3	Transportation and Marketing 14.9

THE major types of fish and shellfish harvested from North American waters and used for food include the following:

1. Groundfish (haddock, cod, whiting, flounder, and ocean perch), lobster, clams, scallops, snow crab, shrimp, capelin, herring, and sardines from New England and Atlantic Canada.
2. Oysters, clams, scallops, striped bass, and blue crab from the Middle and South Atlantic.
3. Shrimp, oysters, red snapper, clams, and mullet from along the Gulf Coast.
4. Lake herring, chubs, carp, buffalofish, catfish, yellow perch, and yellow pike from the Mississippi Valley and the Great Lakes region.
5. Alaska pollock, tuna, halibut, salmon, Pacific cod, various species of flatfish, king and dungeness crab, scallops, shrimp, and oysters from the Pacific Coast and Alaska.
6. Catfish, salmon, trout, oysters, and mussels from aquaculture operations in various parts of North America.

The major industrial fish used for fish meal and oil is menhaden from the Atlantic and Gulf coasts. In addition, the parts of fish not used for human consumption that are removed in fish processing plants are often used to manufacture fish meal and oil.

Fish meal and oil are the principal components of the feed used in the aquaculture of trout and salmon. Meal also is a component of the diets of poultry and pigs. Fish oil is used in margarine, paints, and in the tanning industry. It is also being refined for pharmaceutical purposes.

This chapter deals with the preservation and processing of fresh and frozen fishery products, the care of fresh fish aboard the vessel and ashore, the technology of fish freezing, and present commercial trends in the freezing, frozen storage, and distribution of seafood.

FRESH FISHERY PRODUCTS

CARE ABOARD VESSELS

After they are brought aboard a vessel, fish must be promptly and properly cared for to assure maximum quality. Trawl-caught fish on the New England and Canadian Atlantic coasts, such as haddock and cod, are usually eviscerated, washed, and then iced down in the pens of the vessel's hold. The offshore Canadian fleet, Iceland, the United Kingdom, and other European countries have been icing the fish in boxes for optimum quality. Because of their small size, other groundfish, such as ocean perch, whiting, and flounder, are not eviscerated and are not always washed. Instead, they are iced down directly in the hold of the vessel.

Crustaceans, such as lobsters and many species of crabs, are usually kept alive on the vessel without refrigeration. Warmwater shrimp are beheaded, washed, and stored in ice in the hold of the vessel; on some vessels, however, the catch is frozen either in refrigerated brine or in plate freezers. Coldwater shrimp are stored whole in ice or in chilled sea water, or they may be cooked in brine, chilled, and stored in containers surrounded with ice.

Freshwater fish in the Great Lakes and Mississippi River areas are caught in trap nets, haul seines, or gill nets and sorted according to species into 50- or 100-lb boxes, which are kept on the deck of the vessel. In most cases, fishermen carry ice aboard their vessels, and the fish are landed the day they are caught.

Freshwater fish in the lakes of Canada are iced down in the summertime and stored at collecting stations on the lakes, where they are picked up by a collecting boat with a refrigerated hold. Winter-caught Canadian freshwater fish are usually weather frozen on the ice immediately after catching and are marketed as frozen fish.

Line-caught fish of the Pacific Northwest, such as halibut, which are caught largely by bottom long-line gear, and salmon caught by trolling gear, are eviscerated, washed, and iced in the pens of the vessel. Pacific salmon caught by seines and gill nets for cannery consumption are usually stored whole for several days, either aboard the vessel or ashore in tanks of sea water refrigerated to 30°F. A small but significant volume of halibut is held similarly in refrigerated sea water aboard vessel. Tuna caught offshore by seiners or clipper vessels are usually brine-frozen at sea. However, tuna caught inshore by the smaller trollers or seiners are often iced in the round or refrigerated with a brine spray.

Fish raised by aquaculture farms are usually harvested and sold as required for the fresh fish market. They are usually shipped in containers in which they are surrounded by ice.

Icing of Fish

Fish lose quality because of bacterial or enzymatic activity or both. Reduction of storage temperature retards these activities significantly, thus delaying spoilage and autolytic deterioration.

Low temperatures are particularly effective in delaying growth of psychrophilic bacteria, which are primarily responsible for the spoilage of nonfatty fish. The shelf life of species such as haddock and cod is doubled for each 7 to 10°F lowering of storage temperature within the range of 60 to 30°F.

To be effective, ice must be clean when used aboard the vessel. Bacteriological tests on ice in the hold of a fishing vessel showed bacterial counts as high as 5 billion per gram of ice. These results indicate that chlorinated or potable water should be used in making the ice at the ice plant, ice should be stored under sanitary conditions, and unused ice should be discarded from the vessel at the end of each trip.

Both flake ice and crushed block ice are used aboard fishing vessels, although flake ice is more common because it is cheaper to produce and easier to handle mechanically.

The preparation of this chapter is assigned to TC 11.1, Meat, Fish, and Poultry Products.

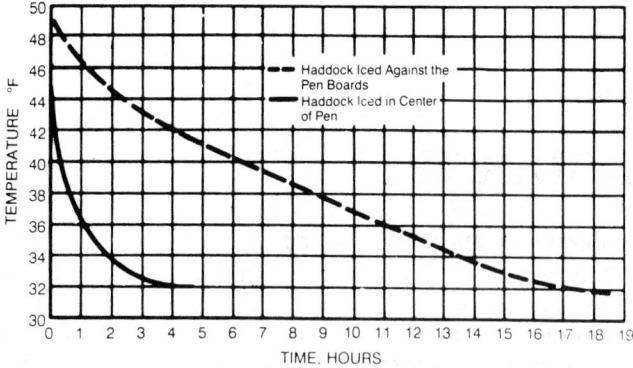

Fig. 1 Cooling Rate of Properly and Improperly Iced Haddock

The amount of ice used aboard vessels varies with the particular fishery and vessel; however, it is essential to provide sufficient ice around the fish to obtain a proper cooling rate (see Figure 1). A common ratio of ice to fish used in bulk icing on New England trawlers is one part ice to three parts fish. Experiments on English trawlers in boxing fish at sea with one part ice to two parts fish demonstrated improved quality in the landed fish, and as ice has become more plentiful and less costly relative to the value of fish, the ratio of ice to fish continues to increase. Mechanical refrigeration is employed in some vessels to retard melting of ice en route to the fishing grounds; however, the hold temperature must be controlled after fish are taken to allow melting of the ice for effective cooling of the fish.

Salt Water Icing

Iced fish storage temperatures must be maintained close to the freezing point of fish. To obtain lower ice temperatures, the freezing point may be depressed by adding salt to the water from which the ice is made. Adequate amounts of ice made from a 3% solution of sodium chloride brine will maintain a storage environment of about 30°F. Tests conducted on the storage of haddock in salt water ice aboard a fishing vessel showed that, under parallel conditions, fish iced with salt water ice cooled faster and to a lower temperature than fish iced in plain ice. However, the salt water ice melted faster than the plain ice because of its lower latent heat and the greater temperature differential of the salt water ice. Therefore, once the salt water ice melted, the fish stored in this ice rose to a higher temperature than those stored in plain ice. Since it is not always possible to renew ice on fish at sea, sufficient quantities of salt water ice must be used initially to make up for the faster melting rate.

In making ice from water containing a preservative, rapid freezing or the use of a stabilizing dispersant, or both, is essential to prevent migration of the additive to the center of the ice block. This problem is not encountered in flake ice, because flake ice machines are designed to freeze water rapidly into thin layers of ice, thus fixing additive within the ice flakes. Chapter 33 describes the manufacture of flake ice in more detail.

Use of Preservatives

In the United States and Canada, the use of antibiotics in ice or in dips for treatment of whole or gutted fish, shucked scallops, and unpeeled shrimp is not permitted by regulation.

Storage of Fish in Refrigerated Sea Water

Refrigerated sea water (RSW) is used commercially for preserving fish. On the Pacific Coast, substantial quantities of net-caught salmon are stored in RSW aboard barges and cannery tenders for delivery to the canneries. On the East and Gulf coasts, RSW installations on fishing vessels are used for chilling and holding menhaden and industrial species needed for production of meal, oil, and pet food. On the east and west coasts of Canada, RSW installations are used for chilling and holding herring and capelin, which are processed on shore for their roe. Other more limited applications of RSW include holding Pacific halibut and Gulf shrimp aboard a vessel; chilling and holding Maine sardines in shore tanks for canning, and short-term holding of Pacific groundfish in shore tanks for later filleting.

With groundfish and shrimp, RSW works well for short-term storage (2 to 4 days), but it is not suitable for longer periods because of the excessive salt uptake, accelerated rancidity, poorer texture, and increased bacterial spoilage that may result. These problems can be partially overcome by introducing CO_2 gas into the RSW; the storage life of some species of fish can be increased by about one week by holding in RSW saturated with CO_2. Additional benefits of RSW are the reduction of handling that results from the bulk storage of the fish and the reduction of pressure on the fish as a result of buoyancy, faster cooling, and lower storage temperature.

In many RSW systems, the refrigeration effect is provided by ammonia flowing through external chillers or pipe coils located within the tanks. Best results have been achieved with external chillers.

Boxing at Sea

There are many advantages to using containers or boxes instead of bulk storage aboard fishing vessels. Known as *boxing at sea*, the use of containers reduces the pressure effects on the fish while they are stowed in the vessel's hold. Because significant reductions in handling during and subsequent to unloading are possible, mechanical damage and product temperature rise may be virtually eliminated and handling costs reduced. Fish can be sorted into boxes by size and species as soon as they are caught. They lend themselves more readily to mechanized handling such as machine filleting, because they are generally firmer and of more nearly uniform shape; fillet yields are generally better than they are with bulk-stored fish.

Boxing at sea is not generally practiced in the United States, except by some inshore vessels. The principal problems associated with converting a fishing vessel from bulk storage to boxed storage are the increased labor required by the crew for handling the boxes, reduced hold capacity, and a relatively large investment for boxes. Many fisheries have difficulties in working out the logistics for assuring the prompt return of properly cleaned boxes to the vessel. Most of these problems have been solved in European fleets, the Canadian offshore fleet, and the South American lake fishing fleets. The use of nonreturnable containers for boxing at sea simplifies logistics and reduces initial capital outlay; it has proved justifiable in some U.S. fisheries.

Reusable containers for boxing at sea are usually made of plastic. Careful icing is necessary to minimize the surface area of fish in contact with the box. Plastic boxes provide more heat transfer resistance than aluminum boxes in vessels with uninsulated fish holds and for in-plant storage prior to processing.

All fish boxes must be equipped with drains, preferably directed outside the boxes on the bottom of a stack.

SHORE PLANT PROCEDURE AND MARKETING

Proper use of ice and adherence to good sanitary practices ensure maintenance of iced fish freshness during unloading from the vessel, at the shore plant, during processing, and throughout the distribution chain. Fish landed in good quality will spoil rapidly if these practices are not carried out.

Fish unloaded from the vessel usually are graded by the buyer for species, size, and minimum quality specification. A price is based in part on the quality in relation to market requirements. Fish also may be inspected by local and federal regulatory agencies for wholesomeness and sanitary condition. Organoleptic

Fishery Products

Table 1 Organoleptic Criteria of Quality Fish

Factor	Good Quality	Poor Quality
Eyes	Bright, transparent, often protruding	Cloudy, often pink, sunken
Odor	Sweet, fishy, similar to seaweed	Stale, sour, presence of sulfides, amines
Color	Bright, characteristic of species, sometimes pearl essence at correct light angles	Faded, dull
Texture	Firm, may be in rigor, elastic to finger pressure	Soft, flabby, little resilience, presence of fluid
Belly	Walls intact, vent pink, normal shape	Often ruptured, bloated, vent brown, protruding
Organs (including gills)	Intact, bright, easily recognizable	Soft to liquid, grey homogeneous mass
Muscle tissue	White or characteristic of species and type	White flesh pink to gray, spreading of blood color around backbone

criteria are most important for evaluating quality; however, there is a growing acceptance, particularly in Canada and some European countries, of objective chemical and physical tests as indexes of quality loss or spoilage. Organoleptic quality criteria vary somewhat among species, but the information in Table 1 can be used as a general guide in judging the quality of whole fish.

In New England and the Canadian Atlantic Provinces, groundfish unloaded from the vessel may be placed in boxes and trucked to the shore plant or conveyed directly from the hold or deck to the shore plant. Pneumatic conveyors are sometimes used to unload groundfish and shrimp from vessels, although their use is being phased out in the groundfish industry. Single- or double-wall insulated boxes are normally used for transporting fish. Wooden boxes are rarely used because they are a source of microbiological contamination. Ice should be applied generously to each box of fish, even if the period prior to processing is only a few hours. Fish in the plant awaiting processing for longer than those few hours should be iced heavily and stored in insulated containers or in single-wall boxes in a chill room refrigerated to 35 °F. If refrigerated facilities are not available, the boxes of fish should be kept in a cool section of the plant that is clean, sanitary, and has adequate drainage.

Large boxes of plastic, resin-coated plywood, or reinforced fiberglass that hold up to 1,000 lb of fish and ice, are used by some plants in preference to icing fish overnight on the floor. These *tote* boxes are moved and stacked by forklift, can be used for trucking fish to other plants, and make better use of plant floor space. Generally, fish awaiting processing should not be kept longer than overnight.

Fresh fish are marketed in different forms: fillets, whole fish, dressed-head on, dressed-headed (head removed), or, in some instances, steaks. The method of preparing fish for marketing depends largely on the species of fish and on consumer preference. For example, groundfish such as cod and haddock usually are marketed as fillets or as dressed-headed fish. Freshwater fish such as catfish and bullheads are usually dressed and skinned; lake trout are not skinned, but are merely dressed; and lake herring are marketed in dressed, round, or filleted form.

PACKAGING FRESH FISH

Most fresh fish is packaged in institutional containers of 5 to 35 lb capacity at the point of processing. Polyethylene trays, steel cans, aluminum trays, plastic-coated solid boxes, wax-impregnated corrugated fiberboard boxes, foamed polystyrene boxes, and polyethylene bags are used.

Fresh fish is often packaged when it still contains process heat from wash water. In these cases, it is advantageous to use a packaging material that is a good heat conductor. The fresh fish industry makes little use of controlled prechilling equipment in packaging systems. As a result, product temperatures may never reach the optimum level subsequent to packaging. Traditionally, institutional fresh fish travels packed in wet ice; in this case, it may cool to the proper level in transit even if process heat is initially present. However, there is a trend toward the use of leaktight shipping containers for fresh fish, because modern transportation equipment is not designed to handle wet shipments. Also, some customers want to avoid the cost of transporting ice and demand a product that is uniformly chilled to within the temperature range of 32 to 36 °F when it reaches their door. Shippers who make use of leaktight shipping containers will have to upgrade their product temperature control systems to assure that the fish reaches ice temperature prior to packaging. Rapid prechilling systems that result in crust freezing can be applied to some fresh seafood products, but this practice must be used with discretion since partial freezing produces deleterious effects on quality.

Some general requirements for institutional containers for holding products such as fillets, steaks, and shucked shellfish are: (1) sufficient rigidity to prevent pressure exerted on the product, even when containers are stacked or heavily covered with ice; and (2) measures to prevent ice-melt water from contaminating the product. Some containers have drains permitting the drip associated with the fish itself to run off. Others are sealed and may be gastight, which increases shelf life. One problem associated with sealed containers is the emission of a strong odor when the package is first opened. Although this odor may be foul, it soon dissipates and has no adverse effect on quality. Dressed or whole fish may be placed in direct contact with ice in a gastight container.

Leaktight shipping containers are used with nonrefrigerated transportation systems, such as air freight, and consequently require insulation. Foamed polystyrene is particularly suited to this application. For typical air freight shipments, the most economical thickness of insulation is between 1 and 2 in. To maintain product temperature in transit, shippers use either dry ice, packaged wet ice, packaged gel refrigerant, or wet ice with absorbent padding in the bottom of the container. Foamed polystyrene containers may be of molded construction, or of the composite type in which foam inserts and a plastic liner are used with a corrugated fiberboard box.

At the retail level, fresh fish may be handled in two ways. Stores with service counters display fish in unpackaged form. In some markets that do not have service counters, however, fish must be packaged prior to displaying for sale. Both types of outlets receive the product in institutional containers. If the fish is prepackaged at the market, high labor and packaging costs may be incurred, and the temperature of the product is likely to rise. Often, relatively warm fish is placed in a foam tray, wrapped, and displayed in a meat case, the temperature of which may be 40 °F or more. This drastically reduces the shelf life of the fish. Centralized prepackaging at the point of initial processing appears to have many important advantages over the present system. A number of retail chains have their suppliers prepackage the product under controlled temperature and sanitation conditions.

FRESH FISH STORAGE

The maximum storage life of fish varies with the species. In general, the storage life of East and West Coast fish properly iced and stored in refrigerated rooms at 35 °F is 10 to 15 days, with a maximum of 15 days. This depends on the condition of the fish when it is unloaded from the boat. Generally, freshwater fish properly iced in boxes and stored in refrigerated rooms may be held for only seven days. Both figures are from the time the fish is landed and processed to the time of consumption.

Cold storage facilities for fresh fish should be maintained at about 35°F and over 90% relative humidity. Air velocity should be limited to control ice loss. Temperatures less than 32°F retard ice melting and can result in excessive fish temperatures. This is particularly important when storing round fish such as herring, which generate heat from autolytic processes.

Floors should have adequate drainage with ample slope toward drains. All inside surfaces of the cold storage room should be easy to clean and have the capability to withstand the corrosive effects of frequent washings with antimicrobial compounds.

Radiopasteurization of Fresh Seafood

Ionizing radiation can double or triple the normal shelf life of refrigerated, unfrozen fish and shellfish stored at 33°F (see Table 2). No off-odors or adverse nutritional or other changes are imparted to the product as a result of the radiation treatment. However, irradiation of fish is not common.

Table 2 Optimal Radiation Dose Levels and Shelf Life at 33°F for Some Species of Fish and Shellfish

Species	Optimal Irradiation Dose—Rads Air Packed	Shelf Life, Weeks
Oysters—shucked, raw	200,000	3-4
Shrimp	150,000	4
Smoked chub	100,000	6
Yellow perch	300,000	4
Petrale sole	200,000	2-3 (4-5 when vac pac)
Pacific halibut	200,000	2 (4 when vac pac)
King crab meat	200,000	4-6
Dungeness crab meat	200,000	3-6
English sole	200,000-300,000	4-5
Soft-shell clam meats	450,000	4
Haddock	150,000-250,000	3-4
Pollock	150,000	4
Cod	150,000	4-5
Ocean perch	250,000	4
Mackerel	250,000	4-5
Lobster meat	150,000	4

Modified Atmosphere Packaging

A product environment with modified levels of nitrogen, carbon dioxide, and oxygen can curtail the growth of bacterial spoilage and extend shelf life of fresh fish. For example, whole haddock stored in a 25% carbon dioxide atmosphere from the time it is caught keeps about twice as long as in air. However, a modified atmosphere does not inhibit all microbes, and spoilage bacteria, because of their great number, usually restrict the growth of the few pathogenic bacteria present. Therefore, obvious spoilage is a safeguard against eating fish that may have dangerous levels of pathogenic bacteria.

Because modified atmosphere packaging can be a safety hazard, it is being introduced slowly in several countries, under close monitoring by regulatory agencies. This type of packaging requires a complete knowledge of regulations and a good control system that maintains proper temperature and sanitation levels.

FROZEN FISHERY PRODUCTS

The production of frozen fishery products varies with geographical location and includes primarily the production of groundfish fillets, scallops, breaded precooked fish sticks, breaded raw fish portions, fish roe, and bait and animal food in the northeastern states and in Atlantic Canada; round or dressed halibut and salmon, halibut and salmon steaks, groundfish fillets, surimi, herring roe, and bait and animal food in the northwestern states and in British Columbia; halibut, groundfish fillets, crab, salmon, and surimi in Alaska; shrimp, oysters, crabs, and other shellfish and crustacea in the Gulf of Mexico and South Atlantic states; and round or dressed fish in the areas bordering on the Great Lakes.

The fish obtained from these areas differ considerably in both physical and chemical composition. For example, cod or haddock are readily adaptable to freezing and have a comparatively long storage life, while other fatty species, such as mackerel, tend to become rancid during frozen storage and, therefore, have a relatively short storage life. The differences in composition of many species of fish and in marketing requirements necessitate consideration of the specific product with regard to quality maintenance and methods of packaging, freezing, cold storage, and handling.

Temperature is the most important factor limiting the storage life of frozen fish. At temperatures below freezing, bacterial activity as a cause of spoilage is limited. However, even fish frozen within a few hours of catching and stored at −20°F will deteriorate very slowly until it becomes unattractive to look at and unpleasant to eat.

Fish proteins are permanently altered during freezing and cold storage. This denaturation occurs quickly at temperatures not far below freezing, and even at 0°F fish deteriorates rapidly. Badly stored fish are easily recognized; the thawed product is opaque, white, and dull, and juice can easily be squeezed from it. While the properly stored product is firm and elastic, poorly stored fish is spongy and, in very bad cases, the flesh may break up. Instead of the succulent curdiness of cooked fresh fish, cooked samples at first have a wet and sloppy consistency, and on further chewing become dry and fibrous.

Among other factors that determine how quickly quality deteriorates in cold storage are the initial quality and composition of the fish, protection of the fish from dehydration, the freezing method, and the environment during storage and transport. These factors are reflected in four principal phases of frozen fish production and handling—packaging, freezing, cold storage, and transportation.

PACKAGING

Materials for packaging frozen fish are similar to those for other frozen foods. A package should: (1) be attractive and appeal to the consumer, (2) protect the product, (3) allow rapid, efficient freezing and ease of handling, and (4) be cost effective.

Package Considerations in Freezing

Refrigeration equipment and packaging materials are frequently purchased without considering the effect of the package size on freezing rate and efficiency. For example, a thin consumer package results in a faster rate of product freezing, lower total freezing costs, higher handling costs, and higher packaging material costs; a thicker institutional-type package results in a slower rate of product freezing, higher freezing costs, lower handling costs, and lower packaging material costs.

Tests indicate that the time required to freeze packaged fish fillets in a plate freezer is directly proportional to the square of the package thickness. Thus, if it takes 3 h to freeze packaged fish fillets 2 in. thick, it will take about 4.7 h to freeze packaged fish fillets 2.5 in. thick. The insulating effect of the packaging material, the fit of the product in the package, and the total surface area of the package must be considered. A packing material with low moisture-vapor permeability has an insulating effect, which increases freezing time and cost.

The rate of heat transfer through packaging material is inversely proportional to its thickness; therefore, a packaging material

Fishery Products

should be used which is (1) thin enough to produce rapid freezing and adequate moisture-vapor barrier in frozen storage, and (2) thick enough to withstand heavy abuse. Aluminum foil cartons and packages offer an advantage in this regard.

Proper fit of package to product is essential; otherwise, the insulating effect of the airspace formed will reduce the freezing rate of the product and increase freezing cost. The surface area of the package is also important because of its relation to the size of the freezer shelves or plates. Maximum use of freezer space can be obtained by designing the package so that it fits the freezer properly. However, these factors often cannot be changed and still meet customer requirements for a specific package.

Package Considerations for Frozen Storage

Fish products lose considerable moisture and become tough and fibrous during frozen storage unless a package with low moisture-vapor permeability is specified. The package in contact with the product must also be resistant to oils or moisture exuded from the product, or rancidity of the oils and softening of the package material will occur. The package must fit the product tightly to minimize airspaces and thereby reduce moisture migration from the product to the inside surfaces of the package.

Types of Packages

Packaging materials consist of paperboard cartons coated with various waterproofing materials, or cartons laminated with moisture-vapor-resistant films and heat-sealable overwrapping materials with a low moisture-vapor permeability. Paperboard cartons are usually made of a bleached sulfate stock, coated with a suitable fortified wax, polyethylene, or other plastic material, or laminated with aluminum foil or other moisture-vapor-resistant film. Fortified wax-coated paperboard cartons are used for most fish products. In some cases, where the added protection and consumer appeal warrant the extra cost, paperboard cartons laminated with aluminum foil are recommended for specialty items.

Overwrapping materials should be highly resistant to moisture transmission, inexpensive, heat sealable, adaptable to machinery application, and attractive in appearance. Various types of hot melt coated waxed paper, cellophane, polyethylene, and aluminum foil are available in different forms and laminant combinations that make possible the selection of an overwrapping material best suited to each product.

Consumer packages. These usually hold less than 1 lb and are generally printed, bleached paperboard, coated with polyethylene, and closed with adhesive. Fish sticks and portions, shrimp, scallops, crabmeat, precooked dinners, and entrees are packaged in this way. In the case of dinners and entrees, rigid plastic pressboard or aluminum trays are used inside the printed paperboard package. Rigid plastic or pressboard packages are becoming more common because they are better for microwave cooking. The packaging of these products is normally mechanized.

Materials such as polyethylene combined with cellophane, polyvinylidene chloride, or polyester and combinations of other plastic materials are used with high-speed automatic packaging machines to package shrimp, dressed fish, fish fillets, fish portions, and fish steaks prior to freezing. In some instances, tearing of the wrapping material by fins protruding from the fish has been a problem. Otherwise, this method of packaging is satisfactory and affords the product considerable protection against dehydration and rancidity at a comparatively low cost. This packaging method has also created new markets for merchandising frozen fish products. Boil-in-bag type pouches made of polyester-polyethylene and combinations of foil, polyethylene, and paper are used for packaging shrimp, fish fillets, and entrees. These packages are also suitable for microwave cooking.

Institutional packages. The 5-lb and larger cartons used in the institutional trade are almost exclusively constructed of bleached paperboard that has been waxed or polyethylene coated. Folding cartons with self-locking covers, full-telescoping covers, or glued closures are used. Often the cartons are packaged inside a corrugated master carton or are shrink-wrapped in polyethylene film.

Products such as fish fillets and steaks are individually wrapped in cellophane or other moisture-vapor-resistant film, then packed in the carton. Fish, such as headed and dressed whiting and scallop meats, are packed into the carton and covered with a sheet of cellophane. The cover is then put on and the package is frozen upside down in the freezer. Raw, unbreaded products, such as shrimp, scallops, fillets, and steaks, are sometimes individually quick frozen (IQF) prior to packaging. When IQF, they can be glazed to enhance moisture retention. This method is preferred over freezing after packaging, because it leads to a product that is more convenient to handle and sometimes obviates the need to thaw the fish prior to cooking. For all institutional frozen fish, the trend is toward printed paperboard folding cartons coated with moisture-vapor-resistant materials instead of waxed paper or cellophane overwrap. Some frozen fish products and seafood entrees destined for institutional markets are packaged in aluminum trays or in rigid plastic trays so they may be heated within the package.

FREEZING METHODS

Product characteristics, such as size and shape, freezing method, and the rate of freezing, affect the quality, appearance, and cost of production.

Quick freezing of fish offers the following advantages:

- Chills the product rapidly, preventing bacterial spoilage
- Facilitates rapid handling of large quantities of product
- Makes use of conveyors and automatic devices practical, thus materially reducing handling costs
- Promotes maximum use of the space occupied by the freezer
- Produces a packaged product of uniform appearance, with a minimum of voids or bulges

Thus, quick freezing is predominantly employed in the fishing industry today, although sharp (slow) freezing is still used to a limited extent. See also Chapters 29 and 30 of the 1989 ASHRAE *Handbook—Fundamentals* and Chapter 9 of this volume.

Sharp Freezing of Fish

Fish frozen by this method are placed on shelves of refrigerated pipe coils in which ammonia, R-12, or chilled brine is circulated to provide the necessary refrigeration effect. Stainless steel or aluminum pans or plates are often placed over the coils to form flat surfaces on which the fish rest. The heat transfer between the refrigerant in the pipes and the fish at the point of contact is relatively high. However, the overall heat transfer rate is quite low because only a small surface area of the product is in contact with the pipes and is cooled largely by the natural circulation of air within the freezer. This type of freezing has largely disappeared from North American fish-processing plants.

Blast Freezing of Fish

Blast freezers for fishery products are generally small rooms or tunnels in which cold air is circulated by one or more fans over an evaporator and around the product to be frozen contained on racks or shelves. A refrigerant such as ammonia, a halocarbon, or brine flowing through a pipe coil evaporator furnishes the necessary refrigeration effect.

Static pressure in these rooms is considerable, and air velocities average between 500 and 1500 fpm, with 1200 fpm being common. Air velocities between 500 and 1000 fpm give the most economical freezing. Lower air velocities result in slow product freezing, and higher velocities increase unit freezing costs considerably.

Some factories have continuous blast freezers, in which conveyors move fish continuously through a blast room or tunnel.

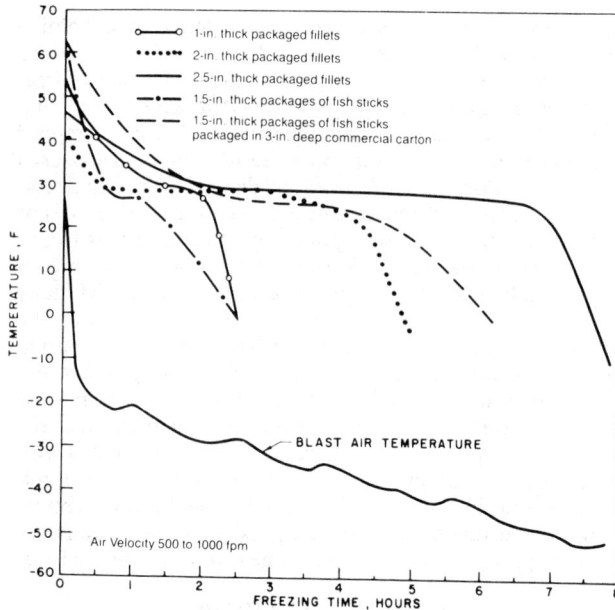

Fig. 2 Freezing Time of Fish Fillets and Fish Sticks in a Tunnel-Type Blast Freezer (Air Velocity 500 to 1000 fpm)

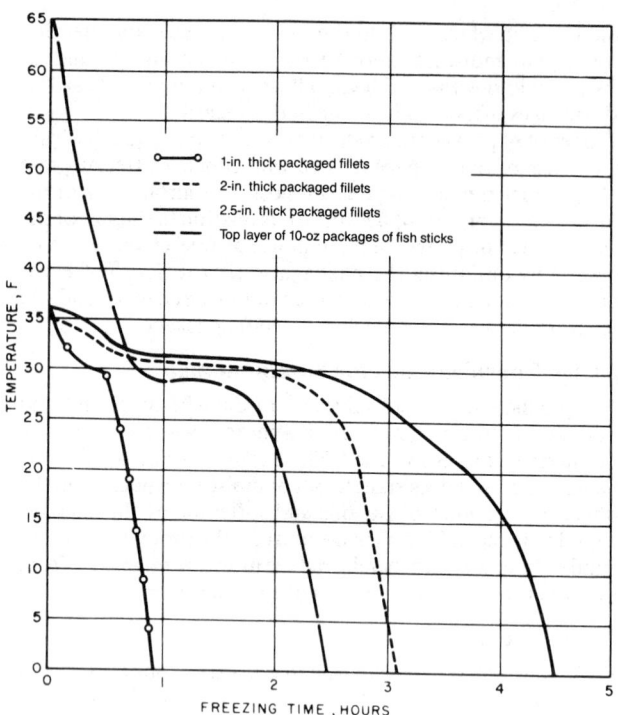

Fig. 3 Freezing Time of Fish Fillets and Fish Sticks in a Plate Freezer

These freezers are built in a number of configurations, including a single pass through the tunnel, multiple passes, spiral belts, and moving trays or carpets. The configuration and type of conveyor belt or freezing surface depend on the type and quantity of the product to be frozen, the space available to install the equipment, and the capital and operating costs of the freezer.

Batch loaded blast freezers are used for freezing shrimp, fish fillets, steaks, scallops, or breaded precooked products packed in institutional packages; round, dressed, or panned fish; and shrimp, clams, or oysters packed in metal cans.

Conveyor-type blast freezers are widely used to freeze products prior to packaging. These products include all types of breaded, precooked seafoods; IQF fillets, loins, tails, and steaks, scallops, and shrimp; and raw, breaded fish portions. In the case of portions, which are sliced or sawed from blocks, the function of the blast freezer is to harden the batter and breading prior to packaging and to lower the temperature of the frozen fish again to storage temperatures if it has been tempered for slicing.

Dehydration of product or freezer burn may occur in freezing unpackaged whole or dressed fish in blast freezers unless the velocity of air is kept to about 500 fpm and the period of exposure to the air is controlled. Consumer packages of fish fillets or fish-fillet blocks requiring close dimensional tolerances undergo bulging and distortion during freezing unless restrained. In blast rooms or tunnels, where the product is frozen on trucks, the use of specially designed freezer trucks enabling distribution of pressure on the surfaces of the package will remedy this condition. It is difficult to control the expansion of the product on conveyor installations. The freezing times for various sizes of packaged fishery products are shown in Figure 2.

Plate Freezing of Fish

In the multiplate freezer, freezing is accomplished by refrigerant flowing through connected passageways in the horizontal movable plates stacked vertically within an insulated cabinet or in an insulated room. The plate freezer is used extensively in the freezing of fishery products packaged in consumer cartons and in 5- and 10-lb institutional-type cartons. Fish to be plate frozen should be properly packaged to keep airspaces in the package to an absolute minimum. Spacers should be used between the plates during freezing to prevent crushing or bulging of the package. For most products, the thickness of the spacers should be about 0.03 to 0.06 in. less than that of the package.

Where very close package tolerances are required, as in the manufacture of fish fillet blocks, a metal frame or tray is used to hold the packages of fish during freezing. The frame or tray is generally the same width as the package and the length of one or two blocks. It must be rigid enough to prevent bulging and to hold the fish block to close dimensional tolerances. This is sometimes accomplished with rigid spacers in order to limit the weight and cost of the tray.

Fish blocks are available in two common sizes: (1) 16.5 lb (19 by 10 by 2.5 in.) and (2) 18.5 lb (19 by 11.5 by 2.5 in.). Other blocks are sized for special applications. The fish may be packed in the block with the long dimension of the fillets generally running the length of the frame (long-pack) or the width of the frame (cross-pack). The orientation depends on the eventual cutting pattern and type of cutting used to convert the block into a finished product.

A tray is not necessary for other packaged seafoods such as shrimp, fillets, fish sticks, or scallops, where close package tolerances are not as essential. Therefore, an automatic continuous plate freezer with properly sized spacers can be satisfactorily used for these products.

The plate freezer provides rapid and efficient freezing of packaged fish products. The freezing time and energy required for freezing packaged fish sticks is greater than that for freezing packaged fish fillets, because heat transfer is slowed by the airspace within the package. The ratio of energy requirements to freezing unit mass of product increases with thickness. The freezing rates of consumer and institutional size packages of fish fillets and fish sticks are shown in Figure 3.

Immersion Freezing of Fish

Immersion in low-temperature brine was one of the first methods used for quick-freezing fishery products. A number of

Fishery Products

of direct immersion freezing machines were developed for freezing whole or panned fish. These machines were generally unsuitable for freezing packaged fish products, which make up the bulk of frozen fish production and have been replaced by methods employing air cooling, contact with refrigerated plates or shelves, or combinations of these methods.

Immersion freezing is used primarily for the freezing of tuna at sea and, to a lesser extent, for freezing shrimp, salmon, and dungeness crab. Extensive research has been conducted on brine freezing of groundfish aboard the vessel, but this method is not in commercial use.

An important consideration in immersion freezing of fish is selection of a suitable freezing medium. The medium should be nontoxic, acceptable to public health regulatory agencies, easy to renew, inexpensive, and have a low freezing temperature and viscosity. It is difficult to obtain a freezing medium that meets all these requirements. Sodium chloride brine and a mixture of glucose and salt in water are acceptable media. The glucose reduces salt penetration into the fish and provides a protective glaze.

Immersion freezing in R-12 spray has been used for handling IQF products such as raw, peeled and cooked shrimp. Its advantage is that it produces the IQF shrimp with the use of very little labor and no dehydration. However, this method of freezing releases some CFC into the atmosphere, so its use will probably be discontinued in the near future.

Liquid nitrogen spray and CO_2 are coming into wider use for IQF seafood products such as shrimp. The quality of fish frozen by these methods is good and, although the cost per unit mass is high, there is virtually no weight loss from dehydration, and there are space and equipment savings. In freezing fish in liquid nitrogen, the fish should not be immersed directly in the liquid nitrogen since this will cause the flesh to shatter and rupture.

Immersion freezing of shrimp. Several companies are now producing shrimp immersion freezing systems that work quite satisfactorily in the Gulf industry. Compact refrigeration systems for blast or contact freezing of shrimp at sea have also been installed on fishing vessels and are proving satisfactory.

Immersion freezing of tuna. Most tuna harvested in the United States is brine-frozen aboard the fishing vessel. Freezing at sea enables the vessel to make extended voyages and return to port with a full payload of high quality fish.

Tuna are frozen in brine wells, which are lined with galvanized pipe coils on the inside. Direct expansion of ammonia into the evaporator coils provides the necessary refrigeration effect. The wells are so designed that tuna can first be precooled and washed with refrigerated sea water and then frozen in an added sodium chloride brine. After the fish are frozen, the brine is pumped overboard and the tuna kept in 10°F dry storage. Prior to unloading, the fish are thawed in 30°F brine on the vessel. In some cases, the fish are thawed in tanks at the cannery. Therefore, if the fish are thawed ashore, thawing on the vessel is not required beyond the stage needed to separate those fused together in the vessel's wells.

Sometimes tuna are held in the wells for a long period prior to freezing or are frozen at a very slow rate because of high well temperatures caused by overloading, insufficient refrigeration capacity, or inadequate brine circulation. These practices have a detrimental effect on product quality, especially on the smaller fish which are more subject to salt penetration and quality changes. Tuna that is not promptly and properly frozen may undergo excessive changes, absorb excessive quantities of salt, and may even be bacteriologically spoiled when landed. Some freezing times for tuna of various sizes are shown in Figure 4.

Freezing Fish at Sea

Freezing fish at sea has found increasing commercial application in leading fishery nations such as Japan, the Soviet Union, the United Kingdom, Norway, Spain, Portugal, Poland, Iceland, and the United States. Including freezer trawlers, factory ships, and refrigerated transports in fisheries, hundreds of large freezer vessels are operating throughout the world. U.S. factory freezer trawlers, factory surimi trawlers, and floating factory ships supplied by catcher vessels operate off Alaska. These vessels process mainly Alaskan pollock, cod, and flounder, although they do process other species.

Freezing groundfish at sea has not developed in the northeastern United States largely because fresh fish commands a better price than does frozen fish. For the same reason, east coast United States producers avoid putting their product into frozen packs if they can sell it fresh. Hence, much of the frozen fish used in the United States, with the exception of Alaskan fish, is imported from other countries.

Where used, the factory vessel is equipped to catch, process, and freeze the fish at sea and to use the waste material in the manufacture of fish meal and oil. A large European factory vessel measures 280 ft in length, displaces 3700 tons, and is equipped to stay at sea for about 80 days without being refueled. About 65 to 100 people are required to operate the vessel and to process and handle the fish. On most vessels of this type, contact-plate freezers are used. The freezers can freeze about 30 tons of fish per day, and the total capacity of the frozen fish hold may be as high as 750 tons.

Because the factory trawler stays at sea for long periods, it can fully use its space for storing fish. However, because there is a limited amount of labor available on these vessels, the packs are generally restricted to the less labor-intensive types.

The freezer trawler was designed to resolve the disadvantages associated with factory freezer vessels. It is a smaller vessel equipped to freeze fish in bulk for later thawing and processing ashore. Freezer trawlers use vertical plate freezers to freeze dressed fish in blocks of about 100 lb.

Some countries use freezer trawlers to supplement the raw material for shore-based processing plants producing frozen fish products. This allows the trawlers to fill their holds in distant waters and transport the fish to home base, where the fish becomes frozen raw material that is held in storage until required for processing. In some cases, trawlers have been designed as dual fisheries, *i.e.,* fishing and freezing groundfish blocks during part

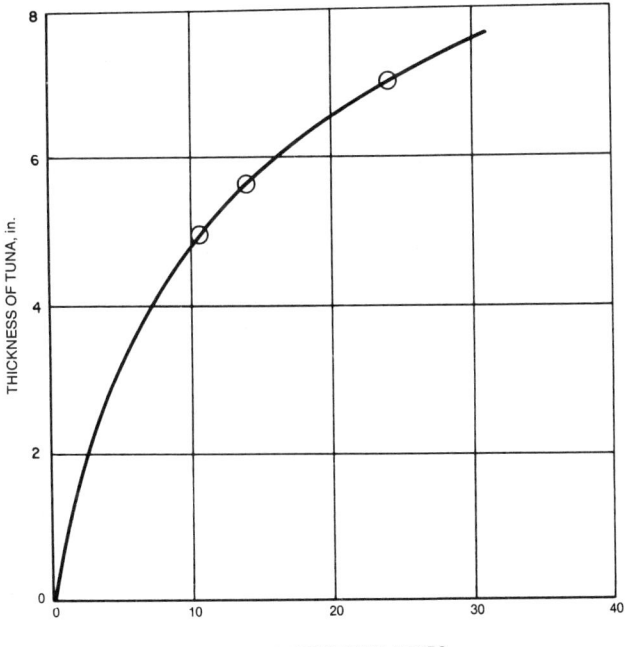

Fig. 4 Freezing Time for Tuna Immersed in Brine

of the year and catching, processing, and freezing Northern shrimp for the rest of the year.

STORAGE OF FROZEN FISH

Fishery products may undergo undesirable changes in flavor, odor, appearance, and texture during frozen storage. These changes are attributable to dehydration (moisture loss) of the fish, oxidation of the oils or pigments, and enzyme activity in the flesh. The rate at which these changes occur depends on: (1) the composition of the species of fish; (2) the level and constancy of storage room temperature and humidity; and (3) the protection afforded the product through the use of suitable packaging materials and glazing compounds.

Composition

The composition of a particular species of fish affects its frozen storage life considerably. Fish having a high oil content, such as some species of salmon, tuna, mackerel, and herring, have a comparatively short frozen storage life because of the development of rancidity as a result of the oxidation of the oils and pigments in the flesh. Certain fish, such as sablefish, are quite resistant to oxidative deterioration in frozen storage, despite their high oil content. The development of rancidity is less pronounced in fish with a low oil content. Therefore, lean fish such as haddock and cod, if handled properly, can be kept in frozen storage for many months without serious loss of quality. The relative susceptibility of various species of fish to oxidative changes during frozen storage is shown in Table 3.

Table 3 Relative Susceptibility of Representative Species of Fish to Oxidative Changes in Frozen Storage

Severe	Moderate	Minor	Very Slight
Pink salmon	Chum salmon	Cod	Yellow pike
Rockfish	Coho salmon	Haddock	Yellow perch
Lake chub	King salmon	Flounder	Crab
Whiting	Halibut	and sole	Lobster
	Ocean perch	Sablefish	
	Herring	Oysters	
	Mackerel		
	Tuna		
	Lake herring		
	Sheepshead		
	Lake trout		

Temperature

The loss of quality of frozen fish due to storage depends primarily on storage temperature and the length of time in storage. Fish stored at −20°F will have a shelf life of more than a year. In Canada, the Department of Fisheries recommends a storage temperature of −15°F or lower. Storage above −10°F, even for a short period, will result in rapid loss of quality. Time-temperature tolerance studies show that frozen seafoods have memory; that is, each time they are subjected to high temperatures or poor handling practices, the loss in quality is recorded. When the product is finally thawed, the total effect of each exposure to high temperatures or other mistreatments is reflected in the quality of the product at the consumer level. Continuous storage at temperatures lower than −15°F reduces oxidation, dehydration, and enzymatic changes, resulting in longer product shelf life. Frozen seafoods should be kept at temperatures as close to −15°F as possible from the time they are frozen until they reach the consumer. The shelf life of frozen fish products stored at different temperatures is given in Table 4. Note the increase in shelf life at the lowest temperatures.

For many years, it was thought too costly to operate refrigerated warehouses at temperatures lower than −10°F. However, improvements in the design and operation of refrigeration equipment have made operation of refrigerated warehouses at this temperature or lower economically possible. The production of surimi by West Coast-based factory ships has resulted in the construction of ultra-cold rooms for its storage. Japanese standards call for this product to be kept at −22°F.

Table 4 Review of Effect of Storage Temperature on Shelf Life of Frozen Fishery Products

Product	Temp., °F	Shelf Life, Months	Product	Temp., °F	Shelf Life, Months
Packaged Haddock Fillets	10 0 −20	4 to 5 11 to 12 Longer than 12	Glazed Whole Halibut	10 0 −10 −20	3 6 9 12
Packaged Cod Fillets	10 0 −10	5 6 10 to 11	Whole Blue Fin Tuna	10 0 to −5 −20	4 8 12
Packaged[a] Pollock Fillets	20 10 0 −10 −20	1 2 8 11 24	Glazed Whole Herring	0 −17	6 9
Packaged Ocean Perch Fillets	15 10 0 −10	1½ to 2 3½ to 4 6 to 8 9 to 10	Packaged Mackerel Fillets	15 0 −10	2 3 3 to 5
Packaged Striped Bass Fillets	15 0	4 9			

[a]Prepared from 1-day old iced fish.

Humidity

A high relative humidity in the cold storage room tends to reduce the evaporation of moisture from the product. The relative humidity of air in the refrigerated room is directly affected by the temperature difference between room cooling coils and room temperature. A large temperature differential results in decreased relative humidity and an accelerated rate of moisture withdrawal from the frozen product. A small temperature difference between the air and evaporator cooling coils results in high relative humidity and reduced moisture loss from the product.

The relative humidity in commercial cold storages is 10 to 20% higher than that of an empty cold storage because of constant evaporation of moisture from the product. In a cold storage operating at 0°F, with a 70% rh and pipe coil temperature of −10°F, the moisture-vapor pressure of the air within the package and in direct contact with the frozen fish would be 0.0185 psia. The air in the cold storage would have a vapor pressure of 0.0132 psia, and the moisture-vapor pressure at the coils would be 0.0108 psia. These differences in moisture-vapor pressure will result in considerable moisture loss from the product unless it is adequately protected by suitable packaging materials or glazing compounds. The evaporator coils in the freezer should be sized properly so that the desired high relative humidities can be obtained. However, because of material costs and space limitations, a temperature difference of 10°F between evaporator coils and room air is the most practical.

Packaging and Glazing

Adequate packaging of fishery products is important to prevent product dehydration and consequent quality loss. The packaging, which, in most instances, occurs prior to freezing, has been described. Individual fish, frozen in the round or dressed, cannot usually be suitably packaged; therefore, they must be protected by a suitable glazing compound.

A glaze acts as a protective coating against the two main causes of deterioration during storage, i.e., dehydration and oxidation.

Fishery Products

Table 5 Storage Conditions and Storage Life for Frozen Fish

Fish	Recommended Protection[a]	Storage Life (0°F), Months
Chub, pink salmon	Ice glazing and packaging	4-6
Mackerel, sea herring, pollock, chub, smelts	Ice glazing and packaging	5-9
Pacific sardines, tuna	Packaging	4-6
Buffalofish, flounders, halibut, ocean perch, rockfish, sablefish, red, sockeye, silver or coho salmon, whiting, shrimp	Packaging	7-12
Haddock, blue pike, cod, hake, lingcod	Packaging	Over 12

[a]All packaging should be with moisture-resistant films.

It acts against dehydration by preventing moisture from leaving the product, and against oxidation by mechanically preventing air contact with the product. It may also act by chemical means to minimize these changes if the glaze carries a suitable antioxidant.

Maximum storage life of fishery products can be obtained by employing the following procedures:

- Select only high quality fish for freezing
- Use moisture-vapor-resistant packaging materials and fit package tightly around product
- Freeze fish immediately after processing or packaging
- Glaze frozen fish prior to packaging, and round, unpackaged fish prior to cold storage
- Put fish in frozen storage immediately after freezing and glazing, if required
- Store frozen fish at temperatures of −15°F or lower
- Renew glaze on round, unpackaged fish as required during frozen storage

The recommended protection and expected storage life for various species of fish at 0°F are shown in Table 5.

Space Requirements

Packaged products such as fillets and steaks are usually packed in cardboard master cartons for storage and shipment. These master cartons are stacked on pallets and transferred to various areas of the cold storage room by forklift. The master cartons are strong enough to support one or two pallet loads placed on the shelf of each rack in the cold storage. In cold storages without racks, cartons should only be stacked to a height that does not cause crushing of the bottom cartons. Cartons for products in packages that contain a lot of air, such as IQF fillets, must be stronger than those for solid packages of fish to resist crushing during storage.

Whole or dressed fish frozen in blocks in metal pans, such as mackerel, chub, or whiting, are removed from the pans after freezing, glazed, and then packaged in wooden boxes lined with wax-impregnated paper or in cardboard cartons.

Round fish stored in wooden boxes can be easily reglazed at periodic intervals during frozen storage. The space requirements for the storage of fishery products are shown in Table 6.

Thawing of frozen fish. Frozen fishery products can be thawed by using circulatory air or water. In thawing, the fish should not be allowed to rise above refrigerated temperatures; otherwise rapid deterioration may occur. Thawing is a slower and more difficult process than freezing when done to ensure that quality is maintained. Each application should be carefully designed.

TRANSPORTATION AND MARKETING

Conditions of temperature and humidity recommended for frozen storage should also be applied during transportation and marketing to minimize product quality loss. Shipment in nonrefrigerated or improperly refrigerated carriers, exposure to high ambient temperatures during transfer from one environment to another, improper loading of common carriers or display cases, equipment failure, and other poor practices lead to increased product temperature and consequently, quality loss.

Frozen fish are transported under mechanical refrigeration in trucks, railroad cars, or ships. Many of these vehicles are capable of maintaining temperatures of 0°F or lower. Additional information on equipment used in the transportation and marketing of frozen fish and other foods is given in Chapters 25 to 32.

To minimize quality loss during transportation and marketing, the following procedures should be adhered to:

1. Transport frozen fish in refrigerated carriers (mechanical or dry ice systems) with ample capacity to maintain 0°F over long distances.
2. Precool refrigerated carriers to at least 10°F before loading.
3. Remove frozen products from the warehouse only when the carrier is ready to be loaded. Load directly into the refrigerated carrier and do not allow the product to sit on the dock.
4. Check the frozen fish temperature with a thermometer before loading.
5. Do not stack frozen fish directly against floors or walls of the carrier. Provide floor and wall racks or strips to permit air circulation around the entire load.
6. Continuously record the temperature of the refrigerated carrier during transit. Use an alarm to warn of equipment failure.
7. Measure the temperature of the product when it is removed from the common carrier at its destination.
8. If products are shipped in an insulated container, apply sufficient dry ice to maintain 0°F or lower temperatures for the duration of the trip.

Table 6 Space Requirements for Frozen Fishery Products

Commodity	Product Package	Container for Storage	Space Required, lb/ft³
Fish sticks, breaded shrimp, breaded scallops	8 or 10 oz	Corrugated master containers	25-30
Fish fillets, fish steaks, small dressed fish	1, 5, or 10 lb	Corrugated master containers	50-60
Shrimp	2.5 and 5 lb	Corrugated master containers	35
Panned, frozen fish (mackerel, herring, chub)	None	Wooden or fiberboard boxes	35
Round halibut	None	Wooden box	30-35
		Stacked loose	38
Round groundfish (cod, etc.)	None	Stacked loose	32
Round salmon	None	Stacked loose	33-35

9. Maintain food delivery or breakup rooms at 0 to 10°F. Do not hold products in breakup rooms any longer than necessary.
10. When received at the retail store, place the product in a 0°F storage room immediately.
11. Hold display cases in retail stores at 0°F or lower.
12. Do not overload display cases, especially above the frost line.
13. Record the temperature of the display cases. Provide an alarm to warn of excessive rise in temperature.
14. Because of the accelerated deterioration of frozen fish products in the distribution and retail chain, hold products in these areas for as short a period as possible.

BIBLIOGRAPHY

Barnett, H.J, R.W. Nelson, P.J. Hunter, S. Bauer, and H. Groninger. 1971. Studies on the use of carbon dioxide dissoved in refrigerated brine for the preservation of whole fish. *Fishery Bulletin* 69(2).

Charm, S.E. and P. Moody. 1966. Bound water in haddock muscle ASHRAE *Journal* 8(4):39.

Dassow, J.A. and D.T. Miyauchi. 1965. Radiation preservation of fish and shellfish of the Northeast Pacific and Gulf of Mexico; Ronsivalli, L.J., M.A. Steinberg, and H.L. Seagran. Radiation preservation of foods. National Academy of Science Publication No. 1273. Washington, D.C.

Feiger, E.A. and C.W. du Bois. 1952. Conditions affecting the quality of frozen shrimp. *Refrigerating Engineering* (September):225.

Holston, J. and S.R. Pottinger. 1954. Some factors affecting the sodium chloride content of haddock during brine freezing and water thawing. *Food Technology* 8(9):409.

Nelson, R.W. 1963. Storage life of individually frozen Pacific oyster meats glazed with plain water or with solutions of ascorbic acid or corn syrup solids. *Commercial Fisheries Review* 25(4):1.

Peters, J.A. 1964. Time-temperature tolerance of frozen seafood. ASHRAE *Journal* 6(8):72.

Peters, J.A., E.H. Cohen, and F.J. King. 1963. Effect of chilled storage on the frozen storage life of whiting *Food Technology* 17(6):109.

Peters, J.A. and J.W. Slavin: Comparative keeping quality, cooling rates, and storage temperatures of haddock held in fresh water ice and salt water ice. *Commercial Fisheries Review* 20(1):6.

Ronsivalli, L.J. and J.W. Slavin. 1965. Pasteurization of fishery products with gamma rays from a cobalt 60 source. *Commercial Fishery Review* 27(10):1.

Stansby, M.E., ed. 1976. *Industrial Fishery Technology*, 2nd ed. Robert E. Krieger Publishing Co., Huntington, NY.

Tressler, D.K, W.B. van Arsdel, and M.J. Copley, eds. 1968. *The freezing preservation of foods*, 4th ed. Avi Publishing Co., Westport, CT.

Wagner, R.L, A.F. Bezanson, J.A. Peters. Fresh fish shipments in the BCF insulated leakproof container *Commercial Fisheries Review* 31(8 and 9):41.

CHAPTER 15

DAIRY PRODUCTS

Milk Production and Processing	15.1
Butter Manufacture	15.6
Cheese Manufacture	15.10
Frozen Dairy Desserts	15.13
UHT Temperature Sterilization and Aseptic Packaging	15.19
Evaporated, Sweetened, Condensed, and Dry Milk	15.22

RAW milk is either processed for beverage milks, creams, and related milk products for marketing, or is used for the manufacture of dairy products. Dairy plant operations include receiving of raw milk; purchase of equipment, supplies, and services; processing of milk and milk products; manufacture of frozen dairy desserts, butter, cheeses, and cultured products; packaging; maintenance of equipment and other facilities; quality control; sales and distribution; engineering; and research.

Farm cooling tanks and most dairy processing equipment manufactured in the United States meet the requirements of the 3-A Sanitary Standards (IAMFES). These standards set forth the minimum design criteria acceptable for composition and surface finishes of materials in contact with the product, construction features such as minimum inside radii, accessibility for inspection and manual cleaning, insulation of nonrefrigerated holding and transport tanks, and other factors which may adversely affect the quality and safety of the product or the ease of cleaning and sanitizing the equipment. Also available are *3-A Accepted Practices*, which deal with construction, installation, operation, and testing of certain systems rather than individual items of equipment.

The *3-A Sanitary Standards* and *Practices* are developed by the 3-A Standards Committees which are composed of conferees representing state and local sanitarians, the U.S. Public Health Service, dairy processors, and equipment manufacturers. Compliance with the *3-A Sanitary Standards* is voluntary, but a manufacturer who complies and has approval of the 3-A Symbol Council may affix to his equipment a plate bearing the 3-A Symbol, which indicates to regulatory inspectors and purchasers that the equipment meets the pertinent sanitary standards.

MILK PRODUCTION AND PROCESSING

Handling Milk at the Dairy

Most dairy farms have bulk milk tanks to receive, cool, and hold the milk. Tank capacity ranges from 200 to 5000 gal, with a few larger tanks. As the cows are mechanically milked, the milk flows through sanitary pipelines to an insulated stainless steel bulk tank. An electric agitator stirs the milk, and mechanical refrigeration begins to cool it even during milking.

When operated with a condensing unit of the minimum capacity given on the nameplate, a tank must have sufficient refrigerated surface at the first milking to cool 50% of its capacity in an everyday pickup tank, or 25% of its capacity in a tank for every-other-day pickup, from 90 to 50°F within the first hour, and from 50 to 40°F within the next hour. During the second and subsequent milking, there must be sufficient refrigerating capacity to prevent the temperature of the blended milk from rising above 45°F. The nameplate must state the maximum rate at which the milk may be added and still meet the cooling requirements of the *3-A Sanitary Standards*.

Automatic controls maintain the desired temperature within a preset range in conjunction with the agitation. Some dairies continuously record the milk temperature in the tank, which is required in some states. Since the milk is picked up from the farm tank daily or every other day, the milk from the additional milkings generally flows into the reservoir cooled from the previous one. Some large dairy farms may use a plate or tubular heat exchanger to rapidly cool the milk. Cooled milk may be stored in a silo tank (a vertical cylinder 10 ft or more in height).

Milk in the farm tank is pumped into a stainless steel tank on a truck for delivery to the dairy plant or receiving station. The tanks are well insulated to alleviate the need for refrigeration of the milk during transportation. Temperature rise when testing the tank full of water should not be more than 2°F in 18 h, when the average temperature difference between the water and the atmosphere surrounding the tank is 30°F.

The most common grades of raw milk are Grade A and Manufacturing Grade. The former is that used for market milk and related products such as cream. Surplus Grade A milk is used for ice cream and/or manufactured products. To produce Grade A milk the dairy farmer must meet state and federal standards. In addition to the state requirements, a few municipal governments also have raw milk regulations.

For milk produced under the provisions of the Grade A Pasteurized Milk Ordinance recommended by the U.S. Public Health Service, the dairy farmer must have healthy cows, adequate facilities (barn, milkhouse, and equipment), maintain satisfactory sanitation of these facilities, and have milk with a bacteria count of less than 100,000 per mL. The milk should not contain pesticides, antibiotics, sanitizers, and so forth. However, current methods detect even minute traces, and total purity is difficult, if not impossible. Milk should be free of objectionable flavors and odors.

The preparation of this chapter is assigned to TC 11.3, Dairy Products.

Table 1 U.S. Requirements for Milkfat and Nonfat Solids in Milks and Creams

Product	Legal Minimum Milkfat, %			Legal Minimum Nonfat Solids, %		
	Federal	Range	Most Often	Federal	Range	Most Often
Whole milk	3.25	3.0–3.8	3.25	8.25	8.0–8.7	8.25
Lowfat milk	0.5	0.5–1.9	2.0	8.25	8.25–10.0	8.25
Skim milk	0.5[a]	0.1–0.5	0.5[a]	—	8.25–9.0	8.25
Flavored milk	—	2.8–3.8	3.25	—	7.5–10.0	8.25
Half-and-half	10.5	10.0–18.0[a]	10.5	—	—	—
Light (coffee) cream	18.0	16.0–30.0[a]	18.0	—	—	—
Light whipping cream	30.0	30.0–36.0[a]	30.0	—	—	—
Heavy cream	36.0	36.0–40.0	36.0	—	—	—
Sour cream	18.0	16.0–20.0	18.0	—	—	—

[a] Maximum

Receiving and Storage of Milk

A milk plant receives, standardizes, processes, packages, and merchandizes milk products that are safe and nutritious for human consumption. Most dairy plants either receive milk in bulk from a producer organization or arrange for pickup directly from dairy farms. The milk level in a farm tank is measured with a dip stick or a direct-reading gage, and the volume is converted to weight. Fat test and weight are common measures used to base payment to the farmer. A few organizations and the state of California include the percent of nonfat solids of protein content.

Plants determine the amount of milk received by weighing the tanker, metering the milk as it is pumped from the tanker to a storage tank, and using load cells on the storage tank or other methods associated with the amount in the storage tank.

Milk is generally received more rapidly than it is processed, so ample storage capacity is needed. A holdover supply of raw milk at the plant may be needed for startup before arrival of the first tankers in the morning. Storage may be required for nonprocessing days and emergencies. Storage tanks vary in size from 1000 to 50,000 gal. The tanks have a stainless steel lining and are well insulated.

The *3-A Sanitary Standards* for silo-type storage tanks specify that the insulating material shall be of a nature and an amount sufficient to prevent freezing, or an average 18-h temperature change of no more than 3 °F in the tank filled with water when the average temperature differential between the water and the surrounding air is 30 °F. Insulation material on a tank should have a total R-value at least equivalent to 2 in. of cork (approximately 6.7 Btu/h·ft²·°F), while tanks installed partially or wholly outside of a building should have insulation equivalent to the total R-value of 3 in. of cork (about 10 Btu/h·ft²·°F). For horizontal storage tanks, the allowable temperature change under the same conditions is 2 °F.

Agitation is essential to maintain uniform milkfat distribution. Milk held in such large tanks as the silo type is continuously agitated with a slow-speed propeller driven by a gearhead electric motor or with filtered compressed air. The tank may or may not have refrigeration, depending on the temperature of the milk flowing into it and the maximum holding time. Some plants pass the milk through a plate cooler to maintain 40 °F or less on all milk directed into the storage tanks.

If cooling is provided for milk in a storage tank, it may be by refrigeration of the surface around the lining. This cooling surface may be an annular space from a plate welded to the outside of the lining for direct refrigerant cooling or circulation of chilled water or glycol solution. Another system provides a distributing pipe at the top for the chilled liquid to flow down the lining and drain from the bottom. Direct refrigerant cooling must be carefully applied to prevent milk from freezing on the lining. This limits the evaporator temperature to approximately 25 to 28 °F.

Separation and Clarification

Before pasteurizing, milk and cream are standardized and blended to control the milkfat content with legal and practical limits. Nonfat solids may also need to be adjusted for some products; some states require added nonfat solids, especially for low-fat milk such as 2% (fat) milk. Table 1 shows the approximate legal milkfat and nonfat solids requirements for milks and creams in the United States.

One means of obtaining the desired fat standard is by separating a portion of the milk. The required amount of cream or skim milk is returned to the milk. Milk with an excessive fat content may be processed through a standardizer-clarifier that removes fat to a predetermined percentage and clarifies it at the same time. This machine can be adjusted to remove 0.1 to 2.0% of the fat in milk. To increase the nonfat solids, condensed skim milk or low-heat nonfat dry milk may be added.

Milk separators are enclosed and fed with a pump. Separators designed to separate cold milk, usually not below 40 °F, have increased capacity and efficiency as the milk temperature is increased. Capacity of a separator is doubled as milk temperature is raised from 40 to 90 °F. The efficiency of fat removal with a cold milk separator decreases as temperature decreases below 40 °F. The maximum efficiency for fat removal is attained at approximately 45 to 50 °F or above. Milk is usually separated at 70 to 90 °F, but not above 100 °F in warm milk separators. If raw, warmed milk or cream is to be held for more than 20 min before pasteurizing, it should be immediately recooled to 40 °F or below after separation. The pump supplying milk to the separator should be adjusted to pump the milk at the desired rate without causing a partial churning action.

At an early stage between receiving and before pasteurizing, the milk or the resulting skim milk and cream should be filtered or clarified. An optimum time to effectively filter is during the transfer from the pickup tanker into the plant equipment. A clarifier removes extraneous matter and leucocytes, thus improving the appearance of homogenized milks.

Pasteurization and Homogenization

There are two systems of pasteurization—batch and continuous. The minimum feasible continuous operation is about 2000 lb/h. Therefore, batch pasteurization is used for relatively small quantities of liquid milk products. The product is heated in a stainless steel-lined vat to not less than 145 °F and held at that temperature or above for not less than 30 min. The Grade A Pasteurized Milk Ordinance requires that means be provided and used in batch or vat pasteurizers to keep the atmosphere above the product at a temperature not less than 5 °F higher than the minimum required temperature of pasteurization during the holding period. Whole milk, low fat milk, half-and-half, and coffee cream are cooled, usually in the vat, to 130 °F and then

Dairy Products

homogenized. Cooling is continued in a heat exchanger (*e.g.*, a plate or tubular unit) to 40 °F or lower and then packaged.

Plate coolers may have two sections, one using plant water and the second using chilled water. The temperature of the product leaving the cooler depends on the flow rates and temperature of the cooling medium. Pasteurizing vats are heated with hot water or with steam vapor in contact with the outer surface of the lining. One heating method consists of spraying the heated water around the top of the lining. It flows to the bottom where it drains into a sump, is reheated by steam injection, and again is pumped through the spray distributor. Steam-regulating valves control the temperature of the hot water.

Most pasteurizing vats are constructed and installed so that the plant's cold water is used for initial cooling of the product after pasteurization. For final vat cooling, refrigerated water is recirculated through the jacket of the vat to attain a product temperature of 40 °F or less. Cooling time to 40 °F should be less than 1 h.

Brine is rarely used in milk plants as the final cooling medium; it corrodes milk equipment and presents the danger of freezing the milk, especially with plate heat exchangers. Flow of the chilled water for final cooling should be counter-current to the product flow in heat exchangers. The rate of flow should be adjusted so that the temperature increase is not more than 10 °F during one pass through the equipment. A ratio of chilled water to product is usually about 4:1.

High temperature, short time pasteurization (HTST) is a continuous process in which the milk is heated to a temperature of not less than 161 °F and held at this temperature for at least 15 s. The complete pasteurizing system usually consists of a series of heat exchange plates contained in a press, a milk balance tank, one or more milk pumps, a holder tube, flow diversion valve, automatic controls, and sources of hot and chilled water for heating and cooling the milk. Homogenizers are used in many HTST systems used to process Grade A products. The heat exchanger plates are arranged so that milk to be heated or cooled flows between two plates, and the heat exchange medium flows in the opposite direction between alternate pairs of plates.

Ports in the plates are arranged to direct the flow where desired, and gaskets are arranged so that any leakage will be to the outside of the press. Terminal plates are inserted to divide the press into three sections (heating, regenerating, and cooling) and arranged with ports for inlet and outlet of milk, hot water, or steam for heating, and chilled water for cooling. To provide a sufficient heat-exchange surface for the temperature change desired in a section, the milk flow is arranged for several passes through each section. The capacity of the pasteurizer can be increased by arranging several streams for each pass made by the milk. The capacity range of a complete HTST pasteurizer is 100 to about 100,000 lb/h. A few shell-and-tube and triple-tube HTST units are in use, but the plate type is by far the most prevalent.

Figure 1 shows one example of a flow diagram for an HTST plate pasteurizing system. Raw product is first introduced into a constant level (or balance) tank from a storage tank or receiving line by either gravity or a pump. A uniform level is maintained in this tank by means of a float-operated valve or similar device. A booster pump is often used to direct the flow through the regeneration section. The product may be clarified and/or homogenized or simply directly pumped to the heating section by means of a timing pump. From the heating section, the product continues through a holding tube to the flow diversion valve. If the product is at or above the preset temperature, it passes back through the opposite sides of the plates in the regeneration section and then through the final cooling section. The flow diversion valve is set at 161 °F or above; if the product is below this minimum temperature, it is diverted back into the balance tank for repasteurization. The exchange of heat in the regeneration section causes the cold raw milk to be heated by the cooled pasteurized milk. The pasteurized milk pressure must be maintained at least 1 psig above the raw milk pressure in the regeneration section. The flow rate of both products is the same, and the temperature change is about the same.

Most HTST heat exchangers have 80 to 90% regeneration. The cost of additional equipment to achieve more than 90% regeneration should be compared with savings in the increased regeneration to determine feasibility. The percentage of regeneration may be calculated as follows:

$$\frac{138\,°F\,(\text{regeneration}) - 40\,°F\,(\text{raw product})}{161\,°F\,(\text{pasteurization}) - 40\,°F\,(\text{raw product})} = \frac{98}{121} = 81\%$$

The temperature of a product going into the cooling section can be calculated if the percent regeneration is known and the raw product and pasteurizing temperatures are determined. If they are 80%, 45 °F, and 161 °F, respectively:

$$(161 - 45) \times 0.80 = 92.8\,°F$$
$$161 - 92.8 = 68.2\,°F$$

The product should be cooled to at least 40 °F, preferably even lower, to compensate for the increase while in the sanitary pipelines or package (including filling, sealing, casing, and transfer into cold storage). The average temperature increase of milk between the time of discharge from the cooling section of the HTST unit and arrival at the cold storage in various quart containers is as follows: glass bottles, 8 °F; preformed paperboard cartons, 6 °F; formed paperboard, 5 °F; and semirigid plastic, 4 °F.

Many plate-type pasteurizing systems are equipped with a cooling section containing propylene glycol solution to cool the milk or milk product to temperatures lower than are practical by circulating only chilled water. This requires an additional section in the plate heat exchanger, a glycol chiller, a pump for circulating the glycol solution, and a product temperature actuated control to regulate the flow of glycol solution and prevent freezing of the product. Milk is usually cooled this way to approximately 34 °F, then packaged. The lower temperature allows the milk to absorb heat from the containers and still maintain a low enough temperature for excellent shelf life. Milk should not be cooled to temperatures between 33.5 °F and freezing because of the tendency toward increased foaming in this range. The propylene glycol is usually chilled to approximately 28 °F for circulation through the milk cooling section.

The product flow rate through the pasteurizer may be more or less than the filling rate of the packaging equipment. Pasteurized

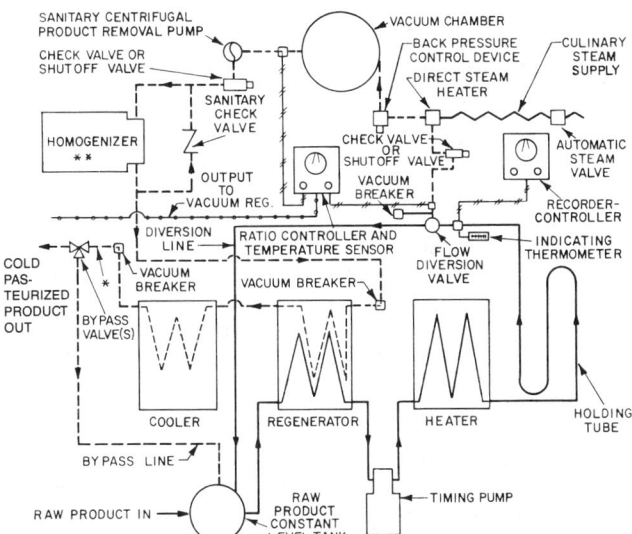

Fig. 1 Flow Diagram of Plate HTST Pasteurizer with Vacuum Chamber

product storage tanks are generally used to hold the product until it is packaged.

The number of plates in the pasteurizing unit is determined by the volume of product needed per unit of time, the desired percentage of regeneration, and the temperature differentials between the product and the heating and cooling media. The heating section will usually have ample surface so that the temperature of the hot water entering the section will be not more than 2 to 6 °F warmer than the pasteurizing, or outlet, temperature of the product. On larger units, steam may be used for the heater section instead of hot water. The cooling section is usually sized so that the temperature of the pasteurized product leaving the section is about 5 °F warmer than the entering temperature of the chilled water.

The holder tube size and length is selected so that not less than 15 s will elapse for the product to flow from one end of the tube to the other. An automatic, power-actuated, flow-diversion valve, controlled by a temperature recorder-controller, is located at the outlet end of the holder tube and diverts the flow back to the raw product constant level tank as long as the product is below the minimum set pasteurizing temperature. The product timing pump is a variable speed, positive displacement, rotary type that can be sealed by the local government milk plant inspector at a maximum speed and volume. This assures a product hold of not less than 15 s in the holder tube.

As a means of reducing undesirable flavors and odors in milk (usually caused by specific types of dairy cattle feed), some plants use a vacuum process in addition to the usual pasteurization. In this process, milk from the flow diversion valve passes through a direct steam injector or steam infusion chamber where it is heated with culinary steam to a temperature of 180 to 200 °F. The milk is then immediately sprayed into a vacuum chamber, where it is cooled by evaporation to the pasteurizing temperature and promptly pumped to the regeneration section of the pasteurizing unit. The vacuum in the evaporating chamber is automatically controlled so that the same amount of moisture will be removed as was added by steam condensate. Noncondensable gases are removed by the vacuum pump, and vapor from the vacuum chamber is condensed in a heat exchanger which is cooled by the plant water.

The vacuum chamber can be installed with any type of HTST pasteurizer. In some plants, after preheating in the HTST system, the product is further heated by direct steam infusion or injection. It then is deaerated in the vacuum chamber. The product is pumped from the chamber by a timing pump through final heating, holder, flow diversion valve, and regenerative and cooling sections. Homogenization may occur either immediately after preheating for pasteurization or after the product passes through the flow diversion valve. If the product is heated by direct steam injection and deaerated, the preferred practice is to homogenize after deaeration.

Where volatile weed and feed taints in the milk are mild, some processors use only a vacuum treatment for reduction of the off-flavor. The main objection to vacuum treatment alone is that, to be effective, the vacuum must be low enough to cause some evaporation, and the moisture so removed constitutes a loss of product. With this type of treatment, the vacuum chamber may be installed immediately after preheating, where it effectively deaerates the milk prior to heating or immediately after the flow diversion valve where it is more effective in removing the volatile taints.

Nearly all milk processed in the United States is homogenized to improve stability of the milkfat emulsion, thus preventing creaming (concentration of the buoyant milkfat in the upper portion of the containerized milk) during normal shelf life. The homogenizer is a high-pressure, reciprocating pump with 3 to 7 pistons, fitted with a special homogenizing valve. There are several types of homogenizing valves in use, all of which cause the molten fat globules in the stream of milk to be subjected to enough shear to be divided into several smaller globules. Homogenizing valves may either be single or two in series.

For effective homogenization of whole milk, the fat globules should be 2 μm or less in diameter. The usual temperature range is from 130 to 180 °F, and the higher the temperature within this range, the lower the pressure required for satisfactory homogenization. The homogenizing pressure for a single-stage homogenizing valve ranges from about 1200 to 2500 psi for milk; for a two-stage valve, from 1200 to 2000 psi on the first stage plus 300 to 700 psi on the second, depending on the design of the valve and the product temperature and composition. To conserve energy, use the lowest homogenizing pressure consistent with satisfactory homogenization; the higher the pressure, the greater the power requirements.

Packaging of Milk Products

Cold product from the pasteurizer cooling section flows to the packaging machine and/or a surge tank 1000 to 10,000 gal or larger. These tanks are stainless steel, well insulated, and have agitation and usually refrigeration.

Milk and related products are packaged for distribution in paperboard, plastic, or glass containers in one-half pint, 10-ounce, pint, quart, one-half gallon, and gallon sizes. Milk in plastic bags or metal cans in sizes ranging from 2.5 to 12 gal may be used for dispenser machines. Fillers vary in design. Gravity flow is used, but positive piston displacement is more common. Filling speeds range from roughly 16 to 250 units/min, but vary with container size. Some fillers handle only one size, while others may be adjusted to automatically fill and seal one of several size containers. Paperboard cartons are usually formed ahead of filling, but may be preformed prior to delivery to the plant. Semirigid plastic containers may be blow-molded in-plant ahead of the filler or preformed. Plastic pouches (called bags) arrive at the plant ready for filling and sealing. The filling of dispenser cans and bags is a semi-manual operation.

The paperboard carton for milk consists of a 16-mil thick kraft paperboard from virgin paper with a 1-mil polyethylene film laminated onto the inside and a 0.75-mil film onto the outside. Gas or electric heaters supply heat for sealing while pressure is applied.

Blow-molded plastic milk containers are fabricated from high density polyethylene resin. The resin temperature for blow-forming varies from 340 to 425 °F. The molded gallon weighs approximately 60 to 70 g, and the one-half gallon, about 45 g. Most equipment uses direct expansion water chillers or another cooling medium to cool the mold head and clutch. The refrigeration demand is sufficiently large to require the cooling load to be included in planning a plastic blow-molded operation. The blow-molded equipment manufacturer should be contacted for the refrigeration requirements of a specific machine.

Packages containing the product may be placed into cases mechanically. Stackers place the cases 5 or 6 high, and conveyors transfer the stacks into the cold storage area.

Most milk processing activities from receiving to storage can be automated. This necessitates the installation and operation of control panels and a digital computer. Automation uses a meter-based system that controls the separation, fat and/or nonfat solids content, and ingredient addition for a variety of common products. If the initial fat tests fed into the computer are correct, the accuracy of the fat content of the standardized product is ±0.01%. Added ingredients (*e.g.*, flavoring and sweetening products) must be in liquid form for a computerized operation.

Several systems to automatically clean the equipment in place (CIP) are used in milk processing plants. These may involve the holding and reuse of the detergent solution or the preparation of a fresh solution (single-use) each day. The means of programming the automatic control of each cleaning and sanitizing step also

Dairy Products

varies. Tanks, vats, and other large equipment can be cleaned by using spray balls and similar devices that assure complete coverage of soiled surfaces. Tubing, HTST units, and equipment with relatively low volume may be cleaned by the full-flood system. The solutions should have a velocity of not less than 5 ft/s and must be in contact with all soiled surfaces. Surfaces used for heating milk products, such as in batch or HTST pasteurization, are more difficult to clean than the other equipment surfaces. Other surfaces difficult to clean are those in contact with high-fat products, products containing added solids and/or sweeteners, and highly viscous products. The usual cleaning steps for this equipment are a warm water rinse, hot acid solution wash, rinse, hot alkali solution wash, and rinse. The variables of time, temperature, concentration, and velocity may need to be adjusted for effective cleaning. Just before use, the product surfaces should be sanitized with chemical solution, hot water, or steam.

Milk Storage and Distribution

Cases containing packaged products are conveyed into a cold storage room or directly to delivery trucks for wholesale or retail distribution. The temperature of the storage area should be 33 to 40 °F, and for improved keeping quality, the product temperature in the container on arrival in storage should be 40 °F or less.

The refrigeration calculation for the cold storage area should include losses through insulation, conveyor passes, and doors; heat added due to temperature of products, containers, cases, lights, conveyor drive motors, washdown water, and personnel; and lift truck operation. Relative humidity in a storage area is generally high and should be considered when selecting coils or diffuser units for maintaining the cold temperature. Frost collects rapidly on the coils, so automatic defrosting during the off-period is common. Warm gas pumped from the compressor discharge into the coils for a short time causes coils to warm and melt frost or ice. Defrosting via electric resistance heaters is also common.

Warming the room air to melt frost is generally considered too slow; in addition, the product temperature may rise.

The floor space required for cold storage depends on product volume, height of stacked cases, kind of package (glass requires more space than paperboard), whether mechanized or manual handling, and the number of processing days per week. A 5-day processing week requires a capacity for holding product supply for 2 days. A very general estimate is that 100 lb of milk product in paperboard cartons can be stored per square foot of area. Approximately one-third more area should be allowed for aisles.

Milk product may be transferred by conveyor from storage room to dock for loading onto delivery trucks. In-floor drag-chain conveyors are commonly used, especially for retail trucks. Refrigeration losses are reduced if the load-out doorway has the protrusion of cushioning material to contact the doorway frame of the truck as it is backed to the dock.

Distribution trucks need refrigeration to protect quality and extend the keeping quality of milk products. Refrigeration capacity must be sufficient to maintain Grade A products at 45 °F or less. Many plants use insulated truck bodies with integral refrigerating systems powered by the truck engine or one that can be plugged into a remote electric power source when it is parked. In some facilities, cold plates in the truck body are connected to a coolant source in the parking space. These refrigerated trucks can also be loaded when convenient and held over at the connecting station until the next morning.

Water Chilling Equipment

A large portion of the refrigeration capacity in a milk plant is required for water chilling. Chilled water is usually used for cooling pasteurized milk and other dairy products and may be used for cooling incoming milk and holding or lowering the temperature in milk storage tanks. Chilled water is refrigerated at a central source and recirculated by pumps through the various pieces of dairy equipment. Refrigeration of the water may be by flash cooling equipment or by circulation through an ice bank. Some plants use both systems.

The flash-type water chiller may be arranged as: a bank of coils having the circulated water sprayed over the coils; banks of coils or plates arranged so water flows down over the coils by gravity from distributing pipes or troughs; a vertical open shell-and-tube chiller with the water flowing down the inside of the vertical tubes; or refrigerated coils submerged in a water tank with rapid circulation of the water around the coils. A closed shell-and-tube type or shell-and-coil type of chiller should not be used because of the low water temperature required and the danger of freezing the water tubes.

Flash-type water chilling equipment may be used for cooling pasteurized milk and other uniform continuous cooling loads required at the same time. Also, high peak cooling loads can be handled by an ice-bank type of chilling system. To maintain a uniform chilled water supply at about 34 °F in a flash system, sufficient water should be circulated so the temperature rise of the water returning will not be greater than 10 °F. Refrigerant temperatures in the flash chiller depend on unit design (usually about 28 °F).

The ice-bank type of water chiller should be evaluated against the flash-type water chiller for each application, considering both initial capital and operating costs. The ice-bank system permits the use of a refrigerating system having considerably less capacity than is required for the peak cooling load by building ice on the refrigerated surface during the time that chilled water is not required. When chilled water is required, the melting ice adds cooling capacity to that supplied by the refrigerating system.

The ice-bank water chiller consists of an insulated tank containing refrigerated coils or plates submerged in the water and spaced so that ice can build up on the refrigerated surface to a depth of about 2.5 in. Agitators and baffles keep water moving over the surface of the ice. The agitator may contain a motor-driven propeller to circulate water around the baffles and over the coils at a rate of five times the amount of water being pumped for product cooling, or agitation may be by means of compressed air. For air agitation, perforated pipes along the bottom of the tank extend under each stand of pipe coils so the air bubbles rising cause the water to move over the surface of the ice.

An ice-bank water chiller with a 5-in. diameter ice buildup on 1.25-in. steel pipe (7 lb of ice per foot of pipe) has 1000 Btu of cooling capacity per foot of pipe available from the melting ice. With good agitation, the water for process cooling leaves the tank at a temperature of 33 to 34 °F as long as ice remains on the refrigerated surface. Warm water returning to the tank should discharge near the tank bottom or at the inlet to the agitator. Water for process cooling should be taken from near the top of the tank. The refrigerant temperature in the ice-bank system varies from 5 to 15 °F; it is usually controlled by a dual-pressure back pressure regulator. The lower setting (5 to 15 °F) is maintained during ice building, while an automatic ice thickness control switches the regulator to the higher setting (40 °F or so) to terminate further ice building when maximum ice has been built.

In selecting or designing an ice-bank water chilling system, sufficient refrigerated surface and ice buildup capacity should be provided to handle the entire chilled water requirements for a full day. The refrigerating effect supplied by the refrigerating system while chilled water is being supplied to processing equipment should not be included in calculating capacity. This selection provides a surplus in capacity to take care of losses in insulation and circulating piping, allows for unusual peak operating conditions, and provides for some expansion.

Half-and-Half and Cream

Half-and-half is standardized to 10.5 to 12% milkfat and in most states to about the same percent nonfat milk solids. Coffee cream should be standardized to 18 to 20% milkfat. Both are pasteurized, homogenized, cooled, and packaged similarly to milk. Milkfat content of whipping cream is adjusted to 30 to 35%. Care must be taken during processing to preserve the whipping properties; this includes the omission of homogenization.

Buttermilk, Yogurt, and Sour Cream

Retail buttermilk is not from the butter churn but is rather a cultured product. To reduce the microorganisms to a low level and improve the body of the resulting buttermilk, skim milk is pasteurized at 180°F or higher for 0.5 to 1 h and cooled to 70 to 72°F. One percent of a lactic acid culture (starter) specifically for buttermilk is added and the mixture incubated until firmly coagulated by the correct lactic acid production (pH 4.5). The product is cooled to 40°F or less with gentle agitation to inhibit serum separation subsequent to packaging and distribution. Salt and/or milkfat (0.5 to 1.0%) in the form of cream or small fat granules may be added. Package equipment and containers are the same as for milk. Pasteurizing, setting, incubating, and cooling are usually accomplished in the same vat. Rapid cooling is necessary, so chilled water is used. If a 500 gal vat is used, as much as 25 to 30 tons of refrigeration may be needed. Some plants have been able to cool buttermilk with a plate heat exchanger without causing a serum separation problem (wheying off).

Cultured half-and-half and cultured sour cream are manufactured similarly to the method for cultured buttermilk. Rennet may be added at a rate of 0.5 mL (diluted in water) per 10 gal cream. Care must be exercised to use an active lactic culture and to prevent postpasteurization contamination by bacteriophage, bacteria, yeast, or molds. An alternate method consists of packaging the inoculated cream, incubating, and then cooling by placing packages in a refrigerated room.

Skim milk may be used, or milkfat standardized to within the range of 1 to 5%, and a 0.1 to 0.2% stabilizer may be added to yogurt. Either vat pasteurization at 150 to 200°F for 0.5 to 1 h or HTST at 185 to 285°F for 15 to 30 s can be used. For yogurt to have optimum body, the milk homogenization is at 130 to 150°F and 500 to 2000 psi. After cooling to between 110 and 100°F, the product is inoculated with a yogurt culture. Incubation for 1.5 to 2.0 h is necessary; the product is then cooled to about 90°F, packaged, incubated 2 to 3 h (acidity 0.80 to 0.85%), and chilled to 40°F or below in the package. Varying yogurt cultures and yogurt manufacturing procedures should be selected on the basis of consumer preferences. Numerous flavorings are used (fruit is quite common), and sugar is usually added. The flavoring material may be added at the same time as the culture, after incubation, or ahead of packaging. In some dairy plants, a fruit (or sauce) is placed into the package before filling with yogurt.

BUTTER MANUFACTURE

Much of the butter production is in combination butter-powder plants. These plants get the excess milk production after current market needs for milk products, frozen dairy desserts, and, to some extent, cheeses are met. Consequently, seasonal variation in the volume of butter manufactured is large; spring is the period of highest volume, fall the lowest.

Separation and Pasteurization

After separation of the milk, the cream with 30 to 40% fat is either pumped to the pasteurizer or cooled to 45°F and held for later pasteurization. Cream from cold milk separation does not need to be recooled except for extended storage. Cream is received, weighed, sampled, and, in some plants, graded according to flavor and acidity. It is pumped to a refrigerated storage vat and cooled to 45°F if held for a short period or overnight. Cream with developed acidity is warmed to 80 to 90°F, and neutralized to 0.12 to 0.15% titratable acidity just prior to pasteurization. If the acidity is above 0.40%, it is neutralized with a soda-type compound in aqueous solution to about 0.30% and then to the final acidity with aqueous lime solution. Sodium neutralizers are $NaHCO_3$, Na_2CO_3, and $NaOH$. Limes include $Ca(OH)_2$, MgO, and CaO.

Batch pasteurization is usually at 155 to 175°F for 0.5 h, depending on intended storage temperature and time. HTST continuous pasteurization is at 185 to 250°F for at least 15 s. HTST systems may be plate or tubular. After pasteurization, the cream is immediately cooled. The temperature depends on the time that the cream will be held before churning, whether or not it is ripened, the season (higher in winter due to fat composition), and the churning method. The range is 40 to 55°F. The ripening process consists of adding a flavor-producing lactic starter to tempered cream and holding until acidity has developed to 0.25 to 0.30%. The cream is cooled to prevent further acid development and warmed to the churning temperature just before churning. Ripening cream is not a common practice in the United Sates, but is customary in some European countries such as Denmark. First, tap water is used to reduce the temperature to between 80 and 100°F. Refrigerated water or brine is then used to reduce the temperature to the desired level. The cream may be cooled by passing the cooling medium through a revolving coil in the vat or through the vat jacket, or by using a plate or tubular cooler.

If the temperature of 1000 lb of cream is to be reduced by refrigerated water from 104 to 39°F and the specific heat is 0.85 Btu/(lb·°F), the heat units to be removed would be:

$$1000 (104 - 39) 0.85 = 55{,}250 \text{ Btu}$$

This heat can be removed by 55,250/144 = 384 lb of ice at 32°F plus 10% for mechanical loss.

The temperature of the refrigerated water commonly used for cooling cream is 33 to 34°F. The ice-bank system is efficient for this purpose. Brine is not currently used. About 265 gal of cream can be cooled from 100 to 40°F in a vat using refrigerated water in an hour.

After a vat of cream has been cooled to the desired temperature, the temperature increases during the following 3 h. It may increase several degrees depending on the rapidity with which the cream was cooled, the temperature to which it was cooled, the richness of the cream, and the properties of the fat. The rise in temperature is due to liberation of heat when fat is changed from a liquid to a crystal form.

Rishoi (1951) presented data in Figure 2 that show the thermal behavior of cream heated to 167°F followed by rapid cooling to 86°F and to 50.7°F, as compared with cream heated to 122°F and cooled rapidly to 88.5°F and to 53.6°F. The curves indicate that when cream is cooled to a temperature at which the fat remains liquid, the cooling rate is normal, but when the cream is cooled to a temperature at which some fractions of the fat have crystallized, a spontaneous temperature rise takes place after cooling.

Rishoi also determined the amount of heat liberated by the part of the milkfat that crystallizes in the temperature range of 85 to 33°F. The results are shown in Table 2 and Figure 3.

Table 2 shows that, at a temperature below 50°F, about one-half of the liberated heat evolved in less than 15 s. The heat liberated during fat crystallization constitutes a considerable portion of the refrigeration load required to cool fat-rich cream. Rishoi states:

"If we assume an operation of cooling cream containing 40% fat from about 150 to 40°F, heat of crystallization evolved represents about 14% of the total heat to be removed. In plastic cream containing 80% fat it represents about 30% and in pure milkfat oil about 40%."

Dairy Products

Table 2 Heat Liberated from Fat in Cream Cooled Rapidly from about 86°F to Various Temperatures

Calculated temperature for zero time, °F:	33.4	39.6	53.1	58.0	63.4	80.4	85.6[a]
First observed temperature:	36.2	43.7	54.5	58.6	63.8	80.4	85.6
Final equilibrium temperature:	39.3	46.1	58.0	60.7	65.3	82.6	85.6
Lapsed time, min	\multicolumn{7}{c}{**Btu Liberated per Pound**}						
0.25	18.3	16.7	7.75	2.9	2.3	0	0
15	23.6	20.0	12.8	7.2	5.2	0.35	0
30	25.5	23.9	18.0	12.2	9.0	2.1	0
60	30.4	27.5	21.4	14.0	10.2	2.7	0
120	32.4	29.3	25.0	14.7	10.6	2.9	0
180	32.9	30.2	26.6	16.0	10.6	4.3	0
240	34.0	30.9	27.2	17.5	10.6	4.9	
300	33.6	31.8	27.9	18.4		5.0	
360	33.6	32.5	27.2	18.8		5.0	

Percent heat liberated at zero time compared with that at equilibrium: 54.5, 51.3, 27.7, 20.7, 21.7.
Percent total heat liberated compared with that liberated at about 32°F: 100.0, 95.7, 82.0, 55.0, 31.0, 12.5, 0.
Iodine values of three samples of butter produced while these tests were in progress were: 28.00, 28.55, and 28.24.
[a] Cooled in an ice water bath.

Churning

To maintain the yellow color of butter from cream that came from cows on green pasture in spring and early summer, yellow coloring is added to the cream in the amount needed to produce the color that is obtained naturally during other periods of the year. After cooling, pasteurized cream should be held a minimum of 2 h and preferably overnight. It is tempered to the desired batch churning temperature, which varies with the season and feed of the cows but ranges from 45°F in early summer to 56°F in winter, to maintain a churning time 0.5 to 0.75 h. Lower churning time results in soft butter that is more difficult (or impossible) to work into a uniform composition.

Today most butter is churned by continuous churns, but batch units remain in use, especially in smaller butter factories. Batch churns are usually made of stainless steel, although a few aluminum ones are still in use. They are cylinder, cube, cone, or double cone in shape. The inside surface of metal churns is sandblasted during fabrication to reduce or prevent butter from sticking to the surface. Metal churns may have accessories to draw a partial vacuum or introduce an inert gas, *e.g.*, nitrogen, under pressure. Working the butter under a partial vacuum reduces the air in the

Fig. 2 Thermal Behavior of Cream Heated to 167°F Followed by Rapid Cooling to 86°F and to 50.7°F; Comparison with Cream Heated to 122°F, then Rapid Cooling to 88.5°F and to 53.6°F

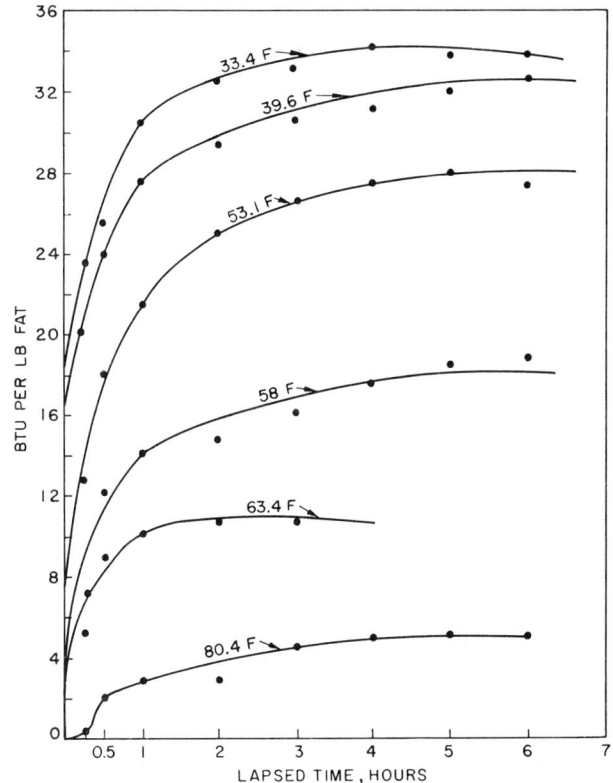

Fig. 3 Heat Liberated from Fat in Cream Cooled Rapidly from Approximately 86°F to Various Temperatures
(Rishoi 1951)

butter. Churns have two or more speeds, with the faster rate for churning. The speed should provide maximum agitation of the cream, usually within 0.25 to 0.5 r/s.

When churning, temperature is adjusted and the churn is filled 40 to 48% of capacity. The churn is revolved until the granules break out and attain a diameter of 5 mm or slightly larger. The buttermilk is drained and should have no more than 1% milkfat. The butter may or may not be washed. The purpose of washing is to remove buttermilk and temper the butter granules if they are too soft for adequate working. Wash water temperature is adjusted to 0 to 10°F below churning temperature. The preferred procedure is to spray wash water over granules until it appears clear from the churn drain vent. The vent is then closed, and water is added to the churn until the volume of butter and water is approximately equal to the amount of the former cream. The churn is revolved slowly 12 to 15 times and drained or held for an additional 5 to 15 min for tempering so granules will work into a mass of butter without becoming greasy.

The butter is worked at a slow speed until free moisture is no longer extruded. The free water is drained, and the butter is analyzed for moisture content. The amount of water needed to obtain the desired content (usually 16.0 to 18.0%) is calculated and added. Salt may be added to the butter. The percent is standardized between 1.0 and 2.5% according to customer demand.

Salt may be added in dry form either to a trench formed in the butter or spread over the top of the butter. It also may be added in moistened form using the water required for standardizing the composition to not less than 80.0% fat. Working continues until the granules are completely compacted and the salt and moisture droplets are uniformly incorporated. The moisture droplets should become invisible by normal vision with adequate working. Most churns have ribs or vanes which cause tumbling and folding of the butter as the churn revolves. The butter passes between the narrow slit of shelves attached to the shell and the roll. A leaky butter is inadequately worked, possibly leading to economic losses due to reduction of weight and shorter keeping quality. The average composition of U.S. butter on the market has these ranges:

| Fat | 80.0 to 81.2 | Moisture | 16.0 to 18.0 |
| Salt | 1.0 to 2.5 | Curd, etc. | 0.5 to 1.5 |

Cultured skim milk is added to unsalted butter as part of the moisture and thoroughly mixed in during working. Cultured skim milk increases acid flavor and the diacetyl content associated with butter flavor.

Butter may be removed manually from small churns, but it is usually emptied by a mechanized procedure. One method is to dump the butter from the churn directly into a stainless steel boat on casters or a tray that has been pushed under the churn with the door removed. Butter in boats may be augered to the hopper for printing (forming the butter into retail sizes) or pumping into cartons 60 to 68 lb in size. The bulk cartons are held cold before printing or shipment. Butter may be stored in the boats or trays and tempered until printing. A hydraulic lift may be used for hoisting the trays and dumping the butter into the hopper. Cone-shaped churns with a special pump can be emptied by pumping butter from churn to hopper.

Continuous Churning

The basic steps in two of the continuous buttermaking processes developed in the United States are: fat emulsion in the cream is destabilized and the serum separated from the milkfat; the butter mix is prepared by thoroughly mixing together the correct amount of milkfat, water, salt, and cultured skim milk; this mixture is worked and chilled at the same time; butter is extruded at 38 to 50°F with a smooth body and texture.

Several European continuous churns consist of a single machine that directly converts the cream to butter granules, drains off the

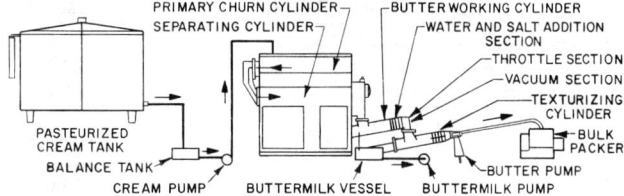

Fig. 4 Flow Diagram of Continuous Butter Manufacture

buttermilk, and washes and works the butter, incorporating the salt in continuous flow. Each brand of continuous churn may vary in equipment design and specific operation details for obtaining the optimum composition and quality control of the finished product. Figure 4 shows a flow diagram of a continuous churn.

In one system, milk is heated to 110°F and separated to cream with 35 to 50% fat and skim milk. The cream is pasteurized at 203°F for 16 s, cooled to a churning temperature of 46 to 55°F, and held for 6 h. The cream enters the balance tank and is pumped to the churning cylinder where it is converted to granules and serum in less than 2 s by vigorous agitation. Buttermilk is drained off and the granules are sprayed with tempered wash water while being agitated.

Next, salt, in the form of 50% brine prepared from microcrystalline sodium chloride, is fed into the product cylinder by a proportioning pump. If needed, yellow coloring may be added to the brine. The high speed agitators work the salt and moisture into the butter in the texturizer section and then extrude it to the hopper for packaging into bulk cartons or retail packages. The cylinders on some designs have a cooling system to maintain the desired temperature of the butter from churning to extrusion. The butterfat content is adjusted by fat test of the cream, churning temperature of the cream, and flow rate of product.

Continuous churns are designed for cleaning-in-place (CIP). The system may be automated or a cream tank may be used to prepare the detergent solution prior to circulation through the churn after the initial rinsing.

Packaging Butter

"Printing" refers to the process of forming (or cutting) butter into retail sizes. Each is then wrapped with parchment or parchment-coated foil. The wrapped prints may be inserted in paperboard cartons or overwrapped in cellophane, glassine, and so forth, and heat-sealed. Common sizes are 0.25, 0.5, and 1 lb. For institutional uses, butter may be extruded into slabs. These are cut into patties, embossed, and each slab of patties wrapped in parchment paper. Most common numbers of patties are 48 to 72 per lb.

Butter keeps better if stored in bulk. If the butter is intended to be stored for several months, the temperature should not be above

Table 3 Specific Heats of Milk and Milk Derivatives
Btu/(lb · °F)

	32°F	59°F	104°F	140°F
Whey	0.978	0.976	0.974	0.972
Skim milk	0.940	0.943	0.952	0.963
Whole milk	0.920	0.938	0.930	0.918
15% Cream	0.750	0.923	0.899	0.900
20% Cream	0.723	0.940	0.880	0.886
30% Cream	0.673	0.983	0.852	0.860
45% Cream	0.606	1.016	0.787	0.793
60% Cream	0.560	1.053	0.721	0.737
Butter	(0.512)[a]	(0.527)[a]	0.556	0.580
Milkfat	(0.445)[a]	(0.467)[a]	0.500	0.530

[a] For butter and milkfat, values in parentheses were obtained by extrapolation, assuming that the specific heat is about the same in the solid and liquid states.

Dairy Products

0 °F, and preferably below −20 °F. For short periods, 32 to 40 °F is satisfactory for bulk or printed butter. Butter should be well protected to prevent absorption of off-odors during storage and weight loss due to evaporation, and to minimize surface oxidation of the fat.

The specific heat of butter and other dairy products at temperatures varying from 32 to 140 °F is given in Table 3. The temperature of the butter when removed from the churn ranges from 56 to 62 °F. Assuming a temperature of 60 °F of the packed butter, the heat that must be removed from 1000 lb to reduce the temperature to 32 °F is:

$$1000 (60 - 32) 0.52 = 14,560 \text{ Btu}$$

It is assumed that the average specific heat at the given range of temperatures is 0.52 Btu/(lb · °F). Heat units that are to be removed from the butter containers and packaging material should be added.

Deterioration of Butter in Storage

The development of an undesirable flavor in butter during storage may be caused by: (1) growth of microorganisms (proteolytic organisms cause putrid and bitter off-flavors); (2) absorption of odors from the atmosphere; (3) fat oxidation; (4) catalytic action by metallic salts; (5) activity of enzymes, principally from microorganisms; and (6) low pH (high acid) of salted butter.

Normally, microorganisms do not grow below 32 °F; if salt-tolerant bacteria are present, their growth will be slow below 32 °F. Growth of microorganisms does not take place at 0 °F or below, but some may survive in the butter held at this temperature. It is important to store butter in a room where the atmosphere is free of odors. Butter will readily absorb odors from the atmosphere or from odoriferous materials with which it comes into contact.

Oxidation causes a stale, tallowy flavor. Chemical changes take place slowly in butter held in cold storage, but are hastened by the presence of metals or metallic oxides.

With almost 100% replacement of tinned copper equipment with stainless steel equipment, a tallowy flavor is not as common as in the past. Factors that favor oxidation are light, high acid, high pH, and metal.

Enzymes present in raw cream are inactivated by current pasteurization temperatures and holding times. The only enzymes that may cause deterioration of butter are those produced by microorganisms that gain entrance to the pasteurized cream and butter or survive pasteurization. The chemical changes caused by enzymes present in butter are retarded by a lowering of the storage temperature.

A fishy flavor may develop in salted butter during cold storage. The development of the defect is favored by high acidity (low pH) of the cream at the time of churning and by metallic salts. With the use of stainless steel equipment and the proper control of the butter's pH, this defect now occurs very rarely. For salted butter to be stored for several months, even at −10 °F, it is advisable to use good quality cream, avoid exposing the milk or cream to strong light, copper, or iron, and to adjust any acidity developed in the cream so that the butter serum will have a pH of 6.8 to 7.0.

Total Refrigeration Load

Some dairy plants that manufacture butter also process and manufacture other products such as ice cream, fluid milk, and cottage cheese. A common refrigeration system is used. The method of determining the refrigeration load is illustrated by the following example:

Example 1. Determine the total refrigeration load for a plant manufacturing butter from 12,600 lb of 30% cream per day in three churnings.

Solution: The refrigeration requirement is obtained in the following steps A through E.

A. If the cream is separated in the plant rather than on the farm, it would have to be cooled down from 90 °F separating temperature to 40 °F for holding until it is processed.

$$\frac{12,600 (90 - 40) 0.85}{144} = 3720 \text{ lb ice}$$

B. After pasteurization, the temperature of the cream is reduced to approximately 100 °F with city water, then down to 40 °F with refrigeration.

$$\frac{12,600 (100 - 40) 0.85}{144} = 4460 \text{ lb ice}$$

C. After churning, the butter wash water (city water) is usually cooled to 45 °F, then used to wash the butter granules. A volume of water equal to the volume of cream churned may be used.

$$\frac{12,600 (60 - 45) 1.00}{144} = 1310 \text{ lb ice}$$

Total ice load	9490 lb ice
Plus 10% mechanical loss	950 lb ice
Total ice required	10,440 lb ice

D. Approximately 4725 lb of butter would be obtained (12,600 lb cream × 30% fat = 3780 lb of fat). If butter contains approximately 80% fat, 3780 divided by 80% equals approximately 4725 lb of butter. The butter temperature going into the refrigerated storage room is usually about 62 °F and must be cooled down to 40 °F during the following 16 h. (For long-term storage, the butter is held at −10 to 0 °F.)

4725 lb (62 − 40) 0.55	= 57,170 Btu
300 lb (metal container) × (75 − 40) 0.12	= 1260 Btu
Total/24 h	58,430 Btu

E. After 24 h or longer, the butter is removed from the cooler to be cut and wrapped in 1 lb or smaller units. During this process, the butter temperature will rise to approximately 55 °F, which constitutes another product load in the cooler when it goes back for storage.

4725 lb (55 − 40) 0.55	= 38,980 Btu
200 lb (paper container) × (75 − 40) 0.33	= 2310 Btu
Total/24 h	41,290 Btu

Total of Steps D and E, Product Load in Cooler

$$\frac{41,290 + 58,430}{16 \text{ h}} = 6230 \text{ Btu/h}$$

On this basis, the following calculations develop:

The cream would be cooled in steps A and B. The butter would then have to be cooled through steps C, D, and E. Refrigerated water is normally used as a cooling medium in steps A, B, and C. The ice-bank system is used to produce the 34 °F water, and the load should be expressed in pounds of ice required that would have to be melted to handle steps A, B, and C. This load would be added to the refrigerated water load from the various other products such as milk, cottage cheese, and so forth, in sizing the ice-bank unit.

Whipped Butter

To whip butter by the batch method, the butter is tempered to 62 to 70 °F, depending on such factors as the season, type of whipper, and so forth. The butter is cut into slabs for placing into the whipping bowl. The whipping mechanism is activated, and air is incorporated until the desired overrun is obtained, usually between 50 and 100%. The whipped butter is packaged mechanically or manually into semirigid plastic containers.

With one continuous system, butter directly from cooler storage is cut into pieces and augered until soft. However, it can be tempered and the augering step omitted. The butter is then pumped into a cylindrical continuous whipper that uses the same principles as those for incorporating air in ice cream. Air or nitrogen is incorporated until the desired overrun is obtained. Another continuous method (used less commercially) is to melt butter or standardize butter oil to the composition of butter with moisture and salt. The fluid product is pumped through a chiller-whipper.

Metered air or nitrogen provides overrun control. Whipped butter is pumped in a soft state to the hopper of the filler and packaged in rigid or semirigid containers, such as plastic. It is chilled and held in storage at 32 to 40 °F.

CHEESE MANUFACTURE

Cheese is becoming increasingly important in the United States. Per capita consumption of cheese has increased approximately 36% during the last decade. Approximately 800 cheeses have been named, but there are only 18 distinctly different types. A few of the more popular types in the United States are cheddar, cottage, Roquefort or blue, cream, ricotta, mozzarella, Swiss, Edam, and Provolone. Such details of manufacture as setting (starter organisms, enzyme, milk or milk product, temperature, and time), cutting, heating (cooking), stirring, draining, pressing, salting, and curing (including temperature and humidity control) are varied to produce a characteristic variety and its optimum quality.

The production of cheddar cheese in the United States far exceeds the other cured varieties; cottage cheese production is much greater than that of the other uncured types. Another trend in the cheese industry is large factories. These plants may have sufficient curing facilities for the total production. If not, the cheese is shipped to central curing plants.

The physical shape of cured cheese varies considerably. The most common are 70-lb cheddar; 60-, 40-, and 20-lb blocks; 35-lb twin; 20-lb daisies; and 12-lb longhorns. There also are a variety of small units such as midgets: 1-, 2-, 5-, and 10-lb loaves. Barrel cheese is common; it is cured in a metal barrel or similar impervious container in units of approximately 500 lb. Cheese may also be cured in rectangular metal containers holding 2000 lb.

The microbiological flora of cured cheese are important in the development of flavor and body. Heating the milk for cheese is general practice. The milk may be pasteurized at the minimum HTST conditions or be given a subpasteurization treatment that results in a positive phosphatase test. This is possible when the milk quality is good (low level of spoilage microorganisms and pathogens). Such treatments of milk give the cheese some of the characteristics of raw-milk cheese in curing, such as production of higher flavors, in a shorter time. Pasteurization to produce phosphatase-negative milk is practiced in the making of soft, unripened varieties of cheese and some of the more perishable of the ripened types such as Camembert, Limburger, and Munster.

The standards and definitions of the Federal Food and Drug Administration and of most state regulatory agencies require that cheese that is not made from phosphatase-negative milk must be cured for not less than 60 days at not less than 35 °F. Raw-milk cheese contains not only lactic acid producing organisms such as *Streptococcus lactis*, which are added to the milk during the cheese making process, but also the heterogeneous mixture of microorganisms present in the raw milk, many of which may produce gas and off-flavors in the cheese. With the advent of milk pasteurization, some control of the bacterial flora of the cheese is possible.

Freshly manufactured cheese of the cured types is rubbery in texture and has little flavor; perhaps the more characteristic flavor is slightly acid. The presence of definite flavor(s) in freshly made cheese indicates poor quality, probably resulting from off-flavored milk. On curing under proper conditions, however, the body of the cheese breaks down, and the nut-like, full-bodied flavor characteristic of aged cheese develops. These changes are accompanied by certain chemical and physical changes during the curing process. The calcium paracaseinate of cheese gradually changes into proteoses, amino acids, and ammonia. These changes are a part of the ripening process and may be controlled by time and temperature of storage. As cheese cures, varying degrees of lipolytic activity also occur. In the case of blue, or Roquefort cheese, this partial fat breakdown contributes substantially to the characteristic flavor.

During curing, the microbiological development produces changes according to the species and strains present. It is possible to predict from the microorganism data some of the usual defects in cheddar. In some cheeses (*e.g.*, Swiss), gas production accompanies the desirable flavor development.

Cheese quality is evaluated on the basis of a score card. Flavor and odor, body and texture, and color and finish are principal factors. They are influenced by milk quality, the skill of manufacturing (including starter preparation), and the control effectiveness of maintaining optimum curing conditions.

Cheddar Cheese

Manufacture. Raw or pasteurized whole milk is tempered to 86 to 88 °F. It is set by adding 0.75 to 1.25% active cheese starter and annatto yellow color, which depends on the market demand. After 15 to 30 min, 99 mL of single-strength rennet per 1000 lb milk is diluted in water 1:40 and slowly added with agitation of milk in the vat. After a quiescent period of 25 to 30 min, the curd should have developed proper firmness. The curd is cut into 0.25 to 0.40 in. cubes. After 15 to 30 min of gentle agitation, the cooking is initiated by heating water in the vat jacket by means of steam for 30 to 40 min. The curd and whey should increase 2 °F per 5 min. Then a temperature of 100 to 102 °F is maintained for approximately 45 min.

The whey is drained and the curd trenched along both sides of the vat, allowing a narrow area free of curd the length of the midsection of the vat. Slabs about 10-in long are cut and inverted at 15-min periods during the cheddaring process. When acidity of the small whey drainage is at a pH of 5.3 to 5.2, the slabs are milled (cut into small pieces) and returned to the vat for salting and stirring, or the curd goes to a machine for the automatic addition of salt and its uniform incorporation into the curd. Weighed curd goes into hoops, which are placed into a press, and 20 psi is applied. After 0.5 to 1 h, the hoops are taken out of the press, the bandage adjusted to remove wrinkles, and then the cheese is pressed overnight at 25 to 30 psi or higher. Cheese may or may not be subjected to a vacuum treatment to improve body by reducing or eliminating air pockets. After the surface is dried, the cheese is coated by dipping into melted paraffin or wrapped with one of several plastic films, or oil with a plastic film, and sealed. Yield is about 10 lb per 100 lb of milk.

As cheese factories have grown larger, a change toward faster and more mechanized methods of making cheddar cheese has evolved. The stirred curd method (whereby the cheddaring step is omitted) is being used by more cheese makers. Deep circular or oblong cheese vats with special, reversible agitators and means for cutting the curd are becoming popular. The curd is pumped from these vats to draining and matting tables with sloped bottoms and low sides, then milled, slated, and hooped. The latest method makes use of a draining-matting conveyor with a porous plastic belt which permits draining as the curd is carried to a second belt for cheddaring. The second belt carries the cheddared curd to the mill. The milled curd is then carried to a finishing table where it is salted, stirred, and moved out for hooping.

Another system, imported from Australia, is used in a number of cheddar cheese factories. This system requires a short method of setting. After the curd is cut and cooked, it is transferred to a series of perforated stainless steel troughs traveling on a conveyor where draining and partial fusion take place. The slabs are then transferred into buckets of a forming conveyor, transferred again to transfer buckets, and finally to compression buckets where cheddaring takes place. Cheddared slabs are discharged to a slatted conveyor which carries them to the mill and then to a final machine where the milled curd is salted, weighed, and hooped.

Curing. Curing temperature and time vary widely among cheddar plants. A temperature of 50 °F cures the cheddar more rapidly

Dairy Products

than lower temperatures. The higher the temperature above 50°F to about 80°F, the more rapid the curing and the more likely that off-flavors will develop. At 50°F, 3 to 4 months are required for a mild to medium cheddar flavor. Six months or more are necessary for an aged (sharp) cheddar cheese. Relative humidity should be roughly 70%. Cheddar intended for processed cheese is cured in many plants at 70°F because of the economy of time. Some experts suggest that the cheddar, after its coating or wrapping, should be held in cold storage at approximately 40°F for about 30 days, then transferred to the 50°F curing room. During cold storage, the curd particles knit together forming a close-bodied cheese. The small amount of residual lactose is slowly converted to lactic acid, along with other changes in optimum curing.

The maximum legal moisture content of cheddar is 39% and the fat must be not less than 50% of total solids. The amount of moisture directly affects the curing rate to some extent within the normal range of 34 to 39%. Cheese having a loose or crumbly body and a high acidity is less likely to cure properly. For best curing, the cheese should have a sodium chloride content of 1.5 to 2.0%. A lower percentage encourages off-flavors to develop, and higher amounts retard flavor development.

Moisture Losses. The loss in weight of cheese during curing is largely attributed to moisture loss. Paraffined cheddar cheese going into cure averages approximately 37% moisture. After a 12-month cure at 40°F, paraffined cheese averages approximately 33% moisture. This loss is a real loss to the cheese manufacturer unless the cheese is sold on the basis of total solids. Control of humidity can have an important role in moisture loss. Figure 5 shows the loss from paraffined longhorns in boxes held at 38°F and 70% rh over 12 months. The conditions were well controlled but the average loss was 7%. The high loss shown on the graph was influenced by the larger surface area in 12-lb longhorns, as compared to 70-lb cheddars. Curing the cheese within a good quality sealed wrapper having a low moisture transmission (but some oxygen and carbon dioxide) largely eliminates loss of moisture.

Provolone (Pasta Filata Types)

Provolone is an Italian plastic-curd-type cheese which represents a large group of pasta-filata-type cheeses. While these cheeses vary widely in size and composition, they are all manufactured by a similar method. After the curd has been matted, like cheddar, it is cut into slabs which are worked and stretched in hot water at 150 to 180°F. The curd is kneaded and stretched in the hot water until it reaches a temperature of about 135°F. The maker then takes the amount necessary for one cheese and folds, rolls, and kneads it by hand to give the cheese its characteristic shape and smooth, closed surface. Molding machines have been developed for large scale operations to eliminate this hand labor. The warm curd of some varieties of pasta filata is placed in molds and submerged in cold water to harden into the desired shape. The hardened cheese is then salted in brine for from one to several days depending on the size and variety. Some pasta filata cheese, such as mozzarella for pizza, is packaged for shipment with wrappers to protect it for the period it is held before use. This cheese may be sealed under vacuum in plastic bags for prolonged holding.

Provolone is salted by submersion in 24% sodium chloride solution at 45°F for 1 to 3 days, depending on the size. It is hung on a light rope to dry and then transferred to the smokehouse and exposed to hickory or other hardwood smoke for 1 to 3 days. The cheese is hung in a curing room for 3 weeks at 55°F and then for 2 to 10 months at 40°F. The size varies as well as the shape, but the most common in the United States is 14 lb and pear-shaped. The moisture content ranges from 37 to 45% and the salt from 2 to 4%. Milkfat usually comprises 46 to 47% of the total solids. The yield is roughly 9.5 lb per 100 lb of milk.

Swiss Cheese

One of the distinguishing characteristics of Swiss cheese is the eye formation during curing. These eyes result from the development of CO_2. Raw or heat-treated milk is tempered to 95°F and pumped to a large kettle or vat. The starter, consisting of 27 mL of *Propioni bacterium shermanii*, 165 mL of *Streptococcus thermophilus,* and 165 mL of *Lactobacillus bulgaricus,* is added per 1100 lb of milk. After mixing, 77 mL of rennet per 1100 lb is diluted 1:40 with water and slowly added with agitation of the milk. Curd is cut when firm (after 25 to 30 min) into very small granules. After 5 min, curd and whey are agitated for 40 min, and then the steam is released into the jacket without water. Curd is heated slowly to 122 to 130°F in 30 to 45 min. Without additional steam, the cooking is continued until curd is firm and has no tendency to stick when a group of particles is squeezed together (0.5 to 1 h and whey pH 6.3). Curd is dipped into hoops (160 lb) and pressed lightly for 6 h, redressing and turning the hoops every 2 h. Pressing is continued overnight. The next step consists of soaking the cheese in brine until it has about 1.5% salt. Table 4 shows temperature and time at which curing occurs.

Roquefort and Blue Cheese

Roquefort and blue cheese require a mold (*Penicillium roqueforti*) to develop the typical flavor. Roquefort is made from ewes' milk in France. Blue cheese in the United States is made from cow's milk. The equipment used for the manufacture and curing of blue cheese is the same as that used for cheddar with a few exceptions. The hoops are 7.5 in. in diameter and 6 in. high. They have no top or bottom covers and are thoroughly perforated with small holes. A manually or pneumatically operated device with 50 needles, which are 6 in. long and 0.125 in. in diameter, is used to punch holes in the blue curd wheels. Apparatus is also needed to feed moisture into the curing room to maintain at least 95% rh without causing a drip onto the cheese.

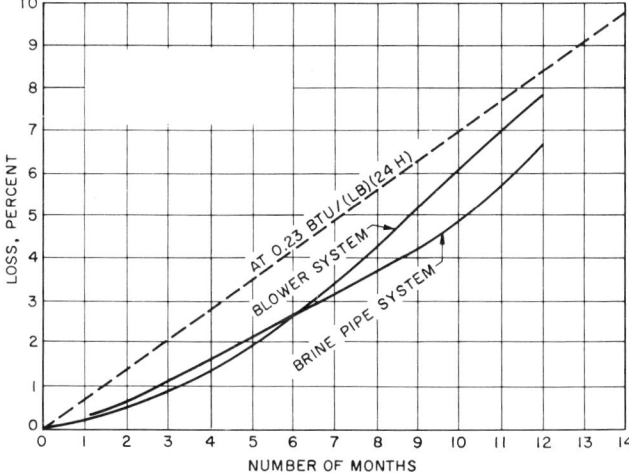

Fig. 5 Shrinkage of Cheese in Storage

Table 4 Swiss Cheese Manufacturing Conditions

Processing Step	Temperature, °F	Relative Humidity, %	Time
Setting	95	—	0.4 to 0.5 h
Cooking	122 to 130	—	1.0 to 1.5 h
Pressing	80 to 85	—	12 to 15 h
Salting (brine)	50 to 52	—	2 to 3 days
Cool room hold	50 to 60	90	10 to 14 days
Warm room hold	70 to 75	80 to 85	3 to 6 weeks
Cool room hold	40 to 45	80 to 85	4 to 10 mos

Table 5 Typical Blue Cheese Manufacturing Conditions

Processing Step	Temperature, °F	Relative Humidity, %	Time
Setting	85 to 86	—	1 h
Acid development	85 to 86	—	1 h
(after cutting)	92	—	120 s
Curd matting	70 to 75	80 to 90	18 to 24 h
Dry salting	60	85	5 days
Curing	50 to 55	>95	30 days
Additional curing	36 to 40	80	60 to 120 days

The milk may be raw or pasteurized and separated. The cream is bleached and may be homogenized at low pressure. Skim milk is added to the cream, and the milk is set with 2 to 3% active lactic starter. After 30 min, 3 to 4 oz of rennet per 1100 lb is diluted with water (1:35) and thoroughly mixed into the milk. When the curd is firm (after 30 min), it is cut into 5/8-in. cubes. Agitation is begun 5 min later. After whey acidity is 0.14% (1 h), the temperature is raised to 92°F and held for 20 min. The whey is drained and trenched. Approximately 2 lb of coarse salt and 1 oz of *P. roqueforti* powder is mixed into each 100 lb of curd.

The curd is transferred to stainless steel perforated cylinders (hoops). These hoops are inverted each 15 min for 2 h on a drain cloth, and curd matting is continued overnight. The hoops are removed and surfaces of the wheels covered with salt. The cheese is placed in a controlled room at 60°F and 85% rh and resalted daily for 4 more days (total 5 days). Small holes are punched through the wheels of cheese from top to bottom of the flat surfaces. The cheese is placed in racks on its curved edge in the curing room and held at 50 to 55°F and not less than 95% humidity. At the end of the month, the cheese surfaces are cleaned; the cheese is wrapped in foil and placed in a 36 to 40°F cold room for 2 to 4 months (Table 5). The surfaces are again scraped clean, and the wheels are wrapped in new foil for distribution.

Originally, Roquefort and blue cheese were cured in caves with high humidity and constant cool temperature. The refrigeration of insulated blue cheese curing rooms to the optimum temperature is not difficult. However, maintaining a uniform relative humidity of not less than 95% without excessive expense seems to be an engineering challenge, at least in some plants.

Cottage Cheese

Cottage cheese is made from skim milk. It is a soft, unripened curd and generally has a cream dressing added to it. There are small and large curd types, and they may or may not have added fruits or vegetables. The cheese plant equipment may consist of receiving apparatus, storage tanks, clarifier-separator, pasteurizer, cheese vats with mechanical agitation, curd pumps, drain drum, blender, filler, conveyors, and such accessory items as refrigerated trucks, laboratory testing facilities, and whey disposal equipment. The largest vats have a 45,000-lb capacity. The basic steps are separation, pasteurization, setting, cutting, cooking, draining and washing, creaming, packaging, and distribution.

Skim milk is pasteurized at the minimum temperature and time of 161°F for 15 s to avoid adversely affecting the curd properties. If substantially higher heat treatment is practiced, the cottage cheese manufacturing procedure must be altered to obtain good body and texture quality and reduce curd loss in the whey. Skim milk is cooled to the setting temperature, which is 86 to 90°F for the short set (5 or 6 h) and 70 to 72°F for the overnight set (12 to 15 h). A medium set is used in a few plants. For the short set, 5 to 8% of a good cultured skim milk (starter) and 1 to 1.5 mL of rennet diluted in water are added per 1000 lb of skim milk. For the long set, 0.25 to 1% starter and 0.5 to 1 mL of diluted rennet per 1000 lb are thoroughly mixed into the skim milk. The use of rennet is optional. The setting temperature is maintained until the curd is ready to cut. The acidity of whey at cutting time depends on the total solids content of the skim milk (0.55% for 8.7% and 0.62% for 10.5%). The pH is typically 4.80, but it may be necessary to adjust for specific make procedures.

The curd is cut into 0.5-in. cubes for large curd and 0.25 in. for small curd cottage cheese. After the cut curd sets for 10 to 15 min, heat is applied to water in the vat jacket to maintain a temperature rise in the curd and whey of 2°F each 5 min. In very large vats, jacket heating is not practical, and superheated culinary steam in small jet streams is used directly in the vat; 20 to 30 min after cutting, very gentle agitation is applied. Heating rate may be increased to 3 or 4°F per 5 min as the curd firms enough to resist shattering. Cooking is completed when the cubes contain no whey pockets and have the desired firmness. The final temperature of curd and whey is usually 120 to 130°F, but some cheesemakers heat to 145°F when making the small type curd.

After cooking is completed, the hot water in both the jacket and the whey is drained. Wash water temperature is adjusted to about 70°F for the first washing and added gently to the vat to reduce curd temperature to 80 to 85°F. After gentle stirring and a brief hold, the water-whey mixture is drained. The temperature of the second wash is adjusted to reduce the temperature of curd to 50 to 55°F and to 40°F with a third wash. Water for the last wash may have 3 to 5 ppm of added chlorine. The curd is trenched for adequate drainage. The dressing is made from low-fat cream, salt, and usually 0.1 to 0.4% stabilizer based on cream weight. Salt averages 1%, and milkfat must be 4% or more in creamed cottage cheese or 2% in low-fat cottage cheese. The dressing is cooled to 40°F and blended into the curd.

A cheese vat can be reused sooner if the cheese pumps quickly convey the curd and whey at the completion of cooking to a special tank for drainage of the whey, washing, and blending of the dressing with the curd. Creamed cottage cheese is transferred mechanically to an automatic packaging machine. One type of filler employs an oscillating cylinder which holds a specific volume. Another type has a piston in a cylinder which discharges a definite volume. The common retail containers are 32-, 16-, 12- and 8-oz sizes of semirigid plastic. Cottage cheese is perishable and must be stored at 40°F or lower for prolonging the keeping quality to 2 or 3 weeks. A good yield is 15.5 lb of curd per 100 lb of skim milk with 9% total solids.

Other Cheeses

Table 6 presents data on a few additional common varieties of cheese in the United States. With the exception of the soft ripened cheeses such as Camembert and Liederkranz, freezing of cheese results in undesirable texture changes. This can be serious, as in the case of cream cheese, where a mealy, pebbly texture results. Other types, such as brick and Limburger, undergo a slight roughening of texture which is undesirable but which still might be acceptable to certain consumers. As a general rule, cheese should not be subjected to temperatures under 29°F.

When cured cheese is held above the melting point of milkfat, it becomes greasy because of the oiling off. The oiling-off point

Table 6 Curing Temperature, Humidity, and Time of Some Cheese Varieties

Variety	Curing Temperature, °F	Relative Humidity, %	Curing Time
Brick	60 to 65	90	60 days
Romano	50 to 60	85	5 to 12 mos
Mozzarella	70	85	24 to 72 h
Edam	50 to 60	85	3 to 4 mos
Parmesan	55 to 60	85 to 90	14 mos
Limburger	50 to 60	90	2 to 3 mos

Dairy Products

Table 7 Temperature Range of Storage, Common Types of Cheese

Cheese	Ideal Temperature, °F	Maximum Temperature, °F
Brick	30 to 34	50
Camembert	30 to 34	50
Cheddar	30 to 34	60
Cottage	30 to 34	45
Cream	32 to 34	45
Limburger	30 to 34	50
Neufchatel	32 to 34	45
Process American	40 to 45	75
Process brick	40 to 45	75
Process Limburger	40 to 45	75
Process Swiss	40 to 45	75
Roquefort	30 to 34	50
Swiss	30 to 34	60
Cheese foods	40 to 45	55

of all types of cheese except process cheese begins at 68 to 70 °F. Consequently, storage should be substantially below the melting point (Table 7). Uncured cheese (*e.g.*, cottage, cream) is highly perishable and thus should not be stored above 45 °F and preferably at 35 °F.

Processing protects cheese from oiling off. By heating the bulk cheese to temperatures of 140 to 180 °F, and through the incorporation of emulsifying salts, a more stable emulsion is formed than in natural or nonprocessed cheese. Process cheese will not oil off even at melting temperatures. Because of the temperatures used in processing, process cheese is essentially a pasteurized product. Microorganisms causing changes in the body and flavor of the cheese during cure are largely destroyed; hence there will be practically no further flavor development. Consequently, the maximum permissible temperature for the storage of process cheese is considerably higher than any of the other types. Table 7 shows the maximum temperatures of storage for cheese of various types.

Refrigeration of Cheese Rooms

Cheeses that are to be dried prior to wrapping or waxing enter the cooler at approximately room temperature. Sufficient refrigerating capacity must be provided to reduce the cheeses to drying room temperature. The product load may be taken as 1 ton for each 12,000 to 15,000 lb per day. Product load in a cheese-drying room is usually small compared to total room load. Extreme accuracy in calculating product load is not warranted.

When determining the peak refrigerating load in a cheese drying room, the factor to be remembered is that peak cheese production may coincide with periods of high ambient temperature. In addition, these rooms normally open directly into the cheese-making room, where both temperature and humidity are quite high. Also, traffic in and out of the drying room may be heavy; therefore, ample allowance for door losses should be made. Two to three air changes per hour are quite possible during the flush season.

To maintain the desired humidity, refrigerating units for the cheese-drying room should be sized to handle the peak summer load with not more than a 20 °F difference between average air temperature and evaporator temperature. Units operated from a central refrigerating system should be equipped with a back pressure regulator.

Temperature control is best obtained through a room thermostat controlling a solenoid valve in the liquid supply to the unit or units. Fans should be allowed to run continuously. The modulating back pressure regulator is not a satisfactory temperature control for a cheese-drying room because it causes undesirable variations in humidity.

Air circulation should only be sufficient to assure uniform temperature and humidity throughout the room. Strong drafts or air currents should be avoided because they cause uneven drying and cracking of the cheeses. The most satisfactory refrigerating units are the ceiling-suspended between-the-rails type. One unit for each 400 to 500 ft^2 of floor area usually assures uniform conditions. One unit should be placed near the door to the room to cool the warm moist air before it has a chance to spread over the ceiling. Otherwise, condensation dripping from the ceiling or mold growth will result.

Humidity control during winter presents certain problems. Since most of the refrigeration load (during the peak season) is due to insulation losses and warm air entering through the door, refrigerating units may not operate enough during cold weather to take up the moisture given up by the cheese, resulting in excess humidity and improper drying of the cheese. Within certain limits, this can be overcome by reducing the speed of the unit fans and lowering the back pressure. If there are several units in the room, the refrigeration may be turned off on some. Fans should be left running to assure uniform conditions throughout the room. If these adjustments are not sufficient, or if automatic control of humidity is desired, it is necessary to use reheat coils in the airstream leaving the units. These may be electric heaters, steam or hot water coils, or hot gas from the refrigerating system. A heating capacity of 15 to 20% of the refrigerating capacity of the units is usually sufficient.

A humidistat may be used to operate the electric heater when the humidity rises above the desired level. The heater should be wired in series, with a second room thermostat set to shut it off if the room temperature becomes excessive. Because of variation in size and shape of drying rooms, it is impossible to generalize on air velocities and capacities. Airflow should be regulated so that the cheese feels moist for the first 24 h, after which it becomes progressively drier and firmer.

Calculation of the refrigeration load for a cheese curing room involves a simple computation of heat to be removed from the cheese at the incoming temperature in order to bring it to curing temperature, using 0.65 Btu/(lb·°F) as the specific heat of cheese. For most varieties, the heat given off during cure is negligible.

While the fermentation of lactose to lactic acid which occurs in cheese is an exothermic reaction, this process is substantially completed in the first week after cheese is made; hence further heat given off during cure is of no significance. Assuming that average conditions for American cheese curing are approximately 45 °F and 70% rh, if 30 to 35 °F refrigerant is used in the cooling system, a humidity of about 70% will be maintained.

FROZEN DAIRY DESSERTS

Ice cream is the most common frozen dairy dessert. Legal guidelines for the composition of frozen dairy desserts generally follow the Federal Standards. The amount of air incorporated during freezing is controlled for the prepackaged products by the standard specifying the minimum density and/or a minimum density of food solids.

The basic dairy components of frozen dairy dessert are milk, cream, and condensed or nonfat dry milk. Some plants also use butter, butter oil, buttermilk (liquid or dry), and dry or concentrated whey of the sweet type. The acid type whey (*e.g.*, from cottage cheese) can be used for sherbets.

Ice Cream

Milkfat content (called butterfat by some standards) is one of the principal factors in the legal standards for ice cream. The fat in other ingredients such as eggs, nuts, cocoa, or chocolate do not satisfy the legal minimum. Federal standards set the minimum milkfat content at 8% for bulky flavored ice-cream mixes (*e.g.*, chocolate) and 10% or above for the other flavors (*e.g.*, vanilla). Manufacturers, however, usually make two or more grades of ice cream, one being competitively priced with the minimum legal fat content, and the others richer in fat, higher in total solids, and lower in overrun for a special trade. Such ice cream may be made with a fat content of 16 or 18%, although most ice cream is made with a fat content ranging from 10 to 12%.

Serum solids content designates the nonfat solids from milk. The chief components of milk serum are lactose, milk proteins (casein, albumin, and globulin), and milk salts (sodium, potassium, calcium, and magnesium as chlorides, citrates, and phosphates). The following average composition percentages for serum solids are useful for general calculations: lactose, 54.5; milk proteins, 37.0; and milk salts, 8.5.

The serum solids in ice cream produce a smoother texture, better body, and better melting characteristics. Because serum solids are relatively inexpensive compared with fat, they are used liberally. The total solids content usually is kept below 40%.

The lower limit in the serum solids content, 6 to 7%, is found in a homemade type of ice cream, where the only dairy ingredients are milk and cream. Ice creams with an unusually high fat content are also kept near this serum solids content so that the total solids content will not be excessive. Most ice cream, however, is made with condensed or nonfat dry milk added to bring the serum solids content within the range of 10 to 11.5%. The upper extreme of 12 to 14% serum solids can be used only where rapid product sales turnover or other special means avoids sandiness.

The sugar content of ice cream is of special interest because of its effect on the freezing point of the mix and its hardening behavior. The extreme range of sugar content encountered in ice cream is from 12 to 18%, with 16% being most representative of the industry. The chief sugar used is sucrose (cane or beet sugar), either in granulated or liquid form. Many manufacturers use dextrose and corn syrup solids to replace part of the sucrose. Some manufacturers prefer sucrose in liquid form, or in a mixture with syrup, because of lower cost and easier handling in tank car lots. In some instances, 50% of the sucrose content has been replaced by other sweetening agents. A more common practice is to replace from one-fourth to one-third of the sucrose by dextrose or corn syrup solids, or a combination of the two.

Practically all ice cream is made with a stabilizer to help maintain a smooth texture, especially under the conditions that prevail in retail cabinets. Manufacturers who do not use stabilizers offset this omission by a combination of factors such as a high fat and solids content, the use of superheated condensed milk to aid in smoothing the texture and imparting body, and a sales program designed to provide rapid turnover. The stabilizing substances most commonly added are carboxymethylcellulose (CMC) and sodium alginate, a product made from giant kelp gathered off the coast of California. Gelatin is used for some ice-cream mixes that are to be batch pasteurized. Other stabilizers are locust or carob bean gum, gum arabic or acacia, gum tragacanth, gum Karaya, psyllium seed gum, and pectin. The amount of stabilizer commonly used in ice cream ranges from 0.20 to 0.35% of the mass of the mix.

Many plants now combine an emulsifier with the stabilizer to produce a smoother and richer product. The emulsifier reduces the surface tension between the water and fat phase to produce a drier-appearing product.

Egg solids in the form of fresh whole eggs, frozen eggs, or powdered whole eggs or egg yolks are used by some manufacturers. While flavor and color may motivate this choice, the most common reason for selecting them is to aid the whipping qualities of the mix. The amount required is about 0.25% egg solids, with 0.50% being about the maximum content for this purpose. To obtain the desired result, the egg yolk should be in the mix at the time it is being homogenized.

In frozen custards or parfait ice cream, the presence of eggs in liberal amounts and the resulting yellow color are identifying characteristics. Federal standards specify a minimum 1.4% egg yolk content for these products.

Milk Ices or Ice Milk

Ice milk commonly contains 3 to 4% fat and 13 to 15% serum solids; practices with respect to sugar and stabilizers are very similar to those for ice cream. The sugar content is somewhat higher in order to build up the total solids content, and the stabilizer content is higher than ice cream, approximately in proportion to the higher water content (lower total solids content) of ice milk. The overrun is approximately 70%.

Soft Ice Milk or Ice Cream

Machines that serve freshly frozen ice cream are common in roadside stands, retail ice-cream stores, and restaurants. These establishments must meet sanitation requirements and have facilities for proper cleaning of the equipment, but very few blend and process the ice-cream mix used. The mix is usually supplied either from a plant specializing in producing ice-cream mix only or from an established ice-cream plant. The mix should be cooled to about 35 °F at the time of delivery, and the ice-cream outlet should have ample refrigerated space in which to store the mix until it is frozen. To be served in a soft condition, this ice cream is usually frozen stiffer than would be customary for a regular plant operation and at 30 to 50% overrun. Some mixes are prepared only for soft serve. They are 1 to 2% greater in serum solids and have 0.5% stabilizer-emulsifier to aid in producing a smooth texture. The overrun is limited to 30 to 40% during freezing to the softserve condition.

Sherbets

Sherbets are fruit (and mint) flavored frozen desserts characterized by their high sugar content and tart flavor. They must contain between 1 and 2% milkfat and not more than 5% by mass of total milk solids. While the milk solids can be supplied by milk, the general practice is to supply them by using ice-cream mix. Typically, a solution of sugars and stabilizer in water is prepared as a base for sherbets of various flavors. To 70 lb of sherbet base, 20 lb of flavoring and 10 lb of ice-cream mix are added. The sugar content of sherbets ranges from 25 to 35%, with 28 to 30% being most common. One example of a sherbet formula is: milk solids, 5%, of which 1.5 is milkfat; sugar, 13%; corn syrup solids, 22%; sherbet stabilizer, 0.3%; and flavoring, acid, and water, 59.7%.

In sherbets, and even more so in ices, a high overrun is not desirable because the resulting product will appear foamy or spongy under serving conditions. The overrun should be kept within 25 to 40%. This fact and the problem of preventing bleeding (the leakage of syrup from the frozen product) emphasize the importance of the choice of stabilizers. If gelatin is selected as the stabilizer in sherbets, the freezing conditions must be managed to avoid an excessive overrun. The gums added to ice cream are commonly used as the stabilizer in sherbets and ices.

Ices

Ices contain no milk solids, but closely resemble sherbets in other respects. To offset the lack of solids from milk, the sugar content of ices is usually slightly higher than in sherbets; it is usually 30 to 32%. A combination of sugars should be used to prevent crusty sugar crystallization, just as in the case of sherbets. The usual procedure is to make an ice-base solution of the sugars and stabilizer, from which different flavored ices may be prepared by adding the flavoring material in the same general manner as mentioned for sherbets.

Ices contain few ingredients with lubricating qualities and often cause extensive wear on scraper blades in the freezer. Frequent resharpening of the blades is necessary. Where a number of

Dairy Products

freezers are available, and the main production is ice cream, it is desirable to confine the freezing of ices and sherbets to a specific freezer or freezers, which should then receive special attention to resharpening.

Making the Ice-Cream Mix

The chosen composition for a typical ice cream would be:

Fat	12.5%	Sugar	15.0%
Serum solids	10.5%	Stabilizer	0.3%

Mixing and Pasteurizing. Generally, the the liquid dairy ingredients are placed in a vat equipped with suitable means of agitation, especially to keep the sugar in suspension until it is dissolved. The dry ingredients are then added, with suitable precautions to prevent lump formation of such products as stabilizers, nonfat dry milk solids, powdered eggs, and cocoa. Gelatin should be added while the temperature is still low to allow time for the gelatin to imbibe water before its dissolving is promoted by heat. Dry ingredients that tend to form lumps may be successfully added by first mixing them with some of the dry sugar so that moisture may penetrate freely. Where vat agitation is not fully adequate, sugar may be withheld until the liquid portion of the mix is partly heated so that promptness of solution avoids settling out.

The mix is pasteurized to destroy any pathogenic organisms, to lower the bacterial count so as to enhance the keeping quality of the mix and comply with bacterial count standards, to dissolve the dry ingredients, and to provide a temperature suitable for efficient homogenization. A pasteurizing treatment of 155 °F maintained for 0.5 h is the minimum allowed. The mix should be homogenized at the pasteurizing temperature. Vat batches should be homogenized in 1 h and preferably less.

Many ice cream plants use continuous pasteurization methods that use plate-type heat exchange equipment for heating and cooling the mix. If some solid ingredients are selected, such as skim milk powder and granulated sugar, a batch is made up in a mixing tank at a temperature of 100 to 140 °F. This preheated mix is then pumped through a heating section of the plate unit where it is heated to a temperature of 175 °F or higher, and held for 25 s while passing through a holding tube. The mix is then homogenized and pumped to the precooling plate section using city, well, or cooling tower water as the cooling medium. Final cooling of the mix may be done in an additional plate section, using chilled water as the cooling medium, or the mix may be cooled through a separate mix cooler. A glycol medium is sometimes used for cooling to temperatures just above the freezing point.

Large plants generally use all liquid ingredients, especially if the production is automated and computerized. The ingredients are blended at 40 to 60 °F. The mix passes through the product-to-product regeneration section of a plate-type heat exchanger with about 70% regeneration during preheating. The mix is HTST heated to not less than 175 °F, homogenized, and held for 25 s. Greater heat treatment is common, and 220 °F for 40 s is not unusual. The final heating may be accomplished with plate equipment, a swept-surface heat exchanger, or a direct steam injector or infusor.

Steam injection and infusion equipment may be followed by vacuum chamber treatment, whereby the mix is flash cooled to 180 to 190 °F by a partial vacuum. It is further cooled through a regenerative plate section and additionally cooled indirectly to 40 °F or less with chilled water. The chief advantage of the vacuum treatment is the flavor improvement of the mix if prepared from mediocre quality raw materials.

Homogenizing the Mix. Homogenization disperses the fat in a very finely divided condition so it will not churn out during freezing. Most of the fat in milk and cream is in globules 3 to 7 μm in diameter. Some of the globules are as large as 12 μm or larger in diameter, especially if there has been some churning incidental to handling. In a properly homogenized mix, globules are seldom over 2 μm in diameter.

Cooling and Holding Mix. Methods of final cooling of ice-cream mix after pasteurization depend on the equipment used and the final mix temperature desired. The mix should be as cold as possible, to about 30 °F minimum for greater capacity and less refrigeration load on the ice cream freezers. Smaller plants generally use the vat holding system of pasteurization with either a Baudelot-type surface cooler or a plate-type heat exchanger, both with precooling and final cooling sections. The precooling may be done with city, well, or cooling tower water, and mix leaving the precooling section is about 10 °F warmer than the entering water temperature. The Baudelot-type cooler may be arranged for final cooling with chilled water, brine, or direct expansion refrigerant. A final mix temperature of 30 to 33 °F can be obtained over the surface cooler using brine or refrigerant. Final mix temperature when using chilled water will be about 40 °F.

For larger ice-cream plants, where low mix temperature is desired and where plate-type pasteurizing equipment is installed, it may be desirable to use separate equipment for the final cooling. Where the mix is preheated to about 140 °F, it will be precooled to about 10 °F warmer than the entering precooling water temperature; final cooling can be done in a remote cooler. An ammonia jacketed, scraped surface chiller is often selected. Where cold liquid mix is used through a continuous pasteurizing, high-heat vacuum system with regeneration at about 70%, the temperature of the mix to the final cooling unit will be 85 °F, assuming 40 °F original mix temperature and 190 °F temperature of mix returning through the regenerating section.

Where plants have ample ice-cream mix holding tank capacity (enabling the mix to be held overnight), part of the final mix cooling may be accomplished by means of a refrigerated surface built into the tanks. Using refrigerated mix holding tanks, the average rate of cooling may be estimated at 1 °F/h. Mixes made up with gelatin as a stabilizer should be aged 24 h to allow time for the full set of the gelatin to develop. Mixes made with sodium alginate or other vegetable-type stabilizers develop maximum viscosity on being cooled, and can be used in the freezer immediately.

Freezing Procedure

The ice-cream freezer freezes the mix to the desired consistency and whips in the desired amount of air in a finely divided condition. The aim is to conduct the freezing and later hardening to obtain the smoothest possible texture.

Freezing an ice-cream mix means, of course, freezing a mixed solution. The solutes that determine the freezing point are the lactose and soluble salts contained in the serum solids and the sugars added as sweetening agents. The other constituents of the mix affect the freezing point only indirectly, by displacing water and affecting the in-water concentration of the solutes mentioned. Leighton (1927) developed a reliable method for computing the freezing points of ice-cream mixes from their known composition. He added the lactose and sucrose content of the mix, expressed their concentration in terms of parts of sugar per 100 parts of water, and determined the freezing-point depression due to the sugars by reference to published data for sucrose. This computation is justified because lactose and sucrose have the same molecular mass.

$$\% \text{ lactose in mix} = 0.545 \, (\% \text{ serum solids})$$

$$\frac{(\% \text{ lactose} + \% \text{ sucrose}) \, 100}{\% \text{ water in mix}} = \frac{\text{parts lactose} + \text{sucrose}}{\text{per 100 parts water}}$$

To the freezing point depression caused by these sugars, he added the depression that will be caused by the soluble milk salts. The depression caused by the salts is computed as follows:

Freezing-point depression caused by salt solids in °F
$= 4.27 \, (\% \text{ serum solids})/\% \text{ water in mix}$

Table 8 Freezing Points of Typical Ice Creams, Sherbet, and Ice

	Composition of the Mix, %					Freezing Point, °F
	Fat	Serum Solids	Sugar	Stabilizer	Water	
Ice cream	8.5	11.5	15	0.4	64.6	27.59
	10.5	11.0	15	0.35	63.15	25.57
	12.5	10.5	15	0.30	61.7	27.55
	14.0	9.5	15	0.28	61.22	27.68
	16.0	8.5	15	0.25	60.25	27.79
	10.5	8.4	{ S 12 / D 4 }	0.40	64.7	27.39
Sherbet	1.2	1.0	{ S 22 / D 8 }	0.50	67.3	25.97
Ice	0	0	{ S 23 / D 9 }	0.50	67.5	25.68

S = Sucrose   D = Dextrose

Table 9 Freezing Behavior of a Typical Ice Cream

Water Frozen to Ice, %	Freezing Point of Unfrozen Portion, °F	Water Frozen to Ice, %	Freezing Point of Unfrozen Portion, °F
0	27.55	40	24.40
5	27.35	45	23.63
10	27.05	50	22.62
15	26.78	55	21.42
20	26.40	60	19.79
25	26.04	70	14.99
30	25.70	80	5.14
35	25.03	90	−22.29

Composition, %: fat, 12.5; serum solids, 10.5; sugar, 15; stabilizer, 0.30; water, 61.7.

Table 8 presents the freezing points of various ice creams and a typical sherbet and an ice, as computed by Leighton's method. The freezing point represents the temperature at which freezing commences. As in the case of all solutions, the unfrozen portion becomes more concentrated as the freezing progresses, and the freezing temperature therefore decreases as the freezing progresses. In a simple solution, containing only one solute, this trend will progress until the unfrozen portion represents a saturated solution of the solute, and thereafter the temperature will remain constant until the freezing has been completed. This temperature is known as the cryohydric point of the solute. In a mixed solution such as ice cream, which contains several sugars and a number of salts, no such point can be recognized.

On the contrary, the sugars remain in solution in a supersaturated state in the unfrozen portion of the product. This is due to the fact that by the time the saturation point has been reached, the temperature is so low and the viscosity so high that essentially a glass state exists. In a mixed solution, however, the temperature required for complete freezing must be somewhat below the cryohydric point of that solute with the lowest cryohydric point. In ice cream, that solute is calcium chloride, contained as a component of serum solids. The cryohydric point of calcium chloride is −59.8°F. Therefore, the ice cream ranges from 0 to 100% frozen within the approximate range of 27.5 to −67°F.

Therefore, the temperature to which the ice cream has been frozen becomes a measure of the degree to which it has been frozen, as illustrated by Table 9. In the table, the freezing points of the unfrozen portions of the third ice cream listed in Table 8 have been computed when 0 to 90% of the original water has been frozen out as ice.

Refrigeration Requirements. Exact calculation of refrigeration requirements is complicated by the number of factors involved. The specific heat of the mix varies with its composition. According to Zhadan (1940), the specific heat of food products may be computed by assuming the following specific heats in Btu/(lb·°F) for the chief components: carbohydrates, 0.34; proteins, 0.37; fats, 0.40; and water, 1.00. Salts are normally not included. Where they are present in significant amounts, as in ice cream (9.5% of the serum solids), a specific heat of 0.20 is accurate. The value given by Zhadan for fats is apparently for solid fats. For milkfat in a liquid condition, Hammer and Johnson (1913) found the specific heat to be 0.52. In addition, their data clearly show that the latent heat of fusion of fats becomes involved. From their data, the latent heat of fusion of milkfat is about 35 Btu/lb.

The change from liquid to solid fat occurs over a wide temperature range, approximately 80 to 40°F; in changing from solid to liquid fat, the range is approximately 50 to 105°F. This wide discrepancy between solidifying and melting behavior is apparently due to the fact that milkfat is a mixture of glycerides, and mutual solubility of the glycerides is involved. In any case, the latent heat of fusion of the fat is involved in cooling the mix from the pasteurizing and homogenizing temperature down to the aging temperature of 38 to 40°F. Instead of making detailed calculations, a specific heat of 0.80 Btu/(lb·°F) is assumed for ice-cream mix, which is high for mixes ranging from 36 to 40% total solids.

In calculating the refrigeration required for freezing and hardening, a single value of a specific heat for frozen ice cream cannot be chosen. As shown in Table 9, any change in temperature in freezing and hardening involves some latent heat of fusion of the water, as well as the sensible heat of the unfrozen mix and the ice. Near the initial freezing point, much more latent heat of fusion is involved per degree temperature change than in well-hardened ice cream, e.g., at −10 to −11°F. For this reason, instead of using an overall value of specific heat, freezing load may be computed as follows:

1. First determine the temperature to which the freezing is to be carried; then determine (by calculations such as those used to develop Table 9) how much water will be converted to ice. The heat to be removed is the product of the heat of fusion of ice and the mass of water frozen.

2. Compute the sensible heat that must be removed in the desired temperature change, by treating the product as a mix; i.e., use the specific heat for ice-cream mix. The temperature change times the lb of product times 0.80 = sensible heat to be removed.

In such a calculation, the water present is treated as though it all remained in a liquid form until the desired temperature had been reached, although ice was forming progressively. Because ice has a specific heat of 0.492 instead of 1.00 as for water, this calculation will err in the direction of generous refrigeration. To offset this, the freezer agitation develops friction heat. Approximately 80% of the energy input in the motor of the freezer is converted to heat in the product. Where the product is frozen to a stiff consistency, power requirements increase, and should be added to the load calculation.

A gallon of ice-cream mix weighs from 9 lb for mixes with a high fat content, to 9.2 lb for mixes with a low fat content and a high content of serum solids and sugar. The weight of a unit volume of ice cream varies with the mix weight and overrun according to the following relationship:

$$\frac{\text{Percentage}}{\text{overrun}} = 100 \frac{(\text{Wt/gal of mix} - \text{Wt/gal of ice cream})}{\text{Wt/gal of ice cream}}$$

Freezing Ice Cream. Both batch and continuous ice-cream freezers are in general use. Both are arranged with a freezer cylinder having either an annular space or coils around the cylinder, where cooling is accomplished by direct refrigerant cool-

Dairy Products

ing, either in a flooded arrangement with an accumulator or controlled by a thermostatic expansion valve. The freezer cylinder has a dasher, which revolves within the cylinder. Sharp metal blades on the dasher scrape the inner surface of the cylinder to remove the frozen film of ice cream as it forms. Some batch freezers use plastic dashers and blades.

Batch freezers range in size from 2 to 40 quarts of ice cream per batch, the smaller sizes being used for retail or soft ice-cream operations, and the 40-quart size used in small commercial ice-cream plants or in large plants for running small special-order quantities. Batch freezers larger than the 40-quart size have not been used extensively since the development of the continuous freezer.

In operation, a measured quantity of mix is placed in the freezer cylinder and the required flavor, fruit, or nuts are added as freezing of the mix progresses. Freezing is continued until the desired consistency is obtained in the judgment of the operator or by the indication of a meter showing an increase in the amperage drawn by the motor as the partly frozen mix becomes stiffer. At the desired point of freezing, the refrigeration is cut off from the freezer cylinder, usually by closing the refrigerant suction valve. The dasher continues operating until enough air has been taken into the mix by the whipping action to produce the desired overrun. The overrun is checked by taking and weighing a sample from the freezer. When the desired overrun is obtained, the entire batch is discharged from the freezer cylinder into cans or cartons, and the machine is then ready for a new batch of mix.

The output of a batch-type freezer varies with the sharpness of the blades, the refrigeration supplied, and the overrun desired. The average maximum output for commerical batch freezers is eight batches per hour. This schedule allows 180 to 240 s to freeze, 120 to 180 s to whip, and about 60 s to empty the ice cream and refill with mix. For this time schedule, it is assumed the ice cream is drawn from the freezer at not over 100% overrun, at a temperature of about 24°F and at a refrigerant temperature around the freezer cylinder of about $-15°F$.

Continuous ice-cream freezers range in size from 40 to 2700 gal/h at 100% overrun. This type of machine is used almost exclusively in commercial ice-cream plants. Where large capacities are required, multiple units are installed with the ice-cream discharge from several machines connected together to supply the requirements of automatic or semiautomatic packaging or filling machines. In operation, the ice-cream mix is continuously pumped to the freezer cylinder by a positive displacement rotary pump. Air pressure within the cylinder is maintained from 20 psig to more than 100 psig, supplied by either a separate air compressor or drawn in with the mix through the mix pump. The mix entering the rear of the freezer cylinder becomes partly frozen and takes on the overrun because of air pressure and the agitation of the dasher and freezer blades as it moves to the front of the cylinder and is discharged.

The output capacity of most continuous freezers can be varied from 50 to 100% rated capacity by regulating the variable speed control supplied for the mix pump. Continuous freezers can be used for nearly every flavor of ice cream, ice milk, sherbets, and ices. Where flavors requiring nuts, whole fruits, or candy pellets are run, the base or unflavored mix is run through the continuous freezer and then passed through a fruit feeder which automatically feeds and mixes the flavor particles into the ice cream. Ice cream can be discharged from continuous freezers at temperatures of 25°F, as required for ice-cream bar (novelty) operations, up to a very stiff consistency at 20°F, as required for automatic filling of small packages.

Special low-temperature ice-cream freezers are available to produce very stiff ice cream for extruded shapes, stickless bars, and sandwiches. Ice-cream temperatures as low as 16°F can be drawn with some mixes. When ample refrigerating effect is supplied, a variation of ice-cream discharge temperature can be obtained by regulating the evaporator temperature around the freezer cylinder by a suction pressure regulating valve. For filling cans and cartons, the average discharge temperature from the continuous freezer is about 22°F, when operating with ammonia in a flooded system at about $-25°F$.

To calculate accurately the refrigeration requirement for freezing the ice-cream mix in the freezer, the weight of the mix per gal and the amount of water should be known. This can be checked by weighing, knowing the percentage of water, or by calculating the weight from the mix formula, as in Example 2.

Example 2. Find the weight of mix for the following composition (by percent): milkfat, 12.0; serum solids, 10.5; sugar, 16.0; stabilizer, 0.25; egg solids, 0.25; and water, 61.0.

Solution: The specific gravity of the mix is

$$\frac{100}{\left(\dfrac{\%\text{ milkfat}}{0.93} + \dfrac{\%\text{ solids, not fat}}{1.58} + \dfrac{\%\text{ water}}{1.00}\right)}$$

The specific gravity of the ice-cream mix listed above would be

$$\frac{100}{\left(\dfrac{12}{0.93} + \dfrac{27}{1.58} + \dfrac{61}{1.00}\right)} = 1.099$$

Wt per gal of mix = wt of 1 gal of water × sp gr mix
$$= 8.355 \times 1.099 = 9.18 \text{ lb}.$$

The overrun in ice cream varies from 60 to 100%, which affects the required refrigeration. For a continuous freezer, the required refrigeration may be calculated as in Example 3.

Example 3. Assume a typical ice-cream mix as listed in Example 2 with 100% overrun. The mix contains 61% water and goes to the freezer at a temperature of 40°F. Freezing would start in this mix at about 27°F, and 48% of the water would be frozen at 22°F.

The weight of mix required to produce 100 gal of ice cream is:

$$\frac{100}{100 + \%\text{ overrun}} \times 100 \text{ gal} \times \text{Wt mix per gal}$$

For the ice cream being considered, the weight of mix required for 100 gal would be:

$$\frac{100}{100 + 100} \times 100 \times 9.18 = 459 \text{ lb}$$

Calculations of capacity required to freeze 100 gal/h of ice cream are as follows:

Sensible heat of mix: $459 \times (40 - 27) \times 0.80 = $ 4770 Btu/h
Latent heat: $459 \times 0.61 \times 0.48 \times 144 = $ 19,350 Btu/h
Sensible heat of slush: $459 \times (27 - 22) \times 0.65 = $ 1490 Btu/h
Heat from motors: $5.5 \text{ hp} \times 2545 = $ 14,000 Btu/h

Total = 39,610 Btu/h
5% losses from freezer and piping (estimated) = 1980 Btu/h

Total refrigeration = 41,590 Btu/h
= 3.47 ton

Under the conditions given, 3.5 tons of cooling capacity per 100 gal/h of 100% overrun ice cream is required.

In continuous freezer operations, the heat gain from motors and losses from freezer and piping would remain about the same at all levels of overrun, but the necessary refrigerating effect would vary with the weight of mix required to produce 100 gal of ice cream, as shown in Table 10.

Table 10 Continuous Freezing Loads for Typical Ice-Cream Mix

Overrun, %	Ammonia Refrigeration at 3 psig Suction Pressure, Tons per 100 gal/h
60	4.04
70	3.88
80	3.74
90	3.61
100	3.50
110	3.39
120	3.30

Hardening Ice Cream. After leaving the freezer, ice cream is in a semisolid state and must be further refrigerated to become solid enough for storage and distribution. The ideal serving temperature for ice cream is about 8°F; it is considered hard at 0°F. To retain a smooth texture in hardened ice cream, the remaining water content must be frozen rapidly, so that the ice crystals formed will be small. For this reason, most hardening rooms are maintained at −20°F and some as low as −30°F. Most modern hardening rooms have forced-air circulation, either from a unit cooler or a remote bank of coils. With the ice-cream containers arranged so that air will circulate around them, the hardening time is about one-half that of rooms having overhead coils or coil shelves and gravity circulation. With forced-air circulation in the hardening room and average plant conditions, ice cream in 2.5- or 5-gal containers (or smaller packages in wire basket containers), all spaced to allow air circulation, will harden in about 10 h. Hardening rooms are usually sized to allow space for a minimum of three times the daily peak production and for a stock of all flavors, with the sizing based on 10 gal/ft^2 of floor area, which includes aisles.

Some larger plants use ice-cream hardening tunnels, which discharge into a low-temperature storage room. Because of the various size packages to be hardened, most tunnels are the air-blast type, operating at temperatures of −30 to −40°F and, in some cases, as low as −50°F. Half-gallon, quart, and pint cartons are usually hardened in these blast tunnels in about 4 h.

Contact-plate hardening machines must continuously and automatically load and unload to introduce the packages from the filler without delay. Horizontal continuous-plate freezers are described in Chapter 29. Compared to hardening tunnels, contact-plate hardeners save space and power and eliminate package bulging. They are limited to packages of uniform thickness having parallel flat sides.

Temperatures in the storage room are held at about −20°F. Space in storage rooms can be estimated at 25 gal/ft^2 stacked solid 6 ft high, including space for aisles. Many storage rooms today use pallet storage and racking systems. Such rooms may be 30 ft tall or more, some using stacker-crane automation.

Refrigeration required to harden ice cream varies with the temperature from the freezer and the overrun. The following example calculates the refrigeration required to harden a typical ice-cream mix.

Example 4. Assume ice cream with 100% overrun enters the hardening room at a temperature of 25°F. At this temperature, approximately 30% of the water would be frozen in the ice-cream freezer with the remainder to be frozen in the hardening room. The weight of one gallon of ice cream at 100% overrun, from a mix weighing 9.18 lb/gal would be 4.59 lb. The mix is assumed to contain 61% water, and the hardening room is at −20°F.

$$\text{Latent heat of hardening: } 4.59 \times 0.61 \times 0.70 \times 144 = 282 \text{ Btu}$$
$$\text{Sensible heat: } 4.59 \times (25 + 20) \times 0.50 = 103 \text{ Btu}$$
$$\text{Total} = 385 \text{ Btu}$$
$$\text{Loss due to heat of container and}$$
$$\text{exposure to outside air, assumed 10\%} = 40 \text{ Btu}$$
$$\text{Total Btu per gallon to harden} = 425 \text{ Btu}$$

Percent overrun, when calculated on the basis of the quantity of ice cream delivered by the freezer or the quantity placed in the hardening room, would affect the refrigeration required, as shown in Table 11.

Table 11 Hardening Loads for Typical Ice-Cream Mix

Overrun, %	Hardening Load, Btu/gal
60	532
70	500
80	470
90	447
100	425
110	405
120	386

Example 5. Calculate the refrigeration load in an ice-cream hardening room, assuming 1000 gal of ice cream at 100% overrun are to be hardened in 10 h in a forced-air circulation room at a temperature of −20°F. The hardening room, for three times this daily output, should have 300 ft^2 of floor area measuring approximately 15 × 20 × 9 ft high. The total insulated surface of 1230 ft^2 requires 4 to 6 in. of urethane or equivalent. For this example, the heat conductance through the insulated surface is selected at 0.04 Btu/(ft$^2 \cdot$h$\cdot$°F). The average ambient temperature is assumed to be 90°F.

$$\text{Heat leakage: } 1230 \times 0.04 \times (90 + 20) = 5410 \text{ Btu/h}$$
$$\text{Heat from fan motor: } 2 \text{ hp} \times 2545/0.85 = 5990 \text{ Btu/h}$$
$$\text{Heat from lights: } 600 \text{ W} \times 3.412 = 2050 \text{ Btu/h}$$
$$\text{Air infiltration and persons in room}$$
$$\text{(approximately 20\% leakage)} = 1080 \text{ Btu/h}$$
$$\text{Hardening 1000 gal ice cream}$$
$$\times 425 \text{ Btu/gal in 10 h} = 42{,}500 \text{ Btu/h}$$
$$\text{Total} = 57{,}030 \text{ Btu/h}$$
$$= 4.75 \text{ tons}$$

Other products such as sherbets, ices, ice milk, and novelties, usually represent a small percentage of the total output of the plant, but should be included in the total requirement of the hardening room.

Ice-Cream Bars and Other Novelties

Ice-cream plants may manufacture and merchandise a limited number of the many novelties. The most common are chocolate-coated ice-cream bars, popsicles, fudgsicles, drumsticks, ice-cream sundae cups, ice-cream sandwiches, and so forth. Small plants freeze most of these products, especially those with sticks, in metal trays each containing 24 molds, which are submerged in a special brine tank with a built-in evaporator surface and brine agitation. The product mix is prepared in a tank and cooled to 35 to 40°F. A controlled quantity of mix is poured into the tray molds or measured in with a dispenser. The tray molds are placed in the brine tank for complete freezing. Brine temperature is −30 to −36°F. The freezing rate should be rapid to result in small ice crystals, but it varies with the product and generally is 15 to 20 min. The frozen product is loosened from the molds by momentary melting of the outer layers of the product in a water bath. It is immediately removed from the molds; each is separately wrapped or put in a novelty bag and promptly placed in frozen storage for distribution.

The refrigeration calculations to freeze 100 dozen popsicles per h at 3 oz per popsicle, based on the mix containing 85% water, are:

Weight of mix estimated:

$$100 \times 12 \times 3/16 \times 1.06 \text{ (sp gr)} = 239 \text{ lb/h}$$
$$\text{Cooling mix: } 239 \times (40 - 27) \times 0.87 = 2700 \text{ Btu/h}$$
$$\text{Freezing: } 239 \times 0.85 \times 144 = 29{,}250 \text{ Btu/h}$$
$$\text{Subcooling: } 239 \times (27 + 30) \times 0.5 = 6810 \text{ Btu/h}$$
$$\text{Cooling trays (50/h)}$$
$$50 \times 8 \text{ lb} \times (60 + 30) \times 0.12 = 4320 \text{ Btu/h}$$
$$\text{Heat from agitator: 1 hp} \times 2545 = 2550 \text{ Btu/h}$$
$$\text{Leakage through tank: } 3 \times 12 \times 3 \text{ ft deep} = 750 \text{ Btu/h}$$
$$\text{Loss, top of tank and piping (assumed)} = 7500 \text{ Btu/h}$$
$$\text{Total} = 53{,}880 \text{ Btu/h}$$
$$= 4.5 \text{ tons}$$

In making ice cream, ice milk, and similar kinds of bars, the mix is processed through the freezer and is extruded in a viscous form at about 22°F. Using similar calculations, the estimated refrigeration load to freeze 100 dozen would be 2.2 tons for 3-oz ice-cream bars with 100% overrun.

The equipment to make and package novelties in large plants is available in several designs and capacities. Some are limited to the manufacture of one or a few kinds of similar novelties. Other

Dairy Products

machines have more versatility; for example, they can be used to make novelties with or without sticks, coated or uncoated, and of numerous sizes, shapes, and flavor combinations. Some of these machines include packaging in a bag or wrap, plus placement and sealing in a carton in units of 6, 8, 12, 14, 18, 24, or 48. In other plants, a separate packaging unit may be required. Some units harden the product by air at a temperature within the range of -35 to $-46\,°F$. Brine, usually calcium chloride, with a specific gravity of 1.275 or more and a temperature of -28 to $-38\,°F$ may be the hardening medium. Capacity of novelty makers varies with the shape and size of the specific product. Common production of a novelty machine is generally within the range of 300 to 3000 dozen or more per hour. Novelty equipment in plants may be semiautomatic or automatic in performance of the necessary functions.

An example of a simple novelty machine is one that has two parallel conveyor chains on which the mold strips are fastened. The molds are conveyed through filling, stick inserting, freezing, and defrosting stages. The extractor conveyor removes the frozen product from the mold cups and carries it to packaging or through the dipping operation; it is then discharged at packaging. In the meantime, the molds go through a wash and rinse and back to be filled. The novelty is either bagged or wrapped by machine, grouped and placed into cartons, and conveyed to cold storage.

Refrigeration Equipment Operation

Countertype freezers are usually designed for use with halocarbon refrigerants and may be arranged in a self-contained cabinet with the condensing unit mounted under the freezer. Nearly all commercial ice-cream plants, particularly the larger plants, use ammonia. Some smaller plants operate continuous ice-cream freezers and refrigerate hardening rooms to acceptable temperatures with single-stage refrigerant compressors. In most cases, these plants operate their compressors at conditions above the maximum compression ratio recommended by the manufacturer.

For economy of operation, within reasonable limits of compression ratio, ice-cream plants normally use multistage compression. For freezing ice cream, producing frozen novelties, and refrigerating an ice-cream hardening room to $-20\,°F$, one or more booster compressors may be used at the same suction pressure, discharging into second-stage compressors, which also handle the mix cooling and ingredient cold storage room loads. If a hardening tunnel is used at a temperature of $-40\,°F$ or below, at least two booster compressors should be used, one operating at the suction pressure required for the tunnel and the other operating at a higher suction pressure for the ice-cream freezers and storage room, with both units discharging into the second-stage compressor system. For plants with hardening tunnels arranged for large volumes, an analysis of operating costs may indicate savings in using three-stage compression with the low-temperature booster used for the tunnel, discharging into the second-stage booster used for freezers and storage, and then the second-stage booster discharging into the third-stage compressor system.

High-temperature loads in an ice-cream plant usually consist of refrigeration for cooling and holding cream, cooling ice-cream mix after pasteurization, cooling for mix holding tanks, and refrigeration for the ingredient cold storage room. If direct refrigerant cooling is used for the high-temperature loads, then compressor selection can be made at about $20\,°F$ saturated suction temperature and combined with the compressor capacity required to handle the booster discharge. About the same high suction temperature can be estimated if ice-cream mix and mix holding tanks are cooled by chilled water from a flash-type water chilling system. If an ice bank chills water used for cooling pasteurized ice-cream mix, it may be desirable to provide a separate compressor to handle this ice bank refrigerating load.

Table 12 Typical Ammonia Compressor Power Requirements at 185 psig Condensing Pressure

Temperature, °F	Pressure, absolute, psi	Power, hp One-stage	Two-stage
-25	16.0	2.8	2.2
-30	13.9	3.0	2.4
-35	12.3	3.5	3.5

Refrigeration is an important cost in an ice-cream plant, and heat transfer has a direct relationship to the effectiveness of the refrigeration system. Barriers to heat transfer are air films, frost and ice, scale, noncondensable gases, abnormal temperature differentials, clogged sprays, slow liquid circulation, poor air circulation, and foreign particles. Additional conditions that adversely affect the ice-cream freezing rate are dull scraper blades, viscous mix, low overrun percentage, high mix temperature, low ice-cream discharge temperature, and the mix composition, especially if it has a high sugar content.

Specific causes of low evaporator efficiency are rapid frost and ice development and an oil film. Automatic defrosting is one answer to the ice and frost problem. Regular oil purges help and an oil separator in the discharge line of the compressors is highly desirable. Neglected condensers lead to high head pressures and higher electrical requirements. Condenser surfaces should be kept clean of mineral deposits and other forms of scale or debris.

In compressor operation, the purging of noncondensable gases requires attention to avoid extra power use. Automatic purgers will handle this problem. The low temperatures required in ice-cream plants result in greater efficiency if two- or possibly three-stage compression systems are used. A comparison of power requirements for one- and two-stage ammonia systems at 185 psig condensing pressure is shown in Table 12.

The reduction of refrigeration losses through doors and conveyor openings and sufficient storage room insulation can contribute materially to better efficiency.

UHT STERILIZATION AND ASEPTIC PACKAGING

Ultrahigh temperature (UHT) *sterilization* of liquid dairy products destroys viability of microorganisms with a minimum adverse effect on sensory and nutritional properties. *Aseptic packaging* (AP) containerizes the sterilized product without recontamination. Sterilization, in the true sense, is the destruction or elimination of all viable microorganisms. In industry, however, the term sterilized may refer to a product that does not deteriorate microbiologically, but in which viable organisms may have survived the sterilization process. In other words, heat treatment renders the product safe for consumption and imparts an acceptable shelf life microbiologically.

Sterilization Methods—Equipment

Retort sterilization of milk products has been a commercial practice for many years. It consists of sterilizing the product after hermetically sealing it in a metal or glass container. The heat treatment is sufficiently severe to cause a definite cooked off-flavor in milk and to decrease the heat-labile nutritional constituents of milk products. UHT-AP has the advantage of causing less cooked flavor, color change, and loss of vitamins while having the same sterilization effect as the retort method.

UHT-AP has been applied to common fluid milk products (whole milk, 2% milk, skim milk, and half-and-half), various creams, flavored milks, evaporated milk, and such frozen dessert mixes as ice cream, ice milk, milk shakes, soft-serve, and sherbets. UHT-sterilized dairy foods include eggnog, salad dressings, sauces, infant preparations, puddings, custards, and nondairy coffee whiteners and toppings.

UHT sterilization is accomplished by rapid heating of the product to the sterilizing temperature, holding the temperature for a definite number of seconds, and then rapid cooling. The methods have been classified as direct steam or indirect heating. Among the advantages of the direct methods are (1) heating is faster, (2) processing intervals between equipment cleanings are longer, and (3) the flow rate is easier to change. Among the indirect methods, advantages are: (1) regeneration potential is greater, (2) potable steam is not necessary, and (3) viscous products and those with small pieces of solids can be processed with the scraped-surface unit.

The direct steam method is subdivided into injection or infusion. In direct injection, steam is forced into the product, preferably in small streamlets, with sufficient turbulence to minimize localized overheating of the milk surfaces that the steam initially contacts. In the infusion system, the product is sprayed into a steam chamber. Advantages of infusion over injection are (1) slightly less steam pressure required (there are exceptions), (2) less localized overheating of a portion of the product, and (3) more flexibility for change of the product flow rate.

The three important indirect systems are tubular, plate, and cylinder with mechanical agitation. In the tubular type, the tube diameter must be small and the velocity of flow high to maximize heat transfer into the product.

The essential components for direct steam injection are storage or balance tank, timing pump, preheater (tubular or plate), steam injector or infuser unit, holding unit, flow-diversion valve, vacuum chamber, aseptic pump, aseptic homogenizer, plate or tubular cooler, and control instruments. The minimum items of equipment for steam infusion are the same, except that the infuser is used to heat the product from the preheat to the sterilization temperature.

The necessary equipment for the indirect systems is similar: storage or balance tank, timing pump, preheater (tubular or plate type, and preferably regenerative), homogenizer, final plate or tubular heater, holding tube or plate, flow-diversion valve, cooler (1 to 3 stages), and control instruments. The mechanically agitated heat exchanger replaces the tubes or plates in the final heating stage. Otherwise, the same items of equipment are used for this system of sterilization.

In addition to the basic equipment, many combinations of essential and supplemental items of UHT equipment are available. For example, one deviation in the indirect system is to use the pump portion of the homogenizer as a timing pump when it is installed after the balance tank. The first stage of homogenization may occur after preheating, and the second stage may occur after precooling. A vacuum chamber may be placed in the line after preheating, for precooling after sterilization, or installed in both locations. A condenser in the vacuum chamber allows the advantages of deaeration without moisture losses that otherwise would occur in the indirect system. In Europe, some indirect systems have a hold of several minutes, after preheating, to reduce the rate of solids accumulation on the final heating surfaces of the tubes or plates. A bactofuge may be included in the line after preheating to reduce a high microbiological content, especially of the bacterial spores.

Self-acting controls and other instrumentation are available to assure automatic operation in nearly every respect. Particularly important is automatic control of the temperatures for preheating, sterilizing, and precooling in the vacuum chamber, and to some extent, of the final temperature before packaging. This may include temperature-sensing elements to control heating and cooling and pressure-sensing elements for operating pneumatic valves. The cleaning cycle may be automated, beginning with a predetermined solids accumulation on specific heating surfaces. Timers regulate the various cleaning and rinsing steps.

In some systems, one or more aseptic surge tanks are installed between the UHT sterilizer and the AP equipment. Surge tanks permit continuation of the sterilizer should the operation of the AP equipment be interrupted, or the continuation of the AP equipment should sterilization be interrupted. It also makes the use of two or more AP units easier than direct flow from the UHT sterilizer to the AP machines.

When aseptic surge tanks are used, they must be constructed to withstand the steam pressure required for equipment sterilization and be provided with a sterile air venting system. Aseptic surge tanks may be unloaded by applying sterile air to force product out to the AP equipment. The pressure for air unloading can be controlled at a constant value, making uniform filling possible even when one of several AP machines is removed from service.

Aseptic surge tanks make it possible to hold bulk product, even for several days, until it is convenient to package it.

Basic Steps. After the formula is prepared and the product standardized, the processing steps are: (1) preheat to 150 to 170 °F by a plate or tubular heat exchanger; (2) heat to a sterilization temperature of 285 to 300 °F; (3) hold for 1 to 20 s at sterilizing temperature; and (4) cool to 40 to 100 °F, depending on product keeping quality needs. Cooling may be by one to three stages; generally two are used. The direct steam method requires at least two cooling stages. The first is flash cooling in a vacuum chamber to 150 to 170 °F to remove moisture equal to the steam injected during sterilization. The second stage reduces the temperature to within 50 to 100 °F. A third stage is required in most plants if the temperature is lowered additionally to 35 to 50 °F.

The products with fat are homogenized to increase stability of the fat emulsion. The direct method requires homogenization after sterilization and precooling. Homogenization may follow preheating or precooling, but usually follows preheating in the indirect method. Efficient homogenization is very important in delaying the formation of a cream layer during storage.

Sterilized plain milks (such as whole, 2%, and skim milk) are most vulnerable to having a cooked off-flavor. Consequently, the aim is to have low sterilization temperature and time consistent with satisfactory keeping quality. The total cumulative heat treatment is directly related to the intensity of the cooked off-flavor. The total processing time from preheating to cooling varies widely among systems. Most operations in the United States range from 30 to 200 s; in European UHT processes, it may be much longer.

Several factors influence the minimum sterilization temperature and time needed to control adverse effects on flavor and physical, chemical, and nutritional changes. The type of product, initial number of spores and their heat resistance, total solids of the product, and pH are the most important factors. Obviously, the relationship is direct for the number and the heat resistance of the spores. Total solids also have a direct relationship, but for an acid pH it is inverse.

Several terms are used in the designation of the influence of the UHT on the microbiological population. *Decimal reduction* refers to a reduction of 90% (*e.g.*, 100 to 10, or one log cycle). An example of a three-decimal reduction is 10,000 to 10. *Decimal reduction time*, or *D value*, is the time required to obtain a 90% decrease. *Sterilizing effect*, or *bactericidal effect*, is the number of decimal reductions obtained and expressed as a logarithmic reduction ($\log_{10}$ initial count minus $\log_{10}$ final count). A sterilizing effect of six means one organism remaining from a million per mL (10^6), and seven would be one remaining in 10 mL (a final count of 10^{-1}).

The *Z value* is the temperature increase required to reduce the D value by one log cycle (90% reduction of microorganisms with the time held constant). The *F value* (thermal death time) is the time required to reduce the number of microorganisms by a stated amount or to a specific number. For example, assuming D value of 36 s for *Bacillus substilis* spores at 250 °F and a need to reduce the spores from 10^6 per mL to < 1 per mL, the thermal treatment time would be $6 \times 36\text{ s} = 216\text{ s}$ (F value).

Dairy Products

Aseptic Packaging

Aseptic fillers are available for coated metal cans, glass bottles, plastic-paperboard-foil cartons, thermoformed plastic containers, blow-molded plastic containers, and plastic pouches. The aseptic can equipment includes a can conveyor and sterilizing compartment, filling chamber, lid sterilizing compartment, sealing unit, and instrument controls. The procedure sterilizes the cans with steam at 550 °F as they are conveyed, fills the cans by continuous flow, simultaneously sterilizes the lids, places the lids on the cans, and seals the lids onto the cans. Pressure control apparatus is not used for entry or exit of cans.

A similar system is used for glass bottles or jars. The jars are conveyed into a turret chamber; air is removed by vacuum; the jars are then sterilized for 2 s with wet steam at 60 psi and moved into the filler. The temperature of the glass equalizes to 120 °F and the filling takes place. Next, the transfer is to the capper for placement of sterile caps, which are screwed onto the jars. The filling and capping space is maintained at 500 °F.

Several aseptic blowmold forming and filling systems have been developed. Each system is different, but the basic steps using molten plastic are: (1) extruded into a parison, (2) extended to the bottom of the mold, (3) mold closed, (4) preblown with compressed air that inflates the plastic film into a bottle shape, (5) parison cut and the neck pinched, (6) final air application, (7) bottle filled and foam removed, (8) top sealed, and (9) mold opened and filled bottle ejected.

The basic steps in the manufacture of aseptic, thermoformed plastic containers are: (1) a sheet of plastic (*e.g.*, polystyrene) is drawn from a roll through the heating compartment and then multistamped into units, which constitute the containers; (2) these units are conveyed to the filler, which is located in a sterile atmosphere, and are filled; (3) a sheet of sterilized foil is heat sealed to the container tops; and (4) each container is separated by scoring and cutting.

One of the two aseptic systems for the plastic-paperboard-foil cartons draws the material from a roll through a concentrated hydrogen peroxide bath to destroy the microorganisms. The peroxide is removed by drawing the sheet between twin rolls, by exposure to ultraviolet light and hot air, or by superheated, sterilized air forced through small slits at high velocity. The packaging material is drawn downward in a vertical, sterile compartment for forming, filling by continuous flow, sealing, separation, and ejection.

In the other plastic-paperboard-foil aseptic system, the prepared, flat blanks are formed and the bottoms are heat sealed. In the next step, the inside surfaces are fogged with hydrogen peroxide. Sterilized hot air dissipates the peroxide. The cartons are conveyed into the aseptic filling and then into top-sealing compartments. The air forced into these two areas is rendered devoid of microorganisms by high efficiency filters.

Operational Problems. Aseptic operational problems are reduced by careful installation of satisfactory equipment. The equipment should comply with 3-A Standards. Milk and milk products that are processed to be commercially sterile and aseptically packaged must also meet the Grade A Pasteurized Milk Ordinance and be processed in accordance with 21 CFR Part 113, "Thermally processed low-acid foods packaged in hermetically sealed containers." Generally, the simplest system, with a minimum of equipment for product contact surfaces and processing time, is desirable. It is specifically important to have as few pumps and nonwelded unions as possible, particularly those with gaskets. The gaskets and O-rings in unions, pumps, and valves are much more difficult to clean and sterilize than are the smooth surfaces of chambers, and tubing. Automatic controls, rather than manual attention, is generally more satisfactory.

Complete cleaning and sterilizing of the processing and packaging equipment are essential. Milk solids accumulate rapidly on heated surfaces; therefore, cleaning is necessary after 0.5 h of processing for the tubular or plate UHT heat exchangers, although cleaning after 3 to 4 h is more common. The cleaning practice for the sterilizer, filler, and accessory equipment usually involves the CIP method for the rinse and alkali cleaning cycle, rinse, acid cleaning cycle, and rinse. Some plants omit the acid-cleaning of the storage tank and packaging equipment except periodically; *e.g.*, once or twice a week. Steam sterilization just before processing is customary. At 8 to 10 psig of wet steam, 1.5 to 2.0 h (or a shorter time at higher steam pressure) may be required. Water sterilized by steam injection or the indirect method can be used for rinsing and for the cleaning solution.

Survival of spores during UHT processing, or subsequent recontamination of the product before the container is sealed, is a constant threat. Inadequate sealing of the container also may be troublesome with certain types of containers. Another source of poststerilization contamination is airborne microorganisms. These may contact the product through inadequate sterilization of air that enters the storage vat for the processed product or through air leaks into the product upsteam of the sterilized product pumps or homogenizer, if a reduced pressure is created. During packaging, air may contaminate the inside of the container or the product itself during filling and sealing.

Quality Control

Poor quality of raw materials must be avoided. The higher the spore count of the product before sterilization, the larger the spore survival number at a constant sterilization temperature and time. Poor quality can also contribute to other product defects (off-flavor, short keeping quality) because of sensory, physical, or chemical changes. Heat stability of the raw product must be considered.

A good quality sterilized product has a pleasing, characteristic flavor and color that are similar to the pasteurized samples. The cooked flavor should be slight, or negligible, with no unpleasant aftertaste. The product should be free of microorganisms and adulterants such as insecticides, herbicides, and peroxide or other container residues. It should have good physical, sensory, and keeping quality.

Deterioration in storage may be evaluated by holding samples at 70, 89, 98, or 113 °F for 1 or 2 weeks. The number of samples for storage testing should be selected statistically and should include samplings of the first and last of each product packaged during the processing day. In order to identify the source of microbiological spoilage, continuous aseptic sampling into standard size containers after sterilization and/or just ahead of packaging may be practiced. Sampling rate should be set to change containers each hour.

The rate of change in storage of sterilized milk products is directly related to the temperature. Commercial practice varies with storage ranging from 35 °F to room temperature, which may go as high as 95 °F or more. In plain milks, the cooked flavor may decrease the first few days, and then remain at its optimum for 2 to 3 weeks at 70 °F before gradually declining. When the milk is held at 70 °F, a slight cream layer becomes noticeable in approximately 2 weeks and slowly continues until much of the fat has risen to the top. Thereafter, the cream layer becomes increasingly difficult to reincorporate or reemulsify.

Viscosity increases slightly the first few weeks at 70 °F and then remains fairly stable for 4 to 5 months. Thereafter, gelation gradually occurs. However, milks vary in stability to gelation depending on such factors as feeds, stage of lactation, preheat treatment, and homogenization pressure. The addition of sodium tetraphosphate to some milks causes gelatin to develop more slowly.

Occasionally, some sterilized milk products develop a sediment on the bottom of the container because of crystallization of complex salt compounds or sugars. Browning can also occur during storage. Usually, the off-flavors develop more rapidly and render the product unsalable before the off-color becomes objectionable.

Heat-Labile Nutrients

The results reported by researchers on the effects of UHT sterilization on the heat-sensitive constituents of milk products lack consistency. The variability may be attributed to the analytical methods and to the difference in total heat treatment among various UHT systems, especially in Europe. In a review, Van Eekelen and Heijne (1965) summarized the effect of UHT sterilization on milk as follows: slight or none for Vitamins A, B_2, and D, carotene, pantothenic acid, nicotinic acid, biotin, and calcium; and no decrease in biological value of the proteins. The decreases were: 3 to 10%, thiamine; 0 to 30%, B_6; 10 to 20%, B_{12}; 25 to 40%, C; 10%, folic acid; 2.4 to 66.7%, lysine; 34%, linoleic acid; and 13%, linolenic acid. Protein digestibility was decreased slightly. A substantial loss of Vitamin C, B_6, and B_{12} occurred during a 90-day storage. Brookes (1968) reported that Puschel found that babies fed sterilized milk averaged a gain of 27 g per day, compared to 20 g for the control group.

EVAPORATED, SWEETENED, CONDENSED, AND DRY MILK

Evaporated Milk

Raw milk intended for processing into evaporated milk should have a heat stability quality with little and preferably no developed acidity. As the milk is received it should be filtered and held cold in a storage tank. The milkfat is standardized to nonfat solids at the ratio of 1:2.2785. The milk is preheated to 200 to 205 °F for 10 to 20 min or 240 to 260 °F for 60 to 360 s to reduce product denaturation during the sterilizaton process. Moisture is removed by batch or (usually) continuous evaporation until the total solids have been concentrated to 2.25 times the original content.

The condensed product is pumped from the evaporator and, with or without additional heating, is homogenized at 2000 to 3000 psi and 120 to 140 °F. It is cooled to 45 °F and held in storage tanks for restandardization to not less than 7.9% milkfat and 25.9% total solids. The product is pumped to the packaging unit for filling the cans made from tin-coated sheet steel. The filled cans are conveyed continuously through a retort, whereby the product is rapidly heated with hot water and steam to 245 °F and held for 15 min to complete the sterilization. Rapid cooling with water to 80 to 90 °F follows. The evaporated milk is agitated while in the retort by the can movement. Application of labels and placement of cans in shipping cartons are done automatically.

Storage at room temperature is common, but deterioration of flavor, body, and color is decreased by lowering the storage temperature to 50 to 60 °F. Relative humidity should be less than 50% to reduce can and label deterioration. The recommended inversion of cases during storage to reduce fat separation is shown in Table 13.

Sweetened Condensed Milk

Sweetened condensed milk is manufactured similarly to evaporated milk in several aspects. One important difference, however, is that added sugar replaces heat sterilization to extend storage life. Filtered cold milk is held in tanks and standardized to 1:2.2942 (fat to nonfat solids). The milk is preheated to 145 to 160 °F, homogenized at 2500 psi, and heating is continued to 180 to 200 °F for 5 to 15 min. or to 240 to 300 °F for 30 s to 5 min The milk is condensed in a vacuum pan to slightly higher than a 2:1 ratio. Liquid sugar (pasteurized) is added at the rate of 18 to 20 lb/100 lb of condensed milk.

As the mixture is pumped from the vacuum pan, it is cooled through a heat exchanger to 86 °F and held in a vat with an agitator. Nuclei for proper lactose crystallization are provided by adding finely-powdered lactose (200-mesh). The product is cooled slowly, taking an hour to reach 75 °F with agitation. Then cooling is continued more rapidly to 60 °F. Improper crystallization forms large crystals, which cause sandiness (gritty texture). The sweetened condensed milk is pumped to a packaging unit for filling into retail cans and sealing. Labeling of cans and placement in cases is mechanized, similar to the process used for evaporated milk. The product is usually stored at room temperature, but the keeping quality is improved if the storage temperature is below 70 °F.

Condensing Equipment. Both batch and continuous equipment are used to reduce the moisture content of fluid milk products. The continuous types have single, double, triple, or more evaporating effects. The improvement in efficiency with multiple effects is shown in Table 14 by the reduction in steam required to evaporate 1 lb of water.

A simple evaporator is the horizontal tube. In this design, the tubes are in the lower section of a vertical chamber. During operation, water vapor is removed from the top and the product, from the bottom of the unit. For the vertical short-tube evaporator, the chamber design may be similar to the horizontal tube. The long-tube vertical unit may be designed to operate with a rising or falling film in the tubes. The latter is common. For the falling film, the product Reynolds number should be greater than 2000 for good heat transfer. Falling-film units may have a high k factor at low temperature differentials, resulting in low steam requirements per mass of water evaporated per area of heating surface. This type (falling film) has a rapid startup and shutdown. Thermocompressing and mechanical compressing evaporators have the advantage of operating efficiently at lower temperatures, thus reducing the adverse effect on heat sensitive constituents. Vapors removed from the product are compressed and used as a source of heat for additional evaporation.

Plate-type evaporators are also in common use. They are similar to plate heat exchangers used for pasteurization in that they have a frame and a number of plates gasketed to carry the product in a passage between two plates and the heating medium in adjacent passages. They differ in that, in addition to ports for product, they have large ports to carry vapor to a vapor separator. Vapors flow from the separator chamber to a condenser similar to those used for other types of evaporators. Plate evaporators require less head space for installation than other types, may be enlarged or decreased in capacity by a change in the number of plates, and offer a very efficient heat exchange surface.

Equipment Operation. Positive pumps of the reciprocating type are often used to obtain 24-in. Hg vacuum in the chamber. Steam jet ejectors may be used for 25-in. Hg vacuum, for one stage; two stages permit 28-in. Hg vacuum; and three stages, 29.8-in. Hg vacuum. Condensers between stages remove the heat and may reduce the amount of vapor for the following stage. Either a cen-

Table 13
Inversion Times for Cases of Evaporated Milk in Storage

Storage Temperature, °F	Time
90	1 month initially and each 15 days
80	1 to 2 months
70	2 to 3 months
60	3 to 6 months

Table 14
Typical Steam Requirements for Evaporating Water from Milk

No. of Evaporating Effects	Steam Required, lb Steam/lb Water
Single	1.30 to 1.00
Double	0.60 to 0.50
Triple	0.40 to 0.35
Quadruple	0.30 to 0.25

Dairy Products

trifugal or reciprocating pump may be used to remove water from the condenser. A barometric leg may also be placed at the bottom of a 34 ft or longer condenser to remove the water by gravity.

Dry Milk and Nonfat Dry Milk

There are two important methods of drying milk—spray and drum. Each has modifications, such as the foam spray and the vacuum drum drying methods. Spray drying exceeds by far the other methods for drying milk, and the largest volume of dried dairy product is skim milk.

In the manufacture of spray-dried nonfat dry milk (NDM), cold milk is preheated to 90°F, separated, and the skim milk for low-heat NDM is pasteurized at 161°F for 15 s or slightly higher and/or longer. It is condensed with caution to restrict total heat denaturation of the serum protein to less than 10%. This requires using a low-temperature evaporator or operating the first effect of a regular double effect evaporator at a reduced temperature. After increasing the total solids to 40 to 45%, the condensed skim milk is continuously pumped from the evaporator through a heat exchanger to increase the temperature to 145°F. The concentrated skim milk is filtered and enters a positive pump operating at 3000 to 4000 psi, which forces the product through a nozzle with a very small orifice, producing a mist-like spray in the drying chamber. Hot air of 290 to 400°F or higher dries the milk spray rapidly. Nonfat dry milk with 2.5 to 4.0% moisture is conveyed from the drier by pneumatic or mechanical means, then cooled, sifted, and packaged. Packages for industrial users are 50- or 100-lb bags.

High heat nonfat dry milk is used principally for bread and other bakery products. The manufacturing procedure is the same as for low heat NDM except: (1) the pasteurization temperature is well above the minimum, *e.g.*, 175°F for 20 s or higher; (2) after pasteurization, the skim milk is heated to 185 to 195°F for 15 to 20 min, condensed; and (3) the concentrate is heated to 160 to 165°F ahead of filtering and then is spray dried, similar to the process for low heat NDM. Storage of low or high heat NDM is usually at room temperature.

Dry Whole Milk. The raw whole milk in storage tanks is standardized at a ratio of fat to nonfat solids of 1:2.769. The milk is preheated to 160°F, filtered or clarified, and homogenized at 160°F and 3000 psi on the first stage and 750 psi on the second stage. The heating continues to 200°F with a 180-s hold. The milk is drawn into the evaporator and the total solids are condensed to 45%. The product is continuously pumped from the evaporator, reheated in a heat exchanger to 160°F, and spray dried to 1.5 to 2.5% moisture. Dry whole milk (DWM) is cooled (not below dew point) and sifted through a 12-mesh screen. For industrial use within 2 or 3 months, the dry whole milk is packaged in 50-lb bags and held at room temperature or, preferably, well below 70°F.

In order to retard oxidation, the dry whole milk may be containerized in large metal drums or in customer-size cans unsealed and subjected to 28-in. Hg vacuum. Less than 2% oxygen in the head space of the package after a week of storage is a common aim. The oxygen desorption from the entrapped content in lactose is slow, and 2 vacuum treatments may be necessary with a 7- to 10-day interval between them. Warm DWM directly from the drier has a faster oxygen desorption rate than after it has cooled. Nitrogen is used to restore atmospheric pressure after each vacuum treatment. After the hold period for the first vacuum treatment, the DWM in the drums is dumped into a hopper, mechanically packaged into retail size metal cans, and given the second vacuum treatment.

Foam spray drying permits the total solids to be increased to 50 to 60% in the evaporator prior to drying. Gas, compressed air, or nitrogen is distributed, by means of a small mixing device, into the condensed product between the high pressure pump and the spray nozzle. A regulator and needle valve are used to adjust the gas flow into the product. Gas usage is approximately 0.5 ft^3/gal of concentrated product. Otherwise, the procedure is the same as for regular drying. Foam spray-dried NDM has poor sinkability but good reconstitutability in water. The density is roughly half that of regular spray-dried NDM. The additional equipment for foam spray drying is limited to a compressor, storage drum, pressure regulator, and a few accessory items. The cost is relatively small, especially if compressed air is used.

Spray driers are made in various shapes and sizes with one or many spray nozzles. The horizontal spray driers may be box shaped or a teardrop design. The vertical spray driers are usually cone or silo shaped.

Heat Transfer Calculations. The typical atomization in United States spray-drying plants is produced by a high pressure pump that forces the liquid through a small orifice in a nozzle designed to give a spreading effect as it emerges from the nozzle. Single-nozzle driers have an orifice opening diameter of 0.107 to 0.177 in. The diameter for multinozzle driers is 0.025 to 0.052 in. In Europe, the spinning disk is the most common means of atomizing in milk drying plants. Droplet sizes of 50 to 250 μm in diameter are usual. Droplet size has an inverse relationship to the rate of drying at a uniform hot air temperature. Larger droplets require a higher air drying temperature and/or longer exposure than the smaller ones.

Other essential steps in spray drying are: (1) moving, filtering, and heating the air; (2) incorporating the hot air with the product droplets; and (3) removing the moisture vapors and separating the moist air from the product particles. After passing through a rough or intermediate filter, the air is heated indirectly by steam coils or directly with a gas flame to 250 to 500°F. During the short drying exposure time, the air temperature drops to 160 to 200°F.

Thermal efficiency is the percentage of the total heat used to evaporate the water during the drying process. The efficiency is improved by recovery of heat from the exhaust air, decreased radiation loss, and high drying air temperature versus a low outlet air temperature. Roughly 2.2 to 3.2 lb of steam are needed to evaporate 1 lb of moisture in the drier.

$$\text{Thermal efficiency} = \frac{(1 - R/100)(t_1 - t_2)}{t_1 - t_0}$$

where

- R = radiation loss, percentage of temperature decrease in drier
- t_1 = inlet air temperature, °F
- t_2 = outlet air temperature, °F
- t_0 = ambient air temperature, °F

Most of the dried particles are separated from the drying air by gravity and fall to the bottom of the drier or the collectors. The fine particles are removed by directing the air-powder mixture through bag filters or a series of cyclone collectors. Air movement in the cyclone is designed to provide a centrifugal force for separation of the product particles. In general, several small-diameter cyclones with a fixed pressure drop will be more efficient for removal of the fines than two large units.

The drier has sensing elements to continuously record the hot air (inlet) temperature and the moist air (outlet) temperature. Adjustments of either of these temperatures during drying is done with a steam valve or gas inlet valve.

Drum Drying

Relatively little skim milk or whole milk has been drum dried in the last few years. Drum-dried products, when reconstituted, have a cooked or scorched flavor compared with the spray-dried products. The heat treatment during drying denatures the protein and results in a high insolubility index. In preparation for drying, the skim milk is separated or the whole milk is standardized to 1:2.769, (*e.g.*, 3.2 fat and 8.86 SNF). The product is filtered or clarified, homogenized after preheating, and pasteurized. If the resulting dry product is intended for bakery purposes, the milk is heated to approximately 185 °F for 10 min. The fluid product may or may not be concentrated by moisture evaporation to not more than 2 to 1. The product is then dried on the drum(s)—skim milk to not more than 4.0% and whole milk to not more than 2.5% moisture. A blade pressed against the drum scrapes off the sheet of dried product. An auger conveys the dry material to the hammer mill, where it is pulverized and sifted through an 8-mesh screen. Drum dried milks are usually packaged at the sifter into 50- to 100-lb Kraft bags with a plastic liner.

A double-drum drier is more common than a single drum for drying milk. Cast iron is used more often in drum construction than stainless steel or alloy steel and chrome plate steel. The knife metal must be softer than the drum. In the double drum unit, the drums are spaced from 0.02 to 0.043 in. apart. End plates on the drums create a reservoir into which the product, at 185 °F, is sprayed the length of the drums. The steam-heated drums boil the product continuously as a thin film adheres to the revolving drums. After about 0.875 of one revolution, the film of product is dry and is scraped off. Drums normally revolve between 10 and 19 rpm. The steam pressure inside the drums is approximately 70 to 90 psi, as indicated by the pressure gage at the inlet of the condensate trap.

The steam pressure is adjusted for drying the product to the desired moisture content. Superheated steam will cause scorching of the product. Condensate inside the drums must be continuously removed, while the exterior vapors from the product are exhausted from the building with a hood and fan system. Capacity, dried product quality, and moisture content depend on many factors. Some important ones are: steam pressure in drums, rotation speed of drums, total solids of product, smoothness of drum surface and sharpness of the knives, properly adjusted gap between the two drums, liquid level in drum reservoir, and product temperature as it enters the reservoir.

REFERENCES

Brookes, H. 1968. New developments in longlife milk and dairy products. *Dairy Industries* (May).

Hammer, B.W. and A. R. Johnson. 1913. The specific heat of milk and milk derivatives. *Research Bulletin* No. 14, Iowa Agricultural Experiment Station.

IAMFES. 3-A *Sanitary Standards*. International Association of Milk, Food, and Environmental Sanitarians, Ames, IA.

Leighton, A. 1927. On the calculation of the freezing point of ice cream mixes and of the quantities of ice separated during the freezing process. *Journal of Dairy Science* 10:300.

Rishoi, A.H. 1951. *Physical characteristics of free and globular milk fat*. American Dairy Science Association, Annual Meetings (June).

Van Eeckelen, M. and J.J.I.G. Heijne. 1965. "Nutritive value of sterilized milk." In *Milk sterilization*. Food and Agricultural Organization of the United Nations, Rome.

Zahadan, V.Z. 1940. *Specific heat of foodstuffs in relation to temperature*. Kholod'naia Prom. 18 (4):32. Cited from Stitt and Kennedy (Russian).

BIBLIOGRAPHY

Arbuckle, W.S. 1972. *Ice cream*, 2nd ed. AVI Publishing Co., Westport, CT.

Farrall, A.W. 1963. *Engineering for dairy and food products*. John Wiley and Sons, Inc., New York.

Griffin, R.C. and S. Sacharow. 1970. *Food packaging*. AVI Publishing Co., Westport, CT.

Hall, C.W. and T.I. Hedrick. 1971. *Drying of milk and milk products*. AVI Publishing Co., Westport, CT.

Henderson, F.L. 1971. *The fluid milk industry*. AVI Publishing Co., Westport, CT.

Judkins, H.F. and H.A. Keener. 1960. *Milk production and processing*. John Wiley and Sons, Inc., New York.

Kosikowski, F.V. 1966. *Cheese and fermented milk foods*. Published by author, Ithaca, NY.

Reed, G.H. 1970. *Refrigeration*. Hart Publishing Co., Inc., New York.

Sanders, G.P. Cheese varieties and descriptions. *Agriculture Handbook* No. 54. USDA, U.S. Government Printing Office, Washington, DC.

Webb, B.W. and E. A. Whittier. 1970. *Byproducts from milk*. AVI Publishing Co., Westport, CT.

Wilcox, G. 1971. *Milk, cream and butter technology*. Noyes Data Corporation, Park Ridge, NJ.

Wilster, G.H. 1964. *Practical cheesemaking*, 10th ed. Oregon State University Bookstore, Corvallis, OR.

CHAPTER 16

DECIDUOUS TREE AND VINE FRUITS

General Produce Considerations 16.1	Peaches and Nectarines 16.8
Storing and Handling of Fruits 16.1	Apricots 16.9
Apples 16.1	Berries 16.9
Pears 16.4	Strawberries 16.9
Grapes 16.5	Figs 16.10
Plums 16.7	Supplements to Refrigeration 16.10
Sweet Cherries 16.8	

THE most obvious losses from marketing fruit crops are caused by mechanical injury, decay, and aging. Losses in moisture, vitamins, sugars, and starches are less obvious, but they adversely affect quality and nutrition. Rough handling and holding at undesirably high or low temperatures increases loss. Loss can be substantially reduced by greater care in handling and by following recommended storage practices.

GENERAL PRODUCE CONSIDERATIONS

Quality and Maturity

Maximum storage life can be obtained only by storing high quality commodities soon after harvest. Different lots of fruits may vary greatly in their storage behavior due to variety, climate, soil and cultural conditions, maturity, and handling practices. When fruits are transported from a distance, grown under unfavorable conditions, or deteriorated, proper storage allowance should be made.

Fresh fruits for storage should be as free as possible from skin breaks, bruises, and decay. Much more decay will develop on bruised areas of apples than on nonbruised areas. A single severe bruise on an apple may increase the moisture loss by as much as 400%.

The amount of incipient decay infection, which influences storage potential of grapes and apples, can be predicted in the early storage period. Only lots with good storage potential should be held for late season marketing.

Maturity of the fruit at harvest time determines the refrigerated storage life and quality of the product. For any given produce there is a maturity best suited for refrigerated storage. Undermature produce will not ripen or develop good quality during or following refrigerated storage; overmature produce deteriorates quickly during storage.

Handling and Harvesting

Rising handling costs have encouraged the use of bulk handling and large storage bins for many vegetables and fruits. Moving, loading, and stacking bins by forklift trucks require care to maintain proper ventilation and refrigeration of the product. Bins should not be so deep that excessive weight damages the produce near the bottom.

Mechanical harvesters for fruits frequently cause some bruising. This damage can materially reduce the refrigerated storage life and quality of the produce.

Transportation

As in storage, losses from deterioration during distribution are affected by temperature, moisture, diseases, and mechanical damage. Gradual aging and deterioration are continuous after harvest. Time in transit may represent a large portion of the postharvest life for some commodities, such as cherries and strawberries. Thus, the environment during this period largely determines produce salability when it reaches the consumer.

To prevent undue warming and condensation of moisture, which promote decay and deterioration, most storages are built on railroad sidings. Canvas tunnels between the railcar and storage help minimize warming and condensation of moisture.

When produce is removed from storage for distribution to wholesale and retail markets, the storage operator can do little to prevent undesirable condensation. Warming the packages until they are above the dewpoint of the air would prevent it, but this takes time and space and is seldom practicable. Deterioration in flavor and condition accelerates after long periods of storage; therefore, the produce should be moved to consumers as quickly as possible.

STORING AND HANDLING OF FRUITS

Details on storage and handling of common fruits are given in the following sections. For more information on storage requirements and physical properties of specific commodities, see Chapter 26.

APPLES

Apples are not only the most important fruit stored on a tonnage basis, but their average storage period is considerably greater than any other fruit. The length of storage may be short for fall varieties and those going into processing, but cold storage is critical to proper handling and marketing.

Storage Temperature

For most varieties, 30°F is the recommended storage temperature. This is 2°F above the highest freezing point (28°F) of most apples and should be safe for modern storage rooms. In older rooms with poor air distribution, a 30°F temperature is unsafe, because of variable temperatures at different locations within the room.

Some apple varieties held at 30°F develop physiological disorders that impair storage life and marketability. However, if storage temperatures are maintained at 39°F, such disorders may not be an economic factor. Unfortunately, such elevation in temperature results in an accelerated rate of deterioration. Since the remedy may have consequences nearly as serious as the physiological disorder, a storage temperature of approximately 36°F is often used.

Storage life of apple varieties depends on harvest maturity, elapsed time, and temperature between harvest and storage, cooling rate in storage, and sometimes cultural factors. The best

---

The preparation of this chapter is assigned to TC 11.5, Fruits, Vegetables and Other Products.

Table 1 Storage Periods for Certain Apple Cultivars and Their Susceptibility to Storage Disorders

Cultivar	Months of Storage Normal	Months of Storage Maximum[a]	Storage Scald Susceptibility	Other Disorders Likely to Occur in Storage[b]
Baldwin	4-5	6-7	Moderate	Bitter pit, brown core
Cortland	3-4	6-7	Very high	Senescent breakdown
Delicious	5-6	8-11	Moderate	Bitter pit, senescent breakdown, soft scald
Empire	4-5	8-9	Slight	—
Golden Delicious	5-6	7-11	Slight	Shriveling, bitter pit, senescent, breakdown
Gravenstein	2	3	Moderate	Bitter pit, Jonathan spot
Granny Smith	5-6	7-9	High	Bitter pit, brown core, senescent breakdown
Idared	5-6	8-9	Slight	Breakdown, Jonathan spot
Jonathan	3-4	6-8	Moderate	Breakdown, Jonathan spot, soft scald
McIntosh	4-5	7-8	Moderate	McIntosh breakdown, brown core
Northern Spy	4-5	7-8	Slight	Senescent breakdown, Spy spot, bitter pit
Rome Beauty	5-6	7-8	Very high	Jonathan spot, soft scald
Spartan	5-6	7-10	Slight	Spartan breakdown, brown core
Stayman Winesap	4-5	7-8	Very high	Senescent breakdown
Winesap	5-6	8-9	High	Senescent breakdown
Yellow Newtown	5-6	8-9	High	Internal browning, senescent breakdown, bitter pit
York Imperial	4-5	6-7	Very high	Cork spot

Source: Hardenburg et al. (1986).
[a] For maximum storage, cultivars must be harvested at optimum maturity, stored under ideal temperature and humidity, and in most cases in the recommended controlled atmosphere. Some fruit may be stored 1-2 months longer than shown.
[b] Water core not listed for Delicious, Jonathan, Winesap, Stayman, and others, as it is present at harvest and does not develop in storage.

storage potential is usually found in apples that are mature but have not yet attained their peak of respiration when harvested. However, the grower is inclined to sacrifice storage quality for the better color often gained in red varieties by holding them longer on the tree. Even if harvesting began at the proper time, the fruit picked last may be at an advanced stage of maturity. Such late picked apples do not have good storage characteristics; neither do those harvested on the immature side, but this is seldom a problem with apples intended for storage before marketing. Harvest at proper maturity, careful handling, and prompt storage after harvest are conducive to long storage life.

Chilling injury is the term commonly applied to disorders that occur at low storage temperatures where freezing is not a factor. The exact mutual relationship of the many types of chilling injury is unknown. The principal disorders classed as chilling injuries in apples are: (1) soft scald; (2) soggy breakdown; (3) brown core; and (4) internal browning. Varieties susceptible to one or more of these disorders are Rome Beauty, Jonathan, Golden Delicious, Grimes Golden, McIntosh, Rhode Island Greening, and Yellow Newtown. In addition to variable susceptibility by variety, there are also yearly variations related to climate, fruit size, and cultural factors.

Table 1 lists the range in storage life at 30°F of several apple varieties susceptible to chilling injury at 30°F when grown under certain climatic or cultural conditions and may be more properly stored at temperatures of 36°F or higher. It also depends greatly on the expected length of storage before marketing and the availability of storage space at different temperatures.

Controlled Atmosphere Storage

Controlled atmosphere (CA) storage offers important gains in extending the market life of certain apple varieties. Chilling injury of some varieties is eliminated by elevating the storage temperature of these varieties to about 40°F and altering the composition of the atmosphere.

Only apples of good quality and high storage potential should be placed in CA storage. Harvest maturity and handling practices are crucial, and only fruits harvested at proper maturity should be considered. In any one district, this limits the apple harvest for CA storage to only a few days. Immature apples or those retained on the tree to gain better color, as is often the case with Delicious and McIntosh, are equally undesirable.

The following practices affect the condition of apples held for both conventional and CA storage:

Maturity. Because there is no reliable maturity index, growers must use personal experience of the variety, area, or orchard to decide when the crop is mature. Availability of labor, size of operation and crop, weather, storage facilities, and intended length of storage also affect the time of harvest.

Packaging. In many sections of the United States and Canada, apples are stored as harvested (*orchard-run*) in bulk bins. No sizing or sorting is done until the fruit is prepared for market. An exception is in the Pacific Northwest, where apples are sized and sorted shortly after harvest, packaged in corrugated containers that hold about 45 lb of fruit, palletized, and returned to storage. Stacking racks or supports must be used if pallets are stacked.

Handling to Storage. Apples should be cooled promptly after picking because they can deteriorate as much in one day at field temperatures as during one week held at proper storage temperatures. Normally, apples are placed in storage and cooled by the room refrigeration equipment to about 32°F in 1 to 3 days. Hydrocooling is sometimes used, but it requires careful disease control. It also interferes with scald inhibitors, which must be applied to the warm fruit.

The rooms for gas storage must be gastight, and there must be provision to remove excess CO_2 from the air. Carbon dioxide may be removed by circulating the air through washers or scrubbers filled with sodium hydroxide, ethanolamine, or plain water, or through a cabinet or room containing bags of hydrated lime.

Deciduous Tree and Vine Fruits

The modified atmosphere may be obtained by filling the room with fruit, sealing it, and allowing the respiration of the fruit to provide the desired proportions. These proportions are maintained by operation of the scrubber and by ventilation.

Table 2 lists approximate temperature and atmospheric requirements for CA storage of several varieties. Since these may vary from region to region, more precise requirements should be determined from authorities within a region. For example, Jonathan, Rome Beauty, and Stayman Winesap are reported best at 2% CO_2 in New York, whereas a somewhat wider range is acceptable in Washington.

Table 2 Requirements for Controlled Atmosphere Storage of Apples

Cultivar	Carbon dioxide, %	Oxygen, %	Temp., °F
Cortland	5	2-3	36
	2-3	2-3	32
Delicious	1-2	1.5-2[a]	31-32
Golden Delicious	1-3	1.5-2[a]	31-32
Granny Smith	1-3	2-3	31-32
Idared	2-3	2-3	31-32
McIntosh	2-3 one month, then 5	2.5-3	36
	2-3 one month, then 5	2	38
Rome Beauty	1-3	2-3	31-32
Stayman Winesap	2-5	2-3	31-32
Yellow Newtown (Calif.)	8	3	40
(Oreg.)	5-6	3	36

Source: Hardenburg *et al.* (1986).
[a] 1.5% oxygen not recommended for Delicious or Golden Delicious in New York, because less than 2% oxygen is injurious.

Disadvantages of gas storage include the difficulty in making the storage room gastight, the danger of suffocation to persons entering the room, and the impossibility of entering the room to examine the fruit without losing the desired atmosphere. Some of these disadvantages have been overcome by a method that passes the air going into the storage room through a generator, which reduces the O_2 level and raises the CO_2 content to desired levels. The process is continuous, so airtight rooms are not essential. Also, the rooms may be opened for inspection and the desired atmosphere restored quickly.

Boxes with polyethylene liners sealed after being packed with the apples have been used to create a modified atmosphere. In many instances, it is possible to closely approach the atmospheric composition of a CA room because of differential permeability of the film to O_2 and CO_2. Unfortunately, the composition of the atmospheres in different containers varies greatly, so that no dependence can be placed on improved storage life. Also, when the apples are marketed, the film liner must be opened in some way to avoid possible harmful effects of low O_2 or high CO_2 as the temperature rises. Although film liners seem impractical as a means of altering atmospheric O_2 and CO_2, unsealed film liners have proven very helpful in controlling excessive moisture loss from such varieties as Golden Delicious. Although often severe in cold storage warehouses, desiccation should not be of sufficient concern to justify film liners for fruit in CA rooms.

Storage Diseases and Deterioration

Storage problems in apples may be caused either by invading microorganisms or by the fruit's own physiological processes. Physiological disorders, although sometimes resembling rots, are related to biochemical processes within the fruit. Susceptiblity to such disorders is often a variety characteristic, but it may be influenced by cultural and climatic factors and storage temperature.

Alternaria Rot. Dark brown to black, firm, fairly dry to dry storage decay centering at wounds, in skin cracks, in core area, or in scald patches—one of the blackest of storage decays. *Control:* Cultural practices that produce apples of good finish and prevent skin diseases and injuries that open way for infection.

Ammonia Gas Discoloration. Circular spots centering at lenticles; dull green on unblushed side and brown to black on blushed side. Injury may disappear from slightly affected fruits. *Control:* Ventilate as soon as possible. Examine fruit for injury at various points in the room because some sections may escape.

Bitter Pit. Many small, sunken bruise-like spots, usually on the calyx half of the fruit. Masses of brown, spongy tissue occur adjacent to surface pits or may be found deeper in the flesh. In storage, spongy tissue near surface loses moisture and tends to become hollow. New areas may appear and develop in storage. *Control:* Apply boron and calcium, as recommended, in the orchard. Follow cultural practices that promote regular bearing and stabilize moisture. Store fruit of proper maturity and cool promptly to 32 °F. Maintain humidity high enough to prevent moisture loss.

Blue Mold Rot (*Penicillium*). Spots of various sizes with decayed tissue that is soft, watery, and can be readily scooped out of the surrounding healthy flesh. Rot usually as deep as wide. Advanced stages have white tufts of mold which turn bluish green as spores are produced under moist conditions. Affected tissue has moldy or musty flavor and odor. Most prevalent type of storage decay of apples. *Control:* Handle carefully to prevent skin breaks. Cool promptly to 32 °F. Use fungicides in wash treatments. Keep picking boxes, packing house, and storage room sanitary. Whitewash walls and ceiling.

Brown Core. No external symptoms. First appears as slight browning or discoloration of core tissue between the seed cavities. Later, part or all of the flesh between the seed cavities and the core line may become brown. Serious in McIntosh and other susceptible varieties stored for long periods at 30 °F. *Control:* Pick at proper stage of maturity. Use controlled atmosphere storage at 38 °F. A disorder with similar symptoms has been reported as a result of exposure to excessive concentrations of carbon dioxide.

Freezing Injury. Watersoaked, rubbery condition of large areas or of entire apple. Vasculars (water-conducting strands) brown. Bruised areas in frozen apples large, with wrinkled gray to light brown surface. Moisture lost rapidly from affected areas. In refrigerator cars, most prevalent on floor, and at doorways; in storage rooms, most injury in bottom layer boxes, near coils, or against walls next to freezer storage. *Control:* Heat car during subfreezing weather. Prevent cold pockets in storage rooms by adequate air circulation. Minimum handling of fruit while frozen. Thaw at 40 to 50 °F. Move thawed fruit into trade channels promptly; do not allow it to become overripe.

Internal Breakdown. Mealy breakdown of internal tissue in overripe fruit. Flesh soft. Surface often duller and darker than normal. Hastened by too high storage temperature, freezing, bruising, or presence of watercore which it often follows. *Control:* Pick before overmature. Cool promptly at temperatures as near as possible for varieties that tolerate that temperature. Watch ripening rate particularly of fruit with watercore.

Internal Browning. No abnormal skin appearance. Sometimes appears only around core; the apple's outer fleshy portion remains normal in appearance. Occasionally only outer flesh is involved; but when internal browning develops in outer fleshy portion it is usually accompanied by browning around the core. Disease develops uniformly throughout tissue. Occurs in firm, soundappearing apples. *Control:* Use controlled atmosphere at 38 °F for Yellow Newtowns and other susceptible varieties.

Jonathan Spot. Slate-brown to black, entirely superficial or very slightly sunken, skin-deep spots in color-bearing cells of skin. In some varieties, spots center at lenticels. *Control:* Refrigerate promptly as this disease is greatly aggravated by delayed storage. Use controlled atmosphere storage.

Lenticel Rots. Bullseye rot (*Neofabrabraea*) most common of group; of importance only from Northwest; spots fairly firm, pale centers, decay mealy, may penetrate nearly as deep as wide. Fisheye rot (*Corticum*), tough leathery spot, often follows scab, decayed tissue stringy. Side rots (*Phialophora*), spots shallow with tender skin, decayed tissue wet, slippery. *Control:* Harvest at prime maturity; store and cool promptly; use forecasting technique for Bullseye rot to determine potential keeping quality.

Scab (*Venturia*). Occasionally active scab spots on fruits at time of storage will enlarge. Fruits may be infected in orchard but show no disease at the time of storage. Disease may subsequently develop in storage as small brown or jet black spots in peel, often without breaking cuticle of fruit. *Control:* Follow recommended orchard spray schedule.

Scald. Diffuse browning and killing of skin of fruit stored for several months. Ordinarily most prevalent on immature fruit or on green portions of fruit. *Control:* Pick apples when well matured. Treat with effective scald-inhibiting chemicals. Scald develops less on fruit in controlled atmospheres.

Soft Scald. Sharply defined or slightly sunken ribbon-like areas in the skin. Affected tissue shallow and rubbery. Most severe on Jonathan, Golden Delicious, and Wealthy. *Control:* Store promptly. Use recommended controlled atmospheres, temperatures, and lengths in storage for each variety.

Soggy Breakdown. Light brown, moist, rubbery, definitely delimited areas in cortex of apple. Not visible on surface. Worst in Grimes Golden, Wealthy, and Golden Delicious. *Control:* Same as for soft scald.

Water Core. Hard, glassy, watersoaked regions in flesh of apple at core or under skin. Decreases in extent during storage but predisposes fruit to internal breakdown. *Control:* Pick as soon as mature. Watch fruit in storage and move before becoming overripe.

PEARS

Bartlett is the most important pear variety, exceeding the total of all others by a wide margin. Other Pacific Coast varieties are Hardy, Comice, Anjou, Bosc, and Winter Nelis. The eastern states have limited varieties due to the severe problem of fire blight, and grow primarily the Kieffer variety. Although most Bartlett, Hardy, and many Winter Nelis pears are canned, cold storage prior to ripening for canning is the usual procedure. A 10-day to 2-week cold storage period for Bartlett pears is commonly used by canners because it improves uniform ripening. Substantial quantities may also be stored for periods approaching maximum storage life of the variety to better use processing facilities.

Maturity at harvest has a very important bearing on subsequent storage life, as it does on the apple. However, unlike apples, pears do not ripen on the tree, nor do most varieties ripen at cold storage temperatures. If harvested too early, they are subject to excessive water loss in storage. If permitted to become overmature on the tree, their storage life is shortened and they may be highly susceptible to scald and core breakdown. Flesh firmness as measured by a pressure tester is perhaps the best measure of potential storage life of pears from any single orchard. For the Bartlett variety, a firmness of 19 to 17 lb, measured with a Magnus-Taylor pressure tester or similar device, using an 0.31 in. plunger head, indicates best storage quality. If average firmness is as low as 15 lb, storage for any prolonged period is hazardous. Pressure test information for each lot of pears going into storage may be very helpful to both the fruit owner and the cold storage operator in determining the storage program.

Careful harvesting and handling are essential to good storage quality. Bruises and skin breaks are likely sites for infection by microorganisms. Varieties such as Winter Nelis and Bosc are highly susceptible to punctures caused by stems broken in the harvesting operation. Comice is also easily damaged because of its very tender skin. Many pears are now being harvested into pallet bins holding about 1000 lb of fruit. Care in dumping fruit from a picking container is important in keeping mechanical damage to a minimum.

For best storage quality, rapid cooling after harvest is essential. Pears ripen rapidly at elevated temperatures but do not soften or change color in the early ripening stages. Therefore, a considerable part of the storage life may be used up without a visible change in the fruit. If cold storage rooms do not have adequate refrigeration and ventilation capacity for the rapid cooling of fruit, precooling in special rooms (or hydrocooling) should be considered prior to placing in the storage room. When warm fruit is placed in a room with cold fruit, the loading arrangements should be such that the temperature of the cold fruit is not elevated.

Pears are very sensitive to temperature and should be stored at 30°F and 90 to 95% rh. Recommendations as low as 29°F have been made, but the risk of freezing injury is great unless the temperature in all parts of the room can be controlled precisely. Pears are not subject to chilling injury as are some apple varieties, so elevated storage temperatures are not required. The stacking arrangement recommended for apples in the cold storage room also applies for pears.

Since pears lose water more readily than most apple varieties, good humidity conditions in the storage room must be maintained. For long storage, 90 to 95% rh is recommended. Perforated film box liners are excellent for moisture loss control.

The approximate storage life of pears at 30°F is shown in Table 3. These values assume an additional time for transportation and marketing. If Bartlett pears for canning are harvested at the best stage of maturity and quickly cooled to 30°F, their safe storage life may be as long as 4 months, since marketing involves only ripening for processing. However, note that quality deteriorates during storage, particularly as the maximum storage life is approached.

Table 3 Storage Life of Pear Varieties at 30°F

Variety	Storage Life, Months
Bartlett, Hardy, and Kieffer	2 to 3
Bosc, Comice, and Seckel	3 to 4
Anjou	6 to 7
Packham	5 to 6
Winter Nelis	7 to 8

After removal from storage, best dessert quality is attained if pears are ripened in a controlled temperature range of about 60 to 70°F. This applies to fruit for the cannery and for fresh use. Ripening at 68 to 72°F is more practical for cannery fruit than lower temperatures, because the shorter time involved reduces overhead costs with no measurable difference in quality.

The practice of controlled atmosphere storage of pears is promising. The storage life of Bartlett pears can be extended to 5 to 6 months at 30°F for fruit of desirable maturity (17 to 20 lb firmness) in an atmosphere containing 2 to 2.5% oxygen and 0.8 to 1% CO_2. Bartlett pears of advanced maturity are intolerant to elevated CO_2 and develop core and flesh browning within a few weeks. They are tolerant to low O_2, but have less storage potential than pears of desirable maturity. Chronological age and pressure test (under 16 lb firmness) is evidence of advanced maturity.

Commercial storage of pears in a controlled atmosphere has not been considered as necessary as it is in the apple industry. Since no low-temperature disorders have been recognized, there has not been a need for further extension of the storage life.

Deciduous Tree and Vine Fruits

Many pears from western states, when packed for storage before marketing, use perforated polyethylene liners in the container. While such liners give excellent protection against water loss, there is no agreement as to their value in modifying the atmosphere within the container.

Storage Diseases and Deterioration

The principal storage disorders in pears are: (1) core breakdown; (2) scald and failure to ripen; and (3) fungus rots.

Core Breakdown. Often accompanies scald. Soft, brown breakdown in core area having acrid, disagreeable odor of acetaldehyde. *Control:* Do not allow pears to become overmature on tree. Cool promptly. Store at 30°F. Ripen fruit between 65 and 75°F.

Core breakdown is associated with overmaturity at harvest. This problem has become more important because of growth regulator sprays used to keep pears from dropping during the harvest season. Pressure test information of late harvested fruit is helpful in locating lots susceptible to core breakdown. Records of the time lapse between harvest and storage are important, since pressure test information may not be a true measure of relative storage quality where storage is delayed. Pear color, particularly in California Bartletts, is a very poor measure of potential storage life because of great variability among pears from different districts.

Scald. Often accompanies core breakdown. Brown to black softening of large areas of skin and tissues immediately beneath skin. Affected areas slough off readily. Acetaldehyde odor and flavor prominent. *Control:* Pick before overmature. Cool promptly. Store only for proper period. Cannot be controlled by oiled paper wraps.

Pear scald is associated with pears that have been stored too long and have lost their capacity to ripen. It is not related to apple scald and cannot be controlled by any supplemental treatments. The problem develops progressively earlier as the temperature is raised above 30°F. Yellowing of the fruit is the principal storage symptom; Bartlett and Bosc are the two most susceptible varieties. Anjou and Comice may not develop scald but do lose their capacity to ripen. Periodic inspection is desirable to be sure that green pear varieties are removed from storage before yellowing progresses to the danger point. Yellow pears may show no scald in storage but may develop scald on removal to a ripening temperature. If pear scald does show in storage, the pears have been kept too long and may be worthless.

Anjou Scald. Anjou pears are often affected with a surface browning more superficial than common scald and distinct from it, resembling apple scald. Anjou scald is controlled by oiled paper wraps and effective scald inhibiting chemicals.

Gray Mold Rot (*Botrytis*). Extensive, firm, dull brown, watersoaked decay with bleached border. Dirty white to gray extensive mycelium forming *nests* of decayed fruits. *Control:* Wrap fruit in copper-impregnated paper. Use fungicide in spray or wax on packing line. Cool promptly to 30°F.

Gray mold rot caused by *Botrytis cinerea* will grow at cold storage temperatures and can be a serious threat to long stored winter varieties such as Anjou and Winter Nelis. Without control measures, the disease may spread from one fruit to another by contact.

Alternaria Rot. Surface is dark brown to black. Decayed tissue is gray to black, dry in center, gelatinous at edge, easily removable as core from surrounding flesh. Found late in storage season, usually at punctures. *Control:* Prevent skin breaks. Remove from storage at first appearance of trouble.

Brown Core. Anjou and Bartlett pears stored in sealed, polyethylene-lined boxes with inadequate permeability may show various degrees of pithy brown core and desiccated air pockets. Prolonged storage in concentrations of 4% or more CO_2 often produces brown core, particularly when pears are harvested at advanced maturity or are cooled slowly after packing. *Control:* Harvest at proper maturity. Cool promptly. Store at 30°F. Use perforated film liners to maintain CO_2 level at 1 to 3%.

Freezing Injury. Bartlett and Anjou pears exposed for 4 to 6 weeks just below their freezing point develop glassy, watersoaked external appearance with tan pithy area around core. Pears frozen sharply may break down completely or show abruptly sunken large pits where slightly bruised while frozen. *Control:* Keep transit and storage temperature above 30°F.

GRAPES

Grapes are widely grown in the United States, but over 90% are grown in California. This state produces grapes of the *Vitis vinifera* species almost exclusively. This species can withstand the rigors of handling, transport, and storage required of table grapes for wide distribution over a long marketing period. Almost all of this fruit is precooled and much of it stored for varying periods before consumption. On the other hand, for fresh use, the fruit of the species *Vitis labrusca* (Eastern type) is largely limited to local market distribution.

Grapes grow relatively slowly, and should be mature before harvest because all of their ripening occurs on the vine. *Mature* here means that stage of physiological development when the fruit appears pleasing to the eye and can be eaten with satisfaction. However, grapes should not be overripe, as this predisposes the fruit to two serious postharvest disorders: (1) weakening of the stem attachment in some varieties, such as Thompson seedless, which causes the berries to separate from the pedicel attachment; and (2) progressively greater susceptibility to invading decay organisms. Danger of fruit decay is increased with exposure to rain or excessively damp weather before harvest (conditions favorable for the inception of field infections by *Botrytis cinerea Pers*).

Cooling and Storage

Grapes are vulnerable to the drying effect of the air because of their relatively large surface to volume ratio, especially that of the stems. Stem condition is an important quality factor and an excellent indicator of the past treatment of the fruit. Stems should be maintained in a fresh green condition not only for appearance but because they become brittle when dry and are apt to break. The stem of a grape cluster, unlike that of other fruits, is the handle by which the fruit is carried; if breakage (shatter) occurs, the fruit is lost for all practical purposes even though the shattered berries may still be in excellent condition. Therefore, careful attention should be paid to those operations that minimize moisture loss.

The rate of water loss is especially high before and during precooling because grapes are normally harvested under hot, dry conditions. Remove field heat promptly after the fruit is picked to minimize the exposure of grapes to low vapor pressure conditions. Volume and temperature of the precooling air, velocity of the air past or through the containers, and accessibility of the fruit to this air are significant factors in the rate of heat removal and are drastically influenced by the location and amount of venting of the containers, alignment of the containers (air channels), and packing materials such as curtains, cluster wraps, and pads.

Two general systems of air handling used for precooling table grapes are (1) the *velocity* or conventional system, and (2) the *pressure* system.

In the velocity system, air is forced through channels parallel to the long axis of the containers. Heat transfer is effected by conduction through the packaging materials and by penetration of the cold air to the fruit from turbulence in and around the vents. Satisfactory precooling rates can be attained if: (1) the containers are aligned so that there are no obstructed channels; (2) the velocity of the air through these channels is at least 100 fpm; and (3) air of not more than 35°F can be supplied at the rate of at least 0.16 cfm/lb of fruit. Very humid air helps to reduce the rate of water loss from the fruit. Large cooling surfaces can be maintained, or atomized water can be added to the airstream. However, the rate of removal of field heat is the significant factor; humidifying techniques that retard precooling are likely a liability.

In the pressure system, a pressure gradient is set up so that there is a positive flow of cold air through the fruit from one vented side of the container to the other. The containers are arranged so that the air must pass *through* the containers before returning to the refrigeration surface. Precooling time may be as little as one-fifth that of the velocity system if the pressure differential across the packages is equivalent to at least 0.25 in. of water and cold air is supplied at the rate of at least 1 cfm/lb of fruit.

Recommended storage temperatures for *Vitis vinifera* (European or California type) grapes are 30 °F. The relative humidity should be from 90 to 95%. Although temperatures as low as 29 °F have not been injurious to well-matured fruit of some varieties, other varieties of low sugar content have been reported damaged by exposure to 31 °F. Grape storage plants in California should provide uniform air circulation in the rooms. Some have precooling rooms where the grapes are cooled to about 39 °F in 6 to 24 h before storage. In some plants, all of the cooling is done in the storage rooms, but only a few have sufficient air movement to cool the fruit as quickly as desired. After the fruit has been precooled, the air velocity should be reduced to a rate that will maintain uniform temperatures throughout the room (no more than 10 to 20 fpm in the channels between the lugs). Ventilation is required only to exhaust sulfur dioxide and air following fumigation.

The greatest change that takes place in grapes in storage is loss of water. The first noticeable effect is drying and browning of stems and pedicels. This effect becomes evident with a loss of only 1 to 2% of the mass of the fruit. When the loss reaches 3 to 5%, the fruit loses its turgidity and softens.

Maintaining a relative humidity of 90 to 95% in grape storage is often a problem especially at the beginning of the storage season when the rooms are being filled with dry lugs. Each lug will absorb 0.33 to 0.67 lb of water over a month's period and, unless moisture is supplied to the room, this water must come from the fruit. An effective method of supplying water to minimize shrinkage is by spray humidification. A fine spray can be obtained that will vaporize readily even at 31 °F with proper balance of water and air pressure using the correct type of nozzle.

Fumigation

Vinifera grapes must be fumigated with sulfur dioxide after they are packed to prevent or retard the spread of decay. The treatment surface sterilizes the fruit, particularly wounds made during handling.

Fumigation with sulfur dioxide in storage prevents new infections of the fruit but does not control infections that have already occurred in the vineyard. Frequently, these have not developed far enough to be detected at harvest and consequently are the primary cause of decay in storage. A method of measuring field infection has been developed and used to forecast decay during storage. The forecast indicates the lots that are sound and can be safely stored and also those that are likely to decay and should be marketed early (Harvey 1955, 1984).

It has become common practice to accumulate packed fruit in the precooler during the daily packing and to fumigate the fruit in the evening. In this way, precooling is not delayed and fumigation can be done after most of the working crew has left. This initial treatment often becomes the responsibility of the refrigeration personnel.

Amount of Sulfur Dioxide. Other commodities should not be treated with the grapes or even held where the fumigant can reach them, as most of them are very easily injured by the gas. Because grapes also can be injured, they should be exposed to the minimum quantity of sulfur dioxide required, which will depend on: (1) the decay potential and condition of the fruit; (2) the amount of fruit to be treated; (3) the type of containers and packing materials; (4) the air velocity and uniformity of air distribution; (5) the size of the room; and (6) losses from leakage and sorption on walls. Under favorable conditions a basic sulfur dioxide concentration of 0.5% by volume for 20 min. is adequate. To keep the concentration at this level, consider the absorptive capacity of the containers and fruit as well as their volume. The dosage can then be calculated from the following equation:

$$W_s = \frac{AB}{100\,E} + CD \qquad (2)$$

where

W_s = quantity of sulfur dioxide required, lb
A = concentration of sulfur dioxide to be used, %
B = free volume of room (total volume minus 0.5 ft^3 for each container), ft^3
C = number of carloads of 28-lb lugs (1000 lugs/car)
D = quantity of sulfur dioxide absorbed by each carload, lb
E = volume occupied by 1 lb of sulfur dioxide gas at 32 °F (5.5 ft^3)

For factor D, 1 lb per car is adequate when the fruit is sound. Air velocities are maintained past both sides of every container at 50 fpm or more (75 to 100 fpm if the fruit has curtains over it or the clusters are wrapped), and the room is relatively gastight with no opportunity for the fumigant to be lost on refrigeration surfaces. Conversely, a higher value of 2 lb car would be used when these factors are less favorable.

Grapes must be fumigated weekly in storage to prevent *Botrytis cinerea* from spreading from infected fruit to adjacent sound fruit. The amount of sulfur dioxide needed depends on the same factors as for the initial treatment. However, a basic concentration of 0.1% for 30 min. is adequate. Also, an absorptive factor in the range of 0.33 to 0.67 lb of sulfur dioxide per carload should be used.

Distribution Procedure. The gas must be distributed quickly and evenly to all parts of the room. This can be done by spacing special nozzles 6 ft apart along the ceiling in the room. If the outlet is placed in front of a fan, there should be one for each fan, or the air from the single fan should be distributed evenly across the room through a plenum.

The same requirements of proper container alignment, adequate fan capacity, and uniform air distribution apply here as for the initial treatment. The lugs should be oriented parallel to the airflow and channels 0.75 to 1.5 in. should be provided on both sides and kept completely unobstructed through the stacked fruit. The fruit should be stacked as near the ceiling as possible or drop curtains provided over the fruit to prevent air from passing over the fruit and thus bypassing the channels. The working distance between pallets should be kept to an absolute minimum to avoid wide channels, and no holes should be left in the wall of lugs when pallets of fruit are withdrawn.

The hot-gas method of delivery may be used if the room requires 10 lb or less of gas. The steel cylinder containing the liquid sulfur dioxide is first connected to the gas inlet and the valve then opened. The cylinder should then be placed in a pot of boiling water to vaporize the fumigant as rapidly as possible. Only about 1 lb/min can be delivered this way.

For larger quantities, the cold-gas method is usually more practical. A riser extends to the bottom of the cylinder through which the liquid sulfur dioxide rises and flows through the delivery line. Every precaution must be taken to provide enough air volume and velocity to vaporize and mix the gas thoroughly with the air before it reaches the fruit. Up to 100 lb of the material can be released in 2 to 3 min. After 30 min., the room should be purged of the gas-

Deciduous Tree and Vine Fruits

laden air until personnel can remain in the space without excessive discomfort.

In plants that are devoted entirely to the storage of grapes, the gas is sometimes released into the air ducts of the plant, thus using the air-cooling system for even distribution and good circulation in the rooms. With a brine spray system of refrigerating the air, a bypass around the spray chamber prevents the gas from contacting the wet metal surfaces, since it readily forms a corrosive acid in combination with water. For the same reason, sulfur dioxide should be cleared from the air of the rooms before the damper is turned and the air is circulated through the spray. It is advisable to check the acidity of the brine frequently to guard against corrosion.

Precautions. Sulfur dioxide has certain properties that demand care in its use as a fumigant in cold storage plants. The concentrations recommended for the fumigation of grapes in storage can cause respiratory spasms and death if the victim cannot escape from the fumes. When working in even weak concentrations of sulfur dioxide, wear goggles to protect against injury to the eyes and a gas mask fitted with canister for acid gases (not the usual canister for ammonia gas). Concentrations as low as 30 to 40 ppm can be detected by smell. It requires several times these concentrations to cause discomfort.

Because a small segment of the population may experience severe allergic reactions to sulfites, the U.S. Environmental Protection Agency has proposed a 10 ppm tolerance for sulfite residues in table grapes (EPA 1989, Harvey *et al.* 1988). Fruit with residues exceeding the tolerance cannot be marketed.

Another precaution about sulfur dioxide which cannot be overemphasized is its injurious effects on other produce. For this reason, care must be taken that only grapes are stored in the room that is to be fumigated and that there are no leaks through wall or halls to adjacent rooms storing other produce.

Periodic inspection of the fruit is recommended to check whether the sulfur dioxide gas is reaching the center of the stacks or whether some grapes are being overtreated. If the pedicels and stems retain a yellow or green color and broken berries show no mold and appear to be dried or seared, the gas has reached the fruit in question and is having the desired effect. Serious bleaching on unbroken grapes means too high a concentration or too long an exposure, and there should be better distribution of the gas, lower concentration, or shorter fumigation periods.

Diseases

Blue Mold Rot (*Penicillium*). Watery, mushy condition. Early production of typical bluish-green spores on berries and stems. Moldy odor and flavor. *Control:* Prevent deterioration in fruit by careful handling and prompt refrigeration, preferably to 32°F. Fumigate with sulfur dioxide in storage.

Cladosporium Rot. Black, firm, shallow decay which produces an olive-green surface mold. Common on stored grapes harvested early in the season. Infections occur on small growth cracks at the blossom end and sides of the grape. *Control:* Precool and store grapes promptly at 32°F. After harvest, fumigate with sulfur dioxide to reduce spread.

Gray Mold Rot (*Botrytis*). Early stage, slip skin with no mold growth. Later, nest of fairly firm decay covered with abundant fine gray mold and grayish-brown, velvety spore masses. *Control:* Cull out decay when packing. Fumigate grapes with sulfur dioxide. For storage, cool grapes rapidly to 30°F. Use forecasting technique to determine safe storage periods. Use short storage period for grapes harvested in rainy periods or after slight freezes.

Rhizopus Rot. Soft, mushy, leaky decay causing staining of lugs. Coarse extensive mycelium and black sporangia develop under moist conditions. *Control:* Prevent skin breaks. Cool promptly to below 50°F.

Sulfur Dioxide Injury. Bleached sunken areas on berry at skin breaks or cap-stem attachment. Decolorized portions have disagreeable astringent flavor. Does not appear in full severity until cool grapes are warmed. *Control:* Apply proper concentration and distribution of gas for recommended period.

Storage Life

The normal storage life of the principal varieties of California table grapes at 30°F is shown in Table 4. Under exceptional conditions, sound fruit will keep longer than indicated; for example, Emperor grapes have been held in good condition for 7 months, and Thompson seedless for 4 months.

The storage life of grapes is affected mostly by the attention given to selecting and preparing the fruit. Grapes should be picked at the best maturity for storage, especially Thompson Seedless and Ohanez. Stems and pedicels should be well developed and the fruit should be firm and mature. Soft and weak fruit should not be stored. The display lug is a satisfactory package for storage since it can be cooled and fumigated easily.

Cooling to 40 to 45°F is advised for grapes that are to be in transit a day or two before reaching storage. Special care should be taken during transit so that decay does not start. It is not good practice to delay fumigation until the grapes reach a distant storage plant, for in the picking and packing of grapes, many berries are injured sufficiently to permit mold to begin unless the fruit is fumigated promptly.

Table 4 Storage Life of California Table Grapes

Variety	Storage Life, Months
Emperor, Ohanez, Ribier	3 to 5
Malaga, Red Malaga, Cornichon	2 to 3
Thompson Seedless, Tokay	1 to 2.5
Muscat, Cardinal	1 to 1.5

For *labrusca* (Eastern type) grapes, a storage temperature of 32°F and humidity of 85% are recommended. Care in packing and handling the fruit, a minimum of delay before storage, and prompt cooling are important for best results with these varieties, as they are with the *vinifera* grapes. The Eastern varieties are not fumigated with sulfur dioxide due to their susceptibility to injury from it. The storage life of the important commercial varieties at 32°F is shown in Table 5.

Table 5 Storage Life of Labrusca Grapes at 32°F

Variety	Storage Life, Weeks
Catawba	5 to 8
Concord, Delaware	4 to 7
Niagara, Moore	3 to 6
Worden	3 to 5

PLUMS

Plums are not suitable for long storage. Among the major shipping varieties, well-matured Santa Rosa, El Dorado, Nubiana, Queen Ann, Laroda, Late Santa Rosa, and Casselman are sometimes stored for short periods. The Italian Prune can be held for no more than 2 weeks before marketing begins.

Plums intended for storage should be harvested at a high soluble solids level for the variety, although doing so may delay the harvest beyond the normal picking date. Harvested fruit should be carefully graded to remove disease, defects, and injuries before packing in the shipping container.

The fruit should be thoroughly cooled before storage. Cooling may be done in the 900 to 1000-lb bulk bins that are used for harvest. The shipping containers are normally vented to aid cooling after packing. While most fruit is air cooled in conventional room coolers, some shippers use forced air to cool fruit quickly in bulk bins or shipping containers.

Plums can usually be stored for 1 month at 30 to 32°F with 90 to 95% rh. Results of storage life tests have been variable, with some lots in certain seasons remaining in good condition even after 4 to 5 months in storage. Other lots and fruit in some seasons have not been held satisfactorily beyond 2 months. Fruit with the highest soluble solids has consistently shown the longest storage life, even when harvested several weeks after the completion of commercial harvest. Some plum varieties benefit from CA storage.

Storage Diseases and Deterioration

Plum deterioration appears as changes in appearance and flavor. A poststorage holding period should be used in judging the condition of stored fruit. Fruit that appears bright and flavorful in storage can show severe deterioration when removed to room temperature for 2 to 3 days.

Some flesh softening and a gradual loss of varietal flavor and tartness occurs even at low storage temperatures. The first visual sign of deterioration is the development of translucence, first around the pit, then extending outward through the fruit. Translucence is followed by the development of progressively more severe flesh browning following the same pattern. The first noticeable loss in flavor is generally associated with the first symptoms of translucence in the tissue. It is necessary to cut through the fruit to judge condition, since fruit held under good storage conditions may appear sound from the outside while being seriously deteriorated internally. See section on Sweet Cherries for diseases. See Cold Storage and Sulfur Dioxide Injuries under Peaches.

SWEET CHERRIES

Harvesting Techniques

Sweet cherries for storage must be harvested with stems attached. Mechanical harvesting takes most of the fruit without stems and consequently the cherries must be processed or otherwise used immediately.

Cooling

Rapid cooling to 30°F is essential if this fruit is to be stored. Hydrocooling has been used successfully, and the wetting is tolerable as long as the fruit remains cold. Fungicidal postharvest sprays or dips are helpful in reducing decay during storage.

Forced air or pressure cooling can be used to quickly cool the fruit without the problem of wetting. Moisture loss and stem drying can be minimized by rapid movement from the field to cooler, rapid cooling, and maintaining low temperatures and high humidity during cooling and storage.

Storage

When sweet cherries are stored, they are normally held in shipping containers, often with polyethylene liners. These liners permit an increase in carbon dioxide gas surrounding the fruit, which tends to reduce decay rates and increase storage time. Cherries should be stored at 30°F and may be held 2 weeks after harvest and still retain enough quality for shipment to market.

Controlled atmospheres with 20 to 25% CO_2 or 0.5 to 2% oxygen help maintain firmness and bright, full color during storage (Hardenburg et al. 1986). Polyethylene liners can extend the market life. The liner must be perforated when removed from storage.

Diseases

Alternaria and Cladosporium Rot. Light brown, dry, firm decay lining skin breaks that can be removed easily from surrounding healthy tissue. Mycelium on area fine and white above and dark green below. *Control:* Sort out cherries with cracks and other skin breaks at packing. Use fungicide in spray or sizer on packing line.

Blue Mold Rot (*Penicillium*). Circular, flat spots covering conical, soft, mushy decay that can be scooped out cleanly from surrounding healthy flesh. White fungus tufts turning to bluish green develop on surface. Musty odor and flavor. *Control:* Prevent skin breaks. Use fungicide in spray or sizer on the packing line. Market promptly. Refrigerate promptly to 32°F.

Brown Rot (*Monilinia*). See Brown Rot of Peaches. *Control:* Follow recommended orchard spray practices. Use fungicide in spray or sizer on packing line. Refrigerate promptly to 32°F. Package cherries in polyethylene bags to reduce desiccation of stem and fruit, preserve color, and reduce decay development.

Gray Mold Rot (*Botrytis*). Light brown, fairly firm, watery decay covered with extensive delicate, dirty-white mycelium. On completely decayed cherries, grayish-brown velvety spores may be found. *Control:* Handle carefully. Use fungicide in spray or sizer on packing line. Refrigerate promptly to 32°F.

Rhizopus Rot. Extensive soft, leaking decay with little change from normal color. Coarse mycelium and black spore heads are prominent under moist conditions. More prevalent in upper-layer packages in refrigerator car. *Control:* Rhizopus develops very slowly at temperatures below 50°F, so storage at recommended temperature keeps decay in check.

PEACHES AND NECTARINES

This discussion relates primarily to peaches but also applies to nectarines in many respects.

Storage Varieties

Peaches do not adapt well to prolonged storage. However, if they are sound and well matured, most freestone varieties can be stored for up to 2 weeks (some freestone and most clingstone for up to 4 weeks) without any noticeable deterioration in flavor, texture, or appearance. Storage life appears to be geared with the harvest season. Early varieties, particularly the freestone peaches now being grown in Florida and the early clingstones grown in the Southeast, have an extremely short storage life and should be used as soon as possible after harvest. However, some late season varieties can be safely stored for up to 6 weeks. In the West, the Rio Oso Gem is consistently stored for 4 to 6 weeks before being marketed.

Harvest Techniques

Peaches for fresh consumption must be in a condition to survive a postharvest holding period of several days to several weeks. The fruit must be sound and bruise-free and must be handled delicately during the harvesting and packing operations. Today, with the widespread use of bulk bins or pallet boxes, hand-picked fruit requires extra careful handling. Hydrodumpers are generally employed for dumping the pallet bins. With proper care, pallet boxes cause less bruising than small field boxes.

Cooling

Cooling peaches to 40°F soon after harvesting is essential to postharvest retention of quality and control of decay. Peaches begin to soften and decay in a few hours without proper temperature management. All peaches shipped out of the Southeast are hydrocooled. Originally, the fruit was cooled in flood-type hydrocoolers as a final operation after it was packed in containers. In the West, most fresh peaches are air cooled in pressure coolers to remove the field heat rapidly for the postharvest holding period. By using forced air or pressure cooling, peaches or nectarines in two-layer plastic tray packs with 6% side vented corrugated con-

Deciduous Tree and Vine Fruits

tainers will cool 80% in about 6 h with an airflow of 0.2 cfm/lb of fruit.

Storage

Peaches are normally stored in corrugated or tray pack shipping containers.

An environment of 31°F and 90 to 95% rh with very low air movement is best for peaches. Under these conditions, peaches can be held from 2 to 6 weeks, depending on variety.

The same storage conditions may be used for nectarines; however, they are somewhat more susceptible to shrivel than are peaches. Air velocity in the storage room should be as low as possible but still maintain proper storage temperatures. Frequent checks should be made of the fruit at the edge of alleyways, for example, to detect the first signs of shrivel.

Good experimental CA results have been obtained with peaches and nectarines held in 1% O_2 with 5% CO_2 at 32°F. Extended storage of 6 to 9 weeks is possible. The fruit ripens or softens with good flavor and is juicy on removal. Low temperature breakdown, which is usually encountered with lengthy storage, is controlled by CA. While CA reduces decay, it does not completely control it; thus, a fungicide is needed for extended storage.

Diseases

Brown Rot (*Monolinia*). Extensive firm, brown, unsunken areas turning dark brown to black in the center and generally covered with yellowishgray spore masses. Skin clings tightly to center of old lesions. *Control:* Follow recommended field and postharvest control measures involving use of heat treatments and fungicides. Refrigerate promptly to as near 32°F as feasible.

Cold Storage Injury. Fruit loses flavor, becomes dry and mealy. Breakdown starting around pit is grayish brown, watersoaked, or mealy. *Control*: Refrigerate promptly to 32°F. Breakdown appears earlier at 38°F. Store for only 2 to 4 weeks, depending on variety.

Pustular Spot (*Coryneum*). Common on peaches from the West, occasionally on Eastern fruit. At first small purplish-red spots, later up to 0.5 in. in diameter, brown, sunken with white center. *Control:* Treat with orchard sprays. Cool harvested fruit to below 45°F.

Rhizopus Rot. Extensive, fairly firm, watery decay with uniformly brown surface color. Skin slips readily from center of lesions. Coarse mycelium; black spherical sporangia develop. *Control:* Store cannery peaches at 32°F before ripening. Prevent skin breaks. Follow recommended field and postharvest control measures. Refrigerate promptly to as near 32°F as feasible.

Sulfur Dioxide Injury. Bleached and pitted areas on fruit surface. After removal from refrigeration, injured areas of peaches are brown, dry, and collapsed. Skin may slough off. *Control:* Avoid sulfur dioxide contact of peaches (and other stone fruits) in storage or in transit with grapes.

Sour Rot. An unfamiliar postharvest disease in peaches was noticed in some packing sheds in the Southeast. First signs of the infection may be peaches that are easily skinned by the brushes and belts on the packing line. Affected peaches then develop softened and sunken brown lesions that eventually become covered with a white or creamy exudation. The infected areas generally emit a vinegar-like, sour odor. *Control:* Chlorination of dump tank water, chlorination of hydrocooling water, and careful culling of all overripe, bruised, and damaged fruit. In short, good shed sanitation and quality control are the keys to eliminating sour rot.

APRICOTS

Apricots are not stored for a prolonged time but may be held for 2 or 3 weeks if they are picked at a maturity firm enough that they will not bruise. Unfortunately, this maturity does not yield good dessert quality fruit. Care must be used in sizing and packing the fruit going into storage, as small surface bruises can become infected with disease-producing organisms. Chapter 26 has further details.

Apricots for short-term storage are harvested in much the same way as freestone peaches, precooled, and placed in storage promptly. Storage temperature should be 32°F with a 90 to 95% rh.

Diseases and Deterioration

See Peaches and Nectarines.

BERRIES

Blackberries, raspberries, and related berries cannot be stored for more than 2 or 3 days even at 31°F with a relative humidity of 90 to 95%. An atmosphere with 20 to 40% CO_2 will increase storage life by 3 or 4 days by inhibiting fungal rots.

As they come from the field, cranberries are stored in field boxes at 36 to 40°F and 90 to 95% rh. They usually are not stored longer than 2 months. Storage at 30 to 32°F causes chilling and physiological breakdown. Modified atmospheres have not extended the storage life of fresh cranberries beyond that of conventional storage.

Diseases

Cladosporium Rot. Surfaces of berries covered with olive to olive-green mold. In raspberries, the mold is most abundant on inside or cup of berry. *Control:* Avoid bruising; pack and ship promptly. Refrigerate to 32°F.

Gray Mold Rot. Causes soft, watery rot. Fruit may be covered with dense, dusty gray growth of fungus which spreads rapidly in package, forming nests. *Control:* Avoid bruising. Refrigerate to 32°F in transit.

Anthracnose (*Gloeosporium sp.*). Berries may be completely rotted and show masses of spores glistening in salmon-colored droplets on fruit. *Control:* Refrigerate to 32°F.

Alternaria Rot. Affected berries remain firm and show gray-white woolly fungal growth from injured cap stem areas. Nesting occurs in tight clusters scattered throughout containers. *Control:* Refrigerate to 32°F.

Chilling Injury. Berries held for 4 or more weeks become tough and rubbery, surfaces are dull in appearance, red in color throughout. *Control*: Hold fruit at 37°F.

Fungus Rots (*Several Fungi*). Limited portions or entire berries are brown, soft, or collapse. Some berries turn into water bags. *Control:* Spray in field. Handle carefully. Reduce temperatures to 38°F after harvest.

STRAWBERRIES

Diseases

Gray Mold Rot (*Botrytis*). Brown, fairly firm, fairly dry decay. Dirty-gray mold and grayish-brown velvety spore masses present. Nesting common. *Control:* Apply recommended fungicides in field. Handle carefully to prevent skin breaks. Cull out all diseased berries. Cool promptly to 40°F or below.

Leather Rot (*Phytophthora*). Large, slightly discolored tough areas with indefinite purplish margins. Vascular system browned, flavor bitter. *Control:* Mulch plants to keep berries from contact with infested soil. Cool promptly to 40°F or below.

Rhizoctonia Rot. Hard dark brown decay on one side of berry, usually small quantities of soil adhering. Develops only a little after harvest. *Control:* Mulch plants to keep berries from contact with infested soil. Cull thoroughly.

Rhizopus Rot. Mushy, leaky collapse of berries associated with coarse black mycelium and sporangia. Extensive red staining of containers from leaking juice. *Control:* Reduce temperature promptly to 40°F or below. Handle carefully to prevent skin breaks.

FIGS

Diseases

Alternaria Spot. White fungal growth on surfaces which soon darkens. As fungus spots enlarge, tissue beneath becomes slightly sunken. *Control:* Cool promptly after harvest, hold at 45°F in transit.

Black Mold Rot (*Aspergillus*). Disease first appears as a dirty white to pink color of the skin and pulp. White mold growth develops within fig. Cavities formed in fruit become lined with black spore masses. *Control:* Store fresh figs at 32°F at relative humidity of 85 to 90%.

SUPPLEMENTS TO REFRIGERATION

Antiseptic Washes. Many fruits are washed before packing to remove dirt and improve appearance. In some cases, hydrocoolers are used to remove field heat. If the water is recirculated, it may become heavily contaminated with decay-producing bacteria and fungi. Chlorine can be added to the water at a level of 50 to 100 ppm to control the buildup of these organisms. Other fungicides may also be used, but they must be legally registered for the specific application.

Protective Packaging. Proper packaging protects against bruising, moisture loss, and spread of disease. Packaging materials may also contain chemicals to control spoilage. Packages must have good stacking strength for palletizing and must also perform under high humidity conditions.

Selective Marketing. The potential storage life of grapes and apples can be predicted within a few weeks after they are stored. Thus, those with a short storage life may be marketed while still in good condition and longer lived products can be stored for late season marketing. Samples taken from each lot placed in storage are kept for a few weeks at temperatures and relative humidities that favor rapid development of decay. Grapes that will not keep long can be detected in about 2 weeks and apples in about 60 days. Since both of these fruits may be stored for several months, knowing their potential storage life can significantly reduce spoilage losses.

Heat Treatment. Heat treatments to reduce decay also kill insects on the surface of the fruits and microorganisms near the surface without leaving a residue. For example, brown rot and rhizopus rot of peaches are reduced by exposing the fruit for 1.5 min. in 130°F water or for 3 min. in 120°F water.

Fungicides. Fungicides may be applied during cleaning, brushing, or waxing of some fruits. Only fungicides registered for the particular fruit and use may be used.

Irradiation. Gamma radiation has effectively controlled decay in some products. High dosages can cause discoloration, softening, or flavor loss. Commercial application of gamma radiation is limited due to the cost and size of equipment needed for the treatment and to uncertainty about the acceptability of irradiated foods to the consumer (Hardenburg *et al.* 1986).

Ultraviolet lamps are sometimes uses to control bacteria and mold in refrigerated storages. While ultraviolet light kills bacteria and fungi that are sufficiently exposed to the direct rays, it does not reduce decay of packaged fruits in storage. Even ultraviolet light directed on fruits as they passed over a grader did not control decay.

REFERENCES

Environmental Protection Agency. 1989. Interim policy for sulfiting agents on grapes; Pesticide tolerance for sulfur dioxide. *Federal Register* 54 (3): 382-85.

Hardenburg, R.E., A.E. Watada, and C. Y. Wang. 1986. *The Commercial Storage of Fruits, Vegetables, and Florist and Nursery Stocks.* USDA *Agriculture Handbook,* No. 66.

Harvey, J.M. 1955. A method of forecasting decay in California storage grapes. *Phytopathology* 45, 229-32.

Harvey, J.M. 1984. *Instructions for forecasting decay in table grapes for storage.* U.S. Department of Agriculture, ARS-7.

Harvey, J.M., C.M. Harris, T.A. Hanke, and P.L. Hartsell. 1988. Sulfur dioxide fumigation of table grapes: Relative sorption of SO_2 by fruit and packages, SO_2 residues, decay, and bleaching. *American Journal of Enology and Viticulture,* Vol. 39, 132-36.

BIBLIOGRAPHY

Ryall, A.L. and W.T. Pentzer. 1982. *Handling, transportation, and storage of fruits and vegetables*, 2nd ed. Vol. 2, Fruits and tree nuts. AVI Publishing Co., Westport, CT, 610 pp. ISBN 0-87055-410-7

Nelson, K.E. 1979. Harvesting and handling California table grapes for market. University of California *Publication* 4095, 67 pp. ISBN 0-931876-33-8

CHAPTER 17

CITRUS FRUITS, BANANAS, AND SUBTROPICAL FRUITS

CITRUS FRUITS..................................... 17.1	Harvesting and Transportation....................... 17.5
Maturity and Quality............................... 17.1	Diseases and Deterioration......................... 17.5
Harvesting and Packaging........................... 17.1	Exposure to Excessive Temperatures................. 17.5
Transportation..................................... 17.2	Wholesale Processing Facilities.................... 17.5
Storage.. 17.3	SUBTROPICAL FRUITS................................. 17.8
Controlled Atmosphere Storage...................... 17.4	Avocados... 17.8
Storage Disorders and Control...................... 17.4	Mangoes.. 17.8
BANANAS.. 17.5	Pineapples... 17.8

CITRUS FRUITS

THIS chapter discusses the harvesting, handling, storage requirements, and possible disorders of fresh market citrus fruits produced in Florida, California, Texas, and Arizona.

MATURITY AND QUALITY

The degree of citrus fruit ripeness at the time of harvest is the most important factor determining eating quality. Oranges and grapefruit do not improve in palatability after harvest. They contain practically no starch, do not undergo marked composition changes after they are picked from the tree (as do apples, pears, and bananas), and their sweetness comes from natural sugars contained when they are picked.

The ripening of citrus fruits is a slow, gradual process closely related to increases in diameter and weight. Citrus fruits must be of high quality when harvested to assure quality during storage and shelf life.

Quality is often associated with the fruit rind's appearance, firmness, thickness, texture, freedom from blemishes, and color. Actually, quality determination should be based on the texture of the flesh, juiciness, content of total solids (principally sugars), total acid, aromatic constituents, and vitamin and mineral content. Age is also important, because immature fruit is usually coarse, very acid or tart, and has an internal texture that is ricey or coarse. Overripe fruit held on the tree too long may become insipid, develop off-flavors, and possess short transit, storage, and shelf life. The importance of having good quality fruit at harvest cannot be overemphasized. The main objective thereafter is to maintain quality and freshness.

HARVESTING AND PACKING

Picking. Citrus fruit is harvested in the United States throughout the year, depending on the growing area and kind of fruit. The approximate commercial shipping seasons for Florida, California-Arizona, and Texas citrus are shown in Figure 1. The picking operations are conducted by trained crews from independent packinghouses or large associations. These organizations schedule picking to meet market demands. The fruit that is not handled through cooperatives is normally sold on the tree to the shippers or processors and is picked at the latter's discretion.

The fruit is carefully removed from the trees either by hand or with special clippers and is then placed in picking bags which are emptied into field boxes. An increasing amount of fruit is handled in bulk, and the pickers put the fruit into pallet boxes or wheeled carts. In some cases, especially when fruit is picked for processing, it is loaded loose into open truck trailers. In Florida, over 90% of the oranges and slightly more than 50% of the grapefruit specialty fruit are processed, while in California and Arizona less than 35% of all citrus fruit is processed.

At the beginning of the season, the fruit is often spot-picked; only the riper, larger, or outside fruit is harvested. Later, the trees are picked clean. In California, lemons usually are picked for size with the aid of sizing rings.

Various labor-saving devices have been tested, including mechanical platforms and positioners, tree shakers with catch frames, and air blasts for fruit removal. Mechanical harvesting, however, is limited to a very small percentage of the total crop. Because of damage incurred, fruit intended for processing only is mechanically harvested. Preharvest sprays have also been developed to improve the color and loosen the fruit to facilitate harvest.

Handling. After the fruit is received at the packinghouse, it is removed from the boxes or bulk containers by emptying it carefully to prevent damage to the fruit. It is then presized to remove fruit that is over- or undersized. Before washing, the fruit may be floated through a soak tank, which usually contains a detergent for cleaning and an antiseptic for decay control.

The washer is generally equipped with transverse brushes that revolve up to 120 rpm. If not applied at the soak tank, soap or antiseptic may be dribbled or foamed on the first series of brushes. The fruit is then rinsed by a fresh water spray.

The fruit then passes under fans which circulate warm air through the moving pieces. When dried, the fruit is polished and waxed, and then passed over roller conveyor grading tables. After grading, it is conveyed to sizing equipment to separate the pieces into the standard sizes being packed; the pieces are then dropped at stations for hand packing or conveyed to automatic or semi-automatic box-filling or bagging machines.

The packinghouse handling of California lemons for fresh market is interrupted by an extended storage period. After washing, the fruit is conveyed to a sorting table for color separation by electronic means or by human eye. Usually, four colors are recognized and are designated as dark green, light green, silver, and yellow. The dark green is a full green; the light green, a partially colored green (a green with color well broken); the silver, fully colored with a green tip (stylar end); the yellow, fully colored and mature with no green showing. Dark green fruit has a normal storage life of from 4 to 6 months; and yellow, 3 to 4 weeks. These periods are approximate, as the storage (or keeping) quality of fruit varies considerably with season and grove. A light concentration of water-wax emulsion is usually applied to lemons before they are put into storage. Lemon storage is more fully discussed in a later section.

After storage, lemons are waxed, then sized and packed. Post-storage washing to remove mold soilage is desirable but requires a washer incorporating very soft roller brushes.

Shipping and storage containers vary considerably for the various types of citrus fruit. The 0.8 bushel fiberboard cartons

---

The preparation of this chapter is assigned to TC 11.5, Fruits, Vegetables, and Other Products.

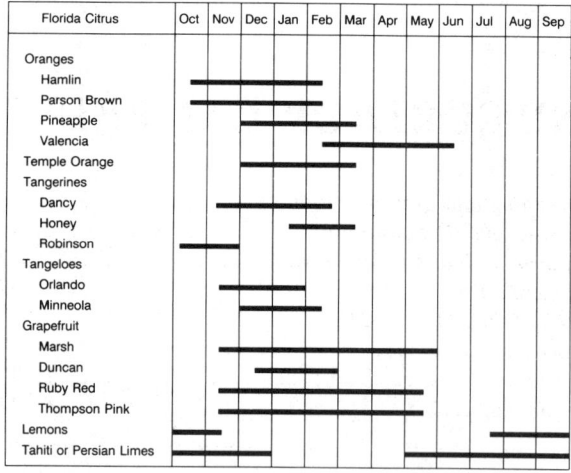
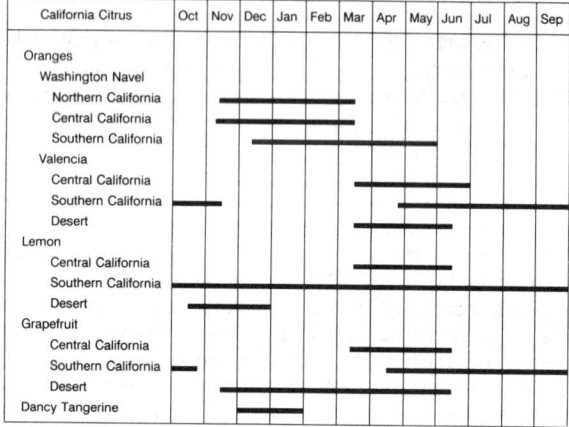
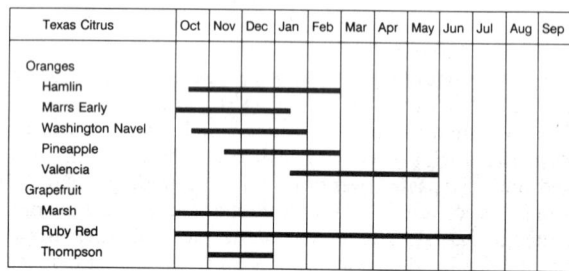

Fig. 1 Approximate Commercial Shipping Season for U.S. Citrus

have become the standard in California and Florida. In addition, over 15% of Florida fresh fruit is consumer-packed in mesh and polyethylene bags which are normally shipped in 40-lb master cartons. After the packages are filled and closed, they are conveyed either to precooling rooms to await shipment or directly to standard refrigerator cars or trucks. The containers are stacked so that air distribution is uniform through the load.

Accelerated coloring or sweating. All varieties of citrus fruit must be mature before they are picked. Color is not always a criterion of maturity. The natural change of color in oranges from dark green to deep orange is a gradual process while the fruit remains on the tree. The fruit remains dark green from its formation until it is nearly full size and approaching maturity; then the color changes may occur very rapidly. The color change is influenced greatly by temperature variations. A few cold nights followed by warm days may completely color oranges that were previously very green. The color changes in lemons and grapefruit are similar, except that the final color is yellow. Unfavorable weather conditions may delay coloring even after maturity.

Up to a certain point, the natural color changes in Valencia oranges follow the trend described, but complete or nearly complete orange color generally develops some time before the fruit is mature. Some regreening of Valencias may occur after the fruit has reached its prime. Navel oranges in California, as well as the Florida varieties of Hamlin, Parson Brown, and Pineapple harvested in late fall and early winter, may be mature and of good eating quality, although the rind is green in color. Grapefruit, lemons, tangerines, tangelos, and other specialty fruit may also be sufficiently mature for eating before they are fully colored. Because the consumer is accustomed to fruit of characteristic color, poorly colored fruit is put through a coloring or degreening process in special rooms, bulk bins, or trailer degreening equipment.

These units are equipped to maintain temperatures and humidities at desired levels. Approximately 5 ppm ethylene in the air is maintained. The concentration of ethylene and the duration of the degreening periods depends on the variety of fruit and the amount of chlorophyl to be removed. During the operation, fresh air is introduced into the room, and a relative humidity of 88 to 92% is maintained. In Florida, temperatures of 82 to 85°F generally are used, while in California temperatures of 65 to 70°F are used. Lemons are usually degreened at 60°F without added ethylene. In California the process is called sweating instead of coloring or degreening.

Oranges, grapefruit, and specialty citrus fruits requiring ethylene treatment are frequently degreened as soon as they are delivered to the packinghouse; but they may receive a fungicide drench prior to degreening. Lemons are washed and graded or color-separated before being degreened.

A high percentage of Florida's early and midseason varieties of oranges and tangelos receive color-added treatment with a certified food dye that causes the rind of pale fruit to take on a brighter and more uniform orange color. This is usually in addition to degreening with ethylene gas. In color-added treatment, the fruit is subjected for 2 to 3 min. to the dye solution, which is maintained at about 120°F. The color-added treatment can be given in an immersion tank filled with vegetable dye solution, or the dye can be flooded on the fruit as it passes on a roller conveyor. The color-added tank is located after the washer and before the wax applicator. Oranges with the desired color at harvest time, as well as tangerines and grapefruit, are bypassed around the dye tank, or the flow of dye may be cut off as the fruit passes over the equipment. Standards for maturity are slightly higher in Florida for oranges given color-added treatment. California oranges are not artificially colored.

Cooling. After the fruit is packed, it is cooled. The efficiency of the cooling rooms depends on the following conditions:

1. Cooling air volume per railcar load: not less than 3000 cfm
2. Relative humidity of supply air: 95% or above
3. Temperature of supply air entering room: not more than 2°F below the selected cooling temperature

The fruit may also be cooled in a refrigerated truck trailer or container after it has been loaded.

In California, air is used to cool oranges but not lemons or grapefruit. In Florida, specialty fruits such as Temple oranges, tangerines, and tangelos may be cooled. Chapter 11 discusses cooling practices and equipment used for various commodities in more detail.

TRANSPORTATION

Fruit packed in piggybacks, trucks, ship vans, or rail cars should be stowed in appropriate modifications of the spaced bonded block to assure good air circulation, a uniform temperature, and a stable load. No dunnage is required. Such stowing provides

Citrus Fruits, Bananas, and Subtropical Fruits

continuous air channels through the interior of the load and improves the likelihood of sound arrival. Trailers and containers that circulate air from the bottom provide uniform temperatures throughout the load with a regular bonded-block stow.

In Florida, the present quarantine treatment for the Caribbean fruit fly, *A. suspensa,* is to subject an export load of citrus to specified temperatures for up to a 17- or 24-day period (Ismail *et al.* 1986). This treatment may be implemented in containers or in a ship's hold.

A uniform sample of 1500 fruit is withdrawn from a shipment before ship or container loading and is held at 80°F or higher. These fruit are then examined after a 10-day incubation period. When an infestation of *A. suspensa* is found, the entire load must undergo the long treatment process. Temperature and time schedules are detailed below:

Temperature, °F	Days
Short Treatment	
33	10
34	12
35	14
36	17
Long Treatment	
33	14
33.5	16
34	17
34.5	19
35	20
35.5	22
36	24

These temperatures are required center pulp temperatures. To avoid chilling injury, a conditioning period of 7 days at 59°F is recommended before initiation of the cold treatment process.

STORAGE

Oranges

Florida- and Texas-grown Valencia oranges can be stored successfully for 8 to 12 weeks at 32 to 34°F with a relative humidity of 85 to 90%. The same requirements apply to Pope's Summer orange, a late-maturing Valencia-type orange. A temperature range of 40 to 44°F for 4 to 6 weeks is suggested for California oranges. March-harvested, Arizona Valencias store best at 48°F, but June-harvested fruits store best at 38°F.

Oranges lose moisture rapidly, so high humidity should be maintained in the storage rooms. For storage longer than the usual transit and distribution periods, 85 to 90% relative humidity is recommended.

Florida and Texas oranges are particularly susceptible to stem end rots. Citrus fruits from all producing areas are subject to blue and green mold rot. These decays develop in the packinghouse, in transit, in storage, and in the market, but can be greatly reduced if fruit is properly treated. Proper temperature is effective in reducing decay. However, once storage fruit is removed to room temperature, decay will develop rapidly.

Storage of oranges is often complicated by the fact that prolonged holding at relatively low temperatures may induce the development of physiological rind disorders not ordinarily encountered at room temperature. Aging, pitting, and watery breakdown are the most prevalent rind disorders induced by low storage temperatures. Generally, California and Arizona oranges are more susceptible to low-temperature rind disorders than Florida oranges.

Successful long storage of oranges requires: harvest at the proper maturity, careful handling of fruit, good packinghouse methods, fungicidal treatments, and prompt storage after harvest.

The rate of respiration of citrus fruits is usually much lower than that of most stone fruits, green vegetables, and somewhat lower than that of apples. Navel oranges have the highest respiration rate, followed by Valencia oranges, grapefruit, and lemons. The heat from respiration is a relatively small part of the heat load. Table 1 shows heat generated through respiration.

Table 1 Heat of Respiration of Citrus Fruits
(Haller *et al.* 1945)

	Btu per ton of fruit per day					
	Oranges			Grapefruit		Lemons
Temp., °F	Florida	California Navels	Valencias	Florida	Calif. Marsh	Calif. Eureka
32	700	900	400	500	500	700
40	1400	1400	1000	1100	800	1100
50	2700	3000	2600	1500	2000	2500
60	4600	5000	2800	2800	2600	3500
70	6600	6000	3900	3500	3900	5000
80	7800	8000	4600	4200	4800	5700

Grapefruit

Florida and Texas grapefruit is frequently placed in storage for 4 to 6 weeks without serious loss from decay and rind breakdown. The recommended temperature is 50°F. A temperature range of 58 to 60°F is recommended for the storage of California and Arizona grapefruit.

A relative humidity of 85 to 90% is usually recommended for the storage rooms in which grapefruit is held. Loss of weight and water occurs rapidly and can be avoided by maintaining the correct humidity and taking the additional precaution of a wax coating.

Decay and rind breakdown are deterrents to long storage of grapefruit and may develop in fruit during storage or following removal from storage. Proper prestorage treatments with fungicides, as discussed in the previous section on disorders and storage temperatures, will greatly reduce these problems. Also, periodic inspections of stored fruits should be made to terminate storage at the least symptom of rind pitting or excessive decay.

Export may require 10 days to 4 weeks of storage in a refrigerated hold and present problems similar to those encountered in refrigerated storage. Marsh Seedless and Ruby Red grapefruit picked before January retain appearance best when stored at 60°F. With riper fruit, 50 to 55°F is better for export shipments. Very ripe fruit harvested in April and May, however, develops excessive decay following storage at 50 to 60°F.

Lemons

Most of the lemon crop is picked during the period of least consumption and stored until consumer demand justifies shipment. Lemons are generally stored near the producing areas rather than the consuming areas.

All lemons, except the relatively small percentage that are ripe when harvested, must be conditioned or cured, and degreened, before shipping. When lemons are stored prior to shipment, the curing and degreening processes proceed during storage. These lemons are usually stored at 58 to 60°F and 86 to 88% rh. Local conditions may suggest slight modifications of these values.

Lemons picked green but intended for immediate marketing, such as most lemons grown in the desert portions of Arizona and California, are degreened and cured from 6 to 10 days at 72 to 78°F and 88 to 90% rh. The thin-skinned Pryor strain of Lisbon lemons degreens in about 6 days, whereas the thick-skinned old-line Lisbon requires as long as 10 days.

Lemon storage rooms must have accurately controlled temperature and relative humidity; the air should be clean and uniformly circulated to all parts of the room. Ventilation should be sufficient to remove harmful metabolic products. Air-conditioning equipment is necessary to provide satisfactory storage conditions, as natural atmospheric conditions are not suitable for the necessary length of time.

A uniform storage temperature of 50 to 60°F is important. Fluctuating or low temperatures cause lemons to develop an undesirable high color or bronzing of the rind. Temperatures 52°F and lower cause a staining or darkening of the membranes dividing the pulp segments and may affect the flavor. Temperatures above 60°F shorten the storage life and are favorable to the growth of decay-producing organisms.

A relative humidity of 86 to 88% is generally considered satisfactory for lemon storage, although a slightly lower humidity may be desirable in some locations. Higher humidities prevent proper curing of the lemons, encourage mold growth on walls and containers, and hasten decay of the fruit; much lower humidities cause excessive shrinkage.

Proper stacking of the fruit containers in storage rooms is important to secure uniform air circulation and temperature control. The stacks should be at least 2 in. apart and the rows, 4 in.; trucking aisles at least 6 ft wide should be provided at intervals.

Specialty Citrus Fruits

In Florida, small amounts of various specialty citrus fruits are grown commercially. These fruits, which are usually eaten out of the hand, include tangerines, tangerine hybrids (Murcott Honey oranges, Temple oranges, tangelos), the King orange, and other mandarin-type fruits.

Careful handling during picking and packing is especially necessary for these fruits. Because of their perishable nature and limited shelf life, these fruits should not be stored longer than required for orderly marketing (2 to 4 weeks). A temperature of 38 to 40°F at 90 to 95% rh is recommended. Adequate precooling and continuous refrigeration during transit are required.

Tahiti or Persian limes, in addition to the fruits just mentioned, are grown in southern Florida. This is the only citrus fruit marketed while it is green in color. The fully ripe (yellow) fruit lacks consumer appeal and is undesirable for fresh market. Limes should be picked while still green, but after the fruit has lost the dimpled appearance around the blossom end. Good quality fruit may be stored satisfactorily for 6 to 8 weeks at temperatures of 48 to 50°F. Mature fruit will gradually turn yellow at this temperature, however. Prevention of desiccation is very important, as is a relative humidity above 85%. Pitting occurs at temperatures below 45°F, while temperatures above those recommended permit the development of stem end rot.

CONTROLLED ATMOSPHERE STORAGE

While some minor benefits have been obtained, tests with oranges, grapefruit, lemons, and limes using modified or controlled atmospheres have given no assurance that storage and market life can be extended. For this reason, controlled atmosphere (CA) storage is not generally recommended for citrus fruits. Atmospheres used for storage of apples and other deciduous fruits are unsatisfactory for citrus fruits and lead to rind injuries, off-flavors, and decay.

The performance of any citrus storage facility depends on: (1) the provision of sufficient capacity for peak loads; (2) evaporator and secondary refrigerating surface sufficient to permit operating at high back pressures, thus preventing low humidities and permitting economical operation; and (3) efficient air distribution, assuring velocities high enough to effect rapid initial cooling and volumes great enough to permit operation during storage with only a small temperature rise between delivery and return air. Chapters 25, 26, and 27 have further information on storage design.

STORAGE DISORDERS AND CONTROL

Postharvest Diseases

Citrus fruits often carry incipient fungus infections when they are harvested. Decay organisms may also enter minor injuries caused during harvesting and handling. The major postharvest diseases (with symptoms) encountered in storage include:

Alternaria Rot. Usually a stylar end in navel oranges as a black, dry, deeply penetrating decay. In other citrus fruits, as a slimy, leaden-brown storage decay of core starting at stem end. *Control:* Provide optimum growing conditions. Harvest oranges before overripe. Do not store tree-ripe lemons. Restrict storage period for other lots known to be weak. Green buttons are an indication of strong fruit. Treatment of lemons at the packinghouse with 2-4-D in wax emulsion or water improves resistance of lemons to disease.

Anthracnose (Colletotrichum). Leathery, dark brown, sunken spots or irregular areas. Internal affected tissues dark gray, fading through pink to normal color. Most serious with degreened early season tangerines, tangerine hybrids, and long-stored oranges and grapefruit. *Control:* Use recommended postharvest fungicide. Avoid long storage; move promptly.

Blue (and Green) Mold Rot (Penicillium). Soft, watery, decolorized lesions which under moist conditions become quickly covered with blue or olive-green powdery spores. *Control:* Prevent skin breaks. Use recommended fungicides in washes. Cool fruit to as near to 32°F as practicable. Use biphenyl in box liners or in fruit wraps.

Brown Rot (Phytophthora). Extensive firm, brown decay having a penetrating rancid odor. Chiefly on fruit from California and Arizona. *Control:* Orchard spraying and good sanitation. Submerge fruit at packing for 2 min. in water at 114°F.

Sour Rot (Geotrichum). Soft, watery rot with sour smell following peel injuries. Similar to early stages of mold rot, except that no powdery spores are formed. Most serious on lemons and mandarin-type fruit. *Control:* Avoid peel injuries at harvest; refrigerate at lowest practical temperature. Approved fungicides are of little or no value.

Stem End Rot (Diplodia; Phomopsis). Pliable, fairly firm, extensive, brown decay starting at stem. Sour, pungent odor. Prevalent in Florida and found occasionally in Arizona and California fruit. *Control:* Treat harvested fruit promptly in recommended fungicides and cool promptly below 50°F.

Several chemical fungicides are approved by EPA for postharvest use on citrus fruits. These include Thiabendazole (TBZ), Orthophenylphenol (OPP or SOPP), and imazalil. These materials are applied after washing and before waxing or are incorporated in the wax coating. Biphenyl, a volatile fungistat, is impregnated in paper wrappers or box liners. It evaporates slowly and inhibits growth of organisms in transit and storage. Under certain contions, it is beneficial to use a combination of these materials since all are not equally effective against the same organism. Strains of the blue and green molds (*Penicillium*), which are resistant to certain fungicides, have developed in citrus storage houses, so care must be taken in selecting a fungicide and the time of application.

Physiological Disturbances

In addition to diseases caused by fungi, some defects are caused by various physiological conditions. These defects can best be avoided by using fruit of prime maturity and by proper handling after harvest. Proper temperature and humidity levels are required during handling, storage, and transit. The physiological disorders (and symptoms) are:

Stem End Rind Breakdown. Small to large sunken, drying, discolored, firm areas in skin around stem button or on the upper

Citrus Fruits, Bananas, and Subtropical Fruits

part of fruit. *Control:* Pick before overmature. Avoid overheating in packinghouse treatments. Wax fruit. Store for limited period only in fairly high relative humidity, 85 to 90%. Follow storage temperatures recommended for variety and growing area.

Freezing Injury. Field freezing is found scattered through boxes. Transit and storage freezing is worse in exposed fruits in bottom-layer boxes or those nearest cooling coils. Affected fruits may show watersoaked areas in rind. Internal tissue disorganized, water-soaked, milky, and with rind flavor. Frozen fruit loses moisture, causing drying, separation of juice vesicles, and buckling of segment walls. The freezing point of citrus fruit is about 28.5 °F.

Internal Decline. In lemons, core tissues near stylar end break down and dry, becoming pink. *Control:* Maintain optimum moisture conditions in grove.

Pitting. This physiological disease is manifested by depressed areas of 0.1 to 0.8 in. diameter in the peel of citrus fruit. Affected tissues collapse and may appear bleached or brown. Pits occur anywhere on the fruit and may coalesce to form large irregular areas. The cause is not fully understood. In general, it is a low-temperature disorder. *Control:* Follow storage temperatures recommended for cultivar and growing area.

BANANAS

This section discusses the harvesting, transportation, processing, and storage of bananas.

HARVESTING AND TRANSPORTATION

Bananas do not ripen satisfactorily on the plant; even if they did, deterioration of ripened fruit is too rapid to allow shipping from tropical growing areas to distant markets. Bananas are harvested when the fruit is mature but unripe, with dark green peels and hard, starchy, inedible pulps. Each banana plant produces a single stem of bananas which contains from 50 to 150 individual fruits (or fingers). The stem is cut from the plant as a unit with fingers attached and transported to nearby boxing stations.

Bananas are removed from the stem, washed, and cut into consumer-sized cluster units of four or more fingers. The clusters are packed in protective fiberboard cartons that contain 40 lb of fruit. The cartons move by rail from the tropical boxing stations to port and then are loaded into the holds of refrigerated ships. On the ship, the fruit is cooled to the optimum carrying temperature, generally 56 to 58 °F, depending on variety.

Bananas are unloaded still green and unripe at seaboard and transported under refrigeration at a holding temperature of 58 °F to interior wholesale distribution centers by both truck and rail car. The objective is to maintain the product in an optimal environment and move it to its destination as quickly as possible to minimize postharvest deterioration.

DISEASES AND DETERIORATION

Bananas are subject to the following diseases and physiological disorders. Proper temperature and moisture during storage and careful handling will slow the aging and development of decay.

Anthracnose (Ripe Rot) (Gloeosporium). Shallow black spots on stems of ripening fruit. Under moist conditions, pink spore masses cover center of spots. Dark discoloration of skin may extend from stem ends over entire fruit. *Control:* Protect fruit from mechanical injury; damage is reduced if fruit is put in corrugated boxes. Schedule ripening so that fruit can be marketed and consumed before appearance of defect.

Black Rot (Ceratocystis). Transmitted from wounds via fibrovascular system of plant. Progresses into crowns and stem ends of fingers. Produces brownish-black areas in peel at fruit ends. As fruit ripens, skin becomes grayish-black in color and water soaked. Pulp rarely affected. *Control:* In the tropics, dip or spray freshly cut tips and bunches with fungicides before boxing. Avoid mechanical injury and maintain sanitation program from tropics to ripening room.

Chilling Injury. Dull gray skin color with increased tendency to darken on slight bruising. Latex in green fruit does not bleed freely and will be clear rather than cloudy. Subsurface peel tissue streaked with brown. Turning or ripe bananas are more susceptible to injury than green fruit. *Control:* Avoid temperatures below 55 °F. Moving air makes chilling more rapid.

Fungus Rots (Several Fungi). Extensive soft rot of scarred, split, or broken fruit. Affected skin and flesh moist and brown to black. Under high humidity, the surface is often covered with mold. *Control:* Handle fruit carefully to avoid bruising and mechanical injury. Cool stored fruit rapidly to 56 °F.

EXPOSURE TO EXCESSIVE TEMPERATURES

Fruit pulp temperatures only a few degrees below optimum holding temperatures, although considerably above the actual freezing point of bananas, can cause chilling injury, as described previously. The severity of chilling injury varies directly with the duration of exposure and indirectly with temperature. It is primarily a peel injury in which certain surface cells of the banana peel are killed. The contents of the dead cells eventually darken (because of oxidation) and give the fruit a dull appearance. Both green and ripe bananas are susceptible to chilling injury; severely chilled green bananas never ripen properly. Fruit pulp temperatures only a few degrees above the optimum holding temperatures can cause the fruit to ripen prematurely in transit.

Once the bananas arrive at wholesale distribution centers, they are unloaded and placed in specially equipped processing rooms for controlled ripening. As soon as the bananas have ripened to an edible state, they are rushed to retail, since ripening cannot be stopped. Even under ideal refrigeration, ripe bananas will have progressed to the point where they are too ripe to be marketed.

WHOLESALE PROCESSING FACILITIES

Wholesale banana facilities are distinguished from general wholesale produce storage facilities by special banana ripening rooms. The ripening room controls initiation and completion of fruit ripening, a natural physiological phenomenon. A typical banana room is shown in Figure 2. The ability to properly ripen bananas is so critically linked to the design of the ripening rooms that major banana importers maintain technical staffs that specialize in banana-room design. Through these technical staffs, free, nonobligatory consultation is provided to wholesalers, architects, engineers, contractors, and anyone else involved in banana ripening facility design, construction, and operation.

A typical banana processing facility consists of a bank of five or more individual ripening rooms. For design purposes, one complete turnover per week is assumed, and therefore the combined capacity of all rooms should approximately equal total weekly volume, allowing for seasonal variations.

Each load is scheduled for optimum ripeness on a particular day. Fruit shipped from this load a day ahead of schedule will be underripe; fruit shipped a day late will be overripe. Therefore, shipping for several days out of one room is not practical. There should be at least as many rooms as there are retail shipout days per week.

Because bananas cannot be processed on a continuous flow basis, individual room capacities are multiples of carlots, usually one half or one carlot. As the capacity of transportation equipment has increased in recent years, so has the design capacity of processing rooms. Generally, one or two large rooms would be cheaper to build than several small rooms with equivalent total capacity. However, minimizing construction cost is not the pertinent consideration; having all bananas in each particular

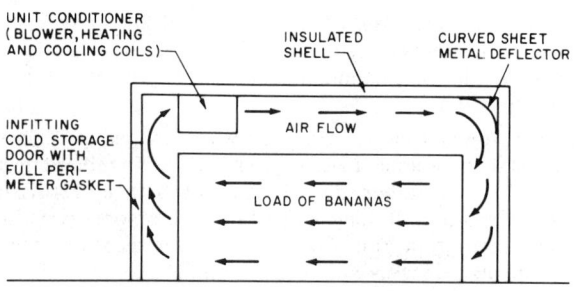

Fig. 2 Banana Room (Side View)

processing room reach optimum ripeness for shipment to retail at the same time is more important.

Airtightness

Banana ripening is initiated by exposing the fruit to ethylene gas, which is introduced into the room from cylinders. The dose is 1 ft^3 of ethylene gas per 1,000 ft^3 of room air space. Ethylene is explosive in air at a concentration between 2.75 and 28.6%. Many ethylene systems gas the fruit automatically over a 24-h period.

To be effective, the gas must be confined to the ripening room for 24 h, so banana rooms must be airtight. Floor drains must be individually trapped to prevent gas leakage. Special care should be taken to seal all penetrations in room walls where refrigerant piping, plumbing lines, and the like, enter rooms. Doors should have single seal gaskets all around and sweep gaskets at the floor line.

Refrigeration

A direct-expansion halocarbon system is recommended for use in banana rooms. Because of ammonia's harmful effect on bananas should leakage occur, direct-expansion ammonia systems should not be used. Malfunctioning refrigeration equipment during processing could cause heavy product losses, so, even with high initial installation costs, each ripening room should have a completely separate system.

For maintenance-free operation in the high humidity environment of processing rooms, evaporator coils should have a fin spacing of 4 fins/in. Coils should be amply sized and capacity rated at a design temperature difference of 15°F with a refrigerant temperature of 40°F. Air temperatures used during processing range from 45 to 65°F. Because of the danger of banana chilling, refrigerant temperatures below 40°F are not recommended. With programmers, suction pressure control devices or hot gas bypass systems must be installed.

Refrigeration Load Calculations

These calculations are based on the same methods used for other fresh fruits. A typical half-carlot-capacity banana room will hold approximately 432 boxes of fruit. Approximate outside dimensions for a three-tier forklift-type room shown in Figure 3 are 30 ft long by 6 ft wide by 22 ft high. Pallets are 48 in. by 40 in. and are stacked 6 pallets deep in each of 3 tiers, totaling 18 pallets per room. Boxes for use in banana rooms are approximately 10 in. high by 16 in. wide by 22 in. long and are stacked 4 boxes per each of 6 layers, totaling 24 boxes per pallet. With 18 pallets per room, 432 boxes of bananas can be stored (18 pallets by 24 boxes). Each box has a net weight of 42 lb and a gross weight of 47 lb.

Transmission load is calculated in the normal manner; the air change load is negligible. The electrical load is based on the continuous operation of multi-kilowatt fan motor(s). The peak heat of respiration is 0.5 Btu/h per pound multiplied by the total net weight of bananas in the ripening room. For product cooling, the specific heat of bananas is 0.8 Btu/(lb·°F) multiplied by the total

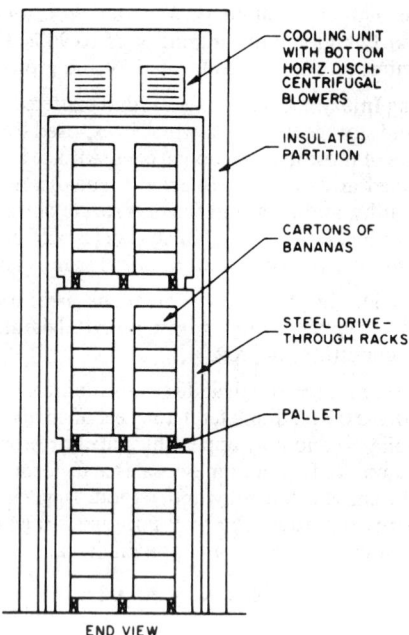

Fig. 3 Three-Tier Forklift Banana Room (End View)

net weight of the bananas, plus the total tare weight of the cartons multiplied by 0.4 Btu/(lb·°F), the specific heat of fiberboard. The total calculated load is thus approximately 80 Btu/h per box. A pulldown rate of 1°F/h is assumed. Total system design capacity is calculated by assuming simultaneous peak respiration and pulldown load.

Heating

Heat is not required during most ripening cycles. However, occasional loads may come in at temperatures below desired levels for treatment with ethylene gas, making heating necessary. Many banana room refrigeration units come with electrical heating elements as an integral part of the unit. If electrical heating strips are used, they should be enclosed in a corrosion-resistant sheath and have a surface temperature of not more than 800°F in dead still air, the temperature limitation being necessary because of proximity to refrigerant coils and the inherent danger should leakage occur. Portable plug-in electric heaters are also used. Heating system capacity should be sufficient to raise load temperature at a rate of 1°F/h. Open-flame gas heaters should never be used in banana rooms because: (1) ethylene gas used during ripening is explosive at certain concentrations; and (2) the necessary room tightness could easily result in the open-flame heaters' consuming the available oxygen within the space, thereby extinguishing the flame and permitting raw gas to enter the room.

Air Circulation

The fruit pulp temperature schedules for 4- to 8-day ripening cycles are shown in Table 2. A temperature variation of only a few degrees will considerably alter the rate of fruit ripening. For even ripening, fruit temperatures must be uniform throughout the room, so comparatively large volumes of air must be continuously circulated throughout the entire load. Centrifugal fans are necessary. They are installed for bottom horizontal discharge, so that the top boxes will not be chilled immediately in front of the unit. Fan air output should be rated at 0.62 in. water external static pres-

Citrus Fruits, Bananas, and Subtropical Fruits

Table 2 Fruit Temperatures for Banana Ripening

Ripening Schedule	Temperature, °F							
	1st Day	2nd Day	3rd Day	4th Day	5th Day	6th Day	7th Day	8th Day
Four days	64	64	62	60				
Five days	62	62	62	62	60			
Six days	62	62	60	60	60	58		
Seven days	60	60	60	60	60	58	58	
Eight days	58	58	58	58	58	58	58	58

sure. Because of heat of respiration, heat must be continually withdrawn from the product even when it is being held at a constant temperature. Therefore, a temperature variation in the load is inevitable, with warmer fruit being downstream relative to the circulated air. Unit conditioners at the front of the room over the door discharge toward the rear of the room. This arrangement leaves riper fruit near the door to be shipped first.

For improved air distribution, a sheet metal (or other suitable material) air deflector curved to a 90° arc is mounted full width on the back room wall. This deflector reduces turbulence and directs the air downward for return through the load.

Air Volume Requirements

Circulated air volume requirements are calculated on the basis of conditions required at the end of the pulldown period. Assume a maximum allowable fruit temperature variation of 2°F, an air temperature drop through the cooling unit of 2°F, and product temperature reduction proceeding at a rate of 0.2°F/h. During the initial pulldown, the air quantity so calculated will give about a 5.5°F drop through the cooling unit. The general equation is:

$$q_t = q_r + q_p = mc_p\Delta t$$

where

- q_t = total heat removed, Btu/h
- q_r = heat of respiration, Btu/h
- q_p = pulldown load, Btu/h
- m = mass of flow rate of air, lb/h
- c_p = specific heat of air, Btu/(lb·°F)
- Δt = temperature change of air, °F

Using values of respiration and specific heat given in the section on load calculations, the value of q_r and q_p can be determined. q_p is calculated on the basis that at the end of the pulldown period, temperature reduction is proceeding at the rate of 0.2°F/h.

$$q_r = 0.5 \times 42 = 21 \text{ Btu/h/box}$$
$$q_p = 0.2[(0.8 \times 42) + (0.4 \times 5)] = 7.12 \text{ Btu/h/box}$$
$$q_t = 21 + 7.12 = 28.12 \text{ Btu/h/box}$$

At equilibrium, the air temperature Δt equals the fruit temperature Δt, and:

$$m = q_t/c_p\Delta t = 28.12/(0.24 \times 2) = 58.58 \text{ lb air per box}$$
$$\text{Volume} = 58.58/(0.075 \times 60) = 13.02 \text{ cfm per box}$$

For a room with 432 boxes, the airflow should be 432 × 13.02 = 5600 cfm at 0.62 in. of water external static pressure.

Humidity

A high relative humidity around the fruit is important during banana ripening. Bananas ripened under low humidity conditions are more susceptible to handling damage. When bananas were ripened on the stem, naked fruit was directly exposed to the moving airstream and automatic room humidifiers were used to prevent excessive fruit dehydration. With the advent of tropical boxing, however, banana room humidifiers are not required.

The fiberboard carton shields the fruit from the moving airstream. In addition, ample sizing of evaporator coils keeps temperature difference across the coil to 10 to 15°F, thereby limiting dehumidification. Both natural transpiration of the fruit and airtight room design also contribute to high room humidity.

Controls

Ripening room air temperatures are varied frequently during banana processing. Temperatures should be controlled by remote bulb-type thermostats, with bulbs for heating and cooling mounted in the return airstream within the ripening room to prevent short cycling of equipment. Thermostats should be mounted on the exterior of the ripening room and have a range of 45 to 70°F, calibrated in 1°F increments, with no more than 2°F differential. The thermostats are best mounted on a control panel having a selector switch providing heating and cooling with continuous fan operation.

Automatic temperature controllers or programmers are being installed in most new facilities. Bananas produce heat continuously, but the rate of heat production varies considerably during the ripening cycle. A generalized heat-of-respiration curve is shown in Figure 4. Although more complex, the removal of heat of respiration from the load can be viewed, for the purpose of analysis, as a simple conduction process. Applying the general conduction equation: $q = kA\Delta t$, where q = heat of respiration; $kA = 1/r$; r = resistance to heat flow of packaging materials; Δt = banana minus air temperatures.

If kA is a constant, Δt must vary as q. Assuming the banana temperature is also constant, the room air temperature must be set lower as q increases during ripening. However, the exact value of q at any particular point in the ripening cycle is unknown. With the conventional, manually adjustable air-sensing thermostat-control system, fruit temperatures are taken manually with a pulp thermometer, and thermostat settings are continually adjusted to follow ripening schedules. This is essentially a trial and error procedure.

By contrast, the temperature programmer has a remote bulb which is placed in a box of bananas in the load. Since the bulb senses fruit temperature directly, heat of respiration is compensated for during ripening. Fruit temperatures are automatically adjusted to follow preset cycles.

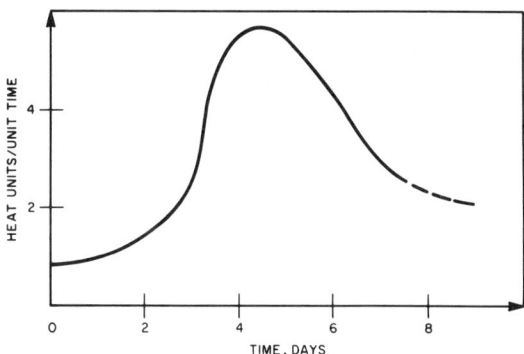

Fig. 4 Heat of Respiration during Banana Ripening

SUBTROPICAL FRUITS

AVOCADOS

Avocado cultivars grown in California are not grown commercially in Florida, and vice versa. Generally, Florida cultivars tend to be larger fruited than those from California. In California, the Fuerte variety accounts for 75% of the annual crop and is available from October through March. Hass (black skin) is available from April through September. In order of importance, Florida cultivars are Booth 8, Lula, Waldin, and Booth 7. Waldin appears on the market in August, followed by Booth 8 in September, and Booth 7 and Lula from October to February.

The best storage temperature for cold-tolerant Florida avocado cultivars, such as Booth 8 and Lula, is 40°F. All Florida avocado cultivars produced during the summer, such as Waldin, are cold-intolerant and store best at 54 to 55°F. A few cultivars, such as Fuerte, store best at 45°F. Cold-tolerant cultivars can be held in storage a month or longer, but storage of cold-intolerant cultivars is usually limited to 2 weeks because of their susceptibility to softening and chilling injury. The best ripening temperature for avocados is 60°F, but temperatures from 55 to 75°F are usually satisfactory. Temperatures above 79°F frequently cause off-flavor, skin discoloration, uneven ripening, and increased decay.

Storage Disorders

Anthracnose (Collecotrichum) are scattered black spots covering firm decayed tissue that can be removed easily from surrounding flesh. Pink spore masses form on the spots under moist conditions. Control requires preventing blemishes and other breaks in the skin.

Chilling injury is typified by small to large sunken pits in the skin, becoming brown or black in color; often accompanied by general browning of the skin and light smoky streaks in the flesh which develop independently.

MANGOES

Important early cultivars, Tommy Atkins and Irwin, mature during June and July, followed by such midseason cultivars as the Kent and Palmer, which mature during July and August. The most important cultivar produced in Florida is the large-fruited Keitt, which is late-season and matures during August and September.

The optimum storage for mangoes is 54 to 55°F for 2 to 3 weeks, although 50°F is adequate for some cultivars for shorter periods. Mangoes are subject to chilling injury at temperatures below 50°F. The best ripening temperatures for mangoes are from 70 to 75°F, but temperatures of 60 to 65°F are also satisfactory under certain conditions. At 60 to 65°F, the fruit develops a bright and most attractive skin color, but the flavor is usually tart and requires an additional 2 to 3 days at 70 to 75°F to attain a sweet flavor. Mangoes ripened at 80°F and higher frequently have a strong flavor and mottled skin.

Storage Disorders

Anthracnose (Collecotrichum) are large scattered black spots in the skin of ripening fruits. Under moist conditions, pink spore masses develop in spots. The disease is controlled on the tree by regular spraying and use of hot water treatment (131°F for 0.12 h) after harvest.

Chilling injury causes pitting of the skin, which sometimes develops a gray cast. Fruit with chilling injury usually does not ripen uniformly. Control requires storing at proper temperatures.

PINEAPPLES

Fresh pineapples are available throughout the year, but a much larger supply is available from March through June. Only three pineapple cultivars are commercially important in the United States: the Smooth Cayenne from Hawaii, and the Red Spanish and Smooth Cayenne from Puerto Rico.

Pineapples harvested at the half-ripe stage can be held for 2 weeks at 45 to 55°F and still have about one week's shelf life. Continuous maintenance of storage temperature is as important as the specific storage temperature. Ripe fruit should be held at 45 to 47°F. Harvesting at the mature-green stage is not recommended because some individual fruits would be so immature that they would fail to ripen. Mature green fruit is especially susceptible to chilling injury at temperatures below 50°F.

Storage Disorders

Black Rot (Ceratocystis). Affected tissues are extensive, soft, and leaky, ranging from normal to jet black in color. To control, treat freshly cut stem parts with benzoic acid-talc dust, prevent bruising, and cool to 50°F.

Brown Rot (Penicillium; Fusarium) has brown, firm decay starting at eyes or cracks; it is common on overripe fruit. To control, provide good growing conditions and move before the fruit is overripe.

Chilling injury is manifested by the fruit taking on a dull hue and developing a water soaking of the flesh and a darkening of the core. To control, hold at recommended temperatures.

REFERENCES

Haller, M.H., D.H. Rose, J.M. Lutz, and P.L. Harding. 1945. Respiration of citrus fruits after harvest. *Journal of Agricultural Resources* 71:327.

Ismail, M.A., T.T. Hatton, D.J. Dezman, and W.R. Miller. 1986. In transit cold treatment of Florida grapefruit shipped to Japan in refrigerated van containers: Problems and recommendations. *Proceedings of the Florida State Horticultural Society* 99:117-121.

BIBLIOGRAPHY

Chace, W.G., Jr., J.J. Smoot, and R.H. Cubbedge. 1979. Storage and transportation of Florida citrus fruits. *Florida Citrus Industry* 51:16

Hardenburg, R.E., A.E. Watada, and C.Y. Wang. 1986. The commercial storage of fruits, vegetables, and florist and nursery stocks. USDA *Handbook* No. 66.

McCornack, A.A., W.F. Wardowski, and G.E. Brown. 1976. Postharvest decay control recommendations for Florida citrus fruit. *Florida Cooperative Extension Service Circular 359-A*.

Smoot, J.J., L.G. Houck, and H.B. Johnson. 1971. Market diseases of citrus and other subtropical fruits. USDA *Handbook*, 398.

Wardowsko, W.F., S. Nagy, and W. Grierson. 1986. *Fresh citrus fruits*. AVI Publishing, Westport, CT.

CHAPTER 18

VEGETABLES

Product Selection and Quality Maintenance 18.1
In-Transit Preservation 18.2
Preservation in Warehouses 18.3
Refrigerated Storage 18.3
Storage of Various Vegetables 18.5

ANNUAL losses (shrinkage) in marketing vegetables (shipping, processing, storage, and retailing) are often mentioned.

Some of these losses are caused, in part, by overly high temperatures during handling, storage, and transport, which increase ripening, decay, and the loss of edible quality and nutrient values. In other cases, freezing or chilling injury from overly low temperatures may be involved. Other serious losses are caused by mechanical injury from careless or rough handling and by shrinkage or wilting because of moisture loss. Many losses can be reduced substantially by following recommended handling, cooling, transport, and storage practices. Improved packaging, refrigerated transport, and awareness of the role of refrigeration in maintaining quality throughout marketing have made it possible to move vegetables to distant cities in field-fresh condition.

PRODUCT SELECTION AND QUALITY MAINTENANCE

The principal hazards to quality retention during marketing include:

1. Metabolic changes associated with respiration, ripening, and aging (composition, texture, color)
2. Moisture loss, with resultant wilting and shriveling
3. Bruising and mechanical injury
4. Parasitic diseases
5. Physiological disorders
6. Freezing and chilling injury
7. Flavor and nutritional changes
8. Growth (sprouting, rooting)

Fresh vegetables are living tissues and have a continuing need for oxygen for respiration. During respiration, stored food such as sugar is converted to heat energy, and the product loses quality and food value. Some of the refrigeration load to maintain commodity temperatures during storage or transportation can be attributed to respiration. For example, a 20,000-lb load of asparagus cooled to 39 °F can produce enough heat of respiration during a cross-country trip to melt 7900 lb of ice.

Vegetables that respire the fastest often give the greatest handling problems because they are the most perishable. Refrigeration is the best method of slowing respiration and other life processes. The respiration rate of many vegetables is given in Chapter 26 of this volume and in Chapter 30 of the 1989 ASHRAE *Handbook—Fundamentals*.

Vegetables are usually covered with microorganisms, which will cause decay given the right conditions. Deterioration because of decay is probably the greatest source of spoilage during marketing. When mechanical injuries break the skin, decay organisms will enter the produce. If it is then exposed to warm temperatures, especially under humid conditions, infection usually increases. Adequate refrigeration is the best method of controlling decay because low temperatures control the growth of most microorganisms.

Many color changes associated with ripening and aging can be delayed by refrigeration. For example, broccoli may show some yellowing in one day on a nonrefrigerated counter, while it remains green at least 3 to 4 days in a refrigerated display.

Refrigeration can retard deterioration caused by chemical and biological reactions. Sweet corn may lose 50% of its initial sugar content in a single day at 70 °F, while only about 5% will be lost in one day at 32 °F (Appleman and Arthur 1919). Also, freshly harvested asparagus will lose 50% of its vitamin C content in one day at 68 °F, whereas it takes 4 days at 50 °F or 12 days at 32 °F to lose this amount (Lipton 1968). With certain exceptions, the best temperature for retarding deterioration resulting from biological processes or from pathogens is the lowest temperature that can be maintained without freezing the commodity. This is about 2 °F above the freezing point of the vegetable.

Loss of moisture with consequent wilting and shriveling is one of the obvious ways in which freshness is lost. *Transpiration* is the loss of water in the vapor state from living tissues. Moisture losses of 3 to 6% are enough to cause a marked loss of quality for many kinds of vegetables. A few commodities may lose 10% or more in moisture and still be marketable, although some trimming may be necessary, such as for stored cabbage. For more on transpiration, see Chapter 30 of the 1989 ASHRAE *Handbook—Fundamentals*.

Postharvest Handling

Care should be exercised in stacking bulk bins in storage so that proper ventilation and refrigeration of the product is maintained. Bins should not be of such depth that excessive weight damages the product near the bottom.

The effect of rough handling of vegetables is cumulative. Several small bruises on a tomato can produce an off-flavor. Bruising also stimulates the rate of ripening of a product such as tomatoes and thereby shortens potential storage and shelf life. Mechanical damage allows increased moisture loss; skinned potatoes may lose 3 to 4 times as much weight as nonskinned ones.

After harvest, most highly perishable vegetables should be removed from the field as rapidly as possible and placed under refrigeration, or they should be graded and packaged for marketing. Since aging and deterioration of vegetables continues after harvest, the marketable life depends greatly on the temperature and care in physical handling. Quality maintenance is aided by the following procedures:

1. Harvest at optimum maturity or quality
2. Handle carefully to avoid mechanical injury
3. Handle rapidly to minimize deterioration
4. Provide protective containers and packaging
5. Use preservative chemical, heat, or modified-atmosphere treatments
6. Enforce good plant sanitation procedures

7. Precool to remove field heat
8. Provide high relative humidity to minimize moisture loss
9. Provide proper refrigeration throughout marketing

Cooling

Rapid cooling of a commodity after harvest, before or after packaging, and before it is stored or moved in transit, prevents deterioration of the more perishable vegetables. The faster field heat is removed after harvest, the longer produce can be maintained in good marketable condition. Cooling slows natural deterioration, including aging and ripening, slows growth of decay organisms (and thereby the development of rot), and reduces wilting, since water losses occur much more slowly at low temperatures than at high temperatures. After cooling, produce should be refrigerated continuously at recommended temperatures. If warming is allowed, much of the benefit of prompt precooling may be lost.

Types of cooling include hydrocooling, vacuum cooling, air cooling, and cooling with contact ice and top ice. These methods are discussed in detail in Chapter 11. The choice of method used depends on factors such as refrigeration sources and costs, volume of product shipped, and product limitations.

IN-TRANSIT PRESERVATION

Good equipment is available to transport perishable commodities to market under refrigeration by rail, truck, piggyback trailers, and containers. A high rh of about 95% is desirable for most vegetables to prevent moisture loss and wilting. Many vegetables benefit from 95 to 100% rh. Humidity in both iced and mechanically refrigerated cars and trailers is usually high. Top ice and package ice, used most often for leafy vegetables, provide refrigeration and added moisture.

Cooling Vehicle and Product

Vehicles used to ship vegetables that require low temperature during transit should have their interiors cooled before loading. With mechanical refrigeration, the units should be operated with the doors closed until the temperature of the interior of the vehicle is reduced to the approximate transit temperature.

Generally, vegetables that require low temperature during transport should be cooled before they are loaded into transport vehicles. Cooling produce in tightly loaded refrigerator cars or trailers is a slow process, and that portion of the load exposed to the cold air discharge may be frozen when the interior of the load is still warm. It is also uneconomical to provide refrigeration capacity in vehicles for cooling.

Packaging, Loading, and Handling

Containers must protect the commodity, permit heat exchange as necessary, and serve as an appropriate merchandising unit with sufficient strength to withstand normal handling. Freight container tariffs describe approved containers and loading procedures.

Containers should be loaded to take advantage of their maximum strength and to permit adequate stripping or use of spacers to hold the load in alignment. Proper vertical alignment of containers is essential to obtain their maximum stacking strength capability, although maximum stacking frequently is incompatible with providing channels for air circulation. Channels for proper air circulation must be maintained, even at the sacrifice of some capacity and resistance of the load to shifting. Ventilation openings in containers, if any, should have the greatest possible exposure to the ventilation channels or flues.

When different types of containers are used in the same load, stacks should be separated so that one type will not damage another. If separation of stacks is not possible, containers made of lighter material, such as fiberboard, should always be loaded on top of any heavier wood containers.

Providing Refrigeration and Air Circulation

Desirable and safe transit temperatures for various vegetables and suggested temperatures to be specified for mechanically refrigerated cars and trailers are given in Table 1. For safety, the thermostat settings suggested for cool season vegetables are usually 2 to 4 °F above the freezing point. The various means for obtaining specific transit temperatures in the United States are provided under the regulations of the Perishable Protective Tariffs issued by the National Freight Committee and the Railway Express Agency.

With the many kinds of refrigeration, heating, and ventilating services now available, the shipper has only to specify the desired transport temperature. Generally, the shipper or the receiver is responsible for selecting the protective service for his commodity in transit. The various protective services are described in detail in USDA *Agriculture Handbooks* No. 669 (Ashby et al. 1987) (truck shipments) and No. 195 (rail shipments).

Table 1 Desirable Transit Temperatures for Various Vegetables

Vegetable	Desirable Transit Temp., °F	Suggested Thermostat Setting[a], °F	Highest Freezing Point[b], °F
Artichokes	32	33	29.8
Asparagus	32-35	35	30.9
Beans, lima	37-41	37	31.0
Beans, snap	40-45	45	30.7
Beets, topped	32	34	30.4
Broccoli	32	34	30.9
Brussels sprouts	32	34	30.6
Cabbage	32	34	30.4
Cantaloupes	36-41	37	29.8
Carrots, topped	32	33	29.5
Cauliflower	32	34	30.6
Celery	32	34	31.1
Corn, sweet	32	34	30.9
Cucumbers	50-55	50	31.1
Eggplant	46-54	50	30.6
Endive and Escarole	32	34	31.8
Greens, leafy	32	34	—
Honeydew melon	45-50	45	30.4
Lettuce	32	34	31.6
Onions, dry	32-39	35	30.6
Onions, green	32	34	30.4
Peas, green	32	34	30.9
Peppers, sweet	45-50	46	30.7
Potatoes:			
Early crop	50-60	50	30.9
Late crop	39-50	40	30.9
For chipping:			
Early crop	64-70	64-70	30.9
Late crop	50-60	50-60	30.9
Radishes	32	34	30.7
Spinach	32	34	31.5
Squash, summer	41-50	41	31.1
Squash, winter	50-55	50	30.5
Sweet potatoes	55-61	55	29.7
Tomatoes:			
Mature green	55-70	55	30.9
Pink	46-50	50	30.6
Watermelons	50-60	50	31.3

Data from USDA *Agricultural Handbook* No. 195 (Redit 1969), USDA *Marketing Research Report* No. 196 (Whiteman 1957), and USDA *Agricultural Handbook* No. 66 (Hardenburg et al. 1986).
[a] For U.S. shipments of vegetables in mechanically refrigerated cars under Rule 710 and in trailers under Rule 800 of the Perishable Protective Tariff.
[b] Highest temperature at which freezing occurs.

Vegetables

Protection from Cold

During the winter, vegetables must be protected from freezing. Mechanically refrigerated freight cars and trucks equipped to handle the full range of both fresh and frozen commodities are designed to provide heat for cold weather protection as well as refrigeration. The heat is supplied by electric heating elements or by reverse cycle operation of the refrigerating unit in which hot gas from the compressor is circulated in the cooling coils. The change from cooling to heating is done automatically by thermostatic controls.

Protection against freezing in transit is a major problem in moving late crop potatoes from storage to market. Cars and trailers should be warmed before loading, protected during loading with canvas door shields or loading tunnels, and loaded properly.

Checking and Cleaning Equipment

If the thermostat is out of adjustment by a few degrees, products may be damaged by freezing or by not receiving sufficient refrigeration. Thermostats should be calibrated at regular intervals to assure that the proper amount of refrigeration is furnished. Trailers and railcars should be cleaned carefully before loading. Debris from previous shipments may contaminate loads and should be swept from floors and floor racks.

Modified Atmospheres in Transit

A variety of systems provide controlled or modified atmospheres in trucks, piggyback trailers, railcars, and seavans. Though the label on each process may differ, all systems alter the levels of oxygen, carbon dioxide, and nitrogen surrounding the produce.

A frequent goal is to lower the oxygen concentration in air to a level of 1 to 5% because this level usually depresses the respiration rate. However, no single modified atmosphere can be expected to benefit more than a few commodities; crop requirements and tolerances are quite specific (Harvey 1965). Also, certain vegetables may tolerate an atmosphere at one temperature but not at another, or they may tolerate a modified atmosphere for only a limited time.

The load compartment must be fairly tight to maintain desirable atmospheres. Modified-atmosphere equipment is predominantly installed in vehicles used to transport chilled perishables in long haul movements (over five days' transit time).

Lettuce is the main vegetable shipped under a modified atmosphere. The physiological disorder known as *russet spotting* is reduced when lettuce is shipped in low oxygen atmospheres with good refrigeration (Stewart and Ceponis 1968). Lettuce is damaged by accumulated carbon dioxide (Stewart et al. 1970), so some fresh hydrated lime is usually placed in the cargo space to absorb carbon dioxide.

Atmosphere systems used in transport either vaporize liquid nitrogen or generate nitrogen by passing heated compressed air through hollow fibers. In either case, nitrogen is vented into the trailer to reduce oxygen levels.

Temperature is critical in determining the benefits of modified atmospheres. If temperatures are higher than those recommended for use with modified atmospheres, decay and other deterioration may be increased rather than reduced. Modified atmospheres should never be used as a substitute for good temperature management.

Use of controlled atmospheres is discussed later in this chapter. For further information on refrigerated transport by truck, railway car, ship, and air, see Chapters 28 through 31.

PRESERVATION IN WAREHOUSES

Wholesale warehouses usually do not have a whole range of controlled temperature rooms to provide optimum storage conditions for each kind of produce, and this is not necessary for short holding. About one-half the produce handled can be stored at 32°F. Enough refrigeration capacity should be available to maintain a year-round temperature of 32°F with 90 to 95% rh. Higher temperatures and lower humidities for more perishable vegetables accelerate quality loss and increase waste. Enough refrigerated coil surface should be provided to allow a differential of only a few degrees between the coil and air temperatures, while still providing adequate refrigeration. A difference of as little as 2°F is desirable to permit optimum humidity conditions.

Desirable air temperature and humidity cannot be maintained if excessive air exchange occurs between the warehouse cold storage room and warmer areas. Operators should consider the use of air curtains or flap doors whenever doors to cold rooms must be opened often or for prolonged periods.

Some vegetables should not be stored at 32°F because of the danger of chilling injury. Other less perishable vegetables, such as dry onions, can be held satisfactorily for short periods at warmer temperatures. For these vegetables, controlled storage conditions during wholesaling are still desirable. Table 2 shows which vegetables should be held at 32°F and which should be held at 50°F during wholesaling.

Table 2 Recommended Temperatures for Maintaining Quality of Fresh Vegetables in Wholesale Warehouses

Store at 32°F and 90 to 95% rh		Store at 50°F 80 to 85% rh
Artichokes	Horseradish	Beans, green
Asparagus	Kohlrabi	Cucumbers
Beans, Lima	Lettuce	Eggplants
Beets	Mushrooms	Garlic, dry
Broccoli	Onions, green	Melons
Brussels sprouts	Parsnips	Okra
Cabbage	Peas, green	Onions, dry
Carrots	Radishes	Peppers
Cauliflower	Rhubarb	Potatoes
Celery	Rutabagas	Pumpkins
Corn, sweet	Spinach	Squash, hard shell
Endive	Squash, summer	Sweet potatoes
Escarole	Turnips	Tomatoes, ripe

Adapted from Bogardus and Lutz (1961).

Mature green tomatoes, for ripening, usually need separate, controlled temperature rooms, where 55 to 70°F temperatures with 85 to 90% rh can be maintained to delay or speed ripening as desired. Information on general aspects of cold storage design and operation can be found in Chapters 25, 26, and 27.

REFRIGERATED STORAGE

The refrigeration requirement of any storage plant must be based on peak refrigeration load. This peak usually occurs when outside temperatures are high and warm produce is being moved into the plant for cooling and storage. The peak refrigeration load depends on the amount of commodity received each day, the temperature of the commodity at the time it is placed under refrigeration, the specific heat of the commodity, and the final temperature attained.

Protective Packaging and Waxing

Vegetables should be stored in containers with adequate cushioning materials, stacking strength, and durability to protect against crushing and to withstand high humidity conditions. Bulging crates should be stacked on their sides or stripped between layers to keep weight off the commodity. Many vegetables are now stored or shipped in fiberboard or corrugated containers, but the weakening of fiberboard materials by moisture absorption at the high humidities in storage is frequently a serious problem.

Manufacturers have improved the strength and reduced the degree of moisture absorption by fiberboard materials. Special treatment of fiberboard permits the use of certain cartons during hydrocooling and with package and top ice. Cartons may be strengthened to withstand stacking by using dividers, wooden corner posts, and full telescoping covers.

Produce consumer-packaged at production points use many types of trays, wraps, and film bags, which may present special storage problems because master containers may lack stacking strength.

Desiccation often can be minimized by using moisture retentive plastic packaging materials. Polyethylene film box liners, pallet covers, and tarpaulins may be helpful to reduce moisture loss from vegetables. Plastic films, if sealed or tightly tied, may restrict the transfer of carbon dioxide, oxygen, and water vapor, leading to harmful concentrations of these respiratory gases; and films restrict heat transfer, which retards the rate of cooling (Hardenburg 1971).

Waxes are applied to rutabagas, cucumbers, mature green tomatoes, and cantaloupes, and to a lesser extent to peppers, turnips, sweet potatoes, and certain other crops. With products such as cucumbers and root crops, waxing reduces moisture loss and thus retards shriveling. With some products, an improved glossy appearance is the main advantage. Thin wax coatings may give little if any protection against moisture loss; coatings which are too heavy may increase decay and breakdown. Waxing alone does not control decay, but waxing and fungicides combined may be beneficial. Waxing is not recommended for potatoes either before or after storage (Hardenburg et al. 1959).

Sprout Inhibitors

Sprout inhibitors are used when cold storage facilities are lacking or if low temperatures might injure the vegetable or affect its processing quality. Sprouting of onions, potatoes, and carrots in storage can be inhibited by spraying the plants a few weeks before harvest with a solution of maleic hydrazide. Potatoes are also sprayed or dipped in a CIPC solution to inhibit sprouting. Vaporized nonanol alcohol is circulated through ducts to suppress potato sprouting in Great Britain (Burton 1958).

Gamma irradiation suppresses sprouting of onions, sweet potatoes, and white potatoes at dosages of 0.05 to 0.15 kGy. Dosages above 0.15 kGy cause breakdown and increase decay in white potatoes (Kader et al. 1984).

Controlled and Modified Atmosphere Storage

Refrigeration is most effective in retarding respiration and lengthening storage life. For some products, reducing the oxygen level in the storage air and/or increasing the carbon dioxide level as a supplement to refrigeration can provide extended storage life. Careful control of the concentration of oxygen and carbon dioxide level is essential. If all of the oxygen is used, produce will suffocate and may develop an alcoholic off-flavor in a few days. Carbon dioxide given off in respiration or from dry ice may accumulate to injurious levels.

Modified or controlled atmospheres (CA) during storage of vegetables have received little application, in contrast to their wide use for apples. Many atmospheres tested on vegetables were injurious or produced only minor benefits (Dewey et al. 1969). If commercial use of CA for vegetables increases, it is likely to be with the use of external generators to create desired atmospheres or with the addition of nitrogen gas or dry ice, rather than by using product respiration in a gastight room.

Danish cabbage has kept better during 4 to 5 months at 32°F in gas mixtures with 2.5 to 5% oxygen and carbon dioxide (with the balance nitrogen) than it has in air (Isenberg and Sayles 1969). Tests have indicated that mature green tomatoes may keep predominantly green for 5 to 6 weeks at 55°F in an atmosphere at 3% oxygen with 97% nitrogen. After removal to air at 64°F, these tomatoes ripened normally with acceptable flavor (Parsons and Anderson 1970).

Asparagus and mushrooms in refrigerated storage have kept better for short periods in atmospheres with 5 to 10% carbon dioxide than in air. The carbon dioxide inhibits soft rot of asparagus (Lipton 1965) and retards cap opening and inhibits mold in mushrooms. Some promising experimental CA results on improved quality retention during short storage also have been obtained for head lettuce, broccoli, brussels sprouts, green onions, and radishes.

Hypobaric storage, or storage at reduced atmospheric pressure, is another supplement to refrigeration that involves principles similar to those in controlled-atmosphere storage. At atmospheric pressures 0.1 or 0.2 of normal, several kinds of produce have an extended storage life. The benefits are attributed both to the low oxygen level maintained and to the continuous removal of ethylene, carbon dioxide, and possibly other metabolically active gases. Their rates of production under hypobaric ventilation are also lower.

Positive hypobaric ventilation is absolutely required to achieve the desired low ethylene concentration within and around produce to retard ripening. The continuous flow of water-saturated air at a low pressure flushes away emanated gases and prevents weight loss.

Injury

Chilling injury may be defined as an injury caused by exposure to low but nonfreezing temperatures, often in a temperature range from 32 to 50°F. At these temperatures, vegetables become weakened because they are unable to carry on normal metabolic processes. Often, vegetables that are chilled look sound when removed from low temperatures. However, symptoms of chilling, such as pitting or other skin blemishes, internal discoloration, or failure to ripen, become evident in a few days at warmer temperatures (Morris and Platenius 1938). Vegetables that have been chilled may be particularly susceptible to decay. *Alternaria* rot is often severe on tomatoes, squash, peppers, and cantaloupes that have been chilled. Tomatoes that have been severely chilled usually ripen slowly and rot rapidly. Figure 1 shows the increasing extent of rot in mature green tomatoes held at chilling temperatures.

Both time and temperature are involved in chilling injury. Damage may occur in a short time if temperatures are considerably below the danger line, but a product may withstand a few degrees in the danger zone for a longer time. Also, the effects of chilling are cumulative. Low temperatures in the field before harvest and in transit add to the total effects of chilling that might occur in storage. A list of vegetables susceptible to chilling injury together with the symptoms and the lowest safe temperature is shown in Table 3.

Table 1 shows the *freezing points* of various vegetables. This is the highest temperature at which ice crystal formation in the tissues has been recorded experimentally. Most vegetables have a freezing point between 28 and 31°F (Whiteman 1957). Different vegetables vary widely in their susceptibility to freezing injury.

The freezing point of the commodity is no indication of the damage to be expected from freezing. For example, tomatoes and parsnips both have freezing points of 30 to 31°F. Parsnips can be frozen and thawed several times without apparent injury, whereas tomatoes are ruined after one freezing. Tissues injured by freezing generally appear water-soaked. Even though some vegetables are somewhat tolerant to freezing, it is desirable to avoid freezing temperatures because they shorten storage life (Parsons and Day 1970).

To minimize damage, fresh commodities should not be handled while frozen. Fast thawing damages tissues, but very slow thawing, such as at 32 to 34°F, permits ice to remain in the tissues too long. Thawing at 39°F is suggested (Lutz 1936).

Vegetables

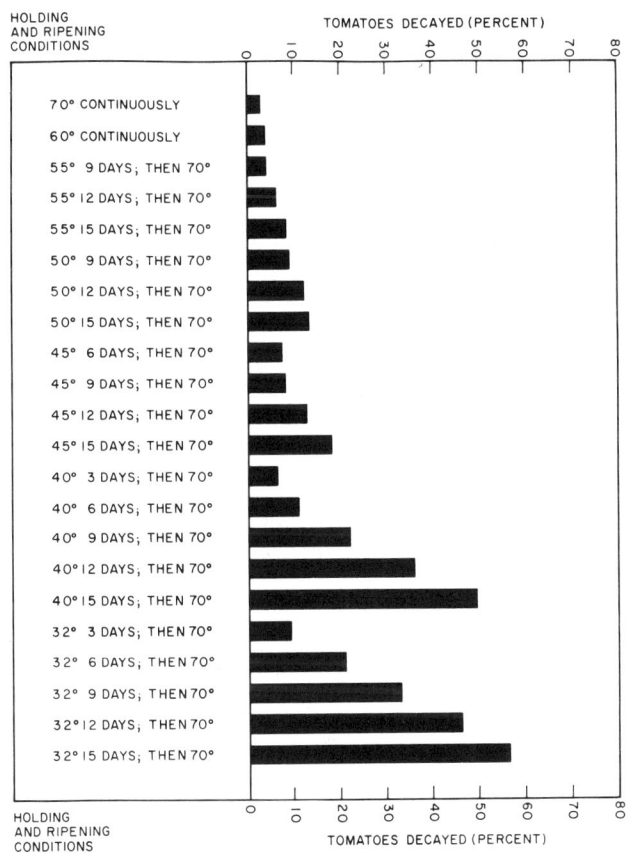

Fig. 1 Effect of Different Temperatures and Holding Periods on Rot, Principally *Alternaria*, in Tomatoes Ripened at 70°F
(McColloch and Worthington 1954)

Table 3 Vegetables Susceptible to Chilling Injury at Moderately Low but Nonfreezing Temperatures

Commodity	Approximate Lowest Safe Temp, °F	Character of Injury between 32°F and Safe Temp.[a]
Beans (snap)	45	Pitting and russeting
Cucumbers	50	Pitting; watersoaked spots, decay
Eggplants	45	Surface scald; *Alternaria* rot
Melons:		
Cantaloupes	a	Pitting; surface decay
Honeydew	45 to 50	Pitting; failure to ripen
Casaba	45 to 50	Pitting; surface decay
Crenshaw and Persian	45 to 50	Pitting; surface decay
Watermelons	40	Pitting; objectionable flavor
Okra	45	Discoloration; watersoaked areas; pitting; decay
Peppers, sweet	45	Sheet pitting, *Alternaria* rot on pods and calyxes
Potatoes	37 to 39	Mahogany browning (Chippewa and Sebago); sweetening[b]
Pumpkins and Hard-Shell Squash	50	Decay, especially *Alternaria* rot
Sweet potatoes	55	Decay; pitting; internal discoloration
Tomatoes:		
Ripe	45 to 50[a]	Watersoaking and softening; decay
Mature green	55	Poor color when ripe; *Alternaria* rot

From USDA *Agricultural Handbook* No. 66 (Hardenburg *et al.* 1986).
[a]See text.
[b]Often these symptoms appear only after removal to warm temperatures, as in marketing.

STORAGE OF VARIOUS VEGETABLES

In the following sections, the temperatures and rh recommended (shown in parentheses) are the optimum for maximum storage in the fresh condition. For short storage, higher temperatures may be satisfactory for some commodities. Temperature requirements represent commodity temperature levels that should be maintained. Much of this information is taken from USDA *Agricultural Handbook* No. 66 (Hardenburg *et al.* 1986).

The quality of each lot of produce should be determined at the time of storage and regular inspections should be made during storage. Such vigilance will permit early detection of disease so that the affected commodity can be moved out of storage before serious loss occurs.

Artichokes, Globe (32°F and 95 to 100% rh)

Globe artichokes are seldom stored, but for temporary holding a temperature of about 32°F is recommended. A high rh of at least 95% will help prevent wilting. This product should keep for 2 weeks in storage if buds are uninjured and wilting is prevented. Perforated polyethylene liners with 23 ¼-in. holes/ft² help retard mosture loss. Hydrocooling or room cooling to 39°F on the day of harvest reduces deterioration.

Gray Mold Rot (*Botrytis*): The most common decay of harvested artichokes. Reddish brown to dark brown firm rot (see Table 4, Note 2). *Control:* Practice sanitation in the field. Refrigerate promptly.

Asparagus (32 to 36°F and 95 to 100% rh)

Asparagus deteriorates very rapidly at temperatures above 36°F and especially at room temperature. It loses sweetness, tenderness, and flavor, and decay develops later. If the storage period is 10 days or less, 32°F is recommended; asparagus is subject to chilling injury if held longer at this temperature. Asparagus is not ordinarily stored except temporarily, but at 36°F with a high relative humidity, it can be kept in salable condition for 3 weeks. However, after a long haul to market, even under refrigeration, it cannot be expected to keep longer than 1 to 2 weeks.

Asparagus should be cooled immediately after cutting. Hydrocooling is the usual method. During transit or storage, the butts of asparagus should be placed on some moist, absorbent material to prevent loss of moisture and to maintain freshness of the spears. Sometimes asparagus bunches are set in shallow pans of water in storage.

Bacterial Soft Rot: Mushy, soft, watersoaked areas on tips and cut ends of asparagus (see Table 4, Note 1). *Control:* Avoid excessive bruising of tips; cool to 39°.

Fusarium Rot: Watersoaked areas changing through yellow to brown, chiefly on asparagus tips; white to pink delicate mold. *Control:* Cool and ship at temperatures of 39°F or lower; handle promptly. Keep tips dry on way to market.

Phytophthora Rot: Large, watersoaked, or brownish lesions at the side of cut asparagus stalks. Lesions later are extensively shriveled. *Control:* Cool to 39°F. Maintain low transit temperatures. Market promptly.

Table 4 Notes on Diseases of General Occurrence

Note 1. Bacterial Soft Rot

Occurs on various vegetables as dark green, greasy or watersoaked soft spots and areas in leaves and stems. Soft, mushy, yellowish spots or soupy areas on stems, roots, and tubers of vegetables. Frequently accompanied by repulsive odor from secondary invaders. *Control:* Use sanitation practices during picking and packing to reduce contamination of harvested product. Use bactericides in postharvest wash treatments. Where possible avoid bruising and injury. Shade harvested produce in the field and reduce temperature promptly to 39 °F or lower for commodities that can withstand low temperatures.

Note 2. Gray Mold Rot (Botrytis)

Decayed tissues are fairly firm to semiwatery. Watersoaked grayish tan to brownish in color. Gray mold and grayish brown, with conspicuous velvety spore masses. *Control:* Use sanitation practices during harvesting and packing. Avoid wounds as much as possible. Use storage and transit temperatures as low as otherwise practicable because decay progresses even at 32 °F.

Note 3. Rhizopus Rot

Decayed tissues are watersoaked, leaky, and softer than those with gray mold rot or watery soft rot. Coarse mycelium and black spore heads develop under moist conditions. Nesting is common. *Control:* Insofar as possible, avoid injury and bruising. Reduce temperature promptly and maintain below 50 °F for commodities that can withstand low temperatures.

Note 4. Watery Soft Rot (Sclerotinia)

Decayed tissues are watersoaked, with slightly pinkish borders, or brownish tan; very soft and watery in later stages, accompanied by development of fine white to dingy cottony mold and black to brown mustardseed-like bodies called *sclerotia*. Nesting is common. *Control:* Use sanitary practices in harvesting and packing. Cull out specimens with discolored or dead portions. Maintain temperature as low as practicable because rot progresses even at 32 °F. Do not store commodities known to have watery soft rot at harvest time.

Beans, Green or Snap (40 to 45 °F and 95% rh)

Green or snap beans are probably best stored at 39 to 45 °F, where they may keep for about 1 week. Even this recommended temperature causes some chilling, but is best for short storage. When stored at 40 °F or below for 3 to 6 days, surface pitting and russet discoloration may appear in a day or two following removal for marketing. The russeting will be especially noticeable in the centers of the containers where condensed moisture remains. *Contact icing is not recommended.* To prevent wilting, the rh should be maintained at 90 to 95%. Beans for processing can be stored up to 10 days at 40 °F. Containers of beans should be stacked to allow abundant air circulation; otherwise, the temperature may rise from the heat of respiration. Beans stored too long or at too high a temperature are subject to decay such as watery soft rot, slimy soft rot, and rhizopus rot.

Anthracnose (Colletotrichum): Circular or oval sunken spots; reddish brown around border with tan centers that frequently bear pink spore mounds. *Control:* Use resistant varieties. Plant disease-free seed. Refrigerate harvested beans promptly to 45 °F.

Bacterial Blight (Pseudomonas): Small, greasy-appearing, watersoaked spots in pod. Older spots show red at the center, with watersoaked area surrounding and penetrating to the seed. *Control:* Keep the field sanitary. Use disease-free seed. Reduce the transit temperature promptly to 45 °F.

Bacterial Soft Rot: See Table 4, Note 1.

Cottony Leak (Pythium): Pods with large, watersoaked spots, accompanied by abundant white cottony mold. *Control:* Sort out diseased pods in packing. Reduce transit temperature promptly to 45 °F.

Freezing Injury: Slight freezing results in watersoaked mottling in surface of exposed pods. Severely frozen beans become completely watersoaked, limp, and dry out rapidly. Snap beans freeze at about 31 °F.

Russeting: Chestnut brown or rusty, diffuse surface discoloration on both sides of pods. *Control:* Permit no surface moisture on warm beans. Cool promptly; avoid temperatures lower than 45 °F.

Soil Rot (Rhizoctonia): Large, reddish brown, sunken, decayed spots on pods. Cream-colored or brown mycelium and irregular, chocolate-colored sclerotia may develop. Nesting is common. *Control:* Maintain transit temperatures of 45 to 50 °F.

Watery Soft Rot: Presence of large black sclerotia in white mold helps to separate this from Cottony Leak (see Table 4, Note 4).

Beans, Lima (37 to 41 °F and 95% rh)

For best quality, lima beans should not be stored. Unshelled lima beans can be stored only about 1 week at 37 to 41 °F. They should be used promptly after removal because the pods discolor rapidly at room temperature. Even with 1 week of refrigerated storage, the pods may develop rusty brown or brown specks, spots, and larger discolored areas which would reduce salability of the beans. The pod discoloration will increase sharply during an additional day at 70 °F following storage.

Beets (32 °F and 98 to 100% rh)

Topped beets are subject to wilting because of the rapid loss of water when the storage atmosphere is too dry. When stored at 32 °F with at least 95% rh, they should keep for 4 to 6 months. Before beets are stored, they should be topped and sorted to remove all diseased beets and those showing mechanical injury. Bunch beets under the same conditions will keep 1 to 2 weeks. Contact icing is recommended. The containers should be well ventilated and stacked to allow air circulation.

Baterial Soft Rot: See Table 4, Note 1.

Broccoli (32 °F and 95 to 100% rh)

Italian or sprouting broccoli is highly perishable and is usually stored for only a brief period as needed for orderly marketing. Good salable condition, fresh green color, and the vitamin C content are best maintained at 32 °F. If it is in good condition and is stored with adequate air circulation and spacing between containers to avoid heating, broccoli should keep satisfactorily 10 to 14 days at 32 °F. Longer storage is undesirable because leaves discolor, buds may drop off, and tissues soften. The respiration rate of freshly harvested broccoli is high, comparable to that of asparagus, beans, and sweet corn. This high rate of respiration should be considered when storing broccoli, especially if it is held without package ice.

Brussels Sprouts (32 °F and 95 to 100% rh)

Brussels sprouts can be stored in good condition for a maximum of 3 to 5 weeks at the recommended temperature of 32 °F. Deterioration, yellowing of the sprouts, and discoloration of the stem are rapid at temperatures of 50 °F and above. Rate of deterioration is twice as fast at 40 °F as at 32 °F. Loss of moisture through transpiration is rather high even if the relative humidity is kept at the recommended level. Film packaging is useful in preventing moisture loss. As with broccoli, sufficient air circulation and spacing between packages is desirable to allow good cooling and to prevent yellowing and decay.

Cabbage (32 °F and 98 to 100% rh)

A large percentage of the late crop of cabbage is stored and sold during the winter and early spring, or until the new crop from southern states appears on the market. If it is stored under proper conditions, late cabbage should keep for 3 to 4 months. The longest keeping varieties belong to the Danish class. Early crop cabbage, especially southern grown, has a limited storage life of 3 to 6 weeks.

Vegetables

Cabbage is successfully held in common storage in northern states, where a fairly uniform inside temperature of 32 to 36 °F can be maintained. An increasing quantity of cabbage is now held in mechanically refrigerated storage, but in some seasons its value does not justify the expense. Use of controlled atmospheres to supplement refrigeration aids quality retention. An atmosphere with 2.5 to 5% oxygen and 2.5 to 5% carbon dioxide can extend the storage life of late cabbage. Cabbage should not be stored with fruits emitting ethylene. Concentrations of 10 to 100 ppm of ethylene cause leaf abscission and loss of green color within 5 weeks at 32 °F.

Pallet bins are used as both field and storage containers, so the cabbage requires no handling from the time of harvest until preparation for shipment. Before the heads are stored, all loose leaves should be trimmed away; only 3 to 6 tight wrapper leaves should be left on the head. Loose leaves interfere with ventilation between heads, which is essential for successful storage. When removed from storage, the heads should be trimmed again to remove loose and damaged leaves.

Chinese cabbage can be stored for 2 to 3 months at 32 °F with 95 to 100% rh.

Alternaria Leaf Spot: Small to large spots bearing brown to black mold. This spotting opens the way for other decays. *Control:* Avoid injuries. Maintain 32 to 34 °F temperature in transit and storage. Practice sanitation in storage rooms.

Bacterial Soft Rot: This slimy decay frequently starts in the pith of a cut stem or in leaf spots caused by other organisms (see Table 4, Note 1).

Black Leaf Speck: Small, sharply sunken, brown or black specks occurring anywhere on outer leaves or in leaves throughout the head. Occurs under refrigeration in transit and storage, and in association with sharp temperature drops. *Control:* No effective control measures are known.

Freezing Injury: Heads frozen slightly may thaw without apparent injury. Freezing injury is found first as brown streaks in the stem, then as light brown watersoaking of the heart leaves and stem. *Control:* Prevent any extended exposure to temperatures below 31 °F.

Watery Soft Rot: See Table 4, Note 4.

Carrots (32 °F and 98 to 100% rh)

Carrots are best stored at 32 °F with a very high rh. Like beets, they are subject to rapid wilting if the humidity is low. For long storage, carrots should be topped and free from cuts and bruises. If they are in good condition when stored, mature carrots should keep 5 to 9 months, if promptly cooled after harvest. Carrots lose moisture readily and wilting results. Humidity should be kept high, but condensation or dripping on the carrots should be avoided, since this is conducive to decay development.

Most carrots for the fresh market are not fully mature. Immature carrots are prepackaged in polyethylene bags, either at the shipping point or in terminal markets. They are usually moved into marketing channels soon after harvest, but they can be stored for a short period to avoid a market glut. If the carrots are cooled quickly and all traces of leaf growth are removed, they can be held 4 to 6 weeks at 32 °F.

In Texas, immature carrots are often stored in clean 50-lb burlap sacks. The sacks of carrots should be stacked in such a manner that at least one surface of each sack is in contact with top ice at all times. Top ice provides some of the necessary refrigeration and prevents dehydration. Bunched carrots may be stored 10 to 14 days at 32 °F. Contact icing is recommended.

Bitterness in carrots, which may develop in storage, is because of abnormal metabolism caused by ethylene given off by apples, pears, and some other fruits and vegetables or from other sources such as internal combustion engines. Bitterness can be prevented by storing carrots away from products that give off ethylene.

Bacterial Soft Rot: See Table 4, Note 1.

Black Rot (Stemphylium): Fairly firm black decay at the crown, on the side, or at the tips of harvested roots. *Control:* Avoid bruising. Store at 32 °F.

Crater Rot (Rhizoctonia): Circular brown craters with white to cream-colored mold in the center. Develops under high humidity in cold storage. *Control:* Field sanitation measures. Avoid surface moisture on storage roots.

Freezing Injury: Roots are flabby, and on cutting show radial cracks in the flesh of the central part and tangential cracks in the outer part. *Control:* Prevent exposure to temperatures below 30 °F.

Gray Mold Rot: See Table 4, Note 2.

Rhizopus Rot: See Table 4, Note 3.

Watery Soft Rot: See Table 4, Note 4.

Cauliflower (32 °F and 95% rh)

Cauliflower may be stored for 3 to 4 weeks at 32 °F with about 95% rh. Successful storage depends on retarding the aging of the head or curd, preventing decay marked by spotting or watersoaking of the white curd, and preventing yellowing and dropping of the leaves. When it is necessary to hold cauliflower temporarily out of cold storage, packing in crushed ice will aid in keeping it fresh. Freezing causes a grayish brown discoloration and softening of the curd, accompanied by a watersoaked condition. Affected tissues are rapidly invaded by soft rot bacteria.

Much of the cauliflower now marketed has the leaves closely trimmed and is prepackaged in perforated cellophane overwraps and packed in fiberboard containers. Vacuum cooling is a fairly efficient method of cooling prepackaged cauliflower. In general, use of controlled atmospheres with cauliflower has not been promising. Atmospheres containing 5% carbon dioxide or higher are injurious to cauliflower, although the damage may not be apparent until after cooking.

Bacterial Soft Rot: See Table 4, Note 1.

Brown Rot (Alternaria): Brown or black spotting of the curd. *Control:* Use seed treatment and field spraying. Keep the curds dry. Maintain low transit temperatures. Store at 32 °F.

Celery (32 °F and 98 to 100% rh)

Celery is a relatively perishable commodity, and for storage of 1 to 2 months, it is essential that a commodity temperature of 32 °F be maintained. The rh should be high enough to prevent wilting (98 to 100%). Considerable heat is given off because of respiration, and for this reason the stacks of crates should be separated and dunnage used to allow cold air circulation under and over the crates and between the bottom crates and the floor. Forced air circulation should be provided; otherwise there may be a 3 to 4 °F temperature differential between the top and the bottom of the room.

Celery can be cooled by forced air cooling, by hydrocooling, or by vacuum cooling. Hydrocooling is the most common cooling method; temperatures should be brought to as near 32 °F as possible. In practice, temperatures are reduced to 40 to 45 °F. Vacuum cooling is widely used for celery packed in corrugated cartons for long-distance shipment.

Bacterial Soft Rot: See Table 4, Note 1.

Black Heart: Brown or black discoloration of tips or all of the heart leaves. Affected celery should not be stored because of rapid development of bacterial soft rot. *Control:* Good cultural practices, with special attention to available calcium. Harvest promptly after celery is mature.

Early Blight (Cercospora): Circular pale yellow spots on leaflets. In advanced stages, spots coalesce and become brown to ashen gray. No spots develop in storage, but affected lots lose moisture and their fresh appearance. *Control:* Control early blight in the field by spraying or dusting.

Freezing Injury: Characteristic loosening of the epidermis can best be demonstrated by twisting the leaf stem. Severe freezing causes celery to become limp and to dry out rapidly. Freezing may cause watery soft rot and bacterial soft rot. The freezing point is about 31°F.

Late Blight (Septoria): Small (1/8 in. or less), yellowish, indefinite spots in the leaflet and elongated spots on the leaf stalk bearing black fruiting bodies of pinpoint size on the surface and surrounding green tissue. Development of blight in storage probably is negligible, but it opens the way for storage decays. *Control:* Control late blight in the field by sanitary measures and fungicides. Store infected lots for short periods only.

Mosaic: Leaflets are mottled; the stalk shows brownish sunken streaks. Badly affected stalks shrivel. *Control:* Eradicate weeds that carry the virus. Grade out all discolored stalks at packing time.

Watery Soft Rot (Pink Rot): This is the principle decay of celery, often severe on field-frozen stock and on celery harvested after prolonged cool, moist weather. In early stages, it often has a pink color. *Control:* Grade out all discolored stalks at packing time. Storage at 32°F will retard but not prevent the disease.

Corn, Sweet (32°F and 95 to 98% rh)

Sweet corn is highly perishable and is seldom stored except to protect an excess supply temporarily. Corn as it usually arrives on the market should not be expected to keep even in 32°F storage for more than 4 to 8 days.

The sugar content, which so largely determines quality in corn and decreases rapidly at ordinary temperatures, decreases less rapidly if the corn is kept at about 32°F. The loss of sugar is about 4 times as rapid at 50°F as it is at 32°F.

Sweet corn should be cooled promptly after harvest. Usually, corn is hydrocooled, but vacuum cooling is also satisfactory if the corn is prewet and top-iced after cooling. Where precooling facilities are not available, corn can be cooled with package and top ice. Sweet corn should not be handled in bulk unless it is copiously iced because of its tendency to heat throughout the pile. Sweet corn ears should be trimmed to remove most shank material before shipment to minimize moisture loss and prevent kernel denting. A loss of 2% moisture from sweet corn can result in objectionable kernel denting.

Cucumbers (50 to 55°F and 95% rh)

Cucumbers can be held only for short periods of 10 to 14 days at 50 to 55°F with a relative humidity of 95%. Cucumbers held at 45°F or below for longer periods develop surface pitting or dark-colored watery areas. These blemishes indicate chilling injury. Such areas soon become infected and decay rapidly on removal of the cucumbers to warmer temperatures. Slight chilling may develop in 2 days at 32°F and severe chilling injury within 6 days at 32°F. The susceptibility of cucumbers to chilling injury does not preclude their exposure to temperatures below 50°F for short intervals, as long as they are used immediately after removal from cold storage. Chilling symptoms develop rapidly only at higher temperatures. Thus, 2 days at 32°F or 4 days at 39°F are harmless under these conditions. Waxing is of some value in reducing weight loss and giving a brighter appearance. Shrink wrapping with polyethylene film can also prevent the loss of turgidity.

At temperatures of 50°F and above, cucumbers ripen rather rapidly, the green color changing to yellow. Ripening is accelerated if they are stored in the same room with ethylene-producing crops for more than a few hours.

Anthracnose (Colletotrichum): Circular, sunken, watersoaked spots that soon produce pink spore masses in the center. Later, the spots turn black. *Control:* Use disease-free seed and fungicidal applications in the field.

Bacterial Soft Rot: See Table 4, Note 1.

Bacterial Spot (Pseudomonas): Small, circular, watersoaked spots, later chalky or moist with gummy exudate. *Control:* If possible, avoid shipping infected cucumbers. Pack them dry and maintain temperatures as near optimum as practicable.

Black Rot (Mycosphaerella): Irregular, brownish, watersoaked spots of varying size, later nearly black. Black fruiting bodies are sometimes present. *Control:* Exclude infected cucumbers from the pack, if possible. Reduce carrying temperatures to about 50°F.

Chilling Injury: Numerous sunken, slightly watersoaked areas in the skin of cucumbers after removal from storage, found on cucumbers stored for longer than a week at temperatures below 45°F. *Control:* Store cucumbers at temperatures between 50 and 55°F, for not longer than 2 weeks.

Cottony Leak (Pythium): Large, greenish, watersoaked lesions. Luxuriant, white, cottony mold over wet decay. *Control:* Exclude infected cucumbers from the pack, if possible. Reduce carrying temperatures to about 50°F.

Freezing Injury: Large areas in cucumbers that are soft, flabby, watersoaked, and wrinkled, especially toward the stem end. *Control:* Prevent exposure to temperatures below 31°F.

Watery Soft Rot: See Table 4, Note 4.

Eggplants (46 to 54°F and 90 to 95% rh)

Eggplants are not adapted to long storage. They cannot be expected to keep satisfactorily even at the optimum temperature of 46 to 54°F for over a week and still retain good condition during retailing. Eggplants are subject to chilling injury at temperatures below 45°F. Surface scald or bronzing and pitting after sand scarring are symptoms of chilling injury. Eggplants that have been chilled are subject to decay by *Alternaria* when they are removed from storage. Exposure to ethylene for 2 or more days hastens deterioration.

Cottony Leak (Pythium): Decayed areas are large, bleached, discolored (tan), wrinkled, moist, and soft; later they exhibit abundant cottony mold. *Control:* Reduce carrying temperatures to about 46°F.

Fruit Rot (Phomopsis): Numerous, somewhat circular brown spots that later coalesce over much of the fruit with pycnidia dotting the older lesions. This is a very common decay of eggplant. *Control:* Use fungicide sprays in the field. Reduce carrying temperature to about 46°F. Move the fruit promptly if decay is evident.

Endive and Escarole (32°F and 95 to 100% rh)

Endive and escarole are leafy vegetables not adapted to long storage. Even at 32°F, which is considered the best storage temperature, they cannot be expected to keep satisfactorily for more than 2 or 3 weeks. They should keep somewhat longer if they are stored with cracked ice in or around the packages. Some desirable blanching usually occurs in endive held in storage.

Garlic, Dry (32°F and 65 to 70% rh)

Garlic should keep at 32°F for 6 to 7 months, if it is in good condition and is well cured when stored. Garlic cloves sprout most rapidly at 40 to 64°F; therefore, prolonged storage at this temperature should be avoided. In California, it is frequently put in common storage, where it can be held for 3 to 4 months or sometimes longer if the building can be kept cool, dry, and well ventilated.

Blue Mold Rot: (Penicillium): Soft, spongy, or powdery dry decay of cloves. Affected cloves finally break down completely into gray or tan powdery masses. *Control:* Prevent bruising; keep garlic dry.

Vegetables

Waxy Breakdown: Yellow or amber waxy translucent breakdown of the outside cloves. *Control:* No control measures have been developed.

Greens, Leafy (32 °F and 95 to 100% rh)

Leafy greens such as collards, chard, and beet and turnip greens are very perishable and should be held as close to 32 °F as possible. At this temperature, they can be held 10 to 14 days. They are commonly shipped with package and top-ice to maintain freshness and are handled like spinach. Kale packed with polyethylene crate liners should keep at least 3 weeks at 32 °F or 1 week at 40 °F. Vitamin content and quality are retained better when wilting is prevented.

Lettuce (32 °F and 95 to 100% rh)

Lettuce is highly perishable. To minimize deterioration, it requires a temperature as close to its freezing point as possible without actually freezing it. Lettuce will keep about twice as long at 32 °F as at 37 °F. If it is in good condition when stored, lettuce should keep 2 to 3 weeks at 32 °F with a high rh. Most lettuce is packed in cartons and vacuum-cooled to about 34 to 36 °F soon after harvest. It should then be immediately loaded into refrigerated cars or trailers for shipment or be placed in cold storage rooms for holding prior to shipping.

An increasing quantity of lettuce is shipped in modified atmospheres to aid quality retention. Modified atmospheres are a supplement to proper transit refrigeration, but are not a substitute for refrigeration. Lettuce is not tolerant of carbon dioxide and is injured by concentrations of 2 to 3% or higher. Romaine and leaf lettuce tolerate slightly higher carbon dioxide levels than head lettuce. Romaine is injured by 10% carbon dioxide, but not by 5% at 32 °F.

Since excess wrapper leaves are usually trimmed off before sale or use, it is suggested that lettuce be trimmed to 2 wrapper leaves before packaging, rather than the usual 5 or 6 to save space and weight. The extra wrapper leaves are not needed to maintain quality.

Bacterial Soft Rot: The most common cause of spoilage in transit and storage. Often, it starts on bruised leaves. This decay normally is the controlling factor in determining the storage life of lettuce and is much less serious at 32 °F than at higher temperatures (see Table 4, Note 1).

Brown Stain: Lesions that are typically tan, brown, or even black, and about ¼ in. wide and ½ in. long, with distinct margins that are darker than the slightly sunken centers. The margins give a halo effect. The lesions develop on head leaves just under the cap leaves. The heart and wrapper leaves are not affected. Brown stain is caused by carbon dioxide accumulation in railcars or trailers from normal product respiration. *Control:* Ventilate to keep carbon dioxide below 2% in transport vehicles by keeping one water drain open. Enclose bags of hydrated lime (in vehicles shipped under a modified atmosphere) to absorb carbon dioxide.

Gray Mold: See Table 4, Note 2.

Pink Rib: Characterized by diffuse pink discoloration near the bases of the midribs of the outer head leaves. In heads with severe symptoms, all but the youngest head leaves may be pink and discoloration may reach into large veins. The cause has not been identified, but shipment in low oxygen atmospheres at undesirably high temperatures (50 °F) can accentuate the disease. It is most common in hard to overmature lettuce. *Control:* Store and ship lettuce at recommended low temperatures.

Russett spotting: This occasionally causes serious losses. Small tan or rust-colored pitlike spots appear mostly on the midrib but possibly develop on other parts of leaves. Exposure to ethylene and to storage or transport temperatures above 37 °F are the main causes of this disorder. Hard lettuce is more susceptible to it than firm lettuce. *Control:* Avoid storing or shipping lettuce with apples, pears, or products that give off ethylene. Precool lettuce adequately to 34 to 37 °F and refrigerate it continuously. Shipment in a low oxygen atmosphere (1 to 8%) gives effective control.

Rusty Brown Discoloration: A serious market disorder of western head lettuce; a diffuse discoloration which tends to follow the veins but also spreads to adjacent tissue. The disorder starts on the outer head leaves but in severe cases may affect all leaves. *Control:* There is no known control method.

Tipburn: Dead, brown areas along the edges and tips of inner leaves. This is considered to be of field origin, but occasionally the severity of the disease increases after harvest. *Control:* Keep the affected stock well cooled and market it promptly after unloading to avoid secondary bacterial rots.

Watery Soft Rots: See Table 4, Note 4.

Melons

Cold storage is hardly used for most kinds of melons except to avoid temporarily adverse market conditions. To avoid injury by chilling, most melons are stored at 45 to 50 °F with 90 to 95% rh.

Persians should keep at this temperature range for up to 2 weeks; *honeydews* for 2 to 3 weeks; and *casabas* for 4 to 6 weeks. It is reported that these melons will be definitely injured in 8 days at temperatures as low as 32 °F. Honeydews are usually given an 18- to 24-h ethylene treatment (5000 ppm) to obtain uniform ripening. Pulp temperature should be 70 °F or above during treatment. Honeydews must be mature when harvested; immature melons fail to ripen even if treated with ethylene.

Cantaloupes harvested at the hard-ripe stage (less than full slip) can be held about 15 days at 36 to 39 °F. Lower temperatures may cause chilling injury. Full-slip hard-ripe cantaloupes can be held for a maximum of 10 to 14 days at 32 to 36 °F. They are more resistant to chilling injury. Cantaloupes are precooled by hydrocooling or forced air cooling before loading, or by top-icing after loading in railcars or trucks.

Watermelons are best stored at 50 to 60 °F and should keep from 2 to 3 weeks. Watermelons decay less at 32 °F than at 40 °F, but they tend to become pitted and have an objectionable flavor after 1 week at 32 °F. At low temperatures, they are subject to various symptoms of chilling injury—loss of flavor and fading of red color. Watermelons should be consumed within 2 to 3 weeks after harvest, primarily because of the gradual loss of crispness.

Alternaria Rot: Irregular, circular, brownish spots, sometimes with concentric rings, later covered with black mold. Often found on melons that have been chilled. *Control:* Avoid chilling temperatures. If cold melons are to be held at room temperature, they should be so stacked that condensed moisture will evaporate readily. Market melons promptly.

Anthracnose (*Colletotrichum*): Numerous greenish, elevated spots with yellow centers, later sunken and covered with moist pink spore masses. *Control:* Apply recommended field control measures.

Chilling Injury: Honeydew and honeyball melons stored for 2 weeks or longer at temperatures of 32 to 34 °F sometimes show large, irregular, water soaked, sticky areas in the rind. *Control:* Store melons at 45 to 50 °F.

Cladosporium Rot: Small black shallow spots later covered with velvety green mold. On cantaloupes, this rot is evident on extensive shallow areas at the stem ends or at points of contact between melons and it can be rubbed off easily. *Control:* Control measures are the same as for *Alternaria rot*.

Fusarium Rot: Brown areas on white melons; white or pink mold over indefinite spots on green melons. Affected tissue is spongy and soft, with white or pink mold. *Control:* Avoid mechanical injuries; reduce carrying temperatures to 45 °F.

Phytophthora Rot: Brown, slightly sunken areas; later water-soaked and covered with a wet, appressed, whitish mold. *Control:*

Cull out the affected fruits during packing. Reduce carrying temperatures to 45 °F.

Rhizopus Rot: The affected melon is soft, but not soupy and leaky as it is in similar decay on other vegetables. Coarse fungus strands may be demonstrated in decayed tissue (see Table 4, Note 3).

Stem End Rot (*Diplodia*): Brown, fairly firm decay usually starting at the stem end and affecting a large part of the watermelon. Black fruiting bodies develop later. *Control:* At the time of loading in cars, recut the stems and treat them with Bordeaux paste or another recommended fungicide.

Mushrooms (32 °F and 95% rh)

Mushrooms are usually processed or sold in a retail market within 24 to 48 h after they are harvested. They keep in good salable condition at 32 °F for 5 days, at 39 °F for 2 days, and at 50 °F or above for about 1 day. A rh of 95% is recommended during storage. While being transported or displayed for retail sale, mushrooms should be refrigerated. Deterioration is marked by brown discoloration of the surfaces, elongation of the stalks, and opening of the veils. Black stems and open veils are correlated with dehydration.

Controlled-atmosphere storage reportedly can prolong the shelf life of mushrooms held at 50 °F, if the oxygen concentration in the atmosphere is 9% or the carbon dioxide concentration is 25 to 50%.

Moisture-retentive film overwraps of caps usually are beneficial in reducing moisture loss.

Okra (45 to 50 °F and 90 to 95% rh)

Okra deteriorates rapidly and it is normally stored only briefly before marketing or processing. It has a very high respiration rate at warm temperatures. Okra in good condition can be kept satisfactorily for 7 to 10 days at 45 to 50 °F. A rh of 90 to 95% is desirable to prevent wilting. At temperatures below 45 °F, okra is subject to chilling injury, which is shown by surface discoloration, pitting, and decay. Holding okra for 3 days at 32 °F may cause pitting. Contact or top-ice causes water spotting in 2 or 3 days at all temperatures.

Fresh okra bruises easily; blackening of the damaged areas occurs within a few hours. A bleaching type of injury may also develop when okra is held in hampers for more than 24 h without refrigeration.

Onions (32 °F and 65 to 70% rh)

A comparatively low rh is essential in the successful storage of dry onions. However, humidities as high as 85% and forced air circulation have given satisfactory results. At higher humidities, at which most other vegetables keep best, onions are disposed to root growth and decay; at too high a temperature, sprouting is encouraged. Storage at 32 °F with 65 to 70% rh is recommended to keep them dormant.

Onions should be adequately cured either in the field, in open sheds, or by artificial means before or in storage. The most common method of curing in northern areas is by forced ventilation in storage. Onions are considered cured when the necks are tight and the outer scales are dried until they rustle. If not cured, onions are likely to decay in storage.

Onions are stored in 50-lb bags, in crates, in pallet boxes that hold about 1000 lb of loose onions, or in bulk bins. Bags of onions are frequently stored on pallets. Bagged onions should be stacked to allow proper air circulation.

In the northern onion-growing states, onions of the globe type are generally held in common storage because average winter temperatures are sufficiently low. They should not be held after early March unless they have been treated with maleic hydrazide in the field to reduce sprout growth.

Refrigerated storage is often used to hold onions for marketing late in the spring. Onions to be held in cold storage should be placed there immediately after curing. A temperature of 32 °F will keep onions dormant and reasonably free from decay, provided the onions are sound and well cured when stored. Sprout growth indicates too high a storage temperature, poorly cured bulbs, or immature bulbs. Root growth indicates the relative humidity is too high.

Globe onions can be held for 6 to 8 months at 32 °F. Mild or Bermuda types can usually be held at 32 °F for only 1 to 2 months. Onions of the Spanish type are often stored; if well matured, they can be held at 32 °F, at least until January or February. In California, onions of the sweet Spanish type are held at 32 °F until April or May.

Onions are damaged by freezing, which appears as watersoaking of the scales when cut after thawing. Onions that have been slightly frozen may recover with little perceptible injury, if allowed to thaw slowly and without handling. When onions are removed from storage in warm weather, they may sweat due to condensation of moisture. This may favor decay. Warming onions gradually (for example to 50 °F over 2 to 3 days) with good air movement should avoid the difficulty. Onions should not be stored with other products that tend to absorb odors.

Onion sets require practically the same temperature and humidity conditions as onions, but because they are smaller in size they tend to pack more solidly. They are handled in approximately 25-lb bags and should be stacked to allow the maximum air circulation.

Green onions (scallions) and green shallots are usually marketed promptly after harvest. They can be stored 3 to 4 weeks at 32 °F with 95% rh. Crushed ice spread over the onions will aid in supplying moisture. Packaging in polyethylene film will also aid in preventing moisture loss. Storage life of green onions at 32 °F can be extended to 8 weeks by packaging them in perforated polyethylene bags or in waxed cartons and holding them in a controlled atmosphere of 1% oxygen with 5% carbon dioxide.

Ammonia Gas Discoloration: Exposure of onions to 1% ammonia in air for 24 h causes the surface of yellow onions to turn brown, red onions to turn deep metallic black, and white onions to turn greenish yellow. *Control:* Ventilate storage rooms as soon as possible after exposure.

Bacterial Soft Rot: This decay often affects one or more scales in the interior of the bulb. Decayed tissue is more mushy than gray mold rot (see Table 4, Note 1).

Black Mold (*Aspergillus*): Black powdery spore masses on the outermost scale or between outer scales. *Control:* Store onions at just above 32 °F and at 65% rh.

Freezing Injury: A watersoaked, grayish yellow appearance of the entire outer fleshy scales results from a slight freezing injury. All scales are affected and become flabby with severe injury. Opaque areas appear in affected scales. *Control:* Prevent exposure to 30 °F temperatures and lower. Thaw frozen onions at 40 °F.

Fusarium Bulb Rot: Semiwatery to dry decay progressing up the scales from the base. Decay usually is covered with dense, low-lying white to pinkish mold. *Control:* Do not store badly affected lots. Pull out infected bulbs in slightly affected lots. Store onions at 32 °F.

Gray Mold Rot: This is the most common type of onion decay; it usually starts at the neck, affecting all scales equally. Decay often is pinkish (see Table 4, Note 2). *Control:* Cure onions thoroughly. Protect them from rain. Store them at just above 32 °F.

Smudge: Black blotches or aggregations of minute black or dark green dots on the outer drying scales of white onions. Under moist conditions, sunken yellow spots develop on fleshy scales. *Control:* Protect onions from rain after harvest. Store them just above 32 °F.

Vegetables

Translucent Scale: The outer 2 or 3 scales are gray and water-soaked, as in freezing. The entire scale may not be discolored; no opaque area is noticeable. Sometimes translucent scale is found in the field. *Control:* No control is known. Store onions at 32°F after curing.

Parsley (32°F and 95 to 100% rh)

Parsley should keep 1 to 2.5 months at 32°F and for a somewhat shorter period at 36 to 39°F. High humidity is essential to prevent desiccation. Package icing is often beneficial.

Bacteria Soft Rot: See Table 4, Note 1.
Watery Soft Rot: See Table 4, Note 4.

Parsnips (32°F and 98 to 100% rh)

Topped parsnips have similar storage requirements to topped carrots and should keep for 2 to 6 months at 32°F. Parsnips held at 32 to 34°F for 2 weeks after harvest attain a sweetness and high quality equal to that of roots subjected to frosts for 2 months in the field. Ventilated polyethylene box or basket liners can aid in preventing moisture loss. Parsnips are not injured by slight freezing while in storage, but they should be protected from hard freezing and should be handled with great care while in a frozen condition. The main storage problems with parsnips are decay, surface browning, and their tendency to shrivel. Refrigeration and high rh will retard deterioration.

Bacterial Soft Rot: See Table 4, Note 1.
Canker (Itersonilia sp.): Organism enters through fine rootlets and through injuries. The surface of the infected parsnip is first brown to reddish; later, it turns black where a depressed canker is formed. *Control:* Follow the recommended field spray program. Practice crop rotation.
Gray Mold Rot: See Table 4, Note 2.
Watery Soft Rot: See Table 4, Note 4.

Peas, Green (32°F and 95 to 98% rh)

Green peas lose part of their sugar rapidly if they are not refrigerated promptly after harvest. They should keep in salable condition 1 to 2 weeks at 32°F. Top icing is beneficial in maintaining freshness. Peas keep better unshelled than shelled.

Bacterial Soft Rot: See Table 4, Note 1.
Gray Mold Rot: See Table 4, Note 2.
Watery Soft Rot: See Table 4, Note 4.

Peas, Southern (40 to 41°F and 95% rh)

Freshly harvested southern peas at the mature-green stage should have a storage life of 6 to 8 days at 40 to 41°F with high rh. Without refrigeration, they remain edible for only about 2 days, the pods yellowing in 3 days and showing extensive decay in 4 to 6 days.

Peppers, Dry Chili or Hot

Chili peppers, after drying to a moisture content of 10 to 15%, are stored in nonrefrigerated warehouses for 6 to 9 months. The moisture content is usually low enough to prevent fungus growth. A relative humidity of 60 to 70% is desirable. Polyethylene-lined bags are recommended to prevent changes in moisture content. Manufacturers of chili pepper products hold part of their supply of the raw material in cold storage at 32 to 50°F, but they prefer to grind the peppers as soon as possible and store them in the manufactured form in airtight containers.

Peppers, Sweet (45 to 55°F and 90 to 95% rh)

Sweet peppers can be stored for a maximum of 2 to 3 weeks at 45 to 55°F. They are subjected to chilling injury if they are stored at temperatures below 45°F. The symptoms of this injury are surface pitting and discoloration near the calyx, which develops in a few hours after removal from storage. At temperatures of 32 to 36°F, peppers usually develop pitting in a few days. When stored at temperatures above 55°F, ripening (red color) and decay development are rapid. Rapid cooling of harvested sweet peppers is essential in reducing marketing losses. It can be done by forced-air cooling, hydrocooling, or vacuum cooling. Peppers are often waxed commercially, which reduces chafing in transit and moisture loss.

Bacterial Soft Rot: See Table 4, Note 1.
Freezing Injury: The outer wall is soft, flabby, watersoaked, and dark green in color. The core and seeds turn brown with severe freezing. Sweet peppers freeze at about 31°F.
Gray Mold Rot: See Table 4, Note 2.
Rhizopus Rot: See Table 4, Note 3.

Potatoes (Temperature, see following; 90 to 95% rh)

The proper storage environment for potatoes might be defined as that environment which will promote the most rapid healing of bruises and cuts, reduce rot penetration to a minimum, allow the least weight and other storage losses to occur, and reduce to a minimum the deleterious quality changes that might occur during storage.

Early-crop potatoes are usually not stored except during congested periods. They are more perishable and cannot be expected to keep as well or as long as late-crop tubers. Refrigerated storage at 40°F following a curing period of 4 or 5 days at 70°F is recommended; or they can be stored for about 2 months at 50°F without curing. If early-crop potatoes are to be used for chipping or French frying, storage at 70°F is recommended. Holding these potatoes in cold storage even at moderate temperatures of 50 to 55°F for only a few days causes excessive accumulation of reducing sugars, which results in production of dark-colored chips.

Late-crop potatoes produced in the northern half of the United States are usually stored. The greater part of the crop is held in nonrefrigerated commercial and farm storages, but some potatoes are held in refrigerated storages. Potatoes in nonrefrigerated storages are usually held in bulk bins 8 to 20 ft deep. Shallower bins are used in milder climates. Some potatoes are stored in pallet boxes. In refrigerated warehouses, potatoes can be stored in sacks, pallet boxes, or bulk.

Late-crop potatoes should be cured immediately after harvest by being held at 50 to 61°F and high relative humidity for about 10 to 14 days to permit suberization and wound periderm formation (healing of cuts and bruises). If properly cured, they should keep in sound dormant condition at 38 to 40°F with 95% rh for 5 to 8 months. A temperature below this is not desirable, except for seed stock for late planting. For this purpose, 37°F is best. At 37°F or below, Irish potatoes tend to become sweet. For ordinary table use, potatoes from 39°F are satisfactory, but they probably will be unsatisfactory for chipping or French frying without being desugared or conditioned at about 64 to 70°F for 1 to 3 weeks prior to use. However, conditioning may be costly, and good results are often uncertain.

Potatoes will remain dormant at 50°F for 2 to 4 months; and since tubers from this temperature are more desirable for both table use and processing than those from 40°F, late-crop potatoes intended for use within 4 months should be stored at 50°F and those for later use at 40°F. All potatoes should be stored in the dark to prevent greening.

A storage temperature of 50 to 55°F is recommended for most cultivars of potatoes intended for chip manufacture. At these temperatures, they usually remain in satisfactory condition if their reducing-sugar content is low enough when they are initially stored. Storage at 61 to 64°F is less desirable because shrinkage, internal and external sprout growth, and decay are greater at these temperatures than at 50 to 55°F. Russet Burbank potatoes for table stock or for chipping are stored at 45°F with 95% rh.

Potatoes usually do not sprout until 2 to 3 months after harvest, even at 50 to 61°F. However, after 2 to 3 months of storage, sprouting can be expected in potatoes stored at temperatures above 39°F and particularly at temperatures around 61°F. Although limited sprouting does not affect potatoes for food purposes, badly sprouted stock shrivels and is difficult to handle and market.

Certain growth-regulating chemicals have been approved by the U.S. Food and Drug Administration to control or reduce sprouting on potatoes. Potatoes treated with chemical sprout inhibitors should not be stored in the same warehouse with seed potatoes. Least shrinkage and best quality potatoes result if 90% or slightly higher rh is maintained. Cunningham *et al.* (1971) recommend 95% or higher rh for late crop potatoes. Condensation on the ceiling and resultant moisture drip is sometimes a problem when very high humidity is maintained.

Ventilation or air circulation in potato storage is needed to provide and maintain optimum temperature and rh throughout storage and the tubers it contains. In northern states, where average outdoor temperatures during storage are low, little circulation or ventilation is needed. Shell or perimeter circulation is extensively used in these areas for seed and table stock potatoes. Forced circulation through the potatoes is required for the higher temperature storage of processing potatoes and for table and seed stock in the warmer parts of the late-crop area. Rapid air circulation may result in lowering the rh of the air immediately surrounding the potatoes; it is conducive to drying and weight loss, which may be desirable if there are disease problems but undesirable with sound potatoes because of increased shrinkage. For late-crop Idaho potatoes, a uniform airflow, which does not have to be continuous, of 45 cfm/lb is recommended. With this ventilation, Russet Burbank potatoes stored at 45°F with 95% rh should keep in good condition for 10 months or longer.

Potatoes should not be kept in the same room with fruits, nuts, eggs, or dairy products because of the objectionable flavor they may impart. Also, potatoes may absorb odors from cheese or from volatile chemicals.

Bacterial Ring Rot: Yellow, soft, cheesy decay of the thin layer of tissue in the vascular ring. The outer 0.25 in. of tuber and the inner part may appear normal. *Control*: Use disease-free seed; store promptly at 40°F.

Bacterial Soft Rot (see Table 4, Note 1): This disease probably causes more loss in the early and intermediate crops than do all other potato diseases combined.

Freezing Injury: If frozen solidly, tubers become soft and cream-colored and exude moisture. Slightly frozen tubers show darkening of the vascular ring and dull gray to black areas in the flesh. Potatoes freeze at about 29 to 31°F.

Fusarium Rot: Brown to black, spongy, and fairly dry; white or pink mold inside cavities in stored potatoes. *Control:* Avoid cutting and bruising during harvesting. After proper curing, maintain well-ventilated storage at 40°F.

Late Blight (*Phytophthora*): Reddish brown to black granular discoloration of the outer 0.12 to 0.25 in. of tuber. The affected tissue is firm to rock hard. *Control:* Apply recommended fungicides in the field. Kill vines prior to harvesting tubers or keep tubers away from blighted vines at harvest. Keep them dry; store at 40°F; market promptly.

Leak (*Pythium*): Large gray to black, moist, decayed area starting at bruises or the stem end of the tuber. The internal tissue is granular and cream-colored at first, turning through reddish brown to inky black. *Control:* Prevent bruising. Refrigerate tubers to 40°F and keep them dry.

Mahogany Browning: Reddish brown patches or blotches in the flesh of tubers. Chippewa and Katahdin varieties are most susceptible. This differs from flesh discoloration caused by freezing in being reddish brown instead of gray. *Control:* Store at 40°F or above because lower temperatures cause the discoloration.

Net Necrosis: Dark brown vascular ring and vascular netting of the flesh, most prominent at the stem end, but extends well toward the bud end; increases during storage. *Control:* Reduce storage temperature promptly to 40°F; the infected tubers show symptoms earlier at higher temperatures.

Scald and Surface Discoloration: On early potatoes, this appears as sunken, injured areas; later, it turns black and sticky and is followed by bacterial rots. *Control:* Move potatoes promptly to market; cool to 40°F.

Southern Bacterial Wilt: Moist, sticky exudation from the vascular ring when the tuber is cut. Sometimes there is advanced, mushy decay in the center of the tuber. *Control:* Avoid shipping infected tubers; market promptly.

Stem End Browning: Dark brown to black vascular tissue, in streaks, extending from 0.4 to 1 in. into the flesh from the stem end; develops during storage. *Control:* After curing, reduce the storage temperature promptly to 40°F. Higher temperatures allow rapid development in susceptible lots.

Tuber Rot (*Alternaria*): Black to purplish, slightly sunken, shallow, irregularly shaped lesions, 0.25 to 1 in. in diameter, developing during storage. *Control:* Apply recommended fungicides in the field. Keep the tubers away from blighted vines as much as possible at harvest time. If the tubers are damp, inspect them frequently during storage and use forced air ventilation to dry up excess moisture.

Pumpkins and Squash

Hard shell winter squash, such as the Hubbards, can be successfully stored for 6 months or longer at 50 to 55°F with a relative humidity of 60 to 75%. Dry storage is needed for quality retention. All specimens should be well-matured, carefully handled, and free from injury or decay when stored. Hubbard and other dark-green skinned squashes should not be stored near apples, as the ethylene from apples may cause the skin to turn orange yellow. Most varieties of *pumpkins* do not keep in storage for as long as the usual storage varieties of squash. Such varieties as Connecticut Field and Cushaw do not keep well and cannot be kept in good condition for more than 2 to 3 months. *Acorn squash* can be stored satisfactorily for 5 to 8 weeks at 50°F. *Butternut squash* should keep at least 2 to 3 months at 50°F with 50% rh.

Summer squash, such as yellow crookneck and giant straightneck, are harvested at an immature stage for best quality. The skin is tender, and these varieties are easily wounded and perishable. They should be refrigerated to about 39°F and moved rapidly to market. They can be held for a few days at 32 to 39°F and a relative humidity of about 90%; if they are held longer than 4 or 5 days, chilling injury causes deterioration. The storage temperature range for summer squash is 41 to 50°F with 95% rh. A temperature of 41°F is best for zucchini squash stored up to 2 weeks.

Black Rot (*Mycosphaerella*): Hard, dry, black decay, dotted with minute black pimplelike fruiting bodies, that occurs at stem ends or sides of the fruit. *Control:* Avoid skin breaks on the fruit; handle promptly.

Dry Rots (*Alternaria; Cladosporium; Fusarium*): Small, deep, dry, decayed areas. The decayed portion is easily lifted out of the surrounding healthy tissue. The surface mold is low-growing, and either greenish black or pinkish white in color. *Control:* Prevent skin breaks. Do not store hard shell squashes below 50°F.

Rhizopus Rot: See Table 4, Note 3.

Radishes (32°F and 95 to 100% rh)

Topped spring radishes after harvest should be precooled quickly, often by hydrocooling, to 41°F or below. If they are then packaged in polyethylene bags, radishes can usually be held 3 to 4 weeks at 32°F and for a somewhat shorter time at 40°F.

Vegetables

Bunched radishes with tops are more perishable. They can be stored at 32°F and a rh of 95 to 100% for 1 to 2 weeks.

Rhubarb (32°F and 95% rh)

Fresh rhubarb stalks wilt and decay rapidly. Rhubarb in good condition can be stored 2 to 4 weeks at 32°F with a 95% rh or above. Moisture loss during holding or storage can be minimized by using nonsealed polyethylene crate liners or by film wrapping of consumer size bunches. Removing and discarding leaf blades at harvest is desirable because it not only reduces the possibility of decay and weight loss but also reduces shipping weight and package size by one-third. Rhubarb is usually marketed with about 3/8 in. of the leaf blade attached to the petiole. Splitting of the petiole will be more serious if the entire leaf is removed.

Fresh rhubarb cut into 1-in. pieces and packaged in 1-lb perforated polyethylene bags can be held 2 to 3 weeks at 32°F with high relative humidity. Splitting of cut ends and curling of these pieces in film bags may be a problem if marketing is at warm temperatures.

Gray Mold Rot (*Botrytis*): Grayish, smoke-colored growths and grayish brown spore masses on stalks. *Control:* Refrigerate to 32°F.

Phytophthora Rots (*Phytophthora*): Watery, greenish brown, sunken lesions starting at the base of the leafstalk, causing brown decay. *Control:* Decay is retarded with transit and storage temperatures below 40°F.

Rutabagas (32°F and 98 to 100% rh)

Rutabagas lose moisture and shrivel readily if they are not stored under high humidity conditions. A hot paraffin wax coating, often given to rutabagas, is effective in preventing wilting and loss of weight and also improves appearance slightly. Too heavy a wax coating may produce severe injury from internal breakdown caused by suboxidation. Rutabagas in good condition, when stored, should keep 4 to 6 months at 32°F.

Freezing Injury: Rare, because the commodity can stand slight freezing without injury. Severe freezing causes watersoaking and light browning of the flesh, a mustard odor, and fermentation. *Control:* Prevent repeated slight freezing or severe freezing.

Gray Mold Rot: See Table 4, Note 2.

Spinach (32°F and 95 to 98% rh)

Spinach is very perishable, and can be stored for only short periods of 10 to 14 days at 32°F, with 95 to 98% rh. It will deteriorate rapidly at higher temperatures. Spinach is commercially vacuum-cooled and forced air-cooled. If it is thoroughly cooled, it can be held for 10 to 14 days at 32°F, without the addition of any package ice prior to storage. When precooling facilities are not available, crushed ice should be placed in each package to provide rapid cooling and to take care of the heat of respiration. Top-ice is also beneficial.

Bacterial Soft Rot: See Table 4, Note 1.

Downy Mildew (*Peronospora*): A field disease, commonly found at the marketing stage as pale yellow irregular areas in the leaves. Downy gray mold is present on the lower surface. *Control:* Control it in the field; market promptly.

White Rust (*Albugo*): Slight yellowing of areas in the leaf above white blisterlike pustules filled with white masses of spores. *Control:* Control it in the field.

Sweet Potatoes (55 to 60°F, 85 to 90% rh)

Most sweet potatoes are stored in nonrefrigerated commercial or farm storages. Preliminary curing at 84°F and 90 to 95% rh for 4 to 7 days is essential for the healing of injuries received in harvesting and handling and in preventing the entrance of decay organisms. After curing, the temperature should be reduced to 55 to 61°F, usually by ventilating the storage, and the relative humidity retained at 85 to 90%. Most varieties will keep satisfactorily for 4 to 7 months under these conditions. Weight loss of 2 to 6% can be expected during curing and about 2% a month during subsequent storage.

Usually, sweet potatoes will not keep satisfactorily if they have been subjected to excessively wet soil conditions just before harvest or chilled before or after harvest by exposure to temperatures of 50°F or below. Short periods at temperatures as low as 50°F need not cause alarm; but after a few days at lower temperatures sweet potatoes may develop discoloration of the flesh, internal breakdown, increased susceptibility to decay, and off-flavors when cooked.

Temperatures above 61°F stimulate development of sprouting (especially at high humidities), pithiness, and internal cork (a virus disease). Refrigeration is frequently used in large sweet potato storages to extend the marketing season into warm weather when ventilation will not maintain low enough temperatures.

Sweet potatoes are usually stored in slatted crates or bushel baskets. Palletization of crates and use of pallet boxes facilitates handling. Sweet potatoes are usually washed and graded and are sometimes waxed before being shipped to market. They may be treated with a fungicide to reduce decay during marketing.

Black Rot (*Ceratocystis*): Greenish black decay, frequently fairly shallow, and sometimes circular in outline at the surface. *Control:* Follow recommended field and postharvest control measures. Heat treatment of seed roots at 106 to 109°F for 24 h will prevent development of black rot.

Chilling Injury: Brown tinged with black discolored areas scattered or associated with vasculars. The interior becomes pithy. Chilling injury is often produced by exposure to lower temperatures for only a few days. Uncured roots are more sensitive than cured ones. *Control:* Store sweet potatoes at 55 to 60°F.

Freezing Injury: Soft, leaky condition of the flesh. The outer layer of the potato is dark brown. *Control:* Do not subject potatoes to low temperatures; sweet potatoes may freeze at 30°F.

Rhizopus Rot: See Table 4, Note 3. *Control:* Cure potatoes for 4 to 7 days at 84°F before storage. Follow recommended field and postharvest control measures.

Tomatoes (Mature Green, 55 to 70°F; Ripe, 45 to 50°F; 90 to 95% rh)

Mature green tomatoes cannot be successfully stored at temperatures that greatly delay ripening, even at a temperature of 55°F, which is considered to be a nonchilling temperature. Tomatoes held for 2 weeks or longer at 55°F may develop an abnormal amount of decay and fail to reach as intense a red color as tomatoes ripened promptly at 64 to 70°F. Temperatures of 64 to 68°F, and a relative humidity of 90 to 95%, are probably used most extensively in commercial ripening of mature green tomatoes. At temperatures above 70°F, decay is generally increased. A temperature range of 57 to 61°F is probably the most desirable for slowing ripening without increasing decay problems. At this temperature, the more mature fruit will ripen enough to be packaged for retailing in 7 to 14 days. Tomatoes should be kept out of cold, wet rooms because, in addition to potential chilling injury, extended refrigeration damages the ability of fruit to develop desirable fresh tomato flavor.

Storage temperatures below 50°F are especially harmful to mature green tomatoes; these chilling temperatures make the fruit susceptible to *Alternaria* decay during subsequent ripening. Increased decay during ripening occurs following 6 days' exposure to 32°F, or 9 days at 39°F (see Figure 1).

Firm ripe tomatoes may be held at 45 to 50°F with a relative humidity of 85 to 90% overnight or over a holiday or weekend.

Tomatoes showing 50 to 75% of the surface colored (the usual ripeness when packed for retailing) cannot be successfully stored for more than 1 week and be expected to have a normal shelf life during retailing. Such fruits should also be held at 45 to 50 °F and 85 to 90% rh. A storage temperature of 50 to 55 °F is recommended for pink-red to firm red tomatoes raised in greenhouses.

When it is necessary to hold firm ripe tomatoes for the longest possible time, consistent with immediate consumption on removal from storage, such as on board a ship or for an overseas military base, they can be held at 32 to 36 °F for up to 3 weeks, with some loss in quality. Mature green, turning, or pink tomatoes should be ripened before storage at this low temperature.

Ethylene gas is sometimes used to hasten and give more even ripening to mature green tomatoes. In ripening rooms, a concentration of one part ethylene per 5000 parts of air daily for 2 to 4 days will usually shorten the ripening period by about 2 days at 64 to 68 °F. Some tomatoes are gassed with ethylene in loaded railcars prior to shipping. Adding ethylene has little or no effect on tomatoes just before or after they have started to turn pink. Tomatoes themselves give off considerable ethylene as they ripen. Interest is increasing in the commercial use of low oxygen atmospheres of 3 to 5% during storage or transport to retard ripening and decay.

Alternaria Rot: Decayed area is brown to black, with or without a definite margin. Lesions are firm; rot extends into the flesh of the fruit. Dense, velvety, olive green or black spore masses frequently grow over affected surfaces. *Control:* Avoid mechanical injuries at packing time. Avoid temperatures below 50 °F in green fruit.

Bacterial Soft Rot: See Table 4, Note 1.

Cladosporium Rot: Thin, brownish blemishes or black shiny spots of shallow decay, later covered by green, velvety mold. *Control:* Take care in harvesting and packing. Ship high quality tomatoes free of field chilling injury under protective services that will provide temperatures of 55 to 68 °F.

Late Blight Rot (*Phytophthora*): Greenish brown to brown, roughened areas with a rusty tan margin. *Control:* Apply recommended field control measures. Cull tomatoes carefully before packing.

Phoma Rot: Slightly sunken, moderately penetrating, black areas at the edge of the stem scar and elsewhere on the fruit. Black pimplelike fruiting bodies develop later. Decayed tissues are firm and brown to black in color. Phoma rot is found in eastern-grown tomatoes. *Control:* Apply field control measures. Exercise care in harvesting and packing. Avoid temperatures below 55 °F.

Rhizopus Rot: See Table 4, Note 3.

Soil Rot (*Rhizoctonia*): Small circular brown spots, frequently with concentric ring markings; later, large, brown, and fairly firm lesions. In advanced stages, under warm conditions, cream-colored or brown mycelium and irregular sclerotia may develop. *Control:* Before packing, sort out tomatoes with early lesions if the disease is prevalent.

Turnips (32 °F and 95% rh)

Turnips in good condition can be expected to keep 4 to 5 months at 32 °F with 90 to 95% rh. At higher temperatures (41 °F and above), decay will develop much more rapidly than at 32 °F. Injured or bruised turnips should not be stored. Store turnips in slatted crates or bins and allow good circulation around containers.

Dipping turnips in hot melted paraffin wax gives them a glossy appearance and is of some value in reducing moisture loss during handling. However, waxing is primarily to aid in marketing and is not recommended prior to long-term storage.

Turnip greens are usually stored for only short periods (10 to 14 days). They should keep about as well as spinach at 32 °F with crushed ice in the packages.

REFERENCES

Agricultural Statistics. 1988. U.S. Department of Agriculture, Washington, D.C.

Appleman, C.O. and J.M. Arthur. 1919. Carbohydrate metabolism in green sweet corn. *Journal of Agricultural Research* 17:137.

Ashby, H.B., R.T. Hinsch, L.A. Risse, W.G. Kindya, W.L. Craig, Jr., and M.T. Turczyn. 1987. Protecting perishable foods during transport by truck. USDA *Agricultural Handbook* No. 669.

Bogardus, R.K., and J.M. Lutz. 1961. Maintaining the fresh quality in produce in wholesale warehouses. *Agricultural Marketing* 6(12):8.

Burton, W.G. 1958. Suppression of potato sprouting in buildings. *Agriculture* 65:299.

Cunningham, H.H., M.V. Zaehringer, and W.C. Sparks. 1971. Storage temperature for maintenance of internal quality in Idaho Russet Burbank potatoes. *American Potato Journal* 48:320.

Dewey, D.H., R.C. Herner, and D.R. Dilley. 1969. Controlled atmospheres for the storage and transport of horticultural crops. *Horticultural Report* No. 9, Michigan State University (July).

Hardenberg, R.E. 1971. Effect of in-package environment on keeping quality of fruits and vegetables. *HortScience* 6(3):198.

Hardenberg, R.E., H. Findlen, and H.W. Hruschka. 1959. Waxing potatoes—Its effect on weight loss, shrivelling, decay, and appearance. *American Potato Journal* 36:434.

Hardenburg, R.E., A.E. Watada, and C.Y. Wang. 1986. The commercial storage of fruits, vegetables, and florist and nursery stocks. USDA *Agriculture Handbook* No. 66.

Harvey, J.M. 1965. Nitrogen—Its strategic role in produce freshness. *Produce Marketing* 8(7):17.

Isenberg, F.M. and R.M. Sayles. 1969. Modified atmosphere storage of Danish cabbage. *Journal of the American Society for Horticultural Science* 94(4):447.

Kader, A.A. et al. 1984. Irradiation of plant products. Comments from CAST 1984-1. Council of Agricultural Science and Technology, Ames, IA. ISSN 0194-4096.

Lipton, W.J. 1965. Post-harvest responses of asparagus spears to high carbon dioxide and low oxygen atmospheres. *Proceedings of the American Society for Horticultural Science* 86:347.

Lipton, W.J. 1968. Effect of temperature on asparagus quality. *Proceedings*, Conference on Transportation of Perishables, Davis, CA, 147.

Lutz, J.M. 1936. The influences of rate of thawing on freezing injury of apples, potatoes and onions. *Proceedings of the American Society for Horticultural Science* 33:227.

McColloch, L.P. and J.T. Worthington. Ways to prevent chilling mature green tomatoes. *PrePack-Age* 7(6):22.

Morris, L.L. and H. Platenius. 1938. Low temperature injury to certain vegetables after harvest. *Proceedings of the American Society for Horticultural Science* 36:609.

Parsons, C.S. and R.E. Anderson. 1970. Progress on controlled-atmosphere storage of tomatoes, peaches and nectarines. *United Fresh Fruit and Vegetables Association Yearbook*, 175.

Parsons, C.S. and R.H. Day. 1970. Freezing injury of root crops—Beets, carrots, parsnips, radishes, and turnips. USDA *Marketing Research Report* No. 866.

Redit, W.H. 1969. Protection of rail shipments of fruits and vegetables. USDA *Handbook* No. 195.

Stewart, J.K. and M.J. Ceponis. 1968. Effects of transit temperatures and modified atmospheres on market quality of lettuce shipped in nitrogen-refrigerated and mechanically refrigerated trailers. USDA *Marketing Research Report* No. 832 (December).

Stewart, J.K., M.J. Ceponis, and L. Beraha. 1970. Modified atmosphere effects on the market quality of lettuce shipped by rail. USDA *Marketing Research Report* No. 863.

Watada, A.E. and L.L. Morris. 1966. Effect of chilling and nonchilling temperatures on snap bean fruits. Proceedings of the American Society for Horticultural Science 89:368.

Whiteman, T.M. 1957. Freezing points of fruits, vegetables and florist stocks. USDA *Marketing Research Report* No. 196 (December).

CHAPTER 19

FRUIT JUICE CONCENTRATES

Processing and Quality Control 19.1
Concentration Methods 19.2
Citrus Juices 19.4
Pineapple and Apple Juices 19.4
Grape and Berry Juices 19.5

MOST frozen juice concentrates are of the four-fold (or three-plus-one) type in which three volumes of water are added to one volume of concentrate for reconstitution. The resulting product is approximately 42° Brix (soluble solids content) in concentration, and on dilution with three volumes of water gives a product of nearly 12° Brix.

Institutional packs in larger containers are frequently of higher concentration. The use of higher concentration reduces the tendency for a concentrate to clarify or form a gel if abnormally high temperatures are encountered during distribution, but there is not a corresponding improvement in flavor stability.

Storage at 0°F or below is essential to retain the delicate flavor, aroma, and naturally-occuring and added vitamins (C and others) over extended periods. Storage tests for up to five years have indicated excellent retention of flavor at 0°F. At 5°F flavor remains stable for a few months, but at 10°F and above, increased rates of deterioration in flavor, cloud loss, and gelation can be expected. Storage at 0°F is essential for orange concentrate and is recommended strongly for all concentrates to preserve fresh flavor.

PROCESSING AND QUALITY CONTROL

Because citrus concentrates predominate, the various steps of processing this fruit will be discussed. Deviations from these methods will be indicated when other individual fruit concentrates are discussed.

Selection, Grading, and Handling of Fresh Fruit

Fruit is selected for proper quality and maturity. Some fresh market fruit that is blemished in appearance but sound in quality is juiced. A major portion of the crop moves directly from the grove to the processing plant. To be mature, the fruit must have the proper Brix-acid ratio, and the juice content and Brix must be above specified values. Fruit should be handled without delay. Samples are taken mechanically as fruit enters the bins, and records of chemical analyses are maintained. Usually, fruit from two or more bins is used simultaneously to improve uniformity. The fruit passes over inspection tables, both before and after temporary storage in bins, and damaged or deteriorated fruit is removed.

Washing and Juice Extraction

Prior to juice extraction, the fruit is wetted by sprays—the wetting agent is dispensed and falls onto the fruit as it travels the brushes. The fruit is rinsed by water sprays near the end of the washer unit. Solutions of chlorine or quaternary NH_3 compounds are used to sanitize the fruit skins, conveyors, and elevators.

Citrus juices are extracted in high-speed mechanical juice extractors. Single machines may handle from 5 to 12 pieces of fruit per second. In some machines, the fruit is halved; then the juice is reamed or squeezed from the half. In other machines, a tube is inserted through the middle of the fruit and pressure is applied which squeezes the juice through fine holes into the tube, sieving away the seeds and large pieces of membrane. After the juice has been extracted, it passes to finishers which remove the seeds, pieces of peel, and excess cell or fruit membrane.

Usually, most of the pulp is separated from the juice by one or two stages of screw-type finishers. Pulp washing, a process in which soluble solids in separated segment and cell walls are recovered by countercurrent extraction with water, is no longer permitted in Florida in the manufacture of frozen concentrated orange juice. Maximum yield of juice or concentrate is controlled by regulation. The juice from the finishers is usually treated in high-speed desludging centrifuges, which remove suspended matter before the juice is transferred to the evaporator. This operation decreases the viscosity of the juice in the evaporator, boosts the efficiency of evaporation, and improves the appearance of the final product.

Heat Treatment and Flavor Fortification

Concentrate, if prepared from good fruit, will remain stable for a long time at 0°F and for nearly a year at 5°F. However, with large-scale production, storage cannot always be kept below 5°F. Concentrates originally of good quality show tendencies to gel or clarify rapidly during storage. Heat treatment, however, will inactivate enzymes responsible for development of these defects during improper storage.

Formerly, frozen concentrated orange juice was overconcentrated and fresh juice added to reduce the concentration to the desired level. This process provided fresh flavor in the final product and was used extensively in frozen concentrates. However, flavor levels could not be standardized by this method alone. Essential oil from orange peel alone does not supply completely balanced fresh flavor, but is used extensively to control flavor intensity. Some peel oil is found in the cut-back juice, but little remains in juice from the evaporator. Peel oil is added to the finished concentrate at the rate of approximately 0.014% by volume in the reconstituted juice. No emulsifier is needed to disperse the oil, as it readily mixes with the concentrate with ordinary stirring. Several variations have been introduced to supplement the peel oil and cut-back fresh juice for flavor fortification. Often the condensate from the first stage of concentration, using evaporator distillers, helps produce an essence which is restored to the concentrate to enhance the flavor.

Packing, Storage, and Distribution

Concentrate from the evaporators is collected in cold-wall tanks equipped with slowly moving agitators. Fresh cut-back juice may be added automatically under the control of an inline electronic refractometer set to yield the desired final Brix value. Cold-pressed peel oil also is added to the desired level and the product cooled to about 35°F. From the tank, the concentrate usually is pumped through swept-surface heat exchangers, where the temperature is

---

The preparation of this chapter is assigned to TC 11.6, Prepared Food Products. This chapter last received a major revision in 1967.

further reduced to about 25 °F, and then passed to the filling and can closing machines.

The concentrate is packed in hermetically sealed cans; 12 oz is the most common size. Some 6-oz, 8-oz, and 16-oz cans are used in consumer packs; larger cans, 32 oz to 1 gal, are favored by the institutional and restaurant trade.

The filled cans pass through freezing tunnels where the temperature is reduced to 0 °F or below. Freezers may be of the cold air-blast type or refrigerated alcohol spray. In the air-blast freezers, air at about −40 °F is forced down through the transport belt. From 45 to 90 min. is allowed for freezing. In the alcohol spray type, the cans are transported on mesh belts or by a walking beam system. The alcohol is maintained at −25 to −40 °F, and from 30 to 45 min. is allowed for freezing.

In calculating cooling requirements, concentrates may be considered as having the specific heat of a sucrose solution of the same Brix value; that is, 0.68 Btu/(lb·°F) for 59.8° Brix concentrate and 0.74 Btu/(lb·°F) for 42° Brix concentrate. Morgan (1951) found the thermal conductivity of a 58.9° Brix orange concentrate to be approximately 0.17 Btu·ft/(h·ft^2·°F), and the conductivity for 42° Brix concentrate was about 0.18 Btu·ft/(h·ft^2·°F). These values apply only to the liquid state, as the heat transfer coefficient increases when ice crystals begin to form. Ice begins to form at about 18 °F in 42° Brix concentrate and at about 9 °F in 58.9° Brix concentrate.

Temperature must be maintained at 0 °F or below during storage, transportation, and distribution. Frequently, temperature recorders or throwaway maximum temperature indicators are placed in the shipment to see that proper conditions are maintained.

Quality Control

While the most important factor in quality control is the use of good sound fruit, the quality must be checked throughout processing and again in the final product. Concentrates are checked for Brix value, Brix-acid ratio, peel-oil content, and other factors. Other tests may be run to see that requirements of a particular brand are met. At periodic intervals, bacteriological samples are taken at various stages in the plant, plated on orange serum agar, and examined after incubation.

When the weather is warm, stringent and frequent cleanups are required. Normally, evaporators are cleaned at least every 24 h, but, where arrangements are favorable, as long as 7 days may elapse between cleanups. Total plate counts in final product are generally maintained below one million organisms per millilitre of reconstituted juice. Counts of a few thousand per millilitre are common. Sanitation is based on maintaining cleanliness (asepsis) rather than cleaning after microorganisms have started to grow (antisepsis). The natural acidity and high sugar content of citrus juice concentrate normally inhibit rapid growth of organisms. Generally, counts tend to decrease in storage. Even at elevated temperatures, conditions are favorable for growth, mainly for yeast and a few *Lactobacilli*.

CONCENTRATION METHODS

The major methods for the production of concentrates are: (1) freezing and mechanical separation; (2) low-temperature vacuum evaporation; and (3) high-speed, high-temperature evaporation.

Freezing and Mechanical Separation

Removal of 144 Btu is required to freeze out a pound of water while about 1000 Btu are required to remove the same amount of water by evaporation. However, this advantage is diminished because refrigeration cycles have a heat efficiency of only about 20% and multi-effect evaporation increases the efficiency of this method of removing water. The outstanding advantages of freeze concentration are that no losses of volatile flavor substances with water vapors occur as in evaporation, and low temperatures eliminate the danger of heat damage. The freeze concentration process does, however, entrain juice solids in the ice. When freezing is very slow, as in natural freezing, this is not much of a problem, but when juice is frozen in continuous slush freezers and subsequently centrifuged, a 5% loss of soluble solids is common. Loss of suspended solids during freeze concentration has been more troublesome than loss of soluble solids. The freeze-concentrated juice is often pale in color because of the loss of carotenoid pigments and weak in flavor because of a loss in suspended essential oils. One solution is to melt the ice and pass the solution into a conventional evaporator. The freeze-concentrated juice is then used to flavor the concentrate from the evaporator. In the process, freeze concentration becomes an extra step and increases cost. Variations in the freeze-concentration process have sometimes been used in the citrus processing industry.

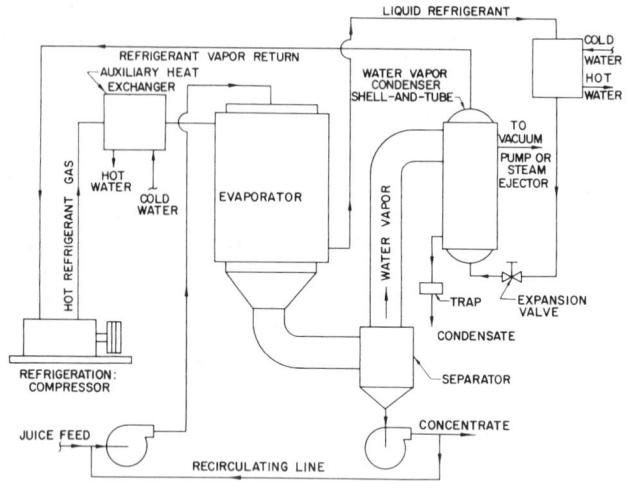

Fig. 1 System of Concentration Using Heat Pump Principle (Direct Refrigerant Contact)

Direct Refrigerant Contact. Figure 1 is a schematic of a concentrator in which hot refrigerant gas supplies the heat for the evaporation of juice. The water vapor is condensed by the evaporating liquid refrigerant in a shell-and-tube condenser, and the water vapor heats the refrigerant. The refrigerant is returned to the suction of the compressor unit and the cycle is repeated. The evaporator, for simplicity, is shown as a single-effect design with means for recirculating the juice. Variations may employ multiple effects or parallel single effects. Vapor and concentrate are separated, usually by an arrangement of cyclones and baffles. The concentrate is pumped out of the system and fed to the blending, chilling, canning, and freezing processes. Pressure in the system can be maintained by a mechanical vacuum pump.

Advantages of this system are:

1. Very little cooling water or steam is required.
2. Equipment associated with cooling water and steam is minimized, saving space and expense.
3. Because the refrigerant vapor directly heats the juice and cools the vapor, a minimum temperature differential from refrigerant to juice can be employed. This requires minimum energy input to the compressor.

Fruit Juice Concentrates

Disadvantages of this system are:

1. Unitary refrigeration cannot be used. The refrigeration systems, controls, and instrumentation must be specifically designed for the particular installation.
2. Refrigerant holdup is relatively high.
3. Evaporating equipment must be constructed to withstand refrigerant pressures. Any leak in the evaporators can cause contamination of the juice with refrigerant.
4. Refrigerant lines are relatively long.

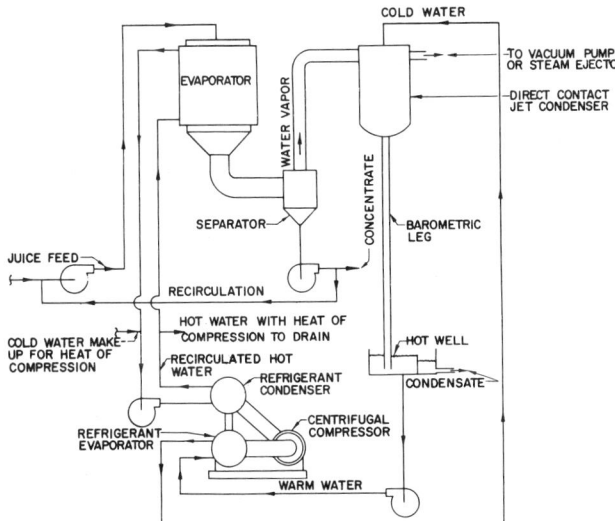

Fig. 2 System of Concentration Using Heat Pump Principle (Indirect Refrigerant Contact)

Indirect Refrigerant Contact. Figure 2 shows a schematic of a concentrator, which, though similar to the one shown in Figure 1, differs in the following ways: (1) the juice is heated by circulated hot water, which in turn is heated by the hot refrigerant gases (high side); and (2) the vapor from the juice is condensed by direct contact with cold water in a barometric condenser, the water having been cooled by the boiling refrigerant (low side). The juice cycle is the same as in Figure 1. The pressure is maintained by a relatively small steam ejector or mechanical vacuum pump. The essential difference is in interposing water as a heat-transfer medium between the hot refrigerant and the juice, and between the vapor and the cold refrigerant.

Advantages of this system are:

1. Centrally located standard unitary refrigeration equipment can be used. Controls and instrumentation can be standard.
2. Refrigerant lines are short and holdup is low.
3. Evaporators need not be designed to hold refrigerant.
4. There is no danger of refrigerant contaminating the juice.
5. Very little steam or water is required.

Disadvantages of this system are:

1. Because water is used as an intermediate heat transfer medium, the hot refrigerant gas must be hotter and the cold refrigerant must be colder than shown in Figure 1. The suction-to-discharge temperature spread on the refrigeration unit is greater and, therefore, the energy input to the compressor per unit mass of water evaporated is greater than for Figure 1.
2. Equipment for handling water streams, which is not required in Figure 1, is required for this process.

Vapor Recompression. Figure 3 shows a schematic for a concentrator in which the vapor from the juice, instead of being condensed, is compressed to a higher temperature and pressure and used to evaporate more juice. The vapor is compressed by a booster steam ejector.

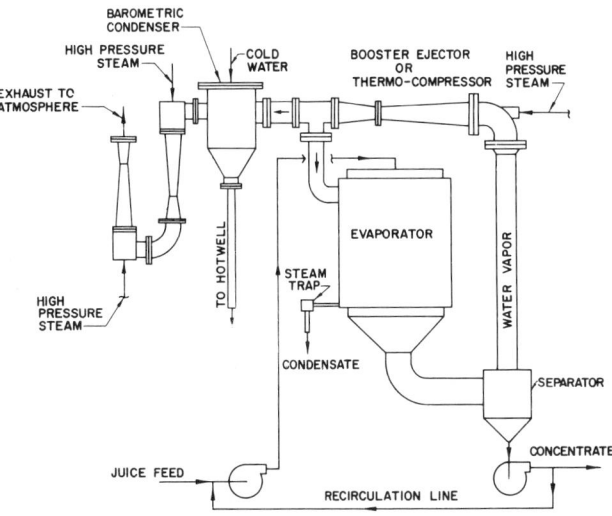

Fig. 3 System of Concentration Using Heat Pump Principle (Direct Vapor Recompression)

The evaporator is heated with steam, and the condensate is pumped out to drain. The juice cycle is similar to that shown in Figures 1 and 2, except that the vapor is recompressed and eventually leaves the system as condensate. The work of compression is supplied by the high-pressure steam fed to the booster ejector.

Advantages of this sytem are:

1. The capital investment for steam ejectors is low compared to that for refrigeration systems.
2. No low-temperature condenser is required.
3. No intermediate water circulation system is required.

Disadvantages of this system are:

1. More steam is required than for either Figure 1 or 2, and its use is relatively inefficient.
2. More cooling water is needed than for either Figure 1 or 2.

High-Temperature, Short-Time Evaporators

The plate evaporator is one type of high-speed, high-temperature evaporator. In large, plate-type heat exchangers, the juice is kept in rapid motion, and heat exchange rates are maintained at comparatively high levels. A typical unit has seven stages plus a vacuum flash cooler arranged so there are four effects. Temperatures in the various stages range from 105 to 205 °F. The units are small for their capacity and lack the vapor-liquid separation chambers of the low-temperature evaporators. Only about 0.30 lb of steam are required to evaporate a pound of water in these units.

Another type of evaporator is an electronic unit which generates frequencies in the 30 MHz range to heat the juice. The product must be kept under pressure at the electrode to prevent bubble formation and localized scorching. Usually product from conventional evaporators feeds the unit. The advantage is in being able to concentrate to 72° Brix and higher. To conserve energy, vapors from evaporators are compressed by a steam booster and used to heat swept-surface heat exchangers. In an example of this type of unit, the concentrate enters the swept-surface heat exchanger at a temperature of 52 °F and leaves at 58 °F. In the electrode chamber, the temperature is further increased to 62 °F. The concentrate then passes through an orifice and is flashed down to the original temperature of 52°F, and is recycled. This process produces a high Brix concentrate of lower viscosity than conventional evaporators.

CITRUS JUICES

Each variety of citrus powder and frozen concentrate has its own processing requirements.

Pure Fruit Juice Powders

Vacuum-dried orange juice powder has been manufactured by introducing concentrate of about 58°Brix into a vacuum chamber where it is dried on a moving stainless steel belt. The dried powder is flavored with a locked-in orange oil prepared by dispersing orange oil in a mixture of molten sugars, extruding, and cooling rapidly. Thus, the oil does not contact the powdered orange juice until water is added to reconstitute the juice.

In the foam-mat drying process a small amount of foam stabilizer is added (0.57 of dry solids content) and the chilled concentrate of about 50°Brix is beaten into a foam, which is laid out in a sheet about 0.12-in. thick on perforated trays. An air blast clears the foam from the holes, and the trays are conveyed, in a stack, up a column where hot air passes up through the perforated trays and reduces the moisture to about 1.25% in 12 min. The product is then chilled to harden and is scraped from the trays. In a variation of the process, a thin layer of the foam is placed on a polished stainless steel belt, dried in a stream of hot air, and finally stripped from the belt in dried form with a doctor blade.

Orange Juice

Most of the orange juice concentrate is made from late-season (Valencia) or mid-season (Pineapple) varieties of oranges. Some early season fruit (Hamlin varieties) may be used, but they are generally pale in color and weak in flavor and must be blended with other concentrates. Concentrate of about 58°Brix is commonly placed in 55-gal barrels with polyethylene liners and called *add-back* to differentiate it from fresh cut-back juice added to restore flavor. Add-back concentrate is thawed and blended with essence and evaporator feed juice to produce the desired concentration and properties for marketing.

Grapefruit Juice

In the production of frozen concentrated grapefruit juice, essentially the same equipment is used (fruit washers, finishers, falling-film evaporators) as in the production of frozen concentrated orange juice. Some adjustments at the extractors are necessary to accommodate grapefruit.

Both sweetened and unsweetened frozen concentrated grapefruit juices are prepared, the sweetened product in greater quantities. The unsweetened product may vary from 38 to 42°Brix. The sweetened product must contain at least 3.47 lb of soluble grapefruit solids per gallon exclusive of added sweetening ingredients. The final Brix may vary from 38 to 48. In Grade-A unsweetened concentrate, the Brix-acid ratio may vary from 9-to-1 to 14-to-1, and in the sweetened product, from 10-to-1 to 13-to-1. While either seedless or seeded varieties can be used, the seeded varieties such as Duncan are generally preferred.

Blended Grapefruit and Orange Juice

USDA *Grade Standards* recommend not less than 50% orange juice in the mixture and as much as 75% orange juice when it is light in color. USDA *Grade Standards* call for 40 to 44°Brix in unsweetened concentrates. In sweetened concentrates, the Brix must be at least 38 before sweetening and 40 to 48 after sweetening. For Grade A the Brix-acid ratios in the packed concentrate may vary from 10-to-1 to 16-to-1, if unsweetened, and from 11-to-1 to 13-to-1 if sweetened.

Tangerine Juice

Differences in the nature of the tangerine require modification in the methods of handling during picking, hauling, and storage at the plant. While grapefruit and oranges withstand considerable rough handling, the tangerine is somewhat flat, irregular in shape, and has a loose, tender skin which is easily broken. If the skin is broken and the fruit bruised, bacteria and yeasts readily attack the fruit, and undesirable enzyme actions occur. For these reasons, they cannot be handled in orange bins but must be handled in boxes or loose in trucks to a depth of not over 2 ft. The Dancy tangerine is the most common variety.

The processes and equipment used in manufacturing the concentrated tangerine juice are practically the same as with oranges. Since the the fruit is smaller, the yield of juice from a given number of extractors is smaller, and about double the amount of extracting equipment is required to furnish juice to keep the evaporators operating at full capacity.

Generally, the values for Brix-acid ratio, peel oil content, and concentration follow those prescribed for the orange product quite closely. A Brix of 44 is common for a three-plus-one concentrate.

PINEAPPLE AND APPLE JUICES

Pineapple Juice

Pineapple juice is prepared from small fruit and the parts of larger pineapples that are not suitable for packing as pieces. The main sources of juice are the cores, the layer of flesh between the shell and cylinder that is cut for the preparation of pineapple slices, and juice that drains from the crushed pineapple. The juice material amounts to about one-third of the weight of the fresh fruit. The juice is extracted by passing through disintegrators and screw presses. It is then centrifuged to remove heavy foreign material and excessive insoluble solids.

Pineapple concentrate is produced from single-strength juice in equipment similar to that used for the production of orange and other fruit juice concentrates. The first step in the concentrating operation is to strip out the volatile flavoring materials. These are separated as about a 100-fold product and added back to the final concentrate. The concentration takes place in multiple-effect evaporators, with the final effect at about 70°F.

Pineapple concentrate is produced either as a 3-to-1 product with a Brix of about 46.5, or as a 4.5-to-1 product with a Brix of about 61. The 3-to-1 concentrate is produced in both a sterile and a frozen form; however, even the sterile product is stored and sold under refrigeration to preserve quality. The 4.5-to-1 concentrate is also produced in both a sterile and frozen form. The concentrate must be stored at 40°F or less if it is to be held for any appreciable length of time.

Bulk pineapple concentrate is principally mixed with citrus concentrate in the production of frozen juice blends. Pineapple concentrate also is used as an ingredient in many types of canned fruit drinks. A popular combination has been a pineapple-grapefruit drink which has reached a volume of production exceeding that of single-strength canned pineapple juice.

Pineapple juice composition varies greatly; Brix varies between 12 and 18, with an average of about 13.5 to 14. Acidity ranges between 0.6 to 1.1, with an average of about 0.8. Brix-acid ratio ranges from 12 to over 20 and usually averages between 16 and 17. Because pineapple concentrate is produced at a standard Brix, variation in composition shows up only in the acidity and Brix-acid ratio.

Apple Juice

Frozen apple juice concentrate is prepared by the conventional vacuum concentration process similar to that used for frozen orange juice concentrate. The volatile components of apple fla-

Fruit Juice Concentrates

vor are recovered in the evaporation process and added to the final concentrate. These procedures take advantage of the fact that most of the volatile flavors are found in the first 10% of the distillate. This portion of the distillate is passed through a fractionating column to obtain the volatile flavors in concentrated form, usually about 150-fold as compared to the fresh juice. The remaining 90% of the original juice is then concentrated under vacuum to somewhat more than the concentration desired for the final product.

For example, 100 gal of apple juice, prepared in the conventional manner, may yield 0.7 gal of 150-fold essence and 24 gal of concentrated juice. When these two fractions are combined, full flavored (four-fold) concentrate is obtained. If greater concentration is desired, the stripped juice is concentrated to a greater extent. (In such instances, however, the juice pectin must be removed to avoid excessive viscosity and gelation of the highly concentrated liquid.) Juice manufactured today is approximately 98% depectinized and 2% non-depectinized. Four-fold apple juice concentrate reconstituted with three volumes of water is not perceptibly different from the original. The keeping quality of frozen apple juice concentrate (depectinized) is essentially unchanged after three years of storage at 0°F.

GRAPE AND BERRY JUICES

Grape Juice

Most of the concentrated grape juice marketed today is prepared from Concord grapes (*Vitis labrusca*). The grapes are harvested when the soluble solids have reached a concentration of 15 to 16% or higher, which varies with maturity and is influenced by cultural and climatic factors. Freezing and frozen storage temperatures are similar to those used for citrus fruit concentrates.

After washing, the grapes are conveyed to the stemmer, which consists of a perforated, slowly revolving (20 rpm) horizontal drum. Inside, several beaters revolve at a much faster speed (200 rpm), knock the berries off the cluster, and partially crush them before they are discharged through the drum perforations. The cluster stems are expelled from the open end of the drum. The crushed fruit is then pumped through a tubular heat exchanger where it is heated to 140 to 145°F for good extraction of the pigments and juice. The hot pulp then may go to hydraulic presses where the juice is removed in the same manner as apple juice. The expressed juice may be clarified in a centrifuge or filter press. If a filter press is used, from 1 to 2% of diatomaceous earth is used to maintain a high filtering rate and provide ample removal of suspended matter. Under normal operating conditions, 190 to 195 gal of juice are obtained from 1 ton of grapes. In some plants screw presses are used for all or part of the crushed grapes, but this increases the suspended matter that must be removed later.

The clarified juice is then pasteurized in tubular or plate-type heat exchangers to a temperature of 180 to 190°F and cooled immediately to 30°F before storage in tanks in refrigerated rooms maintained at 28°F. The cooling of the juice usually is accomplished in two or more steps. In some heat exchange systems, a regeneration cycle is used in the first step whereby the hot juice leaving the pasteurizer preheats entering juice. On occasion, the cooling water discharged from the heat exchangers is piped to the washers to heat the water applied to the incoming grapes.

The method of handling the cooled juice depends on the intended use of the concentrate. Where it is to be used in the later manufacture of jelly, the juice is stored at 28°F for 1 to 6 months to permit settling of the argols, which consist of potassium acid tartrate, tannins, and some colored materials that would give a gritty texture to the jelly or detract from its clarity. In this case, the clear juice is siphoned off the precipitate in the storage tanks and may be refiltered. Where the concentrate is to be sold as a blend formed by mixing sugar and ascorbic acid before canning and freezing, the cold storage tank merely serves as a surge tank, and the juice is pumped to the concentrator within a few hours.

A polishing filter before the concentrator minimizes fouling of the evaporator tubes. The concentrate for either jelly manufacture or blended juice may be stored in tanks at 27°F prior to subsequent processing. Whenever single-strength juice is bulk stored at 27°F prior to concentration, danger exists from spoilage by fermentation. To minimize yeast growth during storage, all pipelines and equipment from the pasteurizer to the cold room should be of a sanitary design for ready and frequent cleaning. The interior surfaces of storage tanks must be relatively smooth or free from crevices, and the tank should be thoroughly cleaned before use.

The juice is concentrated in two steps. In the first, the volatile flavoring materials are stripped from the juice, and the stripped juice is then concentrated to the desired density. The volatile components are removed by heating the single-strength juice to 220 to 230°F for a number of seconds in a heat exchanger, flashing a percentage of the liquid into vapor in a jacketed tube bundle, and then discharging the liquid and vapor through an orifice tangentially into a separator. The separator should be of sufficient size that the vapor velocity is reduced to 10 fps or less for minimal entrainment. From 20 to 30% by weight of the original juice flashes off as a vapor that is led into the base of a fractionating column filled with ceramic saddles or rings. A reflux condenser on the vapor line from the column and a reboiler section at the base of the column provide the necessary reflux ratio. The vent gases from the reflux condenser are then chilled in a heat exchanger, and the condensate containing the essence is collected at a rate equivalent to 1/150 of the volume of entering flavoring material.

An important flavor component of Concord grape juice is methyl anthranilate, which has a boiling point of 512°F and is only slightly soluble in water. This high boiling point and low solubility in water has resulted in losses of methyl anthranilate when the efficiency of the stripping column is low. These losses may be reduced by increasing the vaporizing temperature.

In a typical formulation, grape juice is concentrated to a little over 34° Brix, and essence of fresh cut-back juice is added to reduce it to this concentration. Sucrose is added to 47° Brix and citric acid, until the total acidity is 1.8% calculated as tartaric acid. When diluted with an equal quantity of water, the equivalent of sweetened single-strength juice is obtained; however, for a more palatable beverage, three parts of water are added. The product is labeled as concentrated, sweetened grape juice. The concentrate may be cooled to 20 to 30°F in a heat exchanger or cold-wall tank, filled into cans, sealed, cased, and allowed to freeze in storage below 0°F.

Strawberry and Other Berry Juices

Frozen strawberry juice, as a 7-fold concentrate with separately packed concentrated (100-fold) essence, is used for manufacturing, primarily for jellies. The concentrate provides a more economical way of marketing high quality strawberry juice solids. Concentrates of red raspberry, black raspberry, and blackberry juices are also available in limited quantities.

The process for preparation of strawberry and other berry juice concentrates involves essence recovery in which 12 to 20% of the juice is separated by a stripping process using a steam injection heater. Vapors containing volatile flavors are concentrated to the desired degree in a fractionating column. The juice remaining from the essence recovery step is concentrated under vacuum 3- to 7-fold by volume. A maximum temperature of 100°F for 2.5 h should not be exceeded for strawberry, whereas temperatures up to 130°F may be used in preparing boysenberry concentrate in a batch-type operation.

Preparation of juice for concentration involves chopping or coarse milling of cold, sound berries and mixing with pectic enzymes and filter-aid. After 4 to 5 h at room temperature, juice is expressed with a bag press or rack and cloth press. The cloudy juice is clarified in a filter press. Recovered essences have been concentrated and packaged separately so that the jelly manufacturer can incorporate the essence in the jelly just before filling. This procedure reduces greatly the amount of essence lost by volatilization. The essence also can be incorporated in the concentrate for making a full-flavored product and for shipping as single unit.

Concentrated juice, without essence, can be packed in an enamel-lined container which need only be liquidtight. Concentrated essence should be kept in a carefully sealed can to avoid losses of the highly volatile flavor. Both juice concentrate and essence are kept frozen for proper quality retention.

REFERENCES

Morgan, D. A. 1951. Thermal conductivity in orange concentrate. *Proceedings* of the Florida State Horticultural Society 64:192.

CHAPTER 20

PRECOOKED AND PREPARED FOODS

Physical and Chemical Changes.. 20.1
Types of Precooked Frozen Foods... 20.2
Uniform Storage Temperature... 20.4
Food Service Applications.. 20.5
Standards.. 20.5

PRECOOKED and prepared frozen foods contain dissolved solids such as sugars, salts, fats, proteins, and emulsifying ingredients. They do not begin to freeze until slightly below 32 °F and may not be frozen until 15 °F is reached. Since quality changes resulting from bacteriological, enzymatic, oxidative, and other chemical reactions can adversely affect flavor, aroma, texture, nutritive value, and shelf life, rapid freezing and low-temperature storage are used in most commercial operations.

Precooked and prepared foods, based on the degree of change expected during the freezing, storage, or reheating of the product, are classified as: (1) those frozen, stored, and thawed without marked change, such as applesauce, clear soups, winter squash, baked breads, rolls, cookies and most cakes, and various pies; (2) those altered by freezing, storage, and reheating, but which with certain changes in production or formula can be suitably modified for freezing, including most creamed products, sauces, and gravies; (3) those of excellent initial quality but which deteriorate rapidly at ordinary storage temperatures and should be held at −20 °F or lower, such as fatty fish, shellfish, and many poultry dishes; (4) those so changed by freezing and reheating as to be impractical to freeze, such as custards, cooked egg whites, and salad-type vegetables; and (5) those partially or completely cooked, which, if thawed and reconstituted properly, cannot be differentiated from the unfrozen product, such as seared or cooked steaks and chops.

PHYSICAL AND CHEMICAL CHANGES

Changes occurring when cooked foods are frozen are most often the result of both physical and chemical action. The curdling of custards is partly caused by crystallization of water as ice and partly by a continuing denaturation of egg proteins. Gravies and thickened sauces could curdle for much the same reasons, unless special stabilizers are used to eliminate the effect of fluctuating storage temperatures. The use of proper amylopectin flours or starches and/or gums eliminates curdling. The ordinary starch granule is composed of two fractions, amylose and amylopectin. On defrosting or reheating, the amylose fraction agglomerates, forming tiny clumps which give a curdled appearance. The amylopectin fraction forms long chains which, when reconstituted, hold water without coagulation or clumping; thus, the percentage of amylopectin present determines whether any curdling will take place. Many such amylopectin-modified starches can go through several freeze-thaw cycles without giving the appearance of curdling. Hanson *et al.* (1951) found waxy rice flour to be superior to waxy corn or sorghum flours. When sauces and gravies made from waxy corn or waxy rice flours are compared only in the reheated stage, each yields an equally acceptable product. Frozen sauces and gravies improve stability of frozen precooked foods, possibly by displacing air. Also, sauces surrounding other items in packaged products may have a buffering action.

Changes in some high protein foods can cause coagulation and toughening. Raw egg white is not markedly affected by freezing and thawing, but freezing the cooked product too fast or too slowly makes it tough and rubbery. The water in the elastic gel of the cooked egg white (denatured protein) migrates to increase the size of crystals where nuclei are present. The crystals grow and penetrate the gel, which separates the structure, releasing a part of the elastic tension. The gel contracts due to the migration of the water from within the structure, plus the force exerted by the growth of ice crystals and the release of elastic tension by mechanical cleavage. This contraction is largely irreversible as demonstrated by the liquid-filled spaces remaining after thawing. The remaining structure is tougher, since it contains a considerably higher proportion of protein than the original gel.

Chemically modifying the product by adding starches or precisely controlling the rate of heat extraction during freezing in the case of natural product eliminates the toughness. Here the eggs are separated into white and yolk and later recombined during the cooking process into tubular packages approximately 8 in. long, each portion of which contains a circle of cooked egg yolk surrounded by the proportionate amount of cooked egg white. For further information on eggs, see Chapter 24.

This same denaturation of protein may play a part in the loss of gas from batters during freezing. This loss may also be caused by the separation of ice and the concentration of carbon dioxide in the liquid phase to such a degree that it will not stay in solution. The reaction of the carbonate of soda and acid in the more concentrated aqueous solution, produced by the separation of water as nearly pure ice, also plays a part in the escape of the leavening gas during freezing and thawing. Special leavening agents minimize the loss through control of timing of gas release until after the freezing and reconstituting has begun. All of these reactions, except the one involving baking powder ingredients, cause some loss of gas in yeast doughs. The yeast cells gradually lose their viability over prolonged storage periods. The longer dough is held, the longer it takes to rise after thawing and warming.

Chemical changes of oxidation that cause rancidity may occur during the freezing and storage of some fatty cooked foods. Rancidity in creamed turkey has been detected as soon as it was prepared and, although this level is probably not of commercial significance, the rancidity in precooked products can increase in storage. It has been shown that adding edible antioxidants during the cooking of fatty foods rather than just before packaging increases the effectiveness of the antioxidant in retarding oxidative rancidity.

Cooked meat fat, especially pork, does not turn rancid more quickly than raw fat. The pH has no effect in rancidification of precooked ground pork, and, except at the upper limits of the normal pH range of fresh pork, the precooked product kept better. The cooking process inactivates peroxidizing enzymes, so rancidity usually develops more slowly in the precooked product. The rate of all chemical reactions that cause precooked products to deteriorate is reduced by lowering storage temperatures.

The preparation of this chapter is assigned to TC 11.6, Prepared Food Products.

Assuming a storage life of 12 months at 0 °F, roughly equivalent storage lives at other temperatures are (Tressler and Evers 1950, RRF 1989):

−20 °F	36 months
−10 °F	24 months
0 °F	12 months
10 °F	6 weeks
20 °F	1 week
30 °F	1 day

Factors other than temperature may influence the rate of deterioration. The type of packaging is of great importance; improper packaging may cause freezer burn. Most precooked foods (fish sticks, poultry, and precooked meals) have sufficient airspace in the package so that fluctuating temperatures tend to dehydrate the product, causing quality loss at a rate usually greater than that for the mean temperature. At lower temperatures, the deterioration rate of most precooked frozen foods is about the same at a given uniformly maintained temperature and at the same mean temperature under fluctuating conditions if the maximum temperature is not high enough to partially thaw the product. Widely fluctuating temperatures cause crystal size to grow during storage: this is objectionable in ice cream, sherbets, ices, sauces, gravies, and other emulsions and colloids which may break or separate.

Fruit shortcakes, ice cream toppings, ice creams, and sherbets and ices partially thaw at 15 °F. Products of a high soluble solids content must not be allowed to warm to this temperature.

Defrost temperatures of most commonly prepared frozen foods are well below 30 °F. The relatively high sugar concentrations in fruit pie fillings tend to depress their freezing points more than the soluble proteins and starch in meat products. The type of product also influences the defrosting rate.

Bacteriological and food spoilage must also be considered (see Chapter 10). Precooked foods must be chilled immediately after cooking, frozen rapidly, then stored at well below freezing temperatures (no higher than 0 °F). Particular attention must be paid to the time it takes to pass the critical temperature zone, 104 to 68 °F, in which bacterial growth is rapid. Reheating immediately prior to consumption is necessary for a safe, quality product.

TYPES OF PRECOOKED FROZEN FOODS

Foods Served Hot

Soups and chowders. Tomato and some other soups can be heat processed without marked change in flavor or consistency. Others, although unfavorably changed by canning, are not altered by freezing if they are in concentrated form.

The use of a can instead of a bag-in-carton package for frozen products not only solves the problem of hot-filling and gives a product a lower bacterial content, but it greatly reduces the labor required for packaging through the use of automatic can-filling and closing machines. Another advantage in the hot-filling of cans is an airtight container. The partial vacuum obtained after cooling results in less oxidation and less loss or change of flavor during cooling, freezing, and storage. Methods vary from plant to plant. Three currently popular freezing methods are air-blast, immersion, and cryogenic freezing.

Meat dishes. The popular demand is for pot pies, stews, and roast meat packed with gravy, either alone or as part of a complete frozen meal. In addition, many frozen products contain meat, even though it is not considered a basic ingredient. Whole roasts cooked rare can be vacuum packed and frozen for later defrosting and reheating, if necessary.

Meat stews (and other precooked frozen foods), if allowed to thaw and stand without refrigeration, serve as excellent substrates for the growth of microorganisms. Extreme care in maintaining sanitary conditions in all of the manufacturing operations, and immediate freezing and continued storage at 0 °F or below, are essential.

Antioxidants do not contribute to the keeping quality of stews. Storage time and temperature are the most important factors in the retention of quality in Swiss steaks. Over 9 months' storage, product desirability gradually decreases, although meat stored at −20 °F always scores higher. For further information on frozen meat products, see Chapter 12.

Poultry dishes. Among the principal items are chicken meat for pot pies, deep-fried chicken parts, breasts, legs, thighs, and wings, and cooked turkey meat. *Salmonella, Staphylococci,* and *Streptococci* will be killed during roasting if the temperature at the center of the stuffing reaches 165 °F.

Chicken meat loses flavor and juiciness and becomes dry in texture as a result of biochemical changes taking place in frozen storage. These changes are associated with protein denaturation and proteolysis, which are governed by temperature and time of storage. However, through careful selection of packaging materials or by immersion of the cooked meat in a sauce or gravy, the loss of juiciness and adverse reactions can be greatly controlled. For further information on frozen poultry products, see Chapter 13.

Fish and shellfish. Most of the convenience items consist of breaded fish portions and fish sticks, raw or precooked, which are produced from frozen fish blocks. These products are sold to retail or institutional markets in packages ranging from 10 oz to 5 or 10 lb in capacity or are incorporated into fish dinners. Fish portions have been widely accepted in a fish sandwich sold by several large drive-in restaurant chains.

Frozen-prepared fish products include raw breaded shrimp and scallops, individually quick-frozen (IQF) breaded fish fillets, battered fish products such as fish puffs and fish fries, fish pies, precooked soft-shell clams, deviled crab, clam and fish chowders, and many other products. For further information on frozen fish and shellfish, see Chapter 14.

Vegetables. Some vegetables sold as regular frozen vegetables are really cooked products. Corn on the cob, winter squash, and pumpkin all have been steamed (blanched) enough in preparation for freezing to cook them.

Many precooked vegetables are frozen commercially, either as part of a frozen meal or a prepared dish such as chicken pie, yet the quantities precooked and frozen are small except for potatoes.

Usually vegetables should be slightly undercooked (blanched) so that during cooling, freezing, and reheating, they will not become mushy or lose color and flavor. Undercooking is important in all vegetables except cucumbers, tomatoes, and so forth. The product should be rapidly cooled with a minimum of agitation so the hot product will not continue to cook. Rapid cooling under sanitary conditions prevents multiplication of microorganisms. The vegetable should be packaged, before or after freezing, in airtight, moistureproof, hermetically sealed cartons or other containers, impermeable to oxygen. If small metal packages are used, hot-filling, followed by immediate rapid cooling, is desirable. The packaged product should be rapidly frozen in an air-blast freezer or some other type of quick-freezer. A low storage temperature (0 °F or lower) is essential in the case of loose-frozen products or those not protected with a cream sauce or gravy and not solidly packed.

The use of flexible, boilable, transparent packages (boil-in-bag or boil-in-pouch) containing vegetables in butter sauce, cream, cheese, and various other sauces sealed in vacuum to avoid puffing or swelling, has resulted in uniform heating, retention of flavor, and elimination of errors at point of use. Bulk freezing during the harvest season and subsequent freezer storage, with packaging throughout the year, results in high quality products. Institutional and individual packs allow a variety of ultimate uses. Reconstitution is accomplished in water and by microwave (if a pinpoint

Precooked and Prepared Foods

Table 1 Defrosting and Heating of Frozen Foods

Item	Weight, oz	Packaging	Microwave Heating, s at 800 W	Conventional Heating, s
Green peas	10	Plastic pouch	180 In pouch	840 From second boil
Whole kernel corn	10	Plastic pouch	180 In pouch	840 From second boil
Fordhook lima beans	10	Plastic pouch	420 In covered Pyrex casserole	960 From second boil
Potatoes *au gratin*	11.5	Aluminum dish	180 In covered Pyrex dish with stirring, plus 300 under broiler	3000 In oven at 325 °F
Seafood and/or chicken croquettes	10	4-in. aluminum dish, sauce in plastic pouch	150 For croquettes on serving dish 60 For sauce in pouch	1500-1800 In oven at 350 °F 180 for sauce in pouch in boiling water
Chow mein	16	Polyethylene laminated carton	450 In covered Pyrex with occasional stirring	900-1200 In sauce pan with stirring
Macaroni and cheese	12	Aluminum dish	240 In covered Pyrex dish 240 with stirring, plus 300 under broiler	1800 In oven at 400 °F
Raspberries, strawberries, Mandarin oranges	10	Waxed board with metal ends	180 In original container	1200-1800 In warm water 9000 At room temperature
Swordfish steak	12	Board carton with waxed overwrap	180 In Pyrex pie plate, plus 300-480 under broiler	3600 At room temperature to thaw
Salisbury steak dinner	11	Aluminum tray with foil cover	300-360 In aluminum tray without foil cover	1800-2100 In oven at 375 °F in original container

puncture is made prior to heating to avoid ballooning). For further information on frozen vegetables, see Chapter 18.

Complete meals. Precooked frozen dinners represent the largest sales segment of precooked product in the retail market. The type of heating method used to reconstitute (*i.e.*, reheat) determines to a large extent the selection, formularization, and placement of meal components. Frozen dinners packed for the retail consumer trade are designed for heating in conventional and microwave ovens. The greatest problem in producing meals on a platter is to develop a system of preparing, cooling, freezing, packaging, and reheating the products so that meals will taste freshly cooked and not warmed-over. Each operation must be so timed and coordinated that each product progresses steadily through the plant with a minimum of delay anywhere along the line.

Each tray holding a meal is covered with lightweight aluminum foil. Laminated paper and plastic trays with plastic lids are used for both microwave and conventional ovens. The tray is often slipped into a shallow carton. The packages may then be automatically overwrapped and placed on shelves on a wheeled rack or conveyer for easy movement into a −40 °F air-blast freezer. In some operations, the foil-covered trays are first frozen on shelves on a wheeled rack in a −40 °F air blast. When frozen, the trays are put into the cartons and overwrapped. Sufficient space should be left between the trays or cartons on the rack to permit air circulation around each carton during freezing. Plate freezers, either manually loaded and unloaded or automatic, are often used in this duty. The storage life of meals on trays depends on the component parts.

In a sample examination of various frozen foods (small portions), Decareau (1964) found the required heating time reduced, as shown in Table 1, where package thickness (depth) was comparable for microwave heating. The frozen foods were complete dinners, cooked and prepared seafood, meats, fish, and so forth, defrosted and heated in a microwave oven (frequency, 2450 MHz; power output, 800 W).

Microwave reconstitution and cooking requires materials like paper, plastic, ceramic, or glass that are transparent to microwaves. The uncooked appearance must be compensated for, where applicable, by a pre- or post-microwave treatment, or with browning elements within the microwave oven. Defrosting by microwave may be the only practical application with some foods such as red meat.

In a comparison of frozen beef, stew, fried chicken, macaroni and cheese, cherry pie, and fresh tossed salad, conventionally prepared and ready-prepared, the latter resulted in a dramatic reduction of labor time. In the institutional area, studies indicate that using fully prepared frozen foods results in a minimum saving of 3 to 5% as compared to conventionally prepared foods. In most cases, a randomly selected consumer taste panel preferred the ready-prepared foods, while the professional taste panel preferred the conventionally prepared items.

Freeze-Dried Product

Freeze-drying removes moisture by sublimation from frozen foods without appreciably changing the form, color, or taste of the product. The moisture is reduced to 2% or less, with reconstitution a function of the simple addition of water or any other liquid. Freeze-dried foods receive higher scores when used in prepared mixes or dishes rather than when they are used alone. Institutional sales appear to be a principal market. Such products are feasible for volume production, retaining many of the quality attributes of frozen foods and, when properly packaged, have the further advantage of being storable at ordinary room temperatures for as long as two years without deterioration, although this is usually concomitant with a cost penalty. Rehydration can be rapid and simple. Beef, pork, chicken, seafoods, soups, several mixtures of foods, and coffee are readily available.

Prepared Foods Served at Room Temperature

Appetizers and sandwiches. Properly packaged bread, frozen rapidly and stored for a year at 0 °F, is as fresh as bread held at 70 °F for a single day. When exposed to a dry atmosphere, slices of many kinds of bread quickly become dry and unpalatable. Rapid preparation and sealing of sandwiches, appetizers, and similar products in moistureproof sheetings or packages reduces this problem.

Sandwiches and canapes should be frozen rapidly, otherwise the bread will become stale before it is solidly frozen. When the temperature reaches 0 °F, the staling process almost ceases. To freeze these products rapidly, they must be in relatively thin packages (not shipping containers) in a freezer maintained at −31 to −36 °F, since the depth of the package is the determining factor. The freezer should be designed for rapid heat transfer. The packages of sandwiches, canapes, or hors d'oeuvres should be placed in fiberboard shipping containers after freezing. Many of these products have a limited storage life.

Breads and rolls. Bread becomes stale slowly when it is warm; moderately rapidly at ordinary room temperatures; rapidly when it is chilled between 64 and 20°F; and slowly at 0°F or below. The freezing point of bread containing 36% moisture is 21°F. Frozen bread of high quality can be obtained only if the loaves are chilled and frozen in a very short time; and if the bread is to be held for more than a few weeks, it must be stored considerably below 0°F.

Freezing temperatures should be no higher than −20°F, although −40°F air blast is the recommended method for freezing such products. Products should be cooled to lower the temperature after baking, before freezing them in individually wrapped units and placing them on racks. Air should be kept in constant movement while products are being frozen. Product temperature should be reduced to below 20°F within 6 h, if possible. Under no circumstances should the product remain above 24°F for more than 24 h. Temperature of bread products should be approximately 0°F on removal for shipment and should not rise above 20°F during shipment.

During defrosting, bread and rolls need to pass quickly through the temperature range between 20 and 65°F, at which range bread becomes rapidly stale. A defrosting temperature of 100 to 120°F with an airflow of 200 to 500 fpm works well to defrost wrapped products.

Frozen doughs (unbaked bread, croissants, etc.) for home proofing and baking of bread and rolls are available in both consumer and institutional packs. Because of the freezer-storage effect on yeast cells, these products must be stored at 0°F or below. For further information on frozen breads and rolls, see Chapter 21.

Cakes and cookies. Nearly all kinds of cakes can be frozen and thawed without noticeable change. Researchers still differ as to which kinds freeze and store with less change: batters or prebaked cakes. Few batters produce cakes of good volume, texture, and flavor when baked after 6 months' storage, even at temperatures below 0°F. Many kinds of cakes remain almost perfectly acceptable during storage for this period or longer. Frozen batters have two disadvantages: (1) most have a relatively short storage life; and (2) they require baking and, in some cases, both thawing and baking.

Cakes, both plain and iced, should be packaged in moistureproof sheeting prior to freezing. Moistureproof cellophane, polyethylene, and pliofilm sheetings are satisfactory. The sheeting should be heat sealed, and the cake either placed directly on a tray, or the wrapped cake put into a bleached sulfite carton. In either case, wrapped, or wrapped and packaged, cakes are placed on trays on a wheeled rack, and then, like bread, frozen in a −40°F air-blast or plate freezer.

Because of the relatively low moisture content of most cakes, their freezing rate is faster than that of breads and rolls. Layer cakes become gradually softer and more moist at 0°F or below, while pound cakes become more tender and crumbly. After 4 weeks at 0°F, angel food and chiffon cakes are superior to day-old unfrozen cakes.

Most cookies may be frozen and thawed several times without noticeably affecting quality. Moistureproof packages prevent drying out in storage or loss of crispness because of condensation during thawing.

Nearly all cookies remain fresh for long storage periods if held at low temperatures. Christmas tree cookies, coconut molasses cookies, small bridge cookies, almond macaroons, and chocolate chip cookies retain acceptability for 14 months stored at −15°F. For further information on frozen cakes and cookies, see Chapter 21.

Pies. Frozen, unbaked pies are available in a wide assortment. Meat and poultry pot pies and pizza are also popular. After assembly of the raw pie, modern methods call for cartoning and air-blast or plate freezing at −40°F evaporator temperature until a satisfactory internal temperature is achieved.

The stability of commercially packed apple, cherry, peach, and boysenberry pies is excellent (6 to 18 months) when good quality fruit is used to make the pies. Blueberry pie is the least stable of the common varieties. Changes in filling flavor rather than in crust flavor or texture account for the first flavor changes in fruit pies, but minor appearance changes do occur before any detectable flavor deterioration starts. Surface mold development has been observed on pies stored at 20°F. Frozen unbaked pie shells are also available. For further information on frozen pies, see Chapter 21.

Fruits. Not only may certain fruits be preheated before freezing, but apples, blueberries, raspberries, blackberries, and plums are superior, with the use of proper heating methods and suitable packages. Preheated fruits should be solid packed, without airspaces, in leakproof, moistureproof containers. Storage temperatures of −10°F prevent loss of flavor.

A great variety of prepared and precooked fruit products are frozen, such as pies, sherbets and ices, baby foods, cobblers, fritters, jellies, and spreads, pectinized purees, pie fillings, shortcakes, relishes, toppings or sundae sauces, confections, sauces, and specialty items such as baked apples. Freezing retains the color and flavor of many prepared and precooked fruits far better than does canning. For further information on frozen fruits, see Chapter 16.

Foods Eaten Frozen

This group includes ice creams, sherbets and ices, ice-cream cakes, frozen eclairs, and parfaits. Most of these foods are hardened at −10 to −30°F with or without forced air. They should be stored at 0°F or lower and held at or about 8°F so that they will be spoonable. Room temperature defrosting should be avoided since large ice crystals may form when the product is refrozen. Defrosting time will vary with the size and type of frozen dessert. For further information on ice cream, see Chapter 15.

UNIFORM STORAGE TEMPERATURE

Under fluctuating temperature, if the maximum temperature to which a frozen food is subjected is sufficiently low that the product remains solidly frozen and no liquid separates, the deterioration rate of many foods (except cream sauces and thickened gravies) is about the same as that which occurs when the product is held uniformly at the mean temperature. If the temperature of a food fluctuates between −5 and 5°F, the deterioration rate is the same as what would occur if the food were held at a constant temperature of 0°F.

Fluctuating storage temperatures are considerably more detrimental to white sauces, thickened gravies, cornstarch, and custard puddings than a constant temperature, which is the mean of the maximum and minimum temperatures. However, proper selection of modified starches allow formulations to undergo numerous freeze-thaw cycles before breakdown.

The Association of Food and Drug Officials of the United States (AFDOUS) has developed a frozen food handling code for the proper handling, shipping, warehousing, and retailing of frozen foods. Its purpose is to upgrade the frozen food industry by restricting food handling at a temperature of not more than 0°F to assure high product quality when purchased. Several states have adopted modified versions of this code. Many types of temperature abuse indicators are commercially available. However, they monitor temperature at the container's outside surface and not at the food within the container.

Bacterial counts in frozen, prepared foods are reduced with increased time in frozen storage, and fluctuations in storage temperature within the frozen range accelerate the rate of reduction. Ascorbic acid and thiamine content should decrease during frozen storage, while total solids may increase if dehydration takes place.

Precooked and Prepared Foods

Spice levels mellow during frozen storage; thus, higher initial levels should be formulated.

FOOD SERVICE APPLICATIONS

Individual and multiple portions of precooked entrees and vegetables are also pouch-packed for food-service operations. Constant-temperature water-immersion heaters are available for reconstituting quantities of pouches in institutional operations. High-pressure steamers and microwave ovens may also be used, but greater skill is required.

Bulk packs of prepared entrees for restaurant and institutional use may be heated in quartz plate infrared ovens operating at 650 to 700°F or in high-speed conventional ovens where the requirement for volume is for small continuous feedings, such as a cafeteria serving 6 to 12 people per line.

Plastic wrapped slab packs of frozen entrees, shaped to fit into steam-table pans, are available. Packaging in aluminum foil trays is increasing. Single pans can be quickly reconstituted in infrared ovens. Multiple units can be accommodated in forced air convection ovens specifically designed to reconstitute large amounts of prepared frozen foods.

Infrared ovens are used in restaurants, drive-ins, snack bars, and other types of food service operations. Special infrared ovens designed for airplane galleys can heat 36 foil casserole platters from 0°F to a serving temperature of 160°F or higher in 7 min.

Microwave ovens, while capable of heating prepared foods more rapidly than other devices, are limited in their output and thus not suited to volume applications. Their largest use at present is in vending machines for warming sandwiches or pastries and refrigerated commissary-prepared dinners.

The use of prepared frozen foods by food service operators has grown significantly, since it increases worker productivity and offsets increased labor costs. These products can be most effectively used when they are part of a *food system* in which products, packaging, procurement, storage, heating, holding, and serving procedures, as well as menus, equipment, and utensils, are fully tested, specified, and controlled.

STANDARDS

The U.S. Food and Drug Administration has issued standards of identity for types of raw breaded shrimp and other prepared foods. Product standards have been published by the U.S. Department of Commerce for raw breaded shrimp and fish sticks and for breaded fried fish sticks and fish portions. The U.S. Department of Agriculture has issued standards for breaded onion rings and French fried potatoes. Military specifications exist for precooked frozen meals purchased for in-flight use by the U.S. Air Force. USDA Meat and Poultry Standards Division has established component minimums for more than 200 products. These cover minimally cooked or raw meat contents, as well as key ingredients. The American Frozen Food Institute (AFFI) has prepared a Basic Quality Assurance Program for processors to use as a guide in establishing internal Quality Assurance Programs for frozen precooked and prepared foods.

REFERENCES

Commodity Storage Manual. 1989. Refrigeration Research Foundation, Washington, D.C.

Decareau, R. V. 1964. Microwave defrosting and heating. *Cornell Hotel and Restaurant Administration Quarterly* 5(1):432.

Hanson, H. L., A. Campbell, and H. Lineweaver. 1951. Preparation of stable frozen sauces and gravies. *Food Technology* 5(10):76.

Tressler, D. K. and C. F. Evers. 1950. *The freezing preservation of foods.* Vol. 1, Fresh foods; Vol. 2, Cooked and prepared foods, 3rd ed. Avi Publishing Co., Westport, CT.

CHAPTER 21

BAKERY PRODUCTS

Ingredient Storage 21.1	*Bread Cooling* 21.3
Mixing 21.1	*Slicing and Wrapping* 21.4
Fermentation 21.2	*Bread Freezing* 21.4
Final Proof 21.3	*Freezing Other Bakery Products* 21.5
Baking 21.3	*Retarding Doughs and Batters* 21.5

REFRIGERATION plays an important part in modern bakery production. It stabilizes raw material quality prior to use. Dough temperature control is one of the most important factors governing overall product quality. Refrigerated storage and freezing of finished products improve production scheduling and marketing. Availability of refrigeration can determine operational procedures for large wholesale plants or small retail bakeries.

INGREDIENT STORAGE

Raw materials are generally purchased in bulk quantities except in small operations. Deliveries are made by truck or rail car and stored in bins or tanks with required temperature protection while in transit and storage. Flour is stored in bins at ambient temperatures. Smaller quantities of different types of flour such as clear, rye, and whole wheat, are usually received in bags and stored on pallets. Corn syrup, liquid sugar, lard, and vegetable oil are stored in heated tanks or in enclosed spaces where temperatures are high enough to prevent crystallization or congealing. Holding temperatures for such products are about 125 °F.

Lesser volume and specialized sugars and shortenings are received in drums, bags, or cartons and are stored at conditions that prevent mold or melting and rancidity. Many bakeries use syrups with high levulose content, which may be stored at about 84 °F to maintain fluidity and pumpability. As these syrups are stored at a temperature about one-third lower than conventional syrup, less thermal input is required during storage and the refrigeration load during mixing is significantly reduced.

Yeast in cartons or bulk bags should always be stored in refrigerated spaces at temperatures below 45 °F but above its freezing point, at which value severe cell mortality occurs. Cocoa, milk products, spices, and other raw materials subject to insect infestation, bacterial growth, or unstable oils subject to oxidative rancidity, should be stored under the same conditions as yeast to prevent product loss.

Total plant air conditioning is used more and more in new plant construction except in areas immediately surrounding ovens, in final proofers, and in areas where cooking vessels are located for preparing fruit fillings and hot icings. Total plant cooling was first used in plants producing Danish pastry, puff pastry, and pies and has expanded to new construction facilities for frozen dough operations and for general production. Flour dust in the air should be filtered out because it fouls air passages in the air-conditioning equipment and seriously reduces heat-transfer rates.

MIXING

Bread, buns, and sweet rolls are the most important baked products from the standpoint of production volume.

Mixing is the first active process in such production. Refrigeration is required because of the generation of heat and the necessity to control the dough temperature at the end of the mix.

Metabolism of yeast is materially affected by the temperatures to which the yeast is exposed. During dough mixing, the following heat considerations are encountered: (1) *heat of friction,* by which the electrical energy input of the mixer motor is converted to heat; (2) *specific heat* of each ingredient; and (3) *heat of hydration,* evolved when a dry material absorbs water. If ice is used for temperature control, heat of fusion results. Finally, the temperature of the dough ingredients must be considered. Yeast is dormant at temperatures below 45 °F. It is extremely active in the presence of water and fermentable sugars in a temperature range of 80 to 100 °F, but the cells are killed at about 140 °F and at a lower but sustained pace below its freezing point of 26 °F. Precise temperature control is essential at all stages of storage and production, especially during mixing.

Dough may be formed by either a batch or continuous process. The batch method uses a large bowl with heavy revolving blades or agitator arms which mix flour, water, yeast, and other ingredients into a homogeneous dough.

The two principal types of batch dough mixes are the *straight-dough* process and the *sponge-dough* process. In the straight dough process, all ingredients are mixed at once. In the sponge-dough process, only part of the total amount of flour and water required are mixed with all of the yeast, yeast food, and malt. The resulting mixture, or sponge, is then fermented for a period before it is returned for a final mix with the remaining flour and water, as well as the rest of the ingredients.

The principal heat generated during mixing comes from the heat of hydration as the flour absorbs water and from the heat of friction from the mixer. To absorb this excess heat, and maintain the dough at 79 to 81 °F, the ingredient water is usually supplied to the mix at a temperature of 35 to 39 °F, and the mixer is jacketed to circulate a cooling medium around the bowl.

Liquid sponge ingredients can be incorporated and fermented in special equipment either on a batch or uninterrupted basis. To render the mixture pumpable, more water than the actual amount used in a sponge to be fermented in a trough is incorporated. On completion of the predetermined fermentation time, the sponge (at required pH and titratable acid levels) is chilled through heat exchange equipment from about 90 to 100 °F to about 45 to 65 °F. The cold liquid sponge is then maintained at the required temperature in a storage tank until it is weighed and pumped to the mixer, where it is combined with the remaining ingredients prior to being remixed into a dough. Regular sponges come back for remixing at about 84 °F and often need to use the refrigerated surface on the mixer jacket in conjunction with ice water or ice to achieve the required dough temperature after mixing.

The gluten matrix, which is essential to loaf volume and texture, can be developed only by properly mixing the doughs. Exact temperature control during mixing is essential. Positive dough temperature control is achieved using a cold liquid sponge at 45 to 65 °F, which limits the use of the refrigeration jacket and eliminates using ice in the doughs. In cold weather, remix dough water temperatures of 100 °F or higher are often required. A supple-

---

The preparation of this chapter is assigned to TC 11.6, Prepared Food Products.

mental refrigeration source is used to chill the sponge to retard yeast metabolism, thereby reducing the refrigeration load at the point of final mix.

In the continuous dough mixing process, a brew, or liquid sponge is formed, using 0 to 70% of the flour, 10 to 15% of the sugar, 25 to 50% of the salt, and all of the yeast and yeast food in about 85% of the water. This mixture ferments in a series of tanks and then passes through an incorporator, where the remaining ingredients are added. The resultant thicker batter finally passes through a developer-mixer from which the finished dough, in loaf size, is extruded into the baking pans.

Properly formulated dough can also be deposited on floured belts, rounded up, and then conveyed through conventional makeup and molding equipment. Hot dog and hamburger rolls are also produced in quantity by pumping bun dough from the continuous mix developer directly to the hopper of the makeup equipment. A coating of flour on the outside dough surfaces affects gas development and retention, which in turn leads to a grain/texture more closely resembling that of sponge and dough products. External symmetry and crust characteristics are similarly changed.

Many variety breads, such as raisin and home-style loaves, cannot be made satisfactorily with the continuous mixing process. Thus, some bakeries prepare and ferment the liquid sponge and then pump it into the batch mixer along with the remaining ingredients for conventional finishing.

In continuous mixing, the liquid sponge is made up in an unjacketed tank at a temperature of 70 to 80°F and allowed to ferment for about 1 h, during which time the temperature rises to 84 to 86°F. The second hour of fermentation takes place in a jacketed tank where the brew is held at about 84°F. This second stage consists of alternate transferral from the first stage tank to a pair of tanks to maintain a continuous operation by allowing the brew to be drawn from one tank while fermentation goes on in the alternate tank.

Between final stage tanks and an incorporator, the brew is pumped through a plate heat exchanger where it is cooled to about 70°F. Cooling offsets the heat that will be added by mixing and by adding the remaining ingredients; thus, a predetermined final dough temperature of 79 to 82°F can be obtained.

In many older plants, central ammonia systems are used to cool water for both ingredient use and jacket cooling. The ingredient water is cooled separately in self-contained water chillers that range in capacity from 75 to 650 gpm with 3 to 30 hp condensing units. The jacket cooling is by direct expansion of the refrigerant in the jacket at 30°F.

For batch mixing, individual condensing units are usually located within 15 ft of each mixer. Where a battery of mixers are used in large plants, a compressor room is often located adjacent to the mixing and fermentation rooms. Table 1 lists the sizes of condensing units commonly selected for mixers.

When large batches of dough are handled in the mixers, the required cooling sometimes becomes greater than the available heat transfer surface can produce at 30°F, and the refrigerant temperature must be lowered. A refrigerant temperature below 30°F can, however, cause a thin film of frozen dough to form on the jacketed surface of the mixer. This frozen film effectively insulates the surface and impairs heat transfer from the dough into the refrigerant.

Formulas, batch sizes, mixing times, and almost every other part of the mixing process vary considerably from bakery to bakery and must be determined for each application. These variations usually fall within the following limits:

Final batch dough weights	Up to 2000 lb
Ratio of sponge to final dough weight	50 to 75%
Ratio of flour to final dough weight	50 to 65%
Ratio of water to flour	50 to 65%
Sponge mixing time	360 to 600 s
Final dough mixing time	480 to 720 s
Number of mixes per hour	2 to 5
Continuous mix production rates	Up to 7000 lb/h

The total cooling load is the sum of the heat removal required from each ingredient to bring the homogeneous mass to the desired final temperature plus the removal of the generated heat of hydration and friction. In large batch operations, the sponge and final dough are mixed in different mixers, and refrigeration requirements have to be determined separately for each process. In small operations, a single mixer is used for both sponge and final dough. The cooling load for final dough mix is greater than that for sponge mix and is used to establish refrigeration requirements.

FERMENTATION

After completion of the sponge mix, the sponge is placed in large troughs that are rolled into an enclosed conditioned space for a period of fermentation. This period lasts for 3.5 to 5 h, depending on the dough formula. The sponge comes out of the mixer at a temperature of 75 to 79°F. During the fermentation period, the sponge temperature rises from 8 to 10°F as a result of the heat producing yeast action and the heat of hydration.

To equalize the temperature substantially throughout the dough mass, the room temperature is maintained at the approximate mean sponge temperature of 80°F. Even temperature throughout the batch produces even fermentation action and a uniform product.

Water makes up a large part of the sponge and uncontrolled evaporation causes significant variations in the quality and weight of the bread. A certain amount of evaporation is required to remove excess alcohols and esters that produce too strong a flavor. The rate of evaporation from the surface of the sponge varies with the relative humidity of the ambient air and the airflow rate over the surface. The rate of movement of moisture from the inside of the sponge to the surface does not react similarly to the same external conditions, and the net result can be the drying of the surface and formation of a crust. The crust formation produces an inactive area which causes undeveloped portions. When folded into the dough mass later, these undeveloped portions produce hard, dark streaks in the finished bread.

To control the evaporation rate from the surface of the sponge, the air is maintained at 75% rh, and air movement over the surface is controlled by moving the conditioned air into, through, and out of the process room without producing crust-forming drafts.

In calculating the cooling load, the product is not considered, since the air temperature is maintained at approximately the mean of the various dough temperatures in the room. Transmission heat loss through walls, ceiling, and floor is the principal load source. Infiltration is estimated as 1.5 times the room volume per hour. Usually the light requirements for a fermentation room are about 75 W per 400 ft^2 of floor area. The conditioning units are often placed within the conditioned space, and the full motor heat load must be considered.

Table 1 Size of Condensing Units for Various Mixers

Dough Capacity, lb	R-12 Condensing Unit, hp
800	5
1000	7.5 to 10
1300	10 to 15
1600	15 to 20
2000	20

Bakery Products

The only source of latent heat is the approximate 0.5% weight loss in the sponge. Under full operating load, this could account for a 1.5 °F increase in dew-point temperature. For conditions of 80 °F db and 75% rh, the dew-point temperature would be 71.5 °F, and the supply air would be introduced into the conditioned space at 72 °F dry bulb and 70 °F dew point.

In large rooms, sufficient air volume can be introduced to pick up the sensible heat load with a 8 °F rise in air temperature. In smaller rooms a latent heat load may need to be added by spraying water directly in the room through compressed air atomizing nozzles. Water sprays have been most successful when a relatively large number of nozzles have been spaced around the periphery of the room.

Since a high removal ratio of sensible to latent heat is desirable, the condensing unit is usually specified for operation as close to 60 °F evaporator temperature as is practicable with the temperature and quantity of condensing water available.

FINAL PROOF

After it is formed into loaves and placed in pans, the dough receives its final development or proof before it is baked. It is placed in an insulated enclosure having a controlled atmosphere for 50 to 75 min. To keep the almost exhausted yeast active, the temperature is maintained at 95 to 120 °F, depending on the exact formula, the prior intensity of dough handling, and the character of the baked loaf required.

For proper crust development during the bake, the exposed surface of the dough must be kept pliable by maintaining the relative humidity of the air around 85 to 90%. There are exceptions to humidities in this range in bakeries, where they have found it necessary to compromise on a lower range because of the effect on the pan glazing which is frequently used in place of the older method of greasing the pan.

With dew points at about 92 to 110 °F and ambient conditions going down to at least 65 °F at times, the enclosure must be adequately insulated to keep the inside surface warm enough to prevent the formation of condensation, because mold growth can be a serious problem on warm, moist surfaces. A thermal conductance of $C = 0.12$ Btu/(h·ft²·°F) is adequate for most conditions.

The enclosures most commonly used for the proofing process consist of a series of aisles with doors at both ends. Frequent opening of the doors to move racks of panned dough in and out makes control of the conditioned air circulation an unusually important engineering consideration. The air is recirculated at rates of 90 to 120 changes per hour and is introduced into the room to temper the infiltration air and cause a rolling turbulence throughout the enclosure. This brings the newly entered racks and pans up to room temperature as soon as possible.

The problem of air circulation becomes somewhat simpler when large automatically loaded and unloaded tray-type proofers have only minimal openings for the entrance and exit of the pans, and the load is materially reduced by elimination of the racks moving in and out.

BAKING

The fully developed loaves are baked in the oven at temperatures around 400 to 450 °F for 18 to 30 min, depending on the variety of bread. Two principal types of ovens are used: the tunnel or traveling hearth, which is loaded at one end and unloaded at the opposite end, and the tray, which is loaded and unloaded at the same end. There are three varieties of tray-type ovens. The reel type is used for small production. Single and double-lap traveling tray ovens are used for larger production. The latter type, in which the trays travel back and forth twice, is used principally where there is an advantage in extending the oven upward to save floor space.

Heating is mostly with gas, but oil firing is more economical in some areas. Where gas interruption is a problem, combination oil and gas burners are an alternative. While both indirect- and direct-fired gas systems are used, the indirect firing is generally more popular because of its greater flexibility. Ribbon burners running across the width of the oven are operated as an atmospheric system at pressures of 6 to 8 in. of water.

The burners are located directly beneath the path of the hearth, or trays, and the flames may be adjusted along the length of the burner to equalize the heat distribution across the oven. The oven is divided into zones, with the burners on zone control so that the heat can be varied for different periods of the bake.

Controlled air circulation inside the oven assures uniform heat distribution around the product. To secure a satisfactory golden-toned crust color, the sugar on the surface is caramelized by spraying low-pressure steam on the exposed surface before the crust has begun to bake out. This is accomplished with a series of perforated steam tubes located in the first zone of baking in place of the air recirculating tubes. Heating calculations are based on 450 Btu/lb of bread baked per hour. Steam requirements run approximately 1 bhp per 125 lb of bread baked per hour.

BREAD COOLING

Baked loaves come out of the oven with an internal temperature equalized at 200 to 210 °F because of the evaporative cooling effect of the moisture that is driven off during baking. The crust temperature is closer to 450 °F.

The loaves for are removed from the pans and allowed to cool to an internal temperature of 90 to 95 °F. The cooling of a hygroscopic material takes place in two stages or phases, which are not distinct periods. When the bread first comes out of the oven, the vapor pressure of the moisture in the loaf is high compared to the vapor pressure of the moisture in the surrounding air. Moisture is rapidly replaced with a resultant evaporative cooling effect, which, combined with heat transmission at a relatively high temperature differential from bread to air, causes rapid cooling as indicated by the steepness of a time-temperature reduction curve in the early cooling stage (Figure 1). As the vapor pressure approaches equilibrium, heat transfer is mainly by transmission and the cooling curve flattens rapidly.

While many bakeries still cool bread on racks standing on the open floor for 1.5 to 3 h, depending on air conditions, many large operators cool their bread while it is moving continuously on belt- or tray-type conveyors. Cooling is mostly atmospheric even on these conveyors. However, to insure a uniform final product, cooling is often handled in air-conditioned enclosures with a conventional counterflow movement of air in relation to the product.

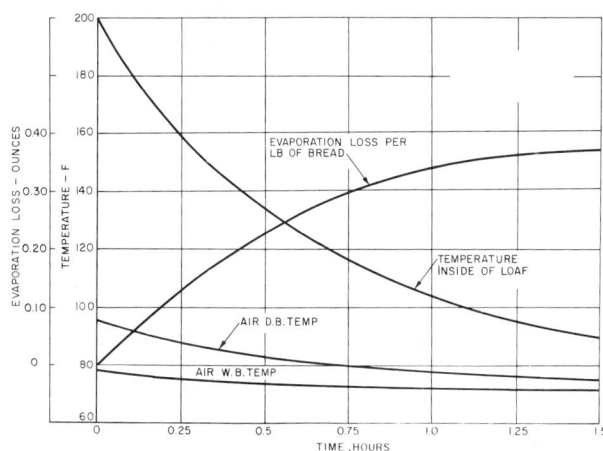

Fig. 1 Moisture Loss and Air Temperature Rise in Counterflow Bread-Cooling Tunnel

An internal temperature of 90 to 95 °F stabilizes moisture in the bread enough to reduce excessive condensation inside the wrapper, which can encourage mold development. Approximately 1.5 h of cooling time is required to bring the internal loaf temperature to 90 °F because of the more or less fixed time factor for movement of heat and moisture from the center of the loaf to the surface (Figure 1).

In counterflow cooling, the optimum temperature for the air introduced into the cooling tunnel is about 75 °F. The entering air at 80 to 85% rh controls the moisture loss from the bread during the latter stage of cooling and improves the keeping quality of the bread.

Attempting to shorten the cooling time by forced cooling with air temperatures below 75 °F is not satisfactory, because the rate of heat and moisture loss from the surface and the movement of the heat and moisture from the interior to the surface as a replacement becomes so unequal that the crust shrivels and cracks.

Cooling time is often shortened when mold inhibitors are used by taking the bread to the slicers at 105 to 110 °F. The product heat load is calculated assuming a sensible heat transfer based on a specific heat of 0.74 Btu/(lb·°F) for bread and a temperature reduction from 180 to 90 °F. For calculating purposes, the evaporation of moisture from the bread may be considered responsible for reducing the bread temperature to 180 °F.

The conditioning of the air can best be accomplished under normal conditions by evaporative cooling in an air washer. For periods when the outdoor wet-bulb temperature exceeds 72 °F, which is the maximum allowable based on 75 °F dry-bulb and 85% saturation, refrigeration is required.

The amount of air required is rather high because the maximum temperature rise of the air in passing over the bread is usually about 20 °F; consequently, the refrigeration load is comparatively high. In localities where wet-bulb temperatures above 72 °F are rare or of short duration, the expense of refrigeration is not considered justified, and the bread is sliced at a higher than normal temperature for this period.

SLICING AND WRAPPING

Bread from the cooler goes through the slicer where high-speed cutting blades, similar to band saw blades, cut cleanly through the bread if it has been properly cooled. If the moisture evaporation rate from the surface and the replacement rate of the surface moisture from the interior of the loaf are not kept in balance, the bread develops a soggy undercrust that fouls the blades, causing the loaf to crush during slicing. A brittle crust may also develop, which leads to excessive crumbing during the slicing. From the slicer, the bread moves automatically into the bagger.

BREAD FREEZING

The bagged bread moves normally into the shipping area for delivery to various markets by local route trucks or by long-distance haulers. Part of the production may go into a quick freezing room and then into cold storage.

Two important problems face bakers in the freezing of bread and other bakery products. The first is connected with the short work week. Most bakeries are inoperative on Saturday; thus, weekend production is much larger than for the earlier part of the week. The problem increases for bakeries on a 5-day week, with Tuesday usually being the second day off. Freezing a portion of each day's production for distribution on the days off would enable a more even daily production schedule.

A second problem is the increasing demand for variety breads and other products. The daily production run of most varieties is comparatively small, so the constant setup change is expensive and time-consuming. Running a week's supply of each variety at one time and freezing to fill daily requirements can reduce operating cost.

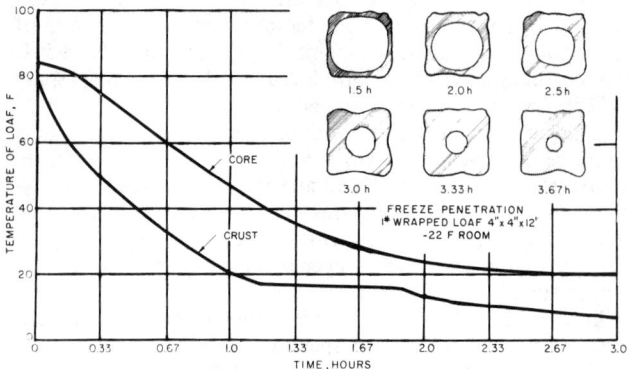

Fig. 2 Core and Crust Temperatures in Freezing Bread

Both these problems are concerned with the staleness of bread, which is a perishable commodity. After baking, starch from the loaf progressively crystallizes and loses moisture. Until a critical point of moisture loss is reached, freshness can be restored by heating and reabsorbing starch crystals. A tight wrap helps keep the moisture content high over a reasonable time. The crystallization of the starch which, when complete, produces the crumbly texture of stale bread, is a spontaneous action that increases in rate, both with decreased moisture and decreased temperature. The moisture loss rate apparently increases with a temperature decrease down to the freezing point. The rate then decreases until the temperature reaches 0 °F, at which point moisture loss seems to be somewhat arrested.

Bread freezes at 16 to 20 °F (Figure 2). For freezing of all cellular structures, it is necessary to cool the bread through the freezing or latent heat removal phase as fast as possible to preserve the cell structure.

Since the moisture loss rate increases with reduced temperature, the bread should be cooled rapidly through the entire range from the initial temperature down to, and through, the freezing points. Successful freezing has been reported in room temperatures of 0, −10, −20, and −30 °F.

In USDA laboratory tests, loaves of wrapped bread placed in cold air blasts of 700 fpm were brought down from 70 to 15 °F core temperatures in the comparative times given:

Freezer Air Temperature, °F	Time, h	Temperature at End of 2 h, °F
−40	2	15
−30	2.25	16
−20	3	19
−10	3.75	21
0	5	22

Another interesting observation showed that with wrapped bread, changes in air velocity from 200 to 1300 fpm caused relatively little change in cooling the item.

Comparative tests between wrapped and unwrapped bread indicate a shortening of cooling time for the unwrapped bread from 10 to 30 min through the range of 0 to −20 °F, showing that the wrapper is an insulator. However, while the wrapper is a deterrent to fast freezing, its value in retaining the moisture content in the product during freezing and thawing makes freezing in the wrapped condition advisable.

Some commercial installations freeze wrapped bread in corrugated shipping cartons. The additional insulation provided by the carton increases the freezing time considerably and causes a

Bakery Products

wide variation in time between variously located loaves. One test found that a corner loaf reached 15 °F in 5.5 h and the center loaf required 9 h.

Most freezers are batch loaded rooms with the bread being placed on the wire shelves of steel racks. One of the principal disadvantages of this arrangement is that the racks must be moved in and out of the room manually. The continuous-type freezers with wire belt or tray-type conveyors permit a steady flow and do not expose personnel to the freezer temperature.

For a −20 °F room temperature, a compound compression refrigeration system is advisable. In one commercial installation, the first stage is handled with an R-12 rotary compressor operating at −5.5 lb vacuum to 37 psig, discharge. The second stage is a reciprocating unit operating at 37 psig absolute suction and 125 psig discharge. The use of a nonfreeze propylene glycol sprayed over the finned evaporator coils prevents frost formation.

In addition to the primary air blowers designed to handle about 10 cfm per pound of bread frozen per h, a series of fans are employed to assure good air turbulence in all parts of the room. Heat load calculations are based on the specific and latent heat values in Table 2.

Table 2 Important Heat Data for Baking Applications

Specific heat	
Baked bread (above freezing)	0.70 Btu/(lb·°F)
Baked bread (below freezing)	0.34 Btu/(lb·°F)
Butter	0.57 Btu/(lb·°F)
Dough	0.60 Btu/(lb·°F)
Flour	0.42 Btu/(lb·°F)
Ingredient mixture	0.40 Btu/(lb·°F)
Lard	0.45 Btu/(lb·°F)
Milk (liquid whole)	0.95 Btu/(lb·°F)
Liquid sponge (50% flour)	0.70 Btu/(lb·°F)
Heat of friction per horsepower of mixer motor	42.4 Btu/min
Heat of hydration of dough or sponge	6.49 Btu/lb
Latent heat of baked bread	46.90 Btu/lb
Specific heat of steel	0.12 Btu/(lb·°F)

After the quick-freeze, the bread is moved into a 0 °F holding room where the temperature evens out throughout the loaf. The bread is often placed in shipping cartons after freezing and stacked tightly on pallets.

Frozen bread must be thawed or defrosted for final use. Slow, uncontrolled defrosting requires only that the frozen item be left to stand, usually in normal atmospheric conditions. For quality control, the defrosting rate is not as critical as the freezing rate. However, for time control, increases in relative humidity and air temperature decrease the defrosting time. Too high a relative humidity causes excessive condensation on the wrapper with some resultant susceptibiltiy to handling damage. For purposes of calculation, at 120 °F air temperature with 50% or less rh the product will defrost in about 1.75 h. Good air movement over the entire product surface at 200 fpm or higher helps to minimize the condensation and make the defrosting rate more uniform.

FREEZING OTHER BAKERY PRODUCTS

Retail bakeries freeze many products to help meet fluctuating demand. Cakes, pies, sweet yeast dough products, soft rolls, and doughnuts are all successfully frozen. A summary of tests and commercial practice indicates that these products are less sensitive to the freezing rate than are bread and rolls. Freezing at temperatures of 0 to 10 °F apparently produces just as satisfactory results as freezing at −10 to −20 °F.

Storage temperatures of 0 °F or lower will keep packaged dinner rolls and yeast-raised and cake doughnuts satisfactorily fresh for 8 weeks. Storage temperatures above 0 °F reduce the satisfactory keeping time, materially affecting their quality. Cinnamon rolls keep satisfactorily for only about 3 weeks, apparently because of the presence of raisins, which absorb moisture from the crumb of the roll. Pound, yellow layer, and chocolate layer cakes can be frozen and held at 10 °F for 3 weeks without materially affecting their quality. Sponge and angel food cakes tend to be much softer as the freezing temperature is reduced to 0 °F.

Layer cakes with icing freeze well, but condensation on exposed icing during thawing ruins the gloss so that these cakes are wrapped before freezing. Unlike cakes that can be satisfactorily frozen after baking, frozen baked pies have an unsatisfactory crust color, and the bottom crust of fruit pies becomes soggy when the pie is thawed.

The freezing of unbaked fruit pies is highly successful. Freezing time has little, if any, effect on product quality, but storage temperatures do have an effect. Frozen pies stored at temperatures above 0 °F develop badly soaked bottom crusts after 2 weeks and the fillings tend to boil out during baking, possibly due to the weakened structure of the crust.

Danish and sweet dough products are frozen baked or unbaked, depending on how quickly they will be required for sale when they are removed from the freezer. Custard and chiffon pie fillings have not had uniformly good results, but some retail bakeries have achieved satisfactory results by carefully selecting starch ingredients. Meringue does not stand up very well, but whipped cream seems to improve with freezing. Cheesecake, pizza, and cookies also freeze well.

Although some products may be of better quality if they are frozen at 0 °F while others are improved by freezing at 10 °F, variety shops must compromise on a single freezer temperature so that the various products can be placed in the same freezer. These freezers are usually maintained between 5 and 10 °F. Freezing time is not a factor because the products are kept in storage in the freezer. The freezers range from large reach-in refrigerators in retail shops to walk-in boxes in wholesale shops.

RETARDING DOUGHS AND BATTERS

Freezing is usually employed where the products are to be held from 3 days to 3 weeks. For shorter holding periods, such as might be required to have freshly baked products all day from one batch mix, a temperature only cold enough to retard fermentation action in the dough is applied. Temperatures of 32 to 40 °F slow up the leavening action sufficiently to allow holding from 3 h to 3 days.

The doughs are sometimes made up into final shape units ready for proofing and baking. Storage of cold slabs of the dough, with the units for baking being made up after thawing, are also used. This method is especially satisfactory for Danish pastry dough and other doughs with rolled-in shortening, such as pie crust. Chilling to retarded temperature seems to improve the flakiness of this type of product. For design calculations, the following values are commonly used: 300 Btu/ft^2 of box area and a 10 °F differential between coil and room temperatures.

In the retarding refrigerator, about 85% rh is required to prevent the product from drying out. Condensation on the product is undesirable. This type of operation usually employs complete batch loading, so refrigeration calculations must be based on the introduction of the batch over a short period of time. The capacity of the refrigeration equipment must be sufficient to absorb the product and carrier heat load in 0.75 to 3 h, depending on cabinet size and handling technique. The products most commonly handled in this manner are Danish pastry, dough for sweet rolls and coffee cake, cookies, layer cake mixes, pie crust mixes, and bun doughs.

The required retarding temperatures are very similar to the temperatures required for storing ingredients, and the refrigerators are usually designed to handle both the ingredient storage and the retarded dough. A common small refrigerator accommodates bun pans on one side and ingredient containers in another compartment.

CHAPTER 22

CANDIES, NUTS, DRIED FRUITS, AND VEGETABLES

CANDY MANUFACTURE 22.1	Coating Kettles or Pans 22.4
Milk and Dark Chocolate 22.1	Packing Rooms 22.4
Enrobing or Hand Dipping 22.2	Refrigeration Plant 22.4
Bar Candy 22.2	STORAGE 22.5
Hard Candy 22.2	Candy 22.5
Hot Rooms 22.3	Nuts 22.7
Cold Rooms 22.3	Dried Fruits and Vegetables 22.7
Cooling Tunnels 22.4	Controlled Atmosphere 22.8

CANDY MANUFACTURE

AIR conditioning is essential for successful candy manufacturing. Proper atmospheric control results in increased production, lower production costs, and better and more uniform product quality.

One or more of several standardized spaces or operations are encountered in every plant. These include: hot rooms; cold rooms; cooling tunnels; coating kettles; packing, enrobing, or dipping rooms; and storage.

Sensible heat must be adsorbed by air-conditioning equipment, which includes the air-distribution system, plates, tables, and cold slabs in tunnels or similar coolers. In calculating the loads, consider such sensible heat sources as people, power, lights, sun effect, transmission losses, infiltration, steam and electric heating apparatus, and the heat of the entering product. Table 1 summarizes the optimum design conditions for refrigeration and air conditioning.

Two of the basic ingredients in candy are sucrose and corn syrup. These easily change from a crystalline form to a fluid, depending on temperature, moisture content, or combinations of the two. The surrounding temperature and humidity must be controlled to prevent moisture gain or loss, which will affect the product's texture and storage life. Temperature should be relatively low, generally below 70 °F. The relative humidity should be 60% or less, depending on the type of sugar used. For chocolate coatings, temperatures of 65 °F or less are desirable, with 50% rh or less.

MILK AND DARK CHOCOLATE

Cocoa butter is either the only or the principal fat in chocolate, constituting 25 to 40% or more of various types. Cocoa butter is a complex mixture of triglycerides of high molecular weight fatty acid, mostly stearic, oleic, and palmitic. Since cocoa butter is present in such large amounts in chocolate, anything affecting cocoa butter affects the chocolate product too.

Because cocoa butter is a mixture of triglycerides, it does not act as a pure compound. Its physical properties, melting point, solidification point, latent heat, and specific heat affect the mixture. Cocoa butter softens over a wide temperature range, starting at about 80 °F and melting at about 94 °F. It has no definite solidification point; this varies from just below its melting point to 80 °F or lower, depending on the amount of cocoa butter and the time it is held at various temperatures. The presence of milk fat in milk chocolate lowers both the melting point and the solidification point of the cocoa butter. High quality milk chocolate remains fluid for easy handling at temperatures as low as 86 to 88 °F. Sweet chocolate remains fluid at temperatures as low as 90 to 92 °F.

Chocolate can be subcooled below its melting point without crystallization. In fact, it does not crystallize in mass but rather in successive stages, as solid solutions of a very unstable crystalline state are formed under certain conditions. The latent heat of crystallization (or fusion) is a direct function of the manner in which the chocolate has been cooled and solidified. Once crystallization has started, it will continue until completion, taking from several hours to several days, depending on its exposure to cooling, particularly to low temperatures (subcooling).

The latent heat of solidification of the grades of chocolate commonly used in candy manufacture varies from approximately 36 to 40 Btu/lb. An average value for its specific heat may be taken as 0.56 Btu/(lb·°F) before solidification and 0.30 Btu/(lb·°F) after solidification. To calculate the cooling load, start with these figures and then add a margin of safety.

Cocoa butter's cooling and solidification properties exist in five polymorphic forms: one stable form and four metastable or labile ones. Cocoa butter usually solidifies first in one of its metastable forms, depending on the rate and temperature at which it solidifies. In solidified cocoa butter, the lower melting labile forms change rapidly to the higher melting forms. The higher melting labile forms change slowly, and seldom completely, to the stable form.

Commercial chocolate blocks are cast in metal molds after the tempering process. During this process, it is desirable to cool the chocolate in the molds as quickly as possible, thus yielding the shortest possible cooling tunnel. However, over-rapid cooling (particularly in large commercial blocks, which are standard 10-lb cakes) may cause checking or cracking which, while not serious to quality, adversely affects appearance. Depositing the chocolate into metal molds at 85 to 90 °F is common.

Dark chocolate should be cooled very slowly at a temperature of 90 to 92 °F; milk chocolate at 86 to 88 °F. Air entering the cooling tunnel, where the goods are unmolded, may be 40 °F. The air may be 62 °F where the goods enter the tunnel. After chocolate is deposited in the mold, it can be moved into a cooling tunnel for a continuous cooling process, or the molds can be stacked up and placed in a cooling room with forced-air circulation. In either case, temperatures of 40 to 50 °F are satisfactory. The discharge room from the cooling tunnel or the room to which the molds are transferred for packing should be maintained at a low enough dew point to prevent condensation on the cooled chocolate. In load calculations for the cooling or cold room, it is necessary to account for transmission and infiltration losses, any load derived from further cooling of the molds, and the sensible and latent heat cooling loads of the chocolate itself.

The tunnel is designed to introduce a 40 °F air countercurrent to the flow of chocolate, so that the coldest air enters the tunnel as the cooled chocolate leaves the tunnel. Thus, as the tunnel air

---

The preparation of this chapter is assigned to TC 11.5, Fruits, Vegetables, and Other Products. The chapter last received a major revision in 1967.

Table 1 Optimum Design Air Conditions

Department or Process	Dry-Bulb[a] Temp.,°F	rh, %
Chocolate pan supply air	55 to 62	55 to 45
Enrober room	80 to 85	30 to 25
Chocolate cooling tunnel supply air	40 to 45	85 to 70[b]
Hand dipper	62	45
Molded goods cooling	40 to 45	85 to 70
Chocolate packing room	65	50
Chocolate finished stock storage	65	50
Centers tempering room	75 to 80	35 to 30
Marshmallow setting room	75 to 78	45 to 40
Grained marshmallow (deposited in starch) drying	110	40
Gum (deposited in starch) drying	125 to 150	25 to 15
Sanded gum drying	100	25 to 40
Gum finished stock storage	50 to 65	65
Sugar pan supply air (engrossing)	85 to 105	30 to 20
Polishing pan supply air	70 to 80	50 to 40
Pan rooms	75 to 80	35 to 30
Nonpareil pan supply air	100 to 120	20
Hard candy cooling tunnel supply air	60 to 70	55 to 40
Hard candy packing	70 to 75	40 to 35
Hard candy storage	50 to 70	40
Caramel rooms	70 to 80	40
Raw Material Storage		
Nuts (insect)	45	60 to 65
Nuts (rancidity)	34 to 38	85 to 80
Eggs	30	85 to 90
Chocolate (flats)	65	50
Butter	20	
Dates, figs, etc.	40 to 45	75 to 65
Corn syrup[c]	90 to 100	
Liquid sugar	75 to 80	40 to 30
Comfort air conditions	75 to 80	60 to 50

Conditions given in this table are intended as a guide and represent values found to be satisfactory for many installations. However, specific cases may vary widely from these values because of such factors as type of product, formulas, cooking process, method of handling, and time. Acceleration or deceleration of any of the foregoing will change the temperature, humidity, or both to some degree.

[a] Temperature and humidity ranges are given in respective order, *i.e.*, first temperature corresponds to first humidity.
[b] Optimum conditions.
[c] Depends on removal system. With higher temperatures, coloration and fluidity are greater.

warms up on its way out, the warmest air leaves the tunnel at the point where the warmest molten chocolate enters. The leaving chocolate is markedly cooler than the entering chocolate, and the subcooling is greatly reduced. This in turn reduces the large temperature difference between the chocolate and the cooling air along the entire tunnel length.

For any particular problem, only testing will determine the length of time the chocolate should remain in the tunnel and the subsequent temperature requirements. Good cooling is generally a function of tunnel length, belt speed, and the actual time the product contacts the cooling medium.

ENROBING OR HAND DIPPING

Chocolate coatings are applied to the center material in one of two ways: either formed by hand or cast in starch or rubber molds and then dipped by hand or enrobed mechanically. The chocolate supply for hand dipping is normally kept in a pan maintained at the lowest possible temperature that will still secure sufficient fluidity for the process. Because this temperature is higher than the dipping room temperature, a heat source, such as electrically heated dipping pans with thermostatic controls, is required. The dipped candy is placed either on trays or belts while the chocolate coating sets.

Setting is controlled by conditioning the air in the dipping room. A dry-bulb temperature of 35 to 40°F best promotes rapid setting and provides a high gloss on the finished goods. However, the temperature in the dipping room is raised for human comfort. The recommended conditions for hand-dipping rooms are 64°F dry-bulb and a relative humidity not exceeding 50 to 55%. The principal aim is to achieve uniform air distribution without objectionable drafts. The loads for this type of installation include transmission, lights, and people, as well as the heat load from the chocolate and the heat used to warm dipping pots.

In high-speed production of bar candy, the chocolate coating is applied in an enrober machine, which consists essentially of a reservoir for the fluid chocolate that is heated and thermostatically controlled. This mass is then pumped to an upper flow pan which allows it to flow in a curtain down to the main reservoir. An open chain-type belt carries the centers through the flowing chocolate curtain where the covering is picked up. At the same time, grooved rolls pick up some chocolate and apply it to the bottom of the centers. The centers should be cooled to 75 to 80°F to assist solidification and retention of the proper amount of coating.

The enrober tunnel sets the balance of the chocolate coating as rapidly as is consistent with high quality and good appearance of the finished goods. The discharge end of the enrober tunnel is normally in the packing room, where the finished candy is then wrapped and packed.

While not an absolute necessity, air conditioning the enrobing room is desirable. Because the coating is exposed to the room atmosphere, the atmosphere should be clean to prevent contamination of the coating with foreign material. It is advisable to maintain enrobing room conditions of 75 to 80°F dry bulb and 50 to 55% rh, *i.e.*, low enough to prevent the centers from warming up and to assist in the setting of the chocolate after it is applied.

The coated pieces are transferred from the enrober to the bottomer slab and then pass into the enrober cooling tunnel. The function of the bottomer slab is to set the bottom coating as rapidly as possible, thus maintaining this coating and forming a firm base for the piece as it passes through the enrober tunnel. A bottomer slab is often a simple plate evaporator fed with chilled water or brine or directly supplied with refrigerant. The belt carrying the candy passes directly over this plate and heat transfer must take place from the candy through the belt to the surfaces of the bottomer slab. A bottomer slab is sometimes located *before* the enrober to obtain a good bottom prior to full coverage.

BAR CANDY

The production of bar candy calls for high-speed semiautomatic operations to keep costs to a minimum. From the kitchen, the center material is delivered to spreaders, which form layers on tables, or is cast in starch molds. Depending on the composition of the center, the hot material may be delivered at temperatures as high as 160 to 180°F. Successive layers of different color or flavor may be placed on top of each other to build up the entire center material. These layers usually consist of nougat, caramel, marshmallow whip, or similar ingredients, to which peanuts, almonds, or other nuts may be added. Since each of the ingredients requires a different cooking process, each separate ingredient is deposited in a separate operation. Thus, a ⅛-in. layer of caramel may be first deposited and then a layer of peanuts, followed by a ¾-in. layer of nougat. Except for nuts, it is necessary to allow time for each successive layer to set before the next is applied. If the candy is spread in slabs, the slab must first be cooled and then cut with rotary knives into pieces the size of the finished center.

HARD CANDY

Manufacturing hard candy with high-speed machinery requires air conditioning to maintain temperature and humidity. The re-

Candies, Nuts, Dried Fruits, and Vegetables

quirements of candy made of cane sugar are somewhat different from that made partly with corn syrup. For example, a dry-bulb temperature of 75 to 80 °F with 40% rh is satisfactory for corn syrup (as the corn syrup percentage increases, the relative humidity must decrease), whereas the same temperature with a 50% rh is necessary for cane sugar.

Where relative humidity is to be maintained at 40% or less, standard dehydrating systems employing such chemicals as lithium chloride, silica gel, or activated alumina should be used. A combination of refrigeration and dehydration is also used.

The amount of air required is a direct function of the sensible heat of the room. Approximate rules indicate that the quantity should be between 1.5 and 2.5 cfm/ft^2 of floor area, with a minimum of 15% outdoor air or 30 cfm per person. Note that the sensible heat in hard candy is at a high temperature to keep it pliable during forming.

Where concentrations of the finished product in containers or tubs are located in the general conditioned area, the quantity of air must be increased to prevent the product sticking to the container.

Unitary air conditioners employing dry coils are satisfactory if they have a sufficient number of rows and adequate surface. A central station apparatus employing cooling and dehumidifying coils or similar design also may be used. Good filtration is essential for air purity as well as for preventing dirt accumulation on cooling coils. Reheat control is needed for constant temperature and humidity conditions. Air distribution should be designed to provide uniform conditions and eliminate the possibility of drafts.

HOT ROOMS

Such products as jellies and gums can best be dried in air-conditioned hot rooms. These products are normally cast into starch molds. The molds are contained in a tray approximately 35 by 15 by 1.5 in. with an extra 0.50 to 0.75 in. blocking at the bottom for air circulation. These trays are then racked up on trucks, with the number of trays per truck (usually 25 to 30) determined by the method of loading. The trucks are then loaded into the hot room where the actual drying is accomplished.

Normal drying temperatures average between 120 and 150 °F dry bulb. While humidity is important, close humidity control is not necessary; even with the maximum outdoor temperature conditions generally encountered when this air is heated, the relative humidity will be low (from 19 to 13%).

Some operators prefer to have manual humidity control, which requires frequent inspection of actual conditions; others prefer automatic humidity control by instruments calibrated to steadily maintain desired dry-bulb temperature and relative humidity in the hot room regardless of the moisture coming out of the candy and the supply air. With full automation, the supply should provide dry air to the unit and also cool the air usually needed in hot weather to purge the hot room after completing the drying cycle.

For proper air distribution in the hot room, it is necessary to arrange for the maximum amount of air to be in contact with the product. Providing space between trays is one means of accomplishing this. The trucks within the hot room must be placed to insure a continuous airflow from truck to truck with the shortest path for the airflow. Space must also be maintained at the entering and leaving air sides to assure flow from the top to the bottom tray for each truck. A large air quantity is required to secure uniformity over the entire product zone.

One method for achieving uniformity is the ejector nozzle system. Basically, this system consists of a supply header fitted with conical ejector nozzles designed to have a tip velocity of 2000 to 5000 fpm with a static pressure behind the nozzle ranging as high as 12 in. of water. Nozzles arranged in this way will induce a flow of air about three times that actually supplied by the nozzles. This ratio gives the most economical balance between air quantity supplied and fan horsepower. By using an ejector system, the primary and induced airstreams mix over the product and space between the top tray; sufficient ceiling height must be allowed for this mixing, which rapidly decreases the necessary differential between the air supply temperature and the actual room temperature. The high airflow thus created decreases the temperature drop across the product. Since the temperature drop is proportional to the heat pickup, a greater airflow will have a lower temperature drop, so that the spread between the air temperatures entering and leaving the product zone is reduced; this promotes uniform drying.

When drying is completed, it is necessary to cool the product rapidly to facilitate unloading. This quick cooling is provided for by a second outdoor air intake which bypasses the heating coil or the air-conditioning unit. At the same time, this bypass intake is put into operation, drop dampers are opened in the bottom of the ejector header, and this air also bypasses the ejector nozzles and flushes out the heat in the room. The use of an exhaust fan is recommended to rapidly remove the heated air, which at this time rises to the ceiling.

The equipment for this type of operation consists of a fan and heating coil located outside the hot room. (No electric motor should be in the room because of the hazard of sparking.) This unit has outdoor air intakes, ejector headers, return air dampers, and dampers for the outdoor air intake. Most operators prefer to use a recording controller to maintain an accurate record of each batch. The controller simply regulates the flow of steam to the heating coil to maintain the desired room temperature. Gradual switches should be provided to control the position of the outdoor and return air dampers, since with a rise in humidity more outdoor air should be taken, and with a drop in humidity more return air should be taken. In some cases, this function can be provided automatically by a humidity control. An end position can be included on the gradual switch for the cooling down period so that when the outdoor air damper is opened wide, the exhaust fan is started.

COLD ROOMS

Many confectionary products—such as marshmallows, certain types of bar centers, and cast cream centers—require chilling and drying but cannot withstand high temperatures. Drying conditions of about 75 °F and 45% rh are required, and the drying period varies from 24 to 48 h. Here an ejector-type system is used for the hot rooms, the difference being that cooling coils are provided in the unit. The load must be carefully calculated to determine the sensible and latent heat quantities from which the actual air quantity can be found, together with air supply temperature and refrigerant temperature.

When properly balanced, controlling the relative humidity becomes an inherent part of the system design. This control system is basically the same as for the hot room, except that its recording regulator must control the admission of steam to the heating coil in winter and regulate the flow of refrigerant to the cooling coil in summer. Flushing dampers and a cooling down cycle are necessary in this type of operation. One precaution must be observed in the equipment for the cold room in connection with starch or sugar dust picked up in the return airstream. During the cooling cycle, the cooling coil normally condenses moisture, and accumulated starch or sugar dust rapidly forms a paste on the cooling coil, reducing its capacity and necessitating frequent maintenance and cleaning of the equipment. Thus, filters should always be used for the outdoor and return air entering the cooling coil.

COOLING TUNNELS

Various candy plant cooling requirements can best be handled in a cooling tunnel, including the cooling of coated centers after leaving the enrober, the cooling of cast chocolate bars, and the cooling of hard candy. These operations are usually set up for a continuous flow of high rate production. The product is normally conveyed on belts either through the enrober or the casting machine and then through a cooling chamber.

A cooling tunnel consists of an insulated box placed around the conveyor so that the product may travel through it in a continuous flow. Cold air is supplied to this enclosure to cool the product. To achieve maximum heat transfer between the air and product, air should flow counter to the material flow. In general, air supply temperatures of 35 to 45 °F with air velocities up to 2500 fpm have been found satisfactory. The high velocity improves the heat transfer and creates a turbulent condition in the airflow which further improves heat-transfer efficiency.

The actual size of the tunnel is determined by the size of the conveyor belt and the air quantity. The air quantity depends on the heat load and the desired rate of cooling. The rise in air temperature through the tunnel should be limited to 15 to 20 °F maximum. One air-conditioning unit, which normally consists of a fan and coil together with the necessary duct connections to and from the unit, is generally used for each tunnel. An outdoor air intake is advisable, since cooling can be frequently accomplished with outdoor air, without operating the refrigeration plant. The tunnel should be made as tight as possible and the entrances and exits for the candy should be reduced to a minimum to limit air loss from the tunnel or air infiltration into the tunnel; in some cases, it is advisable to use a flexible canvas curtain to control airflow. As some loss is unavoidable, it is practicable to take a small amount of outdoor air or air from adjoining spaces to provide a slight excess pressure in the tunnel.

For chocolate-enrobing work, the condition of the air is the paramount factor in securing the best possible luster and most even coating. The best results are obtained with rather slow cooling, but this requires either a low production rate or excessively long tunnels. The final design is a compromise. The coating must be in the proper condition when it is poured over the centers, since improper temperature at this point will result in blushing or loss of luster. Proper temperature, however, is a function of the enrober machine and its operation; no amount of correction in the tunnel will compensate.

A variation of the standard single-pass counterflow tunnel has been used for enrobing. The tunnel is divided horizontally by an uninsulated sheet metal partition. The belt carrying the candy rides directly on this partition, and the return belt is brought back through the lower space below the partition. The cold air is supplied to the lower chamber near the enrober, progresses to the opposite end of the tunnel, is transferred to the top chamber, progresses back to a point near the enrober, and then is returned to the cooling equipment. This tunnel has two important advantages—it chills the return belt so that this belt can act as a bottomer slab to give a quick setting of the base of the coated piece, and the uninsulated partition with the coldest air below assists in this bottoming operation. In this manner, the air supply has already absorbed some of its heat load by the time it is actually introduced to the product-cooling zone. The method approaches the advantage of slow cooling but keeps the tunnel at a minimum length.

COATING KETTLES OR PANS

The application of air conditioning to revolving coating kettles or pans is satisfactory. Originally these pans were merely provided with a supply of warm air, ranging from 80 to 125 °F. This air was then exhausted from the kettle to the room, creating a severe nuisance because of sugar dust blowing out of the kettle. In addition, a portion of the energy required to produce rotation of the kettle is converted to heat in the centers being coated, which causes enough expansion to produce cracking or checking. Thus, a portion of the coating is applied, the product is withdrawn for a seasoning period of up to 24 h, and the material is then returned to the kettles for additional coating.

Some installations overcome most of the sugar-dust problem by using a conditioned supply of air to the kettles and providing positive exhaust from the kettles. The wet- and dry-bulb temperature of the air supplied to the kettles is so controlled that the rate of coating evaporation and the drying are uniform with a high production rate and reduced labor cost. Evaporation of the moisture in the coating material tends to take place at the wet-bulb temperature of the air supplied to the kettle. A large portion of the heat of crystallization entering the product is absorbed, and the centers are not overheated, thus eliminating the need for a seasoning period and permitting continuous operation. With air conditioning applied to coating kettles, the number of rejects caused by splitting, cracking, uneven coating, or doubles can be reduced considerably. Sugar dust recovery is accomplished with a cyclone-type dust collecting device.

PACKING ROOMS

Air conditioning is essential to production in a number of rooms or departments in every candy plant. The average manufacturer spends considerable time and effort in the design and application of packaging materials, since this affects product keeping quality. Important packaging considerations are moistureproof containers, antimoisture insulating for abnormally high humidity conditions, and protection against freezing or extreme heat. Manufacturers often overlook the fact that the air in the packing room surrounds the products when they are placed in one of these moistureproof containers. Controlling this air is essential for proper packing. For instance, products packed in a room at 85 °F dry bulb and 60% rh, which may easily be encountered in a normal summer, would have air surrounding the product at a 69 °F dew point. If this sealed package were subjected to temperatures below 69 °F, the air in the package would become supersaturated and moisture would condense on the surfaces of the container and the product. If the product was then subjected to a higher temperature, the moisture would reevaporate; but in this process, chocolates would lose their luster or show sugar bloom, and marshmallows would develop either a sticky or a grained surface, depending on the formula used. In practice, conditions of 65 °F dry bulb and 50% rh have been found most practicable. This could be improved if the relative humidity were reduced several points. In the case of hard candy, which is intensely hygroscopic, the relative humidity should be reduced to 35 to 40% at 70 to 75 °F.

REFRIGERATION PLANT

Large candy manufacturers often use a central refrigeration plant for cooling water or brine, for circulation throughout the plant to meet various requirements. In some smaller plants, small units are used for each requirement or each group of requirements. Thus, one compressor may be installed on a cooling tunnel, while separate compressors are used in each of the departments for packing, storing, and other operations. This arrangement permits a planned program of expansion or rehabilitation of the plant a portion at a time. In a large plant, this results in a multiplicity of small units and adds considerably to the work of the operation.

The type of equipment used depends on plant size, initial and operating costs, plans for future expansion, and the like (see Chapters 1 through 5). The *Safety Code for Mechanical Refrigeration* (ASHRAE *Standard* 15-1978) should be adhered to in the design of the plant, depending on the refrigerant used.

Candies, Nuts, Dried Fruits, and Vegetables

Some larger plants use centrifugal compressors in a central plant design. The centrifugal machine can be driven by motor or turbine. Most candy plants generate steam at high pressure for use in various cooling operations. Turbines can either use this high-pressure steam, exhausting at a back pressure for lower temperature cooking operations, or can condense exhaust steam. The centrifugal unit is extremely reliable, with low operating cost, ease of maintenance, and long life.

Brine at 28 to 32°F is desirable for central installations. This secondary coolant temperature is low enough to obtain the necessary low dew-point temperatures for many processes, but is high enough to prevent frosting on the finned evaporators. Solutions of alcohol and water, or a weak solution of calcium chloride brine, have been used in some plants. Increased production has caused the lowering of brine temperatures, particularly to provide cooler air supplied to tunnels.

A secondary coolant distribution system makes it possible to connect all the varied services to one source of refrigeration and, by using control devices, maintain dry-bulb and dew-point temperatures more precisely. Cold slabs are supplied for caramels, nougats, and similar products with coolant at this temperature, and its use is extended to comfort cooling in nearby offices and clerical operations.

For smaller manufacturing plants, multiple condensing units using halocarbon refrigerants and direct expansion air-conditioning units are more satisfactory. Similarly, cooling tunnels are equipped with separate cold diffusers and connected to individual halocarbon refrigerant compressors (or a group of units is connected to one single or duplex compressor). These compressors vary in size from 5 to 50 hp and can be located adjacent to the spaces conditioned.

STORAGE

CANDY

Most candies are held for one week to more than a year between manufacture and consumption. Storage may be in the factory, in warehouses during shipping, or in retail outlets. It is important that the candy not lose quality during that time.

Low-temperature storage does not produce undesirable results if: (1) the candies are made of proper ingredients for refrigerated storage; (2) the packages have a moisture barrier; (3) the storage room is held at equilibrium humidity with desirable moisture conditions for preserving the candy; and (4) the candy is brought to room temperature before the packages are opened.

The storage period depends on: (1) the marketing season of the candy; (2) the stability of the candy; and (3) the storage temperature and humidity (see Table 2).

The shelf life of a candy is determined by the stability of its individual ingredients. Common candy ingredients are sugar (including sucrose, dextrose, corn syrups, corn solids, and invert syrups), dark and milk chocolates, nuts (including coconut, peanuts, pecans, almonds, walnuts, and others), fruits (including cherries, dates, raisins, figs, apricots, and strawberries), dried milk and milk products, butter, dried eggs, cream of tartar, gelatin, soybean flour, wheat flour, starch, and artificial colors.

Refrigerated storage of candy ingredients is especially advantageous for seasonal products, such as peanuts, pecans, almonds, cherries, coconut, and chocolate. Ingredients with delicate flavors and colors, such as butter, dried eggs, and dried milk, retain quality more evenly year-round if kept properly refrigerated. Otherwise, ingredients containing fats or proteins may lose considerable flavor or develop off-flavors before being used.

Candies are semiperishable: the finest candies or candy ingredients may be ruined by a few weeks of improper storage. This includes many candy bars, packaged candies, and some choice bulk candies. Unless refrigeration is provided from time of manufacture through the retail outlet, the types of candies offered for sale must be greatly reduced in the summer.

Benefits from refrigerated storage of candies, especially during the summer, are: (1) insects are rendered inactive at temperatures below 48°F; (2) the tendency to become stale or rancid is reduced; (3) candies remain firm as an assurance against sticking to the wrapper or being smashed; (4) loss of color, aroma, and flavor is reduced; and (5) candies can be manufactured year-round and accumulated for periods of heavy sales.

Color

Many colors used in hard candies, hard creams, and bonbons gradually fade when stored at room temperature, especially in the light. However, the most marked effect of storage temperature on color occurs with chocolates. In candies containing high protein and nuts, there is a gradual darkening of color, especially at higher storage temperatures.

Temperatures from 85 to 95°F cause graying of chocolates in only a few hours and darkening of nuts within a month. Graying or fat bloom appears as *cold grease* on the surface and occurs because of crystals of fat on the surface of the chocolate coating. While this is usually associated with old candies, new candies can become gray in one day under adverse storage conditions. Chilling of chocolates following exposure to high temperatures produces graying very quickly, but chilling without previous heat exposure does not.

Sugar blooming of chocolate looks similar to graying or fat blooming and is caused by deposition of crystallized sugar on the surface from condensation of moisture, following removal of the candy from refrigerated storage without proper tempering. Experiments have shown that chocolate tempered by gradually raising the temperature to normal without opening the package did not incur sugar blooming even after storage at 0°F or lower. Sugar bloom may also occur because of storage in overhumid air and migration of moisture from the centers to the surface.

Table 2 Expected Storage Life for Candy

Candy	Moisture Content, %	Relative Humidity, %	Storage Life, Months Storage Temperature, °F			
			68	48	32	0
Sweet chocolate	0.36	40	3	6	9	12
Milk chocolate	0.52	40	2	2	4	8
Lemon drops	0.76	40	2	4	9	12
Chocolate-covered peanuts	0.91	40 to 45	2	4	6	8
Peanut brittle	1.58	40	1	1.5	3	6
Coated nut roll	5.16	45 to 50	1.5	3	6	9
Uncoated peanut roll	5.89	45 to 50	1	2	3	6
Nougat bar	6.14	50	1.5	3	6	9
Hard creams	6.56	50	3	6	12	12
Sugar bonbons	7.53	50	3	6	12	12
Coconut squares	7.70	50	2	3	6	9
Peanut butter taffy kisses	8.20	40	2	3	5	10
Chocolate-covered creams	8.09	50	1	3	6	9
Chocolate-covered soft creams	8.22	50	1.5	3	5	9
Plain caramels	9.04	50	3	6	9	12
Fudge	10.21	65	2.5	5	12	12
Gumdrops	15.11	65	3	6	12	12
Marshmallows	16.00	65	2	3	6	9

Flavor

Keeping candy fresh is one of the chief reasons for refrigerated storage. Most flavors added to candies, including peppermint, lemon, orange, cherry, and grape, are distinctive and stable during storage. Less pronounced flavors such as those of butter, milk, eggs, nuts, and fruits, are more sensitive to high temperatures.

Low temperatures retard the development of staleness and rancidity in fats, preserve flavors in fruit ingredients, and prevent staleness and other off-flavors in candies containing such semiperishable ingredients as milk, eggs, gelatin, nuts, and coconut. Candies containing fruits become strong in flavor when they are stored at room temperature or higher for more than a few weeks; those containing nuts become rancid. There are no specific critical temperatures at which undesirable changes occur, but the lower the temperature, the more slowly they take place.

Texture

Candy becomes increasingly soft at high temperatures and increasingly hard and brittle at low temperatures, reaching an optimum for eating at about 70 °F. These changes in texture are reversible from below 0 to 80 °F, enabling refrigerated candies to be returned satisfactorily to any desired temperature for eating. This is extremely important, since the texture of most candies subjected to very low temperatures (or even shipped in contact with dry ice at about −110 °F) is not permanently changed.

Most candies are manufactured at controlled temperatures. Their texture (except in hard candies and hard creams) is maintained best at temperatures below 68 °F.

Insects

Candies containing fruits, chocolate, nuts, or coconut are favorite hosts for insects. Since fumigation and insect repellants are seldom permissible with candies, refrigeration is used to inactivate insects in candy and candy ingredients.

Common insects become active at about 50 °F and activity increases as the temperature is raised to 100 °F. While common cold storage temperatures do not kill many insects, they do inactivate them at temperatures below 50 °F. Both adults and eggs may exist for months at above freezing temperatures without feeding or propagating. Candies with insect eggs on either the product or wrappers may be refrigerated for long periods with no apparent damage, but when brought back to warm rooms, a serious insect infestation may develop.

Both adults and eggs are killed when infested materials are stored at or near 0 °F. Lower temperatures and long storage periods are lethal to insects. However, storing candies at 0 °F for a few weeks usually destroys all forms of insect life.

Storage Temperature

Regarding their effect on candies, storage temperature is difficult to separate from humidity, but the latter is more important. There are no specific critical refrigerated temperatures at which a certain type of candy must be held. In general, the lower the temperature, the longer the storage life, but the greater the problem of moisture condensation on removal.

Air conditioning (68 to 70 °F). Since the storage of candies begins in the tempering room of the manufacturing plant, 68 °F and 50% rh is desirable to prevent soft sticky pieces from being packaged. Under these conditions, all candies remain firm and there will be little or no graying of chocolates; their original luster can be held.

Hard candies or types containing only sugar ingredients keep in good condition for more than 6 months at 68 to 70 °F. Other types become stale, lose flavor and luster, and darken in color. Under prolonged storage, candy containing nuts or chocolate becomes musty or rancid, and even the colors and flavors of some hard candies may fade. Only *summer candies* should be held for more than a few weeks at 68 to 70 °F, or higher. Unless precautions are taken, the temperature of truck or rail shipments may rise to the melting point of semisoft candies or to the graying point of chocolates. Candies temporarily stored in the sunshine or in warm places in buildings may suffer severely in loss of shape, luster, and color. Some companies use refrigerated trucks and railroad cars for hauling candies. As portable refrigerators become more generally available, their use for candies should increase and thereby supply the missing link of conditioned storage for candies from manufacturer to consumer.

Cool storage (48 to 50 °F). Candies stored at this temperature remain firm and retain good texture and color; only those containing nuts, butter, cream, or other fats become stale or rancid within 4 months. Candies that remain practically fresh for 4 months are fudge, caramels, sugar bonbons, gumdrops, marshmallows, lemon drops, hard creams, and semisweet chocolates; they are wrapped in aluminum foil to give added protection. Candies that become stale at this temperature are peanut butter, taffy kisses, peanut brittle, uncovered peanut rolls, chocolate-covered peanut rolls, and nougat bars.

Cold storage (32 to 34 °F). Most candies can be successfully held in cold storage for at least one year, and many for much longer. Only those containing nuts, coconut, chocolate, or other fatty materials will become stale or rancid.

Freezer storage (0 °F). The need and economic justification for freezing candy is the same as for any other food—better preservation for a longer time. This method of preservation is suitable for candies (1) in which high quality standards must be maintained; (2) in which a longer shelf life is desired than is accomplished from other methods of storage; (3) which are normally manufactured from 6 to 9 months in advance of consumption; and (4) which are especially suitable for retailing as frozen items. Because of their high sugar and low moisture content, little ice formation occurs in candies at 0 °F.

One of the chief reasons for freezing candies is to hold them in an unchanged condition for as long as 9 months, then thaw and sell them as fresh candies. Experience shows that this is not only possible but practical if the manufacturer (1) freezes only those candies that would lose quality when held at a higher temperature; (2) eliminates the few kinds that crack during freezing; (3) packages the candies in moistureproof containers similar to those used for other frozen foods; and (4) thaws the candies in the unopened packages to avoid condensation of moisture on the surface.

Moistureproof packaging. Experiments show that candies for freezing require more protection than those for common storage because the storage period is usually longer and a greater tendency exists for condensation on removal. A single layer of moistureproof material—aluminum foil, polyethylene, saran, mylar, polypropylene, cellophane, glasine, or laminations including one of these—afford adequate protection. Candies not fully protected from desiccation become hard, grainy, and lose flavor.

Protection is provided when the moisture barrier is in contact with the candy in the form of a sealed, individual wrapper. Inner liners for the boxes protect candy, provided they are sealed (which is difficult and seldom accomplished). The usual manner of applying moisture barriers is as overwraps for the boxes, chiefly because overwraps are easiest to apply and seal by machines. Overwraps provide less protection than wraps for individual pieces of candy because of the relatively large amount of air enclosed in the box. Also, boxes with extended edges offer less protection.

Thawing. Frozen candies should be thawed in the unopened packages. While freezing itself affects only a few candies, the manner of removal from storage affects all of them, especially candies unprotected by special coatings or individual wraps.

Candies, Nuts, Dried Fruits, and Vegetables

Improvement. A few candies are improved by freezing. Candies that are improved in freshness, mellowness, or textural smoothness by freezing include those with high moisture content and without protective coatings or individual wrapping. Usually these are candies ordinarily subject to surface drying. Marshmallows, jellies, caramels, fudges, divinities, coconut macaroons, fruit loaves, coconut bonbons, panned Easter eggs, malted milk balls, and chocolate puffs are in this group.

The stability of candies after freezing is good. Candies may be held frozen for 6 months or more, carefully thawed, and then sold as fresh. Some candies are prepared especially for freezing. These are made of low melting point fat, have more flavor and softer texture than most candies, and should be eaten while they are cold.

Humidity Requirements

Sugar ingredients are stable over a wide range of storage temperatures, but they are sensitive to high or low humidity. The initial moisture content of candies largely determines the optimum relative humidity of the storage atmosphere. Candies with a moisture content of 12 to 16%—marshmallows, gumdrops, coconut sticks, jelly beans, and fudge—should be stored at 60 to 65% rh to avoid becoming sticky, runny or moldy, or hard and crusty. Candies with a moisture content of 5 to 9%—most fine candies, nougat bars, nut bars, hard and soft creams, bonbons, and caramels—should be stored at 50 to 55% rh to retain the original weight, finish, and texture. Candies with moisture contents below 2%—milk chocolate bars, chocolate covered nuts, and all kinds of hard candies—should be stored at 45% rh or lower.

The hygroscopicity of the ingredients also determines the relative humidity at which candies must be held to retain the original firmness and finish. Candies with a high proportion of invert syrups, such as taffy kisses, must be kept in a very dry place, even though their moisture content is not extremely low. Other candies containing high proportions of invert syrup, honey, or corn syrup, must be held in a drier atmosphere than the moisture content indicates.

Candies stored at low temperatures have a wider range of critical relative humidities than those are stored at high temperatures. For example, nougat bars stored at 65% rh (10% too high) will become sticky within a few days at room temperature, but at 40°F or lower, stickiness might not show up for many weeks. Similarly, marshmallows stored in a room with 55% rh (10% too low) become dry and crusty within a few days at room temperature, but if stored at 40°F or lower, show little change for several weeks. Thus, refrigeration retards the ill effects of storage under improper humidity conditions. Humectants, such as sorbitol, glycerine, and high conversion corn syrup, are advantageous for maintaining original moisture content of certain candies.

NUTS

Nuts commonly refrigerated include peanuts, walnuts, pecans, almonds, filberts, chestnuts, and imported cashews and Brazil nuts. The advantages of refrigerated storage of nuts are: (1) marketable life is increased as much as ten times; (2) the natural texture, color, and flavor are retained almost perfectly from one season to the next; (3) staleness, rancidity, and molding are retarded for more than two years, depending on the temperature; and (4) insect activity is arrested at temperatures below 48°F. With optimum temperature, humidity, atmospheric conditions, and packaging, good quality nuts may be successfully stored for up to 5 years.

Temperature

Other conditions being equal, the lower the temperature, the longer the storage life of nuts. Life may be extended from 2 to 3 times with each 20°F drop in temperature. The freezing point of nuts, depending on the moisture content, is about 23°F for chestnuts; 14°F for walnuts, pecans, and filberts; and 13°F for peanuts. A normal moisture content (percentage) for stored nuts is: chestnuts, 30; peanuts, 6; walnuts, 4.5; pecans, 4; and filberts, 3.5.

Shelled nuts stored from one harvest season to the next without appreciable loss in quality must be held at 36°F or lower; those held for 6 to 9 months must be kept at 48°F or lower; and all nuts stored for 4 to 6 months should be held below 68°F. The period of storage, at a given temperature, doubles if the nuts are unshelled.

Relative Humidity

While the storage temperature of nuts may range from 68 to −20°F or lower, the relative humidity must range between 65 and 75%. This is to maintain the optimum moisture content for desired texture, color, flavor, and stability. Should the moisture content rise as much as 2% above normal, the nuts (except chestnuts) will darken, become stale, and may become moldy. If the moisture drops more than 2% below normal, the nuts become objectionably hard and brittle.

When the relative humidity is suitable, nuts that are too high or too low in moisture may be safely stored with the assurance that rapid air circulation will bring the moisture content to a safe level. In this sense, storage acts as a conditioning room.

Atmosphere

All nuts (again, except chestnuts) contain 45% or more oil and readily absorb odors and flavors from the atmosphere and surrounding products. Certain gases, particularly ammonia, react with tannin in the seed coats of nuts, causing them to turn black. Therefore, the atmosphere in the nut storage room must be free of all odors. This includes the containers, walls of the room, pallets, and other stored products.

Products that can be safely stored with nuts include dried fruits, candies, rice, and goods packaged in cans, bottles, or barrels. Commodities that should *not* be stored with nuts are onions, meats, cheese, chocolate, fresh fruits, and other products with an odor or high moisture content.

Packaging

The storage life of nuts may be greatly influenced by the choice of package. When nuts are shelled, they become bruised and oil *crawls* over the surface and onto the package in a very thin film. Unless this is retarded by a package that acts as a barrier, contains an antioxidant, or removes air by vacuum, the nuts become stale and rancid. Furthermore, some packaging materials such as pliofilm and polyethylene should be avoided because they impart an undesirable odor to the nuts.

DRIED FRUITS AND VEGETABLES

Dried fruits and vegetables differ from each other in that: (1) fruits contain sugars that render them more hygroscopic, harder to dry, and greater absorbers of moisture during storage; (2) the moisture in dried fruits may range from 32% with sorbate treatment to as low as 2%, while that in vegetables ranges from 7% to a low of 0.3%; (3) dried fruits are acid, more highly colored, and more stable during storage than vegetables, which are nonacid; (4) fruits are generally dried raw with active enzyme and respiration systems, while vegetables are blanched or precooked, with no active enzymes (thus dried fruits are more responsive to storage temperature and humidity conditions than are vegetables); (5) the high sugar and acid content of dried fruits provides an adverse physical environment for bacteria and thus makes their growth almost impossible even though the moisture level is higher than that in dried vegetables.

Dried fruits and vegetables maintain quality longer when stored at low temperatures. Packaging in a nitrogen or carbon dioxide atmosphere and in the presence of a desiccant to achieve further reduction in moisture content has been used to extend storage life of dehydrated vegetables. Staleness (charred- or off-flavor) and other deteriorative changes in dehydrated vegetables are inhibited by packing in nitrogen. In air-packed samples, the rate of staling is reduced at low temperature. Treatment with sulfite to retain color and extend storage life has been widely used for cut fruits and dried vegetables. A light coating of laundry starch applied to diced carrots prior to dehydration has achieved excellent results in retention of color and other quality factors during storage in cellophane at 84 °F without sulfite.

Refrigerated storage at 40 to 50 °F or lower retards and controls insect infestation. Substantial killing occurs with exposure at 32 °F for 6 months or longer, and a temperature of 0 °F kills insects within a few hours. An alternative method is fumigation, which is generally used in commercial practice.

Nonenzymatic browning (browning in products that have been scaled or blanched adequately to inactivate enzymes) is reduced in dehydrated vegetables at low temperature. Cold storage offers protection for several years.

Increasing the temperature of dried apricots accelerates oxygen consumption, carbon dioxide production, disappearance of sulfur dioxide, and darkening.

Molds and yeasts will not grow in dried fruits that have an adequate sulfur dioxide content or less than 25% moisture. At 32 °F, relative humidity is less important than at higher temperatures.

Other methods of dehydration include the following:

Dehydrofreezing. In this process, the raw, prepared product is dried to about 50% in weight, followed by freezing. This method has yielded excellent fruit and vegetable products when storage was at 0 °F or lower. Concentration of juices by low temperature and high vacuum and preservation of the concentrate by freezing (Chapter 19) is another application of the process.

Freeze-drying. In this process, the prepared product is frozen quickly to produce small ice crystals, followed by sublimation of the moisture with the use of vacuum and low heat, in a manner to preserve porosity of the product. This method is successfully used with products of high value, high protein, low fat, and low sugar content. While refrigerated storage is not necessary to prevent spoilage of freeze-dried food, it is necessary to preserve maximum flavor and natural color.

Dried Fruit Storage

Refrigeration is beneficial in augmenting drying as a means of preserving fruits. The optimum conditions for holding most dried fruits are at about 55% rh and just above each fruit's individual freezing temperature. Since the sugar content of these products is high, the freezing point varies from about 22 to 26 °F. Refrigerated storage is beneficial in retaining natural flavor, ascorbic acid, carotene, and sulfur dioxide, and in controlling browning, insects, rancidity, and molding. Other than for insect control, low humidity is more important than low temperatures for storing dried fruit. Packaged, sulfured cut fruits will keep adequately at higher humidities.

While most dried fruits are adversely affected by softening and injured by freezing, dates are held best by freezing. Before storage, dried fruits should be brought to the desired moisture content.

When practical, dried fruits should be packed in moistureproof containers made of metal or foil, which not only assure a constant moisture content in storage, but prevent injury from moisture condensation on removal from storage. The permeability of the packaging medium the dried fruits are sealed in is extremely important in relation to adverse effects of storage humidity. Dried fruit storage recommendations are:

Raisins. At 32 to 40 °F and at 50 to 60% rh, sugaring is prevented for one year, provided the moisture content of the dried fruit is not unusually high. Raisins contain 15 to 18% moisture; for extremely long storage, the lowest possible moisture content should be maintained.

Figs. These may be held for a year at 32 to 40 °F and at 50 to 60% rh. A temperature of 55 °F or lower prevents darkening for more than 5 months, and low humidity controls sugaring.

Prunes. These may be held for a year at 32 to 40 °F with a relative humidity from 50 to 60%. For storage of 4 to 5 months, a relative humidity of 75 to 80% is not detrimental.

Apples. At 32 to 40 °F with a relative humidity of 55 to 65%, apples retain excellent color and texture for more than one year. A relative humidity of 70 to 80% is not objectionable at 32 °F, but at 40 °F enough moisture may be gained to cause the fruit to mold within 8 months. Browning develops gradually at temperatures of 40 °F and above. Even at 40 °F, browning will develop but at a slower rate.

Pears. Same as for apples.

Peaches. Sun-dried freestone peaches are harder to store than most dried fruits. Therefore, the temperature should be held close to 32 °F with a rh of 55 to 65%. At 40 °F and moderate humidity, the moisture pickup will cause molding and browning to develop rapidly.

Clingstone peaches (dehydrated after steam scalding) should be stored at 32 to 40 °F and at 55 to 75% rh. Sun-dried peaches will tolerate a slightly higher humidity than dehydrated peaches.

Apricots. Dried apricots are easy to keep in refrigerated storage at 32 to 40 °F with a 55 to 65% rh. They remain in excellent condition for more than one year. At 40 °F with moderate humidity there is enough gain in moisture content to cause molding.

Dates (sucrose or hard type). For storage of 6 months or less, dates may be held at 32 °F with 70 to 75% rh, but for longer storage they should be stored at 24 to 26 °F. Usually it is more convenient to store them at 32 °F, at which temperature they can be stored for over a year.

Soft or invert sugar-type dates may be held for 6 months at 28 to 32 °F, but if stored for 9 to 12 months they should be stored from 0 to 10 °F. Uncured dates should be stored at 0 to 10 °F.

Dried Vegetable Storage

Few specific recommendations for dehydrated vegetable storage temperatures are available. Decrease in storage temperature retards deterioration, but cold storage is considered necessary only for long storage periods. Among the advantages are: (1) control of insects at 45 °F or lower; (2) preservation of natural colors; and (3) retention of initial flavors and vitamins.

Since most dried vegetables have very low moisture content, are well packaged, and are usually surrounded by an atmosphere of nitrogen or rarefied air, refrigeration is less essential than it is for undried vegetables.

CONTROLLED ATMOSPHERE

Low oxygen in the storage atmosphere suppresses the growth of insects and molds, retards rancidity and staleness, and reduces oxidative changes in flavors, odors, and colors. In small packages, oxygen can be reduced by vacuum; in storage and large shipping containers, oxygen can best be flushed out with nitrogen. Excellent results have been obtained by substituting up to 98% of the atmosphere with nitrogen. Nitrogen is preferred to carbon dioxide, ethylene, or other gases for storage of low moisture products; it greatly extends the shelf life even with refrigeration.

CHAPTER 23

BEVERAGE PROCESSES

BREWERIES 23.1	Heat Treatment of Red Musts 23.8
Chemical Aspects 23.1	Juice Cooling 23.9
Processing 23.3	Heat Treatment of Juices 23.9
Storage 23.6	Fermentation Temperature Control 23.9
Pasteurization 23.6	Potassium Bitartrate Crystallization 23.9
Carbon Dioxide 23.6	Storage Temperature Control 23.9
Heat Balance 23.7	Chill Proofing Brandies 23.10
Common Refrigeration Systems 23.7	CARBONATED BEVERAGES 23.10
Vinegar Production 23.8	Water Coolers 23.10
WINE MAKING 23.8	Size of Plant 23.11
Must Cooling 23.8	Liquid Carbon Dioxide Storage 23.11

THIS chapter discusses the processes and use of refrigeration in breweries, wineries, and carbonated beverage plants.

BREWERIES

Malt is the primary raw ingredient used in the brewing of beer. While adjuncts such as corn grits and rice contribute considerably to the composition of the extract, they do not possess the necessary enzymatic components required for the preparation of the wort. Malting is the initial stage in preparing raw grain to make it suitable for mashing. Traditionally this operation was carried out in the brewery, but it is now almost entirely the function of a separate industry.

Various grains such as wheat, oats, rye, and barley can be malted. Barley is by far the predominate one used in the preparation of mash because it has a favorable ratio of protein to starch; it also has the proper enzyme systems required for conversion, and the barley hull provides an important filter bed during lautering.

Barley is malted by steeping the raw grain in water of about 40 to 65 °F for 2 or 3 days, until its moisture content increases from about 12 to about 45%. The water is changed frequently, and the grain is aerated. When the kernels germinate, the water is drained from the batch and germination continues. During the growing period, the green malt is turned over continuously to assure uniform growth of the kernels. Either slowly revolving drums turn over the growing malt, or, in the compartment system, slow-moving mechanically driven, plow-like agitators are used.

On reaching the desired stage in its growth, the green malt is transferred to a kiln, where further growth is checked by reducing the moisture content. Kilning is usually done in two stages. First, the moisture content of the malt is reduced to approximately 8 to 14%; then, the heat is increased until the moisture is further reduced to 4%. This heating procedure reduces excessive destruction of enzymes. The desired color and aroma are obtained by controlling the final degree of heat.

After kilning, the malt is cleaned to separate dried rootlets from the grain, which is then stored for future use. The finished malt differs from the original grain in several significant ways. The hard endosperm is modified and now is chalky and friable. The enzymatic power has been greatly increased, especially alpha amylase, which is not present in unmalted barley. The moisture content is reduced, making it more suitable for storing and subsequent crushing. It now has a distinctive flavor and aroma, and the starch and diastase are readily extractable in the brewhouse.

The preparation of this chapter is assigned to TC 11.7, Fermentation Products.

CHEMICAL ASPECTS

Two distinct chemical reactions are used in brewing beer. In the first, which is carried out in the brewhouse and is called mashing, the starches in the grains are hydrolyzed into sugars, and complex proteins are broken down into simpler proteins, polypeptides, and amino acids. These reactions are brought about by first crushing the malt and suspending it in warm (100 to 122 °F) water by means of agitation in the mash tun. A portion of the malt is cooked separately with an adjunct, usually corn grits or rice. After boiling, this mixture is combined with the main mash to form a mixture in a temperature range of 145 to 162 °F. Within this temperature range, the alpha and beta amylases degrade the starch to mono, di, tri, and higher saccarides. By suitably choosing a time and temperature regimen, the brewer controls the amount of fermentable sugars produced. The enzyme diatase (essentially a mixture of the enzymes alpha and beta amylase) which induces this chemical reaction acts as a catalyst and is not consumed. Some of the maltose is subsequently changed by another enzyme, maltase, into a fermentable monosaccharide, glucose.

Mashing is complete when the starches are converted to iodine-negative sugars and dextrins. At this point, the temperature of the mash is raised to 167 to 170 °F to stop the amylalytic action and fix the ratio of fermentable to nonfermentable sugars. The wort is separated from the mash solids using a Lauter Tub, a mash filter, or a Strainmaster. Hot water (168 to 170 °F) is then "sparged" through the grain bed to recover additional extract. Wort and sparge water are added to the brew kettle and boiled with hops, which may be in the form of pellets, extract, or whole blossoms. After boiling, the brew is quickly cooled and transferred to the starting cellar, where yeast is added to induce fermentation. A flow diagram is shown in Figure 1. On this flow diagram, the grains flow downward by gravity. The hot wort from the bottom of the brewhouse is then pumped to the top of the stockhouse, where it is cooled and again proceeds by gravity through fermentation and lagering.

After cooling the wort, yeast and sterile air are injected into it. The yeast is pumped in as a slurry at a rate of 1 to 3 lbs of slurry for each barrel of wort. The wort may be oxygenated with pure oxygen. Normally, however, oil-free compressed air is filtered and treated with ultraviolet light and then added to the wort to saturate it with oxygen to a level of approximately 11 ppm.

Two types of reactions, respiration and fermentation, take place in the fermenter. Respiration takes place under aerobic conditions.

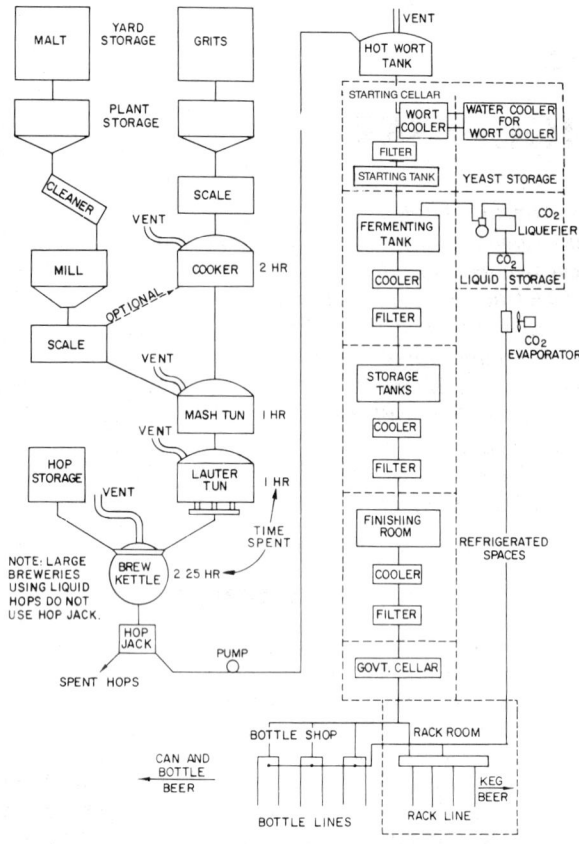

Fig. 1 Brewery Flow Diagram

$$\text{Extract} \xrightarrow[\text{O}_2]{\text{yeast}} CO_2 + H_2O + \text{more yeast}$$

The extent of respiration (and yeast growth) is governed by the oxygen supply as well as nutrients and other metabolic products. When oxygen is depleted, yeast multiplication will continue to increase the number of yeast cells 3 to 5 times, while fermentation begins under the heat of fermentation. Meanwhile, the amount of CO_2 produced by the action of yeast on wort extract in a fermenter will depend on wort composition and the practices of yeast management in each brewery. A representative value for the heat of fermentation is 240 Btu/lb of extract fermented.

The amount of CO_2 produced can be calculated for the typical North American lager wort.

1. $\underset{180}{C_6H_{12}O_6} \longrightarrow \underset{92}{2\ C_2H_5OH} + \underset{88}{2\ CO_2}$

2. $\underset{342}{C_{12}H_{22}O_{11}} + \underset{18}{H_2O} \longrightarrow \underset{184}{4\ C_2H_5OH} + \underset{176}{4\ CO_2}$

3. $\underset{504}{C_{18}H_{32}O_{16}} + \underset{36276}{2\ H_2O} \longrightarrow \underset{264}{6\ C_2H_5OH} + 6\ CO_2$

If fermentable extract is 10, 43, and 12% (1, 2, and 3, respectively) of the total extract, the theoretical yield of CO_2 is 0.512 lb CO_2/lb extract.

Bottom fermentation type yeast, *Saccharomyces uvarium*, is used in fermenting lager beer. Top-fermentation type yeast, *Saccharomyces cerevisiae*, is used in making ale. They are so called because, after fermentation, one settles to the bottom and the other rises to the top. A more significant difference between the two types is that in the top-fermentation type, the fermenting liquid is allowed to attain a higher temperature before a continued rise is checked. The brewing of ale may differ from the brewing of lager in the following ways: (1) for ale, a more highly kilned malt is used; (2) malt forms a greater proportion of the total grist (less adjunct); (3) infusion mashing is used and a wort of higher original density is generally produced; (4) more hops are added during the kettle boil; and (5) a different yeast and temperature of fermentation is employed.

Therefore, ale may have a somewhat higher alcohol content than lager beer and has a fuller and more bitter flavor. With the bottom fermentation yeasts, fermentation is generally carried on between 45 and 65 °F and most commonly between 50 and 60 °F. Ale fermentations are generally carried on at somewhat higher temperatures, often peaking from 70 to 75 °F. In either type, the temperature during fermentation continues to rise, but it is checked by cooling coils or attemperators through which a cooling medium is circulated.

Wort is measured by a saccharometer, a hydrometer calibrated to read the percentage of solids as maltose in solution with water. The standard instrument is the Plato saccharometer, and the reading is referred to as a percentage of solids, or degrees Plato. Table 1 lists the various data deducible from reading a saccharometer.

The same instrument is used to check the progress of fermentation. While it still gives an accurate measure of density of the fermenting liquid, it is no longer a direct indicator of dissolved solids, because the solution now contains alcohol, which is lighter than water. This variation in saccharometer reading is called the apparent extract, which is always greater than the real extract (apparent attenuation is calculated from the hydrometer reading of apparent extract and the original extract). In engineering computations, 81% of the change in apparent extract may be considered a close approximation of the change in real extract. Thus, 81% of the difference between the solids shown in Table 1 for saccharometer readings before and after fermentation represents the mass of maltose fermented. This weight in pounds times 240 Btu

Table 1 Total Solids in Wort

% solids[a]	Specific gravity	Weight per barrel, lb		Specific heat, Btu/lb·°F
		Total	Solids	
0	1.0000	258.7	0.00	1.000
1	1.0039	259.7	2.60	0.993
2	1.0078	260.7	5.21	0.986
3	1.0118	261.8	7.85	0.979
4	1.0157	262.8	10.51	0.972
5	1.0197	263.8	13.19	0.965
6	1.0238	264.9	15.89	0.958
7	1.0278	265.9	18.61	0.951
8	1.0319	267.0	21.36	0.944
9	1.0360	268.0	24.12	0.937
10	1.0402	269.1	26.91	0.930
11	1.0443	270.2	29.72	0.923
12	1.0485	271.2	32.55	0.916
13	1.0528	272.4	35.41	0.909
14	1.0570	273.4	38.28	0.902
15	1.0613	274.6	41.18	0.895
16	1.0657	275.7	44.11	0.888
17	1.0700	276.7	47.06	0.881
18	1.0744	277.9	50.03	0.874
19	1.0788	279.1	53.03	0.867
20	1.0833	280.2	56.05	0.860

Note: One barrel = 31 gal.
[a]Saccharometer readings.

Beverage Processes

gives the heat of fermentation in Btu per barrel (each barrel has a capacity of 31 gal). The difference between the original solids and weight of fermented solids gives the residual solids per barrel. It is assumed that there is no change in the volume because of fermentation. The specific heat of beer may be assumed to be the same as that of the original wort, but the weight per barrel is decreased according to the apparent attenuation.

PROCESSING

Wort Cooling

To prepare the boiling wort from the kettle for fermentation, it must first be cooled to a temperature of 45 to 55 °F. To avoid contamination with foreign organisms, which would adversely affect the subsequent fermentation, cooling must be done as quickly as possible, especially through the temperatures around 100 °F. Besides the primary function of wort cooling, other beneficial effects accrue that are essential to good fermentation. These include precipitation, coagulation of proteins, and oxidation because of natural or induced aeration depending on the type of cooler used.

In the past, the Baudelot type cooler was almost universally used because it is easy to clean and affords the necessary aeration of the wort. However, the traditional open-type Baudelot cooler was replaced by one consisting of a series of swinging leaves encased within a removable enclosure into which sterilized air is introduced for aeration. These have virtually been replaced by the totally enclosed type of heat exchanger. Air for aeration is admitted under pressure into the wort stream, usually at the discharge end of the cooler. The air is first filtered and then irradiated with bactericidal radiation, or it can be sterilized by heating in a double-pipe heat exchanger with steam. Injection of 0.167 ft^3 of air per bbl of wort, which is the amount necessary to saturate the wort, should allow a normal fermentation to occur. The quantity can be increased or diminished accurately as the subsequent fermentation indicates.

The coolant section of wort coolers is usually divided into two or three sections. In the first section, a potable source of water is drawn on. The heated effluent goes to hot water tanks where, after additional heating, it is used for succeeding mashing and sparging brews. Final cooling is done in the last section, either by direct expansion of the refrigerant or by an intermediate coolant such as chilled water or propylene glycol. A third section may be placed between these two sections to recover the warm water, which is stored in a wash-water tank for later use in various washing and cleaning operations around the plant.

Closed coolers save space and reduce the need for expensive cooler room air-conditioning equipment. They also permit a faster cooling rate and provide accurate control of the degree of aeration. The problem of cleaning plate coolers is solved by having a spare set of plates ready for weekly replacement. The cleaning between successive brews is accomplished by circulating cleaning or sterilizing solutions, or both, through the cooler.

In selecting a wort cooler, the following should be considered:

1. The cooling rate should be such that the contents of the kettle can be cooled in less than 2 h.
2. The heat transfer surfaces to be apportioned between the first section using an available water supply and the second section using refrigeration should be such that the most economical use is made of each of these media. Cost of water and its temperature, as well as its availability, should be balanced against the cost of refrigeration. Usual design practice is to cool the wort in the first section to within 10 °F of the available water.
3. Usable heat should be recovered. The effluent from the first section is a good source of preheated water, which can, after additional heating, be used for succeeding brews and as wash water in other parts of the plant. At all times the amount of heat recovered should be consistent with the overall plant heat balance.
4. Meticulous sanitation is of prime importance and its maintenance cost should not be disregarded.

The size of wort coolers is determined by the rate of wort cooling desired, the rate of water flow, and the temperature differences employed. A brew, which may vary in size from 50 to 1000 bbl and over, is ordinarily cooled in less than 1 to 4 h. Open-type coolers are made in stands up to about 20 ft in length. Where more length is needed, two or more stands are operated in parallel.

Open coolers are best operated with a wort flow of from 10 to 11.5 bbl/h per foot of stand. As the flow is increased beyond this rate, an increasingly larger part of the wort is splashed from the top tube of the cooler and dropped directly into the collecting pan below without contacting the cooler surfaces. An increased amount of wort that flows over the surfaces must be subcooled to offset what has been bypassed.

This bypassing does not occur in the plate-type cooler and wort velocities can be increased up to a point where the friction head through the cooler approaches the maximum design pressure of the press and gasketing. The number of passes and streams in each pass afford the designer much latitude in selecting for optimum performance and economical design. This design is based on the specific heat of wort, its initial temperature and the range through which it is to be cooled, the temperature of the available water supply, and the ratio of the quantity of cooling water to wort that is to be used. Design and operating features of a typical plate cooler are as follows:

Specifications

Quantity of wort to be cooled	17,000 lb/h
Temperature of hot wort	210 °F
Temperature of cooled wort	40 °F
Temperature of available water (maximum)	70 °F
Water used, not to exceed	34,000 lb/h
Temperature of water leaving cooler	140 °F
Temperature of wort leaving first section of cooler	80 °F
Temperature of incoming recirculated chilled water	34 °F

Plate Cooler (first section)

Number of plates	40
Heat transfer surface per plate	4.3 ft^2
Heat transfer surface in first section	172 ft^2
Number of passes	5
Number of streams per pass	4
Water flow rate	34,000 lb/h
Wort flow rate	17,000 lb/h

Plate Cooler (second section)

Number of plates	24
Heat transfer surface per plate	4.3 ft^2
Heat transfer surface in second section	103 ft^2
Number of passes	3
Number of streams per pass	4
Chilled water flow rate	52,000 lb/h

A shell-and-tube or plate-type cooler with two stages of cooling can accomplish the wort cooling efficiently. In the first (hot) stage, potable water runs counter-flow to the wort, and the usual discharge temperature is about 168 °F. This hot water is then used in the subsequent brews at various blended temperatures. Excess is used in the brewery's general operations.

In the second stage of wort cooling, a closed system of refrigerated water at about 36 °F through a closed cooler, cools the wort to 50 °F or lower, depending on the brewer. Lower temperature water (33 °F) may be used in open-type units where no danger exists from freezing.

It is possible to cool the wort in one stage, depending on the potable water temperature available and plant refrigeration capacity. This arrangement might be simpler and cheaper, but sewerage cost may be important where water saving is mandatory. Heat exchangers have many benefits including (1) ease of cleaning, (2) uniform temperature control, (3) uniform control of the sterile air injected for aeration, (4) reduction of overall steam requirements for brewing, and (5) a minimum of water wasted. Two-stage cooling has reduced water consumption from 17 to 20 barrels per barrel of beer to 6 to 10 barrels per barrel of beer.

Fermenting Cellar

After cooling, the wort is pitched with yeast and collected in a fermenting tank, where respiration and fermentation occur according to the chemical reaction previously outlined. The daily rate of fermentation varies, depending on the operating procedure adopted in each plant. On the first day, a representative rate might be 2 lb of converted maltose per barrel of wort. The rise in temperature caused by fermentation and by the growth and changing physiology of the yeast increases this rate to 7 lb on the second day. At that time, the maximum desired temperature has been attained and a further rise is checked by an attemperator, so that on the third day another 7 lbs is converted. This rate continues on through the fourth day. Two examples of the fermentation rate follow.

Example 1. Normal Gravity Brewing

Day	°Plato	Real Extract	Extract per bbl, lb	Extract Fermented per bbl, lb
0	11	11.0	29.63	—
1	10	10.2	27.38	2.25
2	8	8.6	22.95	4.43
3	5	6.1	16.12	6.83
4	3	4.5	11.81	4.31
5	2.5	4.1	10.75	1.06
				18.88

Example 2. High Gravity Brewing

Day	°Plato	Real Extract	Extract per bbl, lb	Extract Fermented per bbl, lb
0	16	16.0	43.97	—
1	15	15.2	41.64	2.33
2	12	12.8	34.73	6.91
3	7	8.7	23.11	11.51
4	4	6.3	16.66	6.56
5	3.5	5.9	15.58	1.08
				28.39

After the fourth day, the amount of unconverted maltose remaining in the beer is greatly diminished. Because of the inhibiting effect of alcohol, carbon dioxide, and other products of fermentation on further yeast propagation, the action nearly stops on the fifth day, when only about 3 lb are converted. At this stage, the yeast begins to flocculate (clump together) and either settle to the bottom of the fermenter (bottom yeast) or rise to the top (top yeast). Because of the reduced fermentation rate, the temperature of the beer begins to fall, either as the result of increased attemperation applied to the tank itself or through heat loss from the tank to the surrounding area or a combination of both. Many fermentation programs call for the beer to be cooled to a temperature ranging from 35 to 45°F at this time. This period of rapid cooling aids in settling the yeast. At the completion of this cooling period, the fermentation rate is essentially zero, and the beer is ready to be transferred off the settled yeast. Complete fermentation can generally be effected in about 7 days. The introduction of new types of beers (*i.e.*, reduced calorie, reduced alcohol), and the more general use of high gravity brewing by brewers, has led to a variety of fermentation programs both between brewers making the same product and within the same brewery for different products.

While complete fermentation can be accomplished in less than 7 days, most brewers may take a total from 7 to 14 days or more to accomplish the fermentation and subsequent cooling, depending on original gravity, whether a secondary fermentation is used, and available cooling capacity. Most brewers cool the beer to between 42 and 38°F after end ferment or the final days of quick cooldown in the fermenting tank. Simultaneously, the long rest allows the yeast to settle. Some brewers agitate the beer in cylindrical fermenters, which enables them to ferment the beer faster and then separate out the yeast by centrifuge. Most brewers cool the beer before it goes in RUH storage (for resting and settling between fermentation and final aging) to the desired 29 to 30°F temperature.

Fermenting Cellar Refrigeration

Convection resulting from temperature gradients in the beer provides part of the agitation necessary for heat exchange between the attemperator and the beer. Convection is caused principally by the ebullition set up by the CO_2 bubbles rising to the surface of the liquid. A heat transfer rate ranging from 15 to 30 $Btu/h \cdot ft^2 \cdot °F$ is reasonable for estimating the heat transfer surface required. The heat loss from tank walls and the surface of the liquid may be disregarded in calculating the attemperator coil surface requirements. However, if the room temperature is allowed to drop appreciably below 50°F, the heat dissipated through the metal tank walls becomes important. Depending on the degree of heat dissipation, fermentation may be retarded or even inhibited. In such instances, wooden fermenting tanks or insulated tank walls and bottoms are needed.

The refrigeration requirements are based on the maximum volume of wort in fermentation at one time as illustrated by Example 3.

Example 3. Figure 2 illustrates the volume of wort production based on a 500 bbl per day production rate. The days are represented by the

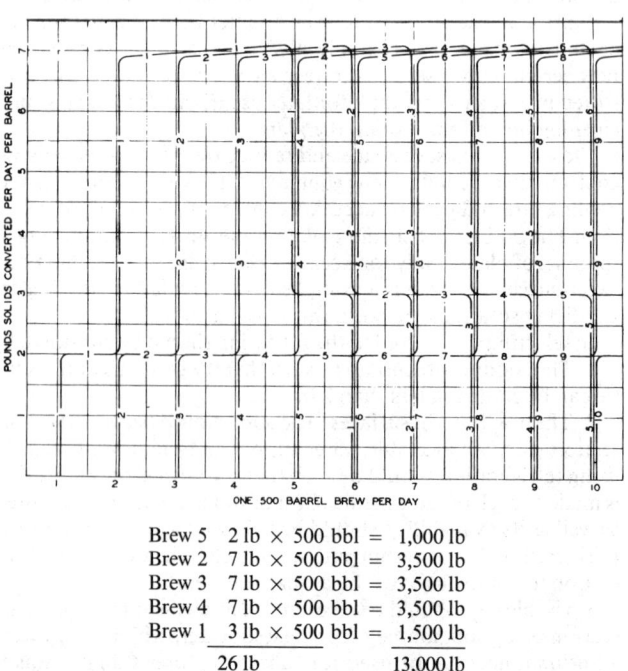

Brew 5	2 lb × 500 bbl =	1,000 lb
Brew 2	7 lb × 500 bbl =	3,500 lb
Brew 3	7 lb × 500 bbl =	3,500 lb
Brew 4	7 lb × 500 bbl =	3,500 lb
Brew 1	3 lb × 500 bbl =	1,500 lb
	26 lb	13,000 lb

Fig. 2. Solids Conversion Rate

Beverage Processes

abscissa, and the pounds of solids converted per day, by the ordinate. The individual brews in fermentation on any particular day are additive. For example, on the fifth day, Brew No. 1 is finishing with a conversion rate of 3 lb for that day; Brew No. 5, which is just beginning the fermentation cycle, is fermenting at the rate of 2 lb; while Brews No. 2, 3, and 4 are at the maximum rate of 7 lb per bbl per day each. The total solids fermented on this day is then 26 lb per bbl for the 2500 bbl in fermentation, totaling 13,000 lb per day of solids converted. Since the heat of fermentation is 280 Btu/lb, the refrigeration load would be:

$$(13{,}000 \times 280)/(24 \times 12{,}000) = 12.6 \text{ tons of refrigeration}$$

Calculations for sizing attemperators must consider: (1) the internal dimensions of the fermenting tank and its capacity; (2) the temperature difference between the coolant and the fermenting beer; (3) the maximum daily sugar conversion rate; and (4) the heat evolved, which is at the rate of 280 Btu/lb of fermentable sugar converted.

Assuming a square fermenting tank 13 ft on a side to hold a brew of 500 bbl, and allowing 1 ft between the tank wall and the attemperator for easy cleaning, a 11 ft² square attemperator can be used, giving 44 perimetrical feet of tubing.

From Figure 2, the maximum daily conversion rate is shown to be 7 lb per barrel. Calculating for 500 bbl per day at 280 Btu/lb of sugar converted,

$$7 \times 500 \times 280 = 980{,}000 \text{ Btu/day or } 40{,}833 \text{ Btu/h}$$

Then assuming 50°F fermenting beer temperature and 20°F brine, a temperature difference of 30°F and a heat transfer rate of 15 Btu/(h·ft²·°F) for the attemperator, the number of square feet of surface required would be:

$$40{,}000/(15 \times 30) = 90.7 \text{ ft}^2$$

Considering 4-in. OD tubing with an external area of 1.05 ft²/ft, the number of lineal feet required would be:

$$90.7/1.05 = 86.4 \text{ ft}$$

Two attemperators, each 11-ft square, would give 88 ft of tubing, which is adequate for the conditions outlined.

The amount of CO_2 generated depends on the original density of the beer and the degree of fermentation. A little more than half the generated CO_2 is collected and liquefied for later use in the brewery for carbonating, for counterpressure in transfer operations, and for the bottle, can, and keg lines.

Adequate fresh air must be provided in an open tank fermenting cellar to safely dilute the CO_2 emanating from the fermenters. The amount permissible should be such that there is no health hazard to the personnel working in these spaces. Concentrations below 0.5% are considered safe. Increasing amounts reduce the efficiency of a worker and 4% concentrations make the work performance for protracted periods untenable. Carbon dioxide tends to settle to the floor, from where it may be withdrawn by scupper connections at the floor level. Fresh air may be introduced through the ceiling through openings located so as not to remove or disturb the layer of protective CO_2 gas on top of the fermenting beer.

Attemperators consist of a partial exterior jacket on the walls of the tank and are supported at about two-thirds the height of the liquid. The jacket is a multipass of three or more baffled passes, or a dimple plate design. A glycol solution or liquid ammonia may be circulated through the cooling jackets.

Stock Cellar

The stock cellar is a refrigerated room containing tanks into which the cooled beer from the fermenter is transferred for aging and settling. The retention of beer in the stock cellar is a matter of the brewer's discretion and may be from one week to several months. The beer is stored in the presence of some yeast remaining from the original fermentation. Under these conditions, slow, subtle, but important changes take place and contribute to the flavor characteristic of the beer, including the coagulation of protein, which might produce haze in the finished product.

While the CO_2 concentration of the air in the stock cellar is less than in the fermenting room, a fresh air supply in sufficient amounts to keep the concentration below 0.5% is necessary. Air-conditioning equipment using chemical dehumidification and refrigeration holds dry conditions such as 32°F and 50% rh in storage areas to reduce mold growth and rusting of the structural steel. To maintain lower CO_2 concentration in tightly closed cellars and to reduce operational cost, heat exchange sinks and thermal (exchange) wheels are used to cool incoming fresh air and exhaust cold stale air.

Air compressor systems commonly use air driers with refrigerated aftercoolers, 32°F glycol coolers, desiccant drying, or both. Drying is necessary if air lines pass through areas below 32°F.

A continuous aging process used in multistory buildings, all gravity flow, is shown in Figure 3. The process is better for larger operations where one principal brand is produced.

Kraeusen Cellar

Instead of carbonating the beer at the finishing step, some brewers prefer to carbonate by the Kraeusen method. In this procedure, the fully fermented beer is moved from the fermenting tank to a tank capable of holding about 20 psig and a small percentage of actively fermenting beer is added. The tank is allowed to vent freely for a period, generally 24 to 48 h, then is closed and the CO_2 pressure is allowed to build up. Since the amount of CO_2 retained in the beer is a function of temperature and pressure, the brewer can achieve the desired carbonation level by controlling either pressure or temperature or both. At the conclusion of the Kraeusen fermentation, which generally lasts a week or more, the beer may be moved to another storage tank. However, the brewer can accomplish the same effect by leaving the beer in the Kraeusen tank and cooling the beer either by space cooling or tank coils or both.

Heat is generated by this secondary fermentation, but the temperature of the liquid does not rise to as high a temperature as it did in the fermenter because the fermentable sugars are available only from the small percentage of actively fermenting beer, added as Kraeusen. Furthermore, the bulk of the liquid may have a lower starting temperature than in primary fermentation. Typically a temperature of 40 to 50°F may be reached at the peak, after which the liquid cools to the ambient temperature of the room. This cooling can be accelerated in the tanks by means of attemperators

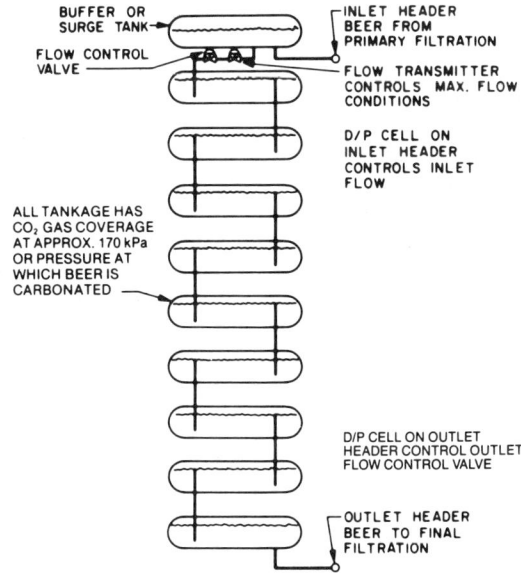

Fig. 3 Continuous Aging Gravity Flow

through which a cooling liquid, such as propylene glycol, is circulated. Since heat is generated during the Kraeusen fermentation, refrigeration load calculations must include removal of this heat either by transfer to the air in the cellar or by tank coils or a combination of both. Furthermore, if the tank is to be used as a storage tank, the calculation must include the necessary heat removal to reduce the beer temperature to the desired level.

Finishing Operations

After flavor maturation and clarification in the storage tanks, the beer is ready for finishing. Finishing includes the processes of carbonation, stabilization, standardization, and clarification.

Carbonation. Any of the following processes are used to raise the CO_2 concentration from 1.2 to 1.7 volumes/volume to about 2.7 volumes/volume:

- Kraeusen
- in-line
- in-tank with stones
- saturator
- aging train

Stabilization. The formation of colloided haze caused by soluble proteins and tannins forming insoluble protein-tannin complexes is reduced by any of the following materials:

- enzymes (papain)
- tannic acid
- tannin absorbents, PVPP, Nylon 66
- protein absorbents, silica gel, bentonite

Standardization. Chilled, deaerated, and carbonated water is added to adjust from high gravity level (14 to 16°Brix) down to normal package levels (10 to 12°Brix or lower for low-calorie beer).

Clarification. In the finishing cellar, the beer is polished by filtration and is then carbonated by any of several different methods. The filtering is usually done through a series of cellulose pulp filters and/or diatomaceous earth filters. The number of filters used depends on the brilliance desired in the finished product. After this final processing, beer is transferred to the government cellar and held until it is required for filling kegs in the racking room, and bottles or cans in the packaging plant. In some breweries, initial clarification is accomplished using centrifuges. This reduces the load on the filtration system allowing higher flow rates and longer filter runs.

STORAGE

Outdoor Storage Tanks

Some breweries use vertical outdoor fermenting and holding tanks (similar to those used in dairies). These tanks have working capacities from 2000 barrels to 10,000 barrels. The geometry of the tanks includes a conical bottom and height to diameter ratios from 1 to 1 up to 5 to 1. The tanks are jacketed and use direct expansion ammonia or propylene glycol as refrigerants. Insulation is usually 4 to 6-in. thick polyurethane foam with an aluminum cladding. They may be built as fermenters or as aging tanks; in many cases, fermenting and aging are completed in the same tank with no beer transfer.

Hop Storage

Hops should be stored at a temperature of 32 to 34°F with 55 to 65% rh, and with very little air motion, to prevent excessive drying. Sweating of the bales should not be permitted, as this would carry off the light aromatic esters and deteriorate the fine hop character. Nothing else should be stored in the hops cellar, since foreign odors may be absorbed by the hops, which can impart off-flavors to the beer.

Yeast Culture Room

In the yeast culture room, yeast is propagated for reseeding and replacing yeast that has lost its viability. Normal fermentation of aerated wort also propagates yeast. The amount of yeast roughly triples during fermentation depending on the degree of aeration. A portion of this yeast is repitched in later fermentation and the balance is discarded as waste yeast (sometimes sold for other purposes). The clean yeast selected for reuse is usually the middle layer of the deposit of yeast remaining in a fermenting tank after the beer is removed.

This yeast is carefully handled to avoid contamination with bacteria and is stored in the yeast room as a liquid slurry (yeast balm) in suitable vats. If open vats are used, a relative humidity of 80% is required to prevent the yeast from hardening on the vat walls. The CO_2 blanket on top of the vats should not be disturbed by excessive air motion. Brewers' practice in handling and recycling yeast varies greatly.

PASTEURIZATION

Plate pasteurizers warm the beer to a temperature sufficient for proper pasteurization (15 s at 160°F or 10 s at 165°F) and then cool the pasteurized product with incoming cold beer. Plate pasteurizers and millipore filtering are used to produce a beer similar to draft beer, which does not require refrigeration to prevent spoilage. It is distributed in bottles and cans, which can be of slightly lighter construction since they do not have to withstand the high pasteurizing pressure created in tunnel-type pasteurizers.

CARBON DIOXIDE

Collection

Carbon dioxide gas, produced as a byproduct of fermentation, can be collected from closed fermenters, compressed, and stored in pressure tanks for later use. It may be used for final carbonation, counterpressure in storage and finishing tanks, transfer, and in bottling and canning of the product. In the past, the CO_2 was stored in the gaseous state at about 250 psig. However, in most medium and large breweries, the gas is collected and, after thorough washing and purification, it is liquefied and stored. Carbon dioxide stored in the liquid state occupies about 0.02 of the volume of an equal weight of gas at the same pressure at room temperature.

As an example, each barrel of wort fermented generates about 13 lb of CO_2 over a period of 5 days, though not at a constant daily rate. Brewers must, therefore, carefully schedule production to provide the necessary CO_2 gas and minimize storage requirements. As a general rule, only about 50 to 60% of the total gas generated is collected. The gas generated at the beginning and at the end of the fermentation cycle is unsuitable because of excessive air content and other impurities and is discarded.

From the fermenting tank, the gas is piped through a foam trap to a gas pressure booster. Surplus gas is discharged to the outside from a water-column safety relief tank, which also protects the fermenting tank from excessive gas pressure. The booster raises the pressure from as low as 1 in. of water to 5 or 6 psig to compensate for friction loss in the long lines to the compressor and to increase compressor capacity.

Compressors, which in the past were of the two-stage type with water injection, are being replaced by nonlubricated compressors using carbon or Teflon rings. Lubricated screw and reciprocating compressors are in use today for food and beverage grade CO_2 production in commercial CO_2 plants. These may be single two-stage compressors, or two compressors comprising individual high and low stages. A complete collection system, shown diagrammatically in Figures 4 and 5, also consists of a suction and foam trap; rotary boosters where required; a scrubber, deodorizer, and

Beverage Processes

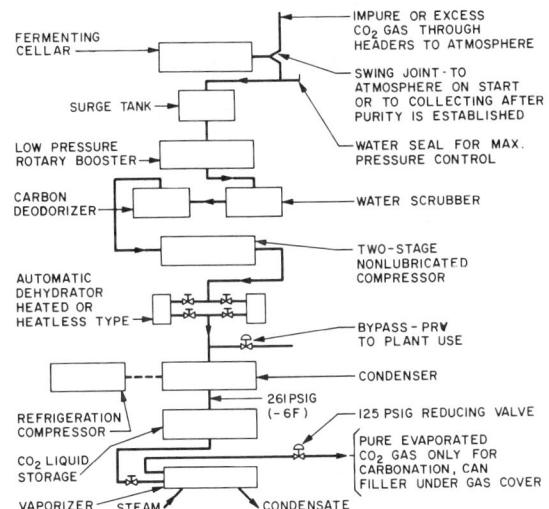

Fig. 4 Typical Arrangement of CO_2 Collecting System

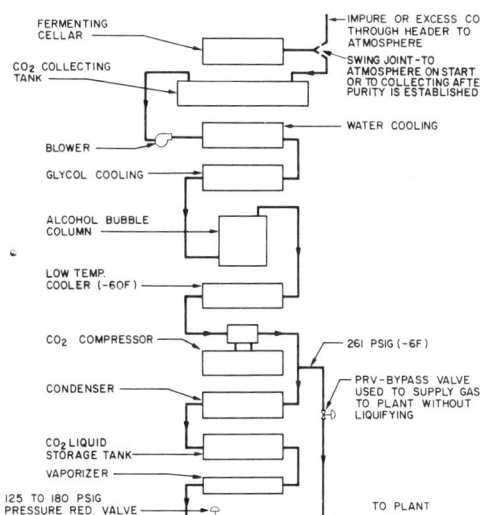

Fig. 5 Special CO_2 Collecting and Liquefaction System

compressor(s); intercoolers and aftercoolers; a dehumidifying tower; a condenser (with refrigeration from plant cascade system or separate compressor); liquid storage tanks; and vaporizers, all interconnected and automatically controlled.

Liquefaction

From tables of the properties of CO_2, the condensing pressures at several temperatures are: $-20\,°F$, 200 psig; $-14\,°F$, 225 psig; and $-8\,°F$, 252 psig.

The latent heat at saturation temperature is about 120 Btu/lb. The refrigerant for liquefying the compressed gas (usually ammonia or R-12) should be about $-21\,°F$ to effectively condense the CO_2. Most of the moisture must be removed from the compressed gas; this may be done by passing the gas through a horizontal-flow finned coil located in a $36\,°F$ cellar, which condenses out about 80% of the moisture, and the condensate is drained from the system. Moisture may also be removed effectively with refrigerant cooled precoolers, intercoolers, and aftercoolers. Further removal is accomplished by passing the gas through desiccant driers. The emerging gas has a slightly higher temperature but has a dew point around $-70\,°F$ or lower.

Under these conditions, the gas is liquefied by contacting the liquefying surfaces which stay ice-free because of the low moisture content ($-40\,°F$ dew point) of the gas and thus assure continuous service. The driers are installed in duplicate, with automatic timing for regeneration of the desiccant material. Desiccant driers usually rely on liquid CO_2 as the refrigeration gas. An earlier method used dual sets of double-pipe driers in which moisture was frozen out and retained in the heat exchanger.

Liquefiers are vertical shell-and-tube, inclined double-pipe, or shell-and-tube. The refrigerant side is operated fully flooded, with the refrigerant being supplied from the main plant and a booster compressor discharging into the plant suction main. Carbonating systems have undergone changes with all-closed fermenters collecting, and refrigerated condensing systems, and large liquid CO_2 holding tanks.

Storage and Reevaporation

The condensed CO_2 drains into a storage tank, which is usually designed for 300 psig working pressure, and varies in storage capacity from 10,000 to 120,000 lb each. The vessel is enclosed in an insulated box and is equipped with equalizing connections, safety valves, liquid-level indicator or trycocks, and electric heating units. Tests are regularly conducted for gas purity from samples taken from above the liquid level.

As liquid is withdrawn from the tank, it is introduced into a steam heated liquid evaporator, which is automatically controlled to give the desired superheat to the reevaporated gas. This type of evaporator is gradually replacing other types because of its ability to control the temperature of the reevaporated gas very closely. The evaporated gas is then temporarily stored in buffer tanks until needed in the brewhouse and package plant.

HEAT BALANCE

Most of the steam required for processing, heating water, and general plant heating can be obtained as a byproduct. Because the manufacture of beer is a batch process with various peaks occurring at different times, the study of the best heat balance possible is rather involved. In a given plant, a comprehensive study of all factors is of utmost importance. Further automation has been accomplished by programming the flow of materials in the brewhouse as well as the entire brewing operation. The newest breweries are fully automated.

In brewery locations where electric energy costs are high, cogeneration facilities should be considered. However, in plants that produce over 3 million bbl annually, the steam turbine is often the prominent prime mover. A bleeder-type turbine operating at 400 psig can drive a refrigeration compressor, electrical generator, or both, and steam can be bled from it for process and other needs requiring low-pressure steam. In smaller plants, a less favorable heat balance must be accepted in line with a more economical initial plant investment. Each brewery requires individual study to procure the most economical equipment.

COMMON REFRIGERATION SYSTEMS

Absorption Machine for Heat Balance (especially for air conditioning and cooling [water] for wort cooling). The unit requires a careful heat balance study to determine if it is economical.

Halocarbon Refrigerant Cascade System. Oil-free ammonia as a coolant can be pumped at a 4 to 1 ratio. The system requires pumps, but more often it is expanded on a 1 to 1 ratio if it is part of a compression system. The ratio of motive power to refrigeration at 25 and 185 psig is 1.5 bhp/ton.

Direct Centrifugal. This often is an oil-free ammonia system with a probable 1 to 1 ratio, or it can be recirculated. The unit requires pumps or a pressure transfer system.

Oil Sealed Screw Compressors. Ammonia is circulated at a 1 to 1 ratio in this system. The units require pumps or a pressure transfer system. The ratio of motive power to refrigeration is 1.38 bhp/ton.

Oil-free Compressors, Screw Type. The ratio of motive power to refrigeration is 1.5 bhp/ton.

Large Balanced Opposed Horizontal Double-Acting Reciprocating Compressor. While the system is not oil-free, good oil separation equipment minimizes this problem. It can use recirculation or direct expansion. The ratio of motive power to refrigeration is 1.25 bhp/ton.

Where necessary, a cooling tower may be used to reduce thermal pollution or to conserve water in the pasteurizing phase. Ecology plays an important part in the brewery; stacks are monitored for particulates, effluent is checked, and heat from the kettle vents and others is recovered. Water usage is being more closely regulated, and water-saving equipment, including evaporative condensers, are used in the refrigeration systems.

VINEGAR PRODUCTION

Vinegar is produced from any liquid capable of first being converted to alcohol (such as wine, cider, and malt) and syrups, glucose, molasses, and the like. The process occurs in two stages:

First Stage: Conversion of sugar to alcohol by yeast

$$C_6H_{12}O_6 + \text{yeast} = C_2H_5OH + CO_2$$

Second Stage: Conversion of alcohol to acetic acid by bacterial action

$$CH_3CH_2OH + O_2 + \text{bacteria} = CH_3COOH + H_2O$$

Bacteria are active only at the surface of the liquid where air is available, so two methods are used to increase the air-to-vinegar surface. In the *packed generator*, a vertical cylinder with perforated plate is filled with oak shavings or other inert support material to increase the column surface area. A weak alcohol and vinegar culture is introduced, and the solution is continuously circulated through a sparger arm. Air is introduced through drilled holes in the top of the tank. A heat exchanger removes the heat generated and holds the solution at 86°F. This batch process requires 72 h.

In the *submerged fermentation process*, air bubbles are distributed through the mash in the tank that is nearly filled with cooling coils to hold the temperature to 86°F. This batch process requires 39 h. The vinegar is best concentrated by removing some of the water in the form of ice, which achieves a 12 to 40% increase in acid. A rotator is often used, and the ice is separated in a centrifuge. About 0 to 10°F is required on the evaporating surface to produce the best crystals. The vinegar is then stored for 30 days before filtering. The solution can be concentrated by distillation (as in "distilled white vinegar").

WINE MAKING

The use of refrigeration to control the rates of various physical, chemical, enzymatic, and microbiological reactions in commercial wine making is well established. Periods of high temperature followed by rapid cooling can be used to denature oxidative enzymes and proteins in grape juice, to retain desirable volatile constituents of the grapes and to enhance the extraction of color pigments from the skins of red grapes, to modify the aroma of the juices from certain white grape cultivars, and to inactivate the fungal populations of mold-infected grapes. Reduced temperatures can be used to slow the growth rate of natural yeast and of the enzymatic oxidation of certain phenolic compounds, to assist in the natural settling of grape solids in juices, and to favor the formation of certain byproducts during fermentation. Reduced temperatures can also be used to enhance the nucleation and crystallization of potassium bitartrate from wines, to slow the rate of aging reactions during storage, and to promote the precipitation of wood extractives of limited solubility from aged brandies.

The extent to which refrigeration is used in these applications depends on such factors as the climatic region in which the grapes have been grown, the grape cultivars used, the physical condition of the fruit at harvest, the styles and types of wines being produced, and the discretion of the wine maker. A variety of enological practices and winery equipment can be found between the batch emphasis of small wineries (crushing tens of tons per season) and the continuous emphasis of large wineries (crushing hundreds or thousands of tons per season).

The applications of refrigeration to wine making is classified and considered in the following order: (1) must cooling; (2) heat treatment of red musts; (3) juice cooling; (4) heat treatment of juices; (5) control of fermentation temperature; (6) potassium bitartrate crystallization; (7) control of storage temperatures; and (8) chill proofing of brandies

MUST COOLING

Must cooling is the cooling of crushed grapes prior to the separation of the juice from the skins and the seeds. White musts are often cooled prior to entering a juice-draining system or a skin-contacting tank. This early cooling reduces the rate of oxidation of certain juice components and prevents the onset of spontaneous fermentation by wild and potentially undesirable organisms. It can be employed when grapes are delivered to the winery at excessively high temperatures or when they have been heated to aid in pressing or extracting red color pigments.

In general, tube-in-tube or spiral heat exchangers are used for this application. Tubes of at least 4 in. internal diameter with detachable end sections of large-radius return bends reduce the chance of blockage by any stems that might be left in the must after the crushing-destemming operation. The cooling medium can be chilled water, a glycol solution, or a direct expanding refrigerant. Overall heat transfer coefficients for must cooling range between 70 and 125 Btu/(h·ft²·°F), depending on the proportions of juice and skins. In small wineries, jacketed draining tanks and fermenters are often used to cool musts in a relatively inefficient batch procedure. Overall coefficients in these tanks are around 1.7 to 5.3 Btu/(h·ft²·°F) because the must is stationery and, therefore, rate controlling.

HEAT TREATMENT OF RED MUSTS

Most red grapes have white or greenish-white flesh and juice. The coloring matter (anthocyanins) resides in the skins. Color can be rapidly extracted from these varieties by heat treating the musts so that the pigment-containing cells are disrupted prior to actual fermentation. This is done in several countries when grape skins are low in color or when color extraction during fermentation is poor. Heating the must is necessary to produce the desirable flavor profile of some varieties, most notably Concord, in the manufacture of juices, jellies, and wines. The series of operations is called *thermovinification*. The must is heated to 135 to 167°F, generally by draining off some of the juice, condensing steam, and returning the hot juice to the skins for a given contact time, often for about 30 min. The complete must can be cooled prior to separation and pressing, or the colored juice can be drawn off and cooled prior to fermentation.

Red grapes can also be heat treated to inactivate the more active oxidative enzymes found in red grapes that are infected by the mold, *Botrytis cinerea*, or to improve the action of pectic enzyme preparations added to facilitate the pressing of some cultivars. In all cases, the temperature time pattern employed is a compromise between desirable and undesirable reactions. The two most undesirable reactions are caramelization and accelerated oxidation of the juice. Condensing steam and tube-in-tube exchangers are

Beverage Processes

generally used for these applications, with design coefficients similar to those given previously for must cooling.

JUICE COOLING

Juices separated from the skins of white grapes are usually cooled to between 35 and 70°F to aid in the natural settling of suspended grape solids, to retain volatile components in the juice and to prepare it for a cool fermentation. Tube-in-tube, shell-and-tube, and spiral exchangers and small jacketed tanks are used either with the direct expansion of the refrigerant, an ethylene glycol solution, or with chilled water as the cooling medium. Overall coefficients for juice cooling range between 95 and 150 Btu/(h·ft²·°F) for the exchanger and 4.4 to 8.8 Btu/(h·ft²·°F) for small jacketed tanks. The jacketed small tank values can be improved significantly by juice agitation. Transport and thermal properties of grape juices can be taken to be those of 24% sucrose solutions by mass. Medium-sized and large wineries tend to employ continuous-flow juice cooling of jacketed tanks.

HEAT TREATMENT OF JUICES

Juices from sound grapes can be exposed to a high temperature for a short time treatment (HTST) to denature grape proteins, reduce the number of unwanted microorganisms, and, some winemakers believe, enhance the varietal aroma of certain juices. The denaturation of proteins reduces the need for their removal from the finished wine by absorptive clays such as bentonite. However, the treatment can cause turbid juices and wines, presumably because the pectin and polysaccharide fraction is modified. Clarified juices from mold-infected grapes can be heat treated to denature oxidative enzymes and to inactivate the molds.

A typical program would include rapidly raising the juice temperature to 194°F, holding it for 2 s, and then rapidly cooling it to 60°F. Plate heat exchangers are used for this treatment because of their thin film paths and high overall coefficients. Grape pulp and seeds can block flow through this equipment if they are not completely removed beforehand.

FERMENTATION TEMPERATURE CONTROL

The anaerobic conversion of grape sugars to ethanol and carbon dioxide by yeast cells is exothermic even though the yeast captures a significant quantity of the overall energy change in the form of high energy phosphate bonds. Bouffard (1895) found experimental values of the heat of reaction range between 79.3 and 95.3 Btu/mole, with the value of 94.4 Btu/mole being generally accepted for fermentation calculations.

One gallon of juice at 16.37 lb/ft³ sucrose (24°Brix) will produce approximately 60 gal of carbon dioxide during fermentation. Allowing for the enthalpy lost by this gas, with its saturation levels of water and ethanol vapors, the corrected heat release value is 90.9 Btu/mole at 59°F and 83.6 Btu/mole at 77°F. The adiabatic temperature rise of the 16.37 lb/ft³ juice would then be 87°F, based on the 59°F value, and 82°F based on the 77°F value. Whether or not fermentation will approach these adiabatic conditions depends on the difference between the rate of heat generation by fermentation and the rate of heat removal by the cooling system. For constant temperature fermentations, the most common type of temperature control, these rates must be equal. Red wine fermentations are generally controlled at temperatures between 75 and 90°F, while white wines are fermented at 50, 60, or 68°F depending on the cultivar used and the type of wine produced. More rapid fermentations of red wines are used in the cooling load calculations of individual fermenters. A more involved calculation, which allows for both red and white fermentations staggered in time, is necessary for the overall daily fermentation loads.

At 68°F, red wines have average fermentation rates in the range of 2.5 to 3.1 lb/ft³ per day, which correspond to heat release rates of approximately 21 to 26 Btu/ft³ per hour. The peak fermentation rate is generally 1.5 times the average, leading to values of 31.5 to 39.0 Btu/ft³ per hour. This value, multiplied by the volume of must fermenting, provides the maximum rate of heat generation. The heat transfer area of the jacket or external exchanger can then be calculated from the average coolant temperature and the overall coefficient.

The largest volume of a fermenter of given proportions, whose fermentation can be controlled by jacket cooling alone, is a function of the maximum fermentation rate and the coolant temperature. The limitation occurs because the volume (and hence the heat generation rate) increases with the diameter cubed, while the jacket area (and hence the cooling rate) only increases with the diameter squared. Similarly, the temperature rise in small fermenters cooled only by the ambient air depends on the fermenters' volume and shape and on the ambient air temperature (Boulton 1979a).

Boulton (1979b) developed a kinetic model for wine fermentations that predicts the daily or hourly cooling requirements of a winery. Many different fermentation temperatures, volumes, and starting times can be incorporated in the model to predict future demands and schedule off-peak electricity use for optimal control of refrigeration compressors.

POTASSIUM BITARTRATE CRYSTALLIZATION

Freshly pressed grape juices are usually saturated solutions of potassium bitartrate. The solubility of this salt decreases as alcohol accumulates so that newly fermented grape wines are generally supersaturated. The extent of supersaturation and even the solubility of this salt depends on the type of wine. Young red wines, for example, are able to hold almost twice the potassium content at the same tartaric acid level as young white wines. Temperature, ethanol concentration, and pH also affect the saturation level. Since salt solubility decreases with temperature, wines are usually cold stabilized so that any crystallization occurs in the tank rather than in the bottle of the finished wine that is to be chilled. In the past, this was done by holding the wine at close to its congealing temperature (usually 25 to 21°F for table wines) for 2 to 3 weeks. Crystallization at these temperatures can be accelerated dramatically by introducing nuclei, either potassium bitartrate powder of other neutral particles, and agitating. Nuclei for crystallization are particularly important in stainless steel tanks, since they do not offer convenient sites for rapid growth as do older wooden cooperage. The holding times can be reduced to between 1 and 4 h by these methods. Several continuous and semicontinuous processes have been developed, most incorporating an interchange of the cold exit stream with the warmer incoming wine (Riese and Boulton 1980). Dessert wines can be stabilized in the same manner, except that the congealing temperatures are usually in the range of 12 to 7°F. Stabilization usually takes place some time after the harvest period, and the suction temperatures of the refrigeration compressors must be adjusted in favor of low coolant temperatures rather than refrigeration capacity.

STORAGE TEMPERATURE CONTROL

The control of storage temperature is perhaps the most important aspect of the post-fermentation handling of wines, particularly generic white wines. As wine is transferred from a fermentor to a storage vessel, it becomes at least partially saturated with oxygen. The rates at which oxidative browning reactions (and the associated development of acetaldehyde) advance depends on the wine, its pH and free sulfur dioxide level, and the storage temperature. Studies by Berg and Akiyoshi (1956) on the oxidation of white wines indicate that for temperatures below ambient, the rate was reduced to one-fifth its value for each 18°F reduction in temperature. Similar studies by Ramey and Ough of the hydrolysis of carboxylic esters produced during low-temperature fermentations indicate that the rate was more than halved for each 18°F reduction in temperature. These data suggest that, on average, the esters have half-lives of 380, 600, and 940 days when wine is stored at 60, 50, and 40 °F, respectively. As a result, wines should generally be stored at temperatures between 40 and 50 °F if oxidation and ester hydrolysis are to be reduced to acceptable levels.

The importance of cold bulk wine storage will likely increase as vintners strive to reduce the amount of sulfur dioxide used. The cooling requirements during storage can be calculated from the dimensions and materials of construction of the vessels and the thickness of the insulation used.

CHILL PROOFING BRANDIES

Aged brandies contain polysaccharide fractions extracted from the wood of the barrels, when the proof is in the range of 100 to 120 (50 to 60% v/v ethanol). When the proof is reduced to 80 for bottling, some of these components with limited solubility become unstable and precipitate, often as a dispersed haze. These components are removed by rapidly chilling the diluted brandy to a temperature in the 0 to 32 °F range and filtering it while cold. This is generally done by using a plate heat exchanger and a pad filter. The outgoing filtered brandy is then used to precool the incoming stream in order to reduce the cooling load. Calculations can be made by using the properties of equivalent ethanol-water mixtures, paying particular attention to the viscosity effects on the heat transfer coefficient.

CARBONATED BEVERAGES

Refrigeration equipment is required in most carbonated beverage bottling plants to cool the finished beverage. The refrigeration load varies with plant and production conditions; small plants may use about 150 tons of refrigeration and large plants may require over 1500 tons. The product water is deaerated before cooling, which aids carbonation and prolongs product life. Cooling the finished beverage (1) facilitates carbonation and obtains maximum stability of the carbonated beverage during filling, which become critical under high-speed operations; (2) promotes uniformity of the finished beverage; and (3) permits reducing the pressure at which the beverage is filled into the container package.

To prepare a product that will have the proper life and show the desired sparkle and release of carbon dioxide gas when the finished beverage is served, the water and syrup are proportioned, cooled, and supersaturated by injecting carbon dioxide gas under pressure in an apparatus known as a carbonator-cooler. The amount of gas dissolved varies inversely with the temperature of the water and directly with the carbon dioxide gas pressure—the lower the temperature of the product, the less the gas pressure needed for a given carbonation. Cold carbonated product is more stable and less likely to lose its gas (to foam) when filling the beverage container.

For example, at 60 °F and atmospheric pressure, a given volume of product will absorb an equal volume of carbon dioxide gas. If the carbon dioxide gas is supplied to the product under a pressure of approximately 15 psig, it will absorb two volumes. For each additional 15 psig, one additional volume of gas is absorbed by the product. Reducing the temperature of the product to 32 °F increases the absorption rate to 1.7 volumes. Therefore, at 32 °F product temperature, each increase of 15 psig in CO_2 pressure results in the absorption of an additional 1.7 volumes instead of one volume as when the product temperature is at 60 °F. Table 2 lists the volume of carbon dioxide dissolved per volume of water at various temperatures. Beverages of low carbonation fall below 3.5 volumes of carbon dioxide per volume of water, while the higher carbonation may run up to 5.0 volumes.

WATER COOLERS

Both instantaneous or storage-type water coolers, based on the rate of cooling and the volume of water stored, are available. The instantaneous cooler is easier to sanitize, operates more economically, and requires less space than the storage type. However, the instantaneous units must be sized carefully to match the refrigeration demands of the beverage production facility.

Water may be cooled by a Baudelot-tank unit, cascade tray, shell-and-coil, shell-and-tube, or carbonator-cooler. The Baudelot or descending film cooler consists of stamped, corrugated, stainless steel plates which resemble a washboard and in which the corrugations form channels for the refrigerant flow. The water trickles down the surface of the plates in a thin film from an overhead pan with distributing holes and is collected in a lower tank.

In the tank-type cooler, refrigerant expansion coils are immersed in a tank of water. Other arrangements baffle the water for long flow. In the cascade tray unit, the expansion coils are laid in a series of inclined trays. The water flows over the coils by gravity, usually in a zigzag path.

In the internal-coil or shell-and-coil unit, water flows through a closed pipe coil immersed in the refrigerant. This closed system cooler normally operates under pressure and can be used to cool water after it has been carbonated.

All these coolers have direct expansion or direct heat exchange between the evaporating refrigerant and the water to be cooled. Brine or propylene glycol secondary coolants may be used either for special design or as a precaution against the possibility of refrigerant leakage into the water. In some plants, this system has been adapted to brine already cooled in some other part of the plant for another purpose.

In combination units, the cooling surfaces are built directly onto the carbonator. Hence the name carbonator-cooler. Here the surfaces both cool and present a large area to the carbon dioxide atmosphere. The designs are a variation of the descending film cooler, and the capacity of the cooling surface or units is based on manufacturers' specifications.

Product Temperature

In the U-tube water cooler, a series of pipes immersed in a shell containing water carries the refrigerant. The pipes are bent in a horizontal U to provide a smooth passage for the refrigerant. Both the carbonator cooler and the U-tube cooler may operate flooded or as direct expansion chillers with the refrigerant inside the pipes or plates.

Table 2 Volume of CO_2 Gas Absorbed in One Volume of Water

Temperature °F	Pressure in Bottle, psig										
	0	10	20	30	40	50	60	70	80	90	100
32	1.71	2.9	4.0	5.2	6.3	7.4	8.6	9.7	10.9	12.2	13.4
40	1.45	2.4	3.4	4.3	5.3	6.3	7.3	8.3	9.2	10.3	11.3
50	1.19	2.0	2.8	3.6	4.4	5.2	6.0	6.8	7.6	8.5	9.5
60	1.00	1.7	2.3	3.0	3.7	4.3	5.0	5.7	6.3	7.1	7.8
70	0.85	1.4	2.0	2.5	3.1	3.7	4.2	4.8	5.4	6.1	6.6
80	0.73	1.2	1.7	2.2	2.7	3.2	3.6	4.1	4.6	5.2	5.7
90	0.63	1.0	1.5	1.9	2.3	2.7	3.2	3.6	4.0	4.5	4.9
100	0.56	0.9	1.3	1.7	2.0	2.4	2.8	3.2	3.5	3.9	4.3

Beverage Processes

Sanitation

It is most important to keep the beverage free from contamination, either by foreign substances or organisms picked up from the atmosphere, by infections, or by metals dissolved in transit. For these reasons, coolers are designed for ease of cleaning and freedom from water stagnation, and they are constructed of corrosion-resistant nontoxic metal (preferably stainless steel).

Smaller cooling units can usually be placed anywhere in the plant, while larger units may be built into a refrigerated room. Water should be protected from contamination by covering the cooling surfaces and reservoirs. All parts should be metal, because whenever wood or organic material comes in contact with water, the possibility for the growth of microorganisms increases. If units built into a refrigerated room are shown on plant tours, inspection windows should be provided rather than allowing visitors to enter the room. The coolers should be locked to prevent tampering.

Refrigeration Plant

Halogenated hydrocarbons or ammonia refrigerants are commonly used for these coolers. Compressors vary also with the particular unit in question, and, as with other commercial units, they vary from the two-cylinder vertical units to the larger, multicylinder V-style compressors.

The refrigeration plant should be centralized in larger production facilities. With the multiplicity of product sizes, production speed variations, and other factors affecting the refrigeration load, an automatically controlled central plant conserves energy, reduces electrical power costs, and improves opportunities for a preventive maintenance program.

Small plants commonly use main or well water for condensers. Water use restrictions have encouraged the use of evaporative condensers and cooling towers, however. Increased economy can be gained by using the spent water from empty can and bottle rinsers as makeup water for the evaporative condensers and cooling towers.

The temperature to which the product must be cooled depends on the type of filling machinery used. These may be divided into three general classes: (1) those that use water at supply temperature or less; (2) those that operate with water at 45 to 55 °F, and (3) those that require water of 40 °F or lower.

Filling machines that use water at the temperature of the water supply must hold pressure against the carbonated water continuously during the filling process. Claims for this system state that extremely stable carbonated water is not needed since the pressure maintained prevents loss of carbonation. This filler will, nevertheless, operate at lower filling pressures if the water is cooled, and a more uniform product is likely to result. The exact temperature to which the water should be cooled depends on the requirements of the particular plant.

Those units that operate with beverages cooled from 45 to 55 °F are the intermediate type. Because of their valve design, they can release the pressure on the filled bottles before crowning without losing carbonation. They have proved fully successful and can be made to operate at reduced speed with somewhat warmer water if necessary.

The third class of filler requires a very stable carbonated product, which is achieved by cooling the product to 40 °F or lower. Only with low product temperature and properly designed valves can the best results, high speed, and proper filling be obtained. With warmer product, irregularity of the finished product is likely to occur. In each installation, it is important to determine the type of filling equipment first.

Refrigeration Load

The refrigeration load is determined by the amount of water being cooled per unit of time. Since most cooling units are of the instantaneous type, they must furnish the desired output of cold water continuously without relying on storage reserve.

To determine the water demand of the given equipment, the maximum fluid output of the filler should be determined. This is usually found when filling litre bottles with carbonated water. Under these circumstances, the common commercial units require from 60 to 200 gpm of water.

Knowing the water temperature from the municipal source, the temperature to which the water is to be cooled, and the water demand, the refrigeration load can be determined by the following approximation:

$$q_t = Qc_p(t_s - t_c)/24$$

where

q_t = cooling load, tons
Q = water flow rate, gpm
t_s = supply water temperature, °F
t_c = cold water temperature, °F
c_p = specific heat of water = 1 Btu/lb·°F

In the computation of the load, one of the most troublesome values to determine is the highest temperature the main water can be expected to reach. This temperature usually occurs in the hottest summer period. Moreover, allowance should be made for additional warming of the water going through the water treatment equipment in the bottling plant. The extent to which water will rise in temperature as it progresses to the filling equipment depends on the insulation of the equipment and pipes. Newer apparatus is well insulated. In poor installations, or those in which the insulation has become wet or damaged, the gain may reach and even exceed 10 °F. Therefore, this factor must be judged for each individual plant installation.

SIZE OF PLANT

The output of each plant depends on the filling capacity of the plant equipment. Small individual units turn out approximately 600 cases of 24 bottles each of splits (approximately half-pint capacity) per hour; intermediate units turn out up to 1200 cases per hour; and high-speed fully automatic machines begin at approximately 1200 cases per hour and go through several increases in size onto the largest units, which approach 5000 cases per hour.

Operation of these machines, which also determines the demand on refrigeration machinery, usually exceeds 8 h per day, especially during summer months when market demands are highest.

An arbitrary classification of plants may be: small plants that produce under 1,250,000 cases per year; intermediate plants that produce about 2,500,000 cases per year; and large plants that produce 15,000,000 or more cases per year. The latter require installation of multiple filling lines.

The usual area of distribution of the finished beverages is within the metropolitan area of the city in which the plant is located. Some plants have built such a reputation for their goods that they ship to warehouses several hundred miles distant. Local distribution is made from there. A few nationally known products are shipped long distances from producing plants to specialized markets.

Cans and nonreturnable bottles are commonly warmed to a temperature exceeding the dew-point temperature in the warehouse to prevent condensation and resulting package damage. Transportation of the beverages does not always require refrigeration, although it is advisable to protect bottled goods against excessive temperature and direct sunlight. At the point of consumption, the beverage is customarily cooled to temperatures of approximately 40 °F.

LIQUID CARBON DIOXIDE STORAGE

Carbon dioxide for carbonating water and beverages is furnished compressed and liquefied in steel cylinders designed for pressures up to 1500 psi. Liquefied carbon dioxide may also be delivered to the plant in trucks. The liquid is piped from the trucks to large storage-converter tanks equipped with mechanical refrigeration and electrical heating. The typical unit is maintained at internal temperatures not in excess of 0°F, so that the equilibrium pressure of the carbon dioxide does not exceed 300 psig and the storage tanks need not be built for excessively high pressures. Full controlled equipment heats or refrigerates, and safety relief valves discharge sufficient carbon dioxide to relieve excess pressure.

REFERENCES

Berg, H.W. and M. Akiyoshi. 1956. Some factors involved in the browning of white wines. *American Journal of Enology* 7:1.

Bouffard, A. 1895. Determination de la chaleur degagee dans la fermentation alcoolique. Comptes rendus hebdomadaires des seances, Academie des sciences, Paris 121:357. *Progres agricole et viticole* 24:345.

Boulton, R. 1979. The heat transfer characteristics of wine fermentors. *American Journal of Enology and Viticulture* 30:152.

Boulton, R. 1979. A kinetic model for the control of wine fermentations. Biotechnology and Bioengineering Symposium No. 9, 167.

MBAA. 1981. *The practical brewer*, 2nd ed. Master Brewer's Association of the Americas, Madison, WI.

Ramey, D.R. and C.S. Ough. 1980. Volatile ester hydrolysis or formation during storage of model solutions and wine. *Agriculture and Food Chemistry* 28:928.

Riese, H. and R. Boulton. 1980. Speeding up cold stabilization. *Wines and Vines* 61(11):68.

CHAPTER 24

EGGS AND EGG PRODUCTS

Shell Eggs 24.1
Egg Products 24.3
Refrigeration Design Data 24.4
E-3-A Sanitary Standards 24.5

REFRIGERATION of shell eggs is the most effective and practical means for preservation of quality and is widely practiced at farm holding rooms, processing plants, and in marketing channels. Monthly production of shell eggs fluctuates less than 10% above or below the monthly average, and demand is fairly constant. Thus, the warehousing of eggs is a matter of backlogging stocks for handling and grading and for adjusting supply to the current demand, decreasing long-term cold storage to a negligible quantity. Liquid egg products (about 40% of all egg products) must be consumed almost immediately because of their short storage life. The other egg products are packed either in frozen or dried form, and may be stored for a considerable time.

SHELL EGGS

Quality Grades and Weight Classes

The U.S. Egg Products Inspection Act of 1970 requires that all eggs moving in interstate commerce be graded for size and quality. No cracked, dirty, or loss eggs may be sold except direct to consumers by producers in limited amounts or to an authorized breaking plant. (Loss eggs may not be used for human consumption, but may be used in pet foods.) Shell eggs only for intrastate commerce are not regulated by the USDA, but most states have egg-grading laws or regulations very similar or identical to those of the USDA.

The USDA has established specifications for quality of individual shell eggs (Table 1), as well as weight classes, specifications, and tolerances for consumer (Table 2), procurement (Table 3), and wholesale packs (Table 4). Practically all eggs produced on commercial poultry farms are processed mechanically. They are washed, candled, frequently oiled, and then packed.

Refrigeration Conditions

The recommended refrigeration conditions for shell eggs to prevent quality loss during short- and long-term storage are as follows:

Temp, °F	rh, %	Storage Period
50 to 60	70 to 80	2 to 3 weeks
29 to 31	85 to 92	5 to 6 months

Eggs for long-term storage should be held as cold as possible, just above the freezing point of the egg, about 27°F. However, such low temperatures are seldom used because most eggs are consumed within a short period, and low temperature may cause sweating. Sweating occurs when eggs are removed from refrigerated storage below the dew point and moisture condenses on the shell surface. Sweating has been a primary cause of subsequent microbial spoilage. For short-term storage (2 to 4 weeks), during normal marketing, eggs can be held at higher temperatures, but not over 60°F. When eggs are removed from refrigerated storage, it may be necessary to temper them to a higher temperature to prevent sweating.

Relative humidity in egg storage rooms must be maintained to prevent moisture loss with a subsequent loss of egg weight. Even for oiled eggs, not lower than 70 to 80% rh is necessary for long-term storage. Too high a relative humidity causes mold growth, which can penetrate the pores of the shell and contaminate the egg contents. Mold will grow on eggs above 90 to 94% rh. Moving air inhibits this growth, so eggs for long-term storage in still-air rooms should be held at lower than 90% rh, but can be held at up to 92% rh if the air is moving at about 40 fpm.

Supplements to Refrigeration

Supplementing the storage air with 0.5 to 0.6% CO_2 helps maintain the CO_2 content of eggs, which is important in preserving the quality of the egg white. It also retards development of molds, off-odors, and off-flavors.

Ozone has been used as a supplement to refrigeration to prevent mold on eggs. A continuous concentration of 1.5 ppm in the aisles maintains 0.6 ppm of ozone in the center of the cases. This concentration prevents mold growth at 31°F and 90% rh if the eggs, cases, and liners are clean when stored. Concentrations as high as 3.5 ppm for several months do not cause injury, but 10 ppm for 5 months causes off-flavors.

Keeping Quality

Shell eggs deteriorate in three distinct ways: (1) changes from chemical reactions; (2) decomposition by bacteria and molds; and (3) changes because of absorption of flavors and odors from the environment. Some changes that take place during storage are caused by chemical action and temperature effect. As the egg ages, the pH of the white increases, the thick white thins, and the yolk membrane thins; ultimately, the white becomes quite watery, although total protein content changes very little. High temperature accelerates these changes. Some coincidental loss in flavor usually occurs, although this change develops more slowly.

Eggs contaminated with certain microorganisms spoil quickly. Some common end results are black, red, or green rot eggs with crusted yolks, moldy eggs, and the like. However, eggs occasionally

---

The preparation of this chapter is assigned to TC 11.1, Meat, Fish, and Poultry Products.

Table 1 Summary of U.S. Standards for Quality of Shell Eggs

Quality Factor	Specifications of Each Quality Factor			
	AA Quality	A Quality	B Quality	C Quality
Shell	Clean, unbroken, and practically normal	Clean, unbroken, and practically normal	Clean to slightly stained, unbroken, may be slightly abnormal	Clean to moderately stained, unbroken, may be abnormal
Air Cell	Not over 0.12 in. in depth	Not over 0.19 in. in depth	Not over 0.37 in. in depth [May show unlimited movement and may be free or bubbly]	May be over 0.37 in. in depth
Yolk	Outline slightly defined, practically free from defects	Outline fairly well defined, practically free from defects	Outline well defined, slightly enlarged, and flattened; may show definite but not serious defects	Enlarged and flattened; shows clearly visible germ development but no blood; may show other serious defects; outline plainly visible
White	Clear and firm	Clear and reasonably firm	Clear but may be slightly weak	May be weak and watery; small blood clots may be present[a]

Condensed from the Federal Register, CFR 7, Part 56, effective July 15, 1967.
[a] If they are small, aggregating not more than 0.12 in. in diameter.

Table 2 U.S. Weight Classes for Consumer Grades

Size or Weight Class	Minimum Net Weight per Dozen, oz	Minimum Net Weight per 30-Dozen Case, lb	Minimum Weight for Individual Eggs at Rate per Dozen, oz
Jumbo	30	56.0	29
Extra large	27	50.5	26
Large	24	45.0	23
Medium	21	39.5	20
Small	18	34.0	17
Peewee	15	28.0	—

become heavily contaminated without any outward manifestations of spoilage. Clean, fresh eggs are seldom contaminated internally. However, if they become wet, particularly under conditions of fluctuating temperatures, bacteria and/or molds may cause spoilage.

Many factors affect egg odor and flavor. Among these factors are the feed and breeding of the hen, microbiological infection, chemical changes during storage, and absorption of odors from the environment. The latter is important in refrigerated warehousing. Particularly to be avoided are strong odors such as those resulting from ammonia gas leaking from the refrigeration system and from certain fruits and vegetables (such as citrus fruits, apples, onions, potatoes, and cabbages).

Control of Quality

Maintaining appearance. Egg quality is evaluated by shell appearance, air-cell size, and the thickness of the yolk and white. Lowering storage temperature and shell oiling slow down the escape of carbon dioxide and moisture and prevent shrinkage and thinning of the white. White mineral oil is sprayed on the shell to give partial protection in most commercial operations. Dipping the eggs in oil completely covers the egg, giving maximum protection, but this is not a common practice.

Bacteriological spoilage. The storage of dirty or improperly cleaned eggs is the most common source of bacteriological spoilage. A large percentage of dirty eggs are contaminated with bacteria. Improper washing, such as using water colder than the eggs or water with a high iron content, increases the contamination but removes the evidence of dirt. Most improperly cleaned eggs invariably spoil during long-term storage. Safe washing procedures for storage eggs require extremely high sanitary standards. The hazards of sweating eggs have already been mentioned.

Table 3 Weight Classes for U.S. Procurement Grades

Weight Classes	Average Net Weight on Lot Basis 30-Dozen Case, lb	Minimum Weight Individual Case, lb	Minimum Weight of Individual Eggs at Net Rate per Dozen, oz	Maximum Avg. % of Individual Eggs below Minimum Wt. Lot Average[a]
Extra large	50.5	50.0	26	3.33
Large	45.0	44.5	23	3.33
Medium	39.5	39.0	20	3.33
Small	34.0	33.5	17	3.33

[a] Individual cases may contain not over 10% of individual eggs below minimum weights specified in any weight class but such eggs shall weigh not less than the minimum specified next lower weight class.

Table 4 Weight Classes for U.S. Wholesale Grades

Weight Classes	Avg. Net Wt. on a Lot Basis,[a] lb	Min. Net Wt. Individual Case Basis,[b] lb	Min. Wt. at Rate per Dozen Wt., oz	Wt. Variation Tolerance Not More Than 10%, by Count of Individual Eggs, oz
Extra large	50.5	50.0	26	Under 26 but not under 24
Large	45.0	44.0	23	Under 23 but not under 21
Medium	39.5	39.0	20	Under 20 but not under 18
Small	34.0	None		None

[a] Lot means any quantity of 30 dozen or more eggs.
[b] Case means standard 30-dozen egg case as used in the U.S. commercial practice.

Eggs and Egg Products

Table 5 Minimum Cooling and Temperature Requirements for Liquid Egg Products

Product	Liquid (Other Than Salt Product) Held 8 h or Less, °F	Liquid (Other Than Salt Product) Held in Excess of 8 h, °F	Liquid Salt Product, °F	Temperatures Within 2 h after Pasteurization, °F	Temperatures Within 3 h after Stabilization, °F
Whites (not to be stabilized)	55 or lower	45 or lower		45 or lower	
Whites (to be stabilized)	70 or lower	55 or lower			55 or lower[a]
All other product (except product with 10% or more salt added)	45 or lower	40 or lower		If held 8 h or less, 45 °F or lower. If held in excess of 8 h, 40 °F or lower	
Liquid egg product (10% or more salt added)			If to be held 30 h or less, 65 °F or lower. If to be held in excess of 30 h, 45 °F or lower	65 °F[b] or lower	

[a] Stabilized liquid whites shall be dried as soon as possible after removal of glucose. The storage of stabilized liquid whites shall be limited to that necessary to provide a continuous operation.
[b] The cooling process shall be continued to assure that any salt product to be held in excess of 24 h is cooled and maintained at 45 °F or lower.

Mold spoilage. Mold growth on eggs can be controlled easily for long periods by keeping the eggs in air below 85% rh. However, such eggs lose an appreciable amount of water under these conditions (commonly termed shrinkage), and commercially this lowers their market value. On the other hand, most molds grow luxuriantly on eggs if the relative humidity at the shell surface is more than 96% (where shrinkage is minimized). A slight growth, commonly called whiskers, occurs at 90 to 94% rh. If eggs are kept at these humidities for extended periods, the mold will eventually penetrate the eggs and cause spoilage.

The optimum humidity for egg storage, as ordinarily determined in the aisle of the storage room, depends on many factors, including air circulation, dunnage, moisture content of cases, fillers and flats, and fluctuation of temperature. It is safe to use 85 to 86% rh when no artificial circulation of air is used, and 87 to 92% rh with positive circulation. Fillers, flats, and cases are sometimes treated with chemicals to minimize mold growth.

EGG PRODUCTS

Among the more important liquid and frozen egg products are plain yolks, sugared yolks, salted yolks, whites, whole egg, and yolk-white blends containing sugar or syrup. The solids content of commercial yolk varies depending on the white content but is not less than 43%; whole eggs in their natural composition are not less than 24.7% solids. Egg solids products include albumen, yolk, whole egg, and blends of yolk, white, and additives.

Egg Product Quality

Criteria usually used in evaluating egg product quality are: odor, color (for yolk), bacteria count, solids and fat content (for yolk and whole egg), yolk content (for whites), and performance. All users generally want a product with a normal odor, low bacteria count (usually less than 10,000 per gram), and satisfactory performance for the many ways in which it is used. For noodles, the user generally wants as high a solids content and color as possible. Bakers are particular about performance: the whites will give trouble in angel food cakes if excessive yolk is present. Performance also is critical in candy (using whites) and salad dressing (using yolks).

Processing

The raw material of egg products is commonly farm-run eggs, although a large volume of eggs with defective shells or other superficial defects is also used. Prior to breaking, eggs are washed to remove dirt and surface microbiological contamination. The principles of proper egg washing are simple: proper use of equipment that includes continuous addition of clean water and a final sanitizing rinse, with the entire process under good temperature control.

Commercial mechanical breaking machines, after breaking the eggs, separate the contents or not, depending on whether whole eggs or yolks and whites are being produced. The products are made homogeneous by mixing or by a special milling process. The USDA-required holding temperatures for egg products are shown in Table 5. A schematic outline of the various unit processes involved in the manufacture of egg solids (dried egg products) is shown in Figure 1.

Pasteurization

Practically, all egg products are required to be pasteurized and free from pathogens before being shipped to the market. The minimum required temperatures and times for pasteurization of each type of egg product as specified by the USDA are listed in Table 6.

Egg white solids may be made *Salmonellae* negative by heat treatments. Spray-dried albumen is heated in closed containers so that the temperature throughout the material is not less than 130 °F for not less than 7 days. For pan-dried albumen, the requirements are 125 °F for 5 days. For the dried whites to be labeled *pasteurized,* the USDA requires that each lot be sampled, cultured, and found to contain no viable *Salmonellae*.

Dehydration

Dehydration of eggs is accomplished by pan, foam, or spray drying. Freeze drying has been applied to some special egg products such as scrambled eggs. Whites, and occasionally yolk and whole egg, are desugared prior to drying, to improve storage life. The final products are about 2% moisture and are usually packaged in fiber drums lined with water-vaporproof liners. (Military specifications usually call for gas-packing in tin cans.)

Storage Life

Liquid egg products are extremely perishable and should be kept at 34 °F or lower at all times after production until used. Storage temperatures of 0 °F or below guarantee quality in frozen eggs for long periods. Products tested after 10 years' storage at 0 °F have been found to be entirely satisfactory. Dried egg white (desugared egg white solids) does not require refrigeration, although dried whole egg and yolk products benefit from some refrigeration. For long-term storage, temperatures of 30 to 40 °F materially aid in retaining flavor and performance characteristics. The relative humidity should be low enough to prevent moisture pickup by fiber drums when they are used.

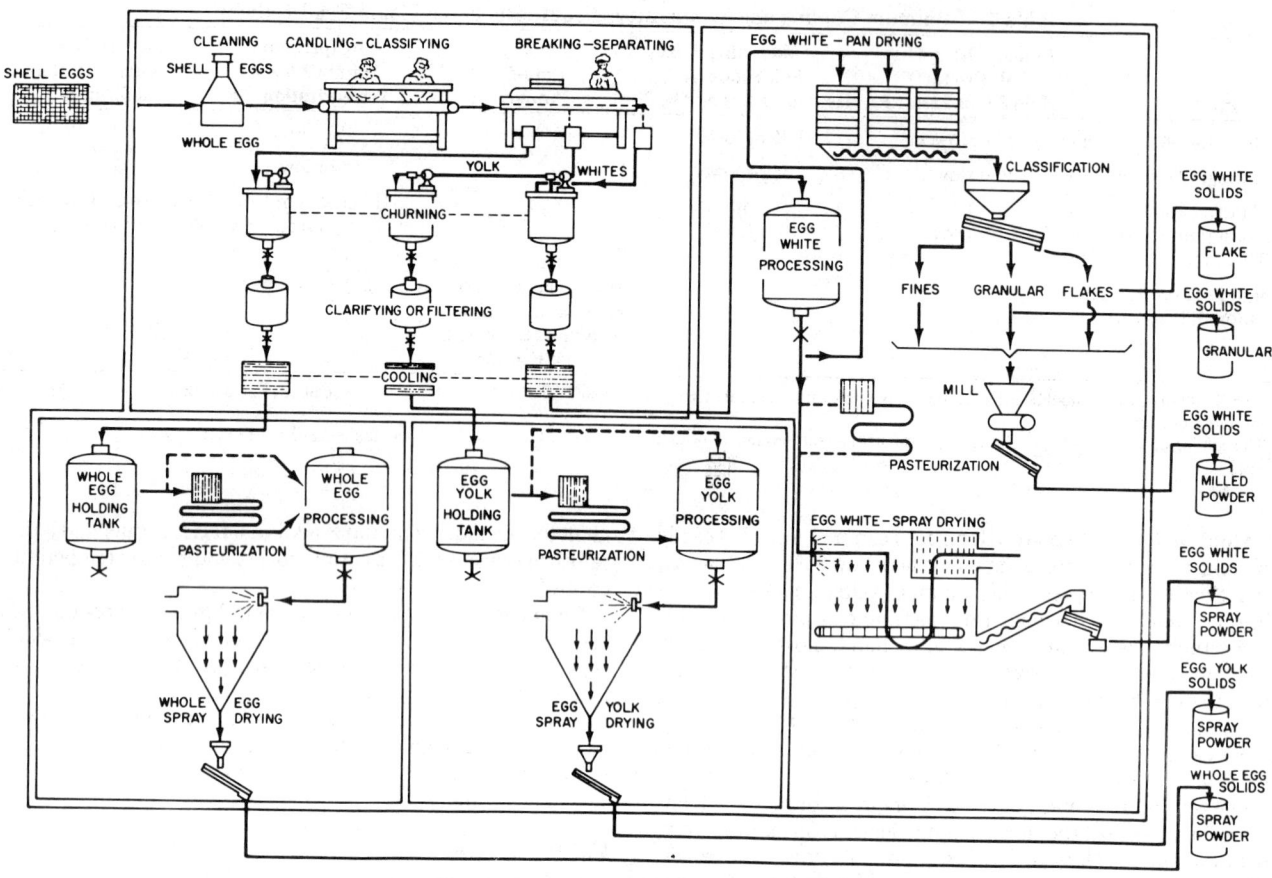

Fig. 1 Steps in Egg Product Processing

Defrosting

Frozen eggs may be defrosted in shallow tanks with 40 to 50 °F running water. They may be defrosted more rapidly by removing them from the can after a few hours of the above mentioned treatment and putting them through a stainless ice crusher into a coil vat or forewarmer. Water at 70 to 110 °F circulated through the rotating coils provides the necessary heat to complete thawing without damage to the product. Both methods are used commercially. Defrosted eggs should be used promptly. In no case should they be allowed to warm to above 45 to 50 °F where bacterial spoilage will occur very rapidly.

REFRIGERATION DESIGN DATA

Shell Egg Storage

Typical conditions affecting the refrigeration load are:

Storage temperature	30 °F
Relative humidity	80 to 90%
Average weight of egg case	50 lb
Standard case size:	
Width by height by length, in.	12 × 13 × 26
Insulation losses	Normal
Air infiltration losses	Normal
Lighting per 1 ft² floor	1 W
Occupant per 400 ft² floor	1
Max. allowable air velocity	
across eggs (forced circulation)	40 fpm

Evaporator Design

The evaporator system for egg coolers can be industrial air conditioners, bare pipe coils, or a combination of both. In small plants, any halogenated hydrocarbon commonly used as a refrigerant would be acceptable as a refrigerant. In large cold storage applications, the refrigerant may be direct expansion ammonia. Using a secondary coolant as the source of refrigerant, although costly, avoids the hazard of an ammonia leak. Large plants usually use this type of refrigeration, with finned or bare pipe coils, for holding duty. In a storage room, space should be provided on all sides for air circulation.

Egg Freezing

Freezing is usually accomplished by air blast at temperatures down to −4 °F directed at staggered piles of cans. Freezing time

Table 6 Pasteurization Requirements

Liquid Egg Products	Minimum Temperature, °F	Minimum Hold Time, s
Albumen	134	210
(without use of chemicals)	132	312
Whole egg	140	210
Whole egg blends (less than 2% added nonegg ingredients)	142	210
Fortified whole eggs and blends (24 to 38% egg solids, 2 to 12% nonegg ingredients)	144	210
	142	372
Salt whole egg (with 2% or more salt added)	146	210
	144	372
Sugar whole egg (2-12% sugar added)	142	210
	140	372
Plain yolk	142	210
	140	372
Sugar yolk (2% or more sugar added)	146	210
	144	372
Salt yolk (2-12% salt added)	146	210
	144	372

Eggs and Egg Products

Table 7 Heat Control of Egg Products

Item	Percent Solids	Freezing Point, °F	Specific Heat, Btu/lb·°F		Latent Heat of Freezing[a], Btu/lb
			Above Freezing	Below Freezing	
Water	0	32	1.00	0.5	144
Whole eggs	25	31	0.88	0.5	108
Whites	12	31	0.94	0.5	127
Yolks	44	31	0.78	0.5	81
Sugared yolks	50	25	0.75	0.5	72
Salted yolks	50	1	0.75	0.5	64

From USDA *Egg Pasteurization Manual,* ARS 74-48, February 1968.

[a] These data together with the specific heat values can be used to calculate the energy required to lower (or raise) the temperature of eggs including the freezing step. For example, the Btu that must be removed to lower the temperature of 30 lb of whole eggs from 40°F to 0°F are obtained as follows:

$$30 \times 0.88 (40 - 31) + (30 \times 108) + 30 \times 0.5(31 - 0) = 3943 \text{ Btu}$$

varies from 16 to 60 h. Very rapid freezing may be accomplished by a number of means, none of which has any overriding advantages and none of which is in common use. If very rapid freezing is practiced, the lid of the can must be kept from being forced off by the very rapid rise in the liquid core (because of expansion) against the lid. Salted yolks should not be frozen or held at under −10°F to reduce gelation.

Heat content values of egg products, which define the requirements for heating, cooling, and freezing, were estimated by using the following equations:

Specific heat above freezing (Btu/lb·°F) =
[(% water) + 0.55 (% solids)]/100

Specific heat of frozen egg = 0.5 Btu/(lb·°F)

Latent heat of freezing (Btu/lb) =
(% water/100) [144 − 0.5(32 − freezing point)]

The last equation assumes that all water in the product becomes ice during freezing. This is approximately true when the product is cooled to 10 to 20°F below the freezing point. The latent heat values in Table 7 therefore provide the maximum refrigeration requirements for freezing.

E-3-A SANITARY STANDARDS

Egg-3-A Sanitary Standards and accepted practices are formulated by the cooperative effort of U.S. Public Health Service, U.S. Department of Agriculture, Poultry and Egg Institute of America, Dairy Industry Committee, International Association of Milk, Food, and Environmental Sanitarians, and Dairy and Food Industries Supply Association. The Standards are published by the *Journal of Milk and Food Technology.*

CHAPTER 25

REFRIGERATED WAREHOUSE DESIGN

Initial Building Considerations 25.1	*Insulation Techniques* 25.8
Building Design 25.2	*Insulation Thickness* 25.9
Specialized Storage Designs 25.3	*Applying Insulation* 25.9
Construction Methods 25.4	*Other Considerations* 25.11
Refrigeration System Selection 25.6	

A REFRIGERATED warehouse is any building or section used for storage controlled conditions, with refrigeration. Two basic storage facilities are: (1) coolers which protect commodities at temperatures usually above 32°F, and (2) low-temperature rooms, operating under 32°F to prevent spoilage. Frozen foods should be kept below 0°F, preferably at −10 to −20°F.

The conditions within a closed refrigerator chamber must be maintained to preserve the stored product. This refers particularly to seasonal, shelf life, and long-term storage. Specific items for consideration are:

- Uniform temperatures
- Length of air blow and impingement on stored product
- Effect of relative humidity
- Effect of air movement on employees
- Controlled ventilation, if necessary

The Association of Food and Drug Officials (AFDO) developed a guide for establishing standards for all phases of handling frozen foods. The code treats receiving, handling, freezing, storage, and transporting frozen foods and calls for sanitary measures, as well as temperature requirements of 0°F and lower for holding frozen foods. These standards must be recognized in the design and operation of refrigerated storage warehouses.

Regulations of the Occupational Safety and Health Act (OSHA), Environmental Protection Agency (EPA), and U.S. Department of Agriculture (USDA) and other standards must also be incorporated in warehouse operations and procedures.

Refrigerated warehouses may be operated for and by an individual company, for public use, or for both. Important locations for warehouses, public or private, are: (1) point of processing; (2) intermediate points for general or long-term storage; and (3) final distributor or distribution point.

The five categories for the classification of refrigerated storages for preservation of food quality are:

- Coolers at 27 to 28°F
- Coolers at temperatures of 32°F and above
- Controlled atmosphere for long-term storage of fruits and vegetables
- Low-temperature storage rooms for general frozen products, usually maintained at −10 to −20°F
- Low-temperature storages at −10 to −20°F, with a surplus of refrigeration for freezing products received above 0°F

Refer to Chapters 20 and 21 in the 1989 ASHRAE *Handbook—Fundamentals* for further information.

The preparation of this chapter is assigned to TC 10.5, Refrigerated Distribution and Storage Facilities.

INITIAL BUILDING CONSIDERATIONS

Private refrigerated space is usually adjacent to, or in the same building with, other operations of the owner.

Public space should be located to serve a producing area, a transit storage point, a large consuming area, or various combinations to develop a good average occupancy. Also, it should have:

- Convenient location for producers, shippers, and distributors, considering the present tendency toward decentralization and avoidance of congested areas
- Good switching facilities and service with minimum switching charges from all trunk lines to plant tracks
- Easy access from main highway truck routes as well as local trucking, but avoiding location on congested streets
- Location with ample land for trucks, truck movement, and plant utility space
- Location with a reasonable land cost
- Adequate power and water supply
- Provisions for surface, waste, and sanitary water disposal
- Consideration to zoning limitations and fire protection
- Location that avoids residential areas where noise of outside operating equipment, such as fans and engine-driven equipment on refrigerated vehicles, would be objectionable
- External appearance may be a factor in the community
- Taxes and insurance to be investigated
- Plant security
- Favorable undersoil bearing conditions and good surface drainage

Plants are often located away from congested areas or even outside city limits where the cost of increased trucking distance is offset by better plant layout possibilities, a better road network, better or lower priced labor supply, or other economies of operation.

Size Determination

Building orientation and size of a cold storage warehouse are determined by the following factors:

- Is the receipt and shipment of goods to be primarily by rail or truck? Shipping practices will affect the platform areas and internal traffic pattern.
- What relative percentage of merchandise is for cooler and freezer storage? Products requiring specially controlled conditions, such as fresh fruits and vegetables, may justify several individual rooms. Seafood, butter, and nuts also require special treatment. However, overall occupancy may be reduced because of seasonal conditions.
- What percentage is anticipated for long-term storage? Products that are stored long-term can usually be stacked more densely.

- Will the product be primarily in small or large lots? The drive-through rack system or a combination of pallet racks and a mezzanine have proved effective in achieving efficient operation and effective space use.
- How will the product be palletized? A dense product such as meat, tinned fruit, drums of concentrate, batteries, and cases of canned goods can be stacked very efficiently. Palletized containers and special pallet baskets or boxes effectively hold meat, fish, and loose products.

  The slip sheet system which requires no pallets, eliminates the waste space of the pallet and can be used effectively for some products.

  Pallet stacking irons make it feasible to use the full height of the storage and palletize any closed or boxed merchandise.
- Will rental space be provided for tenants? Rental space usually requires special personnel and office facilities. An isolated platform area for the tenant operations is also desirable. These areas are usually leased on a unit area basis, and plans are worked into the main building layout.

Stacking Arrangement

The height of refrigerated spaces varies between 28 and 32 ft clear space between the floor and structural steel to allow fork truck operation. Pallet rack systems use the greater height. The practical height for stacking pallets without racks is 15 to 18 ft. The clear space above the pallet stacks is used for air distribution and sprinkler lines. Overhead space is inexpensive, and since the refrigeration requirement for the extra height is insignificant, a minimum of 20 ft clear height is desirable. Greater heights are practical if automated or mechanized material-handling equipment is contemplated.

High stacking may influence insurance rates. The floor area in a warehouse where a diversity of merchandise is to be stored can be calculated on the basis of 8 to 10 lbs per gross cubic ft of volume to allow about 40% for aisles and space above the pallet stacks.

For special purpose or production warehouses, products can be stacked with less aisle and open space, to result in an allowance factor of about 20%. Docks should be refrigerated, particularly in humid and warm climates. The relatively high cost for doors, door cushion closures, and refrigeration influences the economic dock size and number of doors. A commercial warehouse usually requires more truck dock space than specialized storage.

BUILDING DESIGN

Most refrigerated warehouses are single-story structures. Small columns on wide centers permit palletized storage with minimal lost space. This type of building usually provides additional highway truck unloading space. The single-story design has the following characteristics that must be considered: (1) horizontal traffic distances, which to some extent offset the vertical travel required in a multistory building; (2) difficulty of using the stacking height with many commodities or with small lot storage and movement of goods; (3) the necessity for treatment of the floor below freezers to give economical protection against possible ground heaving; and (4) high land cost for capacity of the building. A one-story design for moderate or low stacking heights has a high cubage cost because of the high ratio of roof and insulated areas, added land cost, and refrigeration load.

One-Story Configuration

Figure 1 shows a one-story layout incorporating facilities that comply with current practices. The following essential items and functions are included:

- Refrigerated shipping docks with seal-cushion closures on the doors
- Conditioned vestibules and automatic doors
- Batten curtains or strip doors
- Low-temperature storage held at −10°F or lower
- Pallet-rack systems to facilitate the handling of small lots and to comply with regulations of First-In-First-Out, which is required for some products
- A blast freezer or separate sharp freezer room for isolation of products being frozen
- Cooler or convertible space
- Space for brokerage offices and for distributors
- Space for empty pallet storage and repair
- Space for shop and battery charging
- Automatic sprinklers in accordance with NFPA regulations

A modified one-story design is sometimes used to reduce horizontal distances and land costs. The room height is equivalent to two high-ceiling stories. A balcony or mezzanine in part of the high-ceiling area, two low-ceiling areas adjacent to the one high-ceiling area, or one moderate height and one high-ceiling floor may provide economical low-stacking storage. A basement could be provided below the freezer area for utilities and coolers. Another alternative is to locate nonproductive services (including offices and the mechanical room) on a second-floor level, usually over the truck platform work area to permit full use of the ground floor for production work and storage. However, potential vibration of the second floor from reciprocating equipment must be considered.

One story, or modification of the one story, gives the maximum capacity per unit of investment with a minimum of overall operating expense, including cost of investment, refrigeration, and labor. Mechanization must be considered as well.

Designs that give minimum overall costs restrict office facilities and utility areas to a minimum. Dock area is of major importance to assure efficiency in the operation of loading and unloading merchandise.

Shipping and Receiving Docks

Regulations on temperature control during all steps of product handling have led to designing the trucking dock as a refrigerated anteroom to the cold storage area. Typically, loading and unloading of transport vehicles is handled by separate work crews.

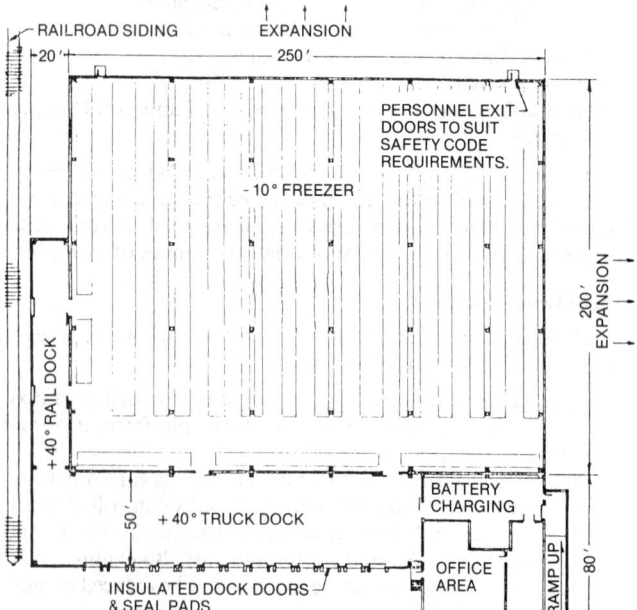

Fig. 1 Typical Plan of One-Story Warehouse

Refrigerated Warehouse Design

One crew moves the product in and out of the vehicles, and a warehouse crew moves the product in and out of the refrigerated storage. This procedure may allow the merchandise to accumulate on the shipping dock. Maintaining the dock at 35 to 45°F offers the following advantages:

- Reduced refrigeration load in the low-temperature storage area, where energy demand per unit capacity of refrigeration is high
- Less ice or frost forms in the low-temperature storage because less warm air infiltrates into the area
- Refrigerated products held on the dock maintain a favorable temperature
- Packaging remains in good condition because it stays dry. Warehouse personnel are more comfortable because temperature differences are less
- Requires less maintenance because condensation on powered lift trucks and other equipment is reduced
- Reduces or eliminates the need for anterooms or vestibules

Utility Space

Space for a general office, locker room, and machine room is needed. A superintendent's and warehouse records office should be located near the center of operations and a checker's office should be in view of the dock and traffic arrangement. Rented space should be isolated from warehouse operations.

The machine room should include ample space for refrigerating equipment and maintenance, adequate ventilation, standby capacity for emergency ventilation, and adequate segregation from other areas. Separate exits are required by most building codes. A maintenance shop and space for parking, charging, and servicing warehouse equipment should be located adjacent to the machine room. Electrically operated materials handling equipment is used to eliminate inherent safety hazards of combustion-type equipment.

SPECIALIZED STORAGE DESIGNS

Warehouse handling methods and storage requirements dictate design. Automated materials handling within the storage, particularly for high stack piling, may be an integral part of the structural system or require special structural treatments. Controlled atmospheres and minimal air circulation require special building designs and mechanical equipment to achieve required conditions. Drive-in and/or drive-through rack systems can improve product inventory control and can be used in combination with stacker cranes, special narrow-aisle high stacker cranes, and automatic conveyors in various phases of automation. In general, warehouses may be classified as follows:

- Public refrigerated warehouse designed to handle all commodities with several chambers to control temperatures from 28 to 60°F with humidity control and to −10°F otherwise.
- Refrigerated warehouse for case and breakup distribution, automated to varying degree, with racks arranged with the appropriate number of pallet spaces in acceptable pallet faces, to accommodate the projected usage.
- Warehouse designed for a processing operation with a block storage for frozen ingredients and rack storage for palletized outshipment of processed merchandise. An economic adaptation is to adjoin a commercial warehouse to a processing plant. The warehouse would receive inventory of frozen ingredients and deliver to the processor on order. It would also receive processed merchandise, take inventory, and make up outshipments for commercial carrier pickup and delivery. Process specialization for the manufacturer and experienced efficient operation of the warehouse staff are advantages.
- Public warehouse serving several production manufacturers, storing and inventorying products in lots and assembling out shipments for commercial carrier pickup and delivery.
- Specialized mechanized warehouse with stacker cranes, racks, infeed and outfeed conveyors, and special conveyor vestibules. In the design of such a facility, which may be 60 to 100 ft high, the cooling units must be mounted in the highest internal area to pick up moisture infiltration. They must also be mounted to avoid dripping on the stored product and mechanical equipment below.

Controlled Atmosphere Storage Rooms

Controlled atmosphere (CA) storage rooms may be required for specialized storages, particularly those handling apples. In addition to refrigeration apparatus to control temperature, these storages also include special gastight seals to facilitate the maintenance of an atmosphere that is lower in oxygen and higher in nitrogen and carbon dioxide than normal atmosphere. These storage rooms require apparatus to control the CO_2 concentration and, in many cases, use nitrogen-generating or oxygen-consuming devices.

The desired atmosphere must be experimentally determined for the commodity as produced in the specific area that the storage is to serve.

Information is available for certain commodities. Commercial application of CA storage has been limited to fresh fruits and vegetables that respire in storage, consuming O_2 and producing CO_2. The storages may be classified as either: (1) having product-generated atmospheres, where the room is sufficiently well sealed so that the normal oxygen consumption of the fruit can consume the oxygen in the storage plus the O_2 that leaks into the room; or (2) externally generated atmospheres where nitrogen generators or O_2 consumers assist the normal action of the fruit. The second type of system can cope with a poorly sealed room, but the cost of operation may be quite high. However, even with the external gas generator type system, a well-sealed room is desired.

In most cases, a CO_2 scrubber is required. The only exception is fruit storages where the total of the desired O_2 and CO_2 content is 21%. In such cases, a normal balance between oxygen consumption and CO_2 exists, and no CO_2 scrubbing is required. Carbon dioxide may be removed by: (1) passing the room air over dry lime that is replaced periodically; (2) passing the air through wet caustic solutions where the NaOH is replaced periodically; (3) water scrubbers where the CO_2 is absorbed from the room air by a water spray and then desorbed from the water by passing outdoor air through the water in a separate compartment; (4) monethanolamine scrubbers that may have the solution regenerated periodically by a manual process or regenerated continuously by automatic equipment; or (5) dry adsorbents automatically regenerated on a cyclic basis.

Among the systems of room sealing are: (1) lining the walls and ceiling of the room with 28-gage galvanized steel and connecting this into a floor-sealing system; (2) specially faced plywood or plywood with an impervious sealing system applied to the inside face; (3) sprayed urethane carefully applied and finished with thermal (fire) barriers.

To be sufficiently tight for use as a hermetic room, the space should meet the following test: after being pressurized to 1 in. water gauge, the rate of pressure loss is observed and if, at the end of one hour, 0.1 to 0.2 in. remains and the test has been conducted under uniform temperature and barometric conditions. A satisfactory external gas generator type of room may lose pressure at double the above rate. The test prescribed for the room using product-generated atmosphere is roughly equivalent to one air change of the empty room in a 30-day period.

Extreme care in all details of construction is required to obtain a seal that passes this test. Doors are sealed and have sills that can be bolted down and sealed. Electrical conduits and special seals around all pipe and hanger penetrations allow some movement, but the seal remains intact. Structural penetrations through the seal must be avoided, and the structure must be stable. CA rooms in multifloor frame buildings, where the structure deflects appreciably under load, are extremely difficult to seal. Most gas seals are applied in the wrong place, *i.e.*, at the cold side of the insulation, so they can be readily maintained and points of leakage can be detected. Some moisture entrapment is to be expected, and insulation materials must be carefully selected so that minimal damage will result from this incorrect placement of a vapor barrier.

In some installations, cold air with a dewpoint lower than the inside surface temperature is circulated through a space between the insulation and the gas seal for dryness. For additional information on CA storage, see Chapter 18.

Jacketed Storages

Mechanically cooled walls, floors, and ceilings may be economical for long-term storage duty, such as bin storage, CA storage, and for products that deteriorate with active air movement. Embedded pipe coils or airspaces through which refrigerated air is recirculated can provide the cold surface. With this method, leakage is absorbed in the jacket and prevented from entering the refrigerated space.

The following must be considered in the initial design of the storage:

- Initial cooldown of the product, which can impose short-term peak loads
- Service loads when loading product in and out
- Odor contamination from products that are not compatible
- Product heat of respiration in cooler spaces

Supplementary refrigeration or conditioning units in the refrigerated space, which would be operated only as required, can usually alleviate these problems.

Automated Warehouses

Automated warehouses are usually tall fixed-rack arrangements with stacker cranes under full automatic, semiautomatic, or hand control. The control systems can be tied into a computer system to retain a complete inventory of product and location.

Some of the advantages are:

- First-in and first-out inventory can be controlled.
- The enclosure structure is high, requiring a minimum of floor space and favorable cubage cost.
- Product damage and pilfering are minimized.
- Direct material handling costs are minimized.

CONSTRUCTION METHODS

Cold storages, more than most construction, require correct design, quality materials, good workmanship, and close supervision. Design should ensure that proper installation can be accomplished under various adverse job-site conditions. Materials must be compatible with each other. Installation must be made by careful workers directed by an experienced, well-trained superintendent. Close cooperation between the general, roofing, insulation, and other contractors increases the possibilities of a successful installation

Construction methods can be classified as either: (1) insulated structural panel; (2) mechanically applied insulation; or (3) adhesive or spray-applied foam systems. These construction techniques seal the insulation within an air and moisture-tight envelope, which must not be violated by major structural components.

Three ways are used to achieve an uninterrupted envelope. The first and simplest is total encapsulation of the structural system by an exterior vapor barrier under the floor, on the outside of the walls, and over the roof deck (Figure 2). This method offers the least number of penetrations through the vapor barrier, as well as the lowest cost. The second method is an entirely interior system where the vapor barrier envelope is placed within the room and insulation is added to the walls, floors, and suspended ceiling (Figure 3). This technique is used where walls and ceilings must be washed, where an existing structure is converted to low-temperature space, or for smaller rooms located within large coolers or unrefrigerated facilities. The third method is inside-outside construction (Figure 4), involving an exterior curtain wall of masonry or similar material tied to an interior structural system. Adequate space allows the vapor barrier insulation system to turn up over a roof deck and be incorporated into a built-up roofing system, which serves as the vapor barrier. This construction method is a viable alternative, although it offers more interruptions in the vapor barrier than the exterior system.

The total exterior vapor barrier system (Figure 2) is best because of the advantages of fewest penetrations and lowest cost. Each area of widely varying temperature should be divided into separate envelopes to retard heat and moisture flow between them (Figure 5).

Space Adjacent to Envelope

Condensation at the envelope is usually caused by high humidity and inadequate ventilation. Most often, this occurs within a dead airspace, such as a ceiling plenum or hollow masonry unit, through-metal structure, or beam cavity. All closed airspaces should be eliminated, except those large enough to be ventilated adequately. Ceiling plenums, for instance, are best ventilated by mechanical vents, which move air above the envelope.

The insulation envelope should not be penetrated if possible. Insulate and vapor-seal all steel beams, columns, and large pipes that project through the insulation with a 4-ft wrap of insulation. Conduit, small pipes, and rods should be insulated for a distance

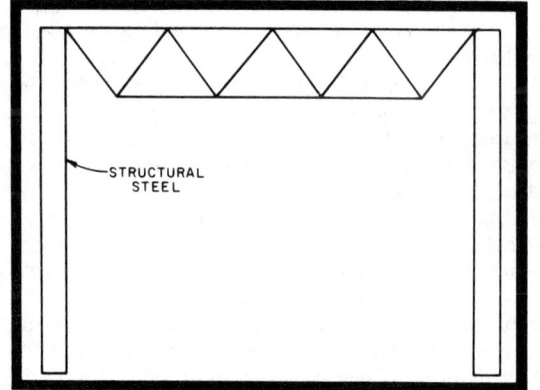

Fig. 2 Total Exterior Vapor Barrier System

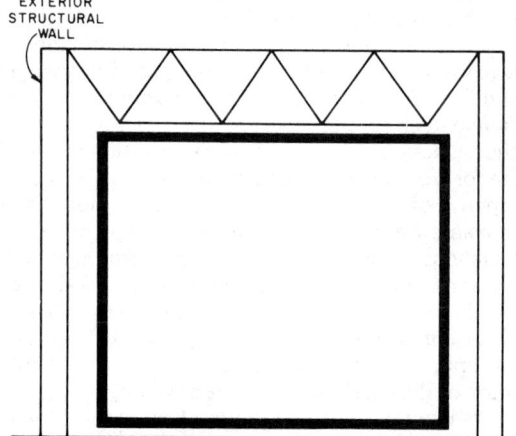

Fig. 3 Entirely Interior Barrier System

Refrigerated Warehouse Design

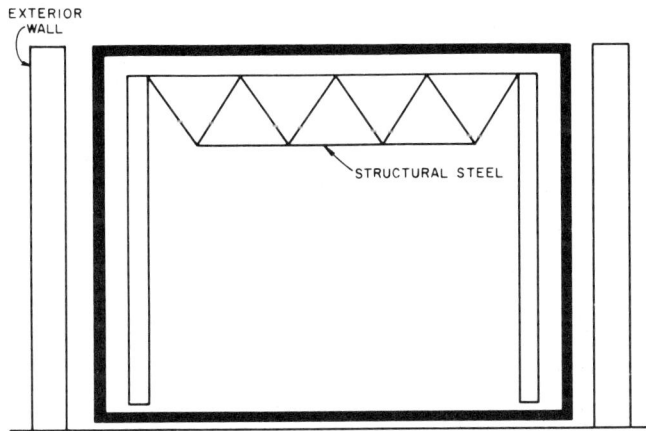

Fig. 4 Interior-Exterior Barrier System

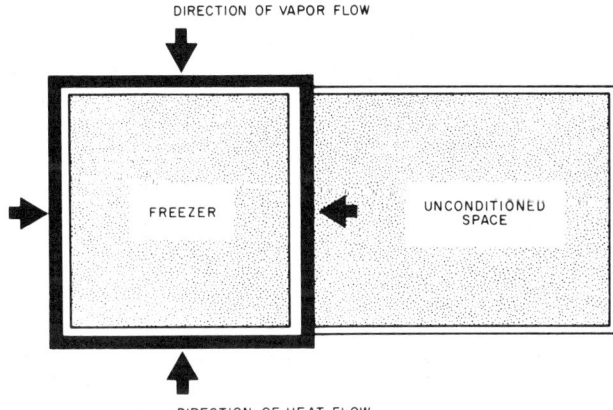

Fig. 5 Separate Barrier System for Each Area of Significantly Different Temperature

four times the regular wall insulation thickness. (Conduits and small pipes should be vapor-sealed on the inside to prevent moisture flow.) In both cases, the thickness of insulation on the projection should be half that on the regular wall or ceiling. Fill any voids within metal projections. Locate the wrap insulation on the warm side where practicable and seal the metal projection on the warm side.

Air/Vapor Treatment at Junctures

Air and vapor leakage at wall/ceiling junctures is probably the predominant construction problem in cold storage facilities.

When a cold room of interior-exterior design (Figure 4) is lowered to operating temperature, the structural elements (roof deck and insulation) contract and can pull the ceiling away from the wall. Because of negative pressure in the space of the wall/ceiling juncture, warm moist air can leak into the room and form frost and ice. Therefore, proper design of the air vapor seal is critical.

An air/vapor flashing sheet system is best for preventing leakage. A good corner flashing sheet must be flexible, tough, airtight, and vaportight. Proper use of flexible insulation at overlaps, mastic adhesive, and a good mastic sealer ensure leak-free performance (Figure 6). To remain airtight and vaportight during the life of the facility, a properly constructed wall/roof junction should:

- Be flexible enough to withstand building movements that may occur at operating temperatures
- Allow for thermal contraction of the insulation as the room is pulled down to operating temperature
- Be constructed with a minimum of penetrations that might cause leaks (Wall ties and structural steel that extends through the corner flashing sheet may eventually leak no matter how well sealed it is during construction; keep these to a minimum, and make them accessible for maintenance.)
- Corner flashing sheet properly lapped and sealed with adhesive and mechanically fastened to the wall vapor barrier
- Corner flashing sealed to roof without openings

The interior-exterior design is likely to be unsuccessful because of extreme difficulty in maintaining an air- or vaportight environment.

The practices outlined for the wall/roof junction apply for other insulation junctures. The insulation manufacturer and designer must coordinate the details of the corner flashing design together.

Poor design and shoddy installation cause moist air leakage into the facility, resulting in frost and ice formation, energy loss, poor appearance, loss of useful storage space, and eventually expensive repairs.

Floor Ventilation

Refrigerated facilities held above freezing need no special underfloor treatment. A below-the-floor vapor retarder is needed in facilities held below freezing, however. In these facilities the subsoil eventually freezes, and any moisture in this soil will also freeze and cause frost heaving. In moderate climates, underfloor tubes vented to ambient air are sufficient to prevent heaving. Artificial heating, either by air circulated through underfloor ducts or glycol circulated through plastic pipe, is needed in more severe climates. Electric heating cables installed under the floor can also be used to prevent frost formation. The choice of heating method depends on energy cost, reliability, and maintenance requirements.

Surface Preparation

When an adhesive is used, the surface against which the insulating material is to be applied should be smooth and dustfree. Where room temperatures will be below freezing, masonry walls should be leveled and sealed with cement back plaster. Smooth poured concrete surfaces may not require back plastering.

No special surface preparation is needed for a mechanical fastening system, assuming that the surface is reasonably smooth and in good repair.

The surface must be warm and dry for a sprayed foam system. Any cracks or construction joints must be prepared to prevent projection through the sprayed insulation envelope. All loose grout and dust must be removed to assure a good bond between the

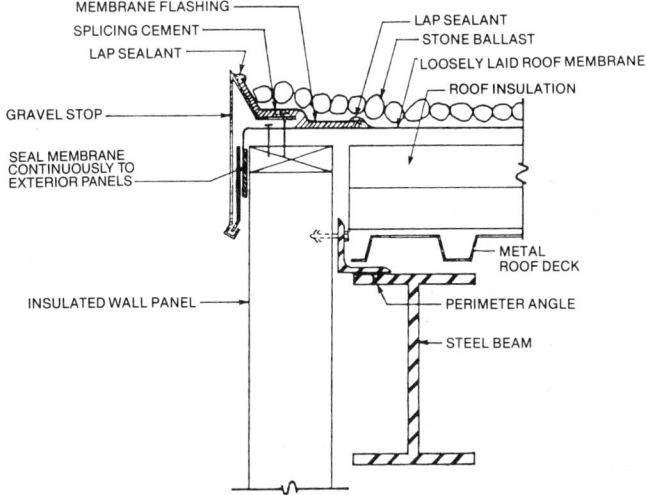

Fig. 6 Flashing Sheet Barrier System at Wall/Roof Juncture

sprayed foam insulation and the surface. Very smooth surfaces may require special bonding agents.

No special surface preparation is needed for panels used as a building lining, assuming the surfaces are sound and reasonably smooth.

Finishes

Insulated structural panels with metal exterior and metal or reinforced plastic interior faces are popular for both coolers and freezers. Their use keeps moisture from the insulation and leaves only the joints between panels as potential areas of moisture penetration.

For sanitary washdowns, a scrubbable finish is sometimes required. Such finishes generally have low permeance, and when applied on the inside surface of the insulation, a lower permeance treatment is required on the outside of the insulation.

All insulated walls and ceilings should have an interior finish. The finish should be pervious to moisture vapor and not serve as a vapor retarder, except for panel construction. The permeance of the in-place interior finish should be significantly greater than the permeance of the in-place vapor barrier.

To select an interior finish to meet in-use requirements of the installation, the following factors should be considered: (1) fire resistance; (2) washdown requirements; (3) mechanical damage; and (4) moisture and gas permeance. All interior walls of insulated spaces should be protected by bumpers and curbs whenever there is a possibility of damage to the finish.

Hung Ceilings

Inexpensive suspended insulated ceilings perform well when they are surfaced on the top or warm side with a vapor barrier flashed to the wall insulation vapor barrier. The space above the ceiling must be ventilated a minimum of six air changes per hour to minimize the possibility of condensation. Roof-mounted exhaust fans and uniformly spaced vents around the perimeter of the plenum may be used. The mechanical ventilators should be thermostat and/or humidistat controlled to turn off when: (1) outside temperature is below 50°F, and (2) outside humidity is below 60% rh. At these conditions, there is little possibility of condensation, and the ventilation only reduces the insulating effect of the dead airspace. Hung ceilings should be designed for light foot traffic for inspection and maintenance.

Floor Drains

Floor drains should be avoided if possible, particularly in freezers. If necessary, they should have short, squat dimensions, and be placed high enough on top of floor insulation in the room to allow the drain and piping to be installed with minimum weakening of the insulation envelope.

Electrical Wiring

Electrical wiring should be brought into a refrigerated room through one location, piercing the wall vapor barrier and insulation only once. Plastic-coated cable is recommended for this service where codes permit. If codes require conduit, the last fitting on the warm side of the run should be of the explosion-proof type, sealed to prevent water vapor from entering the cold conduit. The light fixtures in the room should not be vapor-sealed but should allow free passage of moisture. Care should be taken to obtain a vapor seal between the outside of the electrical service and the cold room vapor barrier.

Tracking

Cold room meat tracking, wherever possible, should be erected and supported within the insulated structure, entirely independent of the building itself. This places all the weight of the tracking on the cold room floor, eliminates flexure on the roof structure or overhead members, and simplifies maintenance.

Cold Storage Doors

Doors should be strong and, at the same time, light enough for easy opening and closing. Hardware should be of good quality, so that it can be set to compress the gasket uniformly against the casing.

In-fitting doors are not recommended for rooms operating below freezing, unless they are provided with heaters, and should not be used at temperatures below 0°F with or without heaters.

Hardware

All metal hardware, either within the construction or exposed to conditions that will rust or corrode the base metal, should be heavily galvanized, plated, or otherwise protected.

Refrigerated Docks

The type of warehouse, whether it is for distribution, intransit storage, or seasonal storage, will vary the loading dock requirements. Shipping docks and corridors should provide liberal space for (1) movement of goods to and from storage, (2) storage of pallets and idle equipment, (3) battery charging, (4) sorting, and (5) inspecting. The dock should be 30 ft or wider.

Floor heights of refrigerated vehicles vary widely, but are often higher than unrefrigerated vehicles. Rail dock heights and building clearances should be verified by the railroad serving the plant. Truck dock heights must comply with the requirements of fleet owners and clients, as well as the requirements for local delivery trucks. A 54 in. dock height above the rail is typical for refrigerated rail cars. Trucks generally require a 54 in. height above the pavement, although local delivery trucks may be much lower. Some reefer trucks are up to 58 in. above grade. Adjustable ramps at some of the truck spots will partly compensate for height variation. Three to five railroad car spots per million cubic feet of storage should be planned. If dimensions permit, 7 to 10 truck spots per million cubic feet should be provided in a public warehouse.

Refrigerated docks maintained at temperatures of 35 to 45°F require about 5 tons of refrigeration per 1000 ft^2 of floor area. Cushion-closure seals around the truck doorways reduce infiltration of outside air. An inflatable or telescoping enclosure can be extended to seal the space between a railcar and the dock. Insulated doors for docks must be mounted on the inside walls.

REFRIGERATION SYSTEM SELECTION

The selection of the refrigeration system for a refrigerated warehouse must be established in the early stages of planning. If it is a single-purpose, low-temperature storage building, almost any type of system can be applied. However, if commodities to be stored require different temperatures and humidities, a system must be selected that may require the use of several isolated rooms and conditions.

Using factory-built package unitary equipment has merit for the smallest structures and also for a multiroom facility that requires a variety of storage conditions. Conversely, the central compressor room has been the accepted standard for larger installations, especially where energy conservation is important.

Direct expansion refrigeration, either a flooded or pumped recirculation system serving fan coil units, is a dependable choice for a central compressor room. Screw compressors, programmable logic controllers, and microprocessor controls complement the reciprocating refrigeration equipment.

Load Determination

The refrigeration loads of warehouses of the same capacity vary widely. Many factors, including building design, indoor and outdoor temperatures, and most important, the type and flow of goods expected, plus the daily freezing capacity, contribute to the

Refrigerated Warehouse Design

Table 1 Refrigeration Design Load Factors for a Typical 100,000 ft² Single-Floor Freezer

Items of Refrigeration Load	Long-term Storage Cooling Capacity		Short-term Storage Cooling Capacity		Distribution Operation Cooling Capacity	
	Tons	Percent	Tons	Percent	Tons	Percent
Transmission losses	98	49	98	43	98	36
Infiltration	10	5	20	9	40	15
Internal operation loads	50	25	56	24	62	22
Cooling of goods received	7	3	15	6	30	11
Other factors	35	18	41	18	45	16
Total design capacity	200	100	230	100	275	100

load. Therefore, no simple design rules apply. Experience from comparable buildings and operations is valuable, but an analysis of any projected operation should be made. Compressor and room cooling equipment should be designed for maximum daily requirements, which will be well above any monthly average.

The factors to be considered include:

- Heat transmission through insulated enclosures
- Heat and vapor infiltration load from warm air passing into refrigerated space
- Heat from pumps or fans circulating refrigerated brine or air, power equipment, personnel working in refrigerated space, and heat from lights
- Heat removed from goods in reducing them from receiving to storage temperatures
- Heat produced by goods in storage
- Heat to be removed in freezing goods received unfrozen
- Other loads, such as office air conditioning, car precooling, or special operations inside the building
- Refrigerated shipping docks
- Heat released from automatic defrost units from the fan motors and defrosting
- Blast freezing or process freezing extra

Heat leakage or transmission load can be calculated using the known overall heat-transfer coefficient of various portions of the insulating envelope, the area of each portion, and the temperature difference between the lowest design cold room temperature and the highest average air temperature for three to five consecutive days at the building location. For floors on ground, the average yearly temperature should be used instead of the maximum

Heat-infiltration load varies greatly with the following variables: size of room, number of openings to warm areas, protection on openings, traffic through openings, cold and warm air temperatures, and humidities. Basis for calculation should be on experience, remembering that most of the load usually occurs during the day operations. Chapter 27 presents a complete analysis of load calculations.

A summation of the average proportional effect of the load factors, as itemized previously and influenced by the type of warehouse design and usage, is shown in Table 1 as a percentage of total load.

Heat from goods received for storage can be approximated from the quantity expected daily and knowing the source. Generally, 10 to 20°F of temperature reduction can be expected, but for some newly processed items and for fruits and vegetables direct from harvesting, 60°F or more temperature reduction may be required. For general public cold storage, the load should run 4 to 8 tons cooling capacity per million cubic ft, to allow for items received direct from harvest in a producing area.

Heat is produced by many commodities in cooler storage, principally fruits and vegetables. Heat of respiration is a sizable factor, even at 32°F, and is a continuing load throughout the storage period. Refrigeration loads should be calculated for maximum expected occupancy of such commodities.

The refrigerating load to be provided for freezing will vary from zero for the purely distribution warehouse, to the major portion of the total load for a warehouse near a producing area. The freezing load will depend on the commodity, the temperature at which it is received, and the method of freezing. More refrigeration is required for blast freezing than for still freezing without forced air circulation.

Air Units and Coils

Fan and Coil Units. These units may have either direct expansion, flooded, or recirculating liquid evaporators with either primary or finned coil surfaces or a brine spray coolant. Storage temperature, packaging method, type of product, etc. must be considered when selecting a unit. The coil surface area, temperature difference between the refrigerant coil and return air, and airflow volume depend on the particular application.

Automatic Defrosting. Properly engineered and installed systems can be automatically defrosted successfully with hot gas, water, electric, or continuously sprayed brine. The sprayed brine system has the advantage of producing the full refrigeration capacity at all times; however it does require a supply- and return-pipe system with a means of boiling off the absorbed condensed moisture.

Water-defrost systems require large supply and drain water lines, which must be designed for quick and effective draining in freezers on the defrost cycle. Hot gas and electric systems on freezer duty must be provided with heated drain pan and drain lines, which are pitched to assure quick and complete drain-off of melted frost. The drain lines are usually insulated with a noncombustible material.

Some unit designs are enclosed in insulated casings and have automatic dampers to recirculate the warmed defrosting air within the casing. These design features provide higher efficiency and keep defrost vapors from escaping into the refrigerated space and connecting ductwork.

Regardless of the type of air units selected, air distribution is very important. For uniform temperature and air motion, room coils widely distributed over the ceiling give good results. When using air units, the air should be distributed as evenly as possible.

The cold side of the ceiling insulation should be vapor sealed adjacent to fan coil units hung directly under the insulated ceilings. During the defrost cycle, vapor and warm air at the front and rear of the units will rise and condense out as frost or free water on the underside of the insulated ceiling. If moisture penetrates the ceiling insulation, the freeze-thaw cycle will destroy the insulation. Under these circumstances, the ceiling insulation in this area should be protected with a vaportight sheet such as aluminum or vaportight glassboard.

Multiple installations. To distribute air without ductwork, multiple installations of coil and fan units have been used. The quantity of air circulated per unit capacity of refrigeration is considerably higher than that required for sprayed-unit systems. Thus, proper application is important. For single-story buildings, sprayed units installed in penthouses with duct distribution of the air have been used to make full use of floor space in the storage area (Figure 7). Either prefabricated refrigeration systems or cooling units connected to a central refrigeration plant can be incorporated in penthouse design.

Unitary cooling units are located in a penthouse, with distributing ductwork projected through the penthouse floor and under the insulated ceiling below. Return air passes up through the penthouse floor grillage. This system avoids the interference of fan coil units hung below the ceiling in the refrigerated chamber.

Defrost water drains through piping passing through the penthouse insulated walls and onto the main storage roof. Refrigerant mains and electrical conduit can be run over the roof on suitable supports to the central compressor room or packaged refrigeration units on the adjacent roof. Temperature control thermostats and electrical equipment can be housed in the penthouse.

A personnel access door into the penthouse is required for convenient equipment and service. The inside insulated penthouse walls and ceiling must be vaportight to keep condensation from deteriorating the insulation. Some primary advantages are:

- Cooling units, catwalks, and piping do not interfere with product storage space and are not subjected to physical damage from stacking truck operations.
- Service to all cooling equipment and controls can be handled by a single individual from essential floor or roof deck location.
- Maintenance and service costs are minimized.

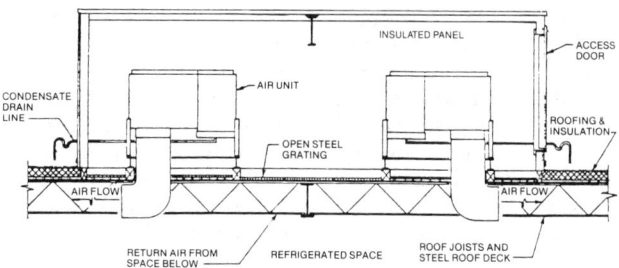

Fig. 7 Penthouse Application of Cooling Units

Freezing Facilities

The public refrigerated warehouse may be located in an area where products from processors are available for quick freezing and storage. Air-blast freezing is the usual method. The equipment must be of adequate capacity to accommodate the anticipated load. Variables in freezing characteristics of the products make using two or more freezing cells with individual fans and coils advisable.

These freezing cells can be constructed within the freezer storage space and oriented near the shipping platform to ensure efficient material handling. To ensure air movement through the stacked product and resultant efficient freezing, special racks or pallets with spacers for stacking the product in the blast freezer chamber should be provided.

Types of Refrigeration Systems and Equipment

Direct refrigeration systems, rather than brine systems, are used almost exclusively. The refrigerant is either flooded, using surge drums, controlled by thermal valves, or recirculated with a pump from an accumulator system. The latter system is advantageous for heavy concentrated loads, such as blast freezers, and is usually preferred for room unit coolers with forced air circulation.

The indirect brine system has advantages in simplicity of operation, ease of control, avoidance of possible leakage of refrigerant into storage areas, and other factors that tend to offset the savings of the direct system. Brine is particularly desirable for convection-coil installation and for widespread systems. However, the high initial and operating cost has almost eliminated use of the brine system in new construction, although still in use.

The refrigerant generally used for central plant systems has been ammonia. The initial and operating cost of halocarbon equipment is slightly higher than with ammonia equipment, but in some cases, this additional investment is justified to meet local conditions.

For large plants with 300 tons cooling capacity or greater, the screw-type compressor may have a somewhat lower initial cost, and maintenance should be less than with reciprocating compressors.

In ammonia systems, oil accumulates in vessels normally located outside the product storage rooms. These vessels are equipped with properly valved (and sometimes heated) oil collection pots to remove oil from the system.

Automation in the form of individual room temperature control, automatic defrosting, and safety shutdown in case of equipment malfunction, is common in most warehouses. A central panel monitors mechanical and electrical equipment and announces trouble areas immediately.

Refrigeration plant equipment should be designed for maximum load of the plant. All equipment should be in multiple units to allow for economical operation at light loads and continuity of operation in case of failure of one unit.

For temperatures below 0°F, two-stage compression is generally used. Compound compressors having capacity control on each stage may be used; but for variable loads of the refrigerated warehouse, separate high- and low-stage (or booster) compressors are better. Each should have capacity control, or the units should be of different capacity for balancing both high- and low-stage operating conditions. If blast freezers are included, pipe connections should be arranged so sufficient booster capacity for the blast freezers can be operated at a lower suction pressure, while the other booster capacity is at higher suction pressure for the freezer room load.

In a two-stage system, the liquid refrigerant should be precooled at the high-stage suction pressure to reduce the low-stage load by 10 to 20%. A good, automatic purger to remove air and other noncondensable gases is essential. Ammonia plants should have oil collection vessels at all points where oil may accumulate in the system. Traps should be placed on the compressor discharge, condensers, intercoolers, and evaporators so oil may be removed periodically without loss of ammonia. Chapters 1 through 8 have further information on refrigeration systems.

The use of commercial, air-cooled, package or factory-assembled units is common, especially in smaller plants. These units have the advantages of lower initial cost, smaller space requirements, and no need for a special machine room or operating engineer. But they use more energy and have higher operating costs, maintenance costs, and a shorter life expectancy for components (usually compressors) than central refrigeration systems.

INSULATION TECHNIQUES

The two main functions of an insulation envelope are to reduce economically the refrigeration requirements for the refrigerated space and to prevent condensation on the exterior.

Vapor Barrier Systems

The primary concern in the design of a low-temperature facility is the vapor barrier system, which should be 100% effective. *The success or failure of an insulation envelope is due entirely to the vapor barrier systems used to prevent water vapor transmission into and through the insulation.*

Once water vapor passes a vapor barrier, a series of events, all of which are detrimental to the refrigerated warehouse and its operation begins. For example, during warm and humid weather water vapor migrating through the insulation is chilled to its dewpoint somewhere in the insulation and condenses to water. In the

Refrigerated Warehouse Design

case of a cooler, the insulation will remain wet, while in a freezer, ice forms. In both cases, the value of the insulation will gradually decrease, and it will eventually be destroyed.

Part of the water vapor entering the insulation, after condensing or freezing, will revaporize or sublime. This vapor may again condense as it advances from the warm side to the cold side of the insulation. Additional infiltration moisture follows, and the insulation gradually becomes less effective because of moisture buildup. Some moisture passes into the refrigerated space and collects on the refrigerating coils, but it is not usually sufficient to dry out the insulation unless the vapor barrier discontinuity is located and corrected.

In walls with insufficient insulation, the dewpoint of the migrating water vapor may, during certain periods, be reached at the inside wall surface and cause condensation and freezing. This can also happen to a wall that originally had adequate insulation but, through water or ice formation in the insulation, lost its insulating value. In either case, the ice deposited on the wall will gradually push the insulation and protective covering away from the wall until the whole insulation structure collapses.

Proper installation of vapor barriers and the sealing of joints between the vapor barrier material or the continuity of the vapor barrier from one surface to another, *i.e.*, wall-to-roof, wall-to-floor, or wall-to-ceiling is extremely important. The failure of vapor barrier systems is almost entirely because of the workmanship during application of the systems for a warehouse.

Types of Insulation

Rigid Insulation. Insulation materials, such as polystyrene, polyisocyanurate, polyurethane, and phenolic material, have proven satisfactory when installed with the proper vapor barrier and finished with materials that provide protection from fire and form a sanitary surface. Consideration should be given to the selection of the proper insulation material based primarily on economics on an installed basis including the finish, sanitation, and fire protection.

Panel Insulation. The use of prefabricated insulated panels for insulated wall and roof construction is widely accepted. These panels are assembled around the building structural frame or can be installed as liner panels in an existing facility. Economics normally prohibits the use of liner panels. These panels can be insulated at the factory with either polystyrene or urethane. Other insulations do not lend themselves to panelized construction.

The basic advantage, besides economics, in using insulated panel construction is that repair and maintenance are simplified because the outer skin also serves as a vapor barrier and is conveniently available. This is of great benefit if the structure is to be enlarged in the future. Proper vapor barrier tie-ins then become practical.

Formed-in-Place Insulation. This application method is gaining acceptance as a result of the developments in polyurethane insulation and equipment for installing this insulation. Portable blending machines are available with a spray or frothing nozzle connected by hoses to pumps and material tanks or with prepacked pressurized material tanks. The partially expanded mixture in the case of froth foaming is fed from the nozzle into the insulation wall, floor, or ceiling cavities to fill the insulation space provided for monolithic insulation construction without joints. Froth foams set in approximately 2 min. Spray foams, however, are normally from a fast-reacting urethane system, which rises and sets in about 10 to 15 s. Expansion pressures are considerably higher for sprayed urethane than for froth urethane.

INSULATION THICKNESS

The R-value of insulation required varies with the temperature held in the refrigerated space and the conditions surrounding the room. Generally, the R-values in Table 2 are recommended for

Table 2 Recommended R-Values[a]

Type of Facility	Temperature Range, °F	Floors R	Walls/ Suspended Ceilings R	Roofs R
Cooler[b]	40 to 50	Perimeter Insulation Only	25	30-35
Chill Cooler[b]	25 to 35	20	24-32	35-40
Holding Freezer	−10 to −20	27-32	35-40	45-50
Blast Freezer[c]	−40 to −50	30-40	45-50	50-60

[a]Because of the wide range in the cost of energy and insulation materials on the thermal performance basis, a recommended R-value is given as a guide in each of the respective areas of construction. For more exact values, consult the designer and/or insulation supplier. R-value units are $h \cdot ft^2 \cdot °F/Btu$.
[b]If a cooler has the possibility of being converted to a freezer in the future, the owner should consider insulating the facility with the higher R-values from the freezer section.
[c]R-values shown are for a blast freezer built within an unconditioned space. If built within a cooler or freezer, consult the designer and/or insulation supplier.

different types of facilities. The range in R-values results from variation in cost of energy, insulation materials, and climatic conditions. For more exact values, consult a designer and/or insulation supplier. R-values less than those shown should not be used.

APPLYING INSULATION

The method and materials used to apply insulating materials to walls, floors, ceilings, roofs, and doors need careful consideration. Four accepted construction techniques are described in this section.

Roofs

Insulating materials should preferably be placed on the roof or floor above the refrigerated space, not under the ceiling. If this type of construction is not feasible and the insulation must be installed under a concrete ceiling or hung from a structural framework, the vapor barrier, insulation, and finish materials should be mechanically supported from the structure above, and should not rely on adhesive application only. Suspending a wood or metal deck from the roof structure and applying insulation and a vapor barrier to the top of the deck is another way of hanging ceiling insulation. Prefabricated panels may also be suspended under the roof. Methods of application and the support structure vary with type of insulation; but skill of application and attention to positive air and vapor seals are essential to continued effectiveness.

Suspended insulated ceilings, whether built-up or prefabricated, require adequate ventilation to maintain the plenum space near ambient conditions to minimize condensation and deterioration of vapor barrier materials. A minimum of six air changes per hour is adequate for most cases. The need for permanent sealing in insulating hanger rods, columns, conduit, and other penetrations is also important.

When installing insulation on top of metal decking or concrete structural slabs for a building larger than 100 by 100 ft, the structural designer usually includes roofing expansion joints. Since the refrigerated space is not normally subject to temperature variations, structural framing is usually designed with no expansion or contraction joints if the structural framing is entirely enclosed within the insulation envelope.

An examination of the coefficients of linear expansion for typical roof construction materials illustrates the need for careful attention to this phase of the building design. The data given in Table 3 should be verified for specific products.

While asphalt built-up roofs have been used, loosely laid membrane roofing has become popular and has proven relatively maintenance free (Figure 6).

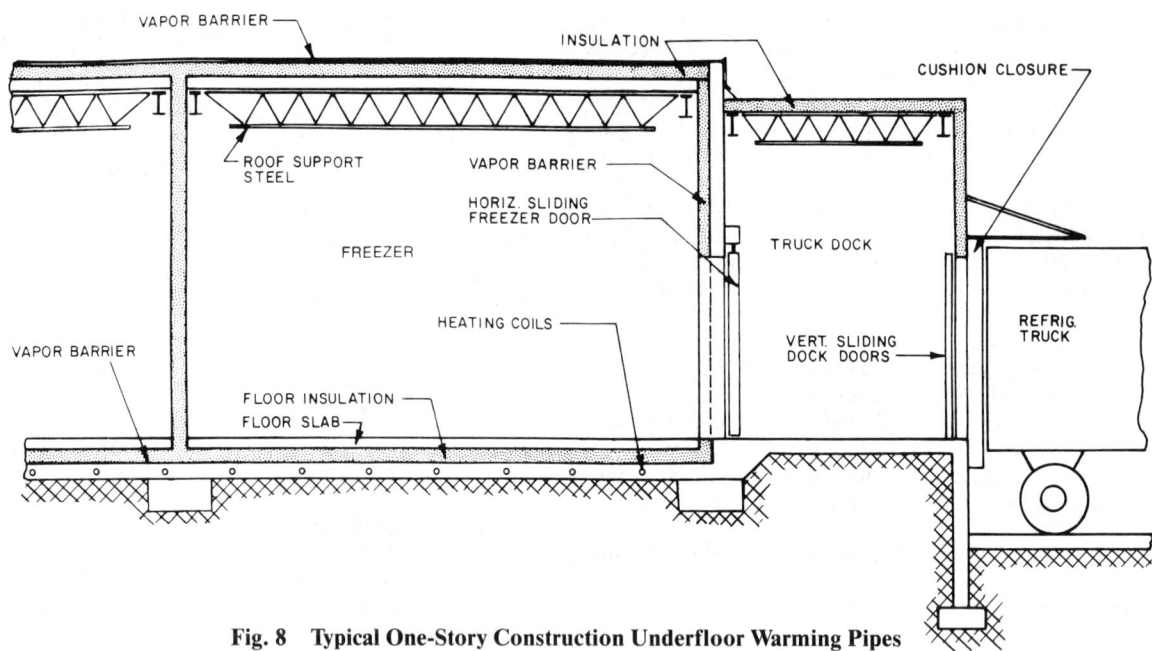

Fig. 8 Typical One-Story Construction Underfloor Warming Pipes

Table 3 Coefficients of Linear Expansion

Material	Linear Thermal Expansion, $10^{-6}/°F$
Aluminum	12
Brick	3.4
Wood (pine)	3.0
Rigid glass foam	4.6
Polystyrene, bead board	15
Polystyrene, foam	40
Polyurethane, foam	60
Corkboard	80
Concrete	5.5
Steel	6.7
Tile	3.6
Asphalt, roof grade	120
Roofing, asphalt impregnated	
Cross	15
Length	25
Coal tar pitch	90
Expanded perlite board	
Cross	900
Length	2000

Walls

Wall construction must be designed so that as few structural members as possible penetrate the insulation envelope. Insulated panels applied to the outside of the structural frame prevent conduction through the framing. Structural framing must be independent of the exterior wall where masonry or concrete wall construction is used. The exterior wall cannot be used as a bearing wall unless a hung insulated ceiling is used instead of the envelope method of insulation construction. Usually, tie rods tie the top of the masonry wall to the interior structural steel. These tie rods require special attention to seal the vapor barrier properly.

Where interior insulated partitions are required, a double column arrangement at the partition prevents structural members from penetrating the wall insulation. Envelope construction should always be used where possible for satisfactory operation and long life of the insulation structure.

Governing codes for fire prevention and sanitation must be followed before a finish is selected for wall insulation. For conventional insulation materials other than prefabricated panels, a vapor retarding system should be selected for walls of the membrane type, *i.e.*, free from the structural wall and flexible enough to withstand building movement.

Abrasion-resistant membranes, such as 10-mil thick black polyethylene film with a minimum of joints, are suitable vapor retarders. These membranes may be mechanically attached to the wall without adhesives. Rigid insulation can then be installed dry and finished with plaster or sheet finishes, as the specific facility requires. Contraction of the interior finish is of more concern than expansion, since temperatures are usually held far below installation ambient temperatures.

Floors

Some freezer buildings have been constructed without floor insulation, and some operate without difficulty. However, the possibility of failure is so great that this practice is seldom recommended.

Underfloor ice formation with resultant heaving of floors and columns can be prevented by adding heat to the soil or fill under the insulation. Air ducts, electric heating elements, or pipes through which a liquid is recirculated have been used.

The air duct system works well for smaller storages. For a larger storage, it should be supplemented with fans and a source of heat when the length of pipe is more than 100 ft. The end openings should be screened to keep out rodents and insects, as well as any material that might close off the air passages. The ducts must drain to release condensed moisture.

The electrical systems are simple to install and maintain if the heating elements are run in conduit or pipe so they may be replaced. Adequate insulation should be used, since this has a direct influence on the energy consumed. Perforated pipes should not be used.

The pipe grid system as shown in Figure 8 is usually best, since it can be designed and installed to warm where needed and can later be regulated to suit varying conditions. Extensions of this system can be placed in vestibules and corridors to reduce the objectionable conditions of ice and wet floors. A source of heat for this system can be obtained by an exchanger in the refrigeration system, steam, or gas engine exhaust. The temperature of the recirculated fluid is controlled at 50 to 70°F, depending on design requirements. Almost universally, the pipes are made of plastic.

The pipe grid system is usually placed in the base concrete slab directly under the insulation. A vapor barrier should be placed

Refrigerated Warehouse Design

below the pipe to prevent corrosion of metal pipe. The fluid should be an antifreeze solution such as glycol, with the proper inhibitor.

The amount of warming for any system can be calculated and is about the same for medium and large refrigerated spaces regardless of ambient conditions. The flow of heat from the earth serves as a safety factor and is about 1.3 Btu/(h·ft^2) of floor. The calculated heat input requirement is the floor insulation leakage based on the temperature difference between the 40°F underfloor earth and room temperature (*e.g.*, 50°F temperature difference for a −10°F storage room).

Doors

The selection and application of cold-storage doors are a fundamental part of cold-storage warehouse design and have a strong bearing on the overall economy of warehouse operation. The trend is to have fewer and better doors. Manufacturers offer many types of doors supplied with the proper thickness of insulation for the intended use. Four basic types of doors are: swinging, horizontal sliding, vertical sliding, and double acting. Door manufacturers' catalogs give detailed illustrations of each. Doors used only for personnel cause few problems. In general, a standard swinging personnel door, 3 ft wide by 6 ft, 6 in. high, designed for the temperatures and humidity involved, is adequate.

The proper door for heavy traffic areas should provide maximum traffic capacity, minimum loss of refrigeration, and minimum maintenance.

Factors to consider when selecting cold storage doors are:

- Automatic doors are a primary requirement with lift truck and automatic conveyor material-handling systems.
- Careless lift truck operators are a hazard to door effectiveness and operation. Guards can be provided but are effective only when the door is open. Photoelectric and ultrasonic beams across the doorway or proximity loop control on both sides of the doorway can provide additional protection by monitoring objects in the door openings or approaches. These systems can also insure that the door is closed after each use.
- The selection of automatic door systems to suit traffic requirements and the building structure justifies experienced technical guidance.
- To ensure continuous door performance, the work area near the doors must be supervised, and the doors must have planned maintenance.
- Cooled or refrigerated shipping platforms increase door efficiency and reduce door maintenance because of lower humidity and temperature difference. Icing of the door is lessened, and fogging in trafficways is reduced.

Air curtains and plastic or rubber strip curtains, when used with doors, give varied effectiveness. Often the curtain suggests to the lift truck driver that it is a substitute for the door, so the door is left open with a concurrent loss of refrigerated air.

Swing or slide. A door with hinges on the right side, when facing the side on which the operating hardware is mounted, is called a right-hand swing. When looking at a sliding door from the side of the wall on which it is mounted, a door sliding to the right to open is called a right-slide door.

Vertical sliding doors. These doors, which are hand or motor operated with counterbalanced spring or weights, are used on truck receiving and shipping openings.

Refrigerated room doors. Doors for pallet material handling are usually automatic horizontal sliding doors, either single-slide or biparting.

Metal or plastic cladding. Light metal cladding or a reinforced plastic skin protect most doors. Areas of abuse must be further protected by heavy metal, either partial or full height.

Heat. To prevent ice formation with subsequent faulty door operation, doors are available with automatic electrical heat— not only in the sides, head, and sill of a door or door frame, but also in switches and cover hoods of power-operated units. The use of such heating elements is necessary on all four edges of double-acting doors in low-temperature rooms. Safe devices that meet electrical codes must be used.

Guard posts and bumpers. Power-operated doors require protection from abuse. Bumpers embedded in the floor on both sides of the wall and on each side of the passageway help preserve the life of the door. Correctly placed guard posts protect sliding doors from traffic damage.

Buck and anchorage. Effective door operation is impossible without good buck and anchorage provisions. The recommendations of the door manufacturers should be coordinated with wall construction.

Door location. Doors should be located to suit material handling safely. Irregular aisles and blind spots in trafficways near doors should be avoided.

Door size. A hinged insulated door opening should provide at least 1 ft clearance on both sides of a pallet. Thus, 6 ft should be the minimum size for a 4-ft pallet load. Double-acting doors should be 8 ft wide. Specific conditions at a particular doorway can require variations from this recommendation. A standard height of 10 ft accommodates all of high stacking lift trucks.

Sill. A concrete sill minimizes the rise at the door sill. A thermal break should be provided in the floor slab at or near the plane of the front of the wall.

Power doors. Horizontal sliding doors are standard when electric operation is provided. The two-leaf biparting unit keeps opening and closing time to a minimum. Also, since leading edges of both leaves have safety edges, personnel, doors, trucks, and product are protected. The door is out of the way when open and protected from damage. The door should remain open a minimum amount of time. A combination of pull cord to open and a time-delay relay, proximity-loop control, or photoelectric cell for closing have been used. Major door damage may be reduced by the proper location of pull-cord switches. They also must be protected from moisture and frost with heat or baffles. Automatic doors should have a preventive maintenance program to check gaskets, door alignment, electrical switches, safety edges. Safety releases on locking devices are necessary to prevent entrapment of personnel.

Fire rated doors. Available in both swing and slide types, fire rated doors are also insulated. As refrigerated buildings have increased in size and the contents have increased in value, insurance companies and fire authorities are requiring fire walls and doors.

Vestibules. Door openings large enough to accommodate lift trucks with high masts, two-pallet-high loads on the forks, and tractor drawn trailers are of such size to cause appreciable refrigeration loss. Infiltration of moisture is objectionable, since it forms as condensate or frost on stacked merchandise and within the warehouse structure. Door heights up to 10 and 12 ft are frequently required, especially where drive-through racks are used. These conditions can be particularly serious when doors are located in opposite walls of a refrigerated space and crossflow of air is possible. It is important to reduce this infiltration by enclosed refrigerated loading docks or, in some instances, by one-way traffic vestibules.

OTHER CONSIDERATIONS

Temperature Pulldown

Because of the low temperatures within freezer facilities, contraction of structural members in this space will be substantially greater than in any surrounding ambient or cooler facilities.

Therefore contraction joints must be properly designed to prevent structural damage during facility pulldown.

The first stage of temperature reduction should be from ambient down to 35 °F at whatever rate of reduction the refrigeration system can achieve.

The room should then be held at that temperature until it is dry. Finishes are especially subject to damage when temperatures are lowered rapidly. Portland cement plaster should be fully cured before the room is refrigerated.

If there is a possibility that the room is airtight (most likely for small rooms, 20 ft by 20 ft max.), swinging doors should be partially open during pulldown to relieve the internal vacuum caused by cooling of the air, or vents should be provided.

The concrete slab will contract during pulldown, causing slab-wall joints, contraction joints, and other construction joints to open. At the end of the holding period (*i.e.*, at 35 °F), any necessary caulking should be done.

An average time for achieving this condition is 72 h. However, there are other indicators that may be used, such as watching the rate of frost formation on the coils or measuring the rate of moisture removal by capturing the condensation during defrost.

After the refrigerated room is dry, the temperature can then be reduced again at whatever rate is achievable with the refrigeration equipment until the operating temperature is reached. Rates of 10 °F per day have been used in the past, but if care has been taken to remove all the construction moisture in the previous steps, faster rates are possible without damage.

Material Handling Equipment

Frozen food public warehouses are a high-volume, year-round operation with fast-moving order pick areas backed by in-transit bulk storage. Distribution warehouses may carry 300 to 3000 items or as many as 30,000 lots. Palletized loads stored either in bulk or on racks are transported by lift trucks or high-rise S/R (storage/retrieval) machines in a 0 to −20 °F environment. Standard, battery-driven forklifts that can lift up to 25 ft can service one-deep, two-deep reach-in, drive-in, drive-through, or gravity-flow storage racks. Special forklifts can lift up to 60 ft.

Automated S/R machine storage operations make better use of storage volume, require fewer personnel, and reduce the refrigeration load because the building has less roof and floor area. This equipment operates in a height range of 23 to 100 ft to service one-deep, two-deep reach-in, two- to twelve-deep rollpin, or gravity flow pallet storage racks. On-line computers and bar code identification allow a system to automatically control the retrieval, transfer, and delivery of products. In addition, these systems can record product location and inventory and load several delivery trucks simultaneously from one order pick conveyor and sorting device.

A refrigerated plant may have two or more material handling systems where the storage area contains fast- and slow-moving reserve storage, plus slow-moving order pick. Fast-moving items may be delivered and order picked by a conventional lift truck pallet operation with up to 30 ft stacking heights. In the fast-moving order pick section, the storage room internal height is raised to 20 ft to accommodate S/R machines, reserve pallet storage, order-pick slots, multilevel palletizing, and the infeed, discharge, and order-pick conveyors. Mezzanines may be considered to gain maximum access to the order-pick slots. Intermediate-level fire protection sprinklers may be required in the high rack or mezzanine areas above 14 ft high.

Fire Protection

Ordinary wet sprinkler systems can be applied to refrigerated spaces that are above freezing temperature. In lower temperature spaces, they cannot be applied without modification. Two types of modified systems are: the brine system and the dry-pipe system. The brine system has been replaced with the dry-pipe system because of brine damage to stored merchandise.

In the dry-pipe system, sprinkler pipes in the refrigerated spaces are filled with dry compressed air. The water is sealed off from the compressed air by a special dry-pipe valve. The air pressure, which is controlled, keeps this valve closed as long as no sprinkler failure occurs. When the air pressure is relieved, the water pressure forces the valve open and feeds water into the sprinkler system.

If a sprinkler head malfunctions or is mechanically damaged, entering water cannot be drained and will freeze in rooms below freezing. If this occurs, the affected piping must be removed, thawed, dried, replaced, and retested. Safety precautions can be designed to prevent water from entering the system, except when heat is released by a fire. This emergency protection can be provided by installing heat-sensing devices in the sprinkler dry-valve system and a special automatic water shutoff valve, which is normally open but will instantly close whenever sprinkler system air is released without heat action of one or more of the heat-sensing devices.

The design of a dry sprinkler system operating in areas below 32 °F requires special knowledge and should not be undertaken without expert guidance. Freezer storage with rack storages 30 ft high or more may require special design, and the initial design should be shown to the insuring company.

Local regulations may require ceiling isolation smoke curtains and smoke vents near the roof in large refrigerated chambers. These features allow smoke to escape and help the fire fighting crew locate the fire. If the building does not have a sprinkler system, central reporting or warning systems are available for hazardous areas.

Maintenance

Buildings can change in dimension due to settling, temperature change, and other factors; therefore, cold-storage facilities should be inspected regularly to spot problems early so that preventive maintenance can be performed in time to avoid serious damage.

Inspection and maintenance procedures fall into two categories: *basic system*—floor, wall, and roof/ceiling systems; and *apertures*—doors, frames, and other access to cold storage rooms.

Basic System

- Stack pallets at a sufficient distance (18 in.) from walls or ceiling to permit air circulation.
- Examine walls and ceiling at random every month for frost buildup. If buildup persists, locate the break in the vapor barrier.
- For insulated ceilings below a plenum, inspect the plenum areas for possible roof leaks or condensation.
- If condensation or leaks are detected, make repairs immediately.

Apertures

- Remind personnel to close doors quickly to reduce frosting in rooms.
- Check the rollers and door travel periodically to be sure that the seal at the door edge is effective. If leaks are detected, adjust the door to restore a moisture- and airtight condition.
- Check doors and door edges to detect damage from forklifts or other traffic. Repair any damage immediately to prevent door icing or motor overload caused by excessive friction.
- Lubricate doors according to the maintenance schedule from the door manufacturer to insure free movement and complete closure.
- Periodically check seals around openings in the walls or ceiling for ducts, piping, and wiring.

CHAPTER 26

COMMODITY STORAGE REQUIREMENTS

Refrigerated Storage 26.1
Refrigerated Storage Plant Operation 26.6
Storage of Frozen Foods 26.7
Specific Products 26.7
Density of Stored Commodities 26.9

THIS chapter presents information on the essential average storage requirements of most of the important perishable foods that enter the market on a commercial scale. Also included is a short discussion on the storage of furs and fabrics. The information is derived from scientific experimentation and the best commercial practice. The data are based on the storage of fresh quality commodities shortly after harvest. For products transported from a distance or deteriorated products, appropriate allowances should be made.

The recommended temperatures are the optimum values for long storage and are actual commodity temperatures, not air temperatures. For short storage, higher temperatures are often satisfactory. Conversely, products subject to chilling injury can sometimes be held at a lower temperature for a short time without injury. Exceptions are bananas, cranberries, cucumbers, eggplant, melons, okra, pumpkins and squash, white and sweet potatoes, and tomatoes. Adhere to minimum recommended temperatures for these products.

The storage life recommendations are based on usual commercial practice, not necessarily including special treatments, which could, in certain instances, extend storage life considerably.

The values for water content and freezing points are given by actual laboratory determinations, but they are only approximate because of the great variability in plant and animal tissues and the products thereof. The optimum storage temperature for many foods is just above their freezing point. Knowledge of freezing points is useful to the refrigerated storage industry in determining how to handle various commodities in storage. The highest temperature at which freezing may occur is generally given. In previous editions, the average freezing point was given, which may be somewhat lower. The highest freezing point may be a better guide for commodities damaged by freezing.

Values of the foods' water content are useful to the refrigerating engineer as a basis for calculating specific heats and the latent heat of freezing. Specific heat is usually calculated by Siebel's formula, as follows:

$$c_p = 0.008a + 0.20 \text{ (above freezing)}$$
$$c_p = 0.003a + 0.20 \text{ (below freezing)}$$

where c_p signifies the specific heat of a substance in Btu/(lb·°F) containing a, the percent of water; 0.20 is the value in Btu/(lb·°F) representing the specific heat of the solid constituents of the substance.

The composition of foods varies to some extent, depending on water content, where grown, and other factors. In tables compiled by different editors, slightly different formulas may have been used in calculations to determine both specific and latent heat. The values given in these tables can be used for refrigeration load calculations. When in doubt, use the higher values. For more precise work, determine the specific heat and other values for the specific sample in question.

Fresh fruits and vegetables are live products, and their heat of respiration must be considered as a part of the refrigeration load in storage handling. The approximate rate of heat evolution for various commodities is given in Chapter 30 of the 1989 ASHRAE *Handbook—Fundamentals*.

REFRIGERATED STORAGE

Cooling

Rapid removal of field heat and cooling the product to the storage temperature substantially increases the storage life. Because deterioration occurs much faster at warm than at low temperatures, the faster that field heat is removed after harvest, the longer produce can be maintained in good marketable condition in storage. Chapter 11 covers cooling methods in detail.

Storage Requirements and Deterioration

Storage requirements for specific fruits and vegetables are so varied that no generalizations can be made. Table 1 presents the recommended storage requirements for various products. Some products require a curing period before storage. Other products, such as Irish potatoes, require different storage conditions depending on the intended use.

Causes of deterioration of meat, fish, poultry, eggs, and dairy products are given in Chapter 10 and in the commodity chapters for those products.

Deterioration. The environment in which harvested produce is placed may greatly influence not only the respiration rate but other changes and products formed in related chemical reactions. In fruits, these changes are described as ripening. In many fruits, such as bananas and pears, the process of ripening is required to develop the maximum edible quality. However, as ripening continues, deterioration begins and the fruit softens, loses flavor, and eventually undergoes tissue breakdown.

In addition to deterioration after harvest by biochemical changes within the product, desiccation and diseases caused by microorganisms are also important.

Deterioration rate is greatly influenced by temperature and is reduced as temperature is lowered. The specific relationships between temperature and deterioration rate vary considerably among commodities and diseases. However, a generalization, assuming a deterioration rate of 1 for a fruit at 30 °F, would be as shown in Table 2. The best temperature to slow down deterioration is the lowest temperature that can safely be maintained without freezing the commodity, which is 1 to 2 °F above the freezing point of the fruit or vegetable.

Some produce will not tolerate low storage temperatures. Severe physiological disorders that develop because of exposure to low

---

The preparation of this chapter is assigned to TC 10.5, Refrigerated Distribution and Storage Facilities. This chapter last received a major revision in 1974.

Table 1 Storage Requirements and Properties of Perishable Products

Commodity	Storage Temperature, °F	Relative Humidity, %	Approximate Storage Life[a]	Water Content, %	Highest Freezing, °F	Specific Heat above Freezing,[b] Btu/lb·°F	Specific Heat below Freezing,[b] Btu/lb·°F	Latent Heat,[c] Btu/lb
Vegetables								
Artichokes								
Globe	32	95 to 100	2 weeks	84	29.9	0.87	0.45	120
Jerusalem	32	90 to 95	5 months	80	27.5[b]	0.83	0.44	114
Asparagus	32 to 36	95 to 100	2 to 3 weeks	93	30.9	0.94	0.48	133
Beans								
Snap or Green	40 to 45	95	7 to 10 days	89	30.7	0.91	0.47	127
Lima	37 to 40	95	3 to 5 days	67	30.0	0.73	0.40	94
Dried	50	70	6 to 8 months	11		0.32	0.23	
Beets								
Roots	32	95 to 100	4 to 6 months	88	30.4	0.90	0.46	126
Bunch	32	95	10 to 14 days		31.3			
Broccoli	32	95 to 100	10 to 14 days	90	30.9	0.92	0.47	130
Brussels Sprouts	32	95 to 100	3 to 5 weeks	85	30.6	0.88	0.46	122
Cabbage, late	32	98 to 100	5 to 6 months	92	30.4	0.94	0.47	132
Carrots								
Topped-immature	32	98 to 100	4 to 6 weeks	88	29.5	0.90	0.46	126
Topped-mature	32	98 to 100	5 to 9 months	88	29.5	0.90	0.46	126
Cauliflower	32	95	2 to 4 weeks	92	30.6	0.93	0.47	132
Celeriac	32	95 to 100	3 to 4 months	88	30.4	0.91	0.46	126
Celery	32	98 to 100	1 to 2 months	94	31.1	0.95	0.48	135
Collards	32	95	10 to 14 days	87	30.6	0.90	0.46	125
Corn, Sweet	32	95 to 98	4 to 8 days	74	30.9	0.79	0.42	106
Cucumbers	50 to 55	95	10 to 14 days	96	31.1	0.97	0.49	137
Eggplant	46 to 54	90 to 95	7 to 10 days	93	30.6	0.94	0.48	133
Endive (Escarole)	32	95 to 100	2 to 3 weeks	93	31.9	0.94	0.48	133
Frozen Vegetables	−10 to 0		6 to 12 months					
Garlic, dry	32	65 to 70	6 to 7 months	61	30.6	0.69	0.40	89
Greens, leafy	32	95 to 100	10 to 14 days	93	31.5	0.94	0.48	133
Horseradish	30 to 32	95 to 100	10 to 12 months	75	28.7	0.78	0.42	104
Kale	32	95	3 to 4 weeks	87	31.1	0.89	0.46	125
Kohlrabi	32	95	2 to 4 weeks	90	30.2	0.92	0.47	129
Leeks, green	32	95	1 to 3 months	85	30.7	0.88	0.46	122
Lettuce, head	32 to 34	95 to 100	2 to 3 weeks	95	31.7	0.96	0.48	136
Mushrooms	32	95	3 to 4 days	91	30.4	0.93	0.47	130
Okra	45 to 55	90 to 95	7 to 10 days	90	28.7	0.92	0.46	129
Onions								
Green	32	95 to 100	3 to 4 weeks	89	30.4	0.91	0.47	127
Dry, and onion sets	32	65 to 75	1 to 8 months	88	30.6	0.90	0.46	126
Parsley	32	95 to 100	1 to 2 months	85	30.0	0.88	0.45	122
Parsnips	32	98 to 100	4 to 6 months	79	30.4	0.84	0.44	112
Peas								
Green	32	95	1 to 3 weeks	74	30.9	0.79	0.42	106
Dried	50	70	6 to 8 months	12		0.30	0.24	
Peppers								
Dried	32 to 50	60 to 70	6 months	12		0.30	0.24	17
Sweet	45 to 50	90 to 95	2 to 3 weeks	92	30.7	0.94	0.47	132
Potatoes								
Early	50 to 55	90		81	30.9	0.85	0.44	116
Main crop	38 to 50	90 to 95	5 to 8 months	78	30.9	0.82	0.43	111
Sweet	55 to 61	85 to 90	4 to 7 months	69	29.7	0.76	0.41	99
Pumpkins	50 to 55	50 to 75	2 to 3 months	91	30.6	0.92	0.47	130
Radishes								
Spring	32	95	3 to 4 weeks	95	30.7	0.95	0.48	134
Winter	32	95 to 100	2 to 4 months	95	30.7	0.95	0.48	134
Rhubarb	32	95	2 to 4 weeks	95	30.3	0.95	0.48	134
Rutabagas	32	98 to 100	4 to 6 months	89	30.0	0.91	0.47	127
Salsify	32	98 to 100	2 to 4 months	79	30.0	0.83	0.44	113
Seed, vegetable	32 to 50	50 to 65	10 to 12 months	7 to 15		0.29	0.23	16
Spinach	32	95 to 98	10 to 14 days	93	31.5	0.94	0.48	133
Squash								
Acorn	45 to 50	70 to 75	5 to 8 weeks		30.6			
Summer	41 to 50	95	5 to 14 days	94	31.1	0.95	0.48	135
Winter	50 to 55	50 to 75	4 to 6 months	85	30.6	0.88	0.45	122
Tomatoes								
Mature green	55 to 70	90 to 95	1 to 3 weeks	93	31.0	0.94	0.48	133
Firm, ripe	45 to 50	90 to 95	4 to 7 days	94	31.1	0.95	0.48	134
Turnips								
Roots	32	95	4 to 5 months	92	30.0	0.93	0.47	132
Greens	32	95	10 to 14 days	90	31.6	0.92	0.47	129
Watercress	32	95	3 to 4 days	93	31.4	0.94	0.48	133
Yams	61	85 to 90	3 to 6 months	74		0.79	0.42	105

Commodity Storage Requirements

Table 1 Storage Requirements and Properties of Perishable Products (*Continued*)

Commodity	Storage Temperature, °F	Relative Humidity, %	Approximate Storage Life[a]	Water Content, %	Highest Freezing, °F	Specific Heat above Freezing,[b] Btu/lb·°F	Specific Heat below Freezing,[b] Btu/lb·°F	Latent Heat,[c] Btu/lb
Fruits and Melons								
Apples	30 to 40	90 to 95	3 to 8 months	84	30.0	0.87	0.45	121
Apples, dried	32 to 41	55 to 60	5 to 8 months	24		0.42	0.27	
Apricots	32	90 to 95	1 to 2 weeks	85	30.0	0.88	0.46	122
Avocados	40 to 55	85 to 90	2 to 4 weeks	65	31.5	0.72	0.40	94
Bananas		85 to 95	a	75	30.6	0.80	0.42	108
Blackberries	31 to 32	90 to 95	3 days	85	30.6	0.88	0.46	122
Blueberries	31 to 32	90 to 95	2 weeks	82	29.7	0.86	0.45	118
Cantaloupes	36 to 40	95	5 to 15 days	92	29.8	0.93	0.48	132
Cherries								
Sour	31 to 32	90 to 95	3 to 7 days	84	29.0	0.87	0.45	121
Sweet	30 to 31	90 to 95	2 to 3 weeks	80	28.8	0.84	0.44	114
Casaba Melons	45 to 50	85 to 95	4 to 6 weeks	93	30.0	0.94	0.48	133
Cranberries	36 to 40	90 to 95	2 to 4 months	87	30.4	0.90	0.46	124
Currants	31 to 32	90 to 95	10 to 14 days	85	30.2	0.88	0.45	122
Dates, cured	0 to 32	75 or less	6 to 12 months	20	3.7	0.36	0.26	29
Dewberries	31 to 32	90 to 95	3 days	85	29.7	0.88	0.45	122
Figs								
Dried	32 to 40	50 to 60	9 to 12 months	23		0.39	0.27	34
Fresh	31 to 32	85 to 90	7 to 10 days	78	27.6	0.82	0.43	112
Frozen fruits	−10 to 0	90 to 95	6 to 12 months					
Gooseberries	31 to 32	90 to 95	2 to 4 weeks	89	30.0	0.90	0.46	127
Grapefruit	50 to 60	85 to 90	6 to 10 weeks	89	30.0	0.90	0.46	127
Grapes								
American	31 to 32	85 to 90	2 to 8 weeks	82	29.7	0.86	0.45	118
Vinifera	31	90 to 95	3 to 6 months	82	28.1	0.86	0.45	118
Guavas	41 to 50	90	2 to 3 weeks	83		0.86	0.45	119
Honeydew Melons	45 to 50	90 to 95	3 to 4 weeks	93	30.4	0.94	0.48	133
Lemons	32 to 50[d]	85 to 90	1 to 6 months	89	29.4	0.91	0.46	127
Limes	48 to 50	85 to 90	6 to 8 weeks	86	29.1	0.89	0.46	123
Mangoes	55	85 to 90	2 to 3 weeks	81	30.4	0.85	0.44	117
Nectarines	31 to 32	90	2 to 4 weeks	82	30.4	0.86	0.44	118
Olives, fresh	41 to 50	85 to 90	4 to 6 weeks	75	29.4	0.80	0.42	108
Oranges	32 to 48	85 to 90	3 to 12 weeks	87	30.6	0.90	0.46	124
Papayas	45	85 to 90	1 to 3 weeks	91	30.6	0.82	0.47	130
Peaches	31 to 32	90 to 95	2 to 4 weeks	89	30.4	0.91	0.46	127
Peaches, dried	32 to 41	55 to 60	5 to 8 months	25		0.43	0.28	
Pears	29 to 31	90 to 95	2 to 7 months	83	29.2	0.86	0.45	118
Persian Melons	45 to 50	90 to 95	2 weeks	93	30.6	0.94	0.48	133
Persimmons	30	90	3 to 4 months	78	28.1	0.84	0.48	112
Pineapples, ripe	45	85 to 90	2 to 4 weeks	85	30.2	0.88	0.45	122
Plums	31 to 32	90 to 95	2 to 4 weeks	86	30.6	0.88	0.45	123
Pomegranates	41	90 to 95	2 to 3 months	82	26.6	0.86	0.44	118
Prunes								
Fresh	31 to 32	90 to 95	2 to 4 weeks	86	30.5	0.88	0.45	123
Dried	32 to 41	55 to 60	5 to 8 months	28		0.46	0.28	
Quinces	31 to 32	90	2 to 3 months	85	28.4	0.88	0.45	122
Raisins				18		0.38	0.25	
Raspberries								
Black	31 to 32	90 to 95	2 to 3 days	81	30.0	0.84	0.44	117
Red	31 to 32	90 to 95	2 to 3 days	84	30.9	0.87	0.45	120
Strawberries	31 to 32	90 to 95	5 to 7 days	90	30.6	0.92	0.47	129
Tangerines	40	90 to 95	2 to 4 weeks	87	30.0	0.90	0.46	125
Watermelons	50 to 60	90	2 to 3 weeks	93	31.3	0.97	0.48	133
Seafood (Fish)								
Haddock, Cod, Perch	31 to 34	95 to 100	12 days	81	28	0.85	0.44	117
Hake, Whiting	32 to 34	95 to 100	10 days	81	28	0.85	0.44	117
Halibut	31 to 34	95 to 100	18 days	75	28	0.80	0.42	107
Herring								
Kippered	32 to 36	80 to 90	10 days	61	28	0.70	0.38	87
Smoked	32 to 36	80 to 90	10 days	64	28	0.72	0.39	92
Mackerel	32 to 34	95 to 100	6 to 8 days	65	28	0.73	0.40	93
Menhaden	34 to 41	95 to 100	4 to 5 days	62	28	0.71	0.39	89
Salmon	31 to 34	95 to 100	18 days	64	28	0.72	0.39	92
Tuna	32 to 36	95 to 100	14 days	70	28	0.77	0.40	100
Frozen fish	−20 to −4	90 to 95	6 to 12 months					
Seafood (Shellfish)[a]								
Scallop meat	32 to 34	95 to 100	12 days	80	28	0.84	0.44	114
Shrimp	31 to 34	95 to 100	12 to 14 days	76	28	0.81	0.43	109
Lobster, American	41 to 50	In sea water	Indefinitely	79	28	0.83	0.44	113

Table 1 Storage Requirements and Properties of Perishable Products (*Continued*)

Commodity	Storage Temperature, °F	Relative Humidity, %	Approximate Storage Life[a]	Water Content, %	Highest Freezing, °F	Specific Heat above Freezing,[b] Btu/lb·°F	Specific Heat below Freezing,[b] Btu/lb·°F	Latent Heat,[c] Btu/lb
Seafood (Shellfish)[a] (*Continued*)								
Oysters, Clams (meat & liquid)	32 to 36	100	5 to 8 days	87	28	0.89	0.46	125
Oyster in shell	41 to 50	95 to 100	5 days	80	27	0.84	0.44	115
Frozen shellfish	−20 to −4	90 to 95	3 to 8 months					
Meat (Beef)								
Beef, fresh, average	32 to 34	88 to 92	1 to 6 weeks	62 to 77	28 to 29[f]	0.70 to 0.84	0.39 to 0.43	89 to 110
Beef carcass								
Choice, 60% lean	32 to 39	85 to 90	1 to 3 weeks	49	29	0.61	0.35	70
Prime, 54% lean	32 to 34	85	1 to 3 weeks	45	28	0.58	0.34	64
Sirloin cut (choice)	32 to 34	85	1 to 3 weeks	56		0.66	0.37	80
Round cut (choice)	32 to 34	85	1 to 3 weeks	67		0.50	0.40	96
Dried, chipped	50 to 59	15	6 to 8 weeks	48		0.60	0.34	69
Liver	32	90	5 days	70	29	0.77	0.41	100
Veal, 81% lean	32 to 34	90	1 to 7 days	66		0.74	0.40	94
Beef, frozen	−10 to 0	90 to 95	6 to 12 months					
Meat (Pork)								
Pork, fresh, average	32 to 34	85 to 90	3 to 7 days	32 to 44	28 to 29[f]	0.48 to 0.57	0.30 to 0.33	46 to 63
Carcass, 47% lean	32 to 34	85 to 90	3 to 5 days	37		0.52	0.31	53
Bellies, 35% lean	32 to 34	85	3 to 5 days	30		0.47	0.29	43
Backfat, 100% fat	32 to 34	85	3 to 7 days	8		0.30	0.22	
Shoulder, 67% lean	32 to 34	85	3 to 5 days	49	28[f]	0.61	0.35	70
Pork, frozen	−10 to 0	90 to 95	4 to 8 months					
Ham								
74% lean	32 to 34	80 to 85	3 to 5 days	56	29[f]	0.66	0.37	80
Light cure	37 to 41	80 to 85	1 to 2 weeks	57		0.67	0.37	82
Country cure	50 to 59	65 to 70	3 to 5 months	42		0.56	0.33	60
Frozen	−10 to 0	90 to 95	6 to 8 months					
Bacon								
Medium fat class	37 to 41	80 to 85	2 to 3 weeks	19		0.38	0.26	27
Cured, farm style	61 to 64	85	4 to 6 months	13 to 20		0.34 to 0.39	0.24 to 0.26	19 to 29
Cured, packer style	34 to 39	85	2 to 6 weeks					
Frozen	−10 to 0	90 to 95	2 to 4 months					
Sausage								
Links or bulk	32 to 34	85	1 to 7 days	38		0.53	0.31	54
Country, smoked	32	85	1 to 3 weeks	50	25	0.62	0.35	72
Frankfurters, average	32	85	1 to 3 weeks	56	29	0.66	0.37	80
Polish style	32	85	1 to 3 weeks	54		0.65	0.36	77
Meat (Lamb)								
Fresh, average	32 to 34	85 to 90	5 to 12 days	60 to 70	28 to 29[f]	0.69 to 0.77	0.38 to 0.41	86 to 100
Choice, 67% lean	32	85	5 to 12 days	61	28	0.70	0.38	87
Leg, choice, 83% lean	32	85	5 to 12 days	65		0.73	0.40	93
Frozen	−10 to 0	90 to 95	8 to 12 months					
Meat (Poultry)								
Poultry, fresh, average	28 to 32	95 to 100	1 to 4 weeks	74	27	0.80	0.42	106
Chicken, all classes	28 to 32	95 to 100	1 to 4 weeks	74	27	0.80	0.42	106
Turkey, all classes	28 to 32	95 to 100	1 to 4 weeks	64	27	0.72	0.39	92
Duck	28 to 32	95 to 100	1 to 4 weeks	69	27	0.76	0.41	99
Poultry, frozen	−10 to 0	90 to 95	12 months					
Meat (Miscellaneous)								
Rabbits, fresh	32 to 34	90 to 95	1 to 5 days	68		0.75	0.40	97
Dairy Products								
Butter	32	75 to 85	1 month	16	−4 to 31	0.36	0.25	23
Butter, frozen	−10	70 to 85	12 months					
Cheese, Cheddar,								
long storage	32 to 34	65	12 months	37	8	0.52	0.31	53
short storage	40	65	6 months	37	8	0.52	0.31	53
processed	40	65	12 months	39	19	0.50	0.31	56
grated	40	65	12 months	31		0.45	0.29	44
Ice cream, 10% fat	−20 to −15		3 to 23 months	63	21	0.70	0.39	86
Milk								
Whole, pasteurized Grade A	32 to 34		2 to 4 months	87	31	0.93	0.46	125

Commodity Storage Requirements

Table 1 Storage Requirements and Properties of Perishable Products (*Concluded*)

Commodity	Storage Temperature, °F	Relative Humidity, %	Approximate Storage Life[a]	Water Content, %	Highest Freezing, °F	Specific Heat above Freezing[b] Btu/lb·°F	Specific Heat below Freezing[b] Btu/lb·°F	Latent Heat[c] Btu/lb
Dairy Products (*Continued*)								
Milk (*Continued*)								
Dried, whole	70	Low	6 to 9 months	2		0.26	0.21	28
Dried, nonfat	45 to 70	Low	16 months	3		0.26	0.21	4
Evaporated	40		24 months	74	29.5	0.79	0.42	106
Evaporated, unsweetened	70		12 months	74	29.5	0.79	0.42	106
Condensed, sweetened	40		15 months	27	5	0.42	0.28	40
Whey, dried	70	Low	12 months	5		0.28	0.22	7
Eggs								
Eggs								
Shell	29 to 32[e]	80 to 85	5 to 6 months	66	28[f]	0.73	0.40	96
Shell, farm cooler	50 to 55	70 to 75	2 to 3 weeks	66	28[f]	0.73	0.40	96
Frozen,								
Whole	0		1 year plus	74		0.80	0.42	106
Yolk	0		1 year plus	55		0.65	0.36	79
White	0		1 year plus	88		0.90	0.46	126
Whole egg solids	35 to 40	Low	6 to 12 months	2 to 4		0.22	0.21	4
Yolk solids	35 to 40	Low	6 to 12 months	3 to 5		0.23	0.21	6
Flake albumen solids		Low	1 year plus	12 to 16		0.31	0.24	20
Dry spray albumen solids		Low	1 year plus	5 to 8		0.26	0.22	11
Candy								
Milk chocolate	0 to 34	40	6 to 12 months	1		0.25	0.20	1
Peanut brittle	0 to 34	40	1.5 to 6 months	2		0.26	0.21	3
Fudge	0 to 34	65	5 to 12 months	10		0.32	0.23	14
Marshmallows	0 to 34	65	3 to 9 months	17		0.37	0.25	24
Miscellaneous								
Alfalfa meal	0	70 to 75	1 year plus					
Beer								
Keg	35 to 40		3 to 8 weeks	90	28	0.92	0.47	129
Bottles and cans	35 to 40	65 or below	3 to 6 months	90				
Bread	0		3 to 13 weeks	32 to 37		0.70	0.34	46 to 53
Canned goods	32 to 60	70 or lower	1 year					
Cocoa	32 to 40	50 to 70	1 year plus					
Coconuts	32 to 35	80 to 85	1 to 2 months	47	30.4	0.58	0.34	67
Coffee, green	35 to 37	80 to 85	2 to 4 months	10 to 15		0.32 to 0.35	0.23 to 0.24	14 to 21
Fur and fabrics	34 to 40	45 to 55	Several years					
Honey	50		1 year plus	17		0.35	0.26	26
Hops	28 to 32	50 to 60	Several months					
Lard (without antioxidant)	45	90 to 95	4 to 8 months	0				
	0	90 to 95	12 to 14 months	0				
Maple syrup				33		0.48	0.31	51
Nuts	32 to 50	65 to 75	8 to 12 months	3 to 6		0.22 to 0.25	0.21 to 0.22	4 to 8
Oil, vegetable, salad	70		1 year plus	0				
Oleomargarine	35	60 to 70	1 year plus	16		0.32	0.25	22
Orange juice	30 to 35		3 to 6 weeks	89		0.91	0.47	127
Popcorn, unpopped	32 to 40	85	4 to 6 weeks	10		0.31	0.24	19
Yeast, baker's compressed	31 to 32			71		0.77	0.41	102
Tobacco								
Hogshead	50 to 65	50 to 65	1 year					
Bales	35 to 40	70 to 85	1 to 2 years					
Cigarettes	35 to 46	50 to 55	6 months					
Cigars	35 to 50	60 to 65	2 months					

Note: The text in this chapter or the appropriate commodity chapter gives additional information on many of the commodities listed. The following individuals helped obtain data for certain commodities: R.L. Hiner, L. Feinstein, and A. Kotula, Meat and Poultry; J.W. White and C.O. Willits, Honey and Maple Syrup; E.B. Lambert, Mushrooms; H. Landani, Furs and Fabrics; M.K. Veldhuis, Orange Juice; A.L. Ryall, Plums and Prunes; L.P. McCulloch, Tomatoes; J.W. Slavin and J.A. Peters, Fish; T.I. Hedrich, Dairy Products; and L. Henrickson, Meat.

[a] Storage life is not based on maintaining nutritional value.

[b] Specific heat c_p (in Btu/lb·°F) is calculated by Siebel's formula. For temperatures above freezing, $c_p = 0.008a + 0.20$, where a is the percent water content of the commodity; for temperatures below freezing, $c_p = 0.003a + 0.20$. Siebel's formula is not very accurate in the frozen region, because foods are not simple mixtures of solids, and liquids are not completely frozen, even at $-20°F$.

[c] The latent heat of fusion is the latent heat of fusion of water (143.4 Btu/lb) multiplied by the water content of the commodity.

[d] Lemons stored in production areas for conditioning are held at 55 to 58°F, but sometimes they are held at 32°F.

[e] Eggs with weak albumen freeze just below 30°F.

[f] Average freezing point.

Table 2 Approximate Relationship of Temperature and Deterioration Rate in Fresh Produce

Temperature, °F	Deterioration Rate
68	8 to 10
50	4 to 5
41	3
37	2
32	1.25
30	1

but not freezing temperatures are classed as *chilling injury*. The banana is a classic example of a fruit displaying chilling injury symptoms, and storage temperatures must be elevated accordingly. Certain apple varieties exhibit this characteristic, and prolonged storage must be held at a temperature well above that usually recommended. The degree of susceptibility of an apple variety to chilling may vary with climatic and cultural factors. Products susceptible to chilling injury, its symptoms, and the lowest safe temperature are discussed in Chapters 10, 16, 17, and 18.

Desiccation. Water loss, which causes a product to shrivel, is a physical factor related to the evaporative potential of air, expressed as:

$$p_D = p(100 - \phi)/100 \quad (1)$$

where

- p_D = vapor pressure deficit, which indicates the combined influence of temperature and relative humidity on the evaporative potential of air
- p = vapor pressure of water at a given temperature
- ϕ = relative humidity, percent

For example, comparing the evaporative potential of air in storage rooms at 32 °F and 50 °F dry bulb, with 90% rh in each room, the vapor pressure deficit at 32 °F is 0.018 in. Hg, while at 50 °F, it is 0.036 in. Hg. Thus, if all other factors are equal, commodities tend to lose water twice as fast at 50 °F dry bulb as at 32 °F at the same rh values. For equal water loss at the two temperatures, the rh has to be maintained at 95% at 50 °F in comparison to 90% at 32 °F. These comparisons are not precise because the water in fruits and vegetables contains a sufficient quantity of dissolved sugars and other chemical materials to cause the water to be in equilibrium with water vapor in the air at 98 to 99% rh instead of 100% rh. Lowering the vapor pressure deficit by lowering the air temperature is an excellent means of reducing water loss during storage.

Other important factors in desiccation include product size, the kind of protective surface on the product, and air movement. Of these, the storage operator can control only the last, and this control is greatly influenced by the container, kind of pack, and stacking arrangement (*i.e.*, the ability of the air to move past individual fruits and vegetables).

As a rule, shrivelling does not become a serious market problem until fruits lose about 5% of their weight, but any loss reduces the salable amount. Moisture losses of 3 to 6% are enough to cause a marked loss of quality for many kinds of produce. A few kinds may lose 10% moisture and still be marketable, although some trimming may be necessary, as for stored cabbage.

The vapor pressure deficit cannot be kept at a zero level, but it should be maintained as low as possible. A maximum of about 0.018 in. Hg, which corresponds to 90% rh at 32 °F, is recommended. Some compromise is possible for short storage periods. In many instances, the refrigerated storage operator may find it desirable to add moisture, or in special cases the owner of the produce may find it desirable to use moisture barriers such as film liners.

REFRIGERATED STORAGE PLANT OPERATION

Checking Temperatures and Humidity

To maintain top product quality, temperature in the cold storage room must be accurately maintained. Variations of 2 to 3 °F in the product temperature above or below the desired temperature are too large in most cases. Storage rooms should be equipped either with accurate thermostats or with manual controls that are given frequent attention.

In refrigerated storage rooms, thermometers are usually placed at a height of about 5 ft for convenience in reading. Temperatures should be monitored where they might be undesirably high or low—one or two aisle temperatures is not enough. A record of both product and air temperatures is necessary to determine performance of the storage plant. A thermometer or recording device of good quality is essential.

Temperature in less accessible storage locations, such as the middle of stacks, can be obtained conveniently with distant reading thermometer equipment such as thermocouples or electrical-resistance thermometers.

Storage instructions or recommendations usually specify a relative humidity within 3 to 5% of the desired levels. An ordinary sling psychrometer at temperatures of 32 °F or lower cannot be read that closely. An error of 0.5 °F in reading either the wet- or the dry-bulb thermometer will cause an error of 5% rh. Carefully calibrated thermometers graduated to 0.1 °F with a range of 25 to 40 °F are best adapted for this purpose in fruit storages. A convenient device for measuring humidities consists of a pair of these thermometers, mounted in a short length of metal casing attached to a spring or motor-operated fan that draws air past the thermometers at a speed of 3 fps or faster. The thermometers should be placed so that they will not be heated by the warm fan motor, and they should be read quickly to prevent warming. The advantage of this instrument over the sling psychrometer is that it can be left in the room long enough to get a true wet-bulb reading. This may require 15 min or more if ice is formed on the wet bulb. Under these conditions, a thin coating of ice is preferable to a thick one in getting accurate readings. Hair hygrometers are satisfactory if not subjected to sudden large changes in humidity and temperature and if checked regularly with a psychrometer.

Perhaps a more accurate method of determining the relative humidity is to record, electrically, the dew-point temperature of the air and to use a resistance thermometer of suitable sensitivity to record the ambient or dry-bulb temperature. From these temperature records, the relative humidity can be calculated.

Air Circulation

Air must be circulated to keep refrigerated storage rooms at an even temperature throughout. Commodity temperatures in a storage room may vary because the air temperature rises as the air passes through the room and absorbs heat from the commodity; also, heat leakage may vary in different parts of the storage. In a duct system, the air near the return ducts will be warmer than the air near delivery ducts. In many storages, refrigeration units are installed over the center aisle. Air circulates from the center of the rooms outward to the walls, down through the rows of produce, and back up through the center of the room.

Rapid air circulation is needed most during removal of field heat. Sometimes this is best done in separate cooling rooms that have more refrigerating and air-moving capacity than regular refrigerated storage rooms (see Chapter 11). After field heat is removed, a high air velocity is usually undesirable. Air movement is needed only to remove respiratory heat and heat entering the room through exterior surfaces and doorways. Excessive air movement can increase moisture loss from a product, with a resulting loss of both weight and quality. However, some air circulated by fans or blowers must be uniformly distributed through all parts of the room.

The nature of the container and the manner of stacking are important factors that influence cooling performance. An elaborate system for air distribution is useless if poor stacking prevents airflow. If spacing is irregular, the wider spaces will get a greater

Commodity Storage Requirements

volume of air than narrower ones. If some spaces are partially blocked, dead air zones will occur with resultant higher temperatures.

Sanitation and Air Purification

The refrigerated storage and the product containers must be clean. Fruits and vegetables coming into the storage are generally contaminated with mold spores, which can enter through punctures or breaks in the skin of the product. Commodities so contaminated can foster a rapid development of the spores, which are carried by air currents throughout the storage. Removing decaying raw material from the storage and sanitizing the product container will reduce the problem. If mold contamination is excessive, the storage develops a musty or moldy odor that is quickly picked up by the fruits and vegetables, a fault of many apple varieties and other products held for several months in refrigerated storage. This problem can best be controlled by means of special cleaners, sanitizers, and deodorizers (see Chapters 16, 17, and 18 for details on product diseases).

During several months' storage, even at 31°F, molds may grow on the surface of packages and on the walls and ceilings of rooms under high relative humidity conditions. These surface molds generally will not rot fruits and vegetables. However, because surface molds are unsightly, storage warehouses should have a thorough cleaning at least once a year. Good air circulation alone is of considerable value in minimizing growth of surface molds. If floors and walls become moldy, they can be scrubbed with a cleaner containing sodium hypochlorite or trisodium phosphate, then rinsed, and the room aired. Field boxes and equipment can be cleaned with 0.25% calcium hypochlorite solutions or by exposing to steam for 2 min.

All inspected plants and warehouses operate under regulations with sanitation requirements clearly set forth in inspection service orders. Plants should be constructed to prevent the entrance of insects and rodents. This involves rat-proof building construction and adequate screening. For doors frequently opened, special measures must be taken to prevent the entrance of insects.

The frequency of cleanup operations and the detergents and sanitizing agents used should be specified by the quality control leader. A representative from quality control should inspect all areas after cleanup and determine whether or not a suitable job has been done. Waste and miscellaneous trash accumulation areas must receive special attention around warehouses, as they become breeding places for rodents and insects. In any food warehouse, the successful quality control and sanitation program depends on the cooperation and vigilance of management.

Air may be purified in storage rooms where odors or volatiles may contribute to off-flavors and hasten deterioration. Air may be cleaned with trays or canisters containing 6 to 14 mesh-activated coconut shell carbon. Pinewood volatiles are removed by activated carbon air-purifying units. Some produce volatiles are also removed, but ethylene, a ripening gas, is not removed by activated carbon alone.

Air washing with water to remove volatiles does not retard fruit ripening. Air washing may increase the relative humidity and thus aid in maintaining good fruit appearance by reducing weight loss.

Removal of Produce from Storage

When produce is removed from storage, undue warming and condensation of moisture, which promote decay and deterioration, must be prevented. Because most storages are built on railroad sidings, canvas tunnels should be installed between the car and the storage through which the produce can be conveyed to minimize warming and condensation of moisture.

When produce is removed from storage for distribution to wholesale and retail markets, the storage operator can do little to prevent undesirable condensation. Warming the packages until they are above the dew point of the air would prevent it, but this takes time and space and is seldom practicable. Deterioration in flavor and condition proceeds rapidly after long storage periods. Therefore, the produce should be moved to consumers as rapidly as possible.

STORAGE OF FROZEN FOODS

Frozen foods deteriorate during the period between production and consumption. The extent of deterioration depends on many factors, such as protection afforded by the package. However, the most important factors are storage temperature and storage time.

Bacteria in frozen foods may be killed during freezing and frozen storage, but all the bacteria present are never completely destroyed. When defrosting, frozen foods are still subject to bacterial decomposition.

SPECIFIC PRODUCTS

Beer

Since beer in bottles or cans is either pasteurized or filtered to destroy or remove the living yeast cells, it does not require as low a storage temperature as keg beer. Bottled beer may be stored at ordinary room temperature 70 to 75°F, but for convenience, it is often stored with keg beer at a lower temperature of 35 to 40°F. The bottled product should be protected from strong light, especially direct sunlight. The storage life will vary from 3 to 6 months, depending largely on the method of processing and packaging. Keg beer, usually stored at 35 to 40°F, has a storage life of 3 to 6 weeks.

Canned Foods

Canned foods that are heat processed in hermetically sealed containers do not benefit from refrigerated storage if they are to be stored for no longer than 2 or 3 months at temperatures that rarely exceed 75°F. However, seasonal commodities produced to provide an inventory for an entire year or longer do benefit from reduced temperature and humidity storage, which delays the onset of considerable color, texture, and flavor changes, loss of nutrients, and container corrosion. Notable examples are canned asparagus, cherries, and catsup. Reduced temperature and humidity storage for canned goods is essential in environments where ambient temperatures and humidities regularly exceed 90°F and 70% rh.

Dried Foods

Dried foods and feeds, particularly those expected to supply a high level of protein, such as dehydrated milk or alfalfa meal, should be protected against high temperatures and humidities. Good packaging, such as canning in vacuum, is helpful and can maintain dried food nutritive value and quality for a year or longer if ambient temperatures do not exceed 90°F regularly. When stored in bulk or in bags that are not good water vapor barriers, the storage life at 40% rh and 70°F should be limited to less than one year, and at 90°F to less than 6 months. At 60% rh and 70°F, the storage life should be limited to 6 months, and at 90°F to not more than 3 to 4 months. The only process by which dried foods can maintain quality and nutritive values for a year and longer at elevated temperatures is by packaging in zero oxygen. Similar or better results can be obtained with 0°F storage regardless of adequacy of the package or relative humidity.

Furs and Fabrics

Refrigerated storage effectively protects furs, floor coverings, garments, and other materials containing wool against insect damage. The commonly used refrigerated storage temperatures do not kill the insects, but inactivate them, preventing insect damage while the susceptible items are in storage (Table 3).

Table 3 Temperature and Time Requirements for Killing Moths in Stored Clothing

Storage Temperature, °F	All Eggs Dead After, Days	All Larvae Dead After, Days	All Adults Dead After, Days
−0.4 to 5	1	2	1
5 to 10	2	21[a]	1
10 to 15	4		1
15 to 19			1
20 to 25	21	67	4
25 to 30	21	125[b]	7
30 to 35		283[c]	

Table adapted from USDA *Publication* AMS-57 (1955).
[a] 50 to 95% of larvae may be killed in 2 days.
[b] A few larvae survived this period.
[c] Larvae survived this period.

However, if insects are present, the article is susceptible to damage as soon as it is removed from refrigerated storage.

Articles should be free of any possible infestation before placement in refrigerated storage. Those items that can be cleaned should be so treated. Others can be either fumigated or mothproofed.

Furs and garments should be stored at 34 to 40 °F. A temperature of 40 °F is most widely used commercially. The low temperature not only inactivates fabric insects but also preserves the vitality and luster of furs and the tensile strength of fabrics.

Continuous storage below the 34 to 40 °F range is a wasteful expense as far as protection from insect damage is concerned. Food should not be stored with fur garments. Some storage firms maintain constant temperatures in their fur vaults between 14 and 32 °F and claim excellent results. However, no research indicates that temperatures in this range are required for storing dressed furs or fabrics. Cured raw furs (but not processed) should be stored at −10 to 10 °F with 45 to 60% rh, and they will keep up to 2 years.

Honey

Both extracted (liquid) and comb honey can be held satisfactorily in common dry storage for about a year. The slow darkening and flavor deterioration at ordinary room temperatures becomes objectionable after this time. Although cold storage is not necessary, temperatures below 50 °F will maintain original quality for several years and retard or prevent fermentation. The range between 50 and 65 °F should be avoided, as it promotes granulation; this increases the probability of fermentation of raw (unheated) honey. As storage temperature increases in the 80 to 100 °F range, deterioration is accelerated; temperatures constantly above 85 °F are unsuitable and above 90 °F, quite damaging.

Honey for export is best kept in cold storage, since the half-life for honey diastase at 77 °F is about 17 months.

Raw honey of greater than 20% moisture is always in danger of fermentation; the likelihood is much less at or below 18.6% moisture. Below 17% moisture, raw honey will not ordinarily ferment. Granulation increases the possibility of fermentation of raw honey by increasing the moisture content of the liquid portion. Properly pasteurized honey will not ferment at any moisture content. Granulated honey can be reliquefied by warming to 120 to 140 °F.

Comb honey should not be stored above 60% rh to avoid moisture absorption through the wax, which leads to fermentation.

Finely granulated honey (honey spread, Dyce process honey, and honey cream) must not be stored above about 75 °F. Higher temperatures will, in time, cause partial liquefaction and destroy the texture. Any subsequent regranulation by lower temperatures will produce an undesirable coarse texture. For holding more than 4 months, cold storage is required.

Maple Syrup

Maple syrup keeps indefinitely at room temperatures without darkening or losing flavor, if packed hot (at or within a few degrees of its boiling point) and in clean containers, which are promptly closed airtight and laid on their sides or inverted to self-sterilize the closure, and then cooled. Refrigerated storage is not necessary. However, once opened, the syrup in a bottle, can, or drum may become contaminated by organisms in the air. Mold or yeast spores, which may be present in improperly pasteurized syrup, though unable to germinate in full-density syrup, may grow in the thin syrup on the surface caused by condensed water. Small packages not completely sterile, containing spores, can be kept free of vegetative growth by periodically inverting the containers to redisperse any thin syrup on the surface caused by condensed water. Maple syrup should never be packaged at temperatures below 180 °F. After pasteurizing, cool the syrup as quickly as possible to prevent stack burn, which darkens the syrup and causes a lowering of its grade.

Nursery Stock and Cut Flowers

The temperature and approximate storage life given in Table 4 for cut flowers allow for a reasonable shelf life after removal from storage; therefore, the storage period may be extended beyond that recommended here.

Low temperature (31 to 33 °F) and dry packaging prevent, or at least greatly retard, flower disintegration and extend the storage life. These conditions, while not widely used commercially, are recommended. Proper dry packing requires a moisture-vapor-proof container in which flowers can be sealed. No free water is added because the package prevents almost all water loss.

Flowers held in water should not be crowded in the containers and should be arranged on shelves or racks to allow good air circulation. Forced air circulation should be provided, but flowers must be kept out of a direct draft. Use clean water and clean containers.

Many kinds of nursery stock can also be stored at temperatures of 31 to 35 °F. Open packages and harden flowers before marketing if the blooms have been stored for long periods. Flowers conditioned at about 50 °F following storage regain full turgidity most rapidly. Cut or crush stem ends and then place in water or a food solution at 80 to 100 °F for 6 to 8 h.

Many kinds of cut flowers and greens are injured if stored in the same room with certain fruits, principally apples and pears, which give off gases (such as ethylene) during ripening. These gases cause premature aging of blooms and may defoliate greens. Greens should not be stored in the same room with cut flowers because the greens, acting in the same way as fruit, can hasten bloom deterioration.

Greens, bulbs, and certain nursery stock are usually packaged or crated when stored. Some bulbs and nursery stock are packed in damp moss or similar material, and low temperatures are required to keep them dormant. Polyethylene wraps or box liners are very effective for maintaining quality of strawberry plants, bare-root rose bushes, and certain cuttings and other nursery stock in storage. Strawberry plants can be stored up to 10 months in polyethylene-lined crates at 30 to 32 °F.

Popcorn

Store popcorn at 32 to 40 °F and about 85% rh. This relative humidity yields the optimum popping condition and the desired moisture content of about 13.5%.

Vegetable Seed

Seeds generally benefit from low temperatures and low humidity storage. High temperatures and high humidity favor loss of viability. Most vegetable seeds undergo no significant decrease

Commodity Storage Requirements

Table 4 Storage Conditions for Cut Flowers and Nursery Stock

Commodity	Storage Temperature, °F	Relative Humidity, %	Approximate Storage Life	Method of Holding	Highest Freezing Point, °F
Cut Flowers:					
Calla Lily	40	90 to 95	1 week	Dry pack	
Camellia	45	90 to 95	3 to 6 days	Dry pack	30.6
Carnation	31 to 32	90 to 95	3 to 4 weeks	Dry pack	30.8
Chrysanthemum	31 to 32	90 to 95	3 to 4 weeks	Dry pack	30.5
Daffodil (Narcissus)	32 to 33	90 to 95	1 to 3 weeks	Dry pack	31.8
Dahlia	40	90 to 95	3 to 5 days	Dry pack	
Gardenia	32 to 34	90 to 95	2 weeks	Dry pack	31.0
Gladiolus	36 to 42	90 to 95	1 week	Dry pack	31.4
Iris, tight buds	31 to 32	90 to 95	2 weeks	Dry pack	30.6
Lily, Easter	32 to 35	90 to 95	2 to 3 weeks	Dry pack	31.1
Lily-of-the-valley	31 to 32	90 to 95	2 to 3 weeks	Dry pack	
Orchid	45 to 55	90 to 95	2 weeks	Water	31.4
Peony, tight buds	32 to 35	90 to 95	4 to 6 weeks	Dry pack	30.1
Rose, tight buds	32	90 to 95	2 weeks	Dry pack	31.2
Snapdragon	40 to 42	90 to 95	1 to 2 weeks	Dry pack	30.4
Sweet Peas	31 to 32	90 to 95	2 weeks	Dry pack	30.4
Tulips	31 to 32	90 to 95	2 to 3 weeks	Dry pack	
Greens:					
Asparagus (plumosus)	35 to 40	90 to 95	2 to 3 weeks	Polylined cases	26.0
Fern, dagger and wood	30 to 32	90 to 95	2 to 3 months	Dry pack	28.9
Fern, leatherleaf	34 to 40	90 to 95	1 to 2 months	Dry pack	—
Holly	32	90 to 95	4 to 5 weeks	Dry pack	27.0
Huckleberry	32	90 to 95	1 to 4 weeks	Dry pack	26.7
Laurel	32	90 to 95	2 to 4 weeks	Dry pack	27.6
Magnolia	35 to 40	90 to 95	2 to 4 weeks	Dry pack	27.0
Rhododendron	32	90 to 95	2 to 4 weeks	Dry pack	27.6
Salal	32	90 to 95	2 to 3 weeks	Dry pack	26.8
Bulbs:					
Amaryllis	38 to 45	70 to 75	5 months	Dry	30.8
Caladium	70	70 to 75	2 to 4 months		29.7
Crocus	48 to 63		2 to 3 months		29.7
Dahlia	40 to 48	70 to 75	5 months	Dry	28.7
Gladiolus	45 to 50	70 to 75	5 to 8 months	Dry	28.2
Hyacinth	63 to 68		2 to 5 months		29.3
Iris, Dutch, Spanish	68 to 77	70 to 75	4 months	Dry	29.3
Gloriosa	50 to 63	70 to 75	3 to 4 months	Polyliner	
Candidum	31 to 33	70 to 75	1 to 6 months	Polyliner and peat	
Croft	31 to 33	70 to 75	1 to 6 months	Polyliner and peat	
Longiflorum	31 to 33	70 to 75	1 to 10 months	Polyliner and peat	28.9
Speciosum	31 to 33	70 to 75	1 to 6 months	Polyliner and peat	
Peony	33 to 35	70 to 75	5 months	Dry	
Tuberose	40 to 45	70 to 75	4 to 12 months	Dry	
Tulip	63	70 to 75	2 to 6 months	Dry	27.6
Nursery Stock:					
Trees and shrubs	32 to 36	95	4 to 5 months	a	
Rose bushes	31 to 36	85 to 95	4 to 5 months	Bare rooted with polyliner	
Strawberry plants	30 to 32	80 to 85	8 to 10 months	Bare rooted with polyliner	29.9
Rooted cuttings	31 to 36	85 to 95	—	Polywrap	
Herbaceous perennials	27 to 28	80 to 85	4 to 8 months	a	
	31 to 35	80 to 85	3 to 7 months		
Christmas trees	22 to 32	80 to 85	6 to 7 weeks		

Data from USDA *Agricultural Handbook* No. 66.
[a] For details for various trees, shrubs, and perennials, see *Storage of Nursery Stock* (AAN 1960).

Table 5 Space, Weight, and Density Data for Commodities Stored in Refrigerated Warehouses

Commodity	Type of Package	Outside Dimensions of Package, in.	Average Gross Wt of Pkg, lb	Average Net Wt Mdse, lb	Average Gross Wt Density, lb/ft³	Average Net Wt Density, lb/ft³
Apples	Wood box, Northwestern	19-1/2 × 11 × 12-3/16	50	42	33.1	27.8
	Fiber tray carton	20-1/2 × 12-1/2 × 13-1/4	46-3/4	43	23.8	21.9
	Fiber master carton	22-1/2 × 12-1/2 × 13	44-3/4	41	21.2	19.4
	Fiber bulk carton	19 × 12-1/2 × 13	44-3/4	41	25.0	22.9
	Pallet bin	47 × 47 × 30	1030	900	26.9	23.5
Beef						
Boneless	Fiber carton	28 × 18 × 6	146	140	83.4	80.0
Fores	Loose					22.2
Hinds	Loose					22.2

Table 5 Space, Weight, and Density Data for Commodities Stored in Refrigerated Warehouses (*Continued*)

Commodity	Type of Package	Outside Dimensions of Package, in.	Average Gross Wt of Pkg, lb	Average Net Wt Mdse, lb	Average Gross Wt Density, lb/ft³	Average Net Wt Density, lb/ft³
Celery	Wirebound crates	20-1/4 × 16 × 9-3/4	60	55	32.8	30.0
	Fiber carton	16 × 11 × 10	36	32	35.4	31.4
Cheese	Hoops	16 × 16 × 13	84	78	43.6	40.5
	Wood, export	17 × 17 × 14	87	76	37.1	32.5
Cheese, Swiss	Wheels	32-1/2 × 32-1/2 × 7		171		40.0
Chili Peppers	Bags	45 × 21 × 26	234	229	16.5	16.1
Citus Fruits						
Oranges	Box	12-1/8 × 13-1/4 × 26-1/4	77	69	31.5	28.3
	Bruce box	13 × 11 × 26-1/4	88	83	40.5	38.2
	Pallet, 40 cartons	40 × 48 × 58-1/2	1690	1480	26.0	22.8
California Oranges	Fiber carton	16-3/8 × 10-1/16 × 10-1/2	40	37	38.0	35.2
Florida Oranges	Fiber carton	19-1/4 × 12-1/4 × 8	45	37	41.3	33.9
Lemons	Fiber carton	16-3/8 × 10-1/16 × 10-1/2	40	37	40.0	37.0
Grapefruit	Fiber carton	19-1/4 × 12-1/4 × 8	40	38	36.7	34.9
Coconut, shredded	Bags	38 × 18-1/2 × 8	101	100	31.0	30.7
Cranberries	Fiber carton	15-3/4 × 11-1/4 × 10-1/2	26	24	24.1	22.2
Cream	Tins	12 × 12 × 14	52-3/4	50	45.2	42.9
Dried Fruit	Wood box	15-1/2 × 10 × 6-1/2	26-1/2	25	45.4	42.9
Dates	Fiber carton	14 × 14 × 11	32	30	25.7	24.0
Raisins, prunes, figs, peaches	Fiber carton	15 × 11 × 7	32	30	47.9	44.9
Eggs, Shell	Wood cases	26 × 12 × 13	55	45	23.4	19.1
Eggs, Frozen	Cans	10 × 10 × 12-1/2	32	30	44.2	41.5
Frozen fishery products						
Blocks	4/13-1/2 lb carton	20-3/4 × 12-1/8 × 6-3/4	56	54	57.0	55.0
	4/16 1/2 lb carton	19-3/4 × 10-3/4 × 11-1/4	68	66	49.2	47.8
Filets	12/16 oz carton	12-3/4 × 8-5/8 × 3-13/16	13.5	12	55.8	49.6
	10/5 lb carton	14-1/2 × 10 × 14	52.25	50	44.6	42.7
	5/10 lb carton	14-1/2 × 10 × 14	52.2	50	44.5	42.7
Fish sticks	12/8 oz carton	11 × 8-3/8 × 3-7/8	6.9	6	33.6	29.3
	24/8 oz carton	16-7/16 × 8-5/16 × 4-5/8	13.8	12	37.8	32.9
Panned fish portions	None, glazed	Wooden boxes				35.0
	2, 3, 5, and 6 lb carton	Custom packing				29-33
Round ground fish	None, glazed	Stacked loose				33-35
Round Halibut	None, glazed	Wooden box, loose				30-35
		Stacked loose				38.0
Round Salmon	None, glazed	Stacked loose				33-35
Shrimp	2-1/2 and 5 lb cartons	Custom packing				35.0
Steaks	1, 5, or 10 lb packages	Custom packing				50-60
Frozen Fruits, Juices, Vegetables						
Asparagus	24/12 oz carton	13-1/2 × 11-3/4 × 8-1/4	21	18	27.7	23.8
Beans, Green	36/10 oz carton	12-1/2 × 11 × 8	25-1/2	22-1/2	40.1	35.3
Blueberries	24/12 oz carton	12 × 11-1/2 × 8	20	18	31.3	28.2
Broccoli	24/10 oz carton	12-1/2 × 11-1/2 × 8-1/2	18-1/2	15	26.2	21.2
Citrus concentrates	Fiber carton 48/6 oz	13 × 8-3/4 × 7-1/2	27	26	54.7	52.7
Peaches	24/1 lb carton	13-1/2 × 11-1/4 × 7-1/2	27	24	41.0	36.4
Peas	6/5 lb carton	17 × 11 × 9-1/2	32	30	31.1	28.2
	48/12 oz carton	21-1/2 × 8-1/2 × 12-1/2	38	36	28.7	27.2
Potatoes, french fries	12/16 oz carton					28.6
	24/9 oz carton					24.0
Spinach	24/14 oz carton	12-1/2 × 11 × 8-1/2	24	21	35.5	31.0
Strawberries	30 lb can	12-1/2 × 10 × 10	32	30	44.2	41.5
	24/1 lb carton	13 × 11 × 8	28	24	42.3	36.2
	450 lb barrel	35 × 25 × 25		450		35.5
Grapes, California	Wood lug box	6-1/2 × 15 × 18	31	28	32.4	29.2
Lamb, boneless	Fiber box	20 × 15 × 5	57	53	65.7	61.0
Lard (2/28 lb)	Wood export box	18 × 13-1/4 × 7-3/4	64	56	59.8	52.5
Lettuce, head	Fiber carton	20-1/2 × 13-1/2 × 9-1/2	37-1/2	35	24.7	
	Fiber carton	21-1/2 × 14-1/4 × 10-1/2	45-55	42-52	26.9	25.2
	Pallet, 30 cartons	42 × 50 × 66	1350	1170	16.8	14.6
Milk, condensed	Barrels	35 × 25-1/2 × 25-1/2	670	600	50.9	45.6
Nuts						
Almonds, in shell	Sacks	24 × 15 × 33	91-1/2	90	13.3	13.1
Almonds, shelled	Cases	6-3/4 × 23-1/2 × 11	32	28	31.7	27.7
English Walnuts, in shell	Sacks	25 × 11 × 31	103	100	20.9	20.3
English Walnuts, shelled	Fiber carton	14 × 14 × 10	27	25	23.8	22.0
Peanuts, shelled	Burlap bag	35 × 10 × 15	127	125	39.2	38.6
Pecans, in shell	Burlap bag	35 × 22 × 12	126-1/2	125	23.7	23.4
Pecans, shelled	Fiber carton	13 × 13 × 11	32	30	29.8	27.9

Commodity Storage Requirements

Table 5 Space, Weight, and Density Data for Commodities Stored in Refrigerated Warehouses (*Concluded*)

Commodity	Type of Package	Outside Dimensions of Package, in.	Average Gross Wt of Pkg, lb	Average Net Wt Mdse, lb	Average Gross Wt Density, lb/ft^3	Average Net Wt Density, lb/ft^3
Peaches	3/4 bushel	16-7/8 top dia	41	48	43.9	40.7
	1/2 bushel	14-1/2 top dia	28	25	45.0	40.2
	Wirebound crate	19 × 11-3/4 × 11-1/8	42	38	29.2	26.4
	Wood lug box	18-1/8 × 11-1/2 × 5-3/4	26	23	38.0	33.1
Pears	Wood box	8-1/2 × 11-1/2 × 18	52	48	51.0	47.1
Pears, place pack	Fiber carton	18-1/2 × 12 × 10	52	46	40.5	35.6
Pork						
Bundle bellies	Bundles	23-1/2 × 10-1/2 × 7	57	57	57.0	57.0
Loins (regular)	Wood box	28 × 10 × 10	60	54	37.0	33.3
Loins (boneless)	Fiber box	20 × 15 × 5	57	52	65.7	59.9
Potatoes	Sack	33 × 17-1/2 × 11	101	100	27.5	27.2
Poultry, fresh (eviscerated)						
Fryers, whole, 24-30 to pkg.	Wirebound crate	24 × 10 × 7	65	60	27.5	25.4
Fryer parts	Wirebound crate	17-3/4 × 10 × 12-1/2	54	50	42.1	38.9
Poultry, frozen (eviscerated)						
Ducks, 6 to pkg.	Fiber carton	22 × 16 × 4	32-1/2	31	39.9	38.0
Fowl, 6 to pkg.	Fiber carton	20-3/4 × 18 × 5-1/2	33 1/2	31	28.2	26.1
Fryers, cut up, 12 to pkg.	Fiber carton	17-1/4 × 15-3/4 × 4-1/4	30 1/2	28	45.4	41.7
Roasters, 8 to pkg.	Fiber carton	20-3/4 × 18 × 5-1/2	32 1/2	30	27.3	25.2
Turkeys,						
3-6 lb, 6 to pkg.	Fiber carton	21 × 17 × 6-1/2	30	27	22.5	20.1
6-10 lb, 6 to pkg.	Fiber carton	26 × 21-1/2 × 7	52-1/2	48	23.3	21.2
10-13 lb, 4 to pkg.	Fiber carton	26-1/2 × 16 × 7-1/2	50	46	27.2	25.0
13-16 lb, 4 to pkg.	Fiber carton	29 × 18-1/2 × 9	67-1/2	62	24.2	22.2
16-20 lb, 2 to pkg.	Fiber carton	17 × 16 × 9	39	36	27.7	25.4
20-24 lb, 2 to pkg.	Fiber carton	19 × 16-1/2 × 9-1/2	47-1/2	44	27.6	25.5
Tomatoes						
Florida	Fiber carton	19 × 10-7/8 × 10-3/4	43	40	33.3	31.0
	Wirebound crate	18-3/4 × 11-15/16 × 11-15/16	64	60	41.3	38.7
California	Wood lug box	17-1/2 × 14 × 7-3/4	34	30	30.9	27.3
Texas	Wood lug box	17-1/2 × 14 × 6-5/8	34	30	36.2	31.9
Veal (boneless)	Fiber carton	20 × 15 × 5	57	53	65.7	61.0

Adapted from the table *Density of Commodities Carried in Cold Storage Warehouses*, courtesy of the Missouri Valley Chapter of the National Association of Refrigerated Warehouses, and from other sources.

in germination during one season when stored at 50 °F and 50% rh. Full viability is retained far longer than one year as temperature and humidity are reduced. A temperature of 20 °F and 15 to 25% rh are considered ideal but rarely necessary unless seed viability must be maintained for many years. Aster, pepper, tomato, and lettuce seed stored under these conditions had equal or better viability after 13 years than the fresh seed.

Low moisture content of the seed is important for germination. Hemp seed containing 9.5% moisture was mostly dead after 12 years storage at 50 °F, but lost only 12% viability when moisture content was 5.7%. If stored at about 0 °F, moisture percentage should not exceed 10% for many species. Seed with higher moisture content, when stored at 0 °F, eventually equilibrates at a lower moisture level but may initially suffer frost damage. At higher temperatures, if it is impossible to keep humidity low enough, seeds must be stored in moistureproof containers.

DENSITY OF STORED COMMODITIES

Table 5 is a compilation of data giving the types of containers used for storage, their dimensions, gross weights, net weights, and density. Additional information on gross weights of packed containers and on dimensions and densities of pallet loads of produce is given in USDA *Marketing Research Report* No. 467.

REFERENCES

Bogardus, R.K. 1961. Wholesale fruit and vegetable warehouses—Guides for layout and design. *Marketing Research Report* No. 467.

Hardenburg, R.E., A.E. Watada, and C.Y. Wang. 1986. The commercial storage of fruits, vegetables, and florist and nursery stocks. USDA *Handbook* No. 66.

Mahlstede, P. and W.E. Fletcher. 1960. *Storage of nursery stock*. American Association Nurserymen, Washington, D.C.

Post, K. and C.W. Fisher, Jr. 1952. Commercial storage of cut flowers. Cornell University Extension *Bulletin* No. 853.

RRF. 1989. *Commodity storage manual*. Refrigeration Research Foundation, Bethesda, MD.

Ryall, A.L. and W.J. Lipton. 1978. *Handling, transportation and storage of fruits and vegetables*. Vol. 1, Vegetables and melons. AVI Publishing Co., Inc., Westport, CT.

Ryall, A.L. and W.T. Pentzer. 1982. *Handling, transportation and storage of fruits and vegetables*. Vol. 2, Fruits and tree nuts. AVI Publishing Co., Inc., Westport, CT.

Tressler, D.K. and C.F. Evers. 1957. *The freezing preservation of foods*, 3rd ed. AVI Publishing Co., Inc., Westport, CT.

USDA. 1955. Protecting stored furs from insects. USDA *Publication* AMS-57.

Webb, B.H., A.H. Johnson, and J.A. Alford. 1973. *Fundamentals of dairy chemistry*. AVI Publishing Co., Inc., Westport, CT.

CHAPTER 27

REFRIGERATION LOAD

Transmission Load 27.1
Product Load 27.1
Internal Load 27.3
Infiltration Air Load 27.3
Equipment Load 27.5
Safety Factor 27.6
Total Refrigeration Load 27.6

THE segments of total refrigeration load are: (1) *transmission load*, heat transferred into the refrigerated space through its surface, (2) *product load*, heat removed from and produced by products brought into and kept in the refrigerated space, (3) *internal load*, heat produced by internal sources, *e.g.*, lights, electric motors, and people working in the space, (4) *infiltration air load*, heat gains associated with air entering the refrigerated space, and (5) *Equipment load*.

The first four segments of load constitute the net heat load for which a refrigeration system is to be provided; the fifth segment consists of all heat gains created by the refrigerating equipment in the process of doing so. Thus, net heat load plus equipment heat load is the total refrigeration load for which a compressor must be selected.

Load calculating procedures and data for the first four segments and load determination recommendations for the fifth segment are contained in this chapter. Information needed for the refrigeration of foods can be found elsewhere in this volume.

TRANSMISSION LOAD

Sensible heat gain through walls, floor, and ceiling varies with type of insulation, thickness of insulation, construction, outside wall area, and temperature difference between refrigerated space and ambient air.

The thermal conductance values per °F temperature difference for several cold storage insulations are listed in Table 1. The thickness of insulation referred to in this table is actual insulation thickness, not the overall wall thickness. Comparative values of various insulation materials can be found in Chapter 22 of the 1989 ASHRAE *Handbook—Fundamentals*.

The overall coefficient of heat transfer U of the wall, floor, or ceiling can be derived by the following equation:

$$U = \frac{1}{1/f_i + x/k + 1/f_o} \quad (1)$$

where

U = overall heat transfer coefficient, Btu/h·ft²·°F
x = wall thickness, in.
k = thermal conductivity of wall material, Btu·in/h·ft²·°F
f_i = inside film or surface conductance, Btu/h·ft²·°F
f_o = outside film or surface conductance, Btu/h·ft²·°F

A value of 1.65 Btu/h·ft²·°F for f_i and f_o is frequently used for still air. If the outer surface is exposed to 15 mph wind, f_o is increased to 6 Btu/h·ft²·°F.

With thick walls and low conductivity, the resistance x/k makes U so small that $1/f_i$ and $1/f_o$ have little effect and can be omitted from the calculation. Walls are not often made of one material; therefore, the value x/k represents the composite resistance value of the several materials used in series to the heat flow, and for a wall with flat parallel surfaces of materials 1, 2, 3, and so forth:

$$U = \frac{1}{x_1/k_1 + x_2/k_2 + x_3/k_3} \quad (2)$$

where x values indicate thicknesses and k values indicate conductivities of the materials used. Table 1 shows possible values of x/k.

The effect on the overall thermal performance of the metal surfaces on prefabricated or insulated panels is negligible and should not be considered in calculating the U-value. In this case, the reciprocal of the conductance values in Table 1 can be substituted for x/k in Equations (1) or (2).

After establishing the coefficient of heat transfer U, the heat gain is given by the basic equation:

$$q = UA\Delta t \quad (3)$$

where

q = heat leakage, Btu/h
A = outside area of section, ft²
Δt = difference between outside air temperature and air temperature of the refrigerated space, °F

Table 2 lists minimum insulation thicknesses based on expanded polyurethane as recommended by the refrigeration industry. Equivalent thicknesses of other insulation materials can be found by comparing C values in Table 1. Consideration should be given to conductivity increases with time, (see Chapter 20 of the 1989 ASHRAE *Handbook—Fundamentals*).

Chapter 24 of the 1989 ASHRAE *Handbook—Fundamentals* gives outdoor design temperatures for major cities; values for 1% should be used.

If the refrigerated room is exposed to the sun, additional heat will be added to the heat load. For practical purposes, the temperature can be adjusted to compensate for solar effect. The values given in Table 3 apply throughout the 24-h period and are added to the ambient temperature when calculating wall heat gain.

Latent heat gain due to moisture transmission through walls, floors, and ceilings of modern-construction refrigerated facilities is negligible. Data in Chapter 22 of the 1989 ASHRAE *Handbook—Fundamentals* may be used to calculate this load should moisture-permeable materials be used.

PRODUCT LOAD

The primary sources of refrigeration load from products brought into and kept in the refrigerated space are: (1) heat removal required to reduce product temperature from receiving to storage

---

The preparation of this chapter is assigned to TC 10.8, Refrigeration Load Calculations.

Table 1 Insulation Conductance Values C for Walls, Floor, and Ceiling (Btu/h·ft²·°F)

Insulation Thickness	Polyurethane (Expanded) $k = 0.16$[b]	Polyurethane (Board) $k = 0.18$[b]	Polystyrene (Extruded) $k = 0.20$[b]	Glass Fiber and Polystyrene (Molded Beads) $k = 0.25$[b]	Corkboard $k = 0.30$[b]
in.	Btu/h·ft²·°F	Btu/h·ft²·°F	Btu/h·ft²·°F	Btu/h·ft²·°F	Btu/h·ft²·°F
1	0.160	0.180	0.200	0.250	0.300
2	0.080	0.090	0.100	0.125	0.150
3	0.053	0.060	0.067	0.083	0.100
4	0.040	0.045	0.050	0.063	0.070
5	0.032	0.036	0.040	0.050	0.060
6	0.027	0.030	0.033	0.042	0.050
7	0.023	0.026	0.029	0.036	0.043
8	0.020	0.022	0.025	0.031	0.038
9	0.018	0.020	0.022	0.028	0.033
10	0.016	0.018	0.020	0.025	0.030

Note: Where wood studs are used in walls, multiply above values by 1.1.
[a]Commonly used values of conductance are tabulated. See Chapters 20 and 22 in the 1989 ASHRAE *Handbook—Fundamentals* for variations due to temperature, aging, and other considerations.
[b]Thermal conductivity k is expressed in Btu·in/h·ft²·°F.

Table 2 Minimum Insulation Thickness

Storage Temperature	Expanded Polyurethane Thickness	
	Northern U.S.	Southern U.S.
°F	in.	in.
50 to 60	1	2
40 to 50	2	2
25 to 40	2	3
15 to 25	3	3
0 to 15	3	4
−15 to 0	4	4
−40 to −15	5	5

Table 3 Allowance for Sun Effect

Typical Surface Types	East Wall °F	South Wall °F	West Wall °F	Flat Roof °F
Dark colored surfaces				
Slate roofing	8	5	8	20
Tar roofing				
Black paint				
Medium colored surfaces				
Unpainted wood	6	4	6	15
Brick				
Red tile				
Dark cement				
Red, gray, or green paint				
Light colored surfaces				
White stone	4	2	4	9
Light colored cement				
White paint				

Note: Add °F to the normal temperature difference for heat leakage calculations to compensate for sun effect—do not use for air-conditioning design.

temperature and (2) heat generation by products in storage, mainly fruits and vegetables.

A product placed in a refrigerated room at a higher temperature than the room temperature loses heat as it approaches room temperature. The quantity of heat to be removed can be calculated from knowledge of the product, including state upon entering the refrigerated space, final state, mass, specific heat above and below freezing temperature, and latent heat. When cooling a definite mass of product from one state and temperature to another state and temperature, the following equations apply:

1. Heat removal in cooling from the initial temperature to some lower temperature above freezing:

$$Q_1 = mc_1 (t_1 - t_2) \quad (4)$$

2. Heat removal in cooling from the initial temperature to the freezing point of the product:

$$Q_2 = mc_1 (t_1 - t_f) \quad (5)$$

3. Heat removal to freeze the product:

$$Q_3 = mh_{if} \quad (6)$$

4. Heat removal in cooling from the freezing point to the final temperature below the freezing point:

$$Q_4 = mc_2 (t_f - t_3) \quad (7)$$

where

Q_1, Q_2, Q_3, Q_4 = heat removal, Btu
m = mass of the product, lb
c_1 = specific heat of the product above freezing, Btu/lb·°F
t_1 = initial temperature of the product above freezing, °F
t_2 = lower temperature of the product above freezing, °F
t_f = freezing temperature of the product, °F
h_{if} = latent heat of fusion of the product, Btu/lb
c_2 = specific heat of the product below freezing, Btu/lb·°F
t_3 = final temperature of the product below freezing, °F

Refrigeration system capacity for products brought into refrigerated spaces is determined from the time allotted for heat removal and assumes that the product is exposed in a manner to remove the heat in that time. The calculation is:

$$q = \frac{Q_2 + Q_3 + Q_4}{n} \quad (8)$$

where

q = product cooling load, Btu/h
n = allotted time period, h

Specific heats above and below freezing and latent heats of fusion for many products are given in Chapter 26. A product's latent heat of fusion is related to its water content and can be estimated by multiplying the percent of water in the product (expressed as a decimal) by the latent heat of fusion of water, which is 144 Btu/lb. Most food products freeze in the range of 26 to 31°F. When the exact freezing temperature is not known, assume that it is 28°F.

Example 1. 220 lb of lean beef is to be cooled from 65 to 40°F, then frozen and cooled to 0°F. Specific heat of beef before freezing is 0.77 Btu/lb·°F; after freezing, 0.40 Btu/lb·°F. The latent heat of fusion is 100 Btu/lb.

Refrigeration Load

Solution:

To cool from 65 to 40 °F in a chilled room:
$220 \times 0.77 \times (65 - 40) = 4235$ Btu

To cool from 40 °F to freezing point in freezer:
$220 \times 0.77 \times (40 - 28) = 2033$ Btu

To freeze:
$220 \times 100 = 22{,}000$ Btu

To cool from freezing to storage temperature:
$220 \times 0.40 \times (28 - 0) = 2464$ Btu

Total:
$4235 + 2033 + 22{,}000 + 2464 = 30{,}732$ Btu

Fresh fruits and vegetables respire and release heat during storage. This heat of respiration varies with product and temperature; the colder the product, the less the heat of respiration. Table 2 of Chapter 26 gives the heat evolution rates for various products.

Calculations in *Example 1* do not cover heat gained when necessary to cool product containers brought into the refrigerated space. When pallets, boxes, or other packing materials are a significant portion of the total mass introduced, this heat gain should be calculated.

Equations (4) through (8) provide the total heat gain of product refrigeration to be extracted by the refrigerating apparatus. Any moisture shrinkage involved in the extraction process appears as latent heat gain. The amount of moisture involved is usually provided by the end-user as a percentage of product mass and, with such information, the latent heat component of the total heat gain may be determined by means of steam tables. The sensible heat component is obtained by subtracting the latent heat component from the total heat gain.

INTERNAL LOAD

All electrical energy dissipated in the refrigerated space (from lights, motors, heaters, and other equipment) must be included in the internal heat load. Heat equivalents of electric motors are shown in Table 4. People release heat at varying rates depending on temperature, type of work, clothing, size, and other factors. The average load caused by occupancy of the refrigerated space is shown in Table 5. When people go into the refrigerated space for short durations, they carry a considerable amount of heat over and above that listed in Table 5, and adjustments must be made if traffic load is heavy.

The latent heat component of internal load is usually very small compared to total refrigeration load and is customarily regarded as all-sensible in total load summaries. However, the latent heat component may be great where water is involved in processing or cleaning and should be independently calculated.

Table 4 Heat Equivalent of Electric Motors

Motor hp	Connected Load in Refrigerated Space[a] Btu/hp·h	Motor Losses Outside Refrigerated Space[b] Btu/hp·h	Connected Load Outside Refrigerated Space[c] Btu/hp·h
1/8 to 1/3	4600 ±	2550	2100 ±
1/2 to 3	3800 ±	2550	1300 ±
5 to 20	3300 ±	255]	800 ±

[a] For use when both useful output and motor losses are dissipated within refrigerated space; motors driving fans for forced circulation unit coolers.

[b] For use when motor losses are dissipated outside refrigerated space and useful motor work is expended within refrigerated space; pump on a circulating brine or chilled water system; fan motor outside refrigerated space driving fan circulating air within refrigerated space.

[c] For use when motor heat losses are dissipated within refrigerated space and useful work expended outside of refrigerated space; motor in refrigerated space driving pump or fan located outside of space.

Table 5 Heat Equivalent of Occupancy

Refrigerated Space Temperature, °F	Heat Equivalent/ Person, Btu/h
50	720
40	840
30	950
20	1050
10	1200
0	1300
−10	1400

Note: Heat equivalent may be estimated by $q_p = 1295 - 11.5t\,(°F)$

INFILTRATION AIR LOAD

Infiltration air is the single largest portion of refrigeration load for many refrigeration applications and can amount to more than half the total load for refrigerated warehouses and similar applications.

Infiltration by Air Exchange

The most common manner of infiltration to refrigerated rooms is air exchange at doorways or similar large openings as illustrated by Figures 1 and 2. For the usual circumstance where mass inflow equals mass outflow less condensed moisture, the room must be sealed except for the opening in question. If the cold room is not sealed, direct flow-through, discussed later, may apply.

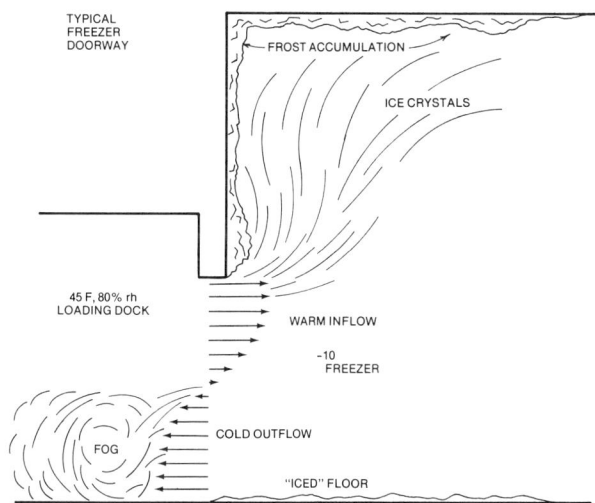

Fig. 1 Flowing Cold and Warm Air Masses that Occur for Typical Open Freezer Doors

The equation for heat gain through doorways from air exchange is as follows:

$$q_t = qD_tD_f \qquad (9)$$

where

q_t = average hourly heat gain for the 24-h or other period, Btu/h
q = sensible and latent refrigeration load for fully established flow, Btu/h
D_t = doorway open-time factor
D_f = doorway flow factor

Gosney and Olama (1975) provide the following air exchange equation for fully established flow:

$$q = 795.6A\,(h_i - h_r)\,\rho_r\,(1 - \rho_i/\rho_r)^{0.5}\,(gH)^{0.5}F_m \qquad (10)$$

where

$795.6 = 3600\text{ s/hr} \times 0.221$
A = doorway area, ft^2
h_i = enthalpy of infiltration air, Btu/lb
h_r = enthalpy of refrigerated air, Btu/lb
ρ_i = density of infiltration air, lb/ft^3

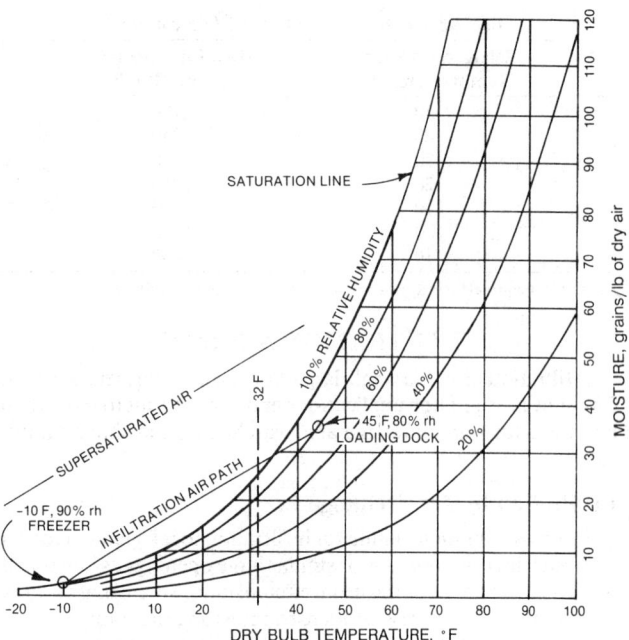

Fig. 2 Psychrometric Depiction of Air Exchange for Typical Freezer Doorway

ρ_r = density of refrigerated air, lb/ft³
g = gravitational constant = 32.174 ft/s²
H = doorway height, ft
F_m = density factor

$$F_m = \left(\frac{2}{1 + (\rho_r/\rho_i)^{1/3}} \right)^{1.5} \tag{11}$$

(Chapter 6 of the 1989 ASHRAE *Handbook—Fundamentals* and the ASHRAE Psychrometrics Chart give air enthalpy and density values.)

Equation (12), when used with with Figure 3, is a simplification of Equation (10):

$$q = 3790 W H^{1.5} (T_s/A)(1/R_s), \text{Btu/h} \tag{12}$$

where

T_s/A = sensible heat load of infiltration air per square foot of doorway opening as read from Figure 3, tons/ft²
W = doorway width, ft
R_s = sensible heat ratio of the infiltration air heat gain, from Tables 6 or 7 (or from a pyschrometric chart)

Table 6 SHR (R_s) for Infiltration from Outdoors to Refrigerated Spaces

Outdoors Temp. °F	rh, %	\-30	\-20	\-10	0	10	20	30	40	50	60
100	50	0.59	0.57	0.55	0.53	0.51	0.49	0.48	0.46	0.45	0.45
	40	0.65	0.63	0.61	0.59	0.57	0.56	0.54	0.53	0.54	0.57
	30	0.71	0.69	0.68	0.66	0.65	0.64	0.63	0.64	0.66	0.76
	20	0.79	0.77	0.76	0.75	0.74	0.74	0.75	0.78	0.87	—
95	60	0.58	0.56	0.54	0.52	0.49	0.47	0.45	0.43	0.42	0.41
	50	0.62	0.60	0.58	0.56	0.54	0.52	0.51	0.49	0.48	0.50
	40	0.67	0.66	0.64	0.62	0.60	0.59	0.57	0.57	0.58	0.64
	30	0.73	0.72	0.71	0.69	0.67	0.66	0.66	0.68	0.72	0.89
90	60	0.61	0.59	0.57	0.55	0.52	0.50	0.48	0.46	0.45	0.45
	50	0.65	0.63	0.61	0.59	0.57	0.55	0.54	0.52	0.52	0.56
	40	0.70	0.68	0.67	0.65	0.63	0.61	0.61	0.61	0.63	0.74
	30	0.76	0.74	0.73	0.71	0.70	0.69	0.70	0.72	0.80	—

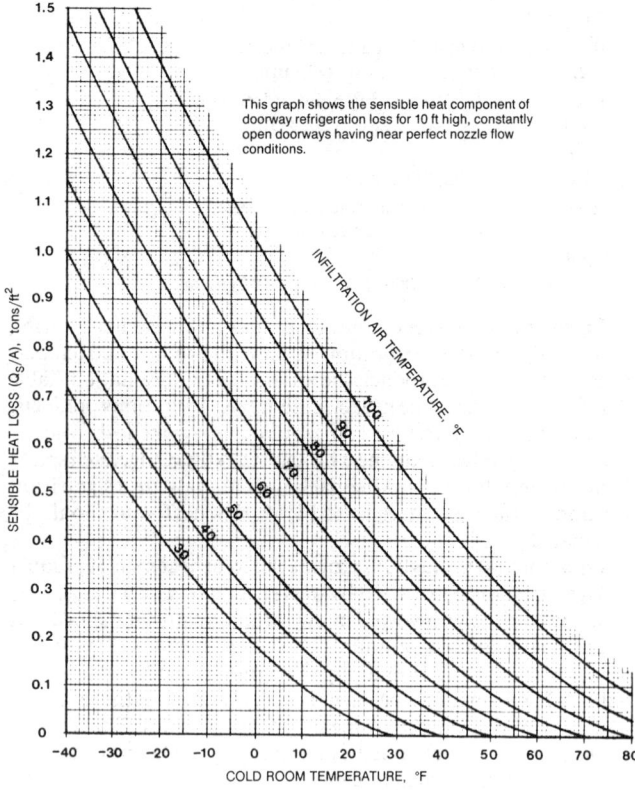

Fig. 3 Sensible Heat Gain by Air Exchange for Continuously Open Door with Fully Established Flow

Table 7 SHR (R_s) for Infiltration from Warmer to Colder Refrigerated Spaces

Warm Space Temp. °F	rh, %	\-40	\-30	\-20	\-10	0	10	20	30	40	50
70	100	0.60	0.58	0.56	0.53	0.50	0.47	0.44	0.41	0.37	0.34
	80	0.66	0.64	0.61	0.59	0.56	0.53	0.50	0.48	0.46	0.44
	60	0.72	0.70	0.68	0.66	0.63	0.61	0.59	0.58	0.59	0.64
	40	0.79	0.78	0.76	0.75	0.73	0.72	0.71	0.73	0.80	—
60	100	0.66	0.64	0.62	0.59	0.56	0.52	0.49	0.45	0.41	0.35
	80	0.71	0.69	0.67	0.64	0.62	0.59	0.56	0.53	0.52	0.53
	60	0.77	0.75	0.73	0.71	0.69	0.67	0.65	0.65	0.70	—
	40	0.83	0.82	0.81	0.79	0.78	0.77	0.78	0.83	—	—
50	100	0.72	0.70	0.67	0.64	0.61	0.57	0.53	0.49	0.43	—
	80	0.76	0.74	0.72	0.70	0.67	0.64	0.61	0.59	0.62	—
	60	0.81	0.80	0.78	0.76	0.74	0.72	0.71	0.75	—	—
	40	0.87	0.86	0.84	0.83	0.82	0.82	0.85	—	—	—
40	100	0.77	0.75	0.72	0.69	0.66	0.62	0.57	0.51	—	—
	80	0.81	0.79	0.77	0.74	0.72	0.69	0.66	0.67	—	—
	60	0.85	0.84	0.82	0.80	0.78	0.77	0.79	0.99	—	—
	40	0.90	0.89	0.88	0.87	0.86	0.88	0.97	—	—	—
30	100	0.82	0.80	0.77	0.74	0.70	0.66	0.59	—	—	—
	80	0.85	0.83	0.81	0.79	0.76	0.73	0.73	—	—	—
	60	0.88	0.87	0.86	0.84	0.83	0.83	0.94	—	—	—
	40	0.92	0.91	0.90	0.90	0.91	0.96	—	—	—	—
20	100	0.86	0.84	0.82	0.79	0.75	0.69	—	—	—	—
	80	0.89	0.87	0.85	0.83	0.81	0.80	—	—	—	—
	60	0.91	0.90	0.89	0.88	0.88	0.95	—	—	—	—
	40	0.94	0.94	0.93	0.94	0.97	—	—	—	—	—
10	100	0.90	0.88	0.86	0.83	0.78	—	—	—	—	—
	80	0.92	0.90	0.89	0.87	0.86	—	—	—	—	—
	60	0.94	0.93	0.92	0.92	0.96	—	—	—	—	—
	40	0.96	0.96	0.96	0.98	—	—	—	—	—	—
0	100	0.92	0.91	0.89	0.85	—	—	—	—	—	—
	80	0.94	0.93	0.92	0.91	—	—	—	—	—	—
	60	0.96	0.95	0.95	0.97	—	—	—	—	—	—
	40	0.97	0.97	0.98	—	—	—	—	—	—	—

Refrigeration Load

R_s values in Tables 6 and 7 are based on 90% rh in the cold room. A small error occurs where these values are used for 80 or 100% rh. This error together with loss of accuracy due to simplification results in a maximum error for Equation (12) of approximately 4%.

For cyclical, irregular, and constant door usage, alone or in combination, the doorway open-time factor D_t can be calculated as follows:

$$D_t = \frac{(P\theta_p + 60\theta_o)}{3600\,\theta_d} \quad (13)$$

where

- D_t = decimal proportion of time doorway is open during the period under consideration
- P = number of doorway passages
- θ_p = door open-close time, s/passage
- θ_o = time door simply stands open, min.
- θ_d = the daily (or other) time period, h

The value of θ_p for motorized pull-cord operated doors is approximately 25 s per passage through the doorway, perhaps a few seconds less in the case of "high-speed" doors. θ_o and θ_d should be provided by the user.

The doorway flow factor D_f is the ratio of actual air exchange to fully established flow. Fully established flow occurs only in the unusual case of an unused doorway standing open to a large room, or to the outdoors, where the cold outflow is not impeded by obstructions (such as stacked pallets within or adjacent to the flow path either within or outside the cold room) and can rapidly escape the vicinity of the doorway. Under these conditions, D_f is 1.0 after 20 to 30 s. However, for the initial 20 to 30-s period that the doorway is open, Gosney and Olama (1975) indicate that D_f is in the 0.50 to 0.60 range.

Pham and Oliver (1983) determined that a frequency of one passage per minute for forklift traffic through an otherwise unrestricted doorway results in a value of $D_f \approx 0.80$, and experience shows a certain constancy between doorway traffic and door-open time, which indicates that an average factor, $D_f = 0.80$, is a good assumption for most cases of infiltration by air exchange whether the doorway is constantly open, cyclically operated, or in-between.

Infiltration by Direct Flow-Through

Building negative pressure is the usual cause of heat gain by direct inflow of warm air to refrigerated rooms where direct flow-through occurs. The corresponding outflow of refrigerated air is to the negative pressure space or, in the case of wind effects, to the lee side of the building.

Warehouses with exhaust fans and production areas with inadequate makeup air are common negative pressure situations. When these spaces open to an adjacent cold room, the result can be serious refrigeration loss by direct flow-through. This outflow cannot occur if inflow openings do not exist in the cold room walls.

If negative pressure is not corrected, plant size and operating cost may increase substantially. Equation (14) for heat gain from infiltration by direct flow-through provides the basis for either correcting the negative pressure or adding to refrigeration capacity.

$$q_t = 60\,VA(h_i - h_r)\,\rho_r D_t \quad (14)$$

where

- q_t = average hourly refrigeration load, Btu/h
- V = average air velocity, ft/min.
- A = opening area, ft^2
- h_i = enthalpy of infiltration air, Btu/lb
- h_r = enthalpy of refrigerated air, Btu/lb
- ρ_r = density of refrigerated air, lb/ft^3
- D_t = proportion of doorway open time

Area A is the smaller of the inflow and outflow openings. If the smaller area consists of truck loading doors in well disciplined loading docks (or staging areas), leakage area around individual doors can vary from a low of 0.3 ft^2 per door to over 1.0 ft^2. For less disciplined or high merchandise movement loading docks, doors will be partially or fully open at times, which is critical in the design.

To evaluate velocity V, the magnitude of negative pressure or other flow-through force must be known. If differential pressure across a doorway is known, velocity can be predicted by converting static head to velocity head. However, attempting to estimate the negative pressure is usually not possible; generally, the only alternative is to assume a commonly encountered velocity. Assumed velocities are essentially the same velocities that occur through exterior doorways opening directly into the negative pressure space in question. The magnitude of these velocities is frequently in the range of 60 to 300 ft/min.

Open Doorway Protection

The following discussion applies only to doorways subject to infiltration by air exchange.

Well-installed and maintained strip doors and push-through type doors can exceed 95% effectiveness against infiltration. Depending upon traffic level and door maintenance, assumed effectiveness E for calculation purposes can range from 0.90 to below 0.80 for freezer applications and from 0.95 to 0.85 for other applications.

"Air-lock" vestibules with strip doors or push-through doors on each end range from 0.95 to 0.85 in effectiveness for freezer applications and from slightly higher than 0.95 to 0.90 for other applications. Manufacturers should provide values of effectiveness and suitable guarantees where air curtains are used.

For any open doorway protective device installed on a doorway subject to two-way air exchange, multiply Equation (9) by $(1-E)$ to determine average hourly heat gain.

The effectiveness of these devices on doorways subject to infiltration by direct flow-through is not readily determined. Depending upon the flow-through pressure differential, its tendency to vary, and the ratio of inflow area to outflow area, effectiveness of the devices can be very low.

Sensible and Latent Heat Components

Upon calculating q_t for infiltration air, the sensible and latent heat components may be obtained by plotting the infiltration air path on the appropriate ASHRAE psychrometric chart, determining the air sensible heat ratio R_s by the chart, and calculating as follows:

$$q_s = q_t R_s (1 - E), \text{ Btu/h sensible heat} \quad (15)$$

$$q_l = q_t (1 - R_s)(1 - E), \text{ Btu/h latent heat} \quad (16)$$

where

$R_s = \dfrac{\Delta h_s}{\Delta h_t}$, as shown on ASHRAE psychrometric charts

EQUIPMENT LOAD

Equipment heat load consists essentially of fan heat where forced air circulation is used, reheat where humidity control is provided, defrosting heat gain where defrosting occurs, and moisture evaporation where the defrosting process is exposed to refrigerated air. To accurately select heat-extracting equipment, a distinction should be made between those equipment heat loads that are felt withing the refrigerated space and those that are introduced directly to the refrigerating fluid.

Equipment heat gain is usually minor at space temperatures above approximately 30°F. Where reheat or other artificial loads are not imposed, total equipment heat gain, including defrosting heat gain where it applies, is about 5% of the four preceding segments of load.

Equipment heat gain becomes major at freezer temperatures. At $-20\,°F$ for example, the theoretical contribution to total refrigeration load due to fan power and coil defrosting alone exceeds, for many systems, 15% of the four preceding loads. This percentage assumes perfect adjustment and operation during defrosting and the absence of light-density frost on the coils. The actual percentage to use may be considerably higher, particularly where room sensible heat ratio (RSHR) is more than a few one-hundreths below 1.00.

SAFETY FACTOR

Generally, a 10% safety factor is applied to the calculated load to allow for possible discrepancies between the design criteria and actual operation. Safety factors should be selected in consultation with the facility user and should be applied individually to the first four heat load segments. A separate safety factor should be added to the coil-defrosting portion of the equipment load for freezer applications that use dry-surface refrigerating coils.

Little data is available to predict heat gain from coil defrosting. For this reason, the experience of existing similar facilities should be sought to obtain an appropriate defrosting safety factor. Similar facilities should have similar room sensible heat ratios.

TOTAL REFRIGERATION LOAD

A correct calculation of the total refrigeration load is necessary to properly size equipment, run a load diversification analysis, operate the system, and estimate operating costs, The load sensible heat ratio should be estimated for installations with dry-surface refrigerating coils and included with the refrigeration load calculations.

BIBLIOGRAPHY

Cole, R.A. 1987. Infiltration load calculations for refrigerated warehouses. *Heating/Piping/Air Conditioning* (April).

Cole, R.A. 1984. Infiltration: A load calculation guide. Proceedings of the International Institute of Ammonia Refrigeration, 6th annual meeting, San Francisco (February).

Cole, R.A. 1989. Refrigeration loads in a freezer due to hot gas defrost, and their associated costs. ASHRAE *Transactions* 95(2):1149-1154.

Dickerson, R.W. 1972. Computing heating and cooling rates of foods. ASHRAE *Symposium Bulletin* No-72-03.

Ellis, W.P. 1964. Interior finishes for cold storage facilities. ASHRAE *Journal* 6(12).

Fisher, D.V. 1960. Cooling rates of apples packed in different bushel containers, stacked at different spacings in cold storage. ASHRAE *Transactions* 66.

Fultz, D.A. 1971. A rating system for evaluating low-temperature construction. ASHRAE *Symposium Bulletin* Wa-71-03.

Gentry, J.P. and R. Guillou. 1966. Fog spray humidification in cold storage. ASHRAE *Journal* 8(10).

Gosney, W.B. and H.A.L. Olama. 1975. Heat and enthalpy gains through cold room doorways. Paper presented before The Institute of Refrigeration at the Faculty of Environmental Science and Technology, The Polytechnic of the South Bank, London (December).

Govan, F.A. 1969. Basic factors to be considered in insulating a refrigerated warehouse. ASHRAE *Symposium Bulletin* DV-69-05.

Hamilton, J.J., D.C. Pearce and N.B. Hutcheon. 1959. What frost action did to a cold storage plant. ASHRAE *Journal* 1(4).

Haugh, C.G., W.J. Standelman and V.E. Sweat. 1972. Prediction of cooling/freezing times for food products. ASHRAE *Symposium Bulletin* NO-72-03.

Heaton, E.K. and J.G. Woodroff. 1970. Humidity and weight loss in cold storage pecans. ASHRAE *Journal* 12(4).

Hendrix, W.A., D.R. Henderson, and H.Z. Jackson. 1989. Infiltration heat gains through cold storage room doorways. ASHRAE *Transactions* 95(2).

Hovanesian, J.D., H.F. Pfost and C.W. Hall. 1960. An analysis of the necessity to insulate floors of cold storage rooms at $35\,°F$. ASHRAE *Transactions* 66.

Jones, B.W., B.T. Beck and J.P. Steele. 1983. Latent loads in low humidity rooms due to moisture. ASHRAE *Transactions* 89(1) Part I.

Karam, H.J. and H. Kreiner. 1962. Reducing heat losses in a commercial refrigerator. ASHRAE *Journal* 5(1).

Kayan, C.F. and J.A. McCague. 1959. Transient refrigeration loads as related to energy-flow concepts. ASHRAE *Journal* 1(3).

Lamiman, D.R. and J.A. Mixon. 1969. Urethane insulation applied to refrigerant buildings. ASHRAE *Journal* 11(5).

Lotz, W.A. 1964. Heat and air transfer in cold storage insulation. ASHRAE *Transactions* 70.

McGarvey, A.R. 1964. Wrap your cold storage room in a vapor barrier. ASHRAE *Journal* 6(12).

Meyer, C.S. 1964. "Inside-out" design developed for low-temperature buildings. ASHRAE *Journal* 5(4).

Pham, O.T. and D.W. Oliver. 1983. Infiltration of air into cold stores. Meat Industry Research Institute of New Zealand, Inc., Hamilton, New Zealand, IIF-IIR, 16th International Congress of Refrigeration, Paris, France.

Pichel, W. 1966. Soil freezing below refrigerated warehouses. ASHRAE *Journal* 8(10).

Popper, K. and F. Nury. 1966. Depot sterilization of cold storage rooms. ASHRAE *Journal* 8(8).

Powell, R.M. 1970. Public refrigerated warehouses. ASHRAE *Journal*, 12(8).

Ruff, A.W. and F.J. Webber. 1969. The panel building for refrigerated warehouse construction. ASHRAE *Symposium Bulletin* DV-69-05.

Seward, R.M. 1970. New thoughts on an old subject—Cold storage insulation. ASHRAE *Journal* 12(7).

Slopa, R.E. 1971. Selection and installation of cold storage doors. ASHRAE *Symposium Bulletin* WA-71-03.

Vevoda, R.F. 1968. Evaluating insulation systems for cold storage facilities. ASHRAE *Journal* 10(1).

Waite, H.J. 1969. Technical requirements for supporting structures. ASHRAE *Symposium Bulletin* DV-69-05.

Wilder, C.M. 1969. Doors for refrigerated warehouses. ASHRAE *Symposium Bulletin* DV-69-05.

# CHAPTER 28

TRUCKS, TRAILERS, AND CONTAINERS

Types of Equipment	28.1
Body Design and Construction	28.1
Auxiliary Equipment	28.3
Types of Refrigerating Systems	28.3
Mechanical Refrigeration Components	28.5
Calculation of Cooling Loads	28.6

THE transport of perishable commodities may be as simple as a direct delivery from farm to market by a small truck, or may require travel by sea, air, rail, highway, or combinations of these modes to satisfy distant markets. For intermodal aspects of container transfer between highway and rail carriage, see Chapter 29; for marine aspects of intermodal container applications, see Chapter 30; and for air transport, see Chapter 31. The Agricultural Research Service (ARS) and the Agricultural Marketing Service (AMS) within the U.S. Department of Agriculture have published many reports on the subject of refrigerated transportion. *Container News* also covers the subject. Back issues of *Refrigerating Engineer* and *Refrigerated Transporter* have covered the topic, although they are no longer published.

TYPES OF EQUIPMENT

Refrigerated transport equipment can be broadly classified by type of operation—highway and intermodal equipment, marine service equipment, or straight trucks. In the United States, each state fixes the maximum weight and dimensions of trucks and trailers that may be operated over its highways. A summary of current state size and weight limits for trucks and truck-trailers is issued by the American Trucking Association (ATA).

Highway and Intermodal Trailers

Refrigerated semitrailers for overland use are up to 53 ft long, 8.5 ft wide, and 14 ft overall height. Where permitted, double or triple trailers may be pulled by one tractor, or a trailer may be pulled by a refrigerated truck. USDA *Agricultural Handbook 669* (Ashby et al. 1987) gives recommended loading methods and temperatures for many perishable products.

Highway trailers, removed from their tractors, are carried piggyback on railroad flat cars (TOFC = trailer on flat car). Trucks and trailers with tractors are driven onto ships (roll-on roll-off). Containers are carried on truck chassis, trailer chassis, on railroad flat cars (COFC = container on flat car), above and below deck on container ships, and, in limited fashion, on cargo aircraft.

Containers differ from standard bodies primarily in the hardware required to facilitate attachment to the chassis and in structural hardware required for lifting and stacking. In a typical trailer-ship transfer operation, for example, the container is lifted from the trailer chassis and placed either on deck or in the hold of the ship. Special fittings permit these containers to be stacked six-high in ship holds (and on shoreside terminals). On-deck carriage of containers may call for stacking two-high under a third dry cargo van. The corner posts must be designed to carry the floor load and added vertical load applied by lashing gear. The American National Standards Institute (ANSI 1979, 1980) and International Standards Organization (ISO 1979a) have developed standards for refrigerated containers for intermodal use.

For long-haul land transportation, either highway or rail, each vehicle body or container requires an independent means for refrigeration. Containers interchanged between marine and land use can be refrigerated with independent systems, or from the ship's central system at sea and with clip-on systems on land. Two or more containers are sometimes connected end-to-end to form a single unit for long-haul operation and separated for local delivery.

Some railroad yards transfer trailer bodies from the trailer chassis to gondola or flat cars in a manner similar to those for the trailer-to-ship operation. Various container transfer machines are used to move, stack, or load containers. Finger-lift, pincer-lift, or cable lift systems each apply characteristic stresses that must be considered in structural design. Many truck and rail carriers apply a standard piggyback technique, *i.e.*, the trailer is driven or lifted onto a railroad flat car and attached to the car for haul by rail.

Trucks

Refrigerated trucks are used primarily for short-haul wholesale delivery within a population center or between population centers. Body styles have been developed to suit the character and distribution needs for perishable products. Trucks and trailers may contain more than one compartment held at different temperatures. Trucks are used in operations that may require more door openings than trailers. Penney and Phillips (1967) measured the air exchange and cooling load caused by door openings of typical refrigerated trucks.

Various devices such as power lift gates and conveyors, which facilitate loading and unloading, are built into the body. Trucks for retail delivery may be furnished for either walk-in or reach-in service. Wheeled racks, which can be rolled into place in the truck, are used to expedite loading.

BODY DESIGN AND CONSTRUCTION

An insulated body, when matched with a suitable refrigeration system, should provide economic temperature control for the commodity being transported or stored. The necessary features are determined by the type of service intended. Trucks may have doors on the two sides rather than at the rear, but most trailers and containers have full opening rear doors large enough to permit forklift trucks to be driven into the vehicle. Many trailers are also equipped with a curbside door. Meat rails are provided if the vehicle is to be used for hanging meat. Temporary or permanent discharge ducts from the refrigerating unit can be used to improve air movement through the cargo.

Insulation

The goal of body builders is to construct a lightweight body of sufficient strength in which the insulation will stay dry and maintain its original insulating value. Insulation should have moderate cost, low density, low thermal conductivity, low moisture permeability and retention, ease of application, uniformity, resistance to breakdown at temperature extremes, and fire resistance. The

---

The preparation of this chapter is assigned to TC 10.6, Transport Refrigeration.

insulation should resist cracking, crumbling, shifting, and packing from the shock, vibration, and flexing of the body structure.

Bodies for low-temperature use (about 0 °F) have been insulated with about 6 in. of an insulating material having a k-factor lower than 0.3 Btu in/h·ft². Legal requirements and other factors limit the outside width, height, and length of vehicles. Therefore, the designer must establish an insulation thickness to obtain optimum cargo space and operating performance. For example, increasing insulation thickness from 3 to 4 in. in a 40-ft long trailer decreases cargo space by 100 ft³, or 5%. Decreased insulation thickness, however, requires an increase in refrigerating or heating capacity. In determining the optimum insulation thickness, the transport body and its refrigeration unit must be considered as one system.

Although glass fiber, expanded polystyrene, and other materials continue to be used as insulation materials, urethane foam now predominates. Such insulation may have a k-factor of 0.18 or lower, even after aging, which permits thinner construction. Loose fill material that settles during use should not be used. Batts or sheets of semirigid materials are usually compressed slightly as a precaution against settling. Chapter 22 of the 1989 ASHRAE *Handbook—Fundamentals* lists characteristics of most available materials.

The heat transferred from outside to inside the vehicle, excluding heat that enters when the door is open for loading or unloading cargo, includes heat transferred (1) through insulation, (2) through structural members, and (3) by air and moisture leakage. Air leakage can affect overall heat transfer significantly, so some buyers specify a maximum rate of air leakage with a given air pressure imposed in the cargo space.

Floor Insulation. Floor loads are frequently supported on rigid insulating material to eliminate floor beams. Most truck, trailer, and container floors must support forklift trucks. Some medium or high-temperature trailers are built without any insulation in the floor, but engine waste heat and radiation from hot road surfaces can raise the temperature of the exposed undersurface as much as 20 °F above the ambient air temperature.

Floor surfaces may be of wood or metal. If wood floors are furnished, they should be tongue-and-groove treated lumber, well-sealed against water penetration. Metal floors should be watertight, and if separate floor racks are not furnished, the floor surface should allow adequate air movement under the load. Several formed or extruded floor surfaces are available, but not all of these permit enough air circulation under the load.

Galvanized or treated steel and aluminum are used widely for floors in ice cream trucks, while aluminum is most often used for trailer floors. Corkboard, expanded polystyrene, and polyurethane foam are used for floor insulation. A metal skirt reaching at least 6 in. up the walls should be bonded to the floor so that water running down the wall or collecting on the floor will not enter the insulation. Drains, if used, must be self-closing. Plywood, aluminum, other metals and certain plastics are used for interior wall surfaces. Glass-reinforced plastic materials are being used for both interior and exterior surfaces.

Reducing Heat Transmission

Often the buyer specifies the maximum allowable heat transfer rate, usually at 100 °F and 50% rh outside and 0 °F inside air temperature. Bodies for low temperature (0 to −20 °F) typically have 3 to 4 in. of polyurethane insulation. For above freezing temperatures, 1 to 2.5 in. of polyurethane is applied. At the option of the customer, the insulation thickness is usually nonuniform on the different surfaces, with the front wall generally thicker for structural reasons. For multistop delivery operations, trailers and trucks are sometimes provided with roll-up insulated doors.

Moisture Penetration

All exterior surfaces of an insulated body must be made as airtight and water vapor-tight as possible. Water vapor will pass through any opening or nonvaporproof barrier in the outer shell and will condense in the insulation itself or in cavities in the insulation space. Some ice cream truck bodies have increased more than 500 lb in weight over a period of use, and large semitrailers in frozen food service have gained up to 1500 lb. Since the k-factor of wet insulation may be much higher than its original value, either the required low temperature may not be obtained or a much greater load is placed on the refrigerating equipment with resultant increased operating and maintenance costs. Use of closed-cell insulation in place of fibrous types minimizes moisture buildup problems.

Air Leakage

Eby and Collister (1955) pointed out the serious heat gain penalty imposed by air entering the insulation space through cracks in the front of the vehicle caused by the motion of the vehicle. They showed that this driving force is 1.21 in. of water at 50 mph, and that a 6-ft length of unsealed seam can permit 1440 ft³/h of air to enter. At conditions of 100 °F ambient temperature at 50% rh and a trailer temperature of 0 °F, the heat gain caused by this amount of infiltration air could be 4600 Btu/h, assuming that this leakage air left the vehicle saturated at 0 °F.

Refrigerated vehicles leak air even when they are stationary, probably because of the stack effect of the temperature difference between inside and outside. This driving force for air infiltration, for a body 8 ft high and a temperature difference of 100 °F, is about 0.030 in. of water (Phillips et al. 1960). Openings with an aggregate area of 1 in² each at the top and bottom of the exterior skin would permit an air leakage of about 120 ft³/h, if they function as thin plate orifices. Observed leakage rates for a nominal 35-ft trailer of more than 700 ft³/h and an accumulation of nearly 1 lb/h of ice at ambient laboratory conditions of 100 °F at 50% rh and 0 °F trailer temperature, illustrate the benefit from improving the vaportightness of the exterior shell.

The need for better sealing is even more significant at road speeds. For example, if a 1-in² opening were located where it was subjected to the 1.21 in. of water ram air pressure at 50 mph, approximately 1150 ft³/h of air could be driven into the insulation space of the vehicle. At ambient temperature and humidity conditions of 100 °F and 50%, the heat gain from this leakage would be about 3800 Btu/h, if the air was cooled to the trailer temperature of 0 °F.

Road tests of nominal 35-ft commercial trailers by Phillips et al. (1960) have shown air leakage rates as high as 1590 ft³/h at 50 mph. Figure 1 shows the heat gain resulting from air leaking into a vehicle with an interior temperature of 0 °F, at various ambient temperatures and relative humidities.

When moist air moves through the insulated space, the water vapor left behind causes trouble in several ways:

- Increased heat gain caused by the latent heat of vaporization and fusion
- Loss of payload because of gain in mass
- Corrosion
- Increased heat gain through insulation
- Coating of coils with frost resulting in loss of refrigerating effect
- More frequent defrosting
- Physical damage to insulation
- Rotting of wood members
- Odor

Additional sealing of the metal skin of a truck or trailer is required to make it leakproof. Three methods are popular: (1) use of foamed-in-place plastic insulations; (2) lining the inside of the exterior skin with a nonpermeable vapor barrier such as aluminum foil coated with a plastic binder, which can be sealed at the joints; or (3) coating the interior surface of the exterior skin with some

Trucks, Trailers, and Containers

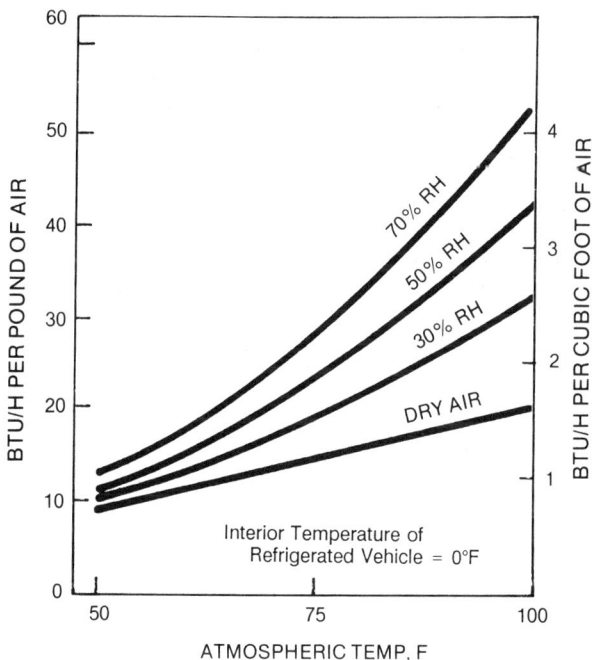

Fig. 1 Heat Load from Air Leakage

type of vapor-sealing compound such as neoprene. The vapor seal must not be destroyed where wiring, piping, or frames penetrate the barrier.

Typical trailer specifications call for air leakage of less than 2 cfm at a negative pressure of 0.5 in. of water in the trailer. The interior skin need not, and probably should not, be vaportight; it must be water resistant so that cleaning does not readily soak the insulating material, however. All doors must fit properly and be gasketed to reduce air and heat leakage. Hardware for these doors must be substantial enough to withstand severe conditions.

The interior walls of trailers, which may be used in vacuum cooling of fresh produce, must be vented, preferably into the cargo space, to prevent pressure damage to the insulated spaces during the vacuum cooling operation.

Insulated bodies with one-piece molded plastic exterior shells are now being produced. Types and sizes range from small retail meat trucks to maximum legal length trailers and containers. High impact and thermal resistance, and good vapor-sealing characteristics are claimed for this type of construction.

Air Circulation

Inadequate air distribution is probably the principal cause of improper refrigeration of cargo during transport. Satisfactory product temperatures will not be maintained unless the load is surrounded by proper air or surface temperatures. Either forced or gravity circulation from plate or pipe coils maintains the product temperature. Forced air units typically discharge air over the stacked cargo either directly from the unit or through ducts directing the air toward the rear of the cargo space.

Products that are to be cooled in transit, or loads that produce heat, must be arranged so that refrigerated air can move not only around the load but through it as well. Nonheat-producing loads, placed in the vehicle at the proper temperature, need only be blanketed with air at the points where heat could enter the product. Most systems using fans can move an adequate air supply over the top of the load. When light density loads approach the ceiling, or when cross bows interfere with the throw of air, supply ducts, either permanent or temporary, facilitate the delivery of air to the rear of the vehicle.

Air channels must be provided on the side walls and inside of the rear doors. Attention must be paid to the movement of air along the floor under the load. Many vehicles have formed metal floors, which allow air to flow the length of the vehicle. Figure 2 shows a trailer with a supply duct and a return air plenum formed by a false front bulkhead extending to the floor racks. Wall strips and floor racks provide channels for airflow.

AUXILIARY EQUIPMENT

Heaters

Most trailer and truck refrigeration units with independent power are designed with a means for heating the cargo space. An augmented hot gas cycle in which the evaporator fan then circulates the heated air is frequently used. Heaters that burn alcohol, kerosene, butane, propane, or charcoal are also used. Because of potential hazards from open flame burners, such heaters must conform to applicable safety regulations.

Modified Atmosphere

Many systems have been developed to control the makeup of the interior environment of the refrigerated vehicle; most limit the amount of oxygen in produce-hauling vehicles and containers. In general, because the time for overland transport is so short, the quality of the products carried under low oxygen is not significantly different from similar products that are adequately cooled and held at proper storage temperature. Some exceptions are claimed for leafy vegetables. A high concentration of carbon dioxide is deleterious to some products, but it has proven beneficial for mold control in others, such as strawberries. Because of the longer transit time, marine shipments should benefit from controlled or modified atmosphere during shipment.

Thermometers

Refrigerated vehicles usually have an indicating thermometer to monitor cargo space temperature. More complex instruments are sometimes used to indicate, control, or perhaps record the cargo space temperature.

TYPES OF REFRIGERATING SYSTEMS

Ventilation

Control of temperature in an insulated vehicle by ventilation with outside air is the least dependable of any system. Obviously, it is limited to areas or times of the year when outdoor conditions are appropriate. Ventilation is usually accomplished by leaving small doors open at the front and rear of the vehicle. Pressure caused by forward motion of the vehicle and the adjustment of doors by the operator control the airflow. Ventilation can be used with mechanical systems to help remove field heat or reduce the concentration of ripening gases generated by a commodity.

Product Subcooling

Some refrigerated commodities are transported without providing refrigerating equipment on the truck or trailer by using the

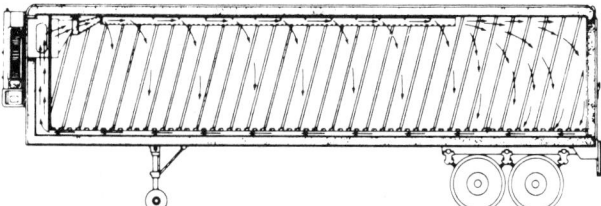

Note false bulkhead which forms return-air plenum taking air from floor.

Fig. 2 Section of Trailer Showing Air Circulation

heat-absorbing capacity of the product itself. Milk, for example, has been transported in tank trucks and trailers, some as large as 5300 gal capacity. Orange juice and many hot liquids are also transported in this way.

For example, 5300 gal of milk will absorb 42,380 Btu/°F temperature rise. If an average tank has a heat leakage of 6500 Btu/h during the trip, the milk could be transported 1600 miles (40 h at an average speed of 40 mph) with a temperature rise of less than 6.5°F and without any refrigeration other than the heat-absorbing capacity of the milk. Orange juice shipped in bulk from Florida to New York, a distance of nearly 1400 miles, has warmed less than 9°F with no refrigeration other than the product itself. Product quality and other factors must be considered, however, before relying on this method for shipments of perishable products.

Water Ice

This means for transport cooling has largely been replaced by other methods. Small amounts are still placed on top of loads to supplement mechanical refrigeration in the removal of field heat.

Dry Ice (Carbon Dioxide)

In the past, dry ice was used to maintain frozen commodities in delivery trucks. Currently it is used for shipping small quantities of specialty items or items requiring very low shipping temperatures. Dry ice sublimates at −109°F and absorbs about 240 Btu/lb in the process.

Liquid Nitrogen or Liquid Carbon Dioxide Spray

Liquid nitrogen has a boiling point of −320°F and with superheating to 0°F has a total refrigerating effect of about 160 Btu/lb. Liquid carbon dioxide with a boiling point of −109°F and superheating to 0°F has a net refrigerating effect of about 120 Btu/lb.

For road operation, the refrigerant is carried in storage vessels located either in or out of the refrigerated space. To eliminate the need for high pressure nitrogen tanks, the storage vessels are insulated and equipped with relief valves set for about 25 psig and vented into the cargo space. Liquid carbon dioxide must be stored at pressures in excess of 60.4 psig, and at temperatures below 87.8°F. For cooling cargo after loading, some terminals use liquid refrigerant sprays fed from dock-mounted storage tanks. In this operation, the distributing tube can be introduced through a partially open door. Caution must be taken to prevent refrigerant from being released when personnel are in the cargo space because of possible harm from freezing or the risk of suffocation. Liquid carbon dioxide systems must be designed to prevent nozzle blockage by snow formation.

Guilfoy (1973) describes a liquid nitrogen clip-on unit for land refrigeration of intermodal containers. During shipboard operation, the containers are refrigerated either by immersion in a temperature-regulated hold or by connecting ducts to a central system.

Eutectic Plates for Holdover

Station Charging. Some retail service trucks use a holdover plate arrangement to maintain low temperatures. A holdover plate consists of a coil for the primary refrigerant mounted inside a thin tank filled with a eutectic solution with a freezing temperature sufficiently low to meet the required conditions. Holdover or eutectic plates vary in size from 18 to 36 in. wide, 40 to 120 in. long, and 1 to 3 in. thick. The refrigerating capacity and weight for a given size plate varies with the thickness. For example, a 30 by 66 in. plate of 2.63 in. thickness has a refrigerating capacity of about 17,000 Btu (at −8 to −9°F eutectic temperature) and weighs about 300 lb, while the same size plate of 1.5 in. thickness has a capacity of about 9000 Btu and weighs about 225 lb. Standard eutectic plates operating at temperatures of −59, −29, −14, −12, −9, −8, −6, 18, 23, and 27°F are available.

Several of these plates are mounted on the walls, ceiling, or both, or are used as shelves or compartment dividers in the vehicle to provide the required refrigerating capacity while the vehicle is away from the garage or station where the plates are refrozen. When the truck is garaged, flexible connections from the stationary refrigerating plant are attached to connecting devices provided on the truck. A charging island, where several trucks can be connected at one time, can be used for fleet operation. Any refrigerant can be used—the most common being Refrigerant 717 or, more recently, Refrigerants 12 and 22. Expansion valves may be mounted on the truck or on the garage wall, and various types of automatic-closure quick connecting devices are available for attaching the flexible lines. In plants using a chilled brine circuit, the plates may be designed to circulate brine through the plates to freeze the eutectic solution.

To provide proper cooling capacity, the plates must be mounted so that air can move freely on both sides, usually 1.5 to 2 in. from walls with top edges 6 to 8 in. from the ceiling. On ceilings, plates should be sloped at least 1 in. per foot across the shortest dimension and should be at least 2 in. below the ceiling at the higher edge. Hangers must be securely fastened to walls, preferably by fastenings incorporated in the studs. Fastenings must not penetrate the outside wall surface. A drip trough may be desirable under each plate, although many vehicles are defrosted by scraping the frost without warming the plates. To protect plates from damage, guard rails should be placed at least 1 in. away from plates, if necessary. Eutectic plate systems usually circulate air by gravity but some have fans with ducts and dampers for temperature control.

Vehicle-Mounted Condensing Unit. Some trucks are equipped with condensing units powered by electric motors. These units operate on plug-in power to refreeze the eutectic plates when the vehicle is stopped. The comparative length of the time for road operation and for refreezing are factors in determining the required size of condensing unit and the amount of eutectic solution (Guilfoy and Mongelli 1971).

Variations of truck-mounted condensing unit systems include those with a separate engine drive, those with a compressor driven by the truck engine, and those with a nose mount or chassis mount condensing unit.

Plastic Plates or Tubes. Several European manufacturers have developed holdover systems that contain the eutectic solution in either rectangular or round cross-sectional tubes. The plastic parts replace relatively heavy steel containers.

Mechanical Refrigeration

Independent Engine or Electric Motor Driven. Many styles of independent engine and/or electric motor driven mechanical refrigerating units are available and constitute the most popular application for both trucks and trailers. The one-piece, plug-type, self-contained unit mounts in an opening in the front wall of the vehicle. The condensing section on the outside and the evaporator section on the inside are separated by an insulated plug that attaches to the vehicle wall and supports the various parts of the refrigerating unit. For trailers, some of the plug units are tall and slender enough to mount between the tractor and trailer; others are short and deep and extend over the tractor roof. For intermodal containers, the most widely used refrigerating unit is electric motor-driven, usually hermetic, with or without a companion engine-driven generator. The mounting plug for the unit essentially forms the entire end wall of the container.

The refrigerating capacity of self-contained, independent engine driven plug-type trailer units in an ambient temperature of 100°F range as follows—20,000 to 40,000 Btu/h at 35°F trailer temperature; 12,000 to 28,000 Btu/h at 0°F trailer temperature; and 6000 to 18,000 Btu/h at −20°F trailer temperature. These units weigh from 800 to 1700 lb. Similar self-contained units, which weigh from 500 to 1000 lb and have a lower cooling capacity, are available for use on trucks.

Trucks, Trailers, and Containers

Some units have two belt-driven compressors, others have one compressor that is either belt- or direct-driven by the auxiliary engine. These auxiliary engines may operate on gasoline from the truck tank, from a separate gasoline or diesel fuel supply on trailers, or on liquefied petroleum gas. In addition to the engine, many refrigeration units are equipped with an electric motor for standby operation. Most units can be used for heating and can be defrosted automatically by the hot gas method. Most units are thermostatically controlled, starting and stopping or reducing speed as refrigeration need requires. Condenser and evaporator fans are driven by the unit engine, and tight-fitting dampers close off the evaporator airflow to permit defrosting without unduly warming the vehicle interior. The one-piece construction facilitates unit replacement in the event of breakdown or removal for maintenance.

Two-piece units are also used for both trucks and trailers. One style uses an engine-driven generator mounted under the trailer to provide electric power to operate one or two independent refrigerating systems contained in a plug-type unit mounted in the front wall. The plug-type unit consists of a self-contained, hermetic compressor system, and the housing under the trailer contains an engine-driven alternator. The compressor can be operated from house current during standby. Another style uses an evaporator section, usually with electric motor-driven fans, mounted in the nose of the vehicle, and an engine-driven condensing unit installed under the body. In this case, permanent piping is run between the condensing unit and evaporator, and a generator on the engine provides electric power to operate the evaporator fan. Some units are designed to cycle on thermostatic control, others operate continuously but change the speed of the engine in response to the need for refrigeration within the body. Some units are designed for automatic defrosting; others are defrosted by manual operation of hot gas valving or electric heaters.

Power Take-Off from Vehicle Engine or Transmission. For most cases, this type of refrigerating system is limited to trucks, although an axle-driven unit has been devised for trailers. Several types of power take-off means are available. One manufacturer has an alternator, belt-driven from the engine crankshaft, which produces a regulated alternating current voltage that is rectified to drive direct current motors for the compressor and fans in the body. temperature control is accomplished by cycling the compressor. Refrigeration is produced only when the truck engine is in operation; full refrigerating capacity is claimed for all engine speeds above 500 rpm.

One manufacturer uses an all-electric, self-contained hermetic unit to provide refrigeration for medium temperature truck bodies. A specially regulated alternator produces an alternating current power supply of variable voltage and frequency at essentially constant current for engine speed from idle operation to 4000 rpm. The alternator is mounted on the truck engine and belt-driven from the engine crankshaft pulley. Standby operation is provided by electrical connection to regularly available alternating current power. A two-speed centrifugally operated gear shift is available to provide more suitable generator speed if the truck engine speed range is too wide.

Power take-off systems of the mechanical type sometimes have the condenser mounted in front of the truck radiator, with flexible tubing to connect to the unit in the cargo space. The most common system uses an engine-mounted compressor, which is belt-driven from the crankshaft pulley. One manufacturer has a flexible shaft drive for the unit compressor and condenser fan, with a belt-driven electric clutch at the forward end of the flexible shaft. In this system, of course, the compressor speed is a function of engine speed.

A variation of the power transmission mechanism for use with a belt-driven flexible shaft includes a standby electric motor to drive the compressor. The flexible shaft drives an electric clutch attached to a universal joint. An overriding clutch idles the electric motor when the compressor is being driven by the flexible shaft. When the electric motor operates, the electric clutch is disengaged. The flexible shaft is driven by belt from the engine crankshaft, and the electric clutch is used for temperature control.

Hydraulic power transmission is a special case of power take-off from the truck transmission or engine. The hydraulic pump is either mounted in the truck engine compartment and belt-driven from the engine crankshaft or is direct-driven from power take-off in the truck transmission. Hydraulic drive systems may be designed to provide a near constant output speed from a variable input speed.

MECHANICAL REFRIGERATION COMPONENTS

The major components of a vapor compression transport are power source, compressor, condenser, evaporator system, and controls.

Power Source

Refrigeration systems for short trucks that require moderate cooling typically use the truck engine for power. A belt on the engine crankshaft drives the compressor, and the battery powers the fans. An optional drive motor can be plugged into power mains. If the cooling demand is higher (as for frozen food, a large truck body, or frequent door openings), the system likely is powered by a small engine (usually diesel with a 25 to 35 in^3 displacement).

Eutectic systems with 5 to 7.5 hp motors may also meet the needs of this application.

Larger diesel engines, up to about 80 in^3, provide power for the largest refrigeration units for large trucks (up to 28 ft) having a high service load or carrying frozen products. Most truck units driven by a self-contained engine are available with an optional electric motor drive.

Trailer refrigeration units are typically driven by diesel engines with 100 to 140 in^3 displacement. Optional electric standby may be available.

Compressor

Small compressors driven by the truck engine are available from the automobile air-conditioning industry. These include in-line or V-piston, rotary vane, and swash-plate piston designs. Larger truck compressors generally have in-line or V-piston configurations with two to four cylinders. Trailer units use piston-type compressors with up to 40 in^3 displacement. Aluminum is used on some designs to reduce weight.

Condensers

Truck engine powered condensers may be mounted (1) in front of the truck engine radiator, (2) above the truck cab with a battery powered fan, and (3) below the truck body. The condenser and fans for larger truck and trailer units are packaged with the compressor, engine, and other components. Condensers are designed for forced convection cooling and generally have aluminum fins and copper tubes.

The high-pressure side of any refrigeration system that operates when the vehicle is in motion must include the following safeguards not normally required of a stationary unit.
- Design all components to withstand shock loads and vibration. Avoid short, rigid lines, which are subject to cold working and are likely to fail.
- Individually anchor heavy objects such as driers, valves, and sight glasses.
- Consider water and dirt in the design.
- Protect all elements from corrosion, especially components for intermodal containers, which are exposed to sea water.
- Protect bearings of all exposed parts.

- Seal shaft seals, particularly against dirt or abrasive material penetration.
- Shield electric motors from splashed or air-carried water.

Evaporator Systems

The evaporator arrangement for eutectic plate systems is described elsewhere in this chapter. Other transport refrigeration systems use plate-fin and tube evaporator coils; fans or blowers provide forced convection airflow. During the defrost cycle, airflow to the cargo is stopped either by stopping the fans or by closing dampers. The heat required for defrost is provided by hot refrigerant gas or electric heaters.

Control Requirements and Devices

Frozen and fresh produce must be maintained close to its optimum temperature for best retention of food value and appearance. The quality of some frozen foods is adequately maintained at 0°F while others, such as ice cream specialty products, may require a temperature of −20°F. Compartmentalized trailers or trucks with two or three temperatures are often used for wholesale delivery of frozen, fresh, or dry commodities in the same vehicle. Certain pharmaceuticals, films, or electronic devices may require humidity control combined with temperature control.

Chapter 26 presents a table of preferred temperatures for long-term storage of perishable products. USDA *Agricultural Handbooks* 669 (Ashby et al. 1987) and 668 (McGregor 1987) discuss many aspects of maintenance of quality of fruits, vegetables, plants, and flowers during transport. The Association of Food and Drug Officials has prepared a model code that outlines storage and transportation temperature limits for frozen foods.

The rigid requirements for maintaining product temperature dictate a suitably designed and maintained truck or trailer with provisions for perimeter airflow, a satisfactory loading pattern, an adequate supply of cooled or heated air from the unit within an acceptable temperature differential, and a precise thermostat with a proper sensor position.

Where the control sensor is located in the return airstream of a refrigerated load, all points upstream of the sensor (*i.e.*, all of the load space) are subject to temperatures lower than the sensor. In this case, the possibility exists for overcooling part of the load. Conversely, if the system is heating a load, all of the space is subject to temperatures higher than the sensor.

In general, little difficulty is experienced with the sensor located in the return air for frozen loads or pre-cooled fresh produce; but the thermostat should be set at a higher set-point temperature to avoid overchilling products requiring cooling in transit. In all cases with return air sensing, the thermostat should be set 4 or 5°F above the product chill or freeze point. For a product with a high specific heat cooling load, the thermostat setting should be increased to 8 to 10°F above the chill or freeze point. If the control sensor is positioned in the supply airstream, the temperature relationship between sensor and the return air is reversed.

The use of various capacity reduction or modulation devices tends to reduce the air temperature differences in the cargo space.

Control Devices. Electronic thermostats used for load temperature control on container units operate over a wide control range with a high degree of accuracy. Controls are multistep and/or modulating for both heating and cooling modes. Auxiliary features include temperature indication, multipoint data logging, and out of range control. Recent developments plan communication by satellite to locate a load or determine the load temperature.

Other operational or safety controls include oil safety devices for engines or compressors, and controls for automatic engine starting and automatic defrost initiation and termination.

Many mechanical refrigeration units are designed to operate from −20 to 80°F in the cargo space and to pulldown from a higher temperature every trip. Various means are used to limit the power required during the transient pulldown period. Such devices include compressor crankcase pressure-limiting controls, compressor unloading, and thermostatic expansion valves with outlet pressure control.

CALCULATION OF COOLING LOADS

Phillips and Penney (1967) describe a method of rating refrigerated trucks. In an earlier study, Phillips *et al.* (1960) describe a rating method for refrigerated trailers. The Truck-Trailer Manufacturers Association (TTMA 1973) developed a method for rating heat transmission in refrigerated vehicles. The American National Standards Institute (ANSI 1980) and International Standards Organization (ISO 1979b) have prepared thermal test standards.

Although suitable standards are not in general use for rating the performance of all types of refrigerating systems used for trucks and trailers, the Air-Conditioning and Refrigeration Institute has developed a standard (ARI 1983) dealing with speed-governed and variable-speed transport refrigeration units using forced circulation air-coolers. Until standards are available for all types of systems, the comparison of the capacity of refrigerating units should be made on the basis of net refrigerating capacity at specified conditions of air temperature at the condenser and evaporator inlet, for example, at 100°F ambient temperature and 0°F trailer interior temperature.

The Refrigerated Transport Foundation (CGTFL 1988) developed a classification system based on the combined thermal performance of both trailers and refrigeration units. The trailer thermal rating is based on a practice recommended by the Truck Trailers Manufacturers Association (TTMA 1973). The refrigeration unit cooling capacity is determined by ARI *Standard* 1110 (1983). By this classification system, the combined trailer and unit can be classified in one or more of four temperature ranges:

F for frozen, 0°F
DF for frozen, −20°F
C35 for chilled temperature, 35°F
C65 for chilled temperature, 65°F

To qualify for the C35 and C65 classification, sufficient heat must be available to achieve the classification at an ambient of 0°F. The classification is voluntary for all parties—the trailer manufacturer, the unit manufacturer, and the carrier or owner. Trailers that are classified display a permanent plate or decal that lists: (1) temperature range(s), (2) trailer U-factor in Btu/h · °F, (3) data on area provided for airflow around the periphery of the loading space, (4) type of bulkhead, (5) presence or not of air distribution ducts on the ceiling, (6) amount of excess unit cooling capacity beyond the steady-state requirements at the coldest range chosen while at 100°F ambient, and (7) airflow available from the unit.

Insulated trailers range from 36 to 53 ft in length, and steady-state heat gain values (U-factor) range from 80 to 240 Btu/h · °F. The refrigeration system must have an additional cooling capacity beyond that needed for steady-state heat transfer to satisfy deterioration factors, solar radiation, air infiltration, door openings, and product loads. Extra capacity also reduces the pulldown time prior to loading, which improves equipment use.

Phillips and Penney (1967) found that solar radiation can increase the cooling load of stationary vehicles more than 20% for several hours during a sunny day; and Penney and Phillips (1967) also noted that the cooling load caused by door usage can be five times the body heat transmission when a truck is used in multistop operation.

Manufacturers of refrigeration units for trucks and trailers have devised computer programs or charts to match the unit to the body for the specific operational details planned by the customer. The input for the calculations include body U-factors, losses from door openings, product data factors, ambient conditions, and refrigeration unit capabilities.

Trucks, Trailers, and Containers

Sample Calculation for Trailer Cooling Load

Assume a trailer is loaded with 38,000 lb of peaches at an average pulp temperature of 54 °F. The load will be delivered 72 h later at an average product temperature of 34 °F. Average ambient temperature is 85 °F.

The trailer U-factor is 155 Btu/h · °F as determined either from data on the trailer or estimated from information on insulation properties, size, and condition of the trailer.

From Chapter 26, the specific heat of peaches (above freezing) is 0.91 Btu/lb · °F.

To estimate the heat of respiration, the average temperature of peaches during transit is assumed to be 40 °F. From Chapter 26, the heat of respiration of peaches at 40 °F is an average of 1700 Btu/(24 h) (ton). A more precise calculation of heat of respiration would recognize that it varies with product temperature, which decreases during the pulldown period.

Total heat load	= Product heat load = Heat of respiration = Heat transmitted
Product heat load	= (Specific heat) (Weight) (Temperature change)
	= (0.91) (38,000) (20)
	= 692,000 Btu
Heat of respiration	= (Heat of respiration factor) (Tons) (Days)
	= (1700) (19) (72/24) = 96,900 Btu
Heat transmitted	= (U-factor) (Hours) (Ambient temperature − Thermostat set temperature)
	= (155) (85 − 36) (72) = 547,000 Btu
Total heat load for 72 h	= 692,000 + 96,900 + 547,000
	= 1,336,000 Btu
Average load/h	= 18,600 Btu/h

If ice was used instead of mechanical refrigeration, the amount of ice melted would be about 1,336,000/144 = 9280 lb. Liquid nitrogen absorbs about 175 Btu/lb during evaporation and warming to 32 °F. So if liquid nitrogen was used in this example the amount of liquid nitrogen required would be 1,336,00/175 = 7630 lb.

The preceding calculation neglects initial cooling requirements for the trailer structure, air in the trailer, and cargo packing materials.

Perishables should be fully cooled prior to loading. However, some cooling is usually required in transit because the product temperature can rise during loading; and, during periods of high production, cooling facilities can be overloaded or insufficient time is permitted to fully cool the center of a product. Cooling during transit can reduce the overall time from harvest to consumption. However, airflow through the load is essential to cool the product in transit successfully. The section on Control Requirements and Devices also discusses product cooling.

REFERENCES

AFDO. Code of recommended practices for the handling of frozen foods. Association of Food and Drug Officials, York, PA.

ANSI. 1979. Requirements for closed van cargo containers. ANSI *Standard* MH5.1.1-1979. American National Standards Institute, New York.

ANSI. 1980. Requirements for thermal containers (refrigerated, heated, and insulated). ANSI *Standard* MH 5.1.2M-80. American National Standards Institute, New York.

ARI. 1983. Mechanical transport refrigeration units. ARI *Standard* 1110-77. Air-Conditioning and Refrigeration Institute, Arlington, VA.

Ashby, H.B., R.T. Hinsch, L.A. Risse, W.G. Kindya, W.L. Craig Jr., and M.T. Turczyn. 1987. Protecting pershable foods during transport by truck. *Agricultural Handbook* 669. U.S. Department of Agriculture, Washington, D.C.

ATA. Summary of size and weight limits. American Trucking Association, Alexandria, VA.

CGTFL. 1988. Refrigerated transportation foundation method classification of controlled temperature vehicles. Recommended practice No. 1-89. California Grape and Tree Fruit League, Fresno, CA.

Eby, C.W. and R.L. Collister. 1955. Insulation in refrigerated transportation body design. *Refrigerating Engineering* (July): 51.

Guilfoy, R.F., Jr. 1973. Refrigeration systems for transporting frozen foods. ASHRAE *Journal* (May):58.

Guilfoy, R.L., Jr. and R.C. Mongelli. 1971. A method for measuring cost and performance of refrigeration systems in local delivery vehicles. ARS *Publication* 52-64. USDA, Agricultural Research Service.

IIR. 1985. Technology advances in refrigerated storage and transport. International Institute of Refrigeration, Paris.

ISO. 1979a. Series I freight containers—external dimensions and ratings, 3rd ed. ISO *Standard* 668:1979. International Standards Organization, Geneva.

ISO. 1979b. Series I freight containers—specifications and testing—part 2: thermal containers, 2nd ed. ISO *Standard* 1496-2:1979. International Standards Organization, Geneva.

McGregor, B. 1987. Tropical products transport handbook. *Agriculture Handbook Number* 668. U.S. Department of Agriculture.

Penney, R.W. and C.W. Phillips. 1967. Refrigeration requirements for truck bodies—effects of door usage. Agricultural Research Service Technical Bulletin No. 1375. USDA.

Phillips, C.W., *et al.* 1960. A rating method for refrigerated trailer bodies hauling perishable foods. Marketing Research Report No. 433. Agricultural Marketing Service, USDA.

Phillips, C.W. and R.W. Penney. 1967. Development of a method for testing and rating refrigerated truck bodies. USDA Agricultural Service, Technical Bulletin No. 1376.

TTMA. 1973. Method for rating heat transmissions of refrigerated vehicles. Recommended Practice No. 38-73. Truck-Trailer Manufacturers' Association, Washington, D.C.

BIBLIOGRAPHY

ASHRAE. 1971. Refrigeration systems for perishable food delivery vehicles. ASHRAE Symposium Bulletin, WA-71-4.

ASHRAE. 1972. Long-haul transportation of respiring perishable commodities in refrigerated containers. ASHRAE Symposium Bulletin NO-72-7.

IIR. 1985. Long distance refrigerated transport—land and sea. International Institute of Refrigeration, Paris.

IIR. 1986. Recommendations for the processing and handling of frozen foods. International Institute of Refrigeration, 177, Boulevard Malesherbes, Paris.

United Nations. 1970. Agreement on the international carriage of perishable foodstuff and the special equipment to be used for such carriage (ATP). Economic Commission for Europe, Inland Transport Committee, United Nations.

CHAPTER 29

RAILROAD REFRIGERATOR CARS

Refrigerator Cars in the United States and Canada 29.1
Protective Services 29.1
Sources of Refrigeration 29.3
Mechanical Refrigeration Equipment 29.3
Special Purpose Refrigerated Cars 29.5
Car Design and Construction 29.5
Heating Equipment 29.7
Refrigerator Cars in Europe and the Soviet Union 29.7

THE development of refrigerator car service by the railroad industry and private refrigerator car lines during the past century has been an important factor in establishing the present nationwide system of distributing perishables. The railroad industry also transports perishables in *piggyback* refrigerated trailers and containers, which are discussed in Chapter 28.

REFRIGERATOR CARS IN THE UNITED STATES AND CANADA

Various types of railroad cars are classified, defined, and given designating letters by the Association of American Railroads (AAR). Refrigerator cars are designated as Class R cars, except those equipped for passenger train service, which falls under Class B (Passenger Equipment Cars) and those designed for transporting solid carbon dioxide, under Class I (Special Cars). The first of the designating letters indicates the car class.

With Class R cars, the second designating letter indicates the type of cooling equipment installed. The letter B indicates that no cooling equipment is provided, C indicates cars using a cryogen, P indicates mechanical refrigeration, and S indicates cars equipped with ice bunkers. The letter L is added to the designating letters of the car equipped with adjustable loading or stowing devices, B to the car designed for use in bulk loading and equipped with interior slope sheets and equipment for loading and unloading, and C to the car equipped with permanently affixed containers. The AAR definitions and letter designations for various types of refrigerator cars are as follows:

RB: Bunkerless refrigerator car with or without ventilating devices and with or without a device for attaching portable heaters. Constructed with insulation in side, ends, floor, and roof to meet maximum UA factor requirement of 250 Btu/°F/h for 50-ft cars and 300 Btu/°F/h for 60-ft cars. Effective for cars ordered new after March 1, 1984.

Note: Cars built or rebuilt prior to March 1, 1984, must have been constructed with a minimum of 3 in. insulation in the sides and ends and 3 in. in floor and roof, based on the insulation requirements given AAR Standard S-2010 or a thickness reduced in proportion to the thermal conductivity of the insulation.

RBL: Car similar in construction to an RB-type car, but equipped with adjustable loading or stowing device.

Note: Cars equipped with interior side rails only, built new, rebuilt, or reclassified on and after January 1, 1966, in order to qualify for the RBL designation, shall have a minimum of 4 usable side rails on each car wall, each extending from doorway to approximately 4 ft from the end of a car.

RC: Refrigerator car similar, to an RB car, using a cryogen to produce temperatures to transport frozen commodities.

RP: Mechanical refrigerator car equipped with or without means of ventilation and provided with apparatus for furnishing protection against heat and/or cold. Apparatus operated by power other than from the car axle.

RPB: Mechanical refrigerator. Similar to RP-type car, but designed primarily for use in bulk potato or similar type loading as cars are equipped with interior slope sheets and conveyors and/or equipment for mechanical loading and unloading.

RPC: Mechanical refrigerator car similar in design to RP car but equipped with permanently affixed container(s).

RPL: Mechanical refrigerator. Similar to RP but equipped additionally with adjustable loading or stowing device.

RS: Bunker refrigerator car equipped with ice bunkers. Designed primarily for use of chunk ice and with or without means of ventilation.

PROTECTIVE SERVICES

Refrigerator cars are exposed to outdoor air conditions that vary from summer desert temperatures to low winter temperatures, and the desired transit temperature for perishable commodities varies from 0°F or slightly lower for frozen foods to as high as 65°F for certain types of bananas and pears. To handle this problem, the National Perishable Freight Committee has established different classes of protective services. These services, together with rules and regulations and charges, are published in the *Perishable Protective Tariff* (Doean 1983).

Perishable freight refers to any commodity susceptible to deterioration or decay and/or that may be protected by cooling, ventilating, or heating. A handbook giving recommended protective services and loading methods for the proper care of perishable agricultural commodities in refrigerator cars is available (Redit 1969). This publication includes a comprehensive bibliography, an explanation of the Perishable Tariff Rules, and methods of calculating the refrigeration loads involved.

Mechanical Protective Service

Cars equipped with mechanical refrigeration units offer temperature control service, with car interior settings from 0 to 70°F at any ambient temperature. These cars cool and heat to maintain the desired interior temperature (see Figure 1).

Mechanical protective service is available both at the standard tariff rate, which includes operating the mechanical system from point of origin to destination, and in modified forms whereby the

---

The preparation of this chapter is assigned to TC 10.6, Transport Refrigeration.

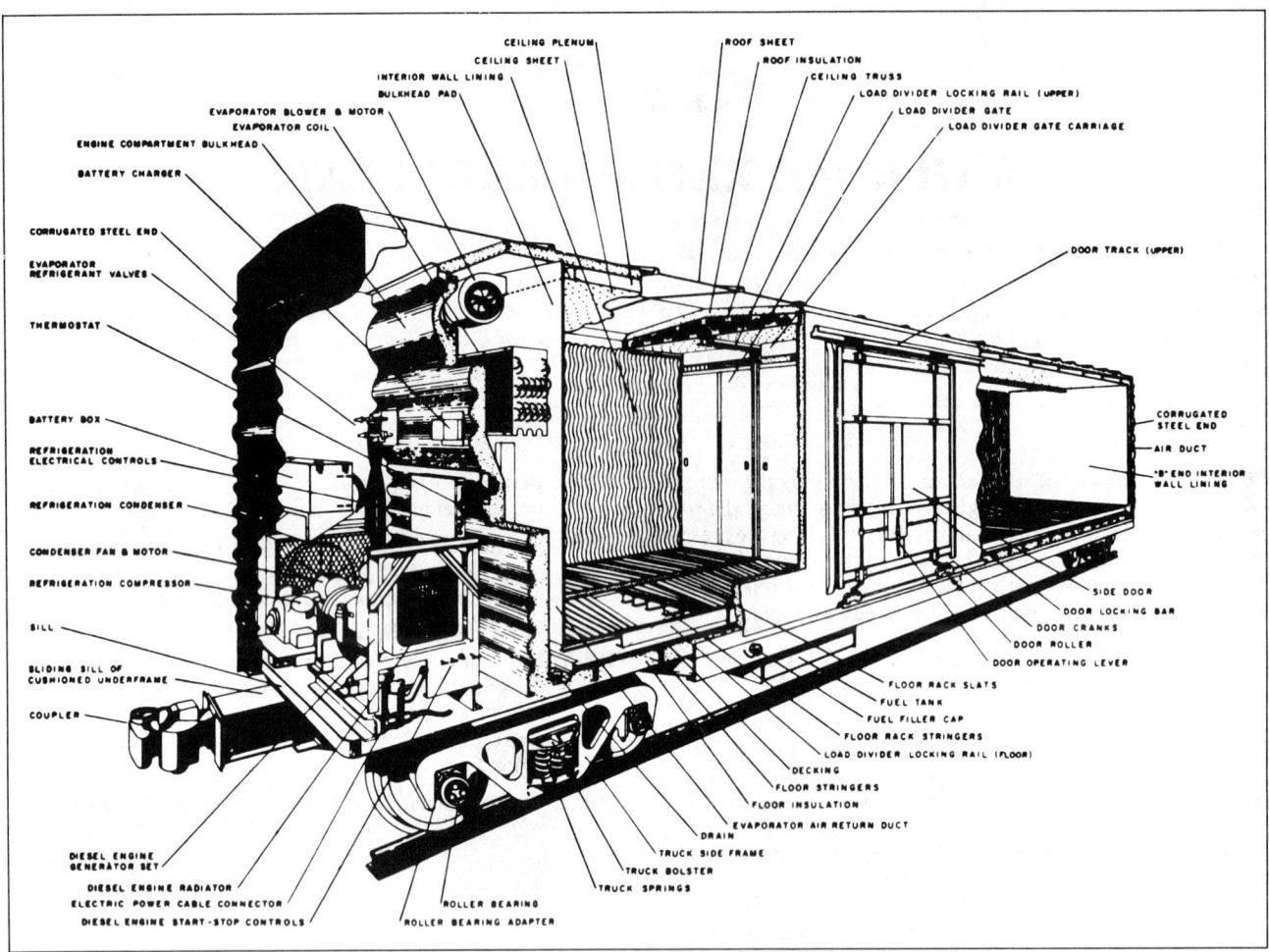

Fig. 1 Typical Refrigeration Components for Mechanical Refrigerator Cars

unit is not started until the indicated car temperature reaches a specified high or low level, or whereby the unit is operated until the desired temperature is reached, after which it is turned off. One such starting or stopping operation is permitted for each trip, especially with commodities that do not require close temperature control, but do require protection when shipped between points with extreme differences in outdoor temperatures.

Refrigeration Services Using Ice

A type of refrigeration service often used when shipping leafy and root vegetables, melons, keg or draught beer, and Christmas trees is *top* or *body* icing; crushed or snow ice is distributed throughout the body of the car (mostly on top of the load) in direct contact with the lading. Top or body icing at the shipment's origin and the placement of ice in loading containers is usually done by the shipper.

The *Perishable Protective Tariff* permits shippers to cool perishable commodities in refrigerator cars at loading stations during or after the loading process. The methods of precooling shipments in cars prior to transit cooling, including contact or top icing and vacuum cooling, are discussed in Chapter 11.

Heater Services

Mechanical refrigerator cars are equipped to provide both heat and cooling to maintain any desired temperatures from 0 to 70°F in any ambient temperature. Therefore, no external devices are needed to provide heater service with mechanical refrigerator cars.

Several types of heater service for protection against frost, freezing, or artificial overheating of shipments during the heater season are available (which varies according to heater service and commodity). With *carrier's protective service,* the carrier furnishes whatever protection it considers necessary. With *shipper's protective service,* heaters are furnished and installed by the shipper and are serviced by the carrier as directed by the shipper. With *shipper's specified service,* heaters are furnished, installed, and serviced by carriers as directed by the shipper.

Protection furnished by the carrier after service has been changed in transit from refrigeration or ventilation is known as *modified carrier's protective service.* Protection supplied without request on a shipment originating outside regular heater territory and west of the Mississippi River is known as *voluntary heater service.* Protection using liquid fuel heaters furnished, installed, and serviced by the carrier with the thermostat set as directed by the shipper and ventilators manipulated at the degree of temperature specified by the shipper is known as *special heater protective service.*

Other Special Services

In addition to the abovementioned protective services, several special services or privileges are provided by the railroad industry for the movement of perishable freight. Shipments of commodities under refrigeration may be stopped in transit and fumigated in the refrigerator car on instructions from the shipper, who may also fumigate the cargo in the cars prior to departure

Railroad Refrigerator Cars

from the loading track. Fumigation is accomplished by the release of certain proprietary gases into the car from pressure tanks.

Sampling or inspection of a shipment in transit may be done by the owner, his representative, or the prospective buyer when authorized by the shipper or owner. When so instructed, a carrier will change, in transit, the type of protective service, the destination of the shipment, or the consignee. About one-half of all refrigerator car shipments are reconsigned in transit. Stopover privileges are permitted to complete loading or to unload partially, and transit privileges are permitted for rehandling, transferring, or reconditioning.

SOURCES OF REFRIGERATION

A refrigerator car fleet consists of ice bunker cars using water ice as their source of refrigeration and mechanical refrigerator cars, which use mechanical compression machines as a refrigeration source. However, the number of ice bunker cars is rapidly diminishing; they are being replaced almost exclusively by mechanically refrigerated cars, most of which are owned by private companies. Those bunker cars still serviceable are being used either in non-ice bunker services, such as ventilation or body ice, or as insulated box cars.

Mechanical refrigeration was determined to be most feasible by using standard compression refrigeration system components, but with an independent electric generator set as its own source of power. This makes the refrigeration system independent of the car movement. By adopting standard commercial components using standard voltages, these systems can also be plugged into carside power outlets to make refrigeration available during loading and unloading without running the engine.

Mechanical Refrigeration

Refrigeration systems are conventional, and the air circulation systems and capacity controls are designed to provide the humidity control required for handling fresh commodities. Powered with their own fuel supply, they are independent and may be operated anywhere at any time. This offers the advantage of providing maximum cooling as soon as the car is loaded; in conventional ice bunker fan cars, the cooling rate was small until the car moved.

The conventional mechanical car is used for transporting frozen foods and fresh perishables in the range from 0 to 70°F. The capacity is sufficient to carry frozen foods in any ambient temperature encountered in rail service; therefore, the same system has more than adequate capacity to hold fresh perishables in the 35°F range, as well as those products at 50°F or higher. Excess refrigeration at the higher temperature levels is controlled by unloading compressor cylinders or by speed variation, or both.

The mechanical car is not effective as a precooler for removing field heat, since the close stacking and tight packaging of commodities in the car prevents the free circulation of air around the individual packages, even though the refrigeration capacity may be available. Therefore, most fresh perishables have the greater portion of their field heat removed before being loaded in these cars; the car is used to obtain the final cooling and desired storage temperature in transit.

Other Sources of Refrigeration

Nitrogen has been used to cool frozen foods to very low temperatures so that the temperature rise for the transit period would allow arrival with temperatures still below 0°F. Initial temperature was predicated on the expected length of the trip, type of packaging, and heat leak of the car.

In other applications, liquid nitrogen is carried in the vehicle in cryogenic vessels at about 20 psig operating pressure; this pressure is used to deliver the nitrogen to the lading compartment through a set of overhead spray nozzles. The refrigeration rate is controlled by a thermostat and a solenoid-operated flow control valve. The principal applications have been with frozen foods and meat.

With an atmospheric boiling point of $-320°F$ for liquid nitrogen, temperature control without overcooling for fresh perishables has been a problem. Early systems depended entirely on the diffusion of the gas throughout the lading space to cool the commodity, but due to the stratification effect at low refrigerant flows, systems now use air-circulating fans to provide more uniform temperatures. These fans are driven by pneumatic motors using the vent gas from the liquid nitrogen storage tanks as a power source.

The principal deterrent to further development of liquid nitrogen systems has been the high cost of the refrigerant. With some applications, such as local delivery of frozen foods, liquid nitrogen systems are less expensive than mechanical refrigeration.

Liquid carbon dioxide has also been tested, and several systems are now commercially available. These may be either the direct-release type or systems using secondary refrigerants flowing in wall or ceiling coils. Application of liquid CO_2 has generally been limited to cars in captive service, such as for meat in regular schedules, since it depends on setting up a supply and distribution system for the liquid refrigerant. Here again, local cost factors determine whether liquid CO_2 will supplant other refrigeration.

MECHANICAL REFRIGERATION EQUIPMENT

With the exception of cars having underslung units, mechanical cars have a louvered engine compartment at one end, which houses most of the mechanical refrigeration equipment. The compartment is separated from the loading space by an insulated bulkhead.

The system is usually divided into major subassemblies, including an engine-generator, a compressor-condenser, evaporator coil and blower, a refrigerant control, and an electrical control panel (see Figure 1). Some of the direct-drive systems have all the components combined into a single unit that is mounted in the machinery compartment on a sliding base for ease in installation, inspection, and replacement. Special consideration has been given in the selection and mounting of all components because of the vibration and impact forces obtained in freight-car service.

The design refrigeration capacity for a mechanical refrigerator car depends primarily on the type of refrigeration service for which the car is intended.

Power and Drive Equipment

Diesel engines have been used almost exclusively to drive the refrigeration systems in mechanical cars.

The generators used with electric-drive systems are connected directly to the engine and provide 220 V, three-phase, 60 Hz current with a rated output from 12 to 25 kW. A majority of the mechanical refrigerator cars have a 20 kW generator connected directly to a 34 hp, two-cycle diesel.

Two-speed engine operation is used to reduce engine maintenance and fuel costs. Full engine speed is used for pulldown and, after the car temperature is within a few degrees of the temperature control setting, low speed operation is maintained. With electric drive systems, 60 Hz operation is obtained at full speed, and 40 Hz operation at low speed. Generator characteristics are designed so that the voltage varies directly as the frequency; thus the generator, battery charging system, electric control, temperature controls, relays, starters, and so forth, all function at both 40 and 60 Hz. Being inductive devices, their impedance varies as the frequency, and they draw approximately the same current at both speeds.

Engines are equipped with shutdown controls that protect against loss of lubricating oil pressure and high coolant temperature.

Refrigeration System Components

Most cars are equipped with single compressor systems using Refrigerant 12. A few systems use dual compressors with two condensers.

The compressors are of several designs, ranging from the two-cylinder, belt-driven type through a six-cylinder, semihermetic type and including multiple cylinder open types and hermetic types. Many of the compressors are equipped with unloaders on all or some of the cylinders to obtain fully unloaded starting conditions. The cylinder unloaders are also used for refrigerant capacity control.

Condensers are of standard air-cooled design, usually constructed of copper tubes and aluminum fins. Air is drawn through louvers in the side of the car across the condenser and discharged out the other side of the car after passing over the engine. Condenser fans are the propeller type, mounted directly on the motor shaft, and delivering about 8,000 cfm with a 2 hp, 220 V, three-phase induction motor.

Evaporators are made of copper tube and aluminum fins and have reinforced tube sheets capable of transmitting the forces from the lading into the bulkhead on which they are mounted in the event of load shifting. Electric tubular heating elements used for car heating and defrosting are placed in the coil in such a way as to effect good heat distribution for proper defrosting, although some systems place the heater assembly beneath the evaporator coil for easy removal. Electric heating elements are also placed in the evaporator drain pan to assist in the removal of meltage formed during defrosting. These are energized only during the defrost cycle.

Some systems use hot gas in lieu of electric heaters for heating and defrosting. Various arrangements use either full reverse cycle operation or modified versions which in the principal refrigerant charge is trapped in the condenser and only gas flows between compressor and evaporator.

Refrigeration systems are generally of the direct expansion type with a thermostatic expansion valve controlling refrigerant flow. In addition, the refrigerant systems contain a solenoid liquid stop valve, filter, dryer, and suction-liquid heat exchanger. The refrigerant controls are usually clustered on a panel and are either mounted on the condensing assembly or on the adjacent bulkhead wall. Connections between these assemblies and the evaporator are made with flexible hoses having union connections to facilitate unit replacements. System components are equipped with shutoff valves enabling the removal and replacement of components without complete loss of refrigerant charge.

Vibration and shock conditions are severe for refrigerant piping and require flexible vibration isolators between compressors and condensers. In addition, all piping is thoroughly secured with braces and clamps to prevent movement or contact with adjacent members. Condensing assemblies are, in turn, shock mounted to the car floor, and all parts of the system are designed for 15-g loading in the direction of travel, 5-g in the vertical direction, and 2-g laterally.

Starting the refrigeration system is accomplished automatically on starting the engine generator set. When the proper voltage is reached, the electrical control circuit is actuated and the condenser and evaporator fans start immediately. The time delay relay is also actuated and, after an approximate 60-s delay, starts the compressor motor. This delay permits the engine to settle down and clear itself after starting before the major load of the compressor is applied.

Electric-drive mechanical refrigeration systems are equipped with a plug-receptacle, which makes it possible to shut the engine down and operate the system from a standby electric power source. Standby operation has been used while correcting engine difficulties and also where the engine noise or exhaust fumes are objectionable during loading and unloading. Fuel for the internal combustion engine is usually carried in fuel tanks located under the car floor.

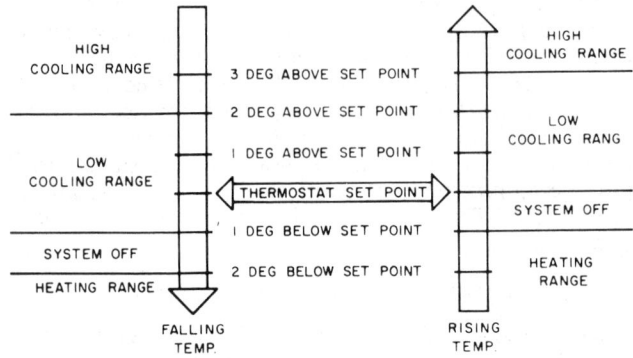

Fig. 2 Thermostat Operation

Electrical controls are contained in a dustproof panel usually mounted on the bulkhead adjacent to the condensing assembly. This panel contains the motor starters and any other electrical devices required by the particular system, such as timers, annunciator lights, and switches. Annunciator lights are generally mounted on the side of the panel visible from the doorway at the side of the car; they indicate whether the system is in a *cooling* or *heating* mode, and, in some cases, an additional light indicates a defrost cycle.

Temperature control is accomplished using a thermostat with a sensing element located in the return air flue to the evaporator coil. The thermostat has multiple switches acting in sequence on a temperature fall to unload the compressor, reduce speed on multispeed systems, stop the compressor, and, if the temperature continues to fall, switch to heat.

Figure 2 shows a typical thermostat operation, with the switch from high cooling to low cooling at 2 °F above set point and the system shut-off just below the thermostat set point. If the temperature continues to drop, the unit will switch to heating at 2 °F below set point. If the low cooling range is insufficient to hold the car temperature, it will rise and switch the unit back to high cooling at 2 to 3 °F above the set point.

Defrosting is automatic and controlled by an electric timer in some systems. Others use a differential pressure switch to measure the pressure drop across the evaporator coil, and the switch is adjusted to initiate the defrost cycle when enough frost has formed to reduce the airflow about one-third. Once tripped, the defrost relay holds the heaters on until a thermostat on top of the coil breaks the circuit and the system returns to normal operation. The termination thermostat setting is usually about 55 °F; in addition, a limiting device set at 150 °F is installed to protect against high temperature in case of malfunction of the regular controls.

During the defrost cycle, the control circuit stops the evaporator fan to prevent any of the heat produced in the coil from reaching the lading compartment. At the termination of defrost, the evaporator fan starts again and runs continuously during cooling, heating, or the off cycle.

All mechanical refrigerator cars are equipped with air-circulating fans, which are usually mounted above the evaporator coil in the bulkhead, between the machinery compartment and the loading area (see Figure 1). Nearly all are of the backward-curved centrifugal type and draw air from the floor through the evaporator coil and discharge it into the ceiling distribution duct. The fans consist of two or three blowers mounted directly on the electric motor shaft or on a separate belt-driven shaft. Fans usually deliver about 4000 cfm and require a 1.5 hp, 220V motor.

Railroad Refrigerator Cars

Temperature Indicating Devices

Two types of temperature-indicating devices have been used in refrigerator cars. Both consist of a liquid-filled metallic bulb located inside the car, connected by a flexible metal capillary tube to a dial indicator mounted outside the car. One type uses mercury under a pressure of 300 to 600 psi, a Bourdon tube to actuate the dial pointer, and a bimetallic strip to compensate for temperature changes in the case. The other type uses toluene under a pressure of 16 to 60 psi and three brass bellows to actuate the dial pointer. One of the bellows responds to the liquid in the bulb and connecting capillary tube. Liquid in a second capillary tube, which is not connected to the bulb but runs parallel to and is the same length as the first tube, actuates a second bellows, which compensates for temperature changes in the tubing. The third bellows compensates for temperature changes in the case.

Since the Canadian railroads use tariff rules for perishable protective services that are based on inside temperature control, nearly all Canadian cars are equipped with temperature indicating devices. The bellows type with two dial pointers (one for top and one for bottom temperatures) is standard for Canadian cars.

SPECIAL PURPOSE REFRIGERATED CARS

Several special-purpose mechanically refrigerated cars have been introduced. One design for bulk shipment is based on the standard mechanical refrigerator-car arrangement (Figure 3). The engine compartment is in one end, but the addition of three subfloor hoppers in the center of the car permit the unloading of bulk products. Mechanically operated hinged endwalls can be moved into a sloped position to facilitate the unloading of bulk products over the flat floor sections. This car can be used as a conventional mechanical refrigerator car when the endwalls are in their vertical position and floor grates are placed over the hoppers.

Another design (Figure 4) is based on a covered hopper car that has three compartments with a full-length hatch and a gravity-type outlet for each. The exterior of the car is covered with a layer of spray-on polyurethane foam insulation protected by a special coating; the roof and other high-wear areas are protected with an additional fiberglass covering. The interior surfaces of the hoppers are covered with a highly abrasive-resistance coating, which has been approved for use with foodstuffs.

Three electric heating elements supply the heat usually needed. If fast preheating is desired, three additional units operate for maximum heat. Noncombustible electric heat preserves oxygen, which is important for lading protection. For cooling, a refrigeration unit provides controlled cold air automatically. The unit is mounted on the end of the car to provide access for service and maintenance and protection against vibration, moisture, and dirt.

In operation, air is drawn off the top of the load and passes through a cooling/heating section at the end of the car where it is brought to the preset temperature. A high-pressure fan forces the conditioned air through the car's side sills, into the bottom of each compartment, and upward through the cargo.

Another conventional mechanical refrigerator car with a slightly different machinery arrangement—a mechanical car with the refrigeration unit in the side door—has been developed. The engine generator set to furnish the electrical power is located beneath the car. The advantage of this system is the ease of changing out the unit for a car that requires only high-temperature applications for fresh perishables.

Another type of refrigerating equipment circulates air that is reverse to that in the conventional unit. In this case, the blower pulls air down through the coil and discharges it under the floor. It passes up through the load and returns to the cooling unit over the top of the load.

CAR DESIGN AND CONSTRUCTION

The basic design and construction of refrigerator cars is very similar to that for railroad box cars because both types of cars must meet all the structural, safety, and performance requirements demanded in railroad freight service. Specifications for standard freight refrigerator cars have been established by the Association of American Railroads and are often revised to incorporate improved standards (AAR 1972).

The principal construction features of refrigerator cars that are different from those found in conventional freight cars include: thermal insulation and water vapor barriers incorporated within the structure of the car, mechanical refrigeration equipment, and various features that facilitate the circulation of air.

Refrigerator car construction features common to the standard box car include: (1) trucks (wheels, axles, bearing, springs, and supporting members); (2) steel underframe; (3) brakes and brake equipment; (4) couplers and draft gears; and (5) a superstructure consisting of structural members, exterior steel sheathing (on the sides, ends, and roof), interior wood or plastic lining and floor, side door openings, and various safety appliances.

Refrigerator cars have cushioned underframes, which absorb and dissipate impact forces through a special cushioning device built into the *center sill* (the longitudinal steel member that is the backbone of the underframe). The center sill of the car is fixed, and a sliding sill within the center sill contains the couplers at each end of the car. The two are connected by the hydraulic cushioning device built into the center sill. Other refrigerator cars have separate cushioning devices in each end of the car that perform a similar function by isolating the coupler impact forces from the main car structure. To permit palletized loading and unloading of refrigerator cars with forklift trucks, cars have door widths from 10 to 10.5 ft to facilitate turning in the doorway area.

Movable steel load dividers (or compartmentizers) brace the load and split the loading compartment into three sections. These load dividers reduce impact damage and segment loads billed to stop for partial loading or unloading.

Most cars now having at least a 4000 cu ft capacity, with load carrying ability of 130,000 lb. A complete list of railway equipment is published quarterly in *The Official Railway Equipment Register*.

Insulation of Car Structure

The usual structural members of rail cars constitute high conductivity heat paths and interfere with the placement of insulation. Various service conditions (including impact and vibration of the car, air pressure caused by the movement of the car at high speeds, body and top icing within the car, periodic washing of the

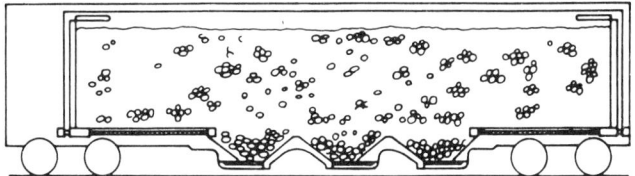

Fig. 3 Mechanical Refrigerator Car for Bulk Loading

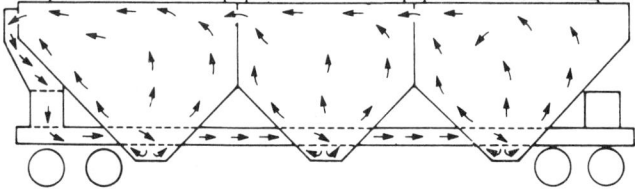

Fig. 4 Covered Hopper Car for Refrigerated Bulk Loading

car interior, and reversal of the direction of water vapor flow with winter and summer seasons) make it difficult to retain the initial resistance to heat flow. Wetting and settling are problems aggravated by these service conditions.

The AAR specifications for freight refrigerator cars establish insulation requirements for newly built cars. The cars are to be insulated by means of rigid, closed-cell polyurethane foam insulation in order to obtain desired UA factor. RP and RS series cars should be designed to obtain a total heat loss rate of 120 Btu/h·°F·h, based on a 50-ft-long car. RB series cars should meet a maximum UA factor requirement of 250 Btu/h·°F·h for 50-ft cars, and 300 Btu/h·°F·h for 60-ft cars.

Membrane-type water vapor barriers usually serve as the covering for blanket insulation. Membrane barriers with one or two reflective surfaces have been placed adjacent to the steel sides in addition to being installed in the roof. Some mechanical cars have a layer of water vapor barrier paper above the top of the ceiling plenum and also on the back side of the inside wood lining.

The use of closed cellular plastic insulation has become standard. The insulation is practically impervious to water and water vapor, and many types possess sufficient compressive strength to withstand the entire floor load in a refrigerator car. Some high conductivity heat paths through the car structure can be avoided because closed cellular plastic insulation provides sufficient strength to eliminate various conventional structural members such as horizontal belt rails and floor stringers.

The insulation may be of two general types—board form or foamed-in-place. Mechanical refrigerator cars use plastic foam insulation, either foam-in-place polyurethanes or prefabricated panels. Foam is added after the interior and exterior structure is in place, or prefabricated sandwich panels incorporating the interior lining and insulation are fastened to the car structure at final assembly.

A typical car design using foamed-in-place insulation is shown in Figure 5. The use of horizontal belt rails has been abandoned; all interior members are placed vertically in the walls or transversely in the floor and roof. This permits the injection and free rise of the foam material so that it will fill all cavities without interference from any structural members. The metal posts are placed on the outside of the car to prevent penetration of the insulation panels.

Mechanical refrigerator cars insulated with the foamed-in-place polyurethanes show a heat leakage only half that of previous insulation and car construction methods. In addition, the use of foam to fill the insulation cavities seals the structure against air leakage and prevents stray airflow through the insulation spaces from the interior air circulation system, the latter contributing heavily to total refrigeration load in many cases. Mechanical refrigerator cars typically have a heat transfer rate of 90 to 100 Btu/(h·°F).

Air Leakage and Other Losses

In addition to conductivity losses through insulation and car structure, a considerable load derives from air infiltration and solar radiation. Doors and drains are principal sources of infiltration. An appreciable amount of air may enter through seams and joints in the car structure. Door gaskets and hardware are carefully designed and applied, as the trend to larger doors increases the exposed areas. Drains under the evaporator drip pan are usually the cuptrap type using drippage water from the cooling coil as the sealant. Drains elsewhere in the car are of the self-closing flapper type.

The interior car lining is also sealed against leakage, since the air circulation system maintains a constant positive pressure in the ceiling plenum and a negative pressure in the air-return flues. All interior lining joints, especially in the evaporator and blower areas, are sealed with tape or some type of mastic sealer.

By improving car sealing methods, the air leakage rate of mechanical cars is usually maintained below 100 ft³/h at 0.5 in. water pressure within the car. This is especially important as more mechanical cars move under modified atmospheric conditions, where excessive infiltration destroys the beneficial effects of the atmosphere.

Cars are finished in light colors, and roofs are usually white or aluminum to obtain the highest reflectivity. To maintain desired reflectivity, cars are regularly cleaned or repainted as increased solar load, because of dirty surfaces, forms a considerable part of the total heat load.

Air Circulation

Floor racks, side and end wall flues, and special ceiling ducts in refrigerator cars facilitate air circulation. Floor racks allow air circulation under the load. Openings in the longitudinal stringers permit side-to-side air circulation and are especially important in cars equipped with side wall flues. Floor racks are designed to carry forklift truck wheel loads of 6000 to 12,500 lb per wheel. Height of floor racks varies from about 4.8 to 7.5 in. Minimum clearance under floor rack slats is 4 in.

Refrigerator cars are equipped with vertical side wall flues that protect loads from direct contact with very warm or cold side walls. Side wall flues are formed by continuous plywood or tongue and groove lining spaced 0.5 to 1.5 in. from the regular inside wall by vertical spacers and open at the top (above the load) and at the bottom (below the floor racks). Cars equipped with mechanical refrigeration have end wall flues in addition to side wall flues. Side

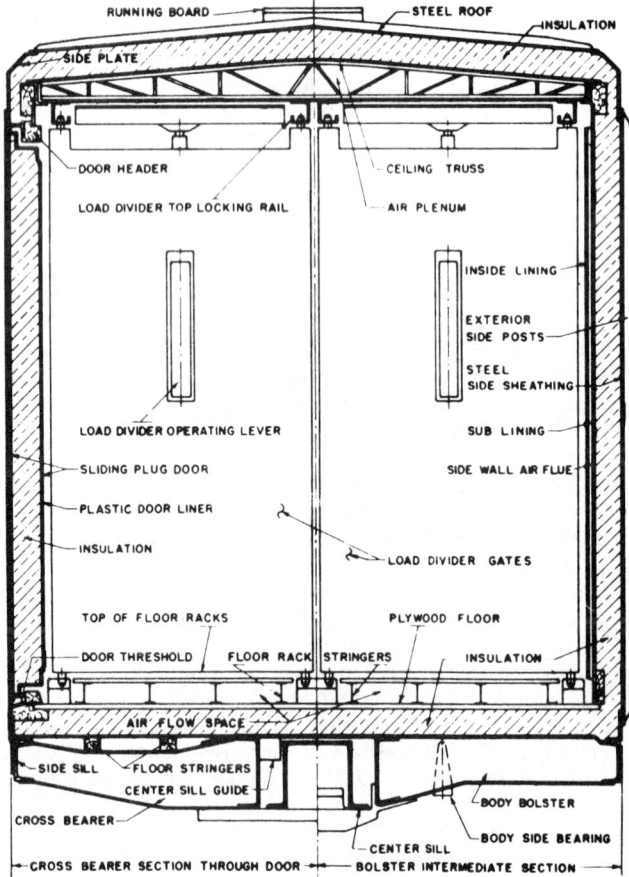

Fig. 5 Cross Section of Typical Mechanical Refrigerator Car Construction

Railroad Refrigerator Cars

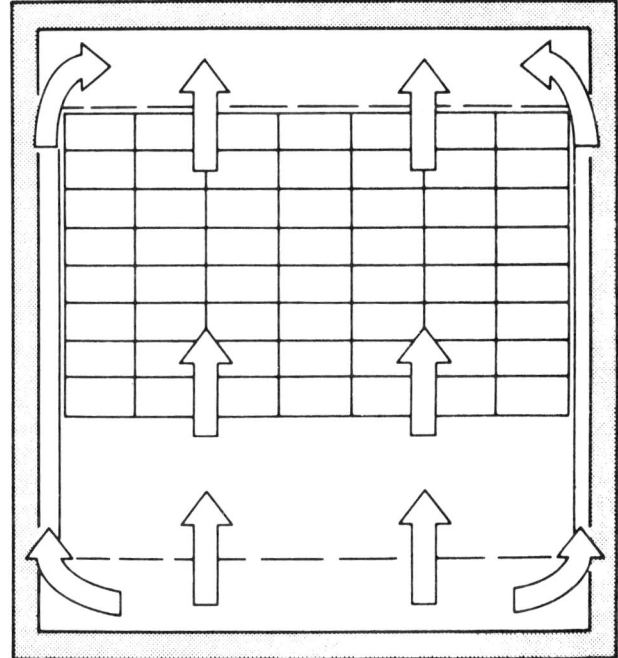

Fig. 6 Cross Section Showing Air Circulation

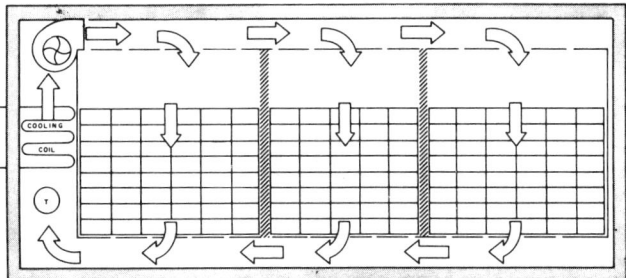

Fig. 7 Conventional Air Circulation System

wall flues in mechanical cars with a ceiling duct extend from the duct to the underside of the floor rack slats; with the complete envelope system, they extend from ceiling plenum to air return duct below the floor.

The principal types of air-circulation systems used in mechanical refrigerator cars are as follows:

Open. Air is discharged directly over the load from the evaporator end of the car. Air passes down through the load and the wall flues and returns to the evaporator via the channels formed by the floor rack stringers.

Plenum. Same as the open system, except the air is discharged into a slotted or perforated ceiling supply duct that distributes the air uniformly over the top of the load.

Envelope. The air is discharged into a ceiling duct. The ceiling duct, wall ducts, and a special floor duct together form a completely closed envelope around the car. The air circulated within the envelope does not enter the loading area, nor does it escape from the envelope even when the side doors are opened.

Figure 6 shows the cross section of the vehicle; some air circulates down the side wall flue and some circulates through the load itself. Both air passes join in the floor duct for return to the cooling unit.

Figure 7 shows the typical air circulation of a mechanical car. The blower draws air through the cooling coil and discharges it into a ceiling duct or air plenum above the load. From here, it is distributed into either the lading compartment or wall flues on either side and the far end of the car. It then goes into the floor duct, which carries the air back to the cooling unit. Just prior to entering the cooling coil, the air flows over the thermostat sensing element, marked *T*. When the load divider gates of the mechanical car are in place, dividing the load into three spaces, the air circulation pattern is essentially the same because airflow is from ceiling to floor, and the load divider gates do not interfere with this circulation pattern.

As the airflow diagrams indicate, air will flow through the load from ceiling to floor. However, the amount that flows through the load, compared with that flowing in the side wall flues, depends entirely on how tight the load is in comparison to the openings in the side wall flues. Wider spaces between containers will increase air circulation through the load itself when it is necessary to remove either field heat or heat of respiration from within the load itself.

In the case of frozen foods (a heavy, dense load), air does not need to circulate through the lading space; all air circulation takes place around the periphery of the load to intercept heat from outside. In this case, little or no air should bypass through the lading itself, and all air must flow around the load to give maximum protection. Therefore, frozen loads are usually packed quite densely.

HEATING EQUIPMENT

Air-circulating fans and heaters were discussed previously in reference to mechanical refrigerator cars. Some specially equipped RB cars also offer heating protection service.

The portable thermostatically controlled alcohol heater (see Figure 8) has now almost completely replaced the portable charcoal heater. This heater was designed in response to the need for adequate temperature control and complete combustion. The heater consists of a 5-gal capacity cylindrical tank, which serves as the heater base, to which are attached the wick, fuel gauge, thermostat, filler cap, chimney assembly, and spring hooks. The alcohol fuel is drawn by capillary action to the top of the compressed glass fiber wick where combustion occurs.

The thermostat assembly includes a bimetal coil, which controls the burning rate by regulating the height of the snuffer plate above the wick surface in response to the thermostat setting. The thermostat control pointer can be set from 30 to 60 °F and up to 70 °F on the latest model, in increments of 2.5 °F and may be locked in a pilot position, where the snuffer plate completely covers the wick. In this position, burning occurs only at the slot in the center of the snuffer plate and this constitutes the pilot flame. The rated heat output with methanol as the fuel is 6000 Btu/h at full burning and 600 Btu/h at pilot burning.

REFRIGERATOR CARS IN EUROPE AND THE SOVIET UNION

One of the major refrigerator car builders is located in East Germany, and it supplies refrigerator rolling stock to Eastern and Western European countries. Two major types of refrigerator rolling stock are used in Europe: refrigerator sections (usually 5-car sections), and autonomous refrigerator cars. These all-purpose types can carry any perishable product in the −10 to 70 °F temperature range, with ambient temperature varying from 113 to −40 °F. Some of the cars are specialized for transporting cooled (not frozen) meat and are equipped with beef rails and hooks.

An autonomous refrigerator car is completely independent, consisting of a fully insulated lading compartment and two engine compartments (one at each end). Each engine compartment has

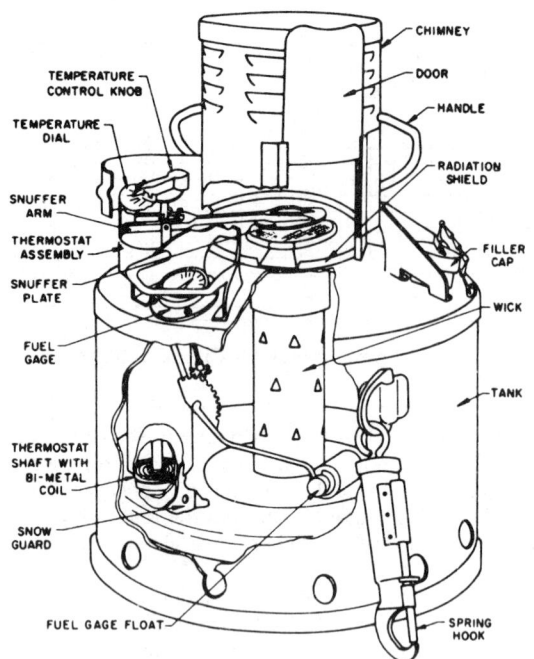

Fig. 8 Portable Thermostatically Controlled Alcohol Heater

a unitized refrigeration and heating unit mounted on top under the car ceiling. This design allows better air circulation and temperature equalization throughout the load. The two independent refrigeration systems also contribute to car reliability. In the event of the one system failing, the second unit can usually provide the necessary cooling or heating capacity to maintain interior temperature within the range allowable for the load. Each unit (running continuouly) is designed to provide 75% of maximum cooling or heating capacity under extreme difference between interior and ambient temperatures.

A diesel-generator in each engine compartment powers the refrigeration/heating unit. All electrical controls and monitoring devices are mounted in dust- and waterproof panels on the engine compartment bulkhead wall.

All cars are equipped with plug-receptacles on the outside of the car, which allow the system to be operated from an outside power source during loading and unloading, or when the car is not running and engine noise and exhaust fumes might be objectionable. Each engine compartment has equipment failure indicating lights and audio signals on the outside. Temperature control boxes provided on every car allow interior temperature readings in transit without entering the load or engine compartment, which are sealed after equipment start-up by authorized personnel. The operation is automatic after initial start-up and is controlled by interior thermostats, which are set at the desirable temperature, depending on the type of load.

A five-car refrigerator section consists of four load refrigerator cars and one "special" car in the middle of the section, which carries the central power plant and the maintenance crew.

Load cars are similar to autonomous refrigerator cars except they have no diesel generators. Instead, electric power for the refrigeration and heating equipment is provided through a central power plant, which usually consists of two diesel generators; each generator is capable of supplying electrical power for up to six fully loaded refrigeration units.

The special car also has a central control panel from which all equipment operation and power distribution are controlled. Equipment failure indicators are also located on the central panel.

Although refrigeration and heating equipment operation is completely automatized (providing there is electrical power), automatics can be overridden by the crew if necessary.

Five-car refrigerator sections are also built in the Soviet Union. They are slightly larger than those used in the rest of Europe. Section arrangement is similar to that in East Germany, but load cars have two refrigeration units, both located at one end of the car. Each unit also has the ability to provide a capacity of up to 75% of the maximum cooling or heating load.

Mechanical Refrigeration Equipment

Autonomous refrigerator cars and refrigerator sections built in East Germany use completely unitized refrigeration units, which can be easily removed and replaced within an hour through an access door located on the car's end wall. The refrigeration unit is of the direct expansion type with thermostatic expansion valve, and it uses a semihermetic two-stage compressor with Refrigerant 12. The condenser is air-cooled by two independent fans driven by separate induction electric motors. Fan operation is controlled by a thermostat, which allows one-, two-, or without fan condenser operation, depending on the load. The evaporator has two blowers which can be turned on without cooling or heating to provide air circulation and ventilation. Electric tubular heating elements, mounted in front of the evaporator are used for car load heating.

Hot gas from the compressor is used for evaporator defrosting. Two solenoid valves are energized alternatively: one on the liquid line to the evaporator, which opens during normal cooling operation and the other, on the defrost line, which opens to supply hot refrigerant to the evaporator during defrosting (the liquid solenoid valve is closed at that time). The defrost cycle is controlled automatically either by an evaporator pressure switch or by a timing device. A hot gas circuit is also used for pressure equalization before compressor start and, in that case, is controlled by time relay.

Pressure gauges mounted on the unit frame and connected to the appropriate points of the system allow vital information, such as suction and discharge pressures and compressor oil pressure, to be read during system operation. Provision is also made for reading compressor intermediate pressure. The electric control panel is mounted on the unit frame in a dustproof box.

Refrigeration units of the five-car sections built in the Soviet Union are similar to the split-type system with two compressor-condenser units, using open compressors located in the car engine compartment and two evaporators, located in the load compartment behind the insulated wall. Each car has only one compartment; it accommodates both refrigeration units (one on top of the other), which can work independently and be controlled automatically depending on the cooling load. Every compressor-condenser unit is generally associated with its own evaporator; however, the piping arrangement and manual bypass valves allow both evaporators to be connected to either refrigeration unit.

Diesel-generators in autonomous refrigerator cars are mounted on a sliding base together with a fuel tank, battery, and electrical control box. The engine compartment side door can be opened wide to allow easy and fast unit replacement. Plug-in electrical conduits are the only disconnections required. The generator provides 220 V, three-phase, 50 Hz (European standard) current. All motors on refrigeration units are of the three-phase induction type. A transformer provides 36 V (dc) current for control circuits.

The power plant for five-car refrigerator section is located in the middle of the section and consists of two diesel generators, which provide 220 V, three-phase, 50 Hz electrical current transmitted to load cars through electrical conduits. Each is designed to supply at least 75% of the current required under maximum load conditions, which makes it possible to operate six out of eight fully loaded refrigeration units. Normally each diesel-generator provides power to an assigned half of the section (two cars, four refrigerator units), but in case one generator fails, a bypass electrical contactor can be activated to allow the second generator to supply power through the entire section. This feature contributes to section reliability and can also save fuel, when load conditions do not require simultaneous operation of both refrigeration units in every car. Fuel tanks are located under the car.

Five-car sections built in East Germany also have a supplementary small diesel generator, similar to the one used in autonomous refrigerator cars. Such a generator can be used to operate auxiliary equipment, such as fuel pumps for pumping diesel fuel from undercar tanks to the dispensing tank under the ceiling of the engine compartment and air compressors (diesel engines on those sections use compressed air start-up). It also allows the car battery to charge while the car is idle for a long time. When the car is in motion, the battery is automatically charged by the axle generator. The small generator is sometimes used when load conditions require only air circulation or ventilation. In this case, small generator capacity is sufficient to run all blower fans in an entire section simultaneously, without using the main diesel generator. Since the diesel fuel consumption for the main generator is much greater than that of the small diesel, this also saves fuel.

The central control panel located next to the engine compartment of the special car allows full control of the refrigeration/heating equipment operation. As five-car sections are operated by a crew (which usually consists of two or three mechanics), the central control panel can be used for remote manual equipment control, overriding automatic operation, in case of the automatic or equipment failure or for reducing the fuel consumption.

The rest of the special car comprises the crew's living quarters, with sleeping compartments, kitchen, bathroom with shower, and dining area. Heat is provided by the boiler (which works on diesel fuel) and hydronic baseboards.

All refrigerator cars have passenger-type trucks for softer rides. Car bodies are completely insulated (except load car engine compartments), with polyurethane foam. Floor racks are provided for better air circulation.

The air-circulation system used on refrigerator cars built in East Germany is a combination of envelope and open types. After passing through the evaporator/heater, conditioned air is discharged by the blowers into a ceiling duct; it then moves to the continuous wall ducts on both sides of the load compartment and enters the load compartment through the floor racks. After passing through the load from bottom to top, air is drawn back to the evaporator. This system, together with the unit's location at both ends of the car, contributes to equal air and temperature distribution throughout the car load.

REFERENCES

AAR. 1972. *Manual of standard and recommended practice.* AAR *Standard* S-2010 Association of American Railroads, Mechanical Division, Washington, D.C.

AAR. *Supplement to the manual of standard and recommended practice.* Association of American Railroads, Mechanical Division, Washington, D.C.

Code of rules for handling perishable freight. *Circular* 40. National Perishable Freight Committee, Chicago.

Doean, J.J. 1983. *Perishable protective tariff* No. 619-B. National Perishable Freight Committee, Chicago.

The car and locomotive cyclopedia. 1984. 5th ed. Simmon-Boardman Books, Inc., Omaha, NE.

The official railway equipment register. R.E.R. Publishing Corp., New York.

USDA. 1969. Protection of rail shipments of fruits and vegetables. USDA *Agricultural Handbook* No. 195, July.

CHAPTER 30

MARINE REFRIGERATION

CARGO REFRIGERATION 30.1	Storage Areas 30.09
Built-In Refrigerators 30.1	Ship Refrigerator Design 30.10
Refrigeration System 30.3	FISHING VESSELS 30.11
Distribution of Refrigeration 30.4	Refrigeration System Design 30.11
Refrigeration Load 30.6	Refrigeration with Ice 30.14
Container Vans 30.7	Refrigeration with Sea Water 30.14
SHIP'S REFRIGERATED STORES 30.8	Tuna Seiners 30.15
Commodities 30.8	

CARGO REFRIGERATION

MARINE transport is an interim operation between preshipment storage of indeterminate duration and early distribution at destination ports. Frequently, the marine transport period is equal to or even exceeds the full high-quality life of the perishables being transported. Good design, therefore, requires that shoreside criteria be applied to the floating cold storage plant.

The increased use of various sizes of cargo containers has affected savings, mostly by providing faster unloading and reloading of the vessels. Most refrigerated cargo can be containerized. The need for controlled atmospheric environments for perishable produce imposes additional limitations over those for dry cargo.

Containers will never completely supplant the ship's built-in cargo refrigerator for such services as those provided by passenger ships, logistic supply ships, all-refrigerated fruit carriers, and special service vessels; these services will continue to require the insulation and refrigeration of the principal structural compartments of a ship. The physical and mechanical aspects of both containerized services, as well as the break-bulk (or individual package) type of operation are discussed in this section.

BUILT-IN REFRIGERATORS

The location and arrangement of insulated compartments and compartment subdivisions within the hull should reflect the trade in which the vessel will serve. The volume of trade, the scheduling of ports of call, and the efficient and speedy handling of produce with minimum exposures are all factors that will influence these arrangements. Perishables should be the last cargo loaded and the first to be discharged.

Arrangement and Utility

Limitations of dimension and arrangement are due to the ship's structure, compliance with floodability compartmentation of the hull, and the fire-resistant regulations to which many ships are subject.

The refrigerators should not be designed exclusively for high-temperature cargo services unless it is certain that the vessel will always remain in that limited trade. Otherwise, all compartments should be readily convertible to any temperature service from below -20 to $60\,°F$, thus allowing flexibility in segregating cargo according to temperature requirements and ports of call.

The preparation of this chapter is assigned to TC 10.6, Transport Refrigeration.

In a vessel in which only a portion is refrigerated and designed for late loading and early discharge, the compartments are usually located under the topmost deck to the hull. The naval architect considers the effect on trim of the vessel in normal loading conditions. Generally, the added weights of the insulation and the perishable cargo are not as influential as the weight of dry cargoes, which are usually of greater density. Often these compartments are empty or lightly loaded during some of the voyage. Trim conditions generally require the refrigerators to be placed forward of the midlength of the vessel or grouped about the middle section, where the moment arms for trim are of lesser consequence. Almost without exception, refrigerators are arranged symmetrically about the ship's longitudinal centerline.

The rooms or areas where the refrigerating control valves are installed should be arranged and located so that the apparatus or controls are accessible to the operating personnel through trunks or passageways at all times. When the boundary bulkheads parallel hatches, they should clear the openings by 3 ft as a safety measure, providing adequate room for handling hatch beams and covers.

The greater the number of subdivisions in the refrigerated compartments, the greater the loss to the ship's revenue spaces, due to the volumes occupied by insulated partitions, cooling apparatus, insulated piping, and access areas. On this basis, the all-refrigerated ship, with only the main structural boundaries insulated, make the most efficient use of a ship's costly enclosures. On the other hand, the conventional all-refrigerated ship, with insulated hatch plugs at the weather deck, has several undesirable features, such as the difficulty of providing uniform compartment temperatures and of refrigerating hatch areas far from the refrigeration apparatus.

With large-volume rooms, the extended period in which the hatch is open to the outside atmosphere for loading and discharge is not conducive to a good environment for frozen or fresh cargoes. Seldom are frozen goods sufficiently cooled to tolerate an extended loading interval. Fresh products often are not precooled before loading, and the accumulation and delayed extraction of respiratory heat in the compartment is detrimental. During long discharge intervals in ports, atmospheric moisture causes sweating or frosting of products, some of which may suffer from this exposure. With partial delivery at a port of call, such exposure may unfavorably affect the remaining cargo. Also, harbor workers may refuse to work in spaces with the refrigerating equipment in operation. Air curtains help substantially to keep heat from entering the refrigerated area.

A more suitable arrangement of the all-refrigerated ship can be made by fitting the insulated compartments in the 'tween decks with access doors in hatch-side bulkheads, and by means of various combinations of insulated plug hatches at the lower hold and 'tween deck levels. Hatch areas may be used for general cargo or treated as separately refrigerated compartments. Many ships are fitted with side port doors at the 'tween deck level, through which cargo is handled by conveyors which are independent of the ship's overall cargo gear.

The central refrigeration machinery plant should be located in, or immediately adjacent to, the main propulsion machinery room, where the ship's responsible watch officers are in constant attendance. A central location for the refrigeration machinery usually results in economy of space and close connections for power and pumping facilities.

Insulation and Construction

Insulation. Moisture-vapor and water-resistant insulation is of particular importance on board ship because of frequent and extreme temperature cycles due to intermittent refrigeration. On termination of refrigeration at discharging ports, insulation will be at lower temperatures than the open room, and often the room surfaces stand dripping wet with atmospheric moisture, which enters through the open door or hatch. Both the warm and the cold side should be moisture-sealed equally, and cold-side breather ports are not recommended. Other common sources of water in ships' refrigerators are melting of ice (packed with vegetables) and defrosting of cool surfaces.

Severe service conditions, which subject the insulation to injury or change by mechanical damage or vibration, and intermittent refrigeration, place exacting requirements on insulation for ships' refrigerators. The ideal shipboard composite insulation should have the following characteristics:

- High insulating value
- Imperviousness to moisture from any source
- Light in weight
- Flexibility and resilience to accommodate ships' stresses and loading
- Good structural strength
- Resistance to infiltrating air
- Resistance to disintegration or deterioration
- Fire resistance or fireproof self-extinguishing requirement
- Odorless
- Not conducive to harboring rodents or vermin
- Reasonable installation cost
- Workability in construction

In the United States, the properties of the insulator and the details of construction must meet approval by the U.S. Coast Guard and the U.S. Public Health Service. For information on insulation materials and moisture barriers, see Chapters 20 and 22 of the 1989 ASHRAE *Handbook—Fundamentals*.

Construction. The three principal parts to the refrigerator boundary include the envelope (or basic structure), the insulating material, and the room lining.

The envelope is usually partly composed of the ship's hull, the watertight decks, or watertight main bulkheads whose members resist the entry of vapor from the warm side. The inboard boundaries outlining the refrigerators should have an equal ability to resist moisture. A continuous steel internal bulkhead with lap seams and welded stiffeners provides a boundary of adequate strength and tightness. Details of design may accommodate dimensioned insulators or facilitate means of fastening these materials. Doorway main bucks of steel channel provide good structure but are usually a source of sweating on low-temperature rooms because of heat gain through the metal. Wooden door bucks minimize sweating but are a retreat from efforts to eliminate concealed wooden structure.

Partitional bulkheads may be of similar detail, but airtight sealing is less important. Some installations are framed with angle-bar grids, between which the insulator is installed. In passenger vessels (over 12 passengers), Coast Guard regulations governing fire-resistant construction restrict wood assembly. Under no circumstances should wood be a part of the floor assembly since it deteriorates rapidly under the prevailing conditions.

The assembled boundary of a ship's refrigerator must withstand heavy floor loads and several wall thrusts of cargo when the vessel rolls or pitches in a heavy sea; it must be able to flex with the hull structure being stressed in any angle. The assembly must resist vibration caused by the propelling machinery, the sea, and the careless handling of cargo. The vapor seal of all surfaces must remain intact.

Only in extreme cases should voids in the insulation assembly be concealed. Filling such volumes with insulating material is cheaper and more effective than constructing internal framing. The exceptions to this rule are in the deep volumes formed by bilge brackets, deck brackets, and open box girders. Solid filling results in more insulation thickness than is needed for a heat barrier in the overhead and the ship's side where beams and frames are deep.

The third part of the boundary assembly is the room lining. This surface must be sturdy enough to withstand the impact of frequent cargo loading and handling. On passenger vessels, U.S. Coast Guard regulations require that the lining be fire-resistant. Tongue-and-groove lumber is considered obsolete for any ship, and other linings are employed. On freight vessels where wood is permissible, exterior grade plywood is sometimes applied. A few installations have been made either of laminated plastic sheets or wood fiber hardboard; both are satisfactory when properly supported. Steel linings are costly, impractical, and difficult to maintain and repair. The favored lining is the cement and fire resistant fiber hardboard panel with aluminum sandwich lamination. When the insulator is secured with adhesives containing volatile solvents, the aluminum laminations should be of perforated metal or mesh.

The U.S. Public Health Service requires all linings to be rat-proofed. Vulnerable linings should have an underlay of 1/4 in., 16-gauge galvanized wire mesh for rat-proofing.

Applying Insulation

In the application of panels with adhesives over block insulators, the butted joints should be separated sufficiently for the adhesive to extrude, provide a moisture-sealed joint, and accommodate movement of the panels by flexing of the ship's structure. The joints may be covered by cargo battens, which are also secured by adhesive and with brass screws. The walls of all refrigerators should be fitted with vertical cargo battens on 15 to 18 in. centers to hold the cargo clear of the insulated wall. This spacing permits circulation of air and prevents the contacting package from assuming the part of a room insulator.

Container vessels of cellular-type construction move the containers along guiding columns; the boundaries of the hold, the hatch coamings excepted, are not subject to mechanical damage. Here the insulation may be of the simplest, low-cost, rat-proofed form without fitted hard panels or cargo battens.

Urethane foam floor insulation requires a restricting surface to confine and distribute the expanding material. In one method, plywood is secured over foam spacer blocks and the material is injected into the space through properly spaced holes. The membrane and the wearing surface covering are laid over the plywood. But wood in the floor is not good practice, nor is the wood a suitable base for a heavy duty wearing surface material. An alternate method calls for laying expanded board or blocks in an approved adhesive. The use of the more resilient corkboard laid in adhesives may be good practice in some applications.

Marine Refrigeration

The greatest weakness in ship refrigerator construction is the floor covering. Research and practice have not brought forth a totally satisfactory covering. Unlike a warehouse, a tightly-packed cargo refrigerator must have floor gratings both to assure the circulation of air and to protect the bottom tier of products from heat leakage. The grating supports carry the cargo weight to concentrated load areas of the floor. The room may be filled with warm general cargo in alternate service, and a thermoplastic covering may be punctured by the supports.

The floor covering must be flexible enough to withstand the flexing of a ship or extreme temperature fluctuations, as well as maintain a moistureproof cover over the insulation. The most satisfactory material is a mastic composed of emulsified asphalt, sand, and cement. This material is applied cold; on setting, it has good load-bearing qualities, is impervious to water, has a small degree of ductility, and may be used in less thickness than concrete. It should always be reinforced, and expansion joints should be included to accommodate shrinkage and adjustment to movements of the ship.

All rigid or semirigid floor coverings should have rubber-base composition expansion joints capable of bonding to the edges of the floor slabs or wall and not subject to shrinking from age. The expansion joints should trace the periphery of the room, the line of all underdeck girder systems, bulkhead offsets to pillars, and similar lines of anticipated ship stress.

Water from ice-packed vegatables and defrost water draining onto the floor covering requires it to be impervious to moisture in the slab as well as the joints. If the floor insulation is wetted, it will deteriorate, lose its efficiency as an insulator, and give off odors. If wet floor insulation freezes, it will lift the deck covering and destroy it. If water penetrates to the steel, the ships' structure will corrode unnoticed.

All floor coverings will crack or become damaged during the ship's life. As a secondary security against water, a membrane should be laid between the insulator and the floor covering. The membrane need not be flashed to the wall if suitable sealing expansion joints are fitted at the juncture of the wall panel and the floor covering.

When an attempt is made to seal insulation with waterproof paper, it should be applied in double thickness, and the laps and perforations should be cement-sealed. Wall and ceiling panels attached with suitable adhesives do not require waterproof paper inner linings.

Floors and Doors

The floor is the weakest point in the ship's refrigerator. The weakest element in the floor is the floor drain or scupper because of the difficulty in bonding the floor covering with the metal drain fitting. Water will often find its way between the covering and the insulation. The conventional floor drain is fitted with a perforated plate flush with the floor covering and hidden by the floor gratings. If the perforations become clogged by debris, water will accumulate at the scupper. A drain fitting near a wall or corner will, on most occasions, create a weak section in the floor covering and will develop cracks running to the wall or across the corner. Figure 1 shows a satisfactory scupper fitting. It is flush with the top of the gratings, has a liftout cover for easy cleaning, and is bonded to the floor covering with expansion joint material. Drains between decks should be omitted wherever practicable.

Refrigerator doors are generally a manufactured product. They should have generously designed steel hardware, and the door and frame should be metal sheathed and have a flat sill and double gasket. Very large or double doors should have additonal dogs to assure proper sealing when closed.

Sliding doors should be installed wherever possible to reduce interferences and conserve adjacent revenue space. When used, brackets should be installed to support portable horizontal spars inside the doors to prevent cargo from falling against the doors in a seaway. Molded glass fiber swinging doors insulated with urethane foams poured in place are available. They are strong and lightweight, are easily handled, and can be fitted with lightweight hardware.

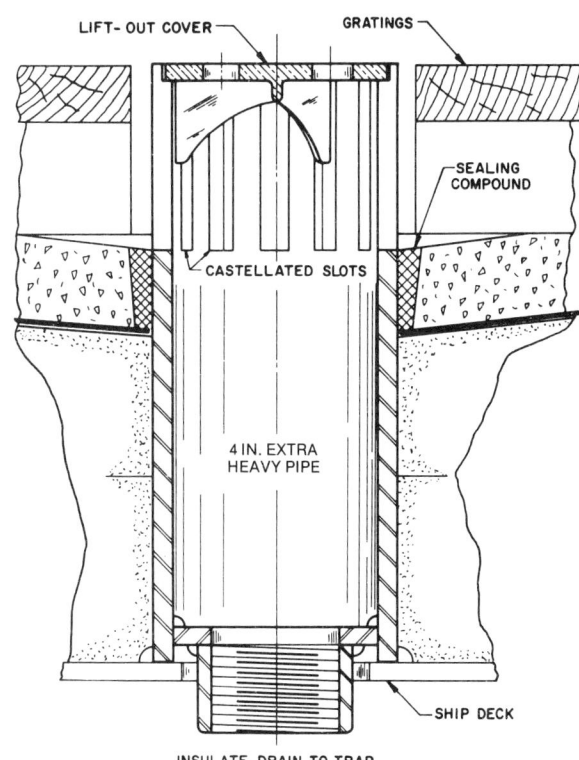

Fig. 1 Floor Drain Fitting

In finishing, wooden surfaces should be varnished rather than shellacked, since the latter material has little protective penetration. Manufactured nonmetallic surfaced materials may be painted or varnished, but if they are nonhygroscopic, their original surface will usually present a good appearance for longer than a painted coating.

Low-temperature apparatus or piping should be inspected carefully during installation. All joints and surfaces should be generously sealed to keep out atmospheric moisture. Special attention should be given to pipe covering ends, valves, and bulkhead penetrations. On subzero services, special composition adhesives should be used. The smallest omission or breach of a seal will allow progressive destruction of the covering.

On shipboard, where piping systems are relatively short, the function of the insulation is more that of preventing sweating or frosting of cold surfaces than preventing heat gain.

REFRIGERATION SYSTEM

Refrigerants 12 and 22 are used in reciprocating compressors, and Refrigerant 11 is used in centrifugal units. R134a is a potential substitute for R12, and R123 for R11. These substitutes and R22 have little or no ozone depletion potential. Refrigerant 22 has temperature-pressure characteristics similar to ammonia, but its use in direct expansion (DX) systems requires special attention to applications issues, such as discharge temperature limits, choice of oil, crankcase heater use, and oil returns. It is suitable for close-connected brine cooling systems under all conditions. Ammonia is widely used on fishing and seafood processing vessels. This section describes halogenated hydrocarbon systems only.

Compressors

Modern ships are outfitted with refrigerators with a minimum design temperature of −20 °F. This range may be reached without excessive compression ratios with R-12 units in compound compression or brine cascading. Centrifugal, reciprocating, and screw compressors are used. Chapter 12 of the 1988 ASHRAE *Handbook—Equipment* describes compressors in detail.

Air and airborne moisture leakage into a refrigerating system is minimized when the low side is maintained at or above atmospheric pressure. However, centrifugal compressors using R-11 and operating at pressures below atmosphere have little trouble from this source because of the effectiveness of shaft seals characteristic of rotative machinery and the continuously operating purging units auxiliary to this type of compressor. The use of brine circulation is mandatory for centrifugal systems using R-11.

Reciprocating R-12 compressors should not be operated in parallel because of the oil's tendency to leave the crankcase with the refrigerant flow and return to flood one unit and starve another. Consequently, each direct expansion evaporator system should be served by its own compressor. Multicompressor units in multiple-evaporator direct-expansion plants are undesirable; paralleling brine systems with fewer condensing units are better.

Capacity control of compressors may be affected by speed variations, automatic cylinder cutouts, unloaders, or bypasses. Intermittent operation or cycling should be reduced to a minimum by speed regulation or unit sizing to provide maximum uniformity in plant conditions.

Centrifugal compressors are essentially large-capacity units with capacities varied by speed controls, bypassing, throttling, or metering of refrigerant. The condenser pressure is controlled by circulating water flow.

Reciprocating units are V-belt driven or direct-connected to an electric motor. Centrifugal compressors may be driven by a step-up gear and motor or by steam turbine direct drive. A steam turbine reduces electric generator demands and often fits well in the main steam plant heat balance.

Compressors should be installed in a fore-and-aft centerline position to reduce the effect of gyroscopic bearing loads when the ship rolls. This position also favors lubrication of the unit.

Reserve compressor capacity and spare part lists are specified by the American Bureau of Shipping. This agency recommends two condensing units, one of which, running continuously, should be capable of maintaining full cargo lading in tropical waters. The rules further state that when refrigerated spaces are of 15,000 ft^3 total capacity or less, consideration should be given to the installation of a single condensing unit with adequate spare parts. An experienced ship operator would consider numerical machine reserves necessary as reserve in refrigeration capacity and would no doubt exceed these minimums. Table 1 suggests reserve capacity of various installations.

**Table 1
Operating and Reserve Capacities of Condensing Units**

No. of Units, 100% Load	Additional or Reserve Unit, %	Total No. of Units
1	100	2
2	50	2
3	33⅓	4
4	25	5
5 or more	20	6 or more

Condensers and Coolers

Halocarbon condensers are of the conventional shell-and-tube design and, since these refrigerants are noncorrosive to all commercial metals commonly used in refrigeration, a wide variety of materials is available. High heat-transfer characteristics and resistance to corrosion/erosion by circulating sea water should be considered in selection and design.

The refrigerant is usually between the tubes and shell of the condenser, and sea water circulates through the tubes in a multiple-pass arrangement. Steel is usually used for shells with cupronickel tubes and cupronickel tube sheets. Bronze water boxes are preferred. If used, cast iron water boxes should have zinc anodes and be coated with a protective material such as epoxy. The tendency to substitute high velocities of circulating water for cooling surface in heat exchange should be resisted to provide long life. Water velocity of 6 fps in the tubes is considered high, and water box velocities above 2 fps are conducive to turbulence and accelerated erosion of tubes. Cooling surfaces are designed for maximum conditions.

Shell-and-tube condensers need double drains if installed level; vertical outlet lines long enough to ensure a liquid seal at the expected angle of list or trim are essential. A single outlet may be used if the condenser is inclined at an angle greater than the expected angle of list or trim. Horizontal receivers must be drained in a similar manner. Vertical receivers with a single outlet are also an option. Receivers may be constructed of welded steel and should be arranged to assure submergence of the liquid outlet under all sea conditions. The receiver should have sufficient capacity to hold a complete charge of the system, plus 20% reserve volume. Means should be provided for operators to observe the liquid level.

Brine cooler specifications are similar to those of the condensers except that the materials must resist the corrosive effects of brine. Steel tubes and tube sheets are generally used. In conventional design, the refrigerant circulates through the tubes to reduce the refrigerant charge to a minimum and provides maximum wetting of cooling surfaces. Large coolers may have two or more independent tube groups in parallel, each with its own expansion valve. Multiple parallel expansion results in improved flexibility, increased reliability, and better control at partial loads.

DISTRIBUTION OF REFRIGERATION

Refrigeration may be distributed by direct-expansion (DX) systems or by brine as the secondary refrigerant. Because the refrigerated compartments are located far from the machinery, direct-expansion systems have many shortcomings. The extended return lines require a large total pressure drop to produce high enough velocities at low loads to return the oil to the machinery. A full load on the same line will reduce the capacity of the compressor cylinders because of the resultant rarefied vapor.

When the evaporators or the return lines in DX systems are lower than the compressor suction, oil traps are possible; these may, under certain conditions, cause alternate starving and slugging of the compressors. Small dual risers (the bottom one of which forms a trap) provide a suitable oil lift when the vapor velocity in vertical return lines falls below 1500 fpm when one compressor is operating. The long low-side lines are generally concealed by insulation or inaccessible for repair at sea or when the ship is loaded. The cargo may be seriously damaged if vibrations, the flexing of the ship, or damages cause the lines to develop leaks. In addition, the extended systems require large charges of refrigerant with many more scattered points where moisture or air could enter the system.

Brine has many advantages over DX systems. The full charge of the primary system of such a plant is small in quantity, and, in case of accident to the piping or the components, escaping refrigerant is confined to the refrigerating machinery room. The short runs of piping result in small total pressure drops and permit an efficient layout with fewer compressor units. This facilitates the return of oil to the compressors.

Brine cushions abrupt thermal changes and eliminates the unwanted sensitivity of control that is characteristic of DX latent heat

Marine Refrigeration

systems. Even though the primary system should be automatically controlled, excellent results in cargo space conditions are obtainable with manual control of return brine flow. Brine systems provide great flexibility in plant operations by the capabilities of compounding or sharing refrigeration loads. Total centralized control is feasible.

Space Cooling

The refrigerators in ships are tightly stowed, without aisles or clearances, and the method of stowage has more influence on cooling than the design of the refrigeration equipment. Marine refrigerators are of countless variations in size, dimension, proportion, and configuration. As a result, conventional tests have little significance as guides to operations.

Because cargoes are so tightly stowed, wall coil refrigeration or gravity air circulation is obsolete. Circulating air removes heat from the cargo, but does accelerate dehydration of vulnerable cargo. A long air path requires a higher velocity airstream than a short path for the same air temperature rise.

One method of refrigeration, used often in banana and other fruit carriers, uses external coil bunkers and fan systems to circulate chilled air through ducts and ported false bulkheads. Typically, brine-cooled serpentine coils of pipe have been used. This system supplies air to one side of the compartment and fans exhaust it from the opposite side. Some installations permit reversal of the direction of the airflow. Refrigeration effectively reaches the entire volume if it is properly stowed and the ports are correctly adjusted. The usual distribution results in low air velocities and relatively low moisture pickup. European builders often employ vertical airstreams between floor and ceiling plenums, but this system has not been adopted by American operators.

Cooling Units—Diffusers

These units are composed of three sections: a bottom air inlet, the middle encasing the cooling coils, and the top housing the motor-driven fans, which are usually in multiple arrangement on a common shaft. Large units may be arranged horizontally in similar sequence. Diffusers are packaged units that are easily and cheaply installed. These units are fitted with distributing ducts that direct the air along the ceiling and down the bulkheads between the cargo battens.

The cooling coils are, in some cases, small tube DX evaporators with closely spaced fins. Each vertically arranged tube pass is fed liquid from the distributing head of a thermoexpansion valve. The outlet consists of a connecting manifold at the base of the coils. When air flows horizontally, the evaporator tubes are horizontal and the connecting manifold vertical. Other types employ prime surface tubes or coils cooled by DX or brine.

The manufacturers of the units will usually specify cooling performance as refrigeration capacity divided by the air-to-refrigerant temperature difference. The ratio provides a ready comparison of competitive designs or proposals.

Cooling units or diffusers are installed behind guards against which the cargo may be piled and behind which access is provided for maintenance. The cargo space lost to these units and their associated ducts increases proportionately as room volume becomes smaller. Fans should have output adjustment in steps to half-rated speed, and the motors should be watertight.

Defrosting Systems

Most marine refrigerator cooling surfaces operate below 32°F, so they rapidly accumulate frost. Prime surface coils require less frequent defrosting than do extended surfaces of equivalent capacity rating. The problem becomes most severe during initial cooling. Maintaining a small temperature difference between the evaporator and room air is the principal means of providing desirable high humidities for fresh perishable cargo.

All cooling surfaces should have means to melt accumulated frost and remove the resulting water. Direct-expansion systems, if not too far from the compressors, may be defrosted by hot gas or, alternatively, with warm water spray. Because hot gas defrosting is complex, most plants defrost with heated fresh or sea water. When using heated water, its temperature should not exceed 90°F to avoid high refrigerant pressures in the evaporator.

Brine-cooled surfaces may be defrosted by passing heated brine through the coils. A steam-heated, thermostatically controlled brine heater may be installed in a central location for the purpose.

In some long voyage chill services, such as for bananas, citrus fruits, and apples, replacement air is necessary to remove excessive quantities of carbon dioxide and other gases. Continuous airflow maintains uniform conditions within the room and is preferred over intermittent airflow. Outside air should first pass through the cooling unit before distribution.

Citrus fruits should have replacement air at the rate of 2 to 3% per minute based on the gross volume of the refrigerator; bananas require as much as 5% replacement air. When replacement ventilation is provided, accessible, efficient, and well-insulated dampers should be fitted to cut off this air supply when frozen food is carried.

Plant Layout and Piping

Plant layout aboard ship should be as simple as possible without sacrificing reliability. In addition, the machinery plant should be closely associated with the main power plant to provide short piping and power connections and facilitate close supervision.

The personnel changes of engineers and refrigeration crew members demand that strangers to the installation be able, on short notice, to trace well-labeled systems and place the plant in operation or maintain it without undue hazards to the machinery or cargo. Even at the expense of valuable revenue space, an uncrowded machinery room should be provided that will give ample space for operations, maintenance, and the repair of apparatus or the ships's structure, and proper clearances should be allowed for the insulation of low-temperature parts.

All machinery should have sturdy foundations, and all parts should be secured against vibrations set up either by themselves or the main propulsion plant. High-speed machinery should be mounted on fore-and-aft centerlines, and all gravity feeds, drains, and tanks should be designed or installed with full consideration of the effects of trim, roll, or pitch of the vessel.

Each refrigeration system must work independently of other systems in the primary refrigerant side. Table 1 lists the recommended number of units for a plant. The entire plant should be cross-connected to back up the load from any system normally assigned to a specific use.

Multiple compartmentalized refrigerator arrangements operating with DX can become a complex central plant, particularly with hot gas defrosting. Remotely located evaporators result in decentralized control with apparatus scattered over the ship.

Halocarbon return piping may be designed on the basis of a 2°F total temperature drop between the evaporator and the compressor. Oil trapping should be avoided. Minimum vapor velocities in horizontal lines of 700 to 800 fpm are recommended for oil return flow. The liquid refrigerant should be subcooled immediately after it leaves the receiver, because flashing is easier to prevent than recondensation. This requires placing the liquid-suction heat interchanger close to the condenser and receiver.

Liquid lines should be arranged to prevent line flashing either from high temperature or from excessive pressure drop in the static head of risers. In situations that require a high lift, a refrigerant-cooled subcooler may be needed in place of a liquid-suction heat interchanger. Liquid lines should be fitted with efficient driers and strainers.

Two-Temperature Brine Plant

Brine refrigerant plants require a condensing unit for each evaporator, but the brine system may be cross connected. Most marine plants operate from subzero to ambient temperatures with any or all compartments being interchangeable. A two-temperature brine plant allows high-temperature brine to refrigerate fresh products with minimum desiccation effect, and low-temperature brine to refrigerate frozen cargoes (see Figure 2). A two-temperature brine plant suggests two circulating piping systems served by two pumps and a standby pump available for either system. The desired brine can be selected by setting valves at the refrigerators.

Brine returns from each cooling unit should run to the central plant return-valve twin manifold where room temperatures may be remotely controlled by brine flow. By controlling return flow, the systems are kept under pressure, and air binding will be at a minimum. The returning brine is next diverted to its respective collector tank. Two types of systems are in use: the *open system*, in which the return brine streams are in view of the operator; and the *closed return system*, in which a static head is maintained in the system by a vented overhead gravity tank. Brine systems use a refrigerant other than R-12.

A simple two-temperature brine system can be made by including either a closed high-temperature brine loop or a cooling unit with a recirculating pump. The selective brine temperature is maintained by a thermostatically operated pilot valve which feeds low-temperature brine to the loop or cooler as required. The central plant thus becomes a one-temperature installation.

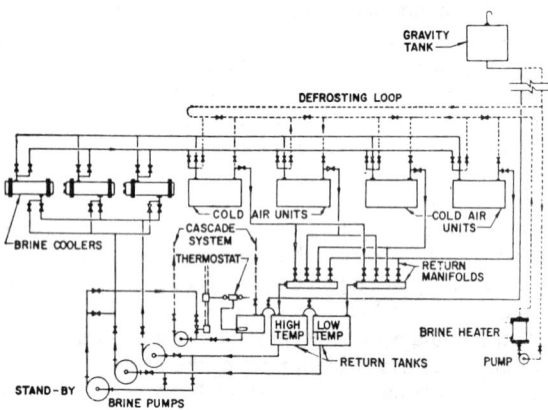

Return tanks may be omitted by vending through system gravity tank.

Fig. 2 Two-Temperature Brine System

With the use of a secondary refrigerant such as brine, an additional temperature differential results between the primary evaporating refrigerant and the compartment air temperature. Cascading of an R-12 refrigerant plant is desirable for room temperature of $-20\,°F$. In this operation, the high-temperature brine circulates through the condenser in lieu of sea water and reduces head pressure and compression ratios over those of single-stage compression.

For defrosting with hot brine, an independent small pipe loop having its own pump and heater is suggested. In this operation, the cold brine valves are closed at the refrigerator, and the cooling coils are placed in series with the pump and heater. Only the brine content of the coil is displaced; the entire pipeline is not heated.

Thermometers

The thermometer is the principal indicator of how a refrigerator plant is functioning. Room temperature depends on the proper placement of the sensing bulb within that space. In rooms operating with wide variation below or above the freezing temperature of the product, the bulb can be placed almost anywhere, but in rooms operating only slightly above the commodity's freezing temperature, care must be taken to place the bulb where it will read the temperature accurately and avoid freezing. The bulb should not be placed in direct contact with warm-side surfaces or refrigerating equipment. Rooms with cold air units should have the bulbs located in the supply airstream, that being the point of critical temperatures for chill cargo. When 32 to 33 °F supply air is distributed, the heat of the load is removed as rapidly as possible with no danger of freezing the product.

Recording thermometers are essential to proper operation and management of cargo refrigerators. Indicating thermometers should have the dial installed outside and near the access doors in each room. All bulbs used for comparative readings should be installed adjacent to each other. Mercury thermometers will, because of their design, be near the refrigerator.

Indicating thermometers inserted in walls through the insulation are not satisfactorily accurate or representative. Indicating thermometers should have the dial installed externally and near the access doors in each room. All bulbs used for comparative readings should be installed adjacent to each other. Switch-operated, electric resistance indicating thermometers should be installed with the dial located at the log desk.

Recording thermometers are essential to proper operation and management of cargo refrigerators. Mercury actuated instruments will, because of system limitations, have the instrument near the refrigerator. Electric resistance recording instruments are best located in the machinery room. In some cases, entire cargoes are electronically monitored and controlled from a central location.

REFRIGERATION LOAD

Specifications

The refrigeration specifications should set forth the extreme operating conditions of loading, ambient and sea temperatures, and rates of pulldown. In the all-purpose installation, each compartment should be designed for the refrigeration of warm, fresh products from the field or orchard; for the overall condition, a percentage division of chill and freezer cargo with simultaneous and total loading should be stated.

Typical conditions include the following:

1. Arrangement and net volume capacities of the refrigerated compartments
2. Thicknesses and kinds of insulation
3. Ambient temperatures
   Weather surfaces — 100 °F
   Adjacent machinery spaces — 100 °F
   Other adjacent spaces — 85 °F
   Sea temperatures — 85 °F
4. Overall stowage factor — 70 ft^3/ton
5. Percentage total loading as chill — 75%
   Percentage total loading as freezer — 25%
6. Receiving temperature, chill cargo — 80 °F
   Receiving temperature, frozen cargo — 25 °F
7. Carrying temperature, chill cargo — 34 °F
   Carrying temperature, frozen cargo — 0 °F
8. Initial period of cargo heat removal (equivalent) — 72 h
9. Replacement air at 85 °F db, 75 °F wb — 3%

Marine Refrigeration

The owner should describe the kind of refrigeration system to be installed and specify the number of compressors and other auxiliary parts or apparatus together with sources of emergency pumping and water facilities. All equipment and installations must be specified as complying with the rules and regulations of the Classification Societies (American Bureau of Shipping, Lloyd's Register, and others), the U.S. Coast Guard, the U.S. Public Health Service, ASHRAE *Standard* 15-1978, *Safety Code for Mechanical Refrigeration*, and ASHRAE *Standard* 26-1985, *Mechanical Refrigeration Installations on Shipboard*.

The specification writer should specify machinery performance only, unless the writer wishes to assume full responsibility for the functioning of the plant and all its parts. The specifier should, however, obtain from the vendors full descriptions, details, capacities, and specifications of the equipment proposed to permit comparative analysis.

Completion tests should be required to determine workmanship and functional performance. Performance guarantees should cover operations under loaded service conditions.

Calculations

The following method of refrigeration load calculation for a general service plant carrying heterogeneous chill cargo may appear simplified, but it is justified by the great range of conditions common to marine installations. Refrigeration loads for freezer cargo may be calculated in a similar manner using the same stowage factor, a specific heat of 0.40 Btu/lb·°F, and an equivalent pulldown period of 72 h. The loads for respiration heat, replacement air, or latent heat of fusion will not be present.

Specialized service in known ambient conditions may be calculated more precisely, but an arbitrary 10% margin should still be added to the results to compensate for aging and unforeseen heat gains.

With general calculations, the following operating conditions should be assumed:

1. *Weather ambient conditions*: Up to 100°F.
2. *Ambient sea conditions*: Up to 85°F.
3. *Conductivity of insulation*: According to standards given for the material, urethane foam with an installed k-value of 0.20 Btu·in/h·ft^2·°F is suggested.
4. *Resistivity of outer boundaries, inner linings, and surface films*: These factors should be ignored because boundaries and linings are usually dense and have high conductivity values.
5. *Infiltration and open door leakage*: For cargo refrigeration installations, such losses at sea are nil. Port exposures during loading and discharge reestablish pulldown conditions.
6. *Ventilation or replacement air*: This factor is often omitted when carrying heterogeneous cargo for short-to-medium length voyages. For specialized service, it may be as much as 300% of the gross room volume per hour.
7. *Electrical energy conversion*: The energy load from fans and brine pumps will be on demand load rather than connected load. An arbitrary value of 3000 Btu/h per brake horsepower may be used.
8. *Product load*: This factor ranges widely for heterogeneous cargo. Its density will vary from 40 to 120 ft^3/ton and its specific heat will vary from 0.22 to 0.95 Btu/lb·°F. It is recommended that the gross weight of the package and the specific heat of the product be used. An average density of 70 ft^3/ton may be used.
9. *Receiving temperature of cargo*: Chill cargo will range from carrying temperature to ambient. Frozen cargo will range from −20 to 28°F.
10. *Carrying temperature of cargo*: Ranges from 32 to 55°F for chill cargo, and at 0 to −20°F for frozen foods.
11. *Respiratory heat of chill cargo*: Meat products, eggs, and dairy products have no respiratory heat. Chapter 30 in the 1989 ASHRAE *Handbook—Fundamentals* lists heat of respiration of many horticultural products at various storage temperatures.

CONTAINER VANS

The early programs for marine refrigerated containerization borrowed from developments in land transport, particularly the highway vehicular concept. Currently called the conventional container, the shortcomings of its precursor have been inherited along with the merits. Highway transport has developed on the premise of an interim transport of relatively short duration between source and delivery by a single carrier. For fresh horticultural products, initial transport is assumed to occur when the products are hardiest and most likely to endure poor environments. These influences are reflected in the conventional marine container. However, the use of interchangeable (or intermodal) transport of containerized products demanding excellent conditions is increasing. For the usual interval of an overseas voyage, which is often added to storage and land haul, good cold storage is essential.

Standards and Assemblies

As surface transport equipment often passes through many countries, international container standards for overall dimensions, fastenings (lashings), handling, and lifting has been published as the *U.S.A. Standard Specifications for Containers* (USASI MH 5.1), sponsored by The American Materials Handling Society, Inc. and The American Society of Mechanical Engineers. The American Bureau of Shipping has issued a *Guide for Certification of Cargo Containers*. Lloyd's Register of Shipping has published rules covering refrigerated cargo containers.

Universal basic standards of dimension are 8 ft wide and 8 ft high. Except for width, which conforms to highway regulations, ship operators have disregarded basic standards, adapting different lengths and heights best suited to a particular operation. These variations make cargo transfers difficult if not impossible. A disregard of universal standards can result in rehandling, with hazardous exposures of sensitive perishable products, at interchange stations.

Most containers are 8.5 ft high, 8 ft wide, and either 20 or 40 ft long. Some 8, 9, or 9.5-ft high containers are also in use. A demand for 40-ft general service containers continues as adequate port terminal facilities are made available and ships' holds are outfitted to accommodate them. However, the problems of air circulation within a conventional refrigerated container rapidly accelerate with increasing length.

True intermodality is further frustrated by the large number of attachable refrigerating machines, all of which are made incompatible by varying characteristics and dimensions. European standards, which are based on centralized marine systems, have reduced this problem. These systems are served by a ship's refrigeration plant and external attachable refrigeration for land haul connects to the air-circulating connections used in sea transport.

Little design choice is left to the owner except for overall length and height, type of insulation, and method of applying refrigeration. Universal practice requires stacking containers six units high. Aluminum shapes, plates, and extrusions are generally used in container assemblies. However, containers are being assembled with plastic embodiment, and others are being built with metal framework and panels of glass fiber and plywood.

Container insulation is mostly urethane foam because of its low density, its closed-cell characteristics, and its high bonding properties. For containers capable of carrying frozen foods, a nominal thickness of 3 in. on the walls and floor and 4 in. in the top is suitable. Considering aging, skin densities, possible imperfections of installation, and the inevitable penetrations and heat bridges, and average k-value of 0.20 Btu·in/h·ft^2·°F is recommended in the calculation of transmitted heat loads. Floors are of extruded aluminum or stainless steel; linings are aluminum or fiberglass. All must have cargo battens or corrugations on the walls to prevent surface contact by the cargo and to assure ventilation.

The methods of assembly, the calculation of refrigeration loads, and the testing procedures for marine containers are the same as for highway vehicles.

The Conex logistics container was initially designed for military dry cargo. Its dimensions and allowable loaded weight promote mobility in handling by forklift trucks, flatbed trucks, rail cars, and helicopter cranes. The larger container is 6 ft-3 in. wide, 6 ft-10.5 in. high, and 8 ft-6 in. long. This size container, if adapted to refrigerated service, is too small and would be needed in too great a quantity to fit each unit with individual attached refrigerating machines. However, it is feasible to fit an insulated container with communicating air ports and an internal fan to circulate the chilled air of a ship hold or warehouse through the cargo. Without auxiliary sources of refrigeration, these containers would have limits in shoreside hauling or holding. For extended shoreside service, attachable machines may be affixed to a closable aperture in the nose of the container.

Conventional Containers

The conventional marine container is fitted with a nose-mounted refrigerating machine powered by a dismountable diesel generator set for land-based operation. To accommodate on-deck stowage alignment with the dry cargo containers, a recess for the refrigerating machine is built into the container. The capacity and effect on evaporator relative humidity are of concern when applied to the long sea voyage.

With the earlier models, cold air blankets the top of the load. The air is recirculated by a fan built as part of the internally extended evaporator. The plant is thermostatically controlled and may include automatic defrosting. In most new containers, the cold air flows between the T-bars that form the deck and up through the cargo. Microprocessor-based control of the incoming air temperature and high airflow can maintain the product temperature within ±1°F throughout the container.

Some units have dual drive, where an internal combustion engine is used in shoreside services and electric motors drive the units on board ship. Others are all electrically driven, the current supplied by chassis-mounted generators or by the ship's electric plant.

Because of the considerable rejected heat of the machines and cargo, conventional containers are carried on the weather decks. Many container refrigeration units are equipped with both a water- and air-cooled condenser. When it is operated in the hold, the water-cooled unit is connected to a cooling water source, and the fan on the air-cooled condenser is turned off.

Developments in Heat Removal Equipment

One development in separable refrigerating units involves an integrated refrigeration plant, with or without an associated internal combustion engine power source. The unit covers the entire transverse area of the container's nose with air-circulating ports that match a ship's resident refrigerating system. The nose of the container is permanently insulated and the refrigeration unit attaches to the container by snap-on toggles.

Liquid nitrogen is economically feasible for many services. In such arrangements, the supply tanks and control apparatus may be carried by the transporting chassis, and the attachments to the spray header and the sensing bulb are separable.

Modified Atmospheres in Transport

As supplements to refrigeration, modified atmospheres slow the metabolism of horticultural products during transport. Sea transport conditions do not allow precise control of the atmosphere. Low-oxygen atmospheres in conventional refrigeration ships' holds are not acceptable because of the hazard to personnel. Individual control of concentrations for shipments in conventional containers is not operationally feasible, but the one-shot application at the point of loading of gastight containers does have promise. Chemical carbon dioxide scrubbers should be enclosed with the cargo when long voyages are anticipated.

Criteria for Marine Carriage

Products intended for export must be of the highest quality and in prime condition. They must be packed in packages that will not be crushed during transport and will allow heat removal; these are responsibilities of the shipper and should be subject to requirements of the marine carrier. All products carried in cargo containers should be cooled to carrying temperature before original departure, and the container should be expertly loaded to allow air movement through the stack, yet remain secure during transport. Uniformity of spacing and product temperature are requisites of good refrigeration for frozen foods as well as fresh horticultural products.

Even with orderly initial stacking within the container, disarrangement may occur en route because of the ship's motion. Vibrations caused by propulsion machinery can cause improperly secured packages to creep. Shifting of packages can lock air passages and effectively retard refrigeration. The shipper's method of stowage, like the quality of the products, is concealed from the marine carrier who assumes the responsibility for good delivery at destination.

Projected Developments

The following developments are projected for cargo refrigeration:

- Increased container size to 48 ft or more
- Remote monitoring and control via powerline communication and satellite link
- Microprocessor control with automated diagnostics and calibration procedures
- Controlled atmosphere systems, which have the ability to sense and control the concentration of some gases

SHIPS' REFRIGERATED STORES

Most vessels carry enough provisions for long voyages without replenishing en route. The refrigerating equipment must operate under extreme ambient conditions. Attention is being given to such factors as the storage of frozen foods, packaging, humidity, air circulation, and the increase of space requirements.

COMMODITIES

Perishable foods can be fresh, dehydrated, canned, smoked, salted, and frozen. For some, refrigeration is necessary; for others it may be omitted if the storage period is not too long. Space aboard ship is costly and limited; many rooms at different temperatures cannot be provided. The benefits of refrigeration and other conditions can be obtained by providing conditions outlined in Table 2 and described in the following sections.

Meats and Poultry

Substantial savings in space and preparation labor and better quality can be obtained with precut, boned, frozen meat, and poultry packed in moisture-vaporproof cartons and wrappers. For this reason, increased 0°F storage space should be anticipated. Fresh meats are less suitable because of their relatively short

Marine Refrigeration

Table 2 Classifications for Ships' Refrigeration Services

Service	Temp., °F	Passenger Vessels	Freight Vessels
Freezer Rooms			
Meats/Poultry	−20	X	X
Frozen foods	−20	X	X
Ice cream	−20	X	X
Fish	−20	X	X
Ice	28	X	
Bread	0	X	
Chill Rooms			
Fresh fruit/Vegetables	34	X	X
Dairy products/Eggs	32	X	X
Thaw rooms	40 to 45	X	X
Wine rooms	48	X	
Bon voyage packages	40	X	
Service Boxes			
Main galley			
Cooks' boxes	40	X	X
Butchers' boxes	40	X	
Bakers' boxes	40	X	
Salad pantry refrigerator	40	X	
Coffee pantry refrigerator	40	X	
Ice cream cabinet	10	X	
Ice chest	See text		
Mess Rooms or Pantries	40	X	X
Deck pantries	40	X	
Wine stewards' box	40	X	
Bars/Fountains	Various	X	
Miscellaneous			
Ice cube freezers	See text	X	
Ice cream freezers	See text	X	
Biologicals	40	X	
Drinking water systems	See text	X	X
Canned foods	See text	X	X
Dehydrated foods	See text	X	X
Ventilated Stores			
Hardy root vegetables	See text	X	X
Flour cereals	See text	X	X

storage life. Also, the space required for fresh meats is two to four times more than that needed for prepackaged meats.

Fish, Ice Cream, and Bread

Good quality fish, properly prepared and packaged, will remain odorless and palatable for a long time.

Commercially prepared ice cream is nearly always available and used to a great extent for both passenger and freight vessels. Ice cream for immediate use should be kept at a slightly higher temperature in an ice cream cabinet in the galley or pantry. Ice cream-making equipment may be desired, in which case provision must be made for hardening the ice cream, ices, and sherbets.

Excellent results can be obtained by purchasing freshly baked bread, sealing it in moistureproof wrappers, and storing it at 0 °F. This supply may be supplemented by bread and other bakery items made on the ship. Frozen bread may be thawed in its wrapper in a few hours.

Fruits and Vegetables

Packaged frozen fruits, fruit juices or concentrates, and vegetables may be stored in any freezer room. All packaged frozen products may be held in a common 0 °F storage space. However, the increased usage, especially on large passenger vessels, may justify separate refrigerated spaces for some or all of the products.

In some cases, fresh-grown product is desired. These items may be stored in a common room, but some compromises with their optimum storage conditions must be expected.

Dairy Products, Ice, and Drinking Water

All dairy products may be stored in a single room, following customary shoreside practice. Strong cheeses with odors that might be adsorbed by other foods should be stored in a tightly enclosed chest or cabinet placed in the dairy refrigerator. Eggs may be processed by oil dipping or heat stabilization to make them less sensitive to unfavorable humidity conditions or odors. Large passenger vessels should be fitted with a separate egg storage room. Butter for reserve supply should come aboard frozen and be kept in a freezer. Frozen homogenized milk has been perfected to a degree that it can be carried for reasonably long periods. Aseptically canned whole milk is now available and may be stored without benefit of refrigeration, but this product has some limitations because of the detectable cooked flavor.

Flake ice machines and automatic ice cube makers are common on passenger and freight vessels. Chilled drinking water is piped to many parts of a ship. The water is cooled in closed-system scuttle-butts, and the necessary circulating lines serve living and machinery spaces where drinking fountains and carafe-filling taps are installed. Remote stations that would require unusually long insulated piping runs are better served by independent domestic-type refrigerated drinking fountains.

STORAGE AREAS

Many borderline perishables, such as potatoes and onions, are satisfactorily stored with ventilation only. Hardy root vegetables are carried on freight vessels not destined for winter zones in slatted bins on a protected weather deck. Flour and cereals must be stored in cool, well-ventilated spaces to minimize conditions conducive to the propagation of weevils and other insects.

Storage Space Requirements

Space requirements for refrigerated ships' stores can be approximated by formulas. However, catering officials and supervising stewards have specific ideas regarding the total volume and subdivisions, and these sometimes vary greatly. A freight ship in ordinary scheduled service seldom exceeds 45 days between replenishment of stores and passenger vessels, considerably less. In addition, deliveries en route are possible. Passenger vessels diverted for cruises can expand into the otherwise unused refrigerated cargo spaces.

The space provided should allow suitable working floor areas for good storekeeping. When possible, stable piling 6 ft high is good practice; and, if the clear height of the room is less than 7 ft, allowances must be made for air circulation. A storage factor of 90 ft³/ton should suffice and allow floor working area. In the absence of a directing caterer or steward for consulation, Equation (1) may be used to estimate the total refrigerated storage space for merchant vessels. Ice storage is not included because of the various methods used in supplying it.

$$V = \left(\frac{NDP}{2000}\right) F \qquad (1)$$

where

V = total volume of refrigerated storage (not including ice), ft³
N = number of crew and passengers
D = number of days between re-storing
P = mass of refrigerated perishables per person per day, lb
   = 10 lb for freighters = 13.5 lb for passenger vessels
F = stowage factor (approximately 90 ft³/ton)

For example, for a freighter on 45-day voyage with a crew of 53 and 12 passengers,

$$V = \left(\frac{65 \times 45 \times 10}{2000}\right) 90 = 1316 \text{ ft}^3$$

With 6-ft high stowage, the net floor area would be 219 ft^2. With 8-ft high ceilings, the gross volume would be 1752 ft^3.

Gross volume represents actual space available for storage of foods up to ceiling height and does not include the space occupied by cooling units, coils, gratings, or other equipment.

Stores' Arrangement and Location

Next to the arrangement of ships to meet their major purpose, the planning of ships' housekeeping facilities is most important. Efficient operation by culinary workers requires not only well-arranged working spaces, but also convenient supply stores. Storerooms are usually located in spaces least suitable for living quarters or revenue-earning volumes and in areas adjacent to the main galley and pantry. The arrangement should provide easy access, which generally places the reserve storage refrigerators on the deck below the galley.

Aboard freight vessels, the refrigerators serve for daily issue as well as reserve storage. Aboard passenger vessels, the reserve storerooms are less frequently entered, and greater use is made of the service or work boxes.

Passenger vessels carry a corps of steward's storekeepers, who should have an issue counter and office located within sight of the exits serving this area. The storerooms should extend to the ship's sides or have passageways reaching to sideport doors through which the stores may be loaded directly into the ship. However, the arrangement of passenger ship stores will likely be compromised because of the interferences of structure, machinery or access hatches, and ventilation trunks.

In addition to the requirements for reserve storage of perishable foods, refrigerators (often referred to as working boxes) must be provided for the galley and pantry crew. On cargo vessels, a large domestic-type refrigerator will suffice. When more space is needed, a commercial walk-in box can be used. On passenger vessels, larger boxes are built-in like reserve refrigerators. The capacity of the passenger ship refrigerators will be governed by the number of passengers carried, the variety of the menu, and the arrangement of the galley and pantries.

Ice cream stored in the reserve boxes is too hard for serving, hence a dry or closed type of serving cabinet that will maintain temperatures from 5 to 10°F must be installed in the pantry. Passenger vessels may require ice cream-making machines as well as bar and soda fountain equipment, the latter being fitted with commercial, independent refrigerating units.

SHIP REFRIGERATOR DESIGN

Marine refrigeration equipment for off-shore vessels should be designed, selected, and applied so that it will function properly under extreme conditions with a minimum dependence on expert servicing.

Refrigerated Room Construction

Free water that might enter the insulation through faulty floor or wall surfaces is the most harmful element to ships' refrigerators. Room linings and floor coverings should be made of the materials and have the surface character that will give life-long resistance to the absorption of water by the insulation and the adherence of moisture on the room's interior surfaces. The construction of reserve and built-in refrigerators should follow details similar to those of the conventionally designed cargo refrigerators described in the Cargo Refrigeration section.

Adequate floor drains of the type that may be cleaned without lifting floor gratings should be provided and located so that, with the probable stowage plan, the scuppers will be accessible for cleaning without the moving of shelving or excessive weights of stores.

All details must be in compliance with the regulations of the U.S. Public Health Service, which also emphasizes rat-proofing. American ships are also subject to strict fire-resistance regulations.

All doors and frames should be of sturdy construction to resist frequent slamming and should have metal sheathing or reinforced glass fiber doors. They should be large enough to facilitate the loading of stores. The locking device should permit release of its fastenings from the inside by a person accidentally locked in.

All rooms should have galvanized iron or stainless steel racks or shelves to meet storekeeping needs, and they should be easily removable from their supports to facilitate rapid and thorough cleaning. The meat room and thaw room should have a single fore-and-aft meat rail for miscellaneous uses and thawing, respectively. Floor gratings or duckboards, fitted to each room, should be of a size and weight to facilitate removal and cleaning of gratings and the room.

The refrigerators should be fitted with waterproof lighting fixtures well guarded from damage by storing operations. The wiring should be bronze basket-weave cables, surface mounted, and well secured. Lighting switches should be mounted inside each room at the door with indicating lights in the outer passageway. Each room should have an efficient audible alarm for the use of any person inadvertently locked inside.

Remote reading thermometers, from which room temperatures can be read in the outside passageway, are essential to good operations. The bulb should be located in a representative location in the room—generally the geometric center at the ceiling. A large passenger installation justifies a duplicate electrical-resistance thermometer, with the instrument located in the refrigeration machinery room.

Service boxes in the galley and pantries should be constructed with a minimum amount of wood. The linings and shelving should be made of materials and have a surface character that facilitates thorough cleaning. Service refrigerators should not have raised door sills, and the floor should present a flush surface that is easily drained and cleaned. The cooling surfaces should be totally accessible for cleaning. Small units should be mounted without floor clearance on elevated bases, or be provided with at least a 8-in. clearance to facilitate scrubbing underneath.

Methods of Refrigeration

Freighters that carry no refrigerated cargo or that normally carry perishables in one direction, will only, in most cases, use direct-expansion systems for stores. Freight ships normally carrying refrigerated cargoes round trip may be arranged to use brine refrigeration for the combined service. Passenger vessels may have brine circulating systems, individual condensing units on each fixture, or a suitable multiplex arrangement.

The location of the condensing units for ships' stores, particularly for galley service, bars, and soda fountains, often presents a plumbing problem with the circulating water. This may make air-cooled units advisable for such applications. When air-cooled units are used, an ample supply of the coolest available air must be assured, and the hot air leaving the condenser must be expelled if the units are located in small rooms. Ventilating fans are mandatory for air-cooled condenser installations. Condensing units should not be installed in hot machinery rooms. Standby units for the more important refrigerating system components are advisable to insure against extended interruptions of service at sea.

Direct-expansion systems. These systems may be classed as central plant, multiplexed, and unit installations. The central plant installation uses two condensing units, one as a standby. If both low- and high-temperature refrigerators are to be served by the single operating unit, it must be selected and based on the lowest refrigerant temperature to be used in the system; this entails some

Marine Refrigeration

sacrifice of compressor capacity. For successful operation, the smallest cooling unit must balance with the condensing unit at a saturated suction temperature above the low-pressure safety cutout switch setting. If this condition is not met, hot gas bypass and desuperheating valves will be needed.

Multiplexed systems. In these systems, the low-temperature fixtures are grouped on one condensing unit, the medium-temperature services on another, and water cooling and high-temperature loads on another. If the refrigerant temperatures required by the various fixtures on one group differ greatly, the higher temperature evaporators should be supplied with evaporator pressure regulators for automatic control. Assuming a series of four refrigerators for ships' foods—one for dairy products, one for fruits and vegetables, one for meats, and a fourth for frozen foods—each would have its own R-12 condensing unit with associated cooling units. This provides maximum protection of perishables in case of mechanical failure. A better arrangement provides one or more standby condensing units. With proper piping and valving, the standby unit can be connected to substitute for any unit that has failed.

Unit installations. These systems apply a single condensing unit to each fixture—an arrangement preferred by some engineers.

Cooling units. Packaged forced-air cooling units with finned tubes speed up and reduce the cost of the installation. Automatic defrosting may be used, where average room temperatures are about 34°F and above. Air should be directed above the food; it should then be allowed to diffuse down and return to the cooling unit for recirculation. One or more of the units may be installed in a room after considering air distribution. If finned-tube, forced-air units are used for temperatures below 34°F, some positive means of defrosting must be provided. Hot gas and electric defrost are successful if properly applied and used.

When wall coils are used, paneled baffles should be installed over them both to prevent the produce from contacting the low-temperature surfaces and to promote air circulation. To cool products in larger spaces, ceiling-mounted fans should be installed for positive air distribution. The fan is generally located in the center of the room. In designing direct-expansion wall coils, the circuits must not be so long that excessive superheat of the expanded refrigerant results. On the other hand, a shortage of cooling surface results in excessive differences between refrigerant and room air temperature, with consequent dehydration of stored products.

Wall coils with an odd number of horizontal passes simplify series connection. Flow should be from top to bottom to minimize oil hang-up. Intermediate risers must be sized to ensure proper velocity for transport of the oil and unevaporated refrigerant leaving that coil section. Each riser may need to be preceded by a close-coupled trap to minimize oil hang-up in the bottom pass. The final riser should be sized as described in Chapter 3. Double risers may be needed to cope with reduced capacity operation.

Suction lines must be connected to overhead mains so oil cannot drain into idle or low-side equipment. Suction lines should run level and above the compressor suction inlets. If a riser must be used, a suction line accumulator may be needed. If a double suction riser is used, a suction line accumulator is essential to avoid oil slugging.

Controls. The most reliable system is the simplest one with a minimum of automatic apparatus. For marine applications, liquid controls should be of the automatic expansion type. Thermostatic valves are suitable for most applications. Rolling, pitching, trim, and list of vessels make float valve operation difficult. Expansion valves, being spring operated, are not affected by these conditions. Capillary tubes are commonly used in unit refrigerators and will function on vessels. If a float-type control is unavoidable, its vertical plane of operation should be fore-and-aft rather than athwartship. Float controls are unsuitable in vessels with propulsion machinery aft or which normally trim aft.

Temperature controls using thermostats, low-pressure controls, evaporator-regulating valves, and solenoid valves perform satisfactorily in ships' service systems. Such equipment, however, must be unaffected by vibration or ships' motion. Mercury bulb controls should not be used.

Direct-expansion controls should be located adjacent to their respective units, be installed clear of damage hazards caused by handling stores, have strong guards, and be fitted with locked covers to prevent unauthorized persons from tampering with them. Controls should be installed outside the refrigerators; on closely coupled systems using a central plant, they may be installed in the machinery room without the benefit of guards.

The calculation of refrigerant loads of a ship's service refrigerator system should follow the same procedure indicated for cargo systems in the section on Cargo Refrigeration. A larger margin for miscellaneous heat gain, such as frequently opened doors, should be allowed. An arbitrary figure of 10 to 25% of the total load may be used.

FISHING VESSELS

All vessels harvesting fish products, from the small open gill net boats used in inland lakes to the large factory processing ships on the high seas, use some form of refrigeration, whether it is ice picked up each day sufficient for the day's catch, or a version of the most advanced blast freezing system.

REFRIGERATION SYSTEM DESIGN

When designing a refrigeration system, the following issues should be considered:

1. The vessel owner and the design engineer must be aware of the monetary value of a fully loaded fish hold. The money saved by selecting and using substandard equipment may be a needless and expensive gamble.
2. The vessel may be several hundred miles from a qualified service technician and have very limited resources on board for emergency repair. In the event of a system failure, effective initial design may maintain temperatures longer, thus preserving the product.
3. Marine refrigeration systems are subjected to severe conditions, including high engine room temperatures, low ambient temperatures, electrolysis (corrosion), impacts, and vibrations. In some cases, these conditions are compounded by little or no maintenance, or even worse, abusive maintenance.
4. The system should be well laid out and designed in such a way as to allow new operators to adapt to the system quickly.
5. All safety and operating controls should be employed. In the event of a component failure, a backup system should be available, or, ideally, built into the system.
6. On completion, the vessel should be provided with all wiring and refrigerant flow diagrams, an operator's manual, and a supply of spare parts.

In the initial planning, the designer must know:

1. For what fisheries the vessel is being equipped and in what area of the world the vessel will operate.
2. In what future fisheries the vessel may be required to work. (At this point, such considerations will probably add little or no cost to the system.) Necessary alterations may be as few as increasing the spacing in the freezing racks.

Hold Preparation

On any vessel presently being refrigerated or being fitted for future refrigeration installations, 6 in. of insulating spray-on urethane is required. Special attention must be given to insulating areas of high heat such as engine rooms, bulkheads, and the underside of the main decks. High heat sources such as the hatch combing, shaft log, and fuel tanks with fuel returns from the engines, must also be insulated. The insulation must be protected to prevent moisture from destroying its insulating quality. In the northern Pacific, laid up fiberglass is commonly used because of its strength, light weight, and versatility. Pen board guides, mounting brackets, and plate racks are at times fiberglassed into the liner, and thus become a very secure part of the vessel. Fiberglass has the advantage of being easily cleaned and sanitized.

Case Study

In this case study, the owner of a 50-ft north Pacific trawler requested plate-type freezing surfaces secured to the front and rear bulkheads and under the deck to control heat gain through insulation, and a freezing rack made up of plates on which fish would be laid during freezing. All equipment uses a single compressor.

Heat Load Calculations. Depending on the sizes of the vessels, the available spaces, and budgets, most vessels employ one or more systems. U.S. law requires the product to be core frozen to $-20°F$ and stored at or below that temperature.

1. Precalculated load from all heat gain sources (not including product load), including a 10% safety factor, is 8000 Btu/h.
2. Saturated suction temperatures: $-40°F$.
3. Required hold temperature: $-20°F$.
4. Temperature difference (TD): $-40 - (-20) = 20°F$.
5. The U-factor, or the amount of heat absorbed by the plate surface, is 2.0 Btu/h·ft²·°F (below 32°F in still air). The U-factor can be increased by increasing airflow over the plates. The heat transfer per square foot of plate surface then equals:

    $$TD \times U = 20 \times 2 = 40 \text{ Btu/h·ft}^2$$

6. Plate heat transfer surface required: $8000/(40) = 200 \text{ ft}^2$
7. On a visit to the vessel, it was found that 22 in. by 48 in. plates would be best suited to this vessel. The number of plates required (assuming both sides of the plate provide cooling in this application) are as follows:

    $22 \times 48 \times 2 \text{ sides}/144 = 14.5 \text{ ft}^2$ cooling surface per plate
    $200/14.5 = 14$ plates required

8. The hatch combing heat load has been considered in the calculation. The hatch is heavily insulated, has a small entry port, and an insulated floor, again with a small entry port into the hold. The hatch should be kept at above 32°F for bulk fresh food storage. It is equipped with two additional plates and an EPR valve (see Figure 3).
9. A typical method of mounting under the deck plate surfaces is shown in Figure 4.

Freezer Cell Loading. A typical freezing cell (Figure 5) is usually located across the front bulkhead or in another suitable space. It is constructed with 4 to 5 in. spaces between the plates and with 2 or 3 spaces at 5 to 7 in. for salmon. Tuna requires spacing of 6 in. and greater. The freezing cell is usually constructed of noncorrosive materials, such as aluminum, with stainless bolts.

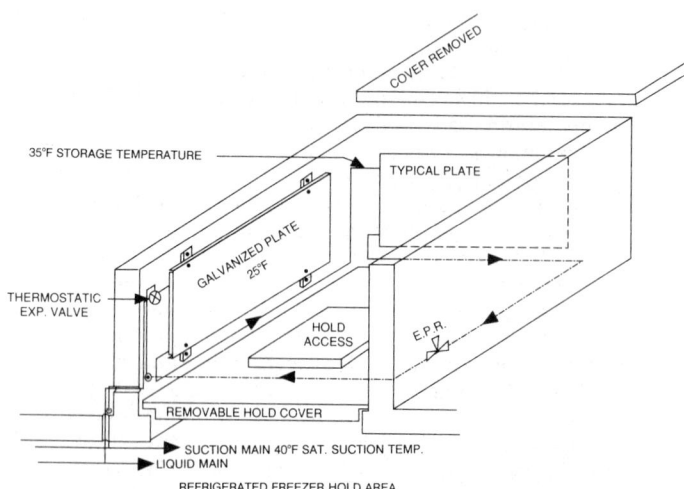

Fig. 3 Cross Section of Product Store

1. The example vessel expects to handle an average catch of 300 lb of salmon daily, and to load its freezing cell twice with 1500 lb per load. On average, 6 to 7 lb of fish per square foot should be used for proper freezing. The surface area required is:

    $$\frac{1500 \text{ lb}}{6 \text{ lb/ft}^2} \times 1.1 \text{ safety factor} = 275 \text{ ft}^2$$

    Note: Only one side of a plate can be used.

2. A 48 in. by 48 in. freezing surface allows easy access to 3 sides and requires a minimum of deck area. It also permits the use of storage pens on both sides. The number of freezing surfaces required are as follows:

    $48 \times 48/144 = 16 \text{ ft}^2$/surface

    No. of freezing surfaces = $275/16 = 17.2$, or 18 freezing surfaces

3. The covering plate not used for loading can be used as a heat transfer surface and must be considered in these calculations. Total heat transfer surface available is:

    $18 \times 16 \times 2 \text{ sides} = 576 \text{ ft}^2$

    Heat absorption capacity = UA (TD)
    $= 576 \times 20 \times 2 = 23,000 \text{ Btu/h}$

4. Total product load for 1500 lb of salmon frozen to $-20°F$ in 12 h is:

    $$q = m [C_1 (t_1 - t_f) + h_{if} + C_2 (t_f - t_2)]/\theta$$

    where
    q = freezing load, Btu/h
    m = mass of product, lb
    C_1 = specific heat of product above freezing
    = 0.86 Btu/lb·°F for salmon
    C_2 = specific heat of product below freezing
    = 0.39 Btu/lb·°F for salmon

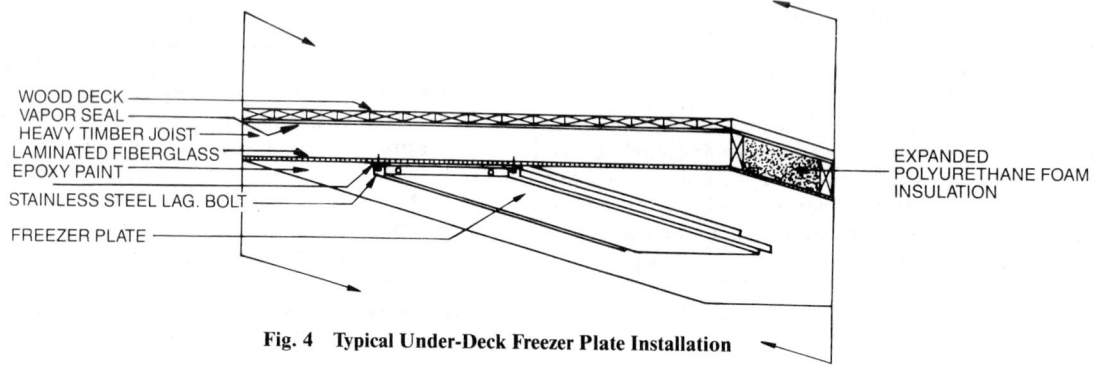

Fig. 4 Typical Under-Deck Freezer Plate Installation

Marine Refrigeration

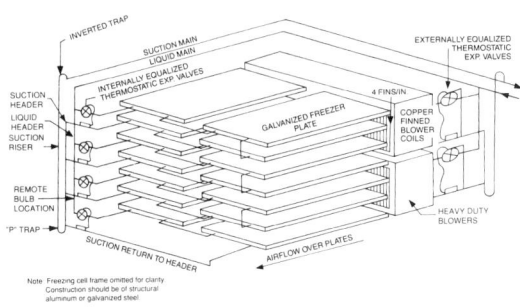

Fig. 5 Typical Marine Freezing Cell

t_1 = initial temperature of product = 50°F
t_2 = final temperature of product = −20°F
h_{if} = latent heat of product
    = 122.4 Btu/lb of salmon
θ = time period, h
q = 1500 {0.86 (50 − 32) + 122.4 + 0.39 [32 − (−20)]}/12
    = 19,770 Btu/h

Assuming a 10% safety factor

q = 19,770 × 1.1 = 21,700 Btu/h

Note that the plate absorption capacity and product bond are closely correlated in still air.

Plate capacity may be increased in two ways. One way is to increase the airflow across the product, which will increase the U-factor of the plate to as high as 4 Btu/h·ft²·°F. The other way is to install a compressor that runs the system at a greater TD. This method has a higher initial cost, however. In addition, problems associated with lower suction temperatures and pressures may arise. In most cases, a fan forcing 200 to 400 cfm over the plate will increase capacity by about 20%. The fan should be turned off when freezing is no longer required.

Total Evaporator Loading. The total evaporator load, then, is the heat absorption capacity of the freezing plates and the heat gain from other sources.

$$q_e = 8,000 + 23,040 = 31,040 \text{ Btu/h}$$

An additional evaporator currently in use for freezing and holding, other than plates, consists of blower coils. As a general rule, these coils have to be custom made due to the size of the loads, hold configurations, and other site-related influences. They should be constructed with not more than 4 fins per inch and with no dissimilar metals to help prevent corrosion. If an all-aluminum coil is used, it should be constructed using marine grade material. Copper-aluminum construction should not be used.

The most effective blast freezing system is a plate freezing cell, sized and constructed as previously described, with hot gas defrost and equipped with a blast coil to force air over the plates. The blast coil should have about 4 fins per inch and deliver an air blast of 1000 cfm/ft² of coil face area. The air blast discharges across one-half of the plate freezing cell and returns across the other half (Figure 5).

Compressor and Drive Selection. The compressor may have two duties: (1) to hold the frozen product, and (2) to freeze the fresh product entering the hold. When not required to freeze fresh product, the evaporator surface assists in pulling the hold down to a selected temperature.

The compressor chosen for this application has a capacity rating of 40,000 Btu/h at −40°F saturated suction temperature (SST) and +90°F saturated discharge temperature (SDT) at 1750 rpm. At 1450 rpm, the rating is 33,000 Btu/h, which is closer to the load of this application and is preferred for longer compressor life. Manufacturers' ratings should be consulted to ensure that the compressor will operate within all specifications (*i.e.*, above minimum rpm to ensure proper lubrication and head and oil cooling, if required).

Mounting Frame. The mounting frame should prevent flexing and vibration. Special consideration should be given to diesel-driven units to prevent harmonic vibration due to their higher compression ratios. The combination of high compression ratios and the great weight of most diesel engines precludes flimsy frame construction. The material used in the frame construction of the example was 3/8 in. × 8 in. channel, well cross braced. This base construction should be considered a minimum for all marine units.

To save space, most marine units are direct driven. The coupling must be aligned to the manufacturer's tolerances, or better, to prevent compressor damage. A good dial gauge and knowledge of its use is required for this procedure.

Since a vessel has many pieces of equipment operating at various speeds and under varying load conditions, harmonic vibrations may be created. These high frequency vibrations cannot be detected in some cases. To prevent damage, all piping and tubing connected to the compressor should have vibration eliminators. Gauges mounted on the compressor should be oil filled and be equipped with anti-pulsation dampers. When possible, all controls, gauges, switch gear, and so forth should be mounted on a remote panel away from the compressor or unit mounting skid. A remote panel gives greater access for servicing the compressor drive system while protecting delicate components. The high noise levels in the engine room require the engineer to use ear protection devices, which makes it necessary to include an alarm system to indicate when a system shuts down. For operator convenience, the alarm, if not audible in the wheelhouse, should be equipped with a flashing indicator light. This light, along with operating pressure gauges and remote reading thermometers, should be mounted in a control console within easy view of the operator.

Condenser Selection. Because condensers are exposed to the heat of the system, cold water, and the corrosive action of salt, zinc anodes need to be used as a sacrificial metal, regardless of what material the condenser is constructed.

On a small vessel, a keel cooler may be fixed to the underside of the vessel, and the refrigerant is piped directly to it. However, this system is not in wide use because, if the vessel hits a submerged object and the keel cooler is ruptured, refrigerant is lost and the system fills with sea water. Secondly, as this system is primarily used for economic reasons, they are at times undersized and do not provide enough heat transfer surface. Keel coolers are inefficient when the vessel is tied up because no water flows past the hull; they are too small to create adequate natural circulation around their surface. This situation causes a higher operating pressure and temperature and, often, a complete system shutdown.

The most widely used condensing system uses cupronickel shell-and-tube receiver condensers or tube-in-tube receiver condensers. Regardless of which condenser is selected, it must have removable heads that are accessible inside the vessel to permit inspection and cleaning. The water system should have at least two pickup points, so if one become plugged, the other may be of use. These intakes should be located so the pumps will not cavitate during rough sea conditions. Easily cleaned sea water filters should be included and the operating head pressure should be controlled with a 3-way water regulating valve.

The receiver-condenser should have sufficient capacity to hold the refrigerant charge. It should be installed in the most stable position available, with a liquid outlet at both ends. If the receiver condenser does not have sufficient holding capacity, both outlets should be piped to a vertical receiver of sufficient size to hold the additional charge. Vertical receivers with a bottom outlet always ensure that a liquid refrigerant column is available to the expansion valves to prevent flash gas regardless of the vessel's attitude. The vertical receiver should be designed into a system with a tube-in-tube condenser with no pump down capacity.

Another condensing system widely used in medium-sized vessels employs an oversized keel cooler to permit full operation at

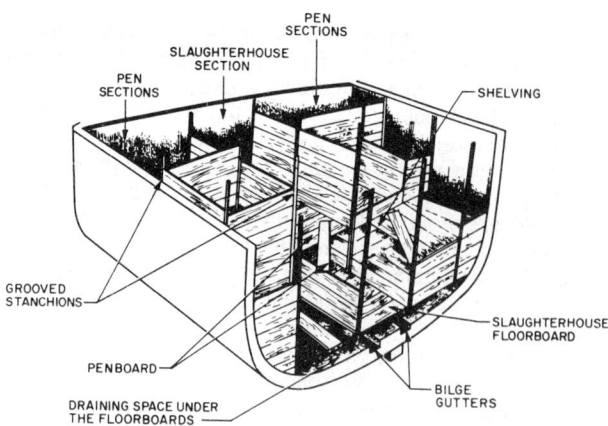

Fig. 6 Typical Layout of Pens in Hold

all times. An antifreeze (propylene glycol) and water solution is circulated through the keel cooler, using a land-based or marine condenser. Operating pressures may be controlled with a 3-way water regulator, thus bypassing excess solution around the condenser back into the keel cooler. Natural water movement allows the refrigeration system to operate at full capacity at all times, even while stationary and even over kelp beds, which may plug the water intakes of salt water cooling systems.

Oil Separators. Oil separators should be used with all low-temperature systems, especially with plate or pipe-style evaporators, because they may oil log if the suction gas velocity is restricted. Crankcase pressure regulators are required to prevent compressor and drive motors from overloading during start-up.

Suction Accumulators. Properly designed suction accumulators prevent compressor wash-outs by liquid refrigerant. The accumulator should be sized to hold at least the low-temperature evaporator operating charge. Accumulators equipped with liquid boil out coils not only help to boil off liquid refrigerant in these accumulators, but also provide some measure of liquid subcooling. Boil out coils are simply several loops of the liquid line inside the accumulator. Boil out accumulators used in conjunction with properly sized liquid suction heat exchangers increase system capacity through liquid subcooling.

Liquid Pumps. Liquid pumps, available in a full range of voltages suitable for marine refrigeration, are used on some vessels. This addition allows the system to operate at lower head pressures, while supplying a full column of liquid refrigerant to the expansion valves, thus preventing flash gas.

Mechanical Subcooling. Mechanical subcooling, which uses an additional refrigeration system to cool the liquid of the main system, provides up to 30% increase in capacity. For R502 subcooling from 100 to 50°F, units and chiller vessels should be sized at 3500 Btu/h per ton of refrigeration capacity. For subcooling below 50°F, 4000 Btu/h per ton should be used. Where an existing system is undersized and cannot be increased for reasons such as space limitations or power limitations of the main system, subcooling becomes an important addition to any system.

Liquid Line Driers. Liquid line driers are usually of the cartridge style and are installed with bypass valves to allow quick and easy changing of the drier blocks. The drier should be placed in a visible location, preferably upstream from the liquid moisture indicator and in an area with minimum vibration. Driers should also be equipped with purge valves.

Other Considerations. Refrigerant flow to the freezing plates must be accurately controlled. There are a few methods to accomplish this. The most common practice is to use a one ton, internally equalized, expansion valve, feeding between 30 to 50 ft^2 of plate surface area. The valve chosen should operate below $-40°F$.

A liquid header of 7/8 in. or larger, to which all expansion valves are attached, ensures a full column of liquid to all valves. Isolation valves should be used wherever possible to isolate various sections of the vessel's refrigerating system. Isolating areas such as the port, starboard, under deck plates, and other sections allows partial use of the system in the event of component failures. Use of these valves should be included in the overall piping plan. All refrigerant lines should be well protected from damage and properly secured to prevent damage through vibration and various vessel movements. If clamps of dissimilar metal to the piping are to be used, the pipe must be protected from direct contact with the clamp, using a good quality rubberized gasket material. This material should be wrapped around the pipe before the clamp is secured to prevent electrolysis that would corrode the copper.

All flow control devices must be of the best possible quality, ensuring that replacement parts are available for future service. All controls, gauges, and thermometers should be waterproof and capable of withstanding strenuous use. All electrical wiring must be completed by a qualified marine electrician, while all refrigeration equipment must be installed by a qualified refrigeration technician.

REFRIGERATION WITH ICE

Ice is commonly used for preservation of groundfish, shrimp, halibut, and most other commercial species. Bin or pen boards are installed to divide the hold as desired (Figure 6). Ice is usually stored in alternate bins so that it is handy for packing around the fish as the fish is loaded into the adjacent bin. The crushed ice varies in size up to lumps as big as a man's fist. As the fish are stowed with crushed ice, each pen is generally divided horizontally by the insertion of boards so that the bottom fish will not be crushed. The compartmentalized sections should not be more than 30 in. high if undesirable crushing and bruising are to be avoided.

The approximate amount of ice required is 1 ton for each 2 tons of fish in summer, and 1 ton to each 3 tons of fish in winter, based on a voyage of about 8 days. Less ice is needed if the ship has supplemental refrigeration.

The method of stowing the fish in ice is very important to the keeping quality. The depth of ice on the floor of the pen should be a minimum of 2 in. at the end of the voyage. This is obtained by having the initial bedding of ice 1 in. thick for each day of the voyage. In stowing, one or two layers of fish are laid on the bedding ice so that the ice is just completely covered. In no case should the layer of fish exceed 12 in. in thickness. The top covering layer of ice is about 9 in. thick, heaped up higher in the center than along the sides. This method of stowing permits the pile to adjust itself to melting and settling and results in good drainage of water and fish slime.

Most small fishing vessels are constructed of wood; the holds are not insulated. The large vessels are of steel construction with insulated holds lined with wood or aluminum. Mechanical refrigeration is used on some vessels to keep the ice from melting so fast and to maintain lower temperatures. The most common mechanical system uses DX cooling coils under the deckhead of the hold and sometimes around the entire shell. Another system uses both a double wall construction with cold air circulated around the entire hold between the insulated hull and deck and an inner metal lining.

REFRIGERATION WITH SEA WATER

In the United States, recirculated refrigerated sea water is commonly used instead of ice for holding industrial fish (i.e., menhaden and salmon) in satisfactory condition. When used for food fish, it is most suitable for larger fish where salt penetration is not

Marine Refrigeration

a problem. The sea water is continuously pumped either around the fish and over baffled cooling coils placed along one side of the insulated tank or through external brine coolers and then sprayed over the fish. Welded reinforced aluminum tanks and aluminum cooling coils have been used successfully, as have ammonia and Refrigerants 12 and 22. The smaller trollers require about 1 ton refrigeration capacity for each 2 tons of fish capacity, while the larger fishing vessels require about 1 ton refrigeration capacity for each 5 to 6 tons fish capacity.

Refrigerated sea water is used to preserve the highly perishable menhaden. In one commercial method of preservation, sea water is chilled at the rate of 5 gpm per ton of refrigeration from 50 to 45 °F with 37.5 °F suction at the compressor, based on cooling 200 tons of menhaden to 35 °F in 24 h at the rate of 3000 fish per ton. The calculated load of 43 tons of refrigeration includes 10% for insulation losses, so a plant of 60 tons capacity provides considerable margin for contingencies. The chilled water is sprayed over the partially filled fish hold and returned from the bottom of the hold to the chiller.

Distant water vessels are usually equipped for freezing and handling the catch at sea because they stay out for periods of 3 to 9 months, making storage with ice unfeasible. Although a large number of different vessels and freezing systems are used, the general types can be classified as either those freezing large whole fish, such as tuna and halibut, and those freezing processed and semiprocessed fishery products, such as fish blocks or groundfish in bulk lots.

The method of freezing is determined by the physical and biochemical characteristics of the fish and the desired end product. For the most part, large fish such as tuna, which are eventually canned and somewhat resistant to salt intake, are conveniently frozen in brine wells where space savings and ease of handling offer convincing benefits. Cod, haddock, hake, and similar demersal or midwater species, which are more delicate than tuna, are usually frozen rapidly either in vertical or horizontal plate freezers as described in the previous section, or in air blast freezers.

TUNA SEINERS

Three main factors distinguish the tuna seine fishery from almost all others. These are the big hauls that may be made at any one time, the great size of the larger fish with its corresponding slow heat removal, and the high temperatures of landed tropical tunas.

A system suitable for freezing tuna must be designed to take account of these factors and also meet other general requirements. Physically, it must be reliable and permit large quantities of tuna to be moved in and out in a short time. It must be compact, accessible for service, and compatible with the other space, weight, and trim requirements of the vessel. The cost of installation and maintenance must be low enough to permit an adequate return on the investment. Finally, the system must be able to chill, hold, freeze, store, and thaw tuna under conditions that assure high quality when delivered to the cannery. Some of these factors are now considered in more detail.

For the vessel. Purse-seining is the predominant catch method used in the eastern Pacific tuna fishery. The vessel capacity may vary from 400 to 2000 tons of fish. For tuna in brine wells, the maximum storage density is approximately 50 lb/ft^3. Maximum storage density for tuna in dry refrigerated holds is approximately 36 lb/ft^3. Enough refrigeration capacity is needed to freeze fish from 70 to 0 °F in 2 to 3 days and to store the frozen fish at 0 °F in dry well.

Operational data. Vessels can range several thousands of miles and may be required to hold fish for 12 weeks or more, although the average trip time is nearer 6 to 8 weeks. The average catch rate varies considerably. Tuna landed for canning vary in size from 5 to 130 lb. The cross-sectional dimensions of these tuna at the point of greatest girth are approximately 7 in. for 26-lb fish and 13 in. for 130-lb fish. Tropical tunas are fished in waters up to 85 °F. The temperatures of the landed fish are higher, reaching as much as 90 °F. Required rate of unloading fish from the vessel is 200 to 400 ton/24 h.

A tuna seiner is over 250 ft long, has a cruising range of 4000 nautical miles, and a hold capacity of over 2000 tons of frozen tuna. To accommodate catches up to 2000 tons, a large seiner may have 20 brine wells (86,000 ft^3) lined with galvanized pipe coils using ammonia direct expansion refrigeration. A total of 400 tons of refrigeration is required for fish storage, ship service, cold storage room, and air conditioning. Each brine well is insulated on all sides with 6 in. foamed-in-place polyurethane. Basic refrigeration for the brine well evaporated coils is supplied by five reciprocating ammonia compressors with capacity reduction controls, horizontal oil separators, and automatic oil return.

The hold is divided into steel wells or tanks arranged on both sides of the shaft alley. The well insulation is about 5 to 6 in. thick. The wells are constructed of 5-in. wooden planking except on sides adjacent to the skin of the ship where 2-in. planking is used. The shaft alley accommodates pipelines, control valves, brine pumps, and other mechanical devices all closely arranged in a very limited space. The tanks are lined with cooling coils and each is equipped with a brine circulating pump, sea water inlet and outlet, and connections to the brine transfer lines. Additional fuel oil is carried in some fish wells, since the permanent fuel tanks cannot carry enough fuel oil for the long trip.

The wells on large seiners are prepared for receiving fish by being filled with fresh sea water and being chilled to 29 °F with the refrigeration system. Since sea water freezes at about 28 °F, it is not practical to cool below 29 °F in the preliminary chilling operation. Before the first well is completely filled with fish, the second well is filled with fresh sea water, cooled down, and made ready for loading. With the refrigeration plant operating and the brine recirculating, the fish are chilled down to 30 °F internal temperature in 24 to 72 h. This preliminary chilling time varies with the operator, some chilling the fish as rapidly as possible and others preferring a three-day precooling period, feeling that the longer time for chilling seals the pores of the fish better and prevents excessive salt penetration, which may increase freezing time when dense brine is used in the later freezing process.

The next operation strengthens the brine so that fish may be frozen at a lower temperature. This is accomplished by dumping salt directly into the well. It requires about 100 lb of salt for each ton of fish capacity of the well. This makes dense brine of the sea water, the dense brine having a specific gravity of 1.126, a specific heat of 0.833 Btu/lb·°F, and a freezing point of 8.2 °F. A cubic foot of the dense brine and fish mixture weighs 70.5 lb, of which 50 lb is fish and 20.5 lb is brine.

Within 2 or 3 days of operation with the dense brine and additional refrigeration, the internal temperature of the fish reaches 20 °F or lower as the brine temperature is maintained at about 18 °F. The fish become rigid because about 75% or more of their water content is frozen. The dense brine solution is then pumped to another well for reuse, although in some cases it may be pumped overboard if badly contaminated with fish slime and blood. After brine freezing, the well is drained of dense brine and the temperature of the fish may be further reduced by the refrigeration coils only.

The fish are maintained in a dry frozen condition at 18 °F or lower until port is almost reached. About one day from port, the circulation of sea water may be started through some of the wells so that the tuna will be sufficiently loosened from each other to start unloading. The fish must be thoroughly thawed before entering the cannery processing line. Steel buckets are lowered into the wells, and the fish are manually thrown into the buckets. Power winches lift the buckets to the wharf. The fish are unloaded into

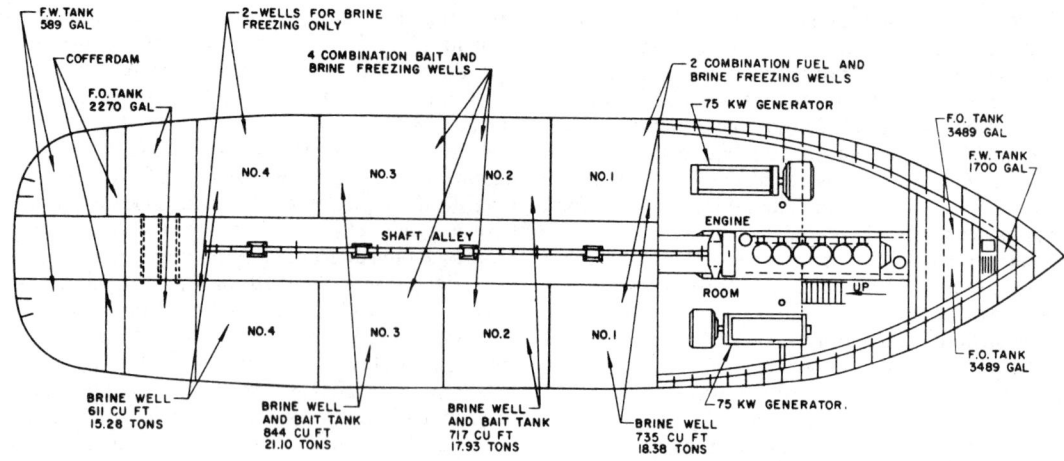

Fig. 7 Plan of Hold for Tuna Clipper

flumes, which carry them to the weighing tank and then into the cannery.

For this type of refrigeration duty, each tuna vessel generally has three or more refrigeration compressors, driven by electric motors, and three different suction lines; one for 29°F brine, one for 18°F brine, and one for holding. Each fish well and tank is connected to each of the three suction lines, and the ammonia compressors are cross-connected so that any or all of them may operate on any of the three different types of loads. Ammonia is generally used because of the desirability of parallel operation of compressors and because oil problems prevent R-12 compressors from being operated interconnected aboard ship.

For a large tuna clipper (see Figure 7), duplicate 10 in. bait pumps, each capable of delivering 2300 gpm of sea water against 20 ft head, and one 3 in. brine transfer pump, capable of handling 300 gpm of sea water against a 50 ft head, are provided. For emergency operation, the brine transfer pump may be cross-connected on the suction and discharge sides with the ship's general service pump.

Each brine tank and each of the three bait tanks, which are also arranged for freezing fish, are provided with a 2 in. brine circulating pump with shutoff valves on inlet and outlet. Each pump circulates 200 gpm of sea water against a 20 ft head. Pumps are mounted in the shaft alley. These pumps draw from the bottoms of the tanks and discharge into the hatches above the tanks through 2.5 in. galvanized piping. This circulation of the brine improves the heat transfer of the cooling coils and makes possible the rapid, uniform cooling and subsequent freezing of the fish.

Freezer Trawler, Factory Vessels, and Mother Ships

On a worldwide basis, the largest quantity of fish frozen at sea comprises such species as cod, hake, pollock, redfish, and other demersal or midwater fish. The fish are caught by bottom or midwater trawls and may be either frozen in bulk on the vessel or processed into fish fillets or fish fillet-type blocks and then frozen. The catch rate varies with the particular fishing operation, but may be as high as 60 to 70 tons per day with averages well below 20 tons per day in most northern fisheries and 35 to 40 tons per day on an overall average. The vessels vary in size from 150 ft for the smaller bulk freezer trawlers, to over 200 ft for factory freezer vessels, and over 300 ft for the larger mother ships.

In bulk freezing aboard vessels in the United Kingdom, the fish are landed on the vessel, eviscerated, and then frozen in bulk in vertical plate freezers using R-22. A typical British bulk freezer trawler will have a length of 215 ft with a capacity of about 1800 gross tons. A vessel with this classification has a 10 to 12 station top loading, side unloading, vertical plate freezer, producing blocks of whole frozen whitefish, 42 in. long by 21 in. wide by 4 in. thick, each block weighing approximately 100 lb. The storage rate is 55 to 60 ft^3 per ton. Total freezing capacity is about 35 tons of fish per day.

In a typical operation, fish are landed on the vessel, sorted and gutted by hand, washed in a rotating cylindrical washer, and conveyed to storage bins alongside the two rows of freezers that run fore and aft on either side of the factory space. The fish are packed between the pairs of freezer plates and are reduced to a temperature of about −5°F in about 3.5 h. Hot trichlorethylene is then circulated through the evaporated plates both to help release the blocks from the freezer and to ensure that the wet fish do not stick to the plates when reloading begins. The discharged frozen blocks of fish are chuted through the insulated hatch in the factory deck to the cold storage room below, which operates at −20°F and has a capacity of about 500 tons. Trawlers of this type make voyages of 30 to 60 days, depending on the catching rate, and fish principally off Greenland, Newfoundland, and Labrador.

Factory trawlers are equipped to catch, fillet, package, and freeze the product on the vessel in a manner very similar to shoreside processing operations. A large vessel with a length of about 250 ft has refrigeration capacity sufficient to freeze 30 tons of fish fillets a day and to store 590 tons at 20°F. After being landed, the fish are stored in ice or chilled sea water and then are headed, filleted, and skinned by machine. The fillets are packed in trays and frozen in blocks, usually in horizontal plate freezers, and then transferred to the cold storage room. This pattern of operations is similar in most factory freezing vessels. Russian, Polish, and East German companies have emphasized use of blast freezers, and some of the larger vessels can freeze at a rate of 100 tons a day. Some vessels may produce a combination of factory-finished products, bulk frozen fish, and iced fish that is the last day's catch.

The mother ship is a processing vessel with no catching capacity. It is the central part of a total fishing complex consisting of a fleet of fishing trawlers and fuel supply vessels that operate at considerable distances from home port. The fishing trawlers, which fish for several days, transfer their catch to the mother ship, where the fish are processed, frozen, and stored in much the same manner as on the factory vessel.

CHAPTER 31

AIR TRANSPORT

Perishable Air Cargo 31.1
Perishable Commodity Requirements 31.2
Design Considerations 31.2
Shipping Containers 31.3
Transit Refrigeration 31.3
Ground Handling 31.4

AIR freight service is provided by all-cargo carriers and passenger airlines, which carry both cargo and passengers. The latter companies also have all-cargo aircraft. Wide-body aircraft have a combination of passenger and cargo mix on the main deck, increasing cargo capacity (Figure 1). All lines maintain regularly scheduled flights so shippers may adequately plan delivery time. Special charter flights are also available from regular terminals and from airports located close to the producing areas. Prospective shippers should contact the airlines serving their locality to obtain specific details for handling perishable shipments.

PERISHABLE AIR CARGO

The use of air transportation for the distribution of perishable commodities has focused attention on the maintenance of their quality in transit. Because speed is one of the primary advantages of air movement, little attention was initially given to providing the proper transit environment. Because commodities that require low temperatures for maximum retention of quality or effectiveness benefit most from air transportation, transit refrigeration became necessary. Because actual refrigeration is not available in cargo aircraft nor in the cargo compartments of passenger aircraft, proper transit temperatures must be provided by adequate precooling of the product and the use of suitable insulated and refrigerated containers.

Fruits and vegetables, flowers and nursery stock, poultry and baby chicks, hatching eggs, meats, seafoods, dairy products, live animals, whole blood, body organs, and drugs (biologicals) are transported by air. Items so shipped are generally of such perishable nature that slower modes of transportation result in excessive deterioration in transit, or air movement is the only possible means of delivery. Certain early season and specialty fruits and vegetables can be flown to distant markets economically because of the high market prices when there is a short supply. Some items, such as cut flowers and papayas, arrive at distant markets in better condition than they would otherwise, so the extra transportation cost is justified. Flowers are shipped on a regular basis from Hawaii to the mainland and from California and Florida to the large midwestern and eastern cities. Air movement of strawberries has increased tremendously, including direct shipments to global destinations. Papayas are shipped from Hawaii almost exclusively by air. The export of perishable agricultural commodities by air is rapidly increasing because of reduced rates generated by new and more efficient aircraft. Payload range comparisons for an all-cargo/all-passenger wide-body jet or a combination are shown in Figure 2.

Fruits and Vegetables

All fresh fruits, vegetables, and cut flowers are alive and remain living throughout their entire salable period. Being alive, they respond to the environment in which they are held and have definite limitations regarding the conditions that can be tolerated. They remain alive by the process of respiration, which breaks down stored foods into energy, carbon dioxide, and water, with the uptake of atmospheric oxygen. Respiration, together with accompanying chemical changes, results in changes of quality and the eventual death of the commodity. These internal changes associated with life cannot be stopped but should be retarded if a high level of quality is to be retained for a prolonged period.

Seafood

Seafood and fish also benefit from the speed of air freight. The abundance of fresh fish at restaurants and markets throughout the country is the result of air shipment.

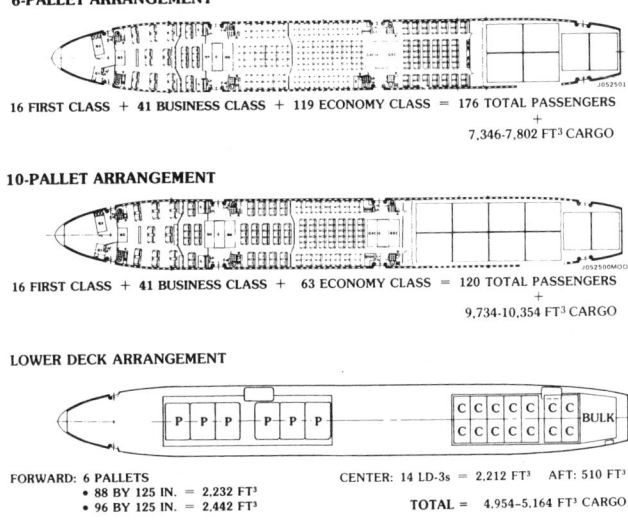

Fig. 1 Flexible Passenger/Cargo Mix

The preparation of this chapter is assigned to TC 10.6, Transport Refrigeration.

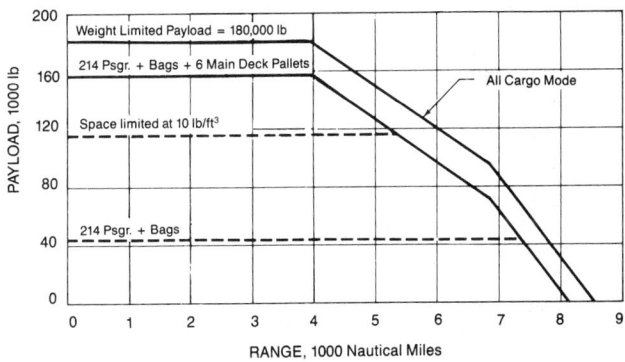

Fig. 2 Payload-Range Comparison for Wide Body Jet

Animals

Federal law requires special handling for livestock (watering, unloading for exercising, and so forth) every few hours. This treatment is expensive, especially if performed at locations without the proper facilities. The faster livestock can be moved, the less costly the loading and reloading procedure. When the greatest segment of the trip is by air, no special unloading and reloading is required. However, some livestock thus shipped require more careful air-conditioning than do human passengers. Whereas human beings complain when they become uncomfortable, some animals suffer actual physical damage, even death, when subjected to temperature changes outside of narrow limits. Baby chicks are a fragile live commodity and are shipped in fairly large numbers by air. Other temperature and ventilation requirements for livestock are shown in Table 1. It suggests, for example, long-haired dogs acclimated to summer temperatures cannot be thrust into a cargo hold at 35 °F and be expected to survive.

PERISHABLE COMMODITY REQUIREMENTS

The justification for the air movement of perishable commodities is based on: (1) the time element; and (2) the delivery of a higher quality product than is possible by other modes of transportation. Better delivered quality makes possible increased returns to the shipper. This not only offsets the added transportation costs but increases consumer demand and acceptance as well. The market quality of perishable items is definitely controlled by a time and temperature relationship. Temperature cannot be ignored even for the few hours now required for transcontinental air movement. Proper temperature and humidity must be maintained at all times.

Pentzer et al. (1958) lists desirable transit environments for most perishable horticultural commodities. Figure 3 shows the result of a test of air shipments of strawberries from California to Chicago in a refrigerated but uninsulated container. The shipments were exposed to high ambient temperatures during ground handling at origin, resulting in fruit temperatures ranging from 50 to 60 °F instead of the desired 32 to 34 °F. These berries were compared with those shipped by rail in 4.5 days with temperatures averaging 38 °F for the transit period. Appearance and decay on delivery were about the same for both lots. Thus, the advantage

Table 1 Livestock Temperature and Ventilation Requirements

A. Recommended maximum temperature:
   95 °F, except 85 °F for small birds and heavy-coated fur-bearing animals, and 100 °F for reptiles

B. Optimum temperature for all confined livestock:
   68 to 75 °F, except 80 to 85 °F for reptiles, lower for some heavy-coated fur-bearing animals

C. Recommended minimum temperatures, °F:
   1. Small monkeys and apes   65
   2. Cattle, horses, and swine   55
   3. Small birds   65
   4. Jungle cats and other tropical mammals   65
   5. Domesticated cats and short-haired dogs   45
   6. Long-haired dogs, foxes, and so forth   35
   7. Hamsters, rabbits, and guinea pigs   55
   8. Reptiles, nonhibernating species such as cobras,
      boa constrictors, and crocodiles   70
   9. Adult poultry   45
   10. Baby chicks in cardboard cartons[a]   70

D. Body temperatures:
   98 to 104 °F for mammals and 102 to 111 °F for birds

E. Heat emission (per pound of body weight):
   2.4 to 8 Btu/h for mammals, 5 to 14 Btu/h for birds

F. Ventilation required:
   For cattle, 3.2 to 3.8 ft^3/h per lb of body weight

[a]Temperature within carton approximately 95 °F.

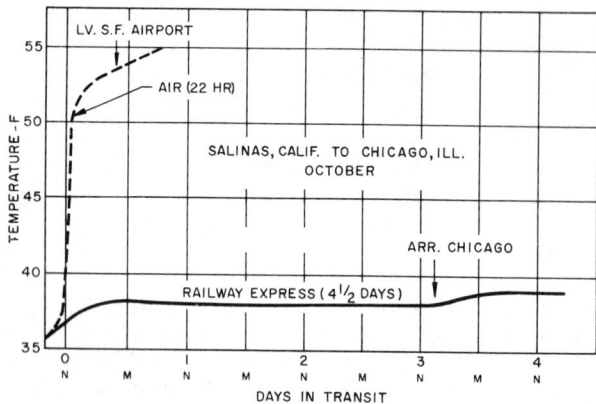

Fig. 3 Temperature of Strawberries Shipped by Air and Rail

of the short 22-h air movement was offset by a loss in quality arising from the unfavorable temperature.

Top quality of many of the most perishable commodities can be significantly reduced by only a few hours' exposure to unfavorably high temperatures. Many drugs (biologicals) and other items, such as whole blood, can be rendered completely ineffective or toxic if not kept at the specified low temperature.

Some flowers, fruits, and vegetables respond favorably to reduced oxygen levels, increased amounts of carbon dioxide, or both, which could be maintained by gastight packaging or containers.

The effect of reduced pressure at high altitudes on fruits, vegetables, and cut flowers is negligible (Barger 1955, Barger and Ryall 1950). Consequently, pressurization of cargo compartments in all-cargo aircraft has not been required for these commodities, although many all-cargo and practically all passenger aircraft maintain pressure in their cargo compartments. Pressurization is, of course, required for shipment of most live animals.

The maintenance of temperatures near freezing is not desirable for all products because some are subject to chilling injury when they are exposed to temperatures well above this point. Chilling injury is most pronounced in tropical products such as bananas, tomatoes, cucumbers, avocados, and orchids. Temperatures above 55 °F are usually safe for cold-sensitive commodities. Other items such as cut flowers require temperatures between 32 and 55 °F.

Certain fruits and vegetables require humidity control. Humidity should be kept between 85 and 95% to prevent wilting and general loss of water. The relative humidity in the cabin of an airplane flying at about 40,000 ft is generally less than 10%. However, the respiration of fruits and vegetables, placed in a closed container with recirculated cooling air, should produce the required humidity level with no additional water added.

Certain vegetables, such as peas, broccoli, lettuce, and sweet corn, have high respiration rates and the heat produced may amount to the equivalent of 250 lb or more of ice meltage per ton of vegetables per day at 60 °F. In designing the refrigeration systems for aircraft containers, the additional evaporator capacity required to handle the heat of respiration should be considered.

DESIGN CONSIDERATIONS

A refrigeration or air-conditioning system for a cargo airplane or airborne cargo containers has conflicting design temperature requirements, depending on the type of cargo to be carried, which makes it difficult to use one optimum refrigeration system for all kinds of cargo. For example, frozen foods should have a temperature of 0 °F or lower, fresh meat and produce 30 to 45 °F, and live animals generally require temperatures in the same comfort range

Air Transport

as for humans. Today, many of the commercial jet cargo planes operate with the main cabin divided between cargo compartments and passenger compartments, and they are supplied by a single air system controlled to the comfort of the human occupants. In this case, perishable cargo must be packed in containers, insulated, and iced or precooled.

The design ambient temperatures that an airplane will experience in flight are given in Chapter 25 of the 1987 ASHRAE *Handbook—Systems and Applications*. A cargo jet cruising close to Mach 0.9, has an increase in skin temperature over ambient of about 50°F. With an all-cargo load, the basic air-conditioning systems are capable of maintaining main cargo compartment temperatures on a design hot day from 40°F at 30,000 ft to 30°F at 40,000 ft.

The air-conditioning systems are equipped with controls to prevent freezing of moisture condensed from the air at low altitudes. With the extremely dry air prevailing at cruise altitudes, an override of these anti-icing controls would permit an even lower cabin temperature although it is doubtful that storage temperatures of frozen goods could be met. Thus, some insulation would still be required in frozen food containers. Further, the airplanes are often required to hold at relatively low altitudes of 20,000 ft or less, because of heavy traffic at the busier airports, for periods of 30 min. or more.

Permanent attachment of a mechanical refrigeration system to a cargo container may not be desirable for several reasons: increased weight; reduction in usable volume; and difficulty in rejecting the condensing unit heat load overboard. These objections are particularly applicable to containers carried in the main cargo hold. On the other hand, permanently attached units would permit refrigeration of just part of the cargo load while the remainder could be held at temperatures in the normal human or animal comfort zone, for example. Temperature control of products requiring widely differing transit and storage temperatures would be more feasible with onboard refrigeration units.

SHIPPING CONTAINERS

Fruits and vegetables are generally shipped in the same containers used for surface transportation: wooden boxes, veneer crates of various types, or fiberboard cartons. Most flower containers are constructed of either plain or corrugated fiberboard materials. Wooden cleats are used for bracing material, generally as dividers or corner braces inside the cargo box. Where lading may be exposed to very cold surfaces, external cleats may be used as spacers to prevent direct contact. Certain flowers such as gardenias and orchids may be packaged in individual cellophane-wrapped boxes or trays and placed in a master container. Any tightly sealed film wrap must be perforated by at least one small hole to permit release of air from the container during ascent to high altitudes.

Containers, built on pallets and shaped to make maximum use of the interior airplane volume, are in use. Containers presently in use with the airlines are described in the IATA Register. Containers for aircraft, except for the belly cargo holds, are not shaped to make maximum use of the interior volume of the airplane. One reason for this is that the individual packages that will fill the containers are generally rectangular in shape anyway. Another reason is to permit easier intermodal transport, *e.g.*, from motor truck to the airplane and vice versa. Because of the size of aircraft loading doors and irregular aircraft cross-sections compared to surface vehicles and vessels, containerization may require this compromise.

Containerization is a system of moving goods in sealed, reusable freight containers too large for manual handling, and which do not have wheels permanently attached. The advantages are: far less cargo damage and pilferage, lower packaging costs, minimized handling, and lower shipping rates. Presently, these containers

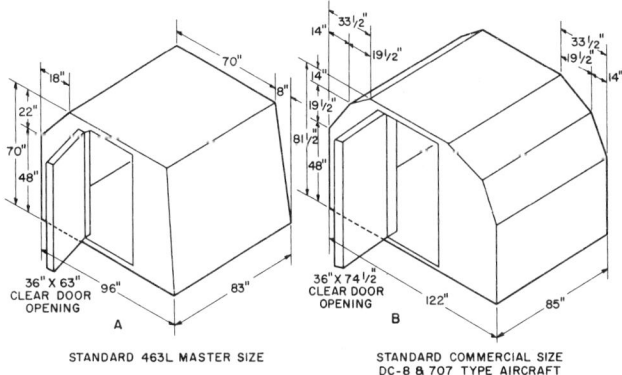

Fig. 4 Insulated Containers Designed to Fit Configuration of Cargo Aircraft

may be loaded at the air freight terminal or loaded at the shipper's facility and transported by flatbed truck-trailer or railroad, or both, to the air terminal.

The critical condition for design of insulation and refrigeration systems (detachable plug-in type or permanently installed) for cargo containers is the time that the container is on the dock in the hot sun waiting for shipment. For this condition, an ambient temperature of 100°F db is assumed. The average outside skin temperature of an unpainted metal container is about 115°F.

Under the conditions just mentioned, the 8 ft by 8 ft by 10 ft container with 0.5 in. of high efficiency insulation (recirculating the air and considering no latent load) requires about 18,000 Btu/h of refrigeration to maintain 35°F inside and about 24,000 Btu/h to maintain 0°F inside. For quick pulldown to these temperatures of the container and fresh perishable contents (assuming prefreezing of frozen products), the capacities should increase by about 50%.

Fresh fish, shrimp, and oysters may be packed in boxes, barrels, or special containers. Proper precaution must be taken to prevent drippage from melted ice into the cargo space. Live lobsters are packed in insulated containers with salt water seaweed. Frozen foods are always packed in insulated containers. Whole blood is shipped in specially developed containers. Insulated bags are also used.

The configurations and dimensions of two insulated containers are shown in Figure 4. Insulated with closed-cell, rigid plastic foam, the containers are a fabricated sandwich structure and are sized to fit conventional pallets and materials handling systems. The heat transfer rate for the entire standard container is 28 Btu/(h·°F) and 32 Btu/(h·°F) for the commercial size. More recent aircraft such as the MD-11 depicted in Figure 1, use pallets 125 in. wide by 64.4, 88, and 96 in. which may be loaded to 64 in. high, and retained by straps. Load capacities are 6700 lb, 10,000 lb, and 11,000 lb, respectively. Containers are LD-3 half width, and LD-6 full width, the latter having twice the capacity and width. Both are 60.4 in. deep and 64 in. high. The LD-3 width is 79 in., volume 158 ft^3, and capacity 3300 lb. A plug-in portable mechanical refrigerating unit may be positioned in the doorway for standby operation. Tight construction permits controlled atmosphere application. A smaller shipping unit, insulated with a foamed plastic, has inside dimensions of 45 in. by 21 in. by 24 in., *i.e.*, a total area of 35.1 ft^2 and a capacity of 13.1 ft^3. The heat transfer rate of the entire container is 2.8 Btu/(h·°F).

TRANSIT REFRIGERATION

Many commodities must receive refrigeration in transit. In most cases, this is accomplished by a refrigerant in the package. Water ice, dry ice, and certain proprietary refrigerants are used. Because no method of transit refrigeration can economically cool a warm commodity to its desired transit temperature, all perishable items must be cooled before shipment. Flowers generally are wrapped in several layers of paper or lightweight insulating material and kept cool by water ice. The ice (solid, chopped, or flaked) is placed in a plastic bag or wrapped in many layers of paper and tied to one of the cleats of the container. In some instances, the block of ice may be chilled to 0°F in a freezer before putting it in the package, thereby obtaining a slightly greater refrigeration capacity. Newspapers are sometimes wadded up and thoroughly wetted and then frozen to 0°F or lower. In all cases, the paper helps absorb the ice meltage water and reduces the chances for leakage into the cargo space.

Some voids should be left in the containers to permit air circulation and uniform cooling. Boxes should be sealed to prevent air exchange. Placing the ice or water for freezing in sealed plastic bags eliminates drippage, but melting ice in open containers increases humidity, which is particularly desirable for cut flowers. A packaged refrigerant with no escape of free liquid must be used with commodities that would be damaged by water. Water ice acts as refrigerant in special-type blood containers.

Dry ice is used extensively with frozen products and fresh strawberries, the amount depending on the type of container and the length of the journey. Sometimes it is placed in with water ice, not only for its own refrigerating value, but also to slow down meltage of the water ice and extend its value to the end of the transit period. Dry ice alone is seldom used for flowers because its very low temperature may cause freezing damage to adjacent blooms if not properly spaced or insulated. The use of large amounts of dry ice may cause a buildup of carbon dioxide gas in concentrations dangerous to humans and animals unless proper ventilation of the cargo compartment is provided.

When the heat transfer rate of insulated shipping containers is known, the amount of refrigeration required can be estimated with reasonable accuracy from

$$Q = HD\Delta t$$

where

Q = total heat transfer, Btu
H = heat transfer rate of entire container, Btu/(h·°F)
Δt = difference between ambient temperature and that at which product is to be carried, °F
D = duration of transit, h

For example, assume the small container previously described holds 15 standard strawberry trays, each holding 13 lb of berries (195 lb total), with the fruit cooled to and carried at 35°F at an ambient temperature of 75°F for a transit time of 24 h:

H = 2.8 Btu/(h·°F)
Δt = 75 − 35 = 40°F
D = 24 h

$$Q = (2.8 \times 40 \times 24) = 2688 \text{ Btu}$$

The heat of respiration generated by the berries at 35°F is about 4000 Btu/(ton·24 h) (see Chapter 26). For 195 lb of berries 24 h in transit:

$$4000 \times (195/2000) = 390 \text{ Btu}$$

The ice required to absorb this heat is

$$(2688 + 390)/144 = 21.4 \text{ lb}$$

The amount of dry ice required would be about 11.9 lb.

These simple calculations can be made only when the thermal conductance or heat-transfer rate of the container is known. It would therefore be of considerable value to all concerned—shipper, carrier, and receiver—to have this factor determined for all insulated shipping containers and clearly displayed. Such ratings have been made on truck-trailer bodies, as discussed in Chapter 28.

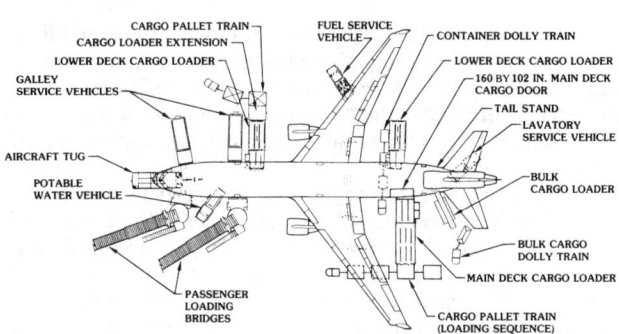

Fig. 5 Ground Service Equipment Arrangement

When package refrigeration is not available, rapid warmup can be retarded by insulated containers or blanket insulation over stacks or pallet loads. This method has been satisfactory with some of the less perishable fruits such as peaches. Temperatures can be maintained for several hours in flight with the proper use of these blankets. Care must be exercised in loading to insure that insulating material is wrapped completely around the cargo and that containers are not in direct contact with hot or cold surfaces. Some cargo compartments on passenger aircraft may be cooled by the air-conditioning system, but the temperature will not be in the optimum range for most perishable commodities.

GROUND HANDLING

All the advantages of speed can be lost if the shipper, carrier, and receiver do not follow good handling practices that keep deterioration to a minimum.

Ground handling can amount to over 70% of the total elapsed time from shipper to receiver. To reduce this ground time, load palletization and special pallet carriers and loaders, in conjunction with improved load-handling systems aboard the aircraft, are used. Air freight terminals are now designed and built to use these new handling techniques. Combination cargo/passenger jets present unique loading techniques. A typical ground service equipment arrangement is shown in Figure 5.

Fast pickup and delivery are also essential. Because of the generally high ground temperatures at shipping point terminals and intermediate points, most perishable agricultural commodities should be cooled as soon after harvesting as possible and delivered to the air terminal in properly refrigerated vehicles, particularly if they are shipped in uninsulated containers. At the terminal, these shipments must be held at proper temperatures if prompt loading from the pickup vehicle is not possible. Holding rooms, refrigerated mechanically or by ice, should be provided. During seasonal loading peaks, refrigerated trucks or trailers may be used as temporary holding rooms. During hot weather, cargo space must be cooled before loading. Portable air-conditioning equipment, such as that used for passenger aircraft, is used.

The airlines have developed rules for handling various perishable commodities. These include the temperatures desired in transit, the amount of seasonal protection needed, loading methods for various types of containers, and other factors involved in proper handling.

REFERENCES

Barger, W.R. 1955. *Altitude Test with Flowers.* USDA, *AMS Report* No. 59.

Barger, W.R. and A.L. Ryall. 1950. *Effects of Changes in Atmospheric Pressure on Fruits and Vegetables.* USDA, *HT&S Report* No. 225 (May).

IATA *Register of Containers and Pallets.* International Air Transport Association, Montreal, Quebec.

Pentzer, W.T., Jr., et al. 1958. *Air Transportation of Fruits, Vegetables and Cut Flowers: Temperature and Humidity Requirements and Perishable Nature* USDA, *AMS Report* No. 280 (October).

CHAPTER 32

RETAIL FOOD STORE REFRIGERATION

Display Refrigerators 32.1	*Condensing Methods* 32.3
Meat Processing 32.2	*Methods of Defrost* 32.5
Walk-in Coolers 32.2	*Refrigerant Lines* 32.6
Refrigerators and Systems 32.3	*Intereffect between Refrigeration and Air Conditioning* .. 32.6
Load versus Ratings 32.3	*Condensing-Unit Noise* 32.6

REFRIGERATION equipment used in self-service retail food stores may be broadly grouped into *display* refrigerators, *storage* refrigerators, *processing* refrigerators, *mechanical* refrigeration machines, and refrigeration. Chapters 35 and 36 of the 1988 ASHRAE *Handbook—Equipment* present this equipment.

DISPLAY REFRIGERATORS

Open, self-service refrigerated display cases for normal and low temperatures are the most widely used refrigeration equipment in food markets. However, glass-door multideck models have rapidly gained in popularity. Many operators combine single-deck and multideck models in most departments where perishables are displayed and sold. Service refrigerators are sometimes used to display delicatessen food and frequently to display fish with crushed ice, instead of or supplemented by mechanical refrigeration. More complex layouts of display refrigerators have developed as new or remodeled stores strive to be distinctively different and more attractive. Refrigerators are allocated in relation to expected sales volume in each department, such as meat or dairy. Thus, floor space is allocated to provide for balanced stocking of merchandise and smooth flow of traffic in relation to expected peak volume periods.

The small store accommodates a wide variety of merchandise in a limited floor space. Thus, display refrigerators installed in small and medium size stores tend to be the glass reach-in type, which can display 50 to 100% more merchandise in the same amount of floor space as single-deck refrigerators. This concentration of large refrigeration loads in a small space makes year-round control of temperature and humidity essential.

Product Temperatures

Display refrigerators are designed to merchandise food to maximum advantage, while providing short-term protection (24 to 72 h) of the food. Temperatures to be maintained are shown in Table 1.

Open refrigerators should not be used to cool the product; the merchandise, when put into the case, should be at or near the proper temperature. Adequate refrigerated storage in another space is therefore needed. Food placed directly in the refrigerator on delivery to the store should come from properly refrigerated trucks. Little or no delay in transferring perishables from storage or trucks to the display refrigerator should be permitted.

Open refrigerators should be loaded properly. The product on display should never be piled so high that it is out of the refrigerated zone or be stacked so densely that circulation of refrigerated air is blocked. The recommendations of the manufacturer should be followed. Proper refrigerator design and loading minimize energy use, maximize the efficiency of the refrigeration equipment, and minimize product loss.

The preparation of this chapter is assigned to TC 10.7, Commercial Food Display and Storage Equipment.

Table 1 Temperatures in Display Refrigerators

Type Fixture	Temperature, °F	
	Minimum[a]	Maximum[b]
Meat, unwrapped		
Display area	35	37
Storage compartment	34	36
Meat, wrapped		
Display area	28	30
Storage compartment	28	30
Produce, display area	35	40
Produce, storage compartment	35	40
Dairy	35	38
Frozen food	[b]	−5
Ice cream, dough, juice	[b]	−12

[a] These temperatures are air temperatures, with thermometer in the refrigerated airstream and not in contact with the product.
[b] Minimum temperatures for frozen foods and ice cream are not critical; maximum temperature is important for proper preservation of product quality.

Store Ambient Effect

Display fixtures are materially affected by temperature, humidity, and movement of surrounding air (Table 2). These fixtures are designed primarily for supermarkets, virtually all of which are either air conditioned or in cool climates. The effect of store ambient on refrigerators is described in Chapter 35 of the 1988 ASHRAE *Handbook—Equipment*.

The application engineer needs to verify that store ambient conditions year-round are within the performance rating limits of the various fixtures selected for the store. The satisfactory performance of display refrigerators usually becomes seriously affected when store conditions exceed 75 °F and 55% relative humidity, which defines a dew point of 57.5 °F and a humidity ratio of 0.0102 lb of moisture per pound of dry air. Food store refrigeration is not designed and cannot be guaranteed to operate above these ambient conditions. The operator often does not understand the significant effect of humidity on performance of refrigeration equipment, particularly low-temperature equipment. The evaporators for low-temperature cases commonly operate at −30 to −40 °F, so reducing the relative humidity in the store by operating the air-conditioning evaporators at about 35 °F can reduce energy consumption significantly.

Moisture from high humidity also degrades performance by:

1. forming excess frost on the evaporator coil, which blocks the airflow
2. increasing the time required for defrosting
3. condensing in and around the refrigerator

Adverse Heat Sources

Lighting. Refrigeration performance and food temperatures are also adversely affected by display lighting and other forms of radiant heat. High-intensity lighting raises product temperatures several degrees, as shown in Figure 1. Light also discolors smoked,

Table 2 Relative Refrigeration Requirements with Varying Store Ambient Conditions

Case Model	70°F Dry Bulb Temperature Relative Humidity, %					78°F Dry Bulb Temperature Relative Humidity, %		
	30	40	55	60	70	50	55	65
Multideck dairy	0.90	0.95	1.00	1.08[a]	1.18[b]	0.99	1.08[a]	1.18[b]
Multideck low temperature	0.90	0.95	1.00	1.08[a]	1.18[a]	0.99	1.08[a]	1.18[b]
Single deck low temperature	0.90	0.95	1.00	1.08[a]	1.15	0.99	1.05	1.15
Single deck red meat	0.90	0.95	1.00	1.08[a]	1.15	0.99	1.05	1.15
Multideck red meat	0.90	0.95	1.00	1.08[a]	1.18[b]	0.99	1.08[a]	1.18
Low temperature reach-in	0.90	0.95	1.00	1.05[a]	1.10[a]	0.99	1.05[a]	1.10

[a] More frequent defrosts required
[b] More frequent defrosts required plus internal condensation (not recommended)

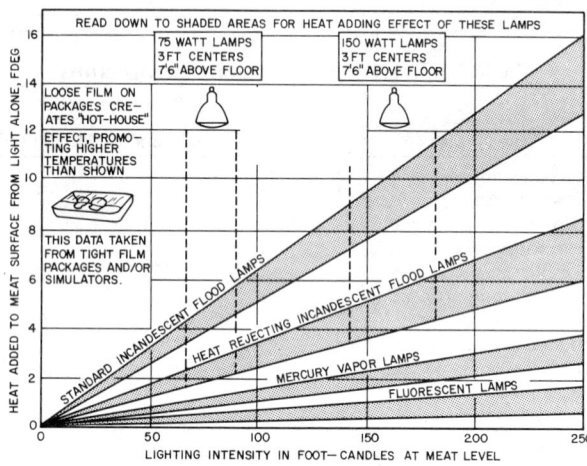

Note: Package warmup may be more than indicated. Standard flood lamps are clear PAR 38 and R-40 types. Blue, pink, or other hues in these types reduce the amount of light, but not the heat load.

Fig. 1 Heating Effect of Lighting on Packaged Meat in Open Display Refrigerators

cured, and table-ready meats. No more than 70 ft-candles of display lighting should be used on smoked, cured, and table-ready meats unless they will be sold within 24 h. Discoloration is directly proportional to both light intensity and length of time displayed.

The heat rejecting lamp is questionable, and conventional incandescent floodlights and supplemental incandescent lighting should never be used to display retail meats.

Ceiling temperatures. An 80°F ceiling can raise the surface temperature of meat in a display case 3 to 5°F. A 100°F ceiling can raise the meat surface temperature 4 to 8°F. Proper ventilation above the ceiling remedies this problem.

Packaging. A loosely wrapped package of meat with an airspace between the film and surface may be 2 to 4°F above the ambient temperature.

Load lines and loading. The following observations indicate the effect of loading:

1. Voids in the display raise the surface temperature of a package in front of a void 2 to 6°F.
2. For all makes and styles of display refrigerators, keeping the top of the product 2 in. below the load line improves the surface temperatures 2 to 6°F.

If meat has been handled with sanitary practices up to the time it is placed in the display case, elevated temperatures can be more tolerable. When there is contamination of the meat surfaces because of dirty knives, meat saws, table tops, and the like, even optimum display temperatures will not prevent premature discoloration and subsequent downgrading of the meat. The elimination of excessive heat from lighting, high ceiling temperatures, and poor display practices should go hand in hand with the institution of strict sanitation practices.

MEAT PROCESSING

In a self-service meat market, the back-room operations of cutting, wrapping, sealing, weighing, and labeling involve precise production control and scheduling to meet varying sales demands. The faster the processing, the less critical the temperature and corresponding refrigeration demand.

The cutting room should not be too dry, but condensation on the meat (providing a medium for bacterial growth) should be avoided by maintaining a reasonable dew-point temperature. Low-velocity fan-coil units are generally used, with precise humidity control being an important concern. Gravity coils are also available and have the advantage of freedom from drafts.

The cutting area generally is cooled to about 45 to 55°F, which is desirable for the personnel, but not low enough for meat storage. Thus, meat should be held in that room only long enough for the cutting operation, packaged as soon as practicable, and displayed or stored at lower temperatures.

The meat-cutting room may be a refrigerated room adjacent to the meat storage cooler or one compartment of a two-compartment cooler. In such a cooler, one compartment is refrigerated at about 28 to 32°F as a meat storage cooler, while the second compartment is refrigerated at about 45°F and used as a cutting and packaging room. Best results, however, are attained when the meat is cut and wrapped in a temperature of 28 to 32°F.

Wrapped Meat Storage

At some point between the cutting-packaging room and the display refrigerator, refrigerated storage for the wrapped cuts of meat must be provided. Without this space, a perfect balance should be maintained between the cutting-packaging rate and the selling rate for each particular cut of meat. Display refrigerators with refrigerated bottom storage compartments, equipped with racks for holding trays of meats, offer one solution to this problem. Wrapped meats stored here are conveniently located near the display shelf. However, the amount of stored meat is not visible, and the inventory cannot be controlled at a glance.

The second method uses a pass-through, reach-in cabinet. This cabinet has both front and rear doors and, when conveniently located between the cutting-packaging room and the display refrigerator, the meats can be passed into the cabinet after wrapping and then withdrawn from the other side for restocking the display refrigerator. Since these pass-through cabinets usually have glass doors, they offer the advantage of making the inventory of wrapped meats visible and, therefore, easily controllable.

The third method uses a section of the backroom walk-in cooler. The cooler is usually equipped with racks into which the trays of meat can be slid. This method of storage also offers visible inventory control although, with many floor plans, it is not possible to locate the cooler conveniently close to the display refrigerators.

WALK-IN COOLERS

Walk-in coolers are required for the storage of products such as meat, produce, dairy products, frozen food, and ice cream. Medium and large stores have separate produce and dairy coolers. Meat coolers are used in all food stores with storage conditions

Retail Food Store Refrigeration

between 28 and 32°F. Meat, fish, and poultry should be stored separately in coolers. Moisture conditions must be confined to a relatively narrow range because an excessive humidity level encourages the growth of bacteria and mold, which leads to sliming. A level that is too low leads to excessive dehydration. Air circulation must be maintained at all times to prevent stagnation, but should not be so rapid as to cause drying out or burning. Forced-air blasts must not be permitted to strike products, so gravity-type coils or low velocity coils are usually selected.

Low-temperature storage capacity equivalent to the total volume of the low-temperature display equipment in the store is satisfactory. Storage requirements can be reduced by the use of central warehousing with frequent deliveries.

Forced-air coils are generally selected for low-temperature coolers because humidity is not critical. For low-temperature coolers, electric heat, hot gas, and latent gas defrosts have been successful. Off-cycle defrosts find application for produce and dairy coolers. Straight-time or time-initiated, time or temperature terminated gas or electric defrosts are generally used for meat coolers.

REFRIGERATORS AND SYSTEMS

All types of perishable foods are sold in food stores, which require a variety of refrigeration systems to best preserve and most dramatically display each product. The variations in refrigerators lead to at least a dozen different suction pressure requirements, starting with a produce processing room (when used) on down to the ice-cream level. Produce prep rooms may approach the suction pressures used in air-conditioning applications. Open ice-cream display cases may have suction pressures as low as −35°F. All other cases and coolers fall in between.

Temperature controls also vary greatly, from a produce prep room (which may operate with a wet coil) requiring no defrost, to the ice-cream case requiring induced heat to defrost the coil periodically. Various solutions include: (1) off-time defrost, (2) hot gas defrost, (3) latent heat defrost, (4) electric defrost, and (5) defrost using ambient air induced into the refrigerator.

Various Design Solutions

The selection of high-side equipment to operate display refrigerators and storage rooms for food stores involves these considerations: (1) refrigeration function; (2) reliability; (3) maintainability; and (4) operating efficiency. Solutions span the simplest one compressor and associated controls on one refrigerator to the complex central refrigeration plant operating all refrigerators in a store.

Condensing-unit systems, which use single compressors with multiple cases or coolers, provide one suitable compromise and are popular. One method places produce, dairy, meat, or frozen food refrigerators on a single compressor, which may have its own condenser or may have a remote condenser.

Another common refrigeration technique couples two units in parallel. Oil distribution, temperature control at more than one suction pressure, and defrosting must be considered. Small parallel units operate like a large condensing unit. The condensers are usually remote air cooled, but they can also be built as part of the unit assembly.

Carrying the sequence one step further leads to a large assembly tying three to six units in parallel. The same needs as defined for two units in parallel apply to the larger styles. However efficiency is sacrificed, when suction pressures with great difference are grouped in one system. More compressors do allow cycling in smaller steps for capacity control. However, capacity control by disabling cylinder banks or controlling compressor speed is better than cycling multiple compressors. Also suction pressures may be varied in a nearly linear and energy efficient manner.

Running all the refrigeration equipment at the low suction pressure required for ice-cream refrigeration is inefficient; thus, for large parallel systems, the ice-cream and meat refrigeration is frequently isolated. Single compressor satellites tied into the parallel compressors are often used for small ice-cream or meat loads, while split suction, parallel equipment is commonly used for larger loads. In both systems different suction pressures are obtained, but all compressors discharge into a common header. Moreover, the refrigerating system must be highly reliable because it must keep operating 24 h a day for ten or more years, protecting the large investment in highly perishable foods.

LOAD VERSUS RATINGS

Food store refrigerator manufacturers publish ratings to match the proper condensing unit with fixture load. For single compressor applications only, the ratings can be stated, for selection convenience, as the capacity the condensing unit must deliver at an arbitrary suction pressure (evaporator temperature). In general, manufacturers of open display refrigerators use ASHRAE *Standard* 72-1983, which specifies a standard rating condition of 75°F and 55% rh, which applies to the sales area. Display refrigerators are a specialty and manufacturers' recommendations must be followed to achieve proper results.

Multiplexed Systems

When dissimilar systems are connected to the same condensing system, the manufacturer should be consulted to determine the true maximum suction pressure at the fixture refrigerant line outlet for each system and the true load that such a system adds to the total. The condensing unit must deliver the total of all the loads at a suction pressure at the machine no higher than the lowest system pressure requirement, less the suction line pressure drop. The other systems should have suction line restriction (by evaporator pressure regulators or some other sure means) to prevent higher temperature evaporators from adding unnecessarily to the load. The regulator then assures that the condensing unit will pull down to the pressure necessary for the coldest evaporator system.

Electronically or mechanically actuated pilot operated EPR valves are used to control evaporator pressures in multiplexed systems. These valves cause little or no measurable pressure drop when they are in the full open position. Another design uses liquid pressure to open and close the EPR valves, so the pressure drop is negligible. The true suction gas temperature leaving display fixtures is often superheated 43 to 77°F. Particularly on low-temperature fixtures, suction line gas temperature increase from heat picked up from the store ambient can be substantial. This increase, which affects condensing unit capacity, must be considered for system design; both capacity effects and the cooling needs of accessible hermetic compressors are involved.

One solution is to place the suction line and liquid line tight together from the refrigerant line outlet, with the pair insulated together a distance of 30 to 60 ft from the fixture outlet. This technique cannot be used with gas defrost, however. Most manufacturers have available suction to liquid heat exchangers, which is another good solution. This technique allows the suction gas to pick up heat from the liquid instead of the store ambient and converts a loss into a gain. The engineer should verify that the liquid entering the fixture is subcooled as planned and that this anticipated capacity did occur. Some case and/or system designs require liquid line insulation. Manufacturer recommendations should be followed.

CONDENSING METHODS

Many commercial refrigeration installations use air-cooled condensers. Alternatively, evaporative condensers or cooling towers may be specified. To obtain the lowest operating costs, equipment should operate at as low a head pressure as possible: 70 psi drop across the expansion valve for R-12, 100 psi for R-22 or R-502.

However, these numbers are arbitrary, because no special care is taken to survive lower condensing temperatures.

Techniques that permit a system to operate satisfactorily with lower condensing temperatures include: (1) insulation of liquid lines and/or receiver tank; (2) subcooling of the liquid refrigerant by design; and (3) receiver connected as a surge tank with appropriate valving. Condensing pressure must still be controlled, at least to the lower limit required by the expansion valve, heat recovery, and gas defrosting. The typical thermostatic expansion valve is capable of feeding with a solid liquid column of refrigerant, assuming a solid liquid column of refrigerant is always supplied to the expansion valve. Balanced port thermostatic expansion valves allow liquid subcooling by mechanical or natural means, which further reduces compression ratios because the valve maintains the proper evaporator feed.

Air-Cooled Machine Room

The arrangement of standard air-cooled condensing units in a separate air-cooled machine room is less prevalent today, but is still used in supermarkets. Fans or motorized dampers controlled by a room thermostat supply air to the room; fans or blowers, also controlled from room temperature, exhaust the air.

A complete air-cooled condensing unit located indoors requires ample, well-distributed ventilation. For winter operation, the exhaust fans are thermostatically controlled. Ventilation requirements vary depending on maximum summer conditions and evaporator temperature, but 750 to 1000 cfm per total condensing unit horsepower has given proper results. Exhaust fans should be spaced for an even distribution of air, allowing one exhaust fan for every 6 to 10 ft of condensing unit (see Figure 2).

Rooftop air intake units should be sized for 750 fpm velocity or less to keep airborne mist from entering the room. About 2 to 2.5 ft^2 of intake louvre area per condensing unit horsepower gives good results. When condensing units are stacked, turning vanes (as shown in Figure 2) assure that upper units receive adequate ventilation. Rooftop intakes are preferred because they are not as sensitive to wind as sidewall intakes, especially in winter in cold climates. In practice, motorized dampers do not seal wall-mounted louvres sufficiently, so they must be boarded up during winter extremes to maintain the compressor room temperature at the desired 70°F. Butterfly dampers installed in upblast exhaust fans, which are controlled by a thermostat in the compressor room, control and prevent the recirculation of warm air from other condensers near the blast fans.

The air baffle prevents intake air from being short-circuited to the exhaust fans (Figure 2). The intake air should not be baffled to flow only through the condensers, because of difficulty in obtaining both proper winter control and proper summer condenser fan operation.

Ventilation fans for air-cooled machine rooms do not normally have a capacity equal to the total of all the individual condenser fans. Therefore, if the air is baffled to flow only through the condenser during maximum ambient temperatures, the condensers will not receive full free air volume when all or nearly all condensing units are in operation. Also, during winter operation, tight baffling of the air-cooled condenser would prevent or complicate recirculation of condenser air, which is essential to maintaining sufficiently high room temperature and condensing temperature for proper refrigeration system performance.

The machine room air entering the condensers should be maintained as close as possible to 70°F operating temperature during the cooler months to maintain adequate head pressures in winter.

Remote Air-Cooled Condenser

The remote condenser may be placed outdoors or indoors to heat portions of the building with rejected heat in winter. Regardless of the arrangement of the remote condenser or the indoor-

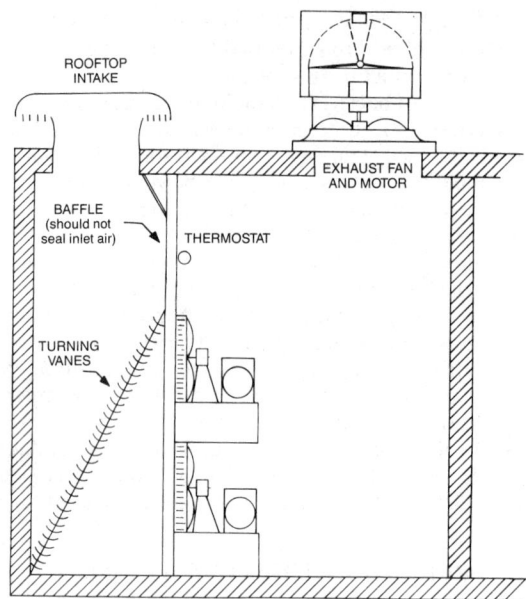

Fig. 2 Typical Air-Cooled Machine Room Layout

outdoor mechanical package, the following design points are relevant.

The air-cooled condenser may be either a single-circuit or a multiple-circuit condenser. The manufacturer's heat rejection factors should be followed to ensure that the proper or desired temperature difference is accommodated.

The head pressure must be controlled on most outdoor condensers. Fan cycle controls work well down to 50°F on condensers with single compressors or with parallel groups of compressors. Below 50°F, condenser flooding (with the refrigerant) can be used alone or with fan controls. Flooding does require a larger refrigerant charge and larger liquid receivers. In contrast, split condensers with solenoid valves in the hot gas lines and bleed-off in the liquid return lines have the capability of reducing the condenser surface during cold weather. Natural subcooling must be carefully integrated into the design, but it can save energy.

Fans are controlled by ambient sensing thermostats, pressure controls, or a combination of both. Sometimes the condenser fan, in conjunction with gravity louvres, is cycled by pressure switches. No refrigerant flooding charge is required by this system.

The sizing of the receiver tank for all types of high-side design, especially for remote condensers, must be considered. Remote condenser installations, particularly when associated with heat recovery and latent heat defrost systems, have substantially higher internal high-side volume than other types of systems. Much of the high side is capable of holding liquid refrigerant, particularly if runs are long and lines are large.

Roof-mounted condensers should have at least 3 ft of space between the roof deck and the bottom of the condenser slab to minimize the radiant heat load from the roof deck to the condenser surface. Also, free airflow to the condenser should not be restricted. Remote condensers should be placed at least 3 ft from any wall, parapet, etc. Two side-by-side condensers should be placed at least 6 ft from each other. In Chapter 14 of the 1989 ASHRAE *Handbook—Fundamentals,* the problems of locating equipment for proper airflow are discussed in detail.

Evaporative Condenser Arrangements

Evaporative condensers are also available as single or multiple-circuit condensers. Manufacturer conversion factors for operating the condensing temperature and wet-bulb temperature must be applied to determine the required size of the evaporative condenser.

In cold climates, the condenser must be installed to assure against freezing during winter. Evaporative condensers demand a regular program of maintenance and water treatment to assure uninterrupted operation. The receiver tank should be capable of storing the extra liquid refrigerant during warm months. Line sizing must be considered to keep a reasonable tank size.

Closed water condenser/evaporative cooler systems see considerable service. In this arrangement, an evaporative condenser cools water instead of refrigerant. This water flows in a closed, chemically stabilized circuit with a regular water-cooled condenser (a two-stage heat transfer system). Heat from the condensing refrigerant transfers to the closed water in the regular water-cooled condenser. The warmed water then passes to the evaporative cooler, where it is cooled by evaporation.

The water-cooled condenser and evaporative cooler must be selected considering the temperature difference (1) from the refrigerant to the circulating water, and (2) from the circulating water to the available wet-bulb temperature. Unless each of these devices is somewhat oversized, the double temperature difference results in higher than normal head pressures. On the other hand, this arrangement causes no corrosion inside the condenser itself, because the water flows in a closed circuit and is chemically stabilized.

The extreme temperature of the entering discharge gas is the prime cause of evaporative condenser contamination and corrosion. The severity and importance of contamination and corrosion can be substantially reduced by using the closed water condensing arrangement. The extent of this corrosion reduction relates directly to the temperature reduction between discharge temperatures experienced even with generously sized evaporative condensers, on the one hand, versus the entering water temperature designed into the closed water circuit on the other.

Water flow in the closed water circuit can be balanced between multiple condensers on the same evaporative cooler circuit with pressure-activated, water-regulating valves. However, low head pressures are usually prevented by temperature control of the closed water circuit. Thus, balancing valves or three-way valves provide satisfactory water distribution control between condensers.

Cooling Tower Arrangements

Few supermarkets use water-cooled condensing units, since the trend is toward air cooling. Nearly all water-cooled condensing units are installed with a watersaving cooling tower because of the shortage or high cost of water and sewage disposal.

The application of water cooling towers for air conditioning is an entirely different engineering problem for food store refrigeration. This is because of the hours of operation required for perishable foods compared to that for space conditioning. It is also because refrigerator operation is required during all four seasons. In particular, cooling towers must survive severe winters in some applications. Year-round control of condensing pressure by thermostatic control of the tower fan must be provided. The control is usually set to turn the fan off when the water temperature drops to a temperature that produces the lowest desired condensing pressure. Water regulating valves are sometimes used in a conventional manner. Dual-speed fan control is also used.

Some engineers use balancing valves for water flow control between condensers and rely on water temperature control to avoid low head pressure. Proper bleedoff and regular water treatment is required to ensure the satisfactory performance and full life of the cooling towers, condensers, water pumps, and piping. Specialists in the field of water treatment should be consulted because each locality has different water and atmospheric conditions. A regular program of water treatment is mandatory.

METHODS OF DEFROST

The most common defrost methods are condensing unit or system off-time, electric heat, latent heat, and air defrost.

Condensing Unit Off-Time

This method simply shuts off the unit and allows it to remain off until the evaporator reaches a temperature that permits defrosting and gives ample time for condensate drainage. Since the heat for this method is obtained from the air circulated in the fixture, the method is quite slow. It is usually limited to fixtures maintaining temperatures of 34 °F or above. Defrost may be controlled by: (1) suction pressure control; (2) time clock initiation and termination; (3) time clock initiation and suction pressure termination; and (4) time clock initiation and temperature termination.

Suction pressure control. This control is adjusted for a cut-in pressure high enough to allow defrosting during the off-cycle. This method is usually used in those fixtures maintaining temperatures from 36 to 43 °F. It provides some cooling effect during the defrost period because air is circulated over melting ice. If an excessive heat or humidity load should occur and cause icing of the evaporator, the evaporator pressure will be lowered to the cutout point of the control, thus initiating a defrost cycle to clear the evaporator.

However, condensing units and/or suction line may, at times, be subjected to low ambient temperatures below the temperature of the evaporator. This prevents the buildup of suction pressure to the cut-in point and allows the condensing unit to remain off for prolonged periods. In such instances, fixture temperatures may become excessively high.

A similar situation can exist if the suction line from a fixture is installed in a trench or conduit with numerous other cold lines. This may also prevent the suction pressure from building up to the cut-in point of the control.

Methods 2, 3, and 4 of controlling off-cycle defrosting use defrost time clocks to break the electrical circuit to the condensing unit initiating a defrost cycle. The difference lies in the manner in which the defrost period is terminated.

Time initiation and termination. A timer initiates and terminates the defrost cycle from clock movement. The length of the defrost cycle must be determined and the clock mechanism set accordingly.

Time initiation and suction pressure termination. This method is similar to the aforementioned, except that the defrost cycle is terminated by suction pressure. The length of the defrost cycle is automatically adjusted to the condition of the evaporator, insofar as frost and ice are concerned. However, to overcome the problem of the suction pressure not rising because of the defrost cut-in pressure previously described, the timer has a fail-safe setting in its circuit to terminate the defrost cycle after a preset time, regardless of suction pressure.

Time initiation and temperature termination. This method is also similar to the time initiation and termination method, except that the defrost cycle is terminated by temperature. The length of the defrost cycle is automatically adjusted to the condition of the evaporator insofar as frost and ice is concerned. A temperature sensor is located on a tube of the evaporator, which always has liquid present, or in the airstream leaving the evaporator. The timer also has a fail-safe setting in its circuit to terminate the defrost cycle after a preset time regardless of the temperature.

Demand and proportional defrost. This system initiates defrost based on demand (need) or proportioned to humidity or dewpoint. Techniques vary from measuring temperature-spread change of the air entering and leaving the coil, to changing the speed of a clock motor based on relative humidity. Other systems employ a device that senses the frost level on the coil.

Electric Defrost

Electric defrost methods usually apply heat externally to the evaporator, and require a longer defrost period than the latent method, usually about 1.5 times longer. The heating element may be in direct contact with the evaporator, depending on conduction for defrost; or it may be located between the evaporator fans and the evaporator, depending on convection or a combination of conduction and convection for defrost. In either instance, a temperature-limiting device should be placed on or near the evaporator to prevent excessive temperature rise if any controlling device fails to operate.

The electric defrost method simplifies the installation of low-temperature fixtures. The controls involved to automate the cycle usually include one or more of these devices: (1) defrost timer, (2) solenoid valve, (3) electrical contactor, and (4) fan delay switch.

Latent Heat Defrost

Latent heat defrost, often called hot gas defrost, uses heat in the discharge gas from the compressor to defrost the evaporators. Occasionally, supplemental electric heaters are added to assure rapid and reliable defrosting. A timer typically terminates the defrost cycle, although temperature termination is sometimes used.

Air Defrost

Air defrost induces or moves air from the store into the refrigerator. A variety of systems are used; some use supplemental electric heat to ensure reliability.

REFRIGERANT LINES

Sizing of refrigerant lines, both liquid and suction, is more critical in the average refrigeration installation because of the typically long horizontal runs and the frequent use of vertical risers. As in all installations, correct liquid line sizes are essential to assure a full feed of liquid to the expansion valve and proper suction-line oil return to the compressor without excessive pressure drop. Oversizing of liquid lines must also be avoided to prevent system pump-down or defrost cycles from operating improperly in single-compressor systems. For gas defrost systems, however, liquid lines may need to be oversized to remove the condensed liquid from a branch with many evaporators in parallel.

Oil separates in the evaporator and moves toward the compressor at a lower velocity than the refrigerant. Unless the suction line is properly installed, the oil can accumulate at low places in the suction line. Oil that collects at low points can cause such problems as compressor slugging, excessive pressure drop, and reduced system capacity. To prevent these problems, suction lines must pitch down toward the compressor, the bottom of all risers must be trapped, and the refrigerant velocity in risers must be maintained at 1500 fpm. To overcome a large pressure drop, the suction lines may be oversized on long runs, but they still must pitch down toward the compressor.

Manufacturers' recommendations and appropriate line sizing charts should be followed to avoid adding heat to either the suction or liquid lines. In large stores, both suction and liquid lines can be profitably insulated, particularly if subcooling is present.

INTEREFFECT BETWEEN REFRIGERATION AND AIR CONDITIONING

Open display equipment often extracts enough heat to reduce the store indoor temperature as much as 16°F below outdoor ambient. The air-conditioning return-duct system or fan kits withdraw chilled air from the floor in front of the cases and discharge it overhead, where it mixes with the general store air.

Heat Reclamation

Heat reclaim condensers and related controls operate as alternates with the normal refrigeration condensers. They can be used in winter to return most of the refrigeration and compressor heat to the store for heat reclamation. They may also be used in mild spring and fall weather when some heating is needed to overcome the cooling effect of the refrigeration system itself. A further use is for humidity control in spring, summer, and fall months. Excess humidity in the store must be avoided because it can increase the display case refrigeration load as much as 50% at the same dry-bulb temperature. Also, heat reclamation can be used to heat water for store use and for space heating. The section on supermarkets in Chapter 18 of the 1987 ASHRAE *Handbook—HVAC Systems and Applications* has more detailed information on the interrelation of the store environment and the refrigeration equipment.

CONDENSING-UNIT NOISE

Air-cooled condensing units located outdoors, either as single units with weather covers or grouped in prefabricated machine rooms, produce sounds that must be evaluated. The largest source of noise is usually moving air from propeller-type condenser fans. Other sources are motor noise from compressor and fan motors, high velocity gas noise, general vibration, and amplification of sound where vibration is transmitted to mounting structures. The last item is most critical when units are roof-mounted.

A fan speed or fan cycle control is helpful in controlling air noise, because then only the amount of air necessary to maintain proper head pressure is generated. Care should be taken not to restrict discharge air. Whenever possible it should be discharged vertically upward.

Resilient mountings for fan motors and small compressors and isolation pads for larger motors and compressors are helpful in reducing noise. Discharge line mufflers are the best answer to high velocity gas noise. The lining of enclosures with sound-absorbing material is of minimal value. Isolation pads can help on roof-mounted units, but even more important is choosing the right location in regard to the supporting structure so that the noise is not amplified.

If sound levels are still excessive after the foregoing controls have been implemented, location becomes the greatest single factor. Distance from a sensitive area is most important in choosing a location; for each time the distance is doubled, the noise level is halved. Direction is also important. Condenser air intakes should face parking lots, open fields, or streets zoned for commercial use. In sensitive areas, avoid ground-level installation close to building walls, as the walls will reflect the sound.

When it is impossible to meet requirements by the foregoing methods, barriers can be used. While a masonry wall is a very effective barrier, it may be objectionable because of cost and weight. If a barrier is used, it must be sealed at the bottom, because any opening will allow sound to escape. Barriers also must not restrict condenser entering air. Keep the open area at the top and sides at least equal to the condenser face area. When noise is a consideration (1) purchase equipment designed to operate as quietly as possible, (2) choose the location carefully, and (3) use barriers when the first two steps do not meet requirements.

CHAPTER 33

ICE MANUFACTURE

Ice Makers	33.1
Thermal Storage	33.4
Ice Storage	33.4
Delivery Systems	33.5
Pneumatic Ice Conveying	33.6
Commercial Ice	33.6
Ice Source Heat Pumps	33.7

MOST commercial ice production is done with ice makers that produce three basic types of fragmentary ice of a type and size required for a particular application. The basic types of fragmentary ice are plate, tubular, and flake. Chapter 54 of the ASHRAE *Handbook and Product Directory—1978 Applications* includes information on block ice manufacturing. Among the many areas where manufactured ice is used are:

- Processing: Fish, meat, poultry, dairy, bakeries, and hydrocooling
- Storage and transporation: Fish, meat, poultry, and dairy
- Manufacturing: Chemicals and pharmaceuticals
- Others: Retail consumer ice; concrete—mixing and curing; and off-peak thermal storage

ICE MAKERS

Flake Ice

Flake ice is produced by applying water on the inside or outside of a refrigerated drum. The drum is either vertical or horizontal and may be either stationary or fixed. Ice removal devices fracture the thin layer of ice produced on the freezing surface of the ice maker, breaking it free from the drum and allowing it to fall into an ice bin, which is generally located below the ice maker.

The thickness of the ice produced by flake ice machines can be varied by adjusting the speed of the rotating part of the machine, varying evaporator temperature, and regulating the water flow on the freezing surface. Production of flake ice is on a continuous basis as contrasted with tube or plate ice, which is made using an intermittent cycle or harvest type of operation. The thickness of ice produced is in the range of 0.04 to 0.12 in. A continuous operation without a harvest cycle results in less refrigeration capacity required to produce a ton of ice than any other type of manufactured ice when similar makeup water and evaporating temperatures are compared. The exact amount of refrigeration required varies by type and design of the flake ice machine. A typical flake ice maker is shown in Figure 1.

All water used by flake ice machines is converted into ice; therefore, there is no waste or spillage. Usually flake ice makers are operated at a lower evaporating temperature than tube or plate ice makers, and the ice is colder when removed from the ice-

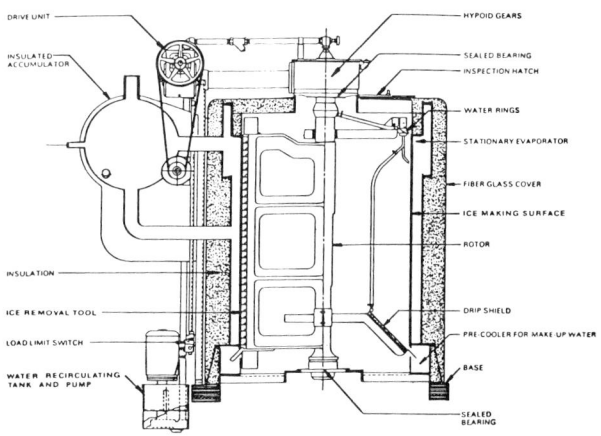

Fig. 1 Flake Ice Maker

making surface. The surface of flake ice is not wetted by thawing during removal from the freezing surface, as is common with other types of ice. Since it is produced at a colder temperature, flake ice is most adaptable to automated storage, particularly when low-temperature ice is desired.

The rapidity of freezing the ice on the freezing surface results in the opaque appearance of flake ice, which is caused by entrained air. For this reason, flake ice is not commonly used for applications where a clear ice appearance is important. For some applications, such as chemical processing and concrete cooling, where rapid cooling is important, flake ice is ideal because the flakes present the maximum amount of cooling surface for a given amount of ice.

When used as ingredient ice in sausage making or other food grinding and mixing, flake ice provides rapid cooling while minimizing mechanical damage to other ingredients and wear on mixing/cutting blades.

Tubular Ice

Tubular ice is produced by freezing a falling film of water either on the outside of a stainless steel tube with evaporating refrigerant on the inside of the tube or freezing water on the inside of tubes surrounded by evaporating refrigerant on the outside.

When ice is produced on the outside of a tube, the freezing cycle is normally from 8 to 15 min., with the final ice thickness from

---

The preparation of this chapter is assigned to TC 10.2, Automatic Icemaking Plants and Skating Rinks.

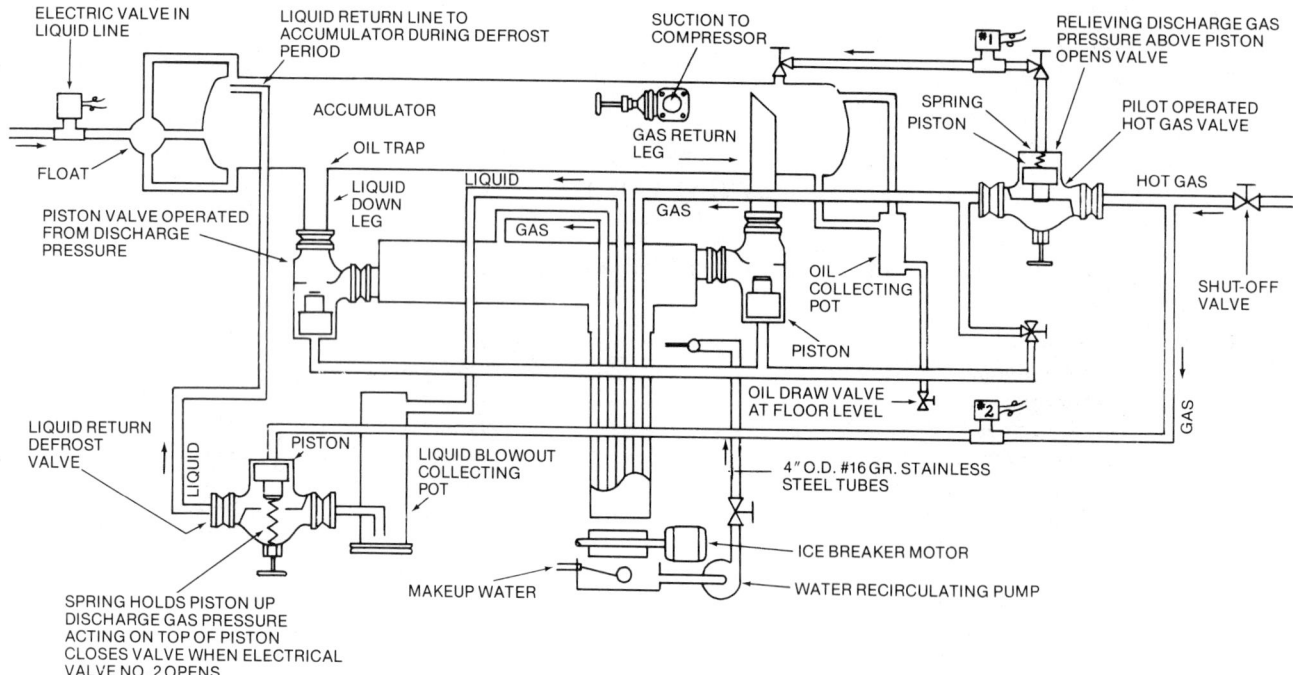

Fig. 2 Making Ice on Outside of Tube

0.2 to over 0.5 in. following the curvature of the tube. The refrigerant temperature inside the tube continually drops from an initial suction temperature of about 25 °F to the terminal suction temperature in the range of 10 to −15 °F. At the end of the freezing cycle, which is controlled by a timer, the circulating water is shut off. Defrost is accomplished by introducing hot discharge gas from the top of a high-pressure receiver. To maintain proper defrost temperatures, typical discharge gas pressure is 160 psia. This drives the liquid refrigerant in the tubes up into an accumulator and melts the inside of the tube of ice, which then slides down the tube through a sizer, mechanical breaker, and down an ice slide through an opening into storage. The defrost cycle is normally about 30 s. The unit returns to the freezing cycle by returning the liquid refrigerant to the tubes from the accumulator.

This type of ice maker operates with refrigerants R-717, R-12, and R-22. Higher capacity units of 10 tons per 24 h and larger usually use R-717. The capacity of the unit increases as the terminal suction pressure decreases. A typical unit with 70 °F makeup water and R-717 as the refrigerant will produce 19.3 tons of ice per 24 h with a terminal suction pressure of 38.5 psia and requires 35.7 tons of refrigeration. This equates to 1.85 tons of refrigeration per ton of ice. The same unit will produce 41.6 tons of ice per 24 h, with a terminal suction pressure of 21 psia and require 80 tons of refrigeration. This equates to 1.92 tons of refrigeration per ton of ice. Figure 2 shows the diagrammatic arrangement of a typical ice maker making ice on the outside of the tubes. Figure 3 shows the physical arrangement for an ice maker that makes ice on the outside of the tubes.

When ice is produced on the inside of a tube, it can be harvested as a cylinder or as crushed ice. The freezing cycle is approximately from 13 to 26 min. The tube is usually 0.9 to 2 in. in diameter, producing a cylinder that can be cut to desired lengths. The refrigerant temperature outside the tube is continually dropping, with an initial temperature of 25 °F and a terminal suction temperature in the range of 11 to −1 °F. At the end of the freezing cycle, which is controlled by a timer, the circulating water is shut off and defrost

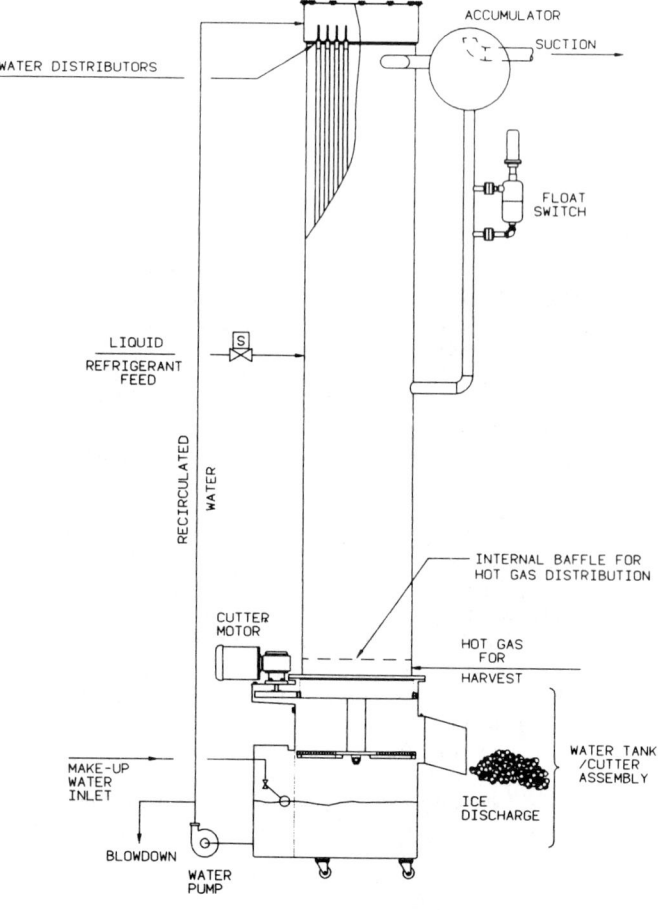

Fig. 3 Tubular Ice Maker

Ice Manufacture

is accomplished by introducing hot discharge gas into the refrigerant in the freezing section. To maintain gas temperature, typical discharge gas pressure is 180 psia. This releases the ice from the tube; the ice descends to a motor-driven cutter plate, which may be adjusted to cut the ice cylinders to the length desired (up to 1.5 in.). At the end of the defrost cycle, the discharge gas valve is closed and the circulating water resumed.

This type of unit can use refrigerants R-717, R-12, and R-22 and the capacity again increases as the terminal suction pressure decreases. A typical unit with 70°F makeup water and R-717 as the refrigerant will produce 43 tons of ice per 24 h with a terminal suction pressure of 40 psia and requires 74.5 tons of refrigeration. This equates to 1.73 tons of refrigeration per ton of ice. The same unit will produce 66 tons of ice per 24 h with a terminal suction pressure of 30 psia and require 135 tons of refrigeration. This equates to 2.04 tons of refrigeration per ton of ice.

Tubular ice makers have the advantage of producing ice at higher suction pressures than other types of ice makers. They can make a relatively thick and clear ice, the curvature of which helps prevent bridging in storage. There is no loss of water during the manufacturing or harvest cycle. Tubular ice makers have a greater height requirement for installation than that of plate or flake ice makers. Provision must be made in the refrigeration system high side to accommodate the volume of refrigerant required for the proper amount of harvest discharge gas. Ice temperatures are similar to those achieved by plate ice makers, but they are generally higher than the temperature capabilities of flake ice makers.

Supply water temperature has a great effect on the capacity of either type of ice maker freezing ice on tubes. If the supply water temperature is reduced from 70 to 40°F, the ice production of the unit will increase approximately 18%. In larger systems, the economics of precooling the water in a separate water cooling system with higher suction pressures should be considered.

Plate Ice

Plate ice makers are commonly defined as those that build ice on a flat, vertical surface. Water is applied above freezing plates and flows by gravity over the freezing plates during the freeze cycle. Liquid refrigerant at a temperature of between −5 and 10°F is contained in internal circuiting inside the plate. The thickness of ice produced is governed by the freezing cycle time. Ice thicknesses in the range of 0.25 to 0.75 in. are quite common, with freeze cycles varying from 12 to 45 min. Figure 4 shows a flow diagram of a plate ice maker using water for harvest. All plate ice makers use a sump and recirculating pump concept whereby an excess of water is applied to the freezing surface. Water not converted to ice on the plates is collected in the sump and is recirculated as precooled water for ice making.

Harvesting of ice from plate ice makers is accomplished by one of two methods. One method involves the application of hot gas to the refrigerant circuit of the plates to warm them to 40 to 50°F, causing the ice surface touching the plate to reach its melting point and thereby release from the plate. The ice falls by gravity to the storage bin below or to a cutter bar or crusher that further reduces the ice to a more uniform size. Plate ice makers using the hot gas method of harvesting are capable of producing ice on two sides of the freezing plate.

The second method of harvesting ice is achieved by flowing warm water on the backside of the plate. In so doing, the refrigerant inside the plate is heated above the ice melting point, and the ice is released. Ice makers of the type using the water warming harvest principle manufacture ice on one side of the plates. Harvest water is prechilled by passing over the plates and collected in the sump and recirculated to become precooled water for the next batch of ice.

For plate ice makers, the freezing time, harvest time, and the related water, pump, and refrigeration are controlled by adjustable

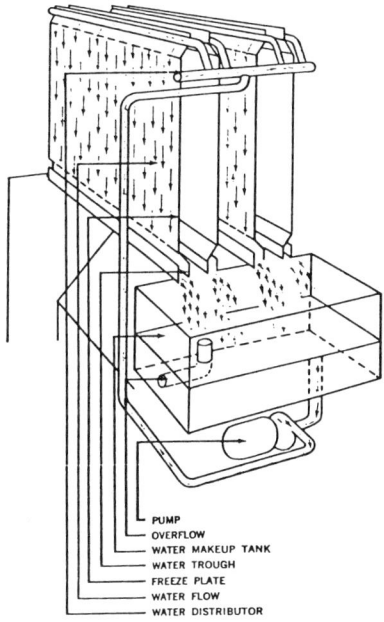

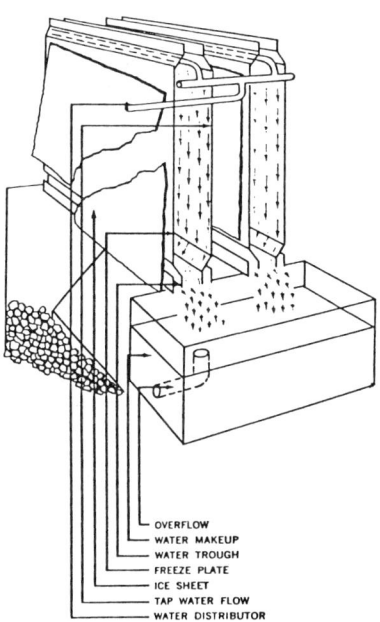

Fig. 4 Plate Ice Maker

electrical timers and relays or electronic devices. Since a wide variety of thicknesses and freezing times is available, plate ice makers can produce clear ice at the longer and slower freezing times. Thus the plate ice maker is commonly used in applications where clear ice is desired.

Because of the harvest cycle involved, plate ice makers require more refrigeration per unit mass of ice production than flake ice makers. This disadvantage is offset by the capability of plate ice makers to operate at higher evaporating temperatures. During the harvest cycle, the suction pressure rises considerably, depending on the design of the ice maker. For this reason, some manufacturers recommend using a separate compressor system for the ice maker. In so doing, a stable suction pressure can be maintained for other refrigeration loads when a common refrigeration system is used for multiple refrigerated requirements; this may occur in large processing plants, refrigerated warehouses, and so forth. Large plate ice makers can be arranged for harvesting of only sections or groups of plates at one time. Proper adjustment of the time spacing for harvesting each section can reduce the fluctuation in suction pressure.

Plate ice makers using the water harvest principle rely on the temperature of the water for harvesting. A minimum of 65 °F for the water is usually recommended to minimize both the harvest cycle time and harvest water consumption in excess of ice making requirements. For installations in cold water areas, or where wintertime inlet water temperatures are low, it is advisable to provide auxiliary means for warming the inlet water to 65 °F.

Ice Builders

Ice builders comprise various types of apparatus in which ice is produced on the refrigerated surfaces of coils or plates submerged in water in an insulated tank. This equipment is commonly known as an ice bank type of water chiller. The ice built on the freezing coils is not used as a manufactured ice product but rather as a means of cooling water circulating through the tank. The ice builder is most often used in applications where high peak and intermittent cooling loads requiring chilled water occur.

Scale Formation

The performance of all ice makers is affected by the characteristics of the inlet water used. Impurities and excessive hardness can cause a scale to be deposited on the freezing surface of the ice maker. The deposit reduces the heat transfer capability of the freezing surface with a resultant reduction in ice making capacity. Deposited scale may also further reduce ice making capacity by causing poor ice removal from the freezing surface during the harvest process. The ice tends to stick on the freezing surface. The rated capacity of all ice makers is based on substantially releasing all the ice from the freezing surface during the removal period. Because the process of freezing water into ice tends to freeze a greater proportion of pure water on the ice maker freezing surface, impurities tend to remain in the excess or recirculated water. A blowdown, or bleedoff, whereby a portion of the recirculated water is bled off and discharged, can be installed. The bleedoff system can control the overconcentration of chemicals and impurities in the recirculated water. Determining the necessity of a bleedoff system and the effectiveness of this concept for controlling scale deposits dependent on local water conditions. Some refrigeration system loss is experienced, since the recirculated water that is bled off to drain has been precooled. Water conditions, water treatment, and related water problems in ice making are covered in detail in Chapter 40 of the 1988 ASHRAE *Handbook—Equipment*, and in Chapter 53 of the 1987 ASHRAE *Handbook—Systems and Applications*.

THERMAL STORAGE

Interest in energy conservation renewed interest in the ice storage concept to provide for thermal storage of cooling capacity for air-conditioning or process applications. Using lower off-peak and weekend power rates, the ice is produced and stored. During the day, stored ice is used to provide the refrigeration of the chilled water system. The design and features of thermal storage equipment are covered in Chapter 46 of the 1987 ASHRAE *Handbook—Systems and Applications*.

ICE STORAGE

Fragmentary ice makers have the capability of producing ice either on a continuous basis or a constant number of harvest cycles per h. The use of the ice is generally not at a constant rate but on a batch basis. Batches may vary from a few hundred lb to many tons per h based on user requirements. The ice must be stored and recovered from storage on demand. Ice storage and storage bin design therefore become important where labor savings, economics, the quantity of ice to be stored, the amount of automation desired, and user delivery requirements are concerned.

Ice makers can produce ice 24 h a day. By making ice during off-shifts and weekends, as well as during work shifts, considerable savings in total ice making and refrigeration system requirements can be achieved. In addition, by using electrical power during off-peak hours, peak loads on the power system are reduced during the day. Many power companies offer reduced rates at off-peak hours.

Ice storages vary in type from short-term to prolonged term storage, with degrees of automation for filling and discharge ranging from manual shoveling to a completely automatic rake system.

Short-term storage generally requires provision for the ice production of one day. The ice maker is mounted over a bin, and ice falls by gravity into the bin. The bin is an insulated, airtight enclosure with one or more insulated doors for access. Ice is removed from the bin by shoveling it into carts. In such a storage, the subcooling effect of the ice generally offsets the heat loss through the insulated bin walls without excessive melting. In most situations, it is not necessary to provide refrigeration units in the ice storage bin where the ice production is being used on a daily basis and where ambient temperatures are reasonable.

Prolonged ice storage requires a refrigerated, airtight, insulated storage bin. Some designs provide for the use of false walls and floor, which produce an envelope effect allowing cold air to circulate completely around the mass of ice in storage.

Time and pressure affect the storage quality of fragmentary ice. Even though a bin is refrigerated to a temperature well below 32 °F, pressure can cause local melting near the bottom. Thus, there is a limit to the size and configuration of a gravity-filled storage bin. The ice falling from an ice maker forms a cone directly underneath the drop in the bin. With slight variation because of the type of ice, the angle of repose is approximately 30°. Fusion of ice under pressure limits the practical ice storage depth from 10 to 12 ft. To use the volume of the bin more efficiently, a leveling screw mounted in the overhead can be used to carry the ice away from the top of the ice cone.

There is also a practical limit to the size of a storage bin in which ice can be manually removed through refrigerator doors. The simplest device used to manually remove ice from the bin is a screw conveyor with trough at floor level, which is equipped with gratings and removable sectional covers. The removable covers protect the screw from ice blockage when the conveyor is not running. The gratings are for the protection of personnel.

Ice Rake System

The ice rake system is used for larger and fully automated storages. These storages generally have a 50 to 300-ton capacity for a single rake system. Depending on plant demands, combinations of rake systems can be developed into an integrated production, storage, and delivery system with capabilities of up to 1000 tons storage with multiple delivery systems by screw conveyor or pneumatic conveying. A range of delivery rates up to 60 tons of ice per hour can be achieved. The advantages of such rake systems provide for the elimination of labor for storing and transporting ice, longer and more effective distribution systems, faster delivery and termination of ice flow, less waste of ice, and elimination of physical contamination.

Storages that incorporate rake systems are of two basic types. One type encloses the ice storage and rake system in an arrangement of steel framework and panels with the complete unit installed in a refrigerated room. The second type involves the construction of an insulated enclosure around the ice bin and rake

Ice Manufacture

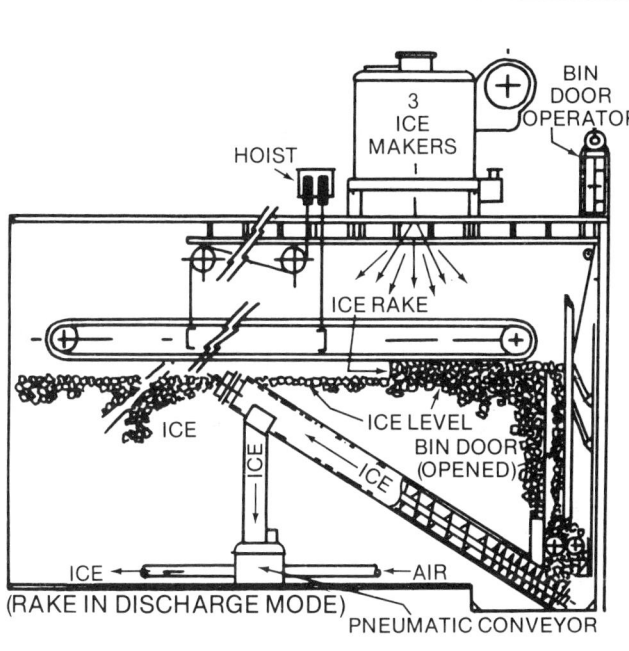

Fig. 5 Ice Rake System

system. This type can be installed in a building or outside, depending on the type of weather protection provided. For either type, the ice makers are mounted over the bin outside of the refrigerated space. Figure 5 shows the arrangement of components for a typical rake system.

Once fragmentary ice comes to rest in the storage bin, it exerts very little side pressure on the walls and a relatively stable mass is formed. The ice will not flow freely once it has come to rest, and a mechanical force is necessary to start the ice moving. The deeper the ice is stored, the greater the pressure on the ice near the bottom. Since pressure generates heat, the ice near the bottom tends to fuse together faster. Thus it is impractical to remove the ice from storage using a system that works from the bottom of the pile. Rake systems work from the top of the ice; they continuously level and fill the storage bin, as well as automatically remove the ice on demand. The systems operate in nonrefrigerated or refrigerated bins. Since most users of large storages also want dry ice for ease of handling, large automated storages are usually refrigerated.

The ice rake itself consists of a structural steel mechanism with drive, which operates similar to the tracks of a crawler tractor. By means of a hoist and timer, the rake is raised or lowered to maintain automatically its position suspended and in close contact with the ice level. Wide scraper-type conveyors, mounted across the tracks the full width of the bin, spread the ice out and drag it toward the back of the bin during the filling mode. To dispense the ice at delivery, the scraper conveyors reverse direction and drag the ice to the opposite end of the bin, dropping it into a screw conveyor mechanism. From this point, the ice is transferred to the external delivery system of screw conveyors.

Some rake systems have features that allow ice deliveries from the bin to be remotely controlled and volumetrically metered. Ice deliveries can be recorded on digital counters at the storage bin, remote stations, and control centers. Accuracy is in the range of ±2%. Another method of metering ice from a rake system storage has the screw conveyor deliver the ice to a weigh belt. As the ice passes along the moving belt, it is electronically weighed and the weight is recorded. Selection of the weigh belt material carrying the ice is critical to prevent ice from sticking to the belt. The weigh belt is often installed in a refrigerated area adjacent to the ice storage.

Another type of ice storage with delivery system capabilities is the *live bottom* type with a multiplicity of screws arranged in various configurations on the bottom of the storage bin. Because of ice fusion, these bins are limited to short-term storage. The limited success of this type of bin depends on the type and quality of fragmentary ice being stored and the ability of the design to overcome particle fusion. Particle fusion results in a condition in which the ice bridges over the top of the screws, and the screws bore holes in the ice rather than empty the bin.

Primarily for the consumer bagged-ice industry, a bin and automatic storage system is used in which the entire floor moves, carrying the ice load into slowly rotating beaters. As the ice breaks loose, it drops to a screw conveyor, which feeds an ice bagger. This type of bin is located in a refrigerated room, and the ice makers are located away from the bin, with the ice generally flowing from ice maker to storage bin by gravity. The ice makers must be shut off so that no ice can flow into the storage during the discharge and bagging process.

The ice silo is used for long or short-term storage with capacities in the range of 20 to 100 tons. The silo tank comprises a cylindrical part and a tapered, conical part leading the ice to the outlet at the bottom of the tank. From this point, the ice is transported by a screw delivery system. A rotating flexible chain arrangement is provided in the silo to assist in ice removal and to partially overcome the fusion problem. The ice maker is mounted over the top of the silo, and the ice falls into the storage. No leveling of ice is required in the bin, since the diameter of the silo is sized to be compatible with the ice maker. Of basically European origin, the silo is relatively successful. The larger the ice storage, the higher the silo, so that proper ice discharge becomes more critical in the design when the fusion effect of ice and the fact that the ice must finally pass through a relatively small opening at the bottom of a tapered zone are considered.

DELIVERY SYSTEMS

The point of ice manufacture is rarely the point of ice usage. Usually it is necessary to move ice from the ice machine or storage bin to some other area where it will be used; thus, a conveying system is required. Most conveyor applications use screws, belts, or pneumatic-type systems. Great care must be taken in selecting the size and type of conveyor to be used, since no matter what type of fragmentary ice is being handled, problems such as fines, freezeup, and ice jams can be encountered with an improperly designed system. Fines, or snow, describes the small particles of ice that chip off the larger pieces during harvesting, crushing, or conveying operations.

Screw and Belt Conveyors

Screw conveyors are the most popular of all the conveyances used for transporting ice. Screw conveyors are manufactured in sizes of 4 in. diameter and up, as well as in various screw pitches. Most ice-conveying operations use 6, 9, and 12 in. diameter screws.

The sizing and drive power requirements of screw conveyors are determined by the ice delivery rate, the inclination of the conveyor, and the conveyor screw pitch. With fragmentary ice, the selection of too small a conveyor will result in excessive conveyor speed or require that the conveyor run too full of ice. These conditions can produce excessive fines.

When screw conveyors transport ice through high ambient inside areas, or outside in the weather, such as in icing fishing vessels, it is advisable to insulate the screw conveyor trough and provide the conveyor with insulated covers. Rain is as much a problem as sunshine for contributing to ice meltage and delivery problems. For this reason, most screw conveyors operating in the weather are provided with sectional and removable covers.

Belt conveyors are often used when excess moisture has to be removed from the ice or to minimize the fines. The belts are of a mesh type to allow snow and excess water to fall through. Stainless steel, galvanized steel, or high-density polyethylene are commonly used for belting.

PNEUMATIC ICE CONVEYING

Pneumatic ice conveying systems have proved desirable, economical, and practical when transporting fragmentary ice distances of 100 ft or more and when multiple delivery stations must be served. A pneumatic system is advantageous when delivery stations are in different directions or at different elevations, when delivery through a pressure hose is needed, or when flexibility is required for future changes or the addition of delivery stations.

The basic principle of conveying ice by a pneumatic system involves the use of a rotary blower, which delivers air to a rotary airlock valve, or conveying valve. Ice is fed into the conveying valve, and the mixture of air and ice is conveyed at high velocity through thin-walled tubing (aluminum, stainless steel, or plastic) by compressed air. The diagrammatic arrangement of pneumatic system components are shown in Figure 6.

Delivery rates are generally between 10 to 40 tons per h, and conveying distances up to 500 ft are common. Delivery distances in excess of this distance can be achieved at reduced delivery rates, with the maximum distance practical being approximately 1000 ft. Conveying pressures range from 4 to 8 psig, the pressure depending on the delivery rate and the maximum distance the ice is to be conveyed. The air velocity required to keep the ice in suspension in the conveying line will vary among the different types of fragmentary ice. Not all types of fragmentary ice can be satisfactorily conveyed by a pneumatic system. The manufacturer of the ice-making equipment should be consulted for recommended line velocities. Storage bins for ice plants using a pneumatic delivery system should be refrigerated to assure cold, free-flowing ice with no moisture. Since the ice remains in the tubing a very short time, tubing insulation is seldom needed or used. The normal tubing used has a diameter of 4 to 8 in. and requires minimum support, making installation easy and economical.

Multiple delivery points are served by automatic Y-type diverter valves, either air or electrically operated. Pneumatically blown ice can be delivered under pressure out of the end of a hose. An alternate method of delivery is by a cyclone receiver, which takes the ice at high velocity, dissipates the air, and drops the ice by gravity. Combinations of hose stations and cyclone delivery stations on the same system are common.

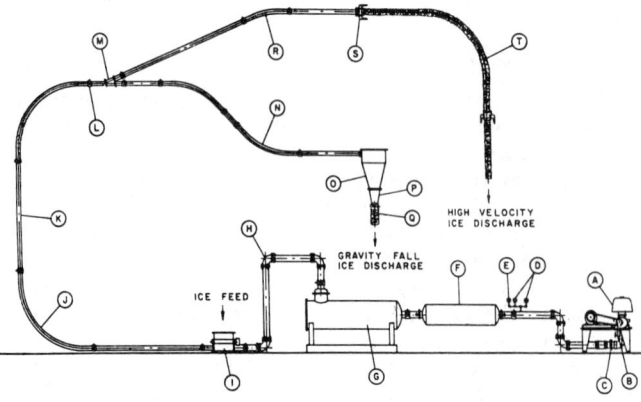

A. Blower
B. Pressure relief valve
C. Check valve
D. Pressure switches
E. Pressure gauge
F. Silencer
G. Heat exchanger (optional)
H. 90° Short radius ell
I. Conveying valve
J. 90° Long radius ell
K. Thin wall tubing
L. Tube coupling
M. Diverter valve
N. 45° Long radius ell
O. Cyclone receiver
P. Hose adapter spout
Q. Flexible hose
R. 22 1/2° Long radius ell
S. Quick coupling disconnect
T. Flexible pressure hose

Fig. 6 Typical Flake Ice Pneumatic Conveying System

When a pneumatic conveying system is used in areas of high ambient and wet-bulb temperatures on systems requiring higher conveying pressures, a heat exchanger is often used to cool and dehumidify the pneumatic air prior to entry to the conveying valve. The heat exchanger is provided with a cooling coil, either refrigerant or chilled-water cooled, a demister or other means of separating moisture from the air, and a condensate trap to expel the entrapped moisture. Geographical location, system pressure, and the quality and use of the ice at the delivery point must be considered when determining whether or not to use a heat exchanger.

Slurry Pumping

A mixture of particle ice and water can be pumped as a slurry. This method has some advantages for transporting ice. Generally, the slurry mix is approximately 50% water and 50% ice. For specialized application, mixtures of up to 80% ice and 20% water can be successfully pumped. Delivery distances of 800 ft have been achieved with delivery rates of 60 tons per h of slurry mix. This concept has been extensively used in the produce industry and has potential in concrete cooling, chemical processing, and other ice or chilled-water related applications. Ice slurry mixes are of particular interest where there is a requirement for low-temperature chilled water at or near 32°F. In converting ice to water, the absorption of latent heat at the usage point enables more cooling to be done with a slurry mix than with straight chilled-water cooling. Pumping volumes and line sizes are minimized, and ice meltage during mixing and pumping does not normally exceed 1 to 3%. The system has a capability of automation for continuous operation.

The basic system for slurry pumping involves a mixing tank in which the ice and water are mixed. The ice is carried by any of the conventional conveying methods from the ice storage bin to the slurry mix tank. Agitators in the tank operate continuously to maintain a mixture uniform in consistency. The slurry mix is then discharged by pumps through pipelines to the usage points. The pumps are of the centrifugal type, modified for pumping slurry. When the icing cycle at the usage points is by intermittent demand, a recirculating system is used to return unused slurry back to the mixing tank. In this way, the slurry is kept moving at all times, and the possibility of ice blockage in the lines is eliminated. The temperature of the slurry solution in the tank is maintained at 32°F.

The fresh produce industry offers a unique application for slurry mixes. Body icing of the fresh produce, which is generally of nonuniform size and configuration, is achieved by applying the slurry mix to the dry packed product. Drain holes are provided in the shipping container to remove the water. Because of its suspension in water, ice is carried to all parts of the container. The ice solidifies as the water drains from the container. The product is then completely surrounded with ice, the voids are filled, and potential hot spots are eliminated. The drained water can be collected and returned to the mixing tank.

COMMERCIAL ICE

Commercial ice is primarily used for human consumption. It is also called *party* or *consumer* ice, and is used for cooling beverages and for other applications in restaurants, hotels, and similar institutions. This type of usage requires packaging at the ice plant for storage and eventual distribution. In bagged form, commercial ice is also made available for sale to the public in grocery stores and automatic, coin-operated vending machines. When the ice is to be used in beverages, ice produced by plate or tube ice makers is preferred because of the clear appearance and the fact that it can be made in greater thicknesses.

A packaging system normally comprises two pieces of equipment: an ice bagger and a bag closer. These components are available from ice packaging equipment manufacturers in various

Ice Manufacture

types and sizes. In the bagging process, the ice is fed from the ice storage bin into the bagging machine by means of a screw or belt conveyor. The bagging machine meters ice into a bag placed below the discharge chute of the bagger. The amount of ice measured into the bag can be determined by weight, volume, or sight approximation, depending on the equipment used.

When packaging by weight, the ice bag is placed on a weighing table located on the bagging machine. Ice is then dispensed into the ice bag until the desired weight of ice is in the bag. At this point, a switch mounted on the weigh scale stops the flow of ice.

The volumetric bagging machine uses a method in which the ice is deposited in a rotating chamber, which is adjustable in volume. After a predetermined volume of ice enters the changer, the ice is discharged into the bag below. Since the shape and size of the ice is not constant, the volumetric chamber is usually set to produce a 3 to 5% overage by volume. Therefore, the proper minimum weight of ice in the bag is assured. The bag closer consists of a mechanical unit that ties and seals the top of the bag with a wire ring, wire twist tie, or plastic clip. Smaller bagging operations do not use a bag closer, and the bags are manually closed with plastic ties, wire rings, staples, and so forth.

In more elaborate systems, the ice bag is automatically removed from the bagging machine and automatically closed before it is dropped onto a conveyor, which carries the bagged ice to the refrigerated storage room. The degree of automation for the bagging and closing operation is determined by the number of bags of ice to be produced per day, the size of the bags, and the cost benefit relationship between automated equipment and reduced labor costs.

Ice bags are made of plastic, most often polyethylene, or heavy moisture resistant paper. Plastic bags are used in most modern plants where the size of the bags is 25 lb or less. Paper bags are used for larger sizes.

Packaged ice requires a refrigerated warehouse or room where the ice is stored prior to distribution. The ice storage is sized to meet the daily production of the plant and the distribution requirements. Generally, a bag ice storage has a capability of storing 3 to 7 days' production. Although the ice will not melt at a storage temperature of below 32°F, it is important that the storage be maintained at a temperature between 10 and 25°F. The lower temperature provides for a subcooling of the ice and avoids meltage during distribution. Depending on the type and quality of the ice being used, the bagged ice can contain some water. The percentage of water can range from zero to 5%. For this reason, provision is made in the storage room refrigeration system for the product load of refreezing the water.

ICE SOURCE HEAT PUMPS

Ice-making systems can be configured to provide heating alone, or heating and cooling, for a building or process. The conversion of water to ice occurs at a relatively high evaporator temperature and coefficient of performance compared to air source heat pumps operating at low ambient temperatures. Systems can provide necessary heating, with the resulting ice either disposed of by melting with low-grade heat, such as solar, or used for useful cooling through daily, weekly, or seasonal storage.

The concept was originally considered mainly for residential heating and cooling. Currently, installations are proving feasible for larger structures, such as office buildings. Energy consumption savings resulting from the coefficient of performance of a conventional heat pump system are achieved. Advantage can also be taken of off-peak night and weekend rates to reduce power costs, and the ice produced is used for building cooling requirements. As a result, a system can be developed that consumes less energy and energy that is consumed at a lower rate.

Ice source heat pumps follow two basic approaches. The first involves using the ice builder principle, with coils in a large tank, as the evaporator component of a heat pump system. The second approach uses a fragmentary type of ice maker as the evaporator of the heat pump system, with ice being stored in a tank as a mixture of ice and water. Many variations, combinations, and adaptations may be developed from these basic systems. The requirement for thermal storage of a large quantity of ice dictates new planning in architectural building design. Heating and cooling system designs for the building are also influenced.

BIBLIOGRAPHY

Dorgan, C.E. 1985. Icemaker heat pumps operation and design. ASHRAE *Transactions* 91(1).

Dorgan, C.E., G.C. Nelson, and W.F. Sharp. 1982. Icemaker heat pump performance—Reedsburg Center. ASHRAE *Transactions* 88(1).

CHAPTER 34

ICE RINKS

Applications 34.1	General Rink Floor Design 34.5
Refrigeration Requirements 34.1	Building, Maintaining, and Planing Ice Surfaces 34.7
Ice Rink Conditions 34.4	Rink Fog and Ceiling Dripping 34.7
Equipment Selection 34.4	Imitation Ice Skating Surfaces 34.8

ANY level sheet of ice made by refrigeration (the term *artificial ice* is sometimes used) is referred to in this chapter as an ice rink regardless of use and whether it is located indoors or outdoors.

The freezing of an ice sheet is usually accomplished by the circulation of a heat-transfer fluid through a network of pipes or tubes located below the surface of the ice. The heat-transfer fluid is predominantly a secondary coolant (brine) such as glycol, methanol, or calcium chloride (see Chapter 18 of the 1989 ASHRAE *Handbook—Fundamentals*).

The primary refrigerant is usually R-22 or ammonia, although R-12 and R-502 are sometimes used. The selection of R-22 as an ice sheet heat transfer fluid, or direct refrigerant, is increasing. Direct ammonia has also been used, but it is now prohibited in public buildings by mechanical refrigeration codes.

APPLICATIONS

Most ice surfaces are used for a variety of sports, although some are constructed for specific purposes and are of specific dimensions. Usual rink sizes are:

Hockey. The accepted North American hockey rink size is 85 by 200 ft. Radius corners of 28 ft are recommended by professional and amateur rules. The Olympic and International hockey rink size is 96 by 196 ft, with 20-ft radius corners. Many rinks are 85 by 185 ft, 80 by 180 ft, and 70 by 170 ft and are considered adequate. In substandard size rinks, a corner radius of not less than 20 ft should be provided to permit the use of mechanical resurfacing equipment.

Curling. Regulation surface for this sport is 14 by 146 ft; however, the width of the ice sheet is often increased to allow space for installation and dividers between the sheets, particularly at the circles. Most are laid out on ice sheets measuring 15 by 150 ft.

Figure Skating. School or compulsory figures are generally done on a patch approximately 16 by 40 ft. Freestyle and dance routines generally require an area of 60 by 120 ft or more.

Speed Skating. Indoor speed skating is done on a hockey-size rink. The olympic-size outdoor speed skating track is a 1400 ft oval, 35 ft wide with 392 ft straightaways and curves with an inner radius of 87.5 ft.

Recreational Skating. Recreational skating can be done on any size or shape rink, as long as it can be efficiently resurfaced. Generally, 30 ft^2 is allowed for each person actually skating; 25 ft^2 per skater is acceptable, except where a large number of pre-teens are skating. An 85 by 200 ft hockey rink with 28-ft radius corners has an area of 16,327 ft^2 and will accommodate a mixed group of about 650 skaters.

The preparation of this chapter is assigned to TC 10.2, Automatic Icemaking Plants and Skating Rinks.

Public Arenas, Auditoriums, and Coliseums

Public arenas, auditoriums, field houses, and so forth, are designed primarily for spectator events. They are alternately used for ice sports, ice shows, and recreational skating, as well as for non-ice events, such as basketball, boxing, tennis, conventions, exhibits, circuses, rodeos, and stock shows. The refrigeration system can be designed so that, with adequate manpower, the ice surface can be produced within 12 to 16 h. However, general practice is to leave the ice sheet in place and hold other events on an insulated floor placed on the ice. Significant time, labor, and energy savings are realized with this approach.

REFRIGERATION REQUIREMENTS

The heat load factors considered in the following section include type of service, length of season, usage, type of enclosure, radiant load from roof and lights, and the geographic location of the rink with associated wet- and dry-bulb temperatures. In the case of outdoor rinks, the sun effect and weather conditions must also be considered.

A fairly accurate forecast of refrigeration requirements can be made based on data from a number of rink installations with the pipes covered by not more than 1 in. of sand or concrete and not more than 1.5 in. of ice—a total of 2.5 in. sand or concrete and ice. The usual practice is to specify 1 ton of refrigeration capacity to freeze and maintain a certain number of square feet of rink surface. These values vary greatly depending on the heat-load factors involved. Table 1 gives the ft^2/ton for a range of applications.

Table 1 Refrigeration Requirements

4 to 5 Winter Months, Above 37° Latitude	
	ft^2/ton
Outdoors, unshaded	85 to 300
Outdoors, covered	125 to 200
Indoors, uncontrolled atmosphere	175 to 300
Indoors, controlled atmosphere	150 to 350
Curling rinks, indoors	200 to 400
Year-Round (Indoors) (Controlled Atmosphere)	
	ft^2/ton
Sports arena	100 to 150
Sports arena, accelerated ice making	50 to 100
Ice recreation center	130 to 175
Figure skating clubs and studios	135 to 185
Curling rinks	150 to 225
Ice shows	75 to 130

Table 2 Ice Rink Heat Loads, Indoor Rinks

Load Source	Approx. Max. Percentage of Total Load, %	Max. Reduction Through Design and Operation, %
Conductive loads:		
Ice resurfacing	12	60
System pump work	15	60
Ground heat	4	80
Header heat gain	2	40
Skaters	4	0
Convective loads:		
Rink air temperature	13	50
Rink humidity	15	40
Radiant loads:		
Ceiling radiation	28	80
Lighting radiation	7	40
Total	100	

Table 3 Ice Rink Heat Loads, Outdoor Rinks

Load Sources	Approx. Max. Percentage of Total Load, %	Max. Reduction Through Design and Operation, %
Conductive loads:		
Ice resurfacing	9	50
System pump work	12	60
Ground heat	2	40
Header heat gain	1	30
Skaters	1	0
Convective loads:		
Air velocity	0 to 15	10
Air temperature	0 to 15	0
Humidity	0 to 15	0
Radiant loads:		
Solar load	10 to 30	60
Total	100	

Heat Loads

The amount of refrigeration used for any particular application depends on the heat load imposed on the ice sheet. Connelly (1976) collected performance data as shown in Tables 2 and 3. The controllability of each of the sources is indicated with an approximate percentage of the maximum possible reduction of the load through effective design and operation.

Conductive Loads. Ground heat gain from both beneath the rink and the edges averages 2 to 4% of the total rink heat load. It is higher when the system is first placed in operation, but decreases as the temperature of the mass under the rink is reduced. Ground heat gain can be reduced with insulation. The effects can be computed from insulation tables (see Chapter 22 of the 1989 ASHRAE *Handbook—Fundamentals*).

Heat gain through the brine mains and headers begins at about 2 to 4% of the total rink heat load, depending on the length of the piping run, surface area, and ambient temperatures. Insulation can control the loss. Natural frost accumulation reduces heat gain without insulation.

The heat gain by the chilled brine as it is pumped to the rink floor through pump work and pipe friction is approximately 10% of the total system heat load. Condenser water pumping and fans are an additional (and substantial) load. These added loads can be controlled by good system design and equipment selection. Additional compressor work is also required to overcome this heat input.

Convective Loads. Heat is transferred through convection by the air to the ice sheet. The load can be estimated using Figures 1 and 2 when the major variables—air temperature, air velocity, and humidity—have been quantified.

Figure 1 shows the relationship between air velocity and film coefficient (conductors) for dry air in the lower part of the graph. The upper part gives the total heat load based on the ambient air temperature and 85% rh, and an ice temperature of 28.5 °F. Total heat load increases at lower ice temperatures used for hockey and curling.

Figure 2 allows the computation of the additional heat load caused by relative humidities from 60 to 90%, compared to a heat

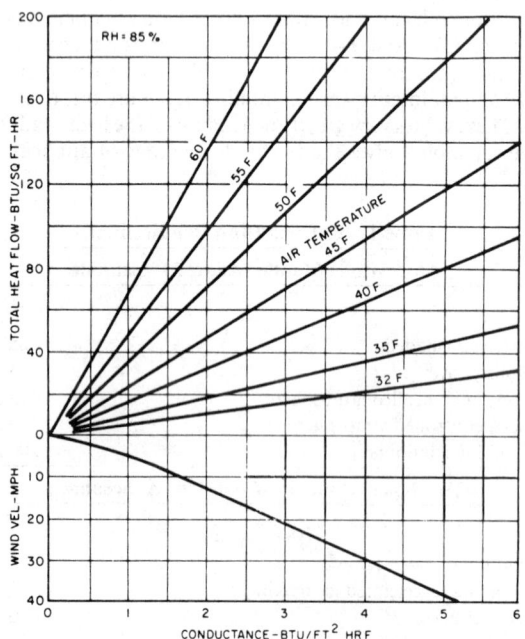

Fig. 1 Heat Transfer from Air to Ice Surface as Affected by Air Velocity and Temperature

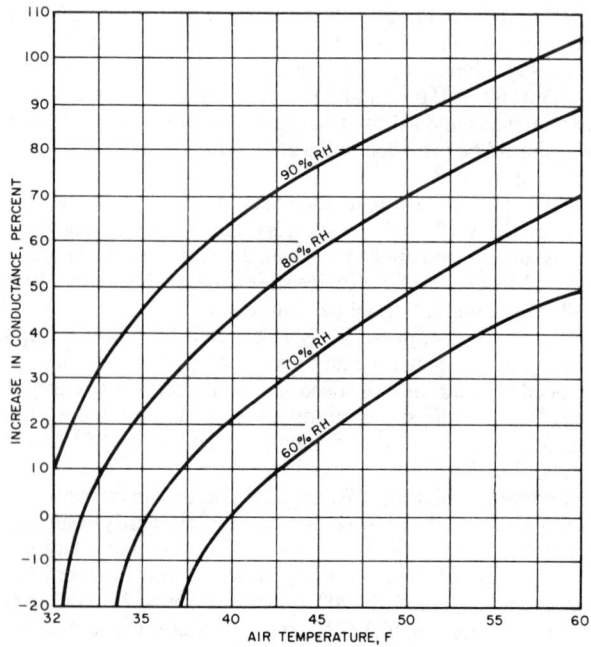

Fig. 2 Effect of Relative Humidity on Heat Transfer from Air to Ice Surface

Ice Rinks

Table 4 Service Factors

Hockey rinks and light public skating	7.5
Heavy public skating	10.0
Curling rinks	5.0
All-year figure skating	2.5
Sports arenas	7.5[a]
Open air rinks (shaded)	15 to 20.0

[a]Does not take into consideration the actual freezing of the ice surface, 0.63 in. thickness in 12 h at 200 Btu/lb.

load based on dry air only. When multiplied by the total ice surface area, either figure represents the total convective heat load.

For example, assume an indoor rink for hockey and light public skating to be 85 by 185 ft, or 15,725 ft². The skating season is to be from October to April with the average monthly maximum wet-bulb temperature obtained from the local meteorological office for the month of October being the greatest at 45 °F.

From Table 4, the service factor for hockey and light public skating is found to be 7.5. Then, using Figure 3 with a 7.5 service factor and 45 °F wet bulb, the heat flow is found to be 37.5 Btu/h · ft².

$$37.5 \times 15{,}725/12{,}000 = 49.14 \text{ tons of refrigeration}$$

Comparing this to the ft²/ton data above,

$$15{,}725/49.14 = 320 \text{ ft}^2/\text{ton}$$

These calculations show that the refrigeration load is a function of the wet-bulb temperature. Where the interior wet-bulb temperature is not controlled by air conditioning or dehumidification, the interior wet-bulb temperature will normally follow the value outdoors except for a lag because of the heat capacity of the building.

In locations where ambient wet-bulb temperatures are high, dehumidification of the building interior should be considered, since this will lower the load on the icemaking plant and also reduce condensation and fog formation. Traditional air conditioning is of little help, since the presence of the ice slab tends to maintain a lower than normal dry-bulb temperature.

Radiant Loads. Indoor ice rinks create a unique condition where a large, relatively cold plane (the ice sheet) is maintained beneath an equally large warm plane (the ceiling). The ceiling is warmed by conductive heat flow from the outside and by the normal stratification of arena air.

Up to about 35% of the heat load on the ice sheet is from radiant sources that flow from either the sun or warm cloud cover in outdoor rinks, or the lights and warm ceiling in indoor and covered rinks. Vertical hanging cloth suspended from east-west horizontal overhead wires has been used successfully to reduce winter sun load. The heat load from lights varies according to the type of lights used. Heat gain by the ice sheet comes from the infrared component of the radiated energy and can be calculated using available tables.

The radiant heat flow can be calculated by the following application of the *Stefan-Boltzmann* equation:

$$q_r = 0.1714 \times 10^{-8} A e (T_c^4 - T_i^4)/C$$

where

q_r = radiant heat load, tons
A = area, ft²
e = emissivity (0.9)
T_c = ceiling temperature, °R
T_i = ice temperature, °R
C = 12,000 Btu/h per ton

Therefore,

$$q_r = 0.1714 \, A e \, [(T_c/100)^4 - (T_i/100)^4]/C$$

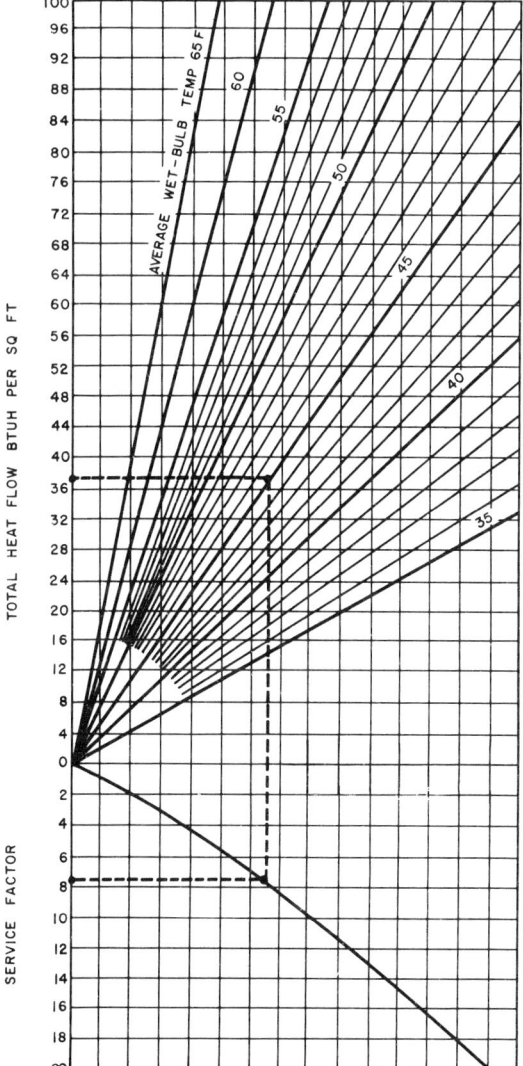

Fig. 3 Total Heat Flow and Service Factors

For example, for 17,000 ft², with 25 °F ice and 60 °F ceiling temperatures:

$$q_r = 0.1714 \, (17{,}000)(0.9)(5.20^4 - 4.85^4)/12{,}000$$
$$= 38.9 \text{ tons}$$

This radiant heat load can be controlled by lowering the temperature of the ceiling, by keeping warm air away from the ceiling, by better roof insulation, and by lowering the emissivity of the ceiling material.

Most ceiling and roof materials and exposed building structural members have an emissivity of about 0.9 (see Chapter 37 of the 1985 ASHRAE *Handbook—Fundamentals* for the emissivity of various materials). Special aluminum paint can lower the emissivity from 0.5 to 0.2. Polished metals, specifically polished aluminum or aluminum foil, have an emissivity of about 0.05.

The use of low-emissivity ceiling material can reduce the heat radiation to the ice sheet significantly. In addition, since the ceiling is not cooled by radiant heat loss, its temperature is maintained above the dewpoint of the rink air, often eliminating condensation, ceiling dripping, and resultant deterioration of materials.

Low-emissivity fabric or tiled ceilings are being incorporated into new and existing facilities to reduce radiation loads, decrease condensation problems, and reduce overall lighting requirements. The economic impact for every application should be analyzed.

The radiant heat gain to the ice, especially in outdoor rinks, can be further controlled by painting the ice approximately 1 in. below the ice surface so as not to interfere with skating and resurfacing operations. Whitewash or slaked lime are common paints. Commercial ice paints are also available. Generally, a material with low solar absorptivity is used in a water base.

ICE RINK CONDITIONS

Indoor commercial rinks operate for 11 months, with a shutdown in the spring for maintenance. However, a 6-month season is considered by many to be more profitable. Outdoor uncovered rinks generally operate from early to mid-November to mid-March above 40° latitude. However, if sufficient refrigeration capacity is provided, ice can be maintained under most heat load conditions.

Indoor rinks are operating successfully in such warm tropical climates as Singapore. Humidity and ceiling radiant loss must be controlled in these special applications to prevent fog, ceiling dripping, and high operating cost.

Steel, frame, brick, concrete, and various forms of plastic have been used to enclose skating rinks. Since a free span of considerable length is necessary, the development of steel truss and rigid frame preengineered systems has greatly reduced the roofing cost for such buildings. Rinks have also been successfully built under air-supported structures for seasonal use over a multipurpose surface.

When rink heating is provided, it is generally in conjunction with a rink dehumidification system whereby some compressor reject heat is recovered. Heated rinks are generally maintained between 50 to 60°F. However, some are maintained above 65°F for greater skater and spectator comfort. In placing radiant heating devices and air-handling equipment, care should be taken not to impose any loads directly on the ice sheet.

Ventilation should be kept to a minimum so that humidity pulled from outside can be kept low. A catalytic exhaust-pipe converter to handle carbon monoxide from a gas-driven resurfacing machine should be provided.

Each rink user group has its own preference for the type of ice used. Hockey players and curlers prefer hard ice, figure skaters prefer softer (i.e., warmer) ice so they can clearly see the tracings of their skates, and recreational skaters prefer even softer ice, which minimizes the buildup of shavings and scraping.

Since it is not practical to measure the ice surface temperature, it is customary to control the ice sheet condition either from a predetermined brine average temperature or from the ice temperature as measured by a thermocouple embedded beneath the ice surface. With approximately 45°F wet bulb and an ice thickness of 1 in., the ice is satisfactory at 22 to 24°F for hockey, at 26°F for figure skating, and at 26 to 28°F for recreational skating. To achieve these values, average brine temperatures approximately 10°F lower than the required ice temperature are sufficient. At higher wet-bulb temperatures or where abnormally high loads are experienced, the temperature of the brine must be lowered to maintain the same ice conditions.

EQUIPMENT SELECTION

Compressors

Two or more compressors should be used in an ice rink system. One compressor should be specified with ample capacity to maintain the ice sheet under normal load conditions. When full capacity is required, the second or additional compressors should pick the load up automatically. In multiple compressor installations, a multistage thermostat and/or a motorized sequence control may be used to sequence the operation of the compressors.

The compressors should be selected to operate at a suction pressure corresponding to a 10°F mean temperature difference between the brine and primary refrigerants in systems using a secondary refrigerant, or the ice and primary refrigerants in direct refrigerant rinks. Total system refrigeration capacity may be calculated by the methods described under the section "Refrigeration Requirements." However, protection should be provided against overload and mechanical failures by dividing the total design load requirements into two or more independent refrigeration systems.

Condensers, Heat Recovery, and Temperature Control

If wells, lakes, or rivers are available, they can be a good source of condenser water cooling with a capacity that is easy to regulate. Since the coolant temperature remains low, it can save energy through operation at low condensing pressures.

Cooling towers used in conjunction with a water-cooled condenser, evaporative condensers, or air cooled condensers are alternates. When selecting a cooling tower or evaporative condenser, not only the maximum expected wet-bulb temperature during the skating season should be considered, but also suitable controls to cover the wide range in capacities and protection against freezeup needed in cold weather. In addition, a water treatment specialist should be consulted.

Air-cooled condensers are used in northern climates, particularly where the rink is used only in the winter. They can be economically sized and require no water, so the possibility of freezeup is eliminated. This type of condenser, however, is not economical for year-round operation, and for seasonal operation it must have wide-range capacity control.

When a cooling tower system is selected, heat from condensers can be used for such energy-saving applications as arena heating, subfloor heating, and snow melting. This is generally done either with the circulation of the condenser cooling water through heat exchangers or with fan coil units. The circulated cooling water can also be used in conjunction with a heat pump as a heat sink/source for heating or cooling various areas within the rink building and for water heating.

Ice rink systems are usually automatic, with thermostatic control either in the brine line sensing the brine return temperature or embedded beneath the ice surface. When embedded within the ice surface, a thermostatic control can be used to cycle the brine pump on and off. In this application, the thermostat must be a close-control type with a differential of no more than 2°F; in all cases, it should be readily adjustable by operating personnel for rink usage and ice thickness.

Rink Piping and Pipe Supports

High flow rate brine systems use standard mild steel pipe 0.75, 1, or 1.25 in. in diameter; thin-walled polyethylene plastic pipe 1 in. in diameter; or UHMW (ultrahigh molecular weight) polyethylene plastic pipe 1 in. in diameter. These are placed at 3.5 or 4-in. centers on the rink floor. A proprietary low flow rate brine system uses 0.25-in. tubing made of flexible plastic with tube spacing averaging 0.75 in. or one dual tube every 1.5 in. Direct refrigerant rinks generally use 0.63 to 0.88-in. steel tubing, which is placed on 3-in. centers for outdoor rinks and 4-in. centers for indoor rinks.

The pipe grid must be maintained as close to level as possible, regardless of the rink piping system used. When a pipe rink surface is of the open type with sand fill around and over the pipes, the conduit usually rests on pressure-treated sleepers set level with the subbase; however, the sleepers can be omitted in a rink that is to be operated year-round. The piping is then spaced with clips, plastic stripping, or punched metal spacers.

Ice Rinks

In permanent concrete floors, the pipe or steel tubes are supported on notched iron supports or welded chair supports. The latter must be used in the case of plastic pipe.

Headers and Expansion Tanks

Brine rinks using large-diameter pipe generally have the piping running lengthwise, with the supply and return headers across one end. Small-diameter tubing rinks generally run crosswise, with the supply and return headers along one side. Direct expansion rinks generally run lengthwise, with the supply header at one end and the return at the opposite end in a balanced system. The header must be sized to assure an even distribution of brine through every pipe. The systems are generally designed with low brine velocities, which do not need brine balancing valves. If at all possible, the brine return header should be placed at the same elevation as the rink piping, with at least two air vents to eliminate the trapping of air.

The balanced flow diagram (see Figure 4) is one of several good designs that assure that the brine on the first and last circuit flows an equal distance. However, with adequately designed headers and piping, the two-header system is more commonly used. To allow for thermal contraction and expansion, headers and rink piping should be free to move. Direct expansion rinks generally use welded steel headers, while brine pipe rinks use steel or ABS plastic, and tubing rinks use PVC plastic.

A closed brine system preferably uses an expansion tank because it reduces the possible contamination of the brine with air and handles the expansion and contraction of the brine volume in the system. The expansion tank should be installed so that it cannot be isolated from the system.

Brine Equipment

The brine circulating pumps must be sized for the particular type of rink and system involved. Large-diameter pipe rinks require 10 to 15 gpm per ton refrigeration to maintain the required 2 to 3 °F temperature differential between incoming and outgoing brine. These operate at approximately 20 to 25 psi. Low flow rate dual tubing or mat rinks use about 2.4 gpm per ton. Differentials of 4 to 6 °F are normal, but 10 to 12 °F differentials can be experienced in high load conditions with no reduction in ice quality. Uniform temperatures are achieved in mat rinks by temperature averaging between closely spaced, adjoining counterflow tubes operating at approximately 40 to 50 psi.

For auditoriums and sports arenas, the rink surface should have provision for deicing in less than 4 h. In this operation, the floor is heated to about 50 °F so that the ice can be peeled off by first breaking the bond between the floor and ice and then breaking the ice and removing it with power tractors.

A standard heat exchanger can be used, with piping so arranged that all the brine can be pumped through the heater, with the brine flowing in the tubes and the steam or water in the shell. Approximately 350 Btu/h·ft² capacity of rink surface is recommended for heating the brine in the cooling system to warm the floor and break the ice bond.

In sports arena rinks designed for frequent deicing, ice may be required every other day, and 0.63 in. ice or more must be frozen in 12 h to allow the ice to temper before being skated on the following day. In such cases, a cold brine accumulator keeps the size of the refrigeration equipment and connected power within reasonable limits. This accumulator is a brine storage tank bypassed from the heating cycle. When refrigeration on the ice surface is not required, a large volume of brine in the accumulator may be cooled to approximately −25 °F and be ready to be pumped into the rink piping when needed. The cold brine accumulator should store sufficient brine to cool the entire cooling system brine volume from 65 to 0 °F. This cold brine tank usually holds more than three times the volume of the cooling system's brine charge.

With the increase in construction and operating costs, the use of accumulators has been declining. In place of accumulators, ice-making equipment is sized to handle the demand loads, and arenas are programmed to eliminate the need for quickly making and removing the ice. Accumulators using phase change materials show promise of reducing size by 6 to 8 times, eliminating high demand charges and allowing total off-peak operation at lower electric rates.

GENERAL RINK FLOOR DESIGN

Generally, five types of rink surface floors are used (Figure 5):

- Open or sand fill type, for plastic or metal piping or tubing
- Permanent, general-purpose type, with piping or tubing embedded in concrete on grade
- All-purpose type, with piping or tubing embedded in concrete with floor slab insulated on grade
- All-purpose floors, supported on piers or walls
- General all-purpose floor; with this type, water table and moisture are severe problems

The open sand fill floor is the least expensive type of rink floor. The cooling pipes rest on wood sleepers over a bed of crushed stone or other fill. The clean washed sand is filled in around the cooling pipes. Curling rink floors, as well as hockey and skating rinks, where first cost is a factor and the building is not intended to be used for other purposes, are usually constructed in this manner. Clay or cinders should never be used in the bed or for fill around the pipes. Tubing rinks do not need supports or sleepers, but tubes are laid on accurately leveled sand.

Rinks using 1-in. plastic pipe or the mat type are usually covered with sand to a depth of 0.5 to 1 in. to provide additional strength to the ice surface and to reduce cracking. Many portable outdoor rinks have used this arrangement for laying the plastic pipes or tubing mats on top of existing sodded areas, black top, or concrete. More permanent installations of outdoor semiportable rinks have used this same arrangement where recreational area is at a premium. Such an installation consists of steel pipes supported on notched steel sleepers, which in turn are supported on concrete piers down to solid ground.

To obtain a better return on investment, most indoor rinks that operate with an ice surface for only a portion of the year have a permanent general-purpose concrete floor so that the floor may be used for other purposes when the skating season is over. The floor withstands the average street load and is usually designed with 1 or 1.25-in. steel or plastic pipe embedded in a steel-reinforced concrete slab 4 to 6 in. thick, depending on the anticipated loading.

In sports arenas, where the ice is removed and the floor made ready for other sports and entertainment, the ice floor must be

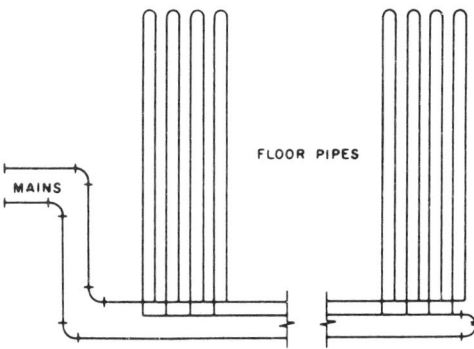

Fig. 4 Balanced Flow System of Distribution

constructed to stand the frequent change from hot to cold, and the refrigerating machinery must be of sufficient capacity to freeze a sheet of ice 0.63 in. thick in 12 h. This type of floor is always insulated.

Subfloor insulation should be installed when quick changeovers are desired, when there is a high moisture content in the subsoil, when the floor is elevated, or when the rink is in continuous use for more than 9 months. This subfloor insulation serves to reduce the load on ice-making equipment and to slow down, but not eliminate, the cooling of the subsoil on surfaces installed on grade.

Drainage

The suitability of the subsoil on which an ice rink is built has a great influence on the rink's success. Complete ice surfaces have had to be rebuilt because of poor drainage and the ultimate heaving of the ice surface. Thus, skating rinks should not be built on swampy or low-lying land unless adequate drainage is provided.

Moist subsoil will freeze in the ground to a depth of 4 ft or more. The frozen water will heave the ice surface when freezing takes place at a depth of 6 in. or more. Heaving creates an uneven skating surface; moves and raises walls, piers, and header trenches; cracks walls and piping; and necessitates the eventual drainage and rebuilding of the rink floor.

Not only should there be a complete drainage system around the footings of the rink to prevent seepage, but there should also be one under the rink surface itself. This is particularly important when a sand fill floor is used; a good system will assure that the ice melted after the skating season will completely drain away and the sand will dry out as quickly as possible.

Subfloor Heating

Subfloor heating by either electrical heating cables or a pipe or tubing recirculating system using warm brine is found in most modern rinks to prevent the danger of heaving. Pipes or tubing are on 12 to 24-in. centers located under 2 to 4 in. of insulation. They are generally installed in sand rinks, which are used year-round, although they may be poured into a concrete base slab with insulation between the base slab and the rink slab (see Figure 5).

Alternatively, the heating pipes may be laid directly in the subfoundation below the rink pipe or insulation. However, an installation not equipped with insulation requires a greater depth between the heating pipes and the ice-making pipes to prevent an increased load.

Neither water nor warm air should be used for subfloor heating. Water, if inadvertently allowed to freeze, cannot be readily melted out. In time, warm air ducts become filled with frost and ice because of high rink humidity and leakage.

Usually the same brine mix used in the ice-making system is used for subfloor heating; it can be heated to the necessary 40 to 50°F in a heat exchanger warmed by compressor waste heat. Subfloor insulation should be of a rigid moistureproof board, such as SM grade polystyrene foam, and be completely enveloped in a polyethylene vapor barrier.

Preparation for Rink Floor

When building on natural ground, regardless of whether a land fill or a permanent general-purpose floor is intended, proper preparation of the bed is important unless the rink is built on elevated sand and gravel subsoil. If the rink is to be built on clay, part clay, or rock subsoil, water should be prevented from collecting in low areas. Either the clay or rock should be excavated or the rink level should be built up with crushed stone and gravel to a height of about 4 ft, after which it should be well rolled. Water should not be used for settling the fill.

In the case of sand fill rinks, quickly draining the melted ice at the end of the skating season will ensure rapid drying of the sand and rink piping and result in a longer life for the steel piping.

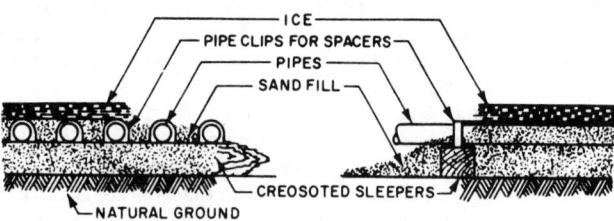
Open or Sand Fill Floor

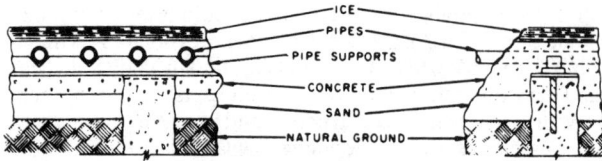
Permanent or General-Purpose Floor

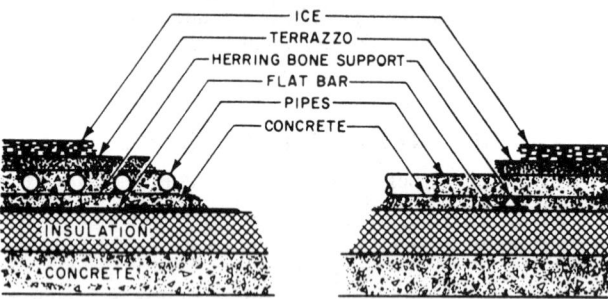
All-Purpose Floor

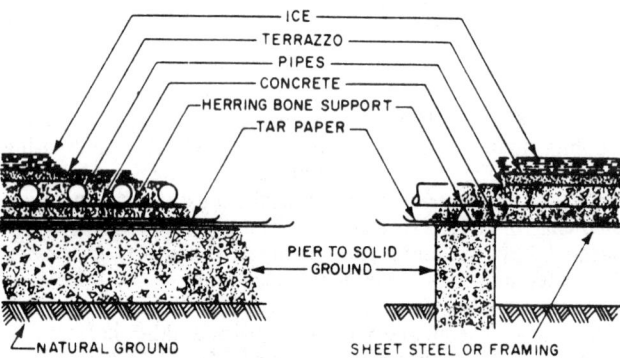
Permanent or All-Purpose Floor on Piers

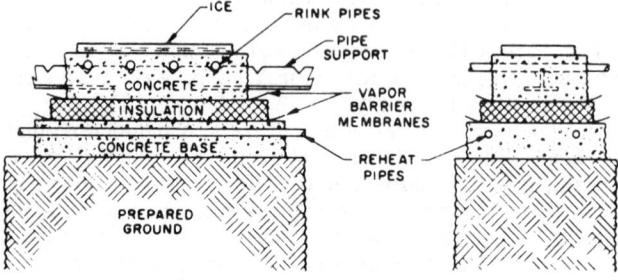
All-Purpose Floor with Reheat

Fig. 5 Ice Rink Floors

Ice Rinks

Cinders should never be used as fill in open sand fill rinks because of the possibility of sulfur in the cinders which, when damp, accelerates corrosion of steel piping.

Care must be taken to assure a level surface over the entire rink with no more than ± 0.13 in. in any 10 ft^2 area and ± 0.25 in. overall.

Permanent General-Purpose Rink Floor

When constructing a permanent general-purpose floor, the same precaution must be taken regarding subsoil as for a sand fill rink. The concrete floor should withstand loads of the average pavement load.

When local conditions make it advisable, the rink floor may be insulated, in which case a level subfloor must be constructed on which the insulation is laid. A concrete subslab may be installed and insulation laid on leveled subsoil.

The concrete mixture recommended is 3000 to 5000 psi concrete with 3-in. slump. Suitable cross-reinforcing is necessary in addition to pipe supports.

Concrete floors with mat-type tubing are poured in two courses. A first course is poured and leveled; the mats are then rolled out and positioned. A 6 by 6 in. wire mesh is laid on top of the mats; then a second course, with grouting between it and the first, is poured on top of the first course, mats, and wire. Air pressure is kept in the tubing to spot any leaks or cuts that may develop. Once started, the pouring of each course of the concrete floor should be continuous.

General purpose rink floors should not be defrosted too frequently. When a rink constructed with a general-purpose floor is to be used during the ice season for purposes that require an ice-free floor, it is preferable to place an insulated portable-section wood floor over the ice for each occasion.

All-Purpose Floors

If a rink floor as used in sports arenas is to withstand both the expansion and contraction of frequent frosting and defrosting and thermal shock because of the circulation of very cold brine, then extra precautions must be taken in its construction, such as provisions for the free movement of the freezing slab with respect to the subfloor.

Header Trench

A well-constructed header trench of sufficient size to house the headers and connections and the subfloor heating system, if applicable, is essential unless the brine headers are cast into the concrete slab as part of the rink. This trench should be equipped with removable covers and be well-drained to facilitate drying out. The headers and piping in the trench are not usually insulated, which allows for periodic inspection and painting of the piping. However, unless a large trench is provided, consideration should be given to insulating the headers on rinks that operate year-round because of the massive buildup of frost. Provision must be made for purging air from the rink piping and header system.

Snow Pit

A snow-melting pit should be provided at a suitable point, usually at one end of the rink. It should be of sufficient size to handle the scraped off snow and the ice accumulated during planing, or it may be made large enough to accommodate the complete ice removal.

Hot water from shell-and-tube condensers, the waste heat recovery system, or some other source must be provided to melt the snow and ice. A large drain with overflow should be provided, as well as a large removable screen to filter out trash.

BUILDING, MAINTAINING, AND PLANING ICE SURFACES

Regardless of the type of rink floor used, when the plant is first placed in operation, the equipment should be operated long enough for a sharp frost to appear on the surface. Then the entire surface should be uniformly covered with a fine spray. This process should be repeated until a 0.5 in. thickness of ice is built on a concrete floor or 0.8 in. thickness on a sand fill floor. It is essential that sand floors be thoroughly wet before freezing because dry sand has poor conductivity. The surface should not be frozen any colder than required after this buildup so as to allow the ice to temper before it is used for skating and also to deter cracking.

To maintain an ice surface, it is customary to scrape off the snow after each skating session or hockey period. In all but the smallest rinks, this is done by a motorized resurfacer. On small rinks, the scraping is done manually with a wide hardened-steel scraper blade. The most satisfactory method of resurfacing the ice between sessions is to use a sprinkler tank filled with hot water, which is wheeled over the ice. The sprinkler has an adjustable valve to control the quantity of water, which is sprayed into a terry cloth bag that wipes the fine snow off the ice surface and fills the crevices cut by the skaters. In this manner the least amount of water is added, reducing the ice buildup and refrigeration load.

By far the most common method is the use of automatic resurfacing machines. Mounted on a four-wheel drive chassis, the machines plane the ice, pick up the snow, and lay down a new ice surface using hot or cold water. Hot water generally gives harder ice, since air bubbles are removed, but higher energy costs have led many rinks to alternate hot and cold water resurfacings. Rink corners should be at least a 20-ft radius for effective use of this equipment. Smaller equipment is available for studio and small-size rinks.

Because of inattentive ice making, improper sprinkling equipment, or deep cutting of the ice during public skating, the ice may become uneven and excessively thick. There may be a fairly slight variation in the ice thickness across the rink, but more serious is the resulting variation in the condition of the ice. In any case, the low spots on the ice must be built up, increasing the thickness and refrigeration requirements.

For example, under assumed conditions, where 18°F brine would be cold enough to hold a 1.5 in. thickness of ice, calculations show that -5°F brine would be required if the ice were permitted to build up to 6 in., with a corresponding decrease in effective refrigeration capacity.

Since ice of 0.5 to 1 in. thickness is satisfactory for skating and is the most economical thickness to freeze and hold, the ice should be periodically planed to maintain the desired thickness.

RINK FOG AND CEILING DRIPPING

During mild weather, particularly in early fall and late spring in the northern United States and Canada, condensation often drips from the roofs and roof supports of rinks (especially curling rinks), because they have insufficient internal heat loads. The condensate dropping on the ice ruins the curling surface and the fog obstructs the view. These conditions cannot be solved by ventilation because the introduction of outside air only aggravates the problem when the weather outside is mild and humid. Insulating the roof also aggravates drip during mild outside weather conditions. Low emissivity ceilings stay warmer and thus reduce condensation and drip.

Under these conditions, to prevent condensation in the roof space and to clear the fog, a six-sheet curling rink, with 12,500 ft^2 of ice, would require the removal of 33 lb of moisture per hour, necessitating about 6 tons of refrigeration.

Units using the brine from the rink piping as a cooling medium avoid frosting by recirculating with a small bypass pump to keep coil inlet brine above 32°F. Reheat coils are often included to add back waste heat from condenser cooling water to counter the cooling effect of the dehumidifier coil. Self-contained, air-cooled, compressor-type packaged dehumidifying units, as well as desiccant drier types with gas or electric regeneration, are available.

IMITATION ICE SKATING SURFACES

A number of different imitation ice-skating surfaces have been marketed; these use semiporous plastic panels dressed with a synthetic lubricant. The coefficient of friction of ice is approximately 3% at 26°F and is even less because of the film of water produced by pressure under the skate. In considering the use of imitation surfaces, the actual friction coefficients of these surfaces—both when freshly lubricated and after a period of usage—should be investigated.

REFERENCE

Connelly, J.J. 1976. ASHRAE Seminar on Ice Rinks (February), Dallas, TX.

CHAPTER 35

CONCRETE DAMS AND SUBSURFACE SOILS

CONCRETE DAMS 35.1	SOIL STABILIZATION 35.4
Methods of Temperature Control 35.1	Thermal Design 35.4
System Selection Parameters 35.3	Passive Cooling 35.4
CONTROL OF SUBSURFACE WATER FLOW 35.3	Active Systems 35.5

REFRIGERATION is one of the more important tools of the heavy construction industry, particularly in the temperature control of large concrete dams. It is also used to stabilize both water-bearing and permanently frozen soil. This chapter briefly describes some of the cooling practices that have been used for these purposes.

CONCRETE DAMS

Without the application of mechanical refrigeration during construction of massive concrete dams, much smaller construction blocks or monoliths would have to be used, which would slow down construction. By removing unwanted heat, refrigeration can speed up construction, improve the quality of the concrete, and lower the overall cost.

METHODS OF TEMPERATURE CONTROL

Temperature control of massive concrete structures can be achieved by (1) selection of the type of cement, (2) replacement of part of the cement with pozzolanic materials, (3) use of embedded cooling coils, or (4) precooling the materials. The measures used depend on the size and type of structure and on the time permitted for its construction.

Cement Selection and Pozzolan Admixtures

The temperature rise that occurs after concrete is placed is due principally to the heat of hydration of the cementing materials. This temperature rise varies directly with cement content per unit volume and, more significantly, with the type of cement. Ordinary Portland cement (Type I) releases about 180 Btu/lb, half of which is typically generated in the first day after the concrete is placed. Type II cement, depending on specifications, may generate slightly less heat. Type IV is a low heat cement that generates less heat and at a slower rate.

Pozzolanic admixtures may be used in lieu of part of the cement. Such pozzolans include fly ash, calcined clays and shales, diatomaceous earths, and volcanic tuffs and pumicites. The heat-generating characteristics of these pozzolans vary, but are generally about one-half that of cement.

When determining the system refrigeration load, heat release data for the cement being used should be obtained from the manufacturer.

Cooling with Embedded Coils

Embedded cooling coils effectively reduce the peak temperature in mass concrete. Further, they lower the temperature of the structure to its final state during construction. This is desirable where volumetric shrinkage of a large mass is necessary during construction, for example, to allow the contraction joint grouting of intermediate abutting structures to be completed.

In the past, thin wall tubing was placed as a grid-like coil on top of each 5 or 7.5 ft lift of concrete in the monoliths. Chilled water was then pumped through the tubing, using a closed loop system to remove the heat. A typical system used 1-in. OD tubing with a flow of about 4 gpm through each embedded coil. While the number of coils in operation at any one time varies with the size of the structure, 150 coils is not uncommon in larger dams. Initially the temperature rise in each coil can be as much as 8 to 10 °F, but it later becomes 3 to 4 °F. An average temperature rise of 6 °F is normal. When sizing the refrigeration equipment, the heat gain of all the circuits is added to the heat gain through the headers and connecting piping.

For a typical system with 150 coils based on a design temperature rise of 6 °F in the embedded coils and a total heat loss of 6 °F elsewhere, the size of the refrigeration plant would be about 300 tons. A flow diagram of a typical embedded coil system is shown in Figure 1.

Cooling with Chilled Water and Ice

The actual temperature of the mix at the time of placement has a greater effect on the overall temperature changes and subsequent contraction of the concrete than any change caused solely by varying the heat-generating characteristics of the cementing materials. Further, placing the concrete at a lower temperature normally results in a smaller overall temperature change than that obtained with embedded coil cooling. Because of these inherent advantages, precooling measures have been applied to most concrete dams.

Glen Canyon Dam illustrates the installation required. The concrete was placed at a maximum placing temperature of 50 °F during summer months when aggregate temperature was about 87 °F, cement temperature was as high as 150 °F, and the river water temperature was about 85 °F. Maximum air temperatures averaged over 100 °F during the summer months. The system selected included cooling aggregates with 35 °F water jets on the way to the storage bins, adding refrigerated mix water at 35 °F, and the addition of flaked ice for part of the cold water mix. Subsequent cooling of the concrete to temperatures varying from 40 °F at the base of the dam to 55 °F at the top was also required. The total connected brake power of the ammonia compressors in the plant was 6200 hp, with a refrigeration capacity equivalent to making 6000 tons of ice per day.

The maximum amount of chilled water that may be added to the concrete mix is determined by subtracting the amount of surface water from the total mix water, which is free water. Frequently, if a chemical admixture is specified, some water must be added to dissolve the admixture—usually about 20% of the total free water. This limits the amount of ice that can be added to the remaining 80% of free water available. After the amount of ice that is needed for cooling is determined, the size of the ice-making equipment can be fixed. When determining the capacity of the equipment, allowances should also be made for cleaning, service time, and storage for the ice during nonproductive times.

When calculating the actual heat removal, the ice is considered to be 32 °F when introduced into the mixer. The chilled water is

---

The preparation of this chapter is assigned to TC 10.1, Custom Engineered Refrigeration Systems.

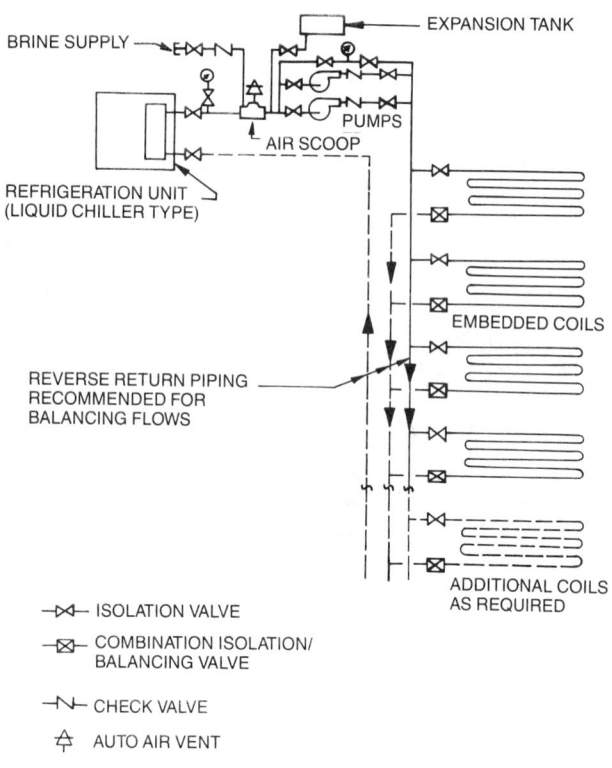

Fig. 1 Flow Diagram—Typical Embedded Coil System

Table 1 Temperature of Various Size Aggregates Cooled by Inundation

Time, min.	Aggregate Size, in.				
	6	3	1.5	0.75	0.375
1	85°F	69°F	49°F	39°F	38°F
2	77	59	41	38 (42)	
5	66	46 (45)	38 (42)		
10	56	40 (42)			
15	50	38			
20	46				
25	44				
30	42				
40	40				
50	39				

Source: Peugh, V.L. and I. Taylor. 1934. "Mathematical theory of cooling concrete aggregates." In *The inundation method of cooling concrete aggregates at Bull Shoals Dam* (House 1949). Numbers in parentheses are from tests by R. McShea.

Note: The temperatures listed are at the center of a cobble with an assumed thermal diffusivity of 0.07 ft^2/h. The aggregate initial temperature is 90°F and the cooling water temperature is 35°F.

assumed to be 40°F entering the mixer, even though the chiller may supply the water at a lower temperature.

Cooling by Inundation

The temperatures specified today cannot be achieved solely by adding ice to the mix. In fact, it is not possible on heavy construction of this type (in view of low cement content and low water-cement ratio specified) to put a sufficient amount of ice in the mix to reduce the mass to the specified temperatures. As a result, inundation (deluging or overflowing) of aggregates in refrigerated water was developed and was one of the first uses of refrigeration in dams.

When aggregates are cooled by inundation with water, generally the three largest sizes are placed in large cylindrical tanks. Normally two tanks are used for each of the three aggregate sizes to provide backup capacity and a constant flow of materials. The cooling tanks, loaders, unloaders, chutes, screens, and conveyor systems from the tanks into the concrete plant should be enclosed and cooled from 45 to 40°F by refrigeration units, with blowers placed at appropriate points in the housing around the tanks and conveyors.

Peugh and Tyler (House 1949) determined the inundation (or soaking) time required by calculations and actual tests. Pilot tests corroborated their computations of cooling times of cobbles as indicated in Table 1.

This study indicated that an immersion period of about 40 min will bring the cobbles down to an average temperature of 40°F. Theoretically, smaller sizes can be brought to the desired temperature in less time. Considerations such as the rate at which cool-

ing water can be pumped, make it unlikely that a cooling period less than 30 min should be considered. Any excess cooling will provide the needed factor of safety. However, the limiting factor on the overall cycle is the cooling time for the largest aggregate, which is nearly 45 min, plus about 15 min for loading and unloading. Backup capacity should be considered for maintaining a constant flow of materials.

Air Blast Cooling

Following the inundation method, air-blast cooling was developed. This method of cooling aggregates does not require particular changes to the materials handling set-up or additional tanks for inundation; the aggregate is cooled by blowing cold air through the aggregate in the batching bins above the concrete mixers. Also, the air cycle used in cooling can be used in heating aggregates during cold weather. Cooling is applied at the final stage of handling aggregates and does not increase the moisture content.

The compartmented bins where air blast cooling is usually accomplished are usually sized so that if any supply breakdowns occur, the mixing plant will not have to shut down before a particular pour can be completed. On this basis, the average concrete octagonal bin above the mixing plant on a large job holds at least 600 yd^3. The size is usually more than adequate to allow time for air cooling. However, certain minimum requirements must be considered. If possible, based on a one-hour loading and cooling schedule, at least 2 h of storage volume should be provided for each size aggregate that is to be cooled by air. The minimum volume should be 1.5 h plus cycling time, based on the tables shown for cooling 6-in. cobble.

Table 2 Bin Compartment Analysis for Determining Refrigeration Loads and Static Pressures

Material Size, in.	Lb/yd^3	% of Total	No. of Bin Compartments	Bin %
Stones				
6 to 3	800	21.33	4	25.00
3 to 1.5	700	18.67	3	18.75
1.5 to 0.75	650	17.33	2	12.50
0.75 to 0.25	600	16.00	3	18.75
Sand	1000	26.67	4	25.00
Total	3750	100	16	100

Note: In practice, these relative sizes will vary from time to time; the amounts shown are for an assumed design mix of the principal classes of concrete. On any given job, several design mixes requiring different amounts for each size are needed.

Concrete Dams and Subsurface Soils

The bin compartment analysis shown in Table 2 may be used as a starting point in determining the air refrigeration loads and static pressures. This type of analysis should give approximately equal storage periods for each size of aggregate used. In practice, after air cycle cooling is calculated, more air volume is needed in the smaller aggregate compartments, and less in the sand and cobble, to obtain maximum cooling. This is because of the higher air resistance in the smaller aggregate sections and the fact that the sand compartment is not as effectively cooled by air.

Computing Air Blast Cooling Loads. To calculate the required cooling, several assumptions must be made as follows:

Assumption 1. Normally, the lowest temperature of air leaving the cooling coils is 38 to 40 °F. While lower air temperatures may be achieved, these should not be counted on, because a temperature lower than 35 °F usually causes rapid frosting of the coils.

Assumption 2. The heat transfer between the aggregate and air will be only 80 to 90% effective. To allow for air temperature rise in the ducts, heat leakage, pressure drop, etc., an empirical factor of 85% may be used. Thus, with 80 °F aggregate and 40 °F (40 °F differential), the effective temperature differential would be $0.85(80-40) = 34$ °F.

Assumption 3. The rise in temperature of air passing through the aggregate compartment normally will not exceed 80% of the difference between the entering aggregate and the entering air temperatures. Thus with 80 °F aggregate and 40 °F air, the temperature rise of the air is $0.80(80-40) = 32$ °F. The maximum temperature of the return air is $40 + 32 = 72$ °F.

Assumption 4. An allowance should be included for heat leakage into the air ducts on the aggregate bin sides—normally about 2% of a total air blast cooling load.

Another important consideration is the static pressure against air flowing through a body of aggregates. The resistance to air flowing through a bin varies as the square of the air volume or velocity as summarized in Table 3. The resistance pressure listed is for a unit height of aggregate. To use the values shown in Table 3, the cross-sectional area of the aggregate compartment, as well as the height of the aggregate column must be known. This information can be obtained from the manufacturers of concrete mixing bins and mixing equipment.

Table 3 Resistance Pressure

Aggregate Size, in.	Velocity, fpm	Resistance Pressure, in. water/ft of ht.
6 to 3	300	0.29
3 to 1	200	0.32
1.5 to 0.75	100	0.27
0.75 to 0.25	60	0.30

Other Cooling Methods

As specifications continued to require lower and lower placing temperatures, direct methods of cooling sand and cement were initiated to obtain the overall heat removal required. Obviously, cement cannot be cooled by inundation and spray methods. Although the sand could be cooled by these methods, the free moisture content of the sand when batched causes trouble in terms of correctly proportioning the mix. Attempts have been made to cool sand by using hollow flight screw conveyors through which chilled water flows. Since cooling of either of these components below the dew point could result in condensation and serious handling problems, this method shows little promise. A vacuum system, which evaporated the surface moisture to reduce the aggregate temperature, has also been used.

SYSTEM SELECTION PARAMETERS

For larger installations, plant selection depends on a number of factors, including the following:

1. Normal pouring rate, yd^3/h (contractor or contract specified)
2. Maximum pouring rate, yd^3/h (contractor or contract specified)
3. Total allowable mixing water, lb/yd^3 (usually contract specified)
4. Required concrete placement temperature (usually contract specified)
5. Concrete temperature when coming from the mixer (to be determined considering materials handling to placement site, time in transit, and storage at placement site)
6. Average ambient temperature and aggregate temperature during period of maximum placement. The average ambient temperature of the aggregates is assumed to be the mean ambient temperature—including night and day—during the period of storage, which is determined by the amount of storage capacity provided by the contractor and the rate of concrete placement. If the minimum live storage is, for example, 100,000 tons of aggregates and the pouring rate is 2000 yd^3/day, the consumption would be approximately 3800 tons of aggregates per day. If the job is working a 6-day week, this would provide 26 days of storage. Unless weather conditions are unusual, the temperature of the rock delivered to the reclaiming tunnel is assumed to equal the average ambient temperature for the 26 days preceding the delivery.
7. Specific heat of materials. The specific heat of the aggregates and sand may vary with project location. Typical values include: sand = 0.106 Btu/lb·°F; aggregates = 0.12 Btu/lb·°F; water = 1.0 Btu/lb·°F; cement = 0.12 Btu/lb·°F.
8. Heat release rate for cement (material specifications)

Where the aggregate cooling range from initial to final temperature of the mix is relatively small (15 to 20 °F), or where the required pour temperature is relatively high (65 °F or more), chilled water plus ice in the mix, or chilled water plus air blast on larger aggregates may handle the entire cooling load. When the overall temperature reduction is greater than this, or when lower pour temperatures are specified, such as 50 °F or less, a combination of all three types of cooling will probably be required because only a limited amount of heat removal can be obtained by one of these methods alone.

Cooling by air blast alone is limited by the entering air temperature, which must be maintained high enough to prevent coil frosting. Cooling by inundation, although it requires large inundation tanks, offers the most positive and sure method of cooling. Ice can also be added to the mix to remove the remainder of the heat. The result is a very satisfactory blending of the aggregates and exact control of the amount of water in the mix.

CONTROL OF SUBSURFACE WATER FLOW

Refrigeration has been used successfully since 1880 to freeze moisture in unstable and water-bearing soil and to stop underground flows of water in pervious material or gravelly stream beds. Other common methods of stopping the flow of water include sheet piling, cement grout, chemicals, and the use of well points. In many cases, freezing has been the last resort after other methods are unsuccessful. In a number of cases, a combination of well points, grouting, and freezing have been used to solve the problem.

Using refrigeration, the common practice is to put down a series of concentric pipes in a line or arch pattern in the path of the subsurface water flow. A wall of earth is frozen by pumping cold brine down the inside pipe and letting it flow back through the annular space between the inside and outside pipe. The growth of frozen soil on the outside of the concentric pipe proceeds until it connects with the frozen cylinder formed on the adjacent pipe. Spacing between pipes can vary, depending on the time available

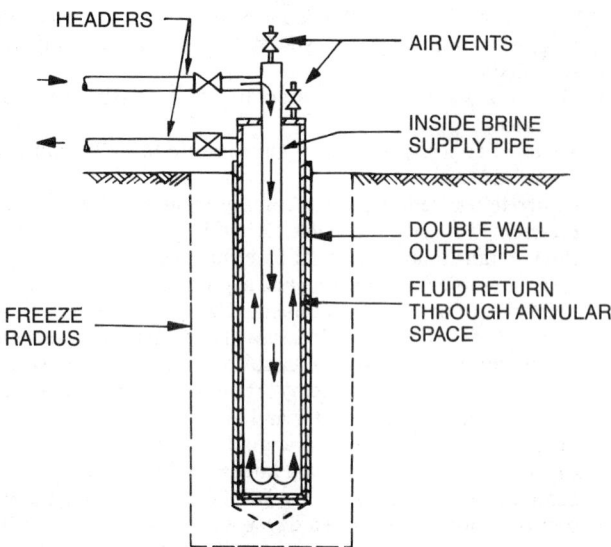

Fig. 2 Typical Freezing Point

to complete the wall of ice, but spacings of 2 to 4 ft on center have been used successfully. Freezing pipes have been successfully used to control subsurface water in both dam and mining projects. A typical freezing pipe and system diagram are shown in Figure 2.

The brine or refrigerant and method of containment should be carefully selected in system design. If a leak develops in the system, groundwater contamination could result, or the soil saturated with the refrigerant may have to be excavated and cleaned, depending on the type of refrigerant used. Double-wall piping, or an environmentally safe refrigerant that will vaporize when exposed to the air, should be considered.

Ammonia is generally used as the basic refrigerating medium. Then brine, chilled by the ammonia, is circulated through the freezing pipes. The brine is commonly calcium chloride ($CaCl_2$), but magnesium chloride ($MgCl_2$) is recommended as it has less tendency to precipitate and clog the piping at low temperatures. To monitor the progress of freezing, thermocouples are normally located throughout the area to be frozen. Brine temperatures may range from +15 to −20°F depending on the state of the freezing and the amount of time available to complete the task.

SOIL STABILIZATION

In northern latitudes where areas of permanently frozen earth, or permafrost, prevail, methods of soil stabilization for building and equipment foundations are required. These methods range from providing a non-frost-susceptible gravel pad to rigid below-grade insulation plus an active or passive refrigeration system to freeze the soil and keep it frozen. Since many of these systems are in remote areas, simplicity and reliability become major factors in system design.

THERMAL DESIGN

Piling Design

Frost heave has long been a problem for the designers of piles and buildings in the arctic or subarctic. Uplift forces as high as 13.5 tons/ft of perimeter must be taken into account when designing nonthermal piling (Long and Yarmak 1982). Designers have used sleeves, greases, waxes, and plastics to reduce the adfreeze bond in the active layer and lower frost heave forces. Almost all the methods for reducing heave forces have proven to be only temporary (Long and Yarmak 1982). Increased pile embedment remains the only sure method of preventing frost heaving of conventional piling. The use of thermopiles can effectively eliminate frost heaving forces (Long and Yarmak 1982). The thermopile freezes the active layer radially from the pile. As a result, active layer temperatures adjacent to the pile remain at the same temperature as the rest of the pile.

Slab-on-Grade Buildings, Outdoor Slabs, and Equipment Pads

When a slab-on-grade building, outdoor slab, or equipment pad is constructed on a permafrost area, the resulting soil thawing or thaw bulb must be considered. To stabilize the structure, slab, or pad, thermopiles, thermoprobes, or an active refrigerated foundation system may be required.

Design Considerations

Soil properties have a pronounced effect on the capacity requirements for passive or active refrigeration systems. The soil within the radius of influence is an integral part of the refrigeration system. Highly conductive soils will increase the radius of influence of the system and allow more heat to be pumped out of the subsoils. Conversely, poor conducting soils will cause the radius of influence to be small; the thermal lag through the soils will be high, and the heat transfer rate, low. Soil moisture contents and soil classifications are valuable for estimating thermal conductivities. Backup capabilities for active and passive systems should be considered to avoid foundation failure.

PASSIVE COOLING

The three processes used by passive systems for heat removal are air convection, liquid convection, and two-phase liquid/vapor convection. All passive refrigeration systems rely on the temperature differential between the soil and the winter air to operate. When the temperature of the soil in contact with the refrigeration system is lower than the air temperature, the system is dormant.

Air Convection Systems

Air convection has been used to provide subgrade cooling below on-grade and pile-supported structures. For a passive air convection system to work, two criteria must be met in the design—the air outlet must be higher than the inlet to promote convection, and the air distribution system must be designed so that the friction imposed by the distribution system is low enough to allow convection to begin. Air convection systems are usually designed to take advantage of the prevailing winds at a specific site. The wind can push air through a distribution system at greater velocities than convection would allow, however, the wind in the arctic frequently carries large quantities of snow. Distribution ducts can be blocked by snow and ice causing failure of the system (Long and Yarmak 1982). Ducting fans are built into many air convection systems for active refrigeration backup.

Liquid Convection Systems

Liquid convection has been used to provide subgrade cooling below on-grade structures. Radiator and heat absorber portions of the system are normally connected by either a double or single pipe with a flow splitter to decrease frictional losses in the system and to provide maximum cooling of the working fluid. Examples of some working fluids trichlorethylene, kerosene, and methanol and water. Frictional losses within the liquid system are high and limit heat transfer rates. Large-diameter pipes may be used to overcome frictional losses within liquid systems so that high heat transfer rates can be achieved. Some liquid convection systems allow an option for mechanical circulation of the working fluid as an active refrigeration backup. Liquid systems must

Concrete Dams and Subsurface Soils

be sealed to avoid leakage into the subsoils. Introduction of the working fluid into the permafrost subsoils could depress the soil freezing point and increase the probability of foundation failure.

Two-Phase Systems

Two-phase liquid/vapor convection systems are the most widely used passive refrigeration systems for permafrost foundations and earth stabilization. A typical two-phase unit is constructed of pipe enclosed at both ends and charged with a passive refrigerant gas. The radiator (aboveground condenser) portion of the unit can have a bare or finned surface, depending on heat transfer requirements (Figure 3). The evaporator portion of the unit can have any configuration as long as a slope remains between the evaporator and the radiator (Figure 4).

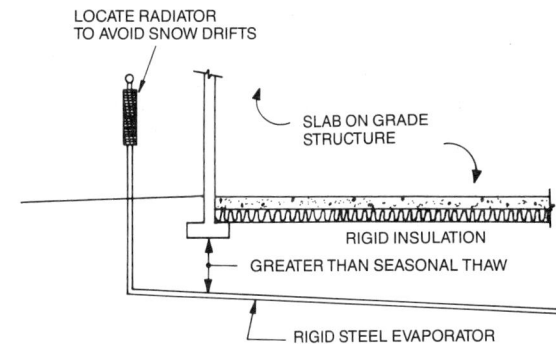

Fig. 4 Typical Thermo-Probe Installation

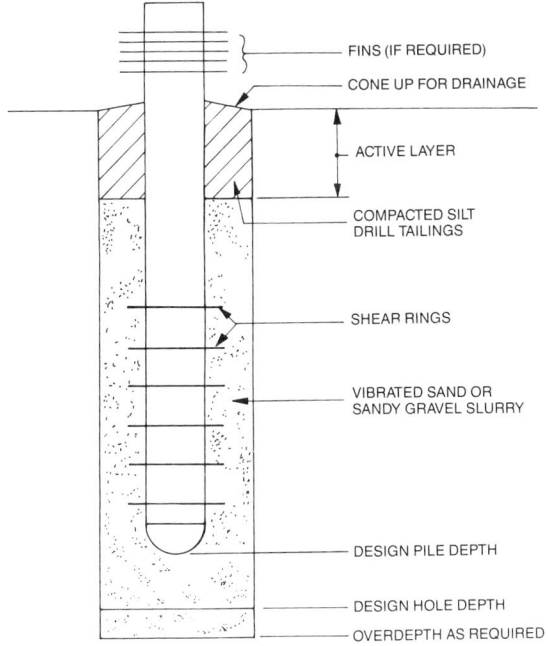

Fig. 3 Thermo Ring Pile Placement

Refrigeration of the subgrade occurs when the radiator is lower in temperature than the soil in contact with the bottom of the evaporator where the liquid portion of the refrigerant is pooled (Yarmak and Long 1982). Condensation occurs in the radiator, initiating evaporation of the refrigerant in the evaporator. The condensate wets the walls of the unit and flows down to the evaporator. Reevaporation of refrigerant condensate with subsequent cooling occurs where the soil in contact with the evaporator is warmer than the soil adjacent to the liquid pool of refrigerant at the bottom of the unit. The entire evaporator unit is then reduced in temperature, cooling the surrounding soil.

A two-phase system will start up with a temperature differential of as little as 0.01°F between the radiator and the evaporator. Liquid and air convection systems may require temperature differentials of 4°F to 15°F before they start up.

Propane, butane, halocarbons, anhydrous ammonia, and carbon dioxide have been used as the refrigerant gas in two-phases systems. Choice of refrigerant gas depends primarily on the allowable internal pressure capabilities of the vessel containing the gas, the quality of available gases, the molecular stability of the gas, and the preference of either the customer or manufacturer of the system. Relatively low-pressure systems using propane, halocarbons, or anhydrous ammonia have been known to gas lock, that is, gases other than the refrigerant gas accumulate in the radiator portion of the unit and prevent the refrigerant gas from condensing. Purging or venting of noncondensable gases may be required on a system following start up.

ACTIVE SYSTEMS

Active ground refrigeration systems are used to keep building or equipment foundations stable when a passive system will not maintain the required stability. The system is normally a network of underground ductwork or piping through which a cooling fluid flows. Heat is removed using a heat exchanger or cooling coil, with refrigeration provided by medium- or low-temperature refrigeration units. System design can include provisions to bypass the refrigeration units during cold weather. Outside air should not be used directly in systems that use air as a cooling fluid due to the moisture introduced into the system in the form of snow and humidity. A defrost system for the cooling coil in a circulating air system should be considered. A typical system installation is shown in Figure 5.

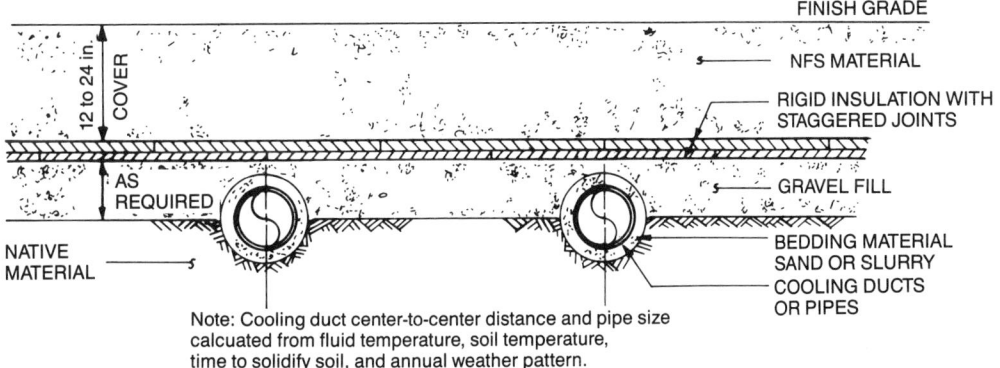

Fig. 5 Active Ground Stabilization System

REFERENCES

House, R.F. 1949. The inundation method of cooling concrete aggregates at Bull Shoals Dam. USA Corps of Engineers, Little Rock, AR.

Long, Erwin L. and Edward Yarmack, Jr. 1982. Permafrost foundations maintained by passive refrigeration. Petroleum Division, Ocean Engineering Division—ASME (March).

Yarmak, Edward, Jr. and Erwin L. Long. 1982. Some considerations regarding the design of two-phase liquid/vapor convection type passive refrigeration systems.

BIBLIOGRAPHY

Kinley, F.B. 1955. Refrigeration for cooling concrete mix. *Air Conditioning, Heating, and Ventilating* (March) 94.

Long, E.L. 1963. The long thermopile, permafrost. Proceedings of International Conference, National Academy of Sciences, Washington, D.C., 487-91.

Townsend, C.L. Control of cracking in mass concrete structures. *Engineering Monograph* No. 34. U.S. Department of Interior, Bureau of Reclamation.

CHAPTER 36

REFRIGERATION IN THE CHEMICAL INDUSTRY

Flow Sheets and Specifications 36.1	Equipment Characteristics 36.3
Refrigeration—Service or Utility 36.1	Design Standards and Codes 36.5
Load Characteristics 36.2	Refrigerants 36.5
Safety Requirements 36.2	Refrigeration Equipment 36.6

CHEMICAL industry refrigeration systems range in capacity from one ton of refrigeration or less to thousands of tons. Temperature levels range from those associated with chilled water through the cryogenic range. The degree of sophistication and interrelation with the chemical process varies from that associated with comfort air conditioning of laboratories or offices to that where the production of reliable refrigeration is vital to product quality or to the safety of the operation.

However, two significant characteristics identify most chemical industry refrigeration systems. Almost exclusively, they are engineered one-of-a-kind systems, and equipment used for normal commercial application may be unacceptable for chemical plant service.

This chapter gives guidance to refrigeration engineers in working with chemical plant designers so they can design an optimum refrigeration system. Refrigeration engineers must be familiar with the chemical process for which the refrigeration facilities are being designed. An understanding of the overall process is also desirable. The accompanying bibliography lists published material on specific processes as well as general articles containing further bibliographical material. Many computer programs are also available that can calculate cooling loads based on the gas chromatographic analysis of a process fluid. These programs accurately define not only the thermodynamic performance of the fluid to be chilled, but also the required heat transfer characteristics of the chiller.

Occasionally, because the process is secret, refrigeration engineers may have limited access to process information. In such cases, chemical plant design engineers must be aware of the restrictions this may place on providing a satisfactory refrigeration system.

FLOW SHEETS AND SPECIFICATIONS

The starting point in attaining a sound knowledge of the chemical process is the flow sheet. Flow sheets serve as a road map to the unit being designed. They include such information as heat and material balances around the major system components and pressures, temperatures, and composition of the various streams within the system. Flow sheets also include refrigeration loads, the temperature level at which refrigeration is to be provided, and the manner in which refrigeration is to be provided to the process (such as via a primary refrigerant or via a secondary coolant). They indicate the nature of the chemicals and processes to be anticipated in the vicinity in which the refrigeration system is to be installed. This information should indicate the need for special safety considerations in the design of the refrigeration system or for construction materials that resist corrosion by process materials or process fumes.

Different portions of a process flow sheet may be developed by different process engineers; consequently, the temperature levels at which refrigeration is specified may vary by only a few degrees. A study of such a situation might reveal that a single temperature is satisfactory for several or even all of the users of refrigeration, which could reduce project cost by eliminating multilevel refrigeration facilities.

Most process flow sheets indicate the design maximum refrigeration load required. The refrigeration engineer should also know the minimum design load. Process loads in the chemical industry tend to fluctuate through a wide range, creating potential operational problems.

Flow sheets also indicate the significance of the refrigeration system to the overall process and the desirability of providing redundant systems, interlocking systems, and so forth. In some cases, refrigeration is mandatory to ensure safe control of a process chemical reaction or to achieve satisfactory product quality control. In other cases, loss or malfunction of the refrigeration system has much less significance.

Other sources of information are also valuable. A properly prepared set of specifications and process data expands on flow sheet information. These generally cover the proposed process design in much more detail than the flow sheets and may also detail the mechanical systems. Information regarding the design philosophy, including continuity of operation, safety hazards, degree of automation, and special startup requirements, is generally found within the specifications. Equipment capacities, design pressures and temperatures, and materials of construction may be included. Specifications for piping, insulation, instrumentation, electrical, pressure vessels and heat exchangers, painting, and so forth are normally issued as part of the design package available to a refrigeration engineer.

It is imperative that refrigeration engineers establish effective communication with chemical process engineers. The refrigeration engineer must know what information to request and what information to give to the chemical process engineer for design optimization. The following sections outline some of the significant characteristics of chemical industry refrigeration systems. A full understanding of these peculiarities is of value in achieving effective communication with chemical plant designers.

REFRIGERATION—SERVICE OR UTILITY

Refrigeration engineers unfamiliar with the chemical industry must understand that unless the chemical process is cryogenic in nature, chemical plant designers probably consider refrigeration merely as a service or utility of the same nature as steam, cooling water, compressed air, and the like. Chemical engineers expect the reliability of the refrigeration to be of the same quality as other services. When a steam valve is opened, chemical engineers expect steam to be available instantly, in whatever quantities demanded. When steam is no longer required, the engineer expects to be able to shut off the steam supply at any time without adversely affecting any other steam user or the source of steam. The same response from the refrigeration system will be expected. This high degree of reliability is usually so strongly implied that no specific mention of it may be made in specifications or process descriptions.

Because refrigeration is frequently considered a service, process designers spend insufficient time analyzing temperature levels,

---

The preparation of this chapter is assigned to TC 10.1, Custom Engineered Refrigeration Systems. This chapter last received a major revision in 1971.

potential load combinations, energy recovery potentials, and the like. The temperature level required for a specific load as indicated on the preliminary flow sheet is often selected from habit. It may have been chosen merely because it was the temperature level at which refrigeration was available on some previous project with which the chemical engineer is familiar. The potential for minimizing the size of the refrigeration system, the total plant investment, or both, by providing refrigeration at a minimum number of temperature levels, is frequently not investigated by process engineers. Likewise, the potential for power recovery is frequently overlooked.

Part of the reason for this attitude is that refrigeration facilities represent only a minor part of the total plant investment. The entire utilities installation for the chemical industry usually falls in the range of 5 to 15% of the total plant investment, with the refrigeration system only a small portion of the utilities investment. Process requirements may be overruling, but process engineers must recognize legitimate process necessities and avoid unnecessary and costly restrictions on the refrigeration system design.

LOAD CHARACTERISTICS

Flow sheet values generally indicate direct process refrigeration requirements and do not include heat gains from the equipment or piping. Flow sheet peaks and average loads generally do not allow for unusual startup conditions or off-normal process operation that may impose unusual refrigeration loads. This information must be gained by a thorough understanding of the process and by discussing the potential effect on refrigeration system design for off-normal process conditions with the process engineer.

Once the true peak loads are established, the duration and frequency of the peaks must be considered. For some simple processes this information is fairly straightforward; however, if the plant is designed for both batch operation and multiproduct manufacture, this can be an enormous task. Computer simulation of such a combination of processes has led to optimization of not only the refrigeration equipment but also the process equipment. Computer simulation assures that the refrigeration machine, secondary coolant storage tanks, and circulating pumps are properly sized to handle both average and peak loads with a minimum investment. Few applications require computer simulation, but a thorough understanding of the relationship of peak loads to average loads and their influence on refrigeration system component sizing is vital to good design.

Sometimes unusually light load conditions must be met. If the process is cyclic and an on-off operation is undesirable, the refrigeration system may run for significant periods under a very light load or even a no-load condition. Light loads often require special design of the system controls, such as multistep unloading and hot gas bypass with reciprocating compressors, a combination of suction throttling and hot gas bypass with centrifugal compressors, and slide valve unloading and hot gas bypass with screw compressors. A secondary coolant system might require a bypass arrangement.

The investment in refrigeration equipment can be kept to a minimum when only a few levels of refrigeration are required. Checking the specified temperatures to be sure they are based on some process requirement may show that fewer temperature levels than shown on the flow sheet are necessary. If multiple levels of refrigeration are required, a compound system should be evaluated. The evaluation must consider the limited ability of a compound system to provide the precise temperatures required of some processes.

Production Philosophy

The flow sheet and specifications will generally indicate whether a chemical process is in continuous or batch operation, but further research may be required to gain an understanding of the required continuity of service for the refrigeration facilities. The chemical industry frequently requires a high degree of continuity; plant production rates are often based on 8000 h or more per year. In general, the refrigeration equipment is worked extremely hard, with no off-peak period because of seasonal changes, and any unscheduled interruption of refrigeration service may create large production losses. In most cases, scheduled maintenance shutdowns are not only infrequent but also highly vulnerable to cancellation or delays because of the press of production requirements.

As a result, reliability is key in the design of chemical industry systems. Equipment that is satisfactory in commercial or light industrial service is frequently unfit where high service rates and minimum availability for maintenance are the rule. In some cases, duplicate systems are justified. More often, multiple part-capacity units are installed so that a refrigeration system breakdown will not create a total process production loss. Major equipment and hardware items that require minimal maintenance or that permit maintenance with the refrigeration system in operation should be selected (for instance, dual lube oil filters or a bypass line that permits temporary reversal of condenser water flow for cleaning). Particular attention should be paid to equipment layout, so that adequate access, tube pullout space, and laydown space are available to minimize refrigeration system maintenance time. In some cases, overhead steel supports for rigging heavy equipment, or permanent monorails, are justifiable.

Flexibility Requirements

The chemical industry constantly develops new processes; consequently, the usual chemical plant undergoes constant modification. On occasion, total processes are rendered obsolete and scrapped before design production rates are ever reached. Thus flexibility should be designed into the refrigeration system so it may be adaptable to some process modification. Designing for optimum flexibility is difficult; however, a study of the potentials or probabilities of process modifications and of the expected life of the process facility will help in making design decisions.

SAFETY REQUIREMENTS

Most chemical processes require special design to ensure safe operation. Many raw materials, intermediates, or finished products are themselves corrosive or toxic or are potential fire or explosion hazards. Frequently, the chemical reactions involved in the process generate extremes of pressure or temperature that must be properly contained for safe operation. Refrigeration engineers must be aware of these potential hazards as well as any abnormal hazards that may develop during startup, unscheduled shutdown, or other upset within the chemical process. When designing modifications or expansions for an existing facility, the possibility that certain construction or maintenance techniques may be safety hazards must be considered.

Corrosion

Refrigeration engineers may become involved in the choice of construction materials when designing heat exchange equipment for cooling of corrosive process materials. The shell, tubes, tube sheets, gaskets, packing, O-rings, seal materials, and components of instrument or control hardware must be properly specified. The potential hazards of leakage between the normal process side and the refrigeration side of heat exchange equipment must be investigated, because an undesirable chemical reaction may occur between the process material and the refrigerant.

An additional corrosive hazard may result from leaks, spills, or upsets within the process area. Safe chemical plant design must anticipate the unusual as well as the usual hazards. For example, if refrigerant piping will run adjacent to a flanged piping system containing a highly corrosive material, special materials for the piping and insulation systems may be justified.

Toxicity

If the refrigeration system indirectly contacts a toxic material via heat exchange equipment, flanges and such elements as gaskets,

Refrigeration in the Chemical Industry

packing, and seals in direct contact with the toxic material must be designed carefully. The possibility of a leak in equipment that might allow refrigerants or secondary coolants to mix with process chemicals and cause a toxic or otherwise dangerous reaction must also be considered. In some cases, the potential for toxic leaks may be so high that a special ventilation system may be required.

Even though containment and ventilation can handle toxic materials under normal process conditions, toxic material may need to be vented in abnormal situations to avoid the hazards of fire or explosion. In such cases, the toxic material is frequently vented or diluted through a tall stack so that ground or operating level concentrations do not reach toxic limits. Refrigeration engineers should evaluate the desirability of locating the refrigeration equipment itself or its controls outside the operating areas. An alternate solution is to enclose either the refrigeration system or its controls with a ventilation system that has a remote air intake.

The hazards of toxicity are not confined to the chemical process itself. Some common refrigeration chemicals are toxic in varying degrees. Because of the chemical industry's interest in safety, refrigeration engineers are required to treat some of the common refrigerants and secondary coolants with much more caution than required for the usual commercial or industrial system.

Fire and Explosion

Plant specifications normally define the area classifications, which determine the need for special enclosures for electrical equipment. For areas designated as being in an explosion-proof environment, standard components are normally grouped in a single large explosion-proof cabinet. Intrinsically safe control systems, which eliminate the need for explosion-proof enclosures at all end devices, are also used. Intrinsically safe shutdown systems eliminate arcing at the end device by using low-voltage signals controlled by a microprocessor. Control equipment can also be mounted in standard or weatherproof cabinets equipped with an inert gas purge system.

Items other than the electrical system may require special consideration in areas of high fire or explosion hazard. The use of flammable materials should be carefully reviewed. Stainless steel instrument tubing should be chosen rather than unprotected plastic tubing, for example. Both insulation materials and insulation finish systems should be chosen to minimize flame spread in the event of a fire. In some extremely hazardous areas, nonflammable refrigerants or secondary coolants may be required, rather than flammable materials that would provide otherwise superior performance.

In areas of high explosion potential, modification of the usual refrigeration system design may be required. Design pressures for refrigeration vessels or piping may be determined by process considerations as well as refrigeration system requirements. Special pressure relief systems, such as rupture disk in series with relief valves, dual full-sized relief valves, or a parallel relief valve and rupture disc with transfer valves between them, are often required.

Refrigeration System Malfunction

A malfunction can itself be a significant safety hazard, whether the upset is caused by an internal failure of the system or by fire, explosion, or some other catastrophe. Most process areas designated as being in a hazardous environment are protected by automatic gas and flame detection systems that shut down the refrigeration system when explosive mixtures or fire are detected in an area. Loss of refrigeration may permit a process reaction to run out of control and cause loss of product, fire, explosion, or the release of toxic material in an area remote from the original source of trouble. Thus, one or several degrees of redundancy may be needed to minimize the consequence of refrigeration system malfunction.

Storage of cold coolant, ice, or cold eutectic, or an alternate emergency supply of cooling water may be necessary to meet emergency peaks. Alternate sources of electric power to the refrigeration system may be desirable. Dual drive capability by either electric motor or steam turbine might even be justified. Uninteruptable power systems may also be used for control systems.

Frequently, special facilities are used to protect against the extreme consequences of refrigeration system malfunction. One example is the quenching of the process reaction via an inhibiting or neutralizing chemical introduced to the process in the event of an unusual pressure or temperature rise. Another simpler and more common example requires closing one or more process valves in the event of a loss of refrigeration.

Maintenance

In an operating chemical plant, because of hazards frequently encountered, the normal maintenance procedures may not be permitted. Because welding, burning, or the use of an open flame is often prohibited throughout large areas of a chemical plant, maintenance flanges or screwed connections are used to permit replacement of piping and equipment. Sometimes extra access space or handling facilities (such as monorails) are provided to permit efficient removal of machinery to an area where welding, burning, and the like, is permitted.

EQUIPMENT CHARACTERISTICS

Automation

A few small chemical processes, generally batch operations, are still designed with a minimum of instrumentation, but, in chemical plant operations, instrumentation plays an important role in achieving a satisfactory process and represents a significant percentage of the plant investment.

For this reason, most chemical plant designers insist on standardization of instrumentation throughout the plant. All instruments, including alarm lights, pressure gages, temperature gages, and thermowells must meet the required standards. These requirements may not include familiar refrigeration components and may create problems in the refrigeration design. Therefore, it is vital that the refrigeration engineer and the chemical or instrumentation engineer agree on the exact instrument requirements early in the design phase.

A second concept frequently adopted by the chemical industry is the control and monitoring of all plant operation from a single central control room. In the chemical industry, it is not unusual to operate a multimillion dollar processing facility with two central control room operators and perhaps a single roving operator. This concept influences the refrigeration system in several ways. Refrigeration system controllers and alarm and shutdown lights are mounted on the central control room (CCR) panel, as are recorders or indicators that display refrigeration system temperatures, pressures, flows, and so forth.

Even with central control room operation, a local panel is required for the display or recording of additional information that can aid in troubleshooting an emergency shutdown. Frequently, startup control is available only at this local panel, to assure that startup is not attempted unless an operator is present to witness the operation of major items of refrigeration equipment. The CCR-mounted hardware would certainly conform to the standards of the process instrumentation to minimize operator confusion either in reading informative devices or in operating control devices. Most plants now use centralized computer or microprocessor type controls. In some cases, the refrigeration unit is controlled from the main computer or microprocessor.

When applying a CCR concept, it is important to determine exactly how much information and control are to be provided at the control room and how much are to be provided locally.

Transmission of unnecessary information to the CCR can be costly, but, if sufficient information is not available, serious process upsets are inevitable. Remember that most process operators are not specifically trained to understand the intricacies of refrigeration machinery. They usually regard the refrigeration system and its control as simply one more piece of what may be an extremely intricate process. The CCR system must permit monitoring of the refrigeration system performance and control of that performance to suit process needs.

Operators require alarms to indicate abnormal conditions for which they can make corrections, either at the CCR or in the field, and to indicate a system breakdown or shutdown. They also require a locally mounted manual shutdown station in the event of an emergency. Devices such as sequencing alarms and lube oil or bearing temperature recorders, which are either troubleshooting aids or can adequately be checked or logged by the roving patrol operator, are best installed locally.

The concept of designing for a minimum of operator attention with personnel that are not refrigeration specialists is yet another reason (in addition to those previously mentioned, such as production and safety requirements) that a high degree of reliability is required for a chemical industry refrigeration system. An engineered system is required, and normal commercial design is frequently inadequate.

Outdoor Construction

Another chemical industry characteristic is outdoor construction. The chemical industry installs sophisticated process and auxiliary facilities outdoors all over the world. These unprotected facilities are expected to operate as designed and with a minimum of difficulty. Whether the problems are those imposed by low temperatures, heavy snows and freezing rain, dust storms, baking heat, or hurricane force winds with salt-laden rains, they are generally unfamiliar to the uninitiated refrigeration engineer. For example, explosion-proof electrical construction is not necessarily weather resistant. Lube oil heaters and a prestartup circulation of heated lube oil may be required for compressors or other rotating machinery. In areas with high winds, special attention must be paid to the detailed installation instructions for insulation jacketing applied to pipe or vessels.

Winter operation of cooling towers may require multiple-cell construction, two-speed or reverse rotation cooling-tower fans, or even facilities for steam heating the cooling-tower basin. Instrument air for transmission of signals or power to pneumatic operators should be dried to a dew point (under pressure conditions) lower than anticipated ambient conditions to avoid condensate or ice from forming in the instrument air lines or in the instruments themselves. In fact, it has become almost standard that the instrument air provided is both oil-free and dried to a low dew point. The effect of ambient temperature, as well as radiation from the sun, should be considered in determining system design pressures, especially when equipment may be idle. Ambient temperatures up to 120 °F and vessel skin temperatures of 165 °F can be experienced in hot climates. To determine whether or not purchased equipment meets the requirements for outdoor installation, a detailed check of vendors' drawings, specifications, descriptive literature, and vendor-procured components is required.

Energy Recovery

Both the installed cost and the operating costs for the refrigeration system of some chemical processes can be reduced by the intelligent use of energy recovery techniques. If the process requires large quantities of low-pressure steam, the use of back pressure turbines to drive the refrigeration compressors could significantly reduce refrigeration system energy costs. On the other hand, if the process generates an excess of low-pressure steam, it is possible that an absorption system could provide an overall saving. For processes with an excess of low-pressure steam only during the summer months when heating requirements are at a minimum, a condensing turbine may be economical if it is sized to operate at the low steam pressure when the excess is available and at a higher pressure during the heating season.

Other energy imbalances occurring within a chemical process can be advantageous. A waste heat boiler installed in a high-temperature gas stream may provide a source for low-cost steam. If the gas stream is at a moderate pressure as well as a high temperature, the possibility of a gas-driven power recovery turbine should be considered.

Another means by which operating costs frequently can be reduced is the reuse of once-through cooling water. Frequently, turbine-driven centrifugal refrigeration machines can use cooling water from the refrigerant condenser to condense the turbine exhaust steam, either in a shell-and-tube or a low-level jet condenser. Another possibility is the reuse of refrigeration system cooling water in process heat exchangers. These examples are typical but not all inclusive. Again, a thorough understanding of the process is a prerequisite for understanding the energy recovery potential, and a flow sheet should be helpful.

Performance Testing

Frequently, a requirement for performance testing of the refrigeration facilities is included within the contract for a chemical industry processing unit. Agreement should be reached as early as possible between the owner and the contractor regarding the exact procedure to be used for testing. If the test is to be run at some condition other than design conditions, both parties must agree on the methods of converting the test results to design conditions. Approximation techniques, such as those outlined in ARI *Standard* 550, are usually unacceptable in the chemical industry. The refrigeration engineer must be assured that adequate facilities for an equitable test are designed into the refrigeration system. This may require additional flowmetering devices or more accurate temperature-measuring devices than are required for normal plant operation.

Insulation Requirements

The service conditions imposed by the chemical industry on both piping and equipment insulation are frequently more exacting than those experienced in the usual commercial or industrial installation. Not only must the initial integrity be as near perfect as possible, but it must also resist the high degree of both physical and chemical abuse that it is likely to incur during its lifetime. To achieve both a minimum permeability and a maximum resistance to abuse, multicomponent finish systems may be required. In some cases, a vapor barrier mastic coating system (which may include reinforcing cloth) is covered with aluminum, stainless steel, or an epoxy-coated carbon steel jacket to protect against physical and chemical abuse. Piping and equipment insulated under ideal shop working conditions must be designed to withstand loading and unloading and erection into position on the job site without damaging the vapor barrier.

Because the fire hazard is generally high and the potential loss of personnel and investment resulting from a serious fire is prohibitive, a strict limit is usually placed on insulation finish systems having a high flame-spread rating, particularly in indoor construction.

The frequent use of stainless steel in the chemical industry for piping and equipment creates another problem with regard to insulation system design. Many stainless steels fail when they are exposed to chlorides. Stress corrosion cracking can occur in a matter of hours. Consequently, chloride-bearing insulation materials must not be used in contact with stainless steels even in minute quantities and should not be used anywhere in the system unless

Refrigeration in the Chemical Industry

a valid vapor barrier is interposed. Chapter 21 of the 1989 ASHRAE *Handbook—Fundamentals* covers the general subject of thermal insulation and water vapor barriers.

DESIGN STANDARDS AND CODES

The bibliography lists common codes and standards that apply to the chemical industry. The grouping is by the area of primary concern. For example, the NBFU and FM bulletins are listed under Fire and Safety, but they contain specific requirements for equipment and piping to meet their codes. Relatively few suppliers of refrigeration equipment regularly manufacture to meet these codes or standards or are familiar with codes relating to their equipment.

Another variation from commercial or industrial design practice is the use of company standards in addition to those previously mentioned. For the usual commercial or industrial plant, the client will at best provide performance specifications and a statement of what is to be done, leaving the preparation of detailed specifications for equipment, piping, ducting, insulation, and painting to the designer. However, in the chemical industry, many such items are covered by company standards, which, though established primarily for use in the process cycles, can yield corporate benefits if they are also used in design of the refrigeration systems. Such benefits include: simplified parts and warehouse stocking, increased degree of interchangeability, simplified maintenance procedures, elimination of control to ensure that nonstandard items purchased for use in the refrigeration system do not get into process or other utility systems, possible reduction of insurance costs, and greater flexibility in reusing the components should that become necessary.

Company standards vary so widely that no listing has been included in the bibliography. A request for all applicable company standards at the start of the design and an effort to use them will avoid costly rework of the design following a review by the client.

Startup and Shutdown

Processes are most hazardous during startup and shutdown. Though present in batch or discontinuous processing units, the problem is usually more severe in continuously operating units for the following reasons:

1. Instrumentation and control must be designed for the normal condition, and the cost of features intended for use only during startup or shutdown often cannot be justified. Frequently, conditions at startup or shutdown fall outside the range of the operating instruments and control, so that manual control is necessary.
2. The same argument holds for much of the process equipment, so that extraordinary measures, such as severely throttled flow, minimum-flow bypasses, and recycling may be needed.
3. The operators go through these conditions only infrequently and may have forgotten the techniques of operation at the time they are most needed.
4. The process conditions at startup and shutdown are usually not recorded on flow sheets or in descriptions because they occur so infrequently and usually vary continually as the units are brought on and off stream. As a consequence, designers have a tendency to overlook them and to concentrate on the conditions in the operating range.

Refrigeration engineers must inquire whether startup or shutdown is likely to impose any special conditions on the refrigeration system. Startup and shutdown is a special burden in this respect, since operators are particularly busy with the processing equipment and cycle during these times and generally cannot monitor or adjust the operation of what they regard as a service system. Therefore, process engineers must be made thoroughly aware of the precise limitations that startup or shutdown of the refrigeration system may impose on process operation.

REFRIGERANTS

Such factors as flammability, toxicity, and compatibility with proposed construction materials may influence the final selection of a refrigerant more than in other applications. Special attention should be paid to the consequences of leakage between the process materials and the refrigerant or the secondary coolant. Chapters 16 and 18 of the 1989 ASHRAE *Handbook — Fundamentals* discuss refrigerants and secondary coolants in detail.

Because of their nontoxic, nonflammable properties, halogenated hydrocarbons have been used predominantly, but recent environmental concerns have reduced their application throughout the chemical industry. Hydrocarbons such as methane, propane, ethane, propylene, ethylene, or ammonia are used in many cases where the process stream involves them as constituents. In the petrochemical industry, the use of these materials within the process is common.

Of the secondary coolants, calcium and sodium chloride brines have been used most often, although glycols and such halocarbons as methylene chloride, trichloroethylene, R-11, and R-12 also have been frequent choices. Again, environmental concerns predicate against the use of R-11 and R-12 in new facilities.

Many of the same factors that influence refrigerant selection must be considered in choosing a secondary coolant. Corrosivity, toxicity, and stability are of special significance in determining suitability for chemical plant service.

Refrigeration Systems

An indirect system, in which brine or chilled water is circulated to air washers, cooling coils, and process heat exchangers from a central refrigeration plant, is much more prevalent in the chemical industry than in the food industry or in residential or light commercial comfort air conditioning. This is particularly true where large capacities or low temperature levels are involved. An indirect system permits centralization of the refrigeration equipment and associated auxiliaries, which may offer significant advantages in operation and maintenance, particularly if remote location of the refrigeration equipment permits design, operation, and maintenance in a nonhazardous location. It also may permit the installation of a minimum number of large units rather than many small units located in remote areas. For low-temperature systems of significant capacity, an indirect brine cooling system installed in the process area close to the process users is common.

Where the number of process heat exchangers requiring cooling and the length of piping can be kept to a minimum, a direct system, which uses the refrigerant in the process heat exchange equipment, often is the optimum design, particularly for small or medium loads. Since an indirect heat exchanger is not required in this case, a higher operating suction pressure and consequent lower operating and investment costs may be possible. Direct systems are also used when a refrigerant is involved in the manufacturing process stream, as in the production of ammonia or many petrochemicals. Here, the length of refrigerant lines, with possible high refrigerant losses because of leakage, is a less significant factor in system selection. Direct systems are usually of the dry expansion or flooded evaporator design; flooded coil systems of the gravity feed or pumped liquid overfeed design are relatively uncommon.

Direct systems for larger capacity, low-temperature, multiple user service have advantages and disadvantages that must be considered for proper system selection. Some of the disadvantages are:

1. Maintaining an extensive refrigeration piping system free of leaks is difficult. Leakage from piping for secondary coolants is frequently less objectionable than the refrigerant gases.

1. Checking for refrigerant leaks or repairing them in certain high explosion hazard process areas can be a problem because halide and electronic leak detectors may not be permitted and burning or welding may not be possible without a plant shutdown. If air or moisture leaks into a system operating at vacuum conditions, extensive icing and corrosion problems can result.
2. Higher piping costs are often involved when all items are considered, including large and expensive vapor and liquid-control valves at individual heat exchangers, particularly for the halocarbon refrigerants. Generous refrigerant knockout separators are necessary at each stage. Although both refrigerant and secondary coolant lines require insulation to prevent capacity losses, sweating, and icing, refrigerant lines, especially on the vapor return to the compressor, can greatly exceed the size of those required for brine recirculation.
3. No system reserve capacity is available as is the case with a secondary coolant, particularly if the latter is designed as a storage system. Process upsets can directly and suddenly increase the load on the refrigeration unit, causing cycling of the equipment and possible damage. For some processes, meeting short, sharp load peaks is of paramount importance to avoid off-standard production or unsafe operating conditions.
4. Constant temperature control is often more difficult or more costly to maintain with direct refrigeration than with a secondary coolant.
5. Initial testing of an extensive direct refrigeration system may be a significant problem. Testing must be done pneumatically rather than hydrostatically to prevent problems associated with water left in the refrigerant system. Pneumatic testing is a hazardous operation and is forbidden in some chemical plants. The alternative of extensive posttesting dehydration is usually both expensive and time consuming.
6. The initial cost for refrigerants is usually much higher in an extensive direct refrigeration system than in a secondary coolant system operating at temperatures most frequently encountered in the chemical industry. In case of system leaks, the costs of makeup coolant are generally less than the costs of makeup primary refrigerant.

Some of the advantages offered by direct refrigeration systems include the following:

1. Careful control of corrosion inhibitors may be necessary to keep secondary coolants stable so that they do not cause extensive equipment damage.
2. Less equipment and maintenance may be required; secondary coolant circulation and control or coolant mixing and makeup facilities are not needed.
3. Power costs are generally lower because of higher suction pressures and, in some designs, because pumps are not required.
4. Damage because of equipment freezing is not likely. Such damage can occur in a secondary coolant system if the coolant condition or the refrigeration plant operation is not properly supervised.

Thus, the broad scope of refrigeration applications within the chemical industry permits the use of virtually any refrigeration system under the proper process conditions. For a more detailed discussion of the most frequently used systems, see Chapters 1 to 5 of this volume.

REFRIGERATION EQUIPMENT

For the most part, the refrigeration equipment used in the chemical industry is identical to, or closely parallels, the equipment used in other industries. The chemical industry is unique, however, in the wide variety of applications, the large temperature ranges covered by these applications, the diversity of equipment usage, and the variation of mechanical specifications required. Where possible, the chemical industry uses standard equipment, but this is frequently impossible because of the particularly rigorous demands of chemical plant service. Therefore, this section only briefly describes the application and modification of refrigeration equipment for chemical plant service. The 1988 ASHRAE *Handbook—Equipment* gives further information on various refrigeration equipment.

Compressors

Refrigeration engineers may find difficulty in applying conventional refrigeration compressors to chemical plant service. Most process engineers are familiar with heavy duty, forged steel, high pressure, single or double throw reciprocating gas compressors; they are uncomfortable with the high speed, cast iron, or steel compressors that are standard to the refrigeration industry. Another difference between commercial and chemical plant usage is the greater use of open-drive equipment in the chemical plant.

The large capacities and low temperatures frequently encountered in chemical plant duty have led to wide use of either centrifugal compressors or high capacity rotary or screw compressors. These large machines vary from standard commercial equipment principally in the amount and complexity of controls or other auxiliaries provided. Load control devices such as multistep unloaders or hot-gas bypass systems are often required to permit a compressor turndown to 10% of full load or, in some cases, to permit no-load operation without either compressor surge problems or on-off operation. Most systems with large multistage centrifugal compressors use economizers to minimize horsepower and suction volume requirements. Compressor lube oil systems are often provided with auxiliary oil pumps, dual oil filters, dual oil coolers, and the like, to permit routine maintenance without shutdown and to minimize shutdown frequency. Compressor control and alarm systems are frequently tied into central control room panel boards to permit monitoring and/or control of compressors.

Compressors for hydrocarbon gas refrigerants find their greatest use within the chemical industry, particularly in the field of petrochemicals. The relatively low cost and ready availability of pure hydrocarbons and hydrocarbon mixtures frequently dictate their use. Many offer the additional advantage of positive pressure operation throughout the entire refrigeration cycle.

Since refrigeration systems within the chemical industry are often required to operate for a year or more without shutdown, standby compression equipment is frequently installed. Even the larger refrigeration loads sometimes require 100% standby protection. Special controls may be required to provide rapid and automatic startup of the standby equipment. The main drive is commonly an electric motor and the standby drive may be either a steam turbine or an internal combustion engine. Provisions must be made via nonelectric drivers or emergency generating equipment to keep all necessary auxiliaries and controls operative during the electrical outage. Oversized crankcase heaters may be required, as well as electric or steam tracing of various lube oil system components.

High in-service requirements, plant standardization, explosion hazards, and corrosive atmospheres all require special controls. Often, the copper instrument tubing normally used on commercial equipment must be replaced with steel or stainless steel tubing more suitable to the proposed plant atmosphere. Lube oil piping must be stainless steel pipe with nonferrous valves, coolers, and filters. This requirement is primarily to minimize expensive delays in initial plant startup that may result when rust or scale within lube oil systems causes damage to bearings or seals in high-speed centrifugal or screw compressors.

Absorption Equipment

Chapter 1 of the 1989 ASHRAE *Handbook—Fundamentals* and Chapter 13 of the 1988 ASHRAE *Handbook—Equipment* discuss absorption equipment in detail. Absorption equipment has seen little use in chemical plants, although this equipment is frequently desirable because plant waste heat is available to operate it.

By special design and reselection of materials, hot streams of many fluids can be used as the energy source instead of using hot water or steam. Direct fired units are available. Hot condensable vapors can also be used as the energy source.

Commercial absorption machine equipment must be modified to permit outdoor operation. Manufacturers' recommendations should be followed regarding changes necessary to prevent freezing on the water side of the equipment and solution crystallization on the absorption side of the equipment, particularly during shutdown conditions.

Condensers

Water-Cooled Condensers. Units for chemical plant service require relatively minor design changes from those provided for commercial installations. Since cooling water is frequently of a lower quality than that normally encountered, special materials of construction may be required throughout the tube side. An example is the necessity to switch from copper to a cupronickel when using cooling water from a tidal source that is high in chlorides. If the cooling water is high in mud or silt content, it is sometimes justifiable to install piping and valving that will permit backflushing the condenser without requiring a refrigeration machine shutdown. Chemical plant requirements normally dictate the use of shell-and-tube condensers of the replaceable tube type. Process engineers may insist on conservative tube side velocities (8 fps or less as a maximum for copper tubes) and replaceable tube bundles. The long hours of operation without opportunity for cleaning and the types of cooling water used frequently require that a higher water-side fouling factor be assumed than on commercial installations.

Air-Cooled Condensers. With increasing restrictions on the use of water for condensing, air-cooled condensing systems have been used in several instances even in the larger centrifugal-type plants. These have usually been installed in warmer locations where the increase in condensing pressure (temperature) over that resulting from the use of cooling tower water or once-through water systems is minimal.

Air-cooled condensers for chemical plant service are normally fabricated to one of the API Standards for forced convection coolers. Care must be taken when specifying these coolers so that the manufacturer understands the type of duty associated with a condensing refrigerant. The service required of an air-cooled condenser in a chemical plant atmosphere dictates either the use of more expensive alloys in the tube construction or conventional materials of greater wall thickness, to give acceptable service life. Air coolers may be more difficult to locate because recirculation of hot discharge air or fouling by hot process exhaust gases must be avoided.

Evaporative Condensers. Evaporative condensers, particularly for smaller refrigerating loads, are used extensively in the chemical industry, and they should become more prevalent as more emphasis is placed on the reduction of thermal contamination of rivers, lakes, and streams. In a few larger installations, the combination of an air cooler and an evaporative condenser operating in series satisfies the condensing requirements of the refrigeration system.

In most cases, the commercial evaporative condenser is totally unsuitable for chemical plant service, but satisfactory results can be obtained if this equipment is carefully specified. The major items of concern are the atmospheric conditions to which such equipment may be exposed and the long in-service requirements of the chemical plant. The chemical plant atmosphere, which may abound in vapors or dusts that are corrosive in themselves, can be an even more serious problem when these vapors and dusts are passed over surfaces that are constantly being wetted. Another problem is that dusts from nearby raw material storages or grinding operations may infiltrate the water recirculating system and plug the spray nozzles. The problems of water treatment and winter freeze protection are usually much more severe in chemical plant service because of the lower quality water that is frequently available and the demand for both year-round operation and a high turndown ratio. Light load operation in freezing weather calls for extreme care in design to avoid freezeup.

Two other areas of commercial evaporative condenser design that must frequently be strengthened for chemical plant duty are the electrical equipment, which must be satisfactory for the plant environment, and the fans, dampers, and recirculating pumps, which must be suitable for long-life low-maintenance service.

Evaporators

The general familiarity of chemical plant design personnel with heat exchanger design and application may sometimes lead them to suggest that refrigeration evaporators for the chemical plant should be designed similarly to evaporators in nonrefrigeration service. While the general laws of heat transfer apply in either case, there are special requirements for evaporators in refrigeration service which are not always present in other types of heat exchanger design. Refrigeration engineers must coordinate the process engineers' experience with the special requirements of a refrigeration evaporator. To do so, the Standards of the Tubular Exchangers Association (TEMA) should be consulted to insure that the end product is familiar to the plant engineer while still performing efficiently as a refrigeration chiller.

Paramount in these special requirements are the proper treatment of oil circulation in the refrigeration evaporator and proper evaluation of liquid submergence as it may affect low-temperature evaporator performance. When the evaporator in chemical plant service is being used with reciprocating and rotary screw compression equipment, continuous oil return from the evaporator must normally be provided. If continuous oil return is not possible, an adequate oil reservoir for the compression equipment, with periodic transfer of oil from the low side of the system, may be needed. On evaporators used with centrifugal compression equipment, continuous oil return from the evaporators is not necessary. In general, centrifugal compressors pump very little oil, so oil contamination of the low side of the system is not as serious as with positive displacement equipment. However, even with centrifugal equipment, the low side evaporators eventually become contaminated with oil, which must be removed. Most centrifugal systems operate for several years before oil accumulation in the evaporator adversely affects evaporator performance. Newer tube surfaces with porous coatings may be more sensitive to the presence of oil in the refrigerant than would be conventional finned surfaces. For newer surfaces, a continuous oil return system may be essential for centrifugal systems.

Flooded shell refrigeration evaporators operating at extremely low temperatures and low suction pressures may build up an excessive liquid head, which can create higher evaporating pressures and temperatures at the bottom of the evaporator than at the top. Spray-type evaporators with pump recirculation of refrigerant eliminate this static head penalty.

Special materials for evaporator tubes and shells of particularly heavy wall thickness are frequently dictated to cool process streams of a highly corrosive nature. Corrosion allowances in evaporator design, which are seldom a factor in the commercial refrigeration field, are often required in chemical service. Ranges of permissible velocities are frequently specified to prevent sludge deposits or erosion at tube ends.

Process side construction suitable for high pressures seldom encountered in usual refrigeration applications is frequently necessary. Choice of process side scale factors must also be made carefully without overstatement.

Differences between process inlet and outlet temperatures of 100 °F or more are not uncommon. For this reason, special consideration must be given to thermal stresses within the refrigerant evaporator. U-tube or floating tube sheet construction is frequently specified in chemical plant service, but minor process side modifications may permit the use of less expensive standard fixed tube sheet design. The refrigerant side of the evaporator may be required to withstand pressures resulting from maximum process temperature or the evaporator must be able to bypass the process stream under certain high-temperature conditions, e.g., in a refrigeration system failure.

Relief devices and safety precautions common to the refrigeration field normally meet chemical plant needs but should be reviewed against individual plant standards and local statutory requirements. Forged steel relief valves are becoming more common as they meet the applicable refinery piping codes. In hazardous service, relief valves are sized for emergency discharge in the event of fire. The effect of chemical vapors on the downstream (outlet) internal parts of relief valves may call for special materials or trapped outlet piping with isolating liquid seals.

Process requirements frequently call for sudden or unexpected load changes on the refrigeration evaporator. Possible thermal shocks, with attendant stresses, must be evaluated, and the evaporator must be designed to meet any such conditions.

Evaporators in chemical plant duty normally require inspection and cleaning on an annual basis. For this reason, they should be located for accessibility and ease of tube replacement. Possible contamination of the process stream or the refrigerant side, because of leakage, should be evaluated. Special means of leak detection from one side of the evaporator to the other may be justified on occasion.

Low-temperature refrigeration in the chemical industry often creates extremely high viscosities on the process side of the equipment. Special evaporator designs may be needed to minimize pressure drops on the process side and to maintain optimum heat-transfer performance. Small tube diameters may not be compatible with the process stream because certain processes may call for extra large tubes. For extremely low temperature and high viscosity duties, evaporators are sometimes provided with rotating internal wall scrapers to assure flow of high viscosity fluids through the evaporators. Similarly, jacketed process vessels are used to cool highly viscous materials, while rotary scrapers keep the vessel walls clean.

For proper process flow, evaporators usually are remote from the other refrigeration equipment to minimize piping and pumping costs. Because remotely located evaporators place special emphasis on proper refrigerant piping practices, secondary coolant systems may be used. Chemical plants frequently use flooded refrigeration systems, which pump refrigerant from the central compressor station to remote evaporators. The use of these systems often reduces the design difficulties in assuring adequate oil return, and special provisions must be made at the central refrigeration station to protect the compressor against liquid carryover in the suction gas. The system must have an adequate accumulator to assure dry gas to the compressor (see Chapter 2).

Standard air-side evaporators may require modification, mainly to solve special corrosion problems in handling air or process gases that attack standard coil materials. Occasionally, process requirements demand coil designs that do not match standard commercial air-side pressure drops, air-side design temperature range, or both. Coils of special depth and special finning may be required and coil casings and fan casings of alloy steel are common.

Instrumentation and Controls

Since the heart of the chemical plant is its instrument control system, it follows that the instrumentation and control is much more advanced in the chemical industry than in commercial or the usual industrial refrigeration applications. As previously discussed, chemical industry refrigeration instrumentation hardware is much more sophisticated in design, particularly in regard to providing increased safety, reliability, and compatibility with process instrumentation devices. This sophistication extends to the application and design of individual hardware items. The chemical industry seldom settles for integral control devices such as self-contained pressure regulators or capillary-actuated thermal-control valves. The usual chemical industry control loop consists of a sensing device, a transmitter, a recorder/controller, a positioner, and an operator, all pneumatically or electrically interconnected. Many plants use central computer and microprocessor controls. Interfacing between the refrigeration system and control system may be necessary.

Cooling Towers and Spray Ponds

In a refrigeration system that uses water-cooled rather than air-cooled or evaporative condensers, heat may be rejected to once-through cooling water, spray ponds, or cooling towers. The chemical industry uses mechanical draft towers almost exclusively. These are generally of the induced draft design and are about evenly divided between crossflow and counterflow operation. Although a familiarity with these items is necessary, chemical plant engineers are usually responsible for their design.

Miscellaneous Equipment

Pumps. Refrigeration system pumps are usually of a high quality centrifugal design, the primary exception being small positive displacement pumps for compressor lube oil systems. In the past, heavy duty design was the rule rather than the exception, and secondary coolant and chilled-water units were usually of a horizontal split-case design, patterned after boiler-house or water plant construction. Chemical process designers have advocated standard chemical plant pump designs, which usually have a vertically split case and an end suction. If the selection is made carefully, this design is successful in many applications, and the resultant savings in pump costs, space requirements, and spare parts stocking requirements make it economically attractive. For pumped materials difficult to contain, such as most refrigerants and many secondary coolants, mechanical shaft seals of various designs are frequently used. As a result of their highly successful use in pumping difficult process fluids, canned or sealless pumps are used in such applications as liquid overfeed systems using halocarbons. Because the pumping of difficult fluids is a common problem, chemical process designers can be of invaluable assistance.

Piping. Chapter 33 of the 1989 ASHRAE *Handbook—Fundamentals* lists pipe sizing methods for refrigerant piping systems. Chapters 1 to 4 also contain valuable information regarding such systems. A brief discussion of piping for secondary coolants is included in Chapter 5.

Several aspects of chemical plant piping systems deserve special mention. First, as a consequence of several factors, including low fluid temperatures, large pipe sizes, congested pipe alley space, and the industry's reluctance to use expansion joints for high duty service and in corrosive atmospheres, piping flexibility problems are much more complex. Expansion joints are frequently prohibited, which increases space requirements dramatically. Secondly, piping and valve standards that apply to both process and service facilities are frequently established by the process designer. The refrigeration engineer who is accustomed to using carbon-steel piping systems with tongue and groove flanging and

Refrigeration in the Chemical Industry

valves may find that plant standards call for a welded nickel-steel system with raised face flanges, spirally-wound stainless steel gaskets, and cast-steel valving.

Most piping construction problems resulting from the difference between expectations of the process engineer and the experience of the refrigeration engineer can be resolved by constructing the system to meet ANSI/ASME *Standard* B31.3, "Chemical Plant and Petroleum Refinery Piping." The ANSI/ASME B16 series of standards that define the flanges and fittings of the process industry should also be followed.

Tanks. Chemical plants use storage tanks for both refrigerants and secondary coolants much more frequently than do most commercial or industrial plants. In chilled water or brine circulation systems, storage tanks often serve a dual purpose: first, to store secondary coolants during operation to provide a reserve capacity and thus smooth out short-term peak requirements; and secondly, to store secondary coolants during a maintenance shutdown of process evaporators. In some cases, brine mix and storage facilities are provided, so that any loss of brine because of leakage or unusual maintenance demands can be quickly replaced, thus minimizing unscheduled process outages. In many cases, refrigerant pumpout compressors and storage receivers can minimize loss of the refrigerant and unscheduled outage time as a result of refrigeration system failures on the refrigerant side.

The chemical industry designs all pressure vessels in accordance with the ASME boiler and pressure vessel codes, in particular Section VIII, Division I, for unfired pressure vessels, regardless of local government regulations requiring such design. In most plants, standards are established regarding such items as pressure relief devices, manhole design, insulation supports, and tank supports. A thorough knowledge of the plant standards to be applied should be gained before specifications and design details are established for refrigeration system tankage.

BIBLIOGRAPHY

Processes

Bourton, K. 1968. *Chemical and process engineering unit operations: A bibliographical guide.* Plenum Press, New York.
Chemical economics handbook. Stanford Research Institute, Menlo Park, CA.
Faith, W.L., D.B. Keyes, and R.L. Clark. 1965. *Industrial chemicals,* 3rd ed. John Wiley and Sons, New York.
GPSA. *Data book.* Gas Processors Suppliers Association, Tulsa, OK.
Haywood, R.W. 1967. *Analysis of engineering cycles.* Pergamon Press, Oxford, England.
Jordan, D.G. 1968. *Chemical process development,* 2 vols. Interscience Publishers, New York.
Modern chemical processes, Vols. I-VII. Reinhold Publishing Corp., New York.
Nelson, W.L. 1958. *Petroleum refinery engineering,* 4th ed. McGraw-Hill Book Co., New York.
Shreve, R.N., J.A. Brink, Jr., and G.T. Austin. 1984. *Shreve's chemical process industries,* 5th ed. McGraw Hill Book Co., New York.
Standen, A., ed. 1969-70. *Kirk-Othmer encyclopedia of chemical technology,* 2nd ed. Interscience Publishers, New York.

New process and product announcements plus plant descriptions and technical developments are regularly carried in the following technical journals:

> *Chemical and Petroleum Engineering (English Translation)*
> *Chemical Engineering*
> *Chemical Engineering and Processing*
> *Chemical Engineering Progress*
> *Chemical Processing*
> *Chemical Week*
> *Cryogenics*
> *European Chemical News*
> *Hydrocarbon Processing*
> *Industrial & Engineering Chemistry*
> *International Oil News: Management Edition*
> *Journal of Petroleum Technology*
> *Oil and Gas Journal*

Check annual indexes under Plants, Processes, Flow Sheets, and specific products.

Chemical Engineering

Canhom, W.G. 1966. Piping codes and the chemical plant. *Chemical Engineering* (September):173.
House, F.F. 1967. Winterizing chemical plants. *Chemical Engineering* (October):199.
Landall, R., ed. 1966. *The chemical plant from process selection to commercial operation.* Reinhold Publishing Corp., New York.
Langhaar, J.W. 1953. Cooling pond may answer your water cooling problem. *Chemical Engineering* (August):194-98.
Lee, E.S. and E.H. Gray. 1967. Optimizing complex chemical plants by mathematical modeling techniques. *Chemical Engineering* (August):131.
Ludwig, E.E. 1964. *Applied process design for chemical and petrochemical plants,* Vol. 2. Gulf Publishing Company, Houston, TX.
Matley, J. 1969. Keys to successful plant startups. *Chemical Engineering* (May):210.
McCabe, W.L. and J.C. Smith. 1967. *Unit operations of chemical engineering.* McGraw-Hill Book Co., New York.
Mead, W.J., ed. 1964. *The encyclopedia of chemical process equipment.* Reinhold Publishing Corp., New York.
Miller, R. 1968. Matching materials to temperatures. *Chemical Engineering* (June):94.
Process control with guide to process industrial elements. 1969. *Chemical Engineering* (June):94.
Shinskey, F.G. 1979. *Process control systems,* 2nd ed. McGraw-Hill Book Co., New York.
Soule, L.M. 1969. Basic concepts of industrial process control. *Chemical Engineering* (September).
Thomas, B.S. 1960. How to calculate heat, water losses (for heated ponds and thickeners). *Chemical Engineering* (August):129-32.
Williams, T.J. Computers and process control. Annual review, January 1969:76 and January 1970:28.

Physical Properties of Materials, Chemical Engineering

API. 1966-68. *Selected values of physical and thermodynamic properties of hydrocarbons and related compounds.* American Petroleum Institute, Tulsa, OK.
Clarke, L. and R.L. Davidson. 1962. *Properties of gases and liquids,* 2nd ed. McGraw-Hill Book Co., New York.
Directory of chemical producers. Stanford Research Institute, Menlo Park, CA.
Dreisbach, R.R. 1961. Physical properties. *Advances in Chemistry Series,* Vol. 3, American Chemical Society.
Engineering Data Book, 8th ed. 1966. Natural Gas Processors Suppliers Association, Tulsa, OK.
Gallant, R.W. 1968. *Physical properties of hydrocarbons.* Gulf Publishing Co., Houston, TX.
Greenwood, D.C. 1961. *Engineering data for product design.* McGraw-Hill Book Co., New York.
Landau, R. and A.S. Cohen. 1966. *Chemical plant from process selection to commercial operation.* Reinhold Publishing Corp., New York.
Perry, R.H. and D.W. Green. 1984. *Chemical engineers handbook,* 6th ed. McGraw-Hill Book Co., New York.
Reid, R.C. and T.K. Sherwood. 1966. *Properties of gases and liquids,* 2nd ed. McGraw-Hill Book Co., New York.
Sax, I.N. 1968. *Dangerous properties of industrial materials,* 3rd ed. Reinhold Publishing Corp., New York.
Technical information. Technical Association of Pulp and Paper Industry, New York.
Weast, R.C. 1988. *Handbook of chemistry and physics.* CRC Press, Boca Raton, FL.

Economics

Bauman, H.C. 1964. *Fundamentals of cost engineering in the chemical industries.* Reinhold Publishing Corp., New York.

Dryden, C.E. and R.H. Furlow. 1966. *Chemical engineering costs*. Ohio State University, Columbus, OH.

Facts and figures for the chemical process industries. 1968, 1969. *Chemical Engineering News*, September, p. 14A; September p. 3A.

Feldman, R.P. 1969. Economics of plant startups. *Chemical Engineering* (November):87.

Guthrie, K.M. 1969. Data techniques for preliminary capital cost estimating. *Chemical Engineering* (March):114.

Hackney, J.W. 1961. How to appraise capital investments. *Chemical Engineering* (May).

Harnby, N. 1967. Cost of overdesign in the chemical industry. *Chemical Engineering* (December):61.

Lobstein, R. 1969. *Guide to chemical plant design*. Noyes Development Corp., Park Ridge, NJ.

Mendez, G. 1968. Cost comparisons for process piping. *Chemical Engineering* (June):255.

Nelson, W.L. 1966. *Guide to refinery operating costs*. Petroleum Publishing Co., Tulsa, OK.

Oriolo, D.J. 1964. Plant evaluation. *Oil and Gas Journal* (February 3, 10, 17, and March 27).

Peters, M.S. and K.D. Timmerhaus. 1980. *Plant design and economics for chemical engineers*, 3rd ed. McGraw-Hill Book Co., New York.

Street, H.H. 1963. Comparing techniques for appraising project alternatives. *Chemical Engineering* (May).

Weaver, J.B. 1961. Profitability measures. *Chemical Engineering News* (September).

Corrosion

Bacchetti, J.A. 1967. Unusual startup problems. *Chemical Engineering Progress* (December):53.

Dillon, C.P. 1968. Corrosion engineer; his role in the process industry. *Materials Protection* (September):32.

Halden, H.A. 1968. Corrosion prevention in chemical plant. *Chemical Process Engineering* (June):75.

Materials of construction. *Chemical Engineering* annually in November issue.

Palmer, J.D. 1969. How process equipment corrodes. *Chemical Engineering Process* (February):47.

Riggs, O.L. and W.P. Banks. 1968. Corrosion problems in phosphoric acid production: Control with anodin protection. *Materials Protection* (May):39.

Rinckhoff, J.B. 1967. Controlling corrosion in wet-gas sulfuric acid plants. *Chemical Engineering* (November):158.

Refrigeration in the Chemical Industry

Ballou, D.F. et al. 1967. Design and cost estimating of mechanical refrigeration systems. *Hydrocarbon Processing* (June):318.

Brunschwiler, W. Low temperature refrigeration plant. *Sulzer Tech. Review* 50(2):53.

Chieffo, A.B. and R.H. McClean. 1968. Fast detection of leaks cuts hydrocarbon losses and pollution. *Chemical Engineering* (July):144.

Codlin, E.M. 1967. *Cryogenics and refrigeration; A bibliographical guide*. MacDonald, London.

Collins, S.C. and R.L. Connaday. 1958. *Expansion machines for low temperature processes*. Oxford University Press.

Dehne, M.F. 1969. Air cooled overhead condensers. *Chemical Engineering Progress* (July):51.

Green, E.S. 1962-63. Some financial aspects of plant selection for industrial refrigeration installations. *Industrial Refrigeration Proceedings*, 89.

Meyre, W. 1969. Multi-purpose refrigeration facility. *Chemical Engineering Progress* (June):91.

Monsanto licked surge loads, compressor failures with circulated evaporators in chlorine plant. 1966. *Air Conditioning, Heating and Refrigeration News* (December):11.

New processes vie for specialty dehydration. 1968. *Chemical Engineering* (May):104.

Scheel, L.F. 1969. Refrigeration; centrifugal or recip? *Hydrocarbon Processing* (March):123.

Spencer, E. 1967. Estimating the size and cost of steam vacuum refrigeration. *Hydrocarbon Processing* (June):136.

Starczewski, J. 1965. Evaporator design. *Modern Refrigeration* (December):1164.

Strother, J.R. 1966. Gas turbines show potential for process refrigeration. *Pulp and Paper* (April):56.

Versagi, F.J. 1966. Life expectancy of equipment. *Air Conditioning, Heating & Refrigeration News* (October):109.

Woolrich, W.R. 1966. *Handbook of refrigerating engineering*, 2 Vols. Avi Publishing Co., Westport, CT.

Zafft, R.W. 1967. How to size and find the cost of absorption refrigeration. *Hydrocarbon Processing* (June):131.

Fire Safety

Adcoch, C.T. and J.D. Waldon. 1967. To minimize the loss in a chemical plant explosion. *Safety Maintenance* (October-November).

Armistead, G., Jr. *Safety in petroleum refining and related industries*. J.G. Simmonds and Co., Inc.

Chemical engineering progress, loss prevention. 1967-69. 3 vols., American Institute of Chemical Engineers, New York.

Engineering for safe operation, 2nd rev. ed. 1966. American Oil Co., Chicago.

Factory Mutual Engineering Division. *Handbook of industrial loss prevention*. McGraw-Hill Book Co., New York.

Fawcett, H.H., ed. 1965. *Safety and accident prevention in chemical operations*. John Wiley and Sons, New York.

Fire hazards in buildings and air handling systems. 1968. *ASHRAE Journal* (September).

Hazard survey of the chemical and allied industries. 1968. American Insurance Association, New York.

Jenett, E. Designing for safety. *Chemical Engineering*, July 20, August 17 and 31, and September 14, 1964.

Plant and design safety. 1965. American Institute of Chemical Engineers, New York.

Pratt, J.D. 1968. *Guide to fire prevention in the chemical industry*. British Chemical Industries Assoc. Ltd.

Preddy, D.L. 1969. Guidelines for safety in chemical plant. *Chemical Engineering* (April):94.

Redding, R.J. 1969. Electrical safety in chemical plant. *Chemical Engineering* (May):98.

Vervalin, C.H., ed. 1964. *Fire protection manual for hydrocarbon processing plants*. Houston, TX.

Bulletins, *Reports*, and *Standards* of the following organizations (see Chapter 42, Codes and Standards, for further details):
    FM—Factory Manual Engineering Division
    NBFU—National Board of Fire Underwriters
    NFPA—National Fire Protection Association

Material, *Equipment*, and *Design Codes and Standards* (see Chapter 42, Codes and Standards, for further details):
    AGMA—American Gear Manufacturers Association
    AIChE—American Institute of Chemical Engineers
    ANSI—American National Standards Institute
    API—American Petroleum Institute
    ASME—American Society of Mechanical Engineers
    ASTM—American Society for Testing and Materials
    CAGI—Compressed Air and Gas Institute
    CMA—Chemical Manufacturers Association
    CTI—Cooling Tower Exchange
    HEI—Heat Exchange Institute
    HI—Hydraulic Institute
    MSS—Manufacturers Standardization Society of the Valve and Fittings Industry
    NACE—National Association of Corrosion Engineers
    NBFU—National Board of Fire Underwriters
    NEMA—National Electrical Manufacturers Association
    PFI—Pipe Fabrication Institute
    TEMA—Tubular Exchanger Manufacturers Association

CHAPTER 37

ENVIRONMENTAL TEST FACILITIES

Cooling Systems 37.1
Heating Systems 37.3
System Control and Instrumentation 37.4
Design Calculations 37.6
Chamber Construction 37.6

ENVIRONMENTAL test facilities are used to simulate an environment or combination of environments under laboratory controlled conditions that duplicate or exaggerate the effects found in actual service. They assist the engineer and scientist in exploring the effects of equipment and in developing equipment for resistance to the many environmental forces.

The acceptance of and demand for environmental simulation facilities result from the following factors: (1) parallel and reproducible tests can be made; (2) equipment being tested can usually be observed and analyzed during testing; and (3) supporting equipment requirements are reduced to a minimum. Field testing and product development costs are reduced, lead time required for completion of product development is shortened, and most desirable reliability features can be incorporated in the original manufacture of the product.

Environmental equipment is used not only to determine the performance of mechanical and electrical equipment, but for certain tests on personnel as well. Personnel testing includes: (1) checking protective equipment and clothing; (2) altitude and space procedures indoctrination; and (3) studying physiological and psychological effects on the human body and mind.

Environmental testing is usually divided into two general classifications—climatic and dynamic. The climatic tests of primary interest include the following:

Temperature Testing. This includes (1) temperature soaks at high and low extremes; (2) temperature shock testing in which the part is subjected to rapid high and low temperature cycling; and (3) programmed cycling in which the parts are subjected to repetitive expansion and contraction stresses and breathing.

Humidity Testing. This may involve simply exposing the equipment to a constant humidity level, or *cycling*, wherein the temperature and relative humidity are varied. This type of testing induces breathing and condensation within the parts tested. Subcooling may also be used to produce icing conditions.

Salt Spray Chambers. These are used for study of the corrosive action of salt vapor on components, usually a constant high humidity test.

Fungus Chambers. These are used to stimulate the growth of fungus on electrical equipment being studied to assure protection against tropical climates.

Miscellaneous Climatic Chambers. These include chambers for the simulation of desert sand and dust with high velocity air movement, sunshine, snow, and rain, as well as chambers for proof testing of explosion-resistant wiring devices.

Altitude and Space Chambers. Altitude chambers have been used for some time in testing aircraft equipment. In recent years, space chambers have been developed for missiles, rockets, and space vehicle development work.

Combined Environment Testing. This type of equipment combines two or more of the above environments in one system, with all the complexity implied. Current thinking is that this type of testing is the only satisfactory means of proving that a piece of equipment will stand up in actual service.

Dynamic or nonclimatic tests include vibration, physical shock, acceleration, mechanical stress, nuclear radiation, cosmic radiation, micrometeorite bombardment, and many other types of stress.

The air-conditioning and refrigerating engineer may be directly concerned with equipment design for the application of most climatic environments. Some suggested approaches to design of equipment for the production and control of various environments will be given; the dynamic tests, however, are beyond the scope of this text. Because of the wide variety of tests that must be produced, detailed descriptions are not possible within this chapter.

Many techniques used in the design of environmental test equipment are the same as those employed for low-temperature metallurgy (Chapter 39), space simulators (Chapter 38), biomedical applications (Chapter 40), and cryogenic equipment (Chapter 38).

Many existing federal specifications outline basic environmental test specifications and the design approach for some of the chambers to be used in this work. The rapid pace of missile and space vehicle development has led to a series of informal special test criteria developed by project contractors. Environmental equipment designers extend the state of their art to develop equipment to meet these needs.

COOLING SYSTEMS

Temperature reduction in test chambers is accomplished by both mechanical refrigeration systems and the use of expendable refrigerants. Engineered refrigeration systems are discussed in detail in Chapter 1. Both approaches can be used directly or in conjunction with secondary heat-transfer fluids.

Primary Refrigerants

Any refrigerant suitable for mechanical refrigeration can be used as the primary refrigerant. Selection requires an evaluation of equipment size, cooling load, type of evaporator, temperatures to be produced, method of condensing, hazards, lubrication, and special operating requirements.

The halocarbon refrigerants are commonly selected because they are not toxic or flammable, and they are stable at elevated temperatures. R-12, R-22, or R-502 is used in single-stage systems and in the high stage of two-stage cascade systems; either R-13 or R-503 is used in the low stage. R-14 is used in the low stage of three-stage cascade systems. R-502 is commonly used in a single stage to an evaporating temperature of $-40\,°F$, and R-502 and R-13 or R-503 refrigerants in a cascade system to $-120\,°F$ evaporating temperature. R-14 is used in the low stage of a three-stage cascade system for lower evaporating temperatures. These choices of refrigerant are common in practice and do not mean that single-stage systems using R-12, R-22, or R-502 for temperatures below $-40\,°F$ are not used or that $-120\,°F$ is the lowest evaporating

---

The preparation of this chapter is assigned to TC 9.2, Industrial Air Conditioning.

temperature possible of a two-stage cascade system. Although single-stage and cascade systems are the most common designs presently used, a two- or three-stage compound system may be selected for certain load applications.

Test chambers are frequently required to operate at high as well as low temperatures. When using primary refrigerants in evaporators, which may be subjected to temperatures above 500°F, thermal isolation and cooling of the evaporator are necessary to prevent oil decomposition or other possible deterioration of the refrigerant circuit. Refrigerant 13 is more stable than either R-12 or R-22. Pumpdown of an R-13 circuit, when operating up to 500°F, is reasonably satisfactory. A secondary refrigerant heat exchange fluid, which is pumped out of the cooling coil at some safe, predetermined temperature, is another method for protection against overheating of a primary refrigerant coil. However, it is more common to use expendable refrigerants for high-temperature equipment, even with the disadvantage of high operating costs at low temperature.

Table 1 Characteristics of Selected Low-Temperature Primary Refrigerants

Refrigerant	Saturation Temperature, °F at pressure indicated		
	10 psia	14.7 psia	200 psia
12	−37.2	−21.6	131.7
22	−55.6	−41.4	96.3
502	−64.3	−50.1	89.8
13	−127	−114.6	−7.8
503	−139.8	−127.6	−12.1
14	−207.7	−198.3	−105.6

Expendable Refrigerants

Expendable refrigerants, such as dry ice, liquid carbon dioxide, liquid nitrogen, and helium, and others discussed in Chapter 38, are suitable for producing low temperatures in environmental chambers. Sublimation of dry ice within the chamber, chilling brine for circulation through heat exchangers, or direct expansion of the liquids within the test space are common expendable refrigerant methods. Liquid nitrogen, helium, and other cryogenic liquids work particularly well below the temperature range of mechanical refrigeration systems.

The advantages of expendable refrigerants in environmental test equipment are reduced initial cost, basic simplicity, reduced weight and size of the chamber, and the ability to produce very rapid pulldown rates to low temperatures. Disadvantages include high operating costs, the need for a reliable source of expendable refrigerants, the serious personnel hazard resulting from the absence of oxygen when the air is displaced in the test space, and the possible detrimental effect of submerging the tested product in the gas or liquid of the expendable refrigerant, which is expanded directly into the test space. Indirect brine systems frequently approach the first cost of mechanical refrigeration systems. Direct expansion of expendable refrigerants in altitude chambers is impracticable.

Chambers using dry ice as the expendable refrigerant have, in most cases, been replaced with direct injection of liquid carbon dioxide or liquid nitrogen, which eliminates the need for a dry ice compartment. Also, the chamber is smaller and the temperature more easily controlled. Only on-off cycling is possible for control of direct injection liquid CO_2 systems because of the triple point of CO_2. Solenoid actuated two-way valves and pneumatically or electrically actuated ball valves have been used with much success. If reduced capacity operation is desirable, the control valve should be time-pulsed or more than one valve should be used with small and large expansion orifices. Solenoid valves with built-in orifices designed for this specific application are available.

Liquid CO_2 is used at two pressures; low-pressure bulk liquid is stored in refrigerated receivers at approximately 0°F and 300 psig. Its latent heat is approximately 120 Btu/lb. High-pressure liquid is stored at room temperature at pressures ranging from 750 to 1000 psig, depending on ambient temperatures. Its latent heat varies from approximately 50 to 75 Btu/lb, depending on initial temperatures. High-pressure liquid is used only for very small or infrequent cooling loads because the latent heat is low, the liquid fills only 60 to 70% of a high-pressure cylinder, and the cost is considerably higher than for low-pressure liquid. For increased efficiency when high-pressure liquid CO_2 is selected, an economizer that precools incoming liquid by heat exchange to the exhaust cold vapors can be used to reduce the CO_2 requirement by 15 to 30%.

Liquid nitrogen can be applied either directly or indirectly to produce temperatures cheaply and easily down to its approximate −321°F boiling point (at 1 atm). Control is accomplished either by on-off cycling of special solenoid actuated valves or by specially designed flow-metering valves. The liquid is stored in Dewar flasks or vacuum-insulated tanks. Transfer is accomplished either by self-pressurizing the storage vessel or by using dry air or nitrogen, properly pressure regulated, to pressurize the flask and force the liquid from a discharge tube.

Piping, valves, heat exchangers, and so forth should be fabricated of nonferrous metals, stainless steel, or high nickel steel. Few plastics are suitable seals in this temperature range. Although its latent heat of evaporation at 1 atm pressure is only 85 Btu/lb, another 56 Btu/lb is available from the superheating of gas to −100°F. Therefore, it has a greater capacity of heat absorption than does low-pressure liquid CO_2 when operating at a −100°F or higher control point. Suitable precautions should be taken against the low-temperature hazards to operating personnel. Oxygen monitoring systems should be located in the test cells to warn personnel if oxygen depletion occurs. Carbon dioxide and nitrogen gases should be vented to the atmosphere after they have been used in the test chamber.

Other cryogenic liquids such as liquid helium and liquid hydrogen provide test temperatures close to absolute zero. These are too expensive for routine work at temperatures which can be obtained by more economical means. Special storage and control valves are required. More detailed information on cryogenic liquids, their handling and piping, may be found in Chapter 38.

Secondary Coolants

Low-temperature heat transfer fluids are used where: (1) control flexibility is better accomplished by their use; (2) large central cooling systems have been chosen; (3) the high temperatures experienced will rule out primary refrigerants because of complex mechanical design problems; (4) expendable refrigerants are used and direct cooling of air or a product is not possible; (5) the wall of a temperature chamber or the shroud of a space simulator is to be conditioned and a close temperature gradient is required; or (6) thermal shock by alternately immersing the test device in liquids of different temperatures is required.

Halocarbons, liquid hydrocarbons, alcohol, some primary refrigerants, silicone fluids, and aqueous glycol solutions are used as secondary coolants. Thermal considerations include viscosity, specific heat, specific gravity, thermal conductivity, freezing and boiling points, and coefficient of expansion. Other important considerations are flammability, toxicity, corrosiveness, vapor pressure, and water solubility (see Table 2).

To minimize evaporation losses, the more volatile fluids are used in closed systems with suitable expansion tanks or accumulators and are frequently pressurized by an inert gas such as nitrogen. However, some secondary refrigerants dissolve in nitrogen. In that case, other methods must be used to pressurize the system such as an expansion tank with a diaphragm separator. Also, certain fluids must be kept refrigerated to prevent boiling off under standby conditions.

Environmental Test Facilities

Table 2 Selected Characteristics of Wide-Range Heat-Transfer Fluids

Property	Santa Barbara Chemical Lexsol 408M	3M Fluorinert FC-77	Liquid Carbonic LQ-1575	R-11	R-1120 Trichloro-ethylene
Boiling Point, °F	320	207	350	75	187
Freezing Point, °F	<−100	−166*	−140	−168	−144
Density at 68°F, lb/ft^3	48.6	111.1	53	92.9	90.5
Specific Heat, Btu/lb·°F at 68°F	0.5	0.25	0.8	0.21	0.23
Viscosity, Centipoise					
68°F	1	1.8	0.9	0.44	0.58
−40°F	3.5	2.2	2.0	0.92	1.17
−60°F	7	3.6	—	1.1	1.4
−80°F	—	7.5	—	1.35	1.6
Thermal Conductivity at 68°F, Btu/h·ft·°F	0.08	0.038	0.07	0.051	0.06

*Pour point

Secondary refrigerants are cooled in insulated sublimation tanks by spraying the liquid over dry ice or directly injecting liquid CO_2 or LN_2 into the refrigerant. Such systems require a pump for recirculation of the secondary refrigerant. Common operating difficulties in sublimation systems include both cavitation in pumps because of the release of dissolved CO_2 or LN_2 gas and foaming in the sublimation tank with resultant carryover of the secondary heat-transfer liquid with the vented gases. Positive displacement pumps and positive suction pressures are recommended.

In a system with a wide temperature range, the pump and shaft seal must be carefully selected. Magnetic drive and canned pumps have no external seals, which eliminates the shaft seal problem.

Secondary coolants can also be cooled by a mechanical refrigeration system or by a heat exchanger cooled by liquid nitrogen. These systems avoid carryover, cavitation, and moisture problems. The evaporator or heat exchanger must be designed to avoid freeze up. In Chapter 18 of the 1989 ASHRAE *Handbook—Fundamentals*, more detail is included and additional coolants and their properties are discussed.

HEATING SYSTEMS

Electric heat is most commonly used in environmental chambers. Prime or extended surface, tubular, or strip heaters are suitable for circulating airstream systems. This method is a conduction or convection type of heating used to duplicate conditions in storage, in transportation, and when a protective housing is provided around equipment. With proper precautions, open nichrome, strip, or coil wire heaters can be used; they offer the advantages of a rapid response because of their low thermal mass. All types of electric heaters require proper insulation and protection against moisture. Generally speaking, temperature test chambers (dry-bulb control) may use open wire resistance heaters, although condensation at low temperatures may produce excessive moisture and corrosion on the heaters.

Some military specifications prohibit direct radiation from chamber heaters to the test object. In this case, heater baffling may be required.

Salt-spray chambers are heated indirectly, usually by circulating heated water or air around the outside of the chamber shell outside the salt-fog atmosphere. Heaters for sand and dust chambers must use sheathed heaters that have good resistance against the erosion of the high-velocity sand and dust. Explosion-proof testing chambers contain an explosive fuel and air mixture. Heaters must have temperature-limiting devices on the sheath with terminals extended to the outside to prevent ignition of the mixture.

Altitude and space chambers present special heat problems. Arcing between terminals under vacuum conditions must be guarded against. General practice brings the terminals to the outside of the vacuum space through a vacuum-tight fitting. The reduced convective heat transfer of a heater under vacuum must be considered in the design to prevent burnout. Thermal and mechanical bonding of sheathed heaters to the exterior wall of the vacuum structure has been successfully used to avoid some of these problems. Internal forced convection heating is usually suitable up to about 50,000 ft of altitude, approximately 0.1 atm.

Indirect heating is also suitable for environmental chambers. Hot water, steam, brines, and oils can be used in coils in various configurations. These heating systems are particularly applicable where close tolerance must be provided within the system because complete modulation is possible. Modulation control is also possible in electrical resistance heating systems by using power proportioning equipment. Hot water recirculating systems and steam systems can be used with relatively standard design approaches. Complete drainage of the coils and piping within the chamber must be assured if the chamber will be operated below freezing. Proper water treatment should be provided.

Many brines and heat-exchange fluids that are not limited to the temperature ranges of water or steam are available. No one fluid for the commonly required wide ranges of environmental test equipment, such as −100 to 500°F, is available at present. Most chemical brines present some problems with viscosity, toxicity, flammability, chemical reaction with the piping, and other limitations. Any heat-exchange fluid must be carefully evaluated for limitations.

Another type of heating system is the radiant type, which is required to simulate such situations as sun exposure where both the heat-flux density and the wave length are important; more commonly, radiant sources are used in equipment exposed to radiation from some high-temperature source. Radiant sources in environmental chambers vary from exposed nichrome wire heaters to specialized equipment such as arc lamps and mercury Xenon lamps used to simulate sunshine.

Since the wavelength of the source is a function of its temperature, the conditions to be simulated must be understood before a suitable radiant heat source can be designed. Where the problem is simply to produce high watt density on the product to be tested, bare nichrome wires, sheathed heaters, carbon and silicon carbide rod heating elements, and similar devices are available. Tubular quartz heat lamps are often suitable for very rapid heat-up conditions. Sunshine at sea level is usually simulated by the use of mercury vapor lamps or combinations of tungsten filament heat lamps to produce the proper wavelength and watt density. Radiant heating systems may be required to produce heat densities in excess of 125 kW/ft^2. Careful design of the chamber structure, proper support of the heating elements, and correct location of the heating elements are required to prevent mechanical and thermal stress problems resulting from large temperature differentials and rapid heat-up and cool-down cycles.

Under altitude simulation conditions, removal of heat from a test object is possible by convection up to about 100,000 ft, 8.3 mm Hg, absolute. To accomplish this, the air mover should be of the largest practical size so that the residual air in the chamber is passed at high velocity through the cooling coils and directed on and around the test object. An envelope of cooled, rarified air, within which temperature gradients will be reasonably small, is thus produced around the test object.

Large gradients may occur throughout the rest of the chamber but will not affect the test results. Airflow should be reduced at low temperatures to minimize the fan brake horsepower addition to the cooling load. Cold wall construction reduces the heat load to be transferred to the cooling coils within the vacuum space, provides surface for radiant transfer, and overcomes some of the losses through the structure; however, high wall temperature gradients may result if direct expansion is used in lieu of a pumped secondary refrigerant.

Evaporators should be maintained at the lowest practical temperature to provide the maximum temperature difference for both radiant and convective heat transfer. In simulated altitudes over 100,000 ft, a radiation heat-transfer system should be designed into the chamber to produce the desired temperature-control conditions on the test part. Treating the walls of the chamber to produce high emissivity is necessary in this case; close control of the entire wall surface is generally advisable. The limitation of radiant heat transfer in the low-temperature range must be considered because the amount of heat transferred from one surface to another is a function of the ratio of the square of the absolute temperatures. With temperatures in the $-100\,°F$ range, the heat-transfer rate is small. The detailed calculation of this effect is described in Chapter 3 of the 1989 ASHRAE *Handbook—Fundamentals*, where the Stefan-Boltzmann equation is discussed.

Air Movement

All types of fans and air movers are used in test chambers. Propeller, axial flow, and centrifugal fans are generally selected. Positive displacement or multistage turbine-type blowers provide the necessary total pressure for ram air simulation under altitude conditions. Drives for internally mounted fans must be located externally when extreme conditions are encountered in the airstream. To eliminate corrosion and minimize heat transmission, stainless steel shafts are commonly used. In almost all instances, a vapor seal is required to eliminate or minimize transmission to or from the ambient air. Internal bearings must withstand the full range of environmental conditions with a reasonable life expectancy. Altitude chamber fan shafts must be equipped with vacuum-tight shaft seals.

Air distribution within the work space of the test equipment is important in producing both uniform gradient conditions and system response. Frequently, close temperature gradients specified for air circulation rate and distribution will complicate the control problem. Air densities may vary almost three-to-one in temperature test chambers and more than fifty-to-one in altitude test chambers. Therefore, motor sizing and speed control of fans require careful attention. The airflow for a specific gradient is directly proportional to the net heat gain or loss in the work space and inversely proportional to the permissible temperature gradient and density of the air.

Usually the air volume must be increased at elevated temperatures because of the lower density. The circuit pressure drop will be approximately proportional to the air density. Therefore, the system balance points should be determined from the fan characteristic curves for various operating range conditions in the chamber.

Humidification

Common means of raising the relative humidity within a test enclosure include directly introducing low-pressure steam, vaporizing the water by electric immersion heaters in an open evaporator, and directly atomizing water sprays alone or in combination with electric heaters to aid vaporization of the spray.

Steam and vapor generators are best for increasing humidity as well as temperatures, since considerable sensible heat is introduced. Control of steam may be modulating or on-off. With the latter, anticipation is suggested to minimize overrun because of too rapid a rise in moisture content. When self-contained steam generators are used, they should include a low water cut-off device, suitable pressure relief valves, and other safety devices. Makeup water should be provided from a distilled or demineralized source.

Atmospheric pressure vapor generators operating at atmospheric pressure are commonly selected when it is desirable to produce test conditions of 95 ± 5% rh at temperatures between 100 and 160°F. Protection against immersion heater burnout is mandatory for good design.

For production of high humidities at temperatures near and below ambient, some form of water spray is recommended because of the adiabatic cooling effect on evaporation. Although water spray systems do not have as rapid a response as steam, this is no disadvantage at lower temperatures where the humidity ratio is low. By controlling the temperature of the sprayed water, both humidification and dehumidification can be achieved with a single system. Heating and cooling means may be located either inside the conditioning spray plenum or externally in the recirculating water circuit. Close control of humidities is possible with such a design approach. Distilled or demineralized water is suggested, since tap water may be contaminated. Recirculating spray systems should be periodically flushed out because of possible airstream contamination of equipment or test parts.

Dehumidification

Low humidity conditions with controlled dew points above freezing are usually produced by mechanical refrigeration cooling and electric reheating because most environmental chambers require mechanical refrigeration for extended dry-bulb temperature range control. For production and control of humidity conditions requiring dew points below freezing and continuous operation, either alternately defrosted dual evaporators or automatically regenerating desiccant dehumidifiers are used. If an enclosure is sufficiently vaportight and there is no internal latent load, an evaporator sufficiently large to handle the frost buildup because of initial latent load may be an acceptable solution.

Ram air coolers that simulate the cooling air supply for electronic airborne equipment at controlled temperature, humidity, and altitude conditions, present special dehumidification problems. If 100% outdoor air is required, it is best handled by a wet coil which removes a large part of the moisture just above freezing, followed by a set of parallel alternately defrosted evaporators. For ram air simulators in a closed-loop arrangement, simpler systems are possible if the cooling of air does not involve dehumidification.

When using certain liquid or solid desiccants in very high or low dry-bulb ranges, mechanical refrigeration cooling may be required for precooling or aftercooling. In such systems, a portion of the chamber air is continuously withdrawn by an auxiliary blower and passed through the drying agent and any necessary precooling or aftercooling coils.

Some tests require the conditioned air dew point to remain sufficiently low to prevent condensation on the test article during dry-bulb control temperature cycling. This requirement can be a problem during rapid heat pull-up and may require a desiccant dehumidifier.

SYSTEM CONTROL AND INSTRUMENTATION

Control Tolerances

Environmental test facilities generally require minimal control tolerances over a wider range than that required for ordinary

Environmental Test Facilities

HVAC systems. Control of temperatures as close as ±0.5 to 1.0°F are commonly specified. Relative humidity control tolerance is at most ±5% and more often ±3% or even less. To meet these requirements, the control and instrumentation system must have an even better accuracy since operating tolerances often exceed control tolerances.

When, in addition, the test chamber is designed to provide a wide range of test conditions, the HVAC equipment must be oversized for most of those conditions. The control system is then required to modulate equipment output to a small percentage of capacity, with all of the difficulties inherent in controlling oversized equipment.

For these reasons, the control devices used with test chambers must be of the highest quality. In general this will be "industrial quality," as used in process control applications. Most systems operate in a modulating mode, using proportional plus integral (PI) or, occasionally, proportional plus integral plus derivative (PID). Proportional Only control is not acceptable due to its inherent deviation from set point. For discussions of control theory, operating modes and typical systems, see Chapter 51 of the 1987 ASHRAE *Handbook—HVAC Systems and Applications* and Haines (1987).

Temperature Controls

Systems for temperature control are of the types discussed in Chapter 51 of the 1987 ASHRAE *Handbook—HVAC Systems and Applications* and Haines (1987). Controllers should operate in PI mode, although PID may be needed when rapid cycling over a wide range of set points is required.

Temperature sensors may be thermocouples, thermisters, or RTDs. Thermisters and RTDs should be specified to have an absolute accuracy of ±0.5°F and a sensitivity of at least 0.1°F. Most good RTD and thermister sensors can meet these requirements. Thermisters must be specified with factory certified performance and a guarantee of not more than 1.0°F drift per year so that they can be readily replaced. Note that thermisters do drift and must, therefore, be recalibrated or replaced at frequent intervals, depending on the tolerance specifications. Wound wire RTDs may be specified with no drift over time. The best quality wound wire RTDs are platinum; these may be obtained with certified accuracies, traceable to industry standards. Platinum RTDs using solid state deposition techniques are also available. These are accurate and have a fast response but are subject to drift over time. Sensors for use with pneumatic controllers should be of the bulb and capillary type—sensitivity and accuracy may not be suitable for close tolerances.

Sensors must be suitable for the temperature ranges required. Consult the sensor manufacturer for this information. Where wide ranges are required, two or more instruments may provide better accuracy over the entire range.

Recorders may be of the strip or circular chart type. They should use the same sensors as are used by the controllers. If separate sensors are used they must be calibrated together. For accurate calibration, laboratory-type mercury thermometers must be used.

Humidity Controls

Accurate humidity sensing is very difficult. Materials that vary dimensionally with changes in relative humidity are not suitable for close control because of hysteresis and drift. Systems using electrolytes and wet-bulb/dry-bulb sensors require essentially continuous maintenance to ensure accuracy. For any system, accuracies at extremely high or below freezing temperatures may be questionable. For reasonably close control (±5% rh), solid-state deposition-type sensors of the capacitance or resistance type are satisfactory and have a fast response. They must be calibrated regularly—the best have a drift of about 1% per year. For very close control, for calibration, and for continuous accurate sensing, the chilled-mirror dew-point sensor is used. This device is available in several packaging arrangements and, if needed, can be obtained with a certification of accuracy traceable to an agency standard. Its response is somewhat slow for use with rapidly changing conditions. In some cases, it has been used to back up and check the solid-state instruments which are actually used for control.

Controllers must operate in the PI mode and recorders should use sensors in common with the controllers.

Pressure Controls

An altitude measurement and control system may consist of a simple manometer for indicating pressure and a hand-throttling or bleed valve to control the level; or the system may be completely automatic, operated by a mechanical or electronic sensing device that actuates modulating bleed controls to automatically regulate the vacuum. The U-tube manometer shows relative pressure between the site barometric pressure and the internal vacuum of the chamber. Absolute vacuum measuring manometers are recommended for altitude chambers to avoid the need to correct for local barometric pressure variations.

Opposed-bellows absolute pressure controllers and electrical instruments using strain gage pressure transducers with bridge circuit electronic controllers are available. These instruments are suitable for altitudes to 200,000 ft. They provide on-off control or proportioning output signals and can control various functions of the vacuum system. Electronic vacuum-sensing devices measure the change in interference across an air gap between two electronic devices because of the reduction in gas molecules in the space. Coupled with amplifiers, this effect initiates action in control circuits at a set point.

One instrument system has an alpha emitter and measures the amount of energy reaching the grid to determine the vacuum level. Another system uses ionization of a filament. A thermocouple measures the change in the heat-exchange rate as a function of vacuum. Most electronic vacuum instruments are available with millivolt output that may be coupled to a standard potentiometer to provide both a record and an output signal for controlling the vacuum system.

The most common way of automatically controlling a vacuum system is to bleed air into the chamber through a modulating valve. This floods the pump to the desired capacity so that good control can be maintained with the proper size bleed valve. Frequently, two or more valves are required to control the pressure in the chamber over a wide range.

Altitude chambers can be controlled by cycling a solenoid valve in the pump suction line or simply turning the pump on and off. The latter measure is not recommended because of the excessive strain on the motor and the drive mechanism. These two methods are not practical for diffusion pumping systems.

Valves and Dampers

Haines (1987) discusses the factors to be addressed in selecting control valves and dampers. The principal point to be considered is the *system gain* due to oversizing of these devices. In test chambers where a wide range of conditions must be simulated, the system gains are aggravated by the need to size the HVAC equipment for the worst conditions, which means that it is oversized for every other condition, sometimes extremely so. Oversizing of the valve or damper inevitably leads to poor control. It follows that valves and dampers should be undersized rather than oversized if close control is to be obtained. Undersizing penalizes the air or water system hydraulics, but may be necessary to obtain the desired results.

Computers as Controllers

Typically, DDC (Direct Digital Control) with computers is used. The computer simply takes over the controller function—sensors programmed to any kind of operating and timed sequence desired. It can also provide visual and printed output of system status, either in real time or as history, including graphic and statistical presentations. Such functions add to the cost of software and software maintenance and should be used only when needed.

Accessories

To provide satisfactory operation, a control system must include other accessories besides sensing elements and controllers. A pneumatic system should include a pressure-reducing valve, a filter-drier, and a surge tank; all are standard with such systems. For systems operating below freezing, nitrogen, or dried compressed air should be considered as the pneumatic medium rather than compressed air.

The selection of safety and other interlock relays and switches should be carefully considered not only to provide safe operation but to control the system and reduce the number of switch settings to be made by the operator. Vacuum switches are frequently used to cut out the heaters or switch them to series wiring whenever an altitude chamber is under vacuum, thus preventing burnout because of low convection transfer. Often there is a temperature limit controller to prevent damage that may occur because of excessive heating or cooling of the test module.

In addition to the conventional control systems for temperature, humidity, and vacuum, the designer may encounter requirements for measuring or controlling many other variables. These variables include: hydrogen-ion concentration (pH); fluid velocity and pressure, dust, and smoke density; explosive fuel and air mixtures; ozone, oxygen, and carbon dioxide levels; and others.

DESIGN CALCULATIONS

Cooling and Heating Loads

Because the rate of temperature change is so important in the operation of test chambers, usually both the steady-state and transient load must be calculated to size the refrigeration and heating system. The steady-state load includes heat transferred through the insulation, framing, windows, sleeves and penetrations, fan shafts and other conducting materials, and the internal live heat loads of lighting, product, fans, and personnel. The sum of these loads that occur simultaneously at the lowest air temperature in the chamber is the minimum design capacity. The equipment needs needs to be sized to handle this design capacity, plus a factor of safety, at the refrigerant evaporating temperature required and as determined by the evaporator size.

Heat transmission losses must also be determined to calculate the total steady-state heating load. The live loads, such as blowers and lights, will decrease the heating load, but usually they are not deducted from the transmission losses. The heaters are then sized like the refrigeration system, except the load is at the high temperature.

For transient conditions, the thermal mass of all items, including chamber air, liner and bracing, insulation and framing, evaporators, heat exchangers, heaters, blower, brackets, shelves or other gear, and the test specimen must be determined. The product of the mean specific heat and the mass gives the average thermal mass. The product of the time rate of temperature change and the thermal mass gives the instantaneous load due to temperature transients. The total load may include transmission and live loads as well.

Under transient conditions, a gradient will exist between the room ambient and the heat source or sink. For small chambers, conservative design assumes that the temperature of all items directly contacting the chamber air changes at the same rate as the air temperature. In cases with very rapid temperature changes and high mass loads, the surface areas, air velocity, and point of temperature control must be considered in designing the necessary refrigerating or heating system. A larger system cooling requirement may be calculated by this method rather than by one based on the total load changing with the air temperature— especially if a linear rate of change is required and the final temperature is low on the capacity curve of the refrigeration equipment. This larger capacity is also required for heating, except electric heating, where the capacity is uniform at all temperatures.

Relative Humidity

Equipment selection for relative humidity control is similar to normal air-conditioning applications, only more complex in that widely varied control conditions must be produced. A typical control range is 35 to 185 °F dry-bulb temperature and from 20 to 98% rh above 35 °F dew point. Very low dew points may be required for special test programs. Sensible cooling loads may be calculated in a manner similar to that for low-temperature loads, but must also account for such additional factors as the sensible heat of steam when used for humidification.

Since dehumidification is normally accomplished by mechanically cooling and reheating the air, the most common approach is to take a portion of the total recirculating air, cool it to the required dew point, then reheat it as required to keep it at the desired dry-bulb temperature level. If the total air volume were cooled and then reheated, rather large dehumidification and reheat capacities would result. When long-term tests are required at controlled dew points near freezing, heat transfer cooling surfaces must be amply sized to operate without freezing.

Where large internal heat gains are present and high humidities are desired, spray contactors producing adiabatic saturation may be used. By controlling the temperature of the recirculating spray, such systems may be used for simultaneous cooling and humidification.

The sizing of heaters for increasing dry-bulb temperature is the same as for normal chamber heating. For humidification in systems not using adiabatic saturators, either atmospheric pressure vapor generators or direct steam injection is practiced. The amount of water that must be added to achieve the desired increase in absolute humidity for the known volume of air within the chamber determines the rate of moisture addition. Both of these methods add sensible heat, as well as latent heat, which must be added to the cooling load to obtain the total load.

CHAMBER CONSTRUCTION

Inner liners of test chambers for high- and low-temperature and humidity are most commonly made of type 302 or 304, heliarc welded stainless steel. The outside liner is made of aluminum or mild steel, welded or sealed to provide a vapor-tight structure for the insulation. For walk-in rooms, prefabricated, urethane insulated, foamed-in-place modular panels are popular. Special care must be taken in the sealing of the panel joints.

A means must be provided to prevent an excessive pressure differential between the inside and outside of the chamber during rapid changes of temperature. Even opening and closing a door, when test space conditions are different from ambient, may change the air temperature and consequent pressure enough to buckle the liner of a well-sealed chamber if it is not properly vented.

The structure must be designed to provide for the considerable expansion and contraction that take place during wide-range temperature cycling. Doors are a particularly difficult problem. Any fan baffles, coil mountings, or other accessories must be carefully planned so that moving parts are not interfered with during temperature cycling runs. Shock loads induced by liquid nitrogen and other refrigerants can cause violent stress concentrations. Humidity chambers must be designed so that any condensate formed on the coils, walls, or floor will drain rapidly from

Environmental Test Facilities

the chamber. This feature is important in programmed humidity chambers in which operation ranges from high- to low-humidity conditions.

Altitude Chambers

Generally, a chamber that simulates altitude to 250,000 ft (0.016 torr) is considered an altitude chamber; above that altitude, the chamber is considered a space simulator. Interior liners of chambers that simulate altitude are usually constructed of aluminum, mild steel, or stainless steel, with exterior structural steel reinforcing designed to withstand a vacuum or 15 psig.

Depending on the desired tests space configuration, chamber liners may be rectangular or cylindrical. Design of the vessel to withstand the exterior pressure should be in accordance with the latest edition of the American Society of Mechanical Engineers' Pressure Vessel Code.

Altitude chambers that control temperature or humidity or both, in combination with altitude simulation, are built with the pressure liner on the inside or outside of the insulation space. The inside liner, whether or not it is the pressure liner, should be made of a corrosion resistant material such as stainless steel. The outside liner may be mild steel. If the outside liner is the pressure liner, it must be of sufficient thickness or reinforced to withstand a vacuum or 15 psi pressure differential. It must also be treated for corrosion resistance due to equalization with the test space, which will introduce various combinations of temperature and (high) humidity into the liner surfaces and insulation.

If the inner liner is the pressure liner, it must be stressed for the temperature range involved as well as the pressure. The inner liner should be the pressure liner if high altitude use is a major consideration, because the insulation space is not exposed to the vacuum. However, because of the liner and reinforcing mass, the cooling and heating load during temperature transition is greater than with an outside pressure liner. With an outside pressure liner, the insulation space must be equalized with the test space to prevent inner liner collapse. Achieving high vacuum on the insulation and insulation space is very difficult unless it is dry; therefore, this fact must be considered during temperature or temperature-humidity runs, or when diving the chamber to atmospheric pressure after an altitude test.

Seam welds of the vacuum vessel should be on the vacuum side and should be of one continuous pass with skip welding, if required, on the outside. This method of welding prevents virtual leaks and allows for leak checking.

Insulation

Chapters 20, 21, and 22 of the 1989 ASHRAE *Handbook—Fundamentals* discuss insulation fundamentals and applications in detail.

Doors

Both overlap and plug-type doors are used, the latter most frequently on both high-humidity chambers (because of internal condensation) and high-temperature chambers, to prevent or minimize warping. They must be rugged and well fitted, particularly for altitude chambers. Vapor sealing with multiple gaskets is important, particularly on smaller chambers, since defrosting means are seldom provided. Suitable gasket materials would be natural rubber, synthetic rubber, and silicone. Plastics or light-gage stainless steel are recommended for thermal breaker strip; pressed wood hardboard materials generally have insufficient resistance to physical damage, high temperatures, or high humidities. Low wattage heater cables, installed under the breaker strip, prevent larger doors from freezing closed. Safety measures are essential for walk-in freezers, including doors easily opened from inside.

For altitude chambers, the door assembly must register against a gasket that will easily vacuum seal. Vacuum gaskets must be properly retained to prevent their being drawn into the chamber.

Windows

For temperatures from -300 to $600\,°F$, special hermetically sealed multi-light window assemblies are manufactured. They are suitable for wide-range humidity tests and, if provided with an inner plane of sufficient strength, for altitude tests as well. Tempered plate is required for the inner panes so that temperature shocks will not cause failure. Because of the pressures that build up, sealed assemblies for higher temperatures require careful design. For temperatures up to $1000\,°F$, windows of vicor glass are required and hermetic sealing is difficult, but work along these lines indicates a possible range of -300 to $1000\,°F$. The number of panes and design for a given task are usually specified by the window manufacturer. A typical -100 to $300\,°F$ window contains six lights of glass enclosing dry gas spaces.

Accessories

Interior lighting is mostly incandescent, in vaportight and sometimes explosionproof fixtures. Ample illumination will be given usually by 6 to 10 W/ft^2 of floor space. Lamps are not recommended for use inside chambers above $600\,°F$.

Power leads may handle normal alternating current or special aircraft voltages and frequencies.

Thermocouples in almost any number may be called for. Bare iron-constantan should not be installed in a high-humidity chamber. For the operator's convenience, thermocouples are frequently connected to plugs and outlets.

Tuning shafts, if small and manually operated, may be coupled to adjustable devices on the item being tested. These shafts should be stainless steel tubes. O-rings on the outside of the chamber may be used for pressure sealing.

Power shafts, provided for driving rotating equipment, must have vacuumtight shaft seals for vacuum chambers.

Sleeves and plugs are required for general purpose testing. Holes of various sizes may be specified. On altitude chambers, threaded or flanged caps must be provided.

Special lines for pneumatically or hydraulically operated equipment may require pressure pipes with special connections at the ends.

Protective panels are used to prevent damage to the chamber if an internal explosion or sudden release of pressure occurs.

Window wipers are installed on windows to remove condensation if vision is required when temperature adjustments are made within the chamber.

Reach-in ports, with or without gauntlets, may be provided for minor work within a test space without opening doors or disturbing internal conditions.

BIBLIOGRAPHY

General Electric Company. 1948. Low temperature refrigeration—Compound systems. SM40-1500 et seq., 20 pages, Out of print.

Haines, R.W. 1987. *Control systems for heating, ventilating and air conditioning*, 4th ed. Van Nostrand Reinhold, New York.

Holladay, W.L. 1950. Low temperature test chamber design. *Refrigerating Engineering* (July):656.

Missimer, D.J. 1956. Cascade refrigeration systems for ultra low temperatures. *Refrigerating Engineering* (February).

Missimer, D.J. 1972. Mechanical system can reach $-140\,°C$. *Research/Development* (July):40.

Missimer, D.J. 1973. Ultra low-temp systems—A practical summary. *Refrigeration Service and Contracting* (December):18.

Missimer, D.J. and W.L. Holladay. 1967. Cascade refrigerating systems—State-of-the-art. *ASHRAE Journal* (April):70.

U.S. Government Printing Office. 1949. Military specifications for environmental testing, aircraft electronic equipment. MIL-T-5422 (Aer). Washington, D.C. (December).

U.S. Government Printing Office. 1952. General specifications for environmental testing, aeronautical associated equipment. MIL-E-5272A. Washington, D.C. (September).

CHAPTER 38

CRYOGENICS

Producing Low Temperatures ... 38.1	Miniature Closed-Cycle Refrigerators ... 38.11
Helium Refrigeration and Liquefaction ... 38.4	Equipment ... 38.14
Hydrogen Liquefaction ... 38.5	Cryogenic Insulation ... 38.16
Liquefied Natural Gas ... 38.7	Storage and Transfer ... 38.19
Nitrogen Refrigeration ... 38.9	Instrumentation ... 38.22
Oxygen Liquefaction ... 38.10	Safety ... 38.23
Argon Extraction ... 38.11	Space Simulators ... 38.26

CRYOGENICS refers to the coldest area known in nature. Figure 1 illustrates this temperature range which has an upper limit arbitrarily defined as −150°F (−250°F by some) and a lower limit of absolute zero. These limits separate it from the temperature range generally used in refrigerating engineering.

One important application of cryogenics is the separation and purification of air into its various components (oxygen, nitrogen, argon, and the rare gases). Oxygen is used in the production of steel and chemicals and as an oxidizing agent in the missile and rocket industry. Oxygen is also widely used to enhance combustion of hazardous waste and conversion of the solid waste into medium heat fuel gas. This stimulated the emergence of the new field of cryogenic engineering.

Other important developments have been the large-scale production of liquid hydrogen; helium extraction from natural gas; storage and transport of liquefied gases such as oxygen, argon, nitrogen, helium, neon, xenon, and hydrogen; liquefaction of natural gas for ocean transport and peak shaving; and many new types of cryogenic refrigeration devices. This chapter introduces the topic of cryogenic engineering.

Cryogenic processes generally range from ambient conditions to the boiling point of the cryogenic fluid. Cryogenic cycles also incorporate two or more pressure levels. These properties must also cover the vapor, vapor-liquid, liquid, and sometimes the solid regions. Therefore, the physical properties of fluids over a great range of temperatures and pressures must be known.

Solubility of contaminants must be known in order to design for their removal. The main physical properties for design purposes are those usually used in unit operations, such as fluid flow, heat transfer, and the like, in addition to those directly related to the Joule-Thomson effect and expansion work. Properties such as density, viscosity, thermal conductivity, heat capacity, enthalpy, entropy, vapor pressure, and vapor-liquid equilibriums are generally obtained in graphical, tabular, or equation form, as a function of temperature and pressure.

Selection of materials for low-temperature application depends on the properties of these materials at the desired temperatures, especially regarding brittleness, elasticity, thermal conductivity, thermal expansion, etc. Chapter 39 briefly discusses materials for low-temperature application.

PRODUCING LOW TEMPERATURES

Figure 2 shows the minimum Carnot cycle work for refrigeration at various temperature levels and the rapid increase in the theoretical work with decreasing temperature. Theoretical work becomes infinite at absolute zero. The actual work for low-temperature refrigeration increases at a more rapid rate than that indicated by the theoretical curve, since the thermodynamic efficiency decreases with decreasing temperatures.

The decrease in efficiency results from the greater heat leak at lower temperatures and from other irreversible processes associated with the refrigerator or liquefier. To attain cryogenic temperatures, correct refrigerating methods and equipment must be selected.

Of the many basic processes by which low temperatures are attained, the most commonly used are the Joule-Thomson expansion process (including the cascade) and the expander or work process. The vapor compression process, so widely applied in conventional refrigeration, is not found in cryogenic applications because the critical temperatures of all cryogenic fluids are below normal ambient temperature. Other processes, such as adiabatic demagnetization and adiabatic desorption, have been successfully applied in specialized laboratory studies.

Joule-Thomson Refrigeration

In the Joule-Thomson process, a gas at high pressure is expanded through a restriction to a low pressure, with a resulting change in gas temperature. If the high-pressure gas is initially below its inversion temperature, the gas temperature decreases as a result of the throttling. By passing such a gas through an effective countercurrent heat exchanger prior to or during the expansion process, it is possible to obtain extremely low temperatures and to partially liquefy the gas.

A typical Joule-Thomson refrigerator is shown in Figure 3. High-pressure gas at pressure p_1 and temperature t_1 enters the

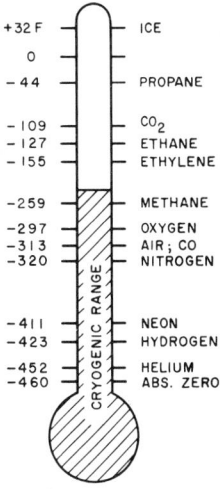

Fig. 1 Cryogenic Thermometer Showing Normal Boiling Temperatures at Atmospheric Pressure

---

The preparation of this chapter is assigned to TC 10.4, Ultralow Temperature Systems and Cryogenics. The chapter last received a major revision in 1971.

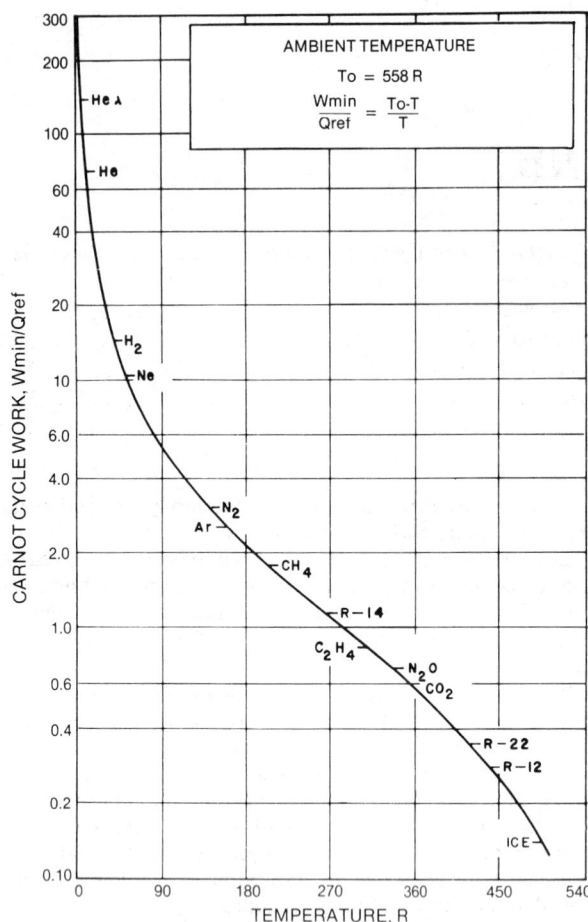

Fig. 2 Effect of Carnot Cycle Work

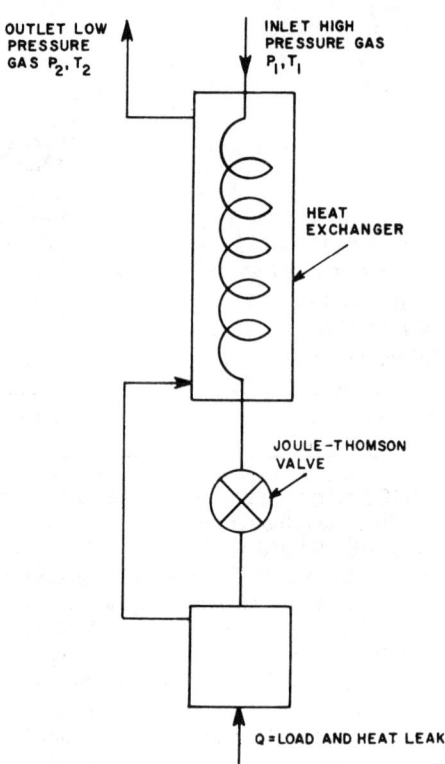

Fig. 3 Simple Joule-Thomson Refrigerator

heat exchanger and is cooled by a countercurrent stream of low-pressure gas. At a low temperature, the gas is expanded to a low pressure through a restriction, shown schematically as a Joule-Thomson valve, and a portion is liquefied. The liquid is vaporized at constant temperature by the heat leak and load, Q. The saturated gas at low pressure returns through the heat exchanger, cooling the incoming high-pressure stream, and leaves the exchanger at temperature t_2 and pressure p_2.

An energy balance around the refrigerator shows that the sum of the heat leak and load, Q, equals the enthalpy difference between the low- and high-pressure gas streams as:

$$Q = H(p_2, t_2) - H(p_1, t_1) \quad (1)$$

Introduction of the enthalpy of the low-pressure gas stream at the temperature t (the enthalpy of the gas stream if it is warmed completely to the temperature of the inlet gas) yields:

$$Q = [H(p_2, t_1) - H(p_1, t_1)] \\ - [H(p_2, t_1) - H(p_2, t_2)] \quad (2)$$

The first term is the isothermal enthalpy change for a Joule-Thomson expansion process. It equals the maximum or theoretical refrigeration available for such a process. The second is an energy term caused by the temperature difference at the warm end of the heat exchanger. Since this term can be expressed in terms of the specific heat of the gas and the temperature difference, Equation (2) may be rewritten as:

$$Q = (\text{Theoretical refrigeration}) - c_p \Delta t \quad (3)$$

in which Δt equals the difference between temperature t_1 and t_2. Since t_1 must be greater than t_2 for a finite heat exchanger, the second term in Equation (3) is positive, thus representing a loss in refrigeration as a result of the heat exchanger inefficiency. Since this loss is proportional to the temperature difference, the most efficient exchangers obtain the maximum amount of refrigeration. Figure 4 shows the theoretical refrigeration of nitrogen as a function of several inlet temperatures and various operating pressures for $p_2 = 1$ atm. These curves show trends typical of gases that exhibit a Joule-Thomson cooling effect; the theoretical refrigeration increases as the pressure increases or the inlet temperature decreases.

The loss in nitrogen refrigeration caused by the warm-end temperature difference of the exchanger equals approximately 1.1 Btu/(h·cfm·°F). For a refrigerator operating at 1000 psi and 80°F, for example, the theoretical refrigeration per unit of gas flow is approximately 14.8 Btu/(h·cfm·°F). From Equation (3), the exchanger temperature difference must be less than 25°F to obtain useful refrigeration or low temperatures. Since the coldest temperature obtained in the heat exchanger is about −321°F, the normal boiling point of nitrogen, the exchanger efficiency, as normally defined, must be greater than

$$\frac{(80 - 25) - (-321)}{80 - (-321)} \times 100 = 94\%$$

Hence, heat exchangers must be extremely efficient to realize useful amounts of refrigeration. In practice, exchanger efficiencies of 98% (warm temperature difference of 8°F for nitrogen systems) or higher are needed.

The Joule-Thomson throttling process is irreversible, and its improper use in a refrigeration process can cause poor process efficiency. A temperature-entropy thermodynamic chart shows that the entropy increase in a throttling process decreases with decreasing temperature and is the least when in the throttling liquid phase. For maximum process efficiency, therefore, the Joule-Thomson throttling valve is always located at the lowest possible temperature.

Cryogenics

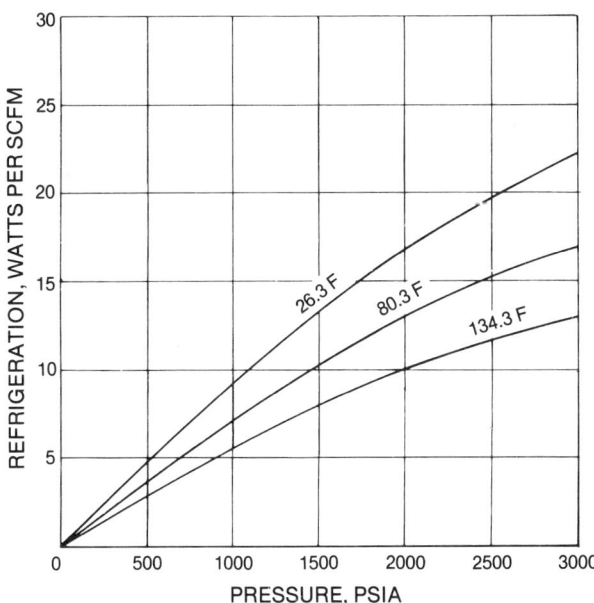

Fig. 4 Theoretical Refrigeration—Nitrogen

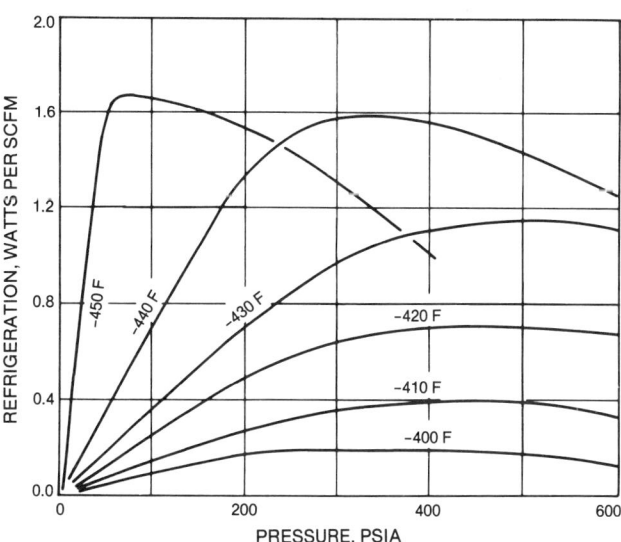

Fig. 6 Theoretical Refrigeration—Helium

Gases such as the fluorinated hydrocarbons, methane, argon, and nitrogen, exhibit cooling Joule-Thomson effects when at normal ambient temperatures and may be used to provide refrigeration at temperatures as low as about $-325\,°F$. However, to reach lower temperatures, the basic Joule-Thomson system must be modified. Neon and hydrogen boil in the -424 to $-406\,°F$ range. However, these gases exhibit an inverse Joule-Thomson effect at normal ambient temperatures: i.e., the gases will warm during an expansion from a high to a low pressure. Since the desired cooling effect is obtained when the gas is cooled to lower temperatures (to the boiling points of nitrogen or argon), as shown in Figure 5, an auxiliary Joule-Thomson circuit may be used to precool them prior to their Joule-Thomson expansion system.

A similar effect is exhibited by helium in Figure 6, which shows the theoretical refrigeration of helium. Note that extremely low temperatures must be obtained for precooling the helium to produce appreciable amounts of refrigeration. The loss in refrigeration caused by the warm-end temperature differences of the exchangers equals about 0.013 Btu/(ft$^3 \cdot$ °F) for neon and helium and for hydrogen below about $-316\,°F$.

A cascade process is one in which gas is precooled by another Joule-Thomson process. Usually the precooling includes phase change with a high-pressure condensing stream in heat exchange with a low-pressure boiling stream. Several stages are usually cascaded together for improved performance. (Figure 13 is an example of a cascade cycle.)

Expander Refrigeration

In the Joule-Thomson process, the high pressure is significant only in that, under these conditions, the nonideality of gases at certain temperature levels may be used to produce low temperatures. A thermodynamically more efficient process results if the high-pressure gas is allowed to do work during its expansion process. In performing work, the energy content of the gas is reduced, with a resulting temperature decrease.

A simple expander or work cycle is shown in Figure 7. High-pressure gas enters the heat exchanger and is cooled by a countercurrent stream of low-pressure gas. At an intermediate temperature, some gas is withdrawn from the heat exchanger and sent to an expansion engine or turbine where it is cooled by expanding work on the machine. The withdrawal temperature is often chosen so that the gas leaving the expander is at low pressure and is a saturated vapor.

The expansion process at constant entropy is ideal. The efficiency of the expander is defined as the ΔH across the expander divided by the ΔH that would result from a constant entropy expansion. The gas not withdrawn to the expander continues to be cooled in the heat exchanger and cools to a temperature much lower than could be obtained in a comparable Joule-Thomson refrigerator. It is expanded in a Joule-Thomson valve and may be partially liquefied. The two gas streams then reunite and, after absorbing energy from the heat load, return through the exchanger.

An energy balance around the system shows that the sum of the heat load and heat leak, Q, equals the sum of the work produced in the expander and the enthalpy difference at the warm end of the exchanger:

$$Q = W + H(\text{l-p outlet}) - H(\text{h-p inlet}) \qquad (4)$$

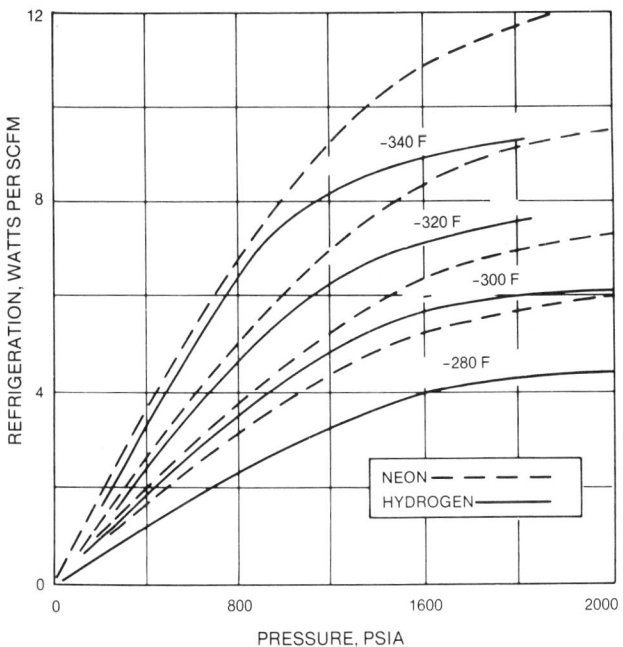

Fig. 5 Theoretical Refrigeration—Neon and Hydrogen

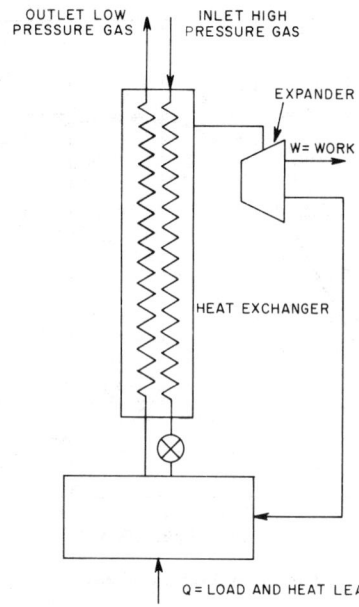

Fig. 7 Simplified Expander Refrigeration Cycle

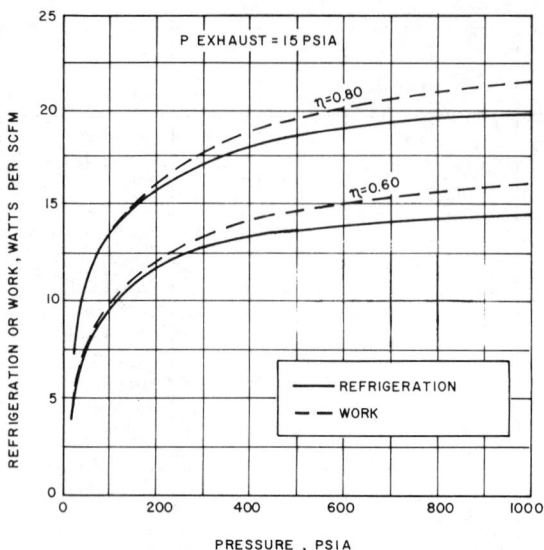

Fig. 8 Work and Refrigeration from Helium Expander Cycle

This equation is comparable to Equation (1), derived for the Joule-Thomson process. The work W represents the refrigeration available through the expansion of the gas. The enthalpy difference may be composed of a Joule-Thomson effect and the warm-end temperature loss (a term that represents the heat exchanger inefficiency), as in the Joule-Thomson process. Hence, in analogy to Equation (3):

$$\text{Refrigeration} = (\text{Work}) + (\text{J-T Effect}) - c_p \Delta t \qquad (5)$$

Thus, the refrigeration obtained from such a cycle will be considerably more than that obtained for the Joule-Thomson cycle. The use of helium represents an extreme example. Helium at ambient temperature exhibits a rise in temperature when expanded from a high pressure to a low pressure by a Joule-Thomson process. Thus, according to the definition used here, the J-T effect is negative. However, the possibility of obtaining work using helium in an expander makes it possible to obtain a finite amount of cooling or refrigeration. Figure 8 shows the refrigeration and work for a helium refrigerator producing refrigeration at about $-320\,°F$ (normal boiling point of nitrogen). At this warm temperature, the helium remains a gas throughout, and all the high-pressure helium is withdrawn to the expander.

The losses contributed by the heat exchanger inefficiencies are assumed to be negligible. The higher value of expansion efficiency is obtained in larger refrigerators, and the lower value is practicable for small laboratory units. The difference between the refrigeration and work curves represents the contribution of the negative Joule-Thomson effect of the helium.

Particular attention must be given to the heat exchanger design. The effects of heat exchanger inefficiency, as manifest in the warm-end temperature differences, apply to this cycle as well as to the Joule-Thomson process. Thus, it is possible to produce only a minute amount of refrigeration, in spite of high expander efficiencies, if the exchangers are poorly designed.

Cycles

The method of adapting the previously described refrigerating means to actual refrigerating or liquefying cycles is determined by many factors, including: (1) the temperature level desired; (2) the ultimate load or product requirement; (3) the product purity; (4) the distribution of auxiliary heat leaks and loads; and (5) the availability of auxiliary refrigeration such as liquid nitrogen. Although basic cycles may be specified, each refrigerator or liquefier must be considered individually because of the varied requirements. In the following sections, the application of the fundamentals to specific requirements are discussed.

HELIUM REFRIGERATION AND LIQUEFACTION

The current applications for helium liquefaction and refrigeration are in space chambers, computers, maser cooling, superconductive application, cryogenic research, and others. Refrigerator capacities range from a fraction of a Btu/h to as high as several tons of refrigeration, at levels ranging from 4.5 to 45 °R. Liquefiers, ranging in size from 2.5 to 25 gal/h of normal saturated liquid at 7.6 °R, have been constructed and operated.

Liquefier Design

The two basic liquefier cycles used in the past are (1) the nitrogen-hydrogen cascade and (2) the precooled expansion engine cycle. The latter is the more popular of the two cycles and ordinarily uses normal boiling liquid nitrogen as the precoolant. Figure 9 is a flow scheme of a typical expander cycle. The requirements for this liquefier, which has a capacity of 13 gal/h (to storage), would be 155 hp and 280 gal per day of liquid nitrogen. A second expander could be used to replace the liquid nitrogen refrigeration requirement.

For a liquefier of greater capacity (26 gal/h and more), the cycle variation as shown in Figure 10 could be used. The utility requirement would be about 1100 hp for a 190 gal/h (to storage) liquefier, plus a liquid nitrogen requirement of 4100 gal per day.

Refrigerator Design

The most common type of refrigerator used with space chambers applies the process shown in Figure 11. This cycle eliminates the need for a precoolant by sizing the expander to make up for the heat leak and the warm-end loss. This requires additional heat exchanger surface area to minimize the losses.

Equipment Design

For larger capacity liquefiers that use liquid nitrogen precoolant, the equipment operating above the boiling nitrogen temperature level may be insulated in steel panel boxes or round

Cryogenics

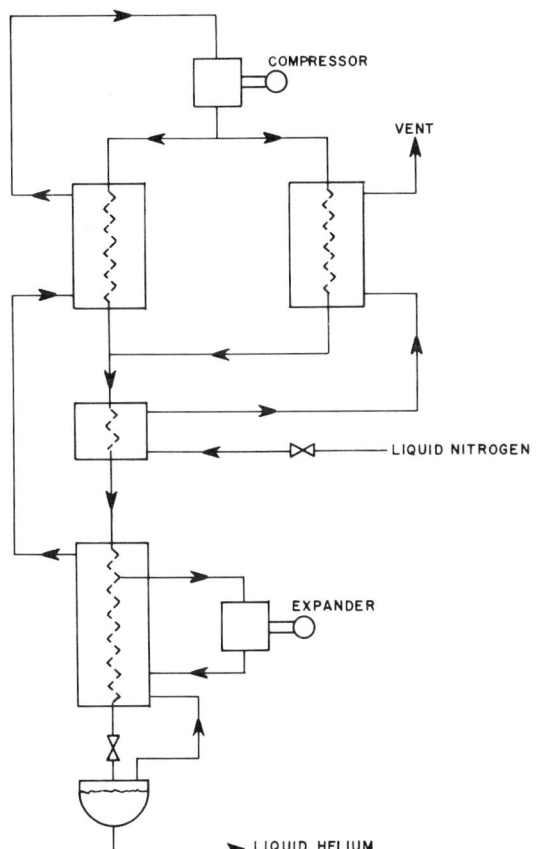

Fig. 9 Small-Scale Helium Liquefier

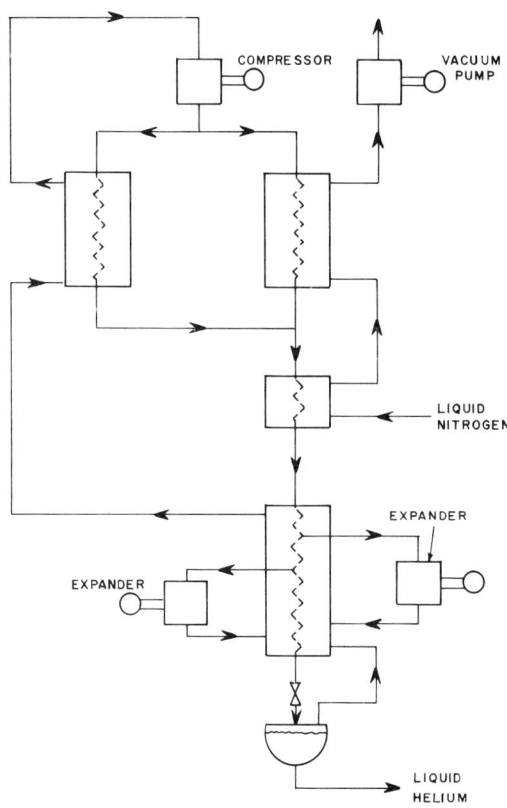

Fig. 10 Large-Scale Helium Liquefier

casings filled with perlite, vermiculite, or rockwool, whereas the lower-temperature equipment should be enclosed by a vacuum-jacketed Dewar, using either multilayer insulation or liquid nitrogen shielding to reduce the heat leak to an economical level.

The main factors that determine the efficiency of a liquefier, once the cycle is chosen, are (1) expander efficiency, (2) transfer losses, and (3) storage heat leak. Each percentage point of expander efficiency will influence the capacity of the liquefier by about 2%. Transfer and storage losses can also influence the plant size by 20 to 30%.

Liquefaction grade helium, which contains less than 50 ppm by volume impurities, requires final purification before reduction below liquid nitrogen temperature. An adsorber located at the nitrogen precoolant level is normally used for this service. The adsorbent must be completely reactivated to remove moisture in order to maintain its capacity for trace impurities in the helium. Following this preliminary reactivation at 350 to 450°F, subsequent normal reactivations may be accomplished by purging the adsorber with ambient temperature helium.

HYDROGEN LIQUEFACTION

Large-scale hydrogen liquefaction ranges in size from 6.5 to 60 ton/day. The problems encountered in the liquefaction of hydrogen are (1) purification, (2) ortho-to-para conversion, (3) precooling and liquefaction, and (4) transfer and storage.

Purification

Hydrogen has been produced from electrolytic cell gas, refinery off-gas, and steam reduction or partial oxidation processes. Hydrogen cell gas has mainly been used for supplying small liquefiers up to 1100 lb per day capacity; for larger plants, other sources are preferred. A partial oxidation process is considered when oil is the only available reactant for the locality. A steam reformer process is preferred when natural gas or naphtha is available. Substantial savings in capital investment can be realized by the use of refinery off-gas.

Purification of hydrogen prior to liquefaction depends on the source. In all cases, the required drying is normally accomplished near ambient temperature, with activated alumina or molecular sieve desiccant. The alumina may be reactivated at a lower temperature and has a lower initial cost, whereas the molecular sieve has the advantages of (1) only slight capacity reduction from heavy hydrocarbon contents in the hydrogen feed, and (2) permitting removal of H_2S and CO_2 along with the moisture in a single step.

Benzene and toluene have been separated from refinery off-gas by an oil scrub system. The other hydrocarbons may be separated in the low-temperature equipment, down to the freezing point of methane (−296.5°F). By cooling the feed gas against liquid nitrogen boiling at about 4 atm, and following it by liquid-gas phase separation, the hydrocarbon concentration may be reduced to a level of about 0.9% by volume. These residual hydrocarbon impurities, along with nitrogen, carbon monoxide, argon, and oxygen, may be removed by one of three methods: (1) adsorption by activated charcoal; (2) absorption by liquid methane, liquid propane, or both; or (3) adsorption by molecular sieve. Adsorption by charcoal is adaptable to small and large plants, and impurities have been reduced to below 1 ppm in a single step. The methane-propane absorption process offers the advantage of less operator attention, whereas the charcoal adsorption method allows for greater flexibility during plant startup and shutdown. The adsorption by molecular seive can reduce impurities to −0.001 ppm and is more applicable to large plants.

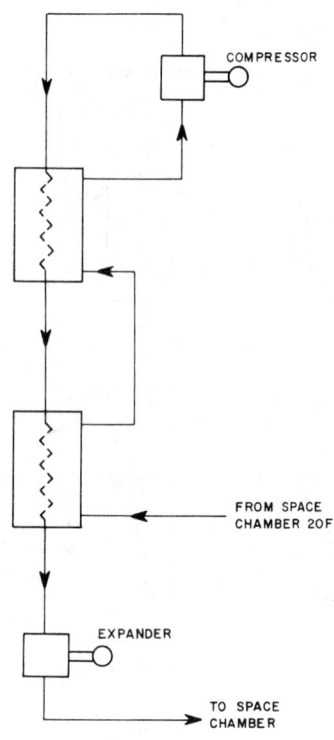

Fig. 11 Helium Refrigerator

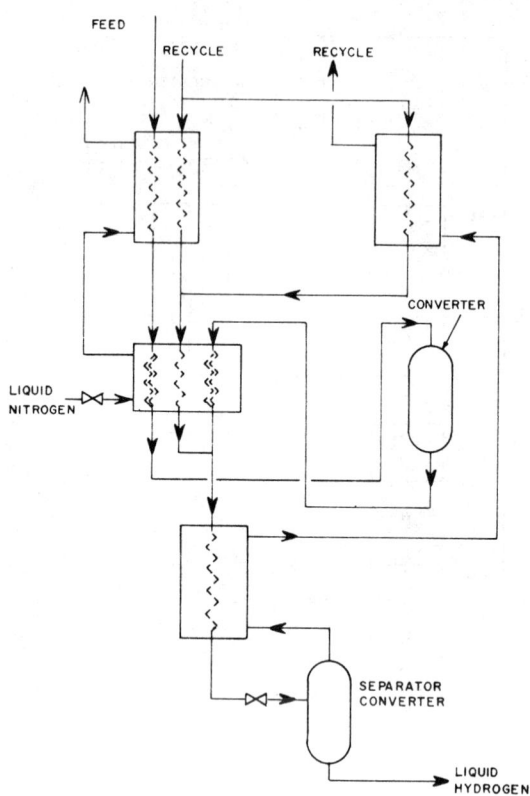

Fig. 12 Hydrogen Liquefier

Ortho-to-Para Conversion

Hydrogen, at ambient temperature, normally exists as 75% ortho (normal hydrogen), whereas, at the normal boiling temperature, its equilibrium concentration is 99.8% para. These forms differ by having parallel (ortho) or opposed (para) nuclear spins in the two atoms forming the hydrogen molecule. In time, the unassisted conversion will take place at the normal liquid temperature by the amount of:

$$1/O_f - 1/O_i = K\theta \qquad (6)$$

where

θ = time, h
K = reaction rate constant, 0.0124/h
O_f = final ortho mol fraction (Grilly 1953)
O_i = initial ortho mol fraction (Grilly 1953)

The reaction is exothermic; 609 Btu of heat are released per pound of hydrogen converted. As shown by Table 1, the boiloff rates in storage would be substantial until a para concentration level of 95 to 99% was reached. This is the level at which the liquid is normally produced.

Table 1 Boiloff Loss in Ortho-to-Para Liquid Hydrogen Conversion

Para Concentration, %	Boiloff Loss, % Per Day
98	0.019
95	0.119
90	0.48
80	1.9
70	4.3
60	7.6
50	11.9
40	17.2
25	26.8

Catalysts increase the conversion rate during the cooling and liquefaction of the hydrogen. Hydrous iron oxide and Cr_2O_3 on an Al_2O_3 gel carrier, and NiO on Al_2O_3 gel have been used. The NiO catalyst is about 90 times as active as the others and produces one of the fastest catalytic reactions known.

One or two stages of conversion at the liquid nitrogen level, followed by staged or continuous conversions down to the liquid hydrogen level, are commonly used in large plants. The staged and continuous conversion methods offer almost equal advantages for increasing cycle efficiencies. Important considerations for the choice between staged and continuous conversion include the process to be used with it and the manner in which the two are integrated to give an overall efficient plant of minimum capital investment.

Precooling and Liquefaction

The generally accepted cycle for small capacity plants is the single or dual pressure, liquid nitrogen precooled, J-T cycle, shown in Figure 12. By staging the conversion at the liquid nitrogen level, the energy consumed is about 5 kWh, and 1 gal of liquid nitrogen is used per gallon of 95% para liquid hydrogen for a plant size in the range of 500 to 1000 lb per day capacity. The process typically uses industrial grade or electrolytic cell gas as the feed, which must be compressed to 1500 psi pressure, dried in a desiccant bed of activated alumina, and cooled to the vacuum liquid nitrogen temperature level of about $-340°F$. This requires a nitrogen evaporative pressure of about 2.7 psia.

Final cleanup and staged para conversion is then accomplished, followed by a final pass against the vacuum nitrogen to remove the heat of conversion and heat leakage. The precooled hydrogen is finally cooled in the J-T heat exchanger against the flashed return vapors and then reduced in pressure to near atmospheric. The flashed gases are separated, and the liquid is converted to above 95% para in the separator-converter; vapors are returned

Cryogenics

to help cool the 1500 psi stream, and the liquid is transferred from the separator into storage.

High-capacity hydrogen liquefiers are commonly equipped with a single or dual pressure, nitrogen precooled, expander cycle. Variations of this type of cycle use up to a 1500 psi hydrogen recycle head pressure. The process will produce liquid hydrogen with an energy consumption of about 3 kWh per gallon into storage, including all refrigeration requirements.

For the liquefiers of small capacity (13 gal/h and less), the cold equipment is contained in one or two vacuum-jacketed Dewars to minimize heat leakage. For large units, the equipment above normal boiling temperature of nitrogen may be contained in a steel panel cold box, using perlite or rockwool insulation. Equipment below boiling nitrogen temperature is contained in vacuum-jacketed Dewars.

LIQUEFIED NATURAL GAS

Liquefying natural gas reduces its specific volume by a ratio of 600:1, which makes handling and storage economically possible despite the added cost of liquefaction and the specially insulated transport and storage equipment. The two principal types of LNG liquefaction plants, peak shaving and base load, are characterized by LNG disposition.

Peak shaving has been applied the most extensively, and the majority of liquefaction plants have been built for this purpose. Peak shaving is a method of smoothing the demand on a gas pipeline by liquefying a portion of the gas during the summer or off-peak season, storing the LNG, and subsequently vaporizing it during peak periods in winter. In this way, fuel requirements are met with a smaller pipeline than would be required without peak shaving.

The base load plant liquefies all of the gas available to it for transportation to another geographical location. These large-capacity plants are located in industrially undeveloped regions where natural gas is available in large quantities. Liquefaction makes natural gas available in areas where natural gas reserves are limited, nonexistent, or being depleted.

Feed Treatment

Natural gas must be treated, usually before liquefaction, to remove components that would freeze and stop the process equipment. The type of feed treatment depends on the source of the natural gas as well as the nature and quantity of the components it contains. Such components commonly consist of carbon dioxide, hydrogen sulfide, water with its associated hydrates, hexane and heavier saturated hydrocarbons, and cyclic compounds such as benzene and toluene.

If the natural gas feed is obtained directly from a pipeline, as in a peak shaving facility, it may already have been purified to a certain degree. Hydrogen sulfide may have been removed, and in the process, some carbon dioxide will also have been removed. The recovery of LPG removes the $C_6 +$ hydrocarbons. Generally, all that will require removal is additional carbon dioxide and water. If the natural gas feed is obtained from the well head, removal of all the components listed must be considered. The extent of removal depends on the composition and temperature of the liquid phases that form during liquefaction. Feed impurities must be reduced in concentration to the extent that the residual amounts are completely soluble in the liquid phases at all points in the process. Typically, carbon dioxide and hydrogen sulfide are reduced to 150 ppm by volume, while water is removed to a dew point of about $-100\,°F$, or 1 ppm at maximum.

The most popular processes to remove acid gases are amine scrubbing, hot potassium carbonate scrubbing, cold methanol scrubbing, and adsorption on molecular sieve. Water can be removed by methanol or glycol systems, but removal on a desiccant such as molecular sieve or activated alumina is most widely applied. Sometimes a single bed of molecular sieve is used to remove hydrogen sulfide, carbon dioxide, and water. The pretreatment system must be considered carefully with respect to the liquefaction system, and is often closely tied into it.

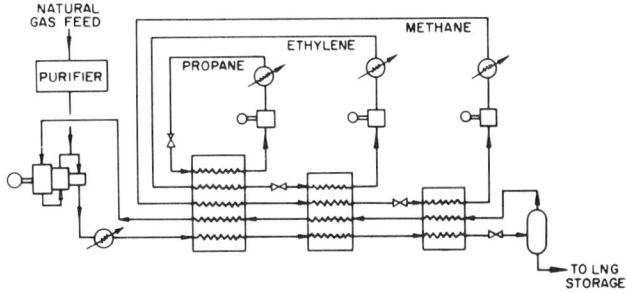
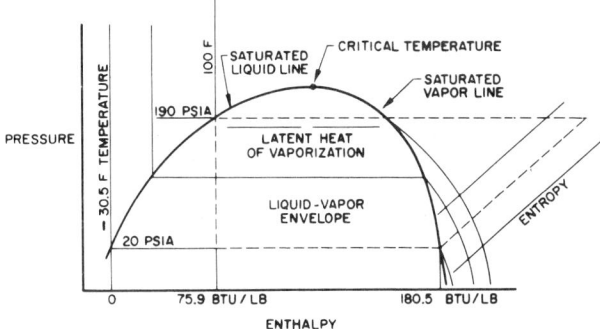

Fig. 13 Cascade Cycle

Liquefaction

Natural gas liquefaction cycles are generally cascade or expander cycles. The cascade cycle provides refrigeration by the Joule-Thomson process or isenthalpic throttling of separate refrigerant streams. The expander cycle provides refrigeration at lower temperature levels by expanding the gas in an engine producing work, ideally, at constant entropy. Figure 13 shows a simplified three-stage cascade cycle containing three independent refrigeration systems. The initial refrigeration is supplied by circulating propane, the intermediate refrigeration by ethylene, and the low-temperature refrigeration by methane.

The pressure-enthalpy chart (Figure 13) indicates the mechanism by which refrigeration is provided by the propane system. In this example, the propane is compressed from a cold section temperature of $-30.5\,°F$ at 20 psia pressure to 190 psia, ideally at constant entropy. The heat of compression and the refrigeration load that vaporizes the propane are rejected to cooling water in the compressor aftercooler at constant condensing pressure, at a saturation temperature of $100\,°F$. The condensed propane is throttled, at constant enthalpy, to 20 psia and cools to its saturation temperature, the initial $-30.5\,°F$. This two-phase stream then flows to the evaporator where it absorbs heat from the other streams (process feed gas, ethylene, and methane) and vaporizes. The pressure-enthalpy chart indicates that, under these conditions, each pound of circulating propane provides refrigeration of $180.5 - 75.9 = 104.6$ Btu.

The ethylene and methane refrigeration systems operate in similar fashion. However, these streams are cooled with the higher boiling refrigerants, in addition to water. The compressor discharge pressure for ethylene is selected so that total liquefaction is obtained at the propane evaporator temperature; the condensing pressure for methane is at the ethylene evaporator temperature.

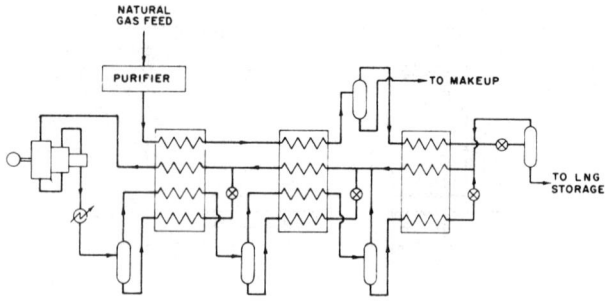

Fig. 14 Mixed Refrigeration Cascade Cycle

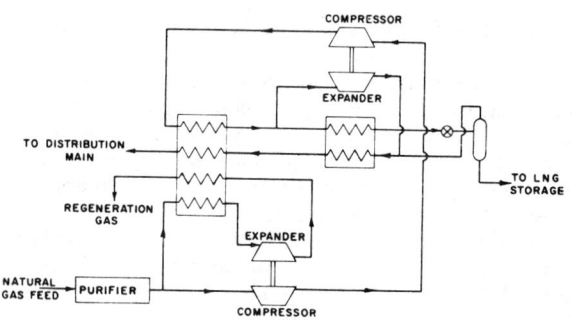

Fig. 15 Dual Expander Cycle

The three-stage cycle shown in Figure 13 is a simplified version of the cascade cycle. Present liquefiers incorporate from five to ten stages of cooling by evaporating each refrigerant at pressures intermediate between the final discharge pressure and the first-stage suction pressure. The overall refrigeration system is more reversible thermodynamically, because it provides refrigeration at several levels at intermediate pressures, thereby reducing the compression power significantly. Other cycle variations integrate the methane refrigerant system with the natural gas feed. One variation replaces the methane refrigeration shown in Figure 13 by an LNG flash separator stage and returns the vapor overhead at a higher pressure, which simplifies the machinery requirements for the facility.

Mixed Refrigeration Cascade Cycle. A variation of the cascade system is the mixed refrigeration or autorefrigerated cascade cycle. A simplified version of this cycle is shown in Figure 14. Instead of using several pure refrigerants in separate closed loops, this process combines all the refrigerants in a single multicomponent stream. Such a refrigerant stream might contain, for example, a mixture of nitrogen, methane, ethane, propane, and butane, with the composition to provide the necessary refrigeration at the required temperature levels.

The compressed refrigerant stream is cooled in the compressor aftercooler where most of the heavier components of the mixture are partially condensed. The liquid and vapor phases are separated and further cooled in separate passes of the first-stage heat exchanger against returning low-pressure refrigerant. The subcooled liquid is throttled into the low-pressure returning refrigerant stream to provide the necessary refrigeration to partially condense the separated vapor streams. This process is repeated in several stages of heat exchange, with each successive liquid fraction consisting of lighter hydrocarbons as the temperature decreases. The compressed natural gas is cooled and liquefied in countercurrent heat exchange with the returning refrigerant stream in a separate pass in each heat exchanger.

With the proper composition of the mixed refrigerant, this process can be made thermodynamically more efficient than a multistage cascade process. This occurs when the refrigeration required by the cooling stream is provided by the warming stream only to the extent required and at a temperature only slightly lower than the cooling stream. The resulting temperature-enthalpy curve for the warming stream is matched closely to the temperature-enthalpy curve for the cooling stream, which represents a condition of low thermodynamic irreversibility.

Additional advantages of the mixed refrigeration cascade cycle are simplicity in equipment and piping, including the use of a single refrigerant loop, a single compressor, reduced plant maintenance, use of the feed gas as the source of components for the refrigerant, and the elimination of facilities for the storage of refrigerant. The refrigerant mixture is usually obtained by separating the necessary components from the feed stream.

These advantages are especially important in remote locations, since the mixed refrigerant system has proved particularly useful in base load LNG plants. It is seldom used in peak shaving plants.

Expander Cycle. In the expander cycle, heat exchange and expander steps are arranged according to the pressure level at which the gas is originally available, the gas composition, and the relative cost and availability of auxiliary refrigeration. Cycle efficiency can be increased by using multiple expansion stages to produce refrigeration at several temperature levels. However, it soon reaches an optimum point, which represents balance between the cost of additional heat transfer surface and expansion engines and the savings obtained through reduced power. Figure 15 shows a dual expander cycle in which refrigeration is produced at two different temperature levels.

The natural gas feed is purified and split into two streams. One stream is precooled before being expanded to produce high temperature level refrigeration; the other is compressed in the two expander-driven booster compressors before precooling to the second expander. The higher pressure drop across the second expander produces refrigeration at a lower temperature level. Refrigeration is recovered from both expanded streams by cooling and liquefying the natural gas feed in countercurrent heat exchange.

The dual expander cycle is used to reduce the investment in plants whose pipeline pressure exceeds that required by the local distribution system. The excess pressure is used to produce the refrigeration needed to liquefy the natural gas, which eliminates the need for the refrigerant cycle compressor.

The type of cycle selected for an LNG plant depends primarily on plant capacity. Large plants invariably use a cascade cycle because the high investment cost imposed by cycle complexity is countered by the low operating cost resulting from high cycle efficiency. Intermediate and small plants usually use an expander cycle because low investment is more important. One small plant has made use of a Joule-Thomson throttling cycle to obtain minimum investment.

LNG Storage

Tanks for LNG storage are among the largest of all cryogenic tankage. In general, storage tanks for a peak shaving facility are larger in relation to the capacity of a liquefaction unit than for a base load shipping or receiving facility. This is because the peak shaving plant must liquefy and store enough gas during its 200 to 275-day off-season period to accommodate peak loads during the remaining season of five to ten times the minimum sendout during off-peak operation. The base load plant must provide no more storage than needed to accept plant liquefaction capacity for the maximum periods that ships are not in port. Tanks with a capacity of 0.5 billion ft^3 at standard conditions are common and one tank is 230 ft in diameter with a 176-ft liquid depth and holds 4 billion ft^3 at standard conditions.

Cryogenics

Three types of tank construction are used: double-wall metal tanks, prestressed concrete, and frozen earth cavity. The double-wall metal tank consists of one container nested within the other with an annular space 4 to 5 ft thick which is filled with insulation, commonly perlite. The inner container holds the LNG and is usually made from 9% nickel steel, although aluminum alloy is also used. The outer tank serves as a vaportight container for the insulation and is made from carbon steel. The insulation space is purged with dry gas to exclude atmospheric moisture. Tank design often includes direct communication between the inner container and the insulation space, so that the boiloff vapors may serve as a purge. This type of tank is erected above ground on a specially prepared site which consists of load-bearing insulation over a select sand fill. Heating coils may be imbedded below the tank to prevent frost heaving.

Concrete tanks are built with walls prestressed in vertical and circumferential directions to place the concrete in compression, thus offsetting tensile stresses induced by operating conditions. Such tanks may be built in a variety of configurations featuring above and below-ground installations with single- and double-wall tanks and external and internal insulation. This type of construction is usually adopted for tanks of large capacity. In-ground installation provides a measure of safety by containing spills in the event of tank leakage or rupture.

In the frozen earth cavity tank, the earth itself serves as the container. The earth is prefrozen by pipes sunk into the ground in a cylindrical pattern through which refrigerant is passed. When a satisfactory impermeable frozen wall is obtained, the cavity is excavated, a suitable insulated concrete or metal roof is added, and the tank is ready for use. Mixed success has been attained with frozen earth cavity tanks because of uncertainties in the ability to know the paths of underground water flows and the inability to prevent cracking in some frozen-earth structures.

The large capacity and thickness of insulation restrict heat inleak. Boiloff rates as low as 0.05% of tank capacity per day are normally obtained for the doublewall metal tanks and many of the prestressed concrete tanks. The frozen earth tank usually exhibits higher boiloff rates on initial fill, followed by slowly decreasing boiloff with passage of time. Boiloff vapors are collected, compressed, and returned to the distribution system.

Sendout Systems

A sendout system consists of pumps, vaporizers, and piping to withdraw LNG from storage and inject the vaporized natural gas into local distribution systems at the proper pressure. Equipment must be sized to deliver sendout requirements imposed by maximum peak loads. Use of sendout equipment at peak shaving units is low because it operates only during peak load seasons. However, reliability is of great importance, and sendout systems are usually provided in duplicate, with each unit capable of delivering maximum sendout. Base load units use sendout equipment almost continuously.

Commonly used vaporizers are of two types. Each relies on natural gas as the source of heat, and each uses an intermediate fluid to transfer the heat from the combustion gases to the LNG. In one type, water is warmed by submerged combustion burners, and the LNG is vaporized in tubes immersed in the water bath. In the second type, isopentane absorbs heat in a direct-fired burner and delivers it in an LNG exchanger. One vaporizer uses sea water for its heat source. The water flows by gravity over tubes in an open rack exchanger, with the LNG flowing inside the tubes.

Pumps are generally multistage, immersed centrifugal units mounted in an insulated vessel outside the main storage tank. Sealed, submerged motor-type pumps have also been used.

Using the Cold LNG

The large amounts of refrigeration available at temperatures as low as $-250°F$ have prompted many to examine potential uses for it. Typically, refrigeration recovery would be practiced at base load rather than at peak shaving units because the large capacity and uniform demand facilitates coordination with the refrigeration user. Obvious uses include food refrigeration and freezing and central air conditioning, although the use pattern of the latter is seasonal and out of phase with fuel demand during the heating season. The safety aspects of using a hydrocarbon stream in connection with air separation would have to be carefully considered. A variation of the air separation plant application is a system of conserving refrigeration. Nitrogen would be liquefied at the sendout site, in exchange with vaporizing LNG, and returned via the methane tanker to the LNG liquefaction plant where it would relinquish its refrigeration, thus substantially reducing the required power.

An olefins petrochemical plant has the necessary scale of operations to make effective use of all the refrigeration available from a typical base load LNG operation. By providing separation equipment, the LNG can also supply the ethane and propane for feedstock to the olefins plant. Such an integration of facilities would provide mutual benefit to both chemical plant operator and gas supplier.

A final example of the integration of LNG vaporization with another facility is provided by magnetohydrodynamic power generation. Refrigeration from the LNG would permit reduction in size of the required magnets.

NITROGEN REFRIGERATION

Large-scale nitrogen refrigeration systems have been built to supply refrigeration to large liquid hydrogen plants. Helium li-

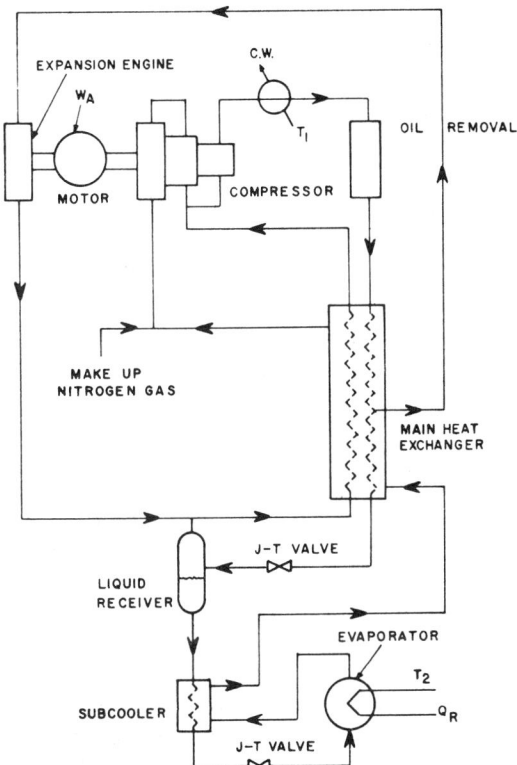

Fig. 16 Nitrogen Refrigeration System

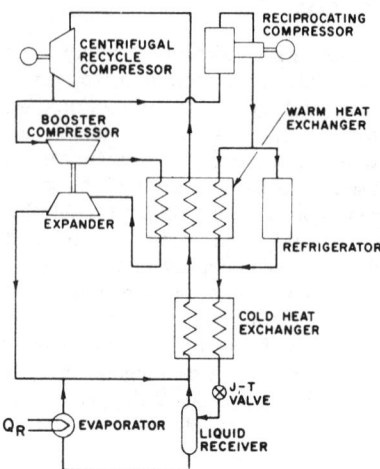

Fig. 17 Dual Pressure Nitrogen Refrigeration System

quefaction plants, liquid nitrogen scrubbing plants for purification of ammonia synthesis gas, argon purification plants, low-temperature plants for the separation of natural gas, and food freezing and shipping systems are examples of cryogenic processes that incorporate nitrogen refrigeration systems. High-pressure nitrogen is also used to extract natural gas from the earth.

Nitrogen refrigeration systems are applied in the temperature range of -344 to $-260\,°F$. The important properties of nitrogen as a refrigerant are as follows:

Molecular weight	28.02
Normal boiling point	$-320.5\,°F$
Latent heat at boiling point	2405 Btu/lb-mol
Liquid density at boiling point	50.4 lb/ft^3
Critical temperature	$-232.78\,°F$
Critical pressure	33.5 atm
Critical density	19.4 lb/ft^3
Melting (triple) point	$-345.9\,°F$

Typical overall requirements of actual power used per quantity of refrigeration obtained range from 26 bhp/ton at $-280\,°F$ to 36 bhp/ton at $-321\,°F$.

The Carnot cycle theoretical work requirement is given to be:

$$W/q = [(t_1 - t_2)/(460 - t_2)] [12{,}000/2545] \quad (7)$$

where
- W = Carnot work, brake hp
- q = refrigeration, tons
- t_1 = cooling water temperature, °F
- t_2 = refrigeration temperature level, °F

Thus, if $t_1 = 85\,°F$ and $t_2 = -315\,°F$,

$$W/q = [85 - (-315)]/[460 + (-315)] [12{,}000/2545]$$
$$= 13 \text{ bhp/ton} \quad (8)$$

Therefore, overall cycle efficiency (E) of a nitrogen refrigerator removing heat at $-315\,°F$ is:

$$E = 100 \frac{W_T/q_R}{W_A/q_R} = 100 \frac{13}{36} = 36\% \quad (9)$$

Significant factors that contribute to the inefficiency are: (1) inefficient mechanical equipment (such as compressors, expanders, and motors); (2) temperature differences and pressure drop required for heat transfer; (3) heat leak from ambient into cold equipment; and (4) piping configuration.

A typical nitrogen refrigeration cycle is shown in Figure 16. This Claude (expansion engine) cycle incorporates a double flash system. Variations of this cycle include multiple refrigeration sources such as a second expansion engine or a high-temperature Joule-Thomson refrigerator to precool a split feed stream. Another variation is shown in Figure 17. This dual pressure arrangement uses a moderate pressure for the stream that passes through the Joule-Thomson valve and is liquefied, and a low pressure for the stream that is recycled through the expansion engine.

The moderate pressure is at about the critical pressure of nitrogen to minimize the adverse thermodynamic effect of the large isothermal heat load of condensing nitrogen. The pressure of the recycle stream is based on optimum performance of highly efficient centrifugal compression and expansion machinery. Although the performance of the cycle is not as efficient as the high-pressure cycles, the lower operating pressures provide advantages in lower initial investment, ease of operation, and decreased maintenance.

Nitrogen required to make up losses from the system can be supplied, if an air separation plant is part of the facility, as in the case of argon purification plants or liquid nitrogen scrub units. If nitrogen from an air separation plant is not available, makeup nitrogen can be supplied by hauled-in liquid nitrogen, or hauled-in gaseous nitrogen, either in conventional cylinders or in tube trailers, or from an inert gas generator. Leakage from the system should be minimized. One of the major sources of leakage is the compressor.

Reciprocating compressors are usually selected for cycles requiring a discharge pressure between 1500 to 3000 psi. Compressor designs that minimize leakage are available. Oil-lubricated reciprocating compressors generally give the best service for compressing dry nitrogen, and highly efficient oil removal devices are necessary to prevent oil from entering the cold section of the system, where it would gradually plug the heat exchangers and prevent continuous operation.

For cycles requiring low to intermediate discharge pressures (<700 psi), centrifugal compressors are widely used, especially for large refrigerators. These machines are well adapted to handling large volumes of gas, are lower in cost than reciprocating compressors, show good operating efficiency, and eliminate oil contamination.

OXYGEN LIQUEFACTION

Liquid oxygen is used in large quantities as an oxidizer with fuels such as kerosene, hydrogen, and others, for rocket and missile propulsion and for steel making. It is also produced in large quantities for commercial use. Delivered in special cryogenic railway tankcars and trailers, the liquid oxygen is vaporized and warmed to near ambient temperature before it is used.

Liquid Oxygen Generator

A condensed flow diagram for a liquid oxygen generator cycle is shown in Figure 18. Filtered air is compressed in a multistage compressor to 2000 to 3000 psi. Between the second and third stages of compression, the air is passed through a caustic scrubbing system to remove carbon dioxide from the airstream.

Air leaving the fifth stage of the compressor passes through a desiccant or molecular sieve filled vessels to remove moisture from the air. After passing through the drier, the airstream is split. One stream passes through the main heat exchangers, which are cooled by effluent gases from the distillation column; it is then expanded through a valve to the pressure in the high-pressure column. The other portion of the airstream is expanded through a reciprocating expander, from compressor discharge pressure to an intermediate pressure, is warmed through coils in the main heat exchanger, and is further expanded in a centrifugal expander, to the pressure of the high-pressure column.

Cryogenics

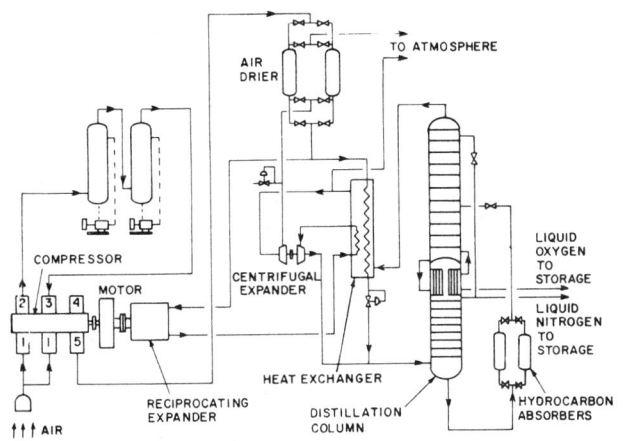

Fig. 18 Simplified Flow Diagram of Liquid Oxygen Generator

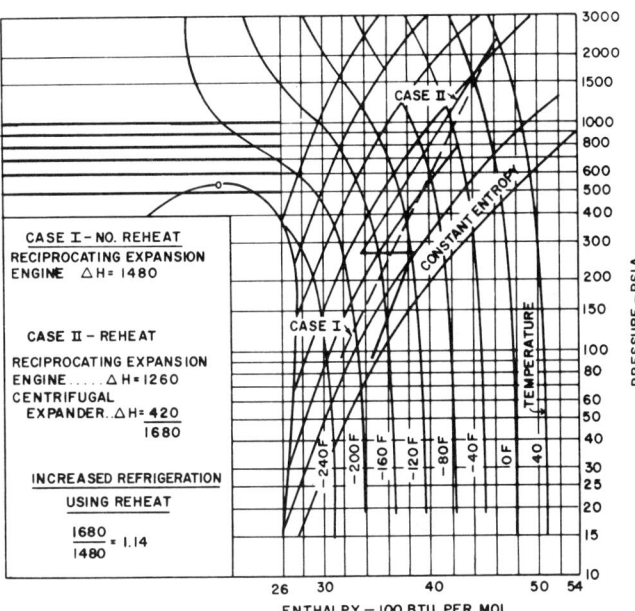

Fig. 19 Effect of Reheat on Expander Refrigeration

A double rectification column is used. Crude liquid oxygen from the high-pressure column is passed through hydrocarbon adsorbers before being expanded to the low-pressure column. Liquid nitrogen product is withdrawn from the stream, which also provides liquid nitrogen reflux to the low-pressure column. Liquid oxygen is withdrawn from the bottom of the low-pressure column. Effluent nitrogen-rich gas from the low-pressure column is warmed in the main heat exchangers countercurrently with incoming air and discharged to atmosphere. A portion of the warm effluent gas loads the centrifugal expander and is then used for reactivating the air drier, hydrocarbon adsorbers, and reciprocating expander oil filters, since dual assemblies of each are provided to ensure continuous operation of the liquid oxygen generator.

Reciprocating and Centrifugal Expanders

The dual reciprocating and centrifugal expander with intermediate reheat of the airstream increases efficiency. This scheme can be effectively applied where airflows are relatively large, since a centrifugal expander works well at lower expansion pressures. The refrigeration increase is the result of increased enthalpy drop, which is caused by the higher inlet temperature to the centrifugal expander.

The reciprocating expander is more efficient when it operates at a higher exhaust pressure because of the smaller losses in the expansion engine ports, valves, and cylinder. This is illustrated in Figure 19, by comparing the conventional arrangement of a single reciprocating engine expanding from compressor discharge pressure to high pressure column pressure. About 14% increase in total expander refrigeration is realized by using this scheme. The reciprocating expansion engine is considerably smaller in size because of the higher density of the exhaust air from the engine.

The overall cycle efficiency of a liquid oxygen generator is about 780 kWh/ton of liquid oxygen plus liquid nitrogen. Because the use of the liquid oxygen may be cyclic with high peak requirements and periods of full storage, generators must be taken off-stream and put back on-stream with a minimum interruption of production. A warm liquid oxygen generator unit requires about 5 h to be placed in full production; if the generator unit is shut down cold, it requires approximately 30 min. to return to production.

Any high-pressure cycle, liquid oxygen generating unit occasionally requires defrost of the main heat exchangers. One unit requires 4 h to defrost and then cool down the main heat exchangers for normal liquid oxygen production again.

Another arrangement used to make commercial liquid oxygen incorporates an expander-type refrigeration unit which operates on the nitrogen gas produced at the top of the high-pressure column. Cold nitrogen vapor withdrawn from the column is warmed to ambient temperature where it is compressed. The aftercooled compressed nitrogen is partially recooled in countercurrent heat exchange with the warming nitrogen and split into two streams. One stream is fed to an expansion engine, where the gas is work-expanded, cooled, and then returned to the column. The other stream continues in heat exchange with the warming nitrogen and is liquefied. This liquid is returned to the column and provides the necessary refrigeration, not only for overcoming the usual process heat influxes but also for withdrawal of the liquid oxygen product from the bottom of the low-pressure column.

Although not so efficient as the preceding process, the latter operates at pressures conveniently handled by centrifugal compression and expansion machinery and offers advantages typical of this equipment, such as low investment, ease of operation, low maintenance, and freedom from contamination of the process streams with hydrocarbon lubricants. It can also be used for easy conversion of a gaseous oxygen plant to a liquid oxygen plant by adding the refrigeration equipment.

ARGON EXTRACTION

Argon can be produced as a byproduct of the oxygen production cycle (Figure 18). Binary separation of argon (an impurity) and oxygen occurs in the section immediately below the upper column in Figure 18. The argon concentration increases on each succeeding tray until a maximum concentration of 6 to 8% is reached. Crude argon is then refined to the desired purity in a separate column.

MINIATURE CLOSED-CYCLE REFRIGERATORS

Cryogenically cooled, solid state devices (masers, infrared detectors, cryotrons, and parametric amplifiers) require miniature refrigerators that are much smaller than those described in the previous sections. The successful adaptation to miniaturization has required simplified basic cycles and an appreciably reduced scale of equipment, although fundamental refrigeration principles apply.

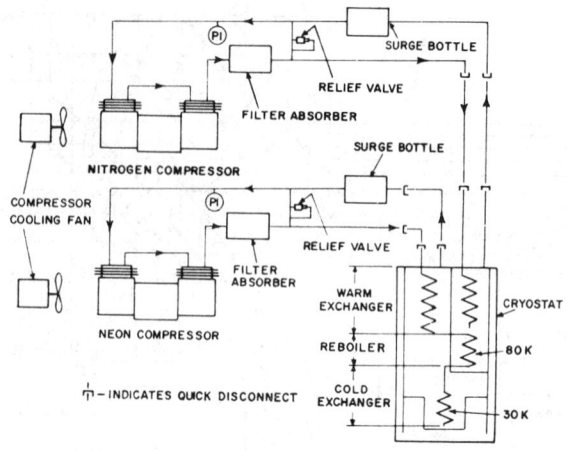

Fig. 20 Cascade Refrigerator (54°R) for Infrared Detector Cooling

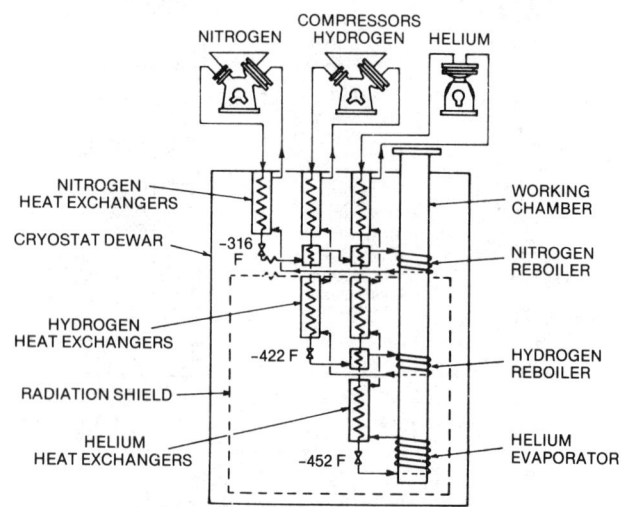

Fig. 21 Cascade Refrigerator (7.9°R) for Maser Cooling

Application of J-T Effect

A typical miniature refrigerator using the Joule-Thomson effect is shown in Figure 20 (Koontz and Lashmet 1963). In this refrigerator, neon is cascaded against a nitrogen circuit, resulting in temperature levels of approximately −406°F. This cycle has been successfully adapted to the cooling of infrared detectors, which require a refrigeration capacity of approximately 0.85 Btu/h. For comparison with the larger refrigerators described earlier, flow rates of approximately 0.33 cfm at standard conditions are used, and the entire system, including compressors, weighs approximately 50 lb.

The extension of the Joule-Thomson principle to produce liquid helium temperatures is shown in Figure 21 (Geist and Lashmet 1963). In this system, nitrogen provides refrigeration at approximately 144°R. Hydrogen, precooled to this temperature, transfers some of the refrigeration to the 38°R temperature level. The final stage of refrigeration is provided by helium, precooled to about 38°R. A typical refrigerator of this type, which is suitable for cooling masers, will develop approximately 4.3 Btu/h of useful refrigeration at 7.9°R. The normal operating conditions for the three circuits are as follows:

Circuit	Discharge Pressure, psia	Suction Pressure, psia	Flow Rate, scfm
Nitrogen	2800 to 3000	15 to 20	3.6 to 4.0
Hydrogen	2000 to 2200	16 to 20	4.3 to 5.0
Helium	250 to 270	15 to 20	3.3 to 3.9

An example of the application to both the expansion engine principle and the Joule-Thomson effect is shown in Figure 22. This particular refrigerator, described by Zeitz and Woolfendon (1963), was designed to produce 0.7 Btu/h of useful refrigeration at 4.5°R. It uses an expander circuit and a Joule-Thomson circuit. All refrigeration is provided by the expander circuit. During steady-state, it operates from 30 to 300 psia, with a flow of 7 cfm. This circuit uses two expansion engines, operating at approximately 126°R and 18°R, respectively. No Joule-Thomson valve is included.

To provide more refrigeration during cooldown, the expander circuit operates between 60 and 600 psia. The expansion engines, which normally operate at 300 rpm, are increased to 600 rpm, and available refrigeration is increased four-fold.

The Joule-Thomson circuit circulates 0.4 cfm between 1.3 and 80 psia. It picks up refrigeration from both the warm expander and the cold expander, or at roughly 126°R and 18°R. Final expansion down to operating temperatures is achieved through a Joule-Thomson exhanger and expansion valve.

Two independent circuits are used to operate the circuit providing the refrigeration in a supercharged condition—60 psia at cooldown and 30 psia at steady-state. If a single circuit were used, all the flow would have to be handled at 1.5 psia, requiring much larger compressors and exchangers and reducing the practicality of this type of system.

Stirling Cycle

In its ideal form, the Stirling cycle is composed of two reversible isothermal processes separated by two reversible constant volume processes. Therefore, it has the theoretical efficiency of a Carnot cycle.

In a real refrigerating machine, the process steps are carried out on a fixed gas charge by the sinusoidal motion of a pair of coupled pistons either in the same cylinder or in a pair of cylinders with a fixed spatial orientation. The pistons are connected to a motor drive through a crank mechanism which preserves a phase relationship between their motion such that the gas charge is shuttled to and from the cylinder through a regenerator. The regenerator acts alternately as a heat source and sink to preserve the temperature difference between the near constant volume compression and expansion processes.

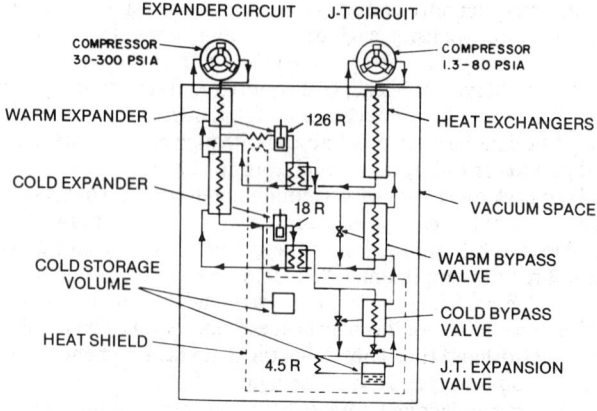

Fig. 22 Expander Refrigerator (4.5°R)

Cryogenics

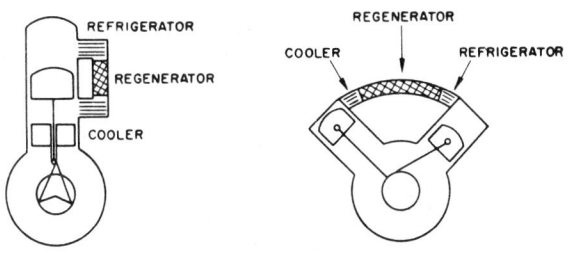

Fig. 23 Two Arrangements of Stirling Refrigerator

Heat exchangers mounted close to both ends of the regenerator provide the near isothermal heat addition and extraction portions of the cycle. Two arrangements for Stirling refrigeration machines are shown in Figure 23. In the coaxial arrangement the expansion and compression volumes are defined by the motion of the two pistons within a single cylinder. This configuration has some advantages of compactness and, as the pressures in the cold and warm working volumes are nearly balanced, the sealing problem is made easier.

Figure 24 shows an arrangement for a Stirling refrigerator to cool at 36°R. While it is possible to produce refrigeration at this temperature by applying a single expansion cycle, a more efficient solution is to use two stages of expansion in a configuration like that shown (Prast 1965). A miniature device patterned after this concept, producing useful refrigeration at 36°R and 3.4 Btu/h at 45°R, requires about 2,000 Btu/h of input power and weighs about 25 lb. The working fluid is helium with a mean working pressure of 125 psig. A no-load cooldown time to 47°R is about 10 min. The item to be cooled is most conveniently mounted at the cold head of the device, but remote cooling through a secondary heat transfer loop has been used.

Because of limiting low-temperature heat capacities of regenerator materials, Stirling refrigerators will not, by themselves, effectively provide refrigeration below about 22°R. Combined with a Joule-Thomson loop, they can cool to liquid helium temperatures. Vibration, induced microphonics in direct-mounted solid state devices, performance degradation from contamination of the regenerators, and piston ring wear are some of the problems associated with these refrigerators.

The Gifford-McMahon cycle (Gifford and Hoffman 1961, Hoffman 1963), a refrigerator made to operate on a modified Stirling cycle, is shown in Figure 25. Like the Stirling device, it uses regenerators to store and release thermal energy alternately while maintaining fixed mean temperature differences between the expansion volumes and the warm-end process fluid. Refrigeration is produced by expansion processes in volumes provided by a piston-cylinder configuration. In this case, two expansion stages in series are used. The expansion processes are made to take place by the motion of the sinusoidally driven piston and by controlled actuation of valves that alternately communicate the expansion spaces to the inlet and discharge side of the compressor. In the configuration shown, the pressure at the top and base of the piston is almost balanced at all times so that the crosshead drive must provide only the forces necessary to overcome friction and pressure losses in the regenerators. Sealing between gas spaces at different temperatures is also made easier by the design feature.

Unlike the straight Stirling refrigerator, the expansion and compression processes are completely separable in both the mechanical and thermodynamic sense. Therefore, more or less conventional compression equipment is used and the expansion engine pod can be located remotely from the compressor. The connection between compressor and engine is provided by warm pres-

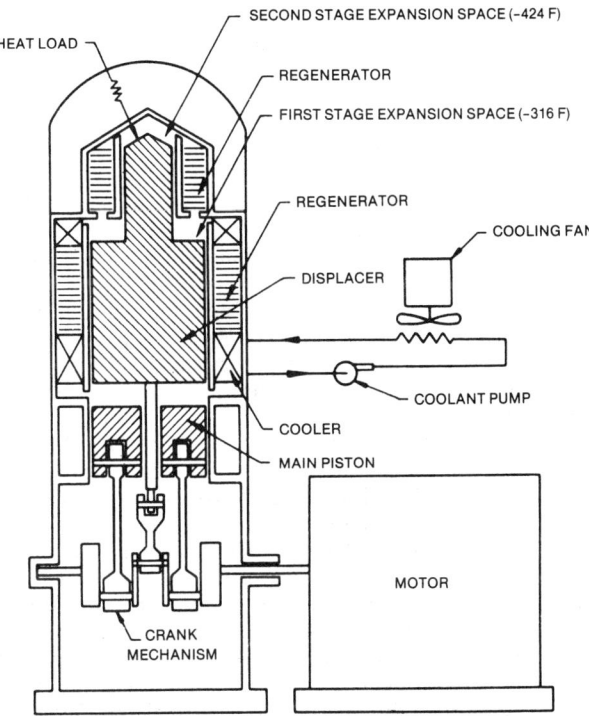

Fig. 24 Two Stage (36 R) Stirling Refrigerator

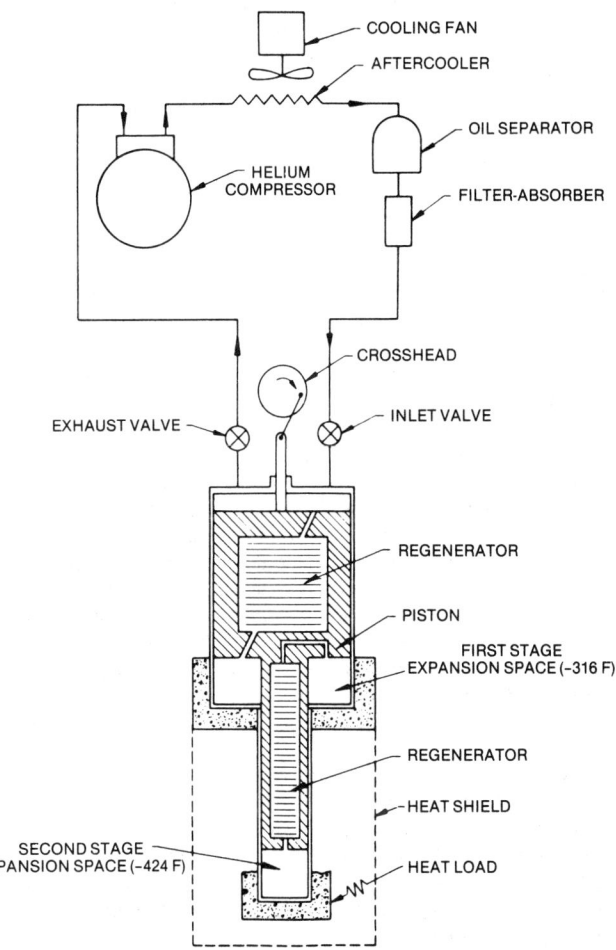

Fig. 25 Two Stage (36 R) Modified Stirling Refrigerator

sure and return lines. The Gifford-McMahon cycle is not ideally reversible, and real refrigeration devices using this cycle do not exhibit the efficiency of the Stirling machine. However, relatively good performance is achieved by using highly effective regenerators and multistage expansion at different temperature levels.

Like the straight Stirling refrigerator, this modified version cannot effectively cool below about 22°R and for the same reason. Similarly, to reach liquid helium temperatures, it is used as a precooler for a Joule-Thomson loop.

EQUIPMENT

The machinery generally used by the cryogenic industry can be grouped as (1) compressors, (2) expanders, and (3) pumps.

Compressors

Compressors may be classified as (1) centrifugal; (2) axial; (3) reciprocating, nonlubricated, or lubricated; (4) diaphragm; and (5) screw or lobe.

A *centrifugal compressor* is used mostly as the main air supply for separation plants. It has a horsepower range from 1200 to 10,000 and a flow capacity between 10,000 and 60,000 scfm. The discharge pressure is usually about 60 to 100 psia. These machines are generally fitted with adjustable inlet guide vanes to facilitate capacity reduction and maintain economical power requirements. Reductions in flow of up to 30% are obtainable without serious power penalties or the danger of surging.

An *axial compressor* is used for the main air supply to the largest air separation plants. Such a compressor typically has a capacity of 125,000 scfm of air at a discharge pressure of about 100 psia and requires 22,000 hp. These machines are usually driven by gas turbines or combination gas and steam turbines. The steam turbine has the additional function of starting the gas turbine. Capacity is adjusted by means of inlet guide vanes, as used with centrifugal compressors.

A *reciprocating compressor* is widely applied in the cryogenic industry. It is used to compress gases to high pressure, *i.e.*, 3000 to 4000 psi. When the flow rates required are 10,000 cfm and under, and high efficiencies are necessary, this machine should be used in preference to the centrifugal type.

The reciprocating compressor is available in lubricated and nonlubricated construction. This refers only to cylinder construction. The main bearings, connection rods, crossheads, and pins are lubricated by hydrocarbon oils in the usual manner on both machines.

Nonlubricated compressors used in oxygen compression have carbon- or bronze-filled TFE piston rings and piston rod packing. The suction and discharge valves are specially constructed for oxygen service. The distance pieces that separate the cylinders from the crankcase are purged with an inert gas, such as nitrogen, to preclude the possibility of high concentrations of oxygen in the area in the event of excessive rod packing leakage. The safety of the operator must be the prime consideration. Consequently, oxygen compressors with shutdown devices that detect high temperature and high pressures in the interstage piping are frequently employed. This type of compressor is usually limited to low-pressure ratios across each stage, so the discharge temperatures are below 325°F.

Another popular type of nonlubricated reciprocating compressor has pistons that depend on labyrinths on the piston cylindrical surface rather than on piston rings to form a seal. These machines are built only in the vertical configuration and, like other nonlubricated machines, are used to compress oxygen because of their inherent safety features, or whenever an uncontaminated gas supply is required.

Another type of nonlubricated reciprocating compressor used in the cryogenic industry is the diaphragm machine. The diaphragm is actuated by a piston connected to the crankshaft by the usual crosshead and connecting rod arrangement. The piston displaces the diaphragm hydraulically. The hydraulic fluid is either a soap and water solution or one of the synthetic oils that are compatible with oxygen, if the machine is used with oxygen. The diaphragm machine is used for high-pressure, low flow rate applications, requiring an uncontaminated gas stream for such service as filling industrial gas bottles.

The screw- and lobe-type compressors are often used in the cryogenic industry. The screw machine is widely applied as a main air source compressor in smaller air separation plants. The spiral rotors are so designed that the air is compressed to discharge pressure within the cavities between the rotors, prior to discharge from the machine. These machines run at rotor speeds of approximately 5000 rpm and are used where airflows are on the order of 8000 to 12,000 scfm and discharge pressures are as high as 100 psia. The lobe compressor is used on low-pressure and on low- to medium-pressure flow rates.

The lubricated reciprocating compressor, as used in the cryogenic industry, is most widely applied as a high-pressure air supply, or on nitrogen recycle service. Machines on this service usually have cylinders lubricated with a synthetic oil of the TCP group, to minimize the possibility of ignition in interstage piping, coolers, or separators. Otherwise, the construction is according to standard air machine practice. To minimize pulsations in the system, close attention is given to the design of the interstage piping. Capacity control on this type of machine is by clearance pockets, automatically operated through a selector switch on the control panel.

Expanders

Expanders may be grouped as (1) reciprocating (nonlubricated or lubricated), and (2) turbine. These are used to expand various gases efficiently from high to low pressures to obtain refrigeration. The gases may be air, nitrogen, natural gas, hydrogen, or helium, depending on the cycle. The expanders obtain high efficiencies by causing the gas to do work on a piston or turbine wheel and thereby realize temperature drops in excess of those obtainable with a J-T valve for the same pressure ratio.

The expanders in modern air separation plants reach high adiabatic efficiencies. For reciprocating types, efficiencies of 80% are normally quoted, and even higher values are quoted for high-capacity turbo machinery.

Varying the inlet valve closing point controls the capacity of *reciprocating expanders* by controlling the amount of gas admitted to the cylinder before expansion takes place. These machines are used to expand pressure gas streams from 3000 psia to pressures as low as 100 psia, depending on the cycle selected by the process designer.

Cylinder lubrication on this type of machine is usually one of the refrigeration compressor oils. Warm water in jackets around the ring-pass section of the cylinder prevents the oil from freezing in the cylinder. The discharge temperatures of an air expander will usually approach, within a few degrees, the saturation temperature at the discharge pressure, which is −275°F at 100 psia.

Nonlubricated reciprocating expansion engines are generally applied where possible oil contamination of the process is intolerable or where extremely low operating temperatures preclude the use of cylinder lubricants. This type of expansion engine is found in hydrogen and helium liquefaction plants and in helium refrigerators. Some nonlube machines are operating in air separation plants where the inlet pressures are moderate, about 1500 psi.

The *turbo-expander* is used when flow conditions are reasonably constant, oil contamination is not permissible, and inlet pressures do not exceed 100 psia. These machines are most frequently the radial-inflow-type turbine. The adiabatic efficiency of the machine can be as high as 90% when inlet nozzles, turbine wheels,

Cryogenics

and blower wheels are accurately matched and process conditions are favorable.

Turbo-expanders are loaded by means of compressors, blowers, generators, and dynamometers to absorb the shaft power produced by expansion of the process gas through the turbine. Compressors and blowers are used either to compress a process stream or simply to compress atmospheric air and discharge back to atmosphere. The generator converts the turbine shaft work to electrical energy, which is returned to the electrical supply system. Dynamometers are usually used in small-scale operations where energy is dissipated by pumping recycled oil through a restriction.

Pumps

Pumps may be grouped as (1) centrifugal and (2) piston. Pumps used in the cryogenic industry differ from those in other industries in materials of construction, design, and mode of operation. A cryogenic centrifugal pump is used as a transfer pump on tank trucks, in in-plant transfer service, and for liquid product pressurization prior to vaporization. The pumps are usually designed for specific duties and are not usually shelf items.

Depending on the duty cycle and discharge pressure, pump casings are made from stainless steel or aluminum. Aluminum is usually selected for impellers. Nonlubricated bearings are bronze, running on a stainless steel sleeve or shaft. Both vertical and horizontal pumps are built. Multistage pumps are usually vertical, and single-stage pumps are horizontal. The reciprocating pumps used in cryogenic applications are generally used for high discharge pressures required for filling industrial gas cylinders and recharger applications. Many single and multicylinder reciprocating pumps are available for pressures up to 3000 psi.

Heat Exchangers

Heat exchangers used in cryogenics must be compact and highly efficient: *i.e.*, effective for the required heat transfer using small stream-to-stream temperature differences. Efficiency stems from the second law of thermodynamics that leads to:

$$W_{lost} = T_o \Delta S_T \tag{10}$$

where

W_{lost} = work required beyond that needed in a thermodynamically ideal process
T_o = ambient temperature
S_T = entropy change of the universe caused by the process under study

In heat exchanger steps, Equation (10) becomes:

$$W_{lost} = \int T_o \frac{T_1 - T_2}{T_1 T_2} dH \tag{11}$$

where

$T_1 - T_2$ = stream-to-stream ΔT at any point in the heat exchanger
T_1, T_2 = hot and cold stream temperatures at any point in the heat exchanger
H = enthalpy level along the heat transfer path

The work lost is especially large where T_1 and T_2 are both small. Thus, any low-temperature ΔT becomes especially costly. The necessity for compactness stems from the need to reduce heat leakage from ambient into all cryogenic process equipment. The requirement of low ΔT also forces the cryogenic engineer to use large surface areas for heat transfer since the heat transfer coefficient cannot easily be changed. Thus, the designer requires a surface area inversely proportional to the ΔT.

Therefore, most cryogenic heat transfer processes take place in heat exchangers of three general types: (1) plate-and-fin exchangers, (2) coiled tube-in-shell exchangers, and (3) regenerators. Each of these devices offers large heat transfer surface area per unit volume; 1000 to 3500 ft^2/ft^3 are normal. Each has particular advantages and limitations.

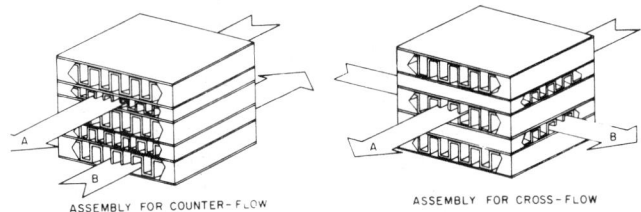

Fig. 26 Plate-and-Fin Heat Exchangers

Plate-and-fin heat exchangers are generally made of aluminum corrugated and flat sheets stacked alternately and brazed en masse by dipping in a molten salt bath. The resulting core is connected with headers to control stream flows, thus producing the final exchanger. Both crossflow and counterflow arrangements are possible, and a wide variety of corrugation designs and flow passage cross sections are available. Figure 26 shows the arrangement schematically. Typical parting-sheet to parting-sheet distances are 0.4 in., and normal heat transfer areas are about 1500 ft^2/ft^3.

Plate-and-fin exchangers have the advantages of high surface-to-volume ratios, relatively low cost, and wide flexibility in multistream flow arrangements and in passages assigned for pressure drop or high heat transfer rate. Disadvantages are that these exchangers are available in limited sizes only, thus requiring frequent manifolding and limiting use up to only moderate pressures, currently about 1000 psi.

The *coiled tube-in-shell heat exchanger* is used extensively in cryogenic systems and is most useful in high-pressure systems where the pressure difference between the tube-side stream and the shell-side stream is large.

The tubing can be either aluminum or copper. Aluminum is most suitable for pressures inside the tube of less than 600 psi. The tubes are headered into tube sheets or, in the case of high-pressure

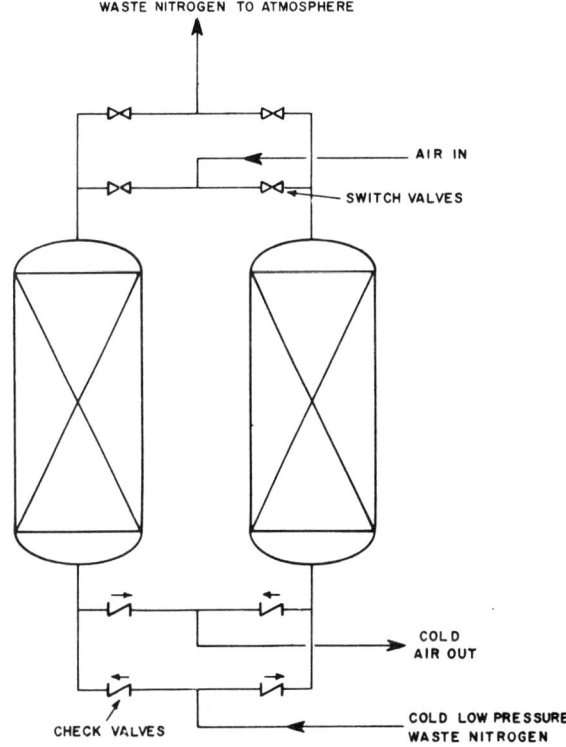

Fig. 27 Regenerator Set

service, pipe headers. The unit is usually designed so that all the tubes in the heat exchanger are the same length, a feature necessary for obtaining uniform distribution.

The cryogenic heat exchanger often operates with a large temperature difference from its warm to its cold end. Thus, there can be significant longitudinal heat flow in the structural members of the exchanger. Generally, this is reduced by building the exchanger relatively long in proportion to its cross section, though this increases the pressure drop and may allow greater heat leakage. Another caution is the use of low-conductivity material for any large member that extends the length of the exchanger. Thus, in the coiled tube-in-shell exchanger, both the mandrel and the outside shell are usually made of stainless steel.

Coiled tube-in-shell heat exchangers are made as small as a finger for use in maser cooling and cryosurgery devices, and as large as 16 ft in diameter and 115 ft long for use in LNG base load plants. This diversity of sizes and the possibility for use at almost any processing pressure are the main advantages; the main disadvantage is relatively high cost.

Regenerators. Figure 27 shows a regenerator set, typical of those used in air separation plants. Each regenerator is a vessel packed with a material of relatively high heat capacity, such as quartz pebbles, and, as shown, two such vessels are needed to replace one heat exchanger. Each vessel is alternately traversed by each gas stream. In one period, the incoming airstream warms the packing in one vessel, while the effluent gas stream cools the packing in the other vessel. After a short interval (up to 10 min.), the gas streams are switched from one vessel to the other by means of a switch and check valve arrangement.

In small cryogenic refrigerator service, such as in the Stirling cycle or the Gifford-McMahon cycle, the regenerator is usually a single vessel packed with copper wool or a similar material. In these regenerators, flow is reversed in a much more rapid cycle, perhaps with each stroke of the compressor piston.

Regenerators have the following advantages as compared to normal heat exchangers:

1. They are relatively easy to manufacture and less costly.
2. They concentrate a very large surface area into a small volume.
3. Well-designed regenerators have low pressure drop characteristics.
4. They allow for impurities, such as the carbon dioxide and water content of atmospheric air, to be deposited on the packing from the gas being cooled during one period, and then allow the same impurities to be evaporated into gas being warmed during the other period, thus eliminating the need for removal of impurities by other means.

Especially in air separation systems, regenerators also have disadvantages that have eliminated them from use in new plants. These include:

1. Their operation depends on the reliable operation of check valves and switch valves, some of which must operate at low temperature.
2. Each switch produces a severe physical blow on all the process piping and machinery.
3. Regenerator operation is inherently cyclical, so the operation of low-temperature expanders and distillation equipment is made difficult.
4. A pure product cannot be withdrawn from a regenerator since each switch contaminates the vessel with a charge of impure gas.
5. Regenerator packing tends to disintegrate in the environment of cyclical changes in temperature and pressure, thus sending dust into the rest of the plant.
6. Regenerators have a high maintenance cost due to cyclic operation.

These disadvantages are not so apparent in small refrigerator applications, so regenerators are extensively used for these processes. Regenerators are usually limited in air separation to heat exchange of relatively low-pressure gases, and where the pressure difference between the two gas streams is small. Regenerator vessels for low-temperature service are usually made of either aluminum, stainless steel, or 9% nickel steel.

Valves

The basic criterion for selecting valves for any service is the efficiency with which they accomplish shutoff or throttling. In low-temperature circuits, the additional consideration of heat leakage must be included. Most process valves are located in the cold box and are constructed of series 300 stainless steel. With these valves, extended bonnets of stainless steel are required to minimize the thermal conductance, thus keeping the packing at near ambient temperatures.

Teflon-impregnated non-asbestos packing is suitable for oxygen service and most cold applications. A thin wall column is constructed around the shaft of liquid-control valves to reduce the heat loss. To reduce leaks, the inaccessible cold valves have no end or bottom flanges. With top-opening valves, the trim can be changed without opening the insulating jacket.

In sizing cold control valves, a designer should calculate the valve capacity to give adequate control for normal cold liquid or gas flow conditions and to allow as much warm defrost flow as is practical. For cryogenic liquid expansion valves, valve sizing should include both liquid and gas calculations, since flashing occurs. It is good practice to install valve positioners on extended stem-control valves.

Vacuum-jacketed valves are used extensively at liquid hydrogen and helium temperatures. Construction considerations include extra long, thin vacuum-jacketed stems to lengthen the heat path, thus reducing the heat leak. Expansion bellows are used to isolate valves from line stress during cooldown.

CRYOGENIC INSULATION

Cryogenic insulations can generally be divided as (1) straight (high) vacuum, (2) filled vacuum (evacuated multilayers, evacuated powders, or evacuated fibers), and (3) nonvacuum (powders, foams, or fibers).

Evacuated systems predominate in those applications concerned with long-term storage of the more volatile fluids, particularly when fluid quantities to be stored are relatively small. Nonvacuum systems predominate in those applications concerned with short-term storage and transfers of the less volatile fluids, storage of large quantities (*e.g.,* field-erected tanks), and production plants.

In general, evacuated multilayer insulations, when properly designed and fabricated, provide the highest impedance to heat flow. They typically trade off some loss of thermal impedance through the mechanism of solid conductance for considerable gain in minimizing gas conduction and radiation. Evacuated powders, typically five to ten times more conductive than evacuated multilayers, enjoy the advantages of cost and tolerance to poor vacuum (*e.g.,* 10^{-2} torr versus 10^{-4} torr for multilayer). Nonvacuum insulations enjoy a further cost advantage, but their considerably higher conductances render them useful primarily for applications in which equipment volume is no problem (*e.g.,* large stationary equipment) and thick insulation can be used. The choice of insulation depends on the particular application.

The following sections give the more important characteristics and properties of various types. Designers can use this information as a guide to selection of insulation for a given use.

Table 2 Effective Emissivity E in Terms of Individual Emissivities of Cold and Hot Walls, e_1 and e_2

	Specular Reflection	Diffuse Reflection
Parallel plates	$\dfrac{e_1 e_2}{e_2 + (1 - e_2)e_1}$	$\dfrac{e_1 e_2}{e_2 + (1 - e_2)e_1}$
Long coaxial cylinders ($L \gg d$)	$\dfrac{e_1 e_2}{e_2 + (1 - e_2)e_1}$	$\dfrac{e_1 e_2}{e_2 + (A_1/A_2)(1 - e_2)e_1}$
Concentric spheres	$\dfrac{e_1 e_2}{e_2 + (1 - e_2)e_1}$	$\dfrac{e_1 e_2}{e_2 + (A_1/A_2)(1 - e_2)e_1}$

Table 3 Selected Minimum Total Emissivities[a]

	Surface Temperature, °R			
Surface	7.2	36	139	540
Copper	0.0050		0.008	0.018
Gold			0.01	0.02
Silver	0.0044		0.008	0.02
Aluminum	0.011		0.018	0.03
Magnesium				0.07
Chromium			0.08	0.08
Nickel			0.022	0.04
Rhodium			0.078	
Lead	0.012		0.036	0.05
Tin	0.012		0.013	0.05
Zinc			0.026	0.05
Brass	0.018			0.035
Stainless steel, 18-8			0.048	0.08
50 Pb 50 Sn solder			0.032	
Glass, paints, carbon				0.9
Silver plate on copper		0.013	0.017	
Nickel plate on copper		0.027	0.033	

[a] These are actually absorptivities for radiation from a source at 540 °R. Normal and hemispherical values are included indiscriminately.

Vacuum Insulation

Straight (High) Vacuum Insulation. Heat transfer through high vacuum insulation predominantly occurs by radiation, but gas conduction may contribute significantly.

The net exchange of radiant energy between two surfaces in Btu/h is:

$$q = E A_1 \sigma (T_2^4 - T_1^4) \tag{12}$$

where

- E = effective emissivity of the two surfaces
- A_1 = inner surface area, ft²
- σ = Stefan-Boltzmann constant, 0.1714×10^{-8} Btu/h·ft²·°R⁴
- T_2 = hot wall temperature, °R
- T_1 = cold wall temperature, °R

Table 2 gives the values of E in terms of individual surface emissivities for three commonly used geometries. Table 3 gives the minimum recorded emissivities for a number of materials used in cryogenics. These data include normal and hemispherical emissivities and absorptivities, since their differences can usually be ignored for engineering calculations.

Existing experimental data on low-temperature emissivity allow the following generalizations:

1. Materials having the lowest emissivities also have the lowest electrical resistances.
2. Emissivity decreases with decreasing temperature.
3. The apparent emissivity of good reflectors is increased by surface contamination.
4. Alloying a metal increases its emissivity.
5. Emissivity is increased by treatments such as mechanical polishing, which result in work-hardening of the surface layer of the metal.
6. Visual appearance (brightness) is not a reliable criterion of reflecting power at long wavelengths.

The best reflecting surfaces are pure, good conductor metals, annealed and cleaned, using a method that avoids strain, such as one involving the use of acids and residue-free organic solvents.

As a rule, the objective when using high-vacuum insulation is to obtain a vacuum of such quality (10^{-6} torr or less) that the heat transfer by residual gas conduction does not seriously contribute to the overall heat transfer. Since this is not always possible, gaseous heat conduction at low pressures, where the mean free path of the gas molecules is large compared to the dimension of the concentric spheres, coaxial cylinders, or parallel plates, may be calculated by using an adaptation of the Knudsen formula.

$$W_{gc} = \frac{\gamma + 1}{\gamma - 1} \alpha \left(\frac{R}{8\pi M T} \right)^{0.5} p(T_2 - T_1) \tag{13}$$

$$\alpha = (\alpha_1 \alpha_2)/[\alpha_2 + \alpha_1(1 - \alpha_2)(A_1/A_2)] \tag{14}$$

where

- W_{gc} = net energy transfer per unit time per unit area of inner surface, Btu/h·ft²
- γ = c_p/c_v, the specific heat ratio of the gas, assumed constant
- R = molar gas constant
- p = pressure, mm Hg
- M = molecular mass of the gas
- T = temperature at the point where p is measured, °R
- α = overall accommodation coefficient
- A = area, and subscripts 1 and 2 refer to the inner and outer surfaces, respectively, ft²

This expression reduces to

$$W_{gc} = C \alpha p (T_2 - T_1) \tag{15}$$

where

$$C = \frac{\gamma + 1}{\gamma - 1} \left(\frac{R}{8\pi M T} \right)^{0.5} \tag{16}$$

Values of this constant for cryogenic applications have been calculated by Corruccini (1957) and are shown in Table 4. The temperature at the pressure gage is assumed to be 80 °F.

Table 4 Constants for the Gas Conduction Equation

Gas	T_2 and T_1, °R	Constant
N_2	⩽ 720	28.0
O_2	⩽ 540	26.0
H_2	540 and 139 or 540 and 162	93.0
H_2	139 and 36	70.1
H	any	49.3
Air	⩽ 720	27.5

Table 5 gives some suggested values for individual accommodation coefficients. The experimental data are meager, and the uncertainty in these values is great. However, for most engineering calculations, these data are adequate.

Table 5 Approximate Accommodation Coefficients

Temperature, °R	He	H_2	Air
540	0.3	0.3	0.8–0.9
139	0.4	0.5	1
36	0.6	1	1

Evacuated Multilayer Insulation. In principle, this insulation consists of many brightly reflective radiation shields, separated by poor conductors, in a vacuum. When properly applied, this insulation has an apparent mean thermal conductivity between 80 and $-424\,°F$ of 2×10^{-5} to 4×10^{-5} Btu/(h·ft·°R).

The radiating shields are usually supported and separated by a spacer with low solid conductance but with relative thermal transparency. In general, the spacers cause a net increase in conductance over the case in which the same number of radiating shields would be floating without spacers. The limit of performance (no spacers) could be evaluated by:

$$Q = [E\sigma A/(n + 1)](T_2^4 - T_1^4) \quad (17)$$

where

- σ = Stefan-Boltzmann constant
- A = area
- E = effective emissivity
- n = number of floating radiation shields
- T_1 and T_2 = absolute temperatures of hot and cold boundaries

The spacer can cause considerable deviation from this limiting performance. Thus, for design purposes, the thermal performance of multilayer insulation is expressed by an apparent mean thermal conductivity k as follows:

$$q_T = [kA(T_2 - T_1)]/b \quad (18)$$

where

- b = thickness of the insulation

Two multilayer systems in wide commercial use are:

1. Thin sheets of mylar that have been vacuum coated with aluminum on one side to produce a low-emissivity surface. In this system, the mylar itself acts as the spacer. The coated mylar is usually crinkled to reduce contact area between successive layers.
2. Thin layers of low-emissivity metal foil (usually aluminum 0.00025-in. thick), separated by a low-conductance spacer (usually glass fiber paper).

Tables 6 and 7 give typical thermal conductivity values for these two materials. Table 7 shows the undesirable effect of mechanical compression on the three aluminum-glass systems. It also illustrates the improved performance as the spacer is made thinner, allowing a larger number of shields per unit thickness of insulation. Cost of these systems rises with the use of thinner spacers. The added cost of the improved performance must be justified by a more critical application.

The following precautions should be taken when using multilayer insulations:

1. Use vacuum lower than 10^{-4} torr for maximum efficiency. Deteriorating effects of elevated pressure are presented by Glaser (1967).
2. Avoid high mechanical loads on the multilayer stack.
3. Use low-emissivity reflecting surfaces.
4. Minimize gaps between adjacent shields at corners or joints because they can add significantly to the total heat transport.

Table 6 Apparent Mean Thermal Conductivity of Aluminized Mylar (Boundary Temperatures 540 and 139°R)

Number of shields per in.	Density, lb/ft³	Conductivity, Btu/h·ft·°R × 10⁶
61	0.0131	28.0
102	0.0219	25.0
150[a]	0.0331	24.0

[a] Optimum density.

Table 7 Apparent Mean Thermal Conductivity of Three Glass Fiber Paper-Aluminum Foil Insulations (Boundary Temperatures 540 and 139°R)

Number of shields per in.	Density, lb/ft³	Conductivity, Btu/h·ft·°R × 10⁶	Type of Insulation
12.7[a]	0.0156	101	Linde SI-10
25.4	0.0312	231	Linde SI-10
71.1[a]	0.0625	20	Linde SI-62
142.2	0.125	58	Linde SI-62
109.2[a]	0.0938	12	Linde SI-92[b]
218.4	0.188	29	Linde SI-92[b]

[a] Optimum density.
[b] Spacer is opacified with aluminum flakes (Grunert et al., 1969).

Evacuated Powder Insulation. The insulating value of powders is increased greatly by removing the interstitial gas. Therefore, when powders are used at low pressures (10^{-2} torr or less), gas conduction is negligible, and heat transport is chiefly by radiation and solid conduction. For some powders, radiation absorption reduces heat transfer more than solid conduction increases it, and these powders thus reduce the overall heat transport. Table 8 shows the apparent mean thermal conductivity between $-323\,°F$ and $80\,°F$ of some selected powders. Figure 28 shows the apparent mean thermal conductivity of several powders as a function of interstitial gas pressure (Fulk 1959).

The amount of heat transport because of radiation through these dielectric powders can be reduced by adding metallic powders. Figure 29 shows the effect of adding metallic powders (Hunter et al. 1960).

The vacuum level required with evacuated powder insulation is much less extreme than with straight vacuum or multilayer insulations. Most powders, however, can adsorb considerable quantities of gases and water. Evacuation to even moderate pressure levels is usually difficult. The accumulation of water can be avoided by suitable powder handling techniques. Adsorbed gases must be driven off at elevated temperatures, or outgassing of noncondensable gas must be accommodated by including adsorbents or getters in the vacuum space. In addition, fine mesh filters must be used to protect the vacuum pumps from these highly abrasive particles.

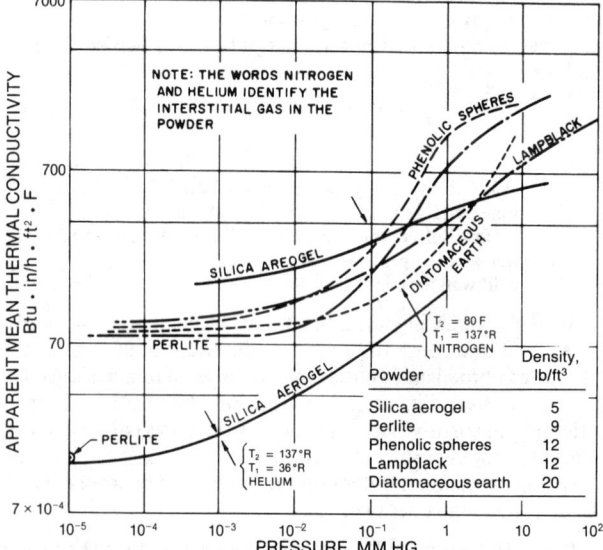

Note: The words Nitrogen and Helium identify the interstitial gas in the powder. T_1 and T_2 are the boundary temperatures.

Fig. 28 Thermal Conductivities of Powders

Table 8 Apparent Mean Thermal Conductivity of Several Powders
(25.4-mm sample thickness; boundary temperatures of 540 and 137°R; and wall emissivities greater than 0.8)

Powder	Remarks	Density, lb/ft^3	Gas Pressure, mm Hg	Interstitial Gas	Conductivity, Btu/(h·ft·°R) × 10^{-4}
		6.2	10^{-4}	—	12[a]
	Chemically	6.2	628[e]	Nitrogen	113
	prepared	6.2	628	Helium	358
	250A	6.2	628	Hydrogen	462
Silica aerogel	+ 10% free silicon dust (by mass)	7.0	10^{-4}	—	10.5
Silica	Flame prepared 150-200A	3.7	10^{-4}	—	12
		3.7	630	Nitrogen	107
	+ 30 mesh	3.7	10^{-4}	—	12
	+ 30 mesh	6.2	10^{-4}	—	10.5
	+ 30 mesh	6.2	628	Nitrogen	192
	+ 30 mesh	6.2	628	Helium	735
	+ 30 mesh	6.2	628	Hydrogen	845
	− 30 + 80 mesh	8.1	10^{-4}	—	7
Perlite, expanded	− 30 + 80 mesh	8.1	628	Nitrogen	188
	− 30 + 80 mesh	8.1	628	Helium	728
	− 30 + 80 mesh	8.1	628	Hydrogen	838
	− 80 mesh	8.7	10^{-4}	—	5.8
	− 80 mesh	8.7	628	Nitrogen	202
	− 80 mesh	8.7	628	Helium	780
	− 80 mesh	8.7	628	Hydrogen	840
	− 30 mesh	8.7	10^{-4}	—	5.8[b,c,d]
Vermiculite	10-14 mesh		760		260 × 10^{-4}
		15.0	10^{-4}	—	9
Diatomaceous earth	1 − 100 μm	15.6	10^{-4}	—	8
		18.0	10^{-4}	—	5.8
Alumina, fused	− 50 + 100 mesh	125	10^{-4}	—	10.5
Alumina, laminar	0.1 − 10 microns	4.4	10^{-4}	—	13.3
	− 20 + 30 mesh 30%	9.4	10^{-4}	—	10.5
Mica, expanded	− 30 + 15 mesh 70%	9.4	628	Nitrogen	290
	− 30 + 15 mesh 70%	9.4	628	Helium	867
Lampblack	—	12.5	10^{-4}	—	7.2
Charcoal peach pits	20-30 mesh	30	10^{-4}	—	10.5
Carbon + 7% Ash[f]	30% 44 μm	12.5	10^{-4}	—	3.5
Calcium silicate (synthetic)	0.02 μm	10.6	10^{-4}	—	4.3
	0.02-0.07 μm	22.5	10^{-4}	—	3.3
	0.02-0.07 μm	22.5	628	Nitrogen	263

[a] 1.2 × 10^{-4} · Btu/h·ft·°R for boundary temperatures of 137 and 36°R
[b] 3.8 × 10^{-4} · Btu/h·ft·°R for boundary temperatures of 540 and 36°R
[c] 2.1 × 10^{-4} · Btu/h·ft·°R for boundary temperatures of 137 and 36°R
[d] 0.46 × 10^{-4} · Btu/h·ft·°R for boundary temperatures of 137 and 7.6°R
[e] Barometric pressure at Boulder, Colorado
[f] Mostly SiO_2 and Al_2O_3

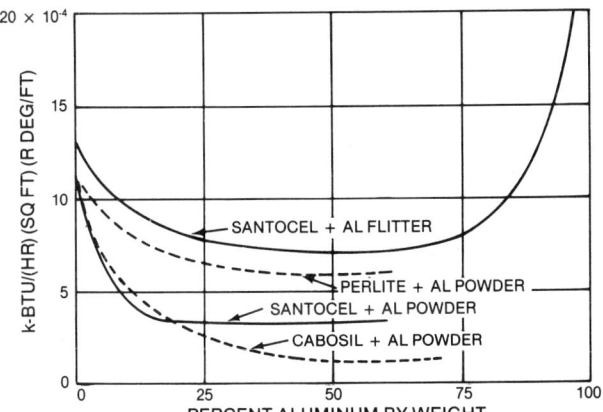

Fig. 29 Apparent Mean Thermal Conductivity Between 540 and 137°R of Evacuated Powders with Added Aluminum

Evacuated fibrous insulations have performance levels comparable to evacuated powders, but usually they are more expensive and require higher vacuum levels to attain their best performance. Both their cost and sensitivity to vacuum level are generally related to fiber size and density; better performance and higher cost are usually associated with small fiber diameter and high density. Typical performance of evacuated fibrous systems is shown in Figure 30.

Nonvacuum Insulations

Nonvacuum insulations may consist of powders, fibers, or foams. Heat conduction in such insulations is primarily because of the interstitial gas and is roughly the thermal conductivity of the gas. For example, the mean thermal conductivity in the range of −323 to 80°F is 0.084 Btu/(h·ft·°F) for a perlite powder containing hydrogen gas at atmospheric pressure. The corresponding value for hydrogen itself is 0.072. Similarly, perlite containing helium is 0.073, and a sample of glass fiber containing helium is 0.077, while the corresponding value for helium gas is 0.063. Again, per-

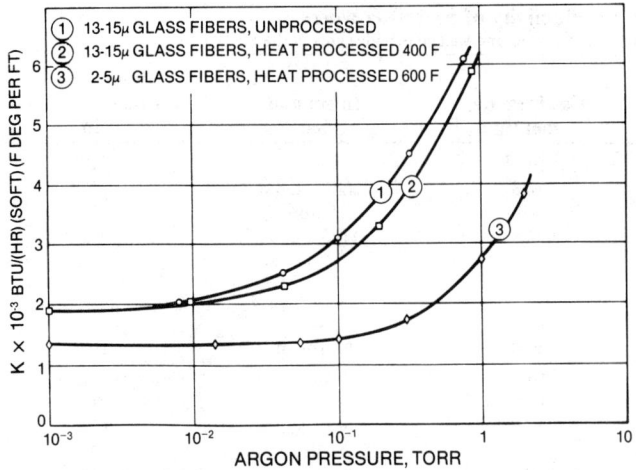

Fig. 30 Thermal Conductivity of Glass Fiber Mats under 750 mm Hg Mechanical Load

lite containing nitrogen at atmospheric pressure is 0.019, the glass fiber containing nitrogen is 0.015, while nitrogen itself is 0.010. In the same temperature range, air-filled Styrofoam is 0.015, while air is 0.010.

Among the powders, silica aerogel has superior insulating value when unevacuated because of its extremely small effective particle size.

Rigid Foam Insulations. Of most interest in cryogenic applications are the more or less closed-cell foams of polystyrene, epoxy, polyurethane, rubber, silica, and glass. The thermal conductivities of some selected samples are shown in Table 9. In general, the thermal conductivity of a foam is determined by the interstitial gas contained in the cells, plus internal radiation and a contribution because of solid conduction. Evacuation of most foams reduces the apparent thermal conductivity, thus showing the partial open-cell nature of most of these materials (Kropschot 1959).

Over time, many CFC-blown foams will be permeated by air, which may increase their conductivity by as much as 30%. If hydrogen or helium is allowed to permeate the cells, the conductivity can be increased by a factor of three or four. Glass and silica foams appear to be the only ones that have fully closed cells. Note that even under evacuation, foam insulations have much higher conductivities than the other types of insulation previously discussed.

Table 9 Apparent Mean Thermal Conductivity of Selected Foams

Foam Type	Density, lb/ft^3	Boundary Temperature, °R	Test Space Pressure, mm Hg	Conductivity, Btu·in / (h·ft^2·°R)
Polystyrene[a]	2.4	540/137	760	0.23
	2.9	540/137	760	0.18
	2.9	137/36	10^{-5}	0.056
Epoxy resin[b]	5.0	540/137	760	0.23
	5.0	540/137	0.01	0.12
	5.0	540/137	0.004	0.090
Polyurethane[c]	5 to	540/137	760	0.23
(isocyanate)	8.5	540/137	0.01	0.083
Rubber[d]	5.0	540/137	760	0.25
Silica[e]	10.0	540/137	760	0.38
Glass[f]	9.0	540/137	760	0.24

[a] Dow Chemical Co. *Styrofoam*
[b] Debell & Richardson *Du Ra Foam*
[c] Nopco Chemical Co. *Lock Foam*
[d] U.S. Rubber
[e] Pittsburgh Corning *Foam Sil*
[f] Pittsburgh Corning *Foam Glass*

STORAGE AND TRANSFER

Cryogenic fluids require more elaborate storage than fluids boiling near ambient temperature. For example, propane and LPG can be stored in carbon steel containers without insulation. At 70 °F, propane has a vapor pressure of 110 psig. More expensive materials of construction are required when the stored fluid is below −40 °F. For fluid temperatures down to about −320 °F, or liquid nitrogen at normal boiling temperature, some type of insulation is required to minimize boiloff losses. Liquid hydrogen and helium below −320 °F requires a vacuum system or its equivalent to prevent air condensation and resultant high boiloff rates.

To minimize losses of fluids boiling below −320 °F that are handled in small quantities, a liquid nitrogen shield is used on containers having high vacuum insulation. In larger sizes, multilayer insulation is a less expensive alternate to the liquid nitrogen shield for liquid hydrogen and helium containers. Table 10 shows a classification for cryogenic storage according to fluid, size, insulation, and application. This classification is only approximate, and economics must be considered in cases where overlap may occur.

Although normal criteria for line sizing apply in the transfer of cryogenic fluids, a designer must also consider cooldown and heat leak losses, the behavior of the construction materials at cryogenic temperatures, and the servicing required to maintain good performance. The selection of insulation for the transfer system is determined on the basis of flluid properties, transfer rate, line lengths, duty, and service. For example, a noninsulated line is normally used for liquid oxygen or nitrogen transfer at high rates, over relatively short distances, and for short durations. For liquid helium transfer, however, multilayer insulation and short transfer lengths are essential.

Dewars

Dewar vessels vary in size from less than a quart of volume to several thousand gallons. The small Dewars for liquid oxygen and nitrogen are vacuum-jacketed and constructed of glass, glass fiber-reinforced epoxy, copper, or stainless steel. Figure 31 shows the basic designs of small-capacity Dewars. Part A of this figure shows the simplest design, which ranges in inside diameter from several inches to about 4 ft and in inside depth from less than a foot to

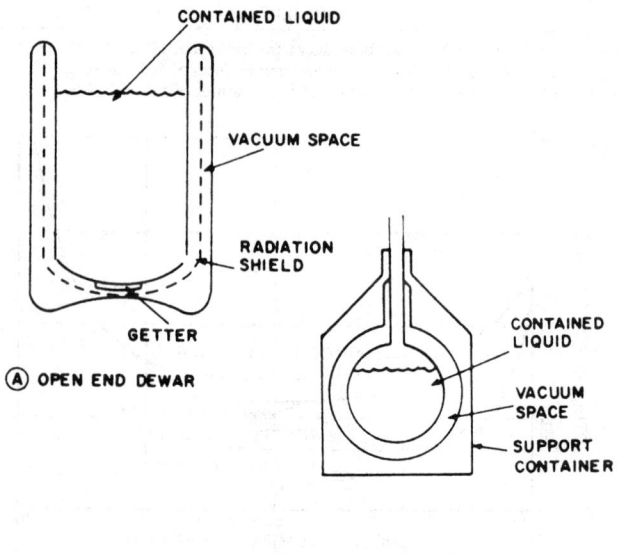

Fig. 31 Storage Dewars for Liquid Oxygen or Nitrogen

Cryogenics

Table 10 Classification of Cryogenic Storage

Fluid	Size, gal	Insulation	Application
Liquid oxygen or nitrogen	Up to 0.026	Vacuum	Use
Liquid oxygen or nitrogen	Up to 0.042	Multilayer	Transport and use
Liquid hydrogen, helium, or neon	Up to 0.040	Vacuum and liquid nitrogen shield or multilayer	Transport and use
Liquid oxygen, nitrogen, or argon	Up to 7000	Evacuated powder	Transport
Liquid oxygen or nitrogen	Up to 13,000	Evacuated powder	Production and use
Liquid oxygen or nitrogen	Greater than 13,000	Powder	Production and use
Liquid argon	Up to 50,000	Evacuated powder	Production and use
Liquid hydrogen	Up to 30,000	Multilayer	Transport
Liquid hydrogen	Up to 850,000	Evacuated powder	Production and use
Liquid helium	Up to 9000	Multilayer	Transport and use
Liquid helium	Up to 30,000	Multilayer	Production
Liquid natural gas	Up to 12,600,000	Powder with nitrogen purge	Production (above ground)
Liquid natural gas	lUp to 10,500,000	Powder or foam with nitrogen purge	Sea transport

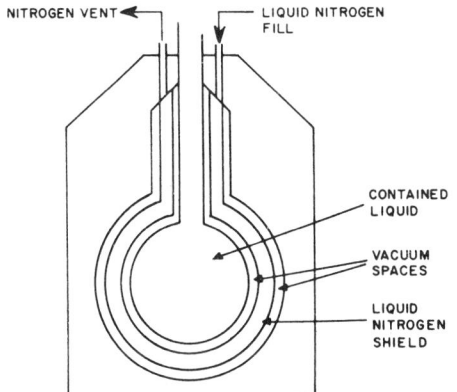

Fig. 32 Storage Dewar for Liquid Hydrogen or Helium

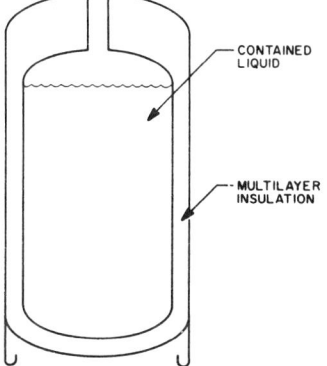

Fig. 33 Shipping Dewar

more than 10 ft. In large sizes, these Dewars are made of stainless steel, and a floating radiation shield is included to decrease the heat leak. Also included in the larger sizes is a getter to maintain the sealed vacuum. This is important because of the outgassing of noncondensables, such as hydrogen, from the metal surfaces.

Dewars are useful for transferring liquid nitrogen and oxygen from larger stationary storage into test apparatus. Their design may also include a flange, which permits bolting them to a top plate enclosure. The Dewar may be used as the enclosure for a cryostat where the low-temperature refrigeration heat exchange surface or test apparatus is supported from the top plate. For the larger capacities (up to 26 gal), having lower boiloff rates, the design shown (the LOX-LIN Dewar design) in Part B of Figure 31 is used. Liquid hydrogen and helium may be stored in the design shown in Figure 32. The liquid nitrogen shielding reduces the radiation heat leak to the hydrogen or helium, which would otherwise boil off at a much higher rate.

Figure 33 shows a multilayer insulated Dewar with a capacity of 26 to 66 gal. These containers have a diameter of 20 in. and a length of 60 to 80 in., and are used for liquid nitrogen, oxygen, natural gas, argon, hydrogen, and helium. Ruggedly constructed with a stainless steel inner shell and a carbon steel outer shell, Dewars are intended for shipment via common carrier or for permanent installation on a vehicle to provide LNG for fuel or LIN for in-transit refrigeration of food. They can be moved directly to a point of use via hand truck and can have a heavy-walled inner shell to supply use-point pressures up to 200 psi.

Stationary Storage

The simplest and least costly container for cryogenic liquids consists merely of an inner shell of nickel steel, stainless steel, or aluminum, enclosed by a thin carbon steel outer shell. The space between the outer and inner shells is filled with a powder or fiber-type insulation, such as perlite or rockwool, to insulate the inner shell contents. This design is commonly used for liquid oxygen and nitrogen storage containers of greater than 13,000 gal capacity and eliminates the need to construct a heavy outer shell to support loads because of an evacuated insulation space. The lower heat leak, evacuated-powder insulated construction is preferred in smaller sizes because its higher surface-to-volume ratio yields higher rates of heat in-leak per unit of contents. Evacuated-powder insulated containers are shop-fabricated to minimize cost. They are thus limited to 13,000 gal by normal shipping clearances, but tanks to 50,000 gal are shippable with special handling.

Liquid hydrogen storage containers range in capacity to almost 1-million gal. Evacuated powder is used on tanks larger than 50,000 gal, which must be field-erected. Smaller, shop-fabricated tanks may also use evacuated powder, unless the more expensive but lower heat-leak multilayer insulation can be justified. The factors favoring multilayer insulation are the lower temperature, higher volatility, and higher cost of liquid hydrogen, compared with liquid oxygen and nitrogen. On the other hand, many liquid hydrogen storage tanks supply gas to a use point, and vent losses can usually be used.

For the storage of liquid helium, the economic justification for high vacuum or multilayer insulation plus liquid nitrogen or vent gas shielding is unquestionable. Liquid helium is not only expensive, but it is more volatile than liquid hydrogen. Only 9.2 Btu are required to vaporize one gallon of liquid helium, compared with 11.4 Btu for liquid hydrogen and 544 Btu for LIN. Since liquid helium cannot normally be shipped and stored without vent losses, multilayer insulated vessels can be designed to recover refrigeration from the vent gas by using this cold gas to chill the insulation and reduce heat leakage. Without this vent gas shielding or high vacuum insulation with liquid nitrogen shielding, liquid helium cannot be retained for more than a few hours.

Transport

A single trailer design is normally adaptable for the transportation of liquid nitrogen, oxygen, or argon. Trailers with capacities up to 7000 gal have been constructed. Highway load restriction has established this as the maximum capacity. For servicing receiver tank pressures up to 250 psi, a specially designed cryogenic centrifugal or turbine pump is used. The pumps may be installed on the trailer or on the ground with each receiver tank, many of which are low pressure (15 psig or less). A simple pressure transfer is used for servicing such tanks, using a built-in self-pressurization coil on the trailer to maintain a pressure differential between the trailer and receiver.

Cryogenic trucks with capacities up to about 2000 gal are widely used to service smaller receiver tanks, using an on-board cryogenic pump. Some trucks may have a second, high-pressure pump and vaporizing system for refilling gas storage systems up to 3000 psi. Railroad tank cars with capacities of up to about 20,000 gal are also in use to transport liquid nitrogen, oxygen, and argon. Trailers, trucks, and tank cars use evacuated-powder insulation.

Liquid hydrogen is transported in trailers up to 14,000 gal capacity, and in railroad tank cars up to 30,000 gal. Multilayer insulation is almost always used. The low mass of liquid hydrogen allows the design of transports with capacity limited by size restrictions, not weight limits. Liquid hydrogen is usually transferred by pressure difference, using a self-pressurization coil. A few trailers are equipped with pumps for servicing higher pressure receiver tanks. These pumps are expensive, because of the low density and high volatility of liquid hydrogen.

Liquid helium is generally shipped in portable Dewars of up to 26 gal capacity. A small number of larger transportable containers and trailers up to 9000 gal capacity have been built and are used for domestic or overseas shipment from producing plant to distribution stations, where smaller Dewars are filled for shipment to the user. Liquid helium is transferred by differential pressure. Usually, the normal pressure buildup in the supply vessel is sufficient. If an external source of helium gas is needed for additional pressure, the liquid-gas interface should not be disturbed. Improper techniques can convert the entire liquid contents to gas in a brief period. Even the best techniques cause a substantial percentage of the transferred liquid to change to gas. When large volumes are transferred, loss recovery techniques are employed to compress gas into storage.

Transfer of Servicing

For the transfer of liquid oxygen and nitrogen, a foam glass block insulation of vacuum-jacketing has been used. For high transfer rates over short durations, uninsulated lines are usually used. The economical breakeven point between no insulation, foam insulation, and vacuum-jacketing depends on transfer rate and quantities, transfer distances, liquid cost, and cooldown and maintenance requirements.

As the multilayer line fabrication becomes less expensive, its application for liquid oxygen and nitrogen service is more accepted in cases of higher product loss costs. Transfer lines for liquid hydrogen service normally use the vacuum-jacketed design or the multilayer insulation in conjunction with the vacuum jacket. Liquid helium requires the use of either multilayer insulation or liquid nitrogen shielding and vacuum-jacketing of lines to accomplish economic transfers.

One of the larger problems with cryogenic vacuum-jacketed transfer line design concerns the contraction of the inner line on cooldown. In the past, stresses have been relieved by expansion loops every 300 ft, by outer line bellows, or by a combination of the two. The use of invar for the inner line material has decreased or eliminated the requirement for the expansion bellows. The inner line is allowed to become stressed between two anchor points, since the invar expansion coefficient is low, relative to the materials previously used.

The optimum line size for the transfer of liquid hydrogen and helium at low pressures is usually the size that gives close to the minimum storage transfer pressure. This is because of the costly storage involved and the higher losses associated with higher transfer pressures. For liquid oxygen and nitrogen, a transfer pump is generally sized to overcome the maximum differential pressure, plus a nominal head for the line pressure drop.

INSTRUMENTATION

Except for in-line devices, cryogenic plants use essentially the same instruments as most chemical processes. Cryogenic fluid in the sensing leads that directly operate controllers, transmitters, and switches is allowed to increase to atmospheric temperatures through warmup loops before entering the elements. Occasionally, when analyzing some gas mixtures from the liquid phase, a separator must be provided to assure no fractional distillation of the components when they change from the liquid to the warm gas phase.

Flow measurement of cryogenic liquids also requires special consideration. For both positive displacement meters and variable head meters, the measurement depends on exact, fixed dimensions. At low temperatures, thermal contraction presents clearance and lubrication problems with positive displacement meters; it also causes an area correction factor of less than 1.0 on variable area meters. Materials such as Monel or 304 stainless steel, which have lower expansion factors, are often selected for this service.

Pressure drop across an orifice plate can cause flashing if the liquid is not sufficiently subcooled before the orifice. Accurate temperature measurements are essential, since the density of some cryogenic fluids varies considerably with temperature. Many low-temperature primary measuring devices, *e.g.*, sonic probes and resistance thermometers, require insertion through vacuum-jacketed vessels. Special electrical connectors that provide a good vacuum seal at liquid helium temperatures are available.

Temperature Measurement

For normal commercial cryogenic temperature measurement, three types of thermometers are widely used: (1) thermocouples; (2) resistance bulbs; and (3) vapor pressure bulbs.

Thermocouples are the most frequently used temperature measuring device in the -350 to $0°F$ range. They are relatively inexpensive and have low heat capacity, thus giving good response characteristics. Their small size lends itself to construction into many shapes and forms, thus allowing their use in many difficult locations. Except in galvanometer-type meters, thermocouples are

Cryogenics

usually connected to the bridge circuit of a potentiometer, which allows readings to be taken at a considerable distance from the installation.

The EMF produced by thermocouples below $-350\,°F$ becomes quite small, making them impracticable below this range, unless special precautions are taken. In addition, inhomogeneities in the wire can cause errors because of local thermocouple effects.

Copper-constantan is used in most thermocouples for moderate cryogenic temperature. With premium grade thermocouple wire, and with a -300 to $-75\,°F$ temperature range, the limit of error is $\pm 1\%$ of the temperature being measured. For low-temperature measurement, No. 20 B&S wire gauge size is recommended. In cold boxes, thermocouples are soldered to the pipe to be measured, with an insulating piece installed to provide a heat barrier at the junction.

Thermocouples using gold, 2.1% atomic cobalt, and copper are also used. This type of thermocouple has the advantage of higher thermoelectric power, but gold tends to be less homogeneous than most other materials. Normally, thermocouples follow standard temperature versus millivolt tables. Calibration deviation can be checked with the bulb and the table.

Resistance thermometers of the older, pure metallic type (mostly platinum) are used to measure temperatures in the cryogenic range of $36\,°R$ to ambient. Platinum is chosen particularly because of its chemical inertness, good workability, high purity, and because of the availability of test data. It is limited at the lower range because of its insensitivity and low resistivity. The sensing element is essentially a coil of fine wire wound around a frame of insulating material and carefully supported so that it will not be subject to the mechanical strain caused by differential thermal expansion.

Platinum resistance thermometers are ideally suited for differential temperature measurement and for temperature control when the span is less than $9\,°R$ and the accuracy requirements are around $0.1\,°R$. The semiconductor resistance thermometer gives good results from 2 to $180\,°R$. Semiconductors have negative temperature coefficients below approximately $27\,°R$, giving them higher sensitivity than platinum, and, used with standard indicators and recorders, need only a precision power source. At low temperatures, the conduction process in semiconductors varies with different temperature ranges. This makes calibration difficult and is therefore used most frequently below the platinum range of $36\,°R$.

The vapor pressure thermal system consists of a pressure measurement element, connected by capillary tubing to a temperature-sensing bulb. These systems are relatively sensitive in the cryogenic ranges and function with good reproducibility. Construction of the bulb depth is critical to prevent convection. Extra care is needed in system design to assure that the lowest temperature is in the bulb. Impurities in the sensing fluid can also produce errors. In measuring liquid hydrogen, a catalyst must be inserted in the bulb to assure a conversion from ortho to para hydrogen. Vapor pressure thermometers are frequently used to check thermocouples and resistance bulbs at the normal boiling points of cryogenic liquids.

Liquid Levels

Liquid levels at low temperatures are often measured with standard instrumentation, but certain techniques must be followed to get satisfactory results. On cryogenic tank storage, the familiar oil tank system of a float connected to a steel line and then to a perforated tape, to drive a local or remote readout device, is a standard application. Suitable materials are used in the parts within the tank. Design must be such that there is no freezeup of the float system where it comes through the tank roof. Levels of storage tanks and of various vessels within the cold box use the common differential pressure meter. The principle used with cryogenic fluids is that both top and bottom meter legs are filled only with gas at the level taps to the meter, and the difference in head is the mass of the fluid.

On low boiling point fluids, the lines are run so that as soon as the liquid leaves the vessel, the heat leak quickly produces a change to the gaseous state at that elevation. With high boiling point liquids, particularly those composed of some higher boiling constituents, fractional distillation can take place at the point of heat leak and cause a buildup of liquid in the lower leg; this will produce an erroneous reading. The density of the vapor of higher boiling point liquids can be significant if the system is operated at higher pressures. Considerable care should therefore be taken when designing and installing liquid level legs. Back-purging with a warm lower boiling point gas is recommended on difficult installations.

Capacitance probes have been used for both single-point (on-off) liquid level measurement and continuous (analog) measurement. Typical construction for a continuous reading system consists of a concentric set of tubes, each of which is a plate of a capacitor. As the temperature considerably affects the dielectric constant of cryogenic fluids, it is necessary to include a dielectric constant cell below the main section of the tube that forms the reference capacitor, to make automatic correction for dielectric changes. The capacitor operates on a 400 Hz current, working on a self-balancing bridge circuit. Readout is obtained on any standard potentiometric indicator or recorder. The simpler point contact probe for on-off service requires no compensating capacitance or balancing bridge circuit.

Ultrasonic probes are basically used for point sensing of cryogenic liquids. An electrical signal from an amplifier external to the probe drives a diaphragm, which is the face of the probe. This motion in the ultrasonic range (above 20,000 Hz) is feedback to the amplifier and operates a control relay sensitive to the amount of oscillation. When the face of the probe becomes immersed in the liquid, the damping effect reduces the feedback oscillation and thus trips the control relay. This system can be used for high or low alarms or to control and measure tank filling and emptying cycles.

Displacer-type liquid level sensing devices have been used in cryogenic liquids. These instruments operate on the principle that the buoyant force acting on a cylindrical displacer reduces the apparent mass of the displacer in proportion to the amount of its submergence in the liquid phase. Standard instruments designed for use in room temperature liquids may be used without modification in heavier cryogens such as liquid nitrogen. For liquid hydrogen or liquid helium applications, the standard instrument should be modified to improve its sensitivity in these low-density fluids. The output of the instrument may be a standard pneumatic or electrical signal or a proportional control signal.

SAFETY

Processing and handling cryogenic fluids requires special safety precautions. Hazards occur because of the temperature levels, the toxic or flammable nature of the fluids, their lack of compatibility with common materials of construction, interaction between the cryogens and impurities found in cryogenic mixtures, and composition changes caused by vaporization of liquid mixtures.

Cryogenic liquids are so cold compared to the human body that contact between these liquids and the skin can cause severe cold burns, which are more painful and harder to treat than hot burns. Unprotected skin contacting very cold surfaces (such as pipelines or vessels containing cryogenic liquids) may stick fast, and the flesh may be torn in removal. When handling these liquids, protective gloves, face shields, and complete leg and arm coverings should be used. Open or porous shoes should not be used, and safety shoes are recommended. Even exposure to cold vapor cloud may be dangerous if intense enough for the body temperature to drop.

Table 11 Properties of Cryogenic Fluids

Gas	Boiling Point, °R	Volume Expansion to Gas[a]	Flammability Limits in Air, Mole %	Toxicity	Specific Gravity, (Air = 1)	Odor
Helium-3	5.8	470/1			0.137	
Helium-4	7.6	757/1			0.137	
Hydrogen	36.7	851/1	4.65-94%		0.06952	
Deuterium	42.5	976/1	yes	Radioactive		
Tritium	45.2	—	yes	Radioactive		
Neon	48.8	1438/1			0.6964	
Nitrogen	139.3	696/1			0.6970	
Carbon monoxide	146.9	—	12.5-74.2	High	0.6978	
Fluorine	153.0	957/1		High	1.312	Sharp
Argon	157.1	842/1			1.38	
Oxygen	162.4	860/1			1.1053	
Methane	201.1	648/1	50-15.0	Slight	0.5544	
Krypton	215.6	693/1			2.818	
Refrigerant 14 (CF_4)	261.4	446/1		Moderate	1.62	
Ozone	190.3	—	yes	High	1.658	Sulfurous
Xenon	295.2	573/1			4.53	
Ethylene	304.7	487/1	2.7-40		0.978	Sweet
Boron trifluoride	310.9	—		High	2.37	Pungent
Nitrous oxide	330.5	666/1		Moderate	1.530	Sweet
Ethane	332.6	436/1	3.0-12.5	Slight	1.047	
Hydrochloric acid	338.4	—		High	1.268	Pungent
Acetylene	340.4	—	2.5-80	Slight	0.9073	Garlic
Fluoroform	340.4	—		Moderate		
1,1-Difluoroethylene	342.0	—	yes	Slight		Faint ether
Refrigerant 13 (CF_3Cl)	344.9	—		Slight		Mild
Carbon dioxide	389.7[b]	641/1		Slight	1.5829	Slight pungent

[a] Gas at 529°R and 1 atm pressure.
[b] Triple point.

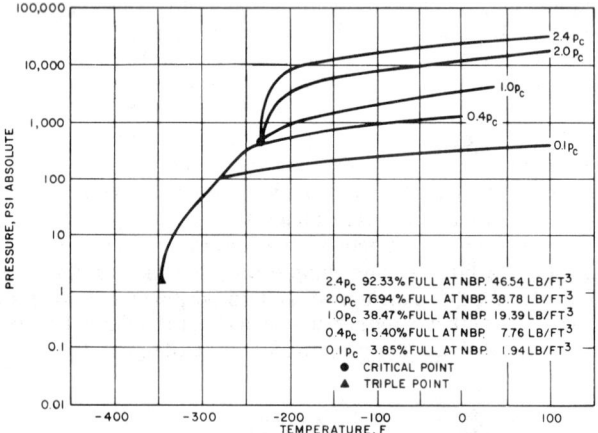

Fig. 34 Pressure Developed on Warming Confined Liquid Nitrogen
(Courtesy of Air Products and Chemicals, Inc., Allentown, PA)

The low temperature levels also ensure that heat leaking into the liquid container will keep the cryogen boiling under whatever pressure is kept in the vessel. If the vessel is not vented, large pressures can build up and may rupture the container as a result of liquid vaporization from heat in-leakage. Figure 34 shows the pressure buildup over a closed tank of liquid nitrogen (Vance and Duke 1962). Similar pressure buildups are found for all cryogenic liquids. Since vapors escaping from a cryogenic vessel are cold, it is possible for a vent line to become clogged with ice, causing pressures to develop even in vented tanks.

Heat leakage and vaporization of cryogenic fluid trapped within valves, fittings, and in sections of piping could cause excessive pressure buildup and rupture of the equipment. For example, liquid hydrogen expands about 800 times when warmed to room temperature. The maximum should be limited through proper equipment design and pressure relief devices such as safety valves and burst disks.

Table 11 lists the toxicity and flammability characteristics of common cryogens. Even though some gases are nontoxic, none of the gases except oxygen will support life. Most cryogens are odorless and hence give no warning of their presence. Thus, vessels filled with cryogenic fluid may be hazardous because of: (1) the toxic effect of the residue vapor; (2) the inability of a worker to tolerate such vapors as nitrogen or helium; or (3) the flammability of the vapor mix (as with hydrogen or methane). Important safety considerations for handling such gases include adequate ventilation, monitoring the atmosphere, a personnel watch, and planned corrective action to be taken in the event that low oxygen levels are encountered.

Combustion in any system requires a fuel, an oxidant, and a means of ignition. In oxygen-containing systems, combustion is possible with almost any fuel, including many metallic and nonmetallic materials suitable for industrial use. This is of special concern to designers and users of vessels, piping, and machinery for handling oxygen. Safety considerations in the design of oxygen systems include attempts to prevent ignition and to prevent the propagation of combustion if initiated. Metallic and nonmetallic materials vary in their susceptibility to ignition and combustion. Many common contaminants such as oil, rust, cloth, paper, tobacco, wood, paint, and metal chips are easily ignited in oxygen and may burn sufficiently to increase the temperature so as to ignite more stable materials such as low carbon steel, stainless steel, or aluminum.

Contaminants in cryogenic systems also result in functional interference with components in the system by particle deposits on moving parts, valve seats, system sensors, and controls. All cryo-

Cryogenics

genic systems must be subjected to specific purging and cleaning procedures.

Prior to the loading of cryogenic vessels and working on such systems, they must be made inert. This can be accomplished by vacuum, pressure, or flowing purge. Vacuum purge is most satisfactory, because it requires fewer operations and eliminates any fuel pockets. It is accomplished by venting the systems to atmosphere, evacuating to a relatively low pressure, repressurizing the system with inert gas to a positive pressure, and again venting to atmospheric pressure. The volume of a gas is inversely proportional to the pressure, and the pressure of a mixture of gases equals the sum of its parts.

If a gaseous H_2 tank is vented to atmospheric pressure, it will contain 1 atm or volume of gas. If the tank is reduced to 10 mm Hg of pressure and then repressurized to 3 atm with an inert gas (N_2), the tank will contain:

$$\frac{10 \text{ mm Hg}}{760 \text{ mm/atm} \times 3 \text{ atm} - 10} \times 100 = 0.44\% \text{ H}_2$$

For hydrogen, since the explosive limit in air ranges from 4 to 75%, the system can be assumed to be inert. Should entry into the vessel be required, the vessels can be thoroughly purged with air, without any combustion occurring. The O_2 content should be raised above 19.5% by volume before entering.

A pressure purge is less satisfactory than a vacuum purge. It requires alternate pressurizing and venting of the system until a safe atmosphere is reached. The method of determining flammable gas content is similar to vacuum purge. Raising the tank pressure to 3 atm with N_2 in the initial cycle gives a VH_2 of 33.3%; the second cycle would be $VH_2 = 33.3/3 = 11.1\%$; and the third, $11.1/3 = 3.7\%$. While a flowing purge is the simplest method, it is the least satisfactory because it provides no positive assurance that a completely inert atmosphere has been attained. Purging configurations should assure turbulent flow of the purging gas; flow rates should be high enough to thoroughly purge all parts of the system. The importance of using safe procedures to purge and render the system inert can be obtained from a study of the flammability limits of gas mixtures.

Figure 35 shows the flammability limits of the system $CH_4-N_2-O_2$ at or near atmospheric pressure (Vance and Duke 1962). The crosshatched area includes all compositions in which the mixture is flammable. At other compositions, combustion does not raise the gas temperature high enough to sustain combustion. The line drawn from 21% O_2/79% N_2 binary to pure CH_4 represents the compositions that can result from mixing CH_4 with air. Note that flammability occurs from about 5 to 15% CH_4 in air. If an LNG tank were to be purged with air, the gas mixture resulting would at first be too rich in CH_4 for combustion. However, as purging continues, flammable mixtures would develop, resulting in a hazardous condition. Data on flammability limits may be found in the literature and interpreted in the light of the behavior shown here.

Organic contaminants, which may enter with a cryogenic liquid, are usually frozen and may float or sink in liquid. Thus, they can collect at low or high points in a system. If this occurs, an explosive mixture can be generated with oxygen when the equipment is warmed up and both the liquid and the deposited solids are vaporized. This has been the cause of many explosions in the cryogenic areas of air separation plants. In these units, hydrocarbons, especially acetylene, from the compressor oil or from the atmosphere can enter the system in trace quantities and must be removed by adsorption or by venting, especially if they are present in concentrations greater than their solubility in the cryogenic liquid.

The release of cryogenic liquids and vapors must be carefully controlled. Although many cryogenic liquids produce vapors of lower molecular weight than air, the lower temperatures result in

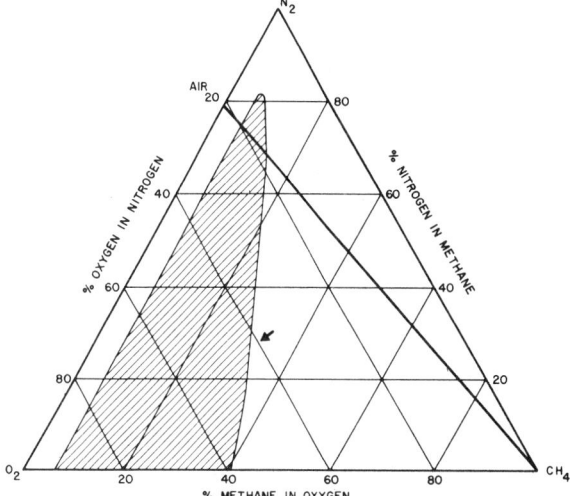

Fig. 35 Flammable Limits of $CH_4-O_2-N_2$
(Courtesy of Air Products and Chemicals, Inc., Allentown, PA)

a more dense fluid. The vapors travel along the ground and collect in low places. Exposure to these vapors is hazardous. Exposure to oxygen vapors could result in clothing or any equipment with oil-lubricated parts becoming oxygen enriched. Incidents of fire have occurred in many such cases. Any personnel exposed to such vapors should ventilate their clothing for at least 20 min. before exposing themselves to a source of ignition. Open flames should not be permitted where oxygen enrichment exists.

Venting for cryogenic vapors should be designed to provide for sufficient dilution of the vapors, making them safe to personnel or operating equipment. Safe distance depends on the vent direction, amount of flow, liquid or gas, wind direction, terrain, and other factors. For oxygen, about 50 ft is a recommended safe distance from the vent to equipment where large quantities are vented.

The disposal of flammable cryogens is best accomplished in a burnoff system in which the liquid or gas is piped to a remote area and burned with air in multiple burner arrangements. Such systems should include pilot ignition methods, warning systems in case of flameout, and means for purging the vent line. Small quantities may be vented unburned from a single vent, which must be released at least 15 ft above a roof peak; the amount depends on the wind direction and velocity, distance from dwellings, and the like.

Should a fire occur at the vent or at other locations, the fuel supply should be shut off. No attempt should be made to put out a hydrogen or LNG fire while it is still being supplied with fuel. If a hydrogen fire is extinguished and the flow of hydrogen is not stopped, a hazardous combustible mixture may start forming at once. The mixture will almost certainly be ignited and there will be an explosion that will cause more damage and restart the fire. Water should be used to keep the metal parts cool until the fire burns itself out.

Spills with liquid hydrogen and LNG (two cryogens produced in bulk) result in a vapor blanket including zones of combustible mixtures that could ignite the entire mass. Both these fluids burn clean; H_2 gives a nearly invisible flame. Compared to a flammability limit of 2 to 9% (by volume) for jet fuel, in air, hydrogen has a flammability limit of 4 to 75% and LNG from 5 to 15%. The ignition of explosive mixtures of H_2 with oxygen or air occurs with low energy input; about one-tenth that of a gasoline-air mixture. All ignition sources should be eliminated, and all equipment and connections should be grounded. Lightning protection in the form of lightning rods, aerial cables, and ground rods, suitably connected, should be provided at all preparation, storage, and use areas for flammable cryogenic fluids.

Knowledge of the mechanical properties of materials at low temperatures is necessary for the design of cryogenic equipment. The low temperatures limit the number of materials that can safely be used, since many materials become brittle as the temperature is lowered. This is especially true of some iron alloys where a transition occurs from ductile to brittle failure modes. The temperature level and range over which this transition occurs depends on the metal composition, its treatment history, and its application conditions. On the other hand, materials without this transition are likely to have greater strength at cryogenic temperatures than at ambient. However, since any system built to operate with cryogens will be warmed to ambient from time to time, the design must permit safe operation at all temperatures from ambient to the lowest temperature possible for the operation.

Seals are a particularly difficult problem where cryogens are to be used. Most materials to be used as sealants, gaskets, and packing act as fuels in an oxygen atmosphere and/or become brittle at low temperatures. Moreover, the wide operating temperature range results in large dimensional changes or stresses in the materials. Even welds cause difficulties because of the composition variations that occur at the edges of the weld and because of the problem of controlling weld heating and cooling rates.

Although plastics have served well in some cryogenic systems, they must be used with care. Some polymers become brittle at low temperatures, whereas others tend to flow when cold. In every case, the problem of relative expansion and contraction rates between the polymeric material and a metal joined to it must be solved. Finally, many materials are attacked chemically by cryogens, especially oxygen and hydrogen. Data should be obtained on the suitability of each material of construction for the temperature range and the specific cryogen to be used before it is used in a cryogenic system.

Engineers entering this field should proceed cautiously, checking the cryogenic literature for guidelines on good design and operating practice. In every case, safe operation will require detailed design and operating consideration of all the possible hazards. The bibliography lists a few sources of information.

SPACE SIMULATORS

A space simulator produces an environment as close to actual space conditions as possible. Normally, the high vacuum and thermal properties of space are the primary simulation goals. Satisfactory methods of producing zero gravity conditions (or the high energy particle and electromagnetic radiation conditions of space) have not been achieved.

A nearly perfect vacuum is required for true space simulation. Table 12, which gives the presently accepted values for pressure at various altitudes, shows that pressures below 10^{-9} mm Hg are necessary to simulate altitudes above 420 mi. The vacuum in interplanetary space is 10^{-12} and 10^{-14} mm Hg.

To achieve such pressures in a space simulation chamber, extremely high pumping speeds, accompanied by a minimization of the gas leak rate into the chamber, must be provided. Materials used in chamber construction must be chosen for low outgassing (release of adsorbed gas under vacuum), since this may provide a major portion of the vacuum pumping load.

Space is a nearly perfect heat sink, equivalent to a black body at a temperature of about 7°R. This condition is simulated in a chamber by surrounding the test object with a cooled, absorptive sink, usually maintained near liquid nitrogen temperature. Radiation is then provided to simulate the sun's radiation, earth albedo, or other effects.

Table 12 Altitude versus Pressure Chart

Altitude, ft	Pressure, mm Hg
240,000	0.0253
260,000	0.00892
280,000	0.00291
300,000	9.5×10^{-4}
350,000	8.5×10^{-5}
400,000	1.6×10^{-5}
500,000	3.5×10^{-6}
600,000	1.5×10^{-6}
800,000	4.0×10^{-7}
1,000,000	1.3×10^{-7}
1,200,000	5.0×10^{-8}
1,500,000	1.4×10^{-8}
2,000,000	2.3×10^{-9}
2,100,000	1.7×10^{-9}
2,200,000	1.2×10^{-9}
2,300,000	8.8×10^{-10}

The satisfactory simulation of solar radiation, including a duplication of the solar spectrum, requires elaborate equipment; *i.e.*, carbon arcs of mercury xenon arc lamps, with complex optical systems for collimation and filtering. For many applications, the use of infrared lamps or tubular heaters placed within the chamber, close to the vehicle, is a satisfactory substitute. These can be programmed in various ways to simulate orbital conditions. When this type of heat source is used, a small reflector is placed behind them to collimate the beam partially. Correction of heat flux density is required because of change in emissivity with wavelength. Earth albedo can also be simulated with these lamps or with warm panels, which can be placed in suitable locations within the chamber.

Although high vacuums are required to simulate space conditions closely, thermal balance studies only require pressures less than about 10^{-5} mm Hg to eliminate heat transfer because of convection. Effects that can be observed at lower pressures include volatilization of various materials, lubrication problems with moving parts, and cold welding, which occurs below 10^{-9} mm Hg.

Chamber Construction

Chambers for space simulation are of welded construction, generally stainless steel or stainless-clad plate. External stiffeners are used to reduce the plate thickness of the tank; these can be mild steel. The interior is polished to reduce the area for outgassing and to reduce the radiant heat transfer. The data in Table 13 show the outgassing rates expected from both mild and stainless steel and the advantage of using stainless steel. Conventional fabrication and welding techniques are used on these chambers, but double welding is avoided to eliminate trapped gas pockets that provide virtual leaks that cannot be detected by ordinary methods. Reliable metallic or elastomer *O* ring seals are used up to large diameters. When double *O* rings are used, the space between them is usually evacuated to provide a guard vacuum for the seal.

When pressures below 10^{-8} or 10^{-9} mm Hg are required, bakeout provisions are included in the form of strip heaters and insulation. Baking the chamber at temperatures of 400 to 750°F removes adsorbed gases and reduces the outgassing rate of the metal walls. This, however, places certain limitations on chamber design, and generally requires cooling for elastomer seals, which are nonbakeable.

Vacuum Pumping

Vacuum pumping is accomplished by mechanical, diffusion, ion, or sublimation pumps.

Table 13 Outgassing Rate for Mild and Stainless Steel
(Giles 1965)

Material and Condition	1 h	10 h	100 h
Mild steel			
Degreased	1.7×10^{-6}	3.2×10^{-7}	6.2×10^{-8}
After 750°F bake	—	3.9×10^{-10}	—
Stainless steel			
As received	6.2×10^{-6}	6.2×10^{-7}	—
Degreased	9.5×10^{-9}	6.2×10^{-9}	2.9×10^{-9}
After 930°F bake	—	5.3×10^{-10}	2.6×10^{-13}

Note: The outgassing rate is in torr·ft³/(s·ft²) where 1 torr = 1 mm Hg. A torr·ft³ is a unit volume of gas corrected to one torr absolute pressure, or

$$\text{Total outgassing rate of material, ft}^3 = \frac{\text{torr} \cdot \text{ft}^3}{\text{s} \cdot \text{ft}^2} \times \frac{\text{Material area, ft}^2}{\text{System press, torr}}$$

Mechanical Vacuum Pumps. Rotary piston-type pumps and vane-type vacuum pumps are most commonly used to produce vacuum conditions within space simulation chambers. Single-stage units blank off at about 10^{-2} mm Hg and two-stage units at 10^{-3} mm Hg.

Systems with gas loads in the micrometre range often employ positive displacement-type blowers as boosters. These blowers, when backed by rotary or vane pumps, give a vacuum usable throughout the range from 10^{-2} to 10^{-3} mm Hg (Dowling et al. 1961).

Mechanical vacuum pumps are displacement machines that use a special oil as the sealant. They tend to collect any condensable gases in the crankcase, thereby limiting the ultimate pressure to the vapor pressure of the contaminants. Gas ballast is frequently used to help discharge moisture from rotary piston pumps. Crankcase heaters may also be useful. When heavy concentrations of condensables are to be pumped, cold traps (condensers) are often used.

Diffusion Pumps. These are a form of booster pump using a special oil and jet assembly. The largest size commonly used on space chambers has a pumping capacity of 1800 cfs. Diffusion pumps are usually installed around the periphery of, and close to, the chamber to maximize pumping speed.

For best results and lowest ultimate pressures, a single refrigerated baffle or series of baffles must be placed between the diffusion pump and the chamber to minimize oil backstreaming. The baffles are generally cooled with water or liquid nitrogen.

Ion and Sublimation Pumps. Ion pumps, or getter-ion pumps, rely on the gas-absorbing power of nascent metal films. This is enhanced by ionization of the pumped gases. The metal film is formed continuously by evaporation or by sputtering. In some cases, a magnetic field is used to increase the path length and improve ionization efficiency. Sublimation pumps rely on an active metal film such as titanium, which reacts chemically with the gases and therefore removes them from the system. These pumps have an advantage over diffusion pumps in that they are free from oil and cannot contaminate the system. Performance improvement of sublimed films may be made by cooling the metal films to liquid nitrogen temperatures (Hunt et al. 1962), thereby increasing the pumping speed, particularly for hydrogen.

Thermal Shrouds

Simulation of space conditions requires that a test object be surrounded with surfaces highly absorptive to thermal radiation and emitting minimal radiation. This is best accomplished with blackened surfaces cooled to cryogenic temperatures. These surfaces are known as cryopanels, heat sinks, or thermal shrouds. Cryogenic surfaces are generally cooled with liquid nitrogen, not only because it is readily available, inexpensive, and safe to use, but because its temperature range gives an adequate simulation of space conditions.

If actual space conditions are to be represented by a temperature of about 7°R, any shroud temperature above this will lead to an increase in the test object temperature above what it would attain in space. This increase is called the simulation error, which is calculated by:

$$\Delta T = \frac{T_s^4}{7 T_o^3} + \frac{T_o}{7}\left(\frac{A_o}{A_s}\right)\left(\frac{1}{E_s} - 1\right) \quad (19)$$

where

ΔT = temperature error, °R
T_o = test object temperature, °R
T_s = shroud temperature, °R
A_o = test object area, ft²
A_s = shroud area, ft²
E_s = shroud emissivity

The first term in the expression, which contains both shroud and test object temperatures, is the error caused by radiant emission of the shroud in excess of what would come from space. The second term, which does not include shroud temperature, is the error caused by reflections from the shroud or by its failure to absorb radiation from the test object. This term would be eliminated if the shroud emissivity were unity.

In a well-designed simulator, the area ratio of the shroud and test object can be made large enough so that, with normally attained values of E_s, the second term will be small with respect to the first term in the equation and can be neglected. For a vehicle temperature of 540°R, the error for a 144°R shroud is 1.2°R; for a 198°R shroud it is 2.5°R. Thus, shroud temperatures can rise considerably above the boiling point of liquid nitrogen before serious errors are introduced. A commonly selected maximum temperature is 198°R, which gives an effective overall temperature of about 173°R. After this temperature is selected, the shroud is designed to maintain temperature under the heat load anticipated in the chamber.

Shrouds are usually of sheet metal construction, with tubes containing liquid nitrogen affixed to the sheet. Tubes may be welded on or formed into the sheet by various processes. A commonly used material is *tube-in-strip*, in which the tubes are formed into the billet before rolling and are inflated by hydraulic pressure after the sheet is rolled. Another method uses a printed pattern of tubes on a metal sheet, which is then bonded to a second sheet and inflated. *Plate-coil* construction is also used to some extent (particularly for stainless steel shrouds) and consists of embossed and welded sheets containing a pattern of tubes.

Liquid Nitrogen Supply Systems

The low-temperature operation of these units limits the choice of materials: aluminum, copper, and stainless steel are the metals normally used. Aluminum and copper have a distinct advantage in thermal conductivity which permits a sizable web between tubes without significant temperature rise. Aluminum presents some fabrication problems over copper but retains its reflectivity much better; this is important in these systems.

Surface treatment of the panels is required to obtain high reflectance to the chamber walls and high absorptivity on the inside. A bright dip etch of aluminum provides a permanently bright surface with an emissivity of 0.1 or less. The other side can be black anodized or painted with certain black paints. The paints have proved more suitable than the black anodize: they have high absorptivity over a wider range of wavelengths and show less tendency to bleach out under ultraviolet radiation. Extended surfaces on the black side will improve the situation to some extent, but will greatly increase both the weight and cost of the shroud.

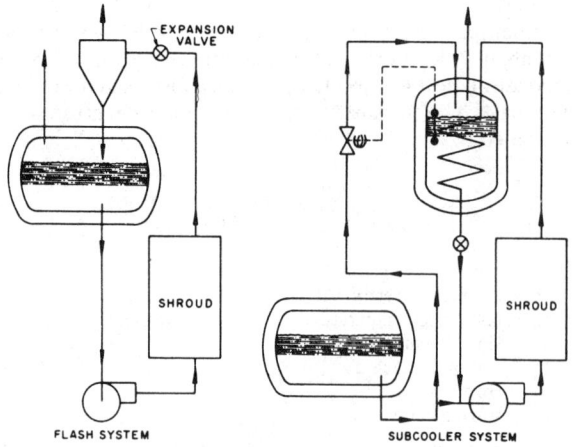

Fig. 36 Liquid Nitrogen Systems

The shroud panels are generally fastened to a support ring or framework inside the chamber, which is suspended by long stainless rods to minimize heat leak. The fastening of the panels to the support structure, and to each other, should be as loose as possible to allow for the relatively large contractions that take place during cooldown.

Liquid nitrogen supply systems for the shrouds generally are of two types, usually described as flash systems or subcooled systems (Figure 36). The flash system pumps liquid directly from the storage tank through the load, then expands it through a valve, either directly into the tank or into a separator from which the remaining liquid is drained into the tank by gravity. The subcooled system pumps the liquid in a closed loop, using a heat exchanger to remove the heat with liquid nirogen boiling at atmospheric pressure.

Both of these systems provide single-phase flow of liquid under pressure through the shroud. This is necessary to prevent hot spots or vapor locking in the tubes and to ensure maximum heat transfer.

The flash-type system is simpler, but the tank must be elevated to provide net positive head for the pumps. Usually a separator, or specially designed tank, is also required to take care of the flash process.

The subcooler uses liquid from a standard tank, with no special provisions, and has the advantage of somewhat smaller pump losses because of the pump suction operating at tank pressure. The tank can be pressurized to 20 to 30 psig, thus reducing the amount of head that must be generated by the pump.

From the pumping system, the liquid is generally fed to a valving system, which divides the liquid stream into the various zones of the shroud. This may be an insulated valve box or separate valves mounted on the chamber with pneumatic operators. In addition to flow control valves, vent valves allow each zone to be cooled individually by venting the vapor; they are then switched on stream when the proper temperature is reached.

Refrigeration and Heat Systems

When temperatures above that of liquid nitrogen are required, other types of systems may be selected. In some cases, a heat transfer fluid or brine is pumped through the shroud. This fluid is heated or cooled externally to produce the desired temperature. Cooling is usually accomplished with single or multistage refrigerators, and temperature is generally limited to about −100°F. In some cases, carbon dioxide is used as a refrigerant for these systems. Heating can be done with electric immersion heaters or with

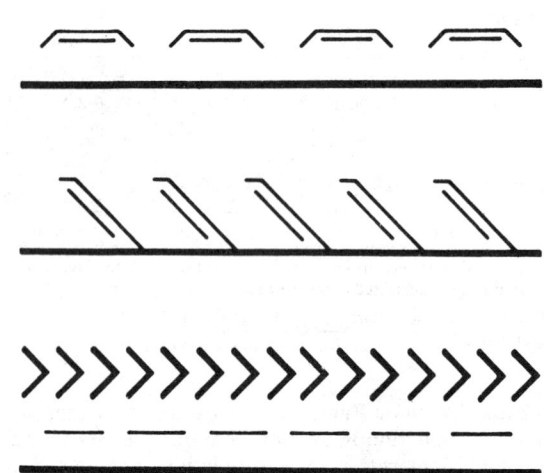

Fig. 37 Cryopumping Arrays (Arrangement of Panels)

steam in a heat exchanger. Such systems can be used to control temperature environments over a fairly wide range.

Another method for covering a wide range of temperatures is the circulation of gas, commonly nitrogen, through the shroud. To increase the mass flow, the gas is circulated in a pressurized loop, generally at 100 to 200 psig. It can then be heated or cooled externally to established controlled temperature levels in the chamber. Liquid nitrogen, which cools the gaseous nitrogen stream, permits operating at all temperatures to −300°F. Heating with steam or electric heaters permits temperature elevation with such a system, but temperatures are generally limited to about 300°F. Electric heating elements sealed in silicone rubber have also been used.

Cryopumping

Lower pressures are possible with cryogenic pumping, which uses a large portion of the chamber walls as a pumping surface, than provided by oil diffusion pumps. Further, large pumping surfaces minimize the possibility of molecules rebounding from the chamber walls and impinging on the object under test. In this sense, cryogenics offers a better simulation of the infinite molecular sink of space. At 36°R, the vapor pressure of the air gases (oxygen and nitrogen) is about 10^{-12} mm Hg, so that surfaces cooled to this temperature act as pumps to pressures as low as 10^{-10} or 10^{-11} mm Hg.

Various panel arrays are used. Although shielding somewhat reduces the pumping speed, these panels are generally shielded from the inside of the chamber, so that they will not receive any direct thermal radiation. In all cases, the shielding panels are part of the cold wall of the chamber and are cooled to liquid nitrogen temperature to minimize the heat load on the panels. The panels are usually reflective and, when possible, are faced with reflective surfaces to further reduce this load.

The pumping speed of these arrays is determined by a quantity defined as the capture probability of the array, a specific property determined by the arrangement and dimensions of the panels and shields. The flat-panel array and chevron-shielded array (top and bottom in Figure 37) have capture efficiencies in the vicinity of 0.20. The angle-fin array (center of Figure 37) has an efficiency of about 0.35 (Barnes and Hood 1961, Pinson and Peck 1962, Hood 1963). The actual speed may be computed by multiplying the orifice speed of 380 cfs/ft^2 by the capture efficiency. Thus, the

Cryogenics

angle fin has a speed in the vicinity of 1300 cfs/ft^2. Exact prediction of these speeds depends on other factors such as test object size, fraction of surface covered by the array, heat sink temperature, and cryopanel temperature, but the above method gives estimates accurate enough for most purposes.

Since the gas molecules and thermal radiation follow the same laws of reflection from the surfaces, the heat load on the panels of the arrays increases as the capture efficiency increases. Since refrigeration to 36°R is quite expensive, a compromise is made between maximum speed and allowable heat load. For most purposes, the angle-fin array has proved to be quite satisfactory, combining adequate shielding with high pumping speed and relative simplicity of construction.

In most space simulation chambers with cryopumping, carbon dioxide, water vapor, and other condensable gases are pumped by the shroud, and oxygen, nitrogen, and argon are handled by the cryopump panels. The remaining hydrogen, helium, and neon are then removed by the diffusion pumps. Thus, any large loads of the latter three gases will result in large pressure increases unless the number of diffusion pumps is quite large.

The refrigeration for cryopumping is generally supplied by helium refrigerators that pump a stream of cold helium gas through the panels. The helium refrigeration cycle used is a simple compressor-expander system. The gas is cooled with a heat exchanger to about 40°R before expansion, is expanded isentropically, and enters the load at about 18 to 27°R.

Either reciprocating or turbine-type expanders may be used. The cold gas returns from the panels at about 36°R and passes through the counterflow heat exchanger where it is warmed to room temperature before returning to the compressor. These systems usually operate at a pressure of about 300 psig and with pressure ratios of 5 to 10. A refrigerator of this type, capable of absorbing 3400 Btu/h of heat at the 36°R level, is sufficient to handle cryopumping arrays of 1000 ft^2 area, which would be found in chambers 30 to 40 ft in diameter.

Further improvement in a cryopumped vacuum chamber can be gained by lowering the panel temperature to 7.6°R with liquid helium. However, the large expense involved with the production and use of liquid helium has limited its use to small vacuum systems.

Cryosorption Pumping

One technique used for the evacuation of space simulation chambers is cryosorption—the physical adsorption of gases on cryogenically cooled surfaces. Materials such as charcoal, zeolite, silica gel, and alumina, when cooled to cryogenic temperatures, possess strong attractive forces for the common gases to be pumped in vacuum chambers. These forces effectively reduce the vapor pressure of a gas at the cryosurface temperature, enabling it to be pumped at its normal boiling point, *e.g.*, nitrogen on a 139°R adsorbent surface or hydrogen on a 36°R surface.

The efficiency of a cryosorption system depends on the maintenance of an adsorbent surface at low temperature. Since an adsorbent is inherently a poor thermal conductor, a thin layer (*e.g.*, 0.12 in.) is bonded in some fashion to a cryogenically cooled metal plate. Cryosorption panels using a molecular sieve as an adsorbent possess high pumping speeds and large capacities for hydrogen when cooled to 36°R, the normal boiling point of hydrogen. Likewise, cryosorption panels cooled to 7.6°R, the normal boiling point of helium, have demonstrated a high pumping speed and capacity for helium (Grenier and Stern 1966). The capacity of a cryosorption panel is extremely important in that the volumetric pumping speed of such a system decreases substantially as saturation is approached.

Regeneration of the panel can be achieved only by bakeout at elevated temperatures (480°F) for periods of up to 24 h, thus desorbing the pumped gas and, presumably, aborting any test under way in the vacuum chamber. Data indicate, however, that the capacity of a panel for hydrogen is sufficient, under normal gas loads, to provide for a test run of several months or more without regeneration (Stern *et al.* 1966).

Since 36°R cryosorption panels can handle only noncondensable gases, principally hydrogen, care must be taken to protect these panels from condensation of the condensable gases and resultant decrease of hydrogen capacity. For example, condensed nitrogen has a detrimental effect on panel operation. For this reason, 36°R baffles must be placed in front of the adsorbent panel to prevent condensation of the condensable gases on its surface.

Design for Human Testing

The emphasis in the space program on manned space flights has resulted in the necessity of placing humans in a simulated space vacuum environment. Whereas vacuum pumping and cryogenic systems for a human-rated chamber are similar to those of a nonman-rated chamber, manned system design must provide adequate safeguards to ensure the safety of an astronaut in a space chamber.

Locks must be provided for astronaut rescue operations, crew exchange, and repair or modification of the spacecraft while it is under space vacuum conditions. Biomedical facilities must be present in the chamber vicinity for monitoring, observation, and emergency rescue.

One problem of these facilities has been the repressurization system, installed for the protection of occupants in case of space suit failure and resultant exposure to chamber vacuum. Repressurization systems admit high-pressure air at a sufficient rate to raise the chamber pressure to 30,000 ft in as little as 5 s. The remainder of the repressurization to standard atmospheric pressure is then achieved at a slower rate (*e.g.*, 25 s). Several areas, such as time, noise level, and fog, are critical for repressurization systems. Extremely dry air must be used for repressurization to prevent excessive fogging and resultant difficulty in finding an injured person.

REFERENCES

Barnes, C.B. and C.B. Hood. 1961. Correlation between pumping speed and cryoplate geometry. *Advances in Cryogenic Engineering* 7:64.

Corruccini, R.J. 1957. Gaseous heat conduction at low pressures and temperatures. *Vacuum* 7:19.

Dowling, D.L., D.B. Herrick, and W.E. Rose. 1961. Evacuating large space chambers with positive displacement blowers. *Vacuum Symposium Transactions*, 1235.

Fulk, M.M. 1959. Evacuated powder insulation for low temperatures. *Progress in Cryogenics* 1:63. Heywood and Co., London.

Geist, J.M. and P.K. Lashmet. 1963. *Advances in Cryogenic Engineering* 8:199. Plenum Press, Inc., New York.

Gifford, W.E. and T.E. Hoffman. 1961. A new refrigeration system for 4.2 K. *Advances in Cryogenic Engineering* 6:82. Plenum Press, Inc., New York.

Giles, S. 1965. Outgassing of systems (Space vacuum technology). *Test Engineering and Management* XIII (April).

Glaser, P.E., *et al.* 1967. Thermal insulation systems. NASA SP-5027.

Grenier, G. and S. Stern. 1966. Cryosorption pumping of helium at 4.2 K. *Journal of Vacuum Science and Technology* 3(6):334.

Grilly, E.R. 1953. *The review of scientific instruments* 24(1):1.

Grunert, W.E., *et al.* 1969. Opacified fibrous insulation. Proceedings of the 4th Thermophysics Conference, AIAA.

Hoffman, T.E. 1963. Reliable, continuous, closed-circuit 4 K refrigeration for a Maser application. *Advances in Cryogenic Engineering* 8:213. Plenum Press, Inc., New York.

Hood, C.B. 1963. Cryopumping speeds in large space simulation chambers. Proceedings of the Institute of Environmental Sciences, 81.

Hunt, A.L., C.D. Damm, and E.C. Popp. 1962. Gettering of residual gases and adsorption of hydrogen on evaporated molybdenum films at liquid-nitrogen temperatures. *Advances in Cryogenic Engineering* 8:110.

Hunter, B.J., *et al.* 1960. Metal powder additives in evacuated powder insulation. *Advances in Cryogenic Engineering* 5:146. Plenum Press, New York.

Koontz, J.K. and P.K. Lashmet. 1963. Paper presented at Annual Meeting of ASME, Philadelphia, November 17-22.

Kropschot, R.H. 1959. Cryogenics insulation materials and techniques. *ASHRAE Journal* (September):48.

Pinson, J.D. and A.W. Peck. 1962. Monte Carlo analysis of high speed pumping systems. *Vacuum Symposium Transactions*, 406.

Prast, G. 1965. A modified Philips-Stirling cycle for very low temperatures. *Advances in Cryogenic Engineering* 10:40. Plenum Press, Inc., New York.

Stern, S., R. Hemstreet, and D. Ruttenbur. 1966. Cryosorption pumping of hydrogen at 20 K. *Journal of Vacuum Science and Technology* 3:165.

Vance, R.W. and W.M. Duke. 1962. *Applied cryogenic engineering*. John Wiley and Sons, New York.

Zeitz, K. and B.K. Woolfendon. 1963. *Advances in Cryogenic Engineering* 8:206. Plenum Press, Inc., New York.

BIBLIOGRAPHY

Coward, H.F. and G.W. Jones. 1965. Limits of flammability of gases and vapors. Bureau of Mines *Bulletin* 503, U.S. Government Printing Office, Washington, D.C.

Hydrogen Safety Manual. 1968. NASA-Lewis Research Center, Cleveland, OH, NASA TM X52454.

Lapin, A. 1972. *Liquid and gaseous oxygen safety review*, 4 vols. NASA CR-120922, Air Products and Chemicals, Inc. (June).

Lewis, B. and G. von Elbe. *Combustion, Flames and Explosions of Gases*, 2nd ed. Academic Press, Inc., New York.

NFPA. 1986. *Fire protection handbook*, 16th ed. National Fire Protection Association. Quincy, MA.

Product Pamphlets. Compressed Gas Association, Inc., New York.

Zabetakis, M.G. 1965. Flammability characteristics of combustible gases and vapors. Bureau of Mines *Bulletin* 627. U.S. Government Printing Office, Washington, D.C.

Zabetakis, M.G. 1967. *Safety with cryogenic fluids.* Department of the Interior, U.S. Bureau of Mines, Plenum Press, New York.

CHAPTER 39

LOW TEMPERATURE METALLURGY

Mechanical Properties 39.1
Material Selection 39.2
Product Improvement 39.4

THIS chapter describes the use of low temperatures for changing properties of metals, the effects of low temperatures on metallic properties, acceptability criteria for design, and the basic structure of construction materials.

MECHANICAL PROPERTIES

Mechanical and physical properties, cost, fabricability, and availability, are some of the important factors that are considered in selecting a material for a low-temperature application. Few generalizations can be made, except that a decrease in temperature increases hardness, strength, and modulus of elasticity. The effect of low temperatures on ductility and toughness varies considerably, however. The ductility of some metals increases with decrease in temperature, while others show an increase in ductility to some limiting low temperature followed by a decrease at lower temperatures. Still others show a decrease in toughness and ductility as the temperature is decreased below that normally encountered in the atmosphere.

Tensile Strength

In Figure 1, the relation of tensile strength to temperature is shown for metals generally selected for structural components. The slopes of the curves indicate that the increase in strength with decrease in temperature varies among the different metals. However, tensile strength is not a particularly good criterion for determining the suitability of a material for low-temperature service because most failures result from embrittlement. Ductility values obtained from the static tensile test may give some clue to degree of embrittlement, but the notched-bar impact test gives a better indication both of how the material performs under dynamic loading and how it reacts to complex multidirectional stress.

Ductility

Ductility, as measured by percentage of elongation in the tensile test, is shown in relation to temperature for several metals in Figure 2. The curves for copper, aluminum (5083), and lead show that ductility of these metals, which have a face-centered cubic crystal structure, increases as the temperature is lowered. However, not all face-centered cubic metals exhibit an increase in ductility at lower temperatures, as illustrated by the AISI 304 stainless steel. Most metals that have a body-centered cubic or hexagonal close packed crystal structure decrease drastically in ductility as the temperature is lowered, similar to zinc and SAE 1010 steel. These metals rupture with little plastic deformation even when tested under static loading.

Notched-Bar Impact

Some question exists as to whether values obtained on notched specimens approximately 0.4 in. square reflect behavior of full-size sections used in engineering structures. Considerable support has been given to the concept that the mode of propagation of fracture (*i.e.*, shear or cleavage) is of greater importance than total

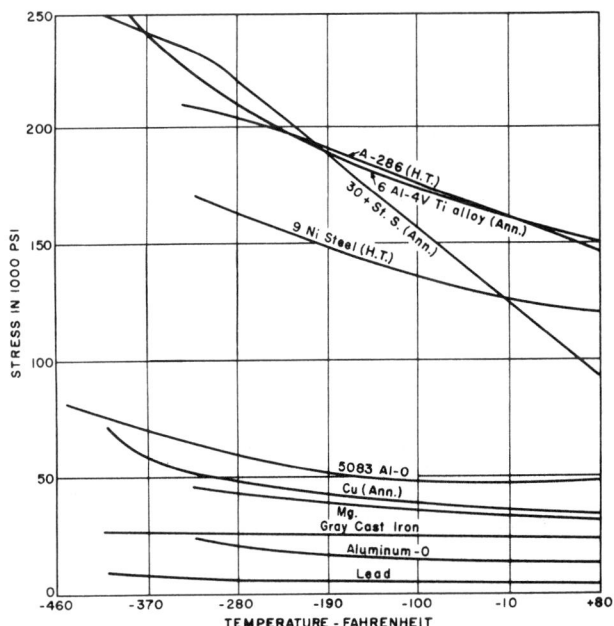

Fig. 1 Relation of Temperature to Tensile Strength of Several Metals

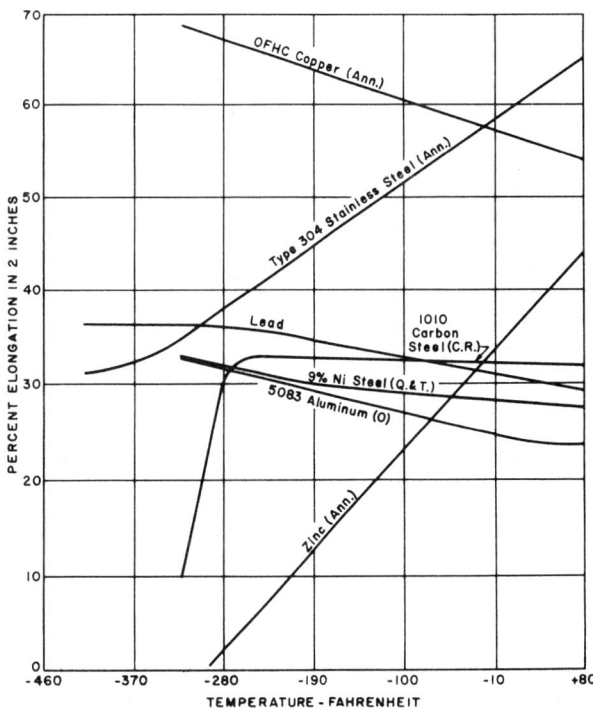

Fig. 2 Elongation Versus Temperature of Several Metals

The prepration of this chapter is assigned to TC 10.4, Ultra-Low Temperature Systems and Cryogenics. This chapter last received a major revision in 1967.

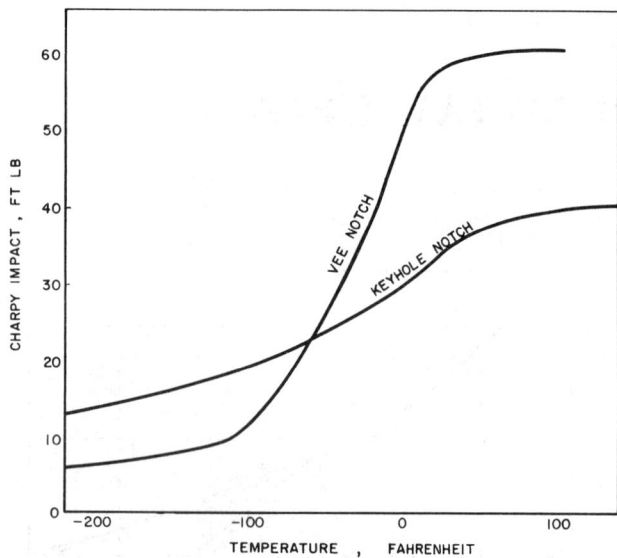

Fig. 3 Effect of Different Notches on Impact Properties of a Normalized and Tempered Cast Nickel Steel

energy to rupture. Another group advocates that the amount of plastic deformation should be the criterion.

The most popular impact tests are the Charpy V and Keyhole notched specimens (see Figure 3) and the NRL drop weight test. The primary purpose of any low-temperature, service evaluation test is to define the temperature or temperature range at which a material fails in a brittle rather than a ductile manner. The temperature distinguishing between ductile and brittle behavior is the *transition temperature*.

Any test can be used to determine the transition temperature of steels, but the transition temperature determined by one test is not necessarily the same as that determined by another. In fact, the type of transition curve (see Figure 2) may change with the test method; and if the criterion is based on a condition of fracture common to all tests, such as 50% cleavage and 50% shear, transition temperature may vary as much as 50°F or more in the same steel when tested by different methods. Temperature, the strain rate, and the degree of restraint all affect the temperature of change from ductile to brittle behavior in any particular steel; as these variables change according to different test methods, differences in transition temperature occur.

Nonferrous alloys commonly used at cryogenic temperatures do not exhibit a transition temperature, as noted in Figure 2. This does not mean that nonferrous materials all have high degrees of resistance to impact. Impact values for many nonferrous alloys are low, but there is no transition temperature below which these materials are subject to brittle failure or are extraordinarily notch-sensitive.

Notch Tensile Strength

Notch tensile testing defines the effect of a notch of particular geometrical configuration. Testing has been done of notches with varying calculated stress concentration factors, so it is sometimes difficult to compare data. However, some advantages of notch tensile testing are:

1. Sheet material can be tested that is not thick enough for accurate notched-bar impact tests.
2. Notched-unnotched ratios less than unity indicate the point at which a notch is effective in producing decreased strengths.
3. This type of test is perhaps more representative of the normal stresses applied to a structure than is the impact test.

Although the notched-unnotched ratio of unity has been used in many instances as an acceptability criterion for a material, relative degrees of brittleness can be tolerated in many situations and ratios of less than unity may be acceptable. No one number separates materials into categories of good or bad for either notch tensile testing or impact testing. Examples of the results of some typical notch tensile tests are shown in Table 1.

Table 1 Typical Notch Tensile Data for Several Materials at −423°F

	Yield Strength, ksi	Tensile Strength, ksi	Elongation, %	Notched-Unnotched Ratio[a]
A286, 0.1-in. thick— solution treated and aged	127	215	30	0.96
Ti-5Al-2.5Sn, 0.063 in., annealed	258	263	2	0.71
Type 310 Stainless Steel, 0.020 in.	261	290	1-12	1.12
Hastelloy B, 0.011 in.	240	290	16	1.09
Inconel X, 0.063 in.	134	283	30	0.85
5052-H38 Aluminum, 0.40 in.	56.5	81.2	37	0.96

[a]K_T for all tests = 6.3.

Fracture Toughness

For fracture to occur, a crack must first be initiated, then propagated. However, in most engineering structures, microscopic cracks are already present and fracture is possible when the applied stress is sufficient to propagate the preexisting microcrack. With this in mind, various fracture toughness tests have been devised. In essence, a crack is introduced into a test specimen and the growth of the crack is recorded as a function of temperature and applied stress. Values of stress needed to fracture the specimen as a function of temperature indicate the sensitivity of the alloy to brittle fracture. The fracture toughness is expressed either as stress factor K_c (critical crack extension force) or G_c (critical stress intensity factor), both of which are adjusted values of the fracture stress. These parameters are regarded as basic material parameters, just as the ultimate stress, and may be used in fracture mechanic calculations. ASTM *Standard* E381 describes this test in detail.

MATERIAL SELECTION

Nonferrous Alloys

Aluminum alloys are characterized by increasing tensile strength, yield strength, and ductility, as temperature is lowered. The impact strength of aluminum alloys may rise, remain at a constant level, or decrease, with decreasing test temperatures. Successful use is being made of most alloys at temperatures at least as low as that of liquid hydrogen (−423°F) and liquid helium (−453°F). No transition temperature exists for aluminum alloys. Typical strength curves for the aluminum-magnesium 5083, 5086, and 5154 alloys are shown in Figure 4. Aluminum alloys that cannot be heat treated can be welded in heavy and light sections; limiting tensile strengths of 400 to 500 psi make these alloys desirable only for moderate stress levels.

Copper and copper alloys behave similarly to aluminum alloys as testing temperatures are decreased. Copper alloys are used in liquefaction plants. For these as well as aluminum alloys, the strength characteristics are inversely proportional to their impact resistance (high strength alloys have low impact resistance). Welding or brazing of heavy sections can present serious fabrication problems.

Low Temperature Metallurgy

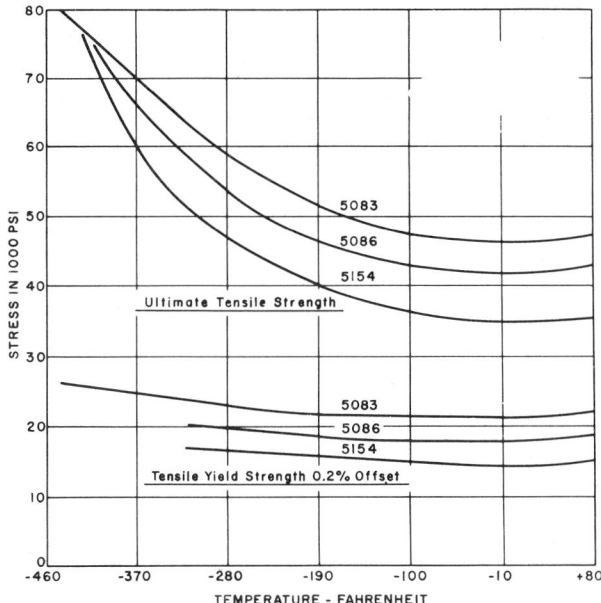

Fig. 4 Typical Tensile Properties of Alloys at Low Temperatures

Nickel and nickel alloys do not exhibit brittle failure tendencies at low temperatures, and high strength alloys such as Inconel X can be used at cryogenic temperatures. Nickel alloys can be welded successfully, but their high cost has decreased their more general use. A 36% nickel-iron alloy has been used where low expansion properties are important.

Materials used in joining, such as the *lead-tin solder alloys*, can be used at low temperatures, but increasing tin contents create transition-type impact curves. Silver-bearing brazing alloys are not generally subject to brittle failure at temperatures at least as low as that of liquid hydrogen ($-423°F$).

Titanium alloys are useful as cryogenic materials, primarily for the aerospace industry, where high strength-weight ratios are important. Some titanium alloys do exhibit transition curves, but the Ti-6 Al-4V alloy has been used extensively at $-320°F$ and the Ti-5Al-2.5 Sn alloy is suitable at liquid hydrogen temperatures ($-423°F$). High fabrication and material costs limit the usefulness of titanium alloys except where extremely high strength-weight ratios are of prime importance. Yield strengths of 100,000 psi coupled with low densities make these alloys attractive.

Ferrous Alloys

The low-temperature properties of steel are dependent on deoxidation practice, heat treatment, and alloy content. The use of aluminum for deoxidation of steel provides a fine-grain structure that is inherently more resistant to brittle failure than is a coarse-grain structure. Steels for low-temperature service are frequently specified to be made to fine-grain practice.

Quenched and tempered steels provide better impact properties than similar steels in the normalized condition; the annealed or soft condition generally provides the worst impact properties. Commercial quenching practices limit the thicknesses for which carbon steel can be fully hardened. Therefore, the use of quenched and tempered steels must be limited to those steels with an alloy content that will allow them to be fully hardened on quenching.

Transition temperatures, below which steels are generally considered to be unusable for cryogenic service, are sometimes ascribed to the transformation of austenite to martensite, with a sharp decrease in impact strength over a rather narrow temperature range. Alloying elements reduce the tendency for transformation to take place.

Nickel is the most effective alloying element for increasing the resistance of steel to low-temperature embrittlement. The temperature at which embrittlement occurs decreases as nickel content increases up to about 15% nickel. With higher percentages of nickel, the impact properties of low carbon steels become practically nonvariant at temperatures below atmospheric (see Figure 5). Manganese, within limits, is also beneficial in steel for increasing the resistance to embrittlement at low temperatures. Other alloys may slightly improve steel's resistance to low-temperature embrittlement, but their effect is generally an indirect one of modifying the form and distribution of the carbides rather than the direct one of having an alloying effect on irons, as is the case with nickel.

Quenched and tempered low alloy steels of the AISI or SAE types are frequently used in light and moderate sections of machinery parts that are to operate at temperatures down to $-150°F$. A type of steel should be selected that will fully harden on quenching; if any welding is performed, it should be done prior to the quenching treatment.

The choice of materials for large pressure vessels and towers has probably received more attention than any other phase of the low-temperature problem. Because of almost universal use of welded construction for these parts, it is essential that a material be used that can be welded by an accepted commercial process and that welds have resistance to low-temperature embrittlement approaching that of the base material. Refer to ASME Section VIII, "Rules for Construction of Pressure Vessels."

In addition, the properties of the base material must not suffer deterioration from the heat of welding or, at least, not to an extent that cannot be corrected by a simple form of heat treatment such as stress relief. This requirement immediately eliminates steels that derive their resistance to low-temperature embrittlement from full heat treatment. As-rolled or normalized aluminum-treated low carbon steel, welded with AWSE 7015 electrodes, is used for comparatively thin wall vessels for temperatures down to $-50°F$. Low carbon, 2.25% nickel steel, welded with AWSE 8105 nickel steel electrodes, is generally used for temperatures down to $-75°F$ or slightly lower, and low carbon, 3.5% nickel steel, welded with AWSE 8015 nickel steel electrodes, has been employed for temperatures down to $-150°F$.

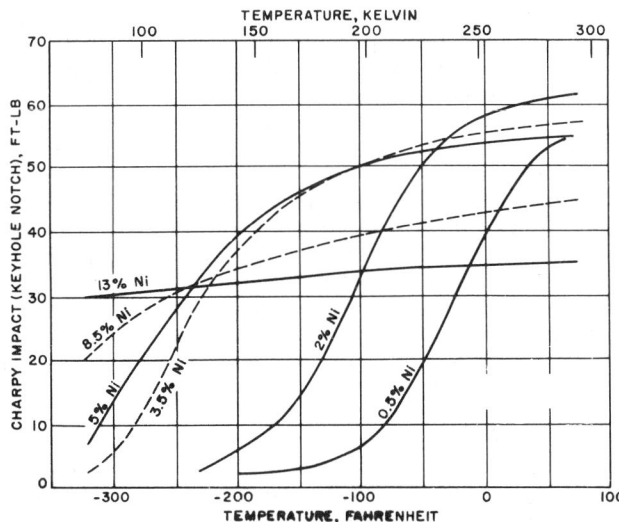

All steels contained 0.10% carbon except the nickel-free steel (0.20% carbon) and the 2% nickel steel (0.15% carbon) (Armstrong and Brophy).

Fig. 5 Effect of Nickel on the Low-Temperature Embrittlement of Normalized Low Carbon Steels

Low carbon quenched and tempered steel plate is used for pressure vessels; but in this case, welding is performed on the heat treated steel plate, and further heat treatment is not usually applied. The limiting low temperature for such vessels is usually dependent on the properties of the weld metal. HY-80 quenched and tempered steel plate has been used in submarine pressure hulls. This alloy steel has a transition temperature below $-120\,°F$ and extremely high yield strengths, in excess of 80,000 psi. It can be welded without subsequent stress relief, but the welds are presently limited to somewhat higher transition temperatures. T1 quenched and tempered steel plate has been used in high strength applications and has acceptable impact properties to $-50\,°F$. This alloy is generally stress relieved after welding.

Low carbon 9% nickel steel is used successfully at temperatures as low as that of liquid nitrogen ($-320\,°F$). In the quenched and tempered condition, this alloy has a yield strength of 75,000 psi and can be welded with inconel-type filler metals without the necessity for additional stress relief.

Chromium-nickel stainless steels have been used more than any other ferrous alloys for use at temperatures to that of liquid helium ($-453\,°F$). No stress relief is generally required after welding, and impact strengths vary only slightly with decreasing temperature. A popular steel for low-temperature service in the chromium-nickel group is type 304, which has a yield strength in excess of 35,000 psi. Where higher strengths are required and welding can be avoided, strain hardened forms of the chromium-nickel stainless steels are available with yield strengths of 100,000 psi.

The *precipitation hardening stainless steels*, such as 17-7PH and 17-4PH, have received some attention in the aircraft and missile industry. Their ability to develop extremely high strength levels may be advantageous for cryogenic purposes down to $-200\,°F$, but they have not been extensively used for these purposes.

Maraging steels have been developed with alloy contents of 18 to 25% nickel. These alloys exhibit high strength at temperatures as low as $-320\,°F$, since they develop up to 250,000 psi yield strength.

Cast iron is usually considered a brittle material at normal atmospheric temperatures, but it is used successfully in many applications, even when some degree of shock resistance is required. Impact measurements on cast iron are usually made on unnotched arbitration test bars 1.2 in. in diameter, and although the impact resistance decreases as the temperature is lowered, a sharp transition temperature has not been observed. The plain cast irons have very low impact resistance at both room and low temperatures, but some of the alloyed cast irons possess a moderate degree of toughness.

The high alloy irons of the Ni-Resist type have appreciably higher toughness, but both the low alloyed and Ni-Resist types show a drop of 25 to 30% in impact values from room temperature down to $-300\,°F$. Since the impact resistance of low alloyed high strength cast irons is approximately 25 to 30 ft lb at room temperature, and that of Ni-Resist, 80 or more ft lb, the impact values at $-300\,°F$ are about 20 ft lb for grey iron and 50 to 90 ft lb for Ni-Resist. These values cannot be compared with impact properties of steels and metals that are measured with notched specimens about 0.4 in. square, as such specimens are unsuitable for measuring differences in inherently brittle metals.

Analysis for Use

The choice of material for a specific use is often a compromise involving several factors. Cost is of major importance. The stress level at which the product will operate may require section sizes in certain of the lower strength materials that are impractical to manufacture or assemble. The operating temperature will further limit acceptable materials. The ability to weld or to stress relieve after welding may have a critical effect on the eventual choice of materials.

The possibilities of corrosion often must be considered. Thermal expansion characteristics have been critical in certain applications. Elastic modulus is certainly of interest in bolting and other applications. Resistance to thermal shock has sometimes been of major importance and, in these cases, alloys with high thermal conductivity are often considered. No general rules can really be established for the choice of material.

PRODUCT IMPROVEMENT

The principal areas in which cold treatment has been practiced to advantage in product improvement of ferrous alloys may be summarized as follows:

1. Increased hardness and wear resistance.
2. Dimensional stability of tools, gauges, and machine parts.
3. Elimination of grinding cracks.
4. Increased cutter tool life.
5. Improved magnetic properties.
6. Ability to store aluminum alloy.

Since austenite is relatively soft and ductile, any appreciable quantity present in the surface structure of a tool or machine part will reduce hardness, which may lead to more rapid wear. Carburizing can increase hardness. Increased hardness values of up to 15 points Rockwell C have been noted in carburized cases of SAE 3310 gears after multiple cold treatment ($-120\,°F$) retempering cycles on a case intentionally carburized to a high surface carbon content.

Where maximum dimensional stability is required in tools, gages, or machine components, cold treatment (or *cold stabilization*) may be used to convert the retained austenite and resultant expansion prior to the final dimensioning of the part.

Dimensional stability is important in gage manufacture and in close tolerance machine parts subject to subsequent low-temperature service conditions in high altitude aircraft. SAE 52100 steel is used in such applications and is particularly sensitive to austenitizing temperature. When austenitized at $1700\,°F$, approximately 13% retained austenite exists in the quenched structure.

Seasoning for dimensional stability is a variation of cold stabilization involving cycling several times over the temperature range found in service. Although conversion of retained austenite may occur, stress relief or equalization of stress is an important factor in producing stability in this application. Formerly, for applications requiring dimensional stability, iron castings were seasoned by storage for months before machining. Today, seasoning consists of cycling cast iron parts from -150 to $700\,°F$, allowing for possible contraction.

The elimination of retained austenite in hardened steel reduces the possibility of grinding cracks normally resulting from thermal stresses caused by the localized heating action of the grinding wheel. This condition is further aggravated by the expansion of retained austenite, which transforms thermally or by the mechanical shock of the wheel and causes conversion to martensite with severe surface stresses.

If high speed steel is properly hardened and then tempered $1050\,°F$ for 2.5 h, very little retained austenite should remain in the structure. With the usual addition of two or more multiple tempering cycles at $1050\,°F$, insufficient retained austenite should remain in the structure to warrant cold treatment. However, cold treatment may serve to reduce the number of required multiple tempering cycles.

High-carbon, high-chrome (1.5% carbon, 12 to 14% chromium, 1% molybdenum), 1% carbon, 5% chromium, 1% molybdenum die steels respond very well to cold treatment. Multiple cycling of high-carbon, high-chrome steels between -120 and $950\,°F$ produces greater stability than is possible by secondary hardening alone in the temper at $950\,°F$.

Low Temperature Metallurgy

Wear resistance can be increased in the 1% carbon, 5% chromium steel by cooling to $-150\,°F$ prior to a $400\,°F$ temper. This is particularly effective whenever the surface carbon content is increased by pack hardening in carburizing compound. This technique of carburizing certain tool steels to improve surface wear resistance is also used with the 0.5% carbon, 1 to 2% tungsten, 1% chromium punch and die steel.

Hardenable magnet steels, such as the 1% carbon types with either 3.5% chromium, 3% cobalt, or 5% tungsten, tend to retain austenite in the hardening process. This nonmagnetic austenite may be converted to magnetic martensite by cooling to $-150\,°F$ as a continuation of the quenching cycle, with a resultant increase of about 5% martensite.

Aluminum alloy rivets made from 2017 and 2024 aluminum will normally start to age at $32\,°F$ within 16 h after the solution treatment. They may be stored for about a week at $0\,°F$ and held indefinitely at $-40\,°F$ without hardening taking place until after the riveting operation. Cold treatment is applied to solution treated castings, forgings, and sheet metal blanks for storage prior to cold working operations. To promote dimensional stability and better machined finishes in precipitation hardenable aluminum and magnesium castings, a combination of cold treatment at $-140\,°F$ followed by overaging $50\,°F$ above the aging temperature has been used. This allows permanent growth to take place prior to the final dimensioning of precision parts and minimizes possible service failures caused by temperature fluctuations over the same range.

Expansion fitting differs from shrink fitting in that an internal part is thermally contracted by cold treatment, inserted into a mating part, and allowed to expand to a tight fit on returning to atmospheric temperature. Shrink fitting, on the other hand, involves the heating of the external part and its subsequent shrinkage when assembled and allowed to cool around an inside part. A combination of both expansion and shrink fitting may be used when greater assembly strength is required.

Shrink disassembly of bushings, sleeves, and hollow inserts is possible if a suitable refrigerated coolant can be allowed to flow through the internal part. Cold treatment in the assembly of parts offers a definite advantage in ease of operation, with freedom from oxidation, distortion, and softening caused by tempering of hardened parts when heated for shrink fitting. A disadvantage of the process is the limited size change possible when cooling with liquid nitrogen to $-320\,°F$ compared to the thermal expansion resulting from heating through several hundred degrees.

In a proprietary process, cold treatment is applied to control the phase change between liquid and solid mercury used as replica material instead of the wax or plastic employed in the lost wax or precision investment casting process. Mercury is rapidly frozen in a master mold by cooling to $-100\,°F$, removed as a solid replica of the part, and then held below $-40\,°F$ during the investment process.

The various means of producing subatmospheric temperatures and the temperatures obtainable are:

Salt and ice	$-6\,°F$
Mechanical refrigeration, single stage	$-50\,°F$
Solid CO_2 and methyl alcohol or acetone	$-100\,°F$
Solid CO_2	$-109\,°F$
Mechanical refrigeration, multiple stage	$-150\,°F$
Liquid air	$-297\,°F$
Liquid nitrogen	$-320\,°F$
Liquid helium	$-453\,°F$

Mechanical refrigeration is generally operated continuously at the desired temperature. Both liquid baths and circulated air chambers are used for cooling parts; however, the advantage of rapid heat transfer in the refrigerated liquid bath is somewhat offset by the greater tendency toward cracking.

CHAPTER 40

BIOMEDICAL APPLICATIONS

Preservation at Temperatures above −79°C 40.1
Preservation at Temperatures below −79°C 40.1
Freeze-Drying 40.2
Clinical Uses of Refrigeration 40.3

REFRIGERATION for biological systems may be categorized as research, preservation at conventional or very low temperatures, freeze-drying, and clinical applications. In clinical or medical applications, refrigeration is used to preserve or to destroy the viability of living biomaterials. Opposed as these end results may be, the techniques are much the same; they differ only in the rates at which the various thermodynamic processes are carried out and in the chemical composition of the media involved.

PRESERVATION AT TEMPERATURES ABOVE −79°C

The most common pieces of medical refrigerating equipment are standard refrigerators, which operate at temperatures slightly above 0°C, and special modifications such as the addition of temperature recorders and alarms. Blood, tissues, and viruses are stored at conventional refrigeration and deep-freezer temperatures. However, whole blood storage is regulated by National Institutes of Health procedures, which allow blood to be stored and unfrozen, at 4°C for a maximum of 21 days for blood collected in ACD (acid-citrate-dextrose) anticoagulant and 28 days for blood collected in CPD (citrate-phosphate-dextrose) anticoagulant. Frozen blood plasma may be kept in a deep freezer at about −20°C for approximately a year. Blood bank refrigerators should be equipped with forced air circulation, a dependable and sensitive thermostat, an alarm system, and a recording thermometer.

Preservation of tissue for transplants is another medical application of refrigeration. Transplants may be satisfactorily preserved by simple ice refrigeration in the container used for transport.

Bone, artery, and other tissue banks make use of conventional refrigeration systems. Bones are preserved by storage at about −20°C or, sometimes, at near dry-ice temperature. The bone is procured, sealed in a sterile tube, and frozen in a mixture of methyl alcohol and solid carbon dioxide. The bone is then placed in an insulated box surrounded by solid carbon dioxide in a freezer. Arterial grafts are stored in a similar manner.

Quick freezing and storage in the −20 to −40°C range preserves tissue and viruses sealed in glass vials or bottles for further study. Some viruses are better preserved at even lower temperatures; a solid CO_2 container, kept in the mechanical cooler, can maintain temperatures in the −70°C range. Dewars containing liquid nitrogen are also used for storage.

Usually the refrigeration system is a two-stage unit with a high- and low-pressure compressor and two evaporators. A low-temperature refrigerant such as R-12, R-22, or sometimes ammonia, is used. Such low temperatures are not easily maintained, and the temperature rises rapidly if mechanical or power failure occurs. Because of the value of these stored materials, an alarm system is essential. Unless the refrigerator is in a corridor or other readily accessible area, an alarm system of local lights or bells, is not sufficient. A central alarm that operates an alarm or bell in the main engineering or maintenance office, supplemented by another bell in the refrigerator, should be installed.

Refrigeration for cadaver preservation is largely limited to medical schools or other institutions where anatomical studies are made. Refrigeration is also needed subsequent to embalming. Because of the difficulty in supplying the demand for cadavers for medical teaching and study, equipment for long-term storage is necessary. Storage of this type requires a large space, sometimes a refrigerated room held at about 2°C.

PRESERVATION AT TEMPERATURES BELOW −79°C

Although the normal freezing point of most biomaterials is around −0.6°C, ultra-low temperatures, *i.e.*, temperatures well below −79°C (the saturation temperature of CO_2 at 101.3 kPa pressure), preserve these materials more satisfactorily than higher temperatures. In fact, many biomaterials may be stored indefinitely at −130°C, the recrystallization point of ice, as the maximum temperature.

Refrigeration at these temperatures can be provided by stored refrigeration in the form of a low-energy fluid such as liquid nitrogen or by active refrigeration provided by a refrigerant operating in a thermodynamic cycle. Since refrigeration loads for preservation at these temperatures fluctuate widely, the stored form of refrigeration is often the most convenient, even though it may not be the most economical. In such cases, materials can be stored in liquid nitrogen or its vapor at temperatures in the range between −170 and −196°C.

Vacuum-insulated Dewars provide storage for a long period between necessary additions of fluid to replace the boiloff quantity; they are commonly used for the storage of spermatozoa, skin, and other biological materials. Solid-state electronics, particularly microprocessors, have facilitated the development of high precision, programmable freezers of this type.

For refrigeration temperatures between those of solid carbon dioxide (−79°C) and liquid nitrogen (−196°C), a system of controlled liquid nitrogen injection may be applied in a manner much like that currently used in refrigerated trailer trucks. Other system designs are cascade cycles, often with hydrocarbons as refrigerants for low-temperature stages.

Freezing the Specimen

For most specimens, freezing and thawing techniques are quite important, particularly for materials other than bone or arteries. Therefore, rates of temperature change, cryoprotective additives, and heat-transfer mechanisms must be considered. One of the most important physical considerations is the cooling rate.

The preparation of this chapter is assigned to TC 10.4, Ultra-Low Temperature Systems and Cryogenics. This chapter last received a major revision in 1976.

Although the exact mechanism of cell destruction is not well understood, very rapid cooling rates and slow cooling rates are shown to have distinct advantages and disadvantages. Mazur *et al.* (1972) proposed that cell death at low cooling rates occurs because of dehydration and the concentration of solutes, while cell death at high cooling rates is a consequence of intracellular ice formation. In between these two extremes is an optimum cooling rate at which cell water is transported across the cell membrane at a rate sufficient to prevent the formation of intracellular ice, yet not so fast as to make dehydration a problem. That is, the cell contents are supercooled, but the extent of supercooling is not great enough to produce intracellular ice.

The introduction of certain chemical compounds such as glycerol, dimethylsulfoxide, and ethylene glycol has minimized the effects of dehydration at low cooling rates while avoiding the problems of intracellular freezing. These compounds, known as *cryophylactic agents*, reduce the amount of water that must form ice to establish chemical equilibrium at reduced temperatures. They also tend to act as water substitutes and prevent solute concentrations from reaching lethal levels.

Of these compounds, glycerol has been shown to be very attractive, particularly in view of its low toxicity. Meryman (1960) states that virtually every body tissue has been successfully frozen and received when protected with 5 to 15% glycerol; however, the extent of recovery on thawing varies widely according to cell type.

To survive freezing, most animal cells must be subjected to low rates of temperature change, usually in the presence of one of the penetrating additives. The rate of penetration of the additive itself can be important in clinical situations. If the extracellular concentration rate of the additive is increased too fast relative to the penetration rate, cells can experience osmotic damage. This problem is particularly acute in the case of glycerol transported across cell membranes at a substantially lower rate than dimethylsulfoxide or ethylene glycol.

Rapid freezing in liquid nitrogen is successful for only a very few classes of living organisms of mammalian cells; only erythrocytes, some bacteria skin, and certain tumor cells will survive rapid freezing. If the freezing is accomplished with the material in direct contact with a cryogenic liquid (usually liquid nitrogen), film boiling rather than nucleate boiling may occur. The use of a liquid that boils at a higher temperature (Cowley *et al.* 1961) or the coating of the specimen with a rough insulating material (Luyet 1961) has been suggested as a means of maintaining low surface temperatures favoring nucleate boiling and its associated high heat fluxes.

Whether the heat removal process must be performed slowly or rapidly, many reports indicate that indefinitely prolonged maintenance of potential cell and tissue viability is achieved by storage at $-100\,°C$ or lower. Among such materials are erythrocytes, bone marrow, tissue cultures, protozoans, bacteria, spermatozoa, corneas, and viruses.

Thawing the Specimen

Thawing the specimen also presents water separation and other problems and may be more lethal than the cooling process. A major problem is the limited driving force for thawing. The protein in mammalian cells normally denatures at temperatures of around $45\,°C$, so that in thawing any volume of cells, the boundary temperatures must remain well below this limit. Thus, as thawing proceeds from the periphery of the specimen toward its center, the rate of heat transfer decreases, and it may become difficult to maintain warming rates near optimum. Attempts to overcome this problem have applied methods of delivering energy on a volumetric basis, such as induction heating, microwave heating, and other methods. For the most part, these methods have suffered from the differential absorption characteristics of the solid and liquid phases present in the thawing specimen. This condition can lead to thermal runaway on the local level, with accompanying high-temperature injury.

Clinical Applications

While an understanding of the physiochemical responses of cells and tissues to freezing and thawing remains incomplete, some notable clinical procedures have applied the principles, such as the longterm preservation of human red blood cells. The American Association of Blood Banks (AABB 1973) developed procedures for the freezing of human erythrocytes on a large scale.

In the case of skin preservation, Theilleux *et al.* (1959) reported 95% survival in skin grafts on animals, using frozen preserved skin. The skin is impregnated in Earle's balanced salt solution, containing 30% glycerol, for 0.5 h. Then, it is deposited on glass coverslips or aluminum strips and frozen in liquid nitrogen; afterward it is stored at dry-ice temperature.

Refrigeration is now widely used in the storage of spermatozoa. Long-term storage is usually accomplished at liquid nitrogen temperature and with the addition of glycerol and other liquids.

Long-term preservation by freezing, with subsequent thaw and transplant, has been the subject of many studies, but the complexities of geometry and structure have allowed only partial success.

The possible cryogenic preservation of corpses has received periodic media attention. The ultimate goal is the eventual thawing and reanimation of the body. Of course, the present state of the art indicates that the probability of success is extremely small.

FREEZE-DRYING

When a biological material is dried from the liquid state, it often shrinks, alters its chemical structure, or becomes relatively insoluble for future reconstitution. Any of these changes may make the produce unusable. However, sublimation drying from the frozen state often allows the gross morphology, the chemical integrity, and the solubility of the material to remain unchanged. Commercially, freeze-drying is used (1) to reduce the mass and volume of preparations and (2) to stabilize materials, which allows them to be stored at ambient temperature.

In theory, only 3 steps are required in freeze-drying. First, the necessary heat of sublimation must be added. Second, the sublimated water vapor must diffuse from the ice crystal surface through the already dried portion of the specimen. Third, the water vapor that reaches the surface of the specimen must be removed to maintain a vapor pressure gradient (Meryman 1960, 1966). Greiff and Rowe (1976) reviewed the application of freeze-drying technology.

Practically every type of biological organism has been freeze-dried. A partial list of biological materials successfully freeze-dried includes:

VIABLE	NONVIABLE
Bacteria (Greaves 1960)	Arteries (Gresham 1964)
Fungi	Bee venom
Molds	Blood plasma
Mushroom spores	Bones (Gresham 1964)
Seeds (Rowe 1960)	Cooked meals (Rowe 1960)
Vaccines	Dura (Savu *et al.* 1959)
Viruses (Greaves 1960)	Enzymes (Savu *et al.* 1959)
Yeasts	Erythrocytes
	Fascia (Greiff and Rowe 1976)
	Human milk (Rowe 1960)
	Proteins
	Rabbit brains
	Skin (Gresham 1964)
	Tumors
	Vegetables
	Whole dead animals (Meryman 1961)

Freezing Apparatus

Materials to be freeze-dried must first be frozen to a temperature at which no liquid phase is present, so as to avoid puffing

Biomedical Applications

or deterioration during drying. Although this temperature differs for every chemical combination, it is usually not necessary to go below −60°C to ensure complete freezing.

Quick freezing is desirable and often necessary. Slow freezing usually produces irreversible biological changes as the water freezes out and always produces large ice crystals that cause difficulties during drying or rehydration. These difficulties do not arise from the small ice crystals formed during rapid freezing, and the biological irreversibilities are reduced. Rowe (1960) describes the methods of prefreezing and chamber freezing in detail.

Drying Apparatus

The usual commercial or laboratory freeze-drying plant incorporates four subsystems: (1) a drying chamber, (2) an ice condenser, (3) a vacuum pump, and (4) a refrigeration unit. The drying chamber may be either tray or manifold. In the tray method, the material to be dried is either frozen on trays or frozen in bottles placed on trays, and the entire chamber is evacuated. In the manifold method, each bottle or ampoule is connected to a manifold, which is then evacuated. Thus, a high vacuum is not required for the entire drying chamber. The arbitrary differentiation between ordinary and extreme dryness is 1% moisture content, relative to the dry mass, which is also the dividing line between one- and two-stage systems.

Most mechanisms of heat addition have been tried at one time or another. Rowe (1960) lists these as: (1) conduction from electrical mats, (2) warm air convection, (3) liquid convection, (4) infrared radiation, and (5) radio-frequency radiation. Greaves (1954) concludes that radiation is theoretically the most efficient means of heat addition, but direct conduction is the method usually practiced commercially because of industrial difficulties with radiant heating. All methods require that a careful balance be maintained between the maximum sublimation rate and the melting point. Abelow and Flosdorf (1957) conducted temperature-time investigations in the drying of biologically sensitive materials.

Since the vapor pressure of water at the drying temperature is the critical factor determining the drying rate, surface temperature should be as high as possible without melting. Because of the critical nature of these drying temperatures, adequate control should be assured by the designer. Some representative values are listed by Neumann (1962) as:

Blood plasma (stabilized)	−15 to −20°C
Blood serum	−9 to −12°C
Coffee extract	−18 to −25°C
Meat	−10 to −15°C
Penicillin	−25 to −40°C
Tissue (for microscopic study)	−25°C
Vegetables	−6 to −8°C
Viruses	−20 to −40°C

Three methods are used to remove water vapor. Chemical desiccants were first used with the drying of blood products, but now are rarely used. Mechanical pumping, the second method, is not frequently used either, since it would require the drying chamber pressure to be lower than the residual vapor pressure of the substance being dried. Condensation of the water vapor on ordinary refrigerator coils or a cryosurface, the third method, is used in almost every current freeze-drying system.

The difference in the water-vapor partial pressures between the drying surface and the atmosphere surrounding the specimen provides the driving force for the water-vapor diffusion process. Most refrigerated systems provide this force by using a vacuum-drying chamber. The standard pumping systems to produce the necessary vacuum—rotary, diffusion, vapor booster, steam ejector, and refrigeration—are well known and commercially available. Rowe (1960) describes their application in freeze-drying systems. Greiff and Rowe (1976) give examples of typical systems.

Storage

Vacuum sealing and filling with inert gases are the two usual methods for storing freeze-dried medical products. The vacuum method appears to be the superior one. However, since glass sealing of ampoules to maintain a vacuum is not as economical as the automatic rubber stoppering of an inert atmosphere, manufacturers generally prefer this less expensive method and accept the limitation of an expiration date on their products. Although freeze-drying is rather expensive, it does become more economical if transportation costs are involved, since freeze-drying eliminates refrigeration costs and 60 to 90% of shipping weight.

CLINICAL USES OF REFRIGERATION

Hypothermia

Originally, *hypothermia* meant the artificial lowering of body temperature for therapeutic purposes, as distinguished from natural hibernation. However, the term is currently taken to mean the state of a warm-blooded animal whose temperature has been deliberately reduced below the norm for that species. Swan and Paton (1961) suggest the following classification of hypothermic temperatures:

Moderate hypothermia	37 to 28°C
Intermediate hypothermia	28 to 20°C
Deep hypothermia	20 to 0°C
Supercooling	below 0°C without ice formation
Freezing	below 0°C with ice formation

Four characteristics of hypothermia are: (1) a pronounced reduction in oxygen consumption, (2) a decrease in the metabolic rate, (3) a reduction in the heartbeat, and (4) the induction of hypotension. Swan and Paton (1961) report the following clinical uses of hypothermia:

Postcardiac arrest encephalopathy
Acute heat trauma
Deliberate temporary interruption of the blood supply
Cardiac and general surgery
Gastro-duodenal hemorrhage
Thyrotoxicosis
Cirrhosis of the liver
Local anesthesia for debilitated patients

Originally, simple body immersion in an ice bath was used to produce hypothermia. Other methods are the temperature control blanket, cold saline intrapleurally, ice bags, and cold air chambers, but all have the disadvantage of taking longer than simple immersion. The more recent method, and one of interest to refrigerating engineers, is to achieve extracorporeal cooling by circulating the blood through a heat exchanger.

Shore (1961) considered the problem as that of cooling a body weighing 70 kg from 37 to 15°C in 0.5 h, or a heat load of 3.5 kW. Basically, the system consists of a shell heat exchanger for the blood, cooled by internally circulated water passing over refrigeration coils. Direct refrigeration of the blood is not used because of the possibility of too rapid chilling. Rewarming is usually accomplished by the reverse of the cooling process: immersion in a warm bath, by a warm air convection chamber, or by the warm water heat exchangers for the blood, as previously mentioned.

Cryogenic Surgery

Essentially, cryogenic surgery is the production of either a temporary blockage of nerve functioning or cellular death within the frozen area by the infliction through controlled freezing of a lesion within the diseased tissue. The controlled destruction of tissue by using cold is applicable at the present time to virtually all of the surgical specialties.

Cryosurgical instrumentation is able to rapidly produce lesions that are sharply delimited and hemostatic. The advancing edge of the permanent lesion is preceded by an area of physiological inhibition and reversibility. Rand *et al.* (1968) prepared a comprehensive text on the application of cryosurgery to the various specialties and the equipment available for these purposes. Cooper *et al.* (1967) pioneered in the development of the *cryosurgical probe,* in which a refrigerant is piped through vacuum-insulated tubing to a small surface, which is uninsulated and serves as the freezing surface.

Refrigerated Microtome

The refrigerated microtome is of special interest to the refrigeration engineer serving medical and biological laboratories because of the required wide temperature range and precise temperature control. Stumpf and Roth (1965) determined that temperatures above $-30\,°C$ facilitate specimen sections thicker than 1 micrometre while temperatures below $-70\,°C$ favor sections of 1 micrometre or less.

Essentially, the refrigerated microtome is a conventionally refrigerated chamber in which microscope slide specimens are sliced, but it may have a liquid nitrogen refrigeration system to produce temperatures from -85 to $-120\,°C$. Also, it normally has a drainage outlet for defrosting the sectioning chamber and a vacuum port to permit freeze-drying of the sliced sections, if desired. Finally, because of the delicate nature of the sectioning, the refrigeration engineer may isolate the compressor to avoid vibrations.

REFERENCES

AABB. 1973. *Red cell freezing.* American Association of Blood Banks, Arlington, VA.

Abelow, I.M. and E.W. Flosdorf. 1957. Improved heat transfer system for freeze-drying. *Chemical Engineering Progress* 53(12):597.

Cooper, I.S. 1967. Cryogenic surgery. *Engineering in the Practice of Medicine,* Chapter 12, p. 122. eds. B.L. Segal and D.C. Kilpatrick. Williams and Wilkens:Baltimore, MD

Cowley, C.W., W.J. Timson, and J.A. Sawdye. 1961. Ultra rapid cooling techniques in the freezing of biological materials. *Biodynamica* 8(170):318.

Greaves, R.I.N. 1954. *Theoretical aspects of drying by vacuum sublimination, biological applications of freezing and drying,* ed. R.J. Harris. New York:Academic Press, Inc.

Greaves, R.I.N. 1960. Preservation of living cells by freeze-drying. *Annals of the New York Academy of Sciences* 85(2):723.

Greiff, D. and T.W.G. Rowe. 1976. Recent advances and applications of freeze-drying technology. *Advances in Cryogenic Engineering* 21:418, New York:Plenum Press.

Gresham, R.B. 1964. Freeze-drying of human tissue for clinical use. *Cryobiology* 1(2):150.

Luyet, B.J. 1961. A method for increasing the cooling rate in refrigeration by immersion in liquid N_2 or in other boiling baths. *Biodynamica* 8(171):331.

Mazur, P., S.P. Liebo, and E.H.Y. Chu. 1972. A two-factor hypothesis of freezing injury. *Experimental Cell Research* 71:345.

Meryman, H.T. 1960. General principles of freezing and freezing injury in cellular materials. *Annals of the New York Academy of Sciences* 85(2):503.

Meryman, H.T. 1960. Principles of freeze-drying . *Annals of the New York Academy of Sciences* 85(2):630.

Meryman, H.T. 1960-61. The preparation of biological museum specimens by freeze-drying. *Curator* 3(1):5; 4(2):153.

Meryman, H.T. 1966. "Freeze drying". In *Cryobiology* 610. New York:Academic Press Inc.

Neumann, K. 1962. Freeze-drying—a modern application of refrigeration. *Linde Report, Scientific Technology* 3:45.

Rand, R.W., A.P. Rinfret, and H. Von Leden, eds. 1968. *Cryosurgery.* Springfield, IL, Charles C. Thomas, publisher.

Rowe, T.W.G. 1960. The theory and practice of freeze-drying. *Annals of the New York Academy of Sciences* 85(2):641.

Savu, A., A. Ciobanu, and G. Adam. 1959. Recherches sur la cryodessication de quelques produits enzymatiques utilises dans l'industrie alimentaire. *Proceedings of the 10th International Congress of Refrigeration* 1:616.

Shore, D.T. 1961. Profound hypothermia. *Journal of Refrigeration* 4(3):50.

Stumpf, W.F. and L.J. Roth. 1965. Frozen sectioning below -60 °C with a refrigerated microtome. *Cryobiology* 1(3):227.

Swan, H. and B.C. Paton. 1961. The current status of hypothermia in cardiovascular surgery. *Progress in Cardiovascular Diseases* 4(3):228.

Theilleux *et al.*1959. Preservation of skin by cold after glycerol treatment. *Proceedings of the 10th International Congress of Refrigeration* 1:621.

BIBLIOGRAPHY

Bulletin of the International Institute of Refrigeration, particularly sections devoted to Commission X (Biology and Medicine).

Cryobiology, Journal of the Society for Cryobiology.

CHAPTER 41

CODES AND STANDARDS

THE Codes and Standards listed in Table 1 represent practices, methods, or standards published by the organizations indicated. They are valuable guides for the practicing engineer in determining test methods, ratings, performance requirements, and limits applying to the equipment used in heating, refrigerating, ventilating, and air conditioning. *Copies can usually be obtained from the organization listed in the Reference column.* These listings represent the most recent information available at the time of publication.

Table 1 Codes and Standards Published by Various Societies and Associations

Subject	Title	Publisher	Reference
Acoustics	Standard Acoustical Terminology (reaffirmed 1976)	ASA	ANSI S1.1-1960
Air Conditioners			
Room	Method of Testing for Rating Room Air Conditioners and Packaged Terminal Air Conditioners	ASHRAE	ANSI/ASHRAE 16-1988
	Methods of Testing for Rating Room Fan-Coil Air Conditioners	ASHRAE	ASHRAE 79-1984
	Room Air Conditioners (1986)	UL	ANSI/UL 484-1986
	Room Air Conditioners	AHAM	ANSI/AHAM RAC 1-1982
	Method of Testing for Rating Room Air Conditioner and Packaged Terminal Air Conditioner Heating Capacity	ASHRAE	ANSI/ASHRAE 58-1986
	Commercial and Residential Central Air Conditioners	CSA	C22.2 No. 119-M1985
Packaged Terminal	Packaged Terminal Air Conditioners	ARI	ARI 310-87
	Packaged Terminal Heat Pumps	ARI	ARI 380-87
Transport	Air Conditioning of Aircraft Cargo (1978)	SAE	SAE AIR806A
	Nomenclature, Aircraft Air-Conditioning Equipment (1978)	SAE	SAE ARP147C
Unitary	Air Conditioners, Central Cooling (1984)	UL	ANSI/UL 465-1984
	Load Calculation for Commercial Summer and Winter Air Conditioning, 4th ed. (1988)	ACCA	ACCA Manual N
	Methods of Testing for Rating Heat Operated Unitary Air-Conditioning Equipment for Cooling	ASHRAE	ASHRAE 40-1986
	Methods of Testing for Rating Unitary Air-Conditioning and Heat Pump Equipment	ASHRAE	ANSI/ASHRAE 37-1978
	Methods of Testing for Seasonal Efficiency of Unitary Air Conditioners and Heat Pumps	ASHRAE	ANSI/ASHRAE 116-1983
	Sound Rating of Outdoor Unitary Equipment	ARI	ARI 270-84
	Application of Sound Rated Outdoor Unitary Equipment	ARI	ARI 275-84
	Unitary Air-Conditioning Equipment	ARI	ARI 210-81
	Unitary Air-Conditioning and Air-Source Heat Pump Equipment	ARI	ARI 210/240-89
	Commercial and Industrial Unitary Air-Conditioning Equipment	ARI	ANSI/ARI 360-86
Air Conditioning	Automotive Air-Conditioning Hose (1989)	SAE	SAE J51 MAY89
	Environmental System Technology (1984)	NEBB	NEBB
	Equipment Selection and System Design Procedures for Commercial Summer and Winter Air Conditioning, 1st. ed. (1977)	ACCA	Manual Q
	Gas-Fired Absorption Summer Air Conditioning Appliances (with 1982 addenda)	AGA	ANSI Z21.40.1-1981
	Load Calculation for Residential Winter and Summer Air Conditioning, 7th ed. (1986)	ACCA	ACCA Manual J
	Equipment Selection and System Design Procedures, 2nd ed. (1984)	ACCA	ACCA Manual D
	Installation Standards for Residential Heating and Air Conditioning Systems (1988)	SMACNA	SMACNA
	HVAC Systems—Duct Design (1981)		
	HVAC Systems—Applications, 1st. ed. (1986)	SMACNA	SMACNA
Transport	Air Conditioning Equipment, General Requirements for Subsonic Airplanes (1961)	SAE	SAE ARP85D
	General Requirements for Helicopter Air Conditioning (1970)	SAE	SAE ARP292B
	Testing of Commercial Airplane Environmental Control Systems (1973)	SAE	SAE ARP217B

Table 1 Codes and Standards Published by Various Societies and Associations (*Continued*)

Subject	Title	Publisher	Reference
Air Curtains	Test Methods for Air Curtain Units	AMCA	AMCA 220-82
	Air Volume Terminals	ARI/ADC	ARI/ADC 880-89
	Selection of Distribution Systems, 1st ed. (1963)	ACCA	ACCA Manual G
	Air Distribution Basics for Residential and Small Commercial Buildings	ACCA	ACCA Manual T
	Residential Equipment Selection	ACCA	ACCA Manual S
	Method of Testing for Rating the Air Flow Performance of Outlets and Inlets	ASHRAE	ASHRAE 70-72
	High Temperature Pneumatic Duct Systems for Aircraft (1981)	SAE	SAE ARP699D
	Metric Units and Conversion Factors	AMCA	AMCA 99-0100-76
	Installation Code for Residential Mechanical Exhaust Systems	CSA	C260.1-1975
	Laboratory Certification Manual	ADC	ADC 1062:LCM-83
	Residential Air Exhaust Equipment (1e)	CSA	C260.2-1976
	Test Code for Grilles, Registers and Diffusers	ADC	ADC 1062:GRD-84
	Standard Methods for Laboratory Air Flow Measurement	ASHRAE	ASHRAE 41.2-1987
Air Ducts and Fittings	Flexible Air Duct Test Code	ADC	ADC FD-72 R1-1979
	Installation of Air Conditioning and Ventilating Systems (1989)	NFPA	ANSI/NFPA 90A-1989
	Installation of Warm Air Heating and Air-Conditioning Systems (1989)	NFPA	ANSI/NFPA 90B-1989
	HVAC Duct Construction Standards—Metal and Flexible, 1st ed. (1985)	SMACNA	SMACNA
	Round Industrial Duct Construction (1977)	SMACNA	SMACNA
	Rectangular Industrial Duct Construction (1980)	SMACNA	SMACNA
	Ducted Electric Heat Guide for Air Handling Systems (1971)	SMACNA	SMACNA
	Marine Rigid and Flexible Air Ducting (1986)	UL	UL 1136
	Thermoplastic Duct (PVC) Construction Manual (1974)	SMACNA	SMACNA
	Marine Rigid and Flexible Air Ducting (1986)	UL	UL 1136
	Residential-Type Air-Conditioning Systems	CSA	B228.1-1968
	Factory-Made Air Ducts and Connectors (1986)	UL	UL 181
Air Filters	Test Performance of Air Filter Units (1987)	UL	UL 900
	Method of Testing Air-Cleaning Devices Used in General Ventilation for Removing Particulate Matter	ASHRAE	ASHRAE 52-76
	Commercial and Industrial Air Filter Equipment	ARI	ARI 850-84
	Air Filter Equipment	ARI	ARI 680-86
	High Efficiency, Particulate, Air Filter Units (1985)	UL	UL 586
	Methods of Test for Atmospheric Dust Spot Efficiency and Synthetic Dust Weight Arrestance	BSI	BS 6540 Part 1
	Method for Sodium Flame Test for Air Filters	BSI	BS 3928
	Electrostatic Air Cleaners (1981)	UL	UL 867
Air-Handling Units	Central Station Air-Handling Units	ARI	ARI 430-89
Boilers	Recommended Design Guidelines for Stoker Firing of Bituminous Coals (1983)	ABMA	ABMA
	Boiler Water Limits and Steam Purity Recommendations for Watertube Boilers (3rd ed., 1982)	ABMA	ABMA
	Boiler Water Requirements and Associated Steam Purity—Commercial Boilers (1981)	ABMA	ABMA
	A Guide to Clean and Efficient Operation of Coal Stoker-Fired Boilers	ABMA	ABMA
	Fluidized Bed Combustion Guidelines	ABMA	ABMA
	Guidelines for Industrial Boiler Performance Improvement	ABMA	ABMA
	Operation and Maintenance Safety Manual	ABMA	ABMA
	Matrix of Recommend Quality Control Requirements	ABMA	ABMA
	Thermal Shock Damage to Hot Water Boilers as a Result of Energy Conservation Measures	ABMA	ABMA
	Boiler and Pressure Vessel Code (11 sections) (1986)	ASME	ASME
	Boiler, Pressure Vessel, and Pressure Piping Code	CSA	B51-M1986
	Heating, Water Supply, and Power Boilers—Electric (1980)	UL	UL834
	Lexicon Boiler and Auxiliary Equipment, 5th ed. (1987)	ABMA	ABMA
Cast-Iron	Testing and Rating Heating Boilers (1989)	HYD I	IBR/SBI
	Ratings for Cast-Iron and Steel Boilers (1989)	HYD I	IBR/SBI
Gas or Oil	Explosion Prevention of Fuel Oil and Natural Gas-Fired Single-Burner Boiler-Furnaces (1987)	NFPA	ANSI/NFPA 85A-1987
	Explosion Prevention of Natural Gas-Fired Multiple-Burner Boiler-Furnaces (1989)	NFPA	ANSI/NFPA 85B-1989
	Gas-Fired Low-Pressure Steam and Hot Water Boilers	AGA	ANSI Z21.13-1987
	Gas Utilization Equipment in Large Boilers (with 1972 and 1976 addenda; R-1983, 1989)	AGA	ANSI Z83.3-1971

Codes and Standards

Table 1 Codes and Standards Published by Various Societies and Associations (*Continued*)

Subject	Title	Publisher	Reference
	Oil Fired Boiler Assemblies (1975)	UL	UL 726
	Commercial-Industrial Gas Heating Equipment (1973)	UL	UL 795
	Prevention of Furnace Explosions in Fuel Oil-Fired Multiple-Burner Boiler-Furnaces (1989)	NFPA	ANSI/NFPA 85D-1989
	Control and Safety Devices for Automatically Fired Boilers	ASME	ASME CSD.1-1988
	Oil Burning Appliance Standards (B140 Series)	CSA	B140.7.1-1976
	Oil-Fired Steam and Hot-Water Boilers for Commercial and Industrial Use (1a)	CSA	B140.7.2-1967
Watertube	Recommended Standard Instrument Connections Manual (1982)	SAMA ABMA	ABMA
Building Codes	BOCA National Building Code, 11th ed. (1990)	BOCA	BOCA
	CABO One- and Two-Family Dwelling Code (1989)	CABO	CABO
	Standard Building Code (1988) (with 1989 revision)	SBCCI	SBCCI
	Uniform Building Code (1988)	ICBO	ICBO
	Uniform Building Code Standards (1988)	ICBO	ICBO
	BOCA National Property Maintenance Code, 3rd ed. (1990)	BOCA	BOCA
	Model Energy Code (1989)	CABO	CABO
	Directory of Building Codes and Regulations, (1988 ed.)	NCSBCS	NCSBCS
Mechanical	BOCA National Mechanical Code, 7th ed. (1990)	BOCA	BOCA
	Safety Code for Elevators and Escalators (plus two yearly supplements)	ASME	ANSI/ASME A17.1-1987
	Uniform Mechanical Code (1988) (with Uniform Mechanical Code Standards)	ICBO/ IAPMO	ICBO/IAPMO
	Standard Mechanical Code (1988) (with 1989 revisions)	SBCCI	SBCCI
	Standard Gas Code (1988) (with 1989 revisions)	SBCCI	SBCCI
Burners	Installation of Domestic Gas Conversion Burners	AGA	ANSI Z21.8-1984
	Domestic Gas Conversion Burners	AGA	ANSI Z21.17-1984
	Commercial-Industrial Gas Heating Equipment (1973)	UL	UL 795
	Oil Burners (1989)	UL	UL 296
	Installation Code for Oil Burning Equipment	CSA	B139-1976
	Supplement No. 1 to B139-1976, Installation Code for Oil Burning Equipment	CSA	B139S1-1982
	General Requirements for Oil Burning Equipment	CAN/CSA	B140.0-M1987
	Vaporizing-Type Oil Burners	CSA	B140.1-1966 (R1980)
	Oil Burners, Atomizing Type	CSA	B140.2.1-1973
	Pressure Atomizing Oil Burner Nozzles	CSA	B140.2.2-1971 (R1980)
	Replacement Burners and Replacement Combustion Heads for Residential Oil Burners	CSA	B140.2.3-M1981
	Guidelines for Burner Adjustments of Commercial Oil-Fired Boilers	ABMA	ABMA
Capillary Tubes	Methods of Testing Flow Capacity of Refrigerant Capillary Tubes	ASHRAE	ASHRAE 28-78
Chillers	Methods of Testing Liquid Chilling Packages	ASHRAE	ASHRAE 30-78
	Absorption Water-Chilling Packages	ARI	ARI 560-82
	Centrifugal Water-Chilling Packages	ARI	ARI 550-88
	Reciprocating Water-Chilling Packages	ARI	ARI 590-86
Chimneys	Chimneys, Fireplaces, and Vents, and Solid Fuel Burning Appliances	NFPA	ANSI/NFPA 211-1988
	Chimneys, Factory-Built, Residential Type and Building Heating Appliance (1983)	UL	UL 103
	Chimneys, Factory-Built, Medium Heat Appliance (1986)	UL	UL 959
Cleanrooms	Procedural Standards for Certified Testing of Cleanrooms (1988)	NEBB	NEBB-1988
Coils	Forced-Circulation Air-Cooling and Air-Heating Coils	ARI	ARI 410-87
	Methods of Testing Forced Circulation Air Cooling and Air Heating Coils	ASHRAE	ASHRAE 33-78
Comfort Conditions	Thermal Environmental Conditions for Human Occupancy	ASHRAE	ANSI/ASHRAE 55-1981
Compressors	Compressors and Exhausters (reaffirmed 1985)	ASME	ANSI PTC 10-74
	Compressed Air and Gas Handbook, 5th ed. (1989)	CAGI	CAGI
	Safety Standard for Compressors for Process Industries	ASME	ASME/ANSI B19.3-1986
	Safety Standard for Air Compressor Systems	ASME	ANSI/ASME B19.1-1985
	Displacement Compressors, Vacuum Pumps and Blowers	ASME	ANSI PTC 9-1974(R1985)
	Compressors and Exhausters	ASME	ANSI PTC 10-1974 (R1985)

Table 1 Codes and Standards Published by Various Societies and Associations (*Continued*)

Subject	Title	Publisher	Reference
Refrigeration	Methods of Testing for Rating Positive Displacement Refrigerant Compressors	ASHRAE	ASHRAE 23-78
	Ammonia Compressor Units	ARI	ARI 510-87
	Hermetic Refrigerant Motor-Compressors (1984)	UL	UL 984
	Positive Displacement Refrigerant Compressors and Condensing Units	ARI	ANSI/ARI 520-85
Computers	Protection of Electronic Computer/Data Processing Equipment	NFPA	ANSI/NFPA 75-1989
Condensers	Water-Cooled Refrigerant Condensers, Remote Type	ARI	ARI 450-87
	Methods of Testing for Rating Remote Mechanical-Draft Air-Cooled Refrigerant Condensers	ASHRAE	ASHRAE 20-70
	Methods of Testing for Rating Water-Cooled Refrigerant Condensers	ASHRAE	ASHRAE 22-78
	Methods of Testing Remote Mechanical-Draft Evaporative Refrigerant Condensers	ASHRAE	ASHRAE 64-74
	Remote Mechanical Draft Air-Cooled Refrigerant Condensers	ARI	ARI 460-87
	Standards for Steam Surface Condensers, 8th ed. (1984)	HEI	HEI
	Addendum I, Standards for Steam Surface Condensers (1989)	HEI	HEI
Condensing Units	Methods of Testing for Rating Positive Displacement Condensing Units	ASHRAE	ASHRAE 14-80
	Refrigeration and Air-Conditioning Condensing and Compressor Units (1987)	UL	ANSI/UL 303-1988
	Commercial and Industrial Unitary Air-Conditioning Condensing Units	ARI	ARI 365-87
Contactors	Definite Purpose Magnetic Contactors	ARI	ARI 780-86
	Definite Purpose Contactors for Limited Duty	ARI	ARI 790-86
Controls	Quick-Disconnect Devices for Use with Gas Fuel	AGA	ANSI Z21.41-1989
	Limit Controls (1989)	UL	UL 353
	Energy Management Control Systems Instrumentation	ASHRAE	ASHRAE 114-1986
	Primary Safety Controls for Gas- and Oil-Fired Appliances (1985)	UL	UL 372
	Temperature-Indicating and Regulating Equipment (1988)	UL	UL 873
	Industrial Control Equipment (1984)	UL	UL 508
	Temperature-Indicating and Regulating Equipment	CSA	C22.2 No. 24-1987
Residential	Automatic Gas Ignition Systems and Components (with 1988 and 1989 addenda)	AGA	ANSI Z21.20-1985
	Temperature Limit Controls for Electric Baseboard Heaters	NEMA	NEMA DC 10-1983 (R 1989)
	Electric Quick-Connect Terminals (1985)	UL	UL 310
	Quick Connect Terminals	NEMA	ANSI/NEMA DC 2-1982 (R 1988)
	Line-Voltage Integrally Mounted Thermostats for Electric Heaters	NEMA	NEMA DC 13-1979 (R 1985)
	Wall-Mounted Room Thermostats	NEMA	NEMA DC3-1984
	Hot-Water Immersion Controls	NEMA	NEMA DC 12-1985
	Warm Air Limit and Fan Controls	NEMA	NEMA DC 4-1986
	Gas Appliance Thermostats (with 1985 and 1988 addenda)	AGA	ANSI Z21.23-1980
	Gas Appliance Pressure Regulators (with 1989 addenda)	AGA	ANSI Z21.18-1987
	Manually-Operated Prego. Electric Spark Gas Ignition Systems and Components	AGA	ANSI Z21.77-1989
	Residential Controls—Class 2 Transformers	NEMA	NEMA DC 20-1986
Coolers			
Air	Methods of Testing Forced Convection and Natural Convection Air Coolers for Refrigeration	ASHRAE	ASHRAE 25-77
	Unit Coolers for Refrigeration	ARI	ARI 420-89
Bottled Beverage	Methods of Testing and Rating Bottled and Canned Beverage Venders and Coolers	ASHRAE	ASHRAE 32-1986
	Refrigerated Vending Machines (1989)	UL	UL 541
Drinking Water	Methods of Testing for Rating Drinking-Water Coolers with Self-Contained Mechanical Refrigeration Systems	ASHRAE	ASHRAE 18-1987
	Drinking Water Coolers (1987)	UL	UL 399
	Drinking Fountains and Self-Contained, Mechanically Refrigerated Drinking Water Coolers	ARI ANSI	ANSI/ARI 1010-84
	Application and Installation of Drinking Water Coolers	ARI	ANSI/ARI 1020-84
	Drinking Water Coolers and Beverage Dispensers	CSA	C22.2 No. 91-1971 (R 1981)
Liquid	Methods of Testing for Rating Liquid Coolers	ASHRAE	ASHRAE 24-78
	Refrigerant-Cooled Liquid Coolers, Remote Type	ARI	ARI 480-87

Codes and Standards

Table 1 Codes and Standards Published by Various Societies and Associations (*Continued*)

Subject	Title	Publisher	Reference
Cooling Towers	Water-Cooling Towers	NFPA	ANSI/NFPA 214-1988
	Atmospheric Water Cooling Equipment	ASME	ANSI/ASME PTC 23-1986
	Acceptance Test Code for Water Cooling Towers: Mechanical Draft, Natural Draft Fan Assisted Types, and Evaluation of Results. Addendum-1, For Thermal Testing of Wet/Dry Cooling Towers (1986)	CTI	CTI ATC-105-1982
	Acceptance Test Code for Spray Cooling Systems (1985)	CTI	CTI ATC-133-1985
	Certification Standard for Commercial Water Cooling Towers (1986)	CTI	CTI STD-201-1986
	Code for Measurement of Sound from Water Cooling Towers	CTI	CTI ATC-128-1981
	Fiberglass-Reinforced Plastic Panels for Application on Industrial Water Cooling Towers	CTI	CTI STD-131-1986
	Nomenclature for Industrial Water-Cooling Towers	CTI	CTI NCL-109-1983
Dehumidifiers	Dehumidifiers	AHAM	ANSI/AHAM DH 1-1986
	Dehumidifiers (3a)	CSA	C22.2 No. 92-1971
	Dehumidifiers (1987)	UL	UL 474
Desiccants	Method of Testing Desiccants for Refrigerant Drying	ASHRAE	ASHRAE 35-1983
Driers	Liquid Line Driers	ARI	ARI 710-86
	Method of Testing Liquid Line Refrigerant Driers	ASHRAE	ASHRAE 63-79
Electrical	National Electric Code (1990)	ANSI/NFPA	ANSI/NFPA 70-1990
	Canadian Electrical Code, Part 1 (15th ed.)	CSA	C22.1-1990
	Compatibility of Electrical Connectors and Wiring (1988)	SAE	SAE AIR1329 A
	Manufacturers' Identification of Electrical Connector Contacts, Terminals and Splices (1982)	SAE	SAE AIR1351 A
	Voltage Ratings for Electrical Power Systems and Equipment	ANSI	ANSI C84.1-1982
Energy	Air Conditioning and Refrigerating Equipment Nameplate Voltages	ARI	ARI 110-80
	Energy Conservation in New Building Design	ASHRAE	ANSI/ASHRAE/IES 90A-1980
	Energy Conservation in Existing Buildings—High Rise Residential	ASHRAE	ANSI/ASHRAE/IES 100.2-1981
	Energy Conservation in Existing Buildings—Commercial	ASHRAE	ANSI/ASHRAE/IES 100.3-1985
	Energy Conservation in Existing Facilities—Industrial	ASHRAE	ANSI/ASHRAE/IES 100.4-1984
	Energy Conservation in Existing Buildings—Institutional	ASHRAE	ANSI/ASHRAE/IES 100.5-1981
	Energy Conservation in Existing Buildings—Public Assembly	ASHRAE	ANSI/ASHRAE/IES 100.6-1981
	Energy Recovery Equipment and Systems, Air-to-Air (1978)	SMACNA	SMACNA
	Energy Conservation Guidelines (1984)	SMACNA	SMACNA
	Model Energy Code (MEC) (1989)	CABO	BOCA/ICBO/SBCCI
	Energy Management Equipment (1984)	UL	UL 916
	Retrofit of Building Energy Systems and Processes (1982)	SMACNA	SMACNA
Exhaust Systems	Method of Testing Performance of Laboratory Fume Hoods	ASHRAE	ASHRAE 110-1985
	Installation of Blower and Exhaust Systems for Dust, Stock, Vapor Removal or Conveying (1990)	NFPA	ANSI/NFPA 91-1990
	Fundamentals Governing the Design and Operation of Local Exhaust Systems	AIHA	ANSI Z9.2-1979
	Safety Code for Design, Construction, and Ventilation of Spray Finishing Operations (reaffirmed 1971)	ANSI	ANSI Z9.3-1985
	Ventilation and Safe Practices of Abrasives Blasting Operations	ANSI	ANSI Z9.4-1985
	Compressors and Exhausters	ASME	ANSI PTC 10-1974 (R 1985)
	Mechanical Flue-Gas Exhausters	CAN	3-B255-M81
	Draft Equipment (1973)	UL	UL 378
Expansion Valves	Method of Testing for Capacity Rating of Thermostatic Refrigerant Expansion Valves	ASHRAE	ANSI/ASHRAE 17-1986
	Thermostatic Refrigerant Expansion Valves	ARI	ARI 750-87
Fan Coil Units	Room Fan-Coil Air Conditioners	ARI	ARI 440-89
	Fan Coil Units and Room Fan Heater Units (1986)	UL	ANSI/UL 883-1986
	Methods of Testing for Rating Room Fan-Coil Air Conditioners	ASHRAE	ASHRAE 79-1984

Table 1 Codes and Standards Published by Various Societies and Associations (*Continued*)

Subject	Title	Publisher	Reference
Fans	Standards Handbook	AMCA	AMCA 99-86
	Fans	ASME	ANSI/ASME PTC 11-1984
	Electric Fans (1977)	UL	UL 507
	Laboratory Methods of Testing Fans for Rating	ASHRAE	ANSI/ASHRAE 51-1985 ANSI/AMCA 210-85
	Methods of Testing Dynamic Characteristics of Propeller Fans—Aerodynamically Excited Fan Vibrations and Critical Speeds	ASHRAE	ANSI/ASHRAE 87.1-1983
	Laboratory Methods of Testing Fans for Rating	AMCA	ANSI/AMCA 210-85
	Drive Arrangements for Centrifugal Fans	AMCA	AMCA 99-2404-78
	Designation for Rotation and Discharge of Centrifugal Fans	AMCA	AMCA 99-2406-83
	Motor Positions for Belt or Chain Drive Centrifugal Fans	AMCA	AMCA 99-2407-66
	Drive Arrangements for Tubular Centrifugal Fans	AMCA	AMCA 99-2410-82
	Site Performance Test Standard Power Plant and Industrial Fans	AMCA	AMCA 803-87
	Fans and Blowers	ARI	ARI 670-85
	Inlet Box Positions for Centrifugal Fans	AMCA	AMCA 99-2405-83
	Fans and Ventilators	CSA	C22.2 No. 113-M1984
Ceiling	AC Electric Fans and Regulators	AMCA	ANSI-IEC Pub. 385
Filters	Flow-Capacity Rating and Application of Suction-Line Filters and Filter Driers	ARI	ARI 730-86
	Grease Filters for Exhaust Ducts (1979)	UL	UL 1046
	Grease Extractors for Exhaust Ducts (1981)	UL	UL 710
Fire Dampers	Fire Dampers and Ceiling Dampers (1979)	UL	UL 555
Fireplaces	Factory-Built Fireplaces (1988)	UL	UL 127
Fire Protection	BOCA National Fire Prevention Code, 8th ed. (1990)	BOCA	BOCA
	National Fire Codes (issued annually)	NFPA	NFPA
	Fire Prevention Code	NFPA	ANSI/NFPA 1-1987
	Fire Protection Handbook, 16th ed.	NFPA	NFPA
	Flammable and Combustible Liquids Code	NFPA	ANSI/NFPA 30-1987
	Heat Responsive Links for Fire Protection Service (1987)	UL	UL 33
	Test Method for Surface Burning Characteristics of Building Materials	ASTM/ NFPA	ASTM E 84-89a
	Fire Doors and Windows	NFPA	ANSI/NFPA 80-1990
	Fire Tests of Building Construction and Materials (1986)	UL	UL 263
	Life Safety	NFPA	ANSI/NFPA 101-1987
	Standard Fire Prevention Code (1988) (with 1989 revisions)	SBCCI	SBCCI
	Standard Method of Fire Tests of Door Assemblies	NFPA	ANSI/NFPA 252-1984
	Uniform Fire Code (1988)	ICBO/ WFCA	ICBO/WFCA
	Uniform Fire Code Standards (1988)	ICBO/ WFCA	ICBO/WFCA
Fireplace Stoves	Fireplace Stoves (1988)	UL	UL 737
Flow Capacity	Method of Testing Flow Capacity of Suction Line Filters and Filter Driers	ASHRAE	ASHRAE 78-1985
Freezers			
Household	Household Refrigerators and Freezers (1983)	UL	UL 250
	Household Refrigerators, Combination Refrigerator-Freezers, and Household Freezers	AHAM	ANSI/AHAM HRF 1-1988
	Capacity Measurement and Energy Consumption Test Methods for Refrigerators, Combination Refrigerator-Freezers, and Household Freezers	CSA	C300 M89
Commercial	Ice Cream Makers (1986)	UL	UL 621
	Soda Fountain and Luncheonette Equipment	NSF	NSF-1
	Dispensing Freezers	NSF	NSF-6
	Commercial Refrigerators and Freezers (1985)	UL	UL 471
	Food Service Refrigerators and Storage Freezers	NSF	NSF-7
Furnaces	Gas-Fired Gravity and Fan Type Direct Vent Wall Furnaces (with 1989 addenda)	AGA	ANSI Z21.44-1988
	Gas-Fired Central Furnaces (except Direct Vent) (with 1988 addenda)	AGA	ANSI Z21.47-1987
	Direct Vent Central Furnaces	AGA	ANSI Z21.64-1988

Codes and Standards

Table 1 Codes and Standards Published by Various Societies and Associations (*Continued*)

Subject	Title	Publisher	Reference
Furnaces (continued)	Gas-Fired Gravity and Fan Type Floor Furnaces (with 1987 and 1988 addenda)	AGA	ANSI Z21.48-1986
	Commercial-Industrial Gas Heating Equipment (1973)	UL	UL 795
	Solid-Fuel and Combination-Fuel Central and Supplementary Furnaces (1981)	UL	UL 391
	Gas-Fired Gravity and Fan Type Vented Wall Furnaces (with 1987 and 1988 addenda)	AGA	ANSI Z21.49-1986
	Methods of Testing for Rating Non-Residential Warm Air Heaters	ASHRAE	ASHRAE 45-1986
	Methods of Testing for Heating Seasonal Efficiency of Central Furnaces and Boilers	ASHRAE	ANSI/ASHRAE 103-1982
	Installation of Oil Burning Equipment	NFPA	NFPA 31-1987
	Oil-Fired Central Furnaces (1986)	UL	UL 727
	Gas-Fired Duct Furnaces (with 1989 addenda)	AGA	ANSI Z83.9-1986
	Oil-Fired Floor Furnaces (1987)	UL	UL 729
	Oil-Fired Wall Furnaces (1987)	UL	UL 730
	Standard Gas Code (1988) (with 1989 revisions)	SBCCI	SBCCI
	Oil Burning Stoves and Water Heaters (2a)	CSA	B140.3-1962 (R1980)
	Oil-Fired Warm Air Furnaces (8a)	CSA	B140.4-1974
	Installation Code for Solid-Fuel-Burning Appliances and Equipment	CAN/CSA	B365-M87
	Solid Fuel-Fired Central Heating Appliances	CAN/CSA	B366.1-M87
	Residential Gas Detectors (1983)	UL	UL 1484
	Electric Central Warm-Air Furnaces	CSA	C22.2 No. 23-1980
Heat Exchangers	Standards of Tubular Exchanger Manufacturers Association, 7th ed. (1988)	TEMA	TEMA
	Liquid Suction Heat Exchangers	ARI	ARI 490-79
	Method of Testing Air-to-Air Heat Exchangers	ASHRAE	ASHRAE 84-78
	Standards for Power Plant Heat Exchangers, 2nd ed. (1989)	HEI	HEI
Heat Pumps	Heat Pumps (1985)	UL	ANSI/UL 559-1985
	Air-Source Unitary Heat Pump Equipment	ARI	ARI 240-81
	Water-Source Heat Pumps	ARI	ANSI/ARI 320-86
	Ground Water-Source Heat Pumps	ARI	ARI 325-85
	Commercial and Industrial Heat Pump Equipment	ARI	ANSI/ARI 340-86
	Central Forced-Air Unitary Heat Pumps with or without Electric Resistance Heat	CSA	C22.2 No. 186.1 M1980
	Add-on Heat Pumps	CSA	C22.2 No. 186.2-M1980
	Performance Standard for Unitary Heat Pumps (1a)	CSA	C273.3-M1977
	Installation Requirements for Air-to-Air Heat Pumps (2a)	CSA	C273.5-1980
	Methods of Testing for Rating Unitary Air-Conditioning and Heat Pump Equipment	ASHRAE	ANSI/ASHRAE 37-1978
Heat Recovery	Energy Recovery Equipment and Systems, Air-to-Air (1978)	SMACNA	SMACNA
	Gas Turbine Heat Recovery Steam Generators	ASME	ANSI/ASME PTC 4.4-1981 (R 1987)
Heaters	Air Heaters	ASME	ANSI PTC 4.3-1974 (R 1985)
	Desuperheater/Water Heaters	ARI	ARI 470-87
	Electric Heaters for Use in Hazardous (Classified) Locations (1985)	UL	UL 823
	Standards for Closed Feedwater Heaters, 4th ed. (1984)	HEI	HEI
	Oil-Fired Air Heaters and Direct-Fired Heaters (1975)	UL	UL 733
	Oil-Fired Room Heaters (1973)	UL	UL 896
	Solid Fuel-Type Room Heaters (1988)	UL	UL 1482
	Gas-Fired Room Heaters, Vol. I, Vented Room Heaters (with 1989 addenda)	AGA	ANSI Z21.11.1-1988
	Gas-Fired Room Heaters, Vol. II, Unvented Room Heaters (with 1984 and 1988 addenda)	AGA	ANSI Z21.11.2-1983
	Gas-Fired Infrared Heaters (with 1989 addenda)	AGA	ANSI Z83.6-1987
	Gas-Fired Construction Heaters (with 1989 addenda)	AGA	ANSI Z83.7-1987
	Direct Gas-Fired Make-Up Air Heaters	AGA	ANSI Z83.4-1989
	Gas-Fired Unvented Commercial and Industrial Heaters (with 1984 and 1989 addenda)	AGA	ANSI Z83.16-1982
	Direct Gas-Fired Industrial Air Heaters	AGA	ANSI Z83.17-1988
	Gas-Fired Pool Heaters (with 1987 and 1988 addenda)	AGA	ANSI Z21.56-1986
	Motor Vehicle Heater Test Procedure (1982)	SAE	SAE J638 JUN82
	Electric Heating Appliances (1987)	UL	UL 499
	Gas-Fired Heating Equipment, Commercial-Industrial (1973)	UL	UL 795
	Fuel-Fired Heaters—Air Heating—for Construction and Industrial Machinery (1989)	SAE	SAE J1024 MAY89

Table 1 Codes and Standards Published by Various Societies and Associations (*Continued*)

Subject	Title	Publisher	Reference
Heaters (continued)	Electric Air Heaters (1980)	UL	UL 1025
	Electric Central Air Heating Equipment (1986)	UL	ANSI/UL 1096-1985
	Direct Gas-Fired Door Heaters (with 1989 addenda)	AGA	ANSI Z83.17-1988
	Space Heaters for Use with Solid Fuels	CSA	B366.2 M1984
	Unvented Kerosene-Fired Room Heaters and Portable Heaters (1982)	UL	UL 647
	Electric Oil Heaters (1980)	UL	UL 574
Heating	Aircraft Electrical Heating Systems (1965) (reaffirmed 1983)	SAE	SAE AIR860
	Environmental System Technology (1984)	NEBB	NEBB
	Installation and Operation of Pulverized Fuel Systems	NFPA	ANSI/NFPA 85F-1988
	Manual for Calculating Heat Loss and Heat Gain for Electric Comfort Conditioning	NEMA	NEMA HE 1-1980 (R1986)
	Heat Loss Calculation Guide (1984)	HYD I	IBR H-21
	Installation Guide for Residential Hydronic Heating Systems, 6th ed. (1988)	HYD I	IBR 200
	Advanced Installation Guide for Hydronic Heating Systems, 2nd ed.	HYD I	IBR 250
	Installation Standards for Residential Heating and Air Conditioning Systems (1988)	SMACNA	SMACNA
	HVAC Systems—Applications, 1st ed. (1986)	SMACNA	SMACNA
	Electric Baseboard Heating Equipment (1987)	UL	UL 1042
	Electric Central Air Heating Equipment (1986)	UL	UL 1096
	Portable Industrial Oil-Fired Heaters	CSA	B140.8-1967 (R1980)
	Portable Kerosene-Fired Heaters	CAN	3-B140.9.3 M86
	Oil-Fired Service Water Heaters and Swimming Pool Heaters (7a)	CSA	B140.12-1976
	Automatic Flue-Pipe Dampers for Use with Oil-Fired Appliances	CSA	B140.14-M1979
	Determining the Required Capacity of Residential Space Heating and Cooling Appliances	CAN/CSA	CAN/CSA-F280-M86
	Performance Standard for Electrical Baseboard Heaters	CSA	C273.2-1971
	Performance Requirements for Electric Heating Line-Voltage Wall Thermostats	CSA	C273.4-M1978
	Electric Air Heaters	CSA	C22.2 No. 46-M1988
	Heater Elements	CSA	C22.2 No. 72-M1984
Humidifiers	Humidifiers (1987)	UL	UL 998
	Central System Humidifiers	ARI	ARI 610-89
	Self-Contained Humidifiers	ARI	ARI 620-89
	Appliance Humidifiers	AHAM	ANSI/AHAM HU 1-1987
Ice Makers	Ice Makers (1984)	UL	UL 563
	Methods of Testing Automatic Ice Makers	ASHRAE	ASHRAE 29-78
	Automatic Commercial Ice Makers	ARI	ARI 810-87
	Split System Automatic Commercial Ice Makers	ARI	ARI 815-87
	Ice Storage Bins	ARI	ARI 820-88
	Automatic Ice-Making Equipment	NSF	NSF-12
Incinerators	Residential Incinerators (1973)	UL	UL 791
	Incinerators, Waste and Linen Handling Systems and Equipment	NFPA	ANSI/NFPA 82-1990
	Incinerator Performance	CSA	Z103-1976
Induction Units	Room Air-Induction Units	ARI	ARI 445-87
Industrial Duct	Round Industrial Duct Construction (1977)	SMACNA	SMACNA
	Rectangular Industrial Duct Construction (1980)	SMACNA	SMACNA
Infrared Sensing Devices	Application of Infrared Sensing Devices to the Assessment of Building Heat Loss Characteristics	ASHRAE	ANSI/ASHRAE 101-1981
Insulation	Commercial and Industrial Insulation Standards	MICA	MICA 1983
	Test Method for Steady-State and Thermal Performance of Building Assemblies by Means of a Guarded Hot Box	ASTM	ASTM C236-87
	Test Method for Steady-State Heat Flux Measurements and Thermal Transmission Properties by Means of the Guarded Hot Plate Apparatus	ASTM	ASTM C177-85
	Test Method for Steady-State Heat Transfer Properties of Horizontal Pipe Insulations	ASTM	ASTM C335-89
	Test Method for Steady-State Heat Flux Measurements and Thermal Transmission Properties by Means of the Heat Flow Meter Apparatus	ASTM	ASTM C518-85
	Specification for Thermal and Acoustical Insulation (Mineral Fiber, Duct Lining Material)	ASTM	ASTM C1071-86

Codes and Standards 41.9

Table 1 Codes and Standards Published by Various Societies and Associations (*Continued*)

Subject	Title	Publisher	Reference
	Thermal Insulation, Mineral Fiber, for Buildings	CSA	A101 M-1983
Louvers	Test Method for Louvers, Dampers, and Shutters	AMCA	AMCA 500-89
Lubricants	Test Method for Carbon-Type Composition of Insulating Oils of Petroleum Origin	ASTM	ASTM D2140-86
	Method for Conversion of Kinematic Viscosity to Saybolt Universal Viscosity or to Saybolt Furol Viscosity	ASTM	ASTM D2161-87
	Method for Calculating Viscosity Index from Kinematic Viscosity at 40 and 100 °C	ASTM	ASTM D2270-86
	Method for Estimation of Molecular Weight of Petroleum Oils from Viscosity Measurements	ASTM	ASTM D2502-87
	Test Method for Molecular Weight of Hydrocarbons by Thermoelectric Measurement of Vapor Pressure	ASTM	ASTM D2503-82 (1987)
	Test Method for Mean Molecular Weight of Mineral Insulating Oils by the Cryoscopic Method	ASTM	ASTM D2224-78 (1983)
	Test Methods for Pour Point of Petroleum Oils	ASTM	ASTM D97-87
	Recommended Practice for Viscosity System for Industrial Fluid Lubricants	ASTM	ASTM D2422-86
	Test Method for Dielectric Breakdown Voltage of Insulating Liquids Using Disk Electrodes	ASTM	ASTM D877-87
	Test Method for Dielectric Breakdown Voltage of Insulating Oils of Petroleum Origin Using VDE Electrodes	ASTM	ASTM D1816-84a
	Method for Separation of Representative Aromatics and Nonaromatics Fractions of High-Boiling Oils by Elution Chromatography	ASTM	ASTM D2549-85
	Method of Testing the Floc Point of Refrigeration Grade Oils	ASHRAE	ANSI/ASHRAE 86-1983
Measurements	Standard Method for Temperature Measurement	ASHRAE	ASHRAE 41.1-1986
	Standard Method for Measurement of Proportion of Oil in Liquid Refrigerant	ASHRAE	ANSI/ASHRAE 41.4-1984
	Standard Method for Measurement of Moist Air Properties	ASHRAE	ANSI/ASHRAE 41.6-1982
	Standard Method for Measurement of Flow of Gas	ASHRAE	ASHRAE 41.7-1984
	Standard Methods of Measurement of Flow of Fluids—Liquids	ASHRAE	ASHRAE 41.8-78
	Standard Methods of Measuring and Expressing Building Energy Performance	ASHRAE	ASHRAE 105-1984
	Procedure for Bench Calibration of Tank Level Gaging Tapes and Sounding Rules	ASME	ANSI B88.2-1974 (R 1981)
	Guide for Dynamic Calibration of Pressure Transducers	ASME	ANSI MC88-1-1972 (R 1978)
	Glossary of Terms Used in the Measurement of Fluid Flow in Pipes	ASME	ANSI/ASME MFC-1M-1979 (R 1986)
	Measurement Uncertainty for Fluid Flow in Closed Conduits	ASME	ANSI/ASME MFC-2M-1983 (R 1988)
	Measurement of Fluid Flow in Pipes Using Orifice, Nozzle, and Venturi	ASME	ASME MFC-3M-1985
	Measurement of Gas Flow by Turbine Meters	ASME	ANSI/ASME MFC-4M-1986
	Measurement of Liquid Flow in Closed Conduits Using Transit-Time Ultrasonic Flowmeters	ASME	ANSI/ASME MFC-5M-1985
	Measurement of Fluid Flow in Pipes Using Vortex Flow Meters	ASME	ASME/ANSI MFC-6M-1987
	Measurement of Gas Flow by Means of Critical Flow Venturi Nozzles	ASME	ASME/ANSI MFC-7M-1987
	Measurement Uncertainty	ASME	ANSI/ASME PTC 19.1-1985
	Pressure Measurement	ASME	ASME/ANSI PTC 19.2-1987
	Temperature Measurement	ANSI	ANSI/ASME PTC 19.3-1985
	Measurement of Rotary Speed	ASME	ANSI PTC 19.13-1974 (R 1986)
	Measurement of Industrial Sound	ASME	ANSI/ASME PTC 36-1985
Mobile Homes and Recreational Vehicles	Plumbing System Components for Manufactured Homes and Recreational Vehicles	NSF	NSF-24
	Recreational Vehicle Parks	NFPA	NFPA 501C-1990
	Shear Resistance Tests for Ceiling Boards for Mobile Homes (1980)	UL	UL 1296
	Roof Trusses for Mobile Homes (1979)	UL	UL 1298

Table 1 Codes and Standards Published by Various Societies and Associations (*Continued*)

Subject	Title	Publisher	Reference
Mobile Homes and Recreational Vechicles (continued)	Gas Burning Heating Appliances for Mobile Homes and Recreational Vehicles (1965)	UL	UL 307B
	Gas Supply Connectors for Manufactured Homes	IAPMO	IAPMO TSC 9-1989
	Liquid-Fuel-Burning Heating Appliances for Mobile Homes and Recreational Vehicles (1978)	UL	UL 307A
	Recreational Vehicle Cooking Gas Appliances (with 1989 addenda)	AGA	ANSI Z21.57-1987
	Gas-Fired Cooking Appliances for Recreational Vehicles (1976)	UL	UL 1075
	Oil-Fired Warm-Air Heating Appliances for Mobile Housing and Recreational Vehicles (2a)	CSA	B140.10-1974 (R1981)
	Mobile Homes	CSA	CAN/CSA-Z240 MH Series-M86
	Recreational Vehicles	CSA	CAN/CSA-Z240 RV Series-M86
	Manufactured Home Installations	NCSBCS	ANSI A225.1-87
	Mobile Home Parks	CSA	Z240.7.1-1972
	Recreational Vehicle Parks	CSA	Z240.7.2-1972
	Roof Jacks for Mobile Homes and Recreational Vehicles (1986)	UL	UL 311
Motors and Generators	Motors and Generators	NEMA	NEMA MG 1-1987
	Impedance Protected Motors (1982)	UL	UL 519
	Electric Motors (1989)	UL	UL 1004
	Energy Efficiency Test Methods for Three-Phase Induction Motors/ Efficiency Quoting Method and Permissible Efficiency Tolerance	CSA	C390-1985
	Steam Generating Units	ASME	ANSI/ASME PTC 4.1-1964 (R 1985)
	Electric Motors and Generators for Use in Hazardous (Classified) Locations (1978)	UL	UL 674
	Code on Nuclear Air and Gas Treatment	ASME	ASME/ANSI AG-1-1988
	Nuclear Power Plant Air Cleaning Units and Components	ASME	ANSI/ASME N509-1980
	Testing and Nuclear Air-Cleaning Systems	ASME	ANSI/ASME N510-1980
Outlets and Inlets	Method of Testing for Rating the Air Flow Performance of Outlets and Inlets	ASHRAE	ASHRAE 70-72
Pipe, Tubing, and Fittings	Power Piping	ASME	ANSI/ASME B31.1-1986
	Plastics Piping Components and Related Materials	NSF	NSF-14
	Scheme for the Identification of Piping Systems	ASME	ANSI/ASME A13.1-1981 (R 1985)
	National Fuel Gas Code	NFPA	ANSI/NFPA 54-1988
		AGA	ANSI Z223.1-1988
	Refrigeration Piping	ASME	ASME/ANSI B31.5-1987
	Refrigeration Tube Fittings (1977)	SAE	ANSI/SAE J513 OCT77
	Specification for Seamless Copper Pipe, Standard Sizes	ASTM	ASTM B42-88
	Specification for Acrylonitrile-Butadiene-Styrene (ABS) Plastic Pipe, Schedules 40 and 80	ASTM	ASTM D1527-89
	Specification for Poly (Vinyl Chloride) (PVC) Plastic Pipe, Schedules 40, 80, and 120	ASTM	ASTM D1785-89
	Specification for Polyethylene (PE) Plastic Pipe, Schedule 40	ASTM	ASTM D2104-89
	Standards of the Expansion Joint Manufacturers Association, Inc., 5th ed. (1980 with 1985 addenda)	EJMA	EJMA
	Rubber Gasketed Fittings for Fire Protection Service (1978)	UL	UL 213
	Tube Fittings for Flammable and Combustible Fluids, Refrigeration Service and Marine Use (1978)	UL	UL 109
Plumbing	BOCA National Plumbing Code, 8th ed. (1990)	BOCA	BOCA
	Standard Plumbing Code (1988) (with 1989 revisions)	SBCCI	SBCCI
	Uniform Plumbing Code (1988)	IAPMO	IAPMO
Pumps	Circulation System Components for Swimming Pools, Spas, or Hot Tubs	NSF	NSF-50
	Pumps for Oil-Burning Appliances (1986)	UL	UL 343
	Motor-Operated Water Pumps (1980)	UL	UL 778
	Electric Swimming Pool Pumps, Filters and Chlorinators (1986)	UL	UL 1081
	Hydraulic Institute Standards, 14th ed. (1983)	HI	HI
	Hydraulic Institute Engineering Data Book, 1st ed. (1979)	HI	HI
	Performance Standard for Liquid Ring Vacuum Pumps, 1st ed. (1987)	HEI	HEI
	Centrifugal Pumps	ASME	ASME PTC 8.2-1990
	Displacement Compressors, Vacuum Pumps and Blowers	ASME	ANSI/ASME PTC 9-1970 (R 1985)

Codes and Standards

Table 1 Codes and Standards Published by Various Societies and Associations (*Continued*)

Subject	Title	Publisher	Reference
Radiation	Testing and Rating Code for Baseboard Radiation, 6th ed. (1981)	HYD I	IBR
	Testing and Rating Code for Finned-Tube Commercial Radiation (1966)	HYD I	IBR
	Ratings for Baseboard and Fin-tube Radiation (1989)	HYD I	IBR
Receivers	Refrigerant Liquid Receivers	ARI	ANSI/ARI 495-85
Refrigerant Containing Components	Refrigerant-Containing Components and Accessories, Non-electrical (1986)	UL	UL 207
Refrigerants	Number Designation of Refrigerants	ASHRAE	ANSI/ASHRAE 34-78
	Refrigeration Oil Description	ASHRAE	ANSI/ASHRAE 99-1987
	Methods of Testing Discharge Line Refrigerant-Oil Separators	ASHRAE	ASHRAE 69-71
	Sealed Glass Tube Method to Test the Chemical Stability of Material for Use Within Refrigerant Systems	ASHRAE	ANSI/ASHRAE 97-1983
	Refrigerant Recovery/Recycling Equipment	UL	UL 1963
	Specifications for Fluorocarbon Refrigerants	ARI	ARI 700-88
Refrigeration	Safety Code for Mechanical Refrigeration	ASHRAE	ANSI/ASHRAE 15-1989
	Capacity Measurement of Field Erected Compression-Type Refrigeration and Air-Conditioning Systems	ASHRAE	ASHRAE 83-1985
	Refrigerated Medical Equipment (1978)	UL	UL 416
	Commercial Refrigerated Equipment	CSA	C22.2 No. 120-1974 (R 1981)
	Equipment, Design and Installation of Ammonia Mechanical Refrigeration Systems	IIAR	ANSI/IIAR 2-1984
Steam Jet	Standards for Steam Jet Vacuum Systems, 4th ed. (1988)	HEI	HEI
	Ejectors	ASME	ASME PTC 24-1976 (R 1982)
Transport	Safety Practices for Mechanical Vapor Compression Refrigeration Equipment or Systems Used to Cool Passenger Compartment of Motor Vehicles (1987)	SAE	SAE J639 JAN87
	Mechanical Refrigeration Installations on Shipboard	ASHRAE	ANSI/ASHRAE 26-1985
	Mechanical Transport Refrigeration Units	ARI	ARI 1110-83
	General Requirements for Application of Vapor Cycle Refrigeration Systems for Aircraft (1973) (reaffirmed 1983)	SAE	SAE ARP731A
Refrigerators	Method of Testing Open Refrigerators for Food Stores	ASHRAE	ANSI/ASHRAE 72-1983
	Methods of Testing Self-Service Closed Refrigerators for Food Stores	ASHRAE	ASHRAE 117-1986
Commercial	Commercial Refrigerators and Freezers (1985)	UL	UL 471
	Food Service Refrigerators and Storage Freezers	NSF	NSF 7
	Food Carts	NSF	NSF 59
	Refrigerating Units (1976)	UL	UL 427
	Refrigeration Unit Coolers (1980)	UL	UL 412
	Soda Fountain and Luncheonette Equipment	NSF	NSF 1
	Food Service Equipment	NSF	NSF-2
Household	Refrigerators Using Gas Fuel (with 1984 and 1989 addenda)	AGA	ANSI Z21.19-1983
	Household Refrigerators and Household Freezers	AHAM	AHAM HRF 1-1988
	Household Refrigerators and Freezers (1983)	UL	UL 250
Roof Ventilators	Power Ventilators (1984)	UL	ANSI/UL 705-1984
Solar Equipment	Method of Testing to Determine the Thermal Performance of Solar Collectors	ASHRAE	ANSI/ASHRAE 93-1986
	Methods of Testing to Determine the Thermal Performance of Solar Domestic Water Heating Systems	ASHRAE	ASHRAE 95-1987
	Methods of Testing to Determine the Thermal Performance of Unglazed Flat-Plate Liquid-Type Solar Collectors	ASHRAE	ANSI/ASHRAE 96-1989
	Method of Testing to Determine the Thermal Performance of Flat-Plate Solar Collectors Containing a Boiling Liquid	ASHRAE	ANSI/ASHRAE 109-1986
	Method of Measuring Solar-Optical Properties of Materials	ASHRAE	ASHRAE 74-73
Solenoid Valves	Solenoid Valves for Liquid Flow Use with Volatile Refrigerants and Water	ARI	ARI 760-87
Sound Measurement	Measurement of Sound from Boiler Units, Bottom-Supported Shop or Field Erected, 3rd ed.	ABMA	ABMA
	Sound Rating of Outdoor Unitary Equipment	ARI	ARI 270-84
	Sound Rating of Non-Ducted Indoor Air-Conditioning Equipment	ARI	ARI 350-86
	Sound Rating of Large Outdoor Refrigerating and Air-Conditioning Equipment	ARI	ARI 370-86

Table 1 Codes and Standards Published by Various Societies and Associations (*Continued*)

Subject	Title	Publisher	Reference
Sound Management (continued)	Rating the Sound Levels and Transmission Loss of Package Terminal Equipment	ARI	ARI 300-88
	Procedural Standards for Measuring Sound and Vibration	NEBB	NEBB-1977
	Sound and Vibration in Environmental Systems	NEBB	NEBB-1977
	Application of Sound Rated Outdoor Unitary Equipment	ARI	ARI 275-84
	Method of Measuring Machinery Sound within Equipment Rooms	ARI	ARI 575-87
	Laboratory Method of Testing In-Duct Sound Power Measurement Procedure for Fans	ASHRAE AMCA	ASHRAE 68-1986 AMCA 300-86
	Specification for Sound Level Meters (reaffirmed 1986)	ASA	ANSI S1.4-1983 ANSI S1.4A-1985
	Method for the Calibration of Microphones (reaffirmed 1986)	ASA	ANSI S1.10-1966 (R 1986)
	Reverberant Room Method for Sound Testing of Fans	AMCA	AMCA 300-85
	Guidelines for the Use of Sound Power Standards and for the Preparation of Noise Test Codes (reaffirmed 1985)	ASA	ASA 10 ANSI S1.30-1979 (R 1985)
	Measurement of Industrial Sound	ASME	ASME/ANSI PTC 36-1985
	Method of Measuring Sound and Vibration of Refrigerant Compressors	ARI	ARI 530-89
Space Heaters	Electric Air Heaters (1980)	UL	UL 1025
Symbols	Graphic Electrical Symbols for Air-Conditioning and Refrigeration Equipment	ARI	ARI 130-88
	Graphic Symbols for Electrical and Electronic Diagrams	ASME	ANSI Y32.2-1975
	Graphic Symbols for Plumbing Fixtures for Diagrams used in Architecture Building Construction	ASME	ANSI Y32.4-1977
	Symbols for Mechanical and Acoustical Elements as used in Schematic Diagrams	ASME	ANSI/ASME Y32.18-1972 (R 1985)
	Graphic Symbols for Pipe Fittings, Valves and Piping	ASME	ASA Z32.2.3-1949 (R 1953)
	Graphic Symbols for Heating, Ventilating, and Air Conditioning	ASME	ASA Z32.2.4-1949 (R 1984)
Testing and Balancing	Procedural Standards for Certified Testing of Cleanrooms (1988)	NEBB	NEBB-1988
	Procedural Standards for Testing, Adjusting, Balancing of Environmental Systems, 4th ed. (1983)	NEBB	NEBB-1983
	HVAC Systems—Testing, Adjusting and Balancing (1983)	SMACNA	SMACNA
	Site, Performance Test Standard-Power Plant and Industrial Fans	AMCA	AMCA 803-87
Terminals, Wiring	Equipment Wiring Terminals for Use with Aluminum and/or Copper Conductors (1988)	UL	UL 486E
Thermal Storage	Method of Testing Active Latent Heat Storage Devices Based on Thermal Performance	ASHRAE	ANSI/ASHRAE 94.1-1985
	Methods of Testing Thermal Storage Devices with Electrical Input and Thermal Output Based on Thermal Performance	ASHRAE	ANSI/ASHRAE 94.2-1981
	Metering and Testing Active Sensible Thermal Energy Storage Devices Based on Thermal Performance	ASHRAE	ANSI/ASHRAE 94.3-1986
Turbines	Steam Turbines for Mechanical Drive Service	NEMA	NEMA SM 23-1985
Unit Heaters	Oil-Fired Unit Heaters (1988)	UL	UL 731
	Gas Unit Heaters (with 1986 and 1989 addenda)	AGA	ANSI Z83.8-1985
Valves	Methods of Testing Nonelectric, Nonpneumatic Thermostatic Radiator Valves	ASHRAE	ASHRAE 102-1983
	Automatic Gas Valves for Gas Appliances (with 1989 addenda)	AGA	ANSI Z21.21-1987
	Manually Operated Gas Valves for Appliances, Applicance Connection Valves, and Hose End Valves	AGA	ANSI Z21.15-1989
	Relief Valves and Automatic Gas Shutoff Devices for Hot Water Supply Systems	AGA	ANSI Z21.22-1986
	Refrigerant Access Valves and Hose Connectors	ARI	ARI 720-88
	Refrigerant Pressure Regulating Valves	ARI	ARI 770-84
	Solenoid Valves for Use with Volatile Refrigerants	ARI	ARI 760-87
	Face-to-Face and End-to-End Dimensions of Ferrous Valves	ASME	ASME/ANSI B16.10-86
	Manually Operated Metallic Gas Valves for Use in Gas Piping Systems up to 125 psig (Sizes 1/2 through 2)	ASME	ANSI B16.33-81
	Valves—Flanged and Buttwelding End	ASME	ANSI B16.34-81

Codes and Standards 41.13

Table 1 Codes and Standards Published by Various Societies and Associations (*Continued*)

Subject	Title	Publisher	Reference
Valves (continued)	Hydrostatic Testing of Control Valves	ASME	ANSI B16.37-80
	Large Metallic Valves for Gas Distribution (Manually Operated, NPS-2 1/2 to 12, 125 psig Maximum)	ASME	ANSI/ASME B16.38-85
	Manually Operated Thermoplastic Gas Shutoffs and Valves in Gas Distribution Systems	ASME	ANSI/ASME B16.40-85
	Safety and Relief Valves	ASME	ANSI/ASME PTC 25.3-1988
	Electrically Operated Valves (1982)	UL	UL 429
	Valves for Anhydrous Ammonia and LP-Gas (Other than Safety Relief) (1980)	UL	UL 125
	Safety Relief Valves for Anhydrous Ammonia and LP-Gas (1984)	UL	UL 132
	Pressure Regulating Valves for LP-Gas (1985)	UL	UL 144
	Valves for Flammable Fluids (1980)	UL	UL 842
Vending Machines	Refrigerated Vending Machines (1989)	UL	UL 541
	Methods of Testing Pre-Mix and Post-Mix Soft Drink Vending and Dispensing Equipment	ASHRAE	ASHRAE 91-76
	Sanitation Ordinance and Code for Vending of Foods and Beverages (1965)	USDA	USDA 546
	Vending Machines for Food and Beverages	NSF	NSF-25
Vent Dampers	Automatic Vent Damper Devices for Use with Gas-Fired Appliances	AGA	ANSI Z21.66-1988
	Vent or Chimney Connector Dampers for Oil-Fired Appliances (1988)	UL	UL 17
Venting	Explosion Prevention Systems	NFPA	ANSI/NFPA 69-86
	Chimneys, Fireplaces, Vents and Solid Fuel Burning Appliances	NFPA	ANSI/NFPA 211-88
	Type L Low-Temperature Venting Systems (1986)	UL	UL 641
	Draft Hoods (with 1983 addenda)	AGA	ANSI Z21.12-1981
	Draft Equipment (1973)	UL	UL 378
	Gas Vents (1986)	UL	UL 441
	National Fuel Gas Code	AGA	ANSI Z223.1-1988
	Guide for Steel Stack Design and Construction (1983)	SMACNA	SMACNA
Ventilation	Removal of Smoke and Grease-Laden Vapors from Commercial Cooking Equipment	NFPA	ANSI/NFPA 96-87
	Parking Structures; Repair Garages	NFPA	ANSI/NFPA 88A-85; 88B-85
	Ventilation for Acceptable Indoor Air Quality	ASHRAE	ASHRAE 62-1989
	Industrial Ventilation	ACGIH	ACGIH
	Food Service Equipment	NSF	NSF-2
Water Heaters	Gas Water Heaters, Vol. I, Storage Water Heaters with Input Ratings of 75,000 Btu per Hour or Less (with 1988 and 1989 addenda)	AGA	ANSI Z21.10.1-1987
	Gas Water Heaters, Vol. III, Storage, with Input Ratings Above 75,000 Btu per Hour, Circulating and Instantaneous Water Heaters (with 1988 and 1989 addenda)	AGA	ANSI Z21.10.3-1987
	Household Electric Storage Tank Water Heaters (1983)	UL	UL 174
	Oil-Fired Storage Tank Water Heaters (1988)	UL	UL 732
	Commercial-Industrial Gas Heating Equipment (1973)	UL	UL 795
	Electric Booster and Commercial Storage Tank Water Heaters (1988)	UL	UL 1453
	Hot Water Generating and Heat Recovery Equipment	NSF	NSF-5
	Construction and Test of Electric Storage Tank Water Heaters	CSA	C22.2 No. 110 M-1981
	Oil Burning Stoves and Water Heaters	CSA	B140.3-1962 (R1980)
	Oil-Fired Service Water Heaters and Swimming Pool Heaters	CSA	B140.12-1976
	CSA Standards on Performance of Electric Storage Tank Water Heaters	CSA	C191-series-M1983
	Methods of Testing to Determine the Thermal Performance of Solar Domestic Water Heating Systems	ASHRAE	ANSI/ASHRAE 95-1987
Woodburning Appliances	Installation Code for Solid Fuel Burning Appliances and Equipment	CSA	CAN/CSA-B365-M87
	Method of Testing for Performance Rating of Woodburning Appliances	ASHRAE	ANSI/ASHRAE 106-1984
	Solid-Fuel-Fired Central Heating Appliances	CSA	CAN/CSA-B366.1-M87
	Space Heaters for Use with Solid Fuels	CSA	B366.2-M1984
	Solid Fuel Type Room Heaters (1988)	UL	UL 1482
	Commercial Cooking and Hot Food Storage Equipment	NSF	NSF-4

ABBREVIATIONS AND ADDRESSES

The Codes and Standards listed in Table 1 can be obtained from the organizations listed in the *Publisher* column.

ABMA	American Boiler Manufacturers Association, Suite 160, 950 N. Glebe Road, Arlington, VA 22203
ACCA	Air Conditioning Contractors of America, 1513 16th Street, NW, Washington, DC 20036
ACGIH	American Conference of Governmental Industrial Hygienists, 6500 Glenway Avenue, Building. D-7, Cincinnati, OH 45211
ADC	Air Diffusion Council, 230 N. Michigan Avenue, Suite 1200, Chicago, IL 60601
AGA	American Gas Association, 1515 Wilson Boulevard, Arlington, VA 22209
AHAM	Association of Home Appliance Manufacturers, 20 N. Wacker Drive, Chicago, IL 60606
AIHA	American Industrial Hygiene Association, 345 White Pond Drive, Akron, OH 44320
AMCA	Air Movement and Control Association, Inc., 30 W. University Drive, Arlington Heights, IL 60004
ANSI	American National Standards Institute, 1430 Broadway, New York, NY 10018
ARI	Air-Conditioning and Refrigeration Institute, 1501 Wilson Boulevard, 6th Floor, Arlington, VA 22209
ASA	Acoustical Society of America, 335 E. 45 Street, New York, NY 10017-3483
ASHRAE	American Society of Heating, Refrigerating and Air-Conditioning Engineers, Inc. 1791 Tullie Circle, NE, Atlanta, GA 30329
ASME	The American Society of Mechanical Engineers, 345 E. 47 Street, New York, NY 10017
	For ordering publications: ASME Marketing Department, Box 2350, Fairfield, NJ 07007-2350
ASTM	American Society for Testing and Materials, 1916 Race Street, Philadelphia, PA 19103
BOCA	Building Officials and Code Administrators International, Inc., 4501 W. Flossmoor Road, Country Club Hills, IL 60478-5795
BSI	British Standards Institution, 2 Park Street, London, W1A 2BS, England
CABO	Council of American Building Officials, 5203 Leesburg Pike, Suite 708, Falls Church, VA 22041
CAGI	Compressed Air and Gas Institute, Suite 1230, Keith Building, 1621 Euclid Avenue, Cleveland, OH 44115
CSA	Canadian Standards Association, 178 Rexdale Boulevard, Rexdale, Ontario M9W 1R3, Canada
CTI	Cooling Tower Institute, P.O. Box 73383, Houston, TX 77273
EJMA	Expansion Joint Manufacturers Association, Inc., 25 N. Broadway, Tarrytown, NY 10591
HEI	Heat Exchange Institute, Suite 1230, Keith Building, 1621 Euclid Avenue, Cleveland, OH 44115
HI	Hydraulic Institute, 712 Lakewood Center N., 14600 Detroit Avenue, Cleveland, OH 44107
HYD I	Hydronics Institute, 35 Russo Place, Berkeley Heights, NJ 07922
IAPMO	International Association of Plumbing and Mechanical Officials, 20001 Walnut Drive South, Walnut, CA 91789
ICBO	International Conference of Building Officials, 5360 S. Workman Mill Road, Whittier, CA 90601
IIAR	International Institute of Ammonia Refrigeration, 111 East Wacker Drive, Chicago, IL 60601
NCSBCS	National Conference of States on Building Codes and Standards, 481 Carlisle Drive, Herndon, VA 22070
NEBB	National Environmental Balancing Bureau, 8224 Old Courthouse Road, Vienna, VA 22180
NEMA	National Electrical Manufacturers Association, 2101 L Street, NW, Suite 300, Washington, D.C. 20037
NFPA	National Fire Protection Association, 1 Batterymarch Park, Quincy, MA 02269-9101
NSF	National Sanitation Foundation, Box 1468, Ann Arbor, MI 48106
SAE	Society of Automotive Engineers, 400 Commonwealth Drive, Warrendale, PA 15096
SBCCI	Southern Building Code Congress International, Inc., 900 Montclair Road, Birmingham, AL 35213
SMACNA	Sheet Metal and Air Conditioning Contractors' National Association, 8224 Old Courthouse Road, Vienna, VA 22182
TEMA	Tubular Exchanger Manufacturers Association, Inc., 25 N. Broadway, Tarrytown, NY 10591
UL	Underwriters Laboratories Inc., 333 Pfingsten Road, Northbrook, IL 60062-2096
WFCA	Western Fire Chiefs Association, Inc., 5360 S. Workman Mill Road, Whittier, CA 90601

ASHRAE HANDBOOK

ADDITIONS AND CORRECTIONS

This section supplements the current handbooks and notes technical errors found in the series. Occasional typographical errors and nonstandard symbol labels will be corrected in future volumes. The authors and editor encourage you to notify them if you find other technical errors. Please send corrections to: Handbook Editor, ASHRAE, 1791 Tullie Circle NE, Atlanta, GA 30329.

1987 HVAC Systems and Applications

p. 7.3. Delete + sign from Equation (1) so it reads as follows:

$$q_r = \sigma F_a F_e [(T_r/100)^4 - (T_p/100)^4] \tag{1}$$

where

$$\sigma = 0.1713 \text{ Btu/h} \cdot \text{ft}^2 \cdot °F \text{ } (5.67 \text{ W/m}^2 \cdot \text{K})$$

p. 12.14. Figure 5 should be replaced with the following:

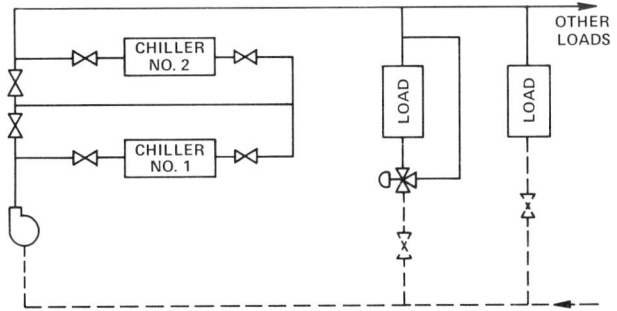

Note: Chiller No. 1 and No. 2 can be operated individually or in series (No. 2 following No. 1)

Fig. 5 Chilled Water System with Constant Flow Rate

p. 13.13, Table 1. Multiply the SI (mm) values for the inside diameters of steel pipe and copper tube by 10.

p. 16.3, LH column, item 3. The value of t_r is estimated as 35 °C in Case 1 and 23 °C in Case 2. Therefore, using Equation (10) for Case 1:

$$h_r = 0.71 \times 4 \times 5.67 \times 10^{-8}[(35 + 15)/2 + 273]^3$$
$$= 4.3 \text{ W/(m}^2 \cdot \text{K)}$$

And for Case 2:

$$h_r = 0.71 \times 4 \times 5.67 \times 10^{-8}[(23 + 10)/2 + 273]^3 = 3.9$$

p. 28.4. Table 2. Soap at 60% rh should read 12.9%, not 122.9%.

p. 32.10. In the 3rd paragraph, the 2nd sentence is repeated.

p. 32.11, Figure 8. Plan views of fan with duct descending to inlet should have a solid, not a broken, line indicating a duct.

p. 52.5. Equation (1) should read:

$$L_p = L_w - 5\log V - 3\log f - 10\log r + 25 \text{ dB} \tag{1}$$
$$L_p = L_w - 5\log V - 3\log f - 10\log r + 12 \text{ dB} \tag{1 SI}$$

p. 52.10, RH column, 3rd paragraph. Add the following text to the paragraph:

Thus, 100 Hz was used in Eq. (6) to calculate the 125 Hz band and 200 Hz for the 250 Hz band in Tables 10 and 11. Similarly, 2500 Hz was used to calculate the 2000 Hz octave band with Eq. (7). The data in the tables were adjusted slightly, so the results from the equations are close to the tabular values, but not exact.

p. 52.11-12, Tables 10 and 11. The line indicating the separation between low- and high-frequency attenuation values at around 1000 Hz was omitted, the line should be added as shown in the partial tables listed below:

Table 10 Octave Band Center Freq.	Table 11 Octave Band Center Freq.
1000 Hz	**1000 Hz**
9.10	19.23
7.58	16.03
.	14.42
.	13.46
.	14.81
.	11.85
2.91	11.11
2.17	8.80
3.45	12.31
2.84	10.26
2.59	9.23
1.65	5.68
2.87	10.66
2.39	8.66
1.90	7.22
1.06	4.04
2.49	9.48
1.88	7.62
.	.
.	.
0.37	0.37

p. 52.14, Figure 8. Change the equations for the characteristic spectrum K, as follows:

In I-P units:
$$K = -36.3 - 10.7 \log (\text{St}) \text{ for St} \leq 25$$
$$K = -1.1 - 35.9 \log (\text{St}) \text{ for St} > 25$$

In SI units:
$$K = 5 - 10.7 \log (\text{St}) \text{ for St} \leq 25$$
$$K = 40.2 - 35.9 \log (\text{St}) \text{ for St} > 25$$

p. 52.40. Add the following reference:

Kuntz, H.L. and R.M. Hoover. 1987. The interrelationships between the physical properties of fibrous duct lining materials and lined duct sound attenuation. ASHRAE *Transactions* 93(2): 449-470.

p. 54.10, RH column. Last sentence should refer to Table 7, not Table 2.

p. 54.15, RH column, Example 8. The required gas input for Solution a, assuming 75% thermal efficiency, is 185,000 Btu/h (54.2 kW).

p. 58.12, Equation (14). Add the following information for the variable B.

p = atmospheric pressure = 14.7 psi (101.325 kPa)

g = acceleration of gravity = 32.174 ft/s^2 (9.806 m/s^2)

R = universal gas constant
 = 53.34 ft·lb$_f$/(lb$_m$·°R)[287 J/(kg·K)]

Therefore $B = 39.7 (1/T_o - 1/T_s)$ where B is in lb/ft^3
or $B = 7.64 (1/T_o - 1/T_s)$ where B is in. of water/ft

p. 58.12, Equation (15). Revise the explanation of C to read:

C = flow coefficient, assumed to equal 0.65

1988 EQUIPMENT

p. 1.3, Table 4a. The values K-2 in the middle section in the +6 and +10 columns should have the footnote d, not e.

p. 1.5, Table 5a. Columns 4 and 5 should read $H_s \times T$ (min), in.

p. 1.10, LH column. The following sentence should be added to the end of the first paragraph:

Ducts that must have an hourly fire resistance rating are usually encased in materials that have the appropriate thermal and durability rating.

p. 1.10, Table 9. The table headings should read: **Galvanized Straps (see Figs. 2 and 3)** and **Rods (see Fig. 3)**.

p. 1.11, Table 10. Footnotes b and c should refer to Figure 3, not Figure 4.

p. 1.11, Table 11. Column 3 should read (see Figure 4), not (see Figure 5).

p. 1.11, Table 11. Delete first sentence in footnote d.

p. 1.15, Table A-5a. Columns 4 and 5 should read $H_s \times T$ (min), mm.

p. 1.18, Table A-7a. Delete the footnote, "*See Table A-7b for notes."

p. 1.19, Table A-7b. Add the following footnote, "*See Table A-7a for notes."

p. 1.19, Table A-9. Change the figure references in the headings from (see Figs. 3 and 4) to (see Figs. 2 and 3) and (see Fig. 4) to (see Fig. 3).

p. 1.19, Table A-9. Change the values 4.11 (under the 6th and 7th columns) to 3.43 and the values 4.11 (under the 8th and 9th columns) to 2.67.

p. 1.20, Table A-10. Change the figure reference in footnote b to read, "For W-dimension, see Figure 3." Change the figure reference in footnote c to (Figure 3).

p.1.20, Table A-11. The 3rd column should read, (see Figure 4), not (see Figure 5).

p. 6.8. In Example 1 replace the calculation for q_{td} with the following:
$$q_{td} = 1248 \times 1.4 \times 3 \times 1(29.4 - 23.9) = 28\,830 \text{ W}$$

p. 24.17, Table 12. Footnote a should refer to Table 11, not Table 7.

p. 30.4. Replace Figure 3 with the following:

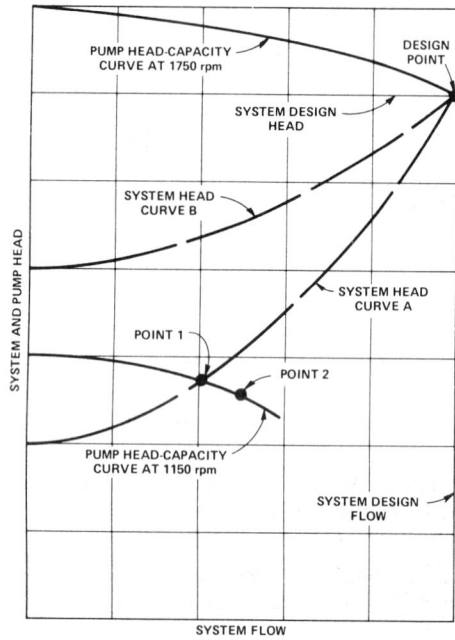

Fig. 3 Application of Affinity Laws

p. 34.18. In the bibliography "Mather" should be spelled "Mathur" in both listings.

1989 FUNDAMENTALS

p. 1.4. Equation (17) should read:

$$\text{COP} = \frac{|Q_2|}{|Q_1| - |Q_2|} \tag{17}$$

p. 3.14. Revise item I.(b)(2) in Table 6 to read as follows:

(2) For very long tubes, when $\left(\dfrac{GD}{\mu}\right)\left(\dfrac{c_p \mu}{k}\right)\left(\dfrac{D}{L}\right) < 20$, Eq. (6) should not be used.

The multiplied variables show a condition for the use of Equation (6) and do not define the convection coefficient.

Additions and Corrections

p. 3.21. Revise reference to read:

Shepard, D.G. 1956. Performance of one-row tube coils with thin plate fins, low velocity forced convection. *Heating, Piping, and Air Conditioning*, 4:137-144.

p. 4.8, Table 2. Revise Equation (13a) to read:
$$G_E = G_v(\rho_l/\rho_v)^{0.5} + G_l \tag{13a}$$

p. 8.3. Equation (16) should read:
$$w = w_{rsw} + 0.06(1 - w_{rsw}) = 0.06 + 0.94 E_{rsw}/E_{max} \tag{16}$$

p. 8.3, The line below Equation (20) should read:

where K_{res} is a proportionality constant (0.065 lb·ft²/Btu).

p. 8.5, RH column, 3rd sentence, para 1. T_r and T_a should be t_r and t_a respectively.

p. 8.5. Equation (41) should read:
$$i_m = \frac{R_{cl} + 1/[f_{cl}(h_r + h_c)]}{R_{cl}/i_{cl} + 1/(h_c f_{cl})} \tag{41}$$

p. 8.5, Equation (42), second line: T_a should read t_a.

p. 8.5, LH column. The unnumbered equation above Equation (47) should read:
$$t_c + w i_m \mathrm{LR} p(t_c) = t_o + w i_m \mathrm{LR} p_a$$

p. 8.5, Figure 2 caption should read:

Fig. 2 Constant Heat Loss Line and its Relationship to t_{oh}, ET*, and t_{ad}

p. 8.5, Figure 2. The equation in the upper left corner should read:
$$t_o + w i_m \mathrm{LR} p_a = \text{constant}$$

p. 8.6, RH column, last para. Delete CLO in line 5.

p. 8.6, RH column, 2nd line from the bottom. The value in parentheses should read (41 lb/h·ft²·°F).

p. 8.7. Equation (61) should read:
$$\alpha = 0.0418 + 0.745/(3600 \, \dot{m} bl + 0.585) \tag{61 SI}$$
$$\alpha = 0.0418 + 0.745/(4.90 \, \dot{m} bl + 0.585) \tag{61 I-P}$$

p. 8.8, Table 1. Under the heading "Walking (on the level)," the speeds should read 2.9 f/s (2 mph); 4.4 f/s (3 mph); 5.9 f/s (4 mph).

p. 8.8, RH column. The first sentence of the paragraph entitled "Mechanical Efficiency," should read:

In the heat balance equation, the rate of work accomplished W must be in the same units as metabolism M and expressed in terms of A_D (i.e., in Btu/h·ft²).

p. 8.9. Equation (67). The variable shown as t_r should read $\bar{t}_r$.

p. 8.9, RH column, paragraph 2. T_{oh} should read t_{oh}.

p. 8.9, Table 3. Section 3 (Walking in still air). The equation should read:
$$h_c = 0.092 \, V^{0.53}$$

p. 8.16. Equation (84) should read:
$$t_{eq} = -0.0818(\mathrm{WCI}) + 91.4 \text{ in } °F \tag{84}$$

p. 8.16. Table 9 should read:

Table 9 Equations for Predicting Thermal Sensation (Y)[a] of Men, Women, and Men and Women Combined[b]

Exposure Period, h	Sex	Regression Equations t = dry-bulb temperature, °F p = vapor pressure, psia
1.0	Male	$Y = 0.122\,t + 1.61\,p - 9.58$
	Female	$Y = 0.151\,t + 1.71\,p - 12.08$
	Combined	$Y = 0.136\,t + 1.71\,p - 10.83$
2.0	Male	$Y = 0.123\,t + 1.86\,p - 9.95$
	Female	$Y = 0.157\,t + 1.45\,p - 12.73$
	Combined	$Y = 0.140\,t + 1.65\,p - 11.34$
3.0	Male	$Y = 0.118\,t + 2.02\,p - 9.72$
	Female	$Y = 0.153\,t + 1.76\,p - 13.51$
	Combined	$Y = 0.135\,t + 1.92\,p - 11.12$

[a] Y values range from -3 to $+3$ where -3 is cold; -2 is cool; -1 is slightly cool; 0 is comfortable; $+1$ is slightly warm; $+2$ is warm; and $+3$ is hot.
[b] For young adult subjects with sedentary activity and wearing clothing with a thermal resistance of approximately 0.5 clo, $\bar{t}_r \cong \bar{t}_a$ and air velocities are < 40 fpm.

p. 8.17. In Equation (87), the second term is added rather than subtracted; i.e. it should read:
$$+ f_{cl} h_c (t_{cl} - t_a)$$

p. 8.18. Equation (92) should read:
$$t_{b,c} = (0.349/18.43)(M - W) + 97.342 \tag{92}$$

p. 8.18. Equation (93) should read:
$$t_{b,h} = (0.625/18.43)(M - W) + 98.004 \tag{93}$$

p. 8.18. Equation (94) should read:
$$\mathbf{TSENS} = \begin{cases} 0.2603\,(t_b - t_{b,c}) & t_b < t_{b,c} \\ 4.7\,\epsilon_{ev}(t_b - t_{b,c})/(t_{b,h} - t_{b,c}) & t_{b,c} \leqslant t_b \leqslant t_{b,h} \\ 4.7\,\epsilon_{ev} + 0.2603\,(t_b - t_{b,h}) & t_{b,h} < t_b \end{cases} \tag{94}$$

p. 8.18. The first line of Equation (95) should read:
$$0.2603(t_b - t_{b,c}) \qquad t_b < t_{b,c}$$

p. 8.19. Equation (96) should read:

$\mathrm{PPD} = 100 - 95 \exp[-(0.03353\,\mathrm{PMV}^4 + 0.2179\,\mathrm{PMV}^2)]$ (96)

p. 8.21, LH column. The beginning of the sentence below Equation (97) should read:

For $V < 10$ insert $V = 10 \ldots$

p. 8.24. Equation (105) should read:

$\mathrm{ERF} = (A_r/A_D)(1.08 + 0.17\,V^{0.5})(t_g - t_a)$ in Btu/h·ft² (105)

p. 8.24, column 2, paragraph 4. The beginning of line 1 should read:

Body heat storage of 315 Btu …

p. 8.28. LH column, Air Contaminants and Health. The first part of the 15th line should read:

in Chapter 23 of the 1987 HVAC Volume.

p. 13.8. Equation (7) should read:
$$\frac{R_1}{R_2} = \frac{390.1 + t_1}{390.1 + t_2} \tag{7}$$

p. 13.9. The explanation for Equation (8) should read:

V = velocity of stream, fps

p. 16.7. Replace Tables 7 and 8 with the following tables.

Table 7 Comparative Performance Per Ton of Refrigeration
Based on 5°F Evaporation and 86°F Condensation

No.	Refrigerant Name	Evaporator pressure, psia	Condensing pressure, psia	Compression, Ratio	Net refrigerating effect, Btu/lb	Refrigerant circulated, lb/min	Liquid Circulated, in³/min	Specific volume of suction gas, ft³/lb	Compressor displacement, cfm	Horsepower, hp	Coefficient of performance	Comp. discharge temp. °F
170	Ethane	236.410	674.710	2.85	69.27	2.88704	289.1266	0.5344	1.543	1.733	2.72	123
744	Carbon Dioxide	332.375	1045.360	3.15	57.75	3.46320	158.5272	0.2639	0.914	1.678	2.81	156
13B1	Bromotrifluoromethane	77.820	264.128	3.39	28.45	7.02901	129.7814	0.3798	2.669	1.134	4.16	104
1270	Propylene	52.704	189.440	3.59	123.15	1.62401	90.7048	2.0487	3.327	1.035	4.56	108
290	Propane	42.37	156.820	3.70	120.30	1.66251	95.0386	2.4589	4.088	1.031	4.57	98
502	22/115 Azeotrope[a]	50.561	191.290	3.78	44.91	4.45305	103.3499	0.8015	3.569	1.067	4.42	98
22	Chlorodifluoromethane	85.926	172.899	2.01	69.90	2.86144	67.6465	1.2394	3.546	1.011	4.67	128
717	Ammonia	34.170	168.795	4.94	474.20	0.42177	19.6087	8.1790	3.450	0.989	4.77	210
500	12/152a Azeotrope[a]	31.064	127.504	4.10	60.64	3.29834	80.1925	1.5022	4.955	1.005	4.69	105
12	Dichlorodifluoromethane	26.505	107.991	4.07	50.25	3.97981	85.2280	1.4649	5.830	0.992	4.75	100
600a	Isobutane	12.924	59.286	4.59	113.00	1.76991	90.0059	6.4189	11.361	1.070	4.41	80
600	Butane	8.176	41.191	5.04	125.55	1.59299	77.7772	10.2058	16.258	0.952	4.95	88
114	Dichlorotetrafluoroethane[b]	6.747	36.493	5.41	43.02	4.64889	89.5631	4.3400	20.176	1.015	4.65	86
11	Trichlorofluoromethane	2.937	18.318	6.24	67.21	2.97592	56.2578	12.2400	36.425	0.939	5.02	110
113	Trichlorotrifluoroethane[b]	1.006	7.884	7.83	52.08	3.84047	68.5997	26.2845	100.945	1.105	4.27	86

[a]See table 1 for composition. [b]Saturated suction except R-113, 114. Enough superheat was added to give saturated discharge.

Table 8 Comparative Refrigerant Performance Per Ton at Various Evaporating and Condensing Temperatures

No.	Refrigerant Name	Suction Temp. °F	Evaporator pressure, psia	Condensing pressure, psia	Compression, ratio	Net refrigerating effect, Btu/lb	Refrigerant circulated, lb/min	Specific volume of suction gas, ft³/lb	Compressor displacement, cfm	Power hp
			A. −130°F Evaporating, −40°F Condensing							
1150	Ethylene	−130	30.887	210.670	6.82	142.01	1.40835	3.8529	5.426	1.756
170	Ethane	−130	13.620	112.790	8.28	156.58	1.27730	8.3575	10.675	1.633
13	Monochlorotrifluoromethane	−130	9.059	88.037	9.72	45.82	4.36529	3.6245	15.822	1.685
			B. −100°F Evaporating, −30°F Condensing							
170	Ethane	−100	31.267	134.730	4.31	157.76	1.26775	3.8671	4.903	1.118
13	Monochlorotrifluoromethane	−100	22.276	106.290	4.77	46.23	4.32581	1.5631	6.762	1.153
22	Chlorodifluoromethane	−100	2.380	19.629	8.25	90.75	2.20397	18.5580	40.901	1.074
			C. −76°F Evaporating, 5°F Condensing							
1150	Ethylene	−76	109.370	416.235	3.81	116.95	1.71021	1.1617	1.987	1.478
170	Ethane	−76	54.634	235.440	4.31	322.65	0.61987	2.2906	1.420	0.566
13	Monochlorotrifluoromethane	−76	40.872	192.135	4.70	39.42	5.07389	0.8801	4.465	1.382
13B1	Bromotrifluoromethane	−76	13.173	77.820	5.91	37.80	5.29128	2.0329	10.757	1.253
290	Propane	−76	6.150	42.367	6.89	147.39	1.35699	14.8560	20.159	1.196
22	Chlorodifluoromethane	−76	5.438	42.963	7.90	84.24	2.37425	8.5925	20.401	1.195
717	Ammonia	−76	3.169	34.162	10.78	535.08	0.37378	75.5710	28.247	1.237
12	Dichlorodifluoromethane	−76	3.277	26.501	8.09	58.61	3.41219	10.2448	34.957	1.191
			D. −40°F Evaporating, 68°F Condensing							
744	Carbon Dioxide	−40	145.770	830.530	5.70	77.22	2.59000	0.6128	1.587	2.208
13B1	Bromotrifluoromethane	−40	31.855	207.854	6.53	28.81	6.94155	0.8915	6.189	1.855
290	Propane	−40	16.099	121.560	7.55	119.33	1.67602	6.0829	10.195	1.670
22	Chlorodifluoromethane	−40	15.268	131.997	8.65	70.65	2.83106	3.2805	9.287	1.606
717	Ammonia	−40	10.376	124.009	11.95	479.49	0.41711	24.9350	10.401	1.577
500	12/152a Azeotrope	−40	10.959	96.948	8.85	60.24	3.31989	3.9895	13.245	1.583
12	Dichlorodifluoromethane	−40	9.304	82.295	8.84	49.44	4.04572	3.8868	15.725	1.596
			E. −10°F Evaporating, 100°F Condensing							
11	Trichlorofluoromethane	−10	1.921	23.637	12.31	62.34	3.20837	18.1590	58.261	1.436
114	Dichlorotetrafluoroethane	5	4.560	46.148	10.12	292.33	0.68417	6.2274	4.261	0.216
12	Dichlorodifluoromethane	−10	19.197	131.720	6.86	44.89	4.45563	1.9803	8.823	1.606
717	Ammonia	−10	23.664	211.400	8.93	453.05	0.44145	11.5260	5.088	1.497
22	Chlorodifluoromethane	−10	31.231	210.670	6.75	64.07	3.12173	1.6757	5.231	1.602
502	22/115 Azeotrope	−10	37.256	230.890	6.20	39.05	5.12177	1.0727	5.494	1.904

Additions and Corrections

Table 8 Comparative Refrigerant Performance Per Ton at Various Evaporating and Condensing Temperatures (*Concluded*)

No.	Refrigerant Name	Suction Temp. °F	Evaporator pressure, psia	Condensing pressure, psia	Compression ratio	Net refrigerating effect, Btu/lb	Refrigerant circulated, lb/min	Specific volume of suction gas, ft³/lb	Compressor displacement, cfm	Power hp
			F. −10°F Evaporating, 100°F Condensing							
11	Trichlorofluoromethane	65	1.921	23.637	12.31	62.34	3.20837	21.2388	68.142	1.679
114	Dichlorotetrafluoroethane	65	4.560	46.148	10.12	292.33	0.68417	7.1379	4.884	0.248
12	Dichlorodifluoromethane	65	19.197	131.720	6.86	44.89	4.45563	2.3597	10.514	1.914
717	Ammonia	65	23.664	211.400	8.93	453.05	0.44145	13.7053	6.050	1.780
22	Chlorodifluoromethane	65	31.231	210.670	6.75	64.07	3.12173	2.0121	6.281	1.924
502	22/115 Azeotrope	65	37.256	230.890	6.20	39.05	5.12177	1.3015	6.666	2.310
			G. −10°F Evaporating + Refrig. Effect to 65°F, 100°F Condensing							
11	Trichlorofluoromethane	65	1.921	23.637	12.31	72.15	2.77204	21.2388	58.875	1.451
114	Dichlorotetrafluoroethane	65	4.560	46.148	10.12	47.22	4.23550	7.1379	30.233	1.534
12	Dichlorodifluoromethane	65	19.197	131.720	6.86	55.83	3.58251	2.3597	8.454	1.539
717	Ammonia	65	23.664	211.400	8.93	493.76	0.40506	13.7053	5.551	1.633
22	Chlorodifluoromethane	65	31.231	210.670	6.75	75.95	2.63326	2.0121	5.298	1.623
502	22/115 Azeotrope	65	37.256	230.890	6.20	51.23	3.90362	1.3015	5.081	1.761
			H. 20°F Evaporating, 80°F Condensing							
290	Propane	20	55.931	144.330	2.58	128.39	1.55775	1.8873	2.940	0.721
22	Chlorodifluoromethane	20	57.786	158.360	2.74	73.12	2.73512	0.9334	2.553	0.707
717	Ammonia	20	48.062	152.690	3.18	485.52	0.41193	5.9202	2.439	0.678
500	12/152a Azeotrope	20	41.936	116.620	2.78	64.15	3.11784	1.1294	3.521	0.702
12	Dichlorodifluoromethane	20	35.765	98.850	2.76	53.22	3.75827	1.1045	4.151	0.701
600a	Isobutane	20	17.916	53.907	3.01	121.45	1.64677	4.7361	7.799	0.706
600	Butane	20	11.557	37.225	3.22	134.18	1.49054	7.3947	11.022	0.686
114	Dichlorotetrafluroethane	30	9.699	32.859	3.39	44.42	4.50207	3.0678	13.812	0.769
			I. 40°F Evaporating, 100°F Condensing							
290	Propane	40	78.782	189.040	2.40	114.96	1.73974	1.3563	2.360	0.750
22	Chlorodifluoromethane	40	83.246	210.670	2.53	68.71	2.91091	0.6557	1.909	0.696
717	Ammonia	40	73.114	211.400	2.89	467.39	0.42791	3.9783	1.702	0.653
500	12/152a Azeotrope	40	60.722	155.790	2.57	60.54	3.30344	0.7920	2.616	0.692
12	Dichlorodifluoromethane	40	51.705	131.720	2.55	50.50	3.96024	0.7784	3.083	0.689
600a	Isobutane	40	26.750	73.364	2.74	115.83	1.72667	3.2564	5.623	0.693
600	Butane	40	17.679	51.683	2.92	129.22	1.54775	4.9754	7.701	0.669
114	Dichlorotetrafluoroethane	52	15.134	46.148	3.05	42.46	4.70998	2.0321	9.571	0.738
11	Trichlorofluoromethane	40	7.050	23.637	3.35	68.50	2.91984	5.4311	15.858	0.636
113	Trichlorotrifluoroethane	47	2.695	10.494	3.89	54.14	3.69433	10.7059	39.551	0.710

Chapter 17. McLinden *et al.* (1989) developed the following data for refrigerants 123 and 134a. See the following reference for further information.

McLinden, M.O. *et al.* 1989. Measurement and formulation of the thermodynamic properties of refrigerants 134a (1,1,1,2-tetrafluoroethane) and 123 (1,1-dichloro-2,2,2-trifluoroethane). ASHRAE Transactions 95(2):263-283.

Refrigerant 123 Saturation Properties

Temp. °F	Pressure psia	Density lb/ft³ liquid	Density lb/ft³ vapor	Enthalpy Btu/lb liquid	Enthalpy Btu/lb vapor	Entropy Btu/lb·°F liquid	Entropy Btu/lb·°F vapor	c_v Btu/lb·°F liquid	c_v Btu/lb·°F vapor	c_p Btu/lb·°F liquid	c_p Btu/lb·°F vapor	σ dyn/cm	Temp. °F
0.	2.0	97.8	0.06	8.2	87.6	0.018	0.190	0.160	0.138	0.202	0.152	20.7	0.
10.	2.6	97.0	0.08	10.2	89.0	0.022	0.190	0.161	0.140	0.205	0.155	20.0	10.
20.	3.5	96.2	0.11	12.3	90.4	0.026	0.189	0.162	0.142	0.209	0.157	19.3	20.
30.	4.5	95.4	0.13	14.4	91.8	0.031	0.189	0.163	0.145	0.213	0.160	18.6	30.
40.	5.8	94.6	0.17	16.6	93.3	0.035	0.189	0.163	0.147	0.218	0.162	17.9	40.
50.	7.3	93.7	0.21	18.8	94.7	0.040	0.189	0.164	0.149	0.222	0.165	17.2	50.
60.	9.2	92.9	0.26	21.0	96.2	0.044	0.189	0.165	0.151	0.226	0.167	16.5	60.
70.	11.4	92.0	0.32	23.3	97.7	0.048	0.189	0.166	0.153	0.231	0.170	15.9	70.
80.	14.1	91.1	0.39	25.6	99.1	0.053	0.189	0.167	0.155	0.235	0.172	15.2	80.
90.	17.2	90.3	0.47	28.0	100.6	0.057	0.189	0.168	0.157	0.240	0.174	14.5	90.
100.	20.8	89.4	0.56	30.4	102.1	0.061	0.189	0.169	0.159	0.244	0.177	13.9	100.
110.	25.0	88.4	0.66	32.9	103.5	0.066	0.190	0.170	0.161	0.248	0.179	13.2	110.
120.	29.8	87.5	0.79	35.4	105.0	0.070	0.190	0.171	0.162	0.252	0.181	12.6	120.
130.	35.3	86.5	0.92	37.9	106.5	0.074	0.191	0.171	0.164	0.256	0.183	12.0	130.
140.	41.5	85.6	1.08	40.5	107.9	0.079	0.191	0.172	0.165	0.259	0.185	11.3	140.
150.	48.5	84.6	1.26	43.1	109.3	0.083	0.192	0.173	0.167	0.263	0.188	10.7	150.
160.	56.5	83.6	1.46	45.8	110.7	0.087	0.192	0.173	0.168	0.265	0.190	10.1	160.
170.	65.3	82.5	1.68	48.5	112.1	0.091	0.193	0.174	0.169	0.268	0.193	9.5	170.
180.	75.2	81.5	1.93	51.2	113.5	0.096	0.193	0.175	0.170	0.271	0.195	8.9	180.
190.	86.1	80.4	2.21	53.9	114.8	0.100	0.194	0.175	0.171	0.273	0.198	8.3	190.
200.	98.2	79.2	2.52	56.6	116.1	0.104	0.194	0.176	0.172	0.275	0.201	7.7	200.

Refrigerant 123 Saturation Properties (*Concluded*)

Temp. °F	Pressure psia	Density lb/ft³ liquid	Density lb/ft³ vapor	Enthalpy Btu/lb liquid	Enthalpy Btu/lb vapor	Entropy Btu/lb·°F liquid	Entropy Btu/lb·°F vapor	c_v Btu/lb·°F liquid	c_v Btu/lb·°F vapor	c_p Btu/lb·°F liquid	c_p Btu/lb·°F vapor	σ dyn/cm	Temp. °F
210.	111.5	78.1	2.87	59.4	117.3	0.108	0.195	0.176	0.173	0.277	0.205	7.1	210.
220.	126.1	76.9	3.26	62.1	118.5	0.112	0.195	0.177	0.174	0.279	0.208	6.5	220.
230.	142.1	75.6	3.69	64.9	119.7	0.116	0.196	0.177	0.175	0.281	0.213	6.0	230.
240.	159.4	74.3	4.18	67.7	120.8	0.120	0.196	0.177	0.177	0.283	0.218	5.4	240.
250.	178.3	73.0	4.72	70.6	121.8	0.124	0.196	0.178	0.178	0.285	0.224	4.8	250.
260.	198.8	71.5	5.33	73.4	122.8	0.128	0.197	0.178	0.179	0.288	0.232	4.3	260.
270.	221.1	70.0	6.01	76.3	123.7	0.132	0.197	0.179	0.181	0.292	0.241	3.8	270.
280.	245.1	68.4	6.78	79.1	124.5	0.136	0.197	0.180	0.183	0.297	0.252	3.3	280.
290.	271.0	66.7	7.66	82.1	125.3	0.140	0.197	0.181	0.186	0.304	0.266	2.8	290.
300.	299.0	64.9	8.67	85.0	125.9	0.143	0.197	0.182	0.189	0.313	0.283	2.3	300.
310.	329.2	62.9	9.83	88.1	126.4	0.147	0.197	0.184	0.193	0.325	0.306	1.9	310.
320.	361.8	60.9	11.19	91.2	126.8	0.151	0.197	0.187	0.197	0.343	0.337	1.4	320.
330.	397.0	58.6	12.79	94.4	127.0	0.155	0.196	0.190	0.203	0.371	0.380	1.0	330.
340.	435.0	56.1	14.73	97.8	127.1	0.159	0.196	0.195	0.210	0.421	0.446	0.7	340.
350.	476.0	53.0	17.19	101.6	126.8	0.164	0.195	0.200	0.218	0.527	0.575	0.3	350.

Refrigerant 134a Saturation Properties

Temp. °F	Pressure psia	Density lb/ft³ liquid	Density lb/ft³ vapor	Enthalpy Btu/lb liquid	Enthalpy Btu/lb vapor	Entropy Btu/lb·°F liquid	Entropy Btu/lb·°F vapor	c_v Btu/lb·°F liquid	c_v Btu/lb·°F vapor	c_p Btu/lb·°F liquid	c_p Btu/lb·°F vapor	σ dyn/cm	Temp. °F
−40.	7.5	88.2	0.17	0.0	96.1	0.000	0.229	0.159	0.154	0.270	0.177	17.7	−40.
−30.	9.9	87.2	0.23	2.7	97.6	0.006	0.227	0.167	0.158	0.276	0.182	16.8	−30.
−20.	12.9	86.2	0.29	5.5	99.1	0.013	0.226	0.174	0.161	0.283	0.186	16.0	−20.
−10.	16.7	85.2	0.37	8.4	100.5	0.019	0.224	0.180	0.165	0.289	0.190	15.1	−10.
0.	21.2	84.2	0.46	11.3	102.0	0.026	0.223	0.186	0.168	0.296	0.195	14.3	0.
10.	26.6	83.1	0.58	14.3	103.5	0.032	0.222	0.190	0.172	0.302	0.200	13.5	10.
20.	33.1	82.0	0.71	17.4	104.9	0.039	0.221	0.194	0.176	0.308	0.206	12.7	20.
30.	40.8	80.9	0.87	20.5	106.3	0.045	0.220	0.198	0.180	0.314	0.211	11.9	30.
40.	49.8	79.7	1.05	23.7	107.7	0.051	0.219	0.201	0.184	0.320	0.217	11.1	40.
50.	60.2	78.6	1.26	26.9	109.1	0.058	0.219	0.203	0.188	0.326	0.224	10.3	50.
60.	72.1	77.4	1.51	30.2	110.4	0.064	0.218	0.205	0.192	0.332	0.231	9.5	60.
70.	85.9	76.2	1.79	33.6	111.7	0.070	0.218	0.207	0.196	0.339	0.238	8.8	70.
80.	101.5	74.9	2.12	37.0	112.9	0.077	0.217	0.208	0.200	0.345	0.246	8.0	80.
90.	119.1	73.6	2.50	40.5	114.1	0.083	0.217	0.209	0.204	0.352	0.255	7.3	90.
100.	138.9	72.2	2.93	44.0	115.2	0.089	0.217	0.210	0.208	0.359	0.266	6.6	100.
110.	161.1	70.7	3.43	47.6	116.3	0.096	0.216	0.210	0.213	0.367	0.277	5.9	110.
120.	185.9	69.2	4.00	51.3	117.2	0.102	0.216	0.211	0.217	0.376	0.291	5.2	120.
130.	213.3	67.6	4.67	55.1	118.1	0.108	0.215	0.212	0.222	0.386	0.308	4.5	130.
140.	243.8	65.8	5.44	59.0	118.8	0.115	0.214	0.213	0.227	0.399	0.328	3.8	140.
150.	277.4	64.0	6.36	62.9	119.3	0.121	0.214	0.214	0.232	0.414	0.355	3.2	150.
160.	314.4	61.9	7.44	67.0	119.7	0.127	0.212	0.216	0.238	0.433	0.391	2.6	160.
170.	355.1	59.7	8.76	71.3	119.7	0.134	0.211	0.218	0.244	0.459	0.442	2.0	170.
180.	399.9	57.2	10.40	75.7	119.4	0.141	0.209	0.222	0.251	0.498	0.520	1.4	180.
190.	449.2	54.2	12.51	80.4	118.6	0.148	0.207	0.227	0.259	0.566	0.650	0.9	190.
200.	503.4	50.6	15.39	85.6	117.0	0.156	0.203	0.233	0.267	0.721	0.918	0.5	200.
210.	563.2	45.0	20.22	92.3	113.6	0.165	0.197	0.245	0.277	1.549	2.129	0.1	210.

p. 17.63, Refrigerant 729 table. The headings for the columns that list pressures should read:

Pressure, psia	
Liquid	Vapor

p. 17.72. Figure 35, Equation 6 should read as follows:

$$t' = \frac{-2E}{D + [D^2 - 4E(C - \log P)]^{0.5}} - 459.72$$

p. 18.2, Table 1. The columns under the heading, "Mass per Unit Volume at 60°F" (columns 6 through 9) should read:

Mass per Unit Volume[b] at 60°F			
CaCl$_2$ lb/gal	Brine lb/gal	CaCl$_2$ lb/ft³	Brine lb/ft³

p. 18.3, Table 2. The columns under the heading, "Mass per Unit Volume at 60°F" (columns 6 through 9) should read:

Mass per Unit Volume[b] at 60°F			
NaCl lb/gal	Brine lb/gal	NaCl lb/ft³	Brine ft³

Additions and Corrections

pp. 19.2, 19.3, 19.4, 19.6. Replace Figures 1, 2, 3, 5, 6, and 7 with the following figures.

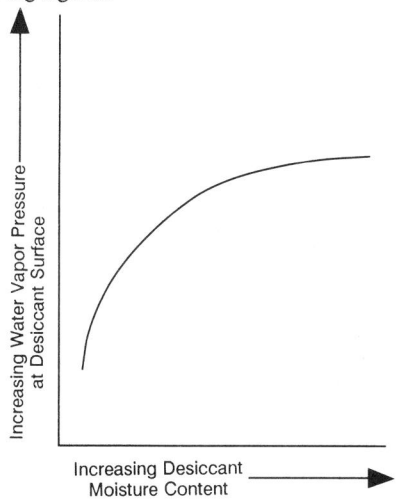

Fig. 1 Desiccant Water Vapor Pressure as a Function of Moisture Content (from *The Dehumidification Handbook* 1982)

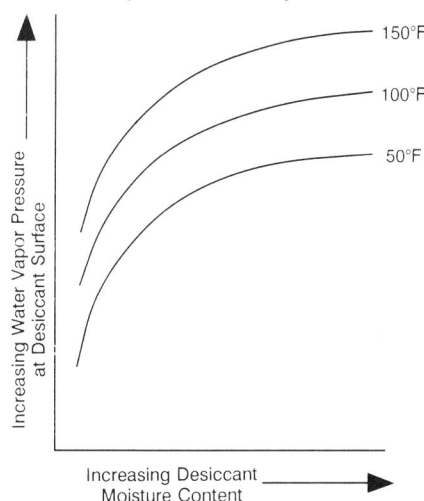

Fig. 2 Desiccant Water Vapor Pressure as a Function of Desiccant Moisture Content and Temperature
(from *The Dehumidification Handbook* 1982)

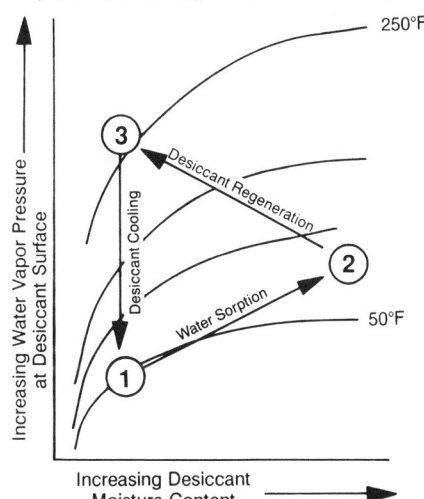

Fig. 3 The Desiccant Cycle
(from *The Dehumidification Handbook* 1982)

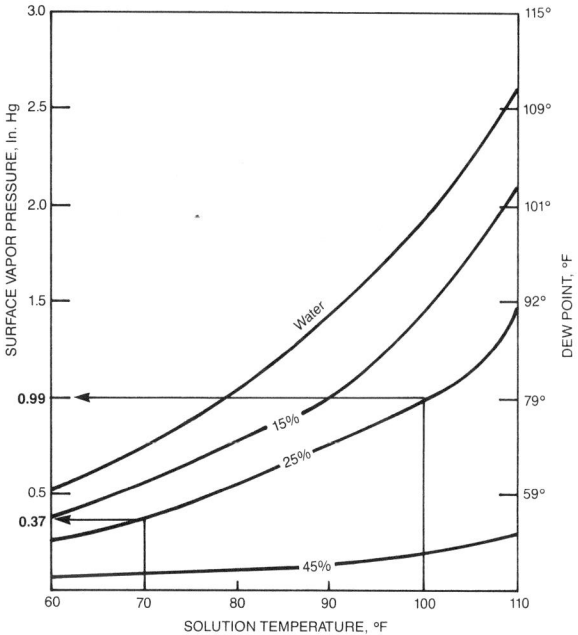

Fig. 5 Surface Vapor Pressure of Water-Lithium Chloride Solutions (from *Lithium Chloride Technical Data* 1988)

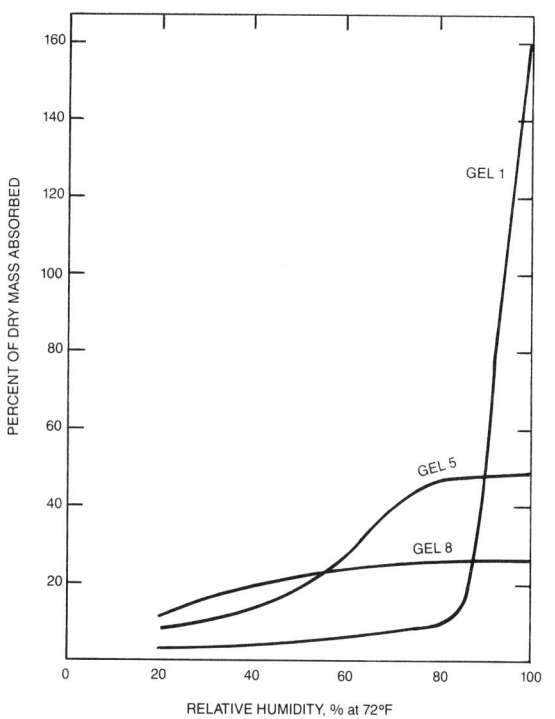

Reference Series Number	Total Surface Area, m^2/g	Average Capillary Diameter, nm	Total Volume of Capillaries, mm^3/g
1	315	21	1700
5	575	3.8	490
8	540	2.2	250

Fig. 6 Adsorption and Structural Characteristics of Some Experimental Silica Gels (from Oscic *et al.* 1982)

A.8   1990 Refrigeration Handbook

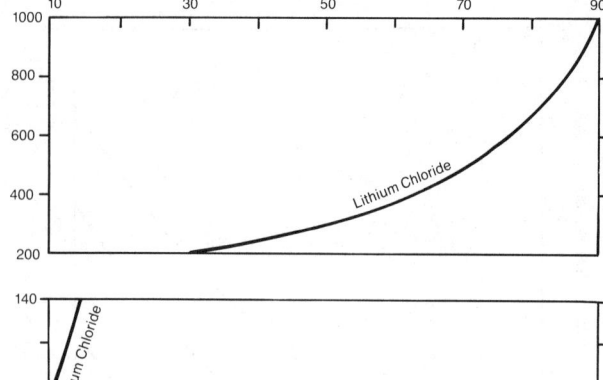

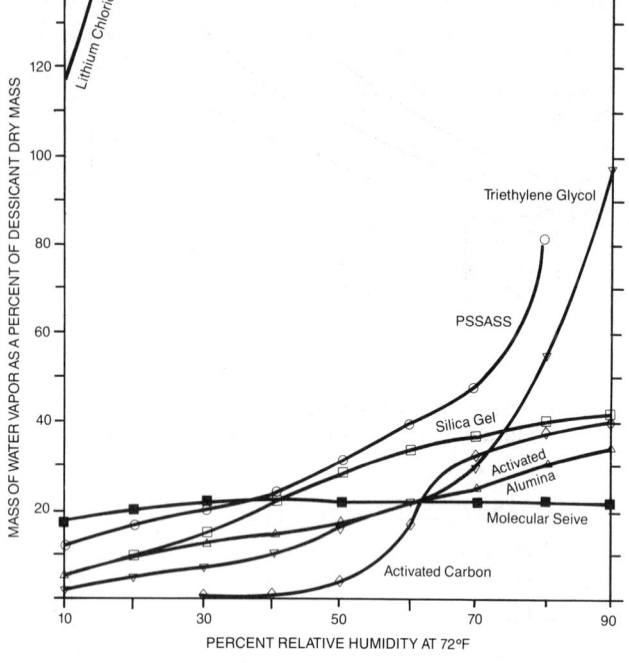

The sources for isotherms presented in the figure include:
PSSASS: Solar Energy Research Institute Report No. PR-255-3308
Lithium chloride: Cargocaire Engineering Division of Munters USA and Kathabar Systems Division of Sommerset Technologies
Triethylene glycol: Dow Chemical Corporation
Silica Gel: Davison Chemical Division of W.R. Grace Co. Inc.
Activated Carbon: Calgon Corporation
Activated Alumina: LaRoche Chemical Company
Molecular Sieve: Davison Chemical Division of W.R. Grace Co.

Fig. 7   Sorption Isotherms of Various Desiccants at 72°F

p. 20.5. Credit Figure 2 to Lotz, W.A. 1969. Facts about thermal insulation. ASHRAE *Journal*, June, p. 83.

p. 20.6, 3rd paragraph. Cross-reference should read: See Table 4, Chapter 22.

p. 20.7, 3rd paragraph, LH column. Reference should be made to Figure 1 of Chapter 22 and to Table 1 of Chapter 22.

p. 20.8. Add a [to the first term in the denominator of Equation (10) so it reads:

$$q_s = \frac{t_{is} - t_{os}}{[r_s \ln(r_1/r_i)]/k_1 + [r_s \ln(r_s/r_1)]/k_2} \quad (10)$$

p. 20.9, LH column. Delete t_o, x, and R_s in the listing of variables for Equations (9) and (10). These equations are based on temperatures at the inner and outer surfaces of the insulation, so they do not incorporate surface resistance values between the pipe and the insulation or between the outer surface and the ambient air.

p. 20.9. Equation (11) should read:

$$q_o = q_s(r_s/r_i) \quad (11)$$

p. 22.7. In Table 4, the Design Value Resistance (*R*) per inch thickness for Cellular polyisocyanurate should be listed as 7.04.

p. 22.8. Under section "Metals," reference should read: See Chapter 37, Table 3

p. 22.13. Some elements of Table 7 should be credited to Lotz, W.A. 1964. Vapor barrier design, neglected key to freezer insulation effectiveness. *Quick Frozen Foods*, November, p. 122.

p. 22.14, Table 7. The headers are correct for "Building paper, felts, roofing papers" and "Liquid-Applied Coating Materials." Change the heading for the "Plastic and Metal Foils and Films" as shown at the bottom of this page.

p. 23.8. Equation (18) should read:

$$Q = C_4 C_v A V \quad (18)$$

where *V* = wind speed, mph

p. 23.8. Equation (19) should read:

$$Q = C_5 K A \, [2g\Delta h_{NPL}(T_i - T_o)/T_i]^{0.5} \quad (19)$$

where T_i = indoor temperature, °R
         T_o = outdoor temperature, °R

Table 7   Typical Water Vapor Permeance and Permeability Values for Common Building Materials[a]

Material	Thickness, in.	Permeance, Perm	Resistance[h], Rep	Permeability, Perm-in.	Resistance/in.[h], Rep/in.
Plastic and Metal Foils and Films[c]					
Aluminum foil	0.001	0.0[d]	∞		
Aluminum foil	0.00035	0.05[d]	20		
Polyethylene	0.002	0.16[d]	6.3		3100
Polyethylene	0.004	0.08[d]	12.5		3100
Polyethylene	0.006	0.06[d]	17		3100
Polyethylene	0.008	0.04[d]	25		3100
Polyethylene	0.010	0.03[d]	33		3100
Polyvinylchloride, unplasticized	0.002	0.68[d]	1.5		
Polyvinylchloride, plasticized	0.004	0.8-1.4[d]	1.3-0.72		
Polyester	0.001	0.73[d]	1.4		
Polyester	0.0032	0.23[d]	4.3		
Polyester	0.0076	0.08[d]	12.5		
Cellulose acetate	0.01	4.6[d]	0.2		
Cellulose acetate	0.125	0.32[d]	3.1		

Additions and Corrections

p. 23.12, RH column. The 7th line of the 7th paragraph should read: The houses in Sweden—which have a residential building airtightness standard of 0.3 air changes per hour at 0.2 in. of water...

p. 23.14, Table 3, RH column. The best estimate of infiltration for Pipes, not caulked or sealed, should be 0.93 in^2 per pipe, not 9.30

p. 23.18, Example 3. The average infiltration should be:

$$I = (9780 \text{ ft}^3/\text{h})/(11,000 \text{ ft}^3) = 0.89 \text{ ACH}$$

p. 25.2, Table 1. The equation for "Infiltration and ventilation air" sensible heat should be:

$$q_s = 0.018 Q \Delta t$$

p. 26.2, RH column. Revise the first two sections of "Calculation Techniques" to read:

LARGE BUILDING CALCULATION TECHNIQUES

Transfer Function Method (Primary Method)

The Transfer Function Method (TFM) (Mitalas 1972) applies a series of weighting factors, or coefficients of Room Transfer Functions, to heat gain and cooling load values from previous hours as well as the current hour; these factors account for the thermal storage effect in converting heat gain to cooling load. Such weighting factors relate specifically to the spatial geometry, configuration, mass, and other characteristics of the space, so as to reflect weighted variations in thermal storage effect on a time basis, rather than a straight-line average.

Transfer Functions. These coefficients relate an output function to the value of one or more driving functions at a given time and at a set period immediately preceding. The transfer function described in this chapter is no different from the thermal response factor used for calculating wall or roof heat conduction and the weighting factor for obtaining cooling load components (ASHRAE 1975). The Bibliography lists reports of various experimental work that has validated the predictive accuracy of the TFM.

While the TFM is preferable to other methods and technically sound for a specific cooling load analysis, its computational complexity requires computer use for effective application in a commercial design environment.

Total Equivalent Temperature Differential Method (Alternative Method)

In the Total Equivalent Temerpature Differential (TETD) Method, various components of space heat gain are added to get an instantaneous total rate of space heat gain. This gain is converted to an instantaneous space cooling load by the Time-Averaging (TA) technique of averaging the radiant portions of the heat gain load components for the current hour with related values from immediately preceding hours. This technique is best solved by computer because of its complexity.

p. 26.5, LH column. The 4th line in the paragraph titled "Adjustments" should read:

listed in these tables, the value of It is approximately 1.15 SHGF.

p. 26.11, Table 9. Revise the values for copiers to read:

p. 26.8, Table 4. Heading under Column C should read:

Motor in, Driven Equipment Out, Btu/h

p. 26.8, Table 4, 1st column. The second 1 under column with the heading, "Motor Nameplate or Rated Horsepower" should read 1.5 (hp).

p. 26.29, Table 26, line 7. The August Sol-Air temperatures for light colored surfaces should be used. In line 11, loads for the roof should be based on the values in line 7.

p. 26.59, Table 64. For "Infiltration," Q should equal ACH (room volume)/60.

p. 26.59. In the notes for Table 64, $t_i - t_o$ = design temperature difference between inside and outside air, °F.

p. 26.60, Table 66. The column titled "Walls and Doors" should also include "Roof."

p. 27.26, RH column. The first equation (Surfaces other than vertical) should read:

$$I_{ds} = CI_{DN}(1 + \cos\Sigma)/2$$

p. 28.2, Equation (1). The units for E are the units in which the fuel is sold, such as gal, therm, kWh, etc.

p. 28.3, Equation (4). The variable shown as g_g should be q_g.

p. 28.3. Equation (5) should read:

$$E_H = \frac{24 \cdot \text{BLC} \cdot \text{DD}_{t_b}}{\bar{\eta}} \qquad (5)$$

where:

E_H = energy consumption, Btu
DD_{t_b} = degree-days calculated to the base temperature
$\bar{\eta}$ = average efficiency of the heating equipment

p. 28.4. The second line below Equation (8) should read, in part, "..., ρ is the density, ..."

p. 28.5, RH column, 3rd para. Replace part of text starting with Equation (14) with the following:

$$\bar{\eta} = \eta_{ss}\text{CF}_{pl}/(1 + \alpha_D) \qquad (14)$$

where:

η_{ss} = steady-state efficiency (rated output/rated input)
α_D = fraction of heat loss from ducts

The term CF_{pl} is a characteristic of the part-load efficiency of the heating equipment, determined by Equations (15) through (19). These equations are based on many annual simulations for the equipment in which RLC is defined as follows:

$$\text{RLC} = \text{BLC}(t_b - t_{od})(1 + \alpha_D)/\text{CHT}$$

where:

t_{od} = outside design temperature
CHT = rated output of equipment

Table 9 Recommended Rate of Heat Gain from Selected Office Equipment

Appliance	Size	Maximum Input		Standby Input		Recommended Rate of Heat Gain	
		Watts	Btu/h	Watts	Btu/h	Watts	Btu/h
Copiers (large)	30-67 copies/min.	5800-22500	1700-6600	5800-22500	900	3100	1700-6600
Copiers	6-30 copies/min.	460-1700	1600-5800	300-900	1000-3100	500-1700	1600-5800

p. 28.6, LH column. Revise the example values used to calculate UA_{inf} as follows:

$$UA_{inf} = 197.9 \text{ cfm} \times 60 \text{ min/h} \times 0.018 \text{ Btu}/(\text{ft}^3 \cdot °F)$$
$$= 213.7 \text{ Btu/h} \cdot °F$$

p. 28.6, RH column. Replace text starting with the sentence that states:

The total gains are now:

$$q_g = q_{internal} + q_{solar} + q_{sol\text{-}air} + q_{ground} \quad \text{[Equation (11)]}$$
$$= 100,000 + 34,300 + 7,000 + 64,500 = 76,800 \text{ Btu/day}$$

The balance temperature is then:

$$t_b = 72 - \frac{76,800 \text{ Btu/day}}{24 \text{h/day} \times 465.6 \text{ Btu/h} \cdot °F} = 65.1°F$$

The average delivered heating efficiency of the system is calculated using Equations (14) and (17) as:

RLC = $465.6[65.1°F - (-4°F)](1 + 0)/50,000$ Btu/h = 0.643
CF_{pl} = $0.9276 + 0.0732(0.643) - 0.0284(0.643)^2$ = 0.963
$\bar{\eta}$ = $0.80 \times 0.963/(1 + 0)$ = 0.770

The seasonal heating fuel consumption is then found from Equation (5) as:

$$E_H = 24 \cdot BLC \cdot DD_{tb}/\bar{\eta}$$
$$= 24 \times 465.6 \times 6031/0.770 = 87.5 \times 10^6 \text{ Btu}$$

p. 28.6, Table 5, first column. The variable for degree days calculated to the base temperature is DD_{tb}.

p. 28.6, Table 5, last column. The seasonal value for q_g is 76.8×1000 Btu/day (not 76.9).

p. 28.8, LH column. Replace part of Example (2) beginning with, "Using Equation (3) and calculating . . ." and continuing to Table 6 with the following:

The balance temperature t_{bc1} is calculated as:

$$t_{bc1} = t_i - q_g/BLC$$
$$q_g = q_{internal} + q_{solar} + q_{sol\text{-}air} + q_i + q_{ground}$$

The internal gain is 100,000 Btu/day as before, q_i is calculated above, and the other terms are calculated using the same procedures as in Example 1 with temperatures and solar gains that are three-month averages for June, July, and August. Then:

$$q_g = (100,000 + 48,300 + 20,300 + 18,800 - 10,500)$$
$$= 177,000 \text{ Btu/d} = 7375 \text{ Btu/h}$$
$$t_{bc1} = 78 - 7375/384 = 58.8°F$$

The second balance temperature t_{bc2} is calculated using Equation (26). Assuming an 8 ft ceiling, the volume of the house is:

$$V = 30 \times 50 \times 8 = 12,000 \text{ ft}^3$$

At 5 air changes per hour the ventilation rate C_f is:

$$C_f = 5 \times 12,000 = 60,000 \text{ ft}^3/\text{h} = 1000 \text{ cfm}$$

Hence, using q_g from above,

$$t_{bc2} = 78 - \frac{7375}{(60 \times 0.018 \times 1000 + 384 - 132)} = 72.5°F$$

p. 28.13, LH column, last equation. The summation runs from $i = 1$ to N EXPOSURE (not N SURF).

p. 28.14, LH column. Change the numbers under the $CLTDS_i$ column in the unnumbered table below Table 10 from 5, 15, 10, 15, 22 (for N, E, S, W, and ROOF exposures) to 0, 6, 21, 6, 3.

p. 28.15, LH column, item (3)(e). The equation for $QTOT_{unocc}$ is:

$$QTOT_{unocc} = 0.0502 \, t - 2.015$$

p. 28.17, LH column, item (5). The equation under Unoccupied hours should read:

$$FREQ_{unocc} = \text{Total bin hours} - FREQ_{occ}$$

p. 28.17, LH column, item (5)b.1. The value for VD is:

VD = design supply air volume = $19,998/12,000 = 1.67$ cfm/ft²

p. 28.19. Equation (30) should read:

$$q_{i,\theta} = h_i(t_{a,\theta} - t_{i,\theta}) + \sum_{k=1}^{n} g_{i,k}(t_{a,\theta} - t_{i,\theta}) + R_i \quad (30)$$

p. 28.23. In the section titled, "Algorithm for VAV System Simulation," all variables of the form m_i should have a dot over the m to correspond to Figure 8.

p. 28.29, Equations (58) and (59). The variable F_R is the collector efficiency factor, while the variable F_r is the collector-heat exchanger efficiency factor.

p. 28.29. Equation (61) should read:

Liquid System: $f = 1.029 \, Y - 0.065 \, X$
$\quad - 0.245 \, Y^2 + 0.0018 \, X^2 + 0.0215 \, Y^3 \quad (61)$

p. 28.32. Equation (68) should read:

$$Q = 88 \, C_v \, VA \quad (68)$$

p. 28.36, Equation (78). The value for σ is 0.1713×10^{-8} Btu/(h·ft²·°R).

p. 28.42, Bibliography. Change the author name to Boyd, R.L.

p. 31.5, LH column, 4th para. The first sentence should read:

Koestel and Tuve (1955) and Reinmann et al. (1959) investigated the effect of air motion on comfort.

p. 31.6. Delete two references to ASHRAE *Standard* 113-88 in the section "Ceiling-Mounted Air Diffuser Systems." This proposed standard is not available at present.

p. 32.5, LH column. Multiply the variable in the first equation by 12 so it reads:

$$f' = 0.11 \left(\frac{12 \, \epsilon}{D_h} + \frac{68}{\text{Re}} \right)^{0.25}$$

p. 32.32, RH column. Values for the 1 splitter vane should be replaced by the following:

Coefficients for elbows with 1 splitter vane (C_o')

R/W	r/W	CR	H/W									
			0.25	0.5	1.0	1.5	2.0	3.0	4.0	5.0	6.0	7.0	8.0
0.05	0.55	0.218	0.52	0.40	0.43	0.49	0.55	0.66	0.75	0.84	0.93	1.0	1.1
0.10	0.60	0.302	0.36	0.27	0.25	0.28	0.30	0.35	0.39	0.42	0.46	0.49	0.52
0.15	0.65	0.361	0.28	0.21	0.18	0.19	0.20	0.22	0.25	0.26	0.28	0.30	0.32
0.20	0.70	0.408	0.22	0.16	0.14	0.14	0.15	0.16	0.17	0.18	0.19	0.20	0.21
0.25	0.75	0.447	0.18	0.13	0.11	0.11	0.11	0.12	0.13	0.14	0.14	0.15	0.15
0.30	0.80	0.480	0.15	0.11	0.09	0.09	0.09	0.10	0.10	0.10	0.11	0.11	0.12
0.35	0.85	0.509	0.13	0.09	0.08	0.07	0.07	0.08	0.08	0.08	0.08	0.09	0.09
0.40	0.90	0.535	0.11	0.08	0.07	0.06	0.06	0.06	0.06	0.07	0.07	0.07	0.07
0.45	0.95	0.557	0.10	0.07	0.06	0.05	0.05	0.05	0.05	0.06	0.06	0.06	0.06
0.50	1.00	0.577	0.09	0.06	0.05	0.05	0.04	0.04	0.04	0.05	0.05	0.05	0.05

p. 32.32, RH column. Illustrations for two and three splitter vanes are reversed for fitting 3-7.

p. 33.22, Equation (16). The variables C and L are defined as follows:

C = factor for viscosity, density, and temperature
$\quad = 0.00354(t + 460)s^{0.848} \mu^{0.152}$

L = pipe length, ft

COMPOSITE INDEX

ASHRAE HANDBOOK SERIES

This index covers the technical data sections of the four current Handbook volumes published by ASHRAE. Listings from each volume are identified by a code letter preceding the page number.

    R = 1990 Refrigeration
    F = 1989 Fundamentals
    E = 1988 Equipment
    H = 1987 HVAC Systems and Applications

The index is alphabetized in a *word-by-word* format; for example, *air diffusers* is listed before *aircraft* and *heat flow* is listed before *heaters*.

Note that the code numbers include the chapter number followed by a decimal point and the page number(s) within the chapter. This index is updated in each Handbook volume.

Abbreviations
  computer programming, F34.1-3
  drawings, F34.1-3
  text, F34.1-3
Absorbents
  liquid, F19.3
Absorbers
  absorption units, E13.3,11
Absorption
  ammonia-water, E13.10-11
  chemical industry, R36.9
  chillers, engine-driven, E32.8
  dehumidification, E7.4-7
  industrial gas cleaning, E11.14-16
  refrigeration cycles, F1.20-26
    ammonia-water, F1.23
    lithium bromide-water, F1.22
  refrigerator, household
    ammonia-water-hydrogen cycle, E13.11
  water-lithium bromide, E13.2-9
Acoustics
  computer applications, H60.7
Activated charcoal
  contaminant control, gaseous, H50.5
Adjusting
  HVAC systems, H57.1-26
ADPI (Air Diffusion Performance Index), F31.8
Adsorbents, F19.3-6
Adsorption
  dehumidification, E7.5-8
  industrial gas cleaning, E11.21-24
  odor removal, F12.5
Air cleaners
  air washers, E4.10
  dry centrifugal collectors, E11.5
  electronic precipitators, E11.10-13
  fabric filters, E11.6-10
  filters, E10.9-11
  gas cleaners, E11.2
  gravity collectors, E11.4
  industrial exhaust systems, H43.10
  momentum collectors, E11.5
  packed scrubbers, E11.16
  particle collection mechanisms, E10.5
  ratings, E10.1
  safety requirements, E10.11
  small forced-air systems, H10.1
  test methods, E10.2-5
    ARI standards, E10.5
    atmospheric dust spot efficiency, E10.3
    DOP penetration, E10.3
    dust holding capacity, E10.3
    environmental, E10.4
    leakage (scan), E10.4
    miscellaneous performance, E10.4
    particle size efficiency, E10.4
    specialized performance, E10.4
    weight arrestance, E10.3
  types, E10.5
    electronic, E10.8
    panel, E10.6
    renewable media, E10.7
    wet-packed scrubbers, E11.15

Air cleaning
  adsorption of vapor emissions, E11.22-24
  electrically augmented scrubbers, E11.25
  gaseous contaminant removal
    absorption efficiency, E11.17-20
    general efficiency comparisons, E11.21
    liquid effects, E11.21
    pressure drop, E11.16
    scrubber packings, E11.15
  incineration of gases and vapors, E11.21
  particulate emission control, E11.2
  performance, E11.2
  wet collectors, E11.13
Air composition
  moist, F6.1
Air conditioners
  packaged terminal, E43.1-4
  residential, H17.1-6
  room, E41.1-6
  unitary, E42.1-8
    mechanical design, E42.7
    refrigeration system design, E42.3-6
    standards, E42.7
    through-the-wall, H5.2
    types, E42.2
    window-mounted, H5.2
Air Conditioning
  aircraft, H25.1-8
  air-and-water systems, H3.1-12
    changeover temperature, H3.4
    description, H3.1
    exterior zone loads, H3.1
    fan-coil, H3.3
    four-pipe, H3.8
    induction, H3.2
    performance, varying load, H3.3
    primary air, H3.3
    refrigeration load, H3.5
    secondary water distribution, H3.9-11
    three-pipe, H3.8
    two-pipe, H3.5-8
  all-air systems, H2.1-18
    air volume, H2.2
    dual duct, H2.13
    dual-path, H2.13-17
    duct systems, H2.1
    evaluation, H2.3
    general, H2.1
    heating, H2.2
    multizone, H2.17
    reheat, H2.3
    single-duct VAV, fan-powered, H2.10
    single-duct VAV, H2.5-8
    single-duct VAV, induction, H2.9
    single-path, H2.3
  all-water systems, H4.1-4
    applications, H4.3
    controls, H4.4
    fan-coil units, H4.1
    four-pipe, H4.3
    piping, H4.4
    two-pipe changeover, H4.2
    ventilation air, H4.3
    water distribution, H4.2

  auditoriums, H20.4
  automobile, H24.1-6
  bars, H19.5-7
  bowling centers, H19.12
  bus, H24.7-9
  central system, H1.2-12
    air distribution, H1.6
    application, H1.2
    components, H1.4-6
  churches, H20.4
  clean spaces, H32.1-8
  commercial buildings, H19.1-16
  communications centers, H19.13
  concert halls, H20.4
  convention centers, H20.6
  cooling load calculations, F26.1-62
  data processing areas, H33.1-8
  domiciliary facilities, H21.1-4
  dormitories, H21.1-4
  educational facilities, H22.1-6
  enclosed stadiums, H20.5
  exhibition centers, H20.6
  gymnasiums, H20.6
  health facilities, H23.1-12
  heat pump systems, applied, H9.1-16
    components, H9.6
    compressors, H9.6
    controls, H9.13
    defrost controls, H9.7
    equipment selection, H9.11-13
    heat reclaim cycle, H9.14
    heat sinks, H9.2-6
    heat sources, H9.2-6
    operating cycles, H9.7-11
    refrigerant controls, H9.6
    thermal storage, H9.7
    types, H9.1-3
  hospitals, H23.1-12
  hotels, H21.1-4
  humid climates, F21.12-14
  industrial, H28.1-10(aboratories, H30.1-16
  motels, H21.1-4
  nightclubs, H19.5-7
  nuclear facilities, H40.1-10
  nursing homes, H23.1-12
  panel heating and cooling systems, H7.1-18
    applications, H7.1
    components, H7.9, 14
    controls, H7.10, 17
    design considerations, H7.7-9, 11
    electrically heated, H7.13
    embedded piping, H7.11
    evaluation, H7.2
    heat transfer, H7.3-7
    metal ceiling panels, H7.11
    resistance, thermal, H7.12
  paper products facilities, H39.1-4
  photographic processing, H36.1-8
  places of assembly, H20.1-10
  printing plants, H34.1-6
  public buildings, H19.1-16
  railroad, H24.9
  residential, H17.1-6
  restaurants, H19.5-7

I.1

retail stores, H18.1-8
schools, H22.1-6
ships, H26.1-8
sports arenas, H20.5
survival shelters, H27.1-22
swimming pools, H20.7
system selection, H1.1-2
temporary exhibit buildings, H20.9
textile processing, H35.1-8
theaters, H20.4
transportation centers, H19.14
underground mines, H42.1-8
unitary systems, H5.1-8
   air-to-air heat pump, H5.3
   characteristics, H5.1
   heat pumps, H5.2, 5-8
   indoor equipment, H5.4
   outdoor equipment, H5.3
   through-the-wall, H5.2
   water-source heat pump, H5.5-8
   window-mounted, H5.2
warehouses, H19.15
wood products facilities, H39.1-4
world fair buildings, H20.9
Air contaminants
aeroallergens, F8.28
measurement, F13.25
respiratory tract, F8.28
Air cooling
evaporative, H56.1-8
mines, underground, H42.2-4
Air diffusers
merchant ships, H26.3
naval surface ships, H26.7
Air diffusion, F32.1-16
ceiling mounted diffusers, F31.6
definitions, F31.1
duct arrangements, F31.14
exhaust inlets, F31.14
outlet
   angle of divergence, F31.9
   centerline velocities, F31.10-12
   effect of walls and ceilings, F31.13
   isothermal jets, F31.8
   isothermal radial jets, F31.13
   jet expansion zones, F31.9
   jet velocity profiles, F31.12
   location, F31.7
   nonisothermal axial flow jets, F31.13
   selection, F31.7
   vertically projected heated and chilled jets, F31.13
performance index (ADPI), F31.8
return air design, optimum, F31.15
return inlets, F31.14
standards for satisfactory performance, F31.5
Air distribution
aircraft air conditioning, H25.3
animal environments, H37.2, 4
banana ripening rooms, R17.8
bowling centers, H19.12
central system, H1.6
communications centers, H19.13
data processing areas, H33.4-6
hospitals, H23.1
laboratories, H30.2, 8
merchant ships, H26.3
naval surface ships, H26.8
places of assembly, H20.2
plant growth facilities, H37.15
railroad air conditioning, H24.10
railway refrigerator cars, R29.8
small forced-air systems, H10.1-8
   air cleaners, H10.1
   airflow requirements, H10.4
   components, H10.1
   controls, H10.2
   design, H10.2
   duct design, H10.4-8
   ducts, H10.1
   equipment selection, H10.3
   grille and register selection, H10.8
   humidifiers, H10.1
   outlets, H10.2
store air conditioning, H18.1-8
swimming pools, H20.8
testing, adjusting, balancing, H57.3-7
textile processing, H35.5
Air exchange
calculating, F23.16-18
infiltration, F23.1
measurement, F23.10
ventilation
   forced, F23.1, 5
   natural, F23.1-9
Air leakage
building components, F23.10-16
commercial buildings, F23.13
controlling, F23.15
exterior doors, F23.14
internal partitions, F23.13
measurement, F23.11
railway refrigerator cars, R29.8
residential, F23.13
Air quality
aircraft air conditioning, H25.2
contaminant control, gaseous, H50.1-8
hospitals, H23.1
indoor, F11.6-8
pollutants, F11.7
rapid transit systems, H29.10
Air transport
commodity requirements, R31.2
design considerations, R31.2
refrigeration, R31.3-4
shipping containers, R31.3
Air washers
cell type, E4.7-8
mass transfer, F5.12
spray type, E4.6-7
water treatment, H53.15
Airborne
contaminants, F11.1-8
particulate matter, F11.2-4
Aircraft air conditioning
air distribution, H25.3
air source, H25.4
cabin pressurization system, H25.7
design conditions, H25.1-3
refrigeration systems, H25.4
temperature control systems, H25.6,8
Airflow
around buildings, F14.1-18
   atmospheric dispersion of exhaust gases, F14.10
   cooling towers, F14.13
   estimating minimum dilution, F14.11
   exhaust stack design, F14.13-16
   flow control, F14.10
   flow patterns, F14.1
   heat rejection equipment, F14.13
   intakes estimating concentrations, F14.11
   internal pressure, F14.10
   meteorological effects, F14.6-9
   modeling, F14.13-17
   testing, F14.13-17
   wind effects on system operation, F14.9
   wind pressures on buildings, F14.4-6
smoke control, H58.1-16
Airports, H19.14
Altitude chambers, R37.7
Ammonia
piping, R4.1-6
system chemistry, R6.1-8
system practices, R4.1-16
-water absorption cycle, F1.24
Ammonia-water solutions
properties, F17.69
Anemometers, F13.14
Animal
air transport, R31.2
laboratories, H30.13
physiology, F9
   cattle, F9.5
   chickens, F9.9
   control systems, F9.2-4
   laboratory animals, F9.11, H30.13
   sheep, F9.6
   swine, F9.8
   turkeys, F9.10
Animal environments, H37.1-8
aircraft ventilation requirements, R31.2
control
   moisture, H37.1
   temperature, H37.1
modification, H37.2
recommended practices, H37.6-8
   beef cattle, H37.6
   dairy cattle, H37.6
   laboratory animals, H37.8
   poultry, H37.8
   swine, H37.7
ventilation design, H37.3-6
Apartments
service water heating, H54.9,11
Attic
temperature, F25.3
ventilation, F21.10
Attics, ventilated, F22.11
Auditoriums, H20.4
concert halls, H20.5
projection booths, H20.4
stages, H20.4
theaters, H20.4
Automobile air conditioning, H24.1-6
components, H24.3-5
controls, H24.6
Bakery products
bread, production of, R21.1-5
thermal properties, R21.5
Balancing, H57.1-19
air distribution systems, H57.3-7
   dual-duct systems, H57.4
   equipment and system check, H57.3
   induction systems, H57.6
   instrumentation, H57.3
   preliminary procedure, H57.3
   reports, H57.6
   variable volume systems, H57.4-6
central system, H1.9
cooling towers, H57.16
definitions, H57.1
energy audit, field survey, H57.17-19
general criteria, H57.1
hydronic systems, H57.7-15
   basic methods, H57.8-11
   chilled water, H57.14
   heat transfer at reduced flow rate, H57.7
   instrumentation, H57.15
   steam distribution, H57.15
measurements, volumetric, H57.2
sound testing, H57.19-22
temperature control verification, H57.17
vibration testing, H57.22-25
Barometers, F13.12
Bars, H19.5-7
Bernoulli equation, F2.2
use of generalized, F2.7
Bin method, F28.10
Biomedical applications, R40
Boilers
modeling, F28.24
residential, H17.3
   piping, H13.25
steam systems, H11.3
water treatment, H53.17
Boiling, F4.1-7
evaporation in tubes, F4.5-7
film, F4.3-4
heat transfer equations, F4.2-3
maximum heat flux, F4.4
nucleate, F4.1-4
point
   liquid, F37.2
   refrigerants, F16.4
   vapor, F37.1
pool, F4.1
Bowling centers, H19.12
Bread (see Bakery products)

Index

Breweries
  carbon dioxide production, R23.6
  chemical aspects, R23.1
  processing
    fermenting cellar, R23.4
    finishing operations, R23.6
    Kraeusen cellar, R23.5
    stock cellar, R23.5
    wort cooling, R23.3
  storage, R23.6
  vinegar production, R23.8
Brines (see also secondary coolants), F18.1-10
Building loss coefficient (BLC), F28.3, 7
Building systems
  insulation, thermal, F20.1-26
Bus air conditioning
  interurban, H24.7
  urban, H24.8
Bus terminals, H29.13-14
Candy
  manufacture, R22.1-5
  storage, R22.5-7
Capillary tubes, E19.21-30
Carbon dioxide
  commercial freezing, use for, R9.5
  modified atmosphere storage, R25.3
  physical properties of, F37.1
  production of, R23.6-8, R16.2, R18.4
  refrigerant properties, F17.36
  volume in water, R23.10
Carbonated beverages
  liquid CO_2 storage, R23.11
  refrigeration, R23.11
Cascade
  ammonia systems, R4.15
  engineered refrigeration systems, R1.3
Cattle
  growth, F9.5
  heat and moisture production, F9.6
  lactation, F9.5
  recommended environment, H37.6
  reproduction, F9.5
Ceiling
  air distributing, E2.5
  diffuser outlets, E2.6
  induction boxes, E2.7
Central plant
  heating and cooling systems, H12.1-16
  chilling plant, H12.13-15
  corrosion, H12.12
  earth temperatures, H12.7-9
  earth thermal conductivity, H12.6
  heat transfer, H12.4
  insulated piping, H12.1-4
  operating guidelines, H12.12
  pipe anchors, H12.10
  pipe system movement, H12.6-10
  piping distribution, H12.1
  water hammer surge, H12.6
Centrifugal
  compressors, E12.26-35
  fans, E3.1-14
  liquid chilling systems, E17.8-13
  pumps, E30.1-10
Charging
  factory, E21.4
Chemical industry refrigeration, R36
  design standards and codes, R36.5, 10
  load characteristics, R36.2
  refrigerants, R36.5
  refrigeration equipment, R36.6-9
  safety requirements, R36.2
Chickens
  growth, F9.9
  heat and moisture production, F9.10
  reproduction, F9.9
Chilled water
  basic systems, H13.6
  testing, adjusting, balancing, H57.14
Chillers
  absorption, engine driven, E32.8
  heat recovery systems, H6.3
  modeling, F28.25
  water coolers, drinking, E38.3
Chilling, liquid systems, E17.1-16
  centrifugal
    auxiliaries, E17.12
    control, E17.11
    equipment, E17.8
    fouling factor, E17.11
    heat recovery, E17.12
    maintenance, E17.13
    noise, E17.11
    operating characteristics, E17.9-11
    operation, E17.13
    performance characteristics, E17.9
    refrigerant selection, E17.9
    selection method, E17.11
    special applications, E17.12
    vibration, E17.11
  general characteristics, E17.1-6
  reciprocating
    control, E17.7
    equipment, E17.6
    fouling factor, E17.7
    operating problems, E17.6
    performance characteristics, E17.6
    selection method, E17.7
    special applications, E17.8
  screw
    auxiliaries, E17.15
    equipment, E17.13
    fouling factor, E17.15
    maintenance, E17.16
    refrigerant selection, E17.14
    selection method, E17.15
    special applications, E17.15
Chimneys
  accessories, E26.21-24
  draft hoods, E26.22
  draft regulators, E26.23
  flue gas extractors, E26.23
  heat exchangers, E26.23
  vent dampers, E26.23
  air supply to fuel-burning equipment, E26.20
  altitude correction, E26.8
  capacity calculation examples, E26.13-18
  codes, E26.27
  design equation, E26.2
  draft, E26.8
  draft fans, E26.24
  fireplace, E26.19
  functions, E26.1
  gas appliance venting, E26.18
  gas temperature, E26.6
  gas velocity, E26.9
  heat transfer, E26.6
  mass flow, E26.4
  materials, E26.20
  resistance coefficients, E26.10-13
  standards, E26.27
  system flow losses, E26.9
  terminations, E26.25
  wind effects, E26.25
Chlorofluorocarbons
  properties, F16.1-15, F17.1-33
  system chemistry, R6
  system practices, R3
Churches, H20.4
Clean spaces
  airflow, H32.2-4
  airborne particles, H32.1
  central equipment, H32.7-8
  exhaust air, H32.6
  humidity control, H32.6
  lighting, H32.6
  makeup air, H32.6
  noise, H32.6
  particulate control, H32.1-2
  pressurization, H32.4
  room construction, H32.7
  temperature control, H32.6
  terminology, H32.1
  vibration, H32.6
Cleaners, air, E10.1-12
Climate (see Weather data)
Clo, F8.3
Clothing, F8.3
Coal, F15.6
Codes, F36.1-14
  boilers, E23.3
  chemical industry, R36.10
  duct construction, E1.1
  engine drives, E32.6
  evaporative condensers, E15.17
  heaters (in-space), E25.6
  motors, E31.2
  service water heating, H54.7
  water-cooled condensers, E15.7
  water coolers, drinking, E38.3
Cogeneration
  absorption units, E13.10
  controls, H8.10-12
  definitions, H8.13
  energy user interconnections, H8.12
  generators, H8.2
  glossary, H8.13
  heat recovery, H8.3
  exhaust gas, H8.7-9
  general, H8.3-5
  jacket, H8.7
  lubricating systems, H8.5
    installation problems, H8.9
  prime movers, H8.1
Coils
  air-heating
    water systems, H15.8
  ammonia systems, R4.9-11
  central system, H1.5-6
  condensers, E15.5, 8, 13
  control, H51.17
  cooling and dehumidifying, F5.13
  merchant ships, H26.3
  naval surface ships, H26.7
  piping, halocarbon refrigerants, R3.17-19
Coils, air cooling
  airflow resistance, E6.6
  construction and arrangement, E6.1
    applications, E6.3
    control, E6.2
    direct expansion, E6.2
    flow arrangement, E6.3
    selection, E6.5
    water, E6.1
  dehumidifying performance, E6.9-14
  heat transfer, E6.6
  maintenance, E6.15
  refrigeration load, E6.14
  sensible cooling performance, E6.7-9
  uses, E6.1
Coils, air heating, E9.1-4
  applications, E9.2
  construction, E9.1
  design, E9.1
  electric, E9.2
  flow arrangement, E9.3
  ratings, E9.3
  selection, E9.3
  steam, E9.1
  water, E9.1
Collectors, solar, H47.10-16, 18
Colleges, H22.1-6
Combustion
  air pollution, F15.13
  altitude compensation, F15.3
  calculations, F15.7-10
    air required, F15.8-11
    efficiency, F15.11
    flue gas produced, F15.10
    sample, F15.10
    seasonal efficiency, F15.12
  condensation, F15.13
  corrosion, F15.13
  engine drives, E32.5
  engine test facilities, H31.6
  flammability limits, F15.2
  fuel classifications, F15.3
  gas burners, E22.3
  gaseous fuels, F15.4

heating values, F15.3
ignition temperature, F15.2
liquid fuels, F15.4
modes, F15.2
odor removal, F12.5
principles, F15.1
products, chimney, E26.4
reactions, F15.1
solid fuels, F15.6
soot, F15.14
Comfort
measurement, F13.9-11
prediction of, F8.16-21
Commercial buildings
air conditioning
airports, H19.14
bars, H19.5-7
bowling centers, H19.12
bus terminals, H19.14
communications centers, H19.13
design concepts, H19.4
design criteria, H19.4
dining and entertainment centers, H19.5-7
general design criteria, H19.1-4
libraries and museums, H19.9-12
load characteristics, H19.4
office buildings, H19.7-9
restaurants, H19.5-7
ship docks, H19.14
transportation centers, H19.14
warehouses, H19.15
Commodity storage, R26
density of packaged commodities, R26.9-11
deterioration, R26.1, 7
properties of perishable products, R26.2-5
storage requirements, R26
Communications centers, H19.13
design concepts, H19.13
load characteristics, H19.13
special considerations, H19.13
Compressors, E12.1-36
air conditioners
room, E41.2
unitary, E42.10
ammonia systems, R4.5-6
automobile air conditioning, H24.3
balancing, E18.1
bus air conditioning, H24.8
centrifugal
application, E12.32
capacity control, E12.32
drivers, E12.33
heat recovery, E12.33
isentropic analysis, E12.28
lubrication, E12.34
maintenance, E12.35
mechanical design, E12.34
noise, E12.33
operation, E12.35
parallel operation, E12.34
performance, E12.31
polytropic analysis, E12.28-30
surging, E12.31
testing, E12.31
theory, E12.27
vibration, E12.32
chemical industry, R36.6
cryogenic, R38.13
engineered refrigeration systems, R1.3
heat pump systems, H9.6, 12
helical rotary (screw), E12.15-26
capacity control, E12.17, 22
double-screw, E12.21-26
economizers, E12.19, 26
lubrication, E12.17
mechanical features, E12.17, 22
oil-injected, E12.18, 25
performance characteristics, E12.19, 26
single screw, E12.16-21
household refrigerators and freezers, E37.8
piping, for halocarbon regrigerants, R3.15-16
positive displacement, E12.1-5
reciprocating, E12.5-11

retail food storage refrigeration, E35.6-9
rotary, E12.11-15
efficiency, E12.13
features, E12.12
large, E12.13
lubrication, E12.13
motor selection, E12.13
performance, E12.12
scroll, E12.14
small, E12.11
trochoidal, E12.15
valves, E12.13
Computer applications, H60.1-16
administrative uses, H60.13
artificial intelligence, H60.15
CAD capabilities, H60.8
communications, H60.9
control, H60.15
design calculations, H60.3-8
hardware options, H60.1-3
microcomputer productivity tools, H60.12
monitoring, H60.15
software, H60.10-12
Computer(s)
abbreviations for programming, F34.1-3
energy estimating programs, F28.26, 38
smoke control analysis, H58.15
Computer rooms, H33.1-8
Condensate
steam systems, H11.7
water treatment, H53.18
Condensation
checklist, F21.11
concealed, E5.2
commercial construction, F21.9
control strategies, F21.8
institutional construction, F21.9
membrane roof systems, F21.9
other considerations, F21.9
residential construction, F21.8
cooled structures, F21.12
dehumidification equipment, E7.1-8
fuel-burning systems, F15.13
heat recovery equipment, E34.14
heated buildings, F21.6
control of excess humidity, F21.7
indoor humidity levels, F21.6
visible, E5.2
floor slab, F21.7
preventing surface, F21.7
zones in United States, F21.8
Condensers
absorption units, E13.3, 11
air conditioners, room, E41.2
air-cooled, E15.8-13
application, E15.9
coil construction, E15.8
control, E15.11
fans, E15.8
heat transfer, E15.9
installation, E15.12
maintenance, E15.12
part of unit, E15.9
rating, E15.9
remote, E15.9
ammonia systems, R4.6-9
automobile air conditioning, H24.3
balancing, E18.2
central system, H1.10
data processing areas, H33.6
engineered refrigeration systems, R1.4
evaporative, E15.13-17
halocarbon refrigerants, R3.13-15, 22
household refrigerators and freezers, E37.7
noise, R32.6
retail food store refrigeration, E35.6, R32.3-5
water-cooled, E15.1-7
circuiting, E15.5
codes, E15.7
construction, E15.7
film coefficient, E15.2
fouling factor, E15.3
heat removed, E15.1

heat transfer, E15.2
liquid subcooling, E15.5
maintenance, E15.7
noncondensable gases, E15.6
operation, E15.7
pressure drop, water, E15.4
shell-and-coil, E15.6
shell-and-tube, E15.5
testing, E15.7
tube-in-tube, E15.6
water systems, H14.7
Condensing
heat transfer coefficients, F4.7-10
inside horizontal tubes, F4.10
inside vertical tubes, F4.9
noncondensable gases, F4.10
outside vertical tubes, F4.9
pressure drop, F4.11
Conduction
equations, F3.3
industrial drying systems, H44.3
resistance, F3.2
steady state, F3.2
Conductivity, thermal
building materials, F23.6-10
design values, F22.6-9
earth, H12.6
food, F30.13-19
industrial insulation, F23.16-20
insulating materials, F23.6-10
liquids, F37.2
soils, F22.21, F23.20, H27.6
solids, F37.3-4
vapor, F37.1
Containers (see Trucks)
air transport, R31.3
Contaminants
animal environments, H37.2
control, gaseous, H50.1-8
activated charcoal, H50.5
air quality, H50.4
applications, H50.1
design considerations, H50.3
design techniques, H50.3
dry sorbent, H50.5
energy considerations, H50.2
equipment, H50.6
methods, H50.5
odor control considerations, H50.3
problem, H50.1
standards, H50.4
system design, H50.7
washing, H50.5
food, R10
refrigerant system, R6, R7
anti-freeze agents, R7.7
dirt, R7.6
metallic, R7.6
moisture, R7.1-6
motor burnout, R7.8-9
noncondensable gases, R7.7
sludge, wax, and tars, R7.6
solvents, R7.7
Contaminants, air, F11.1-10
air composition, F11.1
ambient air standards, F11.6, 8
atmospheric pollen, F11.7
classification, F11.1-2
dusts, F11.1
fogs, F11.2
fumes, F11.1
gases, F11.2
mists, F11.2
smokes, F11.1
vapors, F11.2
combustible dusts, F11.5
flammable gases and vapors, F11.4
indoor air quality, F11.6
industrial, F11.4
measurement, F11.4
microorganisms, F11.8
pollution, F11.5
radioactive, F11.7

Index

Control
- condensation, F21.6-11
- fans, E3.1-14
- humidity, E5.8
- sorption dehumidification, E7.1-8
- sound, F7.1-10
- static pressure, E2.7
- valve loss coefficients, F2.10

Control(s) [see also Control and Control(s), automatic]
- absorption units, E13.5, 11
- air coolers, forced circulation, E8.2
- aircraft air conditioning, H25.6-8
- air-and-water systems, H3.7-10
- air-cooled condensers, E15.11
- air-cycle equipment, E14.2
- all-air systems
  - constant volume, H2.5
  - dual-duct, H2.15
  - multizone, H2.18
  - variable volume, H2.6, 10, 12
- all-water systems, H4.4
- animal environments, H37.1
- automobile air conditioning, H24.6
- banana ripening rooms, R17.7
- beverage coolers, E39.2
- boilers, E23.4
- bus air conditioning, H24.8
- central system, H1.7
- chemical industry, R36.10
- clean spaces
  - air pattern, H32.2
  - humidity, H32.6
  - particulate, H32.1
  - temperature, H32.6
- cogeneration systems, H8.10-12
- contaminant, gaseous, H50.1-8
- data processing areas, H33.6
- drying of farm crops, H38.3
- educational facilities, H22.2
- energy estimating methods, F28.22
- engine, E32.6
- engineered refrigeration systems, R1.9
- environmental test facilities, R37.4
- evaporative cooling systems, H56.3-5
- fire and smoke, H58.1-16
- fuel-burning equipment, H22.13-16
  - combustion, E22.15
  - draft, E22.15
  - operating, E22.13
  - programming, E22.14
  - safety, E22.14
- greenhouses, H37.12
- halocarbon refrigerants, R3.22-25
- heat recovery equipment, E34.6, 12
- heaters (in-space), E25.3, 5
- household refrigerators and freezers, E37.9
- humidifiers, E5.7
- infrared heaters, E29.1-6
- liquid chilling systems, E17.4, 7, 16
- liquid overfeed systems, R2.5
- makeup air units, E27.9
- merchant ships, H26.6
- naval surface ships, H26.8
- odor, vapor adsorption, E11.22
- panel cooling systems, H7.10
- panel heating systems, H7.17
- plant growth facilities, H37.15
- railroad air conditioning, H24.10
- residential air conditioning, H17.5
- small forced-air systems, H10.2
- snow melting systems, H55.14
- solar energy equipment, E44.18-20
- solar energy systems, H47.27
- sound, H52.1-27
- steam systems, H11.14
- store air conditioning, H18.1-8
- testing, adjusting, balancing, H57.17
- thermal storage, H46.3
- turbine, E32.10
- unit heaters, E27.6
- unit ventilators, E27.2
- unitary air conditioning systems, H5.3-8
- vehicular tunnels, H29.7
- vibration, H52.28-39
- water systems, basic, H13.20-23
- water systems, medium, and high temp, H15.8

Control(s), automatic, H51.1-40 [see Control(s)]
- central air handling systems, H51.21-26
  - constant volume, H51.22
  - dual-duct, H51.23
  - hydronic, H51.25
  - makeup air, H51.25
  - multizone, H51.24
  - single zone, H51.24
  - variable volume, H51.21
- central plant, H51.31-35
- central subsystems, H51.12-21
  - cooling coils, H51.18
  - fan control, H51.14-17
  - heating coils, H51.17
  - humidity, H51.19
  - outdoor air quantity, H51.13
  - space considerations, H51.20
- commissioning, H51.38
- components, H51.4-12
  - auxiliary devices, H51.12
  - controlled devices, H51.4
  - controllers, H51.10-12
  - dampers, H51.7
  - operators, H51.5
  - positive positioner, H51.8
  - selection and sizing, H51.6
  - sensors, H51.8-10
  - thermostats, H51.11
  - valves, H51.4
- cooling/heating changeover, H51.37
- design and principles, H51.35
- explosive atmospheres, H51.37
- fundamentals, H51.1-4
  - analog, H51.2
  - block diagrams, H51.2
  - direct digital, H51.2
  - energy sources, H51.4
  - system performance requirements, H51.2
  - terminology, H51.1
  - types of control action, H51.3
- limit controls, H51.37
- lowered night temperature, H51.38
- maintenance, H51.39
- mobile units, H51.37
- operation, H51.39
- retrofitting, H51.39
- safety for duct heaters, H51.37
- space control, H51.26-30
  - constant volume terminal units, H51.28
  - fan-coil units, H51.28
  - radiant panels, H51.30
  - unit ventilators, H51.29
  - VAV terminal units, H51.26-28
- training, H51.39
- tuning, H51.38
- using computers, H51.38

Controlled atmosphere storage
- apples, R16.2-3
- pears, R16.4
- refrigerated warehouses, R25.4
- vegetables, R18.4

Convection
- forced, F3.13-16
  - equations, F3.15-16
  - evaporation in tubes, F4.5-7
  - techniques to augment, F3.15
- industrial drying systems, H44.4
- mass, F5.4-10
- natural, F3.12

Convectors, E28.1-6

Convention centers
- load characteristics, H20.7
- system applicability, H20.7

Conversion factors, F35.1-4

Coolants, secondary, F18, R5
- calcium chloride, F18.2
- corrosion, F18.2
- glycols, F18.4-7
- halocarbons, F18.10
- ice rinks, R34.5
- properties, F18.2-10, R5.2
- sodium chloride, F18.3-5, R5.2
- storage, thermal, H46.9
- system practices, R5.1-7
  - design considerations, R5.3
  - performance comparisons, R5.2
  - selection, R5.1
- water treatment, H53.17

Coolers
- retail food store, R32.3
- walk-in, E36.4, R32.3
- water, drinking, E38.1-6

Coolers, liquid, E16.1-6
- Baudelot, E16.2
- direct expansion, E16.1
- flooded shell-and-tube, E16.1
- freeze prevention, E16.5
- heat transfer
  - coefficients, E16.3
  - fouling factor, E16.3
- insulation, E16.6
- maintenance, E16.5
- oil return, E16.5
- pressure drop, E16.4
- refrigerant flow control, E16.5
- shell-and-coil, E16.3
- vessel design, E16.4-5

Coolers, storage
- retail food store refrigeration, E35.6

Cooling
- energy estimating, F28.7-9
- forced air, R11.4-6
- fruits, R11
- hydrocooling, R11.3-4
- panel systems, H7.1-18
- solar energy systems, H47.21
- time for food, F29.8-12
- vacuum, R11.6-9
- vegetables, R11

Cooling load, F26.1-62
- ceilings, F26.19
- CLTD/CLF calculation procedure, F26.32-48
- appliances, F26.45
- CLTD corrections, F26.37
- CLTD for roofs, F26.34
- CLTD for walls, F26.36
- cooling load factors, F26.44-47
- lighting, F26.44
- people, F26.40
- solar heat gain factor, F26.39-40
- space load through fenestration, F26.38
- wall construction group designation, F26.35
- heat gains
  - appliances, F26.9-11
  - electric motors, F26.9
  - fenestration, F26.5
  - infiltration, F26.12
  - interior surfaces, F26.6
  - latent heat gain, F26.12
  - lighting, F26.7
  - miscellaneous sources, F26.13
  - office equipment, F26.11
  - people, F26.6
  - roofs, F26.3
  - ventilation, F26.12
- residential calculations, F26.54-60
- sol-air temperature, F26.4
- TETD/TA calculation procedure, F26.2, 49-54
- transfer function method, F26.3, 13-32
  - conduction transfer functions, F26.15-19

Cooling towers
- capacity control, E20.8
- central system, H1.10
- design conditions, E20.2
- drift, E20.10
- effect of airflow around buildings, F14.13
- evaporative coolers, E4.10
- field testing, E20.12

size distribution, F11.2-3

fogging, E20.10
free cooling, E20.9
maintenance, E20.11
mass transfer, F5.13
materials, E20.6
performance curves, E20.12
piping, E20.8
principle of operation, E20.1
selection considerations, E20.7
siting, E20.8
sound, E20.10
testing, adjusting, balancing, H57.16
theory, E20.14
    counterflow integration, E20.15
    crossflow integration, E20.16
tower coefficients, E20.17
types
    closed-circuit fluid coolers, E20.6
    coil shed, E20.6
    direct contact, E20.2
    indirect contact, E20.3
    mechanical draft, E20.4
    nonmechanical draft, E20.3
    ponds (spray), E20.5
water-source heat pumps, H5.7
water systems, H14.7
water treatment, E20.11
winter operation, E20.9
Copper plating, R6.6
Copper tube, E33.2, 4
Corrosion, H53.1-20
brine systems, R5.6
brines, F18.2, 7
central plant systems, H12.12
fuel-burning systems, F15.13
geothermal energy systems, H45.9
industrial air conditioning, H28.5
service water heating, H54.6
snow melting systems, H55.6
Costs
capital recovery factors, H49.3
data summary, H49.2
economic analysis techniques, H49.5
equipment service life, H49.7
forced air cooling, R11.5-7
fuels, H49.4
maintenance, H49.5, H59.1
owning
    electrical energy, H49.3
    interest, H49.1
    amortization and depreciation, H49.1
    insurance, H49.1
    return on investment, H49.1
shared savings, H49.6-8
initial, H49.1, 3
Cryogenics
argon extraction, R38.11
cryopumping, R38.27
equipment, R38.13-16
    compressors, R38.13
    expanders, R38.15
    heat exchangers, R38.15
    pumps, R38.15
    regenerators, R38.14
    valves, R38.16
freezers, commercial
    liquid carbon dioxide, R9.5
    liquid nitrogen, R9.5
helium refrigeration and liquefaction, R38.4
hydrogen liquefaction, R38.5
instruments, R38.22
insulation, R38.16-19
liquefied natural gas, R38.6
    liquefaction, R38.7
    storage, R38.9
methods for producing low temp, R38.1
    expander refrigeration, R38.3
    Joule-Thomson refrigeration, R38.1
miniature closed-cycle refrigerators, R38.11
    Joule-Thomson effect, R38.11
    Stirling cycle, R38.12
nitrogen refrigeration, R38.9
oxygen refrigeration, R38.10

properties of fluids, R38.23
safety, R38.23
space simulators, R38.25
    chamber construction, R38.26
    refrigeration, R38.27
    thermal shrouds, R38.26
    vacuum pumping, R38.26
storage and transport, R38.19-21
    Dewars, R38.20
    stationary, R38.20
Cycles
absorption, F1.20-25
Dairy products
asceptic packaging, R15.19-22
butter, R15.6-10
cheese, R15.10-13
contaminants, R10.4
cream, R15.6
display refrigerators, E35.3
ice cream, R15.13-19
    freezing points, R15.16
    making mix, R15.15
    refrigeration requirements, R15.16-18
milk, dry, R15.23-24
milk, evaporated, R15.22
milk, fresh, R15.1-6
milk, sweetened, condensed, R15.22
spoilage, R10.5
storage, R26.4-5
UHT sterilization, R15.19-22
Yogurt, R15.6
Dampers
central system, H1.4
control, H51.7
fire and smoke control, H58.12
loss coefficients, F32.46
sound control, H52.13-16
vehicular facilities, enclosed, H29.15
vent, E26.23
Dams, concrete
cooling, R35.1-3
Data processing areas
air-conditioning systems, H33.3
cooling loads, H33.2
design criteria, H33.1
energy conservation, H33.8
fire protection, H33.7
heat recovery, H33.8
instrumentation, H33.7
return air, H33.6
supply air distribution
water-cooled computer equip., H33.6
Definitions, H32.1
air diffusion, F31.1
cogeneration systems, H8.13
control, H51.1
heat recovery systems, H6.1
measurement, F13.1-32
service water heating, H54.6
steam systems, H11.1
thermodynamics, F1.1
Defrosting
air conditioners, unitary, E42.9
air coolers, forced circulation, E8.2-4
egg products, R24.4
heat pump systems, applied, H9.7, 13
household refrigerators and freezers, E37.6
meat coolers, R12.2-3
refrigerated storage, R1.1
retail food store refrigeration, R32.5, E35.6
Degree day method, F28.2,7
Dehumidification
air washers, E4.8
applications, E7.6-7
coils, E6.1-16
environmental test facilities, R37.4
sorption equipment, E7.1-8
    liquid absorption, E7.2-4
    methods, E7.1
    solid absorption, E7.4
Dehumidifiers
room, E41.6-7
Dehydration

farm crops, F10, H38
industrial equipment, H44
refrigeration systems, E21.1-3
Density
commodities, packaged, R26.9-11
liquids, F37.2
solids, F37.3-4
vapor, F37.1
Desiccants, F19.3-6
adsorbents, solid, F19.4
applications, F19.1
cycle, F19.2
isotherms, F19.5
life, F19.5
refrigerant system, R7.4-5
equilibrium curves, R7.4, 5
    types, F19.3
    water vapor pressure, F19.2
Design conditions
heating, F25.2
indoor, F24.1
outdoor, F24.1-22
Dew point temperature, F6.12
Dewars, R38.20
Diffusion
mass, F5.1-4
molecular, F5.1
Diffusivity, thermal
food, F30.12-14
Dining and entertainment centers, H19.5-7
exhaust hood design, H19.6
load characteristics, H19.5
special considerations, H19.7
    bars, H19.7
    kitchens, H19.7
    nightclubs, H19.7
    restaurants, H19.7
Direct expansion coolers, liquid, E16.1
District cooling systems, H12.1-16
District heating systems, H12.1-16
Domiciliary facilities, H21.1-4
design criteria, H21.1
dormitories, H21.3
energy efficient systems, H21.1
energy neutral systems, H21.2
hotels, H21.3
inefficient energy systems, H21.2
load characteristics, H21.1
multiple-use complexes, H21.4
special considerations, H21.3
total energy systems, H21.2
Door
infiltration, F22.14
Dormitories, H21.1-4
service water heating, H54.8, 11
Draft
chimney, E26.1-28
controls, E22.15
cooling towers, E20.3-4
hoods, E26.22
Draperies, F27.31-33
classification of fabrics, F27.32
double, F27.34
environmental control capabilities, F27.34
shading coefficients, F27.33
Dried fruits and vegetables
storage, R22.7
Driers
refrigerant system, R7.5-6
Drinking water coolers, E38.1-6
central, E38.3-6
    chillers, E38.3
    distribution piping, E38.3
    drinking fountains, E38.4
    pumps, E38.4
    system design, E38.4-6
    unitary, E38.1-3
Drives
engine, E32.1-9
turbine, E32.9-16
Drying
farm crops, F10.4-13, H38.1-10
geothermal energy systems, H45.20

Index

Drying systems, H44.1-8
  agitated bed, H44.6
  applying hygrometry, H44.1
  conduction, H44.3
  convection, H44.4-6
  determining drying time, H44.2
  dielectric, H44.3
  drier calculations, H44.2
  drying mechanism, H44.1
  flash, H44.6
  fluidized bed, H44.6
  freeze drying, H44.6
  microwave, H44.4
  radiant infrared, H44.3
  superheated vapor atmosphere, H44.6
  system selection, H44.3
  ultraviolet radiation, H44.3
  vacuum, H44.6
Dual-duct systems
  air volume, H2.14
  boxes, E2.7
  components, H2.15
  controls, H2.15, H51.23
  evaluation, H2.16
  single-fan, constant volume, H2.13
  supply air conditions, H2.15
  testing, adjusting, balancing, H57.4
  variable volume, H2.13
Duct design, F32
  Bernoulli equation, F32.1
  circular equivalents, F32.8
  computer applications, H60.5
  dynamic losses, F32.7
    local loss coefficients, F32.7, 27-52
  fan/system interface, F32.9
  fitting loss coefficients, F32.27-52
  frictional losses, F32.4-7
    Altschul equation, F32.4
    Darcy-Weisbach equation, F32.4, 9
    chart for round duct, F32.5, 6
    Colebrook equation, F32.4
    correction factors, F32.5
    noncircular ducts, F32.7
    oval ducts, F32.7
    rectangular ducts, F32.7
    roughness factor, F32.5
  head, F32.2
  insulation, F32.12
  leakage, F32.13
  methods, F32.16-25
    equal friction, F32.16
    industrial exhaust system, F33.22-25
    static regain, F32.16
    T-Method, F32.16
  pressure, F32.2
  velocities, F32.15
Duct(s)
  acoustical treatment, E1.6
  all-air systems, H2.1-18
  central system, H1.6
  classification, E1.2
  construction
    commercial, E1.2-11
    for grease- and moisture-laden vapors, E1.10
    industrial, E1.7-9
    residential, E1.2
  ferrous metal, E1.3, 1.9
  fibrous glass, E1.4
  flat-oval, E1.4
  flexible, E1.5
  hangers, E1.7-8
  industrial exhaust systems, H43.6-10
  insulation, F21.17, F32.12
  laboratories, H30.6
  leakage, system, E1.2, F32.13
  materials, E1.2
  plastic, E1.10
  rectangular, E1.3, 1.8
  reinforcement, E1.6-8
  round, E1.3, 1.7
  seismic qualification, E1.12
  small forced-air systems, H10.1-8
  sound control, H52.9-23
  thickness, E1.2-9
  underground, E1.11
  vehicular tunnels, H29.6
  vibration control, H52.36
  welding, E1.12
Dust
  collectors, E11.1-26
Earth
  frozen, stabilization of, R35.4
  temperatures, H27.6
  thermal conductivity, F22.21, H12.6, H27.5
Economics
  commercial freezing methods, R9.6
  forced air cooling, R11.5-7
  maintenance, H59.1
  owning and operating, H49.1-8
  thermal storage, H46.4
  thickness of insulation, F20.10
Educational facilities, H22.1-6
  classroom load profile, H22.3
  colleges, H22.5
  controls, H22.2
  design temperatures, H22.2
  energy considerations, H22.3
  environmental considerations, H22.2
  equipment selection, H22.3
  financial considerations, H22.2
  general design considerations, H22.1
  regulatory considerations, H22.2
  remodeling, H22.5
  schools, H22.5
  sound levels, H22.5
  space considerations, H22.2
  system selection, H22.3
  universities, H22.5-6
  ventilation requirements, H22.1
Efficiency
  adiabatic, F1.9
  air conditioners, room, E41.3
  boilers, E23.3
  dehumidifiers, room, E41.7
  fins, F3.15-18
  furnaces, E24.5, 8-17
  volumetric, F1.9
Eggs
  freezing, R24.4
  products, R24.3
    processing, R24.3
    storage, R24.1
  refrigeration, R24.5
  shell
    quality, R24.1-3
    refrigeration, R24.1
    weight (mass) classes, R24.2
  spoilage, R10.4
  storage requirements, R26.5
  thermal properties, F30.3, 9
Electric
  furnaces, residential, E24.4
  heaters (in-space), E25.4
  infrared heaters, E29.1-6
  infrared radiant heating, H16.1-10
  service water heating, H54.1-20
  snow melting systems, H55
  unit heaters, E27.4
Electrical
  measurement, F13.21-22
Electronic
  air filters, E10.8
Electrostatic precipitation, E11.10
Emissivity
  ratio, F37.3-4
  surface condition, F37.3-4
Energy
  available, F1.3
  balance, F1.2
  geothermal, H45.1-22
  kinetic, F1.1
  potential, F1.1
  simulation
    computer applications, H60.3-5
  thermal, F1.1
Energy conservation
  air conditioners, room, E41.3
  data processing areas, H33.8
  dehumidifiers, room, E41.7
  drying of farm crops, H38.4
  greenhouses, H37.14
  industrial exhaust systems, H43.10
Energy estimating, F28
  bin method, F28.10
  computer programs, F28.26
  cooling, F28.7
  daylighting systems, F28.37
    simulation methods, F28.38
  graphical method, F28.18
  heating, F28.2-7
    modified degree day method, F28.2
    variable-base degree day method, F28.3
  modeling strategies, F28.26
  simulation methods, F28.18-27
    for primary energy conversion systems, F28.23
    heat-balance method, F28.19
    weighting factor method, F28.21
  solar heating/cooling, F28.28-40
    active systems, F28.28
    active/hybrid cooling systems, F28.32-39
    computer programs, F28.38
    evaporative, F28.33
    f-Chart method, F28.28-30
    passive heating, F28.30
    simulation methods, F28.38
Energy management, H48.1-8
  energy use reduction, H48.7
  implementing, H48.3-7
  organizing, H48.2
Energy recovery
  chemical industry, R36.6
Engine drives, E32.1-9
  centrifugal compressors, E32.2
  codes, E32.6
  controls, E32.6-7
  design considerations, E32.3
  dual-service applications, E32.8
  expansion reciprocating, E32.8
  fuels, heating value, E32.2
  gas leak prevention, E32.7
  heat recovery
    absorption chillers, H32.8
    exhaust, E32.7
  heat pumps, E32.2
  installation, E32.3-5
  maintenance, E32.5
  reciprocating compressors, E32.1
  sizing, E32.2
  standards, E32.6
Engine test facilities, H31
Enthalpy
  calculating, F1.6
  food, F30.8-10
  recovery loop, twin tower, E34.8-11
Entropy
  calculating, F1.5
Environmental control, animals, F9
  cattle, F9.5-6
  chickens, F9.9-10
  laboratory animals, F9.11
  physiological control systems, F9.2-4
  sheep, F9.6-7
  swine, F9.8
  turkeys, F9.10-11
Environmental control, humans, F8
Environmental control, plants, F9.10-16
  air composition, F9.16
  humidity, F9.16
  light, F9.12
  night temperatures, F9.15
  photoperiod extension or interruption, F9.14
  pollutants, F9.16
  radiation, F9.12
  supplemental irradiance, F9.12-14
  temperature, F9.14
Environmental test facilities
  altitude chambers, R37.7

chamber construction, R37.6
control, R37.4-6
cooling, R37.1-3
design calculations, R37.6
heating, R37.3
Evaporation
geothermal energy systems, H45.20
in tubes, F4.5-7
    equations, F4.6
Evaporative cooling
animal environments, H37.2
applications, commercial, H56.2
economic considerations, H56.3
equipment, E4.1-10
greenhouses, H37.12
heat recovery equipment, E34.12, 15
indirect precooling, H56.5
indirect/direct (staged) system, H56.5-7
industrial air conditioning, H28.9
mines, underground, H42.8
psychrometrics, H56.3-5
solar systems
    direct, F28.32-36
    indirect, F28.36
survival shelters, H27.19
system analysis, H56.8
system design, H56.1
system load exhaust, H56.8
weather conditions, H56.2
weather data, F24.4
Evaporators
absorption units, E13.3,11
air conditioners, room, E41.2
automobile air conditioning, H24.4
balancing refrigeration components, E18.1
flooded, F4.4-7
household refrigerators and freezers, E37.7
ice makers, E40.2
liquid overfeed systems, R2.6
piping, halocarbon refrigerants, R3.17-19
Exhaust
engine drives, E32.7
engine test facilities, H31.4
industrial air conditioning, H28.9
inlets, F31.14
laboratories, H30.5-15
turbine drives, E32.15
Exhibition centers, H20.6
Expansion
joints, H52.36
pipe, central plant systems, H12.6-10
solar energy systems, E44.10
tanks
    refrigeration systems, R1.7
    secondary coolant, R5.4
    solar energy systems, H47.19
    water systems, basic, H13.13
valves, E19.3-8
water systems, medium and high temp., H15.4-6
Fans
air conditioners, room, E41.3
air-cooled condensers, E15.8
animal environments, H37.5
arrangements, E3.12
central system
    return air, H1.4
    room space requirements, H1.11
    supply air, H1.6
control, E3.13, H51.14-17
draft, E26.24
drying of farm crops, H38.2
duct system interface, F32.9
energy estimating methods, F28.7, 22
furnaces, E24.2
industrial exhaust systems, H43.10
installation, E3.12
isolation, E3.12
laboratories, H30.6
laws, E3.4-5
merchant ships, H26.2
motor selection, E3.13
naval surface ships, H26.7

noise, E3.11
operating principles, E3.4
parallel operation, E3.11
performance curves, E3.10-4
pressurization testing, F23.11
rating, E3.4
selection, E3.10
solar energy systems, H47.20
sound control, H52.7-9
system effects, E3.8-10
    inlet conditions, E3.9
    outlet conditions, E3.8
system pressure relationships, E3.6
testing, E3.4
types, E3.1, 2-3 (table)
vehicular facilities, enclosed, H29.14
Fan-coil
control, H51.28
Farm crops, H38.1-12
drying, F10.4-13
drying equipment, H38.2-7
    batch dryers, H38.3
    column dryers, H38.3
    continuous flow dryers, H38.3
    controls, H38.3
    deep bed drying, H38.4
    dryeration, H38.4
    fans, H38.2
    full-bin drying, H38.4-6
    heaters, H38.3
    layer drying, H38.6
    recirculating/continuous flow bin dryer, H38.7
    recirculation, H38.4
    reducing energy consumption, H38.4
drying specific crops, H38.7-10
    cotton, H38.9
    hay, H38.8
    peanuts, H38.9
    rice, H38.9
    soybeans, H38.7
economics, H38.2
grain quantity, H38.1
storage problems and practices, H38.10-12
    grain aeration, H38.10-12
    moisture migration, H38.10
    seed storage, H38.12
storing, F10.1-4
***f*-Chart**
energy estimating (solar), F28.28-30
Fenestration, F27
component considerations, F27.2
daylight considerations, F27.1
glass, F27.31-33
    daylight utilization, F27.28
    light transmission, F27.28
    occupant acceptance, F27.29
    occupant comfort, F27.29
    safety, F27.29
    sound reduction, F27.29
    strength, F27.29
heat transfer, F27.20-37
    basic principles, F27.21
    glazing material properties, F27.20
    heat balance, F27.23
    shading coefficients, F27.24-37
    solar heat gain factors, F24.5-13, 24-37
    solar intensity (tables), F27.5-13
load-influencing factors, F27.2
    absorbed solar radiation, F27.14
    air infiltration, F27.18
    atmospheric clearness numbers, F27.4
    diffuse solar radiation, F27.14
    direct normal solar intensity, F27.4
    indoor air movement, F27.19
    passive solar gain, F27.15
    shading coefficients, F27.19, 24-37
    solar angle determination, F27.4
    solar radiation, F27.2
    U-values, F27.15-18, 36
plastic
    solar optical properties, F27.30
shading devices, F27.30-37

    double drapery, F27.34
    draperies, F27.31, 33
    exterior, F27.34
    glass block walls, F27.36
    indoor, F27.32
    roof overhangs, F27.35
    shades, F27.30
    shading coefficients, F27.30-37
    skylights, domed, F27.36
    sunshades, F27.34-37
    Venetian blinds, F27.30
    window reveal, F27.36
Fick's Law, F5.1
Filters, air
automobile air conditioning, H24.4
central system, H1.5
clean spaces, H32.7
electronic, E10.9
furnaces, E24.2
hospitals, H23.2
industrial air conditioning, H28.9
industrial gas cleaning, E11.6-10
installation, E10.9-11
laboratories, H30.2, 7
maintenance, E10.9
merchant ships, H26.3
naval surface ships, H26.7
nuclear facilities, H40.4-6
nursing homes, H23.11
panel, E10.6
    dry, extended surface, E10.7
    viscous impingement, E10.6
photographic processing, H36.2
printing plants, H34.5
ratings, E10.1
renewable media, moving-curtain
    viscous impingement, E10.7
    dry media, E10.8
residential air conditioning, H17.5
safety requirements, E10.11
selection, E10.9
swimming pool air conditioning, H20.8
test methods, E10.2-5
    ARI standards, E10.5
    atmospheric dust spot efficiency, E10.3
    DOP penetration, E10.3
    dust holding capacity, E10.3
    environmental, E10.4
    leakage (scan), E10.4
    miscellaneous performance, E10.4
    particle size efficiency, E10.4
    specialized performance, E10.4
    weight arrestance, E10.3
types, E10.6-10
Finned tube
heat transfer, F3.19
units, E28.1-6
Fin(s)
efficiency, F3.15-18
heat transfer, F3.19
thermal contact resistance, F3.19
Fire safety, H58.1-16
data processing areas, H33.7
duct design, F32.12
laboratories, H30.7
survival shelters, H27.19
Fireplace
chimneys, E26.19
Fish
fresh
    icing, R14.1
    packaging, R14.3
    refrigeration, R14.2
    storage, R14.3, R26.3
frozen
    freezing methods, R14.5-7
    packaging, R14.4
    storage, R14.8
    transport, R14.9
spoilage, R10.3
Fishing vessels, R30.11-16
Fittings, E33.1-14
pressure loss

Index

halocarbon refrigerants, R3.3, 9-10
Flowers, cut
  cooling, R11.10
  environmental control, F9.18
  storage, R26.8-9
Fluid flow
  analysis
    Bernoulli equation, use of, F2.7
    conduit friction evaluation, F2.8
    friction factor, F2.9
    Reynolds number, F2.9-10
    section change effects and losses, F2.9
  basic relations, F2.2
    Bernoulli equation, F2.2, 7
    continuity, F2.2
    laminar-viscous, F2.3
    pressure variation across flow, F2.2
    pressure variation along flow, F2.2
    turbulence, F2.3
  fitting losses (table), F2.10, R30.9-10
  fluid properties, F2.1
  incompressible, F2.12
  loss coefficients, F2.10
  measurement, F2.11
  noise, F2.13
  processes
    boundary layer occurrence, F2.4
    cavitation, F2.5
    compressibility effects, F2.7
    drag forces, F2.5
    nonisothermal effects, F2.6
    patterns with separation, F2.4
    wall-friction effects, F2.3
  two-phase, F4.1-16
    boiling, F4.1-7
    condensing, F4.7-12
    convection, forced, F4.6-7
    evaporation, F4.2-7
  unsteady, F2, R35.6-10
Food
  cheese, R15.10-13
  commercial freezing methods, R9
  cooling time, empirical formulas, F29.8-12
  defrosting time, F29.1-16
  fishery products, R14
  freezing time, F29.1-16
    estimating, sample problems, F30.12
    semi-theoretical methods, F29.8
    theoretical methods, F29.1-8
  ice cream, R15.13-19
  meat products, R12
  microbiology, R10
  milk, R15.1-6
  poultry products, R13
  precooked and prepared, R20
    freeze-dried, R20.3
    physical changes, R20.1
    storage temperature, R20.4
    types, R20.2-4
  refrigeration
    candy, R22.3-5
    dairy products, R15
    dried fruits, R22.7
    dried vegetables, R22.7
    eggs and egg products, R24
    fruit juice concentrates, R19
    fruits, fresh, R16, R17
    nuts, R22.7
    retail food store, R32
    vegetables, R18
  retail store refrigeration, E35, R32
  service
    equipment, E36.1-4
    water heating, H54.9-14
  storage requirements, R26
  thermal properties, F30.1-28
    bakery products, R21
    commodity load data, F30.1
    enthalpy, F30.8-11
    freezing point, F30.1-3
    frozen foods, F30.2, 10
    heat of respiration, F30.4-7
    heat transfer coefficient, surface, F30.16-25
    latent heat of fusion, F30.2, 3
    specific heat, F30.2, 3, 11-13
    thermal conductivity, F30.13-19
    thermal diffusivity, F30.12-14
    transpiration, F30.8
    water content, F30.1-3
Fouling factor
  coolers, liquid, E16.3
  liquid chilling systems, E17.7, 11, 15
  water-cooled condensers, E15.3
Four-pipe
  dual-temperature systems, H14.6
Freeze drying, H44.6, R40.2
Freeze prevention
  coolers, liquid, E16.5
  heat recovery equipment, E34.6, 12
  insulation, F21.15
  solar energy equipment, E44.2, 20
  solar energy systems, H47.24
  water systems, basic, H13.23-25
Freezers
  commercial, R9
    air blast, R9.1-4
    contact, R9.4
    cryogenic, R9.5
  household, E37.1-14
  meat products, R12.15
  walk-in, E36.4
Freezing
  bakery products, R21.4-5
  biomedical applications, R40
  commercial methods, R9
  egg products, R24.4
  fish, R14.4-7
  ice cream, R15.13
  meat products, R12.15
  point
    commodities, R26.2-9
    food, F30.2, 3
    liquid, F37.2
    refrigerants, F16.4
  poultry products, R13.4-6
  time, of foods, F29
Friction charts
  ducts, F32.5, 6
  water piping, F33.4, 5
Friction factor, F2.9
Frozen food
  display refrigerators, E35.5
Fruit juice concentrates, R19
  concentration methods, R19.2-4
  processing, R19.1
  storage, R19.1
Fruits, dried
  storage, R22.8
Fruit(s), fresh
  air transport, R31.1
  apples, storage, R16.1, R26.3
  apricots, R16.10, R26.3
  avocados, R17.8, R26.3
  bananas, R17.5-7, R26.3
  berries, R16.9, R26.3
  cherries, sweet, R16.8, R26.3
  citrus, R17.1-5, R26.3
  cooling, R11
    methods, R11.3
    rate, R11.1-2
  display refrigerators, R32.1
  enthalpy, E35.4
  figs, R16.10, R26.3
  grapes, R16.5-7, R26.3
  handling, R16.1, R26.3
  mangoes, R17.8, R26.3
  nectarines, R16.8, R26.3
  peaches, R16.8, R26.3
  pears, R16.4-5, R26.3
  pineapples, R17.8, R26.3
  plums, R16.7, R26.3
  precooling, R11
  refrigeration, R16, R17
    supplements to, R16.10
  storage, R16.1, R26.3
  strawberries, R16.9, R26.3

Fuel oil, F15.3
Fuels, F15.3-8
  central system, H1.9
  engine drives, E32.2
  gaseous, F15.4
  liquid, F15.5
  solid, F15.6
Fuel-burning equipment, E22.1-16
  controls, E22.13-16
    combustion, E22.15
    draft, E22.15
    operating, E22.13
    programming, E22.14
    safety, E22.14
  gas, E22.1-4
    altitude compensation, E22.3
    combustion process, E22.2
    commercial-industrial, E22.2
    residential, E22.1
  oil, E22.4-10
    commercial-industrial, E22.5
    equipment selection, E22.7-10
    residential, E22.4
  solid fuel, E22.10-13
    stoker classifications, E22.10
    stoker types, E22.11
Furnaces
  commercial, E24.17
  floor (in-space), E25.2
  installation practices, E24.18
  residential, E24.1-17
    electric, E24.4
    equipment selection, E24.4-6
    gas, E24.1
    LPG, E24.4
    oil, E24.4
    standards, E24.18
    system design, E24.4-6
    system performance, E24.8-17
    technical data, E24.6-8
  residential air conditioning, H17.2
  wall (in-space), E25.2
Garages, parking
  ventilation, H29.12
Gas
  infrared radiant heating, H16.1-10
  natural, F15.3
    liquefied, R38.6-9
  pipe sizing, F33.22
  turbine
    test cells, H31.3
    drives, E32.14-16
  venting, appliances, E26.18
Gas cleaning, industrial, E11.1-26
Gas-fired
  equipment, E22.1-4
  furnaces
    commercial, E24.17
    installation practices, E24.18
    residential, E24.1-17
    standards, E24.18
  heaters (in-space), E25.1
  infrared heaters, E29.1-6
  makeup air units, E27.9
  service water heating, H54.1-20
  unit heaters, E27.4
Generators
  absorption units, E13.3, 11
  cogeneration systems, H8.2
  water systems, medium and high temp, H15.3
Geothermal energy, H45.1-22
  commercial applications, H45.12-17
    energy storage, H45.16
    sanitary water heating, H45.12
    space cooling, H45.16
    space heating, H45.12
    terminal equipment, H45.14
  description, H45.1
  direct application systems, H45.3-12
    corrosion, H45.9
    equipment, H45.8

general characteristics, H45.4-8
heat exchangers, H45.10-12
materials, H45.8
piping, H45.12
pumps, H45.8-10
environmental aspects, H45.3-5
fluids, H45.2
industrial applications, H45.17-21
life, H45.3
potential development, H45.3
present use, H45.3
residential applications, H45.12-17
temperatures, H45.2
Glass
block walls, F27.36
cooling load factors, F26.41-44
cooling load through, F26.5
daylight use, F27.28
greenhouses, H37.14
heat gain through, F27.1-38
light transmission, F27.28
occupant acceptance, F27.29
occupant comfort, F27.29
safety, F27.29
shading coefficients, F27.24-37
solar optical properties, F27.20
sound reduction, F27.29
strength, F27.29
Grain
drying, F10.4-11
storage, F10.1-4
Greenhouses, H37.9-14
night temperatures, F9.15
supplemental irradiance, F9.14
Growth chambers
supplemental irradiance, F9.14
Hay, F10.10, H38.88
Health
ASHRAE comfort-index, F8.26
environment, F8.26
facilities, H23.1-12
indoor climate, F8.27
outdoor climate, F8.26
qualifying effects, F8.29
thermal environment, F8.27
Heat
latent, of fusion, F30.2,3
of fusion, F37.2
of respiration, F30.4-7
of vaporization, F37.2
production, animals, F9.4-11
specific, F30.2, 3, 11-13, F37.1-3
transfer coefficient, surface, F30.16-25
Heat exchangers
animal environments, H37.6
chimneys, E26.23
geothermal energy systems, H45.10-12
halocarbon refrigerants, R3.17
heat recovery, E34.1-18
solar energy equipment, E44.16-18
solar energy systems, H47.19
steam systems, H11.3
water systems, medium and high temp, H15.8
Heat flow
insulation, F20.4-8
apparent conductivity, F20.4
calculations, F20.8
conductance, surface, F20.6
resistance, thermal, F20.8
Heat gain
appliances, F26.9-11
fenestration, F26.5
infiltration, F26.12
interior surfaces, F26.6
latent heat gain, F26.13
lighting, F26.7
people, F26.6
power, F26.8
roofs, F26.4
ventilation, F26.12
walls, F26.4
Heat loss
flat surfaces, F22.15, 18

pipe, bare steel, F22.15, 18
tube, bare copper, F22.15, 18
Heat pipe
heat exchangers, E34.11-13
Heat pump
applied systems, H9.1-16
components, H9.6
compressors, H9.6
controls, H9.13
defrost controls, H9.7
equipment selection, H9.11-13
heat reclaim cycle, H9.14
heat sinks, H9.2-6
heat sources, H9.2-6
operating cycles, H9.7-11
refrigerant controls, H9.6
thermal storage, H9.7
types, H9.1-3
heat recovery systems, H6.3
ice maker, R33.7
packaged terminal, E43.3
residential air conditioning, H17.2
unitary, E42.8-10
unitary systems
air-to-air, H5.3
through-the-wall, H5.2
water-source, H5-8
window-mounted, H5.2
water heaters, H54.2
water source, E43.4-6, H5.5-8
Heat recovery
absorption units, E13.2
centrifugal compressors, E12.33
cogeneration systems
controls, H8.10-12
definitions, H8.13
energy user interconnections, H8.12
generators, H8.2
glossary, H8.13
heat recovery, H8.3-9
installation problems, H8.9
prime movers, H8.1
data processing areas, H33.8
engine drives, E32.7
heat pump systems, applied, H9.8-14
industrial environment, H41.3
liquid chilling systems, E17.3, 12
retail food store refrigeration, E35.10
steam systems, H11.15
supermarkets, H18.5
systems, H6.1-6
water-source heat pumps, H5.6
Heat recovery, air-to-air, E34
coil energy-recovery loop, E34.7-8
comfort-to-comfort, E34.2
controls, E34.3, 6
corrosion, E34.3
economic evaluation, E34.3
fixed-plate exchangers, E34.13-16
fouling, E34.3
heat pipe heat exchanger, E34.11-13
performance rating, E34.4
process-to-comfort, E34.2
process-to-process, E34.1
rotary energy exchanger, E34.4-7
thermosiphon heat exchangers, E34.16
twin-tower enthalpy recovery loop, E34.8-11
Heat storage
geothermal energy systems, H45.16
water systems, medium and high temp, H15.9
Heat transfer
air-cooled condensers, E15.9
building materials
measurement of, F13.11
central plant systems, H12.4
coils, E6.1-16
air cooling, E6.6-9
dehumidifying, E6.9-14
conduction, F3.1-6
equations, F3.3
steady-state, F3.2

convection
equations, F3.15-16
forced, F3.13
natural, coefficients, F3.12-13
coolers, liquid, E16.3
evaporative condensers, E15.14
extended surfaces
fin efficiency, F3.16
finned-tube, F3.19
thermal contact resistance, F3.19
fluids
low-temperature, R37.3
solar energy systems, H47.18
food
surface coefficients, F30.16-25
overall, F3.2-4
coefficient, F3.2
mean temperature difference, F3.3
radiant, F3.6-12
actual, F3.8
angle factor, F3.9
black body, F3.7
in gases, F3.11
snow melting systems, H55.1
transient, F3.4-6
cylinder, F3.6
slab, F3.5
sphere, F3.7
water systems, basic, H13.2
water-cooled condensers, E15.2
Heat transmission
conductivity
building materials, F22.6-10
industrial insulation, F22.16-20
insulating materials, F22.6-10
soils, F22.21, H12.6, H27.5
design coefficients, F22.1-22
calculating overall, F22.2
definitions, F22.1
heating
basement, F25.5-7
ceiling, F25.5
floor slabs, F25.7
roof, F25.5
industrial insulations
conductivity, F22.16-20
resistance, airspace, F22.2-3
R-values
airspace, F22.2
building materials, F22.6-10
insulating materials, F22.6-10
surface, F22.2
U-factors, component
ceilings, F22.11
doors, F22.12
masonry walls, F22.4
metal panels, F22.5
roofs, F22.11
slab construction, F22.14
windows, F22.12
wood frame walls, F22.3
Heaters
codes, E25.6
electric, E25.4
farm crop drying, H.38.3
gas-fired, E25.2
infrared, E29, H16
oil-fired, E25.3
solid fuel, E25.6
standards, E25.6
testing, E25.7
thermal storage, H46.12-14
unit, E27.3-7
Heating
all-air systems, H2.2
animal environments, H37.6
automobile air conditioning, H24.1
central air conditioning systems, H1.6-9
energy estimating, F28.2-7
environmental test facilities, R37.3-4
equipment
baseboard units, E28.1-6
boilers, E23.1-4

Index

convectors, E28.1-6
finned-tube units, E28.1-6
furnaces, E24.1-18
in-space heaters, E25.1-8
makeup air units, E27.7-10
pipe coils, E28.1-6
radiators, E28.1-6
unit heaters, E27.3-6
geothermal energy systems, H45.12-17
greenhouses, H37.10
industrial air conditioning, H28.7
infrared radiant, H16.1-10
load calculations, F25.1-10
    attic temperature, F25.3
    basement, F25.5-7
    ceiling, F25.5
    crawl space loss, F25.4
    design conditions, F25.1
    estimating temperatures in unheated spaces, F25.2
    floor slabs, F25.7
    general procedure, F25.1
    infiltration loss, F25.8
    internal heat sources, F25.9
    pick-up load, F25.9
    roof, F25.5
    summary, F25.2
    transmission heat loss, F25.5
nuclear facilities, H40.1-10
panel systems, H7.1-18
plant growth facilities, H37.15
railway refrigerator cars, R29.9
solar energy systems, H47.21
thermal storage, H46.20
values (fuels), F15.3-8
water systems
    basic, H13.1-26
    dual-temperature, H14.1-8
    medium and high temp, H15.1-10

Helium
liquefaction, R38.4
refrigerant properties, F17.56

Hoods
draft, E26.22
industrial exhaust systems, H43.1-10
    compound, H43.7
    entry loss, H43.5
    hot process, H43.4
    simple, H43.7
    volumetric flow rates, H43.2-5
laboratory fume, H30.9-10
laminar flow, H30.12

Hospitals
air movement, H23.3
air quality, H23.2
design criteria, H23.3
    administration department, H23.10
    central sterilizing and supply, H23.9
    diagnostic facilities, H23.7
    emergency department, H23.6
    humidity, H23.3
    laboratories, H23.7
    nursery, H23.6
    nursing department, H23.6
    obstetrical department, H23.6
    operating rooms, H23.4
    pressure relationships, H23.4-5
    service department, H23.9
    surgical department, H23.4
    temperature, H23.3
    treatment facilities, H23.7
    ventilation, H23.4
filters, H23.2

Hot water
unit heaters, E27.4

Hotels, H21.1-4

Humid climates
air conditioning, design, F21.13

Humidification
air washers, E4.8
all-air systems, H2.6, 12
control, E5.8
data processing areas, H33.7
enclosure characteristics
    concealed condensation, E5.2
    visible condensation, E5.2
energy considerations, E5.3
environmental conditions
    comfort, E5.1
    disease prevention and treatment, E5.1
    materials storage, E5.1
    process control, E5.1
    static electricity, E5.1
environmental test facilities, R37.4
equipment, E5.4-7
load calculations, E5.3-4
water supply, scaling, E5.7

Humidifiers
central system, H1.6
industrial, E5.6-7
residential, E5.4-5
residential air conditioning, H17.5

Humidity
control, F21.7, H51.19
indoor levels, condensation, F21.6
measurement, F13.23-25
ratio, F6.6
relative, F6.12

Hydrogen
liquefaction, R38.5
refrigerant properties, F17.52-55

Hydronic systems
control, H51.25
residential air conditioning, H17.6
snow melting, H55.1-7
testing, adjusting, balancing, H57.7-15
water treatment, H53.15-18

Hygrometers, F13.24

Ice
manufacture, R33
    delivery systems, R33.6
    flake ice, R33.1
    ice maker heat pumps, R33.8
    ice builders, R33.4
    plate ice, R33.3
    storage, R33.4
    tubular ice, R33.1
storage, thermal, H46.9

Ice cream
manufacture, R15.13-19
display refrigerators, E35.5

Ice makers, E40.1-8, R33.1-4
application, E40.6-8
construction, E40.2
definitions, E40.2
package, E40.3
performance, E40.3
system design, E40.3-6
vending machines, E39.5
water treatment, H53.16

Ice rinks
applications, R34.1
ceiling dripping, R34.8
conditions, R34.4
floor design, R34.5-7
fog, R34.7
heat loads, R34.1-4
refrigeration
    equipment, R34.4
    requirements, R34.1
surface building and maintenance, R34.7

Induction
air-and-water systems, H3.1-12
ceiling boxes, E2.7
testing, adjusting, balancing, H57.6

Industrial
duct construction, E1.7-9
fuel-burning equipment, E22.2-10
gas cleaning, E11.1-26
heat pump systems, applied, H9.10
heat recovery systems, H6.4
humidifiers, E5.6
service water heating, H54.1-20

Industrial air conditioning, H28.1-10
air filtration, H28.9
cooling systems, H28.8
design considerations. H28.5
employee requirements, H28.5
equipment selection, H28.7
heating, H28.6-8
load calculation, H28.6
maintenance, H28.10
process and product requirements, H28.1-5
    air cleanliness, H28.5
    conditioning, H28.1
    corrosion, H28.5
    design temperatures, H28.2-4
    drying, H28.1
    hygroscopic materials, H28.1
    product accuracy and uniformity, H28.4
    product formability, H28.5
    rate of biochemical reactions, H28.4
    rate of chemical reactions, H28.4
    rate of crystallization, H28.4
    regain, H28.1, 4
    static electricity, H28.5
system selection, H28.7

Industrial drying systems, H44.1-8
determining drying time, H44.2
system selection, H44.3
types, H44.3-6

Industrial environment, H41.1-8
dilution ventilation, H41.7
general ventilation, H41.2
    air requirements, H41.2
    heat conservation, H41.3
    heat recovery, H41.3
    locker rooms, H41.4
    need for makeup air, H41.3
    roof ventilators, H41.4
    shower space, H41.4
    toilets, H41.4
heat control, H41.1
local comfort ventilation, H41.5
    design recommendations, H41.6

Industrial exhaust systems
air cleaners, H43.9
air moving devices, H43.10
duct construction, H43.10
duct losses, H43.8
energy recovery, H43.10
exhaust stacks, H43.9
fans, H43.10
fluid mechanics, H43.1
local system components, H43.1-7
    capture velocities, H43.2
    compound hoods, H43.7
    duct considerations, H43.6
    duct size determination, H43.6
    hood entry loss, H43.6
    hood volumetric flow ratios, H43.2-5
    hoods, H43.1
    hot process hoods, H43.44
    lateral systems, H43.5
    simple hoods, H43.7
    special situations, H43.5
maintenance, H43.10
operation, H43.10
system testing, H43.10

Infiltration, F23.7-17
air leakage
    building components, F23.10-16
    commercial buildings, F23.11
    exterior doors, F23.14
    residential, F23.10
calculations
    component leakage, F23.13-16
    empirical models, F23.13
    examples, F23.17
    leakage area, F23.16
    multicell models, F23.13
    single-cell models, F23.13
heat loss calculations
    air change method, F25.8
    crack length method, F25.8
    exposure factors, F25.8
    latent, F25.8
    sensible, F25.8

measurement, F23.10-12
    constant concentration, F23.10
    constant flow, F23.10
    tracer decay method, F23.8
refrigerated storage, R27.3-5

Infrared
heaters, E29.1-6
radiant heating, H16.1-10
snow melting systems, H55.13

Instruments
air contaminants, F13.25
central system, H1.11
combustion analysis, F13.24
data processing areas, H33.7
data recording, F13.28-29
electrical, F13.21-22
    ammeters, F13.21
    power-factor meters, F13.21
    precautions in use, F13.21
    voltmeters, F13.21
    wattmeters, F13.21
error analysis, F13.3
heat transfer (building materials)
    thermal conductivity, F13.11
    wall conductances, F13.11
humidity, F13.23-24
    calibration, F13.24
    dew point hygrometers, F13.24
    dimensional change hygrometers, F13.24
    electrical impedance hygrometers, F13.24
    electrolytic hygrometers, F13.24
    gravimetric hygrometers, F13.24
    psychrometers, F13.23
infrared radiant heating, H16.8
light, F13.26
nuclear radiation, F13.25
    ionization, F13.25
    molecular dissociation, F13.26
pressure, F13.10-13
    absolute, F13.11
    barometers, F13.11
    differential, F13.12
    dynamic, F13.12
    gauges, F13.12
    indirect, F13.12
    low, F13.11
    manometers, F13.12
rotative speed
    stroboscopes, F13.22
    tachometer-generators, F13.22
    tachometers, F13.22
smoke density, F13.25
solar energy systems, H47.25
sound, F13.26
temperature, F13.5-9
    indicating crayons, F13.8
    infrared radiometers, F13.9
    remote temperature sensors, F13.9
    resistance thermometers, F13.8
    static vs. total temperature, F13.9
    thermistors, F13.8
    thermocouples, F13.6-8
    thermometers, F13.6
    winding temperature, F13.8
terminology, F13.1
testing, adjusting, balancing, H57
thermal comfort, F13.9-11
    air humidity, F13.10
    air temperature, F13.9
    air velocity, F13.9
    calculating, F13.10
    integrating instruments, F13.10
    mean radiant temperature, F13.9
velocity, F13.13-15
    airborne tracer techniques, F13.13
    cup anemometers, F13.13
    deflecting vane anemometers, F13.13
    hot-wire anemometer, F13.15
    Pitot tubes, F13.13, 14
    revolving vane anemometers, F13.13
    thermal anemometers, F13.15
    transients, F13.15
    turbulence, F13.15
    vibration, F13.27
volume, F13.15-21
    air change, F13.21
    flow nozzle, F13.16-19
    infiltration, F13.21
    orifice plate, F13.16-19
    positive displacement, F13.20
    turbine flowmeter, F13.20
    variable area flowmeter, F13.19
    Venturi meter, F13.16-19
water systems, medium and high temp, H15.8

Insulation, electrical, R6.2-3

Insulation, thermal
animal environments, H37.6
basic materials, F20.2
central plant systems, H12.1-4
central system, H1.7
clothing, F8.11
conductance, F20.6, F22.6
conductivity, thermal, F22.6, 16
cryogenic, R38.16-20
duct design considerations, F32.12
economic thickness, F20.9-12
factors affecting thermal performance, F20.4-6
hospitals, H23.10
industrial practice, F21.14-19
    ducts, F21.17
    environmental spaces, F21.18
    equipment, F21.16
    land transport vehicles, F21.18
    pipes, F21.14-16
    refrigerated rooms and buildings, F21.17
    tanks, F21.16
    underground, F21.15
physical structure and form, F20.2
preventing surface condensation, F20.18
properties, F20.2-3
railway refrigerator cars, R29.7
refrigerated warehouses, R25.9-12, R27.2
water vapor retarders, F20.9

Jets
air diffusion, F31.1-16

Joule-Thomson refrigeration, R38.1

Kitchens
air conditioning, H19.6
service water heating, H54.14

Laboratories
air balance and flow, H30.8
animal rooms, H30.13, H37.8
    heat generated, H30.13, F9.11
    moisture production, F9.11
    ventilation, H30.13
auxiliary air supply
biological safety cabinets, H30.11-12
containment, H30.14-15
design conditions, H30.2
energy balance interface, H30.14
exhaust systems, H30.5-7
fume hoods, H30.9
laminar flow hoods, H30.12
nuclear facilities, H40.2
recirculation, H30.8
risk assessment, H30.1
special requirements, H30.14
supplementary conditioning, H30.14
supply systems, H30.2
thermal loss, H30.2

Laundries
service water heating, H54.16

Leak detection
ammonia, F16.6
bubble method, F16.5
electronic, F16.5
halide torch, F16.5
sulfur dioxide, F16.6

Lewis relation, F5.9, F8.4

Libraries, H19.9-12
design concepts, H19.10
load characteristics, H19.10

Light
animal physiological control, F9.4
measurement, F13.26
plant environmental control, F9.11

Lighting
greenhouses, H37.12
load, F26.7, 44
plant growth facilities, H37.16-18

Liquid
chilling systems, E17.1-16
coolers, E16.1-6
from condenser to evaporator, E18.5
refrigerant subcoolers
    evaporative condensers, E15.16

Liquid overfeed systems, R2.1-9

Liquid recirculation
ammonia systems, R4.13-14

Lithium bromide-water
properties, F17.69-72
refrigeration cycles, F1.20-26
specific gravity, F16.2
specific heat, F16.2
viscosity, F16.3

Load calculations
air-and-water systems, H3.1, 5
beverage coolers, E39.4
cooling and dehumidifying coils, F5.13
CLTD/CLF calculation procedure, F26.32-48
coils, air cooling and dehumidifying, E6.14
computer applications, H60.3
dams, concrete, R35.3
environmental test facilities, R37.6
heat gains, (see **Heat Gains**)
heating, (see also **Heating**) F25.1-10
humidification, E5.3
industrial air conditioning, H28.6
initial design considerations, F26.3
refrigerated storage, R27
refrigeration
    beef, R12.5-6
    dairy products, R15.9
    marine cargo, R30.6
    trucks, R28.6
residential calculations, F26.54-60
sol-air temperature, F26.4
TETD/TA calculation procedure, F26.2, 49-54
transfer function method, F26.3, 13-32
    conduction transfer functions, F26.15-19

Loss coefficients
control valve, F2.11
duct fittings, F32.27-52
fittings (table), F2.10

LPG
combustion, F15.3-5
furnaces, E24.4, 7

Lubricants
refrigerant-oil properties, R8.4
refrigerant system, R8.1-22

Lubrication
compressors
    centrifugal, E12.34
    helical rotary, E12.17
    reciprocating, E12.9
    rotary, E12.13
engine drives, E32.13

Maintenance
absorption units, E13.8
air cleaners, E10.9
air conditioners, room, E41.5
bus air conditioning, H24.8
centrifugal compressors, E12.35
coils, air cooling, and dehumidifying, E6.15
condensers, E15.7, 12, 17
control, H51.39
coolers, liquid, E16.5
cooling towers, E20.11
definitions, H59.1
economics, H59.1
engine drives, E32.5
heat recovery equipment, E34.6, 10, 13
industrial air conditioning, H28.10
industrial exhaust systems, H43.10
infrared heaters, E29.1-6
life cycle costing, H59.1

Index

liquid chilling systems, E17.5, 13, 16
makeup air units, E27.9
outline for manuals, H59.3
program requirements, H59.3
responsibilities, H59.1-3
solar energy systems, H47.25
store air conditioning, H18.1-8
turbine drives, E32.16
unit heaters, E27.6
unit ventilators, E27.2
Makeup air units, E27.7-10
applications, E27.8
codes and standards, E27.10
controls, E27.9
maintenance, E27.10
ratings, E27.10
selection, E27.8
sound level, E27.10
Marine refrigeration
cargo, R30.1-7
container vans, R30.7-8
fishing vessels, R30.11-16
ships' stores, R30.8-11
Mass transfer
analogy relations, F5.5, 7-10
coefficient, F5.4-6
cylinder, F5.8
eddy diffusion, F5.6
flat plate, F5.7
Lewis relation, F5.9, F8.4
molecular diffusion, F5.1
  analogy to heat transfer, F5.2
  coefficient, F5.2
  Fick's Law, F5.1
  in liquids and solids, F5.4
  of one gas through another, F5.3
simultaneous heat and, F5.10-14
  air washers, F5.11
  basic equations for direct-contact equipment, F5.10
  cooling and dehumidifying coils, F5.13
  cooling towers, F5.13
  enthalpy potential, F5.10
sphere, F5.8
turbulent flow, F5.7
Measurement
air contaminants, F13.25
air exchange, F23.10
air leakage, F23.11
combustion analysis, F13.24
  flue gas, F13.24
data recording, F13.28
electrical, F13.21-22
error analysis, F13.3
flow rate, F2.11
heat transfer (building materials), F13.11
humidity, F13.23-24
infiltration, F23.10-12
light, F13.26
moisture, E21.3-4
  farm crops, F10.3
nuclear radiation, F13.25
odors, F12.1-4
pressure, F13.10-13
rotative speed, F13.22
smoke density, F13.25
sound, F13.26
temperature, F13.5-9
terminology, F13.1
testing, adjusting, balancing, H57.1-26
thermal comfort, F13.9-11
  air humidity, F13.10
  air temperature, F13.9
  air velocity, F13.9
  calculating, F13.10
  integrating instruments, F13.10
  mean radiant temperature, F13.9
velocity, F13.13-15
vibration, F13.27
volume, F13.15-21
Meat
beef
  carcass chilling, R12.1, 3-6
  boxed, R12.7
  cooler layout for, R12.3
  load calculations for, R12.5-6
contaminants, R10.1
display refrigerators, E35.1-3
enthalpy, F30.9
frozen, R12.15
hogs
  chilling, R12.7-9
  pork trimmings, R12.9
processed, R12.11-15
  bacon, R12.12
  lard chilling, R12.14
  sausage dry rooms, R12.13
refrigeration systems for, R12.2, 6
spoilage, R10.2
storage
  modified atmosphere, R12.3
  retail food store, R32.2
variety meats, R12.10
Metallurgy, low temperature, R39
ferrous alloys, R39.3-4
mechanical properties, R39.1-2
nonferrous alloys, R39.2
Milk (see Dairy products)
Mines, underground, H42.1-8
adiabatic compression, H42.1
air conditioning equipment, H42.4-7
air cooling, H42.2-4
dehumidification, H42.2-4
evaporative cooling, H42.8
groundwater, H42.2
mechanical refrigeration, H42.7
spot cooling, H42.7
wall rock heat flow, H42.2
water sprays, H42.8
worker heat stress, H42.1
Mobile homes
residential air conditioning, H17.6
Modeling
airflow around buildings, F14.13-17
energy estimating methods, F28.18-40
Moist air
composition, F6.1
cooling, F6.16
heating, F6.16
thermodynamic properties, F6.1-6
Moisture
condensation, F21.6-12
  attic ventilation, F21.10
  checklist, F21.11
  concealed, F21.8
  control, F21.6-10
  cooled structures, F21.12
  heated buildings, F21.6
  visible, F21.7
  zones in United States, F21.8
content, farm crops, F10.1
in buildings
  air barriers, F21.5
  air leakage, F21.4
  air movement control, F21.5
  basic diffusion equation, F20.14
  condensation tolerance, F21.6
  electrically heated houses, F21.5
  fuel-heated houses, F21.5
  in building materials, F20.12-14
  laboratory testing, F20.16
  permeability, F20.16
  permeance, F20.16
  preventing surface condensation, F20.18
  properties of air/water vapor mixtures, F20.12
  water vapor migration, F20.14, F21.4
  water vapor retarders, F20.15
measurement, E21.3-4
production animals, F9.4-11
refrigerant system, R7.1-6
survival shelters, H27.19
transfer, farm crops, F10.2
Motels, H21.1-4
service water heating, H54.8, 11
Motors
codes, E31.2
compressors, E12.2, 7, 13, 33
control, E31.6-8
efficiency, E31.3
fans, E3.1-14
furnaces, E24.2
heat gain, F26.9
hermetic
  applications, E31.5
  cleanup after burnout, R7.8-9
  insulation, R6.2-4
  non-hermetic, E31.3
  power supply, E31.1
  pump, E30.8
  standards, E31.2
  thermal protection, E31.5
Museums, H19.9-12
classification, F15.4
Natural gas
combustion, H52.3
  calculations, F15.8
  efficiency, F15.11
liquefied, R38.6-9
NC (Noise Criteria) curves, F7.8, H52.3
Nightclubs, H19.5-7
Nitrogen
for commercial freezing, R9
liquefaction, R38.9
properties of, F37.1
refrigerant properties, F17.60
Noise
compressor
  centrifugal, E12.33
  positive displacement, E12.4
cooling towers, E20.10
fans, E3.1-14
liquid chilling systems, E17.11
refrigerant piping, F33.21
Nuclear
radiation measurement, F13.25
Nuclear facilities, H40.1-10
general HVAC design criteria, H40.3-5
nuclear design criteria, H40.1-3
nuclear fuel-processing facilities, H40.6
nuclear power plants, H40.6-10
  HVAC for boiling water reactor, H40.8-10
  HVAC for pressurized water reactor, H40.6-8
plutonium processing plants, H40.5-6
zoning, H40.2
Nursing homes
design concepts and criteria, H23.11
service water heating, H54.8, 11
ventilation, H23.11
Nuts
storage, R22.8, R26.5
Odors, F12.1-6
analytical measurement, F12.3
control, H50.3
removal, F12.4
  adsorption, F12.5, E11.22
  chemical oxidation, F12.5
  combustion, F12.5
  counteraction, F12.5
  masking, F12.5
  scrubbing, F12.5
  ventilation, F12.5
  washing, F12.5
sense of smell, F12.1
sensory measurement, F12.2
sources, F12.4
Office buildings
design concepts, H19.8
load characteristics, H19.7
service water heating, H54.8, 11
special considerations, H19.9
Oil
ammonia systems, R4.3
liquid overfeed systems, R2.3
oil-refrigerant solutions, R8.7-13
properties, R8.4-7
receivers, R4.4
refrigerant reactions, R6.5
refrigerant system, R8

return
  coolers, liquid, E16.5
  in refrigeration systems, R1.6, 8
    halocarbon refrigerants, R3.4-10
  separators, R3.20, R4.4, E19.21
Oil, fuel
  combustion, F15.8-14
  handling, E22.8
  pipe sizing, F33.23
  preparation, E22.9
  storage systems, E22.7
  types, F15.4
Oil-fired
  equipment, E22.4-10
    commercial-industrial, E22.5
    residential, E22.4
    selection, E22.7-10
  furnaces, residential, E24.1-18
  heaters (in-space), E25.3
  infrared heaters, E29.1-6
  makeup air units, E27.9
  service water heating, H54.1-20
  unit heaters, E27.4
One-pipe
  water systems, basic, H13.5
Operating costs, H49.1-8
Operating rooms, H23.4
Outlets
  air diffusion, F31.1-16
  small forced-air systems, H10.2, 8
  supply, E2.1-5
    ceiling diffuser, E2.4
    grille, E2.2
    selection procedure, E2.2
    slot diffuser, E2.3
    sound level, E2.2
    VAV, E2.5
Outpatient surgical facilities, H23.12
Owning costs, H49.1-8
Oxygen
  liquefaction, R38.10
  refrigerant properties, F17.64
Ozone
  modified atmosphere storage, R12.3, R24.1
Packaged terminal air conditioners, H5.1-8
Paper manufacturing, H39.1-4
Peanuts, H38.9
People
  heat gain, F26.6, 43
Permeability
  common building materials, F22.13, 14
Permeance
  common building materials, F22.13, 14
Photographic processing
  film storage, H36.5-7
  filters, H36.2
  manufacture, H36.1
  photographic paper prints, H36.7
  processing and printing, H36.4
Physiological factors, farm crops, F10.1-12
  aeration of grain, F10.10
  deterioration prevention, F10.6-7
  drying
    airflow resistance, F10.8
    barley, F10.12
    cotton, F10.12
    deep bed, F10.9
    grain, F10.11
    hay, F10.10
    peanuts, F10.12
    rice, F10.12
    shelled corn, F10.11
    soybean, F10.12
    theory, F10.4-7
    thin layer, F10.8
    tobacco (curing), F10.13
    wheat, F10.12
  moisture measurement, F10.4-5
  storage, F10.1-3
Physiological principles, humans
  comfort, prediction of F8.16-21
    adaptation, F8.19
    age, F8.19

    asymmetric thermal discomfort, F8.20
    conditions, F8.16
    draft, F8.20
    rhythms, F8.19
    sex, F8.19
    steady state energy balance, F8.17
    two-node transient energy balance, F8.18
    variations, F8.19
    warm or cold floors, F8.21
    zones of comfort and discomfort, F8.18
  effective temperature, F8.13
  energy balance, F8.1-2
  engineering data and measurements, F8.7
    body surface area, F8.7
    clothing insulation, F8.9-11
    environmental parameters, F8.11
    evaporative heat loss, total, F8.10
    heat transfer coefficients, F8.9
    mechanical efficiency, F8.7
    metabolic rate, F8.7, 8
    moisture permeability, F8.9-11
  environmental indices, F8.14-16
    adiabatic equivalent temperature, F8.13
    effective temperature, F8.13
    heat stress, F8.14
    human operative temperature, F8.13
    skin wettedness, F8.14
    wet globe temperature, F8.15
    wet-bulb globe temperature, F8.14
    wind chill, F8.15, 16
  health, F8.26-29
    air contaminants, F8.28
    ASHRAE Comfort-Health Index (CHI), F8.26
    indoor climates, F8.27
    quantifying effects, F8.29
    thermal environment, F8.27
  special environments, F8.22
    extreme cold, F8.25
    high intensity infrared heating, F8.22-24
    hot and humid, F8.24
  thermal interchange, F8.2-6
    effect of clothing, F8.3
    evaporative heat loss from skin, F8.2
    respiratory losses, F8.3
    sensible and latent heat flow, F8.4-6
    sensible heat loss from skin, F8.2
  thermoregulatory control mechanisms
    blood flow, F8.6
    signals and responses, F8.6
    sweating, F8.7
    zones of response, F8.6
Pipe, E33.1-14
  application, E33.8
  bends, E33.9-10
  buried, heat flow in, F22.21
  codes, E33.1
  copper tube, E33.2, 4
  expansion, E33.9-13
  flexibility, E33.9
  hangers, H52.34
  heat loss, bare, F22.18, F23.16-20
  heating coils, E28.1
  insulation, F21.14-16
  insulation thickness, F22.19, 20
  iron, E33.2
  joining methods, E33.2-6
  loops, E33.9
  non-metallic, E33.2
  special systems, E33.7
  standards, E33.1, 6, 7
  steel, E33.1,3
  stress calculations, E33.7
  valves, E33.6
  vibration control, H52.34-36
  wall thickness, E33.7
Pipe sizing, F33.1-26
  chilled and hot water
    air separation, F33.5
    copper tube, F33.5
    friction charts, F33.4, 5
    plastic pipe, F33.5
    pressure drop, F33.6

    steel pipe, F33.4
  gas, F33.22
  general principles, F33.1
    calculating pressure losses, F33.2
    fitting losses, F33.1
    flow rate limitations, F33.2
    valve losses, F33.1
    water hammer, F33.4
  hot and cold service water, F33.6-9
    fixture units, F33.7
    plastic pipe, F33.8
    procedure for cold water systems, F33.8
  oil, F33.23
  refrigerant, F33.19-22
  steam, F33.9-18
    basic chart, F33.12
    charts for high-pressure, F33.11-17
    one pipe gravity, F33.18
    pressure drop, F33.9
    return lines, two-pipe systems, F33.15-18
    tables for low-pressure, F33.10-11
Piping
  all-water systems, H4.4
  ammonia systems, R4.1-16
  central plant systems, H12.1-12
  central system, H1.11
  computer applications, H60.6
  geothermal energy systems, H45.12
  ice rinks, R34.4
  liquid overfeed systems, R2.7
  refrigerant
    ammonia, R4.1-3
    halocarbons, R3.1-14
  refrigeration systems, R1.8
  service water heating, H54.1-20
  solar energy equipment, E44.8-10
  solar energy systems, H47.19
  steam systems, H11.5-7, 12
  system identification, F34.12
  testing, adjusting, balancing, H57.7-15, 25
  unit heaters, E27.6
  water coolers, drinking, E38.3
  water systems, basic, H13.4-7
  water systems, medium and high temp, H15.7
Pitot tube, F13.14
Places of assembly
  auditoriums, H20.4
  churches, H20.4
  common characteristics, H20.1
  concert halls, H20.4
  convention centers, H20.6
  enclosed stadiums, H20.5
  exhibition centers, H20.6
  gymnasiums, H20.6
  sports arenas, H20.5
  swimming pools, H20.7
  system considerations, H20.2
  temporary exhibit buildings, H20.9
  theaters, H20.4
  World fair buildings, H20.9
Plant environment, H37.8-19
  greenhouses, H37.9-14
    carbon dioxide enrichment, H37.12
    energy balance, H37.9
    energy conservation, H37.14
    evaporative cooling, H37.12
    glazing, H37.14
    heating, H37.10
    humidity control, H37.12
    orientation, H37.9
    photoperiod control, H37.13
    radiant energy, H37.12
    sealants, H37.14
    shading, H37.11
    site selection, H37.9
    thermal blankets, H37.14
    ventilation, H37.11
  other facilities, H37.19
  plant growth facilities (chambers), H37.14-18
Plants
  air composition, F9.16
  humidity, F9.16
  irradiance for growth, F9.12-14

Index

light, F9.12
night temperatures
    greenhouse crops, F9.15
photoperiod control, F9.16
pollutants, F9.16
radiation, F9.12
temperature, F9.14

Plastic
ducts, E1.10
greenhouses, H37.14
pipe sizing, water flow, F33.5
solar optical properties, F27.30

Poultry
chilling, R13.1
freezing, R13.4-6
processing, R13
recommended environment, H37.8, F9.9-11
refrigeration, R13.1-4
spoilage, R10.5
storage, R13.6, R26.5
thermal properties, R13.6

Precooling
dams, concrete, R35.2
flowers, cut, R11.10
fruits, R11
methods, R11
    forced air, R11.4-6
    hydrocooling, R11.3-4
    package icing, R11.6
    vacuum, R11.6-9
places of assembly, H20.2
vegetables, R11

Pressure
drying equipment, E7.7-8
measurement, F13.11-13
regulators
    condenser, E19.12
    evaporator, E19.9
    suction, E19.11
smoke control, H58.1-16
steam systems, H11.5
water systems, basic, H13.7-15
water systems, medium and high temp, H15.4-6

Pressure loss
ammonia, R4.1-3
coolers, liquid, E16.4
gaseous contaminant removal, E11.16
halocarbon refrigerants, R3.1-11
refrigerant, F33.19
steam, F33.9
water piping, F33.6
water systems, basic, H13.7

Printing plants
air filtration, H34.5
binding, H34.5
collotype printing, H34.5
control of paper, H34.2
design criteria, H34.1
letterpress, H34.3
lithography, H34.3
    air-conditioning systems, H34.4
    recommended environment, H34.4
platemaking, H34.2
rotogravure, H34.4
salvage systems, H34.5
shipping, H34.5

Properties, mechanical
metallurgy, low temperature, R39

Properties, physical
boiling point, F37.1
conductivity, thermal, F37.1-3
density, F37.1-3
emissivity, F37.3-4
freezing point, F37.2
heat of
    fusion, F37.2
    vaporization, F37.2
liquids (table), F37.2
solids (table), F37.3-4
specific heat, F37.1-3
vapor pressure, F37.2
vapor (table), F37.1

viscosity, F37.1

Psychrometric(s)
air composition, F6.1
charts, F6.14-18
    industrial drying, H44.1
dew point temperature, F6.12
evaporative cooling systems, H56.7
humidity
    ratio, F6.6
    relative, F6.12
perfect gas relationships, F6.12
standard atmosphere, F6.1, 12
thermodynamic properties
    calculation, F6.13
    moist air, F6.1-6
    water (at saturation), F6.6-11
thermodynamic wet-bulb temperature, F6.13
transport properties of moist air, F6.19
typical air-conditioning processes, F6.16-19

Public buildings
air conditioning
    airports, H19.14
    bars, H19.5-7
    bowling centers, H19.12
    bus terminals, H19.14
    communications centers, H19.13
    design concepts and criteria, H19.4
    dining and entertainment centers, H19.5-7
    general design criteria, H19.1-4
    libraries and museums, H19.9-12
    load characteristics, H19.4
    office buildings, H19.7-9
    restaurants, H19.5-7
    ship docks, H19.14
    transportation centers, H19.14
    warehouses, H19.15

Pumps
central system, H1.10
cryopumping, R38.27
energy estimating methods, F28.8
geothermal energy systems, H45.8-10
liquid overfeed systems, R2.1-5
solar energy systems, H47.19
water systems
    basic, H13.8-11
    medium and high temp, H15.6

Pumps, centrifugal
applications, E30.8
arrangement, E30.6
energy conservation, E30.9
installation, E30.10
motive power, E30.8
performance curves, E30.2
selection, E30.6
suction characteristics, E30.3
system characteristics, E30.4
trouble analysis, E30.10
types, E30.1

Radiant
drying systems, H44.3
panel cooling systems, H7.1-18
panel heating systems, H7.1-18
snow melting systems, H55.13

Radiant heating
application considerations, H16.9
comfort design, H16.8
design considerations, H16.3-6
design criteria, H16.2
floor reradiation, H16.6
instrumentation, H16.8

Radiation
solar heat gain factors, F27.5-13
thermal, F3.7-12
    actual, F3.8
    angle factor, F3.9-11
    black body, F3.7
    in gases, F3.11

Radiators, E28.1-6
Radiosity, F3.11
Radon, F11.6
Railroad
air conditioning, H24.9-10
refrigerator cars, R29

Rapid transit systems
design approach, H29.9-11
station air conditioning, H29.8
ventilation, H29.8

RC (Room Criteria) curves, F7.9, H52.3

Receivers
ammonia systems, R4.6-8
halocarbon refrigerants, R3.10
liquid overfeed systems, R2.7-9

Reciprocating
compressors, E12.5-11
liquid chilling systems, E17.6-8

Recirculation systems (liquid overfeed)
ammonia, R4.13-14

Refrigerant control devices, E19.1-30
capillary tubes, E19.21-30
oil separators, E19.21
switches, control, E19.1-3
valves, control
    condensing water regulator, E19.17
    evaporator regulator, E19.9-11
    expansion, constant pressure, E19.8
    expansion, electronic, E19.8
    expansion, thermostatic, E19.3-8
    high side float, E19.13
    low side float, E19.13
    relief devices, E19.19-21
    solenoid, E19.14-17
    suction pressure regulator, E19.11

Refrigerant(s), F16.1-10
ammonia-water, F17.69-70
brine (secondary coolant) systems, R5
effect on construction materials
    elastomers, F16.6, 9
    metals, F16.6
    plastics, F16.6, 10
electrical properties, F16.1, 5
flow control
    balancing, E18.2
    coolers, liquid, E16.5
latent heat, F16.2, 6
leak detection
    ammonia, F16.6
    bubble method, F16.5
    electronic, F16.5
    halide torch, F16.5
    sulfur dioxide, F16.6
line sizing, R1.11
    ammonia, R4.1-3
    halocarbon refrigerants, R3.1-15
liquid overfeed systems
    circulation rate, R2.3, 4 (chart)
    distribution, R2.4
lithium bromide-water, F17.69-72
oil, in systems, R8
performance, F16.2, 8-9
physical properties, F16.1, 4
    boiling point, F16.4
    critical pressure, F16.4
    critical temperature, F16.4
    critical volume, F16.4
    freezing point, F16.4
    refractive index, liquid, F16.4
    surface tension, F16.4
pipe sizing, F33.19-22
piping
    ammonia, R4
    halocarbon refrigerants, R3
pressure drop, E16.4
safety, F16.4, 9
sound velocity, F16.1, 6
standard designation, F16.4
system chemistry, R6
system contaminants
    moisture, R7.1-6
    other, R7.6-8
thermodynamic properties, F17
velocity
    air coolers, forced circulation, E8.3

Refrigerant-oil
solution properties, R8
system chemistry, R6.5

Refrigeration
  absorption cycles, F1.20-26
    ammonia-water, F1.23
    characteristics of refrigerant-absorbent pair, F1.21
    lithium bromide-water, F1.22
    practical, F1.20
    thermodynamic analysis, F1.22
  absorption units, E13.1-12
  air coolers, forced circulation, E8.1-4
  air transport, R31
  air-cycle equipment, E14.1-12
  ammonia system practices, R4
  bakery products, R21
  banana ripening rooms, R17.6
  biomedical applications, R40
  breweries, R23.1-7
  brine (secondary coolant) systems, R5
  candy, R22.5-7
  carbonated beverages, R23.10
  central air conditioning systems, H1.10
  chemical industry, R36.1-10
  commodity storage, R26
  component balancing, E18.1-6
  compression cycles, F1.8-20
    basic vapor, F1.11
    complex vapor, F1.12
    ideal basic vapor, F1.8
    optimum design, F1.13-20
    reversed Carnot, F1.10
    two-stage, F1.13-15
  contaminant control, R7
  coolers, liquid, E16
  cryogenics, R38
  cryopumping, R38.27
  dairy products, R15
  dams, concrete, R35.1
  dried fruits and vegetables, R22.7
  eggs and egg products, R24
  engineered systems, R1
    cascade, R1.3
    compound, R1.2
    economized, R1.2
    multistage, R1.2
    single-stage, R1.2
    liquid overfeed, R2
  environmental test facilities, R37
  factory charging, E21.4
  factory dehydration, E21.1-4
  factory testing, E21.4-8
  fishery products, R14
  food microbiology, R10
  food service equipment, E36.1-4
  freezing, commercial, R9
  fruit juice concentrates, R19
  fruit, R16
  geothermal energy systems, H45.21
  ice manufacture, R33
  ice rinks, R34
  insulation, rooms and buildings, F21.17
  load calculation, R27
  liquid overfeed systems, R2
  lubricants, R8
  marine
    cargo, R30.1-7
    container vans, R30.7-8
    fishing boats, R30.11-16
    ships' stores, R30.8-11
  meat products, R12
  metallurgy, low temperature, R39
  nuts, R22.7
  oil requirements, R8.2
  poultry products, R13
  precooked and prepared foods, frozen, R20.
  precooling, R11
  railway refrigerator cars, R29
  retail food store, E35, R32
  secondary coolant systems, R5
  soils, subsurface, R35.3-5
  space simulation, R38.25-28
  system chemistry, R6
  thermal storage, H46.22
  trucks, R28

  vegetables, R18
  warehouse design, R25.6-9
  wine making, R23.8-10
Refrigerators
  cryogenic, R38.11-13
  household, E37.1-14
    cabinets, E37.2-6
    capillary tube, E37.8
    compressor, E37.8
    condenser, E37.7
    control, temperature, E37.9
    defrosting, E37.6
    evaluation, E37.11
    evaporator, E37.7
    performance characteristics, E37.2
    refrigerating systems, E37.6
    safety requirements, E37.2
    test procedures, E37.12-14
    test rooms, E37.11
  retail food store, E35, R32
Residential
  air conditioning, H17.1-6
    air filters, H17.5
    boilers, H17.3
    capacity selection, H17.4
    central forced-air, H17.6
    controls, H17.5
    furnaces, H17.2
    heat pumps, H17.2
    heating systems, H17.1
    humidifiers, H17.5
    hydronic, H17.6
    mobile homes, H17.5
    multifamily, H17.6
    single-family, H17.2
    solar heating, H17.3
    unitary air conditioners, H17.3
    zonal, H17.3-6
  condensation, F21.8
  cooling load, F26.54-60
  duct construction, E1.2
  gas burners, E22.1
  humidifiers, E5.4
  oil burners, E22.4
  service water heating, H54.1-20
Resistance, thermal, F3.2-4
  design values, F22.6-9
Restaurants, H19.5-7
Retail food store refrigeration
  condensing methods, R32.3-5, E35.6
  defrost methods, R32.5
  display refrigerators, R32.1, E35.1-5
  requirements, R32.2
  walk-in coolers, R32.2
Reynolds number, F2.9-10
Rice, H38.9, F10.12
Roofs
  CLTD tables, F26.34
  insulation, F21.4
  overhang shading, F27.35
  transfer function coefficients, F26.15-19
Rotary
  compressors, E12.11-15
  energy exchanger, E34.4-7
  helical, compressors, E12.16-26
Roughness factor, ducts, F32.5
R-values, F22.6-10
Safety
  air conditioners, room, E41.4
  central system, H1.4
  chemical industry, R36.3
  control, H51.37
  fuel-burning equipment, E22.14
  household refrigerators and freezers, E37.2
  ice makers, E40.5
  nuclear facilities, H40.1
  refrigerants, F16.4, 9
  service water heating, H54.10
  snow melting systems, H55.6
  solar energy systems, H47.25
  solid fuel heaters, E25.7
  textile processing, H35.7
  water systems, medium and high temp, H15.9

Scale
  service water heating, H54.6
  water treatment, H53.12-14
Schools, H22.1-6
Screw compressors (see Compressors, helical rotary)
Scrubbers
  industrial gas cleaning, E11.1-26
Sealed tube text, R6.1
Secondary coolants (see Coolants, secondary)
Service life, equipment, H49.7
Service water heating, H54.1-20
  demand (table), H54.4
  design considerations, H54.6-10
  direct heat transfer equipment, H54.1-2
  distribution systems, H54.3-6
    commercial kitchen, H54.5
    manifolding, H54.6
    return piping, H54.4
    supply piping, H54.3
    two-temperature, H54.5
  indirect heat transfer equipment, H54.2-3
  requirements, H54.8-18
  sizing
    instantaneous, H54.19
    semi-instantaneous, H54.19
  storage equipment sizing, H54.8-18
    apartments, H54.9, 11
    commercial, H54.8-18
    dormitories, H54.8, 11
    food service, H54.9-14
    health clubs, H54.17
    industrial plants, H54.17
    institutional, H54.8-16
    kitchens, H54.14
    laundries, H54.16
    motels, H54.8, 11
    nursing homes, H54.8, 11
    office buildings, H54.8, 11
    ready-mix concrete, H54.17-18
    residential, H54.8-11
    schools, H54.9, 11-16
    showers, H54.13, 15
    sizing examples, H54.12-16
    swimming pools, H54.17
  use (building types), H54.4
Shading
  coefficients, F26.6, F27.19, 24-37
  devices, F27.30-37
Ship(s)
  air conditioning, H26.1-8
  docks, H19.14
  general criteria, H26.1
  merchant ships, H26.1-6
  naval surface ships, H26.6-8
  refrigeration, R30
Shipping docks
  meat products refrigeration, R12.16
  refrigerated warehouses, R25.2, 6
SI unit conversions, F35
Simulation (modeling), F28.18-40
Skylights, domed, F27.36
Smoke
  density measurement, F13.25
Smoke control, H58.1-16
  acceptance testing, H58.15
  computer analysis, H58.15
  dampers, H58.1
  design parameters, H58.9-11
  door-opening forces, H58.7
  duct design, F32.12
  elevators, H58.13
  flow areas effective, H58.8
  principles, H58.4-6
  purging, H58.7
  smoke management, H58.4
  smoke movement, H58.1-4
  stairwells, H58.11-13
  symmetry, H58.9
  zones, H58.14
Snow melting, H55.1-14
  control, H55.14

Index

electric systems, H55.7-14
  cable, H55.9-11
  design data, H55.8
  general design, H55.7
  gutters and downspouts, H55.14
  heat density, H55.7
  infrared, H55.13
  slab design, H55.7
  wire and mat, H55.11-12
hot fluid system design, H55.1-7
operating characteristics, H55.3
snowfall data, H55.6

Soil(s)
subsurface
  control of water flow, 35.3
  stabilization of frozen, R35.4
temperatures, H12.7-9
thermal conductivity, F22.21, H12.6, H27.5

Solar
absorbed radiation, F27.14
absorption units, E13.10
angle, F27.3
diffuse radiation, F27.14
energy estimating, F28.28-40
heat gain factors, F27.5-13, 24-37
heat gain through fenestration, F27
heat pump systems, applied, H9.5
intensity (tables), F27.5-13
optical properties
  glass, F27.20
  plastic, F27.30
passive gain, F27.15
radiation data, F24.4
residential air conditioning, H17.3
air heating systems, E44.1

Solar energy
array design, E44.10-16
  air storage, E44.11
  energy storage, E44.11
  liquid storage, E44.12-14
  piping configuration, E44.10
  shading, E44.11
  short circuiting, E44.15
  sizing, E44.16
  storage tank construction, E44.14
  stratification, E44.15
collectors, E44.3-7
  concentrating collectors, H47.13
  construction, E44.5-7
  flat plate collectors, H47.10
  glazing materials, H47.11
  performance, E44.4, H47.14-16
  plates, H47.12
  test results, E44.7
  testing methods, E44.4
controls, E44.18-20
cooling
  energy estimating, F28.29
  nocturnal radiation and evaporation, H47.21
design, installation, and operation guide, H47.25-26
domestic water heating, H47.16-21
  air systems, H47.17
  collectors, H47.18
  components, H47.17
  controls, H47.20
  drain-back systems, H47.17
  drain-down systems, H47.17
  economic evaluation, H47.21
  expansion tanks, H47.19
  fans, H47.20
  heat exchangers, H47.19
  heat transfer fluids, H47.18
  indirect systems, H47.17
  load requirements, H47.21
  piping, H47.19
  pumps, H47.19
  recirculation systems, H47.16, 18
  system performance evaluation, H47.21
  thermal energy storage, H47.19
  thermosiphon systems, H47.16
  valves, H47.20

freeze protection, E44.2
heat exchangers, E44.16-18
heat gain factors, F26.6
heat storage, H47.16
heating and cooling systems, H47.22-23
liquid heating systems, E44.1
module design, E44.8-10
quality and quantity, H47.1-10
  flat plate collectors, H47.9
  incident angle, H47.3
  longwave atmospheric radiation, H47.9
  solar angle, H47.1
  solar constant, H47.1
  solar irradiation design values, H47.7
  solar irradiation (tables), H47.4, 7
  solar position (tables), H47.4-5
  solar radiation, H47.5-7
  solar spectrum, H47.5
  solar time, H47.2
service water heating, H47.16-21
sizing collector & storage subsystems, H47.23-25

Solid fuel
characteristics, F15.6-7
combustion, F15.8-14
heaters (in-space), E25.6
stokers, E22.10-11

Sol-air temperature, F26.4
Sorbents, F19.3-6
Sorption, E7.2-7
Sound
air outlets, E2.2
characteristics, F7.2-3
clean spaces, H32.6
compressor
  centrifugal, E12.33
  positive displacement, E12.4
control (see Sound control)
converting power to pressure, F7.6
cooling towers, E20.10
determining power, F7.4-5
engine test facilities, H31.6
human response, F7.7-9
liquid chilling systems, E17.11
loudness, F7.3
makeup air units, E27.10
measurement, F13.26
noise, F7.4
power, F7.1
pressure, F7.1
sources, F7.5, 6
testing, adjusting, balancing, H57.19-22
transmission, reducing, F7.6-7
unit heaters, E27.5

Sound control, H52.1-27
airflow noise, H52.13-16
along duct paths, H52.9-13
central system, H1.7
communications centers, H19.13
computer applications, H60.7
design criteria, H52.2-3
design goals, H52.1, 4
ductborne crosstalk, H52.17
educational facilities, H22.5
indoor mechanical systems, H52.4-5
isolation of equipment rooms, H52.24-26
noise generated by air terminals, H52.16-17
noise limit specifications, H52.5
  acoustical design procedures, H52.27
  rooftop equipment, H52.27
places of assembly, H20.3
prediction of fan sound power, H52.7-9
recommended procedures, H52.6
swimming pools, H20.8
textile processing, H35.7
transmission loss, H52.17-23

Soybeans, H38.7
Space heating
water systems, medium and high temp, H15.8
Specific heat of foods, F30.2, 3, 11-13, R26.2-5
Speed (rotative) measurement, F13.23
Sports arenas, H20.5-6

Spot cooling
industrial environment, H41.5
mines, underground, H42.7
Spray ponds, E20.5
Standard atmosphere, F6.1, 12
Standards, R41
air cleaners, E10.2-5
air conditioners, room, E41.4
air conditioners, unitary, E42.7
air quality, H50.4
ambient air quality, F11.6, 8
boilers, E23.3
chemical industry, R36.5, 10
dehumidifiers, room, E41.7
engine drives, E32.6
furnaces, E24.18
heaters (in-space), E25.6
ice makers, E40.5
liquid chilling systems, E17.5
motors, E31.2
packaged terminal air conditioners, E43.3
pipe, tube, fittings, E33.1-14
service water heating, H54.7
vending machines, E39.6
water coolers, drinking, E38.3
water source heat pumps, E43.6
Steam
coils, air heating, E9.1-4
heating systems, H11.1-16
  boiler connections, H11.3
  boilers, H11.3
  condensate removal, H11.7
  definitions, H11.1
  design pressure, H11.5
  distribution, H11.13
  heat exchangers, H11.3
  heat recovery, H11.15
  one-pipe, H11.12
  piping, H11.5-7
  temperature control, H11.14
  terminal equipment, H11.12
  traps, H11.8-10
  two-pipe, H11.13
  valves, pressure reducing, H11.10-12
pipe sizing, F33.9-18
testing, adjusting, balancing, H57.15
turbine drives, E32.9-13
unit heaters, E27.4
Stirling cycle, R38.12
Stokers, E22.10-13
Storage
breweries, R23.5
brine (secondary coolant) systems, R5.3
butter, R15.9
candy, R22.5-7, R26.5
cheese, R15.13, R26.4
carbon dioxide
  breweries, R23.6
  carbonated beverages, R23.11
citrus, R17.3
controlled atmosphere, R16.2, R25.3
cryogenics, R38.20
cut flowers, R26.8-9
farm crops, F10.1-4, H38.10-12
film, R36.5-7
fish
  fresh, R14.3, R26.3
  frozen, R14.8
fruit
  dried, R22.7
  fresh, R16.1-10, R26.3
  juice concentrates, R19
  storage diseases, R16.3-4
furs, R26.7
ice, R33.4-5
ice makers, E40.5-6
meat products, R12, R26.4
milk, fresh, R15.5, R26.4
nursery stock, R26.8-9
nuts, R22.7, R26.5
poultry products, R13.6, R26.5
refrigeration load, R27
service water heating, H54.1-20

solar energy equipment, E44.11-14
thermal, H46.1-24
vegetables
  fresh, R18.3-14, R26.2
  dried, R22.7, R26.2
warehouse design, R25
wines, R23.8

Stores
convenience centers, H18.7
department, H18.6-7
multiple-use complexes, H18.8
regional shopping centers, H18.7-8
small, H18.1
supermarkets, H18.3-5
variety, H18.2

Sunshades, F27.30-37

Survival shelters
apparatus selection, H27.20
climate, H27.5
control of chemical environment, H27.5
evaporative cooling, H27.19
factors affecting environment, H27.1
fire effects, H27.19
moisture in enclosed spaces, H27.19
physiological aspects, H27.1
  time-temperature tolerance, H27.3
  ventilation factors, H27.4
  water requirements, H27.2
soils, H27.5
thermal environment, H27.7
  aboveground shelters, H27.16
  sealed underground shelters, H27.12
  simplified analytical solutions, H27.7
  underground shelters, H27.7
  ventilated underground shelters, H27.13
  ventilated & insulated shelters, H27.8
ventilation, H27.16-18

Swimming pools
air distribution, H20.8
design, H20.7-8
filters, H20.8
noise level, H20.8
service water heating, H54.17

Swine
growth, F9.8
heat and moisture production, F9.8
recommended environment, H37.7
reproduction, F9.8

Symbols, F34
graphical, F34.5-11
letter, F34.1, 4
mathematical, F34.4

Temperature measurement, F13.6-9

Temporary exhibit buildings, H20.9

Terminology
air diffusion, F31.1
control, H51.1
ice makers, E40.2
measurement, F13.1
service water heating, H54.6
testing, adjusting, balancing, H57.1
thermodynamics, F1.1

Testing
air cleaners, E10.2-5
  ARI standards, E10.5
  atmospheric dust spot efficiency, E10.3
  DOP penetration, E10.3
  dust holding capacity, E10.3
  environmental, E10.4
  leakage (scan), E10.4
  miscellaneous performance, E10.4
  particle size efficiency, E10.4
  specialized performance, E10.4
  weight arrestance, E10.3
air conditioners, packaged terminal
  performance, E43.3
air distribution systems, H57.3-7
airflow around buildings, F14.13-17
  critical dilution, F14.14
  full-scale, F14.17
  model test programs, F14.16
  physical modeling, F14.13
  similarity requirements, F14.13

stack exhaust velocity, F14.15
stack height, F14.14
wind tunnel, F14.16
clean space, H32.8
compressors
  centrifugal, E12.31
  positive displacement, E12.5
cooling towers, E20.12, H57.16
desiccants, R7.6
driers, R7.6
engine test facilities, H31.1-6
environmental test facilities, R37.1-7
heat pumps, water source safety, E43.3
heaters (in-space), E25.7
household refrigerators and freezers, E37.11-14
hydronic systems, H57.7-15
industrial exhaust systems, H43.10
leak, E21.4-5
liquid chilling systems, E17.5
moisture, in buildings, F20.16
performance
  complete system, E21.7
  components, E21.7
  compressor, E21.6
refrigerant system, E21.4-8
smoke control acceptance, H58.15
snow melting systems, H55.7
sorption dehumidification, E7.7-8
sound, H57.19-22
vibration, H57.22-25
water-cooled condensers, E15.7

Textile processing
air-conditioning design, H35.4
  air distribution, H35.5
  collector systems, H35.4
  health and safety considerations, H35.7
  integrated systems, H35.4
energy conservation, H35.7
fabric making, H35.3-4
fiber making, H35.1
yarn making, H35.1-3

Thermal storage
applications, H46.16
controls, H46.3
definitions, H46.1
design considerations, H46.2
economics, H46.4
electrically charged devices, H46.12-13
ground-coupled storage, seasonal, H46.14
heat pump systems, applied, H9.7
ice storage, H46.9-11
off-peak cooling, H46.17-20
off-peak heating, H46.20-22
operator training, H46.3
phase change materials, H46.10-12
pumping strategy, H46.2
sensors, H46.3
solar energy systems, E44.11-14, H47.19, 26
storage
  containers, H46.2
  in aquifers, H46.14-16
  location, H46.2
  media, H46.1
  strategies, H46.1
technologies, H46.5-16
water tanks, H46.5-8
water treatment, H46.3

Thermal transmission, F22.1-13
airspace resistance, F22.2
attics, ventilated, F22.11
below-grade construction, F22.13
building envelopes, F22.1
ceilings, F22.11
design values, F22.6-9
doors, F22.12
overall resistance, F22.2
roofs, F22.11
slab-on-grade construction, F22.13
surface, F22.2
U_o concept, F22.12
walls
  masonry, F22.4
  panels containing metal, F22.5

  wood frame, F22.3
  zone method, F22.10
windows, F22.12

Thermodynamic(s)
analysis of absorption cycles, F1.22
basic concepts, F1.1
calculations, F1.5-7
  enthalpy, F11.6
  entropy, F1.5
  internal energy, F1.6
cycle analysis, F1.4
definitions, F1.1
equations of state, F1.4
laws, F1.2
  First, F1.2
  Second, F1.3
properties
  moist air, F6.1-6
  tables, F1.5
  water, F6.6-11
wet-bulb temperature, F6.13

Thermometers, F13.6-8

Thermosiphon
heat exchangers, E34.16

Thermostats, H51.11

Three-pipe
dual-temperature systems, H14.6

Trailers (see Trucks)

Transport
properties, moist air, F6.19

Transportation centers
design concepts, H19.14
load characteristics, H19.14
special considerations, H19.15

Traps
ammonia systems, R4.12
steam, H11.8-9

Trucks, refrigerated, R28

Tube, E33.1-14
heat loss, bare, F22.18
capillary, E19.21-30

Tunnels, vehicular, H29.1-6
allowable CO concentrations, H29.1
CO analyzers and recorders, H29.6
control systems, H29.7
dilution control, H29.1
pressure evaluation, H29.6
straight ducts, H29.6
ventilation requirements, H29.1-3, 5

Turbines, E32.9-16
engine test facilities, H31.3
gas, E32.14-16
steam, H32.9-13

Turkeys
growth, F9.10
heat and moisture production, F9.11
reproduction, F9.10

Two-phase flow, F4.1-14

Two-pipe
chilled water systems, H14.1
dual-temperature systems, H14.2
water systems, basic, H13.6

U-factors, F23.10-15
fenestration, F27.15-18, 36

Vacuum
drying systems, H44.6

Valves, E33.6
air, pressure-reducing, E2.6
ammonia systems, R4.3
compressor
  reciprocating, E12.9
  rotary, E12.13
control, H51.4
  loss coefficients, F2.10, R3.10
halocarbon systems, R3.3, 10
refrigerant control
  check, E19.18
  condenser pressure regulator, E19.12
  condensing water regulator, E19.17
  evaporator regulator, E19.9-11
  expansion, E19.3-8
  high side float, E19.13
  low side float, E19.13

Index

relief devices, E19.19-21
solenoid, E19.14-17
suction pressure regulator, E19.11
solar energy systems, H47.20
steam systems, H11.10-12
Vapor
emission control, E11.22
pressure
desiccant water, F19.2
liquids, F37.2
retarders, water, F21.4-14
transmission, water
permeability, F22.13, 14
permeance, F22.13, 14
Vapor retarders, F20.15-18
refrigerated warehouses, R25.9
Variable volume
algorithm for simulating system, F28.23
control, H51.21, 26-28
dual-duct, H2.14-16
educational facilities, H22.4
laboratories, H30.8
outlets, E2.5
single-duct system, H2.5-8
fan-powered, H2.10-12
induction, H2.9-10
testing, adjusting, balancing, H57.4-6
Vegetables
air transport, R31.1
cooling and freezing times, F29.1-16
display refrigerators, E35.4
dried, storage, R22.8
enthalpy, F30.9
handling, R18.1
heat of respiration, F30.4-7
precooling, R11, R18.2
storage, R18.3-14, R26.2
diseases, R18.5-14
thermal conductivity, F30.17
thermal diffusivity, F30.13
transport, R18.2
Vehicular facilities
bus terminals, H29.11-14
equipment, H29.14-16
rapid transit systems, H29.7-11
tunnels, H29.1-6
Velocity
measurement, F13.13-17
Vending machines, refrigerated, E39.4-6
beverage dispensing, E39.5
food, E39.6
frozen beverage, E39.6
ice makers, E39.5
placement, E39.5
pre- and postmix equipment, E39.4
standards, E39.6
Ventilation
air quality, F23.2
air transport
animal requirements (table), R31.2
aircraft air conditioning, H25.1
all-water systems, H4.1-4
animal
environments, H37.3-6
rooms, H30.13
attic, F21.10
automobile air conditioning, H24.1
bus terminals, H29.12
driving mechanisms, F23.2
combined, F23.5
mechanical systems, F23.5
stack pressure, F23.3
wind pressure, F23.3
educational facilities, H22.1
enclosed vehicular facilities, H29.1-16
engine test facilities, H31.1-6
greenhouses, H37.11
hospitals, H23.4
industrial
environment, H41.1-8
exhaust systems, H43.1-10
infiltration, F23.2, 7-18
calculation, F23.13-17

measurement, F23.8-12
merchant ships, H26.2
natural, F23.1, 6-7
flow, F23.6
guidelines, F23.7
openings, F23.7
naval surface ships, H26.6
nuclear facilities, H40.1-10
nursing homes, H23.11
odor removal, F12.5
parking garages, H29.1-10
places of assembly, H20.1
rapid transit systems, H29.7-11
survival shelters, H27.16-18
underground mines, H42.1-8
unitary air-conditioning systems, H5.3-8
vehicular tunnels, H29.1-6
Ventilators, E27.1-3
Venting
caps, E26.25
gas appliances, E26.18
Vibration
central system, H1.7
clean spaces, H32.6
communications centers, H19.13
compressors
centrifugal, E12.32
positive displacement, E12.4
fans, E3.1-14
fundamentals of isolation, F7.9
liquid chilling systems, E17.11
measurement, F13.26
places of assembly, H20.3
refrigerant piping, F33.21
testing, adjusting, balancing, H57.22-25
Vibration control, H52.28-39
duct vibration, H52.36
floor flexibility, H52.39
instructions to designer/specifier, H52.30
isolation in piping systems, H52.34-36
isolator selection guide, H52.31
resonance, H52.39
seismic protection, H52.37
specific equipment types, H52.32-34
testing isolation systems, H52.38
theory of isolation, H52.29
troubleshooting, H52.37-38
Viscosity
fuel oils, F15.5
oil (refrigerant systems), R8.4-6
vapor, F37.1
Volume
measurement, F13.17-21
Walls
CLTD tables, F26.35-38
transfer function coefficients, F26.15-19
Warehouse(s)
design concepts, H19.16
load characteristics, H19.15
refrigerated, design, R25
construction layout, R25.2
construction methods, R25.4
controlled atmosphere storage, R25.3
doors, R25.11
fire protection, R25.12
initial considerations, R25.1
insulation, R25.9-11
maintenance, R25.13
refrigeration system, R25.6-9
Water
air-and-water systems, H3.1-10
secondary water systems, H3.9-11
all-water air-conditioning systems, H4.1-4
ammonia absorption units, E13.10
chilled water treatment, H53.16
coils, E6.1-16
air heating, E9.1-4
content, food, F30.1
coolers, drinking, E38.1-6
heating
geothermal energy systems, H45.8-10
service (domestic), H54.1-20
solar energy systems, H47.16-22

systems, water treatment, H53.9, 17
humidifier supply, E5.8
lithium bromide absorption units, E13.10
requirements
survival shelters, H27.2
solubility in refrigerants, R7.2
systems, basic, H13.1-26
advantages, H13.15
antifreeze solutions, H13.23-25
boiler room piping, H13.25
chilled, H13.6
combination, H13.6
controls, H13.20-23
design considerations, H13.15
design flow rates, H13.3
design temperature, H13.2
dual-temperature, H13.6
economic design, H13.16
expansion tanks, H13.13
heat sources, H13.17
heat transfer, H13.2
low temperature, H13.18-20
one-pipe, H13.5
pipe sizing, F33.4-9
piping design, H13.4
pressure drop, H 13.7
pressurization, H13.11-15
pump selection, H13.8-11
series loop, H13.5
source heat pumps, H5.5-8
standby losses, H13.26
temperature classification, H13.1
terminal units, H13.17
two-pipe, H13.6
systems, chilled, H14.1-8
brine, H14.1
changeover temperature, H14.3-6
condenser water, H14.7
cooling tower, H14.7
four-pipe, H14.6
once-through, H14.7
three-pipe, H14.6
two-pipe, H14.1
systems, dual-temperature, H14.1-8
changeover temperature, H14.3-6
condenser water, H14.7
cooling tower, H14.7
four-pipe, H14.6
once-through, H14.7
three-pipe, H14.6
two-pipe, H14.2
systems, medium and high temp, H15.1-10
air-heating coils, H15.8
basic, H15.2
characteristics, H15.1
controls, H15.8
design considerations, H15.2
direct-contact heaters, H15.6
direct-fired generators, H15.3
expansion, H15.4-6
heat exchangers, H15.8
heat storage, H15.9
instrumentation, H15.8
piping design, H15.7
pressurization, H15.4-6
pumps, H15.6
safety considerations, H15.9
space heating equipment, H15.8
water treatment, H15.9
tanks
thermal storage, H46.5-8
thermodynamic properties, F6.6-11
treatment
thermal storage, H46.3
vapor transmission, F22.13, 14
Water treatment, H53.1-20
accelerating or intensifying factors, H53.2-3
differential solute concentration, H53.2
dissimilar metal corrosion, H53.2
effects of stress, H53.3
galvanic corrosion, H53.2
moisture, H53.2
oxygen, H53.2

pressure, H53.3
solutes, H53.2
stray current corrosion, H53.3
temperature, H53.3
velocity, H53.3
biological growths, H53.14
cooling towers, E20.11
corrosion control, H53.10-12
definitions, H53.1
evaporative condensers, E15.17
fireside corrosion and deposits, H53.19
preventive and protective measures, H53.4
    cathodic protection, H53.6
    environmental change, H53.7
    materials selection, H53.4
    protective coatings, H53.5
    surface preparation, H53.5
scale control, H53.12-14
selection, H53.15-18
    air washers, H53.15
    boilers, H53.17
    brines, H53.17
    chilled water systems, H53.16
    closed recirculating systems, H53.16
    ice machines, H53.16
    once-through systems, H53.15
    open recirculating systems, H53.15
    return condensate systems, H53.17
    sprayed coil units, H53.15
    water heating systems, H53.17
suspended solids, H53.14
theory, H53.1
underground corrosion, H53.18
water systems, medium and high temp, H15.9

waterside corrosion and deposits, H53.7-10
    cooling systems, H53.9
    heating systems, H53.9
    water characteristics, H53.8
    water-caused troubles, H53.10

Water vapor
air mixture properties, F20.12
migration, F20.14

Water vapor retarders, F20.15-18
condensation, F21.6-14
    attic ventilation, F21.10
    checklist, F21.11
    concealed, F21.7
    control, F21.6-10
    cooled structures, F21.12
    heated buildings, F21.6
    visible, F21.7
    zones in United States, F21.8

Weather data
climatic conditions, F24.4
    Canada, F24.16
    United States, F24.4
    world, F24.18
design factors, F24.3
evaporative cooling, F24.4
hourly weather occurances in U.S. cities, F28.11
interpolation between stations, F24.2
other sources, F24.3
outdoor design conditions, F24.1, 4-22
underground structures, F24.4
weather year (WYEC), F24.3

Welding
sheet metal, E1.12

Wet-bulb temperature, F6.12

Wind
coefficient, F22.2
data sources, F14.7
effect on airflow around buildings, F14.1-18
effects on chimneys, E26.25
tunnel, F14.16

Windows
cooling load through, F26.5
shading
    coefficients, F27.19, 24-37
    devices, F27.30-37
U-values, F27.15-18, 36

Wine
making of, R23.8
temperature control
    fermentation, R23.8
    storage, R23.9

Wood
heaters (in-space), E25.5-7
stoves, E25.6

Wood products facilities, H39.1-4
control rooms, H39.3
general operations, H39.1
    finished product storage, H39.2
    process area air conditioning, H39.1
motor control centers, H39.3
paper test laboratories, H39.3
pulp and paper operations, H39.2-3
system selection, H39.3

World Fair exhibit buildings
design concepts, H20.9-10
system applicability, H20.10

Make the right move... JOIN ASHRAE

10 reasons why...

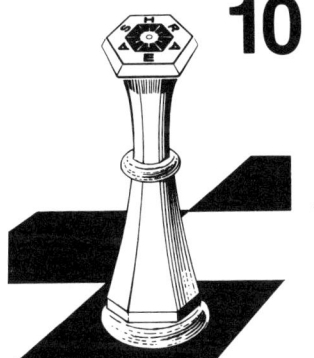

1 ASHRAE Handbooks
Have the industry's most widely referenced book at your side. Members of ASHRAE receive a revised Handbook volume each year. This four volume set—Fundamentals, HVAC, Equipment and Refrigeration—is written by thousands of experts. It offers guidelines for accomplishing tasks and solving problems.

2 ASHRAE Journal
What's new in your industry is reported in the *Journal*. ASHRAE members receive it each month. Feature articles discuss emerging technology. Industry news, governmental action, new products and literature are reviewed.

3 ASHRAE Standards
Participate in the process by which industry tests equipment and sets levels of performance. ASHRAE members are kept informed on standards development through the ASHRAE *Journal*, special mailings and updates at ASHRAE meetings. Members are offered standards at reduced prices.

4 Chapters
Continue your education locally and meet others in your industry. Currently, there are some 150 ASHRAE chapters. Each holds monthly meetings and seminars. Chapters also assist their communities by volunteering technical expertise for sound governmental action and guiding students in engineering careers.

5 ASHRAE Meetings
Exchange ideas with world leaders by attending ASHRAE's Winter and Annual Meetings. ASHRAE *Insights* provides summaries of papers that will be presented—typically several hundred. Papers and discussions are published in ASHRAE *Transactions* and are made available to members at a reduced price. At each of its Winter Meetings, ASHRAE cosponsors the largest HVAC&R equipment exposition in North America.

6 Professional Development Seminars
Take an intensive 2-day course to reduce energy consumption in the buildings and systems you design or operate. Seminars on a variety of energy-related topics are held across North America. Highly qualified faculty and extensive lecture notes set ASHRAE seminars apart from others.

7 ASHRAE Research
No other engineering society supports a research program as extensive as ASHRAE's. As a member you gain access to this research through ASHRAE's Handbooks, *Journal* and *Transactions*. Some research projects call for publication of special bulletins, like the ASHRAE *Manual for Smoke Control* and the *Passive Solar Design Manual*. Members receive research bulletins at reduced prices.

8 International Contact
Become part of a vast international market for information through ASHRAE. Members reside in more than 120 nations. Chapters operate around the world. ASHRAE is associated with societies in more than 20 nations.

9 ASHRAE *Insights*
Learn about ASHRAE activities through *Insights*, a monthly tabloid sent to all members. It reports on committee activities, new publications and upcoming meetings. Here is where you will find advance meeting information.

10 Professionalism
Belonging to ASHRAE is a mark of professional stature. It is a statement that you value continuing education and technological advancement. Through your support of ASHRAE, you help your profession and your industry to benefit humanity.

Who can join ASHRAE?

ASHRAE membership is drawn from many occupations. There are engineers, contractors, researchers, architects and administrators. There are employees of manufacturers, government, design and construction firms and educational institutions. Some ASHRAE members are students in colleges or technical schools.

While most ASHRAE members are directly involved in heating, air conditioning, ventilation or refrigeration, others are employed in industries related to these fields. Anyone who can benefit from participation in ASHRAE can qualify for membership.

To protect the nonproprietary character of ASHRAE, only memberships in individual names are offered.

To receive a membership application, contact:

ASHRAE Membership Department, ASHRAE International Headquarters,
1791 Tullie Circle, N.E., Atlanta, Georgia 30329 U.S.A.   404-636-8400 / Telex: 705343 ASHRAE / FAX: 404-321-5478